SECOND EDITION

Algebra: Beginning and Intermediate

Richard N. Aufmann
Palomar College, California

Joanne S. Lockwood
New Hampshire Community
Technical College, New Hampshire

Houghton Mifflin Company **Boston New York**

Publisher: Richard Stratton
Senior Sponsoring Editor: Lynn Cox
Senior Marketing Manager: Katherine Greig
Marketing Associate: Naveen Hariprasad
Associate Editor: Noel Kamm
Editorial Assistant: Laura Ricci
Senior Project Editor: Nancy Blodget
Editorial Assistant: Sarah Driver
Art and Design Manager: Gary Crespo
Cover Design Manager: Anne S. Katzeff
Photo Editor: Marcy Kagan
Composition Buyer: Chuck Dutton

Cover Image: 2006 © Adrienne O'Toole

Photo credits are listed on page 822, immediately preceding the Appendix.

Printed in the U.S.A.

Library of Congress Catalog Card Number: 2006935492

Instructor's Annotated Edition:
ISBN 10: 0-618-82073-6
ISBN 13: 978-0-618-82073-3

For orders, use student text ISBNs:
ISBN 10: 0-618-82072-8
ISBN 13: 978-0-618-82072-6

3 4 5 6 7 8 9—WEB—11 10 09 08

Contents

2 Solving Equations and Inequalities 83

5 Polynomials 367

6 Factoring 429

7 Rational Expressions 482

8 Rational Exponents and Radicals 547

11　Exponential and Logarithmic Functions　721

R　Review of Beginning Algebra Topics　784

Preface

The second edition of *Algebra: Beginning and Intermediate* examines the fundamental ideas of algebra. The text has been designed not only to meet the needs of the traditional college student, but also to serve the needs of returning students whose mathematical proficiency may have declined during years away from formal education.

In this new edition of *Algebra: Beginning and Intermediate,* we have continued to integrate some of the approaches suggested by AMATYC. Each chapter opens by illustrating and referencing a mathematical application within the chapter. At the end of each section there are "Applying Concepts" exercises that include writing, synthesis, critical thinking, and challenge problems. At the end of each chapter there is a "Focus on Problem Solving" section that introduces students to various problem-solving strategies. This is followed by "Projects & Group Activities," which can be used for cooperative learning activities.

NEW! Changes to the Content in This Edition

In response to user requests, in Section 3 of Chapter 1, the presentation of operations on fractions has been changed from presenting addition and subtraction of rational numbers first, followed by multiplication and division, to presenting multiplication and division of rational numbers first, followed by addition and subtraction.

Chapter 3, Section 2, has been expanded to include not only evaluating functions but also graphing functions and the vertical line test. Section 3.3 now includes an explanation that the x-coordinate of an x-intercept is a zero of a function; students are asked to find zeros of linear functions. This earlier introduction of zeros lays the groundwork for finding zeros of other types of functions later in the text. The first objective in Section 5.1, titled "Evaluate polynomial functions," includes graphs of polynomial functions. Here we revisit the vertical line test by noting that these graphs pass the vertical line test. In Objective 6.5.1, in which we are solving equations by factoring, we revisit the concept of zeros of a function by asking students to find zeros of quadratic functions.

Because the topics of graphing polynomial functions and the vertical line test, which were presented in Chapter 10 of the previous edition, are now presented earlier in the text, Chapter 10 of this text now begins with graphing absolute value functions as an introduction to translations of graphs in Section 1. The earlier presentation and emphasis on functions will provide students with a stronger background for this chapter, the topic of which is "Functions and Relations."

The introductions to both exponential functions in Section 11.1 and logarithmic functions in Section 11.2 have been rewritten. Both topics are presented using applications. This development should lead to greater student understanding of and interest in these concepts.

NEW! Changes to Exercises in This Edition

Two new types of exercises have been added to this edition of the text: Interactive Exercises and Think About It Exercises.

The Interactive Exercises test students' understanding of the basic concepts presented in a lesson. Included when appropriate, these exercises generally appear at the beginning of an objective's exercise set and provide students with guided practice on some of the objective's underlying principles. Sometimes they test students' knowledge of the terms associated with a topic, but more often they are fill-in-the-blank exercises in which students are given part of a solution to a problem and are asked to complete the missing portions. These exercises act as stepping stones to the remaining exercises for the objective. We are confident that these Interactive Exercises will lead to greater student success in mastering the essential skills.

The Think About It Exercises are conceptual in nature. They generally appear near the end of an objective's exercise set and ask the students to think about the objective's concepts, make generalizations, and apply them to more abstract problems. Their focus is on mental mathematics, not calculation or computation, and they are meant to help students synthesize concepts. We are confident that their use will lead to a deeper conceptual understanding of the course content.

Throughout the text, data problems have been updated to reflect current data and trends. These application problems will demonstrate to students the variety of problems in real life that require mathematical analysis. Instructors will find that many of these problems lead to interesting class discussions.

NEW! Enhancements to the Instructional Media Package

HM Testing™ (powered by Diploma®) is a new computerized testing platform that provides instructors with a wide array of algorithmic test items, improved functionality, ease of use, and gradebook functions. HM Testing offers all the tools needed to create, author, deliver, and customize multiple types of tests—including authoring and editing algorithmic questions.

Eduspace®, Houghton Mifflin's Online Learning Tool (powered by Blackboard®), is a web-based learning system that provides instructors and students with powerful course management tools and text-specific content to support all of their online teaching and learning needs. Resources such as HM Assess, algorithmic automatically graded homework exercises, video explanations, an online multimedia eBook, live online tutoring with SMARTHINKING®, and additional study materials all come ready-to-use. Instructors can choose to use the content as is, modify it, or even add their own content.

HM Assess™ is a new diagnostic assessment tool that tests core concepts in mathematics and provides students with access to individual study paths for self-remediation. These paths are carefully designed to offer self-study options and to appeal to a variety of learning styles through review of the concepts, interactive lessons, similar examples, and additional practice exercises. Both course assessments and chapter assessments are available.

Students using Eduspace with HM Assess can:

- Assess their core competencies and remediate themselves via individual study paths.
- Review and reinforce concepts with algorithmically generated exercises.
- View video tutorials.
- Access, complete, and submit homework online.

Acknowledgments

We would like to especially thank users of the previous edition for their helpful suggestions to improve the text. Also, we sincerely appreciate the suggestions and the time and effort of the following people who reviewed this manuscript and/or its previous edition in preparation for the second edition.

Bob Bradshaw, *Ohlone Community College, Fremont, CA*

Tim Chappell, *Penn Valley Community College, Kansas City, MO*

Marilyn Danchanko, *Pennsylvania Highlands Community College, Johnstown, PA*

Demetria N. Gibbs, *Gwinnett Technical College, Lawrenceville, GA*

Edith Hays, *Texas Woman's University, Denton, TX*

Feliks Kostanyan, *DeVry University, Long Island City, NY*

Christine Mirbaha, *The Community College of Baltimore County, Dundalk, MD*

J. Reid Mowrer, *University of New Mexico Los Lunas, Valencia, NM*

Earl T. Odom, Jr., *Bishop State Community College, Mobile, AL*

Elise Price, *Tarrant County College-Southeast Campus, Arlington, TX*

Angelia Reynolds, *Gulf Coast Community College, Panama City, FL*

Mehdi Sadatmousavi, *Pima Community College, Tucson, Arizona*

Lourdes Triana, *Humboldt State University, Arcata, CA*

David E. Turnbull, *Lake Land College, Mattoon, IL*

Jennifer Webb

David White, *The Victoria College, Victoria, TX*

Danny Whited, *Virginia Intermont College, Bristol, VA*

Cynthia Bernice Williams, *Pulaski Technical College, North Little Rock, AR*

STUDENT SUCCESS

Algebra: Beginning and Intermediate is designed to foster student success through an integrated text and media program.

AIM for Success

Welcome to *Algebra: Beginning and Intermediate*. As you begin this course, we know two important facts: (1) We want you to succeed. (2) You want to succeed. To do that requires an effort from each of us. For the next few pages, we are going to show you what is required of you to achieve that success and how you can use the features of this text to be successful.

Motivation

One of the most important keys to success is motivation. We can try to motivate you by offering interesting or important ways mathematics can benefit you. But, in the end, the motivation must come from you. On the first day of class, it is easy to be motivated. Eight weeks into the term, it is harder to keep that motivation.

To stay motivated, there must be outcomes from this course that are worth your time, money, and energy. List some reasons you are taking this course. Do not make a mental list—actually write them out.

Take Note

Motivation alone will not lead to success. For instance, suppose a person who cannot swim is placed in a boat, taken out to the middle of a lake, and then thrown overboard. That person has a lot of motivation but there is a high likelihood the person will drown without some help. Motivation gives us the desire to learn but is not the same as learning.

Although we hope that one of the reasons you listed was an interest in mathematics, we know that many of you are taking this course because it is required to graduate, it is a prerequisite for a course you must take, or because it is required for your major. Although you may not agree that this course is necessary, it is! If you are motivated to graduate or complete the requirements for your major, then use that motivation to succeed in this course. Do not become distracted from your goal to complete your education!

Commitment

To be successful, you must make a commitment to succeed. This means devoting time to math so that you achieve a better understanding of the subject.

List some activities (sports, hobbies, talents such as dance, art, or music) that you enjoy and at which you would like to become better.

ACTIVITY TIME SPENT TIME WISHED SPENT

Thinking about these activities, put the number of hours that you spend each week practicing these activities next to the activity. Next to that number, indicate the number of hours per week you would like to spend on these activities.

Whether you listed surfing or sailing, aerobics or restoring cars, or any other activity you enjoy, note how many hours a week you spend doing it. To succeed in math, you must be willing to commit the same amount of time. Success requires some sacrifice.

The "I Can't Do Math" Syndrome

There may be things you cannot do, such as lift a two-ton boulder. You can, however, do math. It is much easier than lifting the two-ton boulder. When

xxv

page xxv

AIM for Success

This preface helps students develop the study skills necessary to achieve success in college mathematics.

It also provides students with an explanation of how to effectively use the features of the text.

AIM for Success can be used as a lesson on the first day of class or as a student project. A lesson plan, with accompanying PowerPoint slides, is available for instructors on the Online Teaching Center.

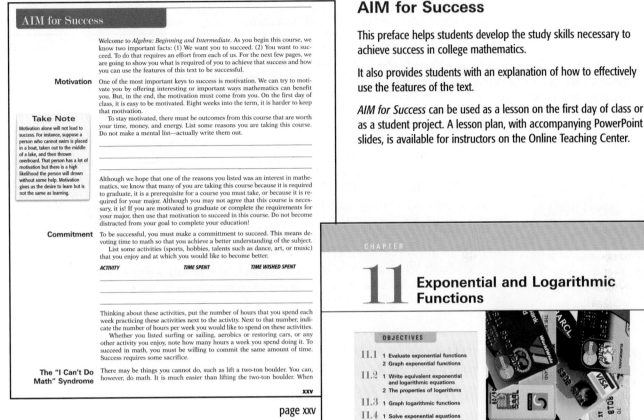

CHAPTER

11 Exponential and Logarithmic Functions

OBJECTIVES

11.1 1 Evaluate exponential functions
2 Graph exponential functions

11.2 1 Write equivalent exponential and logarithmic equations
2 The properties of logarithms

11.3 1 Graph logarithmic functions

11.4 1 Solve exponential equations
2 Solve logarithmic equations

11.5 1 Application problems

Online Study Center

Need help? For online student resources, such as section quizzes, visit the Online Study Center at college.hmco.com/pic/aufmannAB2e.

Having a credit card can be very helpful when making big purchases, but forgetting to pay your bills can cause major problems later on. When applying for things such as student loans, the lender generally wants to know your past credit history. Usually, they will turn to a credit-reporting agency for a full credit report and a credit score, which helps them determine the likelihood that you will repay the loan in a timely manner. One of the most widely used credit scores is the FICO® score. To learn more about credit reports and FICO scores, see the **Projects and Group Activities on page 776.**

721

page 721

Objective-Based Approach

All lessons, exercise sets, tests, and supplements are organized around a carefully constructed hierarchy of objectives. This organization serves as a framework for a complete learning system, allowing students to remediate their learning independently at any point in the process.

Each chapter's objectives are listed on the chapter opener page, which serves as a road-map to student learning.

A numbered objective describes the topic of each lesson.

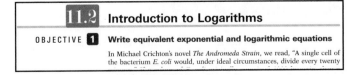

11.2 Introduction to Logarithms

OBJECTIVE **1** Write equivalent exponential and logarithmic equations

In Michael Crichton's novel *The Andromeda Strain*, we read, "A single cell of the bacterium *E. coli* would, under ideal circumstances, divide every twenty

page 732

xv

All exercise sets correspond directly to objectives.

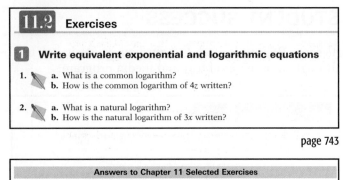

page 743

Answers to the Prep Tests, Chapter Review Exercises, Chapter Tests, and Cumulative Review Exercises refer students back to the original objectives for further study.

Answers to Chapter 11 Selected Exercises

PREP TEST

1. $\frac{1}{9}$ [5.1.2] **2.** 16 [5.1.2] **3.** -3 [5.1.2] **4.** 0; 108 [3.2.1] **5.** -6 [2.1.1]

6. $-2, 8$ [8.1.1] **7.** 6326.60 [1.3.2] **8.** [3.2.2]

page A49

Prep Test

1. Simplify: 3^{-2}

2. Simplify: $\left(\frac{1}{2}\right)^{-4}$

3. Complete: $\frac{1}{8} = 2^?$

4. Evaluate $f(x) = x^4 + x^3$ for $x = -1$ and $x = 3$.

5. Solve: $3x + 7 = x - 5$

6. Solve: $16 = x^2 - 6x$

7. Evaluate $A(1 + i)^n$ for $A = 5000$, $i = 0.04$, and $n = 6$. Round to the nearest hundredth.

8. Graph: $f(x) = x^2 - 1$

page 722

Chapter 11 Summary

Key Words

A function of the form $f(x) = b^x$, where b is a positive real number not equal to 1, is an **exponential function**. The number b is the **base** of the exponential function. [11.1.1, p. 723]

The function $f(x) = e^x$ is called the **natural exponential function**. [11.1.1, p. 724]

For $x > 0, b > 0, b \neq 1, y = \log_b x$ is equivalent to $x = b^y$. Read $\log_b x$ as "the **logarithm** of x, base b." [11.2.1, p. 733]

Examples

$f(x) = 3^x$; 3 is the base of the function.

$f(x) = e^x$; e is an irrational number approximately equal to 2.71828183

$\log_2 8 = 3$ is equivalent to $8 = 2^3$.

page 778

Cumulative Review Exercises

1. Solve: $4 - 2[x - 3(2 - 3x) - 4x] = 2x$

2. Solve $S = 2WH + 2WL + 2LH$ for L.

3. Solve: $|2x - 5| \leq 3$

4. Factor: $4x^{2n} + 7x^n + 3$

page 782

Assessment and Review

Prep Tests assess students' mastery of prerequisite skills for the upcoming chapter.

The Answers found at the back of the book refer them to the appropriate section objective for review.

The **end-of-chapter material** includes:

Chapter Summary

Chapter Review Exercises

Chapter Test

These resources help students prepare for tests.

Cumulative Review Exercises help students maintain skills learned in previous chapters.

Online Assessment and Remediation

Eduspace® (powered by Blackboard®), Houghton Mifflin's text-specific online learning system, now includes **HM Assess™.** This diagnostic assessment tool tests core concepts and provides students with access to *individual study paths* for self-remediation.

Students using HM Assess can:

- Assess their core competencies and remediate themselves based on their individual learning style.
- Review and reinforce concepts with algorithmically generated exercises.
- View video tutorials.
- Take practice quizzes.

Online Practice Problems

Step-by-step guided solutions and fully worked similar examples in **Eduspace**® provide further opportunities for students to actively develop their skills and understanding of the topics.

Live, Online Tutoring

SMARTHINKING® live, online tutoring provides guidance for students as they work through problems without ever giving them the answers.

Online Self Tests

Online Study Center **ACE Self Tests** on the student **Online Study Center** let students practice chapter tests to help them prepare for exams.

AUFMANN INTERACTIVE METHOD (AIM)

Algebra: Beginning and Intermediate uses an interactive style that engages students in trying new skills and reinforcing learning through structured exercises.

An Interactive Approach

EXAMPLE 2 Find the hydrogen ion concentration, H^+, of orange juice that has a pH of 3.6.

Strategy To find the hydrogen ion concentration, replace pH by 3.6 in the equation $pH = -\log(H^+)$ and solve for H^+.

Solution
$$pH = -\log(H^+)$$
$$3.6 = -\log(H^+)$$
$$-3.6 = \log(H^+)$$
$$10^{-3.6} = H^+$$
$$0.00025 \approx H^+$$

The hydrogen ion concentration is approximately 0.00025.

PROBLEM 2 On September 3, 2000, an earthquake measuring 5.2 on the Richter scale struck the Napa Valley, 50 mi north of San Francisco. Find the intensity of the quake in terms of I_0.

Solution See page S36.

- Each section of the text is divided into objectives, and every objective contains one or more sets of matched-pair examples. The first example in each set is worked out.

- The second example, called "Problem," is for the student to work. By solving this problem, the student actively practices concepts as they are presented in the text.

page 768

Problem 2

Strategy ▶ To find the intensity, use the equation for the Richter scale magnitude of an earthquake, $M = \log \dfrac{I}{I_0}$. Replace M by 5.2 and solve for I.

Solution
$$M = \log \frac{I}{I_0}$$
$$5.2 = \log \frac{I}{I_0} \quad \bullet \text{ Replace } M \text{ by } 5.2.$$
$$10^{5.2} = \frac{I}{I_0} \quad \bullet \text{ Write in exponential form.}$$
$$10^{5.2}I_0 = I$$
$$158,489I_0 \approx I$$

The earthquake had an intensity that was approximately 158,489 times the intensity of a zero-level earthquake.

- *Complete worked-out* solutions to these Problems appear at the back of the book for students to check their work.

page S36

NEW! Interactive Exercises

Placed at the beginning of an objective's exercise set (when appropriate), these exercises provide guided practice and test students' understanding of the underlying concepts in a lesson. They also act as stepping-stones to the remaining exercises for the objective.

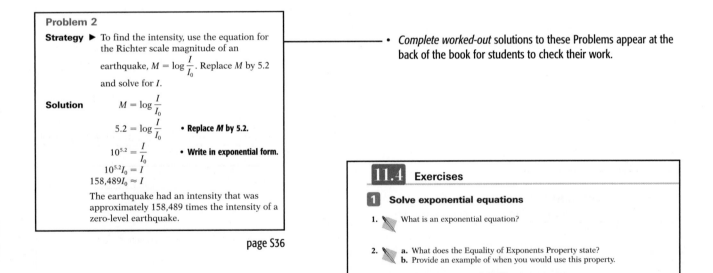

11.4 Exercises

1 Solve exponential equations

1. What is an exponential equation?

2. a. What does the Equality of Exponents Property state?
 b. Provide an example of when you would use this property.

3. Solve: $3^{3x-1} = 9^{2x}$.
 $$3^{3x-1} = 9^{2x} \quad \bullet \text{ Both bases are powers of } \underline{}.$$
 $$3^{3x-1} = (3^{\underline{?}})^{2x} \quad \bullet \text{ Write 9 as a power of 3.}$$
 $$3^{3x-1} = 3^{\underline{?}} \quad \bullet \text{ Simplify the exponent.}$$
 $$\underline{} = \underline{} \quad \bullet \text{ Use the Equality of Exponents Property to equate the exponents.}$$
 $$x = \underline{} \quad \bullet \text{ Solve for } x.$$

4. Solve: $5^x = 18$
 $$5^x = 18 \quad \bullet \text{ The bases are not powers of the same number.}$$
 $$\log 5^x = \log 18 \quad \bullet \text{ Take the } \underline{} \text{ of each side of the equation.}$$
 $$(\underline{}) \log 5 = \log 18 \quad \bullet \text{ Use the Power Property of Logarithms.}$$
 $$x = \frac{\underline{?}}{\underline{?}} \quad \bullet \text{ Solve for } x \text{ by dividing each side of the equation by } \underline{}.$$
 $$x \approx \underline{} \quad \bullet \text{ Use a calculator to evaluate } x \text{ to the nearest ten-thousandth.}$$

page 760

CONCEPTUAL UNDERSTANDING

Algebra: Beginning and Intermediate helps students understand the course concepts through the textbook exposition and feature set.

Features in the Exposition

Take Note

Crichton goes on to say, in the same paragraph, "In this way it can be shown that in a single day, one cell of *E. coli* could produce a super-colony equal in size and weight to the entire planet Earth." In the model we are using, where the bacteria divide once an hour (instead of three times an hour, as in *The Andromeda Strain*), it would take the colony 3 days to become the size of the Earth.

This model would be true if uncontrolled biological growth were possible, but it is not. Other factors, such as food to sustain growth, would not be sufficient to support continued exponential growth.

To determine when there would be 32,000 bacteria in the culture, we would solve the exponential equation $32{,}000 = 1000(2^t)$ for t. Because $A = 1000(2^t)$ is a growth equation and 32,000 is between 8000 and 64,000, t must be between 3 h and 6 h. By trial and error, we find that $t = 5$ h.

$$A = 1000(2^t)$$
$$A = 1000(2^5) \qquad \text{• Replace } t \text{ by 5.}$$
$$A = 1000(32) = 32{,}000$$

Now suppose we want to find how long it would take for the colony to reach 1,000,000 bacteria. From the table, 1,000,000 is between 256,000 and 4,096,000, so this will happen sometime between 8 h and 12 h. If we try 10 h, halfway between 8 and 12, we have

$$A = 1000(2^t)$$
$$A = 1000(2^{10}) \qquad \text{• Replace } t \text{ by 10.}$$
$$A = 1000(1024) = 1{,}024{,}000$$

Because 1,000,000 is less than 1,024,000, the actual value of t is less than 10 h.

We could continue to use trial and error to find t, but it would be more efficient if we could just solve the equation $1{,}000{,}000 = 1000(2^t)$ for t. Using methods described earlier in the text, we have

$$1{,}000{,}000 = 1000(2^t)$$
$$1000 = 2^t \qquad \text{• Divide each side of the equation by 1000.}$$

To solve this equation, it would be helpful to have a function that would give the power of 2 that produces 1000. Around the mid-16th century, mathematicians created such a function, which we now call a *logarithmic function*. We write the solution of $1000 = 2^t$ as $t = \log_2 1000$. This is read "*t equals the logarithm base 2 of 1000*," and it means "*t equals the power of 2 that produces 1000.*"

When logarithms were first introduced, tables were used to find the numerical value of t. Today, a calculator is used. Using a calculator, we can approximate the value of t as 9.97. Thus $9.97 \approx \log_2 1000$ and $1000 \approx 2^{9.97}$.

The ideas presented above are related to the concept of inverse function discussed in the last chapter. Because the exponential function given by $y = b^x$ is a 1–1 function, it has an inverse function. To find that function, we follow the same procedure we used to find the inverse of other functions.

$$y = b^x$$
$$x = b^y \qquad \text{• Interchange } x \text{ and } y.$$

Take Note

Recall that for the equation $2^t = 1000$, we said that "*t* equals the power of 2 that produces 1000" means that "*t* is the logarithm base 2 of 1000." We are doing the same with $x = b^y$.

The equation $x = b^y$ says that y equals the power of b that produces x. That is, y is the logarithm base b of x. The inverse of the exponential function $y = b^x$ is a logarithmic function.

Here is a general definition of logarithm.

Definition of Logarithm

If $x > 0$ and b is a positive constant not equal to 1, then $y = \log_b x$ is equivalent to $b^y = x$.

page 733

- Margin notes titled **Take Note** amplify the concept under discussion or alert students to points requiring special attention.

- Key concepts are presented in orange boxes in order to highlight them and provide for easy reference.

Take Note

The logarithms of most numbers are irrational numbers. Therefore, the value in the display of a calculator is an approximation.

Logarithms base 10 are called **common logarithms.** We usually omit the base, 10, when writing the common logarithm of a number. Therefore, $\log_{10} x$ is written $\log x$. To find the common logarithm of most numbers, a calculator is necessary. A calculator was used to find the value of log 384, shown below.

$$\log 384 \approx 2.584331224$$

When e (the base of the natural exponential function) is used as a base of a logarithm, the logarithm is referred to as the **natural logarithm** and is abbreviated $\ln x$. This is read "el en x." Using a calculator, we find that

$$\ln 23 \approx 3.135494216$$

page 736

- Key terms, in bold, emphasize important vocabulary.

The equation $A = 10\left(\dfrac{1}{2}\right)^{\frac{t}{5}}$ is an example of an **exponential decay equation.**

Compare this equation to the exponential growth equation, and note that for exponential growth, the base of the exponential equation is greater than 1, whereas for exponential decay, the base is between 0 and 1.

A method by which an archeologist can measure the age of a bone is called **carbon dating.** Carbon dating is based on a radioactive isotope of carbon called carbon-14, which has a half-life of approximately 5570 years. The

page 765

- Important points are now highlighted, to help students recognize what is most important and to study more effectively.

CONCEPTUAL UNDERSTANDING (*continued*)

Algebra: Beginning and Intermediate helps students understand the course concepts through the textbook exposition and feature set.

Features in the Exercise Sets

11.1 Concept Review

Determine whether the following statements are always true, sometimes true, or never true.

1. The domain of an exponential function $f(x) = b^x$, $b > 0$, $b \neq 1$, is the set of positive numbers.

2. An exponential function $f(x) = b^x$, $b > 0$, $b \neq 1$, is a 1–1 function.

3. The graph of the exponential function $f(x) = b^x$, $b > 0$, $b \neq 1$, passes through the point $(0, 0)$.

4. For the function $f(x) = b^x$, $b > 0$, $b \neq 1$, the base b is a positive integer.

5. An exponential function $f(x) = b^x$, $b > 0$, $b \neq 1$, has two x-intercepts.

page 728

- **Concept Review** exercises appear at the beginning of each exercise set and are designed to test a student's understanding of the topic.

51. $b > 1$ and m and n are positive numbers such that $m > n$. Is $\log_b m$ less than, equal to, or greater than $\log_b n$?

52. $0 < b < 1$ and $m > 1$. Is $\log_b m$ less than, equal to, or greater than zero?

page 744

- **NEW! Think About It Exercises** are conceptual in nature. They ask the students to think about the objective's concepts, make generalizations, and apply them to more abstract problems. The focus is on mental mathematics, and they are designed to help students synthesize concepts.

11.4 Exercises

1 **Solve exponential equations**

1. What is an exponential equation?

2. **a.** What does the Equality of Exponents Property state?
 b. Provide an example of when you would use this property.

page 760

- **Writing Exercises** require students to verbalize concepts, which is a crucial step in learning them.

11.3 Applying Concepts

31. **Employment** The proficiency of a typist decreases (without practice) over time. An equation that approximates this decrease is given by $S = 60 - 7 \ln(t + 1)$, where S is the typing speed in words per minute and t is the number of months without typing.
 a. Graph the equation.

 b. The point whose approximate coordinates are $(4, 49)$ is on the graph. Write a sentence that describes the meaning of this ordered pair.

32. **Astronomy** Astronomers use the *distance modulus* of a star as a method of determining the star's distance from Earth. The formula is $M = 5 \log s - 5$, where M is the distance modulus and s is the star's distance from Earth in parsecs. (One parsec $\approx 2.1 \times 10^{13}$ miles.)
 a. Graph the equation.

page 754

- **Applying Concepts** are more challenging problems involving multiple concepts that require interpretation, analysis, and critical thinking. They are designed to help students to apply and internalize concepts.

PROBLEM SOLVING

Algebra: Beginning and Intermediate emphasizes applications, problem solving, and critical thinking.

16. Demography The number of centenarians (people aged 100 or older) in the United States has been increasing dramatically. Based on data from the U.S. Census Bureau, and using $x = 0$ to correspond to the year 2000, the equation $y = 66.77(1.0532^x)$ approximates the projected U.S. centenarian population through 2050, where y is in thousands.

 a. According to the model, what is the predicted population of centenarians in 2015? Round to the nearest thousand.

 b. In what year does the model predict that the U.S. centenarian population will first exceed 250,000?

Chemistry For Exercises 17 and 18, use the equation $pH = -\log(H^+)$, where H^+ is the hydrogen ion concentration of a solution. Round to the nearest tenth.

17. Find the pH of milk, for which the hydrogen ion concentration is 3.97×10^{-7}.

18. Find the pH of a baking soda solution, for which the hydrogen ion concentration is 3.98×10^{-9}.

page 771

Integrated Real-Life Applications

Wherever appropriate, the last objective of a section presents applications that require students to use problem-solving strategies, along with the skills covered in that section, to solve practical problems. This carefully integrated applied approach generates student awareness of the value of algebra as a real-life tool. Applications are taken from many disciplines and are emphasized through sourced data.

EXAMPLE 1 The number of words per minute a student can type will increase with practice and can be approximated by the equation $N = 100[1 - (0.9)^t]$, where N is the number of words typed per minute after t days of instruction. In how many days will the student be able to type 60 words per minute?

Strategy To find the number of days, replace N by 60 and solve for t.

Solution
$$N = 100[1 - (0.9)^t]$$
$$60 = 100[1 - (0.9)^t]$$
$$0.6 = 1 - (0.9)^t$$
 • Divide each side of the equation by 100.
$$-0.4 = -(0.9)^t$$
$$0.4 = (0.9)^t$$
 • Subtract 1 from each side of the equation.
$$\log 0.4 = \log(0.9)^t$$
$$\log 0.4 = t \log 0.9$$
$$t = \frac{\log 0.4}{\log 0.9} \approx 8.6967184$$

After approximately 9 days the student will type 60 words per minute.

The graph at the left is the graph of $N = 100[1 - (0.9)^t]$. Note that t is approximately 8.7 when $N = 60$.

PROBLEM 1 An investment of $3000 is placed into an account that earns 12% annual interest compounded monthly. In approximately how many years will the investment be worth twice the original amount? Round to the nearest whole number.

Solution See page S36.

Point of Interest

Saren Sarensen

The pH scale was created by the Danish biochemist Saren Sarensen in 1909 to measure the acidity of water used in the brewing of beer. pH is an abbreviation for pondus hydrogenii, which translates as "potential hydrogen."

The first applications of logarithms (and the main reason why they were developed) were to reduce computational drudgery. Today, with the widespread use of calculators and computers, the computational uses of logarithms have diminished. However, a number of other applications of logarithms have emerged.

A chemist measures the acidity or alkalinity of a solution by the formula $pH = -\log(H^+)$, where H^+ is the concentration of hydrogen ions in the solution. A neutral solution such as distilled water has a pH of 7, acids have a pH less than 7, and alkaline solutions (also called basic solutions) have a pH greater than 7.

Find the pH of vinegar for which $H^+ = 1.26 \times 10^{-3}$. Round to the nearest tenth.

Use the pH equation. $pH = -\log(H^+)$
$H^+ = 1.26 \times 10^{-3}$. $= -\log(1.26 \times 10^{-3})$
 $= -(\log 1.26 + \log 10^{-3})$
 $\approx -[0.1004 + (-3)] = 2.8996$

The pH of vinegar is approximately 2.9.

page 766

Problem-Solving Strategies

A carefully developed approach to problem solving emphasizes the importance of *strategy* when solving problems.

- Students are encouraged to develop their own strategies as part of their solutions to problems.

- Model strategies are always presented as guides for students to follow as they attempt the parallel Problems that accompany each numbered Example.

SECTION 11.5

Problem 1

Strategy ▶ To find the time, solve the compound interest formula for n. Use $P = 6000$, $A = 3000$, and $i = \frac{12\%}{12} = \frac{0.12}{12} = 0.01$.

Solution
$$P = A(1 + i)^n$$
$$6000 = 3000(1 + 0.01)^n$$
$$6000 = 3000(1.01)^n$$
$$2 = (1.01)^n$$
$$\log 2 = \log(1.01)^n$$
$$\log 2 = n \log 1.01$$
$$\frac{\log 2}{\log 1.01} = n$$
$$70 = n$$

70 months ÷ 12 ≈ 5.8 years

In approximately 6 years, the investment will be worth $6000.

page S36

Focus on Problem Solving

Find a Pattern Polya's four recommended problem-solving steps are stated below.

1. Understand the problem.
2. Devise a plan.
3. Carry out the plan.
4. Review your solution

One of the several ways of devising a plan is first to try to find a pattern. Karl Friedrich Gauss supposedly used this method to solve a problem that was given to his math class when he was in elementary school. As the story goes, his teacher wanted to grade some papers while the class worked on a math problem. The problem the teacher gave to the class was to find the sum

$$1 + 2 + 3 + 4 + \cdots + 100$$

Gauss quickly solved the problem by seeing a pattern. Here is what he saw.

Note that

$1 + 100 = 101$
$2 + 99 = 101$
$3 + 98 = 101$
$4 + 97 = 101$

Gauss noted that there were 50 sums of 101. Therefore, the sum of the first 100 natural numbers is

$$1 + 2 + 3 + 4 + \cdots + 97 + 98 + 99 + 100 = 50(101) = 5050$$

Try to solve the following problems by finding a pattern.

1. Find the sum $2 + 4 + 6 + \cdots + 96 + 98 + 100$.

2. Find the sum $1 + 3 + 5 + \cdots + 97 + 99 + 101$.

3. Find another method of finding the sum $1 + 3 + 5 + \cdots + 97 + 99 + 101$ given in the previous exercise.

4. Find the sum $\frac{1}{1 \cdot 2} + \frac{1}{2 \cdot 3} + \frac{1}{3 \cdot 4} + \cdots + \frac{1}{49 \cdot 50}$.

 Hint: $\frac{1}{1 \cdot 2} = \frac{1}{2}, \frac{1}{1 \cdot 2} + \frac{1}{2 \cdot 3} = \frac{2}{3}, \frac{1}{1 \cdot 2} + \frac{1}{2 \cdot 3} + \frac{1}{3 \cdot 4} = \frac{3}{4}$

5. The following problem shows that checking a few cases does not always result in a conjecture that is true for *all* cases. Select any two points on a circle and draw a *chord*, the line connecting the points (see the drawing in the left margin). The chord divides the circle into two regions. Now select three different points and draw chords connecting each of the three points with every other point. The chords divide the circle into four regions. Now select four points and connect each of the points with every other point. Make a conjecture about the relationship between the number

2 points, 2 regions 3 points, 4 regions

4 points, 8 regions 5 points, ? regions

NEW! Think About It Exercises

These exercises focus on concepts, not on calculation or computation, and will help develop students' critical thinking skills.

29. Which function(s) will have the same graph as the graph of the function $f(x) = \log_3 3x$?

 a. $g(x) = 3 \log_3 x$ **b.** $h(x) = 1 + \log_3 x$

 c. $F(x) = \log_3 x^3$ **d.** $G(x) = \dfrac{(\log 3x)}{(\log 3)}$

30. Which function(s) will have the same graph as the graph of the function $f(x) = -\log_b x$?

 a. $g(x) = \log_b(-x)$ **b.** $h(x) = \dfrac{1}{\log_b x}$

 c. $F(x) = \log_b \dfrac{1}{x}$ **d.** $G(x) = \dfrac{(\log b)}{(\log x)}$

Focus on Problem Solving

Students are introduced to various successful problem-solving strategies within the end-of-chapter material. Strategies include:

- drawing a diagram,
- applying solutions to other problems,
- working backwards,
- inductive reasoning,
- and trial and error.

Additional Resources—Get More Value from Your Textbook!

Instructor Resources

Instructor's Annotated Edition (IAE) The Instructor's Annotated Edition is ideal for seasoned instructors, teaching assistants, or adjuncts. Reduced student pages leave room in the margin of the IAE for extensive instructor support materials at point-of-use.

Online Teaching Center This web site offers instructors prep tests and chapter tests, instructor's solutions, quick quizzes, AIM for Success Practice Sheets, topic sequences, and more.

HM Testing™ (Powered by Diploma®) "Testing the way you want it" HM Testing offers all the tools needed to create, author/edit algorithmic questions, customize, and deliver multiple types of tests.

Student Resources

Student Solutions Manual Contains complete solutions to all odd-numbered exercises, and all of the solutions to the end-of-chapter material.

Math Study Skills Workbook, Third Edition, by Paul Nolting Helps students identify their strengths, weaknesses, and personal learning styles in math. Nolting offers study tips, proven test-taking strategies, a homework system, and recommendations for reducing anxiety and improving grades.

Online Study Center This free web site gives students access to ACE practice tests, glossary flashcards, chapter summaries, and other study resources.

Instructional DVDs Hosted by Dana Mosely, these text-specific DVDs cover all sections of the text and provide explanations of key concepts, examples, exercises, and applications in a lecture-based format. DVDs are now closed-captioned for the hearing-impaired.

Eduspace® (powered by Blackboard®) Eduspace is a web-based learning system that provides instructors with powerful course management tools and students with text-specific content to support all of their online teaching and learning needs. In addition to new, enhanced algorithms, Eduspace now includes:

HM Assess™ Houghton Mifflin's new diagnostic tool tests core concepts and provides students with individual study paths for self-remediation. Instructors can use HM Assess to quickly gauge which students are in jeopardy and identify which concepts to spend extra time reviewing.

Visit **www.eduspace.com** for more information.

SMARTHINKING® Live, Online Tutoring SMARTHINKING provides an easy-to-use and effective online, text-specific tutoring service. A dynamic **Whiteboard** and a **Graphing Calculator** function enable students and e-structors to collaborate easily.

Online Course Content for Blackboard®, WebCT®, and eCollege® Deliver program- or text-specific Houghton Mifflin content online using your institution's local course management system. Houghton Mifflin offers homework, tutorials, videos, and other resources formatted for Blackboard, WebCT, eCollege, and other course management systems. Add to an existing online course or create a new one by selecting from a wide range of powerful learning and instructional materials.

For more information, visit **college.hmco.com/pic/aufmannABI2e**
or contact your Houghton Mifflin sales representative.

AIM for Success

Welcome to *Algebra: Beginning and Intermediate*. As you begin this course, we know two important facts: (1) We want you to succeed. (2) You want to succeed. To do that requires an effort from each of us. For the next few pages, we are going to show you what is required of you to achieve that success and how you can use the features of this text to be successful.

Motivation

One of the most important keys to success is motivation. We can try to motivate you by offering interesting or important ways mathematics can benefit you. But, in the end, the motivation must come from you. On the first day of class, it is easy to be motivated. Eight weeks into the term, it is harder to keep that motivation.

To stay motivated, there must be outcomes from this course that are worth your time, money, and energy. List some reasons you are taking this course. Do not make a mental list—actually write them out.

> ### Take Note
>
> Motivation alone will not lead to success. For instance, suppose a person who cannot swim is placed in a boat, taken out to the middle of a lake, and then thrown overboard. That person has a lot of motivation but there is a high likelihood the person will drown without some help. Motivation gives us the desire to learn but is not the same as learning.

Although we hope that one of the reasons you listed was an interest in mathematics, we know that many of you are taking this course because it is required to graduate, it is a prerequisite for a course you must take, or it is required for your major. Although you may not agree that this course is necessary, it is! If you are motivated to graduate or complete the requirements for your major, then use that motivation to succeed in this course. Do not become distracted from your goal to complete your education!

Commitment

To be successful, you must make a commitment to succeed. This means devoting time to math so that you achieve a better understanding of the subject.

List some activities (sports, hobbies, talents such as dance, art, or music) that you enjoy and at which you would like to become better.

ACTIVITY	*TIME SPENT*	*TIME WISHED SPENT*

Thinking about these activities, put the number of hours that you spend each week practicing these activities next to the activity. Next to that number, indicate the number of hours per week you would like to spend on these activities.

Whether you listed surfing or sailing, aerobics or restoring cars, or any other activity you enjoy, note how many hours a week you spend doing it. To succeed in math, you must be willing to commit the same amount of time. Success requires some sacrifice.

The "I Can't Do Math" Syndrome

There may be things you cannot do, such as lift a two-ton boulder. You can, however, do math. It is much easier than lifting the two-ton boulder. When

you first learned the activities you listed above, you probably could not do them well. With practice, you got better. With practice, you will be better at math. Stay focused, motivated, and committed to success.

It is difficult for us to emphasize how important it is to overcome the "I Can't Do Math Syndrome." If you listen to interviews of very successful athletes after a particularly bad performance, you will note that they focus on the positive aspect of what they did, not the negative. Sports psychologists encourage athletes to always be positive—to have a "Can Do" attitude. Develop this attitude toward math.

Strategies for Success

Textbook Reconnaissance Right now, do a 15-minute "textbook reconnaissance" of this book. Here's how:

First, read the table of contents. Do it in three minutes or less. Next, look through the entire book, page by page. Move quickly. Scan titles, look at pictures, notice diagrams.

A textbook reconnaissance shows you where a course is going. It gives you the big picture. That's useful because brains work best when going from the general to the specific. Getting the big picture before you start makes details easier to recall and understand later on.

Your textbook reconnaissance will work even better if, as you scan, you look for ideas or topics that are interesting to you. List three facts, topics, or problems that you found interesting during your textbook reconnaissance.

The idea behind this technique is simple: It's easier to work at learning material if you know it's going to be useful to you.

Not all the topics in this book will be "interesting" to you. But that is true of any subject. Surfers find that on some days the waves are better than others, musicians find some music more appealing than other music, computer gamers find some computer games more interesting than others, and car enthusiasts find some cars more exciting than others. Some car enthusiasts would rather have a completely restored 1957 Chevrolet than a new Ferrari.

Know the Course Requirements To do your best in this course, you must know exactly what your instructor requires. Course requirements may be stated in a *syllabus*, which is a printed outline of the main topics of the course, or they may be presented orally. When they are listed in a syllabus or on other printed pages, keep them in a safe place. When they are presented orally, make sure to take complete notes. In either case, it is important that you understand them completely and follow them exactly. Be sure you know the answer to the following questions.

1. What is your instructor's name?
2. Where is your instructor's office?
3. At what times does your instructor hold office hours?
4. Besides the textbook, what other materials does your instructor require?
5. What is your instructor's attendance policy?
6. If you must be absent from a class meeting, what should you do before returning to class? What should you do when you return to class?
7. What is the instructor's policy regarding collection or grading of homework assignments?

8. What options are available if you are having difficulty with an assignment? Is there a math tutoring center?

9. Is there a math lab at your school? Where is it located? What hours is it open?

10. What is the instructor's policy if you miss a quiz?

11. What is the instructor's policy if you miss an exam?

12. Where can you get help when studying for an exam?

Remember: Your instructor wants to see you succeed. If you need help, ask! Do not fall behind. If you are running a race and fall behind by 100 yards, you may be able to catch up but it will require more effort than had you not fallen behind.

Time Management We know that there are demands on your time. Family, work, friends, and entertainment all compete for your time. We do not want to see you receive poor job evaluations because you are studying math. However, it is also true that we do not want to see you receive poor math test scores because you devoted too much time to work. When several competing and important tasks require your time and energy, the only way to manage the stress of being successful at both is to manage your time efficiently.

Instructors often advise students to spend twice the amount of time outside of class studying as they spend in the classroom. Time management is important if you are to accomplish this goal and succeed in school. The following activity is intended to help you structure your time more efficiently.

List the name of each course you are taking this term, the number of class hours each course meets, and the number of hours you should spend studying each subject outside of class. Then fill in a weekly schedule like the one below. Begin by writing in the hours spent in your classes, the hours spent at work (if you have a job), and any other commitments that are not flexible with respect to the time that you do them. Then begin to write down commitments that are more flexible, including hours spent studying. Remember to reserve time for activities such as meals and exercise. You should also schedule free time.

Take Note

Besides time management, there must be realistic ideas of how much time is available. There are very few people who can *successfully* work full-time and go to school full-time. If you work 40 hours a week, take 15 units, spend the recommended study time given at the right, and sleep 8 hours a day, you will use over 80% of the available hours in a week. That leaves less than 20% of the hours in a week for family, friends, eating, recreation, and other activities.

	Monday	Tuesday	Wednesday	Thursday	Friday	Saturday	Sunday
7–8 a.m.							
8–9 a.m.							
9–10 a.m.							
10–11 a.m.							
11–12 p.m.							
12–1 p.m.							
1–2 p.m.							
2–3 p.m.							
3–4 p.m.							
4–5 p.m.							
5–6 p.m.							
6–7 p.m.							
7–8 p.m.							
8–9 p.m.							
9–10 p.m.							
10–11 p.m.							
11–12 a.m.							

We know that many of you must work. If that is the case, realize that working 10 hours a week at a part-time job is equivalent to taking a three-unit class. If you must work, consider letting your education progress at a slower rate to allow you to be successful at both work and school. There is no rule that says you must finish school in a certain time frame.

Schedule Study Time As we encouraged you to do by filling out the time management form above, schedule a certain time to study. You should think of this time the way you would the time for work or class—that is, reasons for missing study time should be as compelling as reasons for missing work or class. "I just didn't feel like it" is not a good reason to miss your scheduled study time.

Although this may seem like an obvious exercise, list a few reasons you might want to study.

Of course we have no way of knowing the reasons you listed, but from our experience one reason given quite frequently is "To pass the course." There is nothing wrong with that reason. If that is the most important reason for you to study, then use it to stay focused.

One method of keeping to a study schedule is to form a ***study group.*** Look for people who are committed to learning, who pay attention in class, and who are punctual. Ask them to join your group. Choose people with similar educational goals but different methods of learning. You can gain insight from seeing the material from a new perspective. Limit groups to four or five people; larger groups are unwieldy.

There are many ways to conduct a study group. Begin with the following suggestions and see what works best for your group.

1. Test each other by asking questions. Each group member might bring two or three sample test questions to each meeting.

2. Practice teaching each other. Many of us who are teachers learned a lot about our subject when we had to explain it to someone else.

3. Compare class notes. You might ask other students about material in your notes that is difficult for you to understand.

4. Brainstorm test questions.

5. Set an agenda for each meeting. Set approximate time limits for each agenda item and determine a quitting time.

And finally, probably the most important aspect of studying is that it should be done in relatively small chunks. If you can only study three hours a week for this course (probably not enough for most people), do it in blocks of one hour on three separate days, preferably after class. Three hours of studying on a Sunday is not as productive as three hours of paced study.

Text Features That Promote Success

There are 11 chapters in this text. Each chapter is divided into sections, and each section is subdivided into learning objectives. Each learning objective is labeled with a number from 1 to 5.

Preparing for a Chapter Before you begin a new chapter, you should take some time to review previously learned skills. There are two ways to do this. The first is to complete the ***Cumulative Review,*** which occurs after every chapter (except Chapter 1). For instance, turn to page 206. The questions in this review are taken from the previous chapters. The answers for all these exercises can be found on page A10. Turn to that page now and locate the answers for the Chapter 2 Cumulative Review. After the answer to the first exercise, which is 6, you will see the objective reference [1.2.2]. This means that this question was taken from Chapter 1, Section 2, Objective 2. If you missed this question, you should return to that objective and restudy the material.

A second way of preparing for a new chapter is to complete the ***Prep Test.*** This test focuses on the particular skills that will be required for the new chapter. Turn to page 84 to see a Prep Test. The answers for the Prep Test are the first set of answers in the answer section for a chapter. Turn to page A5 to see the answers for the Chapter 2 Prep Test. Note that an objective reference is given for each question. If you answer a question incorrectly, restudy the objective from which the question was taken.

Before the class meeting in which your professor begins a new section, you should read each objective statement for that section. Next, browse through the objective material, being sure to note each word in bold type. These words indicate important concepts that you must know in order to learn the material. Do not worry about trying to understand all the material. Your professor is there to assist you with that endeavor. The purpose of browsing through the material is so that your brain will be prepared to accept and organize the new information when it is presented to you.

Turn to page 2. Write down the title of the first objective in Section 1.1. Under the title of the objective, write down the words in the objective that are in bold print. It is not necessary for you to understand the meaning of these words. You are in this class to learn their meaning.

_____ _____ _____ _____

_____ _____ _____ _____

_____ _____ _____ _____

_____ _____ _____ _____

_____ _____ _____ _____

Math Is Not a Spectator Sport To learn mathematics you must be an active participant. Listening and watching your professor do mathematics is not enough. Mathematics requires that you interact with the lesson you are studying. If you filled in the blanks above, you were being interactive. There are other ways this textbook has been designed to help you be an active learner.

Instructional Examples Blue lines indicate an instructional example that usually has explanatory remarks to the left of the solution. These examples are used for two purposes. The first is to provide additional examples of important concepts. Secondly, these examples examine important techniques or principles that are often used in the solution of other types of problems.

$$y + \frac{1}{2} = \frac{5}{4}$$
$$y + \frac{1}{2} - \frac{1}{2} = \frac{5}{4} - \frac{1}{2}$$
$$y + 0 = \frac{5}{4} - \frac{2}{4}$$
$$y = \frac{3}{4}$$

The solution is $\frac{3}{4}$.

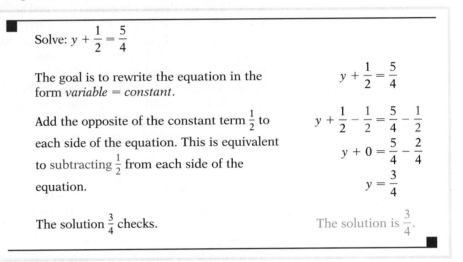

Solve: $y + \frac{1}{2} = \frac{5}{4}$

The goal is to rewrite the equation in the form *variable = constant*.

Add the opposite of the constant term $\frac{1}{2}$ to each side of the equation. This is equivalent to subtracting $\frac{1}{2}$ from each side of the equation.

The solution $\frac{3}{4}$ checks.

$$y + \frac{1}{2} = \frac{5}{4}$$
$$y + \frac{1}{2} - \frac{1}{2} = \frac{5}{4} - \frac{1}{2}$$
$$y + 0 = \frac{5}{4} - \frac{2}{4}$$
$$y = \frac{3}{4}$$

The solution is $\frac{3}{4}$.

page 86

When you complete the example, get a clean sheet of paper. Write down the problem and then try to complete the solution without referring to your notes or the book. When you can do that, move on to the next part of the objective.

Leaf through the book now and write down the page numbers of two other occurrences of an instructional example.

Example/Problem Pairs One of the key instructional features of this text is Example/Problem pairs. Note that each Example is completely worked out and the "Problem" following the example is not. Study the worked-out example carefully by working through each step. Then work the Problem. If you get stuck, refer to the page number following the Problem which directs you to the page on which the Problem is solved—a complete worked-out solution is provided. Try to use the given solution to get a hint for the step you are stuck on. Then try to complete your solution.

Problem 4:

$$-8 = 5 + x$$
$$-8 - 5 = 5 - 5 + x$$
$$-13 = x$$

The solution is -13.

EXAMPLE 4 Solve: $\frac{1}{2} = x + \frac{2}{3}$

Solution $\frac{1}{2} = x + \frac{2}{3}$

$$\frac{1}{2} - \frac{2}{3} = x + \frac{2}{3} - \frac{2}{3}$$ • Subtract $\frac{2}{3}$ from each side of the equation.

$$-\frac{1}{6} = x$$ • Simplify each side of the equation.

The solution is $-\frac{1}{6}$.

PROBLEM 4 Solve: $-8 = 5 + x$

Solution See page S4.

When you have completed your solution, check your work against the solution we provided. (Turn to page S4 to see the solution of Problem 4.) Be aware that frequently there is more than one way to solve a problem. Your answer, however, should be the same as the given answer. If you have any question as to whether your method will "always work," check with your instructor or with someone in the math center.

Browse through the textbook and write down the page numbers where two other Example/Problem pairs occur.

Remember: Be an active participant in your learning process. When you are sitting in class watching and listening to an explanation, you may think that you understand. However, until you actually try to do it, you will have no confirmation of the new knowledge or skill. Most of us have had the experience of sitting in class thinking we knew how to do something only to get home and realize that we didn't.

Word Problems Word problems are difficult because we must read the problem, determine the quantity we must find, think of a method to do that, and then actually solve the problem. In short, we must formulate a *strategy* to solve the problem and then devise a *solution*.

Note in the Example/Problem pair below that both a strategy and a solution are provided for the word problem. The strategy is a written description of how we will solve the problem. In the corresponding Problem, you are asked to formulate a strategy. Do not skip this step, and be sure to write it out.

Take Note

There is a strong connection between reading and being a successful student in math or in any other subject. If you have difficulty reading, consider taking a reading course. Reading is much like other skills. There are certain things you can learn that will make you a better reader.

Problem 9:

Strategy:

To find the percent of questions answers correctly, solve the basic percent equation using $B = 80$ and $A = 72$.

Solution:
$$PB = A$$
$$P(80) = 72$$
$$80P = 72$$
$$\frac{80P}{80} = \frac{72}{80}$$
$$P = 0.9$$

90% of the questions were answered correctly.

Take Note

If a rule has more than one part, be sure to make a notation to that effect.

> **EXAMPLE 9** During a recent year, nearly 1.2 million dogs or litters were registered with the American Kennel Club. The most popular breed was the Labrador retriever, with 172,841 registered. What percent of the registrations were Labrador retrievers? Round to the nearest tenth of a percent. (*Source:* American Kennel Club)
>
> **Strategy** To find the percent, solve the basic percent equation using $B = 1.2$ million $= 1,200,000$ and $A = 172,841$. The percent is unknown.
>
> **Solution**
> $$PB = A$$
> $$P(1,200,000) = 172,841$$
> $$\frac{P(1,200,000)}{1,200,000} = \frac{172,841}{1,200,000}$$
> $$P \approx 0.144$$
> $$P \approx 14.4\%$$
>
> Approximately 14.4% of the registrations were Labrador retrievers.
>
> **PROBLEM 9** A student correctly answered 72 of the 80 questions on an exam. What percent of the questions were answered correctly?
>
> **Solution** See page S5.

page 91

Rule Boxes Pay special attention to rules placed in boxes. These rules give you the reasons certain types of problems are solved the way they are. When you see a rule, try to rewrite the rule in your own words.

Multiplying each side of an equation by the same number (NOT 0) will not change the solution.

Multiplication Property of Equations

Each side of an equation can be multiplied by the same nonzero number without changing the solution of the equation.

page 87

Find and write down two page numbers on which there are rule boxes.

Chapter Exercises When you have completed studying an objective, do the exercises in the exercise set that correspond with that objective. The exercises are labeled with the same number as the objective. Math is a subject that needs to be learned in small sections and practiced continually in order to be mastered. Doing all of the exercises in each exercise set will help you master the problem-solving techniques necessary for success. As you work through the exercises for an objective, check your answers to the odd-numbered exercises with those in the back of the book.

Preparing for a Test There are important features of this text that can be used to prepare for a test.

- Chapter Summary
- Chapter Review Exercises
- Chapter Test

After completing a chapter, read the Chapter Summary. (See page 197 for the Chapter 2 Summary.) This summary highlights the important topics covered in the chapter. The page number following each topic refers you to the page in the text on which you can find more information about the concept.

Following the Chapter Summary are Chapter Review Exercises (see page 202) and a Chapter Test (see page 204). Doing the review exercises is an important way of testing your understanding of the chapter. The answer to each review exercise is given at the back of the book, along with its objective reference. After checking your answers, restudy any objective from which a question you missed was taken. It may be helpful to retry some of the exercises for that objective to reinforce your problem-solving techniques.

The Chapter Test should be used to prepare for an exam. We suggest that you try the Chapter Test a few days before your actual exam. Take the test in a quiet place and try to complete the test in the same amount of time you will be allowed for your exam. When taking the Chapter Test, practice the strategies of successful test takers: (1) scan the entire test to get a feel for the questions; (2) read the directions carefully; (3) work the problems that are easiest for you first; and perhaps most importantly, (4) try to stay calm.

When you have completed the Chapter Test, check your answers. If you missed a question, review the material in that objective and rework some of the exercises from that objective. This will strengthen your ability to perform the skills in that objective.

Your career goal goes here. → Is it difficult to be successful? YES! Successful music groups, artists, professional athletes, chefs, and _____ have to work very hard to achieve their goals. They focus on their goals and ignore distractions. The things we ask you to do to achieve success take time and commitment. We are confident that if you follow our suggestions, you will succeed.

1 Real Numbers and Variable Expressions

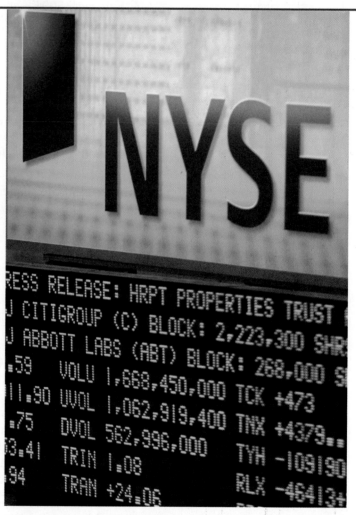

How do investment bankers know which shares of stock to buy or sell? They decide based on each stock's price and whether it is likely to increase or decrease over time. Positive or negative numbers are used to express the amount of change in the price of a share of stock. Stock market analysts calculate the moving average of a stock, which averages the changes in a stock's price over several days. The moving average is used every day to discover trends in the movement of the price and to create investment strategies. **Exercises 156 and 157 on pages 22–23** involve calculating moving averages for different stocks.

DVD **SSM**

Online Study Center

Need help? For online student resources, such as section quizzes, visit the Online Study Center at college.hmco.com/pic/aufmannABI2e.

Prep Test

1. What is 127.1649 rounded to the nearest hundredth?

2. Add: $49{,}147 + 596$

3. Subtract: $5004 - 487$

4. Multiply: 407×28

5. Divide: $456 \div 19$

6. What is the smallest number into which both 8 and 12 divide evenly?

7. What is the greatest number that divides evenly into both 16 and 20?

8. Without using 1, write 21 as a product of two whole numbers.

9. Represent the shaded portion of the figure as a fraction.

1.1 Introduction to Integers

OBJECTIVE 1 Order relations

It seems to be a human characteristic to group similar items. For instance, a botanist places plants with similar characteristics in groups called species. Nutritionists classify foods according to food groups; for example, pasta, crackers, and rice are among the foods in the bread group.

Mathematicians place objects with similar properties in groups called sets. A **set** is a collection of objects. The objects in a set are called **elements** of the set.

The **roster method** of writing sets encloses a list of the elements in braces. The set of sections within an orchestra is written {brass, percussion, strings, woodwinds}.

When the elements of a set are listed, each element is listed only once. For instance, if the list of numbers 1, 2, 3, 2, 3 were placed in a set, the set would be {1, 2, 3}.

The numbers that we use to count objects, such as the number of students in a classroom or the number of people living in an apartment house, are the natural numbers.

Natural numbers = {1, 2, 3, 4, 5, 6, 7, 8, 9, 10, . . .}

The three dots mean that the list of natural numbers continues on and on and that there is no largest natural number.

The natural numbers do not include zero. The number zero is used for describing data such as the number of people who have run a two-minute mile and the number of college students at Providence College who are under the age of 10. The set of whole numbers includes the natural numbers and zero.

$$\textbf{Whole numbers} = \{0, 1, 2, 3, 4, 5, 6, 7, \dots\}$$

The whole numbers do not provide all the numbers that are useful in applications. For instance, a meteorologist also needs numbers below zero.

$$\textbf{Integers} = \{\dots, -5, -4, -3, -2, -1, 0, 1, 2, 3, 4, 5, \dots\}$$

Each integer can be shown on a number line. The integers to the left of zero on the number line are called **negative integers.** The integers to the right of zero are called **positive integers,** or natural numbers. Zero is neither a positive nor a negative integer.

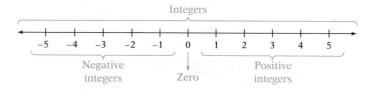

The **graph of an integer** is shown by placing a heavy dot on the number line directly above the number. The graphs of -3 and 4 are shown on the number line below.

Consider the following sentences.

> The quarterback threw the football, and the receiver caught *it*.

> A student purchased a computer and used *it* to write English and history papers.

In the first sentence, *it* is used to mean the football; in the second sentence, *it* means the computer. In language, the word *it* can stand for many different objects. Similarly, in mathematics, a letter of the alphabet can be used to stand for a number. Such a letter is called a **variable.** Variables are used in the following definition of inequality symbols.

Definition of Inequality Symbols

If a and b are two numbers and a is to the left of b on the number line, then a **is less than** b. This is written $a < b$.

If a and b are two numbers and a is to the right of b on the number line, then a **is greater than** b. This is written $a > b$.

Negative 4 is less than negative 1.

$-4 < -1$

5 is greater than 0.

$5 > 0$

There are also inequality symbols for **is less than or equal to** ($\leq$) and **is greater than or equal to** ($\geq$).

$7 \leq 15$ 7 is less than or equal to 15. $6 \leq 6$ 6 is less than or equal to 6.
This is true because $7 < 15$. This is true because $6 = 6$.

EXAMPLE 1 Use the roster method to write the set of negative integers greater than or equal to -6.

 Solution $A = \{-6, -5, -4, -3, -2, -1\}$ • **A set is designated by a capital letter.**

PROBLEM 1 Use the roster method to write the set of positive integers less than 5.

 Solution See page S1.

EXAMPLE 2 Given $A = \{-6, -2, 0\}$, which elements of set A are less than or equal to -2?

 Solution $-6 < -2$ • **Find the order relation between each element of set A and -2.**

$-2 = -2$

$0 > -2$

The elements -6 and -2 are less than or equal to -2.

PROBLEM 2 Given $B = \{-5, -1, 5\}$, which elements of set B are greater than -1?

 Solution See page S1.

OBJECTIVE 2 **Opposites and absolute value**

Two numbers that are the same distance from zero on the number line but are on opposite sides of zero are **opposite numbers**, or **opposites**. The opposite of a number is also called its **additive inverse**.

The opposite of 5 is -5.

The opposite of -5 is 5.

The negative sign can be read "the opposite of."

$$-(2) = -2$$ The opposite of 2 is -2.

$$-(-2) = 2$$ The opposite of -2 is 2.

EXAMPLE 3 Find the opposite number. **A.** 6 **B.** -51

 Solution **A.** -6 **B.** 51

| **PROBLEM 3** Find the opposite number. **A.** -9 **B.** 62

> **Solution** See page S1.

The **absolute value** of a number is its distance from zero on the number line. Therefore, the absolute value of a number is a positive number or zero. The symbol for absolute value is two vertical bars, $|\,|$.

The distance from 0 to 3 is 3. Therefore, the absolute value of 3 is 3.

$$|3| = 3$$

The distance from 0 to -3 is 3. Therefore, the absolute value of -3 is 3.

$$|-3| = 3$$

Point of Interest

The definition of absolute value given in the box is written in what is called rhetorical style. That is, it is written without the use of variables. This is how all mathematics was written prior to the Renaissance. During that period from the 14th to the 16th century, the idea of expressing a variable symbolically was developed. In terms of that symbolism, the definition of absolute value is

$$|x| = \begin{cases} x, & x > 0 \\ 0, & x = 0 \\ -x, & x < 0 \end{cases}$$

Absolute Value

The absolute value of a positive number is the number itself. The absolute value of zero is zero. The absolute value of a negative number is the opposite of the negative number.

| **EXAMPLE 4** Evaluate. **A.** $|-4|$ **B.** $-|-10|$

> **Solution** **A.** $|-4| = 4$
>
> **B.** $-|-10| = -10$ • The absolute value sign does not affect the negative sign in front of the absolute value sign. You can read $-|-10|$ as "the opposite of the absolute value of negative 10."

| **PROBLEM 4** Evaluate. **A.** $|-5|$ **B.** $-|-9|$

> **Solution** See page S1.

1.1 Concept Review

Determine whether the statement is always true, sometimes true, or never true.

1. The absolute value of a number is positive.

2. The absolute value of a number is negative.

3. If x is an integer, then $|x| < -3$.

4. If x is an integer, then $|x| > -2$.

5. The opposite of a number is a positive number.

6. The set of positive integers is $\{0, 1, 2, 3, 4, \ldots\}$.

7. If a is an integer, then $a \leq a$.

8. If a and b are integers and $a > b$, then $a \geq b$.

9. If x is a negative integer, then $|x| = -x$.

10. Suppose a and b are integers and $a < b$. Then $|a| < |b|$.

1.1 Exercises

1 Order relations

1. How do the whole numbers differ from the natural numbers?

2. Explain the difference between the symbols $<$ and $\leq$.

3. Another name for the set of positive integers is the set of ___?___ numbers.

4. The inequality $-5 < -1$ is read "negative five ___?___ negative one."

5. The inequality $0 \geq -4$ is read "zero ___?___ negative four."

6. Which number is greater, -10 or -3?

Place the correct symbol, $<$ or $>$, between the two numbers.

7. -2 -5	**8.** -6 -1	**9.** -16 1	**10.** -2 13
11. 3 -7	**12.** 5 -6	**13.** 0 -3	**14.** 8 0
15. -42 27	**16.** -36 49	**17.** 21 -34	**18.** 53 -46
19. -27 -39	**20.** -51 -20	**21.** -131 101	**22.** 127 -150

23. A number n is to the right of the number 5 on the number line. Which of the following is true?
a. n is positive. **b.** n is negative. **c.** n is 0.
d. n can be positive, negative, or 0.

24. A number n is to the left of the number 5 on the number line. Which of the following is true?
a. n is positive. **b.** n is negative. **c.** n is 0.
d. n can be positive, negative, or 0.

25. Do the inequalities $6 \geq 1$ and $1 \leq 6$ express the same order relation?

26. Use the symbol $\leq$ to rewrite the order relation expressed by the inequality $-2 \geq -5$.

Use the roster method to write the set.

27. the natural numbers less than 9

28. the natural numbers less than or equal to 6

29. the positive integers less than or equal to 8

30. the positive integers less than 4

31. the negative integers greater than -7

32. the negative integers greater than or equal to -5

Solve.

33. Given $A = \{-7, 0, 2, 5\}$, which elements of set A are greater than 2?

34. Given $B = \{-8, 0, 7, 15\}$, which elements of set B are greater than 7?

35. Given $D = \{-23, -18, -8, 0\}$, which elements of set D are less than -8?

36. Given $C = \{-33, -24, -10, 0\}$, which elements of set C are less than -10?

37. Given $E = \{-35, -13, 21, 37\}$, which elements of set E are greater than -10?

38. Given $F = \{-27, -14, 14, 27\}$, which elements of set F are greater than -15?

39. Given $B = \{-52, -46, 0, 39, 58\}$, which elements of set B are less than or equal to 0?

40. Given $A = \{-12, -9, 0, 12, 34\}$, which elements of set A are greater than or equal to 0?

41. Given $C = \{-23, -17, 0, 4, 29\}$, which elements of set C are greater than or equal to -17?

42. Given $D = \{-31, -12, 0, 11, 45\}$, which elements of set D are less than or equal to -12?

43. Given that set A is the positive integers less than 10, which elements of set A are greater than or equal to 5?

44. Given that set B is the positive integers less than or equal to 12, which elements of set B are greater than 6?

45. Given that set D is the negative integers greater than or equal to -10, which elements of set D are less than -4?

46. Given that set C is the negative integers greater than -8, which elements of set C are less than or equal to -3?

2 **Opposites and absolute value**

47. What is the additive inverse of a number?

48. What is the absolute value of a number?

49. The opposite of a negative number is a _____?_____ number.

50. The additive inverse of 7 is _____?_____.

51. The equation $|-5| = 5$ is read "the _____?_____ of negative five is five."

52. Write the statement "the opposite of negative nine is nine" in symbols.

Find the opposite number.

53. 22 **54.** 45 **55.** −31 **56.** −88

57. −168 **58.** −97 **59.** 630 **60.** 450

Evaluate.

61. −(−18) **62.** −(−30) **63.** −(49) **64.** −(67)

65. |16| **66.** |19| **67.** |−12| **68.** |−22|

69. −|29| **70.** −|20| **71.** −|−14| **72.** −|−18|

73. −|0| **74.** |−30| **75.** −|34| **76.** −|−45|

Solve.

77. Given $A = \{-8, -5, -2, 1, 3\}$, find
 a. the opposite of each element of set A.
 b. the absolute value of each element of set A.

78. Given $B = \{-11, -7, -3, 1, 5\}$, find
 a. the opposite of each element of set B.
 b. the absolute value of each element of set B.

79. True or false? The absolute value of a negative number n is greater than n.

80. A number n is positive. In order to make the statement $|n| \underline{\quad ? \quad} n$ a true statement, which symbol must replace the question mark?
 a. > **b.** < **c.** ≤ **d.** =

Place the correct symbol, < or >, between the two numbers.

81. |−83| |58| **82.** |22| |−19| **83.** |43| |−52| **84.** |−71| |−92|

85. |−68| |−42| **86.** |12| |−31| **87.** |−45| |−61| **88.** |−28| |43|

1.1 Applying Concepts

Write the given numbers in order from least to greatest.

89. |−5|, 6, −|−8|, −19

90. −4, |−15|, −|−7|, 0

91. −(−3), −22, |−25|, |−14|

92. −|−26|, −(−8), |−17|, −(5)

Meteorology A meteorologist may report a wind-chill temperature. This is the equivalent temperature, including the effects of wind and temperature, that a person would feel in calm air conditions. The table on the next page gives the wind-chill temperature for various wind speeds and

temperatures. For instance, when the temperature is 5°F and the wind is blowing at 15 mph, the wind-chill temperature is −13°F. Use this table for Exercises 93 and 94.

Wind-Chill Factors															
Wind Speed (in mph)	Thermometer Reading (in degrees Fahrenheit)														
	25	20	15	10	5	0	−5	−10	−15	−20	−25	−30	−35	−40	−45
5	19	13	7	1	−5	−11	−16	−22	−28	−34	−40	−46	−52	−57	−63
10	15	9	3	−4	−10	−16	−22	−28	−35	−41	−47	−53	−59	−66	−72
15	13	6	0	−7	−13	−19	−26	−32	−39	−45	−51	−58	−64	−71	−77
20	11	4	−2	−9	−15	−22	−29	−35	−42	−48	−55	−61	−68	−74	−81
25	9	3	−4	−11	−17	−24	−31	−37	−44	−51	−58	−64	−71	−78	−84
30	8	1	−5	−12	−19	−26	−33	−39	−46	−53	−60	−67	−73	−80	−87
35	7	0	−7	−14	−21	−27	−34	−41	−48	−55	−62	−69	−76	−82	−89
40	6	−1	−8	−15	−22	−29	−36	−43	−50	−57	−64	−71	−78	−84	−91
45	5	−2	−9	−16	−23	−30	−37	−44	−51	−58	−65	−72	−79	−86	−93

93. Which of the following weather conditions feels colder?
 a. a temperature of 5°F with a 20 mph wind or a temperature of 10°F with a 15 mph wind
 b. a temperature of −25°F with a 10 mph wind or a temperature of −15°F with a 20 mph wind

94. Which of the following weather conditions feels warmer?
 a. a temperature of 5°F with a 25 mph wind or a temperature of 10°F with a 10 mph wind
 b. a temperature of −5°F with a 10 mph wind or a temperature of −15°F with a 5 mph wind

Complete.

95. On the number line, the two points that are four units from 0 are _____ and _____ .

96. On the number line, the two points that are six units from 0 are _____ and _____ .

97. On the number line, the two points that are seven units from 4 are _____ and _____ .

98. On the number line, the two points that are five units from −3 are _____ and _____ .

99. If a is a positive number, then $-a$ is a _____ number.

100. If a is a negative number, then $-a$ is a _____ number.

101. An integer that is its own opposite is _____ .

102. True or false? If $a \geq 0$, then $|a| = a$.

103. True or false? If $a \leq 0$, then $|a| = -a$.

104. Give examples of some games that use negative integers in the scoring.

105. Give some examples of English words that are used as variables.

106. Write an essay describing your interpretation of the data presented in the table at the right.

The Twenty-First Century: An Easier Ladder to Climb Number of 30- to 34-year-olds competing for jobs			
	1992	*1999*	*2005*
Men	10.3 million	9.2 million	8.5 million
Women	8.3 million	7.9 million	7.8 million
Total	18.6 million	17.1 million	16.3 million

1.2 Operations with Integers

OBJECTIVE **1** Add integers

A number can be represented anywhere along the number line by an arrow. A positive number is represented by an arrow pointing to the right, and a negative number is represented by an arrow pointing to the left. The size of the number is represented by the length of the arrow.

Addition is the process of finding the total of two numbers. The numbers being added are called **addends**. The total is called the **sum**. Addition of integers can be shown on the number line. To add integers, find the point on the number line corresponding to the first addend. At that point, draw an arrow representing the second addend. The sum is the number directly below the tip of the arrow.

$$4 + 2 = 6$$

$$-4 + (-2) = -6$$

$$-4 + 2 = -2$$

$$4 + (-2) = 2$$

The pattern for addition shown on the preceding number lines is summarized in the following rules for adding integers.

Addition of Integers

Integers with the same sign
To add two numbers with the same sign, add the absolute values of the numbers. Then attach the sign of the addends.

$$2 + 8 = 10$$
$$-2 + (-8) = -10$$

Integers with different signs
To add two numbers with different signs, find the absolute value of each number. Then subtract the smaller of these absolute values from the larger one. Attach the sign of the number with the larger absolute value.

$$-2 + 8 = 6$$
$$2 + (-8) = -6$$

EXAMPLE 1 Add. **A.** $162 + (-247)$ **B.** $-14 + (-47)$

C. $-4 + (-6) + (-8) + 9$

Solution **A.** $162 + (-247) = -85$ • The signs are different. Subtract the absolute values of the numbers $(247 - 162)$. Attach the sign of the number with the larger absolute value.

B. $-14 + (-47) = -61$ • The signs are the same. Add the absolute values of the numbers $(14 + 47)$. Attach the sign of the addends.

C. $-4 + (-6) + (-8) + 9$ • To add more than two numbers, add the
$= -10 + (-8) + 9$ first two numbers. Then add the sum to
$= -18 + 9$ the third number. Continue until all the
$= -9$ numbers have been added.

PROBLEM 1 Add.

A. $-162 + 98$ **B.** $-154 + (-37)$ **C.** $-36 + 17 + (-21)$

Solution See page S1.

OBJECTIVE **2** Subtract integers

Subtraction is the process of finding the difference between two numbers. Subtraction of an integer is defined as addition of the opposite integer.

Subtract $8 - 3$ by using addition of the opposite.

Subtraction → Addition of the Opposite

$$8 \; - \; (+3) \; = \; 8 \; + \; (-3) \; = \; 5$$

└── Opposites ──┘

To subtract one integer from another, add the opposite of the second integer to the first integer.

first number	−	second number	=	first number	+	the opposite of the second number		
40	−	60	=	40	+	(-60)	=	-20
-40	−	60	=	-40	+	(-60)	=	-100
-40	−	(-60)	=	-40	+	60	=	20
40	−	(-60)	=	40	+	60	=	100

EXAMPLE 2 Subtract: $-12 - 8$

Solution $-12 - 8 = -12 + (-8)$ • Rewrite subtraction as addition of the opposite.

$\qquad\qquad = -20$ • Add.

PROBLEM 2 Subtract: $-8 - 14$

Solution See page S1.

When subtraction occurs several times in a problem, rewrite each subtraction as addition of the opposite. Then add.

EXAMPLE 3 Subtract: $-8 - 30 - (-12) - 7 - (-14)$

Solution $-8 - 30 - (-12) - 7 - (-14)$
$\qquad = -8 + (-30) + 12 + (-7) + 14$ • Rewrite each subtraction as addition of the opposite.

$\qquad = -38 + 12 + (-7) + 14$ • Add the first two numbers. Then add the sum to the third number. Continue until all the numbers have been added.
$\qquad = -26 + (-7) + 14$
$\qquad = -33 + 14$
$\qquad = -19$

PROBLEM 3 Subtract: $4 - (-3) - 12 - (-7) - 20$

Solution See page S1.

OBJECTIVE **3**

Multiply integers

Multiplication is the process of finding the product of two numbers.

Several different symbols are used to indicate multiplication. The numbers being multiplied are called **factors;** for instance, 3 and 2 are factors in each of the examples at the right. The result is called the **product.** Note that when parentheses are used and there is no arithmetic operation symbol, the operation is multiplication.

$$3 \times 2 = 6$$
$$3 \cdot 2 = 6$$
$$(3)(2) = 6$$
$$3(2) = 6$$
$$(3)2 = 6$$

Point of Interest

The cross × was first used as a symbol for multiplication in 1631 in a book titled The Key to Mathematics. *In that same year, another book,* Practice of the Analytical Art, *advocated the use of a dot to indicate multiplication.*

When 5 is multiplied by a sequence of decreasing integers, each product decreases by 5.

$$(5)(3) = 15$$
$$(5)(2) = 10$$
$$(5)(1) = 5$$
$$(5)(0) = 0$$

The pattern developed can be continued so that 5 is multiplied by a sequence of negative numbers. The resulting products must be negative in order to maintain the pattern of decreasing by 5.

$$(5)(-1) = -5$$
$$(5)(-2) = -10$$
$$(5)(-3) = -15$$
$$(5)(-4) = -20$$

This illustrates that the product of a positive number and a negative number is negative.

When −5 is multiplied by a sequence of decreasing integers, each product increases by 5.

$$(-5)(3) = -15$$
$$(-5)(2) = -10$$
$$(-5)(1) = -5$$
$$(-5)(0) = 0$$

The pattern developed can be continued so that −5 is multiplied by a sequence of negative numbers. The resulting products must be positive in order to maintain the pattern of increasing by 5.

$$(-5)(-1) = 5$$
$$(-5)(-2) = 10$$
$$(-5)(-3) = 15$$
$$(-5)(-4) = 20$$

This illustrates that the product of two negative numbers is positive.

This pattern for multiplication is summarized in the following rules for multiplying integers.

Multiplication of Integers

Integers with the same sign
To multiply two numbers with the same sign, multiply the absolute values of the numbers. The product is positive.

$$4 \cdot 8 = 32$$
$$(-4)(-8) = 32$$

Integers with different signs
To multiply two numbers with different signs, multiply the absolute values of the numbers. The product is negative.

$$-4 \cdot 8 = -32$$
$$(4)(-8) = -32$$

EXAMPLE 4 Multiply. **A.** $-42 \cdot 62$ **B.** $2(-3)(-5)(-7)$

Solution **A.** $-42 \cdot 62$
$$= -2604$$

• **The signs are different. The product is negative.**

B. $2(-3)(-5)(-7)$
$$= -6(-5)(-7)$$
$$= 30(-7)$$
$$= -210$$

• **To multiply more than two numbers, multiply the first two numbers. Then multiply the product by the third number. Continue until all the numbers have been multiplied.**

PROBLEM 4 Multiply. **A.** $-38 \cdot 51$ **B.** $-7(-8)(9)(-2)$

Solution See page S1.

OBJECTIVE 4 **Divide integers**

For every division problem there is a related multiplication problem.

$$\text{Division: } \frac{8}{2} = 4 \qquad \text{Related multiplication: } 4 \cdot 2 = 8$$

This fact can be used to illustrate the rules for dividing signed numbers.

The quotient of two numbers with the same sign is positive.

$$\frac{12}{3} = 4 \text{ because } 4 \cdot 3 = 12.$$

$$\frac{-12}{-3} = 4 \text{ because } 4(-3) = -12.$$

The quotient of two numbers with different signs is negative.

$$\frac{12}{-3} = -4 \text{ because } -4(-3) = 12.$$

$$\frac{-12}{3} = -4 \text{ because } -4 \cdot 3 = -12.$$

Division of Integers

Integers with the same sign
To divide two numbers with the same sign, divide the absolute values of the numbers. The quotient is positive.

Integers with different signs
To divide two numbers with different signs, divide the absolute values of the numbers. The quotient is negative.

Note that $\frac{-12}{3} = -4$, $\frac{12}{-3} = -4$, and $-\frac{12}{3} = -4$. This suggests the following rule.

If a and b are two integers, and $b \neq 0$, then $\frac{a}{-b} = \frac{-a}{b} = -\frac{a}{b}$.

Read $b \neq 0$ as "b is not equal to 0." The reason why the denominator must not be equal to 0 is explained in the following discussion of zero and one in division.

Zero and One in Division

Zero divided by any number other than zero is zero.

$$\frac{0}{a} = 0, a \neq 0$$ because $0 \cdot a = 0$.

Division by zero is not defined.

$$\frac{4}{0} = ?$$ $? \times 0 = 4$
There is no number whose product with zero is 4.

Any number other than zero divided by itself is 1.

$$\frac{a}{a} = 1, a \neq 0$$ because $1 \cdot a = a$.

Any number divided by 1 is the number.

$$\frac{a}{1} = a$$ because $a \cdot 1 = a$.

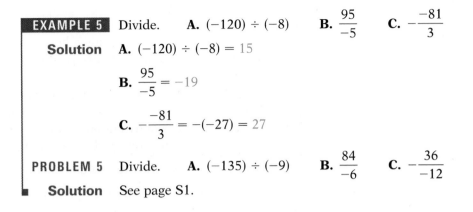

EXAMPLE 5 Divide. **A.** $(-120) \div (-8)$ **B.** $\dfrac{95}{-5}$ **C.** $-\dfrac{-81}{3}$

Solution **A.** $(-120) \div (-8) = 15$

B. $\dfrac{95}{-5} = -19$

C. $-\dfrac{-81}{3} = -(-27) = 27$

PROBLEM 5 Divide. **A.** $(-135) \div (-9)$ **B.** $\dfrac{84}{-6}$ **C.** $-\dfrac{36}{-12}$

Solution See page S1.

OBJECTIVE 5 ## Application problems

In many courses, your course grade depends on the *average* of all your test scores. You compute the average by calculating the sum of all your test scores and then dividing that result by the number of tests. Statisticians call this average an **arithmetic mean.** Besides its application to finding the average of your test scores, the arithmetic mean is used in many other situations.

Stock market analysts calculate the **moving average** of a stock. This is the arithmetic mean of the changes in the value of a stock for a given number of days. To illustrate the procedure, we will calculate the 5-day moving average of a stock. In actual practice, a stock market analyst may use 15 days, 30 days, or some other number.

The table below shows the amount of increase or decrease, in cents, in the closing price of a stock for a 10-day period.

Day 1	Day 2	Day 3	Day 4	Day 5	Day 6	Day 7	Day 8	Day 9	Day 10
+50	−175	+225	0	−275	−75	−50	+50	−475	−50

To calculate the 5-day moving average of this stock, determine the average of the stock for days 1 through 5, days 2 through 6, days 3 through 7, and so on.

Days 1–5	Days 2–6	Days 3–7	Days 4–8	Days 5–9	Days 6–10
+50	−175	+225	0	−275	−75
−175	+225	0	−275	−75	−50
+225	0	−275	−75	−50	+50
0	−275	−75	−50	+50	−475
−275	−75	−50	+50	−475	−50
Sum = −175	Sum = −300	Sum = −175	Sum = −350	Sum = −825	Sum = −600
Average = $\dfrac{-175}{5}$	Average = $\dfrac{-300}{5}$	Average = $\dfrac{-175}{5}$	Average = $\dfrac{-350}{5}$	Average = $\dfrac{-825}{5}$	Average = $\dfrac{-600}{5}$
= −35	= −60	= −35	= −70	= −165	= −120

The 5-day moving average is the list of means: −35, −60, −35, −70, −165, and −120. If the numbers in the list tend to increase, the price of the stock is

showing an upward trend; if they decrease, the price of the stock is showing a downward trend. For the stock shown here, the price of the stock is showing a downward trend. These trends help analysts determine whether to recommend a stock.

EXAMPLE 6 The daily low temperatures, in degrees Celsius, during one week were recorded as follows: $-8°$, $2°$, $0°$, $-7°$, $1°$, $6°$, $-1°$. Find the average daily low temperature for the week.

Strategy To find the average daily low temperature:

▶ Add the seven temperature readings.

▶ Divide the sum by 7.

Solution
$$-8 + 2 + 0 + (-7) + 1 + 6 + (-1)$$
$$= -6 + 0 + (-7) + 1 + 6 + (-1)$$
$$= -6 + (-7) + 1 + 6 + (-1)$$
$$= -13 + 1 + 6 + (-1)$$
$$= -12 + 6 + (-1)$$
$$= -6 + (-1)$$
$$= -7$$

$$-7 \div 7 = -1$$

The average daily low temperature was $-1°C$.

PROBLEM 6 The daily high temperatures, in degrees Celsius, during one week were recorded as follows: $-5°$, $-6°$, $3°$, $0°$, $-4°$, $-7°$, $-2°$. Find the average daily high temperature for the week.

Solution See page S1.

1.2 Concept Review

Determine whether the statement is always true, sometimes true, or never true.

1. The sum of two integers is larger than either of the integers being added.

2. The sum of two nonzero integers with the same sign is positive.

3. The quotient of two integers with different signs is negative.

4. To find the opposite of a number, multiply the number by -1.

5. To find the opposite of an integer, divide the integer by -1.

6. To perform the subtraction $a - b$, add the opposite of a to b.

7. $4(-8) = -4$

8. If x and y are two integers and $y = 0$, then $\frac{x}{y} = 0$.

9. If x and y are two different integers, then $x - y = y - x$.

10. If x is an integer and $4x = 0$, then $x = 0$.

1.2 Exercises

1 Add integers

1. Explain how to add two integers with the same sign.

2. Explain how to add two integers with different signs.

3. In the addition equation $8 + (-3) = 5$, the addends are ___?___ and ___?___, and the sum is ___?___.

4. Use the diagram at the right to complete this addition equation: ___?___ + ___?___ = ___?___.

Add.

5. $-3 + (-8)$	**6.** $-12 + (-1)$	**7.** $-4 + (-5)$	**8.** $-12 + (-12)$
9. $6 + (-9)$	**10.** $4 + (-9)$	**11.** $-6 + 7$	**12.** $-12 + 6$

13. $2 + (-3) + (-4)$ **14.** $7 + (-2) + (-8)$ **15.** $-3 + (-12) + (-15)$

16. $9 + (-6) + (-16)$ **17.** $-17 + (-3) + 29$ **18.** $13 + 62 + (-38)$

19. $-3 + (-8) + 12$ **20.** $-27 + (-42) + (-18)$

21. $13 + (-22) + 4 + (-5)$ **22.** $-14 + (-3) + 7 + (-6)$

23. $-22 + 10 + 2 + (-18)$ **24.** $-6 + (-8) + 13 + (-4)$

25. $-126 + (-247) + (-358) + 339$ **26.** $-651 + (-239) + 524 + 487$

Complete Exercises 27 and 28 without actually finding the sums.

27. Is the sum $812 + (-537)$ positive or negative?

28. Is the sum of -57 and -31 positive or negative?

2 Subtract integers

29. Explain the meaning of the words *minus* and *negative*.

30. Explain how to rewrite $6 - (-9)$ as addition of the opposite.

31. $-10 - 4 = -10 + $ _____?_____ • Rewrite subtraction as addition of the opposite.
 $= $ _____?_____ • Add.

32. $8 - (-5) = 8 + $ _____?_____ • Rewrite subtraction as addition of the opposite.
 $= $ _____?_____ • Add.

Subtract.

33. $16 - 8$ **34.** $12 - 3$ **35.** $7 - 14$

36. $7 - (-2)$ **37.** $3 - (-4)$ **38.** $-6 - (-3)$

39. $-4 - (-2)$ **40.** $6 - (-12)$ **41.** $-12 - 16$

42. $-4 - 3 - 2$ **43.** $4 - 5 - 12$ **44.** $12 - (-7) - 8$

45. $-12 - (-3) - (-15)$ **46.** $4 - 12 - (-8)$ **47.** $13 - 7 - 15$

48. $-6 + 19 - (-31)$ **49.** $-30 - (-65) - 29 - 4$ **50.** $42 - (-82) - 65 - 7$

51. $-16 - 47 - 63 - 12$ **52.** $42 - (-30) - 65 - (-11)$

53. $-47 - (-67) - 13 - 15$ **54.** $-18 - 49 - (-84) - 27$

Complete Exercises 55 and 56 without actually finding the differences.

55. Is the difference $-25 - 52$ positive or negative?

56. Is the difference 8 minus -5 positive or negative?

3 Multiply integers

57. Explain how to multiply two integers with the same sign.

58. Explain how to multiply two integers with different signs.

59. Name the operation in each expression. Justify your answer.
 a. $8(-7)$ **b.** $8 - 7$ **c.** $8 - (-7)$
 d. $-xy$ **e.** $x(-y)$ **f.** $-x - y$

60. Name the operation in each expression. Justify your answer.
 a. $(4)(-6)$ **b.** $4 - (6)$ **c.** $4 - (-6)$
 d. $-ab$ **e.** $a(-b)$ **f.** $-a - b$

61. In the equation $(-10)(7) = -70$, the factors are ___?___ and ___?___, and the product is ___?___.

62. In the equation $15(-3) = -45$, the 15 and -3 are called the ___?___, and -45 is called the ___?___.

63. For the product $(-4)(-12)$, the signs of the factors are the same. The sign of the product is ___?___. The product is ___?___.

64. For the product $(10)(-10)$, the signs of the factors are different. The sign of the product is ___?___. The product is ___?___.

Multiply.

65. $14 \cdot 3$ | **66.** $62 \cdot 9$ | **67.** $5(-4)$

68. $4(-7)$ | **69.** $-8(2)$ | **70.** $-9(3)$

71. $(-5)(-5)$ | **72.** $(-3)(-6)$ | **73.** $(-7)(0)$

74. $-32 \cdot 4$ | **75.** $-24 \cdot 3$ | **76.** $19(-7)$

77. $6(-17)$ | **78.** $-8(-26)$ | **79.** $-4(-35)$

80. $-5(23)$ | **81.** $5 \cdot 7(-2)$ | **82.** $8(-6)(-1)$

83. $(-9)(-9)(2)$ | **84.** $-8(-7)(-4)$ | **85.** $-5(8)(-3)$

86. $(-6)(5)(7)$ | **87.** $-1(4)(-9)$ | **88.** $6(-3)(-2)$

89. $4(-4) \cdot 6(-2)$ | **90.** $-5 \cdot 9(-7) \cdot 3$ | **91.** $-9(4) \cdot 3(1)$

92. $8(8)(-5)(-4)$ | **93.** $-6(-5)(12)(0)$ | **94.** $7(9) \cdot 10(-1)$

95. Is the product of three negative integers positive or negative?

96. Is the product of four positive numbers and three negative numbers positive or negative?

4 Divide integers

97. Write the division $\frac{-15}{3}$ using the division symbol: ___?___ $\div$ ___?___.

98. Write the division $8 \div (-4)$ as a fraction: $\frac{?}{?}$. The quotient is ___?___.

Write the related multiplication problem.

99. $\dfrac{-36}{-12} = 3$ **100.** $\dfrac{28}{-7} = -4$ **101.** $\dfrac{-55}{11} = -5$ **102.** $\dfrac{-20}{-10} = 2$

Divide.

103. $12 \div (-6)$ **104.** $18 \div (-3)$ **105.** $(-72) \div (-9)$

106. $(-64) \div (-8)$ **107.** $0 \div (-6)$ **108.** $-49 \div 0$

109. $45 \div (-5)$ **110.** $-24 \div 4$ **111.** $-36 \div 4$

112. $-56 \div 7$ **113.** $-81 \div (-9)$ **114.** $-40 \div (-5)$

115. $72 \div (-3)$ **116.** $44 \div (-4)$ **117.** $-60 \div 5$

118. $144 \div 9$ **119.** $78 \div (-6)$ **120.** $84 \div (-7)$

121. $-72 \div 4$ **122.** $-80 \div 5$ **123.** $-114 \div (-6)$

124. $-128 \div 4$ **125.** $-130 \div (-5)$ **126.** $(-280) \div 8$

127. $-132 \div (-12)$ **128.** $-156 \div (-13)$ **129.** $-182 \div 14$

130. $-144 \div 12$ **131.** $143 \div 11$ **132.** $168 \div 14$

Complete Exercises 133 and 134 without using a calculator.

133. State whether the quotient $-\dfrac{520}{-13}$ is positive or negative.

134. State whether each quotient is positive, negative, zero, or undefined.
 a. $-61 \div 0$ **b.** $0 \div 85$ **c.** $-172 \div (-4)$ **d.** $-96 \div 4$

5 **Application problems**

135. At 2:00 P.M. the temperature was 85°F. By 10:00 P.M. the temperature had dropped 20°F. Which expression can be used to find the temperature, in degrees Fahrenheit, at 10:00 P.M.?
 a. $85 + 20$ **b.** $85 - 20$ **c.** $20 - 85$ **d.** $85 \div 20$

136. After three tests, a student's test average was 82. After the fourth test, the student's test average was 84. Was the student's grade on the fourth test higher or lower than 82?

137. Temperature Find the temperature after a rise of 9°C from −6°C.

138. Temperature Find the temperature after a rise of 7°C from −18°C.

139. Temperature The high temperature for the day was 10°C. The low temperature was −4°C. Find the difference between the high and low temperatures for the day.

140. Temperature The low temperature for the day was −2°C. The high temperature was 11°C. Find the difference between the high and low temperatures for the day.

141. Chemistry The temperature at which mercury boils is 360°C. Mercury freezes at −39°C. Find the difference between the temperature at which mercury boils and the temperature at which it freezes.

142. Chemistry The temperature at which radon boils is −62°C. Radon freezes at −71°C. Find the difference between the temperature at which radon boils and the temperature at which it freezes.

Geography The elevation, or height, of places on Earth is measured in relation to sea level, or the average level of the ocean's surface. The following table shows height above sea level as a positive number and depth below sea level as a negative number. (*Source: Information Please Almanac*)

Mt. Aconcagua

Continent	Highest Elevation (in meters)		Lowest Elevation (in meters)	
Africa	Mt. Kilimanjaro	5895	Qattara Depression	−133
Asia	Mt. Everest	8844	Dead Sea	−400
Europe	Mt. Elbrus	5634	Caspian Sea	−28
North America	Mt. McKinley	6194	Death Valley	−86
South America	Mt. Aconcagua	6960	Salinas Grandes	−40

143. Use the table to find the difference in elevation between Mt. Elbrus and the Caspian Sea.

144. Use the table to find the difference in elevation between Mt. Aconcagua and Salinas Grandes.

145. Use the table to find the difference in elevation between Mt. Kilimanjaro and the Qattara Depression.

146. Use the table to find the difference in elevation between Mt. McKinley and Death Valley.

147. Use the table to find the difference in elevation between Mt. Everest and the Dead Sea.

148. Temperature The daily low temperatures, in degrees Celsius, during one week were recorded as follows: 4°, −5°, 8°, 0°, −9°, −11°, −8°. Find the average daily low temperature for the week.

149. Temperature The daily high temperatures, in degrees Celsius, during one week were recorded as follows: −8°, −9°, 6°, 7°, −2°, −14°, −1°. Find the average daily high temperature for the week.

150. Temperature On January 22, 1943, the temperature at Spearfish, South Dakota, rose from −4°F to 45°F in two minutes. How many degrees did the temperature rise during those two minutes?

151. Temperature In a 24-hour period in January of 1916, the temperature in Browning, Montana, dropped from 44°F to −56°F. How many degrees did the temperature drop during that time?

Aviation The table at the right shows the average temperatures at different cruising altitudes for airplanes. Use the table for Exercises 152 and 153.

Cruising Altitude	Average Temperature
12,000 ft	16°F
20,000 ft	−12°F
30,000 ft	−48°F
40,000 ft	−70°F
50,000 ft	−70°F

152. What is the difference between the average temperature at 12,000 ft and the average temperature at 40,000 ft?

153. How much colder is the average temperature at 30,000 ft than at 20,000 ft?

154. Education To discourage random guessing on a multiple-choice exam, a professor assigns 5 points for a correct answer, −2 points for an incorrect answer, and 0 points for leaving the question blank. What is the score for a student who had 20 correct answers, had 13 incorrect answers, and left 7 questions blank?

155. Education To discourage random guessing on a multiple-choice exam, a professor assigns 7 points for a correct answer, −3 points for an incorrect answer, and −1 point for leaving the question blank. What is the score for a student who had 17 correct answers, had 8 incorrect answers, and left 2 questions blank?

156. The Stock Market The value of a share of eBay stock on November 15, 2005, was $43.05. The table below shows the amount of increase or decrease, to the nearest 10 cents, from the November 15 closing price of the stock for a 10-day period. Calculate the 5-day moving average for this stock.

Day 1	Day 2	Day 3	Day 4	Day 5	Day 6	Day 7	Day 8	Day 9	Day 10
−50	+130	+90	+150	+60	−10	+10	−130	−90	+30

157. The Stock Market The value of a share of Disney stock on November 15, 2005, was $26.06. The table on the next page shows the amount of

increase or decrease, to the nearest 10 cents, from the November 15 closing price of the stock for a 10-day period. Calculate the 5-day moving average for this stock.

Day 1	Day 2	Day 3	Day 4	Day 5	Day 6	Day 7	Day 8	Day 9	Day 10
−20	+10	−80	−20	−10	+10	0	+10	0	+20

1.2 Applying Concepts

Simplify.

158. $|-7 + 12|$

159. $|13 - (-4)|$

160. $|-13 - (-2)|$

161. $|18 - 21|$

Assuming the pattern is continued, find the next three numbers in the pattern.

162. $-7, -11, -15, -19, \ldots$

163. $16, 11, 6, 1, \ldots$

164. $7, -14, 28, -56, \ldots$

165. $1024, -256, 64, \ldots$

Solve.

166. 32,844 is divisible by 3. By rearranging the digits, find the largest possible number that is still divisible by 3.

167. 4563 is not divisible by 4. By rearranging the digits, find the largest possible number that is divisible by 4.

168. How many three-digit numbers of the form 8__4 are divisible by 3?

In each exercise, determine which statement is false.

169. **a.** $|3 + 4| = |3| + |4|$ **b.** $|3 - 4| = |3| - |4|$ **c.** $|4 + 3| = |4| + |3|$ **d.** $|4 - 3| = |4| - |3|$

170. **a.** $|5 + 2| = |5| + |2|$ **b.** $|5 - 2| = |5| - |2|$ **c.** $|2 + 5| = |2| + |5|$ **d.** $|2 - 5| = |2| - |5|$

Determine which statement is true for all real numbers.

171. **a.** $|x + y| \le |x| + |y|$ **b.** $|x + y| = |x| + |y|$ **c.** $|x + y| \ge |x| + |y|$

172. **a.** $||x| - |y|| \le |x| - |y|$ **b.** $||x| - |y|| = |x| - |y|$ **c.** $||x| - |y|| \ge |x| - |y|$

173. Is the difference between two integers always smaller than either one of the numbers in the difference? If not, give an example for which the difference between two integers is greater than either integer.

174. If $-4x$ equals a positive integer, is x a positive or a negative integer? Explain your answer.

175. Explain why division by zero is not allowed.

 Rational Numbers

OBJECTIVE **1** **Write rational numbers as decimals**

A **rational number** is the quotient of two integers. Therefore, a rational number is a number that can be written in the form $\frac{a}{b}$, where a and b are integers, and b is not zero. A rational number written in this way is commonly called a **fraction.**

$$\frac{a}{b} \begin{array}{l} \leftarrow \text{an integer} \\ \leftarrow \text{a nonzero integer} \end{array}$$

$$\left. \frac{2}{3}, \quad \frac{-4}{9}, \quad \frac{18}{-5}, \quad \frac{4}{1} \right\} \begin{array}{l} \text{Rational} \\ \text{numbers} \end{array}$$

Because an integer can be written as the quotient of the integer and 1, every integer is a rational number.

$$5 = \frac{5}{1} \qquad -3 = \frac{-3}{1}$$

A number written in **decimal notation** is also a rational number.

three-tenths $0.3 = \frac{3}{10}$

thirty-five hundredths $0.35 = \frac{35}{100}$

negative four-tenths $-0.4 = -\frac{4}{10}$

A rational number written as a fraction can be written in decimal notation.

Take Note

The fraction bar can be read "divided by."

$$\frac{5}{8} = 5 \div 8$$

Take Note

Dividing the numerator by the denominator results in a remainder of 0. The decimal 0.625 is a terminating decimal.

EXAMPLE 1 Write $\frac{5}{8}$ as a decimal.

Solution
$$\begin{array}{r} 0.625 \\ 8\overline{)5.000} \\ -4\,8 \\ \hline 20 \\ -16 \\ \hline 40 \\ -40 \\ \hline 0 \end{array}$$
← This is called a **terminating decimal.**

← The remainder is zero.

$$\frac{5}{8} = 0.625$$

PROBLEM 1 Write $\frac{4}{25}$ as a decimal.

Solution See page S1.

EXAMPLE 2 Write $\frac{4}{11}$ as a decimal.

Solution

$$
\begin{array}{r}
0.3636\ldots \quad \leftarrow \text{This is called a \textbf{repeating decimal.}} \\
11\overline{)4.0000} \\
-3\,3 \\
\hline
70 \\
-66 \\
\hline
40 \\
-33 \\
\hline
70 \\
-66 \\
\hline
4 \quad \leftarrow \text{The remainder is never zero.}
\end{array}
$$

$\frac{4}{11} = 0.\overline{36}$ ← The bar over the digits 3 and 6 is used to show that these digits repeat.

PROBLEM 2 Write $\frac{4}{9}$ as a decimal. Place a bar over the repeating digits of the decimal.

Solution See page S1.

See page S1.

> **Take Note**
>
> No matter how far we carry out the division, the remainder is never zero. The decimal $0.\overline{36}$ is a repeating decimal.

Rational numbers can be written as fractions, such as $-\frac{6}{7}$ or $\frac{8}{3}$, where the numerator and denominator are integers. But every rational number also can be written as a repeating decimal (such as 0.25767676. . .) or a terminating decimal (such as 1.73). This was illustrated in Examples 1 and 2.

Numbers that cannot be written either as a repeating decimal or as a terminating decimal are called **irrational numbers.** For example, 2.45445444544445. . . is an irrational number. Two other examples are $\sqrt{2}$ and π.

$$\sqrt{2} = 1.414213562\ldots \quad \pi = 3.141592654\ldots$$

The three dots mean that the digits continue on and on without ever repeating or terminating. Although we cannot write a decimal that is exactly equal to $\sqrt{2}$ or to π, we can give approximations of these numbers. The symbol $\approx$ is read "is approximately equal to." Shown below are $\sqrt{2}$ rounded to the nearest thousandth and π rounded to the nearest hundredth.

$$\sqrt{2} \approx 1.414 \quad \pi \approx 3.14$$

The rational numbers and the irrational numbers taken together are called the **real numbers.**

OBJECTIVE 2

Multiply and divide rational numbers

The sign rules for multiplying and dividing integers apply to multiplication and division of rational numbers. The product of two fractions is the product of the numerators divided by the product of the denominators.

EXAMPLE 3 Multiply: $\dfrac{3}{8} \cdot \dfrac{12}{17}$

 Solution $\dfrac{3}{8} \cdot \dfrac{12}{17} = \dfrac{3 \cdot 12}{8 \cdot 17}$ • **Multiply the numerators. Multiply the denominators.**

$= \dfrac{3 \cdot \overset{1}{\cancel{2}} \cdot \overset{1}{\cancel{2}} \cdot 3}{2 \cdot \underset{1}{\cancel{2}} \cdot \underset{1}{\cancel{2}} \cdot 17}$ • **Write the prime factorization of each factor. Divide by the common factors.**

$= \dfrac{9}{34}$ • **Multiply the numbers remaining in the numerator. Multiply the numbers remaining in the denominator.**

PROBLEM 3 Multiply: $-\dfrac{7}{12} \cdot \dfrac{9}{14}$

 Solution See page S1.

The **reciprocal** of a fraction is the fraction with the numerator and denominator interchanged. For instance, the reciprocal of $\dfrac{2}{3}$ is $\dfrac{3}{2}$, and the reciprocal of $-\dfrac{5}{4}$ is $-\dfrac{4}{5}$. To divide fractions, multiply the dividend by the reciprocal of the divisor.

EXAMPLE 4 Divide: $\dfrac{3}{10} \div \left(-\dfrac{18}{25}\right)$

 Solution $\dfrac{3}{10} \div \left(-\dfrac{18}{25}\right) = -\left(\dfrac{3}{10} \div \dfrac{18}{25}\right)$ • **The signs are different. The quotient is negative.**

$= -\left(\dfrac{3}{10} \cdot \dfrac{25}{18}\right)$ • **Change division to multiplication. Invert the divisor.**

$= -\left(\dfrac{3 \cdot 25}{10 \cdot 18}\right)$ • **Multiply the numerators. Multiply the denominators.**

$= -\left(\dfrac{\overset{1}{\cancel{3}} \cdot \overset{1}{\cancel{5}} \cdot 5}{2 \cdot \underset{1}{\cancel{5}} \cdot 2 \cdot \underset{1}{\cancel{3}} \cdot 3}\right)$

$= -\dfrac{5}{12}$

PROBLEM 4 Divide: $-\dfrac{3}{8} \div \left(-\dfrac{5}{12}\right)$

 Solution See page S1.

> ### Take Note
> The method for dividing fractions is sometimes stated "To divide fractions, invert the divisor and multiply." Inverting the divisor means writing its reciprocal.

To multiply decimals, multiply as in the multiplication of whole numbers. Write the decimal point in the product so that the number of decimal places in the product equals the sum of the decimal places in the factors.

EXAMPLE 5	Multiply: $(-6.89)(0.00035)$

Solution			• **Multiply the absolute values.**

$$\begin{array}{r} 6.89 \\ \times\, 0.00035 \\ \hline 3445 \\ 2067 \quad\;\; \\ \hline 0.0024115 \end{array}$$ 2 decimal places
5 decimal places

7 decimal places

$(-6.89)(0.00035) = -0.0024115$

• **The signs are different. The product is negative.**

PROBLEM 5	Multiply: $(-5.44)(3.8)$
Solution	See page S2.

To divide decimals, move the decimal point in the divisor to make it a whole number. Move the decimal point in the dividend the same number of places to the right. Place the decimal point in the quotient directly over the decimal point in the dividend. Then divide as in the division of whole numbers.

EXAMPLE 6	Divide: $1.32 \div 0.27$. Round to the nearest tenth.

Take Note

Moving the decimal point in the numerator and denominator is the same as multiplying the numerator and the denominator by the same number. For the problem at the right, we have

$$1.32 \div 0.27 = \frac{1.32}{0.27}$$
$$= \frac{1.32}{0.27} \cdot \frac{100}{100}$$
$$= \frac{132}{27}$$

Solution $0.27\overline{)1.32.}$

• **Move the decimal point two places to the right in the divisor and in the dividend. Place the decimal point in the quotient.**

$$\begin{array}{r} 4.88 \approx 4.9 \\ 27\overline{)132.00} \\ -108\quad\;\; \\ \hline 24\,0\;\; \\ -21\,6\;\; \\ \hline 2\,40 \\ -2\,16 \\ \hline 24 \end{array}$$

• **The symbol $\approx$ is used to indicate that the quotient is an approximate value that has been rounded off.**

PROBLEM 6	Divide: $-0.394 \div 1.7$. Round to the nearest hundredth.
Solution	See page S2.

OBJECTIVE 3 **Add and subtract rational numbers**

The sign rules for adding integers apply to addition of rational numbers.

To add or subtract rational numbers written as fractions, first rewrite the fractions as equivalent fractions with a common denominator. The common denominator is the **least common multiple (LCM)** of the denominators. The LCM of the denominators is also called the **lowest common denominator (LCD).**

EXAMPLE 7	Add: $-\dfrac{5}{6} + \dfrac{3}{10}$

Solution	Prime factorization of 6 and 10:	• **Find the LCM of the denominators 6 and 10.**

$6 = 2 \cdot 3 \qquad 10 = 2 \cdot 5$

$\text{LCM} = 2 \cdot 3 \cdot 5 = 30$

$$-\frac{5}{6} + \frac{3}{10} = -\frac{25}{30} + \frac{9}{30}$$

$$= \frac{-25 + 9}{30}$$

$$= \frac{-16}{30}$$

$$= -\frac{8}{15}$$

• Rewrite the fractions as equivalent fractions, using the LCM of the denominators as the common denominator.

• Add the numerators, and place the sum over the common denominator.

• Write the answer in simplest form.

PROBLEM 7 Subtract: $\dfrac{5}{9} - \dfrac{11}{12}$

■ **Solution** See page S2.

EXAMPLE 8 Simplify: $-\dfrac{3}{4} + \dfrac{1}{6} - \dfrac{5}{8}$

Solution $-\dfrac{3}{4} + \dfrac{1}{6} - \dfrac{5}{8} = -\dfrac{18}{24} + \dfrac{4}{24} - \dfrac{15}{24}$ • The LCM of 4, 6, and 8 is 24.

$$= \frac{-18}{24} + \frac{4}{24} + \frac{-15}{24}$$

$$= \frac{-18 + 4 + (-15)}{24}$$

$$= \frac{-29}{24}$$

$$= -\frac{29}{24}$$

PROBLEM 8 Simplify: $-\dfrac{7}{8} - \dfrac{5}{6} + \dfrac{1}{2}$

■ **Solution** See page S2.

To add or subtract decimals, write the numbers so that the decimal points are in a vertical line. Then proceed as in the addition or subtraction of integers. Write the decimal point in the answer directly below the decimal points in the problem.

EXAMPLE 9 Add: $14.02 + 137.6 + 9.852$

Solution
$$\begin{array}{r} 14.02 \\ 137.6 \\ +\ \ 9.852 \\ \hline 161.472 \end{array}$$

• Write the decimals so that the decimal points are in a vertical line.

• Write the decimal point in the sum directly below the decimal points in the problem.

PROBLEM 9 Add: $3.097 + 4.9 + 3.09$

■ **Solution** See page S2.

EXAMPLE 10 Add: $-114.039 + 84.76$

Solution

$$\begin{array}{r} 114.039 \\ -\ \ 84.76 \\ \hline 29.279 \end{array}$$

$-114.039 + 84.76$
$= -29.279$

• The signs are different. Subtract the absolute value of the number with the smaller absolute value from the absolute value of the number with the larger absolute value.

• Attach the sign of the number with the larger absolute value.

PROBLEM 10 Subtract: $16.127 - 67.91$

Solution See page S2.

OBJECTIVE **4** ## Convert among percents, fractions, and decimals

"A population growth rate of 3%," "a manufacturer's discount of 25%," and "an 8% increase in pay" are typical examples of the many ways in which percent is used in applied problems. **Percent** means "parts of 100." Thus 27% means 27 parts of 100.

In applied problems involving a percent, it is usually necessary either to rewrite the percent as a fraction or a decimal, or to rewrite a fraction or a decimal as a percent.

To write 27% as a fraction, remove the percent sign and multiply by $\dfrac{1}{100}$.

$$27\% = 27\left(\frac{1}{100}\right) = \frac{27}{100}$$

To write a percent as a decimal, remove the percent sign and multiply by 0.01.

To write 33% as a decimal, remove the percent sign and multiply by 0.01.

$$33\% \quad = \quad 33(0.01) \quad = \quad 0.33$$

Move the decimal point two places to the left. Then remove the percent sign.

Write 100% as a decimal.

$$100\% = 100(0.01) = 1$$

EXAMPLE 11 Write 130% as a fraction and as a decimal.

Solution $130\% = 130\left(\dfrac{1}{100}\right) = \dfrac{130}{100} = 1\dfrac{3}{10}$

• To write a percent as a fraction, remove the percent sign and multiply by $\dfrac{1}{100}$.

$130\% = 130(0.01) = 1.30$

• To write a percent as a decimal, remove the percent sign and multiply by 0.01.

PROBLEM 11 Write 125% as a fraction and as a decimal.

Solution See page S2.

EXAMPLE 12 Write $33\frac{1}{3}\%$ as a fraction.

Solution $33\frac{1}{3}\% = 33\frac{1}{3}\left(\frac{1}{100}\right) = \frac{100}{3}\left(\frac{1}{100}\right)$ • **Write the mixed number $33\frac{1}{3}$ as the improper fraction $\frac{100}{3}$.**

$= \frac{1}{3}$

PROBLEM 12 Write $16\frac{2}{3}\%$ as a fraction.

Solution See page S2.

EXAMPLE 13 Write 0.25% as a decimal.

Solution $0.25\% = 0.25(0.01) = 0.0025$ • **Remove the percent sign and multiply by 0.01.**

PROBLEM 13 Write 6.08% as a decimal.

Solution See page S2.

A fraction or decimal can be written as a percent by multiplying by 100%. Recall that 100% = 1 and that multiplying a number by 1 does not change the number.

To write $\frac{5}{8}$ as a percent, multiply by 100%.

$$\frac{5}{8} = \frac{5}{8}(100\%) = \frac{500}{8}\% = 62.5\% \text{ or } 62\frac{1}{2}\%$$

To write 0.82 as a percent, multiply by 100%.

$$0.82 \quad = \quad 0.82(100\%) \quad = \quad 82\%$$

Move the decimal point two places to the right. Then write the percent sign.

EXAMPLE 14 Write as a percent. **A.** 0.027 **B.** 1.34

Solution **A.** $0.027 = 0.027(100\%) = 2.7\%$ • **To write a decimal as a percent, multiply by 100%.**

B. $1.34 = 1.34(100\%) = 134\%$

PROBLEM 14 Write as a percent. **A.** 0.043 **B.** 2.57

Solution See page S2.

EXAMPLE 15 Write $\frac{5}{6}$ as a percent. Round to the nearest tenth of a percent.

Solution $\frac{5}{6} = \frac{5}{6}(100\%) = \frac{500}{6}\% \approx 83.3\%$ • **To write a fraction as a percent, multiply by 100%.**

PROBLEM 15 Write $\frac{5}{9}$ as a percent. Round to the nearest tenth of a percent.

Solution See page S2.

EXAMPLE 16 Write $\frac{7}{16}$ as a percent. Write the remainder in fractional form.

Solution $\frac{7}{16} = \frac{7}{16}(100\%) = \frac{700}{16}\% = 43\frac{3}{4}\%$ • **Multiply the fraction by 100%.**

PROBLEM 16 Write $\frac{9}{16}$ as a percent. Write the remainder in fractional form.

Solution See page S2.

 Concept Review

Determine whether the statement is always true, sometimes true, or never true.

1. To multiply two fractions, you must first rewrite the fractions as equivalent fractions with a common denominator.

2. The number $\frac{\pi}{3}$ is an example of a rational number.

3. A rational number can be written as a terminating decimal.

4. The number 2.585585558... is an example of a repeating decimal.

5. An irrational number is a real number.

6. 37%, 0.37, and $\frac{37}{100}$ are three numbers that have the same value.

7. To write a decimal as a percent, multiply the decimal by $\frac{1}{100}$.

8. If a, b, c, and d are natural numbers, then $\frac{a}{b} + \frac{c}{d} = \frac{a + c}{b + d}$.

9. -12 is an example of a number that is both an integer and a rational number.

10. $\frac{7}{9}$ is an example of a number that is both a rational number and an irrational number.

 Exercises

1 **Write rational numbers as decimals**

1. To write $\frac{2}{3}$ as a decimal, divide ____?____ by ____?____. The quotient is 0.6666..., which is a ____?____ decimal.

2. A number such as 0.74744744474444..., whose decimal representation neither ends nor repeats, is an example of an ____?____ number.

Write as a decimal. Place a bar over the repeating digits of a repeating decimal.

3. $\dfrac{1}{3}$ **4.** $\dfrac{2}{3}$ **5.** $\dfrac{1}{4}$ **6.** $\dfrac{3}{4}$ **7.** $\dfrac{2}{5}$

8. $\dfrac{4}{5}$ **9.** $\dfrac{1}{6}$ **10.** $\dfrac{5}{6}$ **11.** $\dfrac{1}{8}$ **12.** $\dfrac{7}{8}$

13. $\dfrac{2}{9}$ **14.** $\dfrac{8}{9}$ **15.** $\dfrac{5}{11}$ **16.** $\dfrac{10}{11}$ **17.** $\dfrac{7}{12}$

18. $\dfrac{11}{12}$ **19.** $\dfrac{4}{15}$ **20.** $\dfrac{8}{15}$ **21.** $\dfrac{7}{16}$ **22.** $\dfrac{15}{16}$

23. $\dfrac{6}{25}$ **24.** $\dfrac{14}{25}$ **25.** $\dfrac{9}{40}$ **26.** $\dfrac{21}{40}$ **27.** $\dfrac{15}{22}$

28. $\dfrac{19}{22}$ **29.** $\dfrac{11}{24}$ **30.** $\dfrac{19}{24}$ **31.** $\dfrac{5}{33}$ **32.** $\dfrac{25}{33}$

33. To write $\dfrac{a}{b}$ as a decimal, we divide a by b (b is not zero). After finding the quotient to 10 decimal places, we still have a nonzero remainder and no repeating pattern of digits. Can we conclude that $\dfrac{a}{b}$ is an irrational number?

34. Is $\dfrac{\sqrt{2}}{2}$ a rational or an irrational number?

2 Multiply and divide rational numbers

35. The product of 1.762 and −8.4 will have ____?____ decimal places.

36. The reciprocal of $\dfrac{4}{9}$ is ____?____. To find the quotient $-\dfrac{2}{3} \div \dfrac{4}{9}$, find the product $-\dfrac{2}{3} \cdot$ ____?____. The quotient $-\dfrac{2}{3} \div \dfrac{4}{9}$ is ____?____.

Simplify.

37. $\dfrac{1}{2}\left(-\dfrac{3}{4}\right)$ **38.** $-\dfrac{2}{9}\left(-\dfrac{3}{14}\right)$ **39.** $\left(-\dfrac{3}{8}\right)\left(-\dfrac{4}{15}\right)$

40. $\dfrac{5}{8}\left(-\dfrac{7}{12}\right)\dfrac{16}{25}$ **41.** $\left(\dfrac{1}{2}\right)\left(-\dfrac{3}{4}\right)\left(-\dfrac{5}{8}\right)$ **42.** $\left(\dfrac{5}{12}\right)\left(-\dfrac{8}{15}\right)\left(-\dfrac{1}{3}\right)$

43. $\dfrac{3}{8} \div \dfrac{1}{4}$ **44.** $\dfrac{5}{6} \div \left(-\dfrac{3}{4}\right)$ **45.** $-\dfrac{5}{12} \div \dfrac{15}{32}$

46. $\dfrac{1}{8} \div \left(-\dfrac{5}{12}\right)$

47. $-\dfrac{4}{9} \div \left(-\dfrac{2}{3}\right)$

48. $-\dfrac{6}{11} \div \dfrac{4}{9}$

49. $(1.2)(3.47)$

50. $(-0.8)(6.2)$

51. $(-1.89)(-2.3)$

52. $(6.9)(-4.2)$

53. $(1.06)(-3.8)$

54. $(-2.7)(-3.5)$

55. $(1.2)(-0.5)(3.7)$

56. $(-2.4)(6.1)(0.9)$

57. $(-0.8)(3.006)(-5.1)$

Complete Exercises 58 and 59 without actually finding the products and quotients.

58. State whether each product or quotient is positive or negative.

 a. $\left(-\dfrac{11}{12}\right)\left(-\dfrac{5}{4}\right)\left(-\dfrac{1}{2}\right)$ **b.** $-1.572 \div -8.4$

59. State whether each product or quotient is greater than one or less than one.

 a. $\dfrac{11}{13} \cdot \dfrac{50}{51}$ **b.** $8.713 \div 7.2$

Simplify. Round to the nearest hundredth.

60. $-24.7 \div 0.09$

61. $-1.27 \div (-1.7)$

62. $9.07 \div (-3.5)$

63. $-354.2086 \div 0.1719$

64. $-2658.3109 \div (-0.0473)$

65. $(-3.92) \div (-45.008)$

3 Add and subtract rational numbers

66. The least common multiple of the denominators of the fractions $\dfrac{5}{8}$, $-\dfrac{1}{6}$, and $\dfrac{2}{9}$ is ____?____.

67. Write the fraction $\dfrac{3}{14}$ as an equivalent fraction with the denominator 28:

$\dfrac{3}{14} = \dfrac{?}{28}$.

Simplify.

68. $\dfrac{3}{8} + \dfrac{5}{8}$

69. $-\dfrac{1}{4} + \dfrac{3}{4}$

70. $\dfrac{7}{8} - \dfrac{3}{8}$

71. $-\dfrac{5}{6} - \dfrac{1}{6}$

72. $-\dfrac{5}{12} - \dfrac{3}{8}$

73. $-\dfrac{5}{6} - \dfrac{5}{9}$

74. $-\dfrac{6}{13} + \dfrac{17}{26}$

75. $-\dfrac{7}{12} + \dfrac{5}{8}$

76. $-\dfrac{5}{8} - \left(-\dfrac{11}{12}\right)$

77. $\dfrac{1}{3} + \dfrac{5}{6} - \dfrac{2}{9}$

78. $\dfrac{1}{2} - \dfrac{2}{3} + \dfrac{1}{6}$

79. $-\dfrac{3}{8} - \dfrac{5}{12} - \dfrac{3}{16}$

80. $-\dfrac{5}{16} + \dfrac{3}{4} - \dfrac{7}{8}$

81. $\dfrac{1}{2} - \dfrac{3}{8} - \left(-\dfrac{1}{4}\right)$

82. $\dfrac{3}{4} - \left(-\dfrac{7}{12}\right) - \dfrac{7}{8}$

83. $\dfrac{1}{3} - \dfrac{1}{4} - \dfrac{1}{5}$

84. $\dfrac{2}{3} - \dfrac{1}{2} + \dfrac{5}{6}$ **85.** $\dfrac{5}{16} + \dfrac{1}{8} - \dfrac{1}{2}$ **86.** $\dfrac{5}{8} - \left(-\dfrac{5}{12}\right) + \dfrac{1}{3}$ **87.** $\dfrac{1}{8} - \dfrac{11}{12} + \dfrac{1}{2}$

88. $-\dfrac{7}{9} + \dfrac{14}{15} + \dfrac{8}{21}$ **89.** $1.09 + 6.2$ **90.** $-32.1 - 6.7$ **91.** $5.13 - 8.179$

92. $-13.092 + 6.9$ **93.** $2.54 - 3.6$ **94.** $5.43 + 7.925$ **95.** $-16.92 - 6.925$

96. $-3.87 + 8.546$ **97.** $6.9027 - 17.692$

98. $2.09 - 6.72 - 5.4$ **99.** $16.4 + 3.09 - 7.93$

100. $-18.39 + 4.9 - 23.7$ **101.** $19 - (-3.72) - 82.75$

102. $-3.07 - (-2.97) - 17.4$ **103.** $-3.09 - 4.6 - 27.3$

104. $317.09 - 46.902 + 583.0714$ **105.** $71.0235 - 86.0974 + 254.309$

Complete Exercises 106 and 107 without actually finding the sums and differences.

106. State whether each sum or difference is positive or negative.
a. $\dfrac{1}{5} - \dfrac{1}{2}$ **b.** $-21.765 + 15.1$ **c.** $0.837 + (-0.24)$ **d.** $-\dfrac{3}{4} + \dfrac{9}{10}$

107. Estimate each sum to the nearest integer.
a. $\dfrac{7}{8} + \dfrac{4}{5}$ **b.** $\dfrac{1}{3} + \left(-\dfrac{1}{2}\right)$ **c.** $-0.125 + 1.25$ **d.** $-1.3 + 0.2$

4 Convert among percents, fractions, and decimals

108. **a.** Explain how to convert a fraction to a percent.
b. Explain how to convert a percent to a fraction.

109. **a.** Explain how to convert a decimal to a percent.
b. Explain how to convert a percent to a decimal.

110. Explain why multiplying a number by 100% does not change the value of the number.

111. To write 80% as a fraction, remove the percent sign and multiply by
____?____ : $80\% = 80 \cdot$ ____?____ $=$ ____?____ .

112. To write 68% as a decimal, remove the percent sign and multiply by ___?___ : 68% = 68 · ___?___ = ___?___ .

113. To write $\frac{3}{10}$ as a percent, multiply by ___?___ : $\frac{3}{10} = \frac{3}{10} \cdot$ ___?___ = ___?___ .

114. To write 1.25 as a percent, multiply by ___?___ : 1.25 = 1.25 · ___?___ = ___?___ .

Write as a fraction and as a decimal.

115. 75% **116.** 40% **117.** 50% **118.** 10%

119. 64% **120.** 88% **121.** 175% **122.** 160%

123. 19% **124.** 87% **125.** 5% **126.** 2%

127. 450% **128.** 380% **129.** 8% **130.** 4%

Write as a fraction.

131. $11\frac{1}{9}\%$ **132.** $37\frac{1}{2}\%$ **133.** $31\frac{1}{4}\%$ **134.** $66\frac{2}{3}\%$

135. $\frac{1}{2}\%$ **136.** $5\frac{3}{4}\%$ **137.** $6\frac{1}{4}\%$ **138.** $83\frac{1}{3}\%$

Write as a decimal.

139. 7.3% **140.** 9.1% **141.** 15.8% **142.** 0.3%

143. 9.15% **144.** 121.2% **145.** 18.23% **146.** 0.15%

Write as a percent.

147. 0.15 **148.** 0.37 **149.** 0.05 **150.** 0.02

151. 0.175 **152.** 0.125 **153.** 1.15 **154.** 2.142

155. 0.008 **156.** 0.004 **157.** 0.065 **158.** 0.083

Write as a percent. Round to the nearest tenth of a percent.

159. $\frac{27}{50}$ **160.** $\frac{83}{100}$ **161.** $\frac{1}{3}$ **162.** $\frac{3}{8}$

163. $\dfrac{4}{9}$ **164.** $\dfrac{9}{20}$ **165.** $2\dfrac{1}{2}$ **166.** $1\dfrac{2}{7}$

Write as a percent. Write the remainder in fractional form.

167. $\dfrac{3}{8}$ **168.** $\dfrac{3}{16}$ **169.** $\dfrac{5}{14}$ **170.** $\dfrac{4}{7}$

171. $1\dfrac{1}{4}$ **172.** $2\dfrac{5}{8}$ **173.** $1\dfrac{5}{9}$ **174.** $1\dfrac{13}{16}$

Complete Exercises 175 and 176 without actually finding the percents.

175. Does $\dfrac{4}{3}$ represent a number greater than 100% or less than 100%?

176. Does 0.055 represent a number greater than 1% or less than 1%?

Employment The graph at the right shows the responses to a survey that asked respondents, "How did you find your most recent job?" Use the graph for Exercises 177 to 179.

177. What fraction of the respondents found their most recent jobs on the Internet?

178. What fraction of the respondents found their most recent jobs through a referral?

179. Did more or fewer than one-quarter of the respondents find their most recent jobs through a newspaper ad?

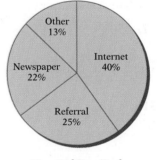

How Did You Find
Your Most Recent Job?

1.3 Applying Concepts

Classify each of the following numbers as a natural number, an integer, a positive integer, a negative integer, a rational number, an irrational number, or a real number. List all that apply.

180. −1 **181.** 28

182. $-\dfrac{9}{34}$ **183.** −7.707

184. $5.2\overline{6}$ **185.** 0.171771777…

Solve.

186. Find the average of $\dfrac{5}{8}$ and $\dfrac{3}{4}$.

187. The Postal Service Postage for first-class mail is \$.39 for the first ounce or fraction of an ounce and \$.24 for each additional ounce or fraction of an ounce. Find the cost of mailing, by first-class mail, a letter that weighs $4\frac{1}{2}$ oz.

Government The table at the right shows the surplus or deficit, in billions of dollars, for the federal budget for selected years from 1945 through 2000. A negative sign (−) indicates a deficit. Use this table for Exercises 188 to 193. (*Source:* U.S. Office of Management and Budget)

Year	Federal Budget Surplus or Deficit (in billions of dollars)
1945	−47.553
1950	−3.119
1955	−2.993
1960	0.301
1965	−1.411
1970	−2.842
1975	−53.242
1980	−73.835
1985	−212.334
1990	−221.194
1995	−163.899
1996	−107.450
1997	−21.940
1998	69.246
1999	79.263
2000	117.305

188. In which year listed was the deficit greatest?

189. Find the difference between the deficits in the years 1980 and 1985.

190. Calculate the difference between the surplus in 1960 and the deficit in 1955.

191. How many times greater was the deficit in 1985 than in 1975? Round to the nearest whole number.

192. What was the average deficit per quarter, in millions of dollars, for the year 1970?

193. Find the average surplus or deficit for the years 1995 through 2000. Round to the nearest million.

194. Use a calculator to determine the decimal representations of $\frac{17}{99}, \frac{45}{99}$, and $\frac{73}{99}$. Make a conjecture as to the decimal representation of $\frac{83}{99}$. Does your conjecture work for $\frac{33}{99}$? What about $\frac{1}{99}$?

195. If the same positive integer is added to both the numerator and the denominator of $\frac{2}{5}$, is the new fraction less than, equal to, or greater than $\frac{2}{5}$?

196. A magic square is one in which the numbers in every row, column, and diagonal sum to the same number. Complete the magic square at the right.

$\frac{2}{3}$		
	$\frac{1}{6}$	$\frac{5}{6}$
		$-\frac{1}{3}$

197. Let x represent the price of a car. If the sales tax is 6% of the price, express the total of the price of the car and the sales tax in terms of x.

198. Let x represent the price of a suit. If the suit is on sale at a discount rate of 30%, express the price of the suit after the discount in terms of x.

199. The price of a pen was 60¢. During a storewide sale, the price was reduced to a different whole number of cents, and the entire stock was sold for $54.59. Find the price of the pen during the sale.

200. Find three natural numbers a, b, and c such that $\frac{1}{a} + \frac{1}{b} + \frac{1}{c}$ is a natural number.

201. If a and b are rational numbers and $a < b$, is it always possible to find a rational number c such that $a < c < b$? If not, explain why. If so, show how to find one.

202. In your own words, define **a.** a rational number, **b.** an irrational number, and **c.** a real number.

203. Explain why you "invert and multiply" when dividing a fraction by a fraction.

204. Explain why you need a common denominator when adding two fractions and why you don't need a common denominator when multiplying two fractions.

1.4 Exponents and the Order of Operations Agreement

OBJECTIVE 1 **Exponential expressions**

Point of Interest

René Descartes (1596–1650) was the first mathematician to use exponential notation extensively as it is used today. However, for some unknown reason, he always used xx for x².

Repeated multiplication of the same factor can be written using an exponent.

$$2 \cdot 2 \cdot 2 \cdot 2 \cdot 2 = 2^5 \leftarrow \text{exponent}$$
$$\underset{\text{base}}{\uparrow\!\!\!\text{---------}}$$

$$a \cdot a \cdot a \cdot a = a^4 \leftarrow \text{exponent}$$
$$\underset{\text{base}}{\uparrow\!\!\!\text{---------}}$$

The **exponent** indicates how many times the factor, called the **base,** occurs in the multiplication. The multiplication $2 \cdot 2 \cdot 2 \cdot 2 \cdot 2$ is in **factored form.** The exponential expression 2^5 is in **exponential form.**

2^1 is read "the first power of two" or just "two." ⟶ Usually the exponent 1 is not written.

2^2 is read "the second power of two" or "two squared."

2^3 is read "the third power of two" or "two cubed."

2^4 is read "the fourth power of two."

2^5 is read "the fifth power of two."

a^5 is read "the fifth power of a."

There is a geometric interpretation of the first three natural-number powers.

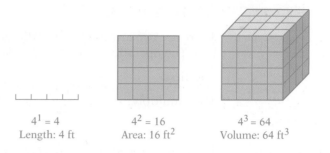

$4^1 = 4$	$4^2 = 16$	$4^3 = 64$
Length: 4 ft	Area: 16 ft^2	Volume: 64 ft^3

To evaluate an exponential expression, write each factor as many times as indicated by the exponent. Then multiply.

$$3^5 = 3 \cdot 3 \cdot 3 \cdot 3 \cdot 3 = 243$$

$$2^3 \cdot 3^2 = (2 \cdot 2 \cdot 2) \cdot (3 \cdot 3) = 8 \cdot 9 = 72$$

Take Note

The -4 is squared only when the negative sign is *inside* the parentheses. In $(-4)^2$, we are squaring -4; in -4^2, we are finding the opposite of 4^2.

EXAMPLE 1 Evaluate $(-4)^2$ and -4^2.

Solution $(-4)^2 = (-4)(-4) = 16$

$-4^2 = -(4 \cdot 4) = -16$

PROBLEM 1 Evaluate $(-5)^3$ and -5^3.

Solution See page S2.

Take Note

The product of an even number of negative factors is positive. The product of an odd number of negative factors is negative.

EXAMPLE 2 Evaluate $(-2)^4$ and $(-2)^5$.

Solution

$$\begin{aligned}(-2)^4 &= (-2)(-2)(-2)(-2) \\ &= 4(-2)(-2) \\ &= -8(-2) \\ &= 16\end{aligned}$$

$$\begin{aligned}(-2)^5 &= (-2)(-2)(-2)(-2)(-2) \\ &= 4(-2)(-2)(-2) \\ &= -8(-2)(-2) \\ &= 16(-2) \\ &= -32\end{aligned}$$

PROBLEM 2 Evaluate $(-3)^3$ and $(-3)^4$.

Solution See page S2.

EXAMPLE 3 Evaluate $(-3)^2 \cdot 2^3$ and $\left(-\dfrac{2}{3}\right)^3$.

Solution $(-3)^2 \cdot 2^3 = (-3)(-3) \cdot (2)(2)(2) = 9 \cdot 8 = 72$

$$\left(-\frac{2}{3}\right)^3 = \left(-\frac{2}{3}\right)\left(-\frac{2}{3}\right)\left(-\frac{2}{3}\right) = -\frac{2 \cdot 2 \cdot 2}{3 \cdot 3 \cdot 3} = -\frac{8}{27}$$

PROBLEM 3 Evaluate $(3^3)(-2)^3$ and $\left(-\dfrac{2}{5}\right)^2$.

Solution See page S2.

OBJECTIVE **2** ## The Order of Operations Agreement

Evaluate $2 + 3 \cdot 5$.

There are two arithmetic operations, addition and multiplication, in this problem. The operations could be performed in different orders.

Add first.	$2 + \underbrace{3 \cdot 5}$	Multiply first.	$2 + \underbrace{3 \cdot 5}$
Then multiply.	$\underbrace{5 \quad \cdot 5}$	Then add.	$\underbrace{2 + \quad 15}$
	25		17

In order to prevent there being more than one answer to the same problem, an Order of Operations Agreement has been established.

The Order of Operations Agreement

Step 1 Perform operations inside grouping symbols. **Grouping symbols** include parentheses (), brackets [], absolute value symbols | |, and the fraction bar.

Step 2 Simplify exponential expressions.

Step 3 Do multiplication and division as they occur from left to right.

Step 4 Do addition and subtraction as they occur from left to right.

EXAMPLE 4 Simplify: $12 - 24(8 - 5) \div 2^2$

Solution $12 - 24(8 - 5) \div 2^2$
$= 12 - 24(3) \div 2^2$ • Perform operations inside grouping symbols.
$= 12 - 24(3) \div 4$ • Simplify exponential expressions.
$= 12 - 72 \div 4$ • Do multiplication and division as they occur from left to right.

$= 12 - 18$
$= -6$ • Do addition and subtraction as they occur from left to right.

PROBLEM 4 Simplify: $36 \div (8 - 5)^2 - (-3)^2 \cdot 2$

Solution See page S2.

One or more of the steps shown in Example 4 may not be needed to simplify an expression. In that case, proceed to the next step in the Order of Operations Agreement.

EXAMPLE 5 Simplify: $\dfrac{4 + 8}{2 + 1} - |3 - 1| + 2$

Solution $\dfrac{4 + 8}{2 + 1} - |3 - 1| + 2$

$= \dfrac{12}{3} - |2| + 2$ • **Perform operations inside grouping symbols (above and below the fraction bar and inside the absolute value symbols).**

$= \dfrac{12}{3} - 2 + 2$ • **Find the absolute value of 2.**

$= 4 - 2 + 2$ • **Do multiplication and division as they occur from left to right.**

$= 2 + 2$ • **Do addition and subtraction as they occur from left to right.**

$= 4$

PROBLEM 5 Simplify: $27 \div 3^2 + (-3)^2 \cdot 4$

Solution See page S2.

When an expression has grouping symbols inside grouping symbols, first perform the operations inside the *inner* grouping symbols by following Steps 2, 3, and 4 of the Order of Operations Agreement. Then perform the operations inside the *outer* grouping symbols by following Steps 2, 3, and 4 in sequence.

EXAMPLE 6 Simplify: $6 \div [4 - (6 - 8)] + 2^2$

Solution $6 \div [4 - (6 - 8)] + 2^2$
$= 6 \div [4 - (-2)] + 2^2$ • **Perform operations inside inner grouping symbols.**

$= 6 \div 6 + 2^2$ • **Perform operations inside outer grouping symbols.**

$= 6 \div 6 + 4$ • **Simplify exponential expressions.**
$= 1 + 4$ • **Do multiplication and division.**
$= 5$ • **Do addition and subtraction.**

PROBLEM 6 Simplify: $4 - 3[4 - 2(6 - 3)] \div 2$

Solution See page S2.

1.4 Concept Review

Determine whether the statement is always true, sometimes true, or never true.

1. $(-5)^2$, -5^2, and $-(5)^2$ all represent the same number.

2. By the Order of Operations Agreement, addition is performed before division.

3. In the expression 3^8, 8 is the base.

4. The expression 9^4 is in exponential form.

5. To evaluate the expression $6 + 7 \cdot 10$ means to determine what one number it equals.

6. When using the Order of Operations Agreement, we consider the absolute value symbol a grouping symbol.

7. Using the Order of Operations Agreement, it is possible to get more than one correct answer to a problem.

8. The Order of Operations Agreement is used for natural numbers, integers, rational numbers, and real numbers.

9. The expression $(2^3)^2$ equals $2^{(3^2)}$.

10. Suppose a and b are positive integers. Then $a^b = b^a$.

1.4 Exercises

1 Exponential expressions

1. In the expression $(-5)^2$, -5 is called the ____?____ and 2 is called the ____?____. To evaluate $(-5)^2$, find the product $(\underline{\quad ? \quad})(\underline{\quad ? \quad}) = \underline{\quad ? \quad}$.

2. The exponential expression 4^3 is read "the ____?____ power of ____?____" or "four ____?____." To evaluate 4^3, find the product $(\underline{\quad ? \quad})(\underline{\quad ? \quad})(\underline{\quad ? \quad}) = \underline{\quad ? \quad}$.

Evaluate.

3. 6^2

4. 7^4

5. -7^2

6. -4^3

7. $(-3)^2$

8. $(-2)^3$

9. $(-3)^4$

10. $(-5)^3$

11. $\left(\dfrac{1}{2}\right)^2$

12. $\left(-\dfrac{3}{4}\right)^3$

13. $(0.3)^2$

14. $(1.5)^3$

15. $\left(\dfrac{2}{3}\right)^2 \cdot 3^3$

16. $\left(-\dfrac{1}{2}\right)^3 \cdot 8$

17. $(0.3)^3 \cdot 2^3$

18. $(0.5)^2 \cdot 3^3$

19. $(-3) \cdot 2^2$

20. $(-5) \cdot 3^4$

21. $(-2) \cdot (-2)^3$

22. $(-2) \cdot (-2)^2$

23. $2^3 \cdot 3^3 \cdot (-4)$

24. $(-3)^3 \cdot 5^2 \cdot 10$

25. $(-7) \cdot 4^2 \cdot 3^2$

26. $(-2) \cdot 2^3 \cdot (-3)^2$

27. $\left(\dfrac{2}{3}\right)^2 \cdot \dfrac{1}{4} \cdot 3^3$

28. $\left(\dfrac{3}{4}\right)^2 \cdot (-4) \cdot 2^3$

29. $8^2 \cdot (-3)^5 \cdot 5$

Complete Exercises 30 and 31 without actually finding the products.

30. Is the fifth power of negative eighteen positive or negative?

31. Is the product $-(3^2)(-5^3)$ positive or negative?

2 The Order of Operations Agreement

32. Why do we need an Order of Operations Agreement?

33. Describe each step in the Order of Operations Agreement.

34. Simplify: $2(3^3)$
$2(3^3) = 2(\underline{\quad ? \quad})$ • **Simplify exponential expressions.**
$= \underline{\quad ? \quad}$ • **Do multiplication and division.**

35. Simplify: $3 - 5(6 - 8)^2$
$3 - 5(6 - 8)^2 = 3 - 5(\underline{\quad ? \quad})^2$ • **Perform operations inside grouping symbols.**
$= 3 - 5(\underline{\quad ? \quad})$ • **Simplify exponential expressions.**
$= 3 - \underline{\quad ? \quad}$ • **Do multiplication and division.**
$= \underline{\quad ? \quad}$ • **Do addition and subtraction.**

Simplify by using the Order of Operations Agreement.

36. $4 - 8 \div 2$

37. $2^2 \cdot 3 - 3$

38. $2(3 - 4) - (-3)^2$

39. $16 - 32 \div 2^3$

40. $24 - 18 \div 3 + 2$

41. $8 - (-3)^2 - (-2)$

42. $16 + 15 \div (-5) - 2$

43. $14 - 2^2 - |4 - 7|$

44. $3 - 2[8 - (3 - 2)]$

45. $-2^2 + 4[16 \div (3 - 5)]$

46. $6 + \dfrac{16 - 4}{2^2 + 2} - 2$

47. $24 \div \dfrac{3^2}{8 - 5} - (-5)$

48. $96 \div 2[12 + (6 - 2)] - 3^3$

49. $4 \cdot [16 - (7 - 1)] \div 10$

50. $16 \div 2 - 4^2 - (-3)^2$

51. $18 \div |9 - 2^3| + (-3)$

52. $16 - 3(8 - 3)^2 \div 5$

53. $4(-8) \div [2(7 - 3)^2]$

54. $\dfrac{(-10) + (-2)}{6^2 - 30} \div |2 - 4|$

55. $16 - 4 \cdot \dfrac{3^3 - 7}{2^3 + 2} - (-2)^2$

56. $0.3(1.7 - 4.8) + (1.2)^2$

57. $(1.65 - 1.05)^2 \div 0.4 + 0.8$

58. $\dfrac{3}{8} \div \left| \dfrac{5}{6} + \dfrac{2}{3} \right|$

59. $\left(\dfrac{3}{4}\right)^2 - \left(\dfrac{1}{2}\right)^3 \div \dfrac{3}{5}$

60. Which expression is equivalent to $9 - 2^2(1 - 5)$?
 a. $7^2(-4)$ **b.** $5(-4)$ **c.** $9 - 4(-4)$ **d.** $9 + 4(-4)$

61. Which expression is equivalent to $15 + 15 \div 3 - 4^2$?
 a. $30 \div 3 - 16$ **b.** $15 + 5 - 16$
 c. $15 + 5 + 16$ **d.** $15 + 15 \div (-1)^2$

1.4 Applying Concepts

62. Using the Order of Operations Agreement, describe how to simplify Exercise 52.

Place the correct symbol, $<$ or $>$, between the two numbers.

63. $(0.9)^3$ 1^5 **64.** $(-3)^3$ $(-2)^5$ **65.** $(-1.1)^2$ $(0.9)^2$

Simplify.

66. $1^2 + 2^2 + 3^2 + 4^2$ **67.** $1^3 + 2^3 + 3^3 + 4^3$

68. $(-1)^3 + (-2)^3 + (-3)^3 + (-4)^3$ **69.** $(-2)^2 + (-4)^2 + (-6)^2 + (-8)^2$

Solve.

70. Find a rational number, r, that satisfies the condition.
 a. $r^2 < r$ **b.** $r^2 = r$ **c.** $r^2 > r$

71. Computers A computer can do 6,000,000 additions in 1 s. To the nearest second, how many seconds will it take the computer to do 10^8 additions?

72. The sum of two natural numbers is 41. Each of the two numbers is the square of a natural number. Find the two numbers.

Determine the ones digit when the expression is evaluated.

73. 34^{202} **74.** 23^{502} **75.** 27^{622}

76. Prepare a report on the Kelvin scale. Include in your report an explanation of how to convert between the Kelvin scale and the Celsius scale.

1.5 Variable Expressions

OBJECTIVE 1

Evaluate variable expressions

Often we discuss a quantity without knowing its exact value, such as the price of gold next month, the cost of a new automobile next year, or the tuition for next semester. In algebra, a letter of the alphabet is used to stand for a quantity that is unknown or one that can change, or *vary*. The letter is called a **variable.** An expression that contains one or more variables is called a **variable expression.**

A variable expression is shown at the right. The expression can be rewritten by writing subtraction as the addition of the opposite.

$$3x^2 - 5y + 2xy - x - 7$$

$$3x^2 + (-5y) + 2xy + (-x) + (-7)$$

Note that the expression has five addends. The **terms** of a variable expression are the addends of the expression. The expression has five terms.

The terms $3x^2$, $-5y$, $2xy$, and $-x$ are **variable terms.**

The term -7 is a **constant term,** or simply a **constant.**

Each variable term is composed of a **numerical coefficient** and a **variable part** (the variable or variables and their exponents).

When the numerical coefficient is 1 or -1, the 1 is usually not written ($x = 1x$ and $-x = -1x$).

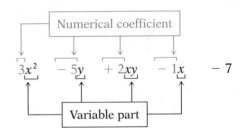

EXAMPLE 1 Name the variable terms of the expression $2a^2 - 5a + 7$.

Solution $2a^2$, $-5a$

PROBLEM 1 Name the constant term of the expression $6n^2 + 3n - 4$.

Solution See page S3.

Variable expressions occur naturally in science. In a physics lab, a student may discover that a weight of 1 pound will stretch a spring $\frac{1}{2}$ inch. A weight of 2 pounds will stretch the spring 1 inch. By experimenting, the student can discover that the distance the spring will stretch is found by multiplying the weight, in pounds, by $\frac{1}{2}$. If we let W represent the number of pounds attached to the spring, the distance, in inches, that the spring stretches can be represented by the variable expression $\frac{1}{2}W$.

With a weight of W pounds, the spring will stretch $\frac{1}{2} \cdot W = \frac{1}{2}W$ inches.

With a weight of 10 pounds, the spring will stretch $\frac{1}{2} \cdot 10 = 5$ inches.

With a weight of 3 pounds, the spring will stretch $\frac{1}{2} \cdot 3 = 1\frac{1}{2}$ inches.

Replacing the variable or variables in a variable expression with numbers and then simplifying the resulting numerical expression is called **evaluating the variable expression.**

EXAMPLE 2 Evaluate $ab - b^2$ when $a = 2$ and $b = -3$.

Solution $ab - b^2$
$2(-3) - (-3)^2$ • **Replace each variable in the expression with the number it represents.**

$= 2(-3) - 9$ • **Use the Order of Operations Agreement to simplify the resulting numerical expression.**
$= -6 - 9$
$= -15$

PROBLEM 2 Evaluate $2xy + y^2$ when $x = -4$ and $y = 2$.

Solution See page S3.

EXAMPLE 3 Evaluate $\dfrac{a^2 - b^2}{a - b}$ when $a = 3$ and $b = -4$.

Solution $\dfrac{a^2 - b^2}{a - b}$

$\dfrac{(3)^2 - (-4)^2}{3 - (-4)}$ • **Replace each variable in the expression with the number it represents.**

$= \dfrac{9 - 16}{3 - (-4)}$ • **Use the Order of Operations Agreement to simplify the resulting numerical expression.**

$= \dfrac{-7}{7}$

$= -1$

PROBLEM 3 Evaluate $\dfrac{a^2 + b^2}{a + b}$ when $a = 5$ and $b = -3$.

Solution See page S3.

EXAMPLE 4 Evaluate $x^2 - 3(x - y) - z^2$ when $x = 2$, $y = -1$, and $z = 3$.

Solution $x^2 - 3(x - y) - z^2$
$(2)^2 - 3[2 - (-1)] - (3)^2$ • **Replace each variable in the expression with the number it represents.**

$= (2)^2 - 3(3) - (3)^2$ • **Use the Order of Operations Agreement to simplify the resulting numerical expression.**
$= 4 - 3(3) - 9$
$= 4 - 9 - 9$
$= -5 - 9$
$= -14$

PROBLEM 4 Evaluate $x^3 - 2(x + y) + z^2$ when $x = 2$, $y = -4$, and $z = -3$.

Solution See page S3.

EXAMPLE 5 The diameter of the base of a right circular cylinder is 5 cm. The height of the cylinder is 8.5 cm. Find the volume of the cylinder. Round to the nearest tenth.

8.5 cm

5 cm

Take Note

For your reference, geometric formulas are printed inside the front cover of this textbook.

Solution $V = \pi r^2 h$ • Use the formula for the volume of a right circular cylinder.

$V = \pi(2.5)^2(8.5)$ • $r = \dfrac{1}{2}d = \dfrac{1}{2}(5) = 2.5$

$V = \pi(6.25)(8.5)$ • Use the π key on your calculator
$V \approx 166.9$ to enter the value of π.

The volume is approximately 166.9 cm^3.

PROBLEM 5 The diameter of the base of a right circular cone is 9 cm. The height of the cone is 9.5 cm. Find the volume of the cone. Round to the nearest tenth.

9.5 cm

9 cm

Solution See page S3.

A graphing calculator can be used to evaluate variable expressions. When the value of each variable is stored in the calculator's memory and a variable expression is then entered on the calculator, the calculator evaluates that variable expression for the values of the variables stored in its memory. See the Appendix for a description of keystroking procedures.

OBJECTIVE 2 **The Properties of Real Numbers**

The Properties of the Real Numbers describe the way operations on numbers can be performed. Here are some of the Properties of the Real Numbers and an example of each.

The Commutative Property of Addition

If a and b are real numbers, then $a + b = b + a$.

Two terms can be added in either order; the sum is the same.

$4 + 3 = 7$ and $3 + 4 = 7$

> ### The Commutative Property of Multiplication
>
> If a and b are real numbers, then $a \cdot b = b \cdot a$.

Two factors can be multiplied in either order; the product is the same.

$$(5)(-2) = -10 \quad \text{and} \quad (-2)(5) = -10$$

> ### The Associative Property of Addition
>
> If a, b, and c are real numbers, then $(a + b) + c = a + (b + c)$.

When three or more terms are added, the terms can be grouped (with parentheses, for example) in any order; the sum is the same.

$$2 + (3 + 4) = 2 + 7 = 9 \quad \text{and} \quad (2 + 3) + 4 = 5 + 4 = 9$$

> ### The Associative Property of Multiplication
>
> If a, b, and c are real numbers, then $(a \cdot b) \cdot c = a \cdot (b \cdot c)$.

When three or more factors are multiplied, the factors can be grouped in any order; the product is the same.

$$(2 \cdot 3) \cdot 4 = 6 \cdot 4 = 24 \quad \text{and} \quad 2 \cdot (3 \cdot 4) = 2 \cdot 12 = 24$$

> ### The Addition Property of Zero
>
> If a is a real number, then $a + 0 = 0 + a = a$.

The sum of a term and zero is the term.

$$4 + 0 = 4 \qquad 0 + 4 = 4$$

> ### The Multiplication Property of Zero
>
> If a is a real number, then $a \cdot 0 = 0 \cdot a = 0$.

The product of a term and zero is zero.

$$(5)(0) = 0 \qquad (0)(5) = 0$$

> ### The Multiplication Property of One
>
> If a is a real number, then $a \cdot 1 = 1 \cdot a = a$.

The product of a term and 1 is the term.

$$6 \cdot 1 = 6 \qquad 1 \cdot 6 = 6$$

> ### The Inverse Property of Addition
>
> If a is a real number, then $a + (-a) = (-a) + a = 0$.

The sum of a number and its opposite is zero.
The opposite of a number is called its **additive inverse.**

$$8 + (-8) = 0 \qquad (-8) + 8 = 0$$

> ### The Inverse Property of Multiplication
>
> If a is a real number, and $a \neq 0$, then $a \cdot \dfrac{1}{a} = \dfrac{1}{a} \cdot a = 1$.

The product of a number and its reciprocal is 1.

$\dfrac{1}{a}$ is the **reciprocal** of a. $\dfrac{1}{a}$ is also called the **multiplicative inverse** of a.

$$7 \cdot \frac{1}{7} = 1 \qquad \frac{1}{7} \cdot 7 = 1$$

> ### The Distributive Property
>
> If a, b, and c are real numbers, then $a(b + c) = ab + ac$ or
> $(b + c)a = ba + ca$.

By the Distributive Property, the term outside the parentheses is multiplied by each term inside the parentheses.

$$2(3 + 4) = 2 \cdot 3 + 2 \cdot 4 \qquad (4 + 5)2 = 4 \cdot 2 + 5 \cdot 2$$
$$2 \cdot 7 = 6 + 8 \qquad\qquad 9 \cdot 2 = 8 + 10$$
$$14 = 14 \qquad\qquad\quad 18 = 18$$

EXAMPLE 6 Complete the statement by using the Commutative Property of Multiplication.

$$(6)(5) = (?)(6)$$

Solution $(6)(5) = (5)(6)$ • **The Commutative Property of Multiplication states that $a \cdot b = b \cdot a$.**

PROBLEM 6 Complete the statement by using the Inverse Property of Addition.

$$7 + ? = 0$$

▪ **Solution** See page S3.

EXAMPLE 7 Identify the property that justifies the statement.

$$2(8 + 5) = 16 + 10$$

Solution The Distributive Property • **The Distributive Property states that $a(b + c) = ab + ac$.**

PROBLEM 7 Identify the property that justifies the statement.

$$5 + (13 + 7) = (5 + 13) + 7$$

▪ **Solution** See page S3.

OBJECTIVE ❸ Simplify variable expressions using the Properties of Addition

Like terms of a variable expression are terms with the same variable part. (Because $x^2 = x \cdot x$, x^2 and x are not like terms.)

Constant terms are like terms. 4 and 9 are like terms.

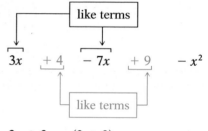

To **combine like terms,** use the Distributive Property $ba + ca = (b + c)a$ to add the coefficients.

$$2x + 3x = (2 + 3)x$$
$$= 5x$$

EXAMPLE 8 Simplify. **A.** $-2y + 3y$ **B.** $5x - 11x$

Solution **A.** $-2y + 3y$

$= (-2 + 3)y$ • **Use the Distributive Property $ba + ca = (b + c)a$.**

$= 1y$ • **Add the coefficients.**

$= y$ • **Use the Multiplication Property of One.**

B. $5x - 11x$

$= [5 + (-11)]x$ • **Use the Distributive Property $ba + ca = (b + c)a$.**

$= -6x$ • **Add the coefficients.**

PROBLEM 8 Simplify. **A.** $9x + 6x$ **B.** $-4y - 7y$

▪ **Solution** See page S3.

In simplifying more complicated expressions, use the Properties of Addition.

The Commutative Property of Addition can be used when adding two like terms. The terms can be added in either order. The sum is the same.

$$2x + (-4x) = -4x + 2x$$
$$[2 + (-4)]x = (-4 + 2)x$$
$$-2x = -2x$$

The Associative Property of Addition is used when adding three or more terms. The terms can be grouped in any order. The sum is the same.

$$3x + 5x + 9x = (3x + 5x) + 9x = 3x + (5x + 9x)$$
$$8x + 9x = 3x + 14x$$
$$17x = 17x$$

By the Addition Property of Zero, the sum of a term and zero is the term.

$$5x + 0 = 0 + 5x = 5x$$

By the Inverse Property of Addition, the sum of a term and its additive inverse is zero.

$$7x + (-7x) = -7x + 7x = 0$$

EXAMPLE 9 Simplify. **A.** $8x + 3y - 8x$ **B.** $4x^2 + 5x - 6x^2 - 2x$

Solution **A.** $8x + 3y - 8x$

$= 3y + 8x - 8x$
- **Use the Commutative Property of Addition to rearrange the terms.**

$= 3y + (8x - 8x)$
- **Use the Associative Property of Addition to group like terms.**

$= 3y + 0$
- **Use the Inverse Property of Addition.**

$= 3y$
- **Use the Addition Property of Zero.**

B. $4x^2 + 5x - 6x^2 - 2x$
$= 4x^2 - 6x^2 + 5x - 2x$
- **Use the Commutative Property of Addition to rearrange the terms.**

$= (4x^2 - 6x^2) + (5x - 2x)$
- **Use the Associative Property of Addition to group like terms.**

$= -2x^2 + 3x$
- **Combine like terms.**

PROBLEM 9 Simplify. **A.** $3a - 2b + 5a$ **B.** $x^2 - 7 + 9x^2 - 14$

Solution See page S3.

OBJECTIVE **4** **Simplify variable expressions using the Properties of Multiplication**

The Properties of Multiplication are used in simplifying variable expressions.

The Associative Property is used when multi-plying three or more factors.
$2(3x) = (2 \cdot 3)x = 6x$

The Commutative Property can be used to change the order in which factors are multiplied.

$(3x) \cdot 2 = 2 \cdot (3x) = 6x$

By the Multiplication Property of One, the product of a term and 1 is the term.

$(8x)(1) = (1)(8x) = 8x$

By the Inverse Property of Multiplication, the product of a term and its reciprocal is 1.

$5x \cdot \dfrac{1}{5x} = \dfrac{1}{5x} \cdot 5x = 1, x \neq 0$

EXAMPLE 10 Simplify. **A.** $2(-x)$ **B.** $\dfrac{3}{2}\left(\dfrac{2x}{3}\right)$ **C.** $(16x)2$

Solution **A.** $2(-x) = 2(-1 \cdot x)$ • $-x = -1x = -1 \cdot x$
$\qquad\qquad\quad = [2 \cdot (-1)]x$ • **Use the Associative Property of Multiplication to group factors.**
$\qquad\qquad\quad = -2x$ • **Multiply.**

B. $\dfrac{3}{2}\left(\dfrac{2x}{3}\right) = \dfrac{3}{2}\left(\dfrac{2}{3}x\right)$ • **Note that** $\dfrac{2x}{3} = \dfrac{2}{3} \cdot \dfrac{x}{1} = \dfrac{2}{3}x.$

$\qquad\qquad\quad = \left(\dfrac{3}{2} \cdot \dfrac{2}{3}\right)x$ • **Use the Associative Property of Multiplication to group factors.**

$\qquad\qquad\quad = 1x$ • **Use the Inverse Property of Multiplication.**
$\qquad\qquad\quad = x$ • **Use the Multiplication Property of One.**

C. $(16x)2 = 2(16x)$ • **Use the Commutative Property of Multiplication to rearrange factors.**

$\qquad\qquad\quad = (2 \cdot 16)x$ • **Use the Associative Property of Multiplication to group factors.**

$\qquad\qquad\quad = 32x$ • **Multiply.**

PROBLEM 10 Simplify. **A.** $-7(-2a)$ **B.** $-\dfrac{5}{6}(-30y^2)$ **C.** $(-5x)(-2)$

Solution See page S3.

OBJECTIVE **5** **Simplify variable expressions using the Distributive Property**

The Distributive Property is used to remove parentheses from a variable expression.

$3(2x - 5)$
$= 3(2x) - 3(5)$
$= 6x - 15$

An extension of the Distributive Property is used when an expression contains more than two terms.

$4(x^2 + 6x - 1)$
$= 4(x^2) + 4(6x) - 4(1)$
$= 4x^2 + 24x - 4$

EXAMPLE 11 Simplify. **A.** $-3(5 + x)$ **B.** $-(2x - 4)$
 C. $(2y - 6)2$ **D.** $5(3x + 7y - z)$

Solution **A.** $-3(5 + x)$
$\qquad\qquad\quad = -3(5) + (-3)x$ • **Use the Distributive Property.**
$\qquad\qquad\quad = -15 - 3x$ • **Multiply.**

B. $-(2x - 4)$
 $= -1(2x - 4)$

 $= -1(2x) - (-1)(4)$
 $= -2x + 4$

- Just as $-x = -1x$,
 $-(2x - 4) = -1(2x - 4)$.
- **Use the Distributive Property.**
- *Note:* When a negative sign immediately precedes the parentheses, remove the parentheses and change the sign of *each* term inside the parentheses.

C. $(2y - 6)2$
 $= (2y)(2) - (6)(2)$
 $= 4y - 12$

- **Use the Distributive Property**
 $(b + c)a = ba + ca$.

D. $5(3x + 7y - z)$
 $= 5(3x) + 5(7y) - 5(z)$
 $= 15x + 35y - 5z$

- **Use the Distributive Property.**

PROBLEM 11 Simplify. **A.** $7(4 + 2y)$ **B.** $-(5x - 12)$
 C. $(3a - 1)5$ **D.** $-3(6a^2 - 8a + 9)$

Solution See page S3.

OBJECTIVE **6** ## Simplify general variable expressions

When simplifying variable expressions, use the Distributive Property to remove parentheses and brackets used as grouping symbols.

EXAMPLE 12 Simplify: $4(x - y) - 2(-3x + 6y)$

Solution $4(x - y) - 2(-3x + 6y)$
 $= 4x - 4y + 6x - 12y$

- **Use the Distributive Property to remove parentheses.**

 $= 10x - 16y$

- **Combine like terms.**

PROBLEM 12 Simplify: $7(x - 2y) - 3(-x - 2y)$

Solution See page S3.

EXAMPLE 13 Simplify: $2x - 3[2x - 3(x + 7)]$

Solution $2x - 3[2x - 3(x + 7)]$
 $= 2x - 3[2x - 3x - 21]$

- **Use the Distributive Property to remove the inner grouping symbols.**

 $= 2x - 3[-x - 21]$

- **Combine like terms inside the grouping symbols.**

 $= 2x + 3x + 63$

- **Use the Distributive Property to remove the brackets.**

 $= 5x + 63$

- **Combine like terms.**

PROBLEM 13 Simplify: $3y - 2[x - 4(2 - 3y)]$

Solution See page S3.

 Concept Review

Determine whether the statement is always true, sometimes true, or never true.

1. The Associative Property of Addition states that two terms can be added in either order and the sum will be the same.

2. When three numbers are multiplied, the numbers can be grouped in any order and the product will be the same.

3. The Multiplication Property of One states that multiplying a number by one does not change the number.

4. The sum of a number and its additive inverse is zero.

5. The product of a number and its multiplicative inverse is one.

6. The terms x and x^2 are like terms because both have a coefficient of 1.

7. $3(x + 4) = 3x + 4$ is an example of the correct application of the Distributive Property.

8. To add like terms, add the coefficients; the variable part remains unchanged.

1.5 **Exercises**

1 **Evaluate variable expressions**

1. To identify the terms of the variable expression $3x^2 - 4x - 7$, first write subtraction as addition of the opposite: $3x^2 + ($___?___$) + ($___?___$)$. The terms of $3x^2 - 4x - 7$ are ___?___, ___?___, and ___?___.

2. Evaluate $mn^2 - m$ when $m = -2$ and $n = 5$.
 $mn^2 - m = (-2)(5^2) - (-2)$ • Replace m with ___?___ and n with ___?___.

 $\quad\quad = (-2)($___?___$) - (-2)$ • Simplify the exponential expression.
 $\quad\quad = $___?___$ - (-2)$ • Multiply.
 $\quad\quad = -50 + $___?___ • Write subtraction as addition of the opposite.

 $\quad\quad = $___?___ • Add.

3. What is the meaning of the phrase "evaluate a variable expression"?

Evaluate the variable expression when $a = 2$, $b = 3$, and $c = -4$.

4. $a^2 - b^2$ 5. $b^2 - c^2$ 6. $(a + b)^2$

7. $a^2 + b^2$ 8. $2a - (c + a)^2$ 9. $(b - a)^2 + 4c$

10. $b^2 - \dfrac{ac}{8}$ **11.** $\dfrac{5ab}{6} - 3cb$ **12.** $(b - 2a)^2 + bc$

Evaluate the variable expression when $a = -2$, $b = 4$, $c = -1$, and $d = 3$.

13. $(b + d)^2 - 4a$ **14.** $(d - a)^2 - 3c$ **15.** $(d - a)^2 \div 5$

16. $(b - c)^2 \div 5$ **17.** $b^2 - 2b + 4$ **18.** $a^2 - 5a - 6$

19. $\dfrac{bd}{a} \div c$ **20.** $\dfrac{2ac}{b} \div (-c)$ **21.** $2(b + c) - 2a$

22. $(b + c)^2 + (a + d)^2$ **23.** $4ac + (2a)^2$ **24.** $3dc - (4c)^2$

Evaluate the variable expression when $a = 2.7$, $b = -1.6$, and $c = -0.8$.

25. $c^2 - ab$ **26.** $(a + b)^2 - c$ **27.** $\dfrac{b^3}{c} - 4a$

Complete Exercises 28 and 29 without a calculator.

28. If $\dfrac{b + c}{abc}$ is evaluated when $a = -25$, $b = 67$, and $c = -82$, will the result be positive or negative?

29. A sphere has a radius of 2 cm.
 a. Can the exact volume of the sphere be V cm^3, where V is a whole number?
 b. Can the exact volume of the sphere be V cm^2, where V is an irrational number?

Solve. Round to the nearest tenth.

30. **Geometry** Find the volume of a sphere that has a radius of 8.5 cm.

31. **Geometry** Find the volume of a right circular cylinder that has a radius of 1.25 in. and a height of 5.25 in.

32. **Geometry** The radius of the base of a right circular cylinder is 3.75 ft. The height of the cylinder is 9.5 ft. Find the surface area of the cylinder.

33. **Geometry** The length of one base of a trapezoid is 17.5 cm, and the length of the other base is 10.25 cm. The height is 6.75 cm. What is the area of the trapezoid?

34. **Geometry** A right circular cone has a height of 2.75 in. The diameter of the base is 1 in. Find the volume of the cone.

35. **Geometry** A right circular cylinder has a height of 12.6 m. The diameter of the base is 7 m. Find the volume of the cylinder.

8.5 cm

10.25 cm

6.75 cm

17.5 cm

12.6 m

7 m

36. The value of z is the value of $a^2 - 2a$ when $a = -3$. Find the value of z^2.

37. The value of a is the value of $3x^2 - 4x - 5$ when $x = -2$. Find the value of $3a - 4$.

38. The value of c is the value of $a^2 + b^2$ when $a = 2$ and $b = -2$. Find the value of $c^2 - 4$.

2 The Properties of Real Numbers

39. The fact that two terms can be added in either order is called the ___?___ Property of Addition.

40. The fact that three or more factors can be multiplied by grouping them in any order is called the ___?___ Property of Multiplication.

41. The Inverse Property of Multiplication tells us that the product of a number and its ___?___ is one.

42. The Inverse Property of Addition tells us that the sum of a number and its ___?___ is zero.

Use the given property to complete the statement.

43. The Commutative Property of Multiplication
$2 \cdot 5 = 5 \cdot ?$

44. The Commutative Property of Addition
$9 + 17 = ? + 9$

45. The Associative Property of Multiplication
$(4 \cdot 5) \cdot 6 = 4 \cdot (? \cdot 6)$

46. The Associative Property of Addition
$(4 + 5) + 6 = ? + (5 + 6)$

47. The Distributive Property
$2(4 + 3) = 8 + ?$

48. The Addition Property of Zero
$? + 0 = -7$

49. The Inverse Property of Addition
$8 + ? = 0$

50. The Inverse Property of Multiplication
$\dfrac{1}{-5}(-5) = ?$

Identify the property that justifies the statement.

51. $-7 + 7 = 0$

52. $(-8)\left(-\dfrac{1}{8}\right) = 1$

53. $23 + 19 = 19 + 23$

54. $-21 + 0 = -21$

55. $3 \cdot 5 = 5 \cdot 3$

56. $-32(0) = 0$

57. $(4 \cdot 3) \cdot 5 = 4 \cdot (3 \cdot 5)$

58. $\dfrac{1}{4}(1) = \dfrac{1}{4}$

3 Simplify variable expressions using the Properties of Addition

59. What are *like terms*? Give an example of two like terms. Give an example of two terms that are not like terms.

60. Explain the meaning of the phrase "simplify a variable expression."

61. Which of the following are like terms?
$3x$, $3x^2$, $5x$, $5xy$

62. Which of the following are like terms?
$-7a$, $-7b$, $-4a$, $-7a^2$

63. Simplify $5a - 8a$ by combining like terms.

$$5a - 8a = 5a + \underline{\quad ? \quad}$$ • **Write subtraction as addition of the opposite.**
$$= [5 + (-8)]a$$ • **Use the** $\underline{\quad ? \quad}$ **Property** $ba + ca = (b + c)a$.
$$= \underline{\quad ? \quad} a$$ • **Add the coefficients.**

64. Simplify $6xy - 6x + 7xy$ by combining like terms.

$$6xy - 6x + 7xy = 6xy + 7xy - 6x$$ • **Use the** $\underline{\quad ? \quad}$ **Property of Addition to rearrange the terms.**
$$= (6 + 7)(\underline{\quad ? \quad}) - 6x$$ • **Use the Distributive Property** $ba + ca = (b + c)a$.
$$= \underline{\quad ? \quad} - 6x$$ • **Add the coefficients.**

Simplify.

65. $-12a + 17a$

66. $-3a + 12a$

67. $5ab - 7ab$

68. $9ab - 3ab$

69. $-12xy + 17xy$

70. $-15xy + 3xy$

71. $-3ab + 3ab$

72. $-7ab + 7ab$

73. $-\dfrac{1}{2}x - \dfrac{1}{3}x$

74. $-\dfrac{2}{5}y + \dfrac{3}{10}y$

75. $\dfrac{3}{8}x^2 - \dfrac{5}{12}x^2$

76. $\dfrac{2}{3}y^2 - \dfrac{4}{9}y^2$

77. $3x + 5x + 3x$

78. $8x + 5x + 7x$

79. $5a - 3a + 5a$

80. $10a - 17a + 3a$

81. $-5x^2 - 12x^2 + 3x^2$

82. $-y^2 - 8y^2 + 7y^2$

83. $7x - 8x + 3y$

84. $8y - 10x + 8x$

85. $7x - 3y + 10x$

86. $8y + 8x - 8y$

87. $3a - 7b - 5a + b$

88. $-5b + 7a - 7b + 12a$

89. $3x - 8y - 10x + 4x$

90. $3y - 12x - 7y + 2y$

91. $x^2 - 7x - 5x^2 + 5x$

92. $3x^2 + 5x - 10x^2 - 10x$

93. State whether the given number will be positive, negative, or zero after the variable expression $15a^2 - 12a + 7 - 21a^2 + 29a - 7$ is simplified.
a. the coefficient of a^2 **b.** the coefficient of a **c.** the constant term

94. Which expressions are equivalent to $-10x - 10y - 10y - 10x$?
a. 0 **b.** $-20y$ **c.** $-20x$ **d.** $-20x - 20y$ **e.** $-20y - 20x$

4 Simplify variable expressions using the Properties of Multiplication

95. Simplify: $4(-12x)$

$4(-12x) = [4(-12)]x$ • Use the ____?____ Property of Multiplication to group factors.

$= \underline{\quad ? \quad}$ • Multiply.

96. Simplify: $\left(\dfrac{1}{2}a^2\right)\left(-\dfrac{3}{2}\right)$

$\left(\dfrac{1}{2}a^2\right)\left(-\dfrac{3}{2}\right) = (\underline{\quad ? \quad})\left(\dfrac{1}{2}a^2\right)$ • Use the Commutative Property of Multiplication to rearrange factors.

$= \left(-\dfrac{3}{2} \cdot \underline{\quad ? \quad}\right)a^2$ • Use the Associative Property of Multiplication to group factors.

$= \underline{\quad ? \quad}a^2$ • Multiply.

Simplify.

97. $4(3x)$

98. $12(5x)$

99. $-3(7a)$

100. $-2(5a)$

101. $-2(-3y)$

102. $-5(-6y)$

103. $(4x)2$

104. $(6x)12$

105. $(3a)(-2)$

106. $(7a)(-4)$

107. $(-3b)(-4)$

108. $(-12b)(-9)$

109. $-5(3x^2)$

110. $-8(7x^2)$

111. $\dfrac{1}{3}(3x^2)$

112. $\dfrac{1}{5}(5a)$

113. $\dfrac{1}{8}(8x)$

114. $-\dfrac{1}{4}(-4a)$

115. $-\dfrac{1}{7}(-7n)$

116. $\left(\dfrac{3x}{4}\right)\left(\dfrac{4}{3}\right)$

117. $\dfrac{12x}{5}\left(\dfrac{5}{12}\right)$

118. $(-6y)\left(-\dfrac{1}{6}\right)$

119. $(-10n)\left(-\dfrac{1}{10}\right)$

120. $\dfrac{1}{3}(9x)$

121. $\dfrac{1}{7}(14x)$

122. $-\dfrac{1}{5}(10x)$

123. $-\dfrac{1}{8}(16x)$

124. $-\dfrac{2}{3}(12a^2)$

125. $-\dfrac{5}{8}(24a^2)$

126. $-\dfrac{1}{2}(-16y)$

127. $-\dfrac{3}{4}(-8y)$

128. $(16y)\left(\dfrac{1}{4}\right)$

129. After $\dfrac{2}{7}x^2$ is multiplied by a proper fraction, is the coefficient of x^2 greater than one or less than one?

130. After $-152m$ is multiplied by a whole number, is the coefficient of m greater than zero or less than zero?

5 **Simplify variable expressions using the Distributive Property**

131. Simplify: $-8(3x - 2)$
$-8(x - 2) = (-8)(3x) - (-8)(2)$ • Use the ___?___ Property.
$\quad = $ ___?___ $x + $ ___?___ • Multiply.

132. Simplify: $(7a - 3)(-2)$
$(7a - 3)(-2) = (7a)($ ___?___ $) - 3($ ___?___ $)$ • Use the Distributive Property.
$\quad = $ ___?___ $ + $ ___?___ • Multiply.

Simplify.

133. $-(x + 2)$ **134.** $-(x + 7)$ **135.** $2(4x - 3)$ **136.** $5(2x - 7)$

137. $-2(a + 7)$ **138.** $-5(a + 16)$ **139.** $-3(2y - 8)$ **140.** $-5(3y - 7)$

141. $(5 - 3b)7$ **142.** $(10 - 7b)2$ **143.** $-3(3 - 5x)$ **144.** $-5(7 - 10x)$

145. $3(5x^2 + 2x)$ **146.** $6(3x^2 + 2x)$ **147.** $-2(-y + 9)$ **148.** $-5(-2x + 7)$

149. $(-3x - 6)5$ **150.** $(-2x + 7)7$ **151.** $2(-3x^2 - 14)$ **152.** $5(-6x^2 - 3)$

153. $-3(2y^2 - 7)$ **154.** $-8(3y^2 - 12)$ **155.** $-(6a^2 - 7b^2)$

156. $3(x^2 + 2x - 6)$ **157.** $4(x^2 - 3x + 5)$ **158.** $-2(y^2 - 2y + 4)$

159. $-3(y^2 - 3y - 7)$ **160.** $2(-a^2 - 2a + 3)$ **161.** $4(-3a^2 - 5a + 7)$

162. After the expression $(17x - 31)$ is multiplied by a negative integer, is the constant term positive or negative?

163. The expression $(5x - 8xy + 7y)$ is multiplied by a positive integer. How many variable terms does the resulting expression have?

6 **Simplify general variable expressions**

164. When simplifying $4(3a - 7) + 2(a - 3)$, the first step is to use the Distributive Property to remove parentheses:
$4(3a - 7) + 2(a - 3) = ($ ___?___ $)a - $ ___?___ $ + ($ ___?___ $)a - $ ___?___

165. When simplifying $12m - (m - 3)$, the first step is to remove the parentheses and change the sign of each term inside the parentheses:
$12m - $ ___?___ $ + $ ___?___

Simplify.

166. $4x - 2(3x + 8)$ **167.** $6a - (5a + 7)$ **168.** $9 - 3(4y + 6)$

169. $10 - (11x - 3)$ **170.** $5n - (7 - 2n)$ **171.** $8 - (12 + 4y)$

172. $3(x + 2) - 5(x - 7)$ **173.** $2(x - 4) - 4(x + 2)$ **174.** $12(y - 2) + 3(7 - 3y)$

175. $6(2y - 7) - 3(3 - 2y)$ **176.** $3(a - b) - 4(a + b)$ **177.** $2(a + 2b) - (a - 3b)$

178. $4[x - 2(x - 3)]$ **179.** $2[x + 2(x + 7)]$ **180.** $-2[3x + 2(4 - x)]$

181. $-5[2x + 3(5 - x)]$ **182.** $-3[2x - (x + 7)]$ **183.** $-2[3x - (5x - 2)]$

184. $2x - 3[x - 2(4 - x)]$ **185.** $-7x + 3[x - 7(3 - 2x)]$ **186.** $-5x - 2[2x - 4(x + 7)] - 6$

187. $4a - 2[2b - (b - 2a)] + 3b$ **188.** $2x + 3(x - 2y) + 5(3x - 7y)$ **189.** $5y - 2(y - 3x) + 2(7x - y)$

190. Which expression is equivalent to $12 - 7(y - 9)$?
 a. $5(y - 9)$ **b.** $12 - 7y - 63$ **c.** $12 - 7y + 63$ **d.** $12 - 7y - 9$

191. Which expression is equivalent to $7[3b + 5(b - 6)]$?
 a. $7[3b + 5b - 6)]$ **b.** $7[3b + 5b - 30)]$ **c.** $7[3b + 5b + 6)]$ **d.** $7[8b - 6]$

1.5 Applying Concepts

Simplify.

192. $C - 0.7C$ **193.** $\dfrac{1}{3}(3x + y) - \dfrac{2}{3}(6x - y)$ **194.** $-\dfrac{1}{4}[2x + 2(y - 6y)]$

Complete.

195. **a.** A number that has no reciprocal is _____.

 b. A number that is its own reciprocal is _____.

196. Determine whether the statement is true or false. If the statement is false, give an example that illustrates that it is false.
 a. Division is a commutative operation.
 b. Division is an associative operation.
 c. Subtraction is an associative operation.
 d. Subtraction is a commutative operation.
 e. Addition is a commutative operation.

197. Give examples of two operations that occur in everyday experience that are not commutative (for example, putting on socks and putting on shoes).

1.6 Translating Verbal Expressions into Variable Expressions

OBJECTIVE 1 **Translate a verbal expression into a variable expression**

One of the major skills required in applied mathematics is the ability to translate a verbal expression into a variable expression. This requires recognizing the verbal phrases that translate into mathematical operations. Here is a partial list of the phrases used to indicate the different mathematical operations.

Point of Interest

The way in which expressions are symbolized has changed over time. Here are some expressions as they may have appeared in the early 16th century.

* **R p. 9** *for* x + 9. *The symbol* **R** *was used for a variable to the first power. The symbol* **p.** *was used for plus.*

* **R m. 3** *for* x − 3. *Again, the symbol* **R** *was used for the variable. The symbol* **m.** *was used for minus.*

* *The square of a variable was designated by* **Q** *and the cube was designated by* **C**. *The expression* x³ + x² *was written* **C p. Q.**

Addition	added to	6 added to y	$y + 6$
	more than	8 more than x	$x + 8$
	the sum of	the sum of x and z	$x + z$
	increased by	t increased by 9	$t + 9$
	the total of	the total of 5 and y	$5 + y$
	plus	b plus 17	$b + 17$
Subtraction	minus	x minus 2	$x - 2$
	less than	7 less than t	$t - 7$
	less	7 less t	$7 - t$
	subtracted from	5 subtracted from d	$d - 5$
	decreased by	m decreased by 3	$m - 3$
	the difference between	the difference between y and 4	$y - 4$
Multiplication	times	10 times t	$10t$
	of	one half of x	$\frac{1}{2}x$
	the product of	the product of y and z	yz
	multiplied by	y multiplied by 11	$11y$
	twice	twice n	$2n$
Division	divided by	x divided by 12	$\frac{x}{12}$
	the quotient of	the quotient of y and z	$\frac{y}{z}$
	the ratio of	the ratio of t to 9	$\frac{t}{9}$
Power	the square of	the square of x	x^2
	the cube of	the cube of a	a^3

Translating a phrase that contains the word *sum, difference, product,* or *quotient* can sometimes cause a problem. In the examples at the right, note where the operation symbol is placed.

the *sum* of x and y $x + y$

the *difference* between x and y $x - y$

the *product* of x and y $x \cdot y$

the *quotient* of x and y $\dfrac{x}{y}$

EXAMPLE 1 Translate into a variable expression.

A. the total of five times b and c

B. the quotient of eight less than n and fourteen

C. thirteen more than the sum of seven and the square of x

Solution **A.** the total of five times b and c

$5b + c$

- Identify words that indicate mathematical operations.
- Use the operations to write the variable expression.

B. the quotient of eight less than n and fourteen

$\dfrac{n - 8}{14}$

- Identify words that indicate mathematical operations.
- Use the operations to write the variable expression.

C. thirteen more than the sum of seven and the square of x

$(7 + x^2) + 13$

PROBLEM 1 Translate into a variable expression.

A. eighteen less than the cube of x

B. y decreased by the sum of z and nine

C. the difference between the square of q and the sum of r and t

Solution See page S3.

In most applications that involve translating phrases into variable expressions, the variable to be used is not given. To translate these phrases, we must assign a variable to an unknown quantity before the variable expression can be written.

Take Note

The expression $n(6 + n^3)$ must have parentheses. If we write $n \cdot 6 + n^3$, then by the Order of Operations Agreement, only the 6 is multiplied by n, but we want n to be multiplied by the *total* of 6 and n^3.

EXAMPLE 2 Translate "a number multiplied by the total of six and the cube of the number" into a variable expression.

Solution the unknown number: n

the cube of the number: n^3
the total of six and the cube of the number: $6 + n^3$

$n(6 + n^3)$

- Assign a variable to one of the unknown quantities.
- Use the assigned variable to write an expression for any other unknown quantity.
- Use the assigned variable to write the variable expression.

PROBLEM 2 Translate "a number added to the product of five and the square of the number" into a variable expression.

Solution See page S3.

EXAMPLE 3 Translate "the quotient of twice a number and the difference between the number and twenty" into a variable expression.

Solution the unknown number: n
twice the number: $2n$
the difference between the number and twenty: $n - 20$

$$\frac{2n}{n - 20}$$

PROBLEM 3 Translate "the product of three and the sum of seven and twice a number" into a variable expression.

Solution See page S3.

OBJECTIVE **2** **Translate a verbal expression into a variable expression and then simplify the resulting expression**

After translating a verbal expression into a variable expression, simplify the variable expression by using the Properties of the Real Numbers.

EXAMPLE 4 Translate and simplify "the total of four times an unknown number and twice the difference between the number and eight."

Solution the unknown number: n

four times the unknown number: $4n$
twice the difference between the number and eight: $2(n - 8)$

$4n + 2(n - 8)$

$= 4n + 2n - 16$
$= 6n - 16$

• Assign a variable to one of the unknown quantities.
• Use the assigned variable to write an expression for any other unknown quantity.
• Use the assigned variable to write the variable expression.
• Simplify the variable expression.

PROBLEM 4 Translate and simplify "a number minus the difference between twice the number and seventeen."

Solution See page S4.

EXAMPLE 5 Translate and simplify "the difference between five-eighths of a number and two-thirds of the same number."

Solution the unknown number: n

five-eighths of the number: $\frac{5}{8}n$

two-thirds of the number: $\frac{2}{3}n$

• Assign a variable to one of the unknown quantities.
• Use the assigned variable to write an expression for any other unknown quantity.

$$\frac{5}{8}n - \frac{2}{3}n$$

- Use the assigned variable to write the variable expression.

$$= \frac{15}{24}n - \frac{16}{24}n$$

- Simplify the variable expression.

$$= -\frac{1}{24}n$$

PROBLEM 5 Translate and simplify "the sum of three-fourths of a number and one-fifth of the same number."

Solution See page S4.

OBJECTIVE **3** **Translate application problems**

Many of the applications of mathematics require that you identify an unknown quantity, assign a variable to that quantity, and then attempt to express another unknown quantity in terms of that variable.

Suppose we know that the sum of two numbers is 10 and that one of the two numbers is 4. We can find the other number by subtracting 4 from 10.

one number: 4
other number: $10 - 4 = 6$
The two numbers are 4 and 6.

Now suppose we know that the sum of two numbers is 10, we don't know either number, and we want to express *both* numbers in terms of the *same* variable.

Let one number be x. Again, we can find the other number by subtracting x from 10.

one number: x
other number: $10 - x$
The two numbers are x and $10 - x$.

Note that the sum of x and $10 - x$ is 10. $x + (10 - x) = x + 10 - x = 10$

EXAMPLE 6 The length of a swimming pool is 20 ft longer than the width. Express the length of the pool in terms of the width.

Solution the width of the pool: W

- Assign a variable to the width of the pool.

the length is 20 more than the width: $W + 20$

- Express the length of the pool in terms of W.

Take Note

Any variable can be used. For example, if the width is y, then the length is $y + 20$.

PROBLEM 6 An older computer takes twice as long to process a set of data as does a newer model. Express the amount of time it takes the older computer to process the data in terms of the amount of time it takes the newer model.

Solution See page S4.

EXAMPLE 7 An investor divided $5000 between two accounts, one a mutual fund and the other a money market fund. Use one variable to express the amounts invested in each account.

Solution the amount invested in the
mutual fund: x

• Assign a variable to the amount invested in one account.

the amount invested in the
money market fund:
$5000 - x$

• Express the amount invested in the other account in terms of x.

PROBLEM 7 A guitar string 6 ft long was cut into two pieces. Use one variable to express the lengths of the two pieces.

Solution See page S4.

1.6 Concept Review

Determine whether the statement is always true, sometimes true, or never true.

1. "Five less than n" can be translated as "$5 - n$."

2. A variable expression contains an equals sign.

3. If the sum of two numbers is 12 and one of the two numbers is x, then the other number can be expressed as $x - 12$.

4. The words *total* and *times* both indicate multiplication.

5. The words *quotient* and *ratio* both indicate division.

6. The expressions $7y - 8$ and $(7y) - 8$ are equivalent.

7. The phrase "five times the sum of x and y" and the phrase "the sum of five times x and y" yield the same variable expression.

1.6 Exercises

1 Translate a verbal expression into a variable expression

For each phrase, identify the words that indicate mathematical operations.

1. the sum of seven and three times m

2. twelve less than the quotient of x and negative two

3. the total of ten and fifteen divided by a number

4. twenty subtracted from the product of eight and the cube of a number

Translate into a variable expression.

5. d less than nineteen

6. the sum of six and c

7. r decreased by twelve

8. w increased by fifty-five

9. a multiplied by twenty-eight

10. y added to sixteen

11. five times the difference between n and seven

12. thirty less than the square of b

13. y less the product of three and y

14. the sum of four-fifths of m and eighteen

15. the product of negative six and b

16. nine increased by the quotient of t and five

17. four divided by the difference between p and six

18. the product of seven and the total of r and eight

19. the quotient of nine less than x and twice x

20. the product of a and the sum of a and thirteen

21. twenty-one less than the product of s and negative four

22. fourteen more than one-half of the square of z

23. the ratio of eight more than d to d

24. the total of nine times the cube of m and the square of m

25. three-eighths of the sum of t and fifteen

26. s decreased by the quotient of s and two

27. w increased by the quotient of seven and w

28. the difference between the square of c and the total of c and fourteen

29. d increased by the difference between sixteen times d and three

30. the product of eight and the total of b and five

31. a number divided by nineteen

32. thirteen less a number

33. forty more than a number

34. three-sevenths of a number

35. the square of the difference between a number and ninety

36. the quotient of twice a number and five

37. the sum of four-ninths of a number and twenty

38. eight subtracted from the product of fifteen and a number

39. the product of a number and ten more than the number

40. six less than the total of a number and the cube of the number

41. fourteen added to the product of seven and a number

42. the quotient of three and the total of four and a number

43. the quotient of twelve and the sum of a number and two

44. eleven plus one-half of a number

45. the ratio of two and the sum of a number and one

46. a number multiplied by the difference between twice the number and nine

47. the difference between sixty and the quotient of a number and fifty

48. the product of nine less than a number and the number

49. the sum of the square of a number and three times the number

50. the quotient of seven more than twice a number and the number

51. the sum of three more than a number and the cube of the number

52. a number decreased by the difference between the cube of the number and ten

53. a number decreased by one-fourth of the square of the number

54. four less than the product of seven and the square of a number

55. twice a number decreased by the quotient of seven and the number

56. eighty decreased by the product of thirteen and a number

57. the cube of a number decreased by the product of twelve and the number

58. the quotient of five and the sum of a number and nineteen

59. Which phrase(s) translate(s) into the variable expression $25n^2 - 9$?

 a. the difference between nine and the product of twenty-five and the square of a number
 b. nine subtracted from the square of twenty-five and a number
 c. nine less than the product of twenty-five and the square of a number

60. Which phrase(s) translate(s) into the variable expression $32 - \dfrac{a}{7}$?

 a. the difference between thirty-two and the quotient of a number and seven
 b. thirty-two decreased by the quotient of a number and seven
 c. thirty-two minus the ratio of a number to seven

2 **Translate a verbal expression into a variable expression and then simplify the resulting expression**

61. The phrase "the total of one-half of a number and three-fourths of the number" can be translated as ($\underline{\quad?\quad}$)n + ($\underline{\quad?\quad}$)n. This expression simplifies to $\underline{\quad?\quad}$.

62. The phrase "the difference between twelve times a number and fifteen times the number" can be translated as ($\underline{\quad?\quad}$)n − ($\underline{\quad?\quad}$)n. This expression simplifies to $\underline{\quad?\quad}$.

Translate into a variable expression. Then simplify the expression.

63. a number increased by the total of the number and ten

64. a number added to the product of five and the number

65. a number decreased by the difference between nine and the number

66. eight more than the sum of a number and eleven

67. the difference between one-fifth of a number and three-eighths of the number

68. a number minus the sum of the number and fourteen

69. four more than the total of a number and nine

70. the sum of one-eighth of a number and one-twelfth of the number

71. twice the sum of three times a number and forty

72. the sum of a number divided by two and the number

73. seven times the product of five and a number

74. sixteen multiplied by one-fourth of a number

75. the total of seventeen times a number and twice the number

76. the difference between nine times a number and twice the number

77. a number plus the product of the number and twelve

78. nineteen more than the difference between a number and five

79. three times the sum of the square of a number and four

80. a number subtracted from the product of the number and seven

81. three-fourths of the sum of sixteen times a number and four

82. the difference between fourteen times a number and the product of the number and seven

83. sixteen decreased by the sum of a number and nine

84. eleven subtracted from the difference between eight and a number

85. five more than the quotient of four times a number and two

86. twenty minus the sum of four-ninths of a number and three

87. six times the total of a number and eight

88. four times the sum of a number and twenty

89. seven minus the sum of a number and two

90. three less than the sum of a number and ten

91. one-third of the sum of a number and six times the number

92. twice the quotient of four times a number and eight

93. the total of eight increased by the cube of a number and twice the cube of the number

94. the sum of five more than the square of a number and twice the square of the number

95. twelve more than a number added to the difference between the number and six

96. a number plus four added to the difference between three and twice the number

97. the sum of a number and nine added to the difference between the number and twenty

98. seven increased by a number added to twice the difference between the number and two

99. fourteen plus the product of three less than a number and ten

100. a number plus the product of the number minus five and seven

3 Translate application problems

101. The sum of two numbers is 25. To express both numbers in terms of the same variable, let x be one number. Then the other number is ___?___.

102. The length of a rectangle is five times the width. To express the length and the width in terms of the same variable, let W be the width. Then the length is ___?___.

103. The width of a rectangle is one-half the length. To express the length and the width in terms of the same variable, let L be the length. Then the width is ___?___.

104. An electrician's bill is $75 for materials and $50 an hour for labor. To express the total amount of the bill in terms of the number of hours of labor, let h be the number of hours of labor. Then the cost of the labor is ___?___, so the total amount for the materials and labor is ___?___.

Write a variable expression.

105. The sum of two numbers is 18. Express the two numbers in terms of the same variable.

106. The sum of two numbers is 20. Express the two numbers in terms of the same variable.

107. Genetics The human genome contains 11,000 more genes than the roundworm genome. Express the number of genes in the human genome in terms of the number of genes in the roundworm genome. (*Source:* Celera, *USA Today* research)

108. Noise Level The noise level of an ambulance siren is 10 decibels louder than that of a car horn. Express the noise level of an ambulance siren in terms of the noise level of a car horn. (*Source:* League for the Hard of Hearing)

109. Space Exploration A survey in *USA Today* reported that almost three-fourths of Americans think that money should be spent on exploration of Mars. Express the number of Americans who think that money should be spent on exploration of Mars in terms of the total number of Americans.

110. Biology According to the American Podiatric Medical Association, the bones in your foot account for one-fourth of all bones in your body. Express the number of bones in your foot in terms of the number of bones in your body.

111. Sports In football, the number of points awarded for a touchdown is three times the number of points awarded for a safety. Express the number of points awarded for a touchdown in terms of the number of points awarded for a safety.

112. WEB Cost of Living A cost-of-living calculator at HOMECARE.com shows that a person living in Manhattan would need twice the salary of a person living in San Diego. Express the salary needed to live in Manhattan in terms of the salary needed to live in San Diego.

113. Astronomy The diameter of Saturn's moon Rhea is 253 miles greater than the diameter of Saturn's moon Dione. Express the diameter of Rhea in terms of the diameter of Dione. (*Source:* NASA)

114. Online Sales A recent survey conducted by Consumer Internet Barometer found that approximately 25% of those people who purchased a product online were extremely satisfied with their experience. Express the number of people who were extremely satisfied with their online purchases in terms of the number of people surveyed.

115. Agriculture In a recent year, Alabama produced one-half the number of pounds of pecans that Texas produced in the same year. Express the amount of pecans produced in Alabama in terms of the amount produced in Texas. (*Source:* National Agricultural Statistics Service)

116. ◕ Taxes According to the Internal Revenue Service, it takes about one-fifth as much time to fill out Schedule B (interest and dividends) as it takes to complete Schedule A (itemized deductions). Express the amount of time it takes to fill out Schedule B in terms of the time it takes to fill out Schedule A.

117. ◕ Demography According to the U.S. Bureau of the Census, the world population in the year 2050 is expected to be twice the world population in 1980. Express the expected world population in 2050 in terms of the world population in 1980.

118. Geometry The length of a rectangle is 5 m more than twice the width. Express the length of the rectangle in terms of the width.

119. Geometry In a triangle, the measure of the smallest angle is 10 degrees less than one-half the measure of the largest angle. Express the measure of the smallest angle in terms of the measure of the largest angle.

120. Sports A halyard 12 ft long is cut into two pieces. Use the same variable to express the lengths of the two pieces.

121. Coins A coin bank contains thirty-five coins in nickels and dimes. Use the same variable to express the number of nickels and the number of dimes in the coin bank.

122. Travel Two cars are traveling in opposite directions and at different rates. Two hours later the cars are 200 mi apart. Express the distance traveled by the faster car in terms of the distance traveled by the slower car.

123. Wages An employee is paid $720 per week plus $27 for each hour of overtime worked. Express the employee's weekly pay in terms of the number of hours of overtime worked.

124. Billing An auto repair bill is $138 for parts and $45 for each hour of labor. Express the amount of the repair bill in terms of the number of hours of labor.

125. Consumerism The cost of a cellular phone service is $19.95 per month plus $.08 per minute for phone calls. Express the monthly cost of the phone service in terms of the number of minutes of phone calls.

1.6 Applying Concepts

126. Chemistry The chemical formula for glucose (sugar) is $C_6H_{12}O_6$. This formula means that there are twelve hydrogen atoms, six carbon atoms, and six oxygen atoms in each molecule of glucose. If x represents the number of atoms of oxygen in a pound of sugar, express the number of hydrogen atoms in the pound of sugar in terms of x.

127. Metalwork A wire whose length is given as x inches is bent into a square. Express the length of a side of the square in terms of x.

128. Pulleys A block-and-tackle system is designed so that pulling one end of a rope five feet will move a weight on the other end a distance of three feet. If x represents the distance the rope is pulled, express the distance the weight moves in terms of x.

129. Translate the expressions $5x + 8$ and $5(x + 8)$ into phrases.

130. In your own words, explain how variables are used.

131. Explain the similarities and differences between the expressions "the difference between x and 5" and "5 less than x."

Focus on Problem Solving

From Concrete to Abstract In your study of algebra, you will find that the problems are less concrete than those you studied in arithmetic. Problems that are concrete provide information pertaining to a specific instance. Algebra is more abstract. Abstract problems are theoretical; they are stated without reference to a specific instance. Let's look at an example of an abstract problem.

How many minutes are there in h hours?

A strategy that can be used to solve this problem is to begin by substituting a number for the variable.

How many minutes are there in 5 hours?

You know that there are 60 minutes in 1 hour. To find the number of minutes in 5 hours, multiply 5 by 60.

$$60 \cdot 5 = 300$$ There are 300 minutes in 5 hours.

Use the same procedure to find the number of minutes in h hours: multiply h by 60.

$$60 \cdot h = 60h$$ There are $60h$ minutes in h hours.

This problem might be taken a step further:

If you walk one mile in x minutes, how far can you walk in h hours?

Consider the same problem using numbers in place of the variables.

If you walk one mile in 20 minutes, how far can you walk in 3 hours?

To solve this problem, you need to calculate the number of minutes in 3 hours (multiply 3 by 60), and divide the result by the number of minutes it takes to walk one mile (20 minutes).

$$\frac{60 \cdot 3}{20} = \frac{180}{20} = 9$$ If you walk one mile in 20 minutes, you can walk 9 miles in 3 hours.

Use the same procedure to solve the related abstract problem. Calculate the number of minutes in h hours (multiply h by 60), and divide the result by the number of minutes it takes to walk one mile (x minutes).

$$\frac{60 \cdot h}{x} = \frac{60h}{x}$$ If you walk one mile in x minutes, you can walk $\frac{60h}{x}$ miles in h hours.

At the heart of the study of algebra is the use of variables. It is the variables in the problems above that make them abstract. But it is variables that allow us to generalize situations and state rules about mathematics.

Try each of the following problems.

1. How many hours are there in d days?

2. You earn d dollars an hour. What are your wages for working h hours?

3. If p is the price of one share of stock, how many shares can you purchase with d dollars?

4. A company pays a television station d dollars to air a commercial lasting s seconds. What is the cost per second?

5. After every v videotape rentals, you are entitled to one free rental. You have rented t tapes, where $t < v$. How many more do you need to rent before you are entitled to a free rental?

6. Your car gets g miles per gallon. How many gallons of gasoline does your car consume traveling t miles?

7. If you drink j ounces of juice each day, how many days will q quarts of the juice last?

8. A TV station has m minutes of commercials each hour. How many ads lasting s seconds each can be sold for each hour of programming?

9. A factory worker can assemble *p* products in *m* minutes. How many products can the factory worker assemble in *h* hours?

10. If one candy bar costs *n* nickels, how many candy bars can be purchased with *q* quarters?

Projects & Group Activities

Investigation into Operations with Even and Odd Integers

An **even integer** is an integer that is divisible by 2. For example, 6, −12, and 480 are even integers. An **odd integer** is an integer that is not evenly divisible by 2. For example, 3, −25, and 971 are odd integers.

Complete each statement with the word *even* or *odd*.

1. If *k* is an odd integer, then *k* + 1 is an _____ integer.

2. If *k* is an odd integer, then *k* − 2 is an _____ integer.

3. If *n* is an integer, then 2*n* is an _____ integer.

4. If *m* and *n* are even integers, then *m* − *n* is an _____ integer.

5. If *m* and *n* are even integers, then *mn* is an _____ integer.

6. If *m* and *n* are odd integers, then *m* + *n* is an _____ integer.

7. If *m* and *n* are odd integers, then *m* − *n* is an _____ integer.

8. If *m* and *n* are odd integers, then *mn* is an _____ integer.

9. If *m* is an even integer and *n* is an odd integer, then *m* − *n* is an _____ integer.

10. If *m* is an even integer and *n* is an odd integer, then *m* + *n* is an _____ integer.

Chapter 1 Summary

Key Words

A **set** is a collection of objects. The objects in the set are called the **elements** of the set. The **roster method** of writing sets encloses a list of the elements in braces. [1.1.1, p. 2]*

Examples

The set of even natural numbers less than 10 is written $A = \{2, 4, 6, 8\}$. The elements in this set are the numbers 2, 4, 6, and 8.

*The numbers in brackets refer to the objective where the Key Word or Essential Rule or Procedure is first introduced. For example, the reference [1.1.1] stands for Chapter 1, Section 1, Objective 1. This notation will be used in all Chapter Summaries throughout the text.

The set of **natural numbers** is {1, 2, 3, 4, 5, 6, 7, ...}. The set of **integers** is {..., −4, −3, −2, −1, 0, 1, 2, 3, 4, ...}.
[1.1.1, pp. 2–3]

A number a **is less than** another number b, written $a < b$, if a is to the left of b on the number line. A number a **is greater than** another number b, written $a > b$, if a is to the right of b on the number line. The symbol $\leq$ means **is less than or equal to.** The symbol $\geq$ means **is greater than or equal to.**
[1.1.1, pp. 3–4]

$-9 < 7$
$4 > -5$

Two numbers that are the same distance from zero on the number line but on opposite sides of zero are **opposite numbers,** or **opposites.** The opposite of a number is also called its **additive inverse.** [1.1.2, p. 4]

5 and −5 are opposites.

The **absolute value** of a number is its distance from zero on the number line. The absolute value of a number is positive or zero. [1.1.2, p. 5]

$|-12| = 12$
$-|7| = -7$

A **rational number** is a number that can be written in the form $\frac{a}{b}$, where a and b are integers and $b \neq 0$. A rational number written in this form is commonly called a **fraction.**
[1.3.1, p. 24]

$\frac{3}{8}, \frac{-9}{11}$, and $\frac{4}{1}$ are rational numbers.

An **irrational number** is a number that has a decimal representation that never terminates or repeats. [1.3.1, p. 25]

$\pi, \sqrt{2}$, and 2.17117111711117... are irrational numbers.

The rational numbers and the irrational numbers taken together are called the **real numbers.** [1.3.1, p. 25]

$\frac{3}{8}, \frac{-9}{11}, \frac{4}{1}, \pi, \sqrt{2}$, and 2.17117111711117... are real numbers.

Percent means "parts of 100." [1.3.4, p. 29]

63% means 63 of 100 equal parts.

An expression of the form a^n is in **exponential form,** where a is the **base** and n is the **exponent.** [1.4.1, p. 38]

6^3 is an exponential expression in which 6 is the base and 3 is the exponent.

A **variable** is a letter that is used to stand for a quantity that is unknown or that can change. A **variable expression** is an expression that contains one or more variables.
[1.5.1, p. 45]

$5x - 4y + 7z$ is a variable expression. It contains the variables x, y, and z.

The **terms of a variable expression** are the addends of the expression. Each term is a **variable term** or a **constant term.**
[1.5.1, p. 45]

The expression $4a^2 - 6b^3 + 7$ has three terms: $4a^2$, $-6b^3$, and 7. $4a^2$ and $-6b^3$ are variable terms. 7 is a constant term.

A variable term is composed of a **numerical coefficient** and a **variable part.** [1.5.1, p. 45]

For the expression $8p^4r$, 8 is the coefficient and p^4r is the variable part.

Replacing the variables in a variable expression with numbers and then simplifying the numerical expression is called **evaluating the variable expression.** [1.5.1, p. 46]

To evaluate $a^2 - 2b$ when $a = -3$ and $b = -5$, simplify the expression $(-3)^2 - 2(-5)$.

The **additive inverse** of a number is the opposite of the number. [1.5.2, p. 49]

The additive inverse of 8 is -8.
The additive inverse of -15 is 15.

The **multiplicative inverse** of a number is the reciprocal of the number. [1.5.2, p. 49]

The multiplicative inverse of $\frac{3}{8}$ is $\frac{8}{3}$.

Like terms of a variable expression are the terms with the same variable part. [1.5.3, p. 50]

For the expression $2st - 3t + 9s - 11st$, the terms $2st$ and $-11st$ are like terms.

Essential Rules and Procedures

To add two integers with the same sign, add the absolute values of the numbers. Then attach the sign of the addends. [1.2.1, p. 11]

$9 + 3 = 12$
$-9 + (-3) = -12$

To add two integers with different signs, find the absolute value of each number. Then subtract the smaller of these absolute values from the larger one. Attach the sign of the number with the greater absolute value. [1.2.1, p. 11]

$9 + (-3) = 6$
$-9 + 3 = -6$

To subtract one integer from another, add the opposite of the second integer to the first integer. [1.2.2, p. 12]

$6 - 11 = 6 + (-11) = -5$
$-2 - (-8) = -2 + 8 = 6$

To multiply two integers with the same sign, multiply the absolute values of the numbers. The product is positive. [1.2.3, p. 13]

$3(5) = 15$
$(-3)(-5) = 15$

To multiply two integers with different signs, multiply the absolute values of the numbers. The product is negative. [1.2.3, p. 13]

$-3(5) = -15$
$3(-5) = -15$

To divide two integers with the same sign, divide the absolute values of the numbers. The quotient is positive. [1.2.4, p. 14]

$14 \div 2 = 7$
$-14 \div (-2) = 7$

To divide two integers with different signs, divide the absolute values of the numbers. The quotient is negative. [1.2.4, p. 14]

$14 \div (-2) = -7$
$-14 \div 2 = -7$

Zero and One in Division [1.2.4, p. 14]

$\frac{0}{a} = 0, a \neq 0$

$\frac{0}{3} = 0$

Division by zero is undefined.

$\frac{9}{0}$ is undefined.

$\frac{a}{a} = 1, a \neq 0$

$\frac{-2}{-2} = 1$

$\frac{a}{1} = a$

$\frac{-6}{1} = -6$

To convert a percent to a decimal, remove the percent sign and multiply by 0.01. [1.3.4, p. 29]

$85\% = 85(0.01) = 0.85$

To convert a percent to a fraction, remove the percent sign and multiply by $\frac{1}{100}$. [1.3.4, p. 29]

$25\% = 25 \cdot \frac{1}{100} = \frac{25}{100} = \frac{1}{4}$

To convert a decimal or a fraction to a percent, multiply by 100%. [1.3.4, p. 30]

$0.5 = 0.5(100\%) = 50\%$

$\frac{3}{8} = \frac{3}{8} \cdot 100\% = \frac{300}{8}\% = 37.5\%$

Order of Operations Agreement [1.4.2, p. 40]

Step 1 Perform operations inside grouping symbols.
Step 2 Simplify exponential expressions.
Step 3 Do multiplication and division as they occur from left to right.
Step 4 Do addition and subtraction as they occur from left to right.

$16 \div (-2)^3 + (7 - 12)$
$= 16 \div (-2)^3 + (-5)$
$= 16 \div (-8) + (-5)$
$= -2 + (-5)$
$= -7$

The Commutative Property of Addition
If a and b are real numbers, then $a + b = b + a$.
[1.5.2, p. 47]

$5 + 2 = 7$ and $2 + 5 = 7$

The Commutative Property of Multiplication
If a and b are real numbers, then $ab = ba$. [1.5.2, p. 48]

$6(-3) = -18$ and $-3(6) = -18$

The Associative Property of Addition
If a, b, and c are real numbers, then $(a + b) + c = a + (b + c)$.
[1.5.2, p. 48]

$-1 + (4 + 7) = -1 + 11 = 10$
$(-1 + 4) + 7 = 3 + 7 = 10$

The Associative Property of Multiplication
If a, b, and c are real numbers, then $(ab)c = a(bc)$.
[1.5.2, p. 48]

$(-2 \cdot 5) \cdot 3 = -10 \cdot 3 = -30$
$-2 \cdot (5 \cdot 3) = -2 \cdot 15 = -30$

The Addition Property of Zero
If a is a real number, then $a + 0 = 0 + a = a$. [1.5.2, p. 48]

$9 + 0 = 9$ $0 + 9 = 9$

The Multiplication Property of Zero
If a is a real number, then $a \cdot 0 = 0 \cdot a = 0$. [1.5.2, p. 48]

$-8(0) = 0$ $0(-8) = 0$

The Multiplication Property of One
If a is a real number, then $1 \cdot a = a \cdot 1 = a$. [1.5.2, p. 49]

$7 \cdot 1 = 7$ $1 \cdot 7 = 7$

The Inverse Property of Addition
If a is a real number, then $a + (-a) = (-a) + a = 0$.
[1.5.2, p. 49]

$4 + (-4) = 0$ $-4 + 4 = 0$

The Inverse Property of Multiplication

If a is a real number and $a \neq 0$, then $a \cdot \dfrac{1}{a} = \dfrac{1}{a} \cdot a = 1$. $6 \cdot \dfrac{1}{6} = 1 \qquad \dfrac{1}{6} \cdot 6 = 1$

[1.5.2, p. 49]

The Distributive Property

If a, b, and c are real numbers, then $a(b + c) = ab + ac$.

[1.5.2, p. 49]

$$5(2 + 3) = 5 \cdot 2 + 5 \cdot 3$$
$$= 10 + 15$$
$$= 25$$

Chapter 1 Review Exercises

1. Use the roster method to write the set of natural numbers less than 7.

2. Write $\dfrac{5}{8}$ as a percent.

3. Evaluate: $-|-4|$

4. Subtract: $16 - (-30) - 42$

5. Divide: $-561 \div (-33)$

6. Write $\dfrac{7}{9}$ as a decimal. Place a bar over the repeating digits of the decimal.

7. Simplify: $(6.02)(-0.89)$

8. Simplify: $\dfrac{-10 + 2}{2 + (-4)} \div 2 + 6$

9. Write 0.672 as a percent.

10. Write $79\dfrac{1}{2}\%$ as a fraction.

11. Multiply: $(-5)(-6)(3)$

12. Subtract: $6.039 - 12.92$

13. Given $A = \{-5, -3, 0\}$, which elements of set A are less than or equal to -3?

14. Write 7% as a decimal.

15. Evaluate: $\dfrac{3}{4} \cdot (4)^2$

16. Place the correct symbol, $<$ or $>$, between the two numbers.

$$-2 \quad -40$$

17. Add: $-\dfrac{2}{5} + \dfrac{7}{15}$

18. Evaluate: $(-3^3) \cdot 2^2$

19. Divide: $-\dfrac{7}{16} \div \dfrac{3}{8}$

20. Simplify: $2^3 \div 4 - 2(2 - 7)$

21. Write $2\dfrac{8}{9}$ as a percent. Write the remainder in fractional form.

22. Given $C = \{-12, -8, -1, 7\}$, find

 a. the opposite of each element of set C.
 b. the absolute value of each element of set C.

23. Divide: $0.2654 \div (-0.023)$
Round to the nearest tenth.

24. Simplify: $(7 - 2)^2 - 5 - 3 \cdot 4$

25. Add: $-12 + 8 + (-4)$

26. Add: $-\dfrac{5}{8} + \dfrac{1}{6}$

27. Add: $-1.329 + 4.89$

28. Subtract: $-5 - 22 - (-13) - 19 - (-6)$

29. Multiply: $\left(\dfrac{1}{3}\right)\left(-\dfrac{4}{5}\right)\left(\dfrac{3}{8}\right)$

30. Place the correct symbol, $<$ or $>$, between the two numbers.

$$-43 \quad -34$$

31. Simplify: $-7y^2 + 6y^2 - (-2y^2)$

32. Simplify: $(12x)\left(\dfrac{1}{4}\right)$

33. Simplify: $5(2x + 4) - 3(x - 6)$

34. Evaluate $a^2 - 3b$ when $a = 2$ and $b = -4$.

35. Complete the statement by using the Inverse Property of Addition.
$-9 + \,? = 0$

36. Simplify: $3[2x - 3(x - 2y)] + 3y$

37. Simplify: $5(3 - 7b)$

38. Simplify: $2x + 3[4 - (3x - 7)]$

39. Identify the property that justifies the statement.
$-4(3) = 3(-4)$

40. Simplify: $3(8 - 2x)$

41. Evaluate $\dfrac{-2ab}{2b - a}$ when $a = -4$ and $b = 6$.

42. Simplify: $(-3)(-12y)$

43. Simplify: $4(3x - 2) - 7(x + 5)$

44. Simplify: $(16x)\left(\dfrac{1}{8}\right)$

45. Simplify: $-3(2x^2 - 7y^2)$

46. Evaluate $3(a - c) - 2ab$ when $a = 2$, $b = 3$, and $c = -4$.

47. Simplify: $2x - 3(x - 2)$

48. Simplify: $2a - (-3b) - 7a - 5b$

49. Simplify: $-5(2x^2 - 3x + 6)$

50. Simplify: $3x - 7y - 12x$

51. Simplify: $3x + (-12y) - 5x - (-7y)$

52. Simplify: $\left(-\dfrac{5}{6}\right)(-36b)$

53. Complete the statement by using the Distributive Property.
$(6 + 3)7 = 42 + \,?$

54. Simplify: $4x^2 + 9x - 6x^2 - 5x$

55. Translate "*b* decreased by the product of seven and *b*" into a variable expression.

56. Translate "ten divided by the difference between *y* and two" into a variable expression.

57. Temperature Find the temperature after a rise of 14°C from −6°C.

58. Temperature The daily low temperatures, in degrees Celsius, for a three-day period were recorded as follows: −8°, 7°, −5°. Find the average low temperature for the three-day period.

59. Temperature Use the table to find the difference between the record high temperature and the record low temperature for January in Bismarck, North Dakota.

Bismarck, North Dakota

	Jan	Feb	Mar	Apr	May	Jun	Jul	Aug	Sep	Oct	Nov	Dec
Record High	62°F (1981)	69°F (1992)	80°F (1967)	93°F (1992)	96°F (1980)	107°F (1988)	109°F (1973)	107°F (1973)	105°F (1959)	95°F (1963)	75°F (1978)	65°F (1979)
Record Low	−44°F (1950)	−43°F (1994)	−28°F (1995)	−12°F (1975)	15°F (1967)	30°F (1969)	35°F (1971)	33°F (1988)	11°F (1974)	−10°F (1991)	−30°F (1985)	−40°F (1983)

Source: www.weather.com

60. Temperature Find the temperature after a rise of 7°C from −13°C.

61. Temperature The temperature on the surface of the planet Venus is 480°C. The temperature on the surface of the planet Pluto is −234°C. Find the difference between the surface temperatures on Venus and Pluto.

62. Translate and simplify "eight times the quotient of twice a number and sixteen."

63. Translate and simplify "the product of four and the sum of two and five times a number."

64. Geometry The length of the base of a triangle is 15 in. more than the height of the triangle. Express the length of the base of the triangle in terms of the height of the triangle.

65. Coffee Blends A coffee merchant made 20 lb of a blend of coffee using only mocha java beans and espresso beans. Use the same variable to express the amounts of mocha java beans and espresso beans in the coffee blend.

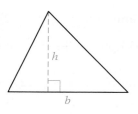

Chapter 1 Test

1. Write 55% as a fraction.

2. Given $B = \{-8, -6, -4, -2\}$, which elements of set B are less than -5?

3. Subtract: $-9 - (-6)$

4. Write $\dfrac{3}{20}$ as a decimal.

5. Multiply: $\dfrac{3}{4}\left(-\dfrac{2}{21}\right)$

6. Divide: $-75 \div 5$

7. Evaluate: $\left(-\dfrac{2}{3}\right)^3 \cdot 3^2$

8. Add: $-7 + (-3) + 12$

9. Use the roster method to write the set of positive integers less than or equal to 6.

10. Write 1.59 as a percent.

11. Evaluate: $|-29|$

12. Place the correct symbol, $<$ or $>$, between the two numbers.
$$-47 \quad -68$$

13. Subtract: $-\dfrac{4}{9} - \dfrac{5}{6}$

14. Divide: $-\dfrac{5}{8} \div \left(-\dfrac{3}{4}\right)$

15. Write $\dfrac{3}{13}$ as a percent. Round to the nearest tenth of a percent.

16. Write 6.2% as a decimal.

17. Subtract: $13 - (-5) - 4$

18. Write $\dfrac{13}{30}$ as a decimal. Place a bar over the repeating digits of the decimal.

19. Multiply: $(-0.9)(2.7)$

20. Add: $15 + (-8) + (-19)$

21. Given $A = \{-17, -6, 5, 9\}$, find

 a. the opposite of each element of set A.
 b. the absolute value of each element of set A.

22. Add: $-18.354 + 6.97$

23. Multiply: $-4(8)(-5)$

24. Simplify: $(9y)4$

25. Simplify: $7x + 5y - 3x - 8y$

26. Simplify: $8n - (6 - 2n)$

27. Identify the property that justifies the statement.
$$\dfrac{3}{8}(1) = \dfrac{3}{8}$$

28. Simplify: $-4(-x + 10)$

29. Simplify: $\dfrac{2}{3}x^2 - \dfrac{7}{12}x^2$

30. Simplify: $(-10x)\left(-\dfrac{2}{5}\right)$

31. Simplify: $(-4y^2 + 8)6$

32. Complete the statement by using the Inverse Property of Addition.
$-19 + ? = 0$

33. Evaluate $\dfrac{-3ab}{2a + b}$ when $a = -1$ and $b = 4$.

34. Simplify: $5(x + y) - 8(x - y)$

35. Simplify: $3(x^2 - 5x + 4)$

36. Evaluate $4(b - a) + bc$ when $a = 2$, $b = -3$, and $c = 4$.

37. Simplify: $6x - 3(y - 7x) + 2(5x - y)$

38. Simplify: $9(-4) \div [2(8 - 5)^2]$

39. Temperature Find the temperature after a rise of 12°C from -8°C.

40. Temperature Use the table to find the average of the record low temperatures in Fairbanks, Alaska, for the first four months of the year.

Fairbanks, Alaska

	Jan	Feb	Mar	Apr	May	Jun	Jul	Aug	Sep	Oct	Nov	Dec
Record High	50°F (1981)	47°F (1987)	56°F (1994)	74°F (1960)	89°F (1960)	96°F (1969)	94°F (1975)	93°F (1994)	84°F (1957)	65°F (1969)	49°F (1997)	44°F (1985)
Record Low	−61°F (1969)	−58°F (1993)	−49°F (1956)	−24°F (1986)	−1°F (1964)	30°F (1950)	35°F (1959)	27°F (1987)	3°F (1992)	−27°F (1975)	−46°F (1990)	−62°F (1961)

Source: www.weather.com

41. Translate "the difference between the sum of a and b and the square of b" into a variable expression.

42. Translate and simplify "twenty times the sum of a number and nine."

43. Translate and simplify "two more than a number added to the difference between the number and three."

44. Astronomy The distance from Neptune to the sun is thirty times the distance from Earth to the sun. Express the distance from Neptune to the sun in terms of the distance from Earth to the sun.

45. Carpentry A nine-foot board is cut into two pieces of different lengths. Use the same variable to express the lengths of the two pieces.

2 Solving Equations and Inequalities

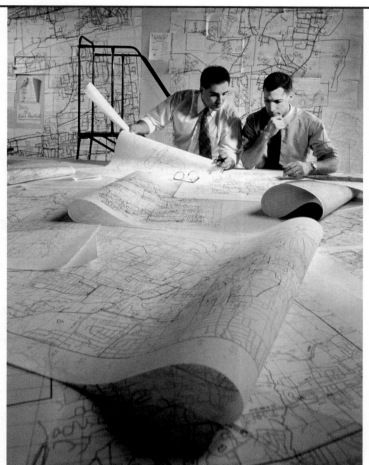

No matter what you do in life, at some point you will be asked to solve a problem that will at first seem very difficult. When this happens, it's usually helpful to take a moment to ask yourself some questions: What do I need to do? How am I going to do it? Just like an architect develops a blueprint to decide how to build a house, you can develop a plan to solve your problem and find the answer. The basic steps involved in problem solving are outlined in the **Focus on Problem Solving on pages 194–195** and used throughout the text. Problem solving is a necessary skill for future success and will only improve with study and practice.

Online Study Center DVD SSM

Need help? For online student resources, such as section quizzes, visit the Online Study Center at college.hmco.com/pic/aufmannABI2e.

Prep Test

1. Subtract: $8 - 12$

2. Multiply: $-\dfrac{3}{4}\left(-\dfrac{4}{3}\right)$

3. Multiply: $-\dfrac{5}{8}(16)$

4. Write 90% as a decimal.

5. Write 0.75 as a percent.

6. Evaluate $3x^2 - 4x - 1$ when $x = -4$.

7. Simplify: $3x - 5 + 7x$

8. Simplify: $8x - 9 - 8x$

9. Simplify: $6x - 3(6 - x)$

2.1 Introduction to Equations

OBJECTIVE 1 Determine whether a given number is a solution of an equation

Point of Interest

One of the most famous equations ever stated is $E = mc^2$. *This equation, stated by Albert Einstein, shows that there is a relationship between mass* m *and energy* E.

An **equation** expresses the equality of two mathematical expressions. The expressions can be either numerical or variable expressions.

$$\left.\begin{array}{l} 9 + 3 = 12 \\ 3x - 2 = 10 \\ y^2 + 4 = 2y - 1 \\ z = 2 \end{array}\right\} \text{Equations}$$

The equation at the right is true if the variable is replaced by 5.

$$x + 8 = 13$$
$$5 + 8 = 13 \qquad \text{A true equation}$$

The equation $x + 8 = 13$ is false if the variable is replaced by 7.

$$7 + 8 = 13 \qquad \text{A false equation}$$

A **solution** of an equation is a number that, when substituted for the variable, results in a true equation. 5 is a solution of the equation $x + 8 = 13$. 7 is not a solution of the equation $x + 8 = 13$.

EXAMPLE 1 Is -3 a solution of the equation $4x + 16 = x^2 - 5$?

Solution

$$4x + 16 = x^2 - 5$$

$4(-3) + 16$	$(-3)^2 - 5$
$-12 + 16$	$9 - 5$

$$4 = 4$$

Yes, -3 is a solution of the equation $4x + 16 = x^2 - 5$.

- Replace the variable by the given number, -3.
- Evaluate the numerical expressions using the Order of Operations Agreement.
- Compare the results. If the results are equal, the given number is a solution. If the results are not equal, the given number is not a solution.

PROBLEM 1 Is $\dfrac{1}{4}$ a solution of $5 - 4x = 8x + 2$?

■ **Solution** See page S4.

EXAMPLE 2 Is -4 a solution of $4 + 5x = x^2 - 2x$?

Solution

$4 + 5x = x^2 - 2x$	
$4 + 5(-4)$	$(-4)^2 - 2(-4)$
$4 + (-20)$	$16 - (-8)$

$$-16 \neq 24$$

- Replace the variable by the given number, -4.
- Evaluate the numerical expressions using the Order of Operations Agreement.
- Compare the results. If the results are equal, the given number is a solution. If the results are not equal, the given number is not a solution.

No, -4 is not a solution of the equation $4 + 5x = x^2 - 2x$.

PROBLEM 2 Is 5 a solution of $10x - x^2 = 3x - 10$?

■ **Solution** See page S4.

OBJECTIVE 2 ## Solve equations of the form $x + a = b$

To **solve an equation** means to find a solution of the equation. The simplest equation to solve is an equation of the form **variable = constant**, because the constant is the solution.

If $x = 5$, then 5 is the solution of the equation because $5 = 5$ is a true equation.

The solution of the equation shown at the right is 7.

$$x + 2 = 9 \qquad 7 + 2 = 9$$

Note that if 4 is added to each side of the equation, the solution is still 7.

$$x + 2 + 4 = 9 + 4$$
$$x + 6 = 13 \qquad 7 + 6 = 13$$

If -5 is added to each side of the equation, the solution is still 7.

$$x + 2 + (-5) = 9 + (-5)$$
$$x - 3 = 4 \qquad 7 - 3 = 4$$

This illustrates the Addition Property of Equations.

Take Note

Think of an equation as a balance scale. If the weights added to each side of the equation are not the same, the pans no longer balance.

> ### Addition Property of Equations
>
> The same number or variable term can be added to each side of an equation without changing the solution of the equation.

This property is used in solving equations. Note the effect of adding, to each side of the equation $x + 2 = 9$, the opposite of the constant term 2. After each side of the equation is simplified, the equation is in the form variable = constant. The solution is the constant.

$$x + 2 = 9$$
$$x + 2 + (-2) = 9 + (-2)$$
$$x + 0 = 7$$
$$x = 7$$

$$\boxed{\text{variable}} = \boxed{\text{constant}}$$

The solution is 7.

In solving an equation, the goal is to rewrite the given equation in the form *variable = constant*. The Addition Property of Equations can be used to rewrite an equation in this form. The Addition Property of Equations is used to remove a term from one side of an equation by adding the opposite of that term to each side of the equation.

EXAMPLE 3 Solve and check: $y - 6 = 9$

Solution

$y - 6 = 9$ • **The goal is to rewrite the equation in the form *variable = constant*.**

$y - 6 + 6 = 9 + 6$ • **Add 6 to each side of the equation (the Addition Property of Equations).**

$y + 0 = 15$ • **Simplify using the Inverse Property of Addition.**

$y = 15$ • **Simplify using the Addition Property of Zero. Now the equation is in the form *variable = constant*.**

Check
$$y - 6 = 9$$
$$15 - 6 \mid 9$$
$$9 = 9$$
• **This is a true equation. The solution checks.**

The solution is 15. • **Write the solution.**

PROBLEM 3 Solve and check: $x - \dfrac{1}{3} = -\dfrac{3}{4}$

Solution See page S4.

Because subtraction is defined in terms of addition, the Addition Property of Equations makes it possible to subtract the same number from each side of an equation without changing the solution of the equation.

Solve: $y + \dfrac{1}{2} = \dfrac{5}{4}$

The goal is to rewrite the equation in the form *variable = constant*.

$$y + \dfrac{1}{2} = \dfrac{5}{4}$$

Add the opposite of the constant term $\dfrac{1}{2}$ to each side of the equation. This is equivalent to subtracting $\dfrac{1}{2}$ from each side of the equation.

$$y + \dfrac{1}{2} - \dfrac{1}{2} = \dfrac{5}{4} - \dfrac{1}{2}$$
$$y + 0 = \dfrac{5}{4} - \dfrac{2}{4}$$
$$y = \dfrac{3}{4}$$

The solution $\dfrac{3}{4}$ checks.

The solution is $\dfrac{3}{4}$.

EXAMPLE 4 Solve: $\dfrac{1}{2} = x + \dfrac{2}{3}$

Solution $\dfrac{1}{2} = x + \dfrac{2}{3}$

$\dfrac{1}{2} - \dfrac{2}{3} = x + \dfrac{2}{3} - \dfrac{2}{3}$ • Subtract $\dfrac{2}{3}$ from each side of the equation.

$-\dfrac{1}{6} = x$ • Simplify each side of the equation.

The solution is $-\dfrac{1}{6}$.

PROBLEM 4 Solve: $-8 = 5 + x$

Solution See page S4.

Note from the solution to Example 4 that an equation can be rewritten in the form *constant = variable*. Whether the equation is written in the form *variable = constant* or in the form *constant = variable*, the solution is the constant.

OBJECTIVE 3 ## Solve equations of the form $ax = b$

The solution of the equation shown at the right is 3.

$$2x = 6 \qquad\qquad 2 \cdot 3 = 6$$

Note that if each side of the equation is multiplied by 5, the solution is still 3.

$$5 \cdot 2x = 5 \cdot 6$$
$$10x = 30 \qquad\qquad 10 \cdot 3 = 30$$

If each side is multiplied by -4, the solution is still 3.

$$(-4) \cdot 2x = (-4) \cdot 6$$
$$-8x = -24 \qquad\qquad -8 \cdot 3 = -24$$

This illustrates the Multiplication Property of Equations.

Multiplication Property of Equations

Each side of an equation can be multiplied by the same nonzero number without changing the solution of the equation.

This property is used in solving equations. Note the effect of multiplying each side of the equation $2x = 6$ by the reciprocal of the coefficient 2. After each side of the equation is simplified, the equation is in the form variable = constant. The solution is the constant.

$$2x = 6$$
$$\dfrac{1}{2} \cdot 2x = \dfrac{1}{2} \cdot 6$$
$$1x = 3$$
$$x = 3$$

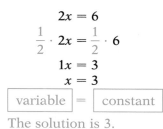

| variable | = | constant |

The solution is 3.

In solving an equation, the goal is to rewrite the given equation in the form variable = constant. The Multiplication Property of Equations can be used to rewrite an equation in this form. The Multiplication Property of Equations is

used to remove a coefficient from a variable term in an equation by multiplying each side of the equation by the reciprocal of the coefficient.

EXAMPLE 5 Solve: $\dfrac{3x}{4} = -9$

Solution

$$\dfrac{3x}{4} = -9$$

$$\dfrac{4}{3} \cdot \dfrac{3}{4}x = \dfrac{4}{3}(-9)$$

$$1x = -12$$

$$x = -12$$

The solution is -12.

• $\dfrac{3x}{4} = \dfrac{3}{4}x$

• Multiply each side of the equation by the reciprocal of the coefficient $\dfrac{3}{4}$ (the Multiplication Property of Equations).
• Simplify using the Inverse Property of Multiplication.
• Simplify using the Multiplication Property of One. Now the equation is in the form *variable = constant*.

• Write the solution.

PROBLEM 5 Solve: $-\dfrac{2x}{5} = 6$

Solution See page S4.

Because division is defined in terms of multiplication, the Multiplication Property of Equations makes it possible to divide each side of an equation by the same number without changing the solution of the equation.

Solve: $8x = 16$

The goal is to rewrite the equation in the form *variable = constant*.

Multiply each side of the equation by the reciprocal of 8. This is equivalent to dividing each side by 8.

The solution 2 checks.

$$8x = 16$$

$$\dfrac{8x}{8} = \dfrac{16}{8}$$

$$x = 2$$

The solution is 2.

When using the Multiplication Property of Equations to solve an equation, multiply each side of the equation by the reciprocal of the coefficient when the coefficient is a fraction. Divide each side of the equation by the coefficient when the coefficient is an integer or a decimal.

EXAMPLE 6 Solve and check: $4x = 6$

Solution

$$4x = 6$$

$$\dfrac{4x}{4} = \dfrac{6}{4}$$

$$x = \dfrac{3}{2}$$

• Divide each side of the equation by 4, the coefficient of *x*.

• Simplify each side of the equation.

Check

$$4x = 6$$

$$4\left(\frac{3}{2}\right) \quad \Big| \quad 6$$

$$6 = 6$$ • **This is a true equation. The solution checks.**

The solution is $\frac{3}{2}$.

PROBLEM 6 Solve and check: $6x = 10$

■ **Solution** See page S4.

Before using one of the Properties of Equations, check to see whether one or both sides of the equation can be simplified. In Example 7, like terms appear on the left side of the equation. The first step in solving this equation is to combine the like terms so that there is only one variable term on the left side of the equation.

EXAMPLE 7 Solve: $5x - 9x = 12$

Solution $5x - 9x = 12$

$-4x = 12$ • **Combine like terms.**

$\dfrac{-4x}{-4} = \dfrac{12}{-4}$ • **Divide each side of the equation by −4.**

$x = -3$

The solution is -3.

PROBLEM 7 Solve: $4x - 8x = 16$

■ **Solution** See page S4.

OBJECTIVE 4 Applications of percent

The solution of a problem that involves a percent requires solving the basic percent equation.

The Basic Percent Equation

Percent · Base = Amount
$$P \quad \cdot \quad B \;\; = \quad A$$

To translate a problem involving a percent into an equation, remember that the word *of* translates into "multiply" and the word *is* translates into "="; the base usually follows the word *of*.

20% of what number is 30?

Given: $P = 20\% = 0.20$
 $A = 30$
Unknown: Base

$$PB = A$$
$$(0.20)B = 30$$
$$\frac{0.20B}{0.20} = \frac{30}{0.20}$$
$$B = 150$$

The number is 150.

What percent of 40 is 30?

Given: $B = 40$
 $A = 30$
Unknown: Percent

We must write the fraction as a percent in order to answer the question.

$$PB = A$$
$$P(40) = 30$$
$$40P = 30$$
$$\frac{40P}{40} = \frac{30}{40}$$
$$P = \frac{3}{4}$$
$$P = 75\%$$

30 is 75% of 40.

Find 25% of 200.

Given: $P = 25\% = 0.25$
 $B = 200$
Unknown: Amount

$$PB = A$$
$$0.25(200) = A$$
$$50 = A$$

25% of 200 is 50.

In most cases, we write a percent as a decimal before solving the basic percent equation. However, some percents are more easily written as a fraction. For example,

$$33\frac{1}{3}\% = \frac{1}{3} \qquad 66\frac{2}{3}\% = \frac{2}{3} \qquad 16\frac{2}{3}\% = \frac{1}{6} \qquad 83\frac{1}{3}\% = \frac{5}{6}$$

EXAMPLE 8 12 is $33\frac{1}{3}\%$ of what number?

Solution $PB = A$

$$\frac{1}{3}B = 12 \qquad \bullet\ 33\frac{1}{3}\% = \frac{1}{3}$$

$$3 \cdot \frac{1}{3}B = 3 \cdot 12$$

$$B = 36$$

The number is 36.

PROBLEM 8 27 is what percent of 60?

Solution See page S4.

Point of Interest

As noted at the right, 172,841 of the dogs or litters registered were Labrador retrievers. Listed below are the next most popular breeds and their registrations.
Golden retrievers: 66,300
German shepherds: 57,660
Dachshunds: 54,773
Beagles: 52,026

EXAMPLE 9 During a recent year, nearly 1.2 million dogs or litters were registered with the American Kennel Club. The most popular breed was the Labrador retriever, with 172,841 registered. What percent of the registrations were Labrador retrievers? Round to the nearest tenth of a percent. (*Source:* American Kennel Club)

Strategy To find the percent, solve the basic percent equation using $B = 1.2$ million $= 1,200,000$ and $A = 172,841$. The percent is unknown.

Solution

$$PB = A$$
$$P(1,200,000) = 172,841$$
$$\frac{P(1,200,000)}{1,200,000} = \frac{172,841}{1,200,000}$$
$$P \approx 0.144$$
$$P \approx 14.4\%$$

Approximately 14.4% of the registrations were Labrador retrievers.

PROBLEM 9 A student correctly answered 72 of the 80 questions on an exam. What percent of the questions were answered correctly?

Solution See page S5.

EXAMPLE 10 According to the Centers for Disease Control and Prevention, 30.8% of the adult population of Kentucky smokes. How many adults in Kentucky smoke? Use a figure of 3,000,000 for the number of adults in Kentucky.

Strategy To find the number of adults who smoke, solve the basic percent equation using $B = 3,000,000$ and $P = 30.8\% = 0.308$. The amount is unknown.

Solution

$$PB = A$$
$$0.308(3,000,000) = A$$
$$924,000 = A$$

Approximately 924,000 adults in Kentucky smoke.

PROBLEM 10 The price of a digital camcorder is $895. A 6% sales tax is added to the price. How much is the sales tax?

Solution See page S5.

The simple interest that an investment earns is given by the equation $\boldsymbol{I = Prt}$, where I is the simple interest, P is the principal, or amount invested, r is the simple interest rate, and t is the time.

Take Note

$18 \text{ months} = \dfrac{18}{12} \text{ years}$
$= 1.5 \text{ years}$

A $1500 investment has an annual simple interest rate of 7%. Find the simple interest earned on the investment after 18 months.

The time is given in months, but the interest rate is an annual rate. Therefore, we must convert 18 months to years.

$18 \text{ months} = 1.5 \text{ years}$

To find the interest earned, solve
the equation $I = Prt$ for I using
$P = 1500$, $r = 0.07$, and $t = 1.5$.

$I = Prt$
$I = 1500(0.07)(1.5)$
$I = 157.5$

The interest earned is $157.50.

EXAMPLE 11 In April, Marshall Wardell was charged $8.72 in interest on an unpaid credit card balance of $545. Find the annual interest rate for this credit card.

Strategy The interest is $8.72. Therefore, $I = 8.72$. The unpaid balance is $545. This is the principal on which interest is calculated. Therefore, $P = 545$. The time is 1 month. Because the *annual* interest rate must be found and the time is given as 1 month, write 1 month as $\frac{1}{12}$ year. Therefore, $t = \frac{1}{12}$. To find the interest rate, solve the equation $I = Prt$ for r.

Solution

$$I = Prt$$

$$8.72 = 545r\left(\frac{1}{12}\right) \qquad \bullet\, I = 8.72, P = 545, t = \frac{1}{12}$$

$$8.72 = \frac{545}{12}r$$

$$\frac{12}{545}(8.72) = \frac{12}{545}\left(\frac{545}{12}r\right)$$

$$0.192 = r$$

The annual interest rate is 19.2%.

PROBLEM 11 Clarissa Adams purchased a $1000 municipal bond that earns an annual simple interest rate of 6.4%. How much must she deposit into a bank account that earns 8% annual simple interest so that the interest earned from each account after one year is the same?

Solution See page S5.

Take Note

Municipal bonds are frequently sold by government agencies to pay for roads, aqueducts, and other essential public needs. An investor who purchases one of these bonds is paid interest on the cost of the bond for a certain number of years. When that time period expires, the cost of the bond is returned to the investor.

The amount of a substance in a solution can be given as a percent of the total solution. For instance, if a certain fruit juice drink is advertised as containing 27% cranberry juice, then 27% of the contents of the bottle must be cranberry juice.

Solving problems involving mixtures is based on the percent mixture equation $Q = Ar$, where Q is the quantity of a substance in the solution, A is the amount of the solution, and r is the percent concentration of the substance.

The formula for a perfume requires that the concentration of jasmine be 1.2% of the total amount of perfume. How many ounces of jasmine are in a 2-ounce bottle of this perfume?

Point of Interest

In the jewelry industry, the amount of gold in a piece of jewelry is measured in karats. *Pure gold is 24 karats. A necklace that is 18 karats is*

$\frac{18}{24} = 0.75 = 75\%$ *gold.*

The amount of perfume is 2 oz. Therefore, $A = 2$. The percent concentration is 1.2%, so $r = 0.012$. To find the number of ounces of jasmine, solve the equation $Q = Ar$ for Q.

$$Q = Ar$$
$$Q = 2(0.012)$$
$$Q = 0.024$$

The bottle contains 0.024 oz of jasmine.

EXAMPLE 12 To make a certain color of blue, 4 oz of cyan must be contained in 1 gal of paint. What is the percent concentration of cyan in the paint?

Strategy The amount of cyan is given in ounces and the amount of paint is given in gallons; we must convert ounces to gallons or gallons to ounces. For this problem, we will convert gallons to ounces: 1 gal = 128 oz.

Solution To find the percent concentration of cyan, solve the equation $Q = Ar$ for r using $Q = 4$ and $A = 128$.

$$Q = Ar$$
$$4 = 128r$$
$$\frac{4}{128} = \frac{128r}{128}$$
$$0.03125 = r$$

The percent concentration of cyan is 3.125%.

PROBLEM 12 The concentration of sugar in Choco-Pops cereal is 25%. If a bowl of this cereal contains 2 oz of sugar, how many ounces of cereal are in the bowl?

Solution See page S5.

2.1 Concept Review

Determine whether the statement is always true, sometimes true, or never true.

1. Both sides of an equation can be multiplied by the same number without changing the solution of the equation.

2. Both sides of an equation can be divided by the same number without changing the solution of the equation.

3. For an equation of the form $ax = b, a \neq 0$, multiplying both sides of the equation by the reciprocal of a will result in an equation of the form $x = constant$.

4. Use the Multiplication Property of Equations to remove a term from one side of an equation.

5. Adding a negative 3 to each side of an equation yields the same result as subtracting 3 from each side of the equation.

6. In the basic percent equation, the amount follows the word *of*.

7. An equation contains an equals sign.

2.1 Exercises

1 Determine whether a given number is a solution of an equation

1. Complete the sentence using the word *expression* or the word *equation*.
 a. $8x = 4x + 10$ is an ___?___ . **b.** $8x - 4x + 10$ is an ___?___ .

2. Determine whether -3 is a solution of the equation $9 - 5x = -3 - 9x$.

$$9 - 5x = -3 - 9x$$

$$\begin{array}{c|c} 9 - 5(-3) & -3 - 9(-3) \\ \hline 9 + \underline{\ \ ?\ \ } & -3 + \underline{\ \ ?\ \ } \\ \underline{\ \ ?\ \ } & \underline{\ \ ?\ \ } \end{array}$$

- Replace the variable by the given number, ___?___ .
- Evaluate the numerical expressions.
- Compare the results.

The results are ___?___ ; therefore, -3 is a solution of the equation $9 - 5x = -3 - 9x$.

3. Is 4 a solution of $2x = 8$?

4. Is 3 a solution of $y + 4 = 7$?

5. Is -1 a solution of $2b - 1 = 3$?

6. Is -2 a solution of $3a - 4 = 10$?

7. Is 1 a solution of $4 - 2m = 3$?

8. Is 2 a solution $7 - 3n = 2$?

9. Is 5 a solution of $2x + 5 = 3x$?

10. Is 4 a solution of $3y - 4 = 2y$?

11. Is 0 a solution of $4a + 5 = 3a + 5$?

12. Is 0 a solution of $4 - 3b = 4 - 5b$?

13. Is -2 a solution of $4 - 2n = n + 10$?

14. Is -3 a solution of $5 - m = 2 - 2m$?

15. Is 3 a solution of $z^2 + 1 = 4 + 3z$?

16. Is 2 a solution of $2x^2 - 1 = 4x - 1$?

17. Is -1 a solution of $y^2 - 1 = 4y + 3$?

18. Is -2 a solution of $m^2 - 4 = m + 3$?

19. Is 4 a solution of $x(x + 1) = x^2 + 5$?

20. Is 3 a solution of $2a(a - 1) = 3a + 3$?

21. Is $-\frac{1}{4}$ a solution of $8t + 1 = -1$?

22. Is $\frac{1}{2}$ a solution of $4y + 1 = 3$?

23. Is $\frac{2}{5}$ a solution of $5m + 1 = 10m - 3$?

24. Is $\frac{3}{4}$ a solution of $8x - 1 = 12x + 3$?

25. Is 2.1 a solution of $x^2 - 4x = x + 1.89$?

26. Is 1.5 a solution of $c^2 - 3c = 4c - 8.25$?

2 **Solve equations of the form** $x + a = b$

27. How is the Addition Property of Equations used to solve an equation?

28. Explain why the goal in solving an equation is to write the equation in the form *variable = constant*.

29. To solve the equation $p - 30 = 57$, use the Addition Property of Equations to add ____?____ to each side of the equation. The solution is ____?____.

30. To solve the equation $18 + n = 25$, ____?____ 18 from each side of the equation. The solution is ____?____.

31. Without solving the equation $x + \frac{13}{15} = -\frac{21}{43}$, determine whether x is less than or greater than $-\frac{21}{43}$. Explain your answer.

32. Without solving the equation $x - \frac{11}{16} = \frac{19}{24}$, determine whether x is less than or greater than $\frac{19}{24}$. Explain your answer.

Solve and check.

33. $x + 5 = 7$

34. $y + 3 = 9$

35. $b - 4 = 11$

36. $z - 6 = 10$

37. $2 + a = 8$

38. $5 + x = 12$

39. $m + 9 = 3$

40. $t + 12 = 10$

41. $n - 5 = -2$

42. $x - 6 = -5$

43. $b + 7 = 7$

44. $y - 5 = -5$

45. $a - 3 = -5$

46. $x - 6 = -3$

47. $z + 9 = 2$

48. $n + 11 = 1$

49. $10 + m = 3$

50. $8 + x = 5$

51. $9 + x = -3$

52. $10 + y = -4$

53. $b - 5 = -3$

54. $t - 6 = -4$

55. $2 = x + 7$

56. $-8 = n + 1$

57. $4 = m - 11$

58. $-6 = y - 5$

59. $12 = 3 + w$

60. $-9 = 5 + x$

61. $4 = -10 + b$

62. $-7 = -2 + x$

63. $m + \frac{2}{3} = -\frac{1}{3}$

64. $c + \frac{3}{4} = -\frac{1}{4}$

65. $x - \frac{1}{2} = \frac{1}{2}$

66. $x - \frac{2}{5} = \frac{3}{5}$

67. $\frac{5}{8} + y = \frac{1}{8}$

68. $\frac{4}{9} + a = -\frac{2}{9}$

69. $m + \frac{1}{2} = -\frac{1}{4}$

70. $b + \frac{1}{6} = -\frac{1}{3}$

71. $x + \frac{2}{3} = \frac{3}{4}$

72. $n + \frac{2}{5} = \frac{2}{3}$

73. $-\frac{5}{6} = x - \frac{1}{4}$

74. $-\frac{1}{4} = c - \frac{2}{3}$

75. $-\frac{1}{21} = m + \frac{2}{3}$

76. $\dfrac{5}{9} = b - \dfrac{1}{3}$

77. $\dfrac{5}{12} = n + \dfrac{3}{4}$

78. $d + 1.3619 = 2.0148$

79. $w + 2.932 = 4.801$

80. $-0.813 + x = -1.096$

81. $-1.926 + t = -1.042$

3 Solve equations of the form $ax = b$

82. How is the Multiplication Property of Equations used to solve an equation?

83. Why, when the Multiplication Property of Equations is used, must the number that multiplies each side of the equation not be zero?

84. To solve the equation $\dfrac{2}{3}w = -18$, use the Multiplication Property of Equations to multiply each side of the equation by ___?___. The solution is ___?___.

85. To solve the equation $56 = 8n$, ___?___ each side of the equation by 8. The solution is ___?___.

86. Without solving the equation $-\dfrac{15}{41}x = -\dfrac{23}{25}$, determine whether x is less than or greater than zero. Explain your answer.

87. Without solving the equation $\dfrac{5}{28}x = -\dfrac{3}{44}$, determine whether x is less than or greater than zero. Explain your answer.

Solve and check.

88. $5x = 15$

89. $4y = 28$

90. $3b = -12$

91. $2a = -14$

92. $-3x = 6$

93. $-5m = 20$

94. $-3x = -27$

95. $-6n = -30$

96. $20 = 4c$

97. $18 = 2t$

98. $-32 = 8w$

99. $-56 = 7x$

100. $8d = 0$

101. $-5x = 0$

102. $36 = 9z$

103. $35 = -5x$

104. $-64 = 8a$

105. $-32 = -4y$

106. $-42 = 6t$

107. $-12m = -144$

108. $\dfrac{x}{3} = 2$

109. $\dfrac{x}{4} = 3$

110. $-\dfrac{y}{2} = 5$

111. $-\dfrac{b}{3} = 6$

112. $\dfrac{n}{7} = -4$

113. $\dfrac{t}{6} = -3$

114. $\dfrac{2}{5}x = 12$

115. $-\dfrac{4}{3}c = -8$

116. $\dfrac{5}{6}y = -20$

117. $-\dfrac{2}{3}d = 8$

118. $-\dfrac{3}{5}m = 12$

119. $-\dfrac{3}{4}x = -24$

120. $\dfrac{2n}{3} = 2$　　　　**121.** $\dfrac{5x}{6} = -10$　　　　**122.** $\dfrac{-3z}{8} = 9$　　　　**123.** $\dfrac{-4x}{5} = -12$

124. $-6 = -\dfrac{2}{3}y$　　　　**125.** $-15 = -\dfrac{3}{5}x$　　　　**126.** $\dfrac{2}{9} = \dfrac{2}{3}y$　　　　**127.** $-\dfrac{6}{7} = -\dfrac{3}{4}b$

128. $-\dfrac{2}{5}m = -\dfrac{6}{7}$　　　　**129.** $5x + 2x = 14$　　　　**130.** $3n + 2n = 20$　　　　**131.** $7d - 4d = 9$

132. $10y - 3y = 21$　　　　**133.** $2x - 5x = 9$　　　　**134.** $\dfrac{x}{1.4} = 3.2$　　　　**135.** $\dfrac{z}{2.9} = -7.8$

136. $3.4a = 7.004$　　　　**137.** $2.3m = 2.415$　　　　**138.** $-3.7x = 7.77$　　　　**139.** $-1.6m = 5.44$

140. In the equation $15x = y$, y is a positive integer. Is the value of x a negative number?

141. In the equation $-6x = y$, y is a negative integer. Is the value of x a negative number?

142. In the equation $-\dfrac{1}{4}x = y$, y is a positive integer. Is the value of x a negative number?

4 Applications of percent

143. Employee A had an annual salary of $52,000, Employee B had an annual salary of $58,000, and Employee C had an annual salary of $56,000 before each employee was given a 5% raise. Which of the three employees now has the highest annual salary? Explain how you arrived at your answer.

144. Each of three employees earned an annual salary of $65,000 before Employee A was given a 3% raise, Employee B was given a 6% raise, and Employee C was given a 4.5% raise. Which of the three employees now has the highest annual salary? Explain how you arrived at your answer.

Solve.

145. 12 is what percent of 50?　　　**146.** What percent of 125 is 50?　　　**147.** Find 18% of 40.

148. What is 25% of 60?　　　**149.** 12% of what is 48?　　　**150.** 45% of what is 9?

151. What is $33\dfrac{1}{3}\%$ of 27?　　　**152.** Find $16\dfrac{2}{3}\%$ of 30.　　　**153.** What percent of 12 is 3?

154. 10 is what percent of 15?　　　**155.** 60% of what is 3?　　　**156.** 75% of what is 6?

157. 12 is what percent of 6?

158. 20 is what percent of 16?

159. $5\frac{1}{4}$% of what is 21?

160. $37\frac{1}{2}$% of what is 15?

161. Find 15.4% of 50.

162. What is 18.5% of 46?

163. Without solving an equation, determine whether 40% of 80 is less than, equal to, or greater than 80% of 40.

164. Without solving an equation, determine whether $\frac{1}{4}$% of 80 is less than, equal to, or greater than 25% of 80.

Complete Exercises 165 to 167 by replacing the question marks with the correct number from the problem situation or with the word *unknown*.

165. **Problem Situation:** It rained on 24 of the 30 days of June. What percent of the days in June were rainy days?

Using the formula $PB = A$, $P = $ ____?____ , $B = $ ____?____ , and $A = $ ____?____ .

166. **Problem Situation:** $7000 invested in a simple interest account earns $420 of interest in 2 years. What is the annual simple interest rate?

Using the formula $I = Prt$, $I = $ ____?____ , $P = $ ____?____ , $r = $ ____?____ , and $t = $ ____?____ .

167. **Problem Situation:** The label on a 32-ounce bottle of lemonade states that it contains 10% real lemon juice. How many ounces of real lemon juice are in the bottle?

Using the formula $Q = Ar$, $Q = $ ____?____ , $A = $ ____?____ , and $r = $ ____?____ .

168. **a.** Convert 30 months to years.
b. Convert 3 months to years.
c. Convert 9 months to years.
d. Convert 15 months to years.

169. **Stadiums** The Arthur Ashe Tennis Stadium in New York has a seating capacity of 22,500 people. Of these seats, 1.11% are wheelchair-accessible seats. How many seats in the stadium are reserved for wheelchair accessibility? Round to the nearest whole number.

170. **Education** The graph at the right shows the sources of revenue in a recent year for public schools in the United States. The total revenue was $419.8 billion. (*Source*: Census Bureau) How many billion dollars more did public schools receive from the state than from the federal government? Round to the nearest billion dollars.

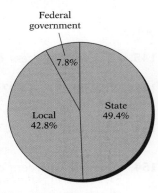

Federal
government

7.8%

Local
42.8%

State
49.4%

**Sources of Revenue for
Public Schools**

171. Natural Resources On average, a person uses 13.2 gal of water per day for showering. This is 17.8% of the total amount of water used per person per day in the average single-family home. Find the total amount of water used per person per day in the average single-family home. Round to the nearest whole number. (*Source:* American Water Works Association)

172. Business During a recent year, 276 billion product coupons were issued by manufacturers. Shoppers redeemed 4.8 billion of these coupons. What percent of the coupons issued were redeemed by customers? Round to the nearest tenth of a percent. (*Source:* NCH NuWorld Consumer Behavior Study, American Coupon Council)

173. Travel According to the annual Summer Vacation Survey conducted by Myvesta, a nonprofit consumer education organization, the average summer vacation costs $2252. If $1850 of this amount is charged on a credit card, what percent of the vacation cost is charged? Round to the nearest tenth of a percent.

174. The Internet A recent survey by Internetworldstats.com found that 973 million people worldwide have access to the Internet. What percent of the world's population has access to the Internet? Use a figure of 6.4 billion people for the world population. Round to the nearest tenth of a percent.

175. The Internet The graph at the right shows the responses to a Yahoo! survey that asked people how many hours per week they typically spend online. How many people were surveyed?

176. Food Industry Of the 63,800 pizzerias in the United States, 2750 are in New York. (*Source*: USA TODAY, February 22, 2005) What percent of the pizzerias in the United States are in New York? Round to the nearest tenth of a percent.

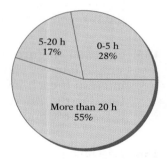

Hours per Week Spent Online

177. Safety Recently the National Safety Council collected data on the leading causes of accidental death. The findings revealed that for people age 20, 30 died from a fall, 47 from fire, 200 from drowning, and 1950 from motor vehicle accidents. What percent of the accidental deaths were not attributed to motor vehicle accidents? Round to the nearest percent.

178. Health According to *Health* magazine, the average American has increased his or her daily consumption of calories from 18 years ago by 11.6%. If the average daily consumption was 1970 calories 18 years ago, what is the average daily consumption today? Round to the nearest whole number.

179. Energy The Energy Information Administration reports that if every U.S. household switched 4 h of lighting per day from incandescent bulbs to compact fluorescent bulbs, we would save 31.7 billion kilowatt-hours of electricity a year, or 33% of the total electricity used for home lighting. What is the total electricity used for home lighting in this country? Round to the nearest tenth of a billion.

180. Investment If Miranda Perry invests $2500 in an account that earns an 8% annual simple interest rate, how much interest will Miranda have earned after 9 months?

181. Investment Kachina Caron invests $1200 in a simple interest account and earns $72 in 8 months. What is the annual simple interest rate?

182. Investment How much money must Andrea invest for two years in an account that earns an annual simple interest rate of 8% if she wants to earn $300 from the investment?

183. Investment Sal Boxer divided a gift of $3000 into two different accounts. He placed $1000 in one account that earned an annual simple interest rate of 7.5%. The remaining money was placed in an account that earned an annual simple interest rate of 8.25%. How much interest did Sal earn from the two accounts after one year?

184. Investment If Americo invests $2500 at an 8% annual simple interest rate and Octavia invests $3000 at a 7% annual simple interest rate, which of the two will earn the greater amount of interest in one year?

185. Investment An investor places $1000 in an account that earns 9% annual simple interest and $1000 in an account that earns 6% annual simple interest. If each investment is left in the account for the same period of time, is the interest rate on the combined investment less than 6%, between 6% and 9%, or greater than 9%?

186. Mixture Problem The concentration of platinum in a necklace is 15%. The necklace weighs 12 g. Find the amount of platinum in the necklace.

187. Mixture Problem A 250-milliliter solution of a fabric dye contains 5 ml of hydrogen peroxide. What is the percent concentration of the hydrogen peroxide?

188. Mixture Problem A carpet is made of a blend of wool and other fibers. If the concentration of wool in the carpet is 75% and the carpet weighs 175 lb, how much wool is in the carpet?

189. Mixture Problem Apple Dan's 32-ounce apple-flavored fruit drink contains 8 oz of apple juice. Forty ounces of a generic brand of apple-flavored fruit drink contains 9 oz of apple juice. Which of the two brands has the greater concentration of apple juice?

190. Mixture Problem Bakers use simple syrup in many of their recipes. Simple syrup is made by combining 500 g of sugar with 500 g of water and mixing it well until the sugar dissolves. What is the percent concentration of sugar in simple syrup?

191. Mixture Problem A pharmacist has 50 g of a topical cream that contains 75% glycerin. How many grams of the cream are not glycerin?

192. Mixture Problem A chemist has 100 ml of a solution that is 9% acetic acid. If the chemist adds 50 ml of pure water to this solution, what is the percent concentration of the resulting mixture?

193. Mixture Problem A 500-gram salt and water solution contains 50 g of salt. This mixture is left in the open air, and 100 g of water evaporates from the solution. What is the percent concentration of salt in the remaining solution?

2.1 Applying Concepts

Solve.

194. $\dfrac{2m + m}{5} = -9$

195. $\dfrac{3y - 8y}{7} = 15$

196. $\dfrac{1}{\dfrac{1}{x}} = 5$

197. $\dfrac{1}{\dfrac{1}{x}} + 8 = -19$

198. $\dfrac{4}{\dfrac{3}{b}} = 8$

199. $\dfrac{5}{\dfrac{7}{a}} - \dfrac{3}{\dfrac{7}{a}} = 6$

200. Geometry Find the value of x in the diagram shown at the right.

201. Consumerism The total cost for a dinner was $97.52. This included a 15% tip calculated on the cost of the dinner after a 6% sales tax. Find the cost of dinner before the tip and tax.

202. Business A retailer decides to increase the original price of each item in the store by 10%. After the price increase, the retailer notices a significant drop in sales and so decides to reduce the current price of each item in the store by 10%. Are the prices back to the original prices? If not, are the prices lower or higher than the original prices?

203. If a quantity increases by 100%, how many times its original value is the new value?

204. Make up an equation of the form $x + a = b$ that has 2 as a solution.

205. Make up an equation of the form $ax = b$ that has -2 as a solution.

General Equations

Solve equations of the form $ax + b = c$

In solving an equation of the form $ax + b = c$, the goal is to rewrite the equation in the form *variable = constant*. This requires applying both the Addition and the Multiplication Properties of Equations.

Solve: $\dfrac{2}{5}x - 3 = -7$

$$\dfrac{2}{5}x - 3 = -7$$

Add 3 to each side of the equation.

$$\dfrac{2}{5}x - 3 + 3 = -7 + 3$$

Simplify.

$$\dfrac{2}{5}x = -4$$

Multiply each side of the equation by the reciprocal of the coefficient $\dfrac{2}{5}$.

$$\dfrac{5}{2} \cdot \dfrac{2}{5}x = \dfrac{5}{2}(-4)$$

Simplify. Now the equation is in the form *variable = constant*.

$$x = -10$$

Check $\dfrac{2}{5}x - 3 = -7$

$$
\begin{array}{c|c}
\dfrac{2}{5}(-10) - 3 & -7 \\
-4 - 3 & -7 \\
-7 & = -7
\end{array}
$$

Write the solution. The solution is -10.

EXAMPLE 1 Solve: $3x - 7 = -5$

Solution
$$3x - 7 = -5$$
$$3x - 7 + 7 = -5 + 7$$ • **Add 7 to each side of the equation.**
$$3x = 2$$ • **Simplify.**
$$\dfrac{3x}{3} = \dfrac{2}{3}$$ • **Divide each side of the equation by 3.**
$$x = \dfrac{2}{3}$$ • **Simplify. Now the equation is in the form** ***variable = constant.***

The solution is $\dfrac{2}{3}$. • **Write the solution.**

PROBLEM 1 Solve: $5x + 7 = 10$

Solution See page S5.

EXAMPLE 2 Solve: $5 = 9 - 2x$

Solution

$5 = 9 - 2x$

$5 - 9 = 9 - 9 - 2x$ • Subtract **9** from each side of the equation.

$-4 = -2x$ • **Simplify.**

$\dfrac{-4}{-2} = \dfrac{-2x}{-2}$ • **Divide each side of the equation by −2.**

$2 = x$ • **Simplify.**

The solution is 2. • **Write the solution.**

PROBLEM 2 Solve: $11 = 11 + 3x$

Solution See page S5.

Note in Example 4 in the Introduction to Equations section that the original equation, $\frac{1}{2} = x + \frac{2}{3}$, contained fractions with denominators of 2 and 3. The least common multiple of 2 and 3 is 6. The least common multiple has the property that both 2 and 3 will divide evenly into it. Therefore, if both sides of the equation are multiplied by 6, both of the denominators will divide evenly into 6. The result is an equation that does not contain any fractions. Multiplying an equation that contains fractions by the LCM of the denominators is called **clearing denominators.** It is an alternative method of solving an equation that contains fractions.

Solve: $\dfrac{1}{2} = x + \dfrac{2}{3}$

Multiply both sides of the equation by 6, the LCM of the denominators.

$$\frac{1}{2} = x + \frac{2}{3}$$

$$6\left(\frac{1}{2}\right) = 6\left(x + \frac{2}{3}\right)$$

Simplify each side of the equation. Use the Distributive Property on the right side of the equation.

$$3 = 6(x) + 6\left(\frac{2}{3}\right)$$

Note that multiplying both sides of the equation by the LCM of the denominators eliminated the fractions.

$$3 = 6x + 4$$

Solve the resulting equation.

$$3 - 4 = 6x + 4 - 4$$

$$-1 = 6x$$

$$\frac{-1}{6} = \frac{6x}{6}$$

$$-\frac{1}{6} = x$$

The solution is $-\frac{1}{6}$.

EXAMPLE 3 Solve: $\dfrac{2}{3} + \dfrac{1}{4}x = -\dfrac{1}{3}$

Solution

$$\dfrac{2}{3} + \dfrac{1}{4}x = -\dfrac{1}{3}$$

$$12\left(\dfrac{2}{3} + \dfrac{1}{4}x\right) = 12\left(-\dfrac{1}{3}\right)$$

$$12\left(\dfrac{2}{3}\right) + 12\left(\dfrac{1}{4}x\right) = -4$$

$$8 + 3x = -4$$
$$8 - 8 + 3x = -4 - 8$$
$$3x = -12$$
$$\dfrac{3x}{3} = \dfrac{-12}{3}$$
$$x = -4$$

- **The equation contains fractions. Find the LCM of the denominators.**
- **The LCM of 3 and 4 is 12. Multiply each side of the equation by 12.**
- **Use the Distributive Property to multiply the left side of the equation by 12.**
- **The equation now contains no fractions.**

The solution is -4.

PROBLEM 3 Solve: $\dfrac{5}{2} - \dfrac{2}{3}x = \dfrac{1}{2}$

Solution See page S5.

OBJECTIVE 2 Solve equations of the form $ax + b = cx + d$

In solving an equation of the form $ax + b = cx + d$, the goal is to rewrite the equation in the form *variable = constant*. Begin by rewriting the equation so that there is only one variable term in the equation. Then rewrite the equation so that there is only one constant term.

Solve: $4x - 5 = 6x + 11$

$$4x - 5 = 6x + 11$$

Subtract $6x$ from each side of the equation.

$$4x - 6x - 5 = 6x - 6x + 11$$

Simplify. Now there is only one variable term in the equation.

$$-2x - 5 = 11$$

Add 5 to each side of the equation.

$$-2x - 5 + 5 = 11 + 5$$

Simplify. Now there is only one constant term in the equation.

$$-2x = 16$$

Divide each side of the equation by -2.

$$\dfrac{-2x}{-2} = \dfrac{16}{-2}$$

Simplify. Now the equation is in the form *variable = constant*.

$$x = -8$$

Check

$$4x - 5 = 6x + 11$$

$$
\begin{array}{c|c}
4(-8) - 5 & 6(-8) + 11 \\
-32 - 5 & -48 + 11 \\
\hline
-37 = & -37
\end{array}
$$

Write the solution. The solution is -8.

Take Note

As shown below, we could have rewritten the equation in the form *constant = variable* by subtracting $4x$ from each side, adding 7 to each side, and then dividing each side by 4.

$$4x - 3 = 8x - 7$$
$$-3 = 4x - 7$$
$$4 = 4x$$
$$1 = x$$

The solution is 1.

EXAMPLE 4 Solve: $4x - 3 = 8x - 7$

Solution

$$4x - 3 = 8x - 7$$

$$4x - 8x - 3 = 8x - 8x - 7$$
• Subtract $8x$ from each side of the equation.

$$-4x - 3 = -7$$
• Simplify. Now there is only one variable term in the equation.

$$-4x - 3 + 3 = -7 + 3$$
• Add 3 to each side of the equation.

$$-4x = -4$$
• Simplify. Now there is only one constant term in the equation.

$$\frac{-4x}{-4} = \frac{-4}{-4}$$
• Divide each side of the equation by -4.

$$x = 1$$
• Simplify. Now the equation is in the form *variable = constant*.

The solution is 1.
• Write the solution.

PROBLEM 4 Solve: $5x + 4 = 6 + 10x$

Solution See page S5.

OBJECTIVE **3** **Solve equations containing parentheses**

When an equation contains parentheses, one of the steps in solving the equation requires the use of the Distributive Property. The Distributive Property is used to remove parentheses from a variable expression. $a(b + c) = ab + ac$

Solve: $4 + 5(2x - 3) = 3(4x - 1)$

$$4 + 5(2x - 3) = 3(4x - 1)$$
Use the Distributive Property to remove parentheses. $4 + 10x - 15 = 12x - 3$

Simplify. $10x - 11 = 12x - 3$

Subtract $12x$ from each side of the equation. $10x - 12x - 11 = 12x - 12x - 3$

Simplify. Now there is only one variable term in the equation. $-2x - 11 = -3$

Add 11 to each side of the equation. $-2x - 11 + 11 = -3 + 11$

<table>
<tr><td>Simplify. Now there is only one constant term in the equation.</td><td>$-2x = 8$</td></tr>
<tr><td>Divide each side of the equation by -2.</td><td>$\dfrac{-2x}{-2} = \dfrac{8}{-2}$</td></tr>
<tr><td>Simplify. Now the equation is in the form <i>variable = constant</i>.</td><td>$x = -4$</td></tr>
</table>

Check

$$4 + 5(2x - 3) = 3(4x - 1)$$

$4 + 5[2(-4) - 3]$	$3[4(-4) - 1]$
$4 + 5(-8 - 3)$	$3(-16 - 1)$
$4 + 5(-11)$	$3(-17)$
$4 - 55$	-51
$-51 =$	-51

Write the solution. The solution is -4.

In Chapter 1, we discussed the use of a graphing calculator to evaluate variable expressions. The same procedure can be used to check the solution of an equation. Consider the example above. After we divide both sides of the equation by -2, the solution appears to be -4. To check this solution, store the value of x, -4, in the calculator. Evaluate the expression on the left side of the original equation: $4 + 5(2x - 3)$. The result is -51. Now evaluate the expression on the right side of the original equation: $3(4x - 1)$. The result is -51. Because the results are equal, the solution -4 checks. See the Appendix for a description of keystroking procedures.

EXAMPLE 5 Solve: $3x - 4(2 - x) = 3(x - 2) - 4$

Solution

$$3x - 4(2 - x) = 3(x - 2) - 4$$
$$3x - 8 + 4x = 3x - 6 - 4$$
- **Use the Distributive Property to remove parentheses.**

$$7x - 8 = 3x - 10$$
- **Simplify.**

$$7x - 3x - 8 = 3x - 3x - 10$$
$$4x - 8 = -10$$
- **Subtract $3x$ from each side of the equation.**

$$4x - 8 + 8 = -10 + 8$$
$$4x = -2$$
- **Add 8 to each side of the equation.**

$$\dfrac{4x}{4} = \dfrac{-2}{4}$$
- **Divide each side of the equation by 4.**

$$x = -\dfrac{1}{2}$$
- **The equation is in the form <i>variable = constant</i>.**

The solution is $-\dfrac{1}{2}$.

PROBLEM 5 Solve: $5x - 4(3 - 2x) = 2(3x - 2) + 6$

Solution See page S6.

EXAMPLE 6 Solve: $3[2 - 4(2x - 1)] = 4x - 10$

Solution $3[2 - 4(2x - 1)] = 4x - 10$
$3[2 - 8x + 4] = 4x - 10$

• **Use the Distributive Property to remove the parentheses.**

$3[6 - 8x] = 4x - 10$

• **Simplify inside the brackets.**

$18 - 24x = 4x - 10$

• **Use the Distributive Property to remove the brackets.**

$18 - 24x - 4x = 4x - 4x - 10$
$18 - 28x = -10$

• **Subtract 4x from each side of the equation.**

$18 - 18 - 28x = -10 - 18$
$-28x = -28$

• **Subtract 18 from each side of the equation.**

$\dfrac{-28x}{-28} = \dfrac{-28}{-28}$

• **Divide each side of the equation by −28.**

$x = 1$

The solution is 1.

PROBLEM 6 Solve: $-2[3x - 5(2x - 3)] = 3x - 8$

Solution See page S6.

OBJECTIVE **4** **Application problems**

EXAMPLE 7 A company uses the equation $V = C - 6000t$ to determine the depreciated value V, after t years, of a milling machine that originally cost C dollars. If a milling machine originally cost $50,000, in how many years will the depreciated value of the machine be $38,000?

Strategy To find the number of years, replace C with 50,000 and V with 38,000 in the given equation, and solve for t.

Solution
$$V = C - 6000t$$
$$38{,}000 = 50{,}000 - 6000t$$
$$38{,}000 - 50{,}000 = 50{,}000 - 50{,}000 - 6000t$$
$$-12{,}000 = -6000t$$
$$\dfrac{-12{,}000}{-6000} = \dfrac{-6000t}{-6000}$$
$$2 = t$$

The depreciated value of the machine will be $38,000 in 2 years.

PROBLEM 7 The value V of an investment of $7500 at an annual simple interest rate of 6% is given by the equation $V = 450t + 7500$, where t is the amount of time, in years, that the money is invested. In how many years will the value of a $7500 investment be $10,200?

Solution See page S6.

Take Note

This system balances because

$$F_1x = F_2(d - x)$$
$$60(6) = 90(10 - 6)$$
$$60(6) = 90(4)$$
$$360 = 360$$

A lever system is shown below. It consists of a lever, or bar; a fulcrum; and two forces, F_1 and F_2. The distance d represents the length of the lever, x represents the distance from F_1 to the fulcrum, and $d - x$ represents the distance from F_2 to the fulcrum.

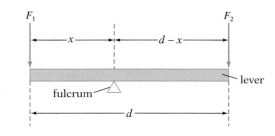

A principle of physics states that when the lever system balances,

$$F_1x = F_2(d - x)$$

EXAMPLE 8 A lever is 10 ft long. A force of 100 lb is applied to one end of the lever, and a force of 400 lb is applied to the other end. When the system balances, how far is the fulcrum from the 100-pound force?

Strategy Draw a diagram of the situation.

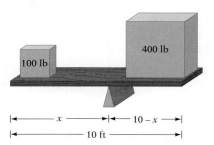

The lever is 10 ft long, so $d = 10$. One force is 100 lb, so $F_1 = 100$. The other force is 400 lb, so $F_2 = 400$. To find the distance of the fulcrum from the 100-pound force, replace the variables F_1, F_2, and d in the lever-system equation with the given values, and solve for x.

Solution
$$F_1x = F_2(d - x)$$
$$100x = 400(10 - x)$$
$$100x = 4000 - 400x$$
$$100x + 400x = 4000 - 400x + 400x$$
$$500x = 4000$$
$$\frac{500x}{500} = \frac{4000}{500}$$
$$x = 8$$

The fulcrum is 8 ft from the 100-pound force.

PROBLEM 8 A lever is 14 ft long. At a distance of 6 ft from the fulcrum, a force of 40 lb is applied. How large a force must be applied to the other end of the lever so that the system will balance?

■ **Solution** See page S6.

Concept Review

Determine whether the statement is always true, sometimes true, or never true.

1. The same variable term can be added to both sides of an equation without changing the solution of the equation.

2. The same variable term can be subtracted from both sides of an equation without changing the solution of the equation.

3. An equation of the form $ax + b = c$ cannot be solved if a is a negative number.

4. The solution of the equation $\frac{x}{3} = 0$ is 0.

5. The solution of the equation $\frac{x}{0} = 3$ is 0.

6. In solving an equation of the form $ax + b = cx + d$, the goal is to rewrite the equation in the form *variable = constant*.

7. In solving an equation of the form $ax + b = cx + d$, subtracting cx from each side of the equation results in an equation with only one variable term in it.

2.2 Exercises

1 Solve equations of the form $ax + b = c$

1. The first step in solving the equation $5 + 8x = 29$ is to subtract ___?___ from each side of the equation. The second step is to divide each side of the equation by ___?___.

2. To clear denominators from the equation $\frac{x}{9} + 2 = \frac{1}{6}$, multiply each side of the equation by ___?___, the least common multiple of the denominators 9 and 6.

Complete Exercises 3 and 4 without actually finding the solutions.

3. Is the solution of the equation $15x + 73 = -347$ positive or negative?

4. Is the solution of the equation $17 = 25 - 40a$ positive or negative?

Solve and check.

5. $3x + 1 = 10$

6. $4y + 3 = 11$

7. $2a - 5 = 7$

8. $5m - 6 = 9$

9. $5 = 4x + 9$

10. $2 = 5b + 12$

11. $13 = 9 + 4z$

12. $7 - c = 9$

13. $2 - x = 11$

14. $4 - 3w = -2$

15. $5 - 6x = -13$

16. $8 - 3t = 2$

17. $-5d + 3 = -12$

18. $-8x - 3 = -19$

19. $-7n - 4 = -25$

20. $-12x + 30 = -6$

21. $-13 = -11y + 9$

22. $2 = 7 - 5a$

23. $3 = 11 - 4n$

24. $-35 = -6b + 1$

25. $-8x + 3 = -29$

26. $-3m - 21 = 0$

27. $7x - 3 = 3$

28. $8y + 3 = 7$

29. $6a + 5 = 9$

30. $3m + 4 = 11$

31. $11 = 15 + 4n$

32. $4 = 2 - 3c$

33. $9 - 4x = 6$

34. $7 - 8z = 0$

35. $1 - 3x = 0$

36. $6y - 5 = -7$

37. $8b - 3 = -9$

38. $5 - 6m = 2$

39. $7 - 9a = 4$

40. $9 = -12c + 5$

41. $10 = -18x + 7$

42. $2y + \dfrac{1}{3} = \dfrac{7}{3}$

43. $3x - \dfrac{5}{6} = \dfrac{13}{6}$

44. $5y + \dfrac{3}{7} = \dfrac{3}{7}$

45. $9x + \dfrac{4}{5} = \dfrac{4}{5}$

46. $4 = 7 - 2w$

47. $7 = 9 - 5a$

48. $-6y + 5 = 13$

49. $-4x + 3 = 9$

50. $\dfrac{1}{2}a - 3 = 1$

51. $\dfrac{1}{3}m - 1 = 5$

52. $\dfrac{2}{5}y + 4 = 6$

53. $\dfrac{3}{4}n + 7 = 13$

54. $-\dfrac{2}{3}x + 1 = 7$

55. $-\dfrac{3}{8}b + 4 = 10$

56. $\dfrac{x}{4} - 6 = 1$

57. $\dfrac{y}{5} - 2 = 3$

58. $\dfrac{2x}{3} - 1 = 5$

59. $\dfrac{3c}{7} - 1 = 8$

60. $4 - \dfrac{3}{4}z = -2$

61. $3 - \dfrac{4}{5}w = -9$

62. $5 + \dfrac{2}{3}y = 3$

63. $17 + \dfrac{5}{8}x = 7$

64. $17 = 7 - \dfrac{5}{6}t$

65. $\dfrac{2}{3} = y - \dfrac{1}{2}$

66. $\dfrac{3}{4} = \dfrac{1}{12}x + 2$

67. $\dfrac{3}{8} = \dfrac{5}{12} - \dfrac{1}{3}b$

68. $\dfrac{2}{3} = \dfrac{3}{4} - \dfrac{1}{2}y$

69. $7 = \dfrac{2x}{5} + 4$

70. $5 - \dfrac{4c}{7} = 8$

71. $7 - \dfrac{5}{9}y = 9$

72. $6a + 3 + 2a = 11$

73. $5y + 9 + 2y = 23$

74. $7x - 4 - 2x = 6$

75. $11z - 3 - 7z = 9$

76. $2x - 6x + 1 = 9$

77. $b - 8b + 1 = -6$

78. $-1 = 5m + 7 - m$

79. $8 = 4n - 6 + 3n$

80. $0.15y + 0.025 = -0.074$

81. $1.2x - 3.44 = 1.3$

82. $3.5 = 3.5 + 0.076x$

83. $-6.5 = 4.3y - 3.06$

Solve.

84. If $2x - 3 = 7$, evaluate $3x + 4$.

85. If $3x + 5 = -4$, evaluate $2x - 5$.

86. If $4 - 5x = -1$, evaluate $x^2 - 3x + 1$.

87. If $2 - 3x = 11$, evaluate $x^2 + 2x - 3$.

2 Solve equations of the form $ax + b = cx + d$

88. To solve the equation $7x - 4 = 2x + 6$, we subtract $2x$ from each side of the equation and add 4 to each side of the equation. The resulting equation is ___?___ = ___?___ . The solution of the equation is ___?___ .

89. To solve $\dfrac{1}{6}x - 7 = \dfrac{3}{4}x$, begin by clearing denominators:

$$(\underline{\quad?\quad})\left(\frac{1}{6}x - 7\right) = (\underline{\quad?\quad})\left(\frac{3}{4}x\right)$$

- Multiply each side of the equation by the LCM of the denominators, 6 and 4.

$$12\left(\frac{1}{6}x\right) - (12)(7) = 9x$$

- Use the ___?___ Property on the left side of the equation. Simplify the right side.

$$\underline{\quad?\quad}x - \underline{\quad?\quad} = 9x$$

- Simplify.

$$2x - \underline{\quad?\quad} - 84 = 9x - \underline{\quad?\quad}$$

- Subtract $2x$ from each side of the equation.

$$-84 = \underline{\quad?\quad}x$$

- Simplify.

$$\frac{-84}{7} = \frac{7x}{7}$$

- ___?___ each side of the equation by ___?___ .

$$\underline{\quad?\quad} = x$$

- Simplify.

Complete Exercises 90 and 91 without actually finding the solutions.

90. Describe the step that will enable you to rewrite the equation $2x - 3 = 7x + 12$ so that it has one variable term with a positive coefficient.

91. If you rewrite the equation $8 - y = y + 6$ so that it has one variable term on the left side of the equation, what will be the coefficient of the variable?

Solve and check.

92. $8x + 5 = 4x + 13$

93. $6y + 2 = y + 17$

94. $7m + 4 = 6m + 7$

95. $11n + 3 = 10n + 11$

96. $5x - 4 = 2x + 5$

97. $9a - 10 = 3a + 2$

98. $12y - 4 = 9y - 7$ **99.** $13b - 1 = 4b - 19$ **100.** $15x - 2 = 4x - 13$

101. $7a - 5 = 2a - 20$ **102.** $3x + 1 = 11 - 2x$ **103.** $n - 2 = 6 - 3n$

104. $2x - 3 = -11 - 2x$ **105.** $4y - 2 = -16 - 3y$ **106.** $2b + 3 = 5b + 12$

107. $m + 4 = 3m + 8$ **108.** $4x - 7 = 5x + 1$ **109.** $6d - 2 = 7d + 5$

110. $4y - 8 = y - 8$ **111.** $5a + 7 = 2a + 7$ **112.** $6 - 5x = 8 - 3x$

113. $10 - 4n = 16 - n$ **114.** $2x - 4 = 6x$ **115.** $2b - 10 = 7b$

116. $8m = 3m + 20$ **117.** $9y = 5y + 16$ **118.** $-x - 4 = -3x - 16$

119. $8 - 4x = 18 - 5x$ **120.** $6 - 10a = 8 - 9a$ **121.** $5 - 7m = 2 - 6m$

122. $8b + 5 = 5b + 7$ **123.** $6y - 1 = 2y + 2$ **124.** $7x - 8 = x - 3$

125. $10x - 3 = 3x - 1$ **126.** $5n + 3 = 2n + 1$ **127.** $8a - 2 = 4a - 5$

128. $\dfrac{1}{5}d = \dfrac{1}{2}d + 3$ **129.** $\dfrac{3}{4}x = \dfrac{1}{12}x + 2$ **130.** $\dfrac{2}{3}a = \dfrac{1}{5}a + 7$

131. $\dfrac{4}{5}c - 7 = \dfrac{1}{10}c$ **132.** $\dfrac{1}{2}y = 10 - \dfrac{1}{3}y$ **133.** $\dfrac{2}{3}b = 2 - \dfrac{5}{6}b$

134. $8.7y = 3.9y + 9.6$ **135.** $4.5x - 5.4 = 2.7x$ **136.** $5.6x = 7.2x - 6.4$

Solve.

137. If $5x = 3x - 8$, evaluate $4x + 2$. **138.** If $7x + 3 = 5x - 7$, evaluate $3x - 2$.

139. If $2 - 6a = 5 - 3a$, evaluate $4a^2 - 2a + 1$. **140.** If $1 - 5c = 4 - 4c$, evaluate $3c^2 - 4c + 2$.

3 Solve equations containing parentheses

141. Use the Distributive Property to remove the parentheses from the equation $4x + 3(x + 6) = 74$: $4x +$ ___?___ $+$ ___?___ $= 74$.

142. Use the Distributive Property to remove the parentheses from the equation $9x - 5(4 - x) = 2(x + 2)$: $9x -$ ___?___ $+$ ___?___ $=$ ___?___ $+$ ___?___.

143. In each of the following equations, y is the same positive integer. For which equations will x have the same solution?
 a. $5 - 2(x - 1) = y$ **b.** $3(x - 1) = y$ **c.** $5 - 2x + 2 = y$
 d. $5 - 2x + 1 = y$

144. In solving the equation $9 - (8 - x) = 5x - 1$, how many times will you use the Distributive Property to remove parentheses?

Solve and check.

145. $5x + 2(x + 1) = 23$

146. $6y + 2(2y + 3) = 16$

147. $9n - 3(2n - 1) = 15$

148. $12x - 2(4x - 6) = 28$

149. $7a - (3a - 4) = 12$

150. $9m - 4(2m - 3) = 11$

151. $5(3 - 2y) + 4y = 3$

152. $4(1 - 3x) + 7x = 9$

153. $10x + 1 = 2(3x + 5) - 1$

154. $5y - 3 = 7 + 4(y - 2)$

155. $4 - 3a = 7 - 2(2a + 5)$

156. $9 - 5x = 12 - (6x + 7)$

157. $3y - 7 = 5(2y - 3) + 4$

158. $2a - 5 = 4(3a + 1) - 2$

159. $5 - (9 - 6x) = 2x - 2$

160. $7 - (5 - 8x) = 4x + 3$

161. $3[2 - 4(y - 1)] = 3(2y + 8)$

162. $5[2 - (2x - 4)] = 2(5 - 3x)$

163. $3a + 2[2 + 3(a - 1)] = 2(3a + 4)$

164. $5 + 3[1 + 2(2x - 3)] = 6(x + 5)$

165. $-2[4 - (3b + 2)] = 5 - 2(3b + 6)$

166. $-4[x - 2(2x - 3)] + 1 = 2x - 3$

167. $0.3x - 2(1.6x) - 8 = 3(1.9x - 4.1)$

168. $0.56 - 0.4(2.1y + 3) = 0.2(2y + 6.1)$

Solve.

169. If $4 - 3a = 7 - 2(2a + 5)$, evaluate $a^2 + 7a$.

170. If $9 - 5x = 12 - (6x + 7)$, evaluate $x^2 - 3x - 2$.

171. If $2z - 5 = 3(4z + 5)$, evaluate $\dfrac{z^2}{z - 2}$.

172. If $3n - 7 = 5(2n + 7)$, evaluate $\dfrac{n^2}{2n - 6}$.

4 Application problems

Physics Black ice is an ice covering on roads that is especially difficult to see and therefore extremely dangerous for motorists. The distance that a car traveling 30 mph on black ice will slide after its brakes are applied is related to the outside temperature by the formula $C = \frac{1}{4}D - 45$, where C is the Celsius temperature and D is the distance in feet that the car will slide.

173. Determine the distance a car will slide on black ice when the outside temperature is $-3°C$.

174. Determine the distance a car will slide on black ice when the outside temperature is $-11°C$.

Physics The pressure at a certain depth in the ocean can be approximated by the equation $P = \frac{1}{2}D + 15$, where P is the pressure in pounds per square inch and D is the depth in feet.

175. Find the depth of a diver when the pressure on the diver is 35 lb/in^2.

176. Find the depth of a diver when the pressure on the diver is 45 lb/in².

Physics The distance s, in feet, that an object will fall in t seconds is given by $s = 16t^2 + vt$, where v is the initial downward velocity of the object in feet per second.

177. Find the initial velocity of an object that falls 80 ft in 2 s.

178. Find the initial velocity of an object that falls 144 ft in 3 s.

Forensic Science Forensic scientists have determined that the equation $H = 2.9L + 78.1$ can be used to approximate the height H, in centimeters, of an adult on the basis of the length L, in centimeters, of that adult's humerus (the bone extending from the shoulder to the elbow).

Humerus

179. Use this formula to approximate the height of an adult whose humerus measures 36 cm.

180. According to this formula, what is the length of the humerus of an adult whose height is 168 cm?

Business The fare, F, to be charged a customer by a taxi company is calculated using the formula $F = 1.50 + 0.95(m - 1)$, where m is the number of miles traveled.

181. A customer is charged $6.25. How many miles was the passenger driven?

182. A passenger is charged $9.10. Find the number of miles the customer was driven.

Employment A person's accurate typing speed can be approximated by the equation $S = \dfrac{W - 5e}{10}$, where S is the accurate typing speed in words per minute, W is the number of words typed in ten minutes, and e is the number of errors made.

183. After taking a 10-minute typing test, a job candidate was told that he had an accurate speed of 35 words per minute. He had typed a total of 390 words. How many errors did he make?

184. A job applicant took a 10-minute typing test and was told that she had an accurate speed of 37 words per minute. If she had typed a total of 400 words, how many errors did she make?

Business A telephone company estimates that the number N of phone calls made per day between two cities of populations P_1 and P_2 that are d miles apart is given by the equation $N = \dfrac{2.51P_1P_2}{d^2}$.

185. Estimate the population P_2, given that P_1 is 48,000, the number of phone calls is 1,100,000, and the distance between the cities is 75 mi. Round to the nearest thousand.

186. Estimate the population P_1, given that P_2 is 125,000, the number of phone calls is 2,500,000, and the distance between the cities is 50 mi. Round to the nearest thousand.

Business To determine the break-even point, or the number of units that must be sold so that no profit or loss occurs, an economist uses the equation $Px = Cx + F$, where P is the selling price per unit, x is the number of units sold, C is the cost to make each unit, and F is the fixed cost.

187. An economist has determined that the selling price per unit for a certain television is $250. The cost to make one such television is $165, and the fixed cost is $29,750. Find the break-even point.

188. A manufacturing engineer determines that the cost per unit for a compact disc is $3.35 and that the fixed cost is $6180. The selling price for the compact disc is $8.50. Find the break-even point.

Lever Systems Use the lever-system equation $F_1x = F_2(d - x)$.

189. In the lever-system equation $F_1x = F_2(d - x)$, what does x represent? What does d represent? What does $d - x$ represent?

190. Two children sit on a seesaw that is 12 ft long. One child weighs 85 lb and the other child weighs 65 lb. If x is the distance of the 65-pound child from the fulcrum when the seesaw balances, state the value of each of the other variables in the lever-system formula $F_1x = F_2(d - x)$: $F_1 = $ ___?___ , $F_2 = $ ___?___ , and $d = $ ___?___ .

For Exercises 191 to 193, use the following information. Two people sit on a seesaw that is 8 ft long. The seesaw balances when the fulcrum is 3 ft from one of the people.

191. How far is the fulcrum from the other person?

192. Which person is heavier, the person who is 3 ft from the fulcrum or the other person?

193. If the two people switch places, will the seesaw still balance?

194. Suppose an 80-pound child is sitting at one end of a 7-foot seesaw, and a 60-pound child is sitting at the other end. The 80-pound child is 3 feet from the fulcrum. Use the lever-system equation to explain why the seesaw balances.

195. Two children are sitting 8 ft apart on a seesaw. One child weighs 60 lb and the second child weighs 50 lb. The fulcrum is 3.5 ft from the child weighing 60 lb. Is the seesaw balanced?

50 lb 60 lb
3.5 ft

8 ft

196. An adult and a child are on a seesaw that is 14 ft long. The adult weighs 175 lb and the child weighs 70 lb. How many feet from the child must the fulcrum be placed so that the seesaw balances?

197. A lever 10 ft long is used to balance a 100-pound rock. The fulcrum is placed 2 ft from the rock. What force must be applied to the other end of the lever to balance the rock?

198. A 50-pound weight is applied to the left end of a seesaw that is 10 ft long. The fulcrum is 4 ft from the 50-pound weight. A weight of 30 lb is then applied to the right end of the seesaw. Is the 30-pound weight adequate to balance the seesaw?

199. In preparation for a stunt, two acrobats are standing on a plank that is 18 ft long. One acrobat weighs 128 lb, and the second acrobat weighs 160 lb. How far from the 128-pound acrobat must the fulcrum be placed so that the acrobats are balanced on the plank?

200. A screwdriver 9 in. long is used as a lever to open a can of paint. The tip of the screwdriver is placed under the lip of the lid with the fulcrum 0.15 in. from the lip. A force of 30 lb is applied to the other end of the screwdriver. Find the force on the lip of the lid.

2.2 Applying Concepts

Solve. If the equation has no solution, write "No solution."

201. $3(2x - 1) - (6x - 4) = -9$

202. $\dfrac{1}{5}(25 - 10b) + 4 = \dfrac{1}{3}(9b - 15) - 6$

203. $3[4(w + 2) - (w + 1)] = 5(2 + w)$

204. $\dfrac{2(5x - 6) - 3(x - 4)}{7} = x + 2$

205. Solve for x: $32{,}166 \div x = 518$ remainder 50

206. One-half of a certain number equals two-thirds of the same number. Find the number.

207. Write an equation of the form $ax + b = cx + d$ that has 4 as a solution.

208. Does the sentence "Solve $3x - 4(x - 1)$" make sense? Why or why not?

209. The equation $x = x + 1$ has no solution, whereas the solution of the equation $2x + 1 = 1$ is zero. Is there a difference between no solution and a solution of zero? Explain your answer.

2.3 Application Problems

OBJECTIVE 1 **Translate a sentence into an equation and solve**

An equation states that two mathematical expressions are equal. Therefore, to translate a sentence into an equation requires recognizing the words or phrases that mean "equals." Besides "equals," these phrases include *is*, *is equal to*, *amounts to*, and *represents*.

Once the sentence is translated into an equation, the equation can be solved by rewriting the equation in the form *variable = constant*.

EXAMPLE 1 Translate "two times the sum of a number and eight equals the sum of four times the number and six" into an equation and solve.

Solution the unknown number: n

two times the sum of a number and eight	equals	the sum of four times the number and six

- Assign a variable to the unknown quantity.
- Find two verbal expressions for the same value.

$$2(n + 8) = 4n + 6$$

- Write a mathematical expression for each verbal expression. Write the equals sign.
- Solve the equation.

$$2n + 16 = 4n + 6$$
$$2n - 4n + 16 = 4n - 4n + 6$$
$$-2n + 16 = 6$$
$$-2n + 16 - 16 = 6 - 16$$
$$-2n = -10$$
$$\frac{-2n}{-2} = \frac{-10}{-2}$$
$$n = 5$$

The number is 5.

PROBLEM 1 Translate "nine less than twice a number is five times the sum of the number and twelve" into an equation and solve.

Solution See page S6.

EXAMPLE 2 The temperature of the sun on the Kelvin scale is 6500 K. This is 4740° less than its temperature on the Fahrenheit scale. Find the Fahrenheit temperature of the sun.

Strategy To find the Fahrenheit temperature, write and solve an equation using F to represent the Fahrenheit temperature.

Solution

6500	is	4740° less than the Fahrenheit temperature

$$6500 = F - 4740$$
$$6500 + 4740 = F - 4740 + 4740$$
$$11,240 = F$$

The temperature of the sun is 11,240°F.

PROBLEM 2 A molecule of octane gas has eight carbon atoms. This represents twice the number of carbon atoms in a butane gas molecule. Find the number of carbon atoms in a butane gas molecule.

Solution See page S6.

EXAMPLE 3 A board 10 ft long is cut into two pieces. Three times the length of the shorter piece is twice the length of the longer piece. Find the length of each piece.

Strategy Draw a diagram. To find the length of each piece, write and solve an equation using x to represent the length of the shorter piece and $10 - x$ to represent the length of the longer piece.

Take Note

We could also let x represent the length of the longer piece and $10 - x$ represent the length of the shorter piece. Then our equation would be

$$3(10 - x) = 2x$$

Show that this equation results in the same solutions.

Solution

three times the length of the shorter piece	is	twice the length of the longer piece

$$3x = 2(10 - x)$$
$$3x = 20 - 2x$$
$$3x + 2x = 20 - 2x + 2x$$
$$5x = 20$$
$$\frac{5x}{5} = \frac{20}{5}$$
$$x = 4$$

• **The length of the shorter piece is 4 ft.**

$$10 - x = 10 - 4 = 6$$

• **Substitute the value of x into the variable expression for the longer piece and evaluate.**

The length of the shorter piece is 4 ft. The length of the longer piece is 6 ft.

PROBLEM 3 A company manufactures 160 bicycles per day. Four times the number of 3-speed bicycles made each day equals 30 less than the number of 10-speed bicycles made each day. Find the number of 10-speed bicycles manufactured each day.

Solution See page S6.

OBJECTIVE 2 Consecutive integer problems

Recall that the integers are the numbers $\ldots, -4, -3, -2, -1, 0, 1, 2, 3, 4, \ldots$.

An **even integer** is an integer that is divisible by 2. Examples of even integers are -8, 0, and 22.

An **odd integer** is an integer that is not divisible by 2. Examples of odd integers are -17, 1, and 39.

Consecutive integers are integers that follow one another in order. Examples of consecutive integers are shown at the right. (Assume the variable n represents an integer.)

11, 12, 13
−8, −7, −6
$n, n + 1, n + 2$

Examples of **consecutive even integers** are shown at the right. (Assume the variable n represents an even integer.)

24, 26, 28
−10, −8, −6
$n, n + 2, n + 4$

Examples of **consecutive odd integers** are shown at the right. (Assume the variable n represents an odd integer.)

19, 21, 23,
−1, 1, 3
$n, n + 2, n + 4$

Solve: The sum of three consecutive odd integers is 51. Find the integers.

> **STRATEGY** **for solving a consecutive integer problem**
>
> ▶ Let a variable represent one of the integers. Express each of the other integers in terms of that variable. Remember that consecutive integers differ by 1. Consecutive even or consecutive odd integers differ by 2.

First odd integer: n
Second odd integer: $n + 2$
Third odd integer: $n + 4$

▶ Determine the relationship among the integers.

The sum of the three odd integers is 51.

$$n + (n + 2) + (n + 4) = 51$$
$$3n + 6 = 51$$
$$3n = 45$$
$$n = 15$$

$n + 2 = 15 + 2 = 17$ • **Substitute the value of n into the variable expressions for the**
$n + 4 = 15 + 4 = 19$ **second and third integers.**

The three consecutive odd integers are 15, 17, and 19.

EXAMPLE 4 Find three consecutive even integers such that three times the second integer is six more than the sum of the first and third integers.

Strategy ▶ First even integer: n
Second even integer: $n + 2$
Third even integer: $n + 4$
▶ Three times the second integer equals six more than the sum of the first and third integers.

Solution $3(n + 2) = n + (n + 4) + 6$
$3n + 6 = 2n + 10$
$n + 6 = 10$
$n = 4$ • **The first even integer is 4.**
$n + 2 = 4 + 2 = 6$ • **Substitute the value of n into the**
$n + 4 = 4 + 4 = 8$ **variable expressions for the second and third integers.**

The three consecutive even integers are 4, 6, and 8.

> **PROBLEM 4** Find three consecutive integers whose sum is -12.
>
> ■ **Solution** See page S7.

OBJECTIVE **3** Value mixture problems

A **value mixture problem** involves combining two ingredients that have different prices into a single blend. For example, a coffee manufacturer may blend two types of coffee into a single blend.

A solution of a value mixture problem is based on the equation $AC = V$, where A is the amount of the ingredient, C is the cost per unit of the ingredient, and V is the value of the ingredient.

The value of 12 lb of coffee costing $5.25 per pound is

$$V = AC$$
$$V = 12(\$5.25)$$
$$V = \$63$$

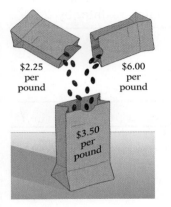

$2.25 per pound

$6.00 per pound

$3.50 per pound

Solve: How many pounds of peanuts that cost $2.25 per pound must be mixed with 40 lb of cashews that cost $6.00 per pound to make a mixture that costs $3.50 per pound?

STRATEGY	for solving a value mixture problem

▶ For each ingredient in the mixture, write a numerical or variable expression for the amount of the ingredient used, the unit cost of the ingredient, and the value of the amount used. For the mixture, write a numerical or variable expression for the amount, the unit cost of the mixture, and the value of the amount. The results can be recorded in a table.

Amount of peanuts: x

	Amount, A	$\cdot$	Unit cost, C	$=$	Value, V
Peanuts	x	$\cdot$	2.25	$=$	$2.25x$
Cashews	40	$\cdot$	6.00	$=$	6.00(40)
Mixture	$x + 40$	$\cdot$	3.50	$=$	3.50($x + 40$)

▶ Determine how the values of the individual ingredients are related. Use the fact that the sum of the values of these ingredients is equal to the value of the mixture.

The sum of the values of the peanuts and the cashews is equal to the value of the mixture.

$$2.25x + 6.00(40) = 3.50(x + 40)$$
$$2.25x + 240 = 3.50x + 140$$
$$-1.25x + 240 = 140$$
$$-1.25x = -100$$
$$x = 80$$

The mixture must contain 80 lb of peanuts.

EXAMPLE 5 How many ounces of a gold alloy that costs $320 an ounce must be mixed with 100 oz of an alloy that costs $100 an ounce to make a mixture that costs $160 an ounce?

Strategy ▶ Ounces of the $320 gold alloy: x

	Amount	Cost	Value
$320 alloy	x	320	$320x$
$100 alloy	100	100	$100(100)$
Mixture	$x + 100$	160	$160(x + 100)$

▶ The sum of the values before mixing equals the value after mixing.

Solution
$$320x + 100(100) = 160(x + 100)$$
$$320x + 10{,}000 = 160x + 16{,}000$$
$$160x + 10{,}000 = 16{,}000$$
$$160x = 6000$$
$$x = 37.5$$

The mixture must contain 37.5 oz of the $320 gold alloy.

PROBLEM 5 A butcher combined hamburger that costs $4.00 per pound with hamburger that costs $2.80 per pound. How many pounds of each were used to make a 75-pound mixture that costs $3.20 per pound?

Solution See page S7.

OBJECTIVE 4 Uniform motion problems

Any object that travels at a constant speed in a straight line is said to be in *uniform motion*. **Uniform motion** means that the speed and direction of an object do not change. For instance, a train traveling at a constant speed of 50 mph on a straight track is in uniform motion.

The solution of a uniform motion problem is based on the equation $d = rt$, where d is the distance traveled, r is the rate of travel, and t is the time spent traveling. For instance, suppose a train travels for 2 h at an average speed of 45 mph. Because the time (2 h) and the rate (45 mph) are known, we can find the distance traveled by solving the equation $d = rt$ for d.

$$d = rt$$
$$d = 45(2) \qquad • r = 45, t = 2$$
$$d = 90$$

The train travels a distance of 90 mi.

A chef leaves a restaurant and drives to his home, which is 16 mi away. If it takes him 20 min to get home, what is the chef's average rate of speed?

Take Note

$20 \text{ min} = \dfrac{20}{60}\text{h}$

$= \dfrac{1}{3}\text{h}$

Because the answer must be in miles per *hour* and the given time is *minutes*, convert 20 min to $\dfrac{1}{3}$ h.

To find the rate of speed, solve the equation $d = rt$ for r using $d = 16$ and $t = \dfrac{1}{3}$.

$$d = rt$$
$$16 = r\left(\dfrac{1}{3}\right)$$
$$16 = \dfrac{1}{3}r$$
$$(3)16 = (3)\dfrac{1}{3}r$$
$$48 = r$$

The average rate of speed is 48 mph.

Take Note

The abbreviation ft/s means feet per second.

If two objects are moving in opposite directions, then the rate at which the distance between them is increasing is the sum of the speeds of the two objects. For instance, in the diagram at the right, two joggers start from the same point and run in opposite directions. The distance between them is changing at a rate of 21 ft/s.

Similarly, if two objects are moving toward each other, the distance between them is decreasing at a rate that is equal to the sum of the speeds. The rate at which the two cyclists at the right are approaching one another is 35 mph.

Two cars start from the same point and move in opposite directions. The car moving west is traveling 50 mph, and the car moving east is traveling 65 mph. In how many hours will the cars be 230 mi apart?

The cars are moving in opposite directions, so the rate at which the distance between them is changing is the sum of the rates of the cars. Thus $r = 115$.

$$50 \text{ mph} + 65 \text{ mph} = 115 \text{ mph}$$

The distance is 230 mi. To find the time, solve $d = rt$ for t.

$$d = rt$$
$$230 = 115t$$
$$\dfrac{230}{115} = \dfrac{115t}{115}$$
$$2 = t$$

The time is 2 h.

The typical cruising speed of a Boeing 777 jet
is 525 mph. However, wind affects the speed
of the plane. For example, when a plane is
flying from California to New York, the wind
is usually in the direction of the plane's flight,
thereby increasing the effective speed of the
plane. If the speed of the wind is 50 mph,
then the effective speed of the plane is the
sum of the plane's speed and the wind's
speed: 525 mph + 50 mph = 575 mph.

If the plane is traveling from New York to
California, the wind is generally in the op-
posite direction of the plane's flight, so it de-
creases the speed of the plane. If the speed
of the wind is 50 mph, then the effective
speed of the plane is the difference between
the plane's speed and the wind's speed:
525 mph − 50 mph = 475 mph.

There are other situations in which these concepts may be applied.

EXAMPLE 6 A Mississippi River sightseeing tour takes tourists from the
tour dock on a 30-mile trip down the river to a Civil War
historical site, where they spend 1 h walking around the
grounds. The tour then returns to the tour dock. If the speed
of the boat is set to travel at 16 mph in calm water and the
rate of the current is 4 mph, find the total time of the trip.

Strategy To determine the total time:

▶ Find the time spent traveling down the river by solving
$d = rt$ for t. The distance is 30 mi. Therefore, $d = 30$.

Because the boat is traveling with the current, the rate of
the boat is the sum of the boat's speed in calm water and
the speed of the current: 16 mph + 4 mph = 20 mph.
Thus $r = 20$ mph.

▶ Find the time spent returning to the tour station by
solving $d = rt$ for t. The distance is 30 mi. Therefore,
$d = 30$. Because the boat is traveling against the current,
the rate of the boat is the difference between the boat's
speed in calm water and the speed of the current:
16 mph − 4 mph = 12 mph. Thus $r = 12$ mph.

▶ Add the time spent traveling down the river, the time
spent at the historical site (1 h), and the time returning.

Solution
$$d = rt$$
$$30 = 20t \qquad \bullet \, d = 30, r = 16 + 4 = 20$$
$$\frac{30}{20} = \frac{20t}{20}$$
$$1.5 = t$$

The time spent traveling down the river is 1.5 h.

$$d = rt$$
$$30 = 12t \qquad \bullet\, d = 30, r = 16 - 4 = 12$$
$$\frac{30}{12} = \frac{12t}{12}$$
$$2.5 = t$$

The time spent returning to the tour station is 2.5 h.

Total time = $1.5 + 1 + 2.5 = 5$

The total time of the trip is 5 h.

PROBLEM 6 A plane that can normally travel at 175 mph in calm air is flying into a headwind of 35 mph. How long will it take the plane to fly 350 mi?

Solution See page S7.

For more difficult uniform motion problems, it may be helpful to record known information in a table. This is illustrated in the following example.

Solve: An executive has an appointment 785 mi from the office. The executive takes a helicopter from the office to the airport and a plane from the airport to the business appointment. The helicopter averages 70 mph, and the plane averages 500 mph. The total time spent traveling is 2 h. Find the distance from the executive's office to the airport.

STRATEGY **for solving a uniform motion problem**

▶ For each object, write a numerical or variable expression for the distance, rate, and time. The results can be recorded in a table.

The total time of travel is 2 h.

Unknown time in the helicopter: t
Time in the plane: $2 - t$

	Rate, r	·	Time, t	=	Distance, d
Helicopter	70	·	t	=	$70t$
Plane	500	·	$2 - t$	=	$500(2 - t)$

▶ Determine how the distances traveled by the individual objects are related. For example, the total distance traveled by both objects may be known, or it may be known that the two objects traveled the same distance.

The total distance traveled is 785 mi.

$$70t + 500(2 - t) = 785$$
$$70t + 1000 - 500t = 785$$
$$-430t + 1000 = 785$$
$$-430t = -215$$
$$t = 0.5$$

|← 70t →|← 500(2 − t) →|

Office Airport Appointment

|←————— 785 mi —————→|

The time spent traveling from the office to the airport in the helicopter is 0.5 h. To find the distance between these two points, substitute 70 for r and 0.5 for t into the equation $rt = d$.

$$rt = d$$
$$70(0.5) = d$$
$$35 = d$$

The distance from the office to the airport is 35 mi.

EXAMPLE 7 A long-distance runner started a course running at an average speed of 6 mph. One and one-half hours later, a cyclist traveled the same course at an average speed of 12 mph. How long after the runner started did the cyclist overtake the runner?

Strategy ▶ Unknown time for the cyclist: t
Time for the runner: $t + 1.5$

	Rate	Time	Distance
Runner	6	$t + 1.5$	$6(t + 1.5)$
Cyclist	12	t	$12t$

▶ The runner and the cyclist traveled the same distance.

Solution $6(t + 1.5) = 12t$
$$6t + 9 = 12t$$
$$9 = 6t$$
$$\frac{3}{2} = t$$ • **The cyclist traveled for 1.5 h.**

$$t + 1.5 = 1.5 + 1.5$$ • **Substitute the value of *t* into the variable**
$$= 3$$ **expression for the runner's time.**

The cyclist overtook the runner 3 h after the runner started.

PROBLEM 7 Two small planes start from the same point and fly in opposite directions. The first plane is flying 30 mph faster than the second plane. In 4 h the planes are 1160 mi apart. Find the rate of each plane.

Solution See page S7.

 Concept Review

Determine whether the statement is always true, sometimes true, or never true.

1. When translating a sentence into an equation, we can use any variable to represent an unknown number.

2. If a rope 19 in. long is cut into two pieces and one of the pieces has length x inches, then the length of the other piece can be expressed as $(x - 19)$ inches.

3. An even integer is a multiple of 2.

4. When $10d$ is used to represent the value of d dimes, 10 is written as the coefficient of d because it is the number of cents in one dime.

5. Given the consecutive odd integers -5 and -3, the next consecutive odd integer is -1.

6. If the first of three consecutive odd integers is n, then the second and third consecutive odd integers are represented as $n + 1$ and $n + 3$.

7. You have a total of 20 coins in nickels and dimes. If n represents the number of nickels you have, then $n - 20$ represents the number of dimes you have.

8. A gold alloy costing $300 per ounce was mixed with a gold alloy costing $200 per ounce. The cost of the mixture was $250 per ounce.

9. Two planes are flying in opposite directions. One plane is flying at a rate of x mph, and the other plane is flying at a rate of y mph. The distance between the planes is increasing at a rate of $(x + y)$ mph.

2.3 Exercises

1 Translate a sentence into an equation and solve

1. When translating a sentence into an equation, we translate the word *is* into the ___?___ sign.

2. The sum of two numbers is twelve. If n represents one of the numbers, then the other number is represented by ___?___.

Translate into an equation and solve.

3. The difference between a number and fifteen is seven. Find the number.

4. The sum of five and a number is three. Find the number.

5. The product of seven and a number is negative twenty-one. Find the number.

6. The quotient of a number and four is two. Find the number.

7. Four less than three times a number is five. Find the number.

8. The difference between five and twice a number is one. Find the number.

9. Four times the sum of twice a number and three is twelve. Find the number.

10. Twenty-one is three times the difference between four times a number and five. Find the number.

11. Twelve is six times the difference between a number and three. Find the number.

12. The difference between six times a number and four times the number is negative fourteen. Find the number.

13. Twice the difference between a number and twenty-five is three times the number. Find the number.

14. Four times a number is three times the difference between thirty-five and the number. Find the number.

15. The sum of two numbers is twenty. Three times the smaller is equal to two times the larger. Find the two numbers.

16. The sum of two numbers is fifteen. One less than three times the smaller is equal to the larger. Find the two numbers.

17. The sum of two numbers is eighteen. The total of three times the smaller and twice the larger is forty-four. Find the two numbers.

18. The sum of two numbers is two. The difference between eight and twice the smaller number is two less than four times the larger. Find the two numbers.

19. The sum of two numbers is fourteen. One number is ten more than the other number. Write two different equations that can be used to find the numbers by first writing an equation in which n represents "the larger number" and then writing an equation in which n represents "the smaller number."

20. The sum of two numbers is seven. Twice one number is four less than the other number. Which of the following equations does *not* represent this situation?
a. $2n = 7 - n - 4$ **b.** $2(7 - x) = x - 4$
c. $2x + 4 = 7 - x$ **d.** $2n - 4 = 7 - n$

Write an equation and solve.

21. Depreciation As a result of depreciation, the value of a car is now $9600. This is three-fifths of its original value. Find the original value of the car.

22. Structures The length of the Royal Gorge Bridge in Colorado is 320 m. This is one-fourth the length of the Golden Gate Bridge. Find the length of the Golden Gate Bridge.

23. Nutrition One slice of cheese pizza contains 290 calories. A medium-size orange has one-fifth that number of calories. How many calories are in a medium-size orange?

24. History John D. Rockefeller died in 1937. At the time of his death, Rockefeller had accumulated a wealth of $1,400 million, which was equal to one-sixty-fifth of the gross national product of the United States at that time. What was the U.S. gross national product in 1937? (*Source: The Wealthy 100: A Ranking of the Richest Americans, Past and Present*)

John D. Rockefeller

25. Agriculture A soil supplement that weighs 18 lb contains iron, potassium, and a mulch. There is fifteen times as much mulch as iron and twice as much potassium as iron. Find the amount of mulch in the soil supplement.

26. Commissions A real estate agent sold two homes and received commissions totaling $6000. The agent's commission on one home was one and one-half times the commission on the second home. Find the agent's commission on each home.

27. Safety Loudness, or the intensity of sound, is measured in decibels. The sound level of a television is about 70 decibels, which is considered a safe hearing level. A food blender runs at 20 decibels higher than a TV, and a jet engine's decibel reading is 40 less than twice that of a blender. At this level, exposure can cause hearing loss. Find the intensity of the sound of a jet engine.

28. Education A university employs a total of 600 teaching assistants and research assistants. There are three times as many teaching assistants as research assistants. Find the number of research assistants employed by the university.

29. Geometry Greek architects considered a rectangle whose length was approximately 1.6 times its width to be the most visually appealing. Find the length and width of a rectangle constructed in this manner if the sum of the length and width is 130 ft.

30. Business A technical information hotline charges a customer $9.00 plus $.50 per minute to answer questions about software. How many minutes did a customer who received a bill for $14.50 use this service?

31. Consumerism The cost to replace a water pump in a sports car was $600. This included $375 for the water pump and $45 per hour for labor. How many hours of labor were required to replace the water pump?

Point of Interest

The low-frequency pulses or whistles made by blue whales have been measured at up to 188 decibels, making them the loudest sounds produced by a living organism.

1.6W

W

32. Utilities The cost of electricity in a certain city is $.08 for each of the first 300 kWh (kilowatt-hours) and $.13 for each kWh over 300 kWh. Find the number of kilowatt-hours used by a family that receives a $51.95 electric bill.

33. Geography Greenland, the largest island in the world, is 21 times larger than Iceland. The combined area of Greenland and Iceland is 880,000 mi². Find the area of Greenland.

34. Metalwork A wire 12 ft long is cut into two pieces. Each piece is bent into the shape of a square. The perimeter of the larger square is twice the perimeter of the smaller square. Find the perimeter of the larger square.

2 Consecutive integer problems

35. Integers that follow one another in order are called ___?___ integers.

36. Two consecutive integers differ by ___?___. Two consecutive even integers differ by ___?___. Two consecutive odd integers differ by ___?___.

37. The sum of three consecutive integers is 54. Find the integers.

38. The sum of three consecutive integers is 75. Find the integers.

39. The sum of three consecutive even integers is 84. Find the integers.

40. The sum of three consecutive even integers is 48. Find the integers.

41. The sum of three consecutive odd integers is 57. Find the integers.

42. The sum of three consecutive odd integers is 81. Find the integers.

43. Find two consecutive even integers such that five times the first integer is equal to four times the second integer.

44. Find two consecutive even integers such that six times the first integer equals three times the second integer.

45. Nine times the first of two consecutive odd integers equals seven times the second. Find the integers.

46. Five times the first of two consecutive odd integers is three times the second. Find the integers.

47. Find three consecutive integers whose sum is negative twenty-four.

48. Find three consecutive even integers whose sum is negative twelve.

49. Three times the smallest of three consecutive even integers is two more than twice the largest. Find the integers.

50. Twice the smallest of three consecutive odd integers is five more than the largest. Find the integers.

51. Find three consecutive odd integers such that three times the middle integer is six more than the sum of the first and third integers.

52. Find three consecutive odd integers such that four times the middle integer is equal to two less than the sum of the first and third integers.

53. Which of the following could *not* be used to represent three consecutive even integers?
 a. $n + 2, n + 4, n + 6$ **b.** $n, n + 4, n + 6$
 c. $n + 1, n + 3, n + 5$ **d.** $n - 2, n, n + 2$

54. If n is an odd integer, which expression represents the third consecutive odd integer after n?
 a. $n + 6$ **b.** $n + 4$ **c.** $n + 3$ **d.** $n + 2$

3 Value mixture problems

55. a. The total value of a 20-pound bag of bird seed that costs $.42 per pound is ____?____.
 b. The cost per ounce of a 24-ounce box of chocolates that has a total value of $16.80 is ____?____.

56. A dried fruit mixture is made from dried cranberries that cost $7.20 per pound and dried apricots that cost $4.50 per pound. Fifteen pounds of the mixture costs $5.40 per pound. Let x be the amount of dried cranberries in the mixture.
 a. Complete the following table.

	Amount, A	·	Unit cost, C	=	Value, V
Dried cranberries	x	·	___?___	=	___?___
Dried apricots	___?___	·	___?___	=	___?___
Mixture	___?___	·	___?___	=	___?___

 b. Use the expressions in the last column of the table in part a to write an equation that can be solved to find the number of pounds of dried cranberries in the mixture: ___?___ + ___?___ = ___?___ .

57. A restaurant chef mixes 20 lb of snow peas costing $1.99 a pound with 14 lb of petite onions costing $1.19 a pound to make a vegetable medley for the evening meal. Find the cost per pound of the mixture.

58. A coffee merchant combines coffee costing $5.50 per pound with coffee costing $3.00 per pound. How many pounds of each should be used to make 40 lb of a blend costing $4.00 per pound?

59. Adult tickets for a play cost $5.00 and children's tickets cost $2.00. For one performance, 460 tickets were sold. Receipts for the performance were $1880. Find the number of adult tickets sold.

60. The tickets for a local production of Gilbert and Sullivan's *H.M.S. Pinafore* cost $6.00 for adults and $3.00 for children and seniors. The total receipts for 505 tickets were $1977. Find the number of adult tickets sold.

61. Fifty liters of pure maple syrup that costs $9.50 per liter are mixed with imitation maple syrup that costs $4.00 per liter. How much imitation maple syrup is needed to make a mixture that costs $5.00 per liter?

62. To make a flour mix, a miller combined soybeans that cost $8.50 per bushel with wheat that costs $4.50 per bushel. How many bushels of each were used to make a mixture of 800 bushels that cost $5.50 per bushel?

63. A hiking instructor is preparing a trail mix for a group of young hikers. She mixes nuts that cost $3.99 a pound with pretzels that cost $1.29 a pound to make a 20-pound mixture that costs $2.37 a pound. Find the number of pounds of nuts used.

64. A silversmith combined pure silver that costs $5.20 an ounce with 50 oz of a silver alloy that costs $2.80 an ounce. How many ounces of the pure silver were used to make an alloy of silver that costs $4.40 an ounce?

65. A tea mixture was made from 30 lb of tea that costs $6.00 per pound and 70 lb of tea that costs $3.20 per pound. Find the cost per pound of the tea mixture.

66. Find the cost per ounce of salad dressing mixture made from 64 oz of olive oil that costs $8.29 and 20 oz of vinegar that costs $1.99. Round to the nearest cent.

67. A fruitstand owner combined cranberry juice that costs $4.20 per gallon with 50 gal of apple juice that costs $2.10 per gallon. How much cranberry juice was used to make cranapple juice that costs $3.00 per gallon?

68. Walnuts that cost $4.05 per kilogram were mixed with cashews that cost $7.25 per kilogram. How many kilograms of each were used to make a 50-kilogram mixture that costs $6.25 per kilogram? Round to the nearest tenth.

69. A coffee merchant mixes a dark roast coffee that costs $10 per pound with a light roast coffee that costs $7 per pound. Which of the following could be true about the cost, *C*, per pound of the mixture? There may be more than one correct answer.
a. $C = \$17$ **b.** $C = \$10$ **c.** $C < \$7$
d. $C > \$7$ **e.** $C > \$10$ **f.** $C < \$10$

70. A snack mix is made from peanuts that cost $3 per pound and caramel popcorn that costs $2.20 per pound. The mixture costs $2.50 per pound. Does the mixture contain more peanuts or more popcorn?

4 Uniform motion problems

71. a. Marvin and Nancy start biking at the same time. Marvin bikes at 12 mph and Nancy bikes at 15 mph. After t hours, Marvin has biked ___?___ miles and Nancy has biked ___?___ miles.

 b. A plane flies at a rate of 380 mph in calm air. The wind is blowing at 20 mph. Flying with the wind, the plane flies ___?___ mph. Flying against the wind, the plane flies ___?___ mph.

72. Sam and Jose leave a movie theatre and go in opposite directions. Sam bikes and Jose walks. Sam bikes four times faster than Jose walks. After a quarter of an hour, Sam and Jose are 3.75 mi apart. Let r represent Jose's rate.

 a. Complete the following table.

	Rate, r	$\cdot$	Time, t	$=$	Distance, d
Sam	___?___	$\cdot$	___?___	$=$	___?___
Jose	r	$\cdot$	___?___	$=$	___?___

 b. Use the expressions in the last column of the table in part a and the distance between Sam and Jose after one-quarter hour to write an equation that can be solved to find Jose's rate: ___?___ + ___?___ = ___?___.

73. Mike and Mindy live 3 mi apart. They leave their houses at the same time and walk toward each other until they meet. Mindy walks faster than Mike.

 a. Is the distance walked by Mike less than, equal to, or greater than the distance walked by Mindy?

 b. Is the time spent walking by Mike less than, equal to, or greater than the time spent walking by Mindy?

 c. What is the total distance traveled by Mike and Mindy?

74. Eric and James ride their bikes from Eric's house to school. James begins biking 10 min before Eric begins. Eric bikes faster than James and catches up with him just as they reach school.

 a. Is the distance biked by James less than, equal to, or greater than the distance biked by Eric?

 b. Is the time spent biking by James less than, equal to, or greater than the time spent biking by Eric?

75. It takes a student 30 min to drive from school to work, a distance of 20 mi. At what average rate of speed does the student drive?

76. A cyclist rides at a rate of 16 mph for 45 min. How far does the cyclist travel in that time?

77. The typical cruising speed for a Boeing 747-400 airplane is 30 mph faster than that for a Boeing 737-800. If a Boeing 747-400 can make a trip of 1680 mi in 3 h, how long would it take a Boeing 737-800 to make the same trip? Round to the nearest tenth.

78. Some hotels in Las Vegas use moving sidewalks to transport guests between various places. Suppose a hotel guest decides to get on a moving sidewalk that is 195 ft long. If the guest walks at a rate of 5 ft/s and the moving sidewalk moves at a rate of 7 ft/s, how many seconds will it take the guest to walk from one end of the moving sidewalk to the other end?

79. Two cyclists start at 1:00 P.M. from opposite ends of a 54-mile race course. The average rate of speed of the first cyclist is 17 mph, and the average rate of speed of the second cyclist is 19 mph. At what time will the two cyclists meet?

80. At 9:00 A.M., Katrina starts down an exercise path and walks at a rate of 4 mph. At 9:30 A.M., Carla leaves from the same place and begins chasing Katrina, catching her at 10:00 A.M. Find Carla's jogging rate.

81. A bicyclist traveling at 18 mph overtakes an in-line skater who is traveling at 10 mph and had a 0.5-hour head start. How far from the starting point does the bicyclist overtake the in-line skater?

82. A helicopter traveling at 120 mph overtakes a speeding car traveling at 90 mph. The car had a 0.5-hour head start. How far from the starting point did the helicopter overtake the car?

83. Two planes are 1380 mi apart and traveling toward each other. One plane is traveling 80 mph faster than the other plane. The planes meet in 1.5 h. Find the speed of each plane.

84. Two jet skiers leave the same dock at the same time and travel in opposite directions. One skier is traveling 14 mph slower than the other skier. In half an hour the skiers are 48 mi apart. Find the rate of the slower skier.

85. A ferry leaves a harbor and travels to a resort island at an average speed of 18 mph. On the return trip, the ferry travels at an average speed of 12 mph because of fog. The total time for the trip is 6 h. How far is the island from the harbor?

86. A commuter plane provides transportation from an international airport to surrounding cities. One commuter plane averaged 210 mph flying to a city and 140 mph returning to the international airport. The total flying time was 4 h. Find the distance between the two airports.

87. Two planes start from the same point and fly in opposite directions. The first plane is flying 50 mph slower than the second plane. In 2.5 h, the planes are 1400 mi apart. Find the rate of each plane.

88. Two hikers start from the same point and hike in opposite directions around a lake whose shoreline is 13 mi long. One hiker walks 0.5 mph faster than the other hiker. How fast does each hiker walk if they meet in 2 h?

89. A student rode a bicycle to the repair shop and then walked home. The student averaged 14 mph riding to the shop and 3.5 mph walking home. The round trip took one hour. How far is it between the student's home and the bicycle shop?

90. An express train leaves Grand Central Station 1 h after a freight train leaves the same station. The express train is traveling 15 mph faster than the freight train. Find the rate at which each train is traveling if the express train overtakes the freight train in 3 h.

91. At noon a train leaves Washington, D.C., headed for Pittsburgh, Pennsylvania, a distance of 260 mi. The train travels at a speed of 60 mph. At 1 P.M. a second train leaves Pittsburgh headed for Washington, D.C., traveling at 40 mph. How long after the train leaves Pittsburgh will the two trains pass each other?

92. A plane leaves an airport at 3 P.M. At 4 P.M. another plane leaves the same airport traveling in the same direction at a speed 150 mph faster than that of the first plane. Four hours after the first plane takes off, the second plane is 250 mi ahead of the first plane. How far did the second plane travel?

2.3 Applying Concepts

93. The amount of liquid in a container triples every minute. The container becomes completely filled at 3:40 P.M. What fractional part of the container is filled at 3:39 P.M.?

94. Wages Four employees are paid at four consecutive levels on a wage scale. The difference between any two consecutive levels is $320 per month. The average of the four employees' monthly wages is $3880. What is the monthly wage of the highest-paid employee?

95. Integer Problem Find four consecutive odd integers whose sum is −48.

96. Integer Problem Find three consecutive odd integers such that the sum of the first and third integers is twice the second integer.

97. Integer Problem Find four consecutive integers such that the sum of the first and fourth integers equals the sum of the second and third integers.

98. Space Travel In 1999, astronomers confirmed the existence of three planets orbiting a star other than the sun. The planets are approximately the size of Jupiter and orbit the star Upsilon Andromeda, which is approximately 260 trillion miles from Earth. How many years after leaving Earth would it take a spacecraft traveling 18 million miles per hour to reach this star? Round to the nearest hundred. (18 million miles per hour is about 1000 times faster than current spacecraft can travel.)

99. Parades If a parade 2 mi long is proceeding at 3 mph, how long will it take a runner, jogging at 6 mph, to travel from the end of the parade to the start of the parade?

100. Uniform Motion Two cars are headed directly toward each other at rates of 40 mph and 60 mph. How many miles apart are they 2 min before impact?

101. Uniform Motion The following problem appears in a math text written around A.D. 1200. Two birds start flying from the tops of two towers 50 ft apart at the same time and at the same rate. One tower is 30 ft high, and the other tower is 40 ft high. The birds reach a grass seed on the ground at exactly the same time. How far is the grass seed from the 40-foot tower?

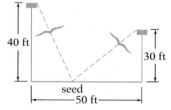

102. Uniform Motion A car travels at an average speed of 30 mph for 1 mi. Is it possible for it to increase its speed during the next mile so that its average speed for the 2 mi is 60 mph?

103. Uniform Motion In December of 1986, pilots Dick Rutan and Jeana Yeager flew the *Voyager* in the first nonstop, nonrefueled flight around the world. They flew east from Edwards Air Force Base in California on December 14, traveled 24,986.727 mi around the world, and returned to Edwards 216 h 3 min 44 s after their departure.
 a. On what date did Rutan and Yeager land at Edwards Air Force Base after their flight?
 b. What was their average speed in miles per hour? Round to the nearest whole number.
 c. Find the circumference of Earth in a reference almanac, and then calculate the approximate distance above Earth that the flight was flown. Round to the nearest whole number.

104. A student jogs 1 mi at a rate of 8 mph and jogs back at a rate of 6 mph. Does it seem reasonable that the average rate is 7 mph? Why or why not? Support your answer.

2.4 Geometry Problems

OBJECTIVE 1 ### Perimeter problems

The **perimeter** of a plane geometric figure is a measure of the distance around the figure. Perimeter is used, for example, in buying fencing for a lawn and in determining how much baseboard is needed for a room.

The perimeter of a triangle is the sum of the lengths of the three sides. Therefore, if a, b, and c represent the lengths of the sides of a triangle, the perimeter, P, of the triangle is given by $P = a + b + c$.

$$P = a + b + c$$

Two special types of triangles are shown below. An **isosceles triangle** has two sides of equal length. The three sides of an **equilateral triangle** are of equal length.

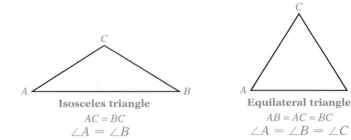

Isosceles triangle
$AC = BC$
$\angle A = \angle B$

Equilateral triangle
$AB = AC = BC$
$\angle A = \angle B = \angle C$

Point of Interest

*Leonardo da Vinci painted the Mona Lisa on a rectangular canvas whose height was approximately 1.6 times its width. Rectangles with these proportions, called **golden rectangles,** were used extensively in Renaissance art.*

The perimeter of a rectangle is the sum of the lengths of the four sides. Let L represent the length and W represent the width of a rectangle. Then the perimeter, P, of the rectangle is given by $P = L + W + L + W$. Combine like terms and the formula is $P = 2L + 2W$.

$$P = 2L + 2W$$

A square is a rectangle in which each side has the same length. Let s represent the length of each side of a square. Then the perimeter, P, of the square is given by $P = s + s + s + s$. Combine like terms and the formula is $P = 4s$.

$$P = 4s$$

Formulas for Perimeters of Geometric Figures	
Perimeter of a triangle:	$P = a + b + c$
Perimeter of a rectangle:	$P = 2L + 2W$
Perimeter of a square:	$P = 4s$

The perimeter of a rectangle is 26 ft. The length of the rectangle is 1 ft more than twice the width. Find the width and length of the rectangle.

Strategy

Let a variable represent the width. Represent the length in terms of that variable. Use the formula for the perimeter of a rectangle.

Width: W
Length: $2W + 1$

Solution

$$P = 2L + 2W$$

$P = 26$. Substitute $2W + 1$ for L. $26 = 2(2W + 1) + 2W$

Use the Distributive Property. $26 = 4W + 2 + 2W$

Combine like terms. $26 = 6W + 2$

Subtract 2 from each side of the equation. $24 = 6W$

Divide each side of the equation by 6. $4 = W$

Find the length of the rectangle by substituting 4 for W in $2W + 1$.

$L = 2W + 1$
$\quad = 2(4) + 1 = 8 + 1 = 9$

The width is 4 ft. The length is 9 ft.

EXAMPLE 1 The perimeter of an isosceles triangle is 25 ft. The length of the third side is 2 ft less than the length of one of the equal sides. Find the measures of the three sides of the triangle.

Strategy ▶ Each equal side: x
The third side: $x - 2$
▶ Use the equation for the perimeter of a triangle.

Solution $P = a + b + c$
$25 = x + x + (x - 2)$
$25 = 3x - 2$
$27 = 3x$
$\;\;9 = x$

$x - 2 = 9 - 2 = 7$ • **Substitute the value of x into the variable expression for the length of the third side.**

Each of the equal sides measures 9 ft.
The third side measures 7 ft.

PROBLEM 1 A carpenter is designing a square patio with a perimeter of 52 ft. What is the length of each side?

Solution See page S7.

OBJECTIVE **2** ## Problems involving angles formed by intersecting lines

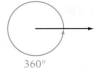

360°

A unit used to measure angles is the **degree**. The symbol for degree is °. ∠ is the symbol for angle.

One complete revolution is 360°. It is probable that the Babylonians chose 360° for a circle because they knew that there are 365 days in one year, and 360 is the closest number to 365 that is divisible by many numbers.

A 90° angle is called a **right angle.** The symbol ∟ represents a right angle. Angle C (∠C) is a right angle.

90°

C

A 180° angle is called a **straight angle.** The angle at the right is a straight angle.

180°

ℓ

An **acute angle** is an angle whose measure is between 0° and 90°. ∠A at the right is an acute angle.

A

An **obtuse angle** is an angle whose measure is between 90° and 180°. ∠B at the right is an obtuse angle.

B

$4x$ $3x$ $5x$

Given the diagram at the left, find x.

Strategy The sum of the measures of the three angles is 360°. To find x, write an equation and solve for x.

Solution $3x + 4x + 5x = 360°$
$12x = 360°$
$x = 30°$

The measure of x is 30°.

Complementary angles are two angles whose measures have the sum 90°.

∠D + ∠E = 70° + 20° = 90°

∠D and ∠E are complementary angles.

70°

D

E 20°

Supplementary angles are two angles whose measures have the sum 180°.

∠F + ∠G = 130° + 50° = 180°

∠F and ∠G are supplementary angles.

130°

F

50°

G

EXAMPLE 2 Find the complement of a 39° angle.

Strategy To find the complement, let x represent the complement of a 39° angle. Use the fact that complementary angles are two angles whose sum is 90° to write an equation. Solve for x.

Solution $39° + x = 90°$
$x = 51°$ • **Subtract 39 from each side of the equation.**

The complement of a 39° angle is a 51° angle.

PROBLEM 2 Find the supplement of a 107° angle.

Solution See page S7.

Parallel lines never meet. The distance between them is always the same. The symbol ∥ means "is parallel to." In the figure at the right, $\ell_1 \| \ell_2$.

Perpendicular lines are intersecting lines that form right angles. The symbol ⊥ means "is perpendicular to." In the figure at the right, $p \perp q$.

Four angles are formed by the intersection of two lines. If the two lines are perpendicular, then each of the four angles is a right angle. If the two lines are not perpendicular, then two of the angles formed are acute angles and two of the angles formed are obtuse angles. The two acute angles are always opposite each other, and the two obtuse angles are always opposite each other.

In the figure at the right, $\angle w$ and $\angle y$ are acute angles, and $\angle x$ and $\angle z$ are obtuse angles.

Two angles that are on opposite sides of the intersection of two lines are called **vertical angles.** Vertical angles have the same measure. $\angle w$ and $\angle y$ are vertical angles. $\angle x$ and $\angle z$ are vertical angles.

Vertical angles have the same measure.

$\angle w = \angle y$
$\angle x = \angle z$

Two angles that share a common side are called **adjacent angles.** For the figure shown above, $\angle x$ and $\angle y$ are adjacent angles, as are $\angle y$ and $\angle z$, $\angle z$ and $\angle w$, and $\angle w$ and $\angle x$. Adjacent angles of intersecting lines are supplementary angles.

Adjacent angles of intersecting lines are supplementary angles.

$\angle x + \angle y = 180°$
$\angle y + \angle z = 180°$
$\angle z + \angle w = 180°$
$\angle w + \angle x = 180°$

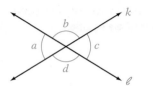

In the diagram at the left, $\angle b = 115°$. Find the measures of angles a, c, and d.

$\angle a$ is supplementary to $\angle b$ because $\angle a$ and $\angle b$ are adjacent angles of intersecting lines.

$$\angle a + \angle b = 180°$$
$$\angle a + 115° = 180°$$
$$\angle a = 65°$$

$\angle c = \angle a$ because $\angle c$ and $\angle a$ are vertical angles. $\angle c = 65°$

$\angle d = \angle b$ because $\angle d$ and $\angle b$ are vertical angles. $\angle d = 115°$

EXAMPLE 3 Find x.

Strategy The angles labeled are adjacent angles of intersecting lines and are therefore supplementary angles. To find x, write an equation and solve for x.

Solution
$$x + (x + 70°) = 180°$$
$$2x + 70° = 180°$$
$$2x = 110°$$
$$x = 55°$$

PROBLEM 3 Find x.

Solution See page S7.

A line that intersects two other lines at different points is called a **transversal.** If the lines cut by a transversal t are parallel lines and the transversal is not perpendicular to the parallel lines, all four acute angles have the same measure and all four obtuse angles have the same measure. In the figure at the right,

$$\angle b = \angle d = \angle x = \angle z$$
$$\angle a = \angle c = \angle w = \angle y$$

Alternate interior angles are two non-adjacent angles that are on opposite sides of the transversal and between the lines. In the figure above, $\angle c$ and $\angle w$ are alternate interior angles, and $\angle d$ and $\angle x$ are alternate interior angles. Alternate interior angles have the same measure.

Alternate interior angles have the same measure.

$$\angle c = \angle w$$
$$\angle d = \angle x$$

Alternate exterior angles are two non-adjacent angles that are on opposite sides of the transversal and outside the parallel lines. In the figure at the left, $\angle a$ and $\angle y$ are alternate exterior angles, and $\angle b$ and $\angle z$ are alternate exterior angles. Alternate exterior angles have the same measure.

Alternate exterior angles have the same measure.

$$\angle a = \angle y$$
$$\angle b = \angle z$$

Corresponding angles are two angles that are on the same side of the transversal and are both acute angles or are both obtuse angles. In the figure above, the following pairs of angles are corresponding angles: $\angle a$ and $\angle w$, $\angle d$ and $\angle z$, $\angle b$ and $\angle x$, and $\angle c$ and $\angle y$. Corresponding angles have the same measure.

Corresponding angles have the same measure.

$$\angle a = \angle w$$
$$\angle d = \angle z$$
$$\angle b = \angle x$$
$$\angle c = \angle y$$

In the diagram at the left, $\ell_1 \parallel \ell_2$ and $\angle f = 58°$. Find the measures of $\angle a$, $\angle c$, and $\angle d$.

$\angle a$ and $\angle f$ are corresponding angles. $\angle a = \angle f = 58°$

$\angle c$ and $\angle f$ are alternate interior angles. $\angle c = \angle f = 58°$

$\angle d$ is supplementary to $\angle a$. $\angle d + \angle a = 180°$
 $\angle d + 58° = 180°$
 $\angle d = 122°$

EXAMPLE 4 Given $\ell_1 \parallel \ell_2$, find x.

Strategy $3x = y$ because corresponding angles have the same measure. $y + (x + 40°) = 180°$ because adjacent angles of intersecting lines are supplementary angles.

Substitute $3x$ for y and solve for x.

Solution $y + (x + 40°) = 180°$
$3x + (x + 40°) = 180°$
$4x + 40° = 180°$
$4x = 140°$
$x = 35°$

PROBLEM 4 Given $\ell_1 \parallel \ell_2$, find x.

Solution See pages S7–S8.

OBJECTIVE **3** **Problems involving the angles of a triangle**

If the lines cut by a transversal are not parallel lines, the three lines will intersect at three points. In the figure at the right, the transversal t intersects lines p and q. The three lines intersect at points A, B, and C. The geometric figure formed by line segments AB, BC, and AC is a **triangle.**

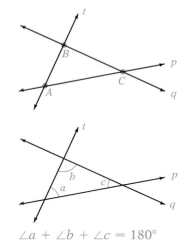

The angles within the region enclosed by the triangle are called **interior angles.** In the figure at the right, angles a, b, and c are interior angles. The sum of the measures of the interior angles is 180°.

$$\angle a + \angle b + \angle c = 180°$$

The Sum of the Measures of the Interior Angles of a Triangle

The sum of the measures of the interior angles of a triangle is 180°.

An angle adjacent to an interior angle is an **exterior angle.** In the figure at the right, angles m and n are exterior angles for angle a. The sum of the measures of an interior angle and a corresponding exterior angle is 180°.

$$\angle a + \angle m = 180°$$
$$\angle a + \angle n = 180°$$

Given that $\angle c = 40°$ and $\angle e = 60°$, find the measure of $\angle d$.

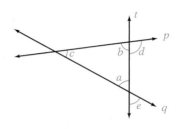

$\angle a$ and $\angle e$ are vertical angles.

$$\angle a = \angle e = 60°$$

The sum of the interior angles is 180°.

$$\angle c + \angle a + \angle b = 180°$$
$$40° + 60° + \angle b = 180°$$
$$100° + \angle b = 180°$$
$$\angle b = 80°$$

∠*b* and ∠*d* are supplementary angles. ∠*b* + ∠*d* = 180°
 80° + ∠*d* = 180°
 ∠*d* = 100°

EXAMPLE 5 Given that ∠*a* = 45° and
∠*x* = 100°, find the measures
of angles *b*, *c*, and *y*.

Strategy ▶ To find the measure of ∠*b*, use the fact that ∠*b* and ∠*x*
are supplementary angles.
▶ To find the measure of ∠*c*, use the fact that the sum of
the measures of the interior angles of a triangle is 180°.
▶ To find the measure of ∠*y*, use the fact that ∠*c* and ∠*y*
are vertical angles.

Solution ∠*b* + ∠*x* = 180°
∠*b* + 100° = 180°
 ∠*b* = 80°

∠*a* + ∠*b* + ∠*c* = 180°
45° + 80° + ∠*c* = 180°
 125° + ∠*c* = 180°
 ∠*c* = 55°

∠*y* = ∠*c* = 55°

PROBLEM 5 Given that ∠*y* = 55°, find the
measures of angles *a*, *b*, and *d*.

Solution See page S8.

EXAMPLE 6 Two angles of a triangle measure 43° and 86°. Find the
measure of the third angle.

Strategy To find the measure of the third angle, use the fact that the
sum of the measures of the interior angles of a triangle is
180°. Write an equation using *x* to represent the measure of
the third angle. Solve the equation for *x*.

Solution *x* + 43° + 86° = 180°
 x + 129° = 180°
 x = 51°

The measure of the third angle is 51°.

PROBLEM 6 One angle of a triangle is a right angle, and one angle
measures 27°. Find the measure of the third angle.

Solution See page S8.

2.4 Concept Review

Determine whether the statement is always true, sometimes true, or never true.

1. The formula for the perimeter of a rectangle is $P = 2L + 2W$, where L represents the length and W represents the width of a rectangle.

2. In the formula for the perimeter of a triangle, $P = a + b + c$, the variables a, b, and c represent the measures of the three angles of a triangle.

3. If two angles of a triangle measure 52° and 73°, then the third angle of the triangle measures 55°.

4. An isosceles triangle has two sides of equal measure and two angles of equal measure.

5. The perimeter of a geometric figure is a measure of the area of the figure.

6. In a right triangle, the sum of the measures of the two acute angles is 90°.

2.4 Exercises

1 Perimeter problems

1. What is the difference between an isosceles triangle and an equilateral triangle?

2. Which of the following units can be used to measure perimeter?
 a. feet
 b. square miles
 c. meters
 d. ounces
 e. cubic inches
 f. centimeters

3. The length of the third side of an isosceles triangle is three units less than twice the length of one of the equal sides. If x is the length of one of the equal sides, an expression for the length of the third side is ___?___.

4. The width of a rectangle is 25% of the length of the rectangle.
 a. If L is the length of the rectangle, then the width of the rectangle is ___?___.
 b. Write an expression for the perimeter of the rectangle:
 2(___?___) + 2(___?___).
 c. Simplify the expression you wrote for the perimeter of the rectangle:
 ___?___.

5. True or false? The formula for the perimeter of a rectangle can be written as $P = 2(L + W)$.

6. The perimeter of an isosceles triangle is 54 ft. Let s be the length of one of the two equal sides. Is it possible for s to be 30 ft?

7. In an isosceles triangle, the third side is 50% of the length of one of the equal sides. Find the length of each side when the perimeter is 125 ft.

8. In an isosceles triangle, the length of one of the equal sides is three times the length of the third side. The perimeter is 21 m. Find the length of each side.

9. The perimeter of a rectangle is 42 m. The length of the rectangle is 3 m less than twice the width. Find the length and width of the rectangle.

10. The width of a rectangle is 25% of the length. The perimeter is 250 cm. Find the length and width of the rectangle.

11. The perimeter of a rectangle is 120 ft. The length of the rectangle is twice the width. Find the length and width of the rectangle.

12. The perimeter of a rectangle is 50 m. The width of the rectangle is 5 m less than the length. Find the length and width of the rectangle.

13. The perimeter of a triangle is 110 cm. One side is twice the second side. The third side is 30 cm more than the second side. Find the length of each side.

14. The perimeter of a triangle is 33 ft. One side of the triangle is 1 ft longer than the second side. The third side is 2 ft longer than the second side. Find the length of each side.

15. The width of the rectangular foundation of a building is 30% of the length. The perimeter of the foundation is 338 ft. Find the length and width of the foundation.

16. The perimeter of a rectangular playground is 440 ft. If the width is 100 ft, what is the length of the playground?

17. A rectangular vegetable garden has a perimeter of 64 ft. The length of the garden is 20 ft. What is the width of the garden?

18. Each of two sides of a triangular banner measures 18 in. If the perimeter of the banner is 46 in., what is the length of the third side of the banner?

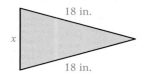

19. The perimeter of a square picture frame is 48 in. Find the length of each side of the frame.

20. A square rug has a perimeter of 32 ft. Find the length of each side of the rug.

2 Problems involving angles formed by intersecting lines

Geometry Problems

21. Name the number of degrees in each of the following.
 a. a right angle
 b. a straight angle
 c. one complete revolution
 d. an acute angle
 e. an obtuse angle
 f. complementary angles
 g. supplementary angles

22. Determine whether enough information is given to solve the problem.
 a. Find the sum of the measures of the two acute angles in a right triangle.
 b. Find the measures of the two acute angles in a right triangle.

23. Let x be the complement of a 53° angle. Then $x + 53 = $ ___?___, so $x = $ ___?___. The complement of a 53° angle is a ___?___ angle.

24. Find x.

$x + x + 20 + 2x = $ ___?___ • A straight angle measures 180°.
$($ ___?___ $)x + 20 = 180$ • Combine like terms.
$4x = $ ___?___ • Subtract 20 from each side of the equation.
$x = $ ___?___ • Divide each side of the equation by 4.

25. Find the complement of a 28° angle.

26. Find the complement of a 46° angle.

27. Find the supplement of a 73° angle.

28. Find the supplement of a 119° angle.

Find the measure of $\angle x$.

29.

30.

31.

Find the measure of ∠*a*.

32.

33.

34.

35.

Find *x*.

36.

37.

38.

39.

40.

41.

Find the measure of ∠*x*.

42.

43.

Find x.

44.

45.

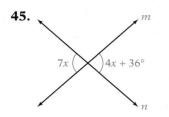

For Exercises 46 to 49, use the diagram below, in which $\ell_1 \| \ell_2$.

46. True or false? $\angle a$ and $\angle b$ have the same measure even if ℓ_1 and ℓ_2 are not parallel.

47. True or false? $\angle a$ and $\angle c$ have the same measure even if ℓ_1 and ℓ_2 are not parallel.

48. True or false? $\angle c$ and $\angle d$ are supplementary only if ℓ_1 and ℓ_2 are parallel.

49. True or false? $\angle c$ and $\angle d$ have the same measure only if t and ℓ_2 are perpendicular.

Given that $\ell_1 \| \ell_2$, find the measures of angles a and b.

50.

51.

52.

53.

Given that $\ell_1 \| \ell_2$, find x.

54.

55.

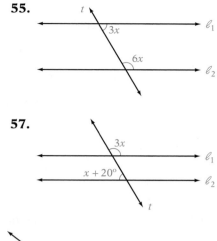

56.

57.

58. Given that $\angle a = 51°$, find the measure of $\angle b$.

59. Given that $\angle a = 38°$, find the measure of $\angle b$.

3 **Problems involving the angles of a triangle**

Geometry Problems

60. The three angles within the region enclosed by a triangle are called the triangle's ____?____ angles. An angle adjacent to one of these angles is an ____?____ angle of the triangle.

61. Given that $\angle x = 50°$ and $\angle c = 85°$, find the measure of $\angle a$.

$\angle b = \angle x = 50°$

$\angle b + \angle c + \angle a = $ ____?____

____?____ + ____?____ + $\angle a = 180°$

____?____ + $\angle a = 180°$

$\angle a = $ ____?____

- ____?____ angles have the same measure.
- The sum of the measures of the interior angles of a triangle is 180°.
- Replace $\angle b$ with 50° and $\angle c$ with 85°.
- Add.
- Subtract 135° from each side of the equation.

62. Given that $\angle a = 95°$ and $\angle b = 70°$, find the measures of angles x and y.

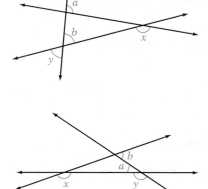

63. Given that $\angle a = 35°$ and $\angle b = 55°$, find the measures of angles x and y.

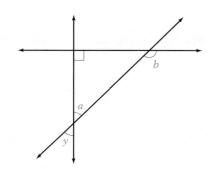

64. Given that $\angle y = 45°$, find the measures of angles a and b.

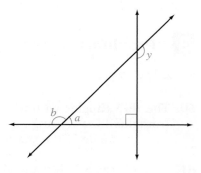

65. Given that $\angle y = 130°$, find the measures of angles a and b.

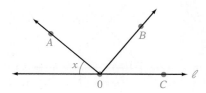

66. Given that $AO \perp OB$, express in terms of x the number of degrees in $\angle BOC$.

67. Given that $AO \perp OB$, express in terms of x the number of degrees in $\angle AOC$.

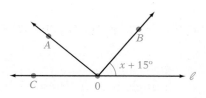

68. One angle in a triangle is a right angle, and one angle measures 30°. What is the measure of the third angle?

69. A triangle has a 45° angle and a right angle. Find the measure of the third angle.

70. Two angles of a triangle measure 42° and 103°. Find the measure of the third angle.

71. Two angles of a triangle measure 62° and 45°. Find the measure of the third angle.

72. A triangle has a 13° angle and a 65° angle. What is the measure of the third angle?

73. A triangle has a 105° angle and a 32° angle. What is the measure of the third angle?

74. In an isosceles triangle, one angle is three times the measure of one of the equal angles. Find the measure of each angle.

75. In an isosceles triangle, one angle is 10° less than three times the measure of one of the equal angles. Find the measure of each angle.

76. In an isosceles triangle, one angle is 16° more than twice the measure of one of the equal angles. Find the measure of each angle.

77. In a triangle, one angle is twice the measure of the second angle. The third angle is three times the measure of the second angle. Find the measure of each angle.

78. True or false? A triangle can have two right angles.

79. True or false? If one angle of a triangle is a right angle, then the other two angles of the triangle are complementary.

80. True or false? A triangle has nine exterior angles.

81. True or false? A triangle that has an exterior angle that is a right angle must also have an interior angle that is a right angle.

2.4 Applying Concepts

82. Geometry A rectangle and an equilateral triangle have the same perimeter. The length of the rectangle is three times the width. Each side of the triangle is 8 cm. Find the length and width of the rectangle.

83. Geometry The length of a rectangle is 1 cm more than twice the width. If the length of the rectangle is decreased by 2 cm and the width is decreased by 1 cm, the perimeter is 20 cm. Find the length and width of the original rectangle.

84. Geometry For the figure at the right, find the sum of the measures of angles x, y, and z.

85. **Geometry** For the figure at the right, explain why $\angle a + \angle b = \angle x$. Write a rule that describes the relationship between an exterior angle of a triangle and the opposite interior angles. Use the rule to write an equation involving angles a, c, and z.

86. Geometry The width of a rectangle is $8x$. The perimeter is $48x$. Find the length of the rectangle in terms of the variable x.

87. Write a report stating the arguments of the proponents of changing to the metric system in the United States or a report stating the arguments of those opposed to switching from the U.S. Customary System of measurement to the metric system.

88. Prepare a report on the use of geometric form in architecture.

<table>
<tr><td>?</td></tr>
<tr><td> $8x$</td></tr>
</table>

2.5 Markup and Discount Problems

OBJECTIVE 1 **Markup problems**

Cost is the price that a business pays for a product. **Selling price** is the price for which a business sells a product to a customer. The difference between selling price and cost is called **markup.** Markup is added to a retailer's cost to cover the expenses of operating a business. Markup is usually expressed as a percent of the retailer's cost. This percent is called the **markup rate.**

The basic markup equations used by a business are

$$\text{Selling price} = \text{Cost} + \text{Markup}$$
$$S \quad = \quad C \; + \quad M$$

$$\text{Markup} = \text{Markup rate} \cdot \text{Cost}$$
$$M \quad = \quad r \quad \cdot \quad C$$

By substituting $r \cdot C$ for M in the first equation, we can also write selling price as

$$S = C + M$$
$$S = C + (r \cdot C)$$
$$S = C + rC$$

The equation $S = C + rC$ is the equation used to solve the markup problems in this section.

EXAMPLE 1 The manager of a clothing store buys a suit for $180 and sells the suit for $252. Find the markup rate.

Strategy Given: $C = \$180$
$S = \$252$
Unknown markup rate: r
Use the equation $S = C + rC$.

Solution

$$S = C + rC$$
$$252 = 180 + 180r$$

• **Substitute the values of C and S into the equation.**

$$252 - 180 = 180 - 180 + 180r$$
$$72 = 180r$$

• **Subtract 180 from each side of the equation.**

$$\frac{72}{180} = \frac{180r}{180}$$

• **Divide each side of the equation by 180.**

$$0.4 = r$$

• **The decimal must be changed to a percent.**

The markup rate is 40%.

PROBLEM 1 The cost to the manager of a sporting goods store for a tennis racket is $120. The selling price of the racket is $180. Find the markup rate.

Solution See page S8.

EXAMPLE 2 The manager of a furniture store uses a markup rate of 45% on all items. The selling price of a chair is $232. Find the cost of the chair.

Strategy Given: $r = 45\% = 0.45$
$S = \$232$
Unknown cost: C
Use the equation $S = C + rC$.

Solution

$$S = C + rC$$
$$232 = C + 0.45C$$
$$232 = 1.45C$$
$$160 = C$$

• **C + 0.45C = 1C + 0.45C = (1 + 0.45)C**

The cost of the chair is $160.

PROBLEM 2 A hardware store employee uses a markup rate of 40% on all items. The selling price of a lawnmower is $266. Find the cost.

Solution See page S8.

O B J E C T I V E **2** **Discount problems**

Discount is the amount by which a retailer reduces the regular price of a product for a promotional sale. Discount is usually expressed as a percent of the regular price. This percent is called the **discount rate** or **markdown rate.**

The basic discount equations used by a business are

$$\text{Sale price} = \text{Regular price} - \text{Discount}$$
$$S \quad = \quad R \quad - \quad D$$

$$\text{Discount} = \text{Discount rate} \cdot \text{Regular price}$$
$$D \quad = \quad r \quad \cdot \quad R$$

By substituting $r \cdot R$ for D in the first equation, we can also write sale price as

$$S = R - D$$
$$S = R - (r \cdot R)$$
$$S = R - rR$$

The equation $S = R - rR$ is the equation used to solve the discount problems in this section.

EXAMPLE 3 In a garden supply store, the regular price of a 100-foot garden hose is $48. During an "after-summer sale," the hose is being sold for $36. Find the discount rate.

Strategy Given: $R = \$48$
$\quad\quad\quad\quad\quad\quad\quad S = \36
Unknown discount rate: r
Use the equation $S = R - rR$.

Solution
$$S = R - rR$$
$$36 = 48 - 48r \quad \bullet \text{ Substitute the values of } \textbf{\textit{R}} \text{ and } \textbf{\textit{S}}$$
$$\text{into the equation.}$$

$$36 - 48 = 48 - 48 - 48r \quad \bullet \text{ Subtract 48 from each side of}$$
$$-12 = -48r \quad\quad\quad\quad\quad \text{the equation.}$$
$$\frac{-12}{-48} = \frac{-48r}{-48} \quad\quad\quad\quad \bullet \text{ Divide each side of the equation}$$
$$\text{by } -48.$$
$$0.25 = r \quad\quad\quad\quad\quad\quad \bullet \text{ The decimal must be changed to}$$
$$\text{a percent.}$$

The discount rate is 25%.

PROBLEM 3 A case of motor oil that regularly sells for $39.80 is on sale for $29.85. What is the discount rate?

Solution See page S8.

EXAMPLE 4 The sale price for a chemical sprayer is $27.30. This price is 35% off the regular price. Find the regular price.

Strategy Given: $S = \$27.30$
$\quad\quad\quad\quad\quad\quad\quad r = 35\% = 0.35$
Unknown regular price: R
Use the equation $S = R - rR$.

Solution
$$S = R - rR$$
$$27.30 = R - 0.35R$$
$$27.30 = 0.65R \quad\quad \bullet \textbf{\textit{R}} - \textbf{0.35}\textbf{\textit{R}} = \textbf{1}\textbf{\textit{R}} - \textbf{0.35}\textbf{\textit{R}} = (\textbf{1} - \textbf{0.35})\textbf{\textit{R}}$$
$$42 = R$$

The regular price is $42.00.

> **PROBLEM 4** The sale price for a telephone is $43.50. This price is 25% off the regular price. Find the regular price.
>
> ■ **Solution** See page S8.

2.5 Concept Review

Determine whether the statement is always true, sometimes true, or never true.

1. In the markup equation $S = C + rC$, the variable r represents the markup rate.

2. Markup is an amount of money, whereas markup rate is a percent.

3. In the discount equation $S = R - rR$, the variable R represents the discount rate.

4. In the markup equation, S represents sale price, and in the discount equation, S represents selling price.

5. $R - 0.45R = 1R - 0.45R = (1 - 0.45)R = 0.55R$ is an example of the application of the Distributive Property.

2.5 Exercises

1 Markup problems

1. Explain the difference between the cost and the selling price of a product.

2. Explain the difference between markup and markup rate.

Complete Exercises 3 and 4 by replacing the question marks with the correct number from the problem situation or with the word *unknown*.

3. **Problem Situation:** A pair of jeans sells for $40. The store uses a markup rate of 25%. Find the cost of the pair of jeans.

 In the formula $S = C + rC$, $S =$ ___?___, $C =$ ___?___, and $r =$ ___?___.

4. **Problem Situation:** A washing machine that costs $620 is sold for $806. Find the markup rate.

 In the formula $S = C + rC$, $S =$ ___?___, $C =$ ___?___, and $r =$ ___?___.

5. A computer software retailer uses a markup rate of 40%. Find the selling price of a computer game that costs the retailer $40.

6. A car dealer advertises a 5% markup over cost. Find the selling price of a car that costs the dealer $26,000.

7. An electronics store uses a markup rate of 58%. Find the selling price of a camcorder that costs the store owner $358.

8. The cost to a landscape architect for a 25-gallon tree is $65. Find the selling price of the tree if the markup rate used by the architect is 30%.

9. A set of golf clubs costing $360 is sold for $630. Find the markup rate on the set of golf clubs.

10. A jeweler purchases a diamond ring for $2399.50. The selling price is $4799. Find the markup rate.

11. A freezer costing $360 is sold for $520. Find the markup rate. Round to the nearest tenth of a percent.

12. A grocer purchases a can of fruit juice for $1.96. The selling price of the fruit juice is $2.45. Find the markup rate.

13. A watch costing $98 is sold for $156.80. Find the markup rate on the watch.

14. A sofa costing $320 is sold for $479. Find the markup rate. Round to the nearest tenth of a percent.

15. The selling price of a compact disc player is $168. The markup rate used by the seller is 40%. Find the cost of the compact disc player.

16. A manufacturer of exercise equipment uses a markup rate of 45%. One of the manufacturer's treadmills has a selling price of $580. Find the cost of the treadmill.

17. A markup rate of 40% was used on a basketball with a selling price of $82.60. Find the cost of the basketball.

18. A digitally recorded compact disc has a selling price of $18.90. The markup rate is 40%. Find the cost of the CD.

19. A markup rate of 25% is used on a computer that has a selling price of $2187.50. Find the cost of the computer.

20. The selling price of a portable CD player is $132. The markup rate used by the retailer is 60%. Find the cost to the retailer.

21. True or false? If a store uses a markup rate of 35%, you can find the store's cost for an item by dividing the selling price of the item by 1 + 0.35, or 1.35.

22. If the markup rate on an item is 100%, what is the relationship between the selling price of the item and the cost of the item?

2 Discount problems

23. Explain the meaning of sale price and discount.

24. Explain the difference between discount and discount rate.

Complete Exercises 25 and 26 by replacing the question marks with the correct number from the problem situation or with the word *unknown*.

25. Problem Situation: The regular price of a camera is $375. The camera is on sale for $318.75. Find the discount rate.

In the formula $S = R - rR$, $S =$ ___?___, $R =$ ___?___, and $r =$ ___?___.

26. Problem Situation: The sale price of a pair of shoes is $44, which is 20% off the regular price. Find the regular price.

In the formula $S = R + rR$, $S =$ ___?___, $R =$ ___?___, and $r =$ ___?___.

27. A tennis racket that regularly sells for $135 is on sale for 25% off the regular price. Find the sale price.

28. A fax machine that regularly sells for $219.99 is on sale for $33\frac{1}{3}$% off the regular price. Find the sale price.

29. A supplier of electrical equipment offers a 5% discount for a purchase that is paid for within 30 days. A transformer regularly sells for $230. Find the discount price of a transformer that is paid for 10 days after the purchase.

30. A clothing wholesaler offers a discount of 10% per shirt when 10 to 20 shirts are purchased and a discount of 15% per shirt when 21 to 50 shirts are purchased. A shirt regularly sells for $27. Find the sale price per shirt when 35 shirts are purchased.

31. A car stereo system that regularly sells for $425 is on sale for $318.75. Find the discount rate.

32. A pair of roller blades that regularly sells for $99.99 is on sale for $79.99. Find the discount rate. Round to the nearest percent.

33. An oak bedroom set with a regular price of $1295 is on sale for $995. Find the discount rate. Round to the nearest tenth of a percent.

34. A stereo set with a regular price of $495 is on sale for $395. Find the markdown rate. Round to the nearest tenth of a percent.

35. A digital camera with a regular price of $325 is on sale for $201.50. Find the markdown rate.

36. A luggage set with a regular price of $178 is on sale for $103.24. Find the discount rate.

37. The sale price of a free-weight home gym is $408, which is 20% off the regular price. Find the regular price.

38. The sale price of a toboggan is $77, which is 30% off the regular price. Find the regular price.

39. A mechanic's tool set is on sale for $180 after a markdown of 40% off the regular price. Find the regular price.

40. The sale price of a modem is $126, which is 30% off the regular price. Find the regular price.

41. A telescope is on sale for $165 after a markdown of 40% off the regular price. Find the regular price.

42. An exercise bike is on sale for $390, having been marked down 25% off the regular price. Find the regular price.

43. True or false? If a store uses a discount rate of 15%, you can find the sale price for an item by multiplying the regular price of the item by $1 - 0.15$, or 0.85.

44. If a discount rate on an item is 50%, which of the following is true?

 a. $S = 2R$ **b.** $R = 2S$ **c.** $S = R$ **d.** $0.50S = R$

2.5 Applying Concepts

45. Markups A pair of shoes that now sells for $63 has been marked up 40%. Find the markup on the pair of shoes.

46. Discounts The sale price of a motorcycle helmet is 25% off the regular price. The discount is $70. Find the sale price.

47. Markups A refrigerator selling for $770 has a markup of $220. Find the markup rate.

48. Discounts The sale price of a copy machine is $765 after a discount of $135. Find the discount rate.

49. Markups The manager of a camera store uses a markup rate of 30%. Find the cost of a camera selling for $299.

50. Discounts The sale price of a television is $180. Find the regular price if the sale price was computed by taking $\frac{1}{3}$ off the regular price, followed by an additional 25% discount on the reduced price.

51. Discounts A customer buys four tires, three at the regular price and one for 20% off the regular price. The four tires cost $304. What was the regular price of a tire?

52. Discounts A lamp, originally priced at under $100, was on sale for 25% off the regular price. When the regular price, a whole number of dollars, was discounted, the discounted price was also a whole number of dollars. Find the largest possible number of dollars in the regular price of the lamp.

53. Discounts A used car is on sale for a discount of 20% off the regular price of $8500. An additional 10% discount on the sale price was offered. Is the result a 30% discount? What is the single discount that would give the same sale price?

54. ✎ Write a report on series trade discounts. Explain how to convert a series discount into a single-discount equivalent.

2.6 Applications: Problems Involving Percent

OBJECTIVE 1 **Investment problems**

The annual simple interest that an investment earns is given by the equation **$Pr = I$**, where P is the principal, or the amount invested, r is the simple interest rate, and I is the simple interest. The solution of an investment problem is based on this equation.

The annual interest rate on a $3000 investment is 9%. The annual simple interest earned on the investment is

$$I = Pr$$
$$I = \$3000(0.09)$$
$$I = \$270$$

Solve: You have a total of $8000 invested in two simple interest accounts. On one account, a money market fund, the annual simple interest rate is 11.5%. On the second account, a bond fund, the annual simple interest rate is 9.75%. The total annual interest earned by the two accounts is $823.75. How much do you have invested in each account?

> **STRATEGY** **for solving a problem involving money deposited in two simple interest accounts**
>
> ▶ For each amount invested, use the equation $Pr = I$. Write a numerical or variable expression for the principal, the interest rate, and the interest earned. The results can be recorded in a table.

The total amount invested is $8000.

Amount invested at 11.5%: x
Amount invested at 9.75%: $8000 - x$

	Principal, P	$\cdot$	Interest rate, r	$=$	Interest earned, I
Amount at 11.5%	x	$\cdot$	0.115	$=$	$0.115x$
Amount at 9.75%	$8000 - x$	$\cdot$	0.0975	$=$	$0.0975(8000 - x)$

▶ Determine how the amounts of interest earned on the individual investments are related. For example, the total interest earned by both accounts may be known, or it may be known that the interest earned on one account is equal to the interest earned on the other account.

The total annual interest earned is $823.75.

$$0.115x + 0.0975(8000 - x) = 823.75$$
$$0.115x + 780 - 0.0975x = 823.75$$
$$0.0175x + 780 = 823.75$$
$$0.0175x = 43.75$$
$$x = 2500$$

The amount invested at 9.75% is $8000 - x$.
Replace x with 2500 and evaluate.

$$8000 - x = 8000 - 2500 = 5500$$

The amount invested at 11.5% is $2500.
The amount invested at 9.75% is $5500.

EXAMPLE 1 An investment of $4000 is made at an annual simple interest rate of 4.9%. How much additional money must be invested at an annual simple interest rate of 7.4% so that the total interest earned is 6.4% of the total investment?

Strategy ▶ Additional amount to be invested at 7.4%: x

	Principal	Rate	Interest
Amount at 4.9%	4000	0.049	$0.049(4000)$
Amount at 7.4%	x	0.074	$0.074x$
Amount at 6.4%	$4000 + x$	0.064	$0.064(4000 + x)$

▶ The sum of the interest earned by the two investments equals the interest earned by the total investment.

Solution
$$0.049(4000) + 0.074x = 0.064(4000 + x)$$
$$196 + 0.074x = 256 + 0.064x$$
$$196 + 0.01x = 256$$
$$0.01x = 60$$
$$x = 6000$$

$6000 must be invested at an annual simple interest rate of 7.4%.

PROBLEM 1	An investment of $3500 is made at an annual simple interest rate of 5.2%. How much additional money must be invested at an annual simple interest rate of 7.5% so that the total interest earned is $575?
Solution	See page S8.

OBJECTIVE 2 **Percent mixture problems**

The amount of a substance in a solution or alloy can be given as a percent of the total solution or alloy. For example, in a 10% hydrogen peroxide solution, 10% of the total solution is hydrogen peroxide. The remaining 90% is water.

The solution of a percent mixture problem is based on the equation **$Ar = Q$,** where A is the amount of solution or alloy, r is the percent of concentration, and Q is the quantity of a substance in the solution or alloy.

The number of grams of silver in 50 g of a 40% silver alloy is

$$Q = Ar$$
$$Q = 50(0.40)$$
$$Q = 20$$

Solve: A chemist mixes an 11% acid solution with a 4% acid solution. How many milliliters of each solution should the chemist use to make a 700-milliliter solution that is 6% acid?

> **Take Note**
>
> As stated at the right, in a 10% hydrogen peroxide solution, 10% of the total solution is hydrogen peroxide. Here we are referring to weight: The solution is 10% hydrogen peroxide by weight. For example, if the solution weighs 100 g, then the hydrogen peroxide in the solution weighs 10 g and the water weighs 90% of 100 g, or 90 g.

> **STRATEGY** **for solving a percent mixture problem**
>
> ▶ For each solution, use the equation $Ar = Q$. Write a numerical or variable expression for the amount of solution, the percent of concentration, and the quantity of the substance in the solution. The results can be recorded in a table.

The total amount of solution is 700 ml.

Amount of 11% solution: x
Amount of 4% solution: $700 - x$

	Amount of solution, A	·	Percent of concentration, r	=	Quantity of substance, Q
11% solution	x	·	0.11	=	0.11x
4% solution	$700 - x$	·	0.04	=	0.04(700 − x)
6% solution	700	·	0.06	=	0.06(700)

> ▶ Determine how the quantities of the substance in the individual solutions are related. Use the fact that the sum of the quantities of the substances being mixed is equal to the quantity of the substance after mixing.

The sum of the amounts of acid in the 11% solution and the 4% solution is equal to the amount of acid in the 6% solution.

$$0.11x + 0.04(700 - x) = 0.06(700)$$
$$0.11x + 28 - 0.04x = 42$$
$$0.07x + 28 = 42$$
$$0.07x = 14$$
$$x = 200$$

The amount of 4% solution is $700 - x$. Replace x by 200 and evaluate.

$$700 - x = 700 - 200 = 500$$

The chemist should use 200 ml of the 11% solution and 500 ml of the 4% solution.

EXAMPLE 2 How many grams of pure acid must be added to 60 g of an 8% acid solution to make a 20% acid solution?

Strategy ▶ Grams of pure acid: x

	Amount	Percent	Quantity
Pure acid (100%)	x	1.00	x
8%	60	0.08	0.08(60)
20%	$x + 60$	0.20	0.20(x + 60)

▶ The sum of the quantities before mixing equals the quantity after mixing.

Solution $x + 0.08(60) = 0.20(x + 60)$
$$x + 4.8 = 0.20x + 12$$
$$0.8x + 4.8 = 12$$
$$0.8x = 7.2$$
$$x = 9$$

To make the 20% acid solution, 9 g of pure acid must be used.

PROBLEM 2 A butcher has some hamburger that is 22% fat and some that is 12% fat. How many pounds of each should be mixed to make 80 lb of hamburger that is 18% fat?

Solution See pages S8–S9.

2.6 Concept Review

Determine whether the statement is always true, sometimes true, or never true.

1. $10,000 is invested in two accounts. If x is invested in one account, then $x - 10,000$ is invested in the other account.

2. The equation $Q = Ar$, where A is the amount of solution or alloy, r is the percent concentration, and Q is the quantity of a substance in the solution or alloy, is used in solving percent mixture problems.

3. If one solution has a 5% concentration and another solution has a 20% concentration, the mixture will have a 25% concentration.

4. An investment of $8000 is divided equally between two accounts with one account paying 6% and the other account paying 8%. The interest received is equivalent to 7% of the total investment.

2.6 Exercises

1 Investment problems

1. Explain the meaning of each variable in the equation $I = Pr$. Give an example of how this equation is used.

2. You invest an amount of money at an annual simple interest rate of 6.25%. You invest a second amount, $500 less than the first, at an annual simple interest rate of 6%. Let x represent the amount invested at 6.25%. Complete the following table.

	Principal, P	$\cdot$	Interest rate, r	$=$	Interest earned, I
Amount at 6.25%	x	$\cdot$	?	$=$	?
Amount at 6%	?	$\cdot$	?	$=$	?

3. The total annual interest earned on the investments in Exercise 2 is $115. Use this information and the information in the table in Exercise 2 to write an equation that can be solved to find the amount of money invested at 6.25%: ____?____ + ____?____ = ____?____ .

4. Suppose $5000 is invested in two simple interest accounts. On one account the annual simple interest rate is 6%, and on the other the annual simple interest rate is 7%. The total annual interest earned on the two accounts is $330. In the context of this situation, explain each term of the equation

$$0.06x + 0.07(5000 - x) = 330$$

5. Joseph Abruzzio decides to divide a gift of $5000 into two different accounts. He deposits $2000 in an IRA account that earns an annual simple interest rate of 5.5%. The remaining money is placed in an account that earns an annual simple interest rate of 7.25%. How much interest will Joseph earn in one year from the two accounts?

6. Kristi invests $1500 at a 7.25% annual simple interest rate, and her sister Kari invests $2000 at a 6.75% annual simple interest rate. Which of the two will earn the greater amount of interest after one year? How much greater?

7. Deon Brown purchases a municipal bond for $2000. The bond earns an annual simple interest rate of 6.4%. How much must Deon invest in an account that earns 8% annual simple interest so that both investments will earn the same amount of interest in one year?

8. A $5000 investment at an annual simple interest rate of 5.2% earned as much interest after one year as another investment in an account that earned 6.5% annual simple interest. How much was invested at 6.5%?

9. Two investments earn a total annual income of $2825. One investment is in a 6.75% annual simple interest certificate of deposit. The other is in a 7.25% tax-free annual simple interest account. The total amount invested is $40,000. Find the amount invested in the certificate of deposit.

10. Two investments earn an annual income of $765. One investment earns an annual simple interest rate of 8.5%, and the other investment earns an annual simple interest rate of 10.2%. The total amount invested is $8000. How much is invested in each account?

11. An investment club invested $5000 at an annual simple interest rate of 8.4%. How much additional money must be invested at an annual simple interest rate of 10.5% so that the total interest earned will be 9% of the total investment?

12. Two investments earn an annual income of $465. One investment is a 5.5% tax-free annual simple interest account, and the other is a 4.5% annual simple interest certificate of deposit. The total amount invested is $9600. How much is invested in each account?

13. Two investments earn an annual income of $575. One investment earns an annual simple interest rate of 8.5%, and the other investment earns an annual simple interest rate of 6.4%. The total amount invested is $8000. How much is invested in each account?

14. An investment club invested $6000 at an annual simple interest rate of 4.0%. How much additional money must be invested at an annual simple interest rate of 6.5% so that the total annual interest earned will be 5% of the total investment?

15. Dee Pinckney made an investment of $6000 at an annual simple interest rate of 5.5%. How much additional money must she invest at an annual

simple interest rate of 10% in order for the total annual interest earned to be 7% of the investment?

16. An account executive deposited $42,000 into two simple interest accounts. On the tax-free account the annual simple interest rate is 3.5%, and on the money market fund the annual simple interest rate is 4.5%. How much should be invested in each account so that both accounts earn the same annual interest?

17. An investment club invested $13,600 into two simple interest accounts. On one account, the annual simple interest rate is 4.2%. On the other, the annual simple interest rate is 6%. How much should be invested in each account so that both accounts earn the same annual interest?

18. Orlando Salavarrio, a financial planner, recommended that 25% of a client's investment be placed in a 4% annual simple interest tax-free account, that 40% be placed in 6% certificates of deposit, and that the remainder be placed in a 9% high-risk investment. The total interest earned from the investments would be $6550. Find the total amount to be invested.

19. The amount of annual interest earned on the x dollars that Will invested in one simple interest account was $0.055x$, and the amount of annual interest earned on the money that Will invested in another simple interest account was $0.072(6000 - x)$.
a. What were the interest rates on the two accounts?
b. What was the total amount of money that Will had invested in the two accounts?

20. Refer to the investments described in Exercise 19. Which of the following could be true about the total amount T of interest earned by Will's two accounts? There may be more than one correct answer.
a. $T < 330$ **b.** $T < 432$ **c.** $T > 432$ **d.** $T > 330$

2 Percent mixture problems

21. The label on a 32-ounce bottle of lemonade says that it contains 10% real lemon juice. The amount of real lemon juice in the bottle is $32(\underline{\;?\;}) = \underline{\;?\;}$ oz.

22. A 25% acid solution is made by adding an amount of 75% acid solution to 50 ml of a 15% acid solution. Let x be the amount of 75% acid solution that is in the mixture.
a. Complete the following table.

	Amount, A	$\cdot$	Percent of concentration, r	$=$	Quantity, Q
15% acid solution	?	$\cdot$	?	$=$	?
75% acid solution	x	$\cdot$	?	$=$	?
Mixture	?	$\cdot$	?	$=$	?

b. Use the expressions in the last column of the table in part a to write an equation that can be solved to find the amount of 75% acid solution that is in the mixture: _____?_____ + _____?_____ = _____?_____.

23. A 32-ounce bottle of Orange Ade contains 8 oz of orange juice. At the same percent concentration, how much orange juice is contained in a 40-ounce bottle of Orange Ade?

24. A chemist has 4 L of a solution that contains 0.36 L of acetic acid. The chemist needs to prepare 6 L of a solution with the same percent concentration of acetic acid. How much acidic acid must the 6 L of the solution contain?

25. A 750-milliliter solution is 4% hydrogen peroxide. How much more hydrogen peroxide is in a 850-milliliter solution that is 5% hydrogen peroxide?

26. One 8-ounce bar of a silver alloy is 30% silver. The percent concentration of silver in a second bar, which weighs 12 oz, is 35%. Find the total amount of silver in the two bars.

27. A jeweler mixed 15 g of a 60% silver alloy with 45 g of a 20% silver alloy. What is the percent concentration of silver in the resulting alloy?

28. A goldsmith mixed 10 g of a 50% gold alloy with 40 g of a 15% gold alloy. What is the percent concentration of the resulting alloy?

29. A silversmith mixed 25 g of a 70% silver alloy with 50 g of a 15% silver alloy. What is the percent concentration of the resulting alloy?

30. A chemist mixed 100 ml of an 8% saline solution with 60 ml of a 5% saline solution. Find the percent concentration of the resulting mixture.

31. How many pounds of a 12% aluminum alloy must be mixed with 400 lb of a 30% aluminum alloy to make a 20% aluminum alloy?

32. How many pounds of a 20% copper alloy must be mixed with 600 lb of a 30% copper alloy to make a 27.5% copper alloy?

33. A hospital staff member mixed a 65% disinfectant solution with a 15% disinfectant solution. How many liters of each were used to make 50 L of a 40% disinfectant solution?

34. A butcher has some hamburger that is 20% fat and some hamburger that is 12% fat. How many pounds of each should be mixed to make 80 lb of hamburger that is 17% fat?

35. How many quarts of water must be added to 5 qt of an 80% antifreeze solution to make a 50% antifreeze solution?

36. Rubbing alcohol is typically diluted with water to 70% strength. If you need 3.5 oz of 45% rubbing alcohol, how many ounces of 70% rubbing alcohol and how much water should you combine?

37. How many ounces of pure water must be added to 60 oz of a 7.5% salt solution to make a 5% salt solution?

38. How much water must be evaporated from 10 gal of a 12% sugar solution in order to obtain a 15% sugar solution?

39. Many fruit drinks are actually only 5% real fruit juice. If you let 2 oz of water evaporate from 12 oz of a drink that is 5% fruit juice, what is the percent concentration of the result?

40. A student mixed 50 ml of a 3% hydrogen peroxide solution with 30 ml of a 12% hydrogen peroxide solution. Find the percent concentration of the resulting mixture. Round to the nearest tenth of a percent.

41. Eighty pounds of a 54% copper alloy is mixed with 200 lb of a 22% copper alloy. Find the percent concentration of the resulting mixture. Round to the nearest tenth of a percent.

42. A druggist mixed 100 cc (cubic centimeters) of a 15% alcohol solution with 50 cc of pure alcohol. Find the percent concentration of the resulting mixture. Round to the nearest tenth of a percent.

43. n ounces of a 30% salt solution is mixed with m ounces of a 50% salt solution. Which relationship between n and m will produce a mixture with a salt concentration of 40%?
a. $m = 2n$ **b.** $m = n$ **c.** $m = 0.4n$

44. n ounces of a 20% acid solution is mixed with m ounces of a 60% salt solution. Which of the following are possible salt concentrations of the mixture? There may be more than one answer.
a. 40% **b.** 80% **c.** 15% **d.** 25% **e.** 55%

2.6 Applying Concepts

45. Investment Problem A financial manager invested 25% of a client's money in bonds paying 9% annual simple interest, 30% in an 8% annual simple interest account, and the remainder in 9.5% corporate bonds. Find the amount invested in each if the total annual interest earned is $1785.

46. Percent Mixture Problem A silversmith mixed 90 g of a 40% silver alloy with 120 g of a 60% silver alloy. Find the percent concentration of the resulting alloy. Round to the nearest tenth of a percent.

47. Value Mixture Problem Find the cost per pound of a tea mixture made from 50 lb of tea that costs $5.50 per pound and 75 lb of tea that costs $4.40 per pound.

48. Percent Mixture Problem How many kilograms of water must be evaporated from 75 kg of a 15% salt solution to produce a 20% salt solution?

49. Percent Mixture Problem A chemist added 20 g of pure acid to a beaker containing an unknown number of grams of pure water. The resulting solution was 25% acid. How many grams of water were in the beaker before the acid was added?

50. Percent Mixture Problem A radiator contains 6 L of a 25% antifreeze solution. How much should be drained and replaced with pure antifreeze to produce a 50% antifreeze solution?

51. Consumer Price Index The consumer price index (CPI), or the "cost-of-living index," measures the change in the price of consumer goods and services. The graph at the right shows the percent increase in the CPI for the years 1998 through 2005.

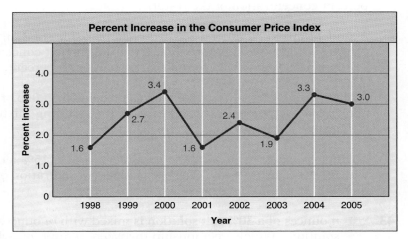

a. During which year shown was the percent increase in the CPI the greatest?

b. During which year was the cost of consumer goods and services highest?

Source: Bureau of Labor Statistics

2.7 Inequalities in One Variable

OBJECTIVE **1** **Set operations and interval notation**

The *roster method* of writing a set encloses the list of the elements of the set in braces. The set of whole numbers, written as $\{0, 1, 2, 3, 4, \ldots\}$, and the set of natural numbers, written as $\{1, 2, 3, 4, \ldots\}$, are **infinite sets.** The pattern of numbers continues without end. It is impossible to list all the elements of an infinite set.

The set of even natural numbers less than 10 is written $\{2, 4, 6, 8\}$. This is an example of a **finite set;** all the elements of the set can be listed.

The **empty set,** or **null set,** is the set that contains no elements. The symbol $\varnothing$ or $\{\,\}$ is used to represent the empty set.

A second method of representing a set is **set-builder notation.** Set-builder notation can be used to describe almost any set, but it is especially useful when writing infinite sets. In set-builder notation, the set of integers greater than -3 is written

$$\{x \mid x > -3, x \in \text{integers}\}$$

and is read "the set of all x such that x is greater than -3 and x is an element of the integers." This is an infinite set. It is impossible to list all the elements of the set, but the set can be described by using set-builder notation.

The set of real numbers less than 5 is written

$$\{x \,|\, x < 5, x \in \text{real numbers}\}$$

and is read "the set of all x such that x is less than 5 and x is an element of the real numbers."

EXAMPLE 1 Use set-builder notation to write the set of integers greater than -8.

Solution $\{x \,|\, x > -8, x \in \text{integers}\}$

PROBLEM 1 Use set-builder notation to write the set of real numbers less than 7.

Solution See page S9.

Just as operations such as addition and multiplication are performed on real numbers, operations are performed on sets. Two operations performed on sets are *union* and *intersection*.

The **union of two sets,** written $A \cup B$, is the set of all elements that belong to either A **or** B. In set-builder notation, this is written

$$A \cup B = \{x \,|\, x \in A \quad \text{or} \quad x \in B\}$$

Given $A = \{2, 3, 4\}$ and $B = \{0, 1, 2, 3\}$, the union of A and B contains all the elements that belong to either A or B. The elements that belong to both sets are listed only once.

$$A \cup B = \{0, 1, 2, 3, 4\}$$

The **intersection of two sets,** written $A \cap B$, is the set of all elements that are common to both A **and** B. In set-builder notation, this is written

$$A \cap B = \{x \,|\, x \in A \quad \text{and} \quad x \in B\}$$

Given $A = \{2, 3, 4\}$ and $B = \{0, 1, 2, 3\}$, the intersection of A and B contains all the elements that are common to both A and B.

$$A \cap B = \{2, 3\}$$

EXAMPLE 2 Find $C \cup D$ given $C = \{1, 5, 9, 13, 17\}$ and $D = \{3, 5, 7, 9, 11\}$.

Solution $C \cup D = \{1, 3, 5, 7, 9, 11, 13, 17\}$

PROBLEM 2 Find $A \cup C$ given $A = \{-2, -1, 0, 1, 2\}$ and $C = \{-5, -1, 0, 1, 5\}$.

Solution See page S9.

EXAMPLE 3 Find $A \cap B$ given $A = \{x \,|\, x \in \text{natural numbers}\}$ and $B = \{x \,|\, x \in \text{negative integers}\}$.

Solution There are no natural numbers that are also negative integers. $A \cap B = \varnothing$

Point of Interest

The symbols $\in$, $\cup$, and $\cap$ were first used by Giuseppe Peano, in Arithmetices Principia, Nova Exposita (The Principle of Mathematics, a New Method of Exposition), published in 1889. The purpose of this book was to deduce the principles of mathematics from pure logic.

PROBLEM 3 Find $E \cap F$ given $E = \{x \mid x \in$ odd integers$\}$ and $F = \{x \mid x \in$ even integers$\}$.

■ **Solution** See page S9.

Set-builder notation and the inequality symbols $>$, $<$, $\geq$, and $\leq$ are used to describe infinite sets of real numbers. These sets can be graphed on the real number line.

The graph of $\{x \mid x > -2, x \in$ real numbers$\}$ is shown below. The set is the real numbers greater than -2. The parenthesis on the graph indicates that -2 is not included in the set.

The graph of $\{x \mid x \geq -2, x \in$ real numbers$\}$ is shown below. The set is the real numbers greater than or equal to -2. The bracket at -2 indicates that -2 is included in the set.

For the remainder of this section, all variables will represent real numbers. Using this convention, the above set is written $\{x \mid x \geq -2\}$.

Take Note

If we asked for the graph of $\{x \mid x \leq 3, x \in$ integers$\}$, the graph would appear as

The domain of a variable affects its graph.

EXAMPLE 4 Graph: $\{x \mid x \leq 3\}$

Solution

• The set is the real numbers less than or equal to 3. Draw a right bracket at 3, and darken the number line to the left of 3.

PROBLEM 4 Graph: $\{x \mid x > -3\}$

■ **Solution** See page S9.

The union of two sets is the set of all elements belonging to either one or the other of the two sets. The set $\{x \mid x \leq -1\} \cup \{x \mid x > 3\}$ is the set of real numbers that are either less than or equal to -1 or greater than 3.

The set is written $\{x \mid x \leq -1 \text{ or } x > 3\}$.

The set $\{x \mid x > 2\} \cup \{x \mid x > 4\}$ is the set of real numbers that are either greater than 2 or greater than 4.

The set is written $\{x \mid x > 2\}$.

The intersection of two sets is the set that contains the elements common to both sets. The set $\{x \mid x > -2\} \cap \{x \mid x < 5\}$ is the set of real numbers that are greater than -2 and less than 5.

The set can be written $\{x \mid x > -2 \text{ and } x < 5\}$. However, it is more commonly written $\{x \mid -2 < x < 5\}$. This set is read "the set of all x such that x is greater than -2 and less than 5."

The set $\{x \mid x < 4\} \cap \{x \mid x < 5\}$ is the set of real numbers that are less than 4 and less than 5.

The set is written $\{x \mid x < 4\}$.

EXAMPLE 5 Graph: $\{x \mid x < 0\} \cap \{x \mid x > -3\}$

Solution The set is $\{x \mid -3 < x < 0\}$.

PROBLEM 5 Graph: $\{x \mid x \geq 1\} \cup \{x \mid x \leq -3\}$

Solution See page S9.

Some sets can also be expressed using **interval notation.** For example, the interval notation $(-3, 2]$ indicates the interval of all real numbers greater than -3 and less than or equal to 2. As on the graph of a set, the left parenthesis indicates that -3 is not included in the set. The right bracket indicates that 2 is included in the set.

An interval is said to be **closed** if it includes both endpoints; it is **open** if it does not include either endpoint. An interval is **half-open** if one endpoint is included and the other is not. In each example given below, -3 and 2 are the **endpoints** of the interval. In each case, the set notation, the interval notation, and the graph of the set are shown.

$\{x \mid -3 < x < 2\}$	$(-3, 2)$ Open interval	
$\{x \mid -3 \leq x \leq 2\}$	$[-3, 2]$ Closed interval	
$\{x \mid -3 \leq x < 2\}$	$[-3, 2)$ Half-open interval	
$\{x \mid -3 < x \leq 2\}$	$(-3, 2]$ Half-open interval	

To indicate an interval that extends forever in one or both directions using interval notation, we use the **infinity symbol** ∞ or the **negative infinity symbol** $-\infty$. The infinity symbol is not a number; it is simply used as a notation to indicate that the interval is unlimited. In interval notation, a parenthesis is always used to the right of an infinity symbol or to the left of a negative infinity symbol, as shown in the following examples.

$\{x \mid x > 1\}$	$(1, \infty)$	
$\{x \mid x \geq 1\}$	$[1, \infty)$	
$\{x \mid x < 1\}$	$(-\infty, 1)$	
$\{x \mid x \leq 1\}$	$(-\infty, 1]$	
$\{x \mid -\infty < x < \infty\}$	$(-\infty, \infty)$	

EXAMPLE 6 Write $\{x \mid 0 < x \le 5\}$ using interval notation.

Solution $(0, 5]$ • **The set is the real numbers greater than 0 and less than or equal to 5.**

PROBLEM 6 Write $\{x \mid -8 \le x < -1\}$ using interval notation.

Solution See page S9.

EXAMPLE 7 Write $(-\infty, 9]$ using set-builder notation.

Solution $\{x \mid x \le 9\}$ • **The set is the real numbers less than or equal to 9.**

PROBLEM 7 Write $(-12, \infty)$ using set-builder notation.

Solution See page S9.

OBJECTIVE 2 Solve inequalities in one variable

The **solution set of an inequality** is a set of numbers, each element of which, when substituted for the variable, results in a true inequality.

The inequality at the right is true if the variable is replaced by 3, -1.98, or $\frac{2}{3}$.

$$x - 1 < 4$$
$$3 - 1 < 4$$
$$-1.98 - 1 < 4$$
$$\frac{2}{3} - 1 < 4$$

There are many values of the variable x that will make the inequality $x - 1 < 4$ true. The solution set of the inequality is any number less than 5. The solution set can be written in set-builder notation as $\{x \mid x < 5\}$.

The graph of the solution set of $x - 1 < 4$ is shown at the right.

In solving an inequality, use the Addition and Multiplication Properties of Inequalities to rewrite the inequality in the form *variable* $<$ *constant* or *variable* $>$ *constant*.

The Addition Property of Inequalities

If $a > b$ and c is a real number, then the inequalities $a > b$ and $a + c > b + c$ have the same solution set.

If $a < b$ and c is a real number, then the inequalities $a < b$ and $a + c < b + c$ have the same solution set.

The Addition Property of Inequalities states that the same number can be added to each side of an inequality without changing the solution set of the inequality. This property is also true for an inequality that contains the symbol $\le$ or $\ge$.

The Addition Property of Inequalities is used to remove a term from one side of an inequality by adding the additive inverse of that term to each side of the inequality. Because subtraction is defined in terms of addition, the same number can be subtracted from each side of an inequality without changing the solution set of the inequality.

Solve: $3x - 4 < 2x - 1$

	$3x - 4 < 2x - 1$
Subtract $2x$ from each side of the inequality.	$x - 4 < -1$
Add 4 to each side of the inequality.	$x < 3$
Write the solution set. The solution set can be written in either set-builder notation or interval notation.	$\{x \mid x < 3\}$ $(-\infty, 3)$

The Multiplication Property of Inequalities is used to remove a coefficient from one side of an inequality so that the inequality can be written in the form *variable* $<$ *constant* or *variable* $>$ *constant*.

<div style="border:1px solid; padding:4px;">

Take Note

$c > 0$ means c is a positive number. Note that the inequality symbols are not changed.

$c < 0$ means c is a negative number. Note that the inequality symbols are reversed.

</div>

The Multiplication Property of Inequalities

Rule 1

If $a > b$ and $c > 0$, then the inequalities $a > b$ and $ac > bc$ have the same solution set.

If $a < b$ and $c > 0$, then the inequalities $a < b$ and $ac < bc$ have the same solution set.

Rule 2

If $a > b$ and $c < 0$, then the inequalities $a > b$ and $ac < bc$ have the same solution set.

If $a < b$ and $c < 0$, then the inequalities $a < b$ and $ac > bc$ have the same solution set.

Rule 1 states that when each side of an inequality is multiplied by a positive number, the inequality symbol remains the same. However, Rule 2 states that when each side of an inequality is multiplied by a negative number, the inequality symbol must be reversed.

Here are some examples of this property.

Rule 1		Rule 2	
$3 > 2$	$2 < 5$	$3 > 2$	$2 < 5$
$3(4) > 2(4)$	$2(4) < 5(4)$	$3(-4) < 2(-4)$	$2(-4) > 5(-4)$
$12 > 8$	$8 < 20$	$-12 < -8$	$-8 > -20$

Because division is defined in terms of multiplication, when each side of an inequality is divided by a positive number, the inequality symbol remains the same. When each side of an inequality is divided by a negative number, the inequality symbol must be reversed.

The Multiplication Property of Inequalities is also true for the symbol $\leq$ or $\geq$.

Take Note

Each side of the inequality is divided *by* a negative number; the inequality symbol must be reversed.

To solve $-3x > 9$ and write the solution set in interval notation, divide each side of the inequality by the coefficient -3, and reverse the inequality symbol.

$$-3x > 9$$
$$\frac{-3x}{-3} < \frac{9}{-3}$$

Simplify.

$$x < -3$$

Write the solution set.

$$(-\infty, -3)$$

EXAMPLE 8 Solve: $x + 3 > 4x + 6$
Write the solution set in set-builder notation.

Solution
$$x + 3 > 4x + 6$$
$$-3x + 3 > 6 \qquad \text{• Subtract } 4x \text{ from each side of the inequality.}$$
$$-3x > 3 \qquad \text{• Subtract 3 from each side of the inequality.}$$
$$x < -1 \qquad \text{• Divide each side of the inequality by } -3, \text{ and reverse the inequality symbol.}$$
$$\{x \mid x < -1\} \qquad \text{• Write the solution set.}$$

PROBLEM 8 Solve: $2x - 1 < 6x + 7$
Write the solution set in set-builder notation.

Solution See page S9.

When an inequality contains parentheses, the first step in solving the inequality is to use the Distributive Property to remove the parentheses.

EXAMPLE 9 Solve: $5(x - 2) \geq 9x - 3(2x - 4)$
Write the solution set in set-builder notation.

Solution
$$5(x - 2) \geq 9x - 3(2x - 4)$$
$$5x - 10 \geq 9x - 6x + 12 \qquad \text{• Use the Distributive Property to remove parentheses.}$$
$$5x - 10 \geq 3x + 12 \qquad \text{• Simplify.}$$
$$2x - 10 \geq 12 \qquad \text{• Subtract } 3x \text{ from each side of the inequality.}$$
$$2x \geq 22 \qquad \text{• Add 10 to each side of the inequality.}$$
$$x \geq 11 \qquad \text{• Divide each side of the inequality by 2.}$$
$$\{x \mid x \geq 11\}$$

PROBLEM 9 Solve: $5x - 2 \leq 4 - 3(x - 2)$
Write the solution set in interval notation.

Solution See page S9.

OBJECTIVE 3 **Solve compound inequalities**

A **compound inequality** is formed by joining two inequalities with a connective word such as "and" or "or." The inequalities shown below are compound inequalities.

$$2x < 4 \quad \text{and} \quad 3x - 2 > -8$$
$$2x + 3 > 5 \quad \text{or} \quad x + 2 < 5$$

The solution set of a compound inequality with the connective word *and* is the set of all elements common to the solution sets of both inequalities. Therefore, it is the *intersection* of the solution sets of the two inequalities.

Solve: $2x < 6$ and $3x + 2 > -4$
Write the solution set in interval notation.

Solve each inequality.	$\begin{array}{ll} 2x < 6 \quad \text{and} & 3x + 2 > -4 \\ x < 3 & 3x > -6 \\ & x > -2 \end{array}$
	$\{x \mid x < 3\} \qquad \{x \mid x > -2\}$
Find the intersection of the solution sets.	$\{x \mid x < 3\} \cap \{x \mid x > -2\} = (-2, 3)$

Solve: $-3 < 2x + 1 < 5$
Write the solution set in set-builder notation.

This inequality is equivalent to the compound inequality $-3 < 2x + 1$ and $2x + 1 < 5$.	$-3 < 2x + 1 < 5$
	$\begin{array}{ll} -3 < 2x + 1 \quad \text{and} & 2x + 1 < 5 \\ -4 < 2x & 2x < 4 \\ -2 < x & x < 2 \end{array}$
	$\{x \mid x > -2\} \qquad \{x \mid x < 2\}$
Find the intersection of the solution sets.	$\{x \mid x > -2\} \cap \{x \mid x < 2\} = \{x \mid -2 < x < 2\}$

There is an alternative method for solving the inequality in the last example.

$$-3 < 2x + 1 < 5$$

Subtract 1 from each of the three parts of the inequality.

$$-3 - 1 < 2x + 1 - 1 < 5 - 1$$
$$-4 < 2x < 4$$

Divide each of the three parts of the inequality by the coefficient 2.

$$\frac{-4}{2} < \frac{2x}{2} < \frac{4}{2}$$
$$-2 < x < 2$$

$$\{x \mid -2 < x < 2\}$$

The solution set of a compound inequality with the connective word *or* is the *union* of the solution sets of the two inequalities.

Solve: $2x + 3 > 7$ or $4x - 1 < 3$
Write the solution set in set-builder notation.

Solve each inequality.	$\begin{array}{ll} 2x + 3 > 7 \quad \text{or} & 4x - 1 < 3 \\ 2x > 4 & 4x < 4 \\ x > 2 & x < 1 \end{array}$
	$\{x \mid x > 2\} \qquad \{x \mid x < 1\}$
Find the union of the solution sets.	$\{x \mid x > 2\} \cup \{x \mid x < 1\} = \{x \mid x > 2 \text{ or } x < 1\}$

EXAMPLE 10 Solve: $1 < 3x - 5 < 4$
Write the solution set in interval notation.

Solution

$$1 < 3x - 5 < 4$$
$$1 + 5 < 3x - 5 + 5 < 4 + 5 \qquad \text{• Add 5 to each of the three parts of the inequality.}$$
$$6 < 3x < 9 \qquad \text{• Simplify.}$$
$$\frac{6}{3} < \frac{3x}{3} < \frac{9}{3} \qquad \text{• Divide each of the three parts of the inequality by 3.}$$
$$2 < x < 3$$
$$(2, 3) \qquad \text{• Write the solution set.}$$

PROBLEM 10 Solve: $-2 \leq 5x + 3 \leq 13$
Write the solution set in interval notation.

Solution See page S9.

EXAMPLE 11 Solve: $11 - 2x > -3$ and $7 - 3x < 4$
Write the solution set in set-builder notation.

Solution

$$\begin{array}{ll} 11 - 2x > -3 \quad \text{and} & 7 - 3x < 4 \\ -2x > -14 & -3x < -3 \\ \quad x < 7 & \quad x > 1 \end{array} \qquad \text{• Solve each inequality.}$$

$$\{x \,|\, x < 7\} \qquad\qquad \{x \,|\, x > 1\}$$

$$\{x \,|\, x < 7\} \cap \{x \,|\, x > 1\} = \{x \,|\, 1 < x < 7\} \qquad \text{• Find the intersection of the solution sets.}$$

PROBLEM 11 Solve: $5 - 4x > 1$ and $6 - 5x < 11$
Write the solution set in interval notation.

Solution See page S9.

EXAMPLE 12 Solve: $3 - 4x > 7$ or $4x + 5 < 9$
Write the solution set in set-builder notation.

Solution

$$\begin{array}{ll} 3 - 4x > 7 \quad \text{or} & 4x + 5 < 9 \\ -4x > 4 & 4x < 4 \\ \quad x < -1 & \quad x < 1 \end{array} \qquad \text{• Solve each inequality.}$$

$$\{x \,|\, x < -1\} \qquad\qquad \{x \,|\, x < 1\}$$

$$\{x \,|\, x < -1\} \cup \{x \,|\, x < 1\} = \{x \,|\, x < 1\} \qquad \text{• Find the union of the solution sets.}$$

PROBLEM 12 Solve: $2 - 3x > 11$ or $5 + 2x > 7$
Write the solution set in set-builder notation.

Solution See page S9.

OBJECTIVE 4 Application problems

EXAMPLE 13 Company A rents cars for $6 a day and 14¢ for every mile driven. Company B rents cars for $12 a day and 8¢ for every mile driven. You want to rent a car for 5 days. How many miles can you drive a Company A car during the 5 days if it is to cost less than a Company B car?

Strategy To find the number of miles, write and solve an inequality using N to represent the number of miles.

Solution Cost of Company A car < cost of Company B car

$$6(5) + 0.14N < 12(5) + 0.08N$$
$$30 + 0.14N < 60 + 0.08N$$
$$30 + 0.06N < 60$$
$$0.06N < 30$$
$$N < 500$$

It is less expensive to rent from Company A if the car is driven less than 500 mi.

PROBLEM 13 The base of a triangle is 12 in., and the height is $(x + 2)$ in. Express as an integer the maximum height of the triangle when the area is less than 50 in².

Solution See page S9.

EXAMPLE 14 Find three consecutive odd integers whose sum is between 27 and 51.

Strategy To find the three consecutive odd integers, write and solve a compound inequality using x to represent the first odd integer.

Solution $\begin{matrix} \text{Lower limit} \\ \text{of the sum} \end{matrix} < \text{sum} < \begin{matrix} \text{upper limit} \\ \text{of the sum} \end{matrix}$

$$27 < x + (x + 2) + (x + 4) < 51$$
$$27 < 3x + 6 < 51$$
$$27 - 6 < 3x + 6 - 6 < 51 - 6$$
$$21 < 3x < 45$$
$$\frac{21}{3} < \frac{3x}{3} < \frac{45}{3}$$
$$7 < x < 15$$

The three integers are 9, 11, and 13; or 11, 13, and 15; or 13, 15, and 17.

PROBLEM 14 An average score of 80 to 89 in a history course receives a B grade. A student has grades of 72, 94, 83, and 70 on four exams. Find the range of scores on the fifth exam that will give the student a B for the course.

Solution See page S9.

2.7 Concept Review

Determine whether the statement is always true, sometimes true, or never true.

1. If $a > b$, then $-a < -b$.

2. If $a < b$ and $a \neq 0$, $b \neq 0$, then $\frac{1}{a} < \frac{1}{b}$.

3. When dividing both sides of an inequality by an integer, we must reverse the inequality symbol.

4. When subtracting the same real number from both sides of an inequality, we must reverse the inequality symbol.

5. Suppose $a < 1$. Then $a^2 < a$.

6. Suppose $a < b < 0$ and $c < d < 0$. Then $ac > bd$.

2.7 Exercises

1 Set operations and interval notation

1. Explain the difference between the union of two sets and the intersection of two sets.

2. Explain the difference between $\{x \mid x < 5\}$ and $\{x \mid x \leq 5\}$.

3. Two ways to write the set of natural numbers less than five are $\{1, 2, 3, 4\}$ and $\{n \mid n < 5, n \in \text{natural numbers}\}$. The first way uses the ___?___ method, and the second way uses ___?___ notation.

4. The symbol for "union" is ___?___. The symbol for "intersection" is ___?___.

5. The symbol ∞ is called the ___?___ symbol.

6. Replace each question mark with *includes* or *does not include* to make the following statement true. The set $[-4, 7)$ ___?___ the number -4 and ___?___ the number 7.

Use set-builder notation to write the set.

7. the integers greater than 4

8. the integers less than -2

9. the real numbers greater than or equal to -2

10. the real numbers less than or equal to 2

11. the real numbers between 0 and 1

12. the real numbers between −2 and 5

13. the real numbers between 1 and 4, inclusive

14. the real numbers between 0 and 2, inclusive

Find $A \cup B$.

15. $A = \{1, 4, 9\}$, $B = \{2, 4, 6\}$

16. $A = \{-1, 0, 1\}$, $B = \{0, 1, 2\}$

17. $A = \{2, 3, 5, 8\}$, $B = \{9, 10\}$

18. $A = \{1, 3, 5, 7\}$, $B = \{2, 4, 6, 8\}$

19. $A = \{-4, -2, 0, 2, 4\}$, $B = \{0, 4, 8\}$

20. $A = \{-3, -2, -1\}$, $B = \{-2, -1, 0, 1\}$

21. $A = \{1, 2, 3, 4, 5\}$, $B = \{3, 4, 5\}$

22. $A = \{2, 4\}$, $B = \{0, 1, 2, 3, 4, 5\}$

Find $A \cap B$.

23. $A = \{6, 12, 18\}$, $B = \{3, 6, 9\}$

24. $A = \{-4, 0, 4\}$, $B = \{-2, 0, 2\}$

25. $A = \{1, 5, 10, 20\}$, $B = \{5, 10, 15, 20\}$

26. $A = \{1, 3, 5, 7, 9\}$, $B = \{1, 9\}$

27. $A = \{1, 2, 4, 8\}$, $B = \{3, 5, 6, 7\}$

28. $A = \{-3, -2, -1, 0\}$, $B = \{1, 2, 3, 4\}$

29. $A = \{2, 4, 6, 8, 10\}$, $B = \{4, 6\}$

30. $A = \{-9, -5, 0, 7\}$, $B = \{-7, -5, 0, 5, 7\}$

Graph.

31. $\{x \mid -1 < x < 5\}$

32. $\{x \mid 1 < x < 3\}$

33. $\{x \mid 0 \le x \le 3\}$

34. $\{x \mid -1 \le x \le 1\}$

35. $\{x \mid x < 2\}$

36. $\{x \mid x < -1\}$

37. $\{x \mid x \ge 1\}$

38. $\{x \mid x \le -2\}$

39. $\{x \mid x > 1\} \cup \{x \mid x < -1\}$

40. $\{x \mid x \le 2\} \cup \{x \mid x > 4\}$

41. $\{x \mid x \le 2\} \cap \{x \mid x \ge 0\}$

42. $\{x \mid x > -1\} \cap \{x \mid x \le 4\}$

43. $\{x \mid x > 1\} \cap \{x \mid x \ge -2\}$

44. $\{x \mid x < 4\} \cap \{x \mid x \le 0\}$

45. $\{x \mid x > 2\} \cup \{x \mid x > 1\}$

46. $\{x \mid x < -2\} \cup \{x \mid x < -4\}$

Write each interval in set-builder notation.

47. $(0, 8)$ **48.** $(-2, 4)$ **49.** $[-5, 7]$ **50.** $[3, 4]$ **51.** $[-3, 6)$

52. $(4, 5]$ **53.** $(-\infty, 4]$ **54.** $(-\infty, -2)$ **55.** $(5, \infty)$ **56.** $[-2, \infty)$

Write each set of real numbers in interval notation.

57. $\{x \mid -2 < x < 4\}$ **58.** $\{x \mid 0 < x < 3\}$ **59.** $\{x \mid -1 \le x \le 5\}$ **60.** $\{x \mid 0 \le x \le 3\}$

61. $\{x \mid x < 1\}$ **62.** $\{x \mid x \le 6\}$ **63.** $\{x \mid -2 \le x < 6\}$ **64.** $\{x \mid x \ge 3\}$

65. $\{x \mid x \in \text{real numbers}\}$ **66.** $\{x \mid x > -1\}$

67. Which set is the empty set?

 a. $\{x \mid x \in \text{integers}\} \cap \{x \mid x \in \text{rational numbers}\}$
 b. $\{-4, -2, 0, 2, 4\} \cup \{-3, -1, 1, 3\}$
 c. $[5, \infty) \cap (0, 5)$

68. Which set is *not* equivalent to the interval $[-1, 6)$?

 a. $\{x \mid -1 \le x < 6\}$
 b. $\{x \mid x \ge -1\} \cup \{x \mid x < 6\}$
 c. $\{x \mid x < 6\} \cap \{x \mid x \ge -1\}$

2 Solve inequalities in one variable

69. State the Addition Property of Inequalities and give numerical examples of its use.

70. State the Multiplication Property of Inequalities and give numerical examples of its use.

71. Which numbers are solutions of the inequality $x + 7 \le -3$?
 a. -17 **b.** 8 **c.** -10 **d.** 0

72. Which numbers are solutions of the inequality $2x - 1 > 5$?
 a. 6 **b.** -4 **c.** 3 **d.** 5

Complete Exercises 73 to 76 by replacing the question mark with *remains the same* or *is reversed*.

73. The inequality $x + 5 > 10$ can be solved by subtracting 5 from each side of the equation. The inequality symbol ___?___.

74. The inequality $\frac{1}{5}x > 10$ can be solved by multiplying each side of the equation by the reciprocal of $\frac{1}{5}$. The inequality symbol ___?___.

75. The inequality $-5x > 10$ can be solved by dividing each side of the equation by -5. The inequality symbol ___?___.

76. The inequality $x - 5 > -10$ can be solved by adding 5 to each side of the equation. The inequality symbol ___?___.

Solve. Write the solution set in set-builder notation.

77. $x - 3 < 2$ **78.** $x + 4 \ge 2$ **79.** $4x \le 8$

80. $6x > 12$ **81.** $-2x > 8$ **82.** $-3x \le -9$

83. $3x - 1 > 2x + 2$ **84.** $5x + 2 \geq 4x - 1$ **85.** $2x - 1 > 7$

86. $4x + 3 \leq -1$ **87.** $6x + 3 > 4x - 1$ **88.** $7x + 4 < 2x - 6$

89. $8x + 1 \geq 2x + 13$ **90.** $5x - 4 < 2x + 5$ **91.** $7 - 2x \geq 1$

92. $3 - 5x \leq 18$ **93.** $4x - 2 < x - 11$ **94.** $6x + 5 \geq x - 10$

Solve. Write the solution set in interval notation.

95. $x + 7 \geq 4x - 8$ **96.** $3x + 1 \leq 7x - 15$

97. $6 - 2(x - 4) \leq 2x + 10$ **98.** $4(2x - 1) > 3x - 2(3x - 5)$

99. $2(1 - 3x) - 4 > 10 + 3(1 - x)$ **100.** $2 - 5(x + 1) \geq 3(x - 1) - 8$

101. $\dfrac{3}{5}x - 2 < \dfrac{3}{10} - x$ **102.** $\dfrac{5}{6}x - \dfrac{1}{6} \leq x - 4$

103. $\dfrac{1}{3}x - \dfrac{3}{2} \geq \dfrac{7}{6} - \dfrac{2}{3}x$ **104.** $\dfrac{7}{12}x - \dfrac{3}{2} < \dfrac{2}{3}x + \dfrac{5}{6}$

105. $\dfrac{1}{2}x - \dfrac{3}{4} > \dfrac{7}{4}x - 2$ **106.** $\dfrac{2 - x}{4} - \dfrac{3}{8} \geq \dfrac{2}{5}x$

107. $2 - 2(7 - 2x) < 3(3 - x)$ **108.** $3 + 2(x + 5) \geq x + 5(x + 1) + 1$

State whether the solution set of an inequality in the given form contains only negative numbers, only positive numbers, or both positive and negative numbers.

109. $x + n > a$, where n and a are both positive and $n < a$

110. $nx > a$, where both n and a are negative

111. $nx > a$, where n is negative and a is positive

112. $x - n > -a$, where n and a are both positive and $n < a$

113. **a.** Which set operation is used when a compound inequality is combined with *or*?
 b. Which set operation is used when a compound inequality is combined with *and*?

114. Explain why writing $-3 > x > 4$ does not make sense.

115. To solve the compound inequality $4x \leq 4$ or $x + 2 > 8$, divide each side of the first inequality by ____?____, and subtract ____?____ from each side of the second inequality: $x \leq$ ____?____ or $x >$ ____?____.

116. The solution set in Exercise 115 can be written in interval notation as $(-\infty,$ ____?____$] \cup (6,$ ____?____$)$.

State whether the inequality describes the empty set, all real numbers, two intervals of real numbers, or one interval of real numbers.

117. $x > -3$ and $x > 2$

118. $x > -3$ or $x < 2$

119. $x < -3$ and $x > 2$

120. $x < -3$ or $x > 2$

3 Solve compound inequalities

Solve. Write the solution set in interval notation.

121. $3x < 6$ and $x + 2 > 1$

122. $x - 3 \leq 1$ and $2x \geq -4$

123. $x + 2 \geq 5$ or $3x \leq 3$

124. $2x < 6$ or $x - 4 > 1$

125. $-2x > -8$ and $-3x < 6$

126. $\frac{1}{2}x > -2$ and $5x < 10$

127. $\frac{1}{3}x < -1$ or $2x > 0$

128. $\frac{2}{3}x > 4$ or $2x < -8$

129. $x + 4 \geq 5$ and $2x \geq 6$

130. $3x < -9$ and $x - 2 < 2$

131. $-5x > 10$ and $x + 1 > 6$

132. $7x < 14$ and $1 - x < 4$

133. $2x - 3 > 1$ and $3x - 1 < 2$

134. $4x + 1 < 5$ and $4x + 7 > -1$

135. $3x + 7 < 10$ or $2x - 1 > 5$

136. $6x - 2 < -14$ or $5x + 1 > 11$

Solve. Write the solution set in set-builder notation.

137. $-5 < 3x + 4 < 16$

138. $5 < 4x - 3 < 21$

139. $0 < 2x - 6 < 4$

140. $-2 < 3x + 7 < 1$

141. $4x - 1 > 11$ or $4x - 1 \leq -11$

142. $3x - 5 > 10$ or $3x - 5 < -10$

143. $2x + 3 \geq 5$ and $3x - 1 > 11$

144. $6x - 2 < 5$ or $7x - 5 < 16$

145. $9x - 2 < 7$ and $3x - 5 > 10$

146. $8x + 2 \le -14$ and $4x - 2 > 10$

147. $3x - 11 < 4$ or $4x + 9 \ge 1$

148. $5x + 12 \ge 2$ or $7x - 1 \le 13$

149. $3 - 2x > 7$ and $5x + 2 > -18$

150. $1 - 3x < 16$ and $1 - 3x > -16$

151. $5 - 4x > 21$ or $7x - 2 > 19$

152. $6x + 5 < -1$ or $1 - 2x < 7$

153. $3 - 7x \le 31$ and $5 - 4x > 1$

154. $9 - x \ge 7$ and $9 - 2x < 3$

155. $\frac{2}{3}x - 4 > 5$ or $x + \frac{1}{2} < 3$

156. $\frac{5}{8}x + 2 < -3$ or $2 - \frac{3}{5}x < -7$

157. $-\frac{3}{8} \le 1 - \frac{1}{4}x \le \frac{7}{2}$

158. $-2 \le \frac{2}{3}x - 1 \le 3$

4 Application problems

159. Translate into an inequality: The minimum value of a number n is 40.

160. Translate into an inequality: A number n is between 150 and 300.

161. Integer Problem Five times the difference between a number and two is greater than the quotient of two times the number and three. Find the smallest integer that will satisfy the inequality.

162. Integer Problem Two times the difference between a number and eight is less than or equal to five times the sum of the number and four. Find the smallest number that will satisfy the inequality.

163. Geometry The length of a rectangle is 2 ft more than four times the width. Express as an integer the maximum width of the rectangle when the perimeter is less than 34 ft.

164. Geometry The length of a rectangle is 5 cm less than twice the width. Express as an integer the maximum width of the rectangle when the perimeter is less than 60 cm.

165. Integer Problem Find four consecutive integers whose sum is between 62 and 78.

166. Integer Problem Find three consecutive even integers whose sum is between 30 and 52.

167. Geometry One side of a triangle is 1 in. longer than the second side. The third side is 2 in. longer than the second side. Find, to the nearest whole number, the length of the second side of the triangle if the perimeter is more than 15 in. and less than 25 in.

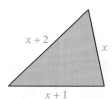

168. Geometry The length of a rectangle is 4 ft more than twice the width. Find, to the nearest whole number, the width of the rectangle if the perimeter is more than 28 ft and less than 40 ft.

169. Consumerism A cellular phone company offers its customers a rate of $49 for up to 200 min per month of cellular phone time, or a rate of $25 per month plus $.40 for each minute of cellular phone time. For how many minutes per month can a customer who chooses the second option use a cellular phone before the charges exceed those of the first option?

170. Consumerism In 2005, the computer service America Online offered its customers the option of paying $23.90 per month for unlimited use. Another option was a rate of $14.95 per month with 10 free hours plus $2.95 per hour thereafter. How many hours per month can you use this second option if it is to cost you less than the first option? Round to the nearest whole number.

171. Consumerism AirTouch advertises text messaging for $6.95 per month for up to 400 messages, and $.10 per message thereafter. A competitor advertises text messaging service for $3.95 per month for up to 400 messages and $.15 per message thereafter. For what number of messages per month is the AirTouch plan less expensive?

172. Consumerism Suppose PayRite Rental Cars rents compact cars for $32 per day with unlimited mileage, and Otto Rentals offers compact cars for $19.99 per day but charges $.19 for each mile beyond 100 mi driven per day. You want to rent a car for one week. How many miles can you drive during the week if Otto Rental is to be less expensive than PayRite?

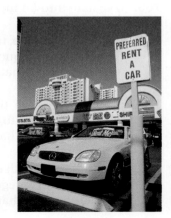

173. Consumerism During a weekday, to call a city 40 mi away from a certain pay phone costs $.70 for the first 3 min and $.15 for each additional minute. If you use a calling card, there is a $.35 fee, and then the rates are $.196 for the first minute and $.126 for each additional minute. How many minutes must a call last if it is to be cheaper to pay with coins than with a calling card?

174. Consumerism Heritage National Bank offers two different checking accounts. The first charges $3 per month plus $.50 per check after the first 10 checks. The second account charges $8 per month with unlimited check writing. How many checks can be written per month if the first account is to be less expensive than the second account?

175. Consumerism Glendale Federal Bank offers a checking account to small businesses. The charge is $8 per month plus $.12 per check after the first 100 checks. A competitor is offering an account for $5 per month

plus $.15 per check after the first 100 checks. If a business chooses the Glendale Federal Bank account, how many checks does the business write monthly if it is assumed that this account will cost less than the competitor's account?

176. Commissions George Stoia earns $1000 per month plus 5% commission on the amount of sales. George's goal is to earn a minimum of $3200 per month. What amount of sales will enable George to earn $3200 or more per month?

177. Automobiles A new car will average at least 22 mpg for city driving and at most 27.5 mpg for highway driving. Find the range of miles that the car can travel on a full tank (19.5 gal) of gasoline.

178. Education An average score of 90 or above in an English course receives an A grade. A student has grades of 85, 88, 90, and 98 on four tests. Find the range of scores on the fifth test that will give the student an A grade.

2.7 Applying Concepts

Use the roster method to list the set of positive integers that are solutions of the inequality.

179. $8x - 7 < 2x + 9$

180. $2x + 9 \geq 5x - 4$

181. $5 + 3(2 + x) > 8 + 4(x - 1)$

182. $6 + 4(2 - x) > 7 + 3(x + 5)$

183. $-3x < 15$ and $x + 2 < 7$

184. $3x - 2 > 1$ and $2x - 3 < 5$

185. $-4 \leq 3x + 8 < 16$

186. $5 < 7x - 3 \leq 24$

187. Temperature The relationship between Celsius temperature and Fahrenheit temperature is given by the formula $F = \frac{9}{5}C + 32$. If the temperature is between 77°F and 86°F, what is the temperature range in degrees Celsius?

188. Integer Problem The average of two negative integers is less than or equal to −15. The smaller integer is 7 less than the larger integer. Find the greatest possible value for the smaller integer.

189. Consumerism The charges for a long-distance telephone call are $1.56 for the first three minutes and $.52 for each additional minute or fraction of a minute. What is the largest whole number of minutes a call can last if it is to cost you less than $5.40?

Absolute Value Equations and Inequalities

2.8

OBJECTIVE **1** **Absolute value equations**

The **absolute value** of a number is its distance from zero on the number line. Distance is always a positive number or zero. Therefore, the absolute value of a number is always a positive number or zero.

The distance from 0 to 3 or from 0 to -3 is 3 units.

$$|3| = 3 \qquad |-3| = 3$$

An equation containing an absolute value symbol is called an **absolute value equation**.

$$\left. \begin{array}{l} |x| = 3 \\ |x + 2| = 8 \\ |3x - 4| = 5x - 9 \end{array} \right\} \begin{array}{l} \text{Absolute} \\ \text{value} \\ \text{equations} \end{array}$$

Absolute Value Equations

If $a \geq 0$ and $|x| = a$, then $x = a$ or $x = -a$.

Given $|x| = 3$, then $x = 3$ or $x = -3$.

Solve: $|x + 2| = 8$

Remove the absolute value sign, and rewrite as two equations.

$$|x + 2| = 8$$
$$x + 2 = 8 \qquad x + 2 = -8$$

Solve each equation.

$$x = 6 \qquad\qquad x = -10$$

Check
$$
\begin{array}{c|c}
|x + 2| = 8 & |x + 2| = 8 \\
\hline
|6 + 2| \;\big|\; 8 & |-10 + 2| \;\big|\; 8 \\
|8| & |-8| \\
8 = 8 & 8 = 8
\end{array}
$$

Write the solution. The solutions are 6 and -10.

EXAMPLE 1 Solve. **A.** $|x| = 15$ **B.** $|2 - x| = 12$
 C. $|2x| = -4$ **D.** $3 - |2x - 4| = -5$

Solution **A.** $|x| = 15$
$$x = 15 \qquad x = -15$$

• Remove the absolute value sign, and rewrite as two equations.

The solutions are 15 and -15.

Take Note

Remember to check the solutions.
For Part B:

$$|2 - x| = 12$$

| $|2 - x| = 12$ | |
|---|---|
| $\|2 - (-10)\|$ | 12 |
| $\|2 + 10\|$ | 12 |
| $\|12\|$ | 12 |
| $12 = 12$ | |

| $|2 - x| = 12$ | |
|---|---|
| $\|2 - 14\|$ | 12 |
| $\|-12\|$ | 12 |
| $12 = 12$ | |

Take Note

$$3 - |2x - 4| = -5$$

| $3 - |2x - 4| = -5$ | |
|---|---|
| $3 - \|2(6) - 4\|$ | -5 |
| $3 - \|12 - 4\|$ | -5 |
| $3 - \|8\|$ | -5 |
| $3 - 8$ | -5 |
| $-5 = -5$ | |

| $3 - |2x - 4| = -5$ | |
|---|---|
| $3 - \|2(-2) - 4\|$ | -5 |
| $3 - \|-4 - 4\|$ | -5 |
| $3 - \|-8\|$ | -5 |
| $3 - 8$ | -5 |
| $-5 = -5$ | |

B. $|2 - x| = 12$

$$2 - x = 12 \qquad 2 - x = -12$$

- Remove the absolute value sign, and rewrite as two equations.

$$-x = 10 \qquad -x = -14$$
$$x = -10 \qquad x = 14$$

- Solve each equation.

The solutions are -10 and 14.

C. $|2x| = -4$

- The absolute value of a number must be nonnegative.

There is no solution to the equation.

D. $3 - |2x - 4| = -5$
$$-|2x - 4| = -8$$
$$|2x - 4| = 8$$

- Solve for the absolute value.
- Multiply each side of the equation by -1.

$$2x - 4 = 8 \qquad 2x - 4 = -8$$

- Remove the absolute value sign, and rewrite as two equations.

$$2x = 12 \qquad 2x = -4$$
$$x = 6 \qquad x = -2$$

- Solve each equation.

The solutions are 6 and -2.

PROBLEM 1 Solve.
 A. $|x| = 25$ **B.** $|2x - 3| = 5$
 C. $|x - 3| = -2$ **D.** $5 - |3x + 5| = 3$

■ **Solution** See page S10.

OBJECTIVE ② Absolute value inequalities

Recall that absolute value represents the distance between two points. For example, the solutions of the absolute value equation $|x - 1| = 3$ are the numbers whose distance from 1 is 3. Therefore, the solutions are -2 and 4.

The solutions of the **absolute value inequality** $|x - 1| < 3$ are the numbers whose distance from 1 is *less than* 3. Therefore, the solutions are the numbers greater than -2 and less than 4. The solution set is $\{x \mid -2 < x < 4\}$. *Note:* In this text, solutions to absolute value inequalities will always be written in set-builder notation.

<div align="center">

Distance Distance
less than 3 less than 3

$-5\ -4\ -3\ -2\ -1\ \ 0\ \ 1\ \ 2\ \ 3\ \ 4\ \ 5$

</div>

Absolute Value Inequalities of the Form $|ax + b| < c$

To solve an absolute value inequality of the form $|ax + b| < c$, solve the equivalent compound inequality $-c < ax + b < c$.

The compound inequality $-c < ax + b < c$ is equivalent to the intersection of the two inequalities $-c < ax + b$ and $ax + b < c$. The intersection of these two inequalities is shown on the following number line.

$$\xleftarrow{} \!\!\!\! \begin{array}{ccc} \\ {}_{-c} & {}_{0} & {}_{c} \end{array} \!\!\!\! \xrightarrow{}$$

From this number line, we see that the solution of $|ax + b| < c$ is all the numbers for which $ax + b$ is greater than $-c$ and less than c.

EXAMPLE 2 Solve: $|4x - 3| < 5$

Solution $|4x - 3| < 5$

- The distance from zero is less than 5 units. $4x - 3$ is between -5 and 5.

$$-5 < 4x - 3 < 5$$

- Solve the equivalent compound in equality.

$$-5 + 3 < 4x - 3 + 3 < 5 + 3$$
$$-2 < 4x < 8$$

- Add 3 to each of the three parts of the inequality.

$$\frac{-2}{4} < \frac{4x}{4} < \frac{8}{4}$$

- Divide each of three parts of the inequality by 4.

$$-\frac{1}{2} < x < 2$$

$$\left\{ x \,\middle|\, -\frac{1}{2} < x < 2 \right\}$$

- Write the solution set.

PROBLEM 2 Solve: $|3x + 2| < 8$

Solution See page S10.

EXAMPLE 3 Solve: $|x - 3| < 0$

Solution $|x - 3| < 0$ • The absolute value of a number must be nonnegative.
$\varnothing$ • The solution set is the empty set.

PROBLEM 3 Solve: $|3x - 7| < 0$

Solution See page S10.

The solutions of the absolute value inequality $|x + 1| > 2$ are the numbers whose distance from -1 is *greater than* 2. Therefore, the solutions are the numbers less than -3 or greater than 1. The solution set is $\{x \,|\, x < -3 \text{ or } x > 1\}$.

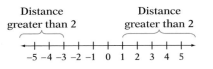

Distance greater than 2 Distance greater than 2

Absolute Value Inequalities of the Form $|ax + b| > c$

To solve an absolute value inequality of the form $|ax + b| > c$, solve the equivalent compound inequality $ax + b < -c$ or $ax + b > c$.

The solution of the compound inequality $ax + b < -c$ or $ax + b > c$ is shown on the following number line.

From this number line we see that the solution of $|ax + b| > c$ is the union of the numbers less than $-c$ and greater than c.

Take Note

Carefully observe the difference between the solution method for $|ax + b| > c$ shown here and that for $|ax + b| < c$ shown in Example 2.

EXAMPLE 4 Solve: $|2x - 1| > 7$

Solution $|2x - 1| > 7$

• The distance from zero is greater than 7. $2x - 1$ is less than -7 or greater than 7.

$$2x - 1 < -7 \quad \text{or} \quad 2x - 1 > 7$$
$$2x < -6 \qquad\qquad 2x > 8$$
$$x < -3 \qquad\qquad x > 4$$

• Solve the equivalent compound inequality.

$$\{x \,|\, x < -3\} \qquad \{x \,|\, x > 4\}$$

$$\{x \,|\, x < -3\} \cup \{x \,|\, x > 4\}$$
$$= \{x \,|\, x < -3 \text{ or } x > 4\}$$

• Find the union of the solution sets.

PROBLEM 4 Solve: $|5x + 3| > 8$

Solution See page S10.

OBJECTIVE 3 Application problems

Piston

The **tolerance of a component,** or part, is the acceptable amount by which the component may vary from a given measurement. For example, the diameter of a piston may vary from the given measurement of 9 cm by 0.001 cm. This is written as 9 cm ± 0.001 cm, which is read "9 centimeters plus or minus 0.001 centimeter." The maximum diameter, or **upper limit,** of the piston is 9 cm + 0.001 cm = 9.001 cm. The minimum diameter, or **lower limit,** of the piston is 9 cm − 0.001 cm = 8.999 cm.

The lower and upper limits of the diameter could also be found by solving the absolute value inequality $|d - 9| \leq 0.001$, where d is the diameter of the piston.

$$|d - 9| \leq 0.001$$

$$-0.001 \leq d - 9 \leq 0.001$$
$$-0.001 + 9 \leq d - 9 + 9 \leq 0.001 + 9$$
$$8.999 \leq d \leq 9.001$$

The lower and upper limits of the diameter are 8.999 cm and 9.001 cm.

EXAMPLE 5 A doctor has prescribed 2 cc of medication for a patient. The tolerance is 0.03 cc. Find the lower and upper limits of the amount of medication to be given.

Strategy Let p represent the prescribed amount of medication, T the tolerance, and m the given amount of medication. Solve the absolute value inequality $|m - p| \leq T$ for m.

Solution $|m - p| \leq T$
$|m - 2| \leq 0.03$ • Substitute the values of p and T into the inequality.

$-0.03 \leq m - 2 \leq 0.03$ • Solve the equivalent compound inequality.
$1.97 \leq m \leq 2.03$

The lower and upper limits of the amount of medication to be given to the patient are 1.97 cc and 2.03 cc.

PROBLEM 5 A machinist must make a bushing that has a tolerance of 0.003 in. The diameter of the bushing is 2.55 in. Find the lower and upper limits of the diameter of the bushing.

Solution See page S10.

2.8 Concept Review

Determine whether the statement is always true, sometimes true, or never true.

1. If $a > 0$ and $|x| = a$, then $x = a$ or $x = -a$.

2. The absolute value inequality $|x| < a$, $a > 0$, is equivalent to the compound inequality $x > a$ or $x < -a$.

3. If $|x + 3| = 12$, then $|x + 3| = 12$ or $|x + 3| = -12$.

4. An absolute value equation has two solutions.

5. If $|x + b| < c$, then $x + b < -c$ or $x + b > c$.

2.8 Exercises

1 Absolute value equations

1. Is 2 a solution of $|x - 8| = 6$?

2. Is -2 a solution of $|2x - 5| = 9$?

3. Is -1 a solution of $|3x - 4| = 7$?

4. Is 1 a solution of $|6x - 1| = -5$?

5. If $|x| = 8$, then $x = $ ___?___ or $x = $ ___?___ .

6. If $|x + 5| = 8$, then $x + 5 = $ ___?___ or $x + 5 = $ ___?___ . Subtract 5 from each side of each equation to find that the two solutions of the equation $|x + 5| = 8$ are ___?___ and ___?___ .

Solve.

7. $|x| = 7$

8. $|a| = 2$

9. $|-t| = 3$

10. $|-a| = 7$

11. $|-t| = -3$

12. $|-y| = -2$

13. $|x + 2| = 3$

14. $|x + 5| = 2$

15. $|y - 5| = 3$

16. $|y - 8| = 4$

17. $|a - 2| = 0$

18. $|a + 7| = 0$

19. $|x - 2| = -4$

20. $|x + 8| = -2$

21. $|2x - 5| = 4$

22. $|4 - 3x| = 4$

23. $|2 - 5x| = 2$

24. $|2x - 3| = 0$

25. $|5x + 5| = 0$

26. $|3x - 2| = -4$

27. $|2x + 5| = -2$

28. $|x - 2| - 2 = 3$

29. $|x - 9| - 3 = 2$

30. $|3a + 2| - 4 = 4$

31. $|8 - y| - 3 = 1$

32. $|2x - 3| + 3 = 3$

33. $|4x - 7| - 5 = -5$

34. $|2x - 3| + 4 = -4$

35. $|3x - 2| + 1 = -1$

36. $|6x - 5| - 2 = 4$

37. $|4b + 3| - 2 = 7$

38. $|3t + 2| + 3 = 4$

39. $|5x - 2| + 5 = 7$

40. $3 - |x - 4| = 5$

41. $2 - |x - 5| = 4$

42. $|2x - 8| + 12 = 2$

43. $|3x - 4| + 8 = 3$

44. $2 + |3x - 4| = 5$

45. $5 + |2x + 1| = 8$

46. $5 - |2x + 1| = 5$

47. $3 - |5x + 3| = 3$

48. $8 - |1 - 3x| = -1$

For Exercises 49 to 52, a and b are positive numbers such that $a < b$. State whether an absolute value equation in the given form has no solution, two negative solutions, two positive solutions, or one positive and one negative solution.

49. $|x - b| = a$

50. $|x - b| = -a$

51. $|x + b| = a$

52. $|x + a| = b$

2 **Absolute value inequalities**

53. If $|x| > 9$, then x ___?___ -9 or x ___?___ 9.

54. If $|x| < 5$, then ___?___ $< x <$ ___?___ .

Solve.

55. $|x| > 3$

56. $|x| < 5$

57. $|x + 1| > 2$

58. $|x - 2| > 1$

59. $|x - 5| \le 1$

60. $|x - 4| \le 3$

61. $|2 - x| \ge 3$

62. $|3 - x| \ge 2$

63. $|2x + 1| < 5$

64. $|3x - 2| < 4$

65. $|5x + 2| > 12$

66. $|7x - 1| > 13$

67. $|4x - 3| \le -2$

68. $|5x + 1| \le -4$

69. $|2x + 7| > -5$

70. $|3x - 1| > -4$

71. $|4 - 3x| \ge 5$

72. $|7 - 2x| > 9$

73. $|5 - 4x| \le 13$

74. $|3 - 7x| < 17$

75. $|6 - 3x| \le 0$

76. $|10 - 5x| \ge 0$

77. $|2 - 9x| > 20$

78. $|5x - 1| < 16$

For Exercises 79 and 80, a and b are positive numbers such that $a < b$. State whether an absolute value inequality in the given form has no solution, all negative solutions, all positive solutions, or both positive and negative solutions.

79. $|x + b| < a$

80. $|x + a| < b$

Application problems

81. The desired dosage of a particular medicine is 50 mg, but it is acceptable for the dosage to vary by 0.5 mg from the desired dosage. The number 0.5 is called the ___?___ for the desired dosage of 50 mg. The value $50 + 0.5 =$ ___?___ mg is called the ___?___ limit for the dosage and the value $50 - 0.5 =$ ___?___ mg is called the ___?___ limit for the dosage.

82. The tolerance for the dosage described in Exercise 81 can be represented by an absolute value inequality by letting D be the actual dosage and stating that the distance between D and 50 must be less than or equal to 0.5: $|D -$ ___?___ $| \le$ ___?___ .

83. **Mechanics** The diameter of a bushing is 1.75 in. The bushing has a tolerance of 0.008 in. Find the lower and upper limits of the diameter of the bushing.

1.75 in.

84. Mechanics A machinist must make a bushing that has a tolerance of 0.004 in. The diameter of the bushing is 3.48 in. Find the lower and upper limits of the diameter of the bushing.

85. Medicine A doctor has prescribed 2.5 cc of medication for a patient. The tolerance is 0.2 cc. Find the lower and upper limits of the amount of medication to be given.

86. Automobiles The diameter of a piston for an automobile is $3\frac{5}{16}$ in. with a tolerance of $\frac{1}{64}$ in. Find the lower and upper limits of the diameter of the piston.

87. Computers A power strip is utilized on a computer to prevent the loss of programming by electrical surges. The power strip is designed to allow 110 volts plus or minus 16.5 volts. Find the lower and upper limits of voltage to the computer.

88. Appliances An electric motor is designed to run on 220 volts plus or minus 25 volts. Find the lower and upper limits of voltage on which the motor will run.

89. Automobiles A piston rod for an automobile is $10\frac{3}{8}$ in. with a tolerance of $\frac{1}{32}$ in. Find the lower and upper limits of the length of the piston rod.

90. A dosage of medicine may safely range from 2.8 ml to 3.2 ml. What is the desired dosage of the medicine? What is the tolerance?

91. The tolerance for the diameter, in inches, of a piston is described by the absolute value inequality $|d - 5| \le 0.01$. What is the desired diameter of the piston? By how much can the actual diameter of the piston vary from the desired diameter?

Electronics The tolerance of the resistors used in electronics is given as a percent of the resistance, which is measured in ohms. For example, if a 20,000-ohm resistor has a tolerance of 4%, then the tolerance is 0.04(20,000) = 800 ohms. The lower and upper limits of the resistor can be found by solving the absolute value inequality $|M - 20,000| \le 800$, where M represents the number of ohms.

92. Find the lower and upper limits of a 29,000-ohm resistor with a 2% tolerance.

93. Find the lower and upper limits of a 15,000-ohm resistor with a 10% tolerance.

94. Find the lower and upper limits of a 25,000-ohm resistor with a 5% tolerance.

95. Find the lower and upper limits of a 56-ohm resistor with a 5% tolerance.

2.8 Applying Concepts

Solve.

96. $\left|\dfrac{2x - 5}{3}\right| = 7$

97. $\left|\dfrac{3x - 2}{4}\right| + 5 = 6$

98. $\left|\dfrac{4x - 2}{3}\right| > 6$

99. $\left|\dfrac{2x - 1}{5}\right| \le 3$

For what values of the variable is the equation true? Write the solution set in set-builder notation.

100. $|x + 3| = x + 3$

101. $|y + 6| = y + 6$

102. $|a - 4| = 4 - a$

103. $|b - 7| = 7 - b$

104. For real numbers x and y, which of the following is always true?
 a. $|x + y| = |x| + |y|$
 b. $|x + y| \le |x| + |y|$
 c. $|x + y| \ge |x| + |y|$

105. Write an absolute value inequality to represent all real numbers within 5 units of 2.

106. Write an absolute value inequality to represent all real numbers within k units of j. (Assume $k > 0$.)

107. Replace the question mark with $\le$, $\ge$, or $=$.
 a. $|x + y|$? $|x| + |y|$
 b. $|x - y|$? $|x| - |y|$
 c. $||x| - |y||$? $|x| + |y|$
 d. $\left|\dfrac{x}{y}\right|$? $\dfrac{|x|}{|y|}$, $y \ne 0$
 e. $|xy|$? $|x||y|$

Focus on Problem Solving

Understand the Problem One of the foremost mathematicians to study problem solving was George Polya (1887–1985). The basic structure that Polya advocated for problem solving has four steps: understand the problem, devise a plan, carry out the plan, and review your solution. The first of these four steps is frequently not given enough attention. There are various exercises that you can try to achieve a good understanding of a problem. Some of these are listed here:

- Try to restate the problem in your own words.
- Determine what is known about this type of problem.
- Determine what information is given.
- Determine what information is unknown.
- Determine whether any of the information given is unnecessary.
- Determine the goal.

To illustrate this aspect of problem solving, consider the following famous riddle.

As I was going to St. Ives,
I met a man with seven wives.
Each wife had seven sacks,
Each sack had seven cats,
Each cat had seven kits.
Kits, cats, sacks, and wives,
How many were going to St. Ives?

To answer the question in the poem, we will ask and answer some of the questions listed above.

1. What is the goal?

 The goal is to determine how many were going to St. Ives. (We know this from reading the last line of the poem.)

2. What information is necessary and what information is unnecessary?

 The first line indicates that the poet was going to St. Ives. The next five lines describe a man the poet *met* on the way. This information is irrelevant.

The answer to the question, then, is 1. Only the poet was going to St. Ives.

Point of Interest

Try this brain teaser: You have two U.S. coins that add up to $.55. One is not a nickel. What are the two coins?

There are many other examples of the importance, in problem solving, of recognizing *irrelevant information*. One that frequently makes the college circuit is posed in the form of a test. The first line of a 100-question test states, "Read the entire test before you begin." The last line of the test reads, "Choose any *one* question to answer." Many people ignore the information given in the first line and just begin the test, only to find out much later that they did a lot more work than necessary.

To illustrate another aspect of Polya's first step in the problem-solving process, consider the problem of finding the area of the oval-shaped region (called an *ellipse*) shown in the diagram at the left. This problem can be solved by doing some research to determine *what information is known about this type of problem*. Mathematicians have found a formula for the area of an ellipse. That formula is $A = \pi ab$, where a and b are as shown in the diagram. Therefore, $A = \pi(6)(4) = 24\pi \approx 75.40$ square units. Without the formula, this problem is difficult to solve. With the formula, it is fairly easy.

For each of the following problems, examine the problem in terms of the first step in Polya's problem-solving method. *Do not solve the problem.*

1. Johanna spent one-third of her allowance on a book. She then spent $5.00 for a sandwich and iced tea. The cost of the iced tea was one-fifth the cost of the sandwich. Find the cost of the iced tea.

2. A flight from Los Angeles to Boston took 6 h. What was the average speed of the plane?

3. A major league baseball is approximately 5 in. in diameter and is covered with cowhide. Approximately how much cowhide is used to cover 10 baseballs?

4. How many donuts are in seven baker's dozen?

5. The smallest prime number is 2. Twice the difference between the eighth and the seventh prime numbers is two more than the smallest prime number. How large is the smallest prime number?

Projects & Group Activities

The Consumer Price Index

WEB

The consumer price index (CPI) is a percent that is written without the percent sign. For instance, a CPI of 160.1 means 160.1%. This number means that an item that cost $100 between 1982 and 1984 (the base years) would cost $160.10 today. Determining the cost is an application of the basic percent equation.

$$PB = A$$

(CPI)(cost in base year) = cost today

$$1.601(100) = 160.1 \qquad \bullet\ \mathbf{160.1\%} = \mathbf{1.601}$$

The table below gives the CPI for various products in October 2005. If you have Internet access, you can obtain current data for the items below plus other items not on this list by visiting the website of the Bureau of Labor Statistics at **www.bls.gov**.

Product	CPI
All items	199.2
Food and beverages	192.5
Housing	198.4
Clothes	122.7
Transportation	184.0
Medical Care	326.2
Energy	204.3
Education: Tuition	454.7

1. Of the items listed, are there any items that cost more than twice as much in 2005 as they cost during the base years? If so, which ones?

2. Of the items listed, are there any items that cost more than one-and-one-half times as much in 2005 as they cost during the base years, but less than twice as much as they cost during the base years? If so, which ones?

3. If the average cost for tuition for one semester at a public college was $4000 in the base years, what was the average cost for tuition for one semester at a public college in 2005?

4. If a new car cost $30,000 in 2005, what would a comparable new car have cost during the base years? Use the transportation category.

5. If rent for an urban apartment in 2005 was $3000, what was the rent for the same apartment during the base years?

6. The CPI for all items in the summer of 1970 was 39.0. What salary in 2005 would have had the same purchasing power as a salary of $10,000 in 1970?

Chapter 2 Summary

Key Words	**Examples**
An **equation** expresses the equality of two mathematical expressions. [2.1.1, p. 84]	$5(3x - 2) = 4x + 7$ is an equation.
A **solution of an equation** is a number that, when substituted for the variable, results in a true equation. [2.1.1, p. 84]	1 is the solution of the equation $6x - 4 = 2$ because $6(1) - 4 = 2$ is a true equation.
To **solve** an equation means to find a solution of the equation. The goal is to rewrite the equation in the form **variable = constant,** because the constant is the solution. [2.1.2, p. 85]	The equation $x = -9$ is in the form *variable = constant*. The constant, -9, is the solution of the equation.
Consecutive integers follow one another in order. [2.3.2, p. 119]	11, 12, 13 are consecutive integers. $-9, -8, -7$ are consecutive integers.
The **perimeter** of a geometric figure is a measure of the distance around the figure. [2.4.1, p. 136]	The perimeter of a triangle is the sum of the lengths of the three sides. The perimeter of a rectangle is the sum of the lengths of the four sides.
An angle is measured in **degrees.** A 90° angle is a **right angle.** A 180° angle is a **straight angle.** One complete revolution is 360°. An **acute angle** is an angle whose measure is between 0° and 90°. An **obtuse angle** is an angle whose measure is between 90° and 180°. An **isosceles triangle** has two sides of equal measure. The three sides of an **equilateral triangle** are of equal measure. **Complementary angles** are two angles whose measures have the sum 90°. **Supplementary angles** are two angles whose measures have the sum 180°. [2.4.1, 2.4.2, pp. 136, 138]	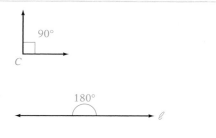
Parallel lines never meet; the distance between them is always the same. **Perpendicular lines** are intersecting lines that form right angles. Two angles that are on opposite sides of the intersection of two lines are **vertical angles;** vertical angles have the same measure. Two angles that share a common side are **adjacent angles;** adjacent angles of intersecting lines are supplementary angles. [2.4.2, p. 139]	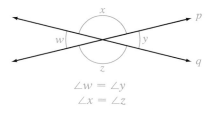

A line that intersects two other lines at two different points is a **transversal.** If the lines cut by a transversal are parallel lines, then pairs of equal angles are formed: **alternate interior angles, alternate exterior angles,** and **corresponding angles.** [2.4.2, pp. 140–141]

$$\angle b = \angle d = \angle x = \angle z$$
$$\angle a = \angle c = \angle w = \angle y$$

Cost is the price that a business pays for a product. **Selling price** is the price for which a business sells a product to a customer. **Markup** is the difference between selling price and cost. **Markup rate** is the markup expressed as a percent of the retailer's cost. [2.5.1, p. 152]

If a business pays $100 for a product and sells that product for $140, then the cost of the product is $100, the selling price is $140, and the markup is $140 − $100 = $40.

Discount is the amount by which a retailer reduces the regular price of a product. **Discount rate** is the discount expressed as a percent of the regular price. [2.5.2, p. 153]

The regular price of a product is $100. The product is now on sale for $75. The discount on the product is $100 − $75 = $25.

The **empty set,** or **null set,** is the set that contains no elements. [2.7.1, p. 168]

The empty set is written as $\varnothing$ or { }.

A **finite set** is a set in which the elements can be counted. [2.7.1, p. 168]

$A = \{1, 2, 3, 4\}$

An **infinite set** is a set in which it is impossible to list all the elements. [2.7.1, p. 168]

$A = \{1, 2, 3, 4, \ldots\}$

The **union** of two sets, written $A \cup B$, is the set that contains all the elements of A and all the elements of B. The elements that are in both set A and set B are listed only once. [2.7.1, p. 169]

If $A = \{2, 3, 4\}$ and $B = \{4, 5, 6\}$, then $A \cup B = \{2, 3, 4, 5, 6\}$.

The **intersection** of two sets, written $A \cap B$, is the set that contains the elements that are common to both A and B. [2.7.1, p. 169]

If $A = \{2, 3, 4\}$ and $B = \{4, 5, 6\}$, then $A \cap B = \{4\}$.

Interval notation is an alternative method of representing a set. An interval is **closed** if it includes both **endpoints.** An interval is **open** if it does not include either endpoint. An interval is **half-open** if one endpoint is included and the other endpoint is not included. A bracket is used to indicate a closed interval, and a parenthesis is used to indicate an open interval. [2.7.1, p. 171]

Closed interval

Open interval

Half-open interval

The **solution set of an inequality** is a set of numbers, each element of which, when substituted in the inequality, results in a true inequality. [2.7.2, p. 172]

Any number greater than 4 is a solution of the inequality $x > 4$.

A **compound inequality** is formed by joining two inequalities with a connective word such as "and" or "or." [2.7.3, p. 174]	$3x > 6$ and $2x + 5 < 7$ $2x + 2 < 3$ or $x + 2 > 4$
An **absolute value equation** is an equation that contains an absolute value symbol. [2.8.1, p. 186]	$\|x - 2\| = 3$
An **absolute value inequality** is an inequality that contains an absolute value symbol. [2.8.2, p. 187]	$\|x - 4\| < 5$ $\|2x - 3\| > 6$

Essential Rules and Procedures

The Addition Property of Equations The same number or variable term can be added to each side of an equation without changing the solution of the equation. [2.1.2, p. 85]	$x + 12 = -19$ $x + 12 - 12 = -19 - 12$ $x = -31$
The Multiplication Property of Equations Each side of an equation can be multiplied by the same non-zero number without changing the solution of the equation. [2.1.3, p. 87]	$-6x = 24$ $\dfrac{-6x}{-6} = \dfrac{24}{-6}$ $x = -4$
To solve an equation of the form $ax + b = c$ **or an equation of the form** $ax + b = cx + d$, use both the Addition and the Multiplication Properties of Equations. [2.2.1 and 2.2.2, pp. 102–105]	$7x - 4 = 3x + 16$ $7x - 3x - 4 = 3x - 3x + 16$ $4x - 4 = 16$ $4x - 4 + 4 = 16 + 4$ $4x = 20$ $\dfrac{4x}{4} = \dfrac{20}{4}$ $x = 5$
To solve an equation containing parentheses, use the Distributive Property to remove parentheses. [2.2.3, pp. 105–107]	$3(2x - 1) = 15$ $6x - 3 = 15$ $6x - 3 + 3 = 15 + 3$ $6x = 18$ $\dfrac{6x}{6} = \dfrac{18}{6}$ $x = 3$
The Basic Percent Equation Percent · base = amount $P \cdot B = A$ [2.1.4, p. 89]	40% of what number is 16? $PB = A$ $0.40B = 16$ $\dfrac{0.40B}{0.40} = \dfrac{16}{0.40}$ $B = 40$ The number is 40.
Consecutive Integers $n, n + 1, n + 2, \ldots$ [2.3.2, p. 119]	The sum of three consecutive integers is 135. $n + (n + 1) + (n + 2) = 135$

Consecutive Even or Consecutive Odd Integers
$n, n + 2, n + 4, \ldots$ [2.3.2, p. 119]

The sum of three consecutive odd integers is 135.

$$n + (n + 2) + (n + 4) = 135$$

Value Mixture Equation
Value of an ingredient
 = amount of the ingredient · cost per unit of the ingredient
$V = AC$
[2.3.3, p. 120]

An herbalist has 30 oz of herbs costing $2 per ounce. How many ounces of herbs costing $1 per ounce should be mixed with the 30 oz to produce a mixture costing $1.60 per ounce?

$$30(2) + 1x = 1.60(30 + x)$$

Uniform Motion Equation
Distance = rate · time
 $d = rt$
[2.3.4, p. 121]

A boat traveled from a harbor to an island at an average speed of 20 mph. The average speed on the return trip was 15 mph. The total trip took 3.5 h. How long did it take to travel to the island?

$$20t = 15(3.5 - t)$$

Formulas for Perimeter
Triangle: $P = a + b + c$
Rectangle: $P = 2L + 2W$
Square: $P = 4s$
[2.4.1, p. 136]

The perimeter of a rectangle is 108 m. The length is 6 m more than three times the width. Find the width of the rectangle.

$$108 = 2(3W + 6) + 2W$$

The perimeter of an isosceles triangle is 34 ft. The length of the third side is 5 ft less than the length of one of the equal sides. Find the length of one of the equal sides.

$$34 = x + x + (x - 5)$$

Sum of the Angles of a Triangle
The sum of the measures of the interior angles of a triangle is 180°.
$\angle A + \angle B + \angle C = 180°$
The sum of an interior angle and a corresponding exterior angle is 180°.
[2.4.3, p. 142]

In a right triangle, the measure of one acute angle is 12° more than the measure of the smallest angle. Find the measure of the smallest angle.

$$x + (x + 12) + 90 = 180$$

Basic Markup Equation
$S = C + rC$, where S is the selling price, C is the cost, and r is the markup rate [2.5.1, p. 152]

The manager of a department buys a pasta maker machine for $70 and sells the machine for $98. Find the markup rate.

$$98 = 70 + 70r$$

Basic Discount Equation
$S = R - rR$, where S is the sale price, R is the regular price, and r is the discount rate [2.5.2, p. 154]

The sale price for a leather bomber jacket is $95. This price is 24% off the regular price. Find the regular price.

$$95 = R - 0.24R$$

Annual Simple Interest Equation

Simple interest = principal · simple interest rate · time

$$I = Prt$$

[2.1.4, p. 91]

An investment of \$4000 is made at an annual simple interest rate of 5%. How much additional money must be invested at an annual simple interest rate of 6.5% so that the total interest earned is \$720?

$$0.05(4000) + 0.065x = 720$$

Percent Mixture Problems

Quantity of a substance in a solution
 = amount of the solution · percent concentration of the substance

$$Q = Ar$$

[2.1.4, 2.6.2, pp. 92, 161]

A silversmith mixed 120 oz of an 80% silver alloy with 240 oz of a 30% silver alloy. Find the percent concentration of the resulting silver alloy.

$$0.80(120) + 0.30(240) = x(360)$$

The Addition Property of Inequalities

If $a > b$, then $a + c > b + c$.

The property is also true for $a < b$. [2.7.2, p. 172]

$$x + 3 > -2$$
$$x + 3 - 3 > -2 - 3$$
$$x > -5$$

The Multiplication Property of Inequalities

Rule 1 If $a > b$ and $c > 0$, then $ac > bc$.
 If $a < b$ and $c > 0$, then $ac < bc$.

$$3x > 2$$
$$\left(\frac{1}{3}\right)(3x) > \left(\frac{1}{3}\right)2$$
$$x > \frac{2}{3}$$

Rule 2 If $a > b$ and $c < 0$, then $ac < bc$.
 If $a < b$ and $c < 0$, then $ac > bc$.

[2.7.2, p. 173]

$$-2x < 5$$
$$\frac{-2x}{-2} > \frac{5}{-2}$$
$$x > -\frac{5}{2}$$

Solutions of Absolute Value Equations

If $a > 0$ and $|x| = a$, then $x = a$ or $x = -a$.

[2.8.1, p. 186]

$$|x - 3| = 7$$

$$x - 3 = 7 \qquad x - 3 = -7$$
$$x = 10 \qquad\quad x = -4$$

Absolute Value Inequalities of the Form $|ax + b| < c$

To solve an absolute value inequality of the form $|ax + b| < c$, $c > 0$, solve the equivalent compound inequality $-c < ax + b < c$. [2.8.2, p. 187]

$$|x - 5| < 9$$
$$-9 < x - 5 < 9$$
$$-9 + 5 < x - 5 + 5 < 9 + 5$$
$$-4 < x < 14$$

Absolute Value Inequalities of the Form $|ax + b| < c$

To solve an absolute value inequality of the form $|ax + b| > c$, solve the equivalent compound inequality $ax + b < -c$ or $ax + b > c$. [2.8.2, p. 188]

$$|x - 5| > 9$$
$$x - 5 < -9 \quad\text{or}\quad x - 5 > 9$$
$$x < -4 \quad\text{or}\quad\quad x > 14$$

Chapter 2 Review Exercises

1. Is 3 a solution of $5x - 2 = 4x + 5$?

2. Solve: $x - 4 = 16$

3. Solve: $8x = -56$

4. Solve: $5x - 6 = 29$

5. Solve: $5x + 3 = 10x - 17$

6. Solve: $3(5x + 2) + 2 = 10x + 5[x - (3x - 1)]$

7. What is 81% of 500?

8. 18 is 72% of what number?

9. 27 is what percent of 40?

10. Solve: $\dfrac{2x - 3}{3} + 2 = \dfrac{2 - 3x}{5}$

11. Find $A \cap B$ given $A = \{0, 1, 2, 3\}$ and $B = \{2, 3, 4, 5\}$.

12. Graph: $\{x \,|\, x < 1\}$

13. Solve: $3x - 7 > -2$
Write the solution set in interval notation.

14. Solve: $2x - 9 < 8x + 15$
Write the solution set in interval notation.

15. Solve: $-5 < 4x - 1 < 7$
Write the solution set in interval notation.

16. Solve: $5x - 2 > 8$ or $3x + 2 < -4$
Write the solution set in interval notation.

17. Solve: $3x < 4$ and $x + 2 > -1$
Write the solution set in set-builder notation.

18. Solve: $3x - 2 > -4$ or $7x - 5 < 3x + 3$
Write the solution set in set-builder notation.

19. Solve: $|2x - 3| = 8$

20. Solve: $|5x + 8| = 0$

21. Solve: $6 + |3x - 3| = 2$

22. Solve: $|2x - 5| \le 3$

23. Solve: $|4x - 5| \ge 3$

24. Solve: $|5x - 4| < -2$

25. **Lever Systems** A lever is 12 ft long. At a distance of 2 ft from the fulcrum, a force of 120 lb is applied. How large a force must be applied to the other end of the lever so that the system will balance? Use the lever-system equation $F_1x = F_2(d - x)$.

26. **Conservation** In Central America and Mexico, 1184 plants and animals are at risk of extinction. This represents approximately 10.7% of all the species at risk of extinction on Earth. Approximately how many plants and animals are at risk of extinction on Earth? (*Source:* World Conservation Union)

27. Physics The pressure at a certain depth in the ocean can be approximated by the equation $P = 15 + \frac{1}{2}D$, where P is the pressure in pounds per square inch and D is the depth in feet. Use this equation to find the depth when the pressure is 55 lb/in².

28. Translate "seven less than twice a number is equal to the number" into an equation and solve.

29. Music A piano wire is 35 in. long. A note can be produced by dividing this wire into two parts so that three times the length of the shorter piece is twice the length of the longer piece. Find the length of the shorter piece.

30. Integer Problem The sum of two numbers is twenty-one. Three times the smaller number is two less than twice the larger number. Find the two numbers.

31. Discounts A ceiling fan that regularly sells for $90 is on sale for $60. Find the discount rate.

32. Geometry In an isosceles triangle, the measure of one of the two equal angles is 15° more than the measure of the third angle. Find the measure of each angle.

33. Markups The manager of a sporting goods store buys packages of cleats for $5.25 each and sells them for $9.45 each. Find the markup rate.

34. Food Mixtures A dairy owner mixes 5 gal of cream that is 30% butterfat with 8 gal of milk that is 4% butterfat. Find the percent concentration of butterfat in the resulting mixture.

35. Geometry The length of a rectangle is four times the width. The perimeter is 200 ft. Find the length and width of the rectangle.

36. Integer Problem The sum of three consecutive odd integers is −45. Find the integers.

37. Geometry Given that $\ell_1 \parallel \ell_2$, find the measures of angles a and b.

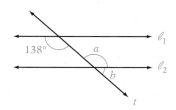

38. Medication A doctor has prescribed 2 cc of medication for a patient. The tolerance is 0.25 cc. Find the lower and upper limits of the amount of medication to be given.

39. Integer Problem Find three consecutive integers such that five times the middle integer is twice the sum of the other two integers.

40. Metallurgy A silversmith combines 40 oz of pure silver that costs $8.00 per ounce with 200 oz of a silver alloy that costs $3.50 per ounce. Find the cost per ounce of the mixture.

41. Food Mixtures A grocer mixed apple juice that costs $3.20 per gallon with 40 gal of cranberry juice that costs $5.50 per gallon. How much apple juice was used to make cranapple juice that costs $4.20 per gallon?

42. Uniform Motion Two planes are 1680 mi apart and traveling toward each other. One plane is traveling 80 mph faster than the other plane. The planes meet in 1.75 h. Find the speed of each plane.

43. Investments Two investments earn an annual income of $635. One investment is earning 10.5% annual simple interest, and the other investment is earning 6.4% annual simple interest. The total investment is $8000. Find the amount invested in each account.

44. Metallurgy An alloy containing 30% tin is mixed with an alloy containing 70% tin. How many pounds of each were used to make 500 lb of an alloy containing 40% tin?

45. Test Scores An average score of 80 to 90 in a psychology class receives a B grade. A student has grades of 92, 66, 72, and 88 on four tests. Find the range of scores on the fifth test that will give the student a B for the course.

Chapter 2 Test

1. Solve: $\dfrac{3}{4}x = -9$

2. Solve: $6 - 5x = 5x + 11$

3. Solve: $3x - 5 = -14$

4. Is -2 a solution of $x^2 - 3x = 2x - 6$?

5. Solve: $5x - 2(4x - 3) = 6x + 9$

6. Solve: $x - 3 = -8$

7. 30% of what is 12?

8. Solve: $4 - 3(x + 2) < 2(2x + 3) - 1$
Write the solution set in interval notation.

9. Solve: $3x - 2 > 4$ or $4 - 5x < 14$
Write the solution set in set-builder notation.

10. Solve: $|3 - 5x| = 12$

11. Solve: $|2x - 1| > 3$

12. Translate "the difference between three times a number and fifteen is twenty-seven" into an equation and solve.

13. Markups The manager of a sports shop uses a markup rate of 50%. The selling price for a set of golf clubs is $300. Find the cost of the golf clubs.

14. Manufacturing A financial manager has determined that the cost per unit to manufacture a calculator is $15 and that the fixed costs per month are $2000. Find the number of calculators produced during a month in which the total cost was $5000. Use the equation $T = U \cdot N + F$, where T is the total cost, U is the cost per unit, N is the number of units produced, and F is the fixed cost.

15. Integer Problem Find three consecutive odd integers such that three times the first integer is one less than the sum of the second and third integers.

16. Geometry In a triangle, the first angle is 15° more than the second angle. The third angle is three times the second angle. Find the measure of each angle.

17. Geometry The width of a rectangle is 12 ft. The length is $(3x + 5)$ ft. Express as an integer the minimum length of the rectangle if the area is greater than 276 ft².

18. Discounts A file cabinet that normally sells for $99 is on sale for 20% off. Find the sale price.

19. Geometry Suppose an obtuse angle of two intersecting lines is 135°. What are the measures of the three other angles?

20. Rental Cars Agency A rents cars for $12 per day and 10¢ for every mile driven. Agency B rents cars for $24 per day with unlimited mileage. How many miles per day can you drive an Agency A car if it is to cost you less than an Agency B car?

21. Medication A doctor prescribed 3 cc of medication for a patient. The tolerance is 0.1 cc. Find the lower and upper limits of the amount of medication to be given.

22. Food Mixtures A butcher combines 100 lb of hamburger that costs $2.60 per pound with 60 lb of hamburger that costs $4.20 per pound. Find the cost of the hamburger mixture.

23. Uniform Motion A jogger runs a distance at a speed of 8 mph and returns the same distance running at a speed of 6 mph. Find the total distance that the jogger runs if the total time running is 1 h and 45 min.

24. Investments An investment of $12,000 is deposited into two simple interest accounts. On one account, the annual simple interest rate is 7.8%. On the other, the annual simple interest rate is 9%. The total interest earned for one year is $1020. How much was invested in each account?

25. Mixture Problem How many ounces of pure water must be added to 60 oz of an 8% salt solution to make a 3% salt solution?

Cumulative Review Exercises

1. Subtract: $-6 - (-20) - 8$

2. Multiply: $(-2)(-6)(-4)$

3. Subtract: $-\dfrac{5}{6} - \left(-\dfrac{7}{16}\right)$

4. Divide: $-2\dfrac{1}{3} \div 1\dfrac{1}{6}$

5. Simplify: $-4^2 \cdot \left(-\dfrac{3}{2}\right)^3$

6. Simplify: $25 - 3 \cdot \dfrac{(5-2)^2}{2^3 + 1} - (-2)$

7. Evaluate $3(a - c) - 2ab$ when $a = 2$, $b = 3$, and $c = -4$.

8. Simplify: $3x - 8x + (-12x)$

9. Simplify: $2a - (-3b) - 7a - 5b$

10. Simplify: $(16x)\left(\dfrac{1}{8}\right)$

11. Simplify: $-4(-9y)$

12. Simplify: $-2(-x^2 - 3x + 2)$

13. Simplify: $-2(x - 3) + 2(4 - x)$

14. Simplify: $-3[2x - 4(x - 3)] + 2$

15. Use the roster method to write the set of negative integers greater than -8.

16. Write $\dfrac{7}{8}$ as a percent. Write the remainder in fractional form.

17. Write 342% as a decimal.

18. Write $62\dfrac{1}{2}\%$ as a fraction.

19. Is -3 a solution of $x^2 + 6x + 9 = x + 3$?

20. Solve: $x - 4 = -9$

21. Solve: $\dfrac{3}{5}x = -15$

22. Solve: $13 - 9x = -14$

23. Solve: $5x - 8 = 12x + 13$

24. Solve: $8x - 3(4x - 5) = -2x - 11$

25. Solve: $-\dfrac{3}{4}x > \dfrac{2}{3}$

26. Solve: $5x - 4 \geq 4x + 8$

27. Solve: $3x + 17 < 5x - 1$

28. Translate "the difference between eight and the quotient of a number and twelve" into a variable expression.

29. Translate and simplify "the sum of a number and two more than the number."

30. Lever Systems A lever is 25 ft long. At a distance of 12 ft from the fulcrum, a force of 26 lb is applied. How large a force must be applied to the other end of the lever so that the system will balance? Use the lever-system equation $F_1x = F_2(d - x)$.

31. Integer Problem Three times the sum of the first and third of three consecutive odd integers is fifteen more than the second integer. Find the first integer.

32. Recreation Tickets for a school play sold for $2.25 for each adult and $.75 for each child. The total receipts for 75 tickets were $128.25. Find the number of adult tickets sold.

33. Uniform Motion Two planes are 1400 mi apart and traveling toward each other. One plane is traveling 120 mph faster than the other plane. The planes meet in 2.5 h. Find the speed of the faster plane.

34. Mixture Problem How many liters of a 12% acid solution must be mixed with 4 L of a 5% acid solution to make an 8% acid solution?

35. Investments An investment advisor invested $10,000 in two accounts. One investment earned 9.8% annual simple interest, and the other investment earned 12.8% annual simple interest. The amount of interest earned in one year was $1085. How much was invested in the 9.8% account?

3

Linear Functions and Inequalities in Two Variables

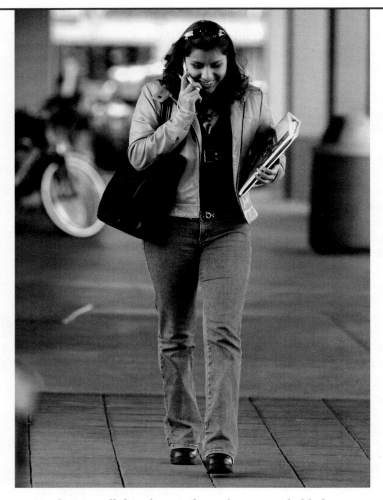

Do you have a cellular phone? If you do, you probably know about the many payment options cellular phone companies offer. Many customers pay a monthly fee plus a small charge for each minute used. The amount charged per minute varies depending on each customer's payment plan. In **Exercise 57 on page 277** you will be asked to write a linear function that models the monthly cost of a cell phone in terms of the number of minutes you use. Being able to use a linear function to model a relationship between two variables is an important skill when it comes to budgeting your phone time—and your money!

OBJECTIVES

3.1
 1 Points on a rectangular coordinate system
 2 Find the length and midpoint of a line segment
 3 Graph a scatter diagram
 4 Average rate of change

3.2
 1 Evaluate a function
 2 Graph of a function
 3 Vertical line test

3.3
 1 Graph a linear function
 2 Graph an equation of the form $Ax + By = C$
 3 Application problems

3.4
 1 Find the slope of a line given two points
 2 Graph a line given a point and the slope

3.5
 1 Find the equation of a line given a point and the slope
 2 Find the equation of a line given two points
 3 Application problems

3.6
 1 Find equations of parallel and perpendicular lines

3.7
 1 Graph the solution set of an inequality in two variables

DVD
 SSM

Online Study Center

Need help? For online student resources, such as section quizzes, visit the Online Study Center at college.hmco.com/pic/aufmannABI2e.

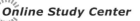

Prep Test

1. Simplify: $-4(x - 3)$

2. Simplify: $\sqrt{(-6)^2 + (-8)^2}$

3. Simplify: $\dfrac{3 - (-5)}{2 - 6}$

4. Evaluate: $-2x + 5$ for $x = -3$.

5. Evaluate $\dfrac{2r}{r - 1}$ for $r = 5$.

6. Evaluate $2p^3 - 3p + 4$ for $p = -1$.

7. Evaluate $\dfrac{x_1 + x_2}{2}$ for $x_1 = 7$ and $x_2 = -5$.

8. Given $3x - 4y = 12$, find the value of x when $y = 0$.

3.1 The Rectangular Coordinate System

OBJECTIVE **1** Points on a rectangular coordinate system

Point of Interest

*A rectangular coordinate system is also called a **Cartesian** coordinate system in honor of Descartes.*

Before the 15th century, geometry and algebra were considered separate branches of mathematics. That changed when René Descartes, a French mathematician who lived from 1596 to 1650, founded **analytic geometry.** In this geometry, a *coordinate system* is used to study relationships between variables.

A **rectangular coordinate system** is formed by two number lines, one horizontal and one vertical, that intersect at the zero point of each line. The point of intersection is called the **origin.** The two lines are called **coordinate axes,** or simply **axes.**

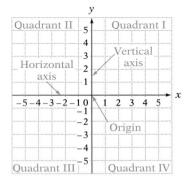

The axes determine a **plane,** which can be thought of as a large, flat sheet of paper. The two axes divide the plane into four regions called **quadrants.** The quadrants are numbered counterclockwise from I to IV.

Each point in the plane can be identified by a pair of numbers called an **ordered pair.** The first number of the pair measures a horizontal distance and is called the **abscissa.** The second number of the pair measures a vertical distance and is called the **ordinate.** The **coordinates** of the point are the numbers in the ordered pair associated with the point. The abscissa is also called the **first coordinate,** or **x-coordinate,** of the ordered pair, and the ordinate is also called the **second coordinate,** or **y-coordinate,** of the ordered pair.

Point
of Interest

Gottfried Leibnitz introduced the words abscissa *and* ordinate. Abscissa *is from Latin meaning "to cut off." Originally, Leibnitz used* abscissa linea, *"cut off a line." The root of* ordinate *is a Latin word used to suggest the sense of order.*

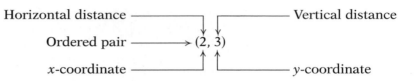

When drawing a rectangular coordinate system, we often label the horizontal axis x and the vertical axis y. In this case, the coordinate system is called an **xy-coordinate system.** To graph or plot a point in the xy-coordinate system, place a dot at the location given by the ordered pair. The **graph of an ordered pair** is the dot drawn at the coordinates of the point in the xy-coordinate system. The points whose coordinates are $(3, 4)$ and $(-2.5, -3)$ are graphed in the figure at the right.

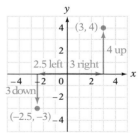

Take Note

The *ordered pair* is an important concept. Remember that there are two numbers (the *pair*) and that the *order* in which they are given is important.

The points whose coordinates are $(3, -1)$ and $(-1, 3)$ are graphed at the right. Note that the graphs are in different locations. The *order* of the coordinates of an ordered pair is important.

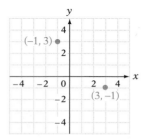

The xy-coordinate system is used to graph equations in *two variables*. Examples of equations in two variables are shown at the right.

$$y = 3x + 7$$
$$y = x^2 - 4x + 3$$
$$x^2 + y^2 = 25$$
$$x = \frac{y}{y^2 + 4}$$

A **solution of an equation in two variables** is an ordered pair (x, y) whose coordinates make the equation a true statement.

Is the ordered pair $(-3, 7)$ a solution of the equation $y = -2x + 1$?

Replace x by -3 and replace y by 7. Then simplify.

$$\begin{array}{c|c} y = -2x + 1 \\ \hline 7 & -2(-3) + 1 \\ 7 & 6 + 1 \\ 7 = 7 \end{array}$$

Compare the results. If the resulting equation is a true statement, the ordered pair is a solution of the equation. If it is not a true statement, the ordered pair is not a solution of the equation.

Yes the ordered pair $(-3, 7)$ is a solution of the equation.

Besides the ordered pair $(-3, 7)$, there are many other ordered-pair solutions of the equation $y = -2x + 1$. For example, $(-5, 11)$, $(0, 1)$, $\left(-\frac{3}{2}, 4\right)$ and $(4, -7)$ are also solutions of the equation.

In general, an equation in two variables has an infinite number of solutions. By choosing any value of x and substituting that value into the equation, we can calculate a corresponding value of y. The resulting ordered-pair solution (x, y) of the equation can be graphed in a rectangular coordinate system.

Graph the solutions (x, y) of $y = x^2 - 1$ when x equals $-2, -1, 0, 1,$ and 2.

Substitute each value of x into the equation and solve for y. Then graph the resulting ordered pairs. It is convenient to record the ordered-pair solutions in a table similar to the one shown below.

x	$y = x^2 - 1$	y	(x, y)
-2	$y = (-2)^2 - 1$	3	$(-2, 3)$
-1	$y = (-1)^2 - 1$	0	$(-1, 0)$
0	$y = 0^2 - 1$	-1	$(0, -1)$
1	$y = 1^2 - 1$	0	$(1, 0)$
2	$y = 2^2 - 1$	3	$(2, 3)$

EXAMPLE 1 Determine the ordered-pair solution of $y = \dfrac{x}{x - 2}$ that corresponds to $x = 4$.

Solution $y = \dfrac{x}{x - 2}$

$y = \dfrac{4}{4 - 2}$ • **Replace x by 4 and solve for y.**

$y = 2$

The ordered-pair solution is $(4, 2)$.

PROBLEM 1 Determine the ordered-pair solution of $y = \dfrac{3x}{x + 1}$ that corresponds to $x = -2$.

Solution See page S10.

EXAMPLE 2 Graph the ordered-pair solutions of $y = x^2 - x$ when $x = -1, 0, 1,$ and 2.

Solution

x	y
-1	2
0	0
1	0
2	2

PROBLEM 2 Graph the ordered-pair solutions of $y = |x + 1|$ when $x = -3, -2, -1, 0,$ and 1.

Solution See page S10.

OBJECTIVE 2

Find the length and midpoint of a line segment

The distance between two points in an xy-coordinate system can be calculated by using the Pythagorean Theorem.

Pythagorean Theorem

If a and b are the lengths of the legs of a right triangle and c is the length of the hypotenuse, then $a^2 + b^2 = c^2$.

Consider the two points and the right triangle shown at the right. The vertical distance between $P_1(x_1, y_1)$ and $P_2(x_2, y_2)$ is $|y_2 - y_1|$. The horizontal distance between the points $P_1(x_1, y_1)$ and $P_2(x_2, y_2)$ is $|x_2 - x_1|$.

The quantity d^2 is calculated by applying the Pythagorean Theorem to the right triangle.

$$d^2 = |x_2 - x_1|^2 + |y_2 - y_1|^2$$

Because the square of a number is always nonnegative, the absolute value signs are not necessary.

$$d^2 = (x_2 - x_1)^2 + (y_2 - y_1)^2$$

The distance d is the square root of d^2.

$$d = \sqrt{(x_2 - x_1)^2 + (y_2 - y_1)^2}$$

The Distance Formula

If $P_1(x_1, y_1)$ and $P_2(x_2, y_2)$ are two points in the plane, then the distance d between the two points is given by

$$d = \sqrt{(x_2 - x_1)^2 + (y_2 - y_1)^2}$$

EXAMPLE 3 Find the exact distance between the points whose coordinates are $(-3, 2)$ and $(4, -1)$.

Solution $d = \sqrt{(x_2 - x_1)^2 + (y_2 - y_1)^2}$ • **Use the distance formula.**

$\qquad\quad = \sqrt{[4 - (-3)]^2 + (-1 - 2)^2}$ • $(x_1, y_1) = (-3, 2)$ and $(x_2, y_2) = (4, -1)$.

$\qquad\quad = \sqrt{7^2 + (-3)^2} = \sqrt{49 + 9}$

$\qquad\quad = \sqrt{58}$

The distance between the points is $\sqrt{58}$.

> **PROBLEM 3** Find the exact distance between the points whose coordinates are (5, −2) and (−4, 3).
>
> **Solution** See page S10.

The **midpoint of a line segment** is equidistant from its endpoints. The coordinates of the midpoint of the line segment P_1P_2 are (x_m, y_m). The intersection of the horizontal line segment through P_1 and the vertical line segment through P_2 is Q, with coordinates (x_2, y_1).

The x-coordinate x_m of the midpoint of the line segment P_1P_2 is the same as the x-coordinate of the midpoint of the line segment P_1Q. It is the average of the x-coordinates of the points P_1 and P_2.

$$x_m = \frac{x_1 + x_2}{2}$$

Similarly, the y-coordinate y_m of the midpoint of the line segment P_1P_2 is the same as the y-coordinate of the midpoint of the line segment P_2Q. It is the average of the y-coordinates of the points P_1 and P_2.

$$y_m = \frac{y_1 + y_2}{2}$$

The Midpoint Formula

If $P_1(x_1, y_1)$ and $P_2(x_2, y_2)$ are the endpoints of a line segment, then the coordinates of the midpoint (x_m, y_m) of the line segment are given by

$$x_m = \frac{x_1 + x_2}{2} \quad \text{and} \quad y_m = \frac{y_1 + y_2}{2}$$

EXAMPLE 4 Find the coordinates of the midpoint of the line segment with endpoints (−5, 4) and (−3, 7).

Solution

$x_m = \dfrac{x_1 + x_2}{2} \qquad\qquad y_m = \dfrac{y_1 + y_2}{2}$ • **Use the midpoint formula.**

$ = \dfrac{-5 + (-3)}{2} \qquad\qquad = \dfrac{4 + 7}{2}$ • $(x_1, y_1) = (-5, 4)$ **and** $(x_2, y_2) = (-3, 7)$.

$ = -4 \qquad\qquad\qquad\quad = \dfrac{11}{2}$

The midpoint is $\left(-4, \dfrac{11}{2}\right)$.

> **PROBLEM 4** Find the coordinates of the midpoint of the line segment with endpoints (−3, −5) and (−2, 3).
>
> **Solution** See page S10.

OBJECTIVE **3** ## Graph a scatter diagram

Discovering a relationship between two variables is an important task in the study of mathematics. These relationships occur in many forms and in a wide variety of applications. Here are some examples.

- A botanist wants to know the relationship between the number of bushels of wheat yielded per acre and the amount of watering per acre.

- An environmental scientist wants to know the relationship between the incidence of skin cancer and the amount of ozone in the atmosphere.

- A business analyst wants to know the relationship between the price of a product and the number of products that are sold at that price.

A researcher may investigate the relationship between two variables by means of *regression analysis*, which is a branch of statistics. The study of the relationship between two variables may begin with a **scatter diagram,** which is a graph of the ordered pairs of the known data.

The following table shows data collected by a university registrar comparing the grade-point averages (GPAs) of graduating high school seniors and their scores on a national test.

GPA, x	3.25	3.00	3.00	3.50	3.50	2.75	2.50	2.50	2.00	2.00	1.50
Score, y	1200	1200	1000	1500	1100	1000	1000	900	800	900	700

The scatter diagram for these data is shown at the right. Each ordered pair represents the GPA and the test score for a student. For example, the ordered pair (2.75, 1000) indicates that one student with a GPA of 2.75 had a test score of 1000.

EXAMPLE 5 The scatter diagram on the following page shows the number of computer majors, in thousands, in the United States and Canada in the fall of each year from 1999 to 2003. (*Source*: Computing Research Association Taulbee Survey)

A. How many students were computer majors in the fall of 2000?

B. In the fall of which year was the number of computer majors less than 19,000?

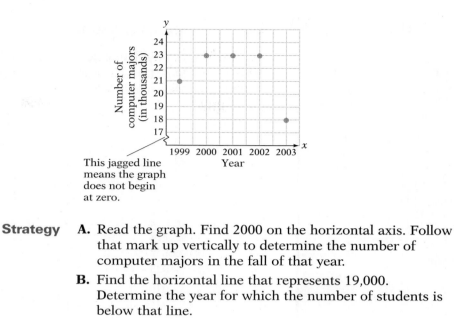

This jagged line means the graph does not begin at zero.

Strategy **A.** Read the graph. Find 2000 on the horizontal axis. Follow that mark up vertically to determine the number of computer majors in the fall of that year.

B. Find the horizontal line that represents 19,000. Determine the year for which the number of students is below that line.

Solution **A.** In the fall of 2000, 23,000 students were computer majors.

B. The number of computer majors was less than 19,000 in 2003.

PROBLEM 5 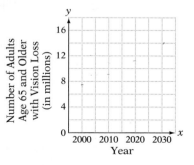 The table below shows the estimated number of adults, in millions, age 65 and older who will have some form of vision loss. Graph the scatter diagram for these data.

Year, x	2000	2010	2020	2030
Adults age 65 and older with vision loss (in millions), y	7.3	8.3	11.2	15.0

Source: Lighthouse National Survey on Vision Loss

Solution See page S11.

OBJECTIVE **4** **Average rate of change**

The table at the left shows the population of California each decade from 1850 to 2000. (*Source:* U.S. Bureau of the Census). These data are graphed in the scatter diagram below.

Population of California	
Year	Population (in millions)
1850	0.1
1860	0.4
1870	0.6
1880	0.9
1890	1.2
1900	1.5
1910	2.4
1920	3.4
1930	5.7
1940	6.9
1950	10.6
1960	15.7
1970	20.0
1980	23.7
1990	29.8
2000	33.9

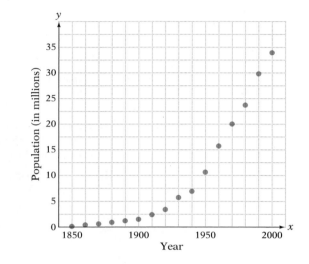

One way to describe how the population of California has changed over time is to calculate the change in population for a given time period. For example, we can find the change in population from 1990 to 2000.

Population in 2000 − population in 1990 = 33.9 million − 29.8 million
= 4.1 million

The amount of change in the population, 4.1 million, does not describe how rapid the growth in population was. To describe the rate of growth, or the *rate of change*, we must consider the number of years over which that growth took place. To determine how rapidly the population grew from 1990 to 2000, divide the change in population by the number of years during which that growth took place.

$$\frac{\text{Change in population}}{\text{Change in years}} = \frac{\text{population in 2000} - \text{population in 1990}}{2000 - 1990}$$

$$= \frac{33.9 - 29.8}{2000 - 1990}$$

$$= \frac{4.1}{10} = 0.41$$

The average rate of change was 0.41 million, or 410,000, people per year. This means that *on average*, from 1990 to 2000, the population of California increased 410,000 people per year.

Consider the two points on the graph that correspond to these years.

(1990, 29.8) and (2000, 33.9)

Note that the average rate of change is calculated by dividing the difference between the *y* values of the two points by the difference between the *x* values of the two points.

Average Rate of Change

The average rate of change of y with respect to x is $\frac{\text{change in } y}{\text{change in } x}$.

To calculate the average rate of change, think

$$\frac{y \text{ value of second point} - y \text{ value of first point}}{x \text{ value of second point} - x \text{ value of first point}}$$

Find the average annual rate of change in the population of California from 1980 to 1990.

In 1980, the population was 23.7 million: (1980, 23.7)
In 1990, the population was 29.8 million: (1990, 29.8)

$$
\begin{aligned}
\text{Average rate of change} &= \frac{\text{change in } y}{\text{change in } x} \\
&= \frac{\text{population in 1990} - \text{population in 1980}}{1990 - 1980} \\
&= \frac{29.8 - 23.7}{1990 - 1980} \\
&= \frac{6.1}{10} = 0.61
\end{aligned}
$$

The average rate of change in the population was 0.61 million, or 610,000, people per year.

Take Note

The rate of change in the population of California is not constant. It varies for different time periods.

From 1990 to 2000, the average rate of change was 410,000 people per year, whereas from 1980 to 1990, the average rate of change was 610,000 people per year. Therefore, the average rate of growth from 1980 to 1990 was greater than the average rate of growth from 1990 to 2000.

EXAMPLE 6 Find the average annual rate of change in California's population from 1850 to 1950.

Solution In 1850, the population was 0.1 million: (1850, 0.1)
In 1950, the population was 10.6 million: (1950, 10.6)

$$
\begin{aligned}
\text{Average rate of change} &= \frac{\text{change in } y}{\text{change in } x} \\
&= \frac{\text{population in 1950} - \text{population in 1850}}{1950 - 1850} \\
&= \frac{10.6 - 0.1}{1950 - 1850} \\
&= \frac{10.5}{100} = 0.105
\end{aligned}
$$

The average rate of change in the population was 105,000 people per year.

Note that if the ordinates and abscissas are subtracted in the reverse order, the result is the same.

$$\text{Average rate of change} = \frac{\text{population in 1850} - \text{population in 1950}}{1850 - 1950}$$

$$= \frac{0.1 - 10.6}{1850 - 1950} = \frac{-10.5}{-100} = 0.105$$

PROBLEM 6 Find the average annual rate of change in California's population from 1900 to 2000.

Solution See page S11.

The table below shows the median salaries in 1990, 1995, 2000, and 2005 for Boston Red Sox players and New York Yankees players. (*Source*: USAToday.com) Use these data for Example 7 and Problem 7.

Year	1990	1995	2000	2005
Median salary, Boston Red Sox	282,500	282,500	2,000,000	2,875,000
Median salary, New York Yankees	437,500	531,000	1,350,000	5,833,334

EXAMPLE 7 Find the average annual rate of change in the median salary of Boston Red Sox players between 1990 and 2005. Round to the nearest thousand dollars.

Solution In 1990, the median salary was 282,500: (1990, 282,500)

In 2005, the median salary was 2,875,000: (2005, 2,875,000)

$$\text{The average rate of change} = \frac{\text{change in } y}{\text{change in } x}$$

$$= \frac{\text{median salary in 2005} - \text{median salary in 1990}}{2005 - 1990}$$

$$= \frac{2,875,000 - 282,500}{2005 - 1990}$$

$$= \frac{2,592,500}{15} \approx 173,000$$

The average rate of change was $173,000 per year.

PROBLEM 7 **A.** Find the average annual rate of change in the median salary of New York Yankees players between 1990 and 2005. Round to the nearest thousand dollars.

B. Was this greater than or less than the average annual rate of change in the median salary of Boston Red Sox players between 1990 and 2005?

Solution See page S11.

3.1 Concept Review

Determine whether the following statements are always true, sometimes true, or never true.

1. An equation in two variables has an infinite number of solutions.

2. A rectangular coordinate system has two real number axes.

3. In the graph of an ordered pair, the first number is the y-coordinate and the second number is the x-coordinate.

4. The solution of an equation with two variables is an ordered pair.

5. The point $(-2, -4)$ is in the fourth quadrant.

6. In quadrant III, the y-coordinate of the point is negative.

3.1 Exercises

1 Points on a rectangular coordinate system

Complete Exercises 1 and 2 by replacing each question mark with the word *left, right, up,* or *down.*

1. To graph the point $(6, -2)$, start at the origin and move 6 units ___?___ and 2 units ___?___.

2. To graph the point $(-3, 5)$, start at the origin and move 3 units ___?___ and 5 units ___?___.

3. Graph the ordered pairs $(0, -1)$, $(2, 0)$, $(3, 2)$, and $(-1, 4)$.

4. Graph the ordered pairs $(-1, -3)$, $(0, -4)$, $(0, 4)$, and $(3, -2)$.

5. Find the coordinates of each point.

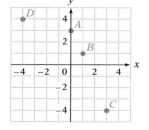

6. Find the coordinates of each point.

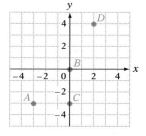

7. Draw a line through all points with an abscissa of 2.

8. Draw a line through all points with an abscissa of −3.

9. Draw a line through all points with an ordinate of −3.

10. Draw a line through all points with an ordinate of 4.

11. Graph the ordered-pair solutions of $y = x^2$ when $x = -2, -1, 0, 1,$ and 2.

12. Graph the ordered-pair solutions of $y = -x^2 + 1$ when $x = -2, -1, 0, 1,$ and 2.

13. Graph the ordered-pair solutions of $y = |x + 1|$ when $x = -5, -3, 0, 3,$ and 5.

14. Graph the ordered-pair solutions of $y = -2|x|$ when $x = -3, -1, 0, 1,$ and 3.

15. Graph the ordered-pair solutions of $y = -x^2 + 2$ when $x = -2, -1, 0, 1,$ and 2.

16. Graph the ordered-pair solutions of $y = -x^2 + 4$ when $x = -3, -1, 0, 1,$ and 3.

17. Graph the ordered-pair solutions of $y = x^3 - 2$ when $x = -1, 0, 1,$ and 2.

18. Graph the ordered-pair solutions of $y = -x^3 + 1$ when $x = -1, 0, 1,$ and $\frac{3}{2}$.

19. Let a and b be positive numbers such that $a < b$. In which quadrant is each point located?

 a. (a, b) **b.** $(-a, b)$ **c.** $(b - a, -b)$ **d.** $(a - b, -b - a)$

20. Let a and b be positive numbers. State whether the two given points lie on the x-axis, the y-axis, a horizontal line other than the x-axis, or a vertical line other than the y-axis.

 a. $(-a, -b)$ and $(a, -b)$ **b.** $(0, a)$ and $(0, -b)$

2 Find the length and midpoint of a line segment

21. For points $P_1(3, -3)$ and $P_2(-1, 4)$, identify each of the following values:
$x_1 = \underline{\quad ? \quad}$, $x_2 = \underline{\quad ? \quad}$, $y_1 = \underline{\quad ? \quad}$, and $y_2 = \underline{\quad ? \quad}$.

22. Find the length of the line segment between the points P_1 and P_2 given in Exercise 21.

$d = \sqrt{(x_2 - x_1)^2 + (y_2 - y_1)^2}$ • Use the $\underline{\quad ? \quad}$ Formula.

$= \sqrt{(\underline{\ ? \ } - \underline{\ ? \ })^2 + (\underline{\ ? \ } - \underline{\ ? \ })^2}$ • Substitute coordinates.

$= \sqrt{(\underline{\ ? \ })^2 + (\underline{\ ? \ })^2}$ • Perform operations inside parentheses.

$= \sqrt{\underline{\ ? \ } + \underline{\ ? \ }}$ • Simplify exponential expressions.

$= \sqrt{\underline{\ ? \ }}$ • Add.

The length of the line segment between $P_1(3, -3)$ and $P_2(-1, 4)$ is $\underline{\quad ? \quad}$.

Find the length and midpoint of the line segment between the given points.

23. $P_1(3, 5)$ and $P_2(4, 1)$

24. $P_1(-2, 3)$ and $P_2(5, -1)$

25. $P_1(0, 3)$ and $P_2(-2, 4)$

26. $P_1(6, -1)$ and $P_2(-3, -2)$

27. $P_1(-3, -5)$ and $P_2(2, -4)$

28. $P_1(-7, -5)$ and $P_2(-2, -1)$

29. $P_1(5, -2)$ and $P_2(-1, 5)$

30. $P_1(3, -5)$ and $P_2(6, 0)$

31. $P_1(5, -5)$ and $P_2(2, -5)$

32. a, b, c, and d are positive numbers such that $a > c$ and $b < d$. In what quadrant is the midpoint of the line segment joining $P_1(a, b)$ and $P_2(-c, -d)$?

33. a, b, c, and d are positive numbers such that $a < c$ and $b < d$. In what quadrant is the midpoint of the line segment joining $P_1(-a, -b)$ and $P_2(c, d)$?

3 Graph a scatter diagram

Solve.

34. **Chemistry** The amount of a substance that can be dissolved in a fixed amount of water usually increases as the temperature of the water increases. Cerium selenate, however, does not behave in this manner. The graph at the right shows the number of grams of cerium selenate that will dissolve in 100 mg of water for various temperatures, in degrees Celsius.
 a. Determine the temperature at which 25 g of cerium selenate will dissolve.
 b. Determine the number of grams of cerium selenate that will dissolve when the temperature is 80°C.

35. **Nutrition** Sodium content in foods is a measure of the salt content in the food. The scatter diagram at the right shows the approximate amount of sodium and the number of calories in various sandwiches offered at a McDonald's restaurant. (*Source:* Gourmet Connection website/Convence Labs)

a. How many calories are in a hamburger?

b. How many milligrams of sodium are in a Big Mac?

36. **Automotive Technology** When water freezes, it expands. This expansion is a potential problem for cars, because the freezing water can cause the engine block of a car to crack. Ethylene glycol is mixed with water to produce an antifreeze that helps prevent the water in the car from freezing. The scatter diagram at the right shows the changes in the freezing point of a solution of water and ethylene glycol that occur as the percent of ethylene glycol is increased.

These jagged lines mean the graph does not begin at zero.

a. At what temperature will a 45% mixture of ethylene glycol and water freeze?

b. What percent of ethylene glycol is required to protect the car to temperatures as low as −30°F?

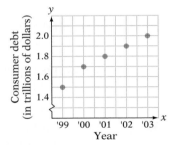

37. **Consumer Debt** The scatter diagram at the right shows how much consumers owed in the years 1999 through 2003. (*Source:* Federal Reserve)

a. What was the consumer debt in 2001?

b. In which year was consumer debt first greater than $1.9 trillion?

c. Looking at the trend of the data, would you expect consumer debt in 2004 to be greater than or less than consumer debt in 2003?

38. Look at the table of data in Exercise 40. The lowest x-value is ___?___ and the highest x-value is ___?___. The lowest y-value is ___?___ and the highest y-value is ___?___. In a scatter diagram of the data, the x-axis must show values from at least ___?___ to ___?___, and the y-axis must show values from at least ___?___ to ___?___.

39. Look at the table of data in Exercise 41. The lowest x-value is ___?___ and the highest x-value is ___?___. The lowest y-value is ___?___ and the highest y-value is ___?___. In a scatter diagram of the data, the x-axis must show values from at least ___?___ to ___?___, and the y-axis must show values from at least ___?___ to ___?___.

40. **Wealth** The table shows the number of U.S. households with a net worth of $1 million or more every three years from 1992 to 2004. Also shown is the millionaire households as a percent of all households in the United States. (*Source:* Edward Wolff, professor of economics, New York University; Federal Reserve) Graph the scatter diagram for these data.

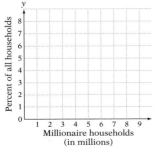

Millionaire households (in millions), x	3.1	3.4	4.1	5.0	8.2
Percent of all households, y	3.2	3.4	4.0	4.7	7.0

41. Telecommunications As shown in the table below, the percent of households with telephones has increased over the years. (*Source: U.S. Census Bureau/Federal Communications Commission*) Draw a scatter diagram for these data.

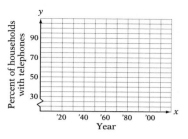

Year, x	'20	'30	'40	'50	'60	'70	'80	'90	'00
Percent, y	35	41	37	62	78	90	93	95	97

4 Average rate of change

42. Temperature On November 5 in midstate New Hampshire, the temperature at 6 A.M. was 34°F. At 2 P.M. on the same day, the temperature was 58°F. Find the average rate of change in temperature per hour.

43. Demography The table at the right shows the population of Texas each decade from 1850 to 2000. (*Source: U.S. Bureau of the Census*)
 a. Find the average annual rate of change in the population of Texas from 1900 to 2000.
 b. Was the average annual rate of change in the population from 1900 to 1950 greater than or less than the average annual rate of change in the population from 1950 to 2000?
 c. Was the average annual rate of change in the population of Texas from 1980 to 1990 greater than or less than the average annual rate of change in the population of California from 1980 to 1990? (See page 217 for the average annual rate of change in the population of California from 1980 to 1990.)

Population of Texas	
Year	Population (in millions)
1850	0.2
1860	0.6
1870	0.8
1880	1.6
1890	2.2
1900	3.0
1910	3.9
1920	4.7
1930	5.8
1940	6.4
1950	7.7
1960	9.6
1970	11.2
1980	14.2
1990	17.0
2000	20.9

For Exercises 44 and 45, use the table in Exercise 43.

44. During which decade was the average annual rate of change in the population the least?

45. During which decade was the average annual rate of change in the population approximately 300,000 people?

46. Cremation In the United States in 1990, there were 2.15 million deaths and 367,975 cremations. In 1999, there were 2.35 million deaths and 595,617 cremations. (*Source: Cremation Association of North America*)
 a. In the 1990s, what was the average annual rate of change in the number of deaths in the United States? Round to the nearest thousand.
 b. In the 1990s, what was the average annual rate of change in the number of cremations in the United States? Round to the nearest thousand.
 c. In the 1990s, which increased at a greater rate, the number of deaths or the number of cremations?

47. Traffic Fatalities The number of pedestrians killed by motor vehicles in 1990 was 6482; in 2000 it was 4727. (*Source: National Highway Transportation Safety Administration*)

a. Find the average annual rate of change in the number of pedestrian fatalities from 1990 to 2000. (*Note:* A negative average rate of change denotes a decrease.)

b. ✎ Why can the answer to part a be considered encouraging?

48. ◗ **Energy Prices** The table at the right shows the average price per gallon of gasoline in the United States each year from 1990 to 2004. (*Source*: Energy Information Administration)

a. Find the average rate of change per year in the price of a gallon of gasoline from 1990 to 2000.

b. Which was greater, the average rate of change per year from 1990 to 1995 or the average rate of change per year from 1995 to 2000? How much greater?

49. ◗ **Trademarks** The table below shows the number of applications for trademark registration in the United States in 1990, 1995, 2000, and 2005. (*Source:* International Trademark Association) One reason for the large number of applications is the increase in popularity of the Internet and the desire to register domain names and websites.

a. Find the average annual rate of change in the number of applications from 1990 to 2000.

b. How much greater was the average annual rate of change in the number of applications from 1995 to 2000 than from 1990 to 1995?

Year	*Price per Gallon*
1990	$1.16
1991	$1.14
1992	$1.13
1993	$1.11
1994	$1.11
1995	$1.15
1996	$1.23
1997	$1.23
1998	$1.06
1999	$1.17
2000	$1.51
2001	$1.46
2002	$1.36
2003	$1.59
2004	$1.88

Year	1990	1995	2000	2005
Applications (in thousands)	120	175	375	324

3.1 Applying Concepts

50. Graph the ordered pairs $\left(x, \frac{1}{x}\right)$, where

$x \in \left\{-2, -1, -\frac{1}{2}, -\frac{1}{3}, \frac{1}{3}, \frac{1}{2}, 1, 2\right\}$.

51. Graph the ordered pairs (x, x^2), where $x \in \{-2, -1, 0, 1, 2\}$.

52. Draw a line passing through every point whose ordinate is the additive inverse of its abscissa.

53. Draw a line passing through every point whose abscissa equals its ordinate.

54. ✎ Describe the graph of all the ordered pairs (x, y) that are 5 units from the origin.

55. ✎ Consider two distinct fixed points in the plane. Describe the graph of all the points (x, y) that are equidistant from these fixed points.

OBJECTIVE **1** **Evaluate a function**

Describing relationships between two quantities is an important aspect of mathematics. One way to describe a relationship between two quantities is in a graph. The graph in Figure 3.1, based on data from IDC, Inc., shows the projected number of users, worldwide, of WiFi subscribers.

Figure 3.1

The data in the graph can be written as ordered pairs: (2002, 1), (2003, 3), (2004, 5), (2005, 10), (2006, 17), and (2007, 25). The ordered pair (2004, 5) means that there were 5 million WiFi subscribers, worldwide, in 2004.

A table is another way of describing a relationship between two quantities. The table at the right shows a grading scale for an exam. This grading scale also can be described with ordered pairs. Some of the possible ordered pairs are (95, A), (72, C), (84, B), (92, A), (61, D), and (88, B).

Exam Score	Grade
90–100	A
80–89	B
70–79	C
60–69	D
0–59	F

A third way to describe a relationship between two quantities is an equation. In the early days of jet travel, the U.S. Air Force wanted to examine the physical effects of high altitudes on pilots in the event that the pilot had to eject from a plane. One of the experiments, on August 16, 1960, involved Joseph Kittinger. He ascended in Excelsior III, a helium-filled balloon, to a height of 102,800 ft above Earth. When he reached this altitude, he jumped from the gondola and began free-falling. The relationship between the distance, in feet, he fell in t seconds after leaving the gondola can be approximated by the equation $s = 16t^2$.

The numbers in the table were generated from the equation $s = 16t^2$. These numbers can be written as ordered pairs, where the first coordinate of the ordered pair is the time Kittinger was falling and the second coordinate is the distance fallen during that time. The ordered pairs are (5, 400), (10, 1600), (15, 3600), (20, 6400), and (25, 10,000). These are only some of the possible ordered pairs. Because there is a known equation relating the two quantities, we can find the distance for a given value of t. Some other possible ordered pairs are (11, 1936) and (16.4, 4303.36).

$16t^2 = s$	
5	400
10	1600
15	3600
20	6400
25	10,000

The three relationships given above have something in common: *ordered pairs*. The relationships can be described as the ordered pairs

(year, number of WiFi subscribers)
(exam score, letter grade)
(time, distance)

In mathematics, a set of ordered pairs is called a *relation*.

Point of Interest

About 28 s after Kittinger left the gondola, he was at an altitude of 90,000 ft and was traveling about 614 mph. This remains the fastest speed a human has ever attained without the aid of a propulsion system. As he continued to descend, air resistance slowed his speed. At 50,000 ft, his speed was about 250 mph. During the first 30 s, the equation $s = 16t^2$ was a good approximation of the distance he fell because air resistance was negligible. At lower altitudes, a different equation that considers air resistance would be more appropriate.

Definition of Relation

A **relation** is any set of ordered pairs.

The **domain** of a relation is the set of first coordinates of the ordered pairs of a relation. The **range** of a relation is the set of second coordinates of the ordered pairs of the relation.

For the year and number of WiFi subscribers relation,

$$\text{Domain} = \{2002, 2003, 2004, 2005, 2006, 2007\}$$
$$\text{Range} = \{1, 3, 5, 10, 17, 25\}$$

For the relation between an exam score and letter grade,

$$\text{Domain} = \{0, 1, 2, 3, 4, \ldots, 97, 98, 99, 100\}$$
$$\text{Range} = \{A, B, C, D, F\}$$

The domain and range for the falling pilot are a little more difficult because there are an infinite number of choices for time, t, the first coordinate of the ordered pair. As mentioned in the first Take Note of this section, the equation $s = 16t^2$ approximates the distance the pilot falls during the first 30 s. We can describe the domain (all of the times between 0 and 30) using set-builder notation or interval notation.

$$\text{Domain} = \{t | 0 \leq t \leq 30\} \quad \text{or} \quad \text{Domain} = [0, 30]$$

For each one of the choices of t, there is a corresponding distance s, the second coordinate of the ordered pair. Therefore, the range is an infinite set. The range is from 0 feet (just starting the descent) to the distance he fell in 30 s. Substituting 30 into $s = 16t^2$ gives $s = 14,400$. Thus

$$\text{Range} = \{s | 0 \leq s \leq 14,400\} \quad \text{or} \quad \text{Range} = [0, 14,400]$$

Although relations are important in mathematics, the concept of *function* is especially useful in applications.

Point of Interest

Functions are used in many applications. For instance, there are functions to calculate π to millions of digits, to predict your chances of winning the lottery, to determine the age of a fossil, and to find a monthly car payment.

Definition of a Function

A **function** is a relation in which no two ordered pairs have the same first coordinate.

To see the difference between a relation and a function, consider the situation of a meteorologist who is taking temperature and percent humidity readings at a weather station at the same time each day. Here are data from nine consecutive days.

Temperature (°F)	72	75	67	71	70	73	68	75	68
Percent humidity	47	50	39	41	36	50	37	44	48

The set of ordered pairs is

$$\{(72, 47), (75, 50), (67, 39), (71, 41), (70, 36),$$
$$(73, 50), (68, 37), (75, 44), (68, 48)\}.$$

The temperature and humidity relation is <u>not</u> a function because the ordered pairs (75, 50) and (75, 44) have the same first coordinate. The ordered pairs (68, 37) and (68, 48) also have the same first coordinate.

For the exam score and letter grade relation discussed earlier, if there were two ordered pairs with the same first coordinate and different second coordinates, it would mean that two students with the same exam score (first coordinate) would receive different letter grades (second coordinate)—for example, (83, B) and (83, F). But this does not happen with a grading scale. Thus there are no two ordered pairs with the same first coordinate. The exam score and grade relation is a function.

<table>
<tr><td>**EXAMPLE 1**</td><td colspan="2">What are the domain and range of the following relation? Is the relation a function?</td></tr>
<tr><td></td><td colspan="2">$\{(2, 4), (3, 6), (4, 8), (5, 10), (4, 6), (6, 10)\}$</td></tr>
<tr><td>**Solution**</td><td>Domain = $[3, 4, 5, 6]$</td><td>• **The domain of the relation is the set of first coordinates of the ordered pairs.**</td></tr>
<tr><td></td><td>Range = $[4, 6, 8, 10]$</td><td>• **The range of the relation is the set of second coordinates of the ordered pairs.**</td></tr>
</table>

Because there are two ordered pairs with the same first coordinate and different second coordinates, (4, 8) and (4, 6), the relation is not a function.

PROBLEM 1 What are the domain and range of the following relation? Is the relation a function?

$$\{(-2, 6), (-1, 3), (0, 0), (1, -3), (2, -6), (3, -9)\}$$

Solution See page S11.

Solution See page S11.

> **Take Note**
>
> Recall that when listing elements of a set, we list duplicate elements only once.

For each element of the domain of a relation, there is a corresponding element in the range of the relation. A possible diagram of the relation in Example 1 is

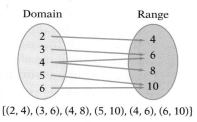

$$\{(2, 4), (3, 6), (4, 8), (5, 10), (4, 6), (6, 10)\}$$

Note from the diagram that there are two arrows from 4 in the domain to different elements, 6 and 8, in the range. This means that there are two ordered pairs, (4, 6) and (4, 6), with the same first coordinate and different second coordinates, and the relation is not a function.

What are the ordered pairs of the relation in the diagram below? Is the relation a function?

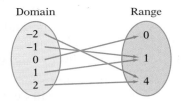

The relation is $\{(-2, 4), (-1, 1), (0, 0), (1, 1), (2, 4)\}$. There are no ordered pairs with the same first coordinate. The relation is a function.

Although a function can be described in terms of ordered pairs, in a table, or by a graph, a major focus of this text will be functions represented by equations in two variables.

For instance, the equation $s = 16t^2$ given earlier describes the relationship between the time, t, Joseph Kittinger fell and the distance, s, he fell. The ordered pairs can be written as (t, s), where $s = 16t^2$. By substituting $16t^2$ for s, we can also write the ordered pairs as $(t, 16t^2)$. Because the distance he falls *depends* on how long he has been falling, s is called the **dependent variable** and t is called the **independent variable.** For the equation $s = 16t^2$, we say that "distance is a function of time."

A function can be thought of as a rule that pairs one number with another number. For instance, the **square function** pairs a real number with its square. This function can be represented by the equation $y = x^2$. This equation states that, given any element x in the domain, the value of y in the range is the square of x. In this case, the independent variable is x and the dependent variable is y. When $x = 6$, $y = 36$ and one ordered pair of the function is $(6, 36)$. When $x = -7$, $y = 49$. The ordered pair $(-7, 49)$ belongs to the function.

Other ordered pairs of the function are $(-2, 4)$, $\left(\dfrac{3}{4}, \dfrac{9}{16}\right)$, and $(-1.1, 1.21)$.

In many cases, you may think of a function as a machine that turns one number into another number. For instance, you can think of the square function machine at the right taking an **input** (an element from the domain) and creating an **output** (an element of the range) that is the square of the input.

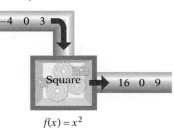

$f(x) = x^2$

Not all equations in two variables define a function. For instance,

$$y^2 = x^2 + 9$$

is not an equation that defines a function. As shown on the next page, the ordered pairs $(4, 5)$ and $(4, -5)$ are solutions of the equation.

> **Take Note**
>
> Sometimes the independent variable is called the input of a function and the dependent variable is called the output.

$y^2 = x^2 + 9$
5^2 $\quad$ $4^2 + 9$
25 $\quad$ $16 + 9$
$25 = 25$

• Let $(x, y) = (4, 5)$.
 Replace x by 4 and y by 5.
• **(4, 5) checks.**

$y^2 = x^2 + 9$
$(-5)^2$ $\quad$ $4^2 + 9$
25 $\quad$ $16 + 9$
$25 = 25$

• Let $(x, y) = (4, -5)$.
 Replace x by 4 and y by 5.
• **(4, −5) checks.**

Consequently, there are two ordered pairs, $(4, 5)$ and $(4, -5)$, with the same first coordinate and *different* second coordinates; the equation does not define a function. The phrase "y is a function of x," or a similar phrase with different variables, is used to describe those equations in two variables that define functions.

Function notation is frequently used for those equations that define functions. Just as x is commonly used as a variable for a number, the letter f is commonly used to name a function. For instance, using the square function, we can write $f(x) = x^2$. The symbol $f(x)$ is read *"the value of f at x"* or *"f of x."* The symbol $f(x)$ is the **value of the function** and represents the value of the dependent variable for a given value of the independent variable.

> This is the value of the function.
> It is the value of the dependent variable.

$$\overbrace{f(x)} = x^2$$

The name of the function is f. $\quad$ This is an algebraic expression that defines the relationship between the dependent variable and the independent variable.

The process of finding the value of $f(x)$ for a given value of x is called **evaluating the function.** For instance, to evaluate $f(x) = x^2$ when x is 4, replace x by 4 and simplify.

$f(x) = x^2$

$f(4) = 4^2$ $\qquad$ • **Replace x by 4. Then simplify.**

$\quad\; = 16$

The *value* of the function is 16 when $x = 4$. An ordered pair of the function is $(4, 16)$.

The letters used to represent a function are somewhat arbitrary. All of the following equations represent the same function. Note that more than one letter can be used.

$$f(x) = x^2 \qquad g(t) = t^2 \qquad P(v) = v^2 \qquad SQ(x) = x^2$$

These represent the square function.

We write $y = f(x)$ to emphasize the relationship between the independent variable, x, and the dependent variable, y. **Remember:** y and $f(x)$ are different symbols for the same number. Also, the <u>name</u> of the function is f; the <u>value</u> of the function at x is $f(x)$. For instance, the equation $s = 16t^2$ discussed at the beginning of this section could be written as $s(t) = 16t^2$. The name of the function is s. The value of the function when $t = 5$ is found by replacing t by 5.

$s(t) = 16t^2$

$s(5) = 16(5)^2$ $\qquad$ • **Replace t by 5.**

$\quad\; = 16(25) = 400$

The value of the function when $t = 5$ s is 400 ft.

Take Note

The notation $f(x)$ does not mean f times x. The letter f is the *name* of the function, and $f(x)$ is the *value* of the function at x.

Take Note

We also could write $s = 16t^2$ in function notation as, for example, $F(t) = 16t^2$, $g(t) = 16t^2$, or $r(t) = 16t^2$. For these equations, the names of the function would be F, g, and r, respectively.

Find the value of $R(v) = v^3 + 3v^2 - 5v - 6$ when $v = -2$.

$R(v) = v^3 + 3v^2 - 5v - 6$

$R(-2) = (-2)^3 + 3(-2)^2 - 5(-2) - 6$ • **Replace v by −2.**

$\qquad = -8 + 12 + 10 - 6$

$\qquad = 8$

The value of $R(v)$ when $v = -2$ is 8.

There are several ways to use a calculator to evaluate a function. The screens at the right, which illustrate the evaluation of $R(-2)$ given above, show one way. See the Appendix Keystroke Guide for other methods of evaluating a function.

```
Plot1  Plot2  Plot3
\Y1 ◼ X^3+3X²−5X−6
\Y2 =    Y1(-2)
\Y3 =              8
\Y4 =
\Y5 =
\Y6 =
\Y7 =
```

EXAMPLE 2 Let $q(r) = 2r^3 + 5r^2 - 6$.

A. Find $q(-3)$.

B. Find the value of q when $r = 2$.

Solution

A. $q(r) = 2r^3 + 5r^2 - 6$

$q(-3) = 2(-3)^3 + 5(-3)^2 - 6$ • **Replace r by −3.**
Then simplify.

$\qquad = 2(-27) + 5(9) - 6 = -54 + 45 - 6$

$q(-3) = -15$

B. To find the value of q when $r = 2$ means to evaluate the function when $r = 2$.

$q(r) = 2r^3 + 5r^2 - 6$

$q(2) = 2(2)^3 + 5(2)^2 - 6$ • **Replace r by 2.**
Then simplify.

$\qquad = 2(8) + 5(4) - 6 = 16 + 20 - 6$

$\qquad = 30$

The value of the function when $r = 4$ is 30.

PROBLEM 2 Let $f(z) = z + |z|$.

A. Find $f(-4)$.

B. Find the value of f when $z = 3$.

Solution See page S11.

It is also possible to evaluate a function at another variable or at a variable expression.

Evaluate $p(x) = 2x^2 + 7x - 9$ when $x = a$.

$$p(x) = 2x^2 + 7x - 9$$
$$p(a) = 2a^2 + 7a - 9 \qquad \bullet \text{ Replace } x \text{ by } a.$$

Evaluate $N(v) = 2v + 3$ when $v = 3 + k$.

$$N(v) = 2v + 3$$
$$N(3 + k) = 2(3 + k) + 3 \qquad \bullet \text{ Replace } v \text{ by } 3 + k.$$
$$= 6 + 2k + 3 \qquad \bullet \text{ Simplify.}$$
$$= 2k + 9$$

EXAMPLE 3 Let $f(x) = 3 - 2x$. Find $f(3h - 2)$.

Solution

$$f(x) = 3 - 2x$$
$$f(3h - 2) = 3 - 2(3h - 2) \qquad \bullet \text{ Replace } x \text{ by } 3h - 2.$$
$$= 3 - 6h + 4 \qquad \bullet \text{ Simplify.}$$
$$= -6h + 7$$

PROBLEM 3 Let $r(s) = 3s - 6$. Find $r(2a + 3)$.

Solution See page S11.

The range of a function contains all the elements that result from evaluating the function for each element of the domain. If the domain contains an infinite number of elements, as in the case of the falling pilot, it may be difficult to find the range. However, if the domain has only a finite number of elements, then the range can be found by evaluating the function for each element of the given domain.

EXAMPLE 4 Find the range of $f(x) = x^2 + 2x - 1$ if the domain is $\{-3, -2, -1, 0, 1\}$.

Solution Evaluate the function for each element of the domain. The range includes the values of $f(-3), f(-2), f(-1), f(0)$, and $f(1)$.

$$f(x) = x^2 + 2x - 1$$
$$f(-3) = (-3)^2 + 2(-3) - 1 = 2$$
$$f(-2) = (-2)^2 + 2(-2) - 1 = -1$$
$$f(-1) = (-1)^2 + 2(-1) - 1 = -2$$
$$f(0) = (0)^2 + 2(0) - 1 = -1$$
$$f(1) = (1)^2 + 2(1) - 1 = 2$$

The range is $\{-2, -1, 2\}$.

Take Note

The order of the elements in a set does not matter. We could have listed the range as $\{-1, 2, -2\}$, but it is customary to list the numbers from least to greatest.

PROBLEM 4 Find the range of $f(x) = x^3 + x$ if the domain is $\{-3, -2, -1, 0, 1, 2\}$.

Solution See page S11.

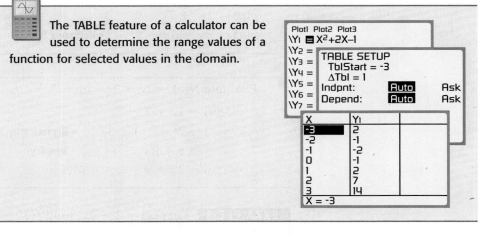

The TABLE feature of a calculator can be used to determine the range values of a function for selected values in the domain.

Given an element in the range of a function, we can find a corresponding element of the domain.

EXAMPLE 5 Given $f(x) = 3x + 1$, find a number c in the domain of f such that $f(c) = 7$. Write the corresponding ordered pair of the function.

Solution
$$f(x) = 3x + 1$$
$$f(c) = 3c + 1 \qquad \text{• Replace } x \text{ by } c.$$
$$7 = 3c + 1 \qquad \text{• } f(c) = 7$$
$$6 = 3c \qquad \text{• Solve for } c.$$
$$2 = c$$

The value of c is 2. The corresponding ordered pair is $(2, 7)$.

PROBLEM 5 Given $f(x) = 2x - 5$, find a number c in the domain of f such that $f(c) = -3$. Write the corresponding ordered pair of the function.

Solution See page S11.

OBJECTIVE **2** **Graph of a function**

The **graph of a function** is the graph of the ordered pairs that belong to the function. Consider $f(x) = 2x - 1$. Evaluating this function when $x = -2$, $-1, 0, 1, 2$, and 3 produces some of the ordered pairs of the function. It is convenient to record the results in a table similar to the one shown above.

x	$y = f(x) = 2x - 1$	(x, y)
-2	$y = f(-2) = 2(-2) - 1 = -5$	$(-2, -5)$
-1	$y = f(-1) = 2(-1) - 1 = -3$	$(-1, -3)$
0	$y = f(0) = 2(0) - 1 = -1$	$(0, -1)$
1	$y = f(1) = 2(1) - 1 = 1$	$(1, 1)$
2	$y = f(2) = 2(2) - 1 = 3$	$(2, 3)$
3	$y = f(2) = 2(3) - 1 = 5$	$(2, 5)$

The graph of the ordered pairs is shown in Figure 1.

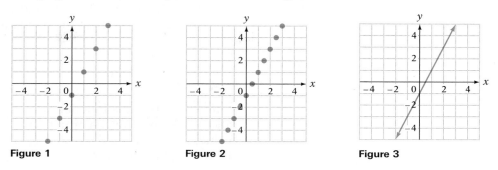

Figure 1 **Figure 2** **Figure 3**

Finding more ordered pairs of the function, such as when $x = -1.5$, -0.5, 0.5, 1.5, and 2.5, results in even more points to graph. See Figure 2. If we continued graphing more and more points, there would be so many dots that the graph would look like the straight line in Figure 3, which is the graph of $f(x) = 2x - 1$.

The graph of $f(x) = 2x - 1$ is shown again at the right. Evaluating f when $x = -1.5$ and $x = 3$, we have

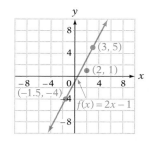

$$f(x) = 2x - 1 \qquad\qquad f(x) = 2x - 1$$
$$f(-1.5) = 2(-1.5) - 1 \qquad f(3) = 2(3) - 1$$
$$f(-1.5) = -4 \qquad\qquad f(3) = 5$$

Thus $(-1.5, -4)$ and $(3, 5)$ are the coordinates of points on the graph. Note also that the point with coordinates $(2, 1)$ is not on the graph and that $f(2) \neq 1$.

Every point on the graph of a function is an ordered pair of the function, and every ordered pair that belongs to the function corresponds to a point on the graph.

EXAMPLE 6 Graph $h(x) = x^2 - 3$ by evaluating the function when $x = -3$, -2, -1, 0, 1, 2, and 3. Plot the resulting ordered pairs. Then connect the points to form the graph.

Solution Evaluate the function for the given values of x. This will produce some ordered pairs of the function. The results can be recorded in a table.

x	$y = h(x) = x^2 - 3$	(x, y)
-3	$h(-3) = (-3)^2 - 3 = 6$	$(-3, 6)$
-2	$h(-2) = (-2)^2 - 3 = 1$	$(-2, 1)$
-1	$h(-1) = (-1)^2 - 3 = -2$	$(-1, -2)$
0	$h(0) = (0)^2 - 3 = -3$	$(0, -3)$
1	$h(1) = (1)^2 - 3 = -2$	$(1, -2)$
2	$h(2) = (2)^2 - 3 = 1$	$(2, 1)$
3	$h(3) = (3)^2 - 3 = 6$	$(3, 6)$

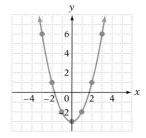

Connect the points to form the graph.

PROBLEM 6 Graph $f(x) = -x^2 - 4x + 2$ by evaluating the function when $x = -5, -4, -3, -2, -1, 0,$ and 1. Plot the resulting ordered pairs. Then connect the points to form the graph.

Solution See page S12.

The graph of an absolute value function will be used to illustrate various concepts throughout this text. We can graph $f(x) = |x|$ by evaluating the function for various values of x, plotting the resulting ordered pairs, and then connecting the points to produce the graph.

| x | $y = f(x) = |x|$ | (x, y) |
|---|---|---|
| -3 | $f(-3) = |-3| = 3$ | $(-3, 3)$ |
| -2 | $f(-2) = |-2| = 2$ | $(-2, 2)$ |
| -1 | $f(-1) = |-1| = 1$ | $(-1, 1)$ |
| 0 | $f(0) = |0| = 0$ | $(0, 0)$ |
| 1 | $f(1) = |1| = 1$ | $(1, 1)$ |
| 2 | $f(2) = |2| = 2$ | $(2, 2)$ |
| 3 | $f(3) = |3| = 3$ | $(3, 3)$ |

The V-shape of this graph is characteristic of the graph of an absolute value function.

EXAMPLE 7 Graph $f(x) = -|x| + 2$ by evaluating the function when $x = -3, -2, -1, 0, 1, 2,$ and 3. Plot the resulting ordered pairs. Then connect the points to form the graph.

Solution Evaluate the function for the given values of x. Then connect the points to form a V-shape graph.

| x | $y = f(x) = -|x| + 2$ | (x, y) |
|---|---|---|
| -3 | $f(-3) = -|-3| + 2 = -1$ | $(-3, -1)$ |
| -2 | $f(-2) = -|-2| + 2 = 0$ | $(-2, 0)$ |
| -1 | $f(-1) = -|-1| + 2 = 1$ | $(-1, 1)$ |
| 0 | $f(0) = -|0| + 2 = 2$ | $(0, 2)$ |
| 1 | $f(1) = -|1| + 2 = 1$ | $(1, 1)$ |
| 2 | $f(2) = -|2| + 2 = 0$ | $(2, 0)$ |
| 3 | $f(3) = -|3| + 2 = -1$ | $(3, -1)$ |

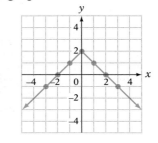

PROBLEM 7 Graph $f(x) = |x - 2|$ by evaluating the function when $x = -1, 0, 1, 2, 3, 4,$ and 5. Plot the resulting ordered pairs. Then connect the points to form the graph.

Solution See page S12.

OBJECTIVE **3** ## Vertical line test

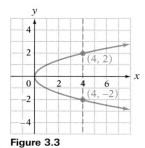

Figure 3.3

A graph is a visual description of a relation. For the relation shown in Figure 3.3, the two ordered pairs (4, 2) and (4, −2) belong to the graph, and those points lie on a vertical line. Because there are two ordered pairs with the same first coordinate, the set of ordered pairs of the graph is not a function. With this observation in mind, we can make use of a quick method to determine whether a graph is the graph of a function.

Vertical Line Test for the Graph of a Function

If every vertical line intersects a graph at most once, then the graph is the graph of a function.

This graphical interpretation of a function is often described by saying that each value in the domain of the function is paired with *exactly one* value in the range of the function.

Take Note

For the second graph on the right, note that there are values of x for which there is only one value of y. For instance, when $x = -5$, $y = 4$. For the relation to be a function, however, <u>every</u> value of x in the domain of the function must have exactly one value of y. If there is even one value of x that has two or more values of y, the condition for a function is not met. This is what the vertical line test states: If there is any place on the graph where a vertical line intersects the graph more than once, the graph is not the graph of a function.

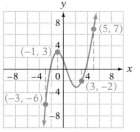

For each x, there is *exactly one* value of y. For instance, when $x = -3$, $y = -6$. This is the graph of a function.

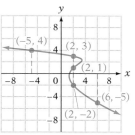

Some values of x can be paired with more than one value of y. For instance, 2 can be paired with −2, 1, and 3. This is not the graph of a function.

EXAMPLE 8 Use the vertical line test to determine whether the graph is the graph of a function.

A.

B.

Solution

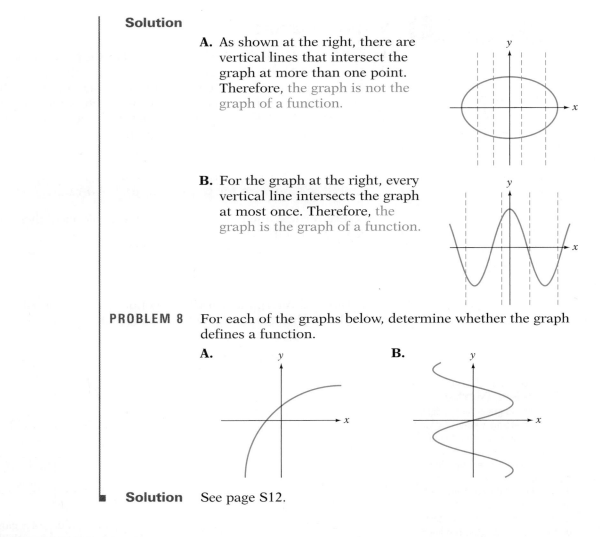

A. As shown at the right, there are vertical lines that intersect the graph at more than one point. Therefore, the graph is not the graph of a function.

B. For the graph at the right, every vertical line intersects the graph at most once. Therefore, the graph is the graph of a function.

PROBLEM 8 For each of the graphs below, determine whether the graph defines a function.

A.

B.

■ **Solution** See page S12.

3.2 Concept Review

Determine whether the following statements are always true, sometimes true, or never true.

1. A set of ordered pairs is called a function.

2. A function is a relation.

3. The domain of a function is the set of real numbers.

4. Zero is excluded from the domain of a function.

5. $f(a)$ represents a value in the range of a function when a is a value in the domain of the function.

6. Two ordered pairs of a function can have the same first coordinate.

3.2 Exercises

1 Evaluate a function

1. Is every function a relation? Is every relation a function? Explain your answer.

2. What is the domain of a function? What is the range of a function?

3. Complete the following sentence using *exactly one* or *more than one*. For a function, each element of the domain is paired with ___?___ element of the range.

4. Complete the following sentences using *can* or *cannot*. If (5, 2) belongs to a function, then (5, 9) ___?___ belong to the function. If (5, 2) belongs to a relation, then (5, 9) ___?___ belong to the relation.

Determine whether the relation is a function. Give the domain and range of the relation.

5. $\{(-3, 1), (-2, 2), (1, 5), (4, -7)\}$

6. $\{(-5, 4), (-2, 3), (0, 1), (3, 2) (7, 11)\}$

7. $\{(1, 5), (2, 5), (3, 5), (4, 5), (5, 5)\}$

8. $\{(1, 0), (10, 1), (100, 2), (1000, 3), (10000, 4)\}$

9. $\left\{\left(-2, -\frac{1}{2}\right), (-1, -1), (1, 1), \left(2, \frac{1}{2}\right), \left(3, \frac{1}{3}\right)\right\}$

10. $\{(0, 0), (1, 1), (4, 2), (9, 3), (1, -1), (4, -2), (9, -3)\}$

11. $\{(2, 3), (4, 5), (6, 7), (8, 9), (6, 8)\}$

12. $\{(-1, 2), (-2, 4), (2, 4), (-2, 5), (4, 9)\}$

13. Does the diagram below represent a function? Explain your answer.

14. 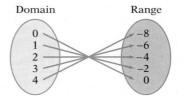 Does the diagram below represent a function? Explain your answer.

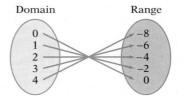

15. Does the diagram below represent a function? Explain your answer.

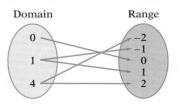

Domain Range

16. Does the diagram below represent a function? Explain your answer.

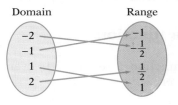

Domain Range

17. Does the diagram below represent a function? Explain your answer.

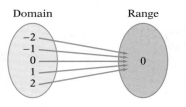

Domain Range

18. Does the diagram below represent a function? Explain your answer.

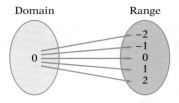

Domain Range

Solve.

19. Shipping The table at the right shows the cost to send an overnight package using United Parcel Service.
 a. Does the table define a function?
 b. Given $x = 3.54$ lb, find y.

Weight, in pounds (x)	Cost (y)
$0 < x \le 1$	$22.75
$1 < x \le 2$	$24.74
$2 < x \le 3$	$27.00
$3 < x \le 4$	$29.25
$4 < x \le 5$	$31.50

20. Shipping The table at the right shows the cost to send an "Express Mail" package using U.S. Postal Service.
 a. Does the table define a function?
 b. Given $x = 0.5$ lb, find y.

Weight, in pounds (x)	Cost (y)
$0 < x \le 0.5$	$13.65
$0.5 < x \le 2$	$17.85
$2 < x \le 3$	$21.05
$3 < x \le 4$	$24.20
$4 < x \le 5$	$27.30

21. What does it mean to evaluate a function? Explain how to evaluate $f(x) = 3x$ when $x = 2$.

22. What is the value of a function?

23. True or False: If f is a function, then it is possible that $f(0) = -2$ and $f(3) = -2$.

24. True or False: If f is a function, then it is possible that $f(4) = 3$ and $f(4) = 2$.

25. Given $f(x) = 5x - 7$, find $f(3)$ by completing the following.

$$f(x) = 5x - 7$$
$$f(\underline{\quad ? \quad}) = 5(\underline{\quad ? \quad}) - 7 \quad \bullet \text{ Insert 3 in the parentheses.}$$
$$f(3) = \underline{\quad ? \quad} \quad \bullet \text{ Simplify.}$$

26. Given $f(x) = x^2 - 3x + 1$, find $f(-2)$ by completing the following.

$$f(x) = x^2 - 3x + 1$$
$$f(\underline{\quad?\quad}) = (\underline{\quad?\quad})^2 - 3(\underline{\quad?\quad}) + 1 \quad \bullet \text{ Insert } -2 \text{ in the parentheses.}$$
$$f(-2) = \underline{\quad?\quad} \qquad\qquad\qquad \bullet \text{ Simplify.}$$

In Exercises 27 to 50, evaluate the function.

27. $f(x) = 4x + 5$
 a. $f(2)$ **b.** $f(-2)$ **c.** $f(0)$

28. $g(x) = -2x - 7$
 a. $g(-3)$ **b.** $g(1)$ **c.** $g(0)$

29. $v(s) = 6 - 3s$
 a. $v(3)$ **b.** $v(-2)$ **c.** $v\left(-\dfrac{2}{3}\right)$

30. $p(z) = 4 - 6z$
 a. $p\left(\dfrac{2}{3}\right)$ **b.** $p(4)$ **c.** $p\left(-\dfrac{5}{6}\right)$

31. $f(x) = -\dfrac{3}{2}x - 2$
 a. $f(4)$ **b.** $f(-2)$ **c.** $f\left(\dfrac{4}{3}\right)$

32. $f(x) = \dfrac{3}{4}x + 1$
 a. $f(8)$ **b.** $f(-4)$ **c.** $f\left(-\dfrac{8}{3}\right)$

33. $p(c) = \dfrac{1}{2}c - \dfrac{3}{4}$
 a. $p\left(\dfrac{7}{2}\right)$ **b.** $p\left(-\dfrac{1}{2}\right)$ **c.** $p\left(\dfrac{1}{2}\right)$

34. $H(x) = \dfrac{2}{3} - \dfrac{5}{6}x$
 a. $H\left(-\dfrac{2}{5}\right)$ **b.** $H(2)$ **c.** $H(0)$

35. $f(x) = 4x - 1$
 a. $f(a + 3)$ **b.** $f(2a)$

36. $f(x) = 3x - 2$
 a. $f(2 - h)$ **b.** $f(-4h)$

37. $f(x) = 2x^2 - 1$
 a. $f(3)$ **b.** $f(-2)$ **c.** $f(0)$

38. $f(x) = 2x^2 + 3$
 a. $f(0)$ **b.** $f(-3)$ **c.** $f(2)$

39. $h(t) = 3t^2 - 4t - 5$
 a. $h(-2)$ **b.** $h(-1)$ **c.** $h(w)$

40. $p(x) = 3 - 4t - 2t^2$
 a. $p(4)$ **b.** $p(-4)$ **c.** $f(0)$

41. $g(x) = x^2 + 2x - 1$
 a. $g(1)$ **b.** $g(-3)$ **c.** $g(a)$

42. $g(x) = 2x^2 - 4x + 1$
 a. $g(-2)$ **b.** $g(4)$ **c.** $g(z)$

43. $p(t) = 4t^2 - 8t + 3$
 a. $p(-2)$ **b.** $p\left(\dfrac{1}{2}\right)$ **c.** $p(-a)$

44. $r(s) = 3 - 6x - 3s^2$
 a. $r(-2)$ **b.** $r\left(-\dfrac{2}{3}\right)$ **c.** $r(-x)$

45. $f(x) = |x - 3|$
 a. $f(-1)$ **b.** $f(5)$ **c.** $f(3)$

46. $h(x) = |2x + 4|$
 a. $h(-2)$ **b.** $h(3)$ **c.** $h(-3)$

47. $C(r) = 3|x| - 2$
 a. $C(-3)$ **b.** $C(4)$ **c.** $C(0)$

48. $y(x) = 3 - |2x|$
 a. $y\left(\dfrac{1}{2}\right)$ **b.** $y\left(-\dfrac{3}{2}\right)$ **c.** $y\left(\dfrac{1}{4}\right)$

49. $K(p) = 5 - 3|x + 2|$
 a. $K(-2)$ **b.** $K(-7)$ **c.** $K(3)$

50. $R(s) = 2|1 - x| - 3$
 a. $R(-7)$ **b.** $R(4)$ **c.** $R(-1)$

51. Find the value of $s(t) = -16t^2 + 48t$ when $t = 3$.

52. Find the value of $T(s) = s^2 - 4s + 1$ when $s = \dfrac{3}{4}$.

53. Find the value of $P(x) = 3x^3 - 4x^2 + 6x - 7$ when $x = 2$.

54. Find the value of $R(s) = s^3 - 2s^2 - 5s + 2$ when $s = -3$.

55. Find the value of $R(p) = \dfrac{3p}{2p - 3}$ when $p = -3$.

56. Find the value of $f(x) = \dfrac{x + 1}{3x - 1}$ when $x = 3$.

Find the range of the function defined by each equation.

57. $f(x) = 3x - 5$; domain $= \{-3, -2, -1, 0, 1, 2\}$

58. $g(x) = 1 - 2x$; domain $= \{-4, -2, 0, 2, 4, 6\}$

59. $r(t) = \dfrac{t}{2}$; domain $= \{-3, -2, -1, 0, 1, 2\}$

60. $v(s) = \dfrac{3}{4}x - 1$; domain $= \{-8, -4, 0, 4, 8\}$

61. $f(x) = x^2 + 3$; domain $= \{-3, -2, -1, 0, 1, 2\}$

62. $r(t) = t^2 - t - 6$; domain $= \{-2, -1, 0, 1, 2, 3\}$

63. $c(n) = n^3 - n - 2$; domain $= \{-3, -2, -1, 0, 1, 2\}$

64. $q(z) = x^3 + 2x^2 - x - 2$; domain $= \{-3, -2, -1, 0, 1, 2\}$

65. Given $f(x) = 2x - 3$, find a number c in the domain of f such that $f(c) = 5$. Write the corresponding ordered pair of the function.

66. Given $f(x) = 3x + 1$, find a number c in the domain of f such that $f(c) = -8$. Write the corresponding ordered pair of the function.

67. Given $f(x) = 1 - 2x$, find a number c in the domain of f such that $f(c) = -7$. Write the corresponding ordered pair of the function.

68. Given $f(x) = -20 - 5x$, find a number c in the domain of f such that $f(c) = -10$. Write the corresponding ordered pair of the function.

69. Given $f(x) = \dfrac{2}{3}x - 2$, find a number c in the domain of f such that $f(c) = 0$. Write the corresponding ordered pair of the function.

70. Given $f(x) = 3x + 3$, find a number c in the domain of f such that $f(c) = 0$. Write the corresponding ordered pair of the function.

2 Graph of a function

71. Which, if any, of the plotted points, *A*, *B*, or *C*, belong to the function?

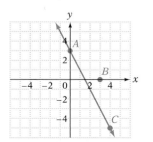

72. Which, if any, of the plotted points, *A*, *B*, or *C*, belong to the function?

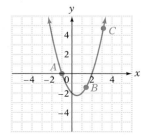

73. Does the ordered pair $(4, -1)$ belong to the function graphed below?

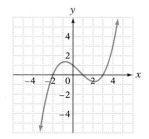

74. Does the ordered pair $(-3, 4)$ belong to the function graphed below?

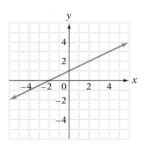

75. If $f(x) = 3x - 4$, does the ordered pair $(-1, -7)$ belong to the function?

76. If $f(x) = x^2 - 4x$, does the ordered pair $(2, 4)$ belong to the function?

Graph each function by evaluating the function at the given values of *x*. Plot the resulting ordered pairs. Then connect the points to form the graph.

77. $f(x) = x - 3$,
$x = -1, 0, 1, 2, 3, 4, 5$

78. $f(x) = -x + 2$,
$x = -2, -1, 0, 1, 2, 3, 4$

79. $g(x) = -2x + 2$,
$x = -2, -1, 0, 1, 2, 3$

80. $g(x) = 2x - 2$,
$x = -2, -1, 0, 1, 2, 3, 4$

81. $h(x) = \dfrac{1}{2}x - 1$,
$x = -6, -4, -2, 0, 2, 4, 6$

82. $h(x) = \dfrac{2}{3}x + 1$,
$x = -6, -3, 0, 3, 6$

83. $f(x) = x^2 - 4$,
 $x = -3, -2, -1, 0, 1, 2, 3$

84. $g(x) = -x^2 + 5$,
 $x = -3, -2, -1, 0, 1, 2, 3$

85. $h(x) = x^2 - 4x$,
 $x = -1, 0, 1, 2, 3, 4, 5$

86. $h(x) = x^2 + 2x$,
 $x = -4, -3, -2, -1, 0, 1, 2$

87. $g(x) = x^2 + 2x - 3$,
 $x = -4, -3, -2, -1, 0, 1, 2$

88. $f(x) = -x^2 + 4x + 2$,
 $x = -1, 0, 1, 2, 3, 4, 5$

89. $f(x) = |x| - 3$,
 $x = -6, -4, -2, 0, 2, 4, 6$

90. $f(x) = -|x| + 4$,
 $x = -6, -4, -2, 0, 2, 4, 6$

91. $h(x) = -2|x| + 5$,
 $x = -3, -2, -1, 0, 1, 2, 3$

92. $h(x) = |2x| - 5$,
 $x = -5, -3, -1, 0, 1, 3, 5$

93. $f(x) = |x - 2| - 3$,
 $x = -6, -4, 0, 2, 4, 5, 6$

94. $g(x) = -|x + 2| + 4$,
 $x = -6, -4, -2, 0, 2, 4, 6$

3 **Vertical line test**

Use the vertical line test to determine whether the graph is the graph of a function.

95.

96.

97.

98.

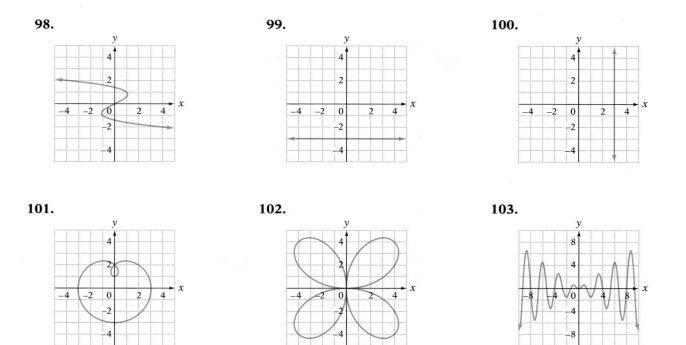

99.

100.

101.

102.

103.

104. Does the graph of the equation pass the vertical line test?

 a. $f(x) = \sqrt{x}$
 b. $g(x) = -\sqrt{x}$
 c. $h(x) = \sqrt{x} + \sqrt{x}$
 d. $F(x) = \sqrt{x} - 2\sqrt{x}$

105. Does the graph of the equation pass the vertical line test?

 a. $f(x) = |x|$
 b. $g(x) = -|x|$
 c. $h(x) = |x| + |x|$
 d. $F(x) = |x| - 2|x|$

3.2 Applying Concepts

106. Given $f(x) = |x|$, find c in the domain of f such that $f(c) = 3$.

107. Given $f(x) = |x - 2|$, find c in the domain of f such that $f(c) = 4$.

108. Given $f(x) = |x| + 4$, is it possible to find a number c such that $f(c) = 2$? Is 2 in the range of f?

109. Given $f(x) = 4x + 7$, write $f(-2 + h) - f(-2)$ in simplest form.

110. Given $P(x) = 3 - 2x$, write $P(3 + h) - P(3)$ in simplest form.

111. Let $F = \{(-3, -6), (-2, -4), (-1, -2), (0, 0), (1, 2), (2, 4), (3, 6)\}$ and $G = \{(-3, 9), (-2, 4), (-1, 1), (0, 0), (1, 1), (2, 4), (3, 9)\}$.
 a. Is F a function?
 b. Is G a function?

c. Write the set of ordered pairs F' that consists of all the ordered pairs of F but with the coordinates reversed. For example, the first two ordered pairs of F' are $(-6, -3)$ and $(-4, -2)$. Is F' a function?

d. Write the set of ordered pairs G' that consists of all the ordered pairs of G but with the coordinates reversed. Is G' a function?

e. Using the results of part **c** and part **d**, if the ordered pairs of a function are reversed, is the resulting set of ordered pairs always a function?

112. **Credit Card Debt** The graph below shows the amount of interest a person with credit card debt of $1000 will pay and the time it will take to repay the debt if that person makes only the minimum monthly payment (and makes no additional charges).

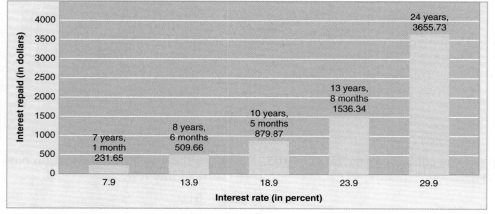

a. How much interest will this person repay if the credit card interest rate is 18.9% (a fairly typical rate)?

b. How many years and months will it take to repay the debt if the interest rate is 23.9%?

113. **Automotive Technology** The distance, s (in feet), that a car will skid on a certain road surface after the brakes are applied is a function of the car's velocity, v (in miles per hour). The function can be approximated by $s = f(v) = 0.017v^2$. How far will a car skid after its brakes are applied if it is traveling 60 mph?

114. **Energy** The power a windmill can generate is a function of the velocity of the wind. The function can be approximated by $P = f(v) = 0.015v^3$, where P is the power in watts and v is the velocity of the wind in meters per second. How much power will be produced by a windmill when the velocity of the wind is 15 m/s?

115. Find the set of ordered pairs (x, y) determined by the equation $y = x^3$, where $x \in \{-2, -1, 0, 1, 2\}$. Does the set of ordered pairs define a function? Why or why not?

116. Find the set of ordered pairs (x, y) determined by the equation $|y| = x$, where $x \in \{0, 1, 2, 3\}$. Does the set of ordered pairs define a function? Why or why not?

Each of the following graphs defines a function. Evaluate the function by estimating the ordinate (which is the value of the function) for the given value of t.

117. Physics The graph in Figure 1 shows the speed, v (in feet per second), at which a parachutist is falling during the first 20 s after jumping out of a plane. Estimate the speed at which the parachutist is falling when:
a. $t = 5$ s **b.** $t = 15$ s

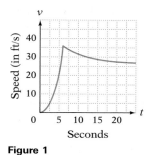

Figure 1

118. Psychology The graph in Figure 2 shows what an industrial psychologist has determined to be the average percent score, P, for an employee taking a performance test t weeks after training begins. Estimate the score an employee would receive on this test when:
a. $t = 4$ weeks **b.** $t = 10$ weeks

Figure 2

119. Physics The graph in Figure 3 shows the temperature, T (in degrees Fahrenheit), of a can of cola t hours after it is placed in a refrigerator. Use the graph to estimate the temperature of the cola when:
a. $t = 10$ h **b.** $t = 15$ h

Figure 3

120. Health Science The graph in Figure 4 shows the decrease in the heart rate, r (in beats per minute), of a runner t minutes after the completion of a race. Use the graph to estimate the heart rate of a runner when:
a. $t = 5$ min **b.** $t = 20$ min

121. Write a report on the longitude-latitude coordinate system used to locate a position on Earth. Include in your report the definition of *nautical mile* and the coordinates of the city in which your school is located.

Figure 4

122. Investigate the relationship between two variables by collecting data, graphing the scatter diagram for these data, and discussing the results. For example, for each person in your classes who commutes to class by car, find out how many miles he or she commutes and the total number of miles on the odometer of the car he or she drives to class. Discuss whether you think there is a relationship between the two variables.

 Linear Functions

OBJECTIVE 1

Graph a linear function

Recall that the ordered pairs of a function can be written as $(x, f(x))$ or (x, y). The **graph of a function** is a graph of the ordered pairs (x, y) that belong to the function. Certain functions have characteristic graphs. A function that can be written in the form $f(x) = mx + b$ (or $y = mx + b$) is called a **linear function** because its graph is a straight line.

Examples of linear functions are shown at the right. Note that the exponent on each variable is 1.

$f(x) = 2x + 5$	$(m = 2, b = 5)$
$P(t) = 3t - 2$	$(m = 3, b = -2)$
$y = -2x$	$(m = -2, b = 0)$
$y = -\dfrac{2}{3}x + 1$	$\left(m = -\dfrac{2}{3}, b = 1\right)$
$g(z) = z - 2$	$(m = 1, b = -2)$

The equation $y = x^2 + 4x + 3$ is not a linear function because it includes a term with a variable squared. The equation $f(x) = \dfrac{3}{x - 2}$ is not a linear function because a variable occurs in the denominator. Another example of an equation that is not a linear function is $y = \sqrt{x} + 4$; this equation contains a variable within a radical and so is not a linear function.

Consider $f(x) = 2x + 1$. Evaluating the linear function when $x = -3, -2, -1, 0, 1,$ and 2 produces some of the ordered pairs of the function. It is convenient to record the results in a table similar to the one at the right. The graph of the ordered pairs is shown in Figure 1 below.

x	$f(x) = 2x + 1$	y	(x, y)
-3	$2(-3) + 1$	-5	$(-3, -5)$
-2	$2(-2) + 1$	-3	$(-2, -3)$
-1	$2(-1) + 1$	-1	$(-1, -1)$
0	$2(0) + 1$	1	$(0, 1)$
1	$2(1) + 1$	3	$(1, 3)$
2	$2(2) + 1$	5	$(2, 5)$

Evaluating the function when x is not an integer produces more ordered pairs to graph, such as $\left(-\dfrac{5}{2}, -4\right)$ and $\left(\dfrac{3}{2}, 4\right)$, as shown in Figure 2 below. Evaluating the function for still other values of x would result in more and more ordered pairs to graph. The result would be so many dots that the graph would look like the straight line shown in Figure 3, which is the graph of $f(x) = 2x + 1$.

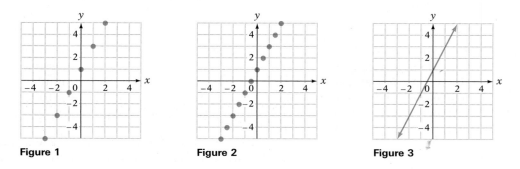

Figure 1 Figure 2 Figure 3

No matter what value of x is chosen, $2x + 1$ is a real number. This means the domain of $f(x) = 2x + 1$ is all real numbers. Therefore, we can use any real number when evaluating the function. Normally, however, values such as π and $\sqrt{5}$ are not used because it is difficult to graph the resulting ordered pairs.

Whether an equation is written as $f(x) = mx + b$ or $y = mx + b$, the equation represents a linear function, and the graph of the equation is a straight line.

Because the graph of a linear function is a straight line, and a straight line is determined by two points, the graph of a linear function can be drawn by finding only two of the ordered pairs of the functions. However, it is recommended that you find at least *three* ordered pairs to ensure accuracy.

Take Note

When the coefficient of x is a fraction, choose values of x that are multiples of the denominator of the fraction. This will result in coordinates that are integers.

EXAMPLE 1 Graph: $f(x) = -\dfrac{3}{2}x - 3$

Solution

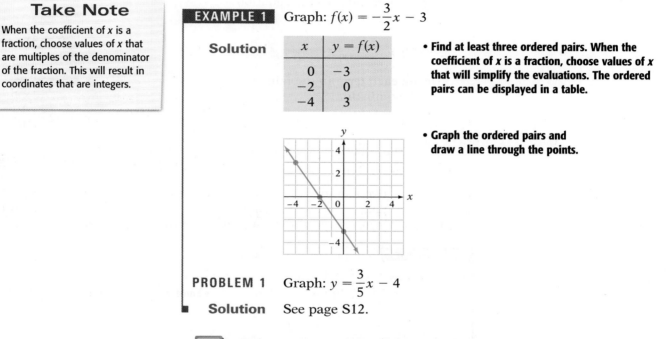

x	$y = f(x)$
0	-3
-2	0
-4	3

• **Find at least three ordered pairs. When the coefficient of x is a fraction, choose values of x that will simplify the evaluations. The ordered pairs can be displayed in a table.**

• **Graph the ordered pairs and draw a line through the points.**

PROBLEM 1 Graph: $y = \dfrac{3}{5}x - 4$

Solution See page S12.

Graphing utilities create graphs by plotting points and then connecting the points to form a curve. Using a graphing utility, enter the equation $y = -\dfrac{3}{2}x - 3$ and verify the graph drawn in Example 1. (Refer to the Appendix on page 823 for suggestions on what domain and range to use.) Trace along the graph and verify that $(0, -3)$, $(-2, 0)$, and $(-4, 3)$ are the coordinates of points on the graph. Now enter the equation $y = \dfrac{3}{5}x - 4$ given in Problem 1. Verify that the ordered pairs you found for this function are the coordinates of points on the graph.

OBJECTIVE 2 ## Graph an equation of the form $Ax + By = C$

A **literal equation** is an equation with more than one variable. Examples of literal equations are $P = 2L + 2W$, $V = LWH$, $d = rt$, and $3x + 2y = 6$.

Linear equations of the form $y = mx + b$ are literal equations. In some cases, a linear equation has the form $Ax + By = C$. In such a case, it may be

convenient to solve the equation for y to get the equation in the form $y = mx + b$. To solve for y, we use the same rules and procedures that we use to solve equations with numerical values.

Write $4x - 3y = 6$ in the form $y = mx + b$.

$$4x - 3y = 6$$

Subtract $4x$ from each side of the equation. $$4x - 4x - 3y = 6 - 4x$$

Simplify. $$-3y = 6 - 4x$$

Divide both sides of the equation by -3. $$\frac{-3y}{-3} = \frac{6 - 4x}{-3}$$

Simplify. $$y = \frac{6 - 4x}{-3}$$

Divide each term in the numerator by the denominator. $$y = \frac{6}{-3} - \frac{4x}{-3}$$

Simplify. $$y = -2 + \frac{4}{3}x$$

Write the equation in the form $y = mx + b$. $$y = \frac{4}{3}x - 2$$

We will show two methods of graphing an equation of the form $Ax + By = C$. In the first method, we solve the equation for y and then follow the same procedure used for graphing an equation of the form $y = mx + b$.

EXAMPLE 2 Graph: $3x + 2y = 6$

Solution $3x + 2y = 6$
$$2y = -3x + 6$$ • Solve the equation for *y*.
$$y = -\frac{3}{2}x + 3$$

x	y
0	3
2	0
4	−3

• Find at least three solutions.

• Graph the ordered pairs in a rectangular coordinate system. Draw a straight line through the points.

PROBLEM 2 Graph: $-3x + 2y = 4$

Solution See page S12.

 Use a graphing utility to graph the equation $3x + 2y = 6$. First solve the equation for y and then enter the equation $y = -\frac{3}{2}x + 3$. Next trace along the graph and verify that $(0, 3)$, $(2, 0)$, and $(4, -3)$ are the coordinates of points on the graph. Follow the same procedure for Problem 2.

An equation in which one of the variables is missing has a graph that is either a horizontal or a vertical line.

The equation $y = -2$ can be written

$$0 \cdot x + y = -2$$

Because $0 \cdot x = 0$ for any value of x, y is -2 for every value of x.

Some of the possible ordered-pair solutions of $y = -2$ are given in the following table. The graph is shown at the right.

x	y
-2	-2
0	-2
3	-2

The equation $y = -2$ represents a function. Some of the ordered pairs of this function are $(-2, -2)$, $(0, -2)$, and $(3, -2)$. In functional notation we would write $f(x) = -2$. This function is an example of a constant function. No matter what value of x is selected, $f(x) = -2$.

Definition of the Constant Function

A function given by $f(x) = b$, where b is a constant, is a **constant function.** The graph of a constant function is a horizontal line passing through $(0, b)$.

■ Graph: $y + 4 = 0$

Solve for y.

$$y + 4 = 0$$
$$y = -4$$

The graph of $y = -4$ is a horizontal line passing through $(0, -4)$.

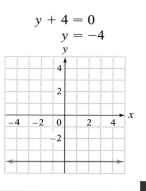

For each value in the domain of a constant function, the value of the function is the same (that is, it is constant). For instance, if $f(x) = 4$, then $f(2) = 4$, $f(3) = 4$, $f(\sqrt{3}) = 4$, $f(\pi) = 4$, and so on. The value of $f(x)$ is 4 for all values of x.

Evaluate $P(t) = -7$ when $t = 6$.

The value of the constant function is the same for all values of the variable.

$$P(t) = -7$$
$$P(6) = -7$$

For the equation $y = -2$, the coefficient of x is zero. For the equation $x = 2$, the coefficient of y is zero. For instance, the equation $x = 2$ can be written

$$x + 0 \cdot y = 2$$

No matter what value of y is chosen, $0 \cdot y = 0$ and therefore x is always 2.

Some of the possible ordered-pair solutions are given in the following table. The graph is shown at the right.

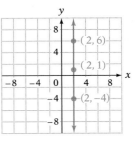

x	y
2	6
2	1
2	-4

Graph of $x = a$

The graph of $x = a$ is a vertical line passing through the point $(a, 0)$.

Recall that a function is a set of ordered pairs in which no two ordered pairs have the same first coordinate and different second coordinates. Because $(2, 6)$, $(2, 1)$, and $(2, -4)$ are ordered-pair solutions of the equation $x = 2$, this equation does not represent a function, and the graph is not the graph of a function.

EXAMPLE 3 Graph: $x = -4$

Solution

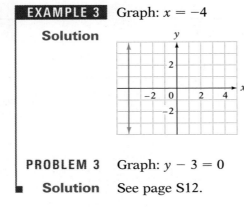

• The graph of an equation of the form $x = a$ is a vertical line passing through the point whose coordinates are $(a, 0)$.

PROBLEM 3 Graph: $y - 3 = 0$

Solution See page S12.

A second method of graphing straight lines uses the *intercepts* of the graph.

The graph of the equation $x - 2y = 4$ is shown at the right. The graph crosses the x-axis at the point $(4, 0)$. This point is called the **x-intercept.** The graph also crosses the y-axis at the point $(0, -2)$. This point is called the **y-intercept.**

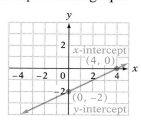

Find the x- and y-intercepts of the graph of the equation $3x + 4y = -12$.

To find the x-intercept, let $y = 0$.
(Any point on the x-axis has x-coordinate 0.)

$$3x + 4y = -12$$
$$3x + 4(0) = -12$$
$$3x = -12$$
$$x = -4$$

The x-intercept is $(-4, 0)$.

To find the y-intercept, let $x = 0$.
(Any point on the y-axis has x-coordinate 0.)

$$3x + 4y = -12$$
$$3(0) + 4y = -12$$
$$4y = -12$$
$$y = -3$$

The y-intercept is $(0, -3)$.

A linear equation can be graphed by finding the x- and y-intercepts and then drawing a line through the two points.

EXAMPLE 4 Graph $4x - y = 4$ by using the x- and y-intercepts.

Solution x-intercept: $4x - y = 4$

$$4x - 0 = 4$$ • To find the **x-intercept, let $y = 0$.**
$$4x = 4$$
$$x = 1$$

The x-intercept is $(1, 0)$.

y-intercept: $4x - y = 4$

$$4(0) - y = 4$$ • To find the **y-intercept, let $x = 0$.**
$$-y = 4$$
$$y = -4$$

The y-intercept is $(0, -4)$.

• **Graph the points (1, 0) and (0, −4). Draw a line through the two points.**

PROBLEM 4 Graph $3x - y = 2$ by using the x- and y-intercepts.

Solution See page S12.

Find the y-intercept of the graph of the equation $y = \frac{2}{3}x + 7$.

$$y = \frac{2}{3}x + 7$$

To find the y-intercept, let $x = 0$.

$$y = \frac{2}{3}(0) + 7$$
$$y = 7$$

The y-intercept is $(0, 7)$.

Note that the y-coordinate of the y-intercept $(0, 7)$ has the same value as the constant term of $y = \frac{2}{3}x + 7$. As shown below, this is always true.

To find the y-intercept, let $x = 0$.

$$y = mx + b$$
$$y = m(0) + b$$
$$y = b$$

Thus the y-intercept is $(0, b)$.

> **y-Intercept of a Straight Line**
>
> For any equation of the form $y = mx + b$, the y-intercept is $(0, b)$.

Find the x- and y-intercepts of $y = \frac{2}{3}x + 2$.

$$y = \frac{2}{3}x + 2$$

To find the x-intercept, let $y = 0$ and solve for x.

$$0 = \frac{2}{3}x + 2$$

$$-2 = \frac{2}{3}x$$

The x-intercept is $(-3, 0)$.

$$-3 = x$$

The y-intercept is $(0, b)$.

$$y = \frac{2}{3}x + 2$$

The y-intercept is $(2, 0)$.

$$b = 2$$

From the example above, for an equation written in the form $y = mx + b$, the x-intercept is found by letting $y = 0$ and then solving for x. If the equation

were written in function notation as $f(x) = \dfrac{2}{3}x - 2$, then to find x-intercept, let $f(x) = 0$ and solve for x. A value of x for which $f(x) = 0$ is called a **zero** of the function.

Find the zero of $f(x) = 2x + 4$.

$f(x) = 2x + 4$

$0 = 2x + 4$ • **To find a zero of a function, let $f(x) = 0$.**

$-4 = 2x$ • **Solve for x.**

$-2 = x$

The zero of $f(x) = 2x + 4$ is -2.

The graph of $f(x) = 2x + 4$ is shown at the right. Note that the x-intercept for the graph is the point $(-2, 0)$. The x-coordinate of the x-intercept is -2, the zero of the function.

EXAMPLE 5 Find the zero of the function defined by $f(x) = -2x - 3$.

Solution $f(x) = -2x - 3$

$0 = -2x - 3$ • **To find a zero of a function, let $f(x) = 0$.**

$2x = -3$ • **Solve for x.**

$x = -\dfrac{3}{2}$

The zero is $-\dfrac{3}{2}$.

PROBLEM 5 Find the zero of the function defined by $f(x) = \dfrac{2}{3}x + 4$.

Solution See page S12.

OBJECTIVE ③ Application problems

There are a variety of applications of linear functions.

On the basis of data from *The Joy of Cooking*, the daily caloric allowance for a woman can be approximated by the equation $C = -7.5A + 2187.5$, where C is the caloric intake and A is the age of the woman.

A. Graph this equation for $25 \le A \le 75$.

B. The point whose coordinates are (45, 1850) is on the graph. Write a sentence that describes the meaning of this ordered pair.

A.

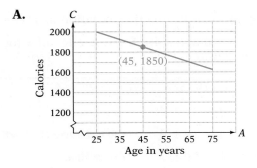

B. The ordered pair (45, 1850) means that the caloric allowance for a 45-year-old woman is 1850 calories per day.

EXAMPLE 6 An electronics technician charges $45 plus $1 per minute to repair defective wiring in a home or apartment. The equation that describes the total cost, *C* (in dollars), to have defective wiring repaired is given by the equation $C = t + 45$, where *t* is the number of minutes the technician works. Graph this equation for $0 \le t \le 60$. The point whose coordinates are (15, 60) is on the graph. Write a sentence that describes the meaning of this ordered pair.

Solution

The ordered pair (15, 60) means that it costs $60 for the technician to work 15 min.

PROBLEM 6 The height, *h* (in inches), of a person and the length, *L* (in inches), of that person's stride while walking are related. The equation $h = \frac{3}{4}L + 50$ approximates the relationship. Graph this equation for $15 \le L \le 40$. The point whose coordinates are (32, 74) is on the graph. Write a sentence that describes the meaning of this ordered pair.

Solution See page S12.

3.3 Concept Review

Determine whether the following statements are always true, sometimes true, or never true.

1. The graph of $y = \frac{1}{x} + 2$ is the graph of a linear function.

2. The graph of $y - 3 = 0$ is a vertical line.

3. The graph of a straight line crosses the x-axis when $y = 0$.

4. The graph of a linear equation is a straight line.

5. The point at which a line crosses the y-axis is the x-intercept.

6. $xy + 2 = 0$ is an example of a linear equation.

3.3 Exercises

1 Graph a linear function

1. Explain how to graph a linear function by plotting points.

2. Give one example of a linear function and one example of a non-linear function.

3. Find three points on the graph of $y = 2x - 5$ by finding the y values that correspond to x values of -1, 0, and 1.
 a. When $x = -1$, $y =$ ___?___. A point on the graph is (___?___, ___?___).
 b. When $x = 0$, $y =$ ___?___. A point on the graph is (___?___, ___?___).
 c. When $x = 1$, $y =$ ___?___. A point on the graph is (___?___, ___?___).

4. To find points on the graph of $y = -\frac{1}{4}x + 1$, it is helpful to choose x values that are divisible by ___?___.

Graph.

5. $y = 3x - 4$ **6.** $y = -2x + 3$ **7.** $y = -\frac{2}{3}x$

8. $y = \dfrac{3}{2}x$ **9.** $y = \dfrac{2}{3}x - 4$ **10.** $y = \dfrac{3}{4}x + 2$

11. $y = -\dfrac{1}{3}x + 2$ **12.** $y = -\dfrac{3}{2}x - 3$ **13.** $y = \dfrac{3}{5}x - 1$

2 **Graph an equation of the form $Ax + By = C$**

Graph.

14. $2x - y = 3$ **15.** $2x + y = -3$ **16.** $2x + 5y = 10$

17. $x - 4y = 8$ **18.** $y = -2$ **19.** $y = 1$

20. $x = -3$ **21.** $x = 4$ **22.** $x - 3y = 0$

23. $2x - 3y = 12$ **24.** $3x - y = -2$ **25.** $3x - 2y = 8$

26. a. The x-intercept of a graph of a linear equation is the point where the graph crosses the __?__. Its __?__-coordinate is 0.
 b. The y-intercept of a graph of a linear equation is the point where the graph crosses the __?__. Its __?__-coordinate is 0.

Find the x- and y-intercepts and graph.

27. $x - 2y = -4$ **28.** $3x + y = 3$ **29.** $2x - 3y = 9$

30. $4x - 2y = 5$

31. $2x - y = 4$

32. $2x + y = 3$

33. $3x + 2y = 5$

34. $4x - 3y = 8$

35. $3x + 2y = 4$

36. $2x - 3y = 4$

37. $3x - 5y = 9$

38. $4x - 3y = 6$

39. 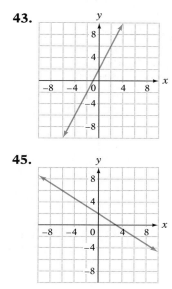 Suppose A and C are negative and B is positive. Is the y-intercept of the graph of $Ax + By = C$ above or below the x-axis?

40. Suppose A, B, and C are negative. Is the x-intercept of the graph of $Ax + By = C$ to the left or to the right of the y-axis?

41. The x-intercept of the graph of $f(x) = -\dfrac{4}{3}x - 4$ is $(-3, 0)$. What is the zero of f?

42. The x-intercept of the graph of $g(x) = -\dfrac{2}{5}x + 4$ is $(10, 0)$. What is the zero of f?

From the graph, determine the zero of the function.

43.

44.

45.

46.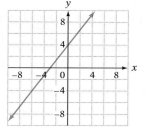

Find the zero of each linear function.

47. $f(x) = 2x - 6$ **48.** $g(x) = 3x + 6$ **49.** $g(x) = -2x + 5$ **50.** $f(x) = -3x - 9$

51. $y(x) = \dfrac{3}{4}x - 6$ **52.** $s(t) = -\dfrac{2}{5}t + 4$ **53.** $f(t) = \dfrac{t}{2} - 3$ **54.** $f(x) = \dfrac{3x}{2} + 3$

55. $g(x) = 3x + 8$ **56.** $h(z) = \dfrac{2}{3}z - 3$ **57.** $f(x) = -\dfrac{3}{4}x + 2$ **58.** $g(x) = \dfrac{5}{2}x + 4$

For Exercises 59 and 60, use this information. The value, V, of an investment of \$1500 at an annual simple interest rate of 7% is given by the equation $V = 105t + 1500$, where t is the amount of time, in years, that the money is invested.

59. a. Find the value of V for $t = 0$: $V = 105(\underline{\quad?\quad}) + 1500 = \underline{\quad?\quad}$.
This means that a point on the graph of the equation $V = 105t + 1500$ is $(0, \underline{\quad?\quad})$.
 b. Find the value of V for $t = 8$: $V = 105(\underline{\quad?\quad}) + 1500 = \underline{\quad?\quad}$. This means that a point on the graph of the equation $V = 105t + 1500$ is $(\underline{\quad?\quad}, \underline{\quad?\quad})$.
 c. The graph of $V = 105t + 1500$ for $0 \leq t \leq 8$ is a straight line joining $(0, \underline{\quad?\quad})$ and $(8, \underline{\quad?\quad})$.

60. The point $(3, 1815)$ is on the graph of $V = 105t + 1500$.
 a. To confirm that this ordered pair is on the graph of $V = 105t + 1500$, substitute 3 for $\underline{\quad?\quad}$ and find the corresponding value of $\underline{\quad?\quad}$:
$V = 105(\underline{\quad?\quad}) + 1500 = \underline{\quad?\quad} + 1500 = \underline{\quad?\quad}$.
 b. The meaning of the ordered pair $(3, 1815)$ in the context of this situation is that after $\underline{\quad?\quad}$ years, the investment is worth $\underline{\quad?\quad}$ dollars.

3 Application problems

Solve.

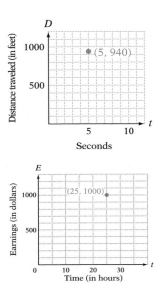

D

Distance traveled (in feet)

1000

● (5, 940)

500

5 10

t

Seconds

E

Earnings (in dollars)

1000

(25, 1000)●

500

0 10 20 30 t

Time (in hours)

61. **Amusement Parks** The Kingda Ka roller coaster at Six Flags Great Adventure in Jackson, N.J., has a maximum speed of 188 ft/s. The equation that describes the total number of feet traveled by the roller coaster in t seconds at this speed is given by $D = 188t$. Use the coordinate axes at the right to graph this equation for $0 \leq t \leq 10$. The point $(5, 940)$ is on this graph. Write a sentence that describes the meaning of this ordered pair.

62. **Sports** A tennis pro is paid \$40 an hour for a private lesson by members of a condominium group. The equation that describes the total earnings, E (in dollars), received by the pro is $E = 40t$, where t is the total number of hours worked. Use the coordinate axes at the right to graph this equation for $0 \leq t \leq 30$. The point $(25, 1000)$ is on the graph. Write a sentence that describes the meaning of this ordered pair.

63. Real Estate A realtor receives $400 per month plus 6% commission on sales. The equation that describes the total monthly income, I (in dollars), of the realtor is $I = 0.06s + 400$, where s is the amount of sales. Use the coordinate axes at the right to graph this equation for $0 \leq s \leq 150{,}000$. The point (60,000, 4000) is on the graph. Write a sentence that describes the meaning of this ordered pair.

64. Human Resources An electronics company pays line workers $9 per hour plus $.05 for each transistor produced. The equation that describes the hourly wage, W (in dollars), is $W = 0.05n + 9$, where n is the number of transistors produced during any given hour. Use the coordinate axes at the right to graph this equation for $0 \leq n \leq 20$. The point (16, 9.80) is on the graph. Write a sentence that describes the meaning of this ordered pair.

65. Food Science A caterer charges a flat rate of $500 plus $.95 for each hot appetizer. The equation that describes the cost, C (in dollars), of catering a dinner party is $C = 0.95n + 500$, where n is the number of hot appetizers. Use the coordinate axes at the right to graph this equation for $0 \leq n \leq 300$. The point (120, 614) is on the graph. Write a sentence that describes the meaning of this ordered pair.

66. Environmental Science A tree service charges $60 plus $35 for each tree removed. The equation that describes the cost, C (in dollars), is $C = 35n + 60$, where n is the number of trees removed. Use the coordinate axes at the right to graph this equation for $0 \leq n \leq 100$. The point (50, 1810) is on the graph. Write a sentence that describes the meaning of this ordered pair.

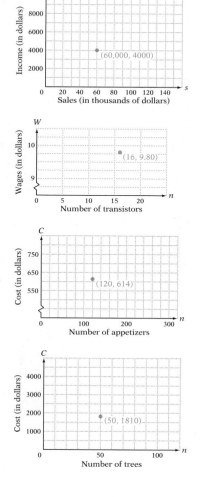

3.3 **Applying Concepts**

67. Business The sale price of an item is a function, s, of the original price, p, where $s(p) = 0.80p$. If an item's original price is $200, what is the sale price of the item?

68. Business The markup of an item is a function, m, of its cost, c, where $m(c) = 0.25c$. If the cost of an item is $150, what is the markup of the item?

 Use a graphing utility to draw the graph of each of the following equations.

69. $f(x) = 1.2x + 2.3$ **70.** $f(x) = 2.4x + 0.5$ **71.** $f(x) = \dfrac{2x}{3} - \dfrac{5}{3}$

72. $y = -\dfrac{3x}{4} - \dfrac{5}{2}$

73. $3x - y = 4$

74. $2x + y = 3$

75. Explain the relationship between the zero of a linear function and the x-intercept of the graph of that function.

76. There is a relationship between the number of times a cricket chirps and the air temperature. The function $f(C) = 7C - 30$, where C is the temperature in degrees Celsius, can be used to approximate the number of times per minute that a cricket chirps. Discuss the domain and range of this function, and explain the significance of its x-intercept.

77. Explain what the graph of an equation represents.

78. Explain why you cannot graph the equation $4x + 3y = 0$ by using just its intercepts.

3.4 Slope of a Straight Line

OBJECTIVE 1 **Find the slope of a line given two points**

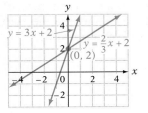

The graphs of $y = 3x + 2$ and $y = \dfrac{2}{3}x + 2$ are shown at the left. Each graph crosses the y-axis at the point $(0, 2)$, but the graphs have different slants. The **slope** of a line is a measure of the slant of the line. The symbol for slope is m.

The slope of a line containing two points is the ratio of the change in the y values between the two points to the change in the x values. The line containing the points whose coordinates are $(-1, -3)$ and $(5, 2)$ is shown below.

The change in the y values is the difference between the y-coordinates of the two points.

$$\text{Change in } y = 2 - (-3) = 5$$

The change in the x values is the difference between the x-coordinates of the two points.

$$\text{Change in } x = 5 - (-1) = 6$$

The slope of the line between the two points is the ratio of the change in y to the change in x.

$$\text{Slope} = m = \frac{\text{change in } y}{\text{change in } x} = \frac{5}{6} \qquad m = \frac{2 - (-3)}{5 - (-1)} = \frac{5}{6}$$

In general, if $P_1(x_1, y_1)$ and $P_2(x_2, y_2)$ are two points on a line, then

Change in $y = y_2 - y_1$ Change in $x = x_2 - x_1$

Using these ideas, we can state a formula for slope.

Slope Formula

The slope of the line containing the two points $P_1(x_1, y_1)$ and $P_2(x_2, y_2)$ is given by

$$m = \frac{y_2 - y_1}{x_2 - x_1}, \quad x_1 \neq x_2$$

Frequently, the Greek letter Δ (delta) is used to designate the change in a variable. Using this notation, we can write the equations for the change in y and the change in x as follows:

Change in $y = y_2 - y_1 = \Delta y$ Change in $x = x_2 - x_1 = \Delta x$

With this notation, the slope formula is written $m = \frac{\Delta y}{\Delta x}$.

In Section 1, Objective 4, of this chapter, we calculated the average rate of change in the population of California for different intervals of time. In each case, the average rate of change was different. Note that the graph of the population data did not lie on a straight line.

The graph of an equation of the form $y = mx + b$ is a straight line, and the average rate of change is constant. We can calculate the average rate of change between any two points on the line, and the rate of change will always be the same. For a linear equation, the rate of change is referred to as slope.

Find the slope of the line containing the points whose coordinates are $(-2, 0)$ and $(4, 5)$.

Let $P_1 = (-2, 0)$ and $P_2 = (4, 5)$. (It does not matter which point is named P_1 and which P_2; the slope will be the same.)

$$m = \frac{y_2 - y_1}{x_2 - x_1} = \frac{5 - 0}{4 - (-2)} = \frac{5}{6}$$

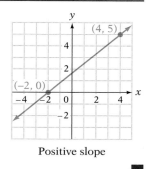

Positive slope

A line that slants upward to the right always has a **positive slope**.

Find the slope of the line containing the points whose coordinates are $(-3, 4)$ and $(4, 2)$.

Let $P_1 = (-3, 4)$ and $P_2 = (4, 2)$.

$$m = \frac{y_2 - y_1}{x_2 - x_1} = \frac{2 - 4}{4 - (-3)} = \frac{-2}{7} = -\frac{2}{7}$$

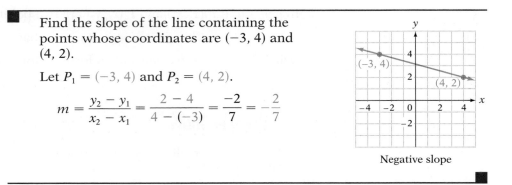

Negative slope

A line that slants downward to the right always has a **negative slope.**

Find the slope of the line containing the points whose coordinates are $(-2, 2)$ and $(4, 2)$.

Let $P_1 = (-2, 2)$ and $P_2 = (4, 2)$.

$$m = \frac{y_2 - y_1}{x_2 - x_1} = \frac{2 - 2}{4 - (-2)} = \frac{0}{6} = 0$$

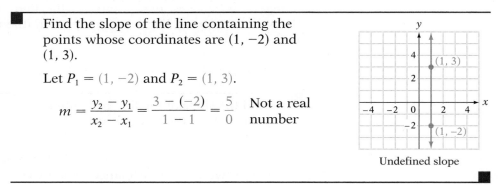

Zero slope

The horizontal line has **zero slope.**

Find the slope of the line containing the points whose coordinates are $(1, -2)$ and $(1, 3)$.

Let $P_1 = (1, -2)$ and $P_2 = (1, 3)$.

$$m = \frac{y_2 - y_1}{x_2 - x_1} = \frac{3 - (-2)}{1 - 1} = \frac{5}{0} \quad \text{Not a real number}$$

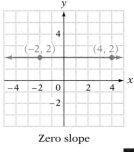

Undefined slope

The slope of a vertical line is **undefined.**

EXAMPLE 1 Find the slope of the line containing the points whose coordinates are $(2, -5)$ and $(-4, 2)$.

Solution $m = \dfrac{y_2 - y_1}{x_2 - x_1} = \dfrac{2 - (-5)}{-4 - 2} = \dfrac{7}{-6} = -\dfrac{7}{6}$ • Let $P_1 = (2, -5)$ and $P_2 = (-4, 2)$.

The slope is $-\dfrac{7}{6}$.

Point
of Interest

One of the motivations for the discovery of calculus was to solve a more complicated version of the distance-rate problem at the right.

You may be familiar with twirling a ball on the end of a string. If you release the string, the ball flies off in a path as shown below.

The question that mathematicians tried to answer was essentially "What is the slope (relative to the horizontal) of the line represented by the arrow?"

Answering questions similar to this led to the development of one aspect of calculus.

PROBLEM 1 Find the slope of the line containing the points whose coordinates are (4, −3) and (2, 7).

Solution See pages S12–S13.

There are many applications of slope. Here are two examples.

The first record for the one-mile run was recorded in 1865 in England. Richard Webster ran the mile in 4 min 36.5 s. His average speed was approximately 19 feet per second.

The graph at the right shows the distance Webster ran during that run. From the graph, note that after 60 s (1 min) he had traveled 1140 ft and that after 180 s (3 min) he had traveled 3420 ft.

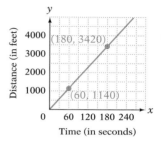

Let the point (60, 1140) be (x_1, y_1) and the point (180, 3420) be (x_2, y_2). The slope of the line between these two points is

$$m = \frac{y_2 - y_1}{x_2 - x_1} = \frac{3420 - 1140}{180 - 60} = \frac{2280}{120} = 19$$

Note that the slope of the line is the same as Webster's average speed, 19 feet per second.

Average speed is related to slope.

The following example is related to agriculture.

The number of farm workers in the United States declined from 10 million in 1950 to 3 million in 2000. (*Source:* U.S. Department of Agriculture)

Let the point (1950, 10) be (x_1, y_1) and the point (2000, 3) be (x_2, y_2). The slope of the line between these two points is

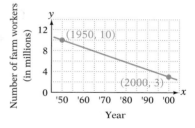

$$m = \frac{y_2 - y_1}{x_2 - y_1}$$

$$= \frac{3 - 10}{2000 - 1950} = \frac{-7}{50} = -0.14$$

Note that if we interpret negative slope as decreasing population, then the slope of the line represents the annual rate at which the number of farm workers was decreasing, 0.14 million or 140,000 workers per year.

In general, any quantity that is expressed by using the word *per* is represented mathematically as slope. In the first example, the slope was 19 feet per second. In the second example, the slope was −0.14 per year.

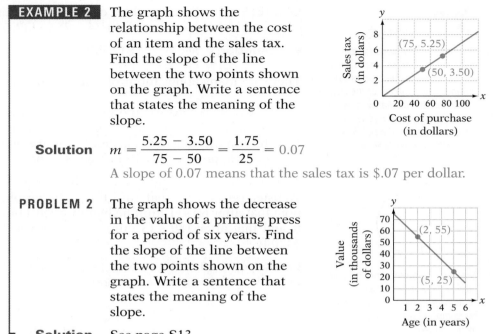

EXAMPLE 2 The graph shows the relationship between the cost of an item and the sales tax. Find the slope of the line between the two points shown on the graph. Write a sentence that states the meaning of the slope.

Solution $m = \dfrac{5.25 - 3.50}{75 - 50} = \dfrac{1.75}{25} = 0.07$

A slope of 0.07 means that the sales tax is \$.07 per dollar.

PROBLEM 2 The graph shows the decrease in the value of a printing press for a period of six years. Find the slope of the line between the two points shown on the graph. Write a sentence that states the meaning of the slope.

Solution See page S13.

OBJECTIVE **2** **Graph a line given a point and the slope**

The graph of the equation $y = -\dfrac{3}{4}x + 4$ is shown at the right. The points $(-4, 7)$ and $(4, 1)$ are on the graph. The slope of the line is

$$m = \frac{7 - 1}{-4 - 4} = \frac{6}{-8} = -\frac{3}{4}$$

Note that the slope of the line has the same value as the coefficient of x. The y-intercept is $(0, 4)$.

Slope-Intercept Form of a Straight Line

The equation $y = mx + b$ is called the **slope-intercept form** of a straight line. The slope of the line is m, the coefficient is x. The y-intercept is $(0, b)$.

When the equation of a straight line is in the form $y = mx + b$, the graph can be drawn by using the slope and the y-intercept. First locate the y-intercept. Use the slope to find a second point on the line. Then draw a line through the two points.

Graph $y = \frac{5}{3}x - 4$ by using the slope and the y-intercept.

The slope is the coefficient of x.

$$m = \frac{5}{3} = \frac{\text{change in } y}{\text{change in } x}$$

The y-intercept is $(0, -4)$.

Beginning at the y-intercept $(0, -4)$, move right 3 units (change in x) and then up 5 units (change in y).

The point whose coordinates $(3, 1)$ is a second point on the graph. Draw a line through the points $(0, -4)$ and $(3, 1)$.

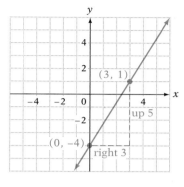

Take Note

When graphing a line by using its slope and y-intercept, *always* start at the y-intercept.

Graph $x + 2y = 4$ by using the slope and the y-intercept.

Solve the equation for y.

$$x + 2y = 4$$
$$2y = -x + 4$$
$$y = -\frac{1}{2}x + 2$$

The slope is $m = -\frac{1}{2} = \frac{-1}{2}$ and the y-intercept is $(0, 2)$.

Beginning at the y-intercept $(0, 2)$, move right 2 units (change in x) and then down 1 unit (change in y).

The point whose coordinates are $(2, 1)$ is a second point on the graph. Draw a line through the points $(0, 2)$ and $(2, 1)$.

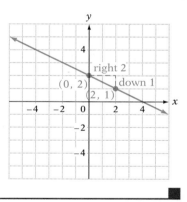

EXAMPLE 3 Graph: $y = -\frac{3}{2}x + 4$ by using the slope and the y-intercept.

Solution y-intercept $= (0, 4)$

$$m = -\frac{3}{2} = \frac{-3}{2}$$

• Locate the **y-intercept**.

• $m = \dfrac{\text{change in } y}{\text{change in } x}$

• Beginning at the y-intercept, **(0, 4)**, move right 2 units and then down 3 units. **(2, 1)** is a second point on the graph.
• Draw a line through the points **(0, 4)** and **(2, 1)**.

> **PROBLEM 3** Graph: $2x + 3y = 6$ by using the slope and the y-intercept.
>
> ■ **Solution** See page S13.

The graph of a line can be drawn when any point on the line and the slope of the line are given.

Graph the line that passes through the point $(-4, -4)$ and has slope 2.

When the slope is an integer, write it as a fraction with denominator 1.

$$m = 2 = \frac{2}{1} = \frac{\text{change in } y}{\text{change in } x}$$

Locate $(-4, -4)$ in the coordinate plane. Beginning at that point, move right 1 unit (change in x) and then up 2 units (change in y).

The point whose coordinates are $(-3, -2)$ is a second point on the graph. Draw a line through the points $(-4, -4)$ and $(-3, -2)$.

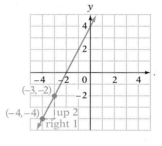

<div style="float:left">

Take Note

This example differs from the others in that a point other than the y-intercept is used. In this case, start at the given point.

</div>

EXAMPLE 4 Graph the line that passes through the point $(-2, 3)$ and has slope $= -\dfrac{4}{3}$.

Solution

- $(x_1, y_1) = (-2, 3)$;

$$m = -\frac{4}{3} = \frac{-4}{3} = \frac{\text{change in } y}{\text{change in } x}$$

- Beginning at the point $(-2, 3)$, move right 3 units and then down 4 units. $(1, -1)$ is a second point on the line.
- Draw a line through the points $(-2, 3)$ and $(1, -1)$.

PROBLEM 4 Graph the line that passes through the point $(-3, -2)$ and has slope 3.

■ **Solution** See page S13.

3.4 Concept Review

Determine whether the following statements are always true, sometimes true, or never true.

1. The slope of a line that slants downward to the left is positive.

2. The y-intercept of a line is the point at which the line crosses the y-axis.

3. The slope of a vertical line is zero.

4. A line whose slope is undefined is parallel to the y-axis.

3.4 Exercises

1 Find the slope of a line given two points

1. To find the slope m of the line containing $P_1(2, -4)$ and $P_2(3, -5)$, use the slope formula $m = \dfrac{y_2 - y_1}{x_2 - x_1}$. Identify each x and y value to substitute into the slope formula.

$y_2 = \underline{\quad ? \quad}$, $y_1 = \underline{\quad ? \quad}$, $x_2 = \underline{\quad ? \quad}$, $x_1 = \underline{\quad ? \quad}$

2. The slope of the line containing $P_1(1, -6)$ and $P_2(5, -10)$ is $\dfrac{-10 - (?)}{5 - (?)} = \underline{\quad ? \quad}$.

This line slants $\underline{\quad ? \quad}$ (upward or downward) to the right.

Find the slope of the line containing the points P_1 and P_2.

3. $P_1(1, 3), P_2(3, 1)$

4. $P_1(2, 3), P_2(5, 1)$

5. $P_1(-1, 4), P_2(2, 5)$

6. $P_1(3, -2), P_2(1, 4)$

7. $P_1(-1, 3), P_2(-4, 5)$

8. $P_1(-1, -2), P_2(-3, 2)$

9. $P_1(0, 3), P_2(4, 0)$

10. $P_1(-2, 0), P_2(0, 3)$

11. $P_1(2, 4), P_2(2, -2)$

12. $P_1(4, 1), P_2(4, -3)$

13. $P_1(2, 5), P_2(-3, -2)$

14. $P_1(4, 1), P_2(-1, -2)$

15. $P_1(2, 3), P_2(-1, 3)$

16. $P_1(3, 4), P_2(0, 4)$

17. $P_1(0, 4), P_2(-2, 5)$

18. $P_1(-2, 3), P_2(-2, 5)$

19. $P_1(-3, -1), P_2(-3, 4)$

20. $P_1(-2, -5), P_2(-4, -1)$

21. **Uniform Motion** The graph below shows the relationship between the distance traveled by a motorist and the time of travel. Find the slope of the line between the two points shown on the graph. Write a sentence that states the meaning of the slope.

22. 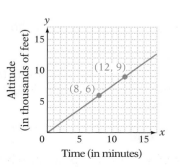 **Aeronautics** The graph below shows how the altitude of an airplane above the runway changes after take-off. Find the slope of the line. Write a sentence that states the meaning of the slope.

23. Automotive Technology The graph below shows how the amount of gas in the tank of a car decreases as the car is driven. Find the slope of the line. Write a sentence that states the meaning of the slope.

24. Meteorology The troposphere extends from the surface of Earth to an elevation of approximately 11 km. The graph below shows the decrease in temperature of the troposphere as altitude increases. Find the slope of the line. Write a sentence that states the meaning of the slope.

25. Sports Lois and Tanya start from the same place on a jogging course. Lois is jogging at 9 km/h, and Tanya is jogging at 6 km/h. The graphs below show the total distance traveled by each jogger and the total distance between Lois and Tanya. Which lines represent which distances?

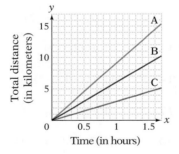

26. Chemistry A chemist is filling two cans from a faucet that releases water at a constant rate. Can 1 has a diameter of 20 mm, and can 2 has a diameter of 30 mm. The depth of the water in each can is measured at 5-second intervals. The graph of the results is shown below. On the graph, which line represents the depth of the water for which can?

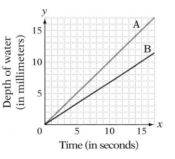

27. Health Science The American National Standards Institute (ANSI) states that the slope for a wheelchair ramp must not exceed $\frac{1}{12}$.

a. Does a ramp that is 6 in. high and 5 ft long meet the requirements of ANSI?

b. Does a ramp that is 12 in. high and 170 in. long meet the requirements of ANSI?

28. Let l be a line passing through points (a, b) and (c, d). If a, b, c and d are all positive with $a > c$ and $b > d$, does l slant upward to the right or downward to the right?

29. The slope of the line with equation $y = 5x - 3$ is ___?___, and its y-intercept is ___?___.

30. The slope of the line with equation $2x + y = 7$ is ___?___, and its y-intercept is ___?___.

2 Graph a line given a point and the slope

Graph by using the slope and the y-intercept.

31. $y = \dfrac{1}{2}x + 2$

32. $y = \dfrac{2}{3}x - 3$

33. $y = -\dfrac{3}{2}x$

34. $y = \dfrac{3}{4}x$

35. $y = -\dfrac{1}{2}x + 2$

36. $y = \dfrac{2}{3}x - 1$

37. $x - 3y = 3$

38. $3x + 2y = 8$

39. $4x + y = 2$

40. Suppose A and B are positive numbers, and C is a negative number. Does the y-intercept of the graph of $Ax + By = C$ lie above or below the x-axis? Does the graph slant upward to the right or downward to the right?

41. Suppose A and C are positive numbers, and B is a negative number. Does the y-intercept of the graph of $Ax + By = C$ lie above or below the x-axis? Does the graph slant upward to the right or downward to the right?

42. Graph the line that passes through the point $(-1, -3)$ and has slope $\dfrac{4}{3}$.

43. Graph the line that passes through the point $(-2, -3)$ and has slope $\dfrac{5}{4}$.

44. Graph the line that passes through the point $(-3, 0)$ and has slope -3.

45. Graph the line that passes through the point $(2, 0)$ and has slope -1.

3.4 Applying Concepts

46. Let $f(x)$ be the digit in the nth decimal place of the repeating digit is $0.\overline{387}$. For example, $f(3) = 7$ because 7 is the digit in the third decimal place. Find $f(14)$.

Complete the sentences.

47. If a line has a slope of 2, then the value of y increases/decreases by _____ as the value of x increases by 1.

48. If a line has a slope of -3, then the value of y increases/decreases by _____ as the value of x increases by 1.

49. If a line has a slope of $\frac{1}{2}$, then the value of y increases/decreases by _____ as the value of x increases by 1.

50. If a line has a slope of $-\frac{2}{3}$, then the value of y increases/decreases by _____ as the value of x increases by 1.

Determine the value of k such that the points whose coordinates are given below lie on the same line.

51. $(3, 2), (4, 6), (5, k)$ <div style="display:inline-block; width:3em"></div> **52.** $(k, 1), (0, -1), (2, -2)$

53. Match each equation with its graph.

 i. $y = -2x + 4$

 ii. $y = 2x - 4$

 iii. $y = 2$

 iv. $2x + 4y = 0$

 v. $y = \frac{1}{2}x + 4$

 vi. $y = -\frac{1}{4}x - 2$

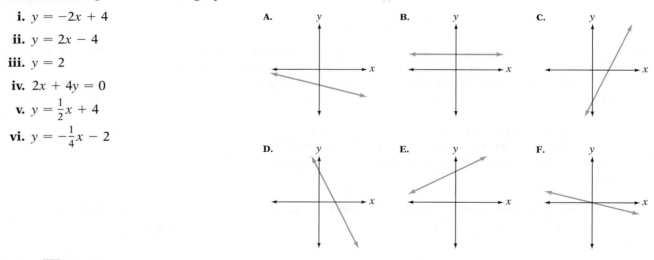

54. Graph $y = 2x + 3$ and $y = 2x - 1$ on the same coordinate system. Explain how the graphs are different and how they are the same. If b is any real number, how is the graph of $y = 2x + b$ related to the two graphs you have drawn?

55. Explain how you can use the slope of a line to determine whether three given points lie on the same line. Then use your procedure to determine whether all of the following points lie on the same line.
 a. $(2, 5), (-1, -1), (3, 7)$ **b.** $(-1, 5), (0, 3), (-3, 4)$

3.5 Finding Equations of Lines

OBJECTIVE 1 **Find the equation of a line given a point and the slope**

When the slope of a line and a point on the line are known, the equation of the line can be determined. If the particular point is the y-intercept, use the slope-intercept form, $y = mx + b$, to find the equation.

Find the equation of the line that contains the point $(0, 3)$ and has slope $\frac{1}{2}$.

The known point is the y-intercept, $(0, 3)$.

Use the slope-intercept form. $y = mx + b$

Replace m with $\frac{1}{2}$, the given slope, and replace b $y = \frac{1}{2}x + 3$
with 3, the y-coordinate of the y-intercept.

The equation of the line is $y = \frac{1}{2}x + 3$.

One method of finding the equation of a line when the slope and any point on the line are known involves using the *point-slope formula*. This formula is derived from the formula for the slope of a line.

Let (x_1, y_1) be the given point on the line, and let (x, y) be another point on the line.

Use the formula for the slope of a line. $\dfrac{y - y_1}{x - x_1} = m$

Multiply both sides of the equation by $(x - x_1)$. $\dfrac{y - y_1}{x - x_1}(x - x_1) = m(x - x_1)$

Then simplify. $y - y_1 = m(x - x_1)$

Point-Slope Formula

Let m be the slope of a line, and let (x_1, y_1) be the coordinates of a point on the line. The equation of the line can be found using the **point-slope formula:**

$$y - y_1 = m(x - x_1)$$

Find the equation of the line that contains the point $(4, -1)$ and has slope $-\frac{3}{4}$.

Use the point-slope formula.

$$y - y_1 = m(x - x_1)$$

Substitute the slope, $-\frac{3}{4}$, and the coordinates of the given point, $(4, -1)$, into the point-slope formula. Then simplify.

$$y - (-1) = -\frac{3}{4}(x - 4)$$

$$y + 1 = -\frac{3}{4}x + 3$$

$$y = -\frac{3}{4}x + 2$$

The equation of the line is $y = -\frac{3}{4}x + 2$.

Find the equation of the line that passes through the point whose coordinates are $(4, 3)$ and whose slope is undefined.

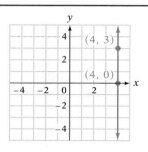

Because the slope is undefined, the point-slope formula cannot be used to find the equation. Instead, recall that when the slope is undefined, the line is vertical and that the equation of a vertical line is $x = a$, where a is the x-coordinate of the x-intercept. Because the line is vertical and passes through $(4, 3)$, the x-intercept is $(4, 0)$.

The equation of the line is $x = 4$.

EXAMPLE 1 Find the equation of the line that contains the point $(-2, 4)$ and has slope 2.

Solution
$$y - y_1 = m(x - x_1)$$
$$y - 4 = 2[x - (-2)]$$
$$y - 4 = 2(x + 2)$$
$$y - 4 = 2x + 4$$
$$y = 2x + 8$$

• Use the point-slope formula.
• Substitute the slope, 2, and the coordinates of the given point, $(-2, 4)$, into the point-slope formula.

The equation of the line is $y = 2x + 8$.

PROBLEM 1 Find the equation of the line that contains the point $(4, -3)$ and has slope -3.

Solution See page S13.

OBJECTIVE 2 Find the equation of a line given two points

The point-slope formula and the formula for slope are used to find the equation of a line when two points are known.

Find the equation of the line containing the points (3, 2) and (−5, 6).

To use the point-slope formula, we must know the slope. Use the formula for slope to determine the slope of the line between the two given points.

Let $(x_1, y_1) = (3, 2)$ and $(x_2, y_2) = (-5, 6)$.

$$m = \frac{y_2 - y_1}{x_2 - x_1} = \frac{6 - 2}{-5 - 3} = \frac{4}{-8} = -\frac{1}{2}$$

Now use the point-slope formula with $m = -\frac{1}{2}$ and $(x_1, y_1) = (3, 2)$.

$$y - y_1 = m(x - x_1)$$
$$y - 2 = -\frac{1}{2}(x - 3)$$
$$y - 2 = -\frac{1}{2}x + \frac{3}{2}$$
$$y = -\frac{1}{2}x + \frac{7}{2}$$

The equation of the line is $y = -\frac{1}{2}x + \frac{7}{2}$.

EXAMPLE 2 Find the equation of the line containing the points (2, 3) and (4, 1).

Solution $m = \dfrac{y_2 - y_1}{x_2 - x_1} = \dfrac{1 - 3}{4 - 2} = \dfrac{-2}{2} = -1$ • **Find the slope. Let $(x_1, y_1) = (2, 3)$ and $(x_2, y_2) = (4, 1)$.**

$$y - y_1 = m(x - x_1)$$
$$y - 3 = -1(x - 2)$$
$$y - 3 = -x + 2$$
$$y = -x + 5$$

• **Substitute the slope and the coordinates of either one of the known points into the point-slope formula.**

The equation of the line is $y = -x + 5$.

PROBLEM 2 Find the equation of the line containing the points (2, 0) and (5, 3).

Solution See page S13.

OBJECTIVE 3 Application problems

Linear functions can be used to model a variety of applications in science and business. For each application, data are collected and the independent and dependent variables are selected. Then a linear function is determined that models the data.

EXAMPLE 3 In 1950, there were 13 million adults 65 years old or older in the United States. Data from the Census Bureau shows that the population of these adults has been increasing at a constant rate of approximately 0.5 million per year. This rate of increase is expected to continue through the year 2010. Find the equation of the line that approximates the population of adults 65 years old or older in terms of the year. Use your equation to approximate the population of these adults in 2005.

Strategy Select the independent and dependent variables. Because we want to determine the population of adults 65 years of age or older, that quantity is the *dependent* variable, y. The year is the *independent* variable.

From the data, the ordered pair (1950, 13), where the y-coordinate is in millions of people, is a point on the line. The slope of the line is the *rate* of increase, 0.5 million per year.

Solution Use the point-slope formula.

$$y - y_1 = m(x - x_1)$$

$$y - 13 = 0.5(x - 1950) \qquad \bullet \ m = 0.5, (x_1, y_1) = (1950, 13)$$
$$y - 13 = 0.5x - 975$$
$$y = 0.5x - 962$$

The linear function is $f(x) = 0.5x - 962$.

$$f(x) = 0.5x - 962$$
$$f(2005) = 0.5(2005) - 962 \qquad \bullet \textbf{ Evaluate the function at 2005 to}$$
$$= 1002.5 - 962 \qquad \qquad \textbf{predict the population in 2005.}$$
$$= 40.5$$

The equation gives an estimate of 40.5 million people 65 years or older in 2005.

PROBLEM 3 Gabriel Daniel Fahrenheit invented the mercury thermometer in 1717. In terms of readings on this thermometer, water freezes at 32°F and boils at 212°F. In 1742 Anders Celsius invented the Celsius temperature scale. On this scale, water freezes at 0°C and boils at 100°C. Find a linear function that can be used to predict the Celsius temperature when the Fahrenheit temperature is known.

Solution See page S13.

3.5 Concept Review

Determine whether the following statements are always true, sometimes true, or never true.

1. Increasing the value of m in the equation $y = mx + b$ increases the slope of the line that is the graph of the equation.

2. Decreasing the value of b in the equation $y = mx + b$ decreases the slope of the line that is the graph of the equation.

3. The point-slope formula for the equation of a line is $y = m(x - x_1)$.

4. $x = a$ is the equation for a vertical line.

5. The line represented by the equation $y = 2x - \frac{1}{2}$ has slope $-\frac{1}{2}$ and x-intercept 2.

6. If $y = mx + b$, then m represents the rate of change of y with respect to x.

7. A horizontal line has no slope.

3.5 Exercises

1 Find the equation of a line given a point and the slope

1. In the equation of the line that has slope $-\dfrac{6}{5}$ and y-intercept $(0, 2)$, m is ___?___ and b is ___?___. The equation is $y = $ ___?___.

2. To find the equation of the line that contains the point $(-4, 5)$ and has slope 2, use the point-slope formula.

$y - y_1 = m(x - x_1)$

$y - 5 = 2(x - (-4))$ • Substitute ___?___ for y_1, ___?___ for m, and ___?___ for x_1.

$y - 5 = 2(x + $ ___?___ $)$ • Simplify inside the parentheses.

$y - 5 = $ ___?___ • Use the Distributive Property on the right side of the equation.

$y = $ ___?___ • Add 5 to each side of the equation.

Find the equation of the line that contains the given point and has the given slope.

3. Point $(0, 5)$, $m = 2$ **4.** Point $(0, 3)$, $m = 1$ **5.** Point $(2, 3)$, $m = \dfrac{1}{2}$

6. Point $(5, 1)$, $m = \dfrac{2}{3}$ **7.** Point $(3, 0)$, $m = -\dfrac{5}{3}$ **8.** Point $(-2, 0)$, $m = \dfrac{3}{2}$

9. Point $(-1, 7)$, $m = -3$ **10.** Point $(-2, 4)$, $m = -4$ **11.** Point $(0, 0)$, $m = \dfrac{1}{2}$

12. Point $(0, 0)$, $m = \dfrac{3}{4}$

13. Point $(2, -3)$, $m = 3$

14. Point $(4, -5)$, $m = 2$

15. Point $(3, 5)$, $m = -\dfrac{2}{3}$

16. Point $(5, 1)$, $m = -\dfrac{4}{5}$

17. Point $(0, -3)$, $m = -1$

18. Point $(2, 0)$, $m = \dfrac{5}{6}$

19. Point $(3, -4)$, slope is undefined

20. Point $(-2, 5)$, slope is undefined

21. Point $(-2, -3)$, $m = 0$

22. Point $(-3, -2)$, $m = 0$

23. Point $(4, -5)$, $m = -2$

24. Point $(-3, 5)$, $m = 3$

25. Point $(-5, -1)$, slope is undefined

26. Point $(0, 4)$, slope is undefined

27. A student found the equation of the line containing the points $(-2, 4)$ and $(1, 1)$ to be $y = x + 6$. Sketch the line containing $(-2, 4)$ and $(1, 1)$ and use your sketch to explain why the student's values for m and b cannot be correct.

28. A student found the equation of the line containing the points $(-4, 4)$ and $(4, 6)$ to be $y = \dfrac{1}{4}x + 5$. Sketch the line containing $(-4, 4)$ and $(4, 6)$ and use your sketch to explain why the student's values for m and b seem reasonable.

2 Find the equation of a line given two points

Find the equation of the line that contains the given points.

29. $P_1(0, 2)$, $P_2(3, 5)$

30. $P_1(0, 4)$, $P_2(1, 5)$

31. $P_1(0, -3)$, $P_2(-4, 5)$

32. $P_1(0, -2)$, $P_2(-3, 4)$

33. $P_1(-1, 3)$, $P_2(2, 4)$

34. $P_1(-1, 1)$, $P_2(4, 4)$

35. $P_1(0, 3)$, $P_2(2, 0)$

36. $P_1(0, 4)$, $P_2(2, 0)$

37. $P_1(-2, -3)$, $P_2(-1, -2)$

38. $P_1(4, 1)$, $P_2(3, -2)$

39. $P_1(2, 3)$, $P_2(5, 5)$

40. $P_1(7, 2)$, $P_2(4, 4)$

41. $P_1(2, 0)$, $P_2(0, -1)$

42. $P_1(0, 4)$, $P_2(-2, 0)$

43. $P_1(3, -4)$, $P_2(-2, -4)$

44. $P_1(-3, 3)$, $P_2(-2, 3)$

45. $P_1(0, 0)$, $P_2(4, 3)$

46. $P_1(2, -5)$, $P_2(0, 0)$

47. $P_1(-2, 5)$, $P_2(-2, -5)$

48. $P_1(3, 2)$, $P_2(3, -4)$

49. $P_1(2, 1)$, $P_2(-2, -3)$

50. $P_1(-3, -2)$, $P_2(1, -4)$

51. $P_1(0, 3)$, $P_2(3, 0)$

52. $P_1(1, -3)$, $P_2(-2, 4)$

53. A plane takes off from an airport that is 500 ft above sea level. The plane climbs at a rate of 1000 ft/min. If $y = mx + b$ is a linear equation that gives the height y (in feet) of the plane above sea level in terms of the time x (in minutes) after take-off, then $m =$ __?__ and an ordered pair that is a solution of the equation is (0, __?__).

54. A plane takes off from an airport that is at sea level and climbs to a height of 12,500 ft in 10 min. If $y = mx + b$ is a linear equation that gives the height y (in feet) of the plane above sea level in terms of the time x (in minutes) after take-off, then two ordered pairs that are solutions of the equation are (0, __?__) and (10, __?__).

3 Application problems

55. Aeronautics The pilot of a Boeing 757 jet takes off from Boston's Logan Airport, which is at sea level, and climbs to a cruising altitude of 32,000 ft at a constant rate of 1200 ft/min. Write a linear equation for the height of the plane in terms of the time after take-off. Use your equation to find the height of the plane 11 min after take-off.

56. Physical Fitness A jogger running at 9 mph burns approximately 14 calories per minute. Write a linear equation for the number of calories burned by the jogger in terms of the number of minutes run. Use your equation to find the number of calories burned after jogging for 32 min.

57. Telecommunications A cellular phone company offers several different options for using a cellular telephone. One option, for people who plan on using the phone only in emergencies, costs the user $4.95 per month plus $.59 per minute for each minute the phone is used. Write a linear equation for the monthly cost of the phone in terms of the number of minutes the phone is used. Use your equation to find the monthly cost of using the cellular phone for 13 min in one month.

58. Aeronautics An Airbus 320 plane takes off from Denver International Airport, which is 5200 ft above sea level, and climbs to 30,000 ft at a constant rate of 1000 ft/min. Write a linear equation for the height of the plane in terms of the time after take-off. Use your equation to find the height of the plane 8 min after take-off.

59. Nutrition There are approximately 126 calories in a 2-ounce serving of lean hamburger and approximately 189 calories in a 3-ounce serving. Write a linear equation for the number of calories in lean hamburger in terms of the size of the serving. Use your equation to estimate the number of calories in a 5-ounce serving of lean hamburger.

60. Chemistry At sea level, the boiling point of water is 100°C. At an altitude of 2 km, the boiling point of water is 93°C. Write a linear equation for the boiling point of water in terms of the altitude above sea level. Use your equation to predict the boiling point of water on the top of Mount Everest, which is approximately 8.85 km above sea level. Round to the nearest degree.

61. **History** In 1927, Charles Lindbergh made history by making the first transatlantic flight from New York to Paris. It took Lindbergh approximately 33.5 h to make the trip. In 1997, the Concorde could make the trip in approximately 3.3 h. Write a linear equation of the time, in hours, it takes to cross the Atlantic in terms of the year. Use your equation to predict how long a flight between the two cities would have taken in 1967. Round your answer to the nearest tenth.

62. **Computers** There were 24 million U.S. homes with computers in 1991. From 1991 to 2003, the average annual rate of growth in homes with computers was 7.42 million homes per year. (*Sources:* U.S. Department of Commerce; U.S. Census Bureau) Write a linear equation for the number of homes with computers in terms of the year. Use your equation to predict the number of homes with computers in 2000.

63. Automotive Technology The gas tank of a certain car contains 16 gal when the driver of the car begins a trip. Each mile driven by the driver decreases the amount of gas in the tank by 0.032 gal. Write a linear equation for the number of gallons of gas in the tank in terms of the number of miles driven. Use your equation to find the number of gallons in the tank after driving 150 mi.

64. Oceanography Whales, dolphins, and porpoises communicate using high-pitched sounds that travel through the water. The speed at which the sound travels depends on many factors, one of which is the depth of the water. At approximately 1000 m below sea level, the speed of sound is 1480 m/s. Below 1000 m, the speed of sound increases at a constant rate of 0.017 m/s for each additional meter below 1000 m. Write a linear equation for the speed of sound in terms of the number of meters below sea level. Use your equation to approximate the speed of sound 2500 m below sea level. Round to the nearest meter per second.

3.5 Applying Concepts

65. What is the x-intercept of the graph of $y = mx + b$?

66. A line contains the points $(4, -1)$ and $(2, 1)$. Find the coordinates of three other points that are on this line.

67. Given that f is a linear function for which $f(1) = 3$ and $f(-1) = 5$, determine $f(4)$.

68. Find the equation of the line that passes through the midpoint of the line segment between $P_1(2, 5)$ and $P_2(-4, 1)$ and has slope -2.

69. If $y = mx + b$, where m is a given constant, how does the graph of this equation change as the value of b changes?

70. Explain the similarities and differences between the point-slope formula and the slope-intercept form of a straight line.

71. Explain why the point-slope formula cannot be used to find the equation of a line that is parallel to the y-axis.

3.6 Parallel and Perpendicular Lines

OBJECTIVE 1 **Find equations of parallel and perpendicular lines**

Two lines that have the same slope do not intersect and are called **parallel lines.**

The slope of each of the lines at the right is $\frac{2}{3}$.

The lines are parallel.

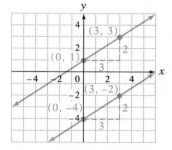

> **Slopes of Parallel Lines**
>
> Two nonvertical lines with slopes of m_1 and m_2 are parallel if and only if $m_1 = m_2$. Any two vertical lines are parallel.

Is the line that contains the points $(-2, 1)$ and $(-5, -1)$ parallel to the line that contains the points $(1, 0)$ and $(4, 2)$?

Find the slope of the line through $(-2, 1)$ and $(-5, -1)$.

$$m_1 = \frac{-1 - 1}{-5 - (-2)} = \frac{-2}{-3} = \frac{2}{3}$$

Find the slope of the line through $(1, 0)$ and $(4, 2)$.

$$m_2 = \frac{2 - 0}{4 - 1} = \frac{2}{3}$$

Because $m_1 = m_2$, the lines are parallel.

EXAMPLE 1 Are the graphs of $y = -\dfrac{3}{2}x - 5$ and $2x + 3y = 6$ parallel?

Solution Write, if necessary, each equation in the form $y = mx + b$, and then compare the values of m, the slopes of the lines. If the values of m are equal, the graphs are parallel.

The equation $y = -\dfrac{3}{2}x - 5$ is in the form $y = mx + b$.

The slope is $-\dfrac{3}{2}$.

Write the equation $2x + 3y = 6$ in the form $y = mx + b$.

$$2x + 3y = 6$$
$$3y = -2x + 6$$
$$y = -\frac{2}{3}x + 2$$

The slope is $-\dfrac{2}{3}$.

$-\dfrac{2}{3} \neq -\dfrac{3}{2}$. The graphs are not parallel.

PROBLEM 1 Are the graphs of $5x + 2y = 2$ and $5x + 2y = -6$ parallel?

Solution See page S13.

Find the equation of the line that contains the point $(2, 3)$ and is parallel to the line $y = \dfrac{1}{2}x - 4$.

The slope of the given line is $\dfrac{1}{2}$. Because parallel lines have the same slope, the slope of the unknown line is also $\dfrac{1}{2}$.

Use the point-slope formula.

$$y - y_1 = m(x - x_1)$$

Substitute $m = \dfrac{1}{2}$ and $(x_1, y_1) = (2, 3)$.

$$y - 3 = \frac{1}{2}(x - 2)$$

Solve for y.

$$y - 3 = \frac{1}{2}x - 1$$

$$y = \frac{1}{2}x + 2$$

The equation of the line is $y = \dfrac{1}{2}x + 2$.

Find the equation of the line that contains the point $(-1, 4)$ and is parallel to the line $2x - 3y = 5$.

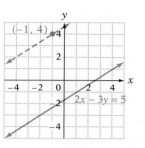

Because the lines are parallel, the slope of the unknown line is the same as the slope of the given line. Solve $2x - 3y = 5$ for y and determine its slope.

$$2x - 3y = 5$$
$$-3y = -2x + 5$$
$$y = \frac{2}{3}x - \frac{5}{3}$$

The slope of the given line is $\frac{2}{3}$. Because the lines are parallel, the slope of the unknown line is also $\frac{2}{3}$.

Use the point-slope formula.

$$y - y_1 = m(x - x_1)$$

Substitute $m = \frac{2}{3}$ and $(x_1, y_1) = (-1, 4)$.

$$y - 4 = \frac{2}{3}[x - (-1)]$$

Solve for y.

$$y - 4 = \frac{2}{3}(x + 1)$$

$$y - 4 = \frac{2}{3}x + \frac{2}{3}$$

$$y = \frac{2}{3}x + \frac{14}{3}$$

The equation of the line is $y = \frac{2}{3}x + \frac{14}{3}$.

EXAMPLE 2 Find the equation of the line that contains the point $(3, -1)$ and is parallel to the line $3x - 2y = 4$.

Solution

$$3x - 2y = 4$$
$$-2y = -3x + 4$$
$$y = \frac{3}{2}x - 2$$

• Solve the equation for y to find the slope.

$$m = \frac{3}{2}$$

$$y - y_1 = m(x - x_1)$$

• Use the point-slope formula.

$$y - (-1) = \frac{3}{2}(x - 3)$$

• $m = \frac{3}{2}$, $(x_1, y_1) = (3, -1)$

$$y + 1 = \frac{3}{2}x - \frac{9}{2}$$

$$y = \frac{3}{2}x - \frac{11}{2}$$

PROBLEM 2 Find the equation of the line that contains the point $(4, 1)$ and is parallel to the line $y = \frac{3}{4}x - 1$.

Solution See page S13.

Two lines that intersect at right angles are **perpendicular lines.**

Any horizontal line is perpendicular to any vertical line. For example, $x = 3$ is perpendicular to $y = -2$, as shown in the graph at the right.

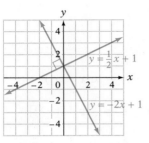

Slopes of Perpendicular Lines

If m_1 and m_2 are the slopes of two lines, neither of which is vertical, then the lines are perpendicular if and only if $m_1 \cdot m_2 = -1$.

A vertical line is perpendicular to a horizontal line.

Solving $m_1 \cdot m_2 = -1$ for m_1 gives $m_1 = -\dfrac{1}{m_2}$. This last equation states that the slopes of perpendicular lines are *negative reciprocals* of each other.

Is the line that contains the points $(4, 2)$ and $(-2, 5)$ perpendicular to the line that contains the points $(-4, 3)$ and $(-3, 5)$?

Find the slope of the line that passes through the points $(4, 2)$ and $(-2, 5)$.

$$m_1 = \frac{5 - 2}{-2 - 4} = \frac{3}{-6} = -\frac{1}{2}$$

Find the slope of the line that passes through the points $(-4, 3)$ and $(-3, 5)$.

$$m_2 = \frac{5 - 3}{-3 - (-4)} = \frac{2}{1} = 2$$

Find the product of the two slopes.

$$m_1 \cdot m_2 = -\frac{1}{2}(2) = -1$$

Because $m_1 \cdot m_2 = -1$, the lines are perpendicular.

Are the graphs of the lines whose equations are $3x + 4y = 8$ and $8x + 6y = 5$ perpendicular?

To determine whether the lines are perpendicular, solve each equation for y and find the slope of each line.

$$3x + 4y = 8$$
$$4y = -3x + 8$$
$$y = -\frac{3}{4}x + 2$$
$$m_1 = -\frac{3}{4}$$

$$8x + 6y = 5$$
$$6y = -8x + 5$$
$$y = -\frac{4}{3}x + \frac{5}{6}$$
$$m_2 = -\frac{4}{3}$$

Then check whether the equation $m_1 \cdot m_2 = -1$ is true.

$$m_1 \cdot m_2 = \left(-\frac{3}{4}\right)\left(-\frac{4}{3}\right) = 1$$

Because $m_1 \cdot m_2 = 1 \neq -1$, the lines are not perpendicular.

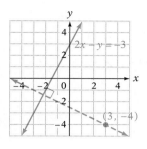

Find the equation of the line that contains the point $(3, -4)$ and is perpendicular to the line $2x - y = -3$.

Determine the slope of the given line by solving the equation for y.

$$2x - y = -3$$
$$-y = -2x - 3$$
$$y = 2x + 3$$

The slope is 2.

The slope of the line perpendicular to the given line is $-\dfrac{1}{2}$, the negative reciprocal of 2.

Use the point-slope formula.

$$y - y_1 = m(x - x_1)$$

Substitute $m = -\dfrac{1}{2}$ and $(x_1, y_1) = (3, -4)$.

$$y - (-4) = -\dfrac{1}{2}(x - 3)$$

Solve for y.

$$y + 4 = -\dfrac{1}{2}x + \dfrac{3}{2}$$
$$y = -\dfrac{1}{2}x - \dfrac{5}{2}$$

The equation of the perpendicular line is $y = -\dfrac{1}{2}x - \dfrac{5}{2}$.

EXAMPLE 3 Are the lines $4x - y = -2$ and $x + 4y = -12$ perpendicular?

Solution $4x - y = -2$ • Solve each equation for y to
$\qquad -y = -4x - 2$ find the slopes.
$\qquad\quad y = 4x + 2 \qquad m_1 = 4$

$x + 4y = -12$
$\quad 4y = -x - 12$
$\qquad y = -\dfrac{1}{4}x - 3 \qquad m_2 = -\dfrac{1}{4}$

$m_1 \cdot m_2 = 4\left(-\dfrac{1}{4}\right) = -1$ • Find the product of the slopes.

Yes, the lines are perpendicular.

PROBLEM 3 Find the equation of the line that contains the point $(-2, 3)$ and is perpendicular to the line $x - 4y = 3$.

Solution See pages S13–S14.

3.6 Concept Review

Determine whether the following statements are always true, sometimes true, or never true.

1. Perpendicular lines have the same y-intercept.

2. Parallel lines have the same slope.

3. The lines given by the equations $y = \frac{3}{2}x + 2$ and $y = -\frac{3}{2}x + 2$ are perpendicular.

4. Two lines are perpendicular if $m_1 \cdot m_2 = 1$.

5. A vertical line is perpendicular to a horizontal line.

6. A line parallel to the y-axis has zero slope.

7. The lines $y = \frac{1}{2}x - 2$ and $y = \frac{1}{2}x - 4$ are parallel.

3.6 Exercises

1 Find equations of parallel and perpendicular lines

1. Given the slopes of two lines, explain how to determine whether the two lines are parallel.

2. Given the slopes of two lines, how can you to determine whether the two lines are perpendicular?

3. The slope of a line is -5. What is the slope of any line parallel to this line?

4. The slope of a line is $\frac{3}{2}$. What is the slope of any line parallel to this line?

5. The slope of a line is 4. What is the slope of any line perpendicular to this line?

6. The slope of a line is $-\frac{4}{5}$. What is the slope of any line perpendicular to this line?

7. Give the slope of any line that is parallel to the line $y = -\frac{1}{3}x + 5$.

8. Give the slope of any line that is perpendicular to the line $y = \frac{3}{5}x + 2$.

9. Is the line $x = -2$ perpendicular to the line $y = 3$?

10. Is the line $y = \frac{1}{2}$ perpendicular to the line $y = -4$?

11. Is the line $x = -3$ parallel to the line $y = \frac{1}{3}$?

12. Is the line $x = 4$ perpendicular to the line $y = -4$?

13. Is the line $y = \frac{2}{3}x - 4$ parallel to the line $y = -\frac{3}{2}x - 4$?

14. Is the line $y = -2x + \frac{2}{3}$ parallel to the line $y = -2x + 3$?

15. Is the line $y = \frac{4}{3}x - 2$ perpendicular to the line $y = -\frac{3}{4}x + 2$?

16. Is the line $y = \frac{1}{2}x + \frac{3}{2}$ perpendicular to the line $y = -\frac{1}{2}x + \frac{3}{2}$?

17. Are the lines $2x + 3y = 2$ and $2x + 3y = -4$ parallel?

18. Are the lines $2x - 4y = 3$ and $2x + 4y = -3$ parallel?

19. Are the lines $x - 4y = 2$ and $4x + y = 8$ perpendicular?

20. Are the lines $4x - 3y = 2$ and $4x + 3y = -7$ perpendicular?

21. Is the line that contains the points $(3, 2)$ and $(1, 6)$ parallel to the line that contains the points $(-1, 3)$ and $(-1, -1)$?

22. Is the line that contains the points $(4, -3)$ and $(2, 5)$ parallel to the line that contains the points $(-2, -3)$ and $(-4, 1)$?

23. Is the line that contains the points $(-3, 2)$ and $(4, -1)$ perpendicular to the line that contains the points $(1, 3)$ and $(-2, -4)$?

24. Is the line that contains the points $(-1, 2)$ and $(3, 4)$ perpendicular to the line that contains the points $(-1, 3)$ and $(-4, 1)$?

25. Is the line that contains the points $(-5, 0)$ and $(0, 2)$ parallel to the line that contains the points $(5, 1)$ and $(0, -1)$?

26. Is the line that contains the points $(3, 5)$ and $(-3, 3)$ perpendicular to the line that contains the points $(2, -5)$ and $(-4, 4)$?

27. Find the equation of the line that contains the point $(-2, -4)$ and is parallel to the line $2x - 3y = 2$.

28. Find the equation of the line that contains the point $(3, 2)$ and is parallel to the line $3x + y = -3$.

29. Find the equation of the line that contains the point $(4, 1)$ and is perpendicular to the line $y = -3x + 4$.

30. Find the equation of the line that contains the point $(2, -5)$ and is perpendicular to the line $y = \frac{5}{2}x - 4$.

31. Find the equation of the line that contains the point $(-1, -3)$ and is perpendicular to the line $3x - 5y = 2$.

32. Find the equation of the line that contains the point $(-1, 3)$ and is perpendicular to the line $2x + 4y = -1$.

33. l is a line with equation $Ax + By = C$ such that A and B are both positive. Does a line parallel to l slant downward to the right or upward to the right?

34. l is a line with equation $Ax + By = C$ such that A is positive and B is negative. Does a line perpendicular to l slant downward to the right or upward to the right?

3.6 Applying Concepts

35. If the graphs of $A_1x + B_1y = C_1$ and $A_2x + B_2y = C_2$ are parallel, express $\dfrac{A_1}{B_1}$ in terms of A_2 and B_2.

36. If the graphs of $A_1x + B_1y = C_1$ and $A_2x + B_2y = C_2$ are perpendicular, express $\dfrac{A_1}{B_1}$ in terms of A_2 and B_2.

For Exercises 37 and 38, suppose a ball is being twirled at the end of a string and the center of rotation is the origin of a coordinate system. If the string breaks, the initial path of the ball is on a line that is perpendicular to the radius of the circle.

37. Suppose the string breaks when the ball is at the point whose coordinates are $P(6, 3)$. Find the equation of the line on which the initial path lies.

38. Suppose the string breaks when the ball is at the point whose coordinates are $P(2, 8)$. Find the equation of the line on which the initial path lies.

39. The graphs of $y = -\frac{1}{2}x + 2$ and $y = \frac{2}{3}x - 5$ intersect at the point whose coordinates are $(6, -1)$. Find the equation of a line whose graph intersects the graphs of the given lines to form a right triangle. (*Hint:* There is more than one answer to this question.)

40. **Geometry** A theorem from geometry states that a line passing through the center of a circle and through a point P on the circle is perpendicular to the tangent line at P. (See the figure at the right.)
 a. If the coordinates of P are $(5, 4)$ and the coordinates of C are $(3, 2)$, what is the equation of the tangent line?
 b. What are the x- and y-intercepts of the tangent line?

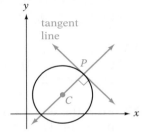

3.7 **Inequalities in Two Variables**

OBJECTIVE **1** **Graph the solution set of an inequality in two variables**

The graph of the linear equation $y = x - 1$ separates the plane into three sets: the set of points on the line, the set of points above the line, and the set of points below the line.

The point whose coordinates are $(2, 1)$ is a solution of $y = x - 1$ and is a point on the line.

The point whose coordinates are $(2, 4)$ is a solution of $y > x - 1$ and is a point above the line.

The point whose coordinates are $(2, -2)$ is a solution of $y < x - 1$ and is a point below the line.

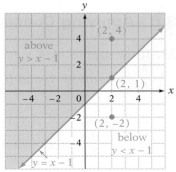

The set of points on the line are the solutions of the equation $y = x - 1$. The set of points above the line are the solutions of the inequality $y > x - 1$. These points form

a **half-plane.** The set of points below the line are solutions of the inequality $y < x - 1$. These points also form a half-plane.

An inequality of the form $y > mx + b$ or $Ax + By > C$ is a **linear inequality in two variables.** (The inequality symbol could be replaced by $\geq$, $<$, or $\leq$.) The solution set of a linear inequality in two variables is a half-plane.

The following illustrates the procedure for graphing the solution set of a linear inequality in two variables.

Take Note

When solving the inequality at the right for y, we divide both sides of the inequality by -4, so the inequality symbol must be reversed.

Graph the solution set of $3x - 4y < 12$.

Solve the inequality for y.

$$3x - 4y < 12$$
$$-4y < -3x + 12$$
$$y > \frac{3}{4}x - 3$$

Change the inequality $y > \frac{3}{4}x - 3$ to the equality $y = \frac{3}{4}x - 3$, and graph the line.

If the inequality contains $\leq$ or $\geq$, the line belongs to the solution set and is shown by a *solid line.* If the inequality contains $<$ or $>$, the line is not part of the solution set and is shown by a *dashed line.*

If the inequality contains $>$ or $\geq$, shade the upper half-plane. If the inequality contains $<$ or $\leq$, shade the lower half-plane.

As a check, use the ordered pair $(0, 0)$ to determine whether the correct region of the plane has been shaded. If $(0, 0)$ is a solution of the inequality, then $(0, 0)$ should be in the shaded region. If $(0, 0)$ is not a solution of the inequality, then $(0, 0)$ should not be in the shaded region.

As shown below, $(0, 0)$ is a solution of the inequality in the example above.

$$y > \frac{3}{4}x - 3$$
$$0 > \frac{3}{4}(0) - 3$$
$$0 > 0 - 3$$
$$0 > -3 \qquad \text{True}$$

Because $(0, 0)$ is a solution of the inequality, $(0, 0)$ should be in the shaded region. The solution set as graphed above is correct.

If the line passes through the point $(0, 0)$, another point, such as $(0, 1)$, must be used as a check.

From the graph of $y > \frac{3}{4}x - 3$, note that for a given value of x, more than one value of y can be paired with the value of x. For instance, $(4, 1)$, $(4, 3)$, $(5, 1)$, and $\left(5, \frac{9}{4}\right)$ are all ordered pairs that belong to the graph.

Because there are ordered pairs with the same first component and different second components, the inequality does not represent a function. The inequality is a relation but not a function.

EXAMPLE 1 Graph the solution set of $x + 2y \le 4$.

Solution $x + 2y \le 4$ • Solve the inequality for y.
$2y \le -x + 4$
$y \le -\frac{1}{2}x + 2$

• Graph $y = -\frac{1}{2}x + 2$ as a solid line. Shade the lower half-plane.

PROBLEM 1 Graph the solution set of $x + 3y > 6$.

Solution See page S14.

EXAMPLE 2 Graph the solution set of $x \ge -1$.

Solution

• Graph $x = -1$ as a solid line.
• The point $(0, 0)$ satisfies the inequality. Shade the half-plane to the right of the line.

PROBLEM 2 Graph the solution set of $y < 2$.

Solution See page S14.

3.7 Concept Review

Determine whether the following statements are always true, sometimes true, or never true.

1. It is possible to write a linear inequality in two variables that has no solutions.

2. The exponents on the variables in a linear inequality in two variables are 1.

3. The graph of a linear inequality is a half-plane.

4. The graph of a linear inequality in two variables represents a function.

5. The solution of a linear inequality in two variables containing $\leq$ or $\geq$ includes the line separating the half-planes.

6. The solution of the inequality $y > x + 2$ is all the points above the line $y = x + 2$.

3.7 Exercises

1 Graph the solution set of an inequality in two variables

1. What is a half-plane?

2. Explain a method that you can use to check that the graph of a linear inequality in two variables has been shaded correctly.

3. Is (0, 0) a solution of $y > 2x - 7$?

4. Is (0, 0) a solution of $y < 5x + 3$?

5. Is (0, 0) a solution of $y \leq -\dfrac{2}{3}x - 8$?

6. Is (0, 0) a solution of $y \geq -\dfrac{3}{4}x + 9$?

Graph the solution set.

7. $y \leq \dfrac{3}{2}x - 3$

8. $y \geq \dfrac{4}{3}x - 4$

9. $y < \dfrac{4}{5}x - 2$

10. $y < \dfrac{3}{5}x - 3$

11. $y < -\dfrac{1}{3}x + 2$

12. $y < -\dfrac{4}{3}x + 3$

13. $x + 3y < 4$

14. $2x - 5y \le 10$

15. $2x + 3y \ge 6$

16. $3x + 2y < 4$

17. $-x + 2y > -8$

18. $-3x + 2y > 2$

19. $y - 4 < 0$

20. $x + 2 \ge 0$

21. $6x + 5y < 15$

22. $3x - 5y < 10$

23. $-5x + 3y \ge -12$

24. $3x + 4y \ge 12$

25. Given that $(0, 0)$ is a point on the graph of the linear inequality $Ax + By > C$, and C is not zero, is C positive or negative?

26. Is $(0, 0)$ a point on the graph of the linear inequality $Ax + By > C$, where C is a positive number?

3.7 Applying Concepts

27. Does the inequality $y < 3x - 1$ represent a function? Explain your answer.

28. Are there any points whose coordinates satisfy both $y \le x + 3$ and $y \ge -\dfrac{1}{2}x + 1$? If so, give the coordinates of three such points. If not, explain why not.

29. Are there any points whose coordinates satisfy both $y \le x - 3$ and $y \ge x + 2$? If so, give the coordinates of three such points. If not, explain why not.

Focus on Problem Solving

Find a Pattern Polya's four recommended problem-solving steps are stated below.

1. Understand the problem.

2. Devise a plan.

3. Carry out the plan.

4. Review your solution

One of the several ways of devising a plan is first to try to find a pattern. Karl Friedrich Gauss supposedly used this method to solve a problem that was given to his math class when he was in elementary school. As the story goes, his teacher wanted to grade some papers while the class worked on a math problem. The problem the teacher gave to the class was to find the sum

$$1 + 2 + 3 + 4 + \cdots + 100$$

Gauss quickly solved the problem by seeing a pattern. Here is what he saw.

$$
\begin{array}{c}
101 \\
101 \\
101 \\
101 \\
1 + 2 + 3 + 4 + \cdots + 97 + 98 + 99 + 100
\end{array}
$$

Note that

$$1 + 100 = 101$$
$$2 + 99 = 101$$
$$3 + 98 = 101$$
$$4 + 97 = 101$$

Gauss noted that there were 50 sums of 101. Therefore, the sum of the first 100 natural numbers is

$$1 + 2 + 3 + 4 + \cdots + 97 + 98 + 99 + 100 = 50(101) = 5050$$

Try to solve the following problems by finding a pattern.

1. Find the sum $2 + 4 + 6 + \cdots + 96 + 98 + 100$.

2. Find the sum $1 + 3 + 5 + \cdots + 97 + 99 + 101$.

3. Find another method of finding the sum $1 + 3 + 5 + \cdots + 97 + 99 + 101$ given in the previous exercise.

4. Find the sum $\dfrac{1}{1 \cdot 2} + \dfrac{1}{2 \cdot 3} + \dfrac{1}{3 \cdot 4} + \cdots + \dfrac{1}{49 \cdot 50}$.

 Hint: $\dfrac{1}{1 \cdot 2} = \dfrac{1}{2}, \dfrac{1}{1 \cdot 2} + \dfrac{1}{2 \cdot 3} = \dfrac{2}{3}, \dfrac{1}{1 \cdot 2} + \dfrac{1}{2 \cdot 3} + \dfrac{1}{3 \cdot 4} = \dfrac{3}{4}$

5. The following problem shows that checking a few cases does not always result in a conjecture that is true for *all* cases. Select any two points on a circle and draw a *chord*, the line connecting the points (see the drawing in the left margin). The chord divides the circle into two regions. Now select three different points and draw chords connecting each of the three points with every other point. The chords divide the circle into four regions. Now select four points and connect each of the points with every other point. Make a conjecture about the relationship between the number

2 points, 2 regions 3 points, 4 regions

4 points, 8 regions 5 points, ? regions

of regions and the number of points on the circle. Does your conjecture work for five points? six points?

6. A *polygonal number* is a number that can be represented by arranging that number of dots in rows to form a geometric figure such as a triangle, square, pentagon, or hexagon. For instance, the first four *triangular* numbers, 3, 6, 10, and 15, are given (and shown) below. What are the next two triangular numbers?

3 6 10 15

Projects & Group Activities

Introduction to Graphing Calculators

Calculator Screen

There are a variety of computer programs and calculators that can graph an equation. A computer or graphing calculator screen is divided into pixels. Depending on the computer or calculator, there are approximately 6000 to 790,000 pixels available on the screen. The greater the number of pixels, the smoother the graph will appear. A portion of a screen is shown at the left. Each little rectangle represents one pixel.

A graphing calculator draws a graph in a manner similar to the way we have shown in this chapter. Values of x are chosen and ordered pairs calculated. Then a graph is drawn through those points by illuminating pixels (an abbreviation for "picture element") on the screen.

Graphing utilities can display only a portion of the xy-plane, called a window. The window [Xmin, Xmax] by [Ymin, Ymax] consists of those points (x, y) that satisfy both of the following:

$$\text{Xmin} \le x \le \text{Xmax} \quad \text{and} \quad \text{Ymin} \le y \le \text{Ymax}$$

The user sets these values before a graph is drawn.

The numbers Xscl and Yscl are the distances between the tick marks that are drawn on the x- and y-axes. If you do not want tick marks on the axes, set Xscl = 0 and Yscl = 0.

The graph at the right is a portion of the graph of $y = \frac{1}{2}x + 1$, as it was drawn with a graphing calculator. The window is

$$\text{Xmin} = -4.7, \text{Xmax} = 4.7$$
$$\text{Ymin} = -3.1, \text{Ymax} = 3.1$$
$$\text{Xscl} = 1 \text{ and } \text{Yscl} = 1$$

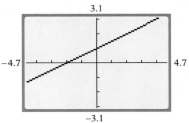

Using interval notation, this is shown as $[-4.7, 4.7]$ by $[-3.1, 3.1]$.

The window $[-4.7, 4.7]$ by $[-3.1, 3.1]$ gives "nice" coordinates in the sense that each time the ▶ or the ◀ is pressed, the change in x is 0.1. The reason for this is that the horizontal distance from the middle of the first pixel to the middle of the last pixel is 94 units. By using Xmin $= -4.7$ and Xmax $= 4.7$, we have[1]

$$\text{Change in } x = \frac{\text{Xmax} - \text{Xmin}}{94} = \frac{4.7 - (-4.7)}{94} = \frac{9.4}{94} = 0.1$$

Similarly, the vertical distance from the middle of the first pixel to the middle of the last pixel is 62 units. Therefore, using Ymin $= -3.1$ and Ymax $= 3.1$ will give nice coordinates in the vertical direction.

Graph each of the following by using a graphing calculator.

Take Note

The Appendix "Keystroke Guide" contains keystroking suggestions to use with this project.

1. $y = 2x + 1$ — For $2x$, you may enter $2 \times x$ or just $2x$. The times sign $\times$ is not necessary on many graphing calculators.

2. $y = -x + 2$ — Many calculators use the $(-)$ key to enter a negative sign.

3. $3x + 2y = 6$ — Solve for y. Then enter the equation.

4. $y = 50x$ — You must adjust the viewing window. Try the window given by $[-4.7, 4.7]$ by $[-250, 250]$ with Yscl $= 50$.

5. $y = \frac{2}{3}x - 3$ — You may enter $\frac{2}{3}x$ as $2x/3$ or $(2/3)x$. Although entering $2/3x$ works on some calculators, it is not recommended.

6. $4x + 3y = 75$ — You must adjust the viewing window.

Chapter 3 Summary

Key Words

A **rectangular coordinate system** is formed by two number lines, one horizontal and one vertical, that intersect at the zero point of each line. The number lines that make up a coordinate system are called the **coordinate axes** or simply the **axes**. The **origin** is the point of intersection of the two coordinate axes. Generally, the horizontal axis is labeled the **x-axis**, and the vertical axis is labeled the **y-axis**. A rectangular coordinate system divides the **plane** determined by the axes into four regions called **quadrants**. [3.1.1, p. 209]

Examples

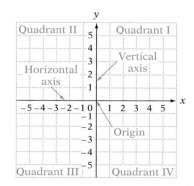

[1]Some calculators have screen widths of 126 pixels. For those calculators, use Xmin $= -6.3$ and Xmax $= 6.3$ to obtain nice coordinates.

An **ordered pair (x, y)** is used to locate a point in a plane. The first number in an ordered pair is called the **abscissa**. The second number in an ordered pair is called the **ordinate**. [3.1.1, p. 209]	(3, 4) is an ordered pair. 3 is the abscissa. 4 is the ordinate.
A **scatter diagram** is a graph of ordered-pair data. [3.1.3, p. 214]	
A **relation** is any set of ordered pairs. [3.2.1, p. 226]	{(2, 3), (2, 4), (3, 4), (5, 7)}
The **domain of a relation** is the set of first coordinates of all the ordered pairs of the relation. The **range of a relation** is the set of second coordinates of all the ordered pairs of the relation. [3.2.1, p. 226]	{(1, 2), (2, 4), (5, 7), (8, 3)} Domain: {1, 2, 5, 8} Range: {2, 3, 4, 7}
A **function** is a set of ordered pairs in which no two ordered pairs have the same first coordinate. [3.2.1, p. 226]	{(2, 3), (3, 5), (5, 7), (6, 9)}
Function notation is used for those equations that represent functions. For the equation at the right, x is the **independent variable** and y is the **dependent variable**. The symbol $f(x)$ is the **value of the function** and represents the value of the dependent variable for a given value of the dependent variable. The process of finding the value of $f(x)$ for a given value of x is called **evaluating the function.** [3.2.1, pp. 228–229]	In function notation, $y = 3x + 7$ is written as $f(x) = 3x + 7$.
The **graph of a function** is the graph of the ordered pairs that belong to the function. A **linear function** is one that can be expressed in the form $f(x) = mx + b$. The graph of a linear function is a straight line. The point at which the graph crosses the x-axis is called the **x-intercept**, and the point at which a graph crosses the y-axis is called the **y-intercept**. [3.2.2, 3.3.1, pp. 232, 246, and 251]	$f(x) = -\dfrac{3}{2}x + 3$
The **slope** of a line is a measure of the slant, or tilt, of a line. The symbol for slope is **m**. A line that slants upward to the right has a **positive slope**, and a line that slants downward to the right has a **negative slope**. [3.4.1, pp. 260–262]	The line $y = 2x - 3$ slants upward to the right. The line $y = -5x + 2$ slants downward to the right.
A horizontal line has **zero slope**. The slope of a vertical line is **undefined**. [3.4.1, p. 262]	$y = 3$ is an equation of a line with zero slope. The slope of the line $x = -3$ is undefined.
An inequality of the form $y > mx + b$ or of the form $Ax + By > C$ is a **linear inequality in two variables**. (The symbol $>$ could be replaced by $\geq$, $<$, or $\leq$.) The solution set of an inequality in two variables is a **half-plane**. [3.7.1, p. 287]	

Essential Rules and Procedures

The **distance formula** $d = \sqrt{(x_2 - x_1)^2 + (y_2 - y_1)^2}$ can be used to find the distance between any two points in a plane. [3.1.2, p. 212]

$(x_1, y_1) = (2, 4), (x_2, y_2) = (-3, 5)$
$d = \sqrt{(-3 - 2)^2 + (5 - 4)^2}$
$ = \sqrt{25 + 1}$
$ = \sqrt{26}$

The **midpoint formula** $(x_m, y_m) = \left(\dfrac{x_1 + x_2}{2}, \dfrac{y_1 + y_2}{2} \right)$ is used to find the midpoint of a line segment. [3.1.2, p. 213]

$(x_1, y_1) = (2, 3), (x_2, y_2) = (-6, -1)$

$x_m = \dfrac{2 + (-6)}{2} = -2,$

$y_m = \dfrac{3 + (-1)}{2} = 1$

$(x_m, y_m) = (-2, 1)$

The **average rate of change** of y with respect to x is $\dfrac{\text{change in } y}{\text{change in } x}$. [3.1.4, p. 217]

In 1998, 1.7 million households used electronic bill payment. This number increased to 9.9 million in 2002. (*Source:* Yankee Group)

$$\frac{9.9 - 1.7}{2002 - 1998} = 2.05$$

The average annual rate of change in the number of households using electronic bill payment from 1998 to 2002 was 2.05 million.

Vertical line test for the graph of a function If every vertical line intersects a graph at most once, then the graph is the graph of a function. [3.2.3, p. 235]

Function

Not a Function

To find the x-intercept of $Ax + By = C$, let $y = 0$. **To find the y-intercept**, let $x = 0$. [3.3.2, p. 251]

$3x + 4y = 12$

Let $y = 0$:
$3x + 4(0) = 12$
$x = 4$

Let $x = 0$:
$3(0) + 4y = 12$
$y = 3$

The x-intercept is (4, 0).

The y-intercept is (0, 3).

A value of x for which $f(x) = 0$ is called a **zero of the function.** [3.3.2, p. 253]

$f(x) = x + 3$

$0 = x + 3$

$-3 = x$

The zero of $f(x) = x + 3$ is -3.

The **slope formula** $m = \dfrac{y_2 - y_1}{x_2 - x_1}$, $x_1 \neq x_2$, is used to find the slope of a line when two points on the line are known. [3.4.1, p. 261]

$(x_1, y_1) = (-3, 2), (x_2, y_2) = (1, 4)$

$$m = \frac{4 - 2}{1 - (-3)} = \frac{2}{4} = \frac{1}{2}$$

The equation $y = mx + b$ is called the **slope-intercept form of a straight line**. [3.4.2, p. 264]

$y = -3x + 2$
The slope is -3; the y-intercept is $(0, 2)$.

The **point-slope formula**, $y - y_1 = m(x - x_1)$, is used to find the equation of a line when the slope and a point on the line are known. [3.5.1, p. 271]

The equation of the line that passes through the point $(4, 2)$ and has slope -3 is

$$\begin{aligned} y - y_1 &= m(x - x_1) \\ y - 2 &= -3(x - 4) \\ y - 2 &= -3x + 12 \\ y &= -3x + 14 \end{aligned}$$

Two lines that have the same slope do not intersect and are called **parallel lines**. [3.6.1, p. 279]

$y = 3x - 4, \quad m_1 = 3$
$y = 3x + 2, \quad m_2 = 3$

Because $m_1 = m_2$, the lines are parallel.

Two lines that intersect at right angles are called **perpendicular lines**. The product of the slopes of two nonvertical perpendicular lines is -1. [3.6.1, p. 282]

$y = \dfrac{1}{2}x - 1, \quad m_1 = \dfrac{1}{2}$

$y = -2x + 2, \quad m_2 = -2$

Because $m_1 \cdot m_2 = -1$, the lines are perpendicular.

Chapter 3 Review Exercises

1. Determine the ordered-pair solution of $y = \dfrac{x}{x - 2}$ that corresponds to $x = 4$.

2. Find the midpoint and length of the line segment with endpoints $(-2, 4)$ and $(3, 5)$.

3. Graph $f(x) = x^2 - 2$ by evaluating the function when $x = -3, -2, -1, 0, 1, 2,$ and 3. Plot the resulting ordered pairs. Then connect the points to form the graph.

4. Graph: $y = -2$

5. Given $P(x) = 3x + 4$, evaluate $P(-2)$ and $P(a)$.

6. What are the domain and range of the relation? Is the relation a function?
$\{(-1, 0), (0, 2), (1, 2), (2, 0), (5, 3)\}$

7. Find the range of $f(x) = x^2 - 2$ if the domain is $\{-2, -1, 0, 1, 2\}$.

8. Given $f(x) = 3x - 1$, find a number c in the domain of f such that $f(c) = 5$. Write the corresponding ordered pair of the function.

9. Find the x- and y-intercepts and graph $4x - 6y = 12$.

10. Graph: $y = -2x + 2$

11. Graph: $2x - 3y = -6$

12. Find the slope of the line that contains the points $(3, -2)$ and $(-1, 2)$.

13. Graph $3x + 2y = -4$ by using the x- and y-intercepts.

14. Graph the line that passes through the point $(-1, 4)$ and has slope $-\frac{1}{3}$.

15. Find the equation of the line that contains the point $(-3, 4)$ and has slope $\frac{5}{2}$.

16. Find the equation of the line that contains the points $(-2, 4)$ and $(4, -3)$.

17. Find the equation of the line that contains the point $(3, -2)$ and is parallel to the line $y = -3x + 4$.

18. Find the equation of the line that contains the point $(-2, -4)$ and is parallel to the line $2x - 3y = 4$.

19. Find the equation of the line that contains the point $(2, 5)$ and is perpendicular to the line $y = -\frac{2}{3}x + 6$.

20. Find the equation of the line that contains the point $(-3, -1)$ and is perpendicular to the line $4x - 2y = 7$.

21. Find the zero of $f(x) = -3x + 12$.

22. Use the vertical line test to determine whether the graph is the graph of a function.

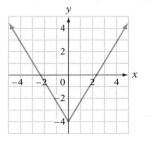

23. Graph the solution set of $y \geq 2x - 3$. **24.** Graph the solution set of $3x - 2y < 6$.

25. 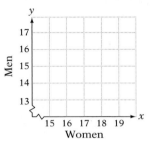 **Life Expectancy** The table below shows the average number of years that women and men were estimated to live beyond age 65 in 1950, 1960, 1970, 1980, 1990, and 2000. (*Source:* 2000 Social Security Trustees Report) Figures are rounded to the nearest whole number. Graph the scatter diagram for these data.

Women, x	15	16	17	18	19	19
Men, y	13	13	13	14	15	16

26. **Demographics** The table below shows the number of foreign-born people living in the United States each decade from 1960 to 2000. (*Source:* U.S. Census Bureau)

a. What was the average annual rate of change in the number of foreign-born people living in the United States from 1980 to 2000? Round to the nearest thousand.

b. Find the average annual rate of change in the number of foreign-born people living in the United States from 1970 to 1980. Was this more than or less than the average annual rate of change in the number of foreign-born people living in the United States from 1980 to 1990?

Year	1960	1970	1980	1990	2000
Number of foreign born (in millions)	9.7	9.6	14.1	19.8	28.4

27. **Uniform Motion** A car is traveling at 55 mph. The equation that describes the distance, d (in miles), traveled is $d = 55t$, where t is the number of hours driven. Use the coordinate axes below to graph this equation for $0 \leq t \leq 6$. The point (4, 220) is on the graph. Write a sentence that describes the meaning of this ordered pair.

28. **Manufacturing** The graph below shows the relationship between the cost of manufacturing calculators and the number of calculators manufactured. Find the slope of the line between the two points shown on the graph. Write a sentence that states the meaning of the slope.

29. Construction A building contractor estimates that the cost to build a new home is $25,000 plus $80 for each square foot of floor space in the house. Determine a linear function that will give the cost to build a house that contains a given number of square feet. Use this model to determine the cost of building a house that contains 2000 ft^2.

Chapter 3 Test

1. Graph $f(x) = -x^2 + 2$ by evaluating the function when $x = -3, -2, -1, 0, 1, 2$, and 3. Plot the resulting ordered pairs. Then connect the points to form the graph.

2. Find the ordered-pair solution of $y = 2x + 6$ that corresponds to $x = -3$.

3. Graph: $y = \dfrac{2}{3}x - 4$

4. Graph: $2x + 3y = -3$

5. Find the equation of the vertical line that contains the point $(-2, 3)$.

6. Find the midpoint and length of the line segment with endpoints $(4, 2)$ and $(-5, 8)$.

7. Find the slope of the line that contains the points $(-2, 3)$ and $(4, 2)$.

8. Given $P(x) = 3x^2 - 2x + 1$, evaluate $P(2)$.

9. Graph $2x - 3y = 6$ by using the x- and y-intercepts.

10. Graph the line that passes through the point $(-2, 3)$ and has slope $-\dfrac{3}{2}$.

11. Find the equation of the line that contains the point $(-5, 2)$ and has slope $\dfrac{2}{5}$.

12. Given $f(x) = 5x - 2$, find a number c in the domain of f such that $f(c) = 3$. Write the corresponding ordered pair of the function.

13. Find the equation of the line that contains the points $(3, -4)$ and $(-2, 3)$.

14. Find the zero of $s(t) = \dfrac{4}{3}t - 8$.

15. What are the domain and range of the relation? Is the relation a function?
$\{(-4, 2), (-2, 2), (0, 0), (3, 5)\}$

16. Find the equation of the line that contains the point $(1, 2)$ and is parallel to the line $y = -\dfrac{3}{2}x - 6$.

17. Find the equation of the line that contains the point $(-2, -3)$ and is perpendicular to the line $y = -\dfrac{1}{2}x - 3$.

18. Graph the solution set of $3x - 4y > 8$.

19. Use the vertical line test to determine whether the graph is the graph of a function.

20. **Business** The director of a baseball camp estimates that 100 students will enroll if the tuition is $250. For each $20 increase in tuition, 6 fewer students will enroll. Determine a linear function that will predict the number of students who will enroll at a given tuition. Use this model to predict enrollment when the tuition is $300.

21. **Depreciation** The graph at the right shows the relationship between the value of a rental house and the depreciation allowed for income tax purposes. Find the slope between the two points shown on the graph. Write a sentence that states the meaning of the slope.

22. 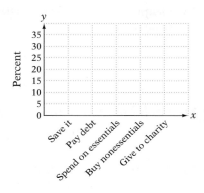 **Extra Income** A survey by Transamerica Center for Retirement Studies asked people what they would do with an extra $1000. The table below shows the responses. Graph the scatter diagram for these data.

Response	Percent of respondents
Save it	31%
Pay debt	24%
Spend on essentials	16%
Spend on nonessentials	20%
Give to charity	2%

Cumulative Review Exercises

1. Identify the property that justifies the statement. $(x + y) \cdot 2 = 2 \cdot (x + y)$

2. Solve: $3 - \dfrac{x}{2} = \dfrac{3}{4}$

3. Solve: $2[y - 2(3 - y) + 4] = 4 - 3y$

4. Solve: $\dfrac{1 - 3x}{2} + \dfrac{7x - 2}{6} = \dfrac{4x + 2}{9}$

5. Solve: $x - 3 < -4$ or $2x + 2 > 3$

6. Solve: $8 - |2x - 1| = 4$

7. Solve: $|3x - 5| < 5$

8. Simplify: $4 - 2(4 - 5)^3 + 2$

9. Evaluate $(a - b)^2 \div ab$ when $a = 4$ and $b = -2$.

10. Graph: $\{x \mid x < -2\} \cup \{x \mid x > 0\}$

11. Solve: $3x - 1 < 4$ and $x - 2 > 2$

12. Given $P(x) = x^2 + 5$, evaluate $P(-3)$.

13. Find the ordered-pair solution of $y = -\dfrac{5}{4}x + 3$ that corresponds to $x = -8$.

14. Find the slope of the line that contains the points $(-1, 3)$ and $(3, -4)$.

15. Find the equation of the line that contains the point $(-1, 5)$ and has slope $\dfrac{3}{2}$.

16. Find the equation of the line that contains the points $(4, -2)$ and $(0, 3)$.

17. Find the equation of the line that contains the point $(2, 4)$ and is parallel to the line $y = -\dfrac{3}{2}x + 2$.

18. Find the equation of the line that contains the point $(4, 0)$ and is perpendicular to the line $3x - 2y = 5$.

19. Graph $3x - 5y = 15$ by using the x- and y-intercepts.

20. Graph the line that passes through the point $(-3, 1)$ and has slope $-\dfrac{3}{2}$.

21. Graph the solution set of $3x - 2y \geq 6$.

22. **Coins** A coin purse contains coins with a value of $1.60. The purse contains nickels and quarters. There are three times as many nickels as quarters. Find the number of nickels in the purse.

23. **Uniform Motion** Two planes are 1800 mi apart and are traveling toward each other. One plane is traveling twice as fast as the other plane. The planes will meet in 3 h. Find the speed of each plane.

24. **Mixture Problem** A grocer combines coffee that costs $8.00 per pound with coffee that costs $3.00 per pound. How many pounds of each should be used to make 80 lb of a blend that costs $5.00 per pound?

25. **Depreciation** The relationship between the depreciated value of a truck for income tax purposes and its age in years is shown in the graph at the right. Write the equation for the line that represents the depreciated value of the truck. Write a sentence that explains the meaning of the slope in the context of this exercise.

4 Systems of Equations and Inequalities

DVD SSM

Online Study Center

Need help? For online student resources, such as section quizzes, visit the Online Study Center at college.hmco.com/pic/aufmannABI2e.

Team sports are a great way to have fun and to get exercise. Informal leagues are formed all over the country for people who want to play a quick game of football or soccer in their free time. Suppose you are the manager of one such soccer league. You have been asked to create a schedule for 12 teams in which each team plays every other team once. How can you calculate how many games must be scheduled? As we learn in the **Focus on Problem Solving on pages 357–358,** one strategy would be to solve an easier problem first. Making connections to other problems you have already solved is an important step toward becoming an excellent problem solver.

Prep Test

1. Simplify: $10\left(\dfrac{3}{5}x + \dfrac{1}{2}y\right)$

2. Evaluate $3x + 2y - z$ for $x = -1$, $y = 4$, and $z = -2$.

3. Given $3x - 2z = 4$, find the value of x when $z = -2$.

4. Solve: $3x + 4(-2x - 5) = -5$

5. Solve:
$0.45x + 0.06(-x + 4000) = 630$

6. Graph: $y = \dfrac{1}{2}x - 4$

7. Graph: $3x - 2y = 6$

8. Graph: $y > -\dfrac{3}{5}x + 1$

4.1 Solving Systems of Linear Equations by Graphing and by the Substitution Method

OBJECTIVE 1 Solve a system of linear equations by graphing

A **system of equations** is two or more equations considered together. The system at the right is a system of two linear equations in two variables. The graphs of the equations are straight lines.

$$3x + 4y = 7$$
$$2x - 3y = 6$$

A **solution of a system of equations in two variables** is an ordered pair that is a solution of each equation of the system.

Is $(3, -2)$ a solution of the system of equations shown at the right?

$$2x - 3y = 12$$
$$5x + 2y = 11$$

Replace x by 3 and y by -2.

$2x - 3y = 12$		$5x + 2y = 11$	
$2(3) - 3(-2)$	12	$5(3) + 2(-2)$	11
$6 - (-6)$		$15 + (-4)$	
	$12 = 12$		$11 = 11$

Yes, because $(3, -2)$ is a solution of each equation, it is a solution of the system of equations.

A solution of a system of linear equations can be found by graphing the lines that represent the equations of the system on the same set of coordinate axes. Three examples of systems of linear equations in two variables are shown below, along with the graphs of the equations of the systems.

System I

$x + 2y = 4$
$2x + y = -1$

System II

$2x + 3y = 6$
$4x + 6y = -12$

System III

$x - 2y = 4$
$2x - 4y = 8$

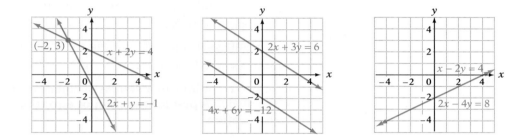

For System I, the two lines intersect at a single point whose coordinates are $(-2, 3)$. Because this point lies on both lines, it is a solution of each equation of the system of equations. We can check this by replacing x by -2 and y by 3. The check is shown below.

Check:

$x + 2y = 4$	$2x + y = -1$	
$-2 + 2(3)$ │ 4	$2(-2) + 3$ │ -1	• **Replace x by -2 and y by 3.**
$-2 + 6$ │ 4	$-4 + 3$ │ -1	
$4 = 4$	$-1 = -1$	

The ordered pair $(-2, 3)$ is a solution of System I.

When the graphs of a system of equations intersect at only one point, the system is called an **independent system of equations.** System I is an independent system of equations.

System II above and the graph of the equations of that system are shown again at the right. Note in this case that the graphs of the lines are parallel and do not intersect. Because the graphs do not intersect, there is no point that is on both lines. Therefore, the system of equations has no solution.

$2x + 3y = 6$
$4x + 6y = -12$

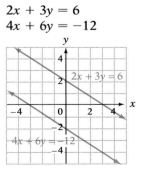

When a system of equations has no solution, it is called an **inconsistent system of equations.** System II is an inconsistent system of equations.

System III from the preceding page and the graph of the equations of that system are shown again at the right. Note that the graph of $x - 2y = 4$ lies directly on top of the graph of $2x - 4y = 8$. Thus the two lines intersect at an infinite number of points. Because the graphs intersect at an infinite number of points, there are an infinite number of solutions of this system of equations. Because each equation represents the same set of points, the solutions of the system of equations can be stated by using the ordered pairs of either one of the equations. Therefore, we can say, "The solutions are the ordered pairs that satisfy $x - 2y = 4$," or we can solve the equation for y and say, "The solutions are the ordered pairs that satisfy $y = \frac{1}{2}x - 2$." We normally state this solution using ordered pairs: "The solutions are the ordered pairs $\left(x, \frac{1}{2}x - 2\right)$."

$$x - 2y = 4$$
$$2x - 4y = 8$$

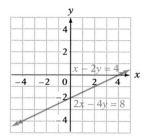

When the two equations in a system of equations represent the same line, the system is called a **dependent system of equations.** System III is a dependent system of equations.

These systems illustrate the three possibilities for a system of linear equations in two variables.

1. The graphs intersect at one point.
 The solution of the system of equations is the ordered pair (x, y) whose coordinates are the point of intersection.
 The system of equations is *independent.*

2. The lines are parallel and never intersect.
 There is no solution of the system of equations.
 The system of equations is *inconsistent.*

3. The graphs are the same line, and they intersect at infinitely many points.
 There are an infinite number of solutions of the system of equations.
 The system of equations is *dependent.*

Solve by graphing: $2x - y = 3$
$\qquad\qquad\qquad\quad 4x - 2y = 6$

Graph each line.
The system of equations is dependent.
Solve one of the equations for y.

$$2x - y = 3$$
$$-y = -2x + 3$$
$$y = 2x - 3$$

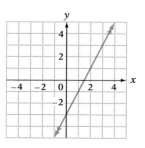

The solutions are the ordered pairs $(x, 2x - 3)$.

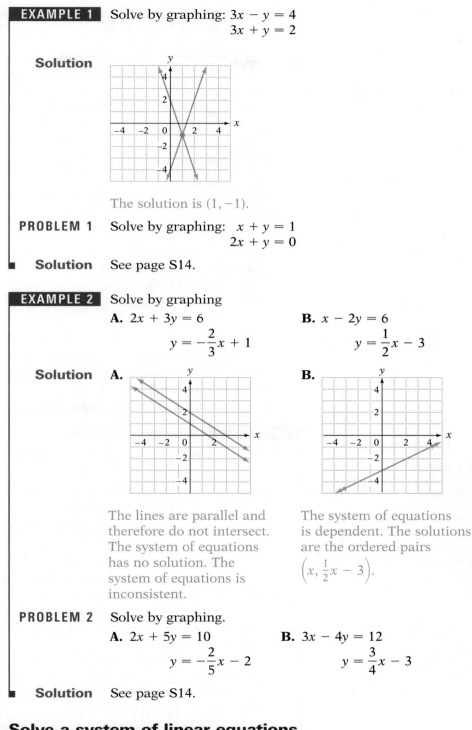

EXAMPLE 1 Solve by graphing: $3x - y = 4$
$3x + y = 2$

Solution

The solution is $(1, -1)$.

PROBLEM 1 Solve by graphing: $x + y = 1$
$2x + y = 0$

■ **Solution** See page S14.

EXAMPLE 2 Solve by graphing

A. $2x + 3y = 6$

$y = -\dfrac{2}{3}x + 1$

B. $x - 2y = 6$

$y = \dfrac{1}{2}x - 3$

Solution **A.** **B.**

The lines are parallel and therefore do not intersect. The system of equations has no solution. The system of equations is inconsistent.

The system of equations is dependent. The solutions are the ordered pairs $\left(x, \dfrac{1}{2}x - 3\right)$.

PROBLEM 2 Solve by graphing.

A. $2x + 5y = 10$

$y = -\dfrac{2}{5}x - 2$

B. $3x - 4y = 12$

$y = \dfrac{3}{4}x - 3$

■ **Solution** See page S14.

OBJECTIVE 2 **Solve a system of linear equations by the substitution method**

A graphical solution of a system of equations is based on approximating the coordinates of a point of intersection. An algebraic method called the **substitution method** can be used to find an exact solution of a system of equations.

To use the substitution method, we must write one of the equations of the system in terms of x or in terms of y.

Solve by the substitution method: $3x + y = 5$
$\qquad\qquad\qquad\qquad\qquad\qquad\qquad 4x + 5y = 3$

(1) $3x + y = 5$
(2) $4x + 5y = 3$

Solve equation (1) for y.
The result is equation (3).

(3) $\qquad\qquad 3x + y = 5$
$\qquad\qquad\qquad\quad y = -3x + 5$

Use equation (2).
Substitute $-3x + 5$ for y and solve for x.

$4x + 5y = 3$
$4x + 5(-3x + 5) = 3$
$4x - 15x + 25 = 3$
$-11x = -22$
$x = 2$

Now substitute the value of x into equation (3) and solve for y.

$y = -3x + 5$
$y = -3(2) + 5$
$y = -6 + 5$
$y = -1$

The solution is $(2, -1)$.

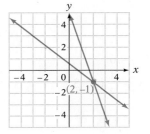

The graph of the system is shown at the left. Note that the graphs intersect at the point whose coordinates are $(2, -1)$, the solution of the system of equations.

EXAMPLE 3 Solve by substitution.

 A. $3x - 3y = 2$ **B.** $9x + 3y = 12$
$\qquad\qquad y = x + 2$ $y = -3x + 4$

Solution **A.** (1) $3x - 3y = 2$
 (2) $\qquad\quad y = x + 2$ • Equation (2) states that $y = x + 2$.
 $3x - 3(x + 2) = 2$ • Substitute $x + 2$ for y in equation (1).
 $3x - 3x - 6 = 2$
 $-6 = 2$

$-6 = 2$ is not a true equation. The system of equations is inconsistent. The system does not have a solution.

 B. (1) $9x + 3y = 12$
 (2) $\qquad\quad y = -3x + 4$ • Equation (2) states that
 $y = -3x + 4$.
 $9x + 3(-3x + 4) = 12$ • Substitute $-3x + 4$ for y in
 $9x - 9x + 12 = 12$ equation (1).
 $12 = 12$

$12 = 12$ is a true equation. The system of equations is dependent. The solutions are the ordered pairs $(x, -3x + 4)$.

PROBLEM 3 Solve by substitution. **A.** $3x - y = 3$ **B.** $6x - 3y = 6$
$\qquad\qquad\qquad\qquad\qquad\qquad\qquad\qquad\qquad\quad 6x + 3y = -4 \qquad\qquad 2x - y = 2$

Solution See page S14.

Take Note

If, after simplifying, there is no variable in the equation and the resulting equation is *not true*, the system of equations is inconsistent. ($-6 = 2$ is not a true equation.) If the resulting equation is *true*, the system of equations is dependent. ($12 = 12$ is a true equation.)

4.1 Concept Review

Determine whether the following statements are always true, sometimes true, or never true.

1. A solution of a system of equations in two variables is an ordered pair that is a solution of every equation in the system.

2. An inconsistent system of equations has no solution.

3. A system of two equations with two unknowns has an infinite number of solutions.

4. A system of two equations with the same slope and different y-intercepts has an infinite number of solutions.

5. When the graphs of a system of equations intersect at only one point, the system of equations is called an independent system of equations.

6. The system $y = \frac{3}{2}x + 2$ and $y = -4x + 2$ has exactly one solution.

4.1 Exercises

1 Solve a system of linear equations by graphing

Determine whether the ordered pair is a solution of the system of equations.

1. $(0, -1)$
$3x - 2y = 2$
$x + 2y = 6$

2. $(2, 1)$
$x + y = 3$
$2x - 3y = 1$

3. $(-3, -5)$
$x + y = -8$
$2x + 5y = -31$

4. $(1, -1)$
$3x - y = 4$
$7x + 2y = -5$

State whether the system of equations is independent, inconsistent, or dependent.

5. **6.** **7.** **8.**

9. Use the graph to solve this system of equations:
$x - 2y = -8$
$3x + y = -3$
The solution is _____?_____ .

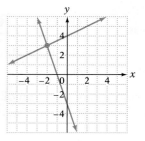

10. An independent system of equations has _____?_____ solution(s), an inconsistent system of equations has _____?_____ solution(s), and a dependent system of equations has _____?_____ solution(s).

Solve by graphing.

11. $x + y = 2$
$x - y = 4$

12. $x + y = 1$
$3x - y = -5$

13. $x - y = -2$
$x + 2y = 10$

14. $2x - y = 5$
$3x + y = 5$

15. $3x - 2y = 6$
$y = 3$

16. $x = 4$
$3x - 2y = 4$

17. $x = 4$
$y = -1$

18. $x + 2 = 0$
$y - 1 = 0$

19. $y = x - 5$
$2x + y = 4$

20. $2x - 5y = 4$
$y = x + 1$

21. $y = \dfrac{1}{2}x - 2$
$x - 2y = 8$

22. $2x + 3y = 6$
$y = -\dfrac{2}{3}x + 1$

23. $2x - 5y = 10$
$y = \dfrac{2}{5}x - 2$

24. $3x - 2y = 6$
$y = \dfrac{3}{2}x - 3$

25. $3x - 4y = 12$
$5x + 4y = -12$

26. $2x - 3y = 6$
$2x - 5y = 10$

In Exercises 27 and 28, A, B, C, and D are nonzero real numbers.

27. Is the following system of equations independent, inconsistent, or dependent?
$y = Ax + C$
$y = Ax + D, D \neq C$

28. Is the following system of equations independent, inconsistent, or dependent?
$x = A$
$y = B$

2 Solve a system of linear equations by the substitution method

29. When you solve a system of equations using the substitution method, how can you tell whether the system of equations is inconsistent?

30. When you solve a system of equations using the substitution method, how can you tell whether the system of equations is dependent?

31. Use this system of equations: (1) $y = 2x - 8$
 (2) $x = 3$

To solve the system by substitution, substitute ___?___ for x in equation (1): $y = 2(3) - 8 =$ ___?___ .

The solution of the system is (___?___ , ___?___).

32. Use this system of equations: (1) $y = x - 1$
 (2) $3x + y = 3$

To solve the system by substitution, substitute ___?___ for y in equation (2): $3x + x - 1 = 3$.

Solving the equation $3x + x - 1 = 3$ for x gives $x =$ ___?___ .

To find y, substitute this value of x into equation (1): $y =$ ___?___ $- 1 =$ ___?___ .

The solution of the system is (___?___ , ___?___).

Solve by substitution.

33. $3x - 2y = 4$
 $x = 2$

34. $y = -2$
 $2x + 3y = 4$

35. $y = 2x - 1$
 $x + 2y = 3$

36. $y = -x + 1$
 $2x - y = 5$

37. $4x - 3y = 5$
 $y = 2x - 3$

38. $3x + 5y = -1$
 $y = 2x - 8$

39. $x = 2y + 4$
 $4x + 3y = -17$

40. $3x - 2y = -11$
 $x = 2y - 9$

41. $5x + 4y = -1$
 $y = 2 - 2x$

42. $3x + 2y = 4$
 $y = 1 - 2x$

43. $7x - 3y = 3$
 $x = 2y + 2$

44. $3x - 4y = 6$
 $x = 3y + 2$

45. $2x + 2y = 7$
 $y = 4x + 1$

46. $3x + 7y = -5$
 $y = 6x - 5$

47. $3x + y = 5$
 $2x + 3y = 8$

48. $4x + y = 9$
 $3x - 4y = 2$

49. $x + 3y = 5$
 $2x + 3y = 4$

50. $x - 4y = 2$
 $2x - 5y = 1$

51. $3x + 4y = 14$
 $2x + y = 1$

52. $5x + 3y = 8$
 $3x + y = 8$

53. $3x + 5y = 0$
 $x - 4y = 0$

54. $2x - 7y = 0$
 $3x + y = 0$

55. $5x - 3y = -2$
 $-x + 2y = -8$

56. $2x + 7y = 1$
 $-x + 4y = 7$

57. $x + 3y = 4$
 $x = 5 - 3y$

58. $6x + 2y = 7$
 $y = -3x + 2$

59. $2x - 4y = 16$
 $-x + 2y = -8$

60. $3x - 12y = -24$
$-x + 4y = 8$

61. $3x - y = 10$
$6x - 2y = 5$

62. $6x - 4y = 3$
$3x - 2y = 9$

63. $y = 3x + 2$
$y = 2x + 3$

64. $y = 3x - 7$
$y = 2x - 5$

65. $x = 2y + 1$
$x = 3y - 1$

66. $x = 4y + 1$
$x = -2y - 5$

67. $y = 5x - 1$
$y = 5 - x$

68. $y = 3 - 2x$
$y = 2 - 3x$

In Exercises 69 and 70, A, B, C, and D are nonzero real numbers.

69. Is the following system of equations independent, inconsistent, or dependent?
$x + By = C$
$x = C - By$

70. Is the following system of equations independent, inconsistent, or dependent?
$Ax + y = C$
$y = D - Ax, D \neq C$

4.1 Applying Concepts

For what values of k will the system of equations be inconsistent?

71. $2x - 2y = 5$
$kx - 2y = 3$

72. $6x - 3y = 4$
$3x - ky = 1$

73. $x = 6y + 6$
$kx - 3y = 6$

74. $x = 2y + 2$
$kx - 8y = 2$

Solve using two variables.

75. Number Problem The sum of two numbers is 44. One number is 8 less than the other number. Find the numbers.

76. Number Problem The sum of two numbers is 76. One number is 12 less than the other number. Find the numbers.

77. Number Problem The sum of two numbers is 19. Five less than twice the first number equals the second number. Find the numbers.

78. Number Problem The sum of two numbers is 22. Three times the first plus 2 equals the second number. Find the numbers.

Solve. (*Hint:* These equations are not linear equations. First rewrite the equations as linear equations by substituting x for $\frac{1}{a}$ and y for $\frac{1}{b}$.)

79. $\dfrac{2}{a} + \dfrac{3}{b} = 4$
$\dfrac{4}{a} + \dfrac{1}{b} = 3$

80. $\dfrac{2}{a} + \dfrac{1}{b} = 1$
$\dfrac{8}{a} - \dfrac{2}{b} = 0$

81. $\dfrac{1}{a} + \dfrac{3}{b} = 2$
$\dfrac{4}{a} - \dfrac{1}{b} = 3$

82. $\dfrac{3}{a} + \dfrac{4}{b} = -1$
$\dfrac{1}{a} + \dfrac{6}{b} = 2$

Use a graphing utility to solve each of the following systems of equations. Round answers to the nearest hundredth.

83. $y = -\dfrac{1}{2}x + 2$
 $y = 2x - 1$

84. $y = 1.2x + 2$
 $y = -1.3x - 3$

85. $y = \sqrt{2}x - 1$
 $y = -\sqrt{3}x + 1$

86. $y = \pi x - \dfrac{2}{3}$
 $y = -x + \dfrac{\pi}{2}$

87. **Demography** Data from the 2000 Census support the following model of the population decline since 1990 of Pittsburgh, Pennsylvania: $y_1 = -0.0036x + 9.559$, where y_1 is the population, in millions, of Pittsburgh in year x. According to data from the same census, Orlando, Florida, has experienced an increase in population that can be modeled by $y_2 = 0.0419x - 82.156$, where y_2 is the population, in millions, of Orlando in year x.

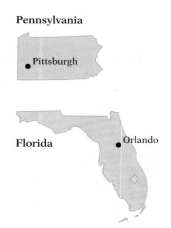

Pennsylvania

Pittsburgh

Florida Orlando

a. Assuming that this model is accurate for future years, will the population of Orlando exceed the population of Pittsburgh in the year 2010?

b. In what year will the population of Orlando first exceed the population of Pittsburgh? Round to the nearest whole number.

c. What does the slope of the line that models Pittsburgh's population mean in the context of this problem?

d. What does the slope of the line that models Orlando's population mean in the context of this problem?

88. Suppose $\begin{array}{l} a_1x + b_1y = c_1 \\ a_2x + b_2y = c_2 \end{array}$ is a system of equations and $\dfrac{a_1}{b_1} = \dfrac{a_2}{b_2}$. Can the system of equations be independent? Explain your answer.

4.2 Solving Systems of Linear Equations by the Addition Method

OBJECTIVE 1 **Solve a system of two linear equations in two variables by the addition method**

The **addition method** is an alternative method for solving a system of equations. This method is based on the Addition Property of Equations. Use the addition method when it is not convenient to solve one equation for one variable in terms of another variable.

Note, for the system of equations at the right, the effect of adding equation (2) to equation (1). Because $-3y$ and $3y$ are additive inverses, adding the equations results in an equation with only one variable.

(1) $5x - 3y = 14$
(2) $2x + 3y = -7$
 $7x + 0y = 7$
 $7x = 7$

The solution of the resulting equation is the first component of the ordered-pair solution of the system.

 $7x = 7$
 $x = 1$

The second component is found by substituting the value of x into equation (1) or (2) and then solving for y. Equation (1) is used here.

$$\begin{aligned}(1)\quad 5x - 3y &= 14\\ 5(1) - 3y &= 14\\ 5 - 3y &= 14\\ -3y &= 9\\ y &= -3\end{aligned}$$

The solution is $(1, -3)$.

Sometimes adding the two equations does not eliminate one of the variables. In this case, use the Multiplication Property of Equations to rewrite one or both of the equations so that when the equations are added, one of the variables is eliminated. To do this, first choose which variable to eliminate. The coefficients of that variable must be additive inverses. Multiply each equation by a constant that will produce coefficients that are additive inverses.

Solve by the addition method:

$$3x + 4y = 2$$
$$2x + 5y = -1$$

We will number the equations (1) and (2).

$$\begin{aligned}(1)\quad 3x + 4y &= 2\\ (2)\quad 2x + 5y &= -1\end{aligned}$$

Eliminate x. Multiply equation (1) by 2 and equation (2) by -3. Note how the constants are selected. The negative sign is used so that the coefficients will be additive inverses.

$$\begin{aligned}2\;&(3x + 4y) = 2(2)\\ -3\;&(2x + 5y) = -3(-1)\end{aligned}$$

The coefficients of the x-terms are additive inverses.

$$\begin{aligned}6x + 8y &= 4\\ -6x - 15y &= 3\end{aligned}$$

Add the equations. Solve for y.

$$\begin{aligned}-7y &= 7\\ y &= -1\end{aligned}$$

Substitute the value of y into one of the equations, and solve for x. Equation (1) is used here.

$$\begin{aligned}(1)\quad 3x + 4y &= 2\\ 3x + 4(-1) &= 2\\ 3x - 4 &= 2\\ 3x &= 6\\ x &= 2\end{aligned}$$

The solution is $(2, -1)$.

Solve by the addition method:

$$\frac{2}{3}x + \frac{1}{2}y = 4$$
$$\frac{1}{4}x - \frac{3}{8}y = -\frac{3}{4}$$

We will number the equations (1) and (2).

$$\begin{aligned}(1)\quad \frac{2}{3}x + \frac{1}{2}y &= 4\\ (2)\quad \frac{1}{4}x - \frac{3}{8}y &= -\frac{3}{4}\end{aligned}$$

Clear the fractions. Multiply each equation by the LCM of the denominators.

$$6\left(\frac{2}{3}x + \frac{1}{2}y\right) = 6(4)$$

$$8\left(\frac{1}{4}x - \frac{3}{8}y\right) = 8\left(-\frac{3}{4}\right)$$

$$4x + 3y = 24$$
$$2x - 3y = -6$$

Eliminate y. Add the equations. Solve for x.

$$6x = 18$$
$$x = 3$$

Substitute the value of x into equation (1), and solve for y.

$$\frac{2}{3}x + \frac{1}{2}y = 4$$

$$\frac{2}{3}(3) + \frac{1}{2}y = 4$$

$$2 + \frac{1}{2}y = 4$$

$$\frac{1}{2}y = 2$$

$$y = 4$$

The solution is (3, 4).

To solve the system of equations at the right, eliminate x. Multiply equation (1) by -2 and add to equation (2).

(1) $3x - 2y = 6$
(2) $6x - 4y = 4$

$$-6x + 4y = -12$$
$$6x - 4y = 4$$

$0 = -8$ is not a true equation.

$$0 = -8$$

The system of equations is inconsistent. The system does not have a solution.

The graph of the equations of the system is shown at the left. Note that the lines are parallel and therefore do not intersect.

EXAMPLE 1 Solve by the addition method.

A. $3x - 2y = 2x + 5$ **B.** $4x - 8y = 36$
 $2x + 3y = -4$ $3x - 6y = 27$

Solution A. (1) $3x - 2y = 2x + 5$
 (2) $2x + 3y = -4$

$$x - 2y = 5$$
$$2x + 3y = -4$$

• Write equation (1) in the form $Ax + By = C$.

$$-2(x - 2y) = -2(5)$$
$$2x + 3y = -4$$

• To eliminate x, multiply each side of equation (1) by -2.

$$-2x + 4y = -10$$
$$2x + 3y = -4$$

$$7y = -14$$
$$y = -2$$

• Add the equations.

$$2x + 3y = -4$$
$$2x + 3(-2) = -4$$
$$2x - 6 = -4$$
$$2x = 2$$
$$x = 1$$

- Replace y in equation (2) by its value.

The solution is $(1, -2)$.

B. (1) $4x - 8y = 36$
(2) $3x - 6y = 27$

$$3(4x - 8y) = 3(36)$$
$$-4(3x - 6y) = -4(27)$$
$$12x - 24y = 108$$
$$-12x + 24y = -108$$
$$0 = 0$$

- To eliminate x, multiply each side of equation (1) by 3 and each side of equation (2) by -4.

- Add the equations.

$0 = 0$ is a true equation. The system of equations is dependent. The solutions are the ordered pairs $\left(x, \frac{1}{2}x - \frac{9}{2}\right)$.

PROBLEM 1 Solve by the addition method.

 A. $2x + 5y = 6$ **B.** $2x + y = 5$
 $3x - 2y = 6x + 2$ $4x + 2y = 6$

Solution See pages S14–S15.

OBJECTIVE 2

Solve a system of three linear equations in three variables by the addition method

An equation of the form $Ax + By + Cz = D$, where A, B, and C are coefficients and D is a constant, is a **linear equation in three variables.** Examples of these equations are shown at the right. The graph of a linear equation in three variables is a plane.

$$2x + 4y - 3z = 7$$
$$x - 6y + z = -3$$

Graphing an equation in three variables requires a third coordinate axis perpendicular to the xy-plane. The third axis is commonly called the z-axis. The result is a three-dimensional coordinate system called the xyz**-coordinate system.** To help visualize a three-dimensional coordinate system, think of a corner of a room: the floor is the xy-plane, one wall is the yz-plane, and the other wall is the xz-plane. A three-dimensional coordinate system is shown at the right.

The graph of a point in an xyz-coordinate system is an **ordered triple** (x, y, z). Graphing an ordered triple requires three moves, the first along the x-axis, the second parallel to the y-axis, and the third parallel to the z-axis. The graphs of the points $(-4, 2, 3)$ and $(3, 4, -2)$ are shown at the right.

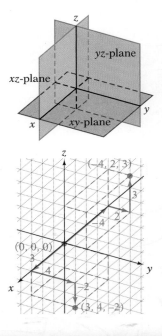

The graph of a linear equation in three variables is a plane. That is, if all the solutions of a linear equation in three variables were plotted in an *xyz*-coordinate system, the graph would look like a large piece of paper extending infinitely. The graph of $x + y + z = 3$ is shown at the right.

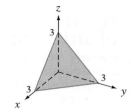

There are different ways in which three planes can be oriented in an *xyz*-coordinate system. The systems of equations represented by the following planes are inconsistent.

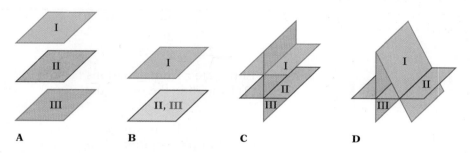

Graphs of Inconsistent Systems of Equations

For a system of three equations in three variables to have a solution, the graphs of the planes must intersect at a single point, they must intersect along a common line, or all equations must have a graph that is the same plane. These situations are shown in the figures that follow.

The three planes shown in Figure E below intersect at a point. A system of equations represented by planes that intersect at a point is independent.

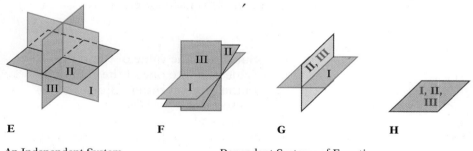

E

An Independent System
of Equations

F G H

Dependent Systems of Equations

The planes shown in Figures F and G above intersect along a common line. The system of equations represented by the planes in Figure H has a graph that is the same plane. The systems of equations represented by these three graphs are dependent.

Just as a solution of an equation in two variables is an ordered pair (x, y), a **solution of an equation in three variables** is an ordered triple (x, y, z). For example, $(2, 1, -3)$ is a solution of the equation $2x - y - 2z = 9$. The ordered triple $(1, 3, 2)$ is not a solution.

Point of Interest

In the early 1980s, Stephen Hoppe became interested in winning Monopoly strategies. Finding these strategies required solving a system that contained 123 equations with 123 variables!

A **system of linear equations in three variables** is shown at the right. A **solution of a system of equations in three variables** is an ordered triple that is a solution of each equation of the system.

$$x - 2y + z = 6$$
$$3x + y - 2z = 2$$
$$2x - 3y + 5z = 1$$

A system of linear equations in three variables can be solved by using the addition method. First, eliminate one variable from any two of the given equations. Then eliminate the same variable from any other two equations. The result will be a system of two equations in two variables. Solve this system by the addition method.

Solve: $x + 4y - z = 10$ (1) $x + 4y - z = 10$
 $3x + 2y + z = 4$ (2) $3x + 2y + z = 4$
 $2x - 3y + 2z = -7$ (3) $2x - 3y + 2z = -7$

Eliminate z from equations (1) and (2) by adding the two equations.

$x + 4y - z = 10$
$3x + 2y + z = 4$
(4) $4x + 6y = 14$

Eliminate z from equations (1) and (3). Multiply equation (1) by 2 and add it to equation (3).

$2x + 8y - 2z = 20$
$2x - 3y + 2z = -7$
(5) $4x + 5y = 13$

Solve the system of two equations in two variables.

(4) $4x + 6y = 14$
(5) $4x + 5y = 13$

Eliminate x. Multiply equation (5) by -1 and add it to equation (4).

$4x + 6y = 14$
$-4x - 5y = -13$
$y = 1$

Substitute the value of y into equation (4) or (5), and solve for x. Equation (4) is used here.

$4x + 6y = 14$
$4x + 6(1) = 14$
$4x + 6 = 14$
$4x = 8$
$x = 2$

Substitute the value of y and the value of x into one of the equations in the original system. Equation (2) is used here.

$3x + 2y + z = 4$
$3(2) + 2(1) + z = 4$
$6 + 2 + z = 4$
$8 + z = 4$
$z = -4$

The solution is $(2, 1, -4)$.

Solve: $2x - 3y - z = 1$ (1) $2x - 3y - z = 1$
 $x + 4y + 3z = 2$ (2) $x + 4y + 3z = 2$
 $4x - 6y - 2z = 5$ (3) $4x - 6y - 2z = 5$

Eliminate x from equations (1) and (2). Multiply equation (2) by -2 and add to equation (1).

$2x - 3y - z = 1$
$-2x - 8y - 6z = -4$
$-11y - 7z = -3$

Eliminate x from equations (1) and (3). Multiply equation (1) by -2 and add to equation (3). This is not a true equation.

$$-4x + 6y + 2z = -2$$
$$4x - 6y - 2z = 5$$
$$0 = 3$$

The system is inconsistent.

EXAMPLE 2 Solve: $3x - y + 2z = 1$
$2x + 3y + 3z = 4$
$x + y - 4z = -9$

Solution

(1) $3x - y + 2z = 1$
(2) $2x + 3y + 3z = 4$
(3) $x + y - 4z = -9$

$3x - y + 2z = 1$
$x + y - 4z = -9$
$4x - 2z = -8$
• Eliminate y. Add equations (1) and (3).

(4) $2x - z = -4$
• Multiply each side of the equation by $\frac{1}{2}$.

$9x - 3y + 6z = 3$
$2x + 3y + 3z = 4$
(5) $11x + 9z = 7$
• Multiply equation (1) by 3 and add to equation (2).

(4) $2x - z = -4$
(5) $11x + 9z = 7$
• Solve the system of two equations.

$18x - 9z = -36$
$11x + 9z = 7$
$29x = -29$
$x = -1$
• Multiply equation (4) by 9 and add to equation (5).

$2x - z = -4$
$2(-1) - z = -4$
$-2 - z = -4$
$-z = -2$
$z = 2$
• Replace x by -1 in equation (4).

$x + y - 4z = -9$
$-1 + y - 4(2) = -9$
$-1 + y - 8 = -9$
$-9 + y = -9$
$y = 0$
• Replace x by -1 and z by 2 in equation (1), (2), or (3). Equation (3) is used here.

The solution is $(-1, 0, 2)$.

PROBLEM 2 Solve: $x - y + z = 6$
$2x + 3y - z = 1$
$x + 2y + 2z = 5$

Solution See pages S15.

4.2 Concept Review

Determine whether the following statements are always true, sometimes true, or never true.

1. The graph of a linear equation in three variables is a plane.

2. A solution of an equation in three variables is an ordered triple (x, y, z).

3. The solution of a system of linear equations in three variables is a straight line.

4. A system of linear equations in three variables that has an infinite number of solutions is an independent system.

5. When you are solving a system of equations by the addition method and the final result is $0 = 5$, the system is inconsistent and has an infinite number of solutions.

4.2 Exercises

1 **Solve a system of two linear equations in two variables by the addition method**

1. If you begin to solve the system of equations $\begin{array}{l} x + y = 6 \\ x - y = 4 \end{array}$ by adding the two equations, the result is $2x = 10$, which means $x = 5$. Does this mean that 5 is a solution of the system of equations? Why or why not?

2. Use this system of equations: (1) $-4x - y = 5$
(2) $x - 2y = 4$

 a. To eliminate x from the system of equations by using the addition method, multiply each side of equation (___?___) by ___?___.
 b. To eliminate y from the system of equations by using the addition method, multiply each side of equation (___?___) by ___?___.

Solve by the addition method.

3. $x - y = 5$
 $x + y = 7$

4. $x + y = 1$
 $2x - y = 5$

5. $3x + y = 4$
 $x + y = 2$

6. $x - 3y = 4$
 $x + 5y = -4$

7. $3x + y = 7$
 $x + 2y = 4$

8. $x - 2y = 7$
 $3x - 2y = 9$

9. $3x - y = 4$
 $6x - 2y = 8$

10. $x - 2y = -3$
 $-2x + 4y = 6$

11. $2x + 5y = 9$
 $4x - 7y = -16$

12. $8x - 3y = 21$
$4x + 5y = -9$

13. $4x - 6y = 5$
$2x - 3y = 7$

14. $3x + 6y = 7$
$2x + 4y = 5$

15. $3x - 5y = 7$
$x - 2y = 3$

16. $3x + 4y = 25$
$2x + y = 10$

17. $3x + 2y = 16$
$2x - 3y = -11$

18. $2x - 5y = 13$
$5x + 3y = 17$

19. $4x + 4y = 5$
$2x - 8y = -5$

20. $3x + 7y = 16$
$4x - 3y = 9$

21. $5x + 4y = 0$
$3x + 7y = 0$

22. $3x - 4y = 0$
$4x - 7y = 0$

23. $3x - 6y = 6$
$9x - 3y = 8$

24. $4x - 8y = 5$
$8x + 2y = 1$

25. $5x + 2y = 2x + 1$
$2x - 3y = 3x + 2$

26. $3x + 3y = y + 1$
$x + 3y = 9 - x$

27. $\dfrac{2}{3}x - \dfrac{1}{2}y = 3$
$\dfrac{1}{3}x - \dfrac{1}{4}y = \dfrac{3}{2}$

28. $\dfrac{3}{4}x + \dfrac{1}{3}y = -\dfrac{1}{2}$
$\dfrac{1}{2}x - \dfrac{5}{6}y = -\dfrac{7}{2}$

29. $\dfrac{2}{5}x - \dfrac{1}{3}y = 1$
$\dfrac{3}{5}x + \dfrac{2}{3}y = 5$

30. $\dfrac{5}{6}x + \dfrac{1}{3}y = \dfrac{4}{3}$
$\dfrac{2}{3}x - \dfrac{1}{2}y = \dfrac{11}{6}$

31. $\dfrac{3}{4}x + \dfrac{2}{5}y = -\dfrac{3}{20}$
$\dfrac{3}{2}x - \dfrac{1}{4}y = \dfrac{3}{4}$

32. $\dfrac{2}{5}x - \dfrac{1}{2}y = \dfrac{13}{2}$
$\dfrac{3}{4}x - \dfrac{1}{5}y = \dfrac{17}{2}$

33. $4x - 5y = 3y + 4$
$2x + 3y = 2x + 1$

34. $5x - 2y = 8x - 1$
$2x + 7y = 4y + 9$

35. $2x + 5y = 5x + 1$
$3x - 2y = 3y + 3$

36. When you solve a system of equations using the addition method, how can you tell whether the system of equations is inconsistent?

37. When you solve a system of equations using the addition method, how can you tell whether the system of equations is dependent?

In Exercises 38 and 39, A, B, C, and D are nonzero real numbers.

38. Is the following system of equations independent, inconsistent, or dependent?
$Ax + By = 0$
$Cx + Dy = 0, D \neq B$

39. Is the following system of equations independent, inconsistent, or dependent?
$Ax + By = C$
$DAx + DBy = DC$

2 Solve a system of three linear equations in three variables by the addition method

40. Verify that the ordered triple $(-1, -2, 3)$ is a solution of the equation $4x - y + 2z = 4$ by substituting -1 for x, -2 for y, and 3 for z.

41. Look at Exercise 42. State two pairs of equations that can be added together to eliminate the variable z.

Solve by the addition method.

42. $x + 2y - z = 1$
$2x - y + z = 6$
$x + 3y - z = 2$

43. $x + 3y + z = 6$
$3x + y - z = -2$
$2x + 2y - z = 1$

44. $2x - y + 2z = 7$
$x + y + z = 2$
$3x - y + z = 6$

45. $x - 2y + z = 6$
$x + 3y + z = 16$
$3x - y - z = 12$

46. $3x + y = 5$
$3y - z = 2$
$x + z = 5$

47. $2y + z = 7$
$2x - z = 3$
$x - y = 3$

48. $x - y + z = 1$
$2x + 3y - z = 3$
$-x + 2y - 4z = 4$

49. $2x + y - 3z = 7$
$x - 2y + 3z = 1$
$3x + 4y - 3z = 13$

50. $2x + 3z = 5$
$3y + 2z = 3$
$3x + 4y = -10$

51. $3x + 4z = 5$
$2y + 3z = 2$
$2x - 5y = 8$

52. $2x + 4y - 2z = 3$
$x + 3y + 4z = 1$
$x + 2y - z = 4$

53. $x - 3y + 2z = 1$
$x - 2y + 3z = 5$
$2x - 6y + 4z = 3$

54. $2x + y - z = 5$
$x + 3y + z = 14$
$3x - y + 2z = 1$

55. $3x - y - 2z = 11$
$2x + y - 2z = 11$
$x + 3y - z = 8$

56. $3x + y - 2z = 2$
$x + 2y + 3z = 13$
$2x - 2y + 5z = 6$

57. $4x + 5y + z = 6$
$2x - y + 2z = 11$
$x + 2y + 2z = 6$

58. $2x - y + z = 6$
$3x + 2y + z = 4$
$x - 2y + 3z = 12$

59. $3x + 2y - 3z = 8$
$2x + 3y + 2z = 10$
$x + y - z = 2$

60. $3x - 2y + 3z = -4$
$2x + y - 3z = 2$
$3x + 4y + 5z = 8$

61. $3x - 3y + 4z = 6$
$4x - 5y + 2z = 10$
$x - 2y + 3z = 4$

62. $3x - y + 2z = 2$
$4x + 2y - 7z = 0$
$2x + 3y - 5z = 7$

63. $2x + 2y + 3z = 13$
$-3x + 4y - z = 5$
$5x - 3y + z = 2$

64. $2x - 3y + 7z = 0$
$x + 4y - 4z = -2$
$3x + 2y + 5z = 1$

65. $5x + 3y - z = 5$
$3x - 2y + 4z = 13$
$4x + 3y + 5z = 22$

4.2 Applying Concepts

Solve. (*Hint:* First multiply both sides of each equation in the system by a multiple of 10 so that the coefficients and constants are integers.)

66. $0.2x - 0.3y = 0.5$
$0.3x - 0.2y = 0.5$

67. $0.4x - 0.9y = -0.1$
$0.3x + 0.2y = 0.8$

68. $1.25x - 0.25y = -1.5$
$1.5x + 2.5y = 1$

69. $2.25x + 1.5y = 3$
$1.75x + 2.25y = 1.25$

70. $1.5x + 2.5y + 1.5z = 8$
$0.5x - 2y - 1.5z = -1$
$2.5x - 1.5y + 2z = 2.5$

71. $1.6x - 0.9y + 0.3z = 2.9$
$1.6x + 0.5y - 0.1z = 3.3$
$0.8x - 0.7y + 0.1z = 1.5$

Solve.

72. The point of intersection of the graphs of the equations $Ax + 3y = 6$ and $2x + By = -4$ is $(3, -2)$. Find A and B.

73. The point of intersection of the graphs of the equations $Ax + 3y + 2z = 8$, $2x + By - 3z = -12$, and $3x - 2y + Cz = 1$ is $(3, -2, 4)$. Find A, B, and C.

74. The distance between a point and a line is the perpendicular distance from the point to the line. Find the distance between the point $(3, 1)$ and the line $y = x$.

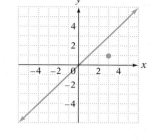

75. **Coin Problem** A coin bank contains only nickels, dimes, and quarters. There is a total of 30 coins in the bank. The value of all the coins is $3.25. Find the number of nickels, dimes, and quarters in the bank. (*Hint:* There is more than one solution.)

The following are not linear systems of equations. However, they can be solved by using a modification of the addition method. Solve each system of equations.

76. $\dfrac{1}{x} - \dfrac{2}{y} = 3$

$\dfrac{2}{x} + \dfrac{3}{y} = -1$

77. $\dfrac{1}{x} + \dfrac{2}{y} = 3$

$\dfrac{1}{x} - \dfrac{3}{y} = -2$

78. $\dfrac{3}{x} + \dfrac{2}{y} = 1$

$\dfrac{2}{x} + \dfrac{4}{y} = -2$

79. $\dfrac{3}{x} - \dfrac{5}{y} = -\dfrac{3}{2}$

$\dfrac{1}{x} - \dfrac{2}{y} = -\dfrac{2}{3}$

80. For Exercise 76, solve each equation of the system for y. Then use a graphing utility to verify the solution you determined algebraically. Note that the graphs are not straight lines.

81. Describe the graph of each of the following equations in an xyz-coordinate system.

a. $x = 3$ **b.** $y = 4$ **c.** $z = 2$ **d.** $y = x$

Solving Systems of Equations by Using Determinants and by Using Matrices

4.3

OBJECTIVE **1** Evaluate determinants

A **matrix** is a rectangular array of numbers. Each number in the matrix is called an **element** of the matrix. The matrix at the right, with three rows and four columns, is called a 3×4 ("3 by 4") matrix.

$$A = \begin{bmatrix} 1 & -3 & 2 & 4 \\ 0 & 4 & -3 & 2 \\ 6 & -5 & 4 & -1 \end{bmatrix}$$

Point of Interest

The word matrix was first used (in a mathematical context) in 1850. The root of this word is the Latin mater. A matrix was thought of as being an object from which something else originates. The idea was that a determinant, discussed at the right, originated (was born) from a matrix. Today, matrices are one of the most widely used tools of applied mathematics.

A matrix of m rows and n columns is said to be of **order $m \times n$**. Matrix A above has order 3×4. The notation a_{ij} refers to the element of a matrix in the ith row and the jth column. For matrix A, $a_{23} = -3$, $a_{31} = 6$, and $a_{13} = 2$.

A **square matrix** is one that has the same number of rows as columns. A 2×2 matrix and a 3×3 matrix are shown at the right.

$$\begin{bmatrix} -1 & 3 \\ 5 & 2 \end{bmatrix} \qquad \begin{bmatrix} 4 & 0 & 1 \\ 5 & -3 & 7 \\ 2 & 1 & 4 \end{bmatrix}$$

Associated with every square matrix is a number called its **determinant.**

Determinant of a 2 × 2 Matrix

The determinant of a 2×2 matrix $\begin{bmatrix} a_{11} & a_{12} \\ a_{21} & a_{22} \end{bmatrix}$ is written $\begin{vmatrix} a_{11} & a_{12} \\ a_{21} & a_{22} \end{vmatrix}$. The value of this determinant is given by the formula

$$\begin{vmatrix} a_{11} & a_{12} \\ a_{21} & a_{22} \end{vmatrix} = a_{11}a_{22} - a_{21}a_{12}$$

Note that vertical bars are used to represent the determinant and that brackets are used to represent the matrix.

Find the value of the determinant $\begin{vmatrix} 3 & 4 \\ -1 & 2 \end{vmatrix}$.

$$\begin{vmatrix} 3 & 4 \\ -1 & 2 \end{vmatrix} = 3 \cdot 2 - (-1)(4) = 6 - (-4) = 10$$

The value of the determinant is 10.

For a square matrix whose order is 3×3 or greater, the value of the determinant of that matrix is found by using 2×2 determinants.

The **minor of an element** in a 3×3 determinant is the 2×2 determinant that is obtained by eliminating the row and column that contain that element.

Find the minor of -3 for the determinant $\begin{vmatrix} 2 & -3 & 4 \\ 0 & 4 & 8 \\ -1 & 3 & 6 \end{vmatrix}$.

The minor of -3 is the 2×2 determinant created by eliminating the row and column that contain -3.

Eliminate the row and column as shown: $\begin{vmatrix} 2 & -3 & 4 \\ 0 & 4 & 8 \\ -1 & 3 & 6 \end{vmatrix}$

The minor of -3 is $\begin{vmatrix} 0 & 8 \\ -1 & 6 \end{vmatrix}$.

Take Note

The only difference between the cofactor and the minor is one of sign. The definition at the right can be stated symbolically as follows: If C_{ij} is the cofactor and M_{ij} is the minor of the determinant element a_{ij}, then $C_{ij} = (-1)^{i+j}M_{ij}$. If $i + j$ is an even number, then $(-1)^{i+j} = 1$ and $C_{ij} = M_{ij}$. If $i + j$ is an odd number, then $(-1)^{i+j} = -1$ and $C_{ij} = -M_{ij}$.

Definition of a Cofactor

The **cofactor** of an element of a determinant is $(-1)^{i+j}$ times the minor of that element, where i is the row number of the element and j is the column number of the element.

For the determinant $\begin{vmatrix} 3 & -2 & 1 \\ 2 & -5 & -4 \\ 0 & 3 & 1 \end{vmatrix}$, find the cofactor of -2 and of -5.

Because -2 is in the first row and the second column, $i = 1$ and $j = 2$. Therefore, $i + j = 1 + 2 = 3$, and $(-1)^{i+j} = (-1)^3 = -1$.

The cofactor of -2 is $(-1)\begin{vmatrix} 2 & -4 \\ 0 & 1 \end{vmatrix}$.

Because -5 is in the second row and the second column, $i = 2$ and $j = 2$. Therefore, $i + j = 2 + 2 = 4$, and $(-1)^{i+j} = (-1)^4 = 1$.

The cofactor of -5 is $1 \cdot \begin{vmatrix} 3 & 1 \\ 0 & 1 \end{vmatrix}$.

Note from this example that the cofactor of an element is -1 times the minor of that element or 1 times the minor of that element, depending on whether the sum $i + j$ is an odd or an even integer.

The value of a 3×3 or larger determinant can be found by **expanding by cofactors** of *any* row or *any* column. The result of expanding by cofactors using the first row of a 3×3 determinant is shown here.

$$\begin{vmatrix} a_{11} & a_{12} & a_{13} \\ a_{21} & a_{22} & a_{23} \\ a_{31} & a_{32} & a_{33} \end{vmatrix} = a_{11}(-1)^{1+1}\begin{vmatrix} a_{22} & a_{23} \\ a_{32} & a_{33} \end{vmatrix} + a_{12}(-1)^{1+2}\begin{vmatrix} a_{21} & a_{23} \\ a_{31} & a_{33} \end{vmatrix} + a_{13}(-1)^{1+3}\begin{vmatrix} a_{21} & a_{22} \\ a_{31} & a_{32} \end{vmatrix}$$

$$= a_{11}\begin{vmatrix} a_{22} & a_{23} \\ a_{32} & a_{33} \end{vmatrix} - a_{12}\begin{vmatrix} a_{21} & a_{23} \\ a_{31} & a_{33} \end{vmatrix} + a_{13}\begin{vmatrix} a_{21} & a_{22} \\ a_{31} & a_{32} \end{vmatrix}$$

Find the value of the determinant $\begin{vmatrix} 2 & -3 & 2 \\ 1 & 3 & -1 \\ 0 & -2 & 2 \end{vmatrix}$.

Expand by cofactors of the first row.

$$\begin{vmatrix} 2 & -3 & 2 \\ 1 & 3 & -1 \\ 0 & -2 & 2 \end{vmatrix} = 2\begin{vmatrix} 3 & -1 \\ -2 & 2 \end{vmatrix} - (-3)\begin{vmatrix} 1 & -1 \\ 0 & 2 \end{vmatrix} + 2\begin{vmatrix} 1 & 3 \\ 0 & -2 \end{vmatrix}$$

$$= 2(6 - 2) - (-3)(2 - 0) + 2(-2 - 0)$$

$$= 2(4) - (-3)(2) + 2(-2) = 8 - (-6) + (-4)$$

$$= 10$$

To illustrate a statement made earlier, the value of this determinant will now be found by expanding by cofactors of the second column.

$$\begin{vmatrix} 2 & -3 & 2 \\ 1 & 3 & -1 \\ 0 & -2 & 2 \end{vmatrix} = -3(-1)^{1+2}\begin{vmatrix} 1 & -1 \\ 0 & 2 \end{vmatrix} + 3(-1)^{2+2}\begin{vmatrix} 2 & 2 \\ 0 & 2 \end{vmatrix} + (-2)(-1)^{3+2}\begin{vmatrix} 2 & 2 \\ 1 & -1 \end{vmatrix}$$

$$= -3(-1)\begin{vmatrix} 1 & -1 \\ 0 & 2 \end{vmatrix} + 3 \cdot 1\begin{vmatrix} 2 & 2 \\ 0 & 2 \end{vmatrix} + (-2)(-1)\begin{vmatrix} 2 & 2 \\ 1 & -1 \end{vmatrix}$$

$$= 3(2 - 0) + 3(4 - 0) + 2(-2 - 2)$$

$$= 3(2) + 3(4) + 2(-4) = 6 + 12 + (-8)$$

$$= 10$$

Note that the value of the determinant is the same whether the first row or the second column is used to expand by cofactors. *Any row or column* can be used to evaluate a determinant by expanding by cofactors.

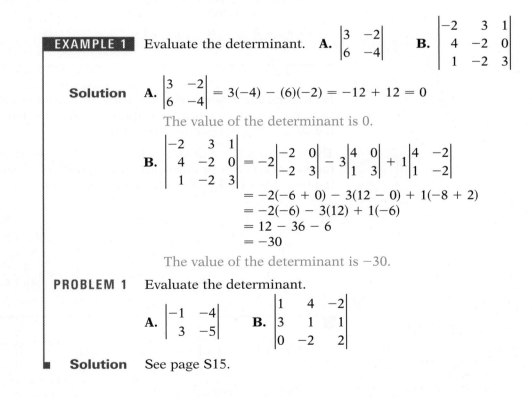

EXAMPLE 1 Evaluate the determinant. **A.** $\begin{vmatrix} 3 & -2 \\ 6 & -4 \end{vmatrix}$ **B.** $\begin{vmatrix} -2 & 3 & 1 \\ 4 & -2 & 0 \\ 1 & -2 & 3 \end{vmatrix}$

Solution **A.** $\begin{vmatrix} 3 & -2 \\ 6 & -4 \end{vmatrix} = 3(-4) - (6)(-2) = -12 + 12 = 0$

The value of the determinant is 0.

B. $\begin{vmatrix} -2 & 3 & 1 \\ 4 & -2 & 0 \\ 1 & -2 & 3 \end{vmatrix} = -2\begin{vmatrix} -2 & 0 \\ -2 & 3 \end{vmatrix} - 3\begin{vmatrix} 4 & 0 \\ 1 & 3 \end{vmatrix} + 1\begin{vmatrix} 4 & -2 \\ 1 & -2 \end{vmatrix}$

$$= -2(-6 + 0) - 3(12 - 0) + 1(-8 + 2)$$

$$= -2(-6) - 3(12) + 1(-6)$$

$$= 12 - 36 - 6$$

$$= -30$$

The value of the determinant is −30.

PROBLEM 1 Evaluate the determinant.

A. $\begin{vmatrix} -1 & -4 \\ 3 & -5 \end{vmatrix}$ **B.** $\begin{vmatrix} 1 & 4 & -2 \\ 3 & 1 & 1 \\ 0 & -2 & 2 \end{vmatrix}$

Solution See page S15.

OBJECTIVE **2** **Solve systems of linear equations by using Cramer's Rule**

The connection between determinants and systems of equations can be understood by solving a general system of linear equations.

Solve: $a_{11}x + a_{12}y = b_1$ (1) $a_{11}x + a_{12}y = b_1$
$\quad\quad\quad a_{21}x + a_{22}y = b_2$ (2) $a_{21}x + a_{22}y = b_2$

Eliminate y. Multiply equation (1)
by a_{22} and equation (2) by $-a_{12}$.

$$a_{11}a_{22}x + a_{12}a_{22}y = b_1a_{22}$$
$$-a_{21}a_{12}x - a_{12}a_{22}y = -b_2a_{12}$$

Add the equations.

$$(a_{11}a_{22} - a_{21}a_{12})x = b_1a_{22} - b_2a_{12}$$

Assuming $a_{11}a_{22} - a_{21}a_{12} \neq 0$, solve
for x.

$$x = \frac{b_1a_{22} - b_2a_{12}}{a_{11}a_{22} - a_{21}a_{12}}$$

The denominator $a_{11}a_{22} - a_{21}a_{12}$ is
the determinant of the coefficients
of x and y. This is called the **coeffi-
cient determinant.**

$$a_{11}a_{22} - a_{21}a_{12} = \begin{vmatrix} a_{11} & a_{12} \\ a_{21} & a_{22} \end{vmatrix}$$

Coefficients of x ———————↑ ↑
Coefficients of y ————————————┘

The numerator for x, $b_1a_{22} - b_2a_{12}$,
is the determinant obtained by
replacing the first column in the
coefficient determinant by the con-
stants b_1 and b_2. This is called the
numerator determinant.

$$b_1a_{22} - b_2a_{12} = \begin{vmatrix} b_1 & a_{12} \\ b_2 & a_{22} \end{vmatrix}$$

Constants of ——————↑
the equations

Following a similar procedure and eliminating x, we can also express the
y-component of the solution in determinant form. These results are summa-
rized in Cramer's Rule.

Cramer's Rule

The solution of the system of equations $\begin{matrix} a_{11}x + a_{12}y = b_1 \\ a_{21}x + a_{22}y = b_2 \end{matrix}$ is given by

$x = \dfrac{D_x}{D}$ and $y = \dfrac{D_y}{D}$, where

$$D = \begin{vmatrix} a_{11} & a_{12} \\ a_{21} & a_{22} \end{vmatrix}, \quad D_x = \begin{vmatrix} b_1 & a_{12} \\ b_2 & a_{22} \end{vmatrix}, \quad D_y = \begin{vmatrix} a_{11} & b_1 \\ a_{21} & b_2 \end{vmatrix}, \text{ and } D \neq 0.$$

Point of Interest

*Cramer's Rule is named after
Gabriel Cramer, who used it in a
book he published in 1750.
However, this rule was published in
1683 by the Japanese
mathematician Seki Kown. That
publication occurred seven years
before Cramer's birth.*

EXAMPLE 2 Solve by using Cramer's Rule: $2x - 3y = 8$
$\quad\quad\quad\quad\quad\quad\quad\quad\quad\quad\quad\quad\quad\quad\quad\quad\quad 5x + 6y = 11$

Solution $D = \begin{vmatrix} 2 & -3 \\ 5 & 6 \end{vmatrix} = 27$ • Find the value of the coefficient determinant.

$D_x = \begin{vmatrix} 8 & -3 \\ 11 & 6 \end{vmatrix} = 81$ • Find the value of each of the numerator determinants.

$D_y = \begin{vmatrix} 2 & 8 \\ 5 & 11 \end{vmatrix} = -18$

$x = \dfrac{D_x}{D} = \dfrac{81}{27} = 3$ • Use Cramer's Rule to write the solutions.

$y = \dfrac{D_y}{D} = \dfrac{-18}{27} = -\dfrac{2}{3}$

The solution is $\left(3, -\dfrac{2}{3}\right)$.

PROBLEM 2 Solve by using Cramer's Rule: $6x - 6y = 5$
$2x - 10y = -1$

■ **Solution** See page S15.

For the system shown at the right, $D = 0$. Therefore, $\dfrac{D_x}{D}$ $6x - 9y = 5$
$4x - 6y = 4$

and $\dfrac{D_y}{D}$ are undefined.

When $D = 0$, the system of equations is dependent if both D_x and D_y are zero. The system of equations is inconsistent if $D = 0$ and either D_x or D_y is not zero.

$$D = \begin{vmatrix} 6 & -9 \\ 4 & -6 \end{vmatrix} = 0$$

A procedure similar to that followed for two equations in two variables can be used to extend Cramer's Rule to three equations in three variables.

Cramer's Rule for Three Equations in Three Variables

$$a_{11}x + a_{12}y + a_{13}z = b_1$$

The solution of the system of equations $a_{21}x + a_{22}y + a_{23}z = b_2$ is given

$$a_{31}x + a_{32}y + a_{33}z = b_3$$

by $x = \dfrac{D_x}{D}$, $y = \dfrac{D_y}{D}$, and $z = \dfrac{D_z}{D}$, where

$$D = \begin{vmatrix} a_{11} & a_{12} & a_{13} \\ a_{21} & a_{22} & a_{23} \\ a_{31} & a_{32} & a_{33} \end{vmatrix}, D_x = \begin{vmatrix} b_1 & a_{12} & a_{13} \\ b_2 & a_{22} & a_{23} \\ b_3 & a_{32} & a_{33} \end{vmatrix}, D_y = \begin{vmatrix} a_{11} & b_1 & a_{13} \\ a_{21} & b_2 & a_{23} \\ a_{31} & b_3 & a_{33} \end{vmatrix},$$

$$D_z = \begin{vmatrix} a_{11} & a_{12} & b_1 \\ a_{21} & a_{22} & b_2 \\ a_{31} & a_{32} & b_3 \end{vmatrix}, \text{ and } D \neq 0.$$

EXAMPLE 3 Solve by using Cramer's Rule: $3x - y + z = 5$
$x + 2y - 2z = -3$
$2x + 3y + z = 4$

Solution $D = \begin{vmatrix} 3 & -1 & 1 \\ 1 & 2 & -2 \\ 2 & 3 & 1 \end{vmatrix} = 28$ • Find the value of the coefficient determinant.

$D_x = \begin{vmatrix} 5 & -1 & 1 \\ -3 & 2 & -2 \\ 4 & 3 & 1 \end{vmatrix} = 28$ • Find the value of each of the numerator determinants.

$D_y = \begin{vmatrix} 3 & 5 & 1 \\ 1 & -3 & -2 \\ 2 & 4 & 1 \end{vmatrix} = 0$

$D_z = \begin{vmatrix} 3 & -1 & 5 \\ 1 & 2 & -3 \\ 2 & 3 & 4 \end{vmatrix} = 56$

$$x = \frac{D_x}{D} = \frac{28}{28} = 1$$ • **Use Cramer's Rule to write the solution.**

$$y = \frac{D_y}{D} = \frac{0}{28} = 0$$

$$z = \frac{D_z}{D} = \frac{56}{28} = 2$$

The solution is $(1, 0, 2)$.

PROBLEM 3 Solve by using Cramer's Rule: $2x - y + z = -1$
$3x + 2y - z = 3$
$x + 3y + z = -2$

■ **Solution** See page S15.

OBJECTIVE 3 Solve systems of linear equations by using matrices

As previously stated in this section, a **matrix** is a rectangular array of numbers. Each number in the matrix is called an **element** of the matrix. The matrix shown below, with 3 rows and 4 columns, is a 3×4 matrix.

$$\begin{bmatrix} 1 & 4 & -3 & 6 \\ -2 & 5 & 2 & 0 \\ -1 & 3 & 7 & -4 \end{bmatrix}$$

The elements $a_{11}, a_{22}, a_{33}, \ldots, a_{nn}$ form the **main diagonal** of a matrix. The elements 1, 5, and 7 form the main diagonal of the matrix above.

By considering only the coefficients and constants for the following system of equations, we can form the corresponding 3×4 **augmented matrix.**

Take Note

For augmented matrices, it is common practice to show a vertical line between the coefficients of the variables and the constant terms.

	System of Equations		**Augmented Matrix**

$$\begin{array}{rcl} 3x - 2y + z &=& 2 \\ x \qquad\quad - 3z &=& -2 \\ 2x - y + 4z &=& 5 \end{array} \qquad \begin{bmatrix} 3 & -2 & 1 & 2 \\ 1 & 0 & -3 & -2 \\ 2 & -1 & 4 & 5 \end{bmatrix}$$

Note that when a term is missing from one of the equations of the system, the coefficient of that term is 0, and a 0 is entered in the matrix.

A system of equations can be written from an augmented matrix.

$$\begin{bmatrix} 2 & -1 & 4 & 1 \\ 1 & 1 & 0 & 3 \\ 3 & -2 & -1 & 5 \end{bmatrix} \qquad \begin{array}{rcl} 2x - y + 4z &=& 1 \\ x + y \qquad &=& 3 \\ 3x - 2y - z &=& 5 \end{array}$$

EXAMPLE 4 Write the augmented matrix for $\begin{array}{rcl} 2x - 3y &=& 4 \\ x + 5y &=& 0 \end{array}$.

Solution The augmented matrix is $\begin{bmatrix} 2 & -3 & 4 \\ 1 & 5 & 0 \end{bmatrix}$.

PROBLEM 4 Write the system of equations that corresponds to the

augmented matrix $\begin{bmatrix} 2 & -3 & 1 & | & 4 \\ 1 & 0 & -2 & | & 3 \\ 0 & 1 & 2 & | & -3 \end{bmatrix}$.

■ **Solution** See page S16.

A system of equations can be solved by writing the system in matrix form and then performing operations on the matrix similar to those performed on the equations of the system. These operations are called **elementary row operations.**

Elementary Row Operations

1. Interchange two rows.

2. Multiply all the elements in a row by the same nonzero number.

3. Replace a row by the sum of that row and a multiple of any other row.

Each of these elementary row operations has as its basis the operations that can be performed on a system of equations. These operations do not change the solution of the system of equations. Here are some examples of each row operation.

1. Interchange two rows.

Take Note

For each example of a row operation at the right, we are using the same system of equations. The solution of this system of equations is (2, 1). You should verify that (2, 1) is also a solution of the new system of equations. This means that these row operations have not changed the solution.

	This notation means to interchange rows 1 and 2.	
Original System		**New System**

$2x - 3y = 1$ $\begin{bmatrix} 2 & -3 & | & 1 \\ 4 & 5 & | & 13 \end{bmatrix}$ $R_1 \leftrightarrow R_2$ $\begin{bmatrix} 4 & 5 & | & 13 \\ 2 & -3 & | & 1 \end{bmatrix}$ $4x + 5y = 13$
$4x + 5y = 13$ $2x - 3y = 1$

The solution is (2, 1). The solution is (2, 1).

2. Multiply all the elements in a row by the same nonzero number.

	This notation means to multiply row 2 by 3.	
Original System		**New System**

$2x - 3y = 1$ $\begin{bmatrix} 2 & -3 & | & 1 \\ 4 & 5 & | & 13 \end{bmatrix}$ $3R_2 \rightarrow$ $\begin{bmatrix} 2 & -3 & | & 1 \\ 12 & 15 & | & 39 \end{bmatrix}$ $2x - 3y = 1$
$4x + 5y = 13$ $12x + 15y = 39$

The solution is (2, 1). The solution is (2, 1).

3. Replace a row by the sum of that row and a multiple of any other row.

This notation means to replace row 2 by the sum of that row and -2 times row 1.

Original System

$2x - 3y = 1$
$4x + 5y = 13$

The solution is $(2, 1)$.

$$\left[\begin{array}{cc|c} 2 & -3 & 1 \\ 4 & 5 & 13 \end{array}\right] \quad -2R_1 + R_2 \rightarrow \quad \left[\begin{array}{cc|c} 2 & -3 & 1 \\ 0 & 11 & 11 \end{array}\right]$$

New System

$2x - 3y = 1$
$11y = 11$

The solution is $(2, 1)$.

Note that we replace the row that follows the addition. See the Take Note at the left.

EXAMPLE 5 Let $A = \left[\begin{array}{ccc|c} 1 & 3 & -4 & 6 \\ 3 & 2 & 0 & -1 \\ -2 & -5 & 3 & 4 \end{array}\right]$. Perform the following elementary row operations on A.

A. $R_1 \leftrightarrow R_3$ **B.** $-2R_3$ **C.** $2R_3 + R_1$

Solution **A.** $R_1 \leftrightarrow R_3$ means to interchange row 1 and row 3.

$$\left[\begin{array}{ccc|c} 1 & 3 & -4 & 6 \\ 3 & 2 & 0 & -1 \\ -2 & -5 & 3 & 4 \end{array}\right] \quad R_1 \leftrightarrow R_3 \quad \left[\begin{array}{ccc|c} -2 & -5 & 3 & 4 \\ 3 & 2 & 0 & -1 \\ 1 & 3 & -4 & 6 \end{array}\right]$$

B. $-2R_3$ means to multiply row 3 by -2.

$$\left[\begin{array}{ccc|c} 1 & 3 & -4 & 6 \\ 3 & 2 & 0 & -1 \\ -2 & -5 & 3 & 4 \end{array}\right] \quad -2R_3 \rightarrow \quad \left[\begin{array}{ccc|c} 1 & 3 & -4 & 6 \\ 3 & 2 & 0 & -1 \\ 4 & 10 & -6 & -8 \end{array}\right]$$

C. $2R_3 + R_1$ means to multiply row 3 by 2 and then add the result to row 1.

$$\left[\begin{array}{ccc|c} 1 & 3 & -4 & 6 \\ 3 & 2 & 0 & -1 \\ -2 & -5 & 3 & 4 \end{array}\right] \quad 2R_3 + R_1 \rightarrow \quad \left[\begin{array}{ccc|c} -3 & -7 & 2 & 14 \\ 3 & 2 & 0 & -1 \\ -2 & -5 & 3 & 4 \end{array}\right]$$

PROBLEM 5 Let $B = \left[\begin{array}{ccc|c} 1 & 8 & -2 & 3 \\ 2 & -3 & 4 & 1 \\ 3 & 5 & -7 & 3 \end{array}\right]$. Perform the following elementary row operations on B.

A. $R_2 \leftrightarrow R_3$ **B.** $3R_2$ **C.** $-3R_1 + R_3$

Solution See page S16.

Elementary row operations are used to solve a system of equations. The goal is to use the elementary row operations to rewrite the augmented matrix with 1's down the main diagonal and 0's to the left of the 1's in all rows except the first. This is called the **row echelon form** of the matrix. Here are some examples of the row echelon form.

$$\begin{bmatrix} 1 & 3 & | & -2 \\ 0 & 1 & | & 3 \end{bmatrix} \qquad \begin{bmatrix} 1 & -2 & 3 & | & 1 \\ 0 & 1 & 2.5 & | & -4 \\ 0 & 0 & 1 & | & 2 \end{bmatrix} \qquad \begin{bmatrix} 1 & 4 & \frac{1}{2} & | & -3 \\ 0 & 1 & 3 & | & 0 \\ 0 & 0 & 1 & | & -\frac{2}{3} \end{bmatrix}$$

We will follow a very definite procedure to rewrite an augmented matrix in row echelon form. For a 2×3 augmented matrix, use elementary row operations to

Step 1: Change a_{11} to a 1.
Step 2: Change a_{21} to a 0. $\quad \begin{bmatrix} a_{11} & a_{12} & | & a_{13} \\ a_{21} & a_{22} & | & a_{23} \end{bmatrix}$
Step 3: Change a_{22} to a 1.

Write the matrix $\begin{bmatrix} 3 & -6 & | & 12 \\ 2 & 1 & | & -3 \end{bmatrix}$ in row echelon form.

Step 1: Change a_{11} to 1. One way to do this is to multiply row 1 by the reciprocal of a_{11}.

$$\begin{bmatrix} 3 & -6 & | & 12 \\ 2 & 1 & | & -3 \end{bmatrix} \;\; \tfrac{1}{3}R_1 \rightarrow \;\; \begin{bmatrix} 1 & -2 & | & 4 \\ 2 & 1 & | & -3 \end{bmatrix}$$

Step 2: Change a_{21} to 0 by multiplying row 1 by the opposite of a_{21} and then adding to row 2.

$$\begin{bmatrix} 1 & -2 & | & 4 \\ 2 & 1 & | & -3 \end{bmatrix} \;\; -2R_1 + R_2 \rightarrow \;\; \begin{bmatrix} 1 & -2 & | & 4 \\ 0 & 5 & | & -11 \end{bmatrix}$$

Step 3: Change a_{22} to 1 by multiplying by the reciprocal of a_{22}.

$$\begin{bmatrix} 1 & -2 & | & 4 \\ 0 & 5 & | & -11 \end{bmatrix} \;\; \tfrac{1}{5}R_2 \rightarrow \;\; \begin{bmatrix} 1 & -2 & | & 4 \\ 0 & 1 & | & -\frac{11}{5} \end{bmatrix}$$

A row echelon form of the matrix is $\begin{bmatrix} 1 & -2 & | & 4 \\ 0 & 1 & | & -\frac{11}{5} \end{bmatrix}$.

The row echelon form of a matrix is not unique and depends on the elementary row operations that are used. For instance, suppose we again start with $\begin{bmatrix} 3 & -6 & | & 12 \\ 2 & 1 & | & -3 \end{bmatrix}$ and follow the elementary row operations below.

$$\begin{bmatrix} 3 & -6 & | & 12 \\ 2 & 1 & | & -3 \end{bmatrix} \; -1R_2 + R_1 \rightarrow \; \begin{bmatrix} 1 & -7 & | & 15 \\ 2 & 1 & | & -3 \end{bmatrix} \; -2R_1 + R_2 \rightarrow \; \begin{bmatrix} 1 & -7 & | & 15 \\ 0 & 15 & | & -33 \end{bmatrix} \; \tfrac{1}{15}R_2 \rightarrow \; \begin{bmatrix} 1 & -7 & | & 15 \\ 0 & 1 & | & -\frac{11}{5} \end{bmatrix}$$

The row echelon forms $\begin{bmatrix} 1 & -7 & | & 15 \\ 0 & 1 & | & -\frac{11}{5} \end{bmatrix}$ and $\begin{bmatrix} 1 & -2 & | & 4 \\ 0 & 1 & | & -\frac{11}{5} \end{bmatrix}$ are different.

Row echelon form is not unique.

The order in which the elements in an augmented 3×4 matrix are changed is as follows:

$$\begin{bmatrix} a_{11} & a_{12} & a_{13} & a_{14} \\ a_{21} & a_{22} & a_{23} & a_{24} \\ a_{31} & a_{32} & a_{33} & a_{34} \end{bmatrix}$$

Step 1: Change a_{11} to a 1.

Step 2: Change a_{21} and a_{31} to 0's.
Step 3: Change a_{22} to a 1.
Step 4: Change a_{32} to a 0.
Step 5: Change a_{33} to a 1.

EXAMPLE 6 Write $\begin{bmatrix} 2 & 1 & 3 & -1 \\ 1 & 3 & 5 & -1 \\ -3 & -1 & 1 & 2 \end{bmatrix}$ in row echelon form.

Solution **Step 1:** Change a_{11} to 1 by interchanging row 1 and row 2.

Note: We could have chosen to multiply row 1 by $\frac{1}{2}$. The sequence of steps to get to row echelon form is not unique.

$$\begin{bmatrix} 2 & 1 & 3 & -1 \\ 1 & 3 & 5 & -1 \\ -3 & -1 & 1 & 2 \end{bmatrix} \quad R_1 \leftrightarrow R_2 \quad \begin{bmatrix} 1 & 3 & 5 & -1 \\ 2 & 1 & 3 & -1 \\ -3 & -1 & 1 & 2 \end{bmatrix}$$

Step 2: Change a_{21} to 0 by multiplying row 1 by the opposite of a_{21} and then adding to row 2.

$$\begin{bmatrix} 1 & 3 & 5 & -1 \\ 2 & 1 & 3 & -1 \\ -3 & -1 & 1 & 2 \end{bmatrix} \quad -2R_1 + R_2 \to \quad \begin{bmatrix} 1 & 3 & 5 & -1 \\ 0 & -5 & -7 & 1 \\ -3 & -1 & 1 & 2 \end{bmatrix}$$

Change a_{31} to 0 by multiplying row 1 by the opposite of a_{31} and then adding to row 3.

$$\begin{bmatrix} 1 & 3 & 5 & -1 \\ 0 & -5 & -7 & 1 \\ -3 & -1 & 1 & 2 \end{bmatrix} \quad 3R_1 + R_3 \to \quad \begin{bmatrix} 1 & 3 & 5 & -1 \\ 0 & -5 & -7 & 1 \\ 0 & 8 & 16 & -1 \end{bmatrix}$$

Step 3: Change a_{22} to 1 by multiplying by the reciprocal of a_{22}.

$$\begin{bmatrix} 1 & 3 & 5 & -1 \\ 0 & -5 & -7 & 1 \\ 0 & 8 & 16 & -1 \end{bmatrix} \quad -\frac{1}{5}R_2 \to \quad \begin{bmatrix} 1 & 3 & 5 & -1 \\ 0 & 1 & \frac{7}{5} & -\frac{1}{5} \\ 0 & 8 & 16 & -1 \end{bmatrix}$$

Step 4: Change a_{32} to 0 by multiplying row 2 by the opposite of a_{32} and then adding to row 3.

$$\begin{bmatrix} 1 & 3 & 5 & -1 \\ 0 & 1 & \frac{7}{5} & -\frac{1}{5} \\ 0 & 8 & 16 & -1 \end{bmatrix} \quad -8R_2 + R_3 \to \quad \begin{bmatrix} 1 & 3 & 5 & -1 \\ 0 & 1 & \frac{7}{5} & -\frac{1}{5} \\ 0 & 0 & \frac{24}{5} & \frac{3}{5} \end{bmatrix}$$

Step 5: Change a_{33} to 1 by multiplying by the reciprocal of a_{33}.

$$\begin{bmatrix} 1 & 3 & 5 & | & -1 \\ 0 & 1 & \frac{7}{5} & | & -\frac{1}{5} \\ 0 & 0 & \frac{24}{5} & | & \frac{3}{5} \end{bmatrix} \quad \frac{5}{24}R_3 \rightarrow \quad \begin{bmatrix} 1 & 3 & 5 & | & -1 \\ 0 & 1 & \frac{7}{5} & | & -\frac{1}{5} \\ 0 & 0 & 1 & | & \frac{1}{8} \end{bmatrix}$$

A row echelon form of the matrix is $\begin{bmatrix} 1 & 3 & 5 & | & -1 \\ 0 & 1 & \frac{7}{5} & | & -\frac{1}{5} \\ 0 & 0 & 1 & | & \frac{1}{8} \end{bmatrix}$.

PROBLEM 6 Write $\begin{bmatrix} 1 & -3 & 2 & | & 1 \\ -4 & 14 & 0 & | & -2 \\ 2 & -5 & -3 & | & 16 \end{bmatrix}$ in row echelon form.

 Solution See page S16.

If an augmented matrix is in row echelon form, the corresponding system of equations can be solved by substitution. For instance, consider the following matrix in row echelon form and the corresponding system of equations.

$$\begin{bmatrix} 1 & -3 & 4 & | & 7 \\ 0 & 1 & 3 & | & -6 \\ 0 & 0 & 1 & | & -1 \end{bmatrix} \qquad \begin{aligned} x - 3y + 4z &= 7 \\ y + 3z &= -6 \\ z &= -1 \end{aligned}$$

Point of Interest

Johann Carl Friedrich Gauss (1777–1855) is considered one of the greatest mathematicians of all time. He contributed not only to mathematics but to astronomy and physics as well. A unit of magnetism called a gauss was named in his honor.

From the last equation of the system above, we have $z = -1$. Substitute this value into the second equation and solve for y. Thus $y = -3$.

$$\begin{aligned} y + 3z &= -6 \\ y + 3(-1) &= -6 \\ y - 3 &= -6 \\ y &= -3 \end{aligned}$$

Substitute $y = -3$ and $z = -1$ in the first equation of the system and solve for x. Thus $x = 2$.

$$\begin{aligned} x - 3y + 4z &= 7 \\ x - 3(-3) + 4(-1) &= 7 \\ x + 9 - 4 &= 7 \\ x &= 2 \end{aligned}$$

The solution of the system of equations is $(2, -3, -1)$.

The process of solving a system of equations by using elementary row operations is called the **Gaussian elimination method.**

EXAMPLE 7 Solve by using the Gaussian elimination method:
$$\begin{aligned} 2x - 5y &= 19 \\ 3x + 4y &= -6 \end{aligned}$$

 Solution Write the augmented matrix and then use elementary row operations to rewrite the matrix in row echelon form.

$$\begin{bmatrix} 2 & -5 & | & 19 \\ 3 & 4 & | & -6 \end{bmatrix} \quad \xrightarrow[\frac{1}{2}R_1]{\text{Change } a_{11} \text{ to 1.}} \quad \begin{bmatrix} 1 & -\frac{5}{2} & | & \frac{19}{2} \\ 3 & 4 & | & -6 \end{bmatrix}$$

• Note that we multiplied R_1 by the reciprocal of a_{11}.

$$\begin{bmatrix} 1 & -\dfrac{5}{2} & \bigg| & \dfrac{19}{2} \\ 3 & 4 & \bigg| & -6 \end{bmatrix} \xrightarrow[\substack{-3R_1 + R_2}]{\text{Change } a_{21} \text{ to } 0.} \begin{bmatrix} 1 & -\dfrac{5}{2} & \bigg| & \dfrac{19}{2} \\ 0 & \dfrac{23}{2} & \bigg| & -\dfrac{69}{2} \end{bmatrix}$$

• Note that we multiply R_1 by the opposite of a_{21}.

$$\begin{bmatrix} 1 & -\dfrac{5}{2} & \bigg| & \dfrac{19}{2} \\ 0 & \dfrac{23}{2} & \bigg| & -\dfrac{69}{2} \end{bmatrix} \xrightarrow[\substack{\frac{2}{23}R_2}]{\text{Change } a_{22} \text{ to } 1.} \begin{bmatrix} 1 & -\dfrac{5}{2} & \bigg| & \dfrac{19}{2} \\ 0 & 1 & \bigg| & -3 \end{bmatrix}$$

• This is row echelon form.

(1) $x - \dfrac{5}{2}y = \dfrac{19}{2}$ • Write the system of equations corresponding to the matrix that is in row echelon form.

(2) $y = -3$

$$x - \dfrac{5}{2}(-3) = \dfrac{19}{2}$$ • Substitute -3 for y in equation (1) and solve for x.

$$x + \dfrac{15}{2} = \dfrac{19}{2}$$

$$x = 2$$

The solution is $(2, -3)$.

PROBLEM 7 Solve by using the Gaussian elimination method:
$4x - 5y = 17$
$3x + 2y = 7$

Solution See page S16.

The Gaussian elimination method can be extended to systems of equations with more than two variables.

EXAMPLE 8 Solve by using the Gaussian elimination method:
$x + 2y - z = 9$
$2x - y + 2z = -1$
$-2x + 3y - 2z = 7$

Solution
$$\begin{bmatrix} 1 & 2 & -1 & \big| & 9 \\ 2 & -1 & 2 & \big| & -1 \\ -2 & 3 & -2 & \big| & 7 \end{bmatrix} \xrightarrow[\substack{-2R_1 + R_2}]{\substack{a_{11} \text{ is } 1. \text{ Change} \\ a_{21} \text{ to } 0.}} \begin{bmatrix} 1 & 2 & -1 & \big| & 9 \\ 0 & -5 & 4 & \big| & -19 \\ -2 & 3 & -2 & \big| & 7 \end{bmatrix}$$

$$\begin{bmatrix} 1 & 2 & -1 & \big| & 9 \\ 0 & -5 & 4 & \big| & -19 \\ -2 & 3 & -2 & \big| & 7 \end{bmatrix} \xrightarrow[\substack{2R_1 + R_3}]{\text{Change } a_{31} \text{ to } 0.} \begin{bmatrix} 1 & 2 & -1 & \big| & 9 \\ 0 & -5 & 4 & \big| & -19 \\ 0 & 7 & -4 & \big| & 25 \end{bmatrix}$$

$$\begin{bmatrix} 1 & 2 & -1 & \big| & 9 \\ 0 & -5 & 4 & \big| & -19 \\ 0 & 7 & -4 & \big| & 25 \end{bmatrix} \xrightarrow[\substack{-\frac{1}{5}R_2}]{\text{Change } a_{22} \text{ to } 1.} \begin{bmatrix} 1 & 2 & -1 & \big| & 9 \\ 0 & 1 & -\dfrac{4}{5} & \big| & \dfrac{19}{5} \\ 0 & 7 & -4 & \big| & 25 \end{bmatrix}$$

$$\begin{bmatrix} 1 & 2 & -1 & \big| & 9 \\ 0 & 1 & -\dfrac{4}{5} & \big| & \dfrac{19}{5} \\ 0 & 7 & -4 & \big| & 25 \end{bmatrix} \xrightarrow[\substack{-7R_2 + R_3}]{\text{Change } a_{32} \text{ to } 0.} \begin{bmatrix} 1 & 2 & -1 & \big| & 9 \\ 0 & 1 & -\dfrac{4}{5} & \big| & \dfrac{19}{5} \\ 0 & 0 & \dfrac{8}{5} & \big| & -\dfrac{8}{5} \end{bmatrix}$$

$$\begin{bmatrix} 1 & 2 & -1 & \bigg| & 9 \\ 0 & 1 & -\frac{4}{5} & \bigg| & \frac{19}{5} \\ 0 & 0 & \frac{8}{5} & \bigg| & -\frac{8}{5} \end{bmatrix} \xrightarrow[\frac{5}{8}R_2]{\text{Change } a_{33} \text{ to } 1.} \begin{bmatrix} 1 & 2 & -1 & \bigg| & 9 \\ 0 & 1 & -\frac{4}{5} & \bigg| & \frac{19}{5} \\ 0 & 0 & 1 & \bigg| & -1 \end{bmatrix}$$

• This is row echelon form.

$$(1) \qquad x + 2y - z = 9$$
$$(2) \qquad y - \frac{4}{5}z = \frac{19}{5}$$
$$(3) \qquad z = -1$$

• Write the system of equations corresponding to the matrix that is in row echelon form.

$$y - \frac{4}{5}(-1) = \frac{19}{5}$$
$$y + \frac{4}{5} = \frac{19}{5}$$
$$y = 3$$

• Substitute −1 for z in equation (2) and solve for y.

$$x + 2y - z = 9$$
$$x + 2(3) - (-1) = 9$$
$$x + 7 = 9$$
$$x = 2$$

• Substitute −1 for z and 3 for y in equation (1) and solve for x.

The solution is $(2, 3, -1)$.

PROBLEM 8 Solve by using the Gaussian elimination method:
$$2x + 3y + 3z = -2$$
$$x + 2y - 3z = 9$$
$$3x - 2y - 4z = 1$$

■ **Solution** See pages S16–S17.

4.3 Concept Review

Determine whether the following statements are always true, sometimes true, or never true.

1. A 3×3 determinant can be evaluated by expanding about any row or column of the matrix.

2. Cramer's Rule can be used to solve a system of linear equations in three variables.

3. A matrix has the same number of rows and columns.

4. Associated with a matrix is a number called its determinant.

5. A system of two linear equations in two variables has no solution, exactly one solution, or infinitely many solutions.

4.3 Exercises

1 Evaluate determinants

For Exercises 1 to 6, use these matrices:

$$A = \begin{bmatrix} -3 & 0 & 1 \\ 2 & 1 & -4 \end{bmatrix} \quad B = \begin{bmatrix} 1 & -1 \\ 5 & 8 \end{bmatrix} \quad C = \begin{bmatrix} 4 & 1 & -7 \\ 0 & -2 & 6 \\ 5 & 0 & -3 \end{bmatrix}$$

1. In matrix B, element b_{21} is ___?___.

2. Matrix A has ___?___ rows and ___?___ columns, so the order of matrix A is ___?___.

3. Because matrix B has the same number of rows as columns, it is called a ___?___ matrix.

4. Only square matrices have determinants. The matrix above that does *not* have a determinant is matrix ___?___.

5. The determinant of matrix B is written $\begin{vmatrix} 1 & -1 \\ 5 & 8 \end{vmatrix}$.

The value of this determinant is $(1)(\underline{\quad?\quad}) - 5(\underline{\quad?\quad}) = \underline{\quad?\quad}$.

6. For the determinant of matrix C, the minor of 6 is $\begin{vmatrix} \underline{\ ?\ } & \underline{\ ?\ } \\ \underline{\ ?\ } & \underline{\ ?\ } \end{vmatrix}$.

Evaluate the determinant.

7. $\begin{vmatrix} 2 & -1 \\ 3 & 4 \end{vmatrix}$

8. $\begin{vmatrix} 5 & 1 \\ -1 & 2 \end{vmatrix}$

9. $\begin{vmatrix} 6 & -2 \\ -3 & 4 \end{vmatrix}$

10. $\begin{vmatrix} -3 & 5 \\ 1 & 7 \end{vmatrix}$

11. $\begin{vmatrix} 3 & 6 \\ 2 & 4 \end{vmatrix}$

12. $\begin{vmatrix} 5 & -10 \\ 1 & -2 \end{vmatrix}$

13. $\begin{vmatrix} 1 & -1 & 2 \\ 3 & 2 & 1 \\ 1 & 0 & 4 \end{vmatrix}$

14. $\begin{vmatrix} 4 & 1 & 3 \\ 2 & -2 & 1 \\ 3 & 1 & 2 \end{vmatrix}$

15. $\begin{vmatrix} 3 & -1 & 2 \\ 0 & 1 & 2 \\ 3 & 2 & -2 \end{vmatrix}$

16. $\begin{vmatrix} 4 & 5 & -2 \\ 3 & -1 & 5 \\ 2 & 1 & 4 \end{vmatrix}$

17. $\begin{vmatrix} 4 & 2 & 6 \\ -2 & 1 & 1 \\ 2 & 1 & 3 \end{vmatrix}$

18. $\begin{vmatrix} 3 & 6 & -3 \\ 4 & -1 & 6 \\ -1 & -2 & 3 \end{vmatrix}$

2 Solve systems of linear equations by using Cramer's Rule

For Exercises 19 to 22, use this system of equations:
$$2x - 3y = 5$$
$$-5x + 4y = -2$$

19. a. The coefficient determinant for the system is $D = \begin{vmatrix} \underline{\ ?\ } & \underline{\ ?\ } \\ \underline{\ ?\ } & \underline{\ ?\ } \end{vmatrix}$.

 b. The value of the coefficient determinant is $D = \underline{\quad?\quad}$.

20. a. The numerator determinant for x is $D_x = \begin{vmatrix} \underline{}\ ? & \underline{}\ ? \\ \underline{}\ ? & \underline{}\ ? \end{vmatrix}$.

 b. The value of the numerator determinant for x is $D_x = \underline{\ ?\ }$.

21. a. The numerator determinant for y is $D_y = \begin{vmatrix} \underline{}\ ? & \underline{}\ ? \\ \underline{}\ ? & \underline{}\ ? \end{vmatrix}$.

 b. The value of the numerator determinant for y is $D_y = \underline{\ ?\ }$.

22. Use your answers to Exercises 19b, 20b, and 21b to find the solution of the system of equations

$$2x - 3y = 5$$
$$-5x + 4y = -2.$$

$$x = \frac{D_x}{D} = \frac{\underline{\ ?}}{\underline{\ ?}} = \underline{\ ?\ } \text{ and}$$

$$y = \frac{D_y}{D} = \frac{\underline{\ ?}}{\underline{\ ?}} = \underline{\ ?\ }, \text{ so the solution is } (\underline{\ ?\ }, \underline{\ ?\ }).$$

Solve by using Cramer's Rule, if possible.

23. $2x - 5y = 26$
 $5x + 3y = 3$

24. $3x + 7y = 15$
 $2x + 5y = 11$

25. $x - 4y = 8$
 $3x + 7y = 5$

26. $5x + 2y = -5$
 $3x + 4y = 11$

27. $2x + 3y = 4$
 $6x - 12y = -5$

28. $5x + 4y = 3$
 $15x - 8y = -21$

29. $2x + 5y = 6$
 $6x - 2y = 1$

30. $7x + 3y = 4$
 $5x - 4y = 9$

31. $-2x + 3y = 7$
 $4x - 6y = 9$

32. $9x + 6y = 7$
 $3x + 2y = 4$

33. $2x - 5y = -2$
 $3x - 7y = -3$

34. $8x + 7y = -3$
 $2x + 2y = 5$

35. $2x - y + 3z = 9$
 $x + 4y + 4z = 5$
 $3x + 2y + 2z = 5$

36. $3x - 2y + z = 2$
 $2x + 3y + 2z = -6$
 $3x - y + z = 0$

37. $3x - y + z = 11$
 $x + 4y - 2z = -12$
 $2x + 2y - z = -3$

38. $x + 2y + 3z = 8$
 $2x - 3y + z = 5$
 $3x - 4y + 2z = 9$

39. $4x - 2y + 6z = 1$
 $3x + 4y + 2z = 1$
 $2x - y + 3z = 2$

40. $x - 3y + 2z = 1$
 $2x + y - 2z = 3$
 $3x - 9y + 6z = -3$

41. $5x - 4y + 2z = 4$
 $3x - 5y + 3z = -4$
 $3x + y - 5z = 12$

42. $2x + 4y + z = 7$
 $x + 3y - z = 1$
 $3x + 2y - 2z = 5$

3 Solve systems of linear equations by using matrices

For Exercises 43 to 49, use this augmented matrix for a system of three equations in three variables x, y, and z:

$$\begin{bmatrix} 1 & -2 & 3 & | & 2 \\ 2 & -3 & 1 & | & 1 \\ 3 & -1 & 2 & | & 9 \end{bmatrix}$$

43. The elements of the main diagonal of the augmented matrix are ___?___, ___?___, and ___?___.

44. The coefficient of y in the second equation of the system of equations associated with the augmented matrix is ___?___.

45. The constant for the third equation of the system of equations associated with the augmented matrix is ___?___.

46. Write the matrix that results from interchanging rows 2 and 3 of the augmented matrix ($R_2 \leftrightarrow R_3$):

$$\begin{bmatrix} 1 & -2 & 3 & | & 2 \\ \underline{?} & \underline{?} & \underline{?} & | & \underline{?} \\ \underline{?} & \underline{?} & \underline{?} & | & \underline{?} \end{bmatrix}$$

47. Write the matrix that results from multiplying row 1 of the augmented matrix by -3 ($-3R_1$):

$$\begin{bmatrix} \underline{?} & \underline{?} & \underline{?} & | & \underline{?} \\ 2 & -3 & 1 & | & 1 \\ 3 & -1 & 2 & | & 9 \end{bmatrix}$$

48. When the augmented matrix is in row echelon form, all the elements of the main diagonal will be ___?___. All the elements to the left of the main diagonal will be ___?___.

49. A row echelon form of the matrix is shown below. The equation that corresponds to the first row is ___?___. The equation that corresponds to the second row is ___?___. The equation that corresponds to the third row is ___?___.

$$\begin{bmatrix} 1 & -2 & 3 & | & 2 \\ 0 & 1 & -5 & | & -3 \\ 0 & 0 & 1 & | & 1 \end{bmatrix}$$

50. 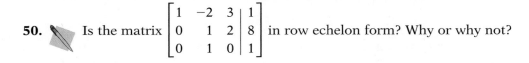 Is the matrix $\begin{bmatrix} 1 & -2 & 3 & | & 1 \\ 0 & 1 & 2 & | & 8 \\ 0 & 1 & 0 & | & 1 \end{bmatrix}$ in row echelon form? Why or why not?

Write each matrix in row echelon form.

51. $\begin{bmatrix} 2 & -4 & | & 1 \\ 3 & -7 & | & -1 \end{bmatrix}$ **52.** $\begin{bmatrix} 4 & 2 & | & -2 \\ 7 & 4 & | & -1 \end{bmatrix}$ **53.** $\begin{bmatrix} 5 & -2 & | & 3 \\ -7 & 3 & | & 1 \end{bmatrix}$ **54.** $\begin{bmatrix} 2 & 5 & | & -4 \\ 3 & 1 & | & 2 \end{bmatrix}$

55. $\begin{bmatrix} 1 & 4 & 1 & | & -2 \\ 3 & 11 & -1 & | & 2 \\ 2 & 3 & 1 & | & 4 \end{bmatrix}$ **56.** $\begin{bmatrix} 1 & 2 & 2 & | & -1 \\ -4 & -10 & -1 & | & 3 \\ 3 & 4 & 2 & | & -2 \end{bmatrix}$ **57.** $\begin{bmatrix} -2 & 6 & -1 & | & 3 \\ 1 & -2 & 2 & | & 1 \\ 3 & -6 & 7 & | & 6 \end{bmatrix}$ **58.** $\begin{bmatrix} 2 & 6 & 10 & 3 \\ 3 & 8 & 15 & 0 \\ 1 & 2 & 3 & -1 \end{bmatrix}$

59. What is the solution of the system of equations that has $\begin{bmatrix} 1 & -1 & 3 & | & -2 \\ 0 & 1 & -1 & | & 1 \\ 0 & 0 & 1 & | & 3 \end{bmatrix}$ as the row echelon form of the augmented matrix for the system of equations?

60. What is the solution of the system of equations that has $\begin{bmatrix} 1 & -3 & 2 & | & 4 \\ 0 & 1 & -2 & | & 3 \\ 0 & 0 & 1 & | & -1 \end{bmatrix}$ as the row echelon form of the augmented matrix for the system of equations?

Solve by using the Gaussian elimination method.

61. $3x + y = 6$
$2x - y = -1$

62. $2x + y = 3$
$x - 4y = 6$

63. $x - 3y = 8$
$3x - y = 0$

64. $2x + 3y = 16$
$x - 4y = -14$

65. $y = 4x - 10$
$2y = 5x - 11$

66. $2y = 4 - 3x$
$y = 1 - 2x$

67. $2x - y = -4$
$y = 2x - 8$

68. $3x - 2y = -8$
$y = \dfrac{3}{2}x - 2$

69. $4x - 3y = -14$
$3x + 4y = 2$

70. $5x + 2y = 3$
$3x + 4y = 13$

71. $5x + 4y + 3z = -9$
$x - 2y + 2z = -6$
$x - y - z = 3$

72. $x - y - z = 0$
$3x - y + 5z = -10$
$x + y - 4z = 12$

73. $5x - 5y + 2z = 8$
 $2x + 3y - z = 0$
 $x + 2y - z = 0$

74. $2x + y - 5z = 3$
 $3x + 2y + z = 15$
 $5x - y - z = 5$

75. $2x + 3y + z = 5$
 $3x + 3y + 3z = 10$
 $4x + 6y + 2z = 5$

76. $x - 2y + 3z = 2$
 $2x + y + 2z = 5$
 $2x - 4y + 6z = -4$

77. $3x + 2y + 3z = 2$
 $6x - 2y + z = 1$
 $3x + 4y + 2z = 3$

78. $2x + 3y - 3z = -1$
 $2x + 3y + 3z = 3$
 $4x - 4y + 3z = 4$

79. $5x - 5y - 5z = 2$
 $5x + 5y - 5z = 6$
 $10x + 10y + 5z = 3$

80. $3x - 2y + 2z = 5$
 $6x + 3y - 4z = -1$
 $3x - y + 2z = 4$

81. $4x + 4y - 3z = 3$
 $8x + 2y + 3z = 0$
 $4x - 4y + 6z = -3$

4.3 Applying Concepts

Solve for x.

82. $\begin{vmatrix} 3 & 2 \\ 4 & x \end{vmatrix} = -11$

83. $\begin{vmatrix} 1 & 0 & 2 \\ 4 & 3 & -1 \\ 0 & 2 & x \end{vmatrix} = -24$

84. $\begin{vmatrix} -2 & 1 & 3 \\ 0 & x & 4 \\ -1 & 2 & -3 \end{vmatrix} = -24$

Complete.

85. If all the elements in one row or one column of a 2 × 2 matrix are zeros, the value of the determinant of the matrix is _____.

86. If all the elements in one row or one column of a 3 × 3 matrix are zeros, the value of the determinant of the matrix is _____.

87. a. The value of the determinant $\begin{vmatrix} x & x & a \\ y & y & b \\ z & z & c \end{vmatrix}$ is _____.

 b. If two columns of a 3 × 3 matrix contain identical elements, the value of the determinant is _____.

88. Show that $\begin{vmatrix} a & b \\ c & d \end{vmatrix} = - \begin{vmatrix} c & d \\ a & b \end{vmatrix}$.

89. **Surveying** Surveyors use a formula to find the area of a plot of land. The *surveyor's area formula* states that if the vertices (x_1, y_1), (x_2, y_2), (x_3, y_3), ..., (x_n, y_n) of a simple polygon are listed counterclockwise around the perimeter, then the area of the polygon is

$$A = \frac{1}{2} \left\{ \begin{vmatrix} x_1 & x_2 \\ y_1 & y_2 \end{vmatrix} + \begin{vmatrix} x_2 & x_3 \\ y_2 & y_3 \end{vmatrix} + \begin{vmatrix} x_3 & x_4 \\ y_3 & y_4 \end{vmatrix} + \cdots + \begin{vmatrix} x_n & x_1 \\ y_n & y_1 \end{vmatrix} \right\}$$

Use the surveyor's area formula to find the area of the polygon with vertices $(9, -3)$, $(26, 6)$, $(18, 21)$, $(16, 10)$, and $(1, 11)$. Measurements given are in feet.

90. Consider the system of equations $\begin{array}{l} 3x - 2y = 12 \\ 2x + 5y = -11 \end{array}$. Graph the equations. Now multiply the second equation by 3, add it to the first equation, and replace the second equation by the new equation. What is the new system of equations? Graph the equations of this system on the same coordinate system. The lines intersect at the same point, so the new system of equations is equivalent to the original system. Explain how this is related to the third elementary row operation.

91. Suppose that applying elementary row operations to an augmented matrix produces the matrix $\begin{bmatrix} 1 & 2 & -3 & | & 5 \\ 0 & 1 & -2 & | & -2 \\ 0 & 0 & 0 & | & -3 \end{bmatrix}$. Explain why the original system of equations has no solution.

4.4 Application Problems

OBJECTIVE 1 Rate-of-wind and water-current problems

Motion problems that involve an object moving with or against a wind or current normally require two variables to solve.

Solve: A motorboat traveling with the current can go 24 mi in 2 h. Against the current, it takes 3 h to go the same distance. Find the rate of the motorboat in calm water and the rate of the current.

> **STRATEGY** for solving rate-of-wind and water-current problems
>
> ▶ Choose one variable to represent the rate of the object in calm conditions and a second variable to represent the rate of the wind or current. Using these variables, express the rate of the object with and against the wind or current. Use the equation $rt = d$ to write expressions for the distance traveled by the object. The results can be recorded in a table.

Rate of the boat in calm water: x
Rate of the current: y

	Rate	·	Time	=	Distance
With current	$x + y$	·	2	=	$2(x + y)$
Against current	$x - y$	·	3	=	$3(x - y)$

▶ Determine how the expressions for the distance are related.

The distance traveled with the current is 24 mi.
The distance traveled against the current is 24 mi.

With the current
$2(x + y) = 24$

Against the current
$3(x - y) = 24$

$$2(x + y) = 24$$
$$3(x - y) = 24$$

Solve the system of equations.

$2(x + y) = 24$ → $\dfrac{1}{2} \cdot 2(x + y) = \dfrac{1}{2} \cdot 24$ → $x + y = 12$

$3(x - y) = 24$ → $\dfrac{1}{3} \cdot 3(x - y) = \dfrac{1}{3} \cdot 24$ → $x - y = 8$

$$2x = 20$$
$$x = 10$$

Replace x by 10 in the equation $x + y = 12$. Solve for y.

$$x + y = 12$$
$$10 + y = 12$$
$$y = 2$$

The rate of the boat in calm water is 10 mph.
The rate of the current is 2 mph.

EXAMPLE 1 Flying with the wind, a plane flew 1000 mi in 5 h. Flying against the wind, the plane could fly only 500 mi in the same amount of time. Find the rate of the plane in calm air and the rate of the wind.

Strategy ▶ Rate of the plane in still air: p
Rate of the wind: w

	Rate	Time	Distance
With wind	$p + w$	5	$5(p + w)$
Against wind	$p - w$	5	$5(p - w)$

▶ The distance traveled with the wind is 1000 mi.
The distance traveled against the wind is 500 mi.

Solution $5(p + w) = 1000$
$5(p - w) = 500$

$p + w = 200$ • **Multiply each side of the equations by $\dfrac{1}{5}$.**
$p - w = 100$

$2p = 300$ • **Add the equations.**
$p = 150$

$$p + w = 200$$
$$150 + w = 200$$
$$w = 50$$

• **Substitute the value of p into one of the equations.**

The rate of the plane in calm air is 150 mph.
The rate of the wind is 50 mph.

PROBLEM 1 A rowing team rowing with the current traveled 18 mi in 2 h. Against the current, the team rowed 10 mi in 2 h. Find the rate of the current.

Solution See pages S17.

OBJECTIVE 2 Application problems

The application problems in this section are varieties of those problems solved earlier in the text. Each of the strategies for the problems in this section will result in a system of equations.

Solve: A store owner purchased twenty 60-watt light bulbs and thirty fluorescent lights for a total cost of $40. A second purchase, at the same prices, included thirty 60-watt bulbs and ten fluorescent lights for a total cost of $25. Find the cost of a 60-watt bulb and of a fluorescent light.

> **STRATEGY** for solving an application problem in two variables
>
> ▶ Choose one variable to represent one of the unknown quantities and a second variable to represent the other unknown quantity. Write numerical or variable expressions for all the remaining quantities. These results can be recorded in two tables, one for each of the conditions.

Cost of 60-watt bulb: b
Cost of a fluorescent light: f

First purchase:

	Amount	·	Unit cost	=	Value
60-watt	20	·	b	=	$20b$
Fluorescent	30	·	f	=	$30f$

Second purchase:

	Amount	·	Unit cost	=	Value
60-watt	30	·	b	=	$30b$
Fluorescent	10	·	f	=	$10f$

> ▶ Determine a system of equations. The strategies presented in Chapter 2 can be used to determine the relationships between the expressions in the tables. Each table will give one equation of the system.

The total of the first purchase was $40. $20b + 30f = 40$
The total of the second purchase was $25. $30b + 10f = 25$

Solve the system of equations.

$20b + 30f = 40$ $\longrightarrow$ $3(20b + 30f) = 3 \cdot 40$ $\longrightarrow$ $60b + 90f = 120$
$30b + 10f = 25$ $-2(30b + 10f) = -2 \cdot 25$ $-60b - 20f = -50$
$70f = 70$
$f = 1$

Replace f by 1 in the equation $20b + 30f = 40$.
Solve for b.

$20b + 30f = 40$
$20b + 30(1) = 40$
$20b + 30 = 40$
$20b = 10$
$b = 0.5$

The cost of a 60-watt bulb was $.50.
The cost of a fluorescent light was $1.00.

EXAMPLE 2 The total value of the nickels and dimes in a coin bank is $2.50. If the nickels were dimes and the dimes were nickels, the total value of the coins would be $3.05. Find the number of dimes and the number of nickels in the bank.

Strategy ▶ Number of nickels in the bank: n
Number of dimes in the bank: d

Coins in the bank now:

Coin	Number	Value	Total value
Nickels	n	5	$5n$
Dimes	d	10	$10d$

Coins in the bank if the nickels were dimes and the dimes were nickels:

Coin	Number	Value	Total value
Nickels	d	5	$5d$
Dimes	n	10	$10n$

▶ The value of the nickels and dimes in the bank is $2.50. The value of the nickels and dimes in the bank would be $3.05.

Solution
(1) $\quad 5n + 10d = 250$
(2) $\quad 10n + 5d = 305$

$10n + 20d = 500$ • **Multiply equation (1) by 2 and equation**
$-10n - 5d = -305$ **(2) by −1.**
$15d = 195$ • **Add the equations.**
$d = 13$

$$5n + 10d = 250$$
$$5n + 10(13) = 250$$
$$5n + 130 = 250$$
$$5n = 120$$
$$n = 24$$

• **Substitute the value of *d* into one of the equations.**

There are 24 nickels and 13 dimes in the bank.

PROBLEM 2 A citrus fruit grower purchased 25 orange trees and 20 grapefruit trees for $290. The next week, at the same prices, the grower bought 20 orange trees and 30 grapefruit trees for $330. Find the cost of an orange tree and the cost of a grapefruit tree.

■ **Solution** See page S17.

Example 3 is an application problem that requires more than two variables. The solution of this problem also illustrates how the substitution method can be used to solve a system of linear equations in three variables.

EXAMPLE 3 An artist is creating a mobile from which three objects will be suspended from a light rod that is 18 in. long as shown at the left. The weight, in ounces, of each object is shown in the diagram. For the mobile to balance, the objects must be positioned so that $w_1d_1 + w_2d_2 = w_3d_3$. The artist wants d_1 to be 1.5 times d_2. Find the distances d_1, d_2, and d_3 so that the mobile will balance.

Strategy There are three unknowns for this problem. Use the information in the problem to write three equations with d_1, d_2, and d_3 as the variables.

The length of the rod is 18 in. Therefore, $d_1 + d_3 = 18$. Using $w_1d_1 + w_2d_2 = w_3d_3$, we have $2d_1 + 3d_2 = 4d_3$. The artist wants d_1 to be 1.5 times d_2. Thus $d_1 = 1.5d_2$.

Solution (1) $d_1 + d_3 = 18$
(2) $2d_1 + 3d_2 = 4d_3$
(3) $d_1 = 1.5d_2$

We will solve the system of equations by substitution. Use equation (3) to replace d_1 in equation (1) and in equation (2).

$$1.5d_2 + d_3 = 18$$
$$2(1.5d_2) + 3d_2 = 4d_3$$

(4) $1.5d_2 + d_3 = 18$
(5) $6d_2 = 4d_3$
(6) $1.5d_2 = d_3$ • **Divide each side of equation (5) by 4.**

Replace d_3 by $1.5d_2$ in equation (4) and solve for d_2.

$$1.5d_2 + 1.5d_2 = 18$$
$$3d_2 = 18$$
$$d_2 = 6$$

From equation (6), $d_3 = 1.5d_2 = 1.5(6) = 9$.

Substituting the value of d_3 into equation (1), we have $d_1 = 9$.

The values are $d_1 = 9$ in., $d_2 = 6$ in., and $d_3 = 9$ in.

PROBLEM 3 A science museum charges $10 for an admission ticket, but members receive a discount of $3, and students are admitted for half price. Last Saturday, 750 tickets were sold for a total of $5400. If 20 more student tickets than full-price tickets were sold, how many of each type of ticket were sold?

■ **Solution** See pages S17–S18.

 Concept Review

Determine whether the following statements are always true, sometimes true, or never true.

1. The rate of a current in a river is x mph, and the rate of a boat in still water is y mph. The rate of a boat going down the river is $x + y$, and the rate going up the river is $x - y$.

2. The speed of a plane is 500 mph. There is a headwind of 50 mph. The speed of the plane relative to an observer on the ground is 550 mph.

3. A contractor bought 50 yd of nylon carpet for x dollars per yard and bought 100 yd of wool carpet for y dollars per yard. The total cost of the carpet is $50x + 100y$.

4. A coin bank contains nickels and dimes. The total value of the coins is $.70. The bank had 5 dimes and 4 nickels.

 Exercises

1 **Rate-of-wind and water-current problems**

1. A boat travels down a river for 3 h (traveling with the current), then turns around and takes 5 h to return (traveling against the current). Let b be the rate of the boat, in miles per hour, in calm water and c be the rate of the current, in miles per hour. Complete the following table.

	Rate, r	·	Time, t	=	Distance, d
With current	?	·	?	=	?
Against current	?	·	?	=	?

2. Suppose the distance traveled down the river by the boat in Exercise 1 is 45 mi. Use the expressions in the last column of the table in Exercise 1 to write a system of equations that can be solved to find the rate of the boat in calm water and the rate of the current.

3. Traveling with the wind, a plane flies m miles in h hours. Traveling against the wind, a plane flies n miles in h hours. Is n less than, equal to, or greater than m?

4. Traveling against the current, it takes a boat h hours to go m miles. Traveling with the current, the boat takes k hours to go m miles. Is h less than, equal to, or greater than k?

5. Flying with the wind, a small plane flew 320 mi in 2 h. Against the wind, the plane could fly only 280 mi in the same amount of time. Find the rate of the plane in calm air and the rate of the wind.

6. A jet plane flying with the wind went 2100 mi in 4 h. Against the wind, the plane could fly only 1760 mi in the same amount of time. Find the rate of the plane in calm air and the rate of the wind.

7. A cabin cruiser traveling with the current went 48 mi in 3 h. Against the current, it took 4 h to travel the same distance. Find the rate of the cabin cruiser in calm water and the rate of the current.

8. A motorboat traveling with the current went 48 mi in 2 h. Against the current, it took 3 h to travel the same distance. Find the rate of the boat in calm water and the rate of the current.

9. Flying with the wind, a pilot flew 450 mi between two cities in 2.5 h. The return trip against the wind took 3 h. Find the rate of the plane in calm air and the rate of the wind.

10. A turbo-prop plane flying with the wind flew 600 mi in 2 h. Flying against the wind, the plane required 3 h to travel the same distance. Find the rate of the wind and the rate of the plane in calm air.

With the wind
$2(p + w) = 600$

Against the wind
$3(p - w) = 600$

11. A motorboat traveling with the current went 88 km in 4 h. Against the current, the boat could go only 64 km in the same amount of time. Find the rate of the boat in calm water and the rate of the current.

12. A rowing team rowing with the current traveled 18 km in 2 h. Rowing against the current, the team rowed 12 km in the same amount of time. Find the rate of the rowing team in calm water and the rate of the current.

13. A plane flying with a tailwind flew 360 mi in 3 h. Against the wind, the plane required 4 h to fly the same distance. Find the rate of the plane in calm air and the rate of the wind.

14. Flying with the wind, a plane flew 1000 mi in 4 h. Against the wind, the plane required 5 h to fly the same distance. Find the rate of the plane in calm air and the rate of the wind.

15. A motorboat traveling with the current went 54 mi in 3 h. Against the current, it took 3.6 h to travel the same distance. Find the rate of the boat in calm water and the rate of the current.

16. A plane traveling with the wind flew 3625 mi in 6.25 h. Against the wind, the plane required 7.25 h to fly the same distance. Find the rate of the plane in calm air and the rate of the wind.

2 Application problems

17. For one project, a builder buys 200 square feet of hardwood flooring and 300 square feet of wall-to-wall carpet. For another project, a builder buys 350 square feet of hardwood flooring and 100 square feet of wall-to-wall carpet. Let h be the cost per square foot of hardwood flooring and w be the cost per square foot of wall-to-wall carpet. Complete the following tables by replacing the question marks.

First Project	Amount	·	Unit cost	=	Value
Hardwood flooring	?	·	?	=	?
Wall-to-wall carpet	?	·	?	=	?

Second Project	Amount	·	Unit cost	=	Value
Hardwood flooring	?	·	?	=	?
Wall-to-wall carpet	?	·	?	=	?

18. Suppose the builder in Exercise 1 spent $2300 on the flooring and carpet for the first project and $2750 on the flooring and carpet for the second project. Use the expressions in the last column of the tables in Exercise 1 to write a system of equations that can be solved to find the cost per square foot of hardwood flooring and the cost per square foot of wall-to-wall carpet.

Complete Exercises 19 and 20 without actually solving the problems.

19. A coffee merchant mixes 3 lb of dark roast coffee with 1 lb of light roast coffee to create a mixture that cost $6.75 per pound. When he mixes 1 lb of the dark roast coffee with 3 lb of the light roast coffee, the mixture costs $3.50 per pound. Is the cost per pound of the dark roast coffee less than, equal to, or greater than the cost per pound of the light roast coffee?

20. The total value of dimes and quarters in a bank is $2.75. If the dimes were quarters and the quarters were dimes, the total value would be $3.20. Is the number of quarters in the bank less than, equal to, or greater than the number of dimes in the bank?

21. **Construction** A carpenter purchased 50 ft of redwood and 90 ft of pine for a total cost of $31.20. A second purchase, at the same prices, included

200 ft of redwood and 100 ft of pine for a total cost of $78. Find the cost per foot of redwood and of pine.

22. **Business** A merchant mixed 10 lb of a cinnamon tea with 5 lb of spice tea. The 15-pound mixture cost $40. A second mixture included 12 lb of the cinnamon tea and 8 lb of the spice tea. The 20-pound mixture cost $54. Find the cost per pound of the cinnamon tea and of the spice tea.

23. **Home Economics** During one month, a homeowner used 400 units of electricity and 120 units of gas for a total cost of $147.20. The next month, 350 units of electricity and 200 units of gas were used for a total cost of $144. Find the cost per unit of gas.

24. **Construction** A contractor buys 20 yd² of nylon carpet and 28 yd² of wool carpet for $1360. A second purchase, at the same prices, includes 15 yd² of nylon carpet and 20 yd² of wool carpet for $990. Find the cost per square yard of the wool carpet.

25. **Coin Problem** The total value of the quarters and dimes in a coin bank is $6.90. If the quarters were dimes and the dimes were quarters, the total value of the coins would be $7.80. Find the number of quarters in the bank.

26. **Coin Problem** A coin bank contains only nickels and dimes. The total value of the coins in the bank is $2.50. If the nickels were dimes and the dimes were nickels, the total value of the coins would be $3.20. Find the number of nickels in the bank.

27. **Manufacturing** A company manufactures both mountain bikes and trail bikes. The cost of materials for a mountain bike is $70, and the cost of materials for a trail bike is $50. The cost of labor to manufacture a mountain bike is $80, and the cost of labor to manufacture a trail bike is $40. During a week in which the company has budgeted $2500 for materials and $2600 for labor, how many mountain bikes does the company plan to manufacture?

28. **Manufacturing** A company manufactures both liquid crystal display (LCD) and cathode ray tube (CRT) color monitors. The cost of materials for a CRT monitor is $50, whereas the cost of materials for an LCD monitor is $150. The cost of labor to manufacture a CRT monitor is $80, whereas the cost of labor to manufacture an LCD monitor is $130. During a week when the company has budgeted $9600 for materials and $8760 for labor, how many LCD monitors does the company plan to manufacture?

29. **Health Science** A pharmacist has two vitamin-supplement powders. The first powder is 25% vitamin B_1 and 15% vitamin B_2. The second is 15% vitamin B_1 and 20% vitamin B_2. How many milligrams of each of the two powders should the pharmacist use to make a mixture that contains 117.5 mg of vitamin B_1 and 120 mg of vitamin B_2?

30. Chemistry A chemist has two alloys, one of which is 10% gold and 15% lead and the other of which is 30% gold and 40% lead. How many grams of each of the two alloys should be used to make an alloy that contains 60 g of gold and 88 g of lead?

31. Business On Monday, a computer manufacturing company sent out three shipments. The first order, which contained a bill for $114,000, was for 4 Model II, 6 Model VI, and 10 Model IX computers. The second shipment, which contained a bill for $72,000, was for 8 Model II, 3 Model VI, and 5 Model IX computers. The third shipment, which contained a bill for $81,000, was for 2 Model II, 9 Model VI, and 5 Model IX computers. What does the manufacturer charge for a Model VI computer?

32. Health Care A relief organization supplies blankets, cots, and lanterns to victims of fires, floods, and other natural disasters. One week the organization purchased 15 blankets, 5 cots, and 10 lanterns for a total cost of $1250. The next week, at the same prices, the organization purchased 20 blankets, 10 cots, and 15 lanterns for a total cost of $2000. The next week, at the same prices, the organization purchased 10 blankets, 15 cots, and 5 lanterns for a total cost of $1625. Find the cost of one blanket, the cost of one cot, and the cost of one lantern.

33. Investments An investor has a total of $18,000 deposited in three different accounts, which earn annual interest of 9%, 7%, and 5%. The amount deposited in the 9% account is twice the amount in the 5% account. If the three accounts earn total annual interest of $1340, how much money is deposited in each account?

34. Investments An investor has a total of $15,000 deposited in three different accounts, which earn annual interest of 9%, 6%, and 4%. The amount deposited in the 6% account is $2000 more than the amount in the 4% account. If the three accounts earn total annual interest of $980, how much money is deposited in each account?

35. Art A sculptor is creating a mobile from which three objects will be suspended from a light rod that is 15 in. long as shown at the right. The weight, in ounces, of each object is shown in the diagram. For the mobile to balance, the objects must be positioned so that $w_1d_1 = w_2d_2 + w_3d_3$. The artist wants d_3 to be three times d_2. Find the distances d_1, d_2, and d_3 so that the mobile will balance.

36. Art A mobile is made by suspending three objects from a light rod that is 20 in. long as shown at the right. The weight, in ounces, of each object is shown in the diagram. For the mobile to balance, the objects must be positioned so that $w_1d_1 + w_2d_2 = w_3d_3$. The artist wants d_3 to be twice d_2. Find the distances d_1, d_2, and d_3 so that the mobile will balance.

37. Coin Problem A coin bank contains only nickels, dimes, and quarters. The value of the 19 coins in the bank is $2. There are twice as many nickels as dimes. Find the number of each type of coin in the bank.

38. Coin Problem The total value of the nickels, dimes, and quarters in a coin bank is $4. If the nickels were dimes and the dimes were nickels, the total value of the coins would be $3.75. If the quarters were dimes and the dimes were quarters, the total value of the coins would be $6.25. Find the number of nickels, dimes, and quarters in the bank.

39. Investments A financial planner invested $33,000 of a client's money, part at 9%, part at 12%, and the remainder at 8%. The total annual income from these three investments was $3290. The amount invested at 12% was $5000 less than the combined amount invested at 9% and 8%. Find the amount invested at each rate.

40. Education The total number of elementary school students, secondary school students, and students enrolled in colleges in the United States is 71 million. The number enrolled in colleges is equal to the number enrolled in secondary school. The number enrolled in elementary school is 7 million more than the total number of students in secondary schools and colleges. (*Source:* www.infoplease.com) Find the number of students in elementary school, the number in secondary school, and the number enrolled in colleges in the United States.

4.4 Applying Concepts

41. Geometry Two angles are complementary. The larger angle is 9° more than eight times the measure of the smaller angle. Find the measure of the two angles. (Complementary angles are two angles whose sum is 90°.)

42. Geometry Two angles are supplementary. The larger angle is 40° more than three times the measure of the smaller angle. Find the measure of the two angles. (Supplementary angles are two angles whose sum is 180°.)

$x + y = 180°$

43. Coin Problem The total value of the nickels, dimes, and quarters in a coin bank is $3.50. If the nickels were dimes and the dimes were nickels, the total value of the coins would be $4.25. If the quarters were dimes and the dimes were quarters, the total value of the coins would be $4.25. Find the number of nickels, dimes, and quarters in the bank.

4.5 Solving Systems of Linear Inequalities

OBJECTIVE **1**

Graph the solution set of a system of linear inequalities

Two or more inequalities considered together are called a **system of inequalities.** The **solution set of a system of inequalities** is the intersection of the solution sets of the individual inequalities. To graph the solution set of a system of inequalities, first graph the solution set of each inequality. The solution set of the system of inequalities is the region of the plane represented by the intersection of the two shaded areas.

Take Note

You can use a test point to check that the correct region has been denoted as the solution set. We can see from the graph that the point (2, 4) is in the solution set, and as shown below, it is a solution of each inequality in the system. This indicates that the solution set as graphed is correct.

$$2x - y \le 3$$
$$2(2) - (4) \le 3$$
$$0 \le 3 \quad \text{True}$$

$$3x + 2y > 8$$
$$3(2) + 2(4) > 8$$
$$14 > 8 \quad \text{True}$$

Graph the solution set: $2x - y \le 3$
$$3x + 2y > 8$$

Solve each equation for y.

$$2x - y \le 3 \qquad\qquad 3x + 2y > 8$$
$$-y \le -2x + 3 \qquad\qquad 2y > -3x + 8$$
$$y \ge 2x - 3 \qquad\qquad y > -\frac{3}{2}x + 4$$

Graph $y = 2x - 3$ as a solid line. Because the inequality is $\ge$, shade above the line.

Graph $y = -\frac{3}{2}x + 4$ as a dashed line.

Because the inequality is $>$, shade above the line.

The solution set of the system is the region of the plane represented by the intersection of the solution sets of the individual inequalities.

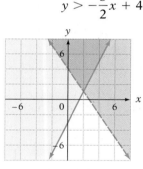

Point of Interest

Large systems of linear inequalities containing over 100 inequalities have been used to solve application problems in such diverse areas as providing health care and hardening a nuclear missile silo.

Graph the solution set: $-x + 2y \ge 4$
$$x - 2y \ge 6$$

Solve each equation for y.

$$-x + 2y \ge 4 \qquad\qquad x - 2y \ge 6$$
$$2y \ge x + 4 \qquad\qquad -2y \ge -x + 6$$
$$y \ge \frac{1}{2}x + 2 \qquad\qquad y \le \frac{1}{2}x - 3$$

Shade above the solid line $y = \frac{1}{2}x + 2$.

Shade below the solid line $y = \frac{1}{2}x - 3$.

Because the solution sets of the two inequalities do not intersect, the solution set of the system is the empty set.

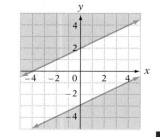

EXAMPLE 1 Graph the solution set.

A. $y \geq x - 1$ B. $2x + 3y > 9$
 $y < -2x$
 $y < -\dfrac{2}{3}x + 1$

Solution A. Shade above the solid line $y = x - 1$.

Shade below the dashed line $y = -2x$.

The solution set of the system is the intersection of the solution sets of the individual inequalities.

B. Solve $2x + 3y > 9$ for y.

$$2x + 3y > 9$$
$$3y > -2x + 9$$
$$y > -\frac{2}{3}x + 3$$

Shade above the dashed line $y = -\dfrac{2}{3}x + 3$.

Shade below the dashed line $y = -\dfrac{2}{3}x + 1$.

The intersection of the system is the empty set because the solution sets of the two inequalities do not intersect.

PROBLEM 1 Graph the solution set.

A. $y \geq 2x - 3$ B. $3x + 4y > 12$
 $y > -3x$
 $y < \dfrac{3}{4}x - 1$

Solution See page S18.

4.5 Concept Review

Determine whether the following statements are always true, sometimes true, or never true.

1. The solution set of a system of linear inequalities is an infinite set.

2. The solution set of a system of linear inequalities is the empty set.

3. The solution set of a system of linear inequalities is the intersection of the solution sets of the individual inequalities.

4.5 Exercises

1 Graph the solution set of a system of linear inequalities

1. Which ordered pair is a solution of the system of inequalities?

$2x - y < 4$
$x - 3y \geq 6$

a. $(5, 1)$ **b.** $(-3, -5)$

2. Which ordered pair is a solution of the system of inequalities?

$3x - 2y \geq 6$
$x + y < 5$

a. $(-2, 3)$ **b.** $(3, -2)$

3. Use the graph showing the four regions of the plane that are determined by the intersecting lines $y = 2x + 2$ and $y = -x + 5$. Complete parts a, b, and c by replacing the question marks with region letters from the graph.

 a. Points in the solution set of the inequality $y \leq 2x + 2$ are located in regions ____?____ and ____?____.

 b. Points in the solution set of the inequality $y \geq -x + 5$ are located in regions ____?____ and ____?____.

 c. Points in the solution set of the system of inequalities

 $y < 2x + 2$
 $y > -x + 5$

 are located in region ____?____.

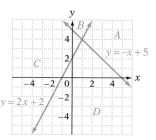

4. Use the graph showing the vertical line through $(2, 0)$ and the horizontal line through $(0, 4)$. Complete parts a and b by replacing the question marks with *greater than* or *less than*.

 a. Points to the right of the vertical line have *x*-coordinates ____?____ 2.

 b. Points below the horizontal line have *y*-coordinates ____?____ 4.

 c. Write the system of inequalities for which the shaded region is the solution set.

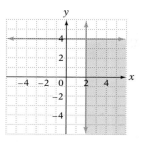

Graph the solution set.

5. $y \leq x - 3$
 $y \leq -x + 5$

6. $y > 2x - 4$
 $y < -x + 5$

7. $y > 3x - 3$
 $y \geq -2x + 2$

8. $y \leq -\dfrac{1}{2}x + 3$
 $y \geq x - 3$

9. $2x + y \geq -2$
 $6x + 3y \leq 6$

10. $x + y \geq 5$
 $3x + 3y \leq 6$

11. $3x - 2y < 6$
 $y \leq 3$

12. $x \leq 2$
 $3x + 2y > 4$

13. $y > 2x - 6$
 $x + y < 0$

14. $x < 3$
$y < -2$

15. $x + 1 \geq 0$
$y - 3 \leq 0$

16. $5x - 2y \geq 10$
$3x + 2y \geq 6$

17. $2x + y \geq 4$
$3x - 2y < 6$

18. $3x - 4y < 12$
$x + 2y < 6$

19. $x - 2y \leq 6$
$2x + 3y \leq 6$

20. $x - 3y > 6$
$2x + y > 5$

21. $x - 2y \leq 4$
$3x + 2y \leq 8$

22. $3x - 2y < 0$
$5x + 3y > 9$

For Exercises 23 to 26, a and b are positive numbers such that $a > b$. Describe the solution set of each system of inequalities.

23. $x + y < a$
$x + y < b$

24. $x + y > a$
$x + y < b$

25. $x + y < a$
$x + y > b$

26. $x + y > a$
$x + y > b$

4.5 Applying Concepts

Graph the solution set.

27. $2x + 3y \leq 15$
$3x - y \leq 6$
$y \geq 0$

28. $x + y \leq 6$
$x - y \leq 2$
$x \geq 0$

29. $x - y \leq 5$
$2x - y \geq 6$
$y \geq 0$

30. $x - 3y \leq 6$
$5x - 2y \geq 4$
$y \geq 0$

31. $2x - y \leq 4$
$3x + y < 1$
$y \leq 0$

32. $x - y \leq 4$
$2x + 3y > 6$
$x \geq 0$

Focus on Problem Solving

Solve an Easier Problem

We begin this problem-solving feature with a restatement of the four steps of Polya's recommended problem-solving model.

1. Understand the problem.
2. Devise a plan.
3. Carry out the plan.
4. Review your solution.

One of the several methods of devising a plan is to try to solve an easier problem. Suppose you are in charge of your softball league, which consists of 15 teams. You must devise a schedule in which each team plays every other team once. How many games must be scheduled?

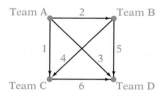

To solve this problem, we will attempt an easier problem first. Suppose that your league contains only a small number of teams. For instance, if there were only 1 team, you would schedule 0 games. If there were 2 teams, you would schedule 1 game. If there were 3 teams, you would schedule 3 games. The diagram at the left shows that 6 games must be scheduled when there are 4 teams in the league.

Here is a table of our results so far. (Remember that making a table is another strategy to be used in problem solving.)

Number of Teams	Number of Games	Possible Pattern
1	0	0
2	1	1
3	3	1 + 2
4	6	1 + 2 + 3

1. Draw a diagram with 5 dots to represent the teams. Draw lines from each dot to a second dot, and determine the number of games required.

2. What is the apparent pattern for the number of games required?

3. Assuming that the pattern continues, how many games must be scheduled for the 15 teams of the original problem?

After solving a problem, good problem solvers ask whether it is possible to solve the problem in a different manner. Here is a possible alternative method of solving the scheduling problem.

Begin with one of the 15 teams (say team A) and ask, "How many games must this team play?" Because there are 14 teams left to play, you must schedule 14 games. Now move to team B. It is already scheduled to play team A, and it does not play itself, so there are 13 teams left for it to play. Consequently, you must schedule 14 + 13 games.

4. Continue this reasoning for the remaining teams and determine the number of games that must be scheduled. Does this answer correspond to the answer you obtained with the first method?

5. Visit **www.allsports.com** to get a listing of the professional sports teams and how they are separated into leagues and divisions. To appreciate how complicated the game schedules can be, try to create a partial schedule where each team in one of the conferences in the National Football League—for instance, the NFC East—plays each other team in that conference twice. An actual schedule would require scheduling an additional eight games for each team in the conference with selected teams in the other conferences.

6. Making connections to other problems you have solved is an important step toward becoming an excellent problem solver. How is the answer to the scheduling of teams related to the triangular numbers discussed in the Focus on Problem Solving in the chapter titled "Linear Functions and Inequalities in Two Variables"?

Projects & Group Activities

Using a Graphing Calculator to Solve a System of Equations

A graphing calculator can be used to solve a system of equations. For this procedure to work on most calculators, it is necessary that the point of intersection be on the screen. This means that you may have to experiment with Xmin, Xmax, Ymin, and Ymax values until the graphs intersect on the screen.

To solve a system of equations graphically, solve each equation for y. Then graph the equations of the system. Their point of intersection is the solution.

For instance, to solve the system of equations

$$4x - 3y = 7$$
$$5x + 4y = 2$$

first solve each equation for y.

$$4x - 3y = 7 \implies y = \frac{4}{3}x - \frac{7}{3}$$

$$5x + 4y = 2 \implies y = -\frac{5}{4}x + \frac{1}{2}$$

The keystrokes needed to solve this system using a TI-83 follow. We are using a viewing window of $[-4.7, 4.7]$ by $[-3.1, 3.1]$. The approximate solution is $(1.096774, -0.870968)$.

Once the calculator has drawn the graphs, use the TRACE feature and move the cursor to the approximate point of intersection. This will give you an approximate solution of the system of equations. A more accurate solution can be found by using the following keystrokes.

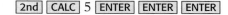

Some of the exercises in the first section of this chapter asked you to solve a system of equations by graphing. Try those exercises again, this time using your graphing calculator.

Chapter 4 Summary

Key Words	*Examples*
A **system of equations** is two or more equations considered together. A **solution of a system of equations in two variables** is an ordered pair that is a solution of each equation of the system. [4.1.1, p. 304]	The solution of the system $$x + y = 2$$ $$x - y = 4$$ is the ordered pair $(3, -1)$. $(3, -1)$ is the only ordered pair that is a solution of both equations.
When the graphs of a system of equations intersect at only one point, the system is called an **independent system of equations**. [4.1.1, p. 305]	$3x - 2y = 4$ $5x + 2y = 6$ $8x\quad\ \ = 10$ • **Add the equations.**
When the graphs of a system of equations coincide, the system is called a **dependent system of equations**. [4.1.1, p. 306]	$-3x + 2y = 4$ $3x - 2y = -4$ $0 = 0$ • **Add the equations.**
When the graphs of a system of equations do not intersect, the system has no solution and is called an **inconsistent system of equations**. [4.1.1, p. 305]	$2x - 3y = -2$ $-2x + 3y = 5$ $0 = 3$ • **Add the equations.**
An equation of the form $Ax + By + Cz = D$ is called a **linear equation in three variables**. [4.2.2, p. 316]	$3x + 2y - 5z = 12$ is a linear equation in three variables.
A **solution of a system of equations in three variables** is an ordered triple that is a solution of each equation of the system. [4.2.2, p. 318]	The ordered triple $(1, 2, 1)$ is a solution of the system of equations $$3x + y - 3z = 2$$ $$-x + 2y + 3z = 6$$ $$2x + 2y - 2z = 4$$
A **matrix** is a rectangular array of numbers. [4.3.1, p. 323]	$\begin{bmatrix} 2 & 3 & 6 \\ -1 & 2 & 4 \end{bmatrix}$ is a 2×3 matrix.
A **square matrix** has the same number of rows as columns. [4.3.1, p. 324]	$\begin{bmatrix} 2 & 3 \\ 4 & -1 \end{bmatrix}$ is a square matrix.
A **determinant** is a number associated with a square matrix. [4.3.1, p. 324]	$\begin{vmatrix} 2 & -3 \\ -4 & 5 \end{vmatrix} = 2(5) - (-4)(-3)$ $= 10 - 12$ $= -2$

The **minor of an element** in a 3 × 3 determinant is the 2 × 2 determinant obtained by eliminating the row and column that contain the element. [4.3.1, p. 324]

$$\begin{vmatrix} 2 & 1 & 3 \\ 4 & 6 & 2 \\ 1 & 8 & 3 \end{vmatrix}$$ The minor of 4 is $\begin{vmatrix} 1 & 3 \\ 8 & 3 \end{vmatrix}$.

The **cofactor of an element of a determinant** is $(-1)^{i+j}$ times the minor of the element, where i is the row number of the element and j is its column number. [4.3.1, p. 325]

In the determinant above, 4 is from the second row and the first column. The cofactor of 4 is

$$(-1)^{2+1} \begin{vmatrix} 1 & 3 \\ 8 & 3 \end{vmatrix} = - \begin{vmatrix} 1 & 3 \\ 8 & 3 \end{vmatrix}$$

The evaluation of the determinant of a 3 × 3 or larger matrix is accomplished by **expanding by cofactors**. [4.3.1, p. 325]

Expand by cofactors of the first column.

$$\begin{vmatrix} 2 & 1 & 3 \\ 4 & 6 & 2 \\ 1 & 8 & 3 \end{vmatrix} = 2 \begin{vmatrix} 6 & 2 \\ 8 & 3 \end{vmatrix} - 4 \begin{vmatrix} 1 & 3 \\ 8 & 3 \end{vmatrix} + 1 \begin{vmatrix} 1 & 3 \\ 6 & 2 \end{vmatrix}$$

An **augmented matrix** is one consisting of the coefficients and constants of a system of equations. [4.3.3, p. 329]

The augmented matrix associated with the system

$$4x - 3y + z = 7$$
$$x + y - 5z = 14$$
$$2x + 3z = -4$$

is

$$\begin{bmatrix} 4 & -3 & 1 & | & 7 \\ 1 & 1 & -5 & | & 14 \\ 2 & 0 & 3 & | & -4 \end{bmatrix}$$

The process of solving a system of equations by using elementary row operations is called the **Gaussian elimination method**. [4.3.3, p. 334]

Inequalities considered together are called a **system of inequalities**. The **solution set of a system of inequalities** is the intersection of the solution sets of the individual inequalities. [4.5.1, p. 353]

$$x + y > 3$$
$$x - y > -2$$

Essential Rules and Procedures

A system of equations can be solved by:

a. Graphing [4.1.1, pp. 305–307]

$$y = \frac{1}{2}x + 2$$

$$y = \frac{5}{2}x - 2$$

b. The substitution method [4.1.2, pp. 307–309]

$$(1)\ 2x - 3y = 4$$
$$(2)\ \ \ \ \ \ \ \ y = -x + 2$$

Substitute the value of y into equation (1).

$$2x - 3(-x + 2) = 4$$

c. The addition method [4.2.1 and 4.2.2, pp. 313–316]

$$-2x + 3y = 7$$
$$\ \ 2x - 5y = 2$$
$$\ \ \ \ \ \ \ \ -2y = 9 \quad \bullet \text{ Add the equations.}$$

Cramer's Rule [4.3.2, pp. 327–328]

For two variables: $a_{11}x + a_{12}y = b_1$
$\qquad\qquad\qquad a_{21}x + a_{22}y = b_2$

$$2x - y = 6$$
$$x + 3y = 4$$

$x = \dfrac{D_x}{D}, y = \dfrac{D_y}{D}$, where $D = \begin{vmatrix} a_{11} & a_{12} \\ a_{21} & a_{22} \end{vmatrix}$, $D_x = \begin{vmatrix} b_1 & a_{12} \\ b_2 & a_{22} \end{vmatrix}$,

$D_y = \begin{vmatrix} a_{11} & b_1 \\ a_{21} & b_2 \end{vmatrix}$, and $D \neq 0$.

$$D = \begin{vmatrix} 2 & -1 \\ 1 & 3 \end{vmatrix} = 7, D_x = \begin{vmatrix} 6 & -1 \\ 4 & 3 \end{vmatrix} = 22,$$

$$D_y = \begin{vmatrix} 2 & 6 \\ 1 & 4 \end{vmatrix} = 2$$

$$x = \frac{D_x}{D} = \frac{22}{7}, y = \frac{D_y}{D} = \frac{2}{7}$$

For three variables: $a_{11}x + a_{12}y + a_{13}z = b_1$
$\qquad\qquad\qquad\quad a_{21}x + a_{22}y + a_{23}z = b_2$
$\qquad\qquad\qquad\quad a_{31}x + a_{32}y + a_{33}z = b_3$

$$x - y + z = 2$$
$$2x + y - 2z = -2$$
$$x - 2y + 3z = 6$$

$x = \dfrac{D_x}{D}, y = \dfrac{D_y}{D}$, and $z = \dfrac{D_z}{D}$, where $D = \begin{vmatrix} a_{11} & a_{12} & a_{13} \\ a_{21} & a_{22} & a_{23} \\ a_{31} & a_{32} & a_{33} \end{vmatrix}$,

$$D = \begin{vmatrix} 1 & -1 & 1 \\ 2 & 1 & -2 \\ 1 & -2 & 3 \end{vmatrix} = 2,$$

$D_x = \begin{vmatrix} b_1 & a_{12} & a_{13} \\ b_2 & a_{22} & a_{23} \\ b_3 & a_{32} & a_{33} \end{vmatrix}, D_y = \begin{vmatrix} a_{11} & b_1 & a_{13} \\ a_{21} & b_2 & a_{23} \\ a_{31} & b_3 & a_{33} \end{vmatrix}, D_z = \begin{vmatrix} a_{11} & a_{12} & b_1 \\ a_{21} & a_{22} & b_2 \\ a_{31} & a_{32} & b_3 \end{vmatrix}$, and

$D \neq 0$

$$D_x = \begin{vmatrix} 2 & -1 & 1 \\ -2 & 1 & -2 \\ 6 & -2 & 3 \end{vmatrix} = 2,$$

$$D_y = \begin{vmatrix} 1 & 2 & 1 \\ 2 & -2 & -2 \\ 1 & 6 & 3 \end{vmatrix} = 4,$$

$$D_z = \begin{vmatrix} 1 & -1 & 2 \\ 2 & 1 & -2 \\ 1 & -2 & 6 \end{vmatrix} = 6$$

$$x = \frac{D_x}{D} = \frac{2}{2} = 1, y = \frac{D_y}{D} = \frac{4}{2} = 2,$$

$$z = \frac{D_z}{D} = \frac{6}{2} = 3$$

Elementary row operations to rewrite a matrix in **row echelon form**:
1. Interchange two rows.
2. Multiply all the elements in a row by the same nonzero number.
3. Replace a row by the sum of that row and a multiple of any other row. [4.3.3, p. 330]

$$\begin{bmatrix} 1 & 4 & -2 & | & 3 \\ 0 & 1 & 3 & | & -1 \\ 0 & 0 & 1 & | & 6 \end{bmatrix} \quad \text{Row echelon form}$$

Chapter 4 Review Exercises

1. Solve by substitution: $2x - 6y = 15$
$x = 3y + 8$

2. Solve by substitution: $3x + 12y = 18$
$x + 4y = 6$

3. Solve by the addition method: $3x + 2y = 2$
$x + y = 3$

4. Solve by the addition method: $5x - 15y = 30$
$x - 3y = 6$

5. Solve by the addition method: $3x + y = 13$
$2y + 3z = 5$
$x + 2z = 11$

6. Solve by the addition method:
$3x - 4y - 2z = 17$
$4x - 3y + 5z = 5$
$5x - 5y + 3z = 14$

7. Evaluate the determinant: $\begin{vmatrix} 6 & 1 \\ 2 & 5 \end{vmatrix}$

8. Evaluate the determinant: $\begin{vmatrix} 1 & 5 & -2 \\ -2 & 1 & 4 \\ 4 & 3 & -8 \end{vmatrix}$

9. Solve by using Cramer's Rule: $2x - y = 7$
$3x + 2y = 7$

10. Solve by using Cramer's Rule: $3x - 4y = 10$
$2x + 5y = 15$

11. Solve by using Cramer's Rule: $x + y + z = 0$
$x + 2y + 3z = 5$
$2x + y + 2z = 3$

12. Solve by using Cramer's Rule: $x + 3y + z = 6$
$2x + y - z = 12$
$x + 2y - z = 13$

13. Solve by the addition method: $x - 2y + z = 7$
$3x - z = -1$
$3y + z = 1$

14. Solve by using Cramer's Rule: $3x - 2y = 2$
$-2x + 3y = 1$

15. Solve by using the Gaussian elimination method:
$2x - 2y - 6z = 1$
$4x + 2y + 3z = 1$
$2x - 3y - 3z = 3$

16. Evaluate the determinant: $\begin{vmatrix} 3 & -2 & 5 \\ 4 & 6 & 3 \\ 1 & 2 & 1 \end{vmatrix}$

17. Solve by using Cramer's Rule: $4x - 3y = 17$
$3x - 2y = 12$

18. Solve by using the Gaussian elimination method:
$3x + 2y - z = -1$
$x + 2y + 3z = -1$
$3x + 4y + 6z = 0$

19. Solve by graphing: $x + y = 3$
$3x - 2y = -6$

20. Solve by graphing: $2x - y = 4$
$y = 2x - 4$

21. Graph the solution set: $x + 3y < 6$
$2x - y > 4$

22. Graph the solution set: $2x + 4y \geq 8$
$x + y \leq 3$

23. **Uniform Motion** A cabin cruiser traveling with the current went 60 mi in 3 h. Against the current, it took 5 h to travel the same distance. Find the rate of the cabin cruiser in calm water and the rate of the current.

24. **Uniform Motion** A plane flying with the wind flew 600 mi in 3 h. Flying against the wind, the plane required 4 h to travel the same distance. Find the rate of the plane in calm air and the rate of the wind.

25. **Recreation** At a movie theater, admission tickets are $5 for children and $8 for adults. The receipts for one Friday evening were $2500. The next day there were three times as many children as the preceding evening and only half as many adults as the night before, yet the receipts were still $2500. Find the number of children who attended the movie Friday evening.

26. **Investments** An investor has a total of $25,000 deposited in three different accounts, which earn annual interest rates of 8%, 6%, and 4%. The amount deposited in the 8% account is twice the amount in the 6% account. If the three accounts earn total annual interest of $1520, how much money is deposited in each account?

Chapter 4 Test

1. Solve by substitution: $3x + 2y = 4$
$x = 2y - 1$

2. Solve by substitution: $5x + 2y = -23$
$2x + y = -10$

3. Solve by substitution: $y = 3x - 7$
$y = -2x + 3$

4. Solve by using the Gaussian elimination method:
$3x + 4y = -2$
$2x + 5y = 1$

5. Solve by the addition method: $4x - 6y = 5$
$$6x - 9y = 4$$

6. Solve by the addition method:
$$3x - y = 2x + y - 1$$
$$5x + 2y = y + 6$$

7. Solve by the addition method:
$$2x + 4y - z = 3$$
$$x + 2y + z = 5$$
$$4x + 8y - 2z = 7$$

8. Solve by using the Gaussian elimination method:
$$x - y - z = 5$$
$$2x + z = 2$$
$$3y - 2z = 1$$

9. Evaluate the determinant: $\begin{vmatrix} 3 & -1 \\ -2 & 4 \end{vmatrix}$

10. Evaluate the determinant: $\begin{vmatrix} 1 & -2 & 3 \\ 3 & 1 & 1 \\ 2 & -1 & -2 \end{vmatrix}$

11. Solve by using Cramer's Rule: $x - y = 3$
$$2x + y = -4$$

12. Solve by using Cramer's Rule:
$$x - y + z = 2$$
$$2x - y - z = 1$$
$$x + 2y - 3z = -4$$

13. Solve by using Cramer's Rule:
$$3x + 2y + 2z = 2$$
$$x - 2y - z = 1$$
$$2x - 3y - 3z = -3$$

14. Solve by graphing: $2x - 3y = -6$
$$2x - y = 2$$

15. Solve by graphing: $x - 2y = -5$
$$3x + 4y = -15$$

16. Graph the solution set: $2x - y < 3$
$$4x + 3y < 11$$

17. Graph the solution set: $x + y > 2$
$$2x - y < -1$$

18. Use a graphing utility to solve the system. Round to the nearest hundredth.
$$2x + 3y = 7$$
$$x - 2y = -5$$

19. **Uniform Motion** A plane flying with the wind went 350 mi in 2 h. The return trip, flying against the wind, took 2.8 h. Find the rate of the plane in calm air and the rate of the wind.

20. **Purchasing** A clothing manufacturer purchased 60 yd of cotton and 90 yd of wool for a total cost of $1800. Another purchase, at the same prices, included 80 yd of cotton and 20 yd of wool for a total cost of $1000. Find the cost per yard of the cotton and of the wool.

Cumulative Review Exercises

1. Solve: $\dfrac{3}{2}x - \dfrac{3}{8} + \dfrac{1}{4}x = \dfrac{7}{12}x - \dfrac{5}{6}$

2. Find the equation of the line that contains the points $(2, -1)$ and $(3, 4)$.

3. Simplify: $3[x - 2(5 - 2x) - 4x] + 6$

4. Evaluate $a + bc \div 2$ when $a = 4$, $b = 8$, and $c = -2$.

5. Solve: $2x - 3 < 9$ or $5x - 1 < 4$

6. Solve: $|x - 2| - 4 < 2$

7. Solve: $|2x - 3| > 5$

8. Given $f(x) = 3x^3 - 2x^2 + 1$, evaluate $f(-3)$.

9. Find the range of $f(x) = 3x^2 - 2x$ if the domain is $\{-2, -1, 0, 1, 2\}$.

10. Given $F(x) = x^2 - 3$, find $F(2)$.

11. Given $f(x) = 3x - 4$, write $f(2 + h) - f(2)$ in simplest form.

12. Graph: $\{x \mid x \le 2\} \cap \{x \mid x > -3\}$

13. Find the equation of the line that contains the point $(-2, 3)$ and has slope $-\dfrac{2}{3}$.

14. Find the equation of the line that contains the point $(-1, 2)$ and is perpendicular to the line $2x - 3y = 7$.

15. Find the distance between the points $(-4, 2)$ and $(2, 0)$.

16. Find the midpoint of the line connecting the points $(-4, 3)$ and $(3, 5)$.

17. Graph $2x - 5y = 10$ by using the slope and y-intercept.

18. Graph the solution set of the inequality $3x - 4y \ge 8$.

19. Solve by substitution: $3x - 2y = 7$
$\qquad\qquad\qquad\qquad\quad y = 2x - 1$

20. Solve by the addition method: $3x + 2z = 1$
$\qquad\qquad\qquad\qquad\qquad\qquad 2y - z = 1$
$\qquad\qquad\qquad\qquad\qquad\qquad x + 2y = 1$

21. Evaluate the determinant: $\begin{vmatrix} 2 & -5 & 1 \\ 3 & 1 & 2 \\ 6 & -1 & 4 \end{vmatrix}$

22. Solve by graphing: $5x - 2y = 10$
$\qquad\qquad\qquad\qquad\quad 3x + 2y = 6$

23. Solve by using Cramer's Rule: $4x - 3y = 17$
$\qquad\qquad\qquad\qquad\qquad\qquad 3x - 2y = 12$

24. Graph the solution set: $3x - 2y \ge 4$
$\qquad\qquad\qquad\qquad\qquad x + y < 3$

25. Coins A coin purse contains 40 coins in nickels, dimes, and quarters. There are three times as many dimes as quarters. The total value of the coins is $4.10. How many nickels are in the coin purse?

26. Mixture Problem How many milliliters of pure water must be added to 100 ml of a 4% salt solution to make a 2.5% salt solution?

27. Uniform Motion Flying with the wind, a small plane required 2 h to fly 150 mi. Against the wind, it took 3 h to fly the same distance. Find the rate of the wind.

28. Purchasing A restaurant manager buys 100 lb of hamburger and 50 lb of steak for a total cost of $490. A second purchase, at the same prices, includes 150 lb of hamburger and 100 lb of steak. The total cost is $860. Find the price of one pound of steak.

29. Electronics Find the lower and upper limits of a 12,000-ohm resistor with a 15% tolerance.

30. Compensation The graph below shows the relationship between the monthly income, in dollars, and sales, in thousands of dollars, of an account executive. Find the slope of the line between the two points shown on the graph. Write a sentence that states the meaning of the slope.

5 Polynomials

DVD SSM *Online Study Center*

Need help? For online student resources, such as section quizzes, visit the Online Study Center at college.hmco.com/pic/aufmannABI2e.

On May 17, 2002, University of Hawaii astronomers discovered 11 new moons orbiting Jupiter. As telescopes become more powerful and NASA sends more spacecraft out into the universe, we are able to see faraway objects in more detail. For instance, these new moons of Jupiter are approximately 629,000,000 miles away from Earth. Distances this great are generally expressed in scientific notation, as illustrated in the **Projects & Group Activities on page 421.**

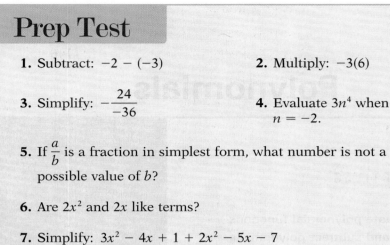

Prep Test

1. Subtract: $-2 - (-3)$

2. Multiply: $-3(6)$

3. Simplify: $-\dfrac{24}{-36}$

4. Evaluate $3n^4$ when $n = -2$.

5. If $\dfrac{a}{b}$ is a fraction in simplest form, what number is not a possible value of b?

6. Are $2x^2$ and $2x$ like terms?

7. Simplify: $3x^2 - 4x + 1 + 2x^2 - 5x - 7$

8. Simplify: $-4y + 4y$

9. Simplify: $-3(2x - 8)$

10. Simplify: $3xy - 4y - 2(5xy - 7y)$

5.1 Introduction to Polynomials

OBJECTIVE 1 Evaluate polynomial functions

A **monomial** is a number, a variable, or a product of numbers and variables. For instance,

7	b	$\dfrac{2}{3}a$	$12xy^2$
A number	A variable	A product of a number and a variable	A product of a number and variables

The expression $3\sqrt{x}$ is not a monomial because $\sqrt{x}$ cannot be written as a product of variables. The expression $\dfrac{2x}{y^2}$ is not a monomial because it is a *quotient* of variables.

The **degree of a monomial** is the sum of the exponents of the variables.

x	degree 1 $(x = x^1)$
$3x^2$	degree 2
$4x^2y$	degree 3
$6x^3y^4z^2$	degree 9
x^n	degree n

The degree of a nonzero constant term is zero.

6	degree 0

A **polynomial** is a variable expression in which the terms are monomials.

A polynomial of one term is a **monomial.**	$5x$
A polynomial of two terms is a **binomial.**	$5x^2y + 6x$
A polynomial of three terms is a **trinomial.**	$3x^2 + 9xy - 5y$

Polynomials with more than three terms do not have special names.

The **degree of a polynomial** is the greatest of the degrees of any of its terms.

$3x + 2$	degree 1
$3x^2 + 2x - 4$	degree 2
$4x^3y^2 + 6x^4$	degree 5
$3x^{2n} - 5x^n - 2$	degree $2n$

The terms of a polynomial in one variable are usually arranged so that the exponents of the variable decrease from left to right. This is called **descending order.**

$2x^2 - x + 8$

$3y^3 - 3y^2 + y - 12$

For a polynomial in more than one variable, descending order may refer to any one of the variables.

The polynomial at the right is shown first in descending order of the x variable and then in descending order of the y variable.

$2x^2 + 3xy + 5y^2$

$5y^2 + 3xy + 2x^2$

The **linear function** given by $f(x) = mx + b$ is an example of a polynomial function. It is a polynomial function of degree 1. A second-degree polynomial function, called a **quadratic function,** is given by the equation $f(x) = ax^2 + bx + c$, $a \neq 0$. A third-degree polynomial function is called a **cubic function.** In general, a **polynomial function** is an expression whose terms are monomials.

To **evaluate a polynomial function,** replace the variable by its value and simplify.

Given $P(x) = x^3 - 3x^2 + 4$, evaluate $P(-3)$.

Substitute -3 for x and simplify.

$$P(x) = x^3 - 3x^2 + 4$$
$$P(-3) = (-3)^3 - 3(-3)^2 + 4$$
$$P(-3) = -27 - 27 + 4$$
$$P(-3) = -50$$

EXAMPLE 1 To overcome the resistance of the tires on the road and the wind, the horsepower (hp), P, required by a cyclist to keep a certain bicycle moving at v mph is given by $P(v) = 0.00003v^3 + 0.00211v$. How much horsepower must the cyclist supply to keep this bicycle moving at 20 mph?

Strategy To find the horsepower, evaluate the function when $v = 20$.

Solution
$$P(v) = 0.00003v^3 + 0.00211v$$
$$P(20) = 0.00003(20)^3 + 0.00211(20) \qquad \bullet \text{ Replace } v \text{ by 20.}$$
$$= 0.2822 \qquad \bullet \text{ Simplify.}$$

The cyclist must supply 0.2822 hp.

PROBLEM 1 The velocity of air that is expelled during a cough can be modeled by $v(r) = 600r^2 - 1000r^3$, where v is the velocity of the air in centimeters per second and r is the radius of the trachea in centimeters. What is the velocity of expelled air during a cough when the radius of the trachea is 0.4 cm?

Solution See page S18.

The **leading coefficient** of a polynomial function is the coefficient of the variable with the largest exponent. The **constant term** is the term without a variable.

Find the leading coefficient, the constant term, and the degree of the polynomial function $P(x) = 7x^4 - 3x^2 + 2x - 4$.

The leading coefficient is 7, the constant term is -4, and the degree is 4.

EXAMPLE 2 Find the leading coefficient, the constant term, and the degree of the polynomial $P(x) = 5x^6 - 4x^5 - 3x^2 + 7$.

Solution The leading coefficient is 5, the constant term is 7, and the degree is 6.

PROBLEM 2 Find the leading coefficient, the constant term, and the degree of the polynomial $R(x) = -3x^4 + 3x^3 + 3x^2 - 2x - 12$.

Solution See page S18.

The following three functions do not represent polynomial functions.

$f(x) = 3x^2 + 2x^{-1}$ A polynomial function does not have a variable raised to a negative power.

$g(x) = 2\sqrt{x} - 3$ A polynomial function does not have a variable expression within a radical.

$h(x) = \dfrac{x}{x - 1}$ A polynomial function does not have a variable in the denominator of a fraction.

EXAMPLE 3 State whether the function is a polynomial function.
 A. $P(x) = 3x^{1/2} + 2x^2 - 3$ **B.** $T(x) = 3\sqrt{x} - 2x^2 - 3x + 2$
 C. $R(x) = 14x^3 - \pi x^2 + 3x + 2$

Solution **A.** A polynomial function does not have a variable raised to a fractional power. This is not a polynomial function.

 B. A polynomial function does not have a variable expression within a radical. This is not a polynomial function.

 C. This is a polynomial function.

PROBLEM 3 State whether the function is a polynomial function.
 A. $R(x) = 5x^{14} - 5$ **B.** $V(x) = -x^{-1} + 2x - 7$
 C. $P(x) = 2x - 2\sqrt{x} - 3$

Solution See page S18.

The graph of a linear function is a straight line and can be found by plotting just two points. The graph of a polynomial function of degree greater than one is a curve. Consequently, many points may have to be found before an accurate graph can be drawn.

Evaluating the quadratic function given by the equation $f(x) = x^2 - x - 6$ when $x = -3, -2, -1, 0, 1, 2, 3$, and 4 gives the points shown in Figure 1. For instance, $f(-3) = 6$, so $(-3, 6)$ is graphed; $f(2) = -4$, so $(2, -4)$ is graphed; and $f(4) = 6$, so $(4, 6)$ is graphed. Evaluating the function when x is not an integer, such as $x = -\dfrac{3}{2}$ and $x = \dfrac{5}{2}$, produces more points to graph, as shown in Figure 2. Connecting the points with a smooth curve results in Figure 3, which is the graph of f.

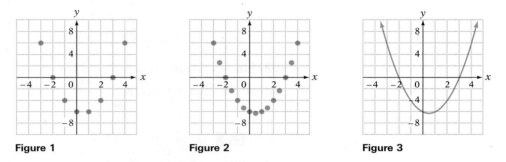

Figure 1 Figure 2 Figure 3

Here is an example of graphing a cubic function, $P(x) = x^3 - 2x^2 - 5x + 6$. Evaluating the function when $x = -2, -1, 0, 1, 2, 3$, and 4 gives the graph in Figure 4. Evaluating at some noninteger values gives the graph in Figure 5. Finally, connecting the dots with a smooth curve gives the graph in Figure 6.

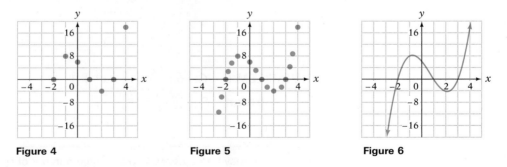

Figure 4 Figure 5 Figure 6

The graphs of $f(x) = x^2 - x - 6$ (Figure 3) and $P(x) = x^3 - 2x^2 - 5x + 6$ (Figure 6) are repeated below. Note that by the vertical line test, the graphs are the graphs of functions.

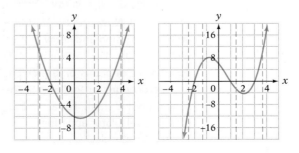

Recall that the domain of a function is the set of first coordinates of the ordered pairs of the function. Because a polynomial function can be evaluated for any real number, the domain of a polynomial function is the set of real numbers.

EXAMPLE 4 Graph: $f(x) = x^2 - 2$

Solution

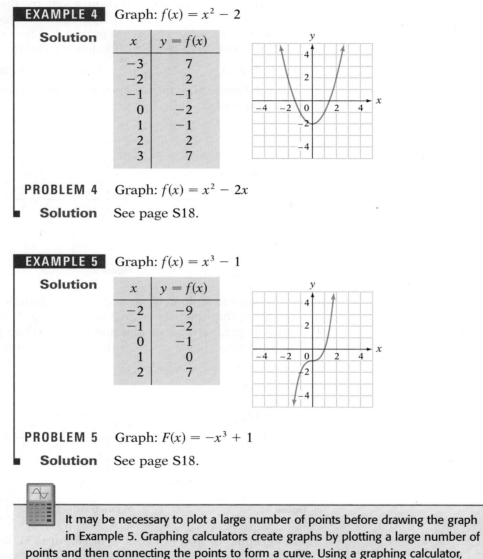

x	$y = f(x)$
-3	7
-2	2
-1	-1
0	-2
1	-1
2	2
3	7

PROBLEM 4 Graph: $f(x) = x^2 - 2x$

Solution See page S18.

EXAMPLE 5 Graph: $f(x) = x^3 - 1$

Solution

x	$y = f(x)$
-2	-9
-1	-2
0	-1
1	0
2	7

PROBLEM 5 Graph: $F(x) = -x^3 + 1$

Solution See page S18.

It may be necessary to plot a large number of points before drawing the graph in Example 5. Graphing calculators create graphs by plotting a large number of points and then connecting the points to form a curve. Using a graphing calculator, enter the equation $y = x^3 - 1$. Graph the equation using XMIN = -4.7, XMAX = 4.7, YMIN = -10, YMAX = 10. Now use the TRACE feature to move around the graph and verify that $(-2, -9)$, $(-1, -2)$, $(0, -1)$, $(1, 0)$, and $(2, 7)$ are the coordinates of points on the graph. Follow the same procedure for Problem 5. The point $(2, -7)$ is shown on the graph. For assistance, refer to the graphing calculator appendix.

OBJECTIVE **2** **Add and subtract polynomials**

Polynomials can be added by combining like terms. Either a vertical or a horizontal format can be used.

Add $(3x^2 + 2x - 7) + (7x^3 - 3 + 4x^2)$. Use a horizontal format.

Use the Commutative and Associative Properties of Addition to rearrange and group like terms.

$(3x^2 + 2x - 7) + (7x^3 - 3 + 4x^2)$
$= 7x^3 + (3x^2 + 4x^2) + 2x + (-7 - 3)$

Combine like terms.

$= 7x^3 + 7x^2 + 2x - 10$

Add $(4x^2 + 5x - 3) + (7x^3 - 7x + 1) + (2x - 3x^2 + 4x^3 + 1)$. Use a vertical format.

Arrange the terms of each polynomial in descending order with like terms in the same column. Add the terms in each column.

$$
\begin{array}{r}
4x^2 + 5x - 3 \\
7x^3 \qquad - 7x + 1 \\
4x^3 - 3x^2 + 2x + 1 \\
\hline
11x^3 + \ x^2 \qquad - 1
\end{array}
$$

The additive inverse of the polynomial $x^2 + 5x - 4$ is $-(x^2 + 5x - 4)$.

To simplify the additive inverse of a polynomial, change the sign of every term inside the parentheses.

$-(x^2 + 5x - 4) = -x^2 - 5x + 4$

To subtract two polynomials, add the additive inverse of the second polynomial to the first.

Take Note

This is the same definition used for subtraction of integers: Subtraction is addition of the opposite.

Subtract $(3x^2 - 7xy + y^2) - (-4x^2 + 7xy - 3y^2)$. Use a horizontal format.

Rewrite the subtraction as the addition of the additive inverse.

$(3x^2 - 7xy + y^2) - (-4x^2 + 7xy - 3y^2)$
$= (3x^2 - 7xy + y^2) + (4x^2 - 7xy + 3y^2)$

Combine like terms.

$= 7x^2 - 14xy + 4y^2$

Subtract $(6x^3 - 3x + 7) - (3x^2 - 5x + 12)$. Use a vertical format.

Rewrite the subtraction as the addition of the additive inverse.

$$(6x^3 - 3x + 7) - (3x^2 - 5x + 12)$$
$$= (6x^3 - 3x + 7) + (-3x^2 + 5x - 12)$$

Arrange the terms of each polynomial in descending order with like terms in the same column. Add the terms in each column.

$$\begin{array}{r} 6x^3 \qquad - 3x + 7 \\ -3x^2 + 5x - 12 \\ \hline 6x^3 - 3x^2 + 2x - 5 \end{array}$$

EXAMPLE 6 Add: $(3x^2 - 2x + 5) + (2 - 3x + 4x^3) + (x - 3x^2 - x^3 - 4)$
Use a vertical format.

Solution
$$\begin{array}{r} 3x^2 - 2x + 5 \\ 4x^3 \qquad\quad - 3x + 2 \\ -x^3 - 3x^2 + x - 4 \\ \hline 3x^3 \qquad\quad - 4x + 3 \end{array}$$

• **Arrange the terms of each polynomial in descending order with like terms in the same column.**

• **Combine like terms in each column.**

PROBLEM 6 Add: $(5x^2 + 3x - 1) + (2x^2 + 4x - 6) + (8 - 7x + x^2)$
Use a vertical format.

Solution See page S18.

EXAMPLE 7 Subtract: $(3x^2 - 2x + 4) - (7x^2 + 3x - 12)$
Use a vertical format.

Solution $(3x^2 - 2x + 4) - (7x^2 + 3x - 12)$
$(3x^2 - 2x + 4) + (-7x^2 - 3x + 12)$

• **Rewrite the subtraction as the addition of the additive inverse.**

$$\begin{array}{r} 3x^2 - 2x + 4 \\ -7x^2 - 3x + 12 \\ \hline -4x^2 - 5x + 16 \end{array}$$

• **Arrange the polynomials in descending order in a vertical format.**

• **Combine like terms in each column.**

PROBLEM 7 Subtract: $(-5x^2 + 2x - 3) - (6x^2 + 3x - 7)$
Use a vertical format.

Solution See page S18.

Function notation may be used when adding or subtracting polynomials. For instance, if $P(x) = 3x^2 - 2x + 4$ and $R(x) = -5x^3 + 4x + 7$, then

$$P(x) + R(x) = (3x^2 - 2x + 4) + (-5x^3 + 4x + 7)$$
$$= -5x^3 + 3x^2 + 2x + 11$$

$$P(x) - R(x) = (3x^2 - 2x + 4) - (-5x^3 + 4x + 7)$$
$$= (3x^2 - 2x + 4) + (5x^3 - 4x - 7)$$
$$= 5x^3 + 3x^2 - 6x - 3$$

EXAMPLE 8 Given $P(x) = -3x^2 + 2x - 6$ and $R(x) = 4x^3 - 3x + 4$, find $S(x) = P(x) + R(x)$.

Solution $S(x) = P(x) + R(x)$
$S(x) = (-3x^2 + 2x - 6) + (4x^3 - 3x + 4)$
$= 4x^3 - 3x^2 - x - 2$

PROBLEM 8 Given $P(x) = 4x^3 - 3x^2 + 2$ and $R(x) = -2x^2 + 2x - 3$, find $D(x) = P(x) - R(x)$.

Solution See page S18.

5.1 Concept Review

Determine whether the statement is always true, sometimes true, or never true.

1. The degree of a polynomial is the greatest of the degrees of its individual terms.

2. The leading coefficient of $4 - 2x - 3x^2$ is 4.

3. Terms with the same variable parts can be added.

4. The additive inverse of a polynomial is a polynomial with every sign changed to the opposite sign.

5. The sum of two trinomials is a trinomial.

6. $f(x) = 3x^2 - 2x + x^{-1}$ is a polynomial function.

5.1 Exercises

1 Evaluate polynomial functions

1. Explain how to determine the degree of a polynomial in one variable. Give two examples of a third-degree polynomial.

2. Explain how to evaluate a polynomial.

For Exercises 3 and 4, use the polynomial function $P(x) = -3x^2 - 2x + 10$.

3. The degree of P is ___?___. The coefficient -3 is called the ___?___ coefficient. The term 10 is called the ___?___ term.

4. **a.** To evaluate $P(-2)$, substitute ___?___ for ___?___ in the expression for $P(x)$: $P(\underline{\quad?\quad}) = -3(\underline{\quad?\quad})^2 - 2(\underline{\quad?\quad}) + 10$
$= \underline{\quad?\quad} + \underline{\quad?\quad} + 10 = \underline{\quad?\quad}$.
 b. Use the results of part a to name a point that is on the graph of $y = P(x)$.

State whether the function is a polynomial function. For those that are polynomial functions, identify **a.** the leading coefficient, **b.** the constant term, and **c.** the degree.

5. $P(x) = -x^2 + 3x + 8$

6. $P(x) = 3x^4 - 3x - 7$

7. $R(x) = \dfrac{x}{x + 1}$

8. $R(x) = \dfrac{3x^2 - 2x + 1}{x}$

9. $f(x) = \sqrt{x} - x^2 + 2$

10. $f(x) = x^2 - \sqrt{x + 2} - 8$

11. $g(x) = 3x^5 - 2x^2 + \pi$

12. $g(x) = -4x^5 - 3x^2 + x - \sqrt{7}$

13. $P(x) = 3x^2 - 5x^3 + 2$

14. $P(x) = x^2 - 5x^4 - x^6$

15. $R(x) = 14$

16. $R(x) = \dfrac{1}{x} + 2$

17. Given $P(x) = 3x^2 - 2x - 8$, evaluate $P(3)$.

18. Given $P(x) = -3x^2 - 5x + 8$, evaluate $P(-5)$.

19. Given $R(x) = 2x^3 - 3x^2 + 4x - 2$, evaluate $R(2)$.

20. Given $R(x) = -x^3 + 2x^2 - 3x + 4$, evaluate $R(-1)$.

21. Given $f(x) = x^4 - 2x^2 - 10$, evaluate $f(-1)$.

22. Given $f(x) = x^5 - 2x^3 + 4x$, evaluate $f(2)$.

Graph.

23. $P(x) = x^2 - 1$

24. $P(x) = 2x^2 + 3$

25. $R(x) = x^3 + 2$

26. $R(x) = x^4 + 1$

27. $f(x) = x^3 - 2x$

28. $f(x) = x^2 - x - 2$

29. Oceanography The length of a water wave close to land depends on several factors, one of which is the depth of the water. However, the length, L (in meters), of deep-water waves (those far away from land) can be approximated by $L(s) = 0.641s^2$, where s is the speed of the wave in meters per second. Find the length of a deep-water wave that has a speed of 6 m/s. Round to the nearest tenth of a meter.

30. Physics The distance, s (in feet), that an object on the moon will fall in t seconds is given by $s(t) = 2.735t^2$. How far will an object fall on the moon in 3 s?

31. Sports The total number of softball games, T, that must be scheduled in a league that has n teams such that each team plays every other team twice is given by $T(n) = n^2 - n$. What is the total number of games that must be scheduled for a league that has 8 teams?

32. Geometry A diagonal of a polygon is a line from a vertex to a nonadjacent vertex. Seven of the possible diagonals for a decagon (10-sided figure) are shown at the right. The total number, T, of diagonals for an n-sided polygon is given by $T(n) = \frac{1}{2}n^2 - \frac{3}{2}n$. Find the total number of diagonals for a decagon.

33. Food Science Baked Alaska is a dessert that is made by putting a 1-inch meringue coating around a hemisphere of ice cream. The amount of meringue that is needed depends on the radius of the hemisphere of ice cream and is given by $M(r) = 6.14r^2 + 6.14r + 2.094$, where $M(r)$ is the volume of meringue needed (in cubic inches) and r is the radius of the ice cream in inches. Find the amount of meringue that is needed for a baked Alaska that has a mound of ice cream with a 6-inch radius. Round to the nearest whole number.

For Exercises 34 and 35, use the following graphs.

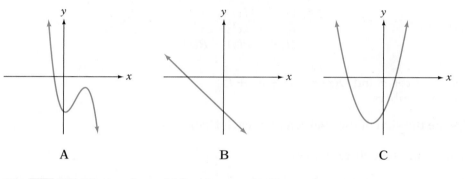

A B C

34. Which graph could be the graph of $y = x^2 + 2x - 3$?

35. Which graph could be the graph of $y = -x^3 + 3x^2 - x - 3$?

36. Suppose $f(x) = x^2 + 1$ and $g(x) = x^3 + x$. Is $f(x) > g(x)$ or is $f(x) < g(x)$ when $0 < x < 1$? Explain your answer.

2 Add and subtract polynomials

37. The additive inverse of $8x^2 + 3x - 5$ is $-(8x^2 + 3x - 5) = $ ___?___.

38. $(4x^2 + 6x - 1) - (3x^2 - 5x + 2) = (4x^2 + 6x - 1) + ($ ___?___ $)$.

Simplify. Use a vertical format.

39. $(5x^2 + 2x - 7) + (x^2 - 8x + 12)$

40. $(3x^2 - 2x + 7) + (-3x^2 + 2x - 12)$

41. $(x^2 + 8 - 3x) - (2x^2 - 3x + 7)$

42. $(2x^2 + 3x - 7) - (5x^2 - 1 - 8x)$

For Exercises 43 to 46, simplify using a horizontal format.

43. $(3y^2 - 7y) + (2y^2 - 8y + 2)$

44. $(-2y^2 - 4y - 12) + (5y^2 - 5y)$

45. $(2a^2 - 3a - 7) - (-5a^2 - 2a - 9)$

46. $(3a^2 - 9a) - (-5a^2 + 7a - 6)$

47. Given $P(x) = 3x^3 - 4x^2 - x + 1$ and $R(x) = 2x^3 + 5x - 8$, find $P(x) + R(x)$.

48. Given $P(x) = 5x^3 - 3x - 7$ and $R(x) = 2x^3 - 3x^2 + 8$, find $P(x) - R(x)$.

49. Given $P(x) = x^{2n} + 7x^n - 3$ and $R(x) = -x^{2n} + 2x^n + 8$, find $P(x) + R(x)$.

50. Given $P(x) = 2x^{2n} - x^n - 1$ and $R(x) = 5x^{2n} + 7x^n + 1$, find $P(x) - R(x)$.

51. Given $P(x) = 3x^4 - 3x^3 - x^2$ and $R(x) = 3x^3 - 7x^2 + 2x$, find $S(x) = P(x) + R(x)$.

52. Given $P(x) = 3x^4 - 2x + 1$ and $R(x) = 3x^5 - 5x - 8$, find $S(x) = P(x) + R(x)$.

53. Given $P(x) = x^2 + 2x + 1$ and $R(x) = 2x^3 - 3x^2 + 2x - 7$, find $D(x) = P(x) - R(x)$.

54. Given $P(x) = 2x^4 - 2x^2 + 1$ and $R(x) = 3x^3 - 2x^2 + 3x + 8$, find $D(x) = P(x) - R(x)$.

For Exercises 55 and 56, $P(x) = -ax^2 - bx + c$ and $Q(x) = dx^2 - ex + f$, where a, b, c, d, e, and f are all positive numbers.

55. If $a > d, b > e$, and $c > f$, state the sign of the coefficient of each term of $Q(x) - P(x)$.
 a. the x^2 term **b.** the x term **c.** the constant term

56. If $a > d, b < e$, and $c > f$, state the sign of the coefficient of each term of $P(x) - Q(x)$.
 a. the x^2 term **b.** the x term **c.** the constant term

5.1 Applying Concepts

Two polynomials are equal if the coefficients of like powers are equal. For example,

$$6x^3 - 7x^2 + 2x - 8 = 2x - 7x^2 - 8 + 6x^3$$

In Exercises 57 and 58, use the definition of equality of polynomials to find the value of k that makes the equation an identity.

57. $(2x^3 + 3x^2 + kx + 5) - (x^3 + x^2 - 5x - 2) = x^3 + 2x^2 + 3x + 7$

58. $(6x^3 + kx^2 - 2x - 1) - (4x^3 - 3x^2 + 1) = 2x^3 - x^2 - 2x - 2$

59. If $P(x)$ is a third-degree polynomial and $Q(x)$ is a fourth-degree polynomial, what can be said about the degree of $P(x) + Q(x)$? Give some examples of polynomials that support your answer.

60. If $P(x)$ is a fifth-degree polynomial and $Q(x)$ is a fourth-degree polynomial, what can be said about the degree of $P(x) - Q(x)$? Give some examples of polynomials that support your answer.

61. Sports The deflection, D (in inches), of a beam that is uniformly loaded is given by the polynomial function $D(x) = 0.005x^4 - 0.1x^3 + 0.5x^2$, where x is the distance in feet from one end of the beam. See the figure at the right. The maximum deflection occurs when x is the midpoint of the beam. Determine the maximum deflection for the beam in the diagram.

62. If $P(2) = 3$ and $P(x) = 2x^3 - 4x^2 - 2x + c$, find the value of c.

63. If $P(-1) = -3$ and $P(x) = 4x^4 - 3x^2 + 6x + c$, find the value of c.

64. Graph $f(x) = x^2$, $g(x) = (x - 3)^2$, and $h(x) = (x + 4)^2$ on the same coordinate grid. From the graphs, make a conjecture about the shape and location of $k(x) = (x - 2)^2$. Test your conjecture by graphing k.

65. Graph $f(x) = x^2$, $g(x) = x^2 - 3$, and $h(x) = x^2 + 4$ on the same coordinate grid. From the graphs, make a conjecture about the shape and location of $k(x) = x^2 - 2$. Test your conjecture by graphing k.

66. Engineering Construction of the Golden Gate Bridge, which is a suspension bridge, was completed in 1937. The length of the main span of the bridge is 4200 ft. The height, in feet, of the suspension cables above the roadway varies from 0 ft at the center of the bridge to 525 ft at the towers that support the cables.

Distance from the center of the bridge	Height of the cables above the roadway
0 ft	0 ft
1050 ft	150 ft
2100 ft	525 ft

The function that approximately models the data is $f(x) = \dfrac{1}{8820}x^2 + 25$, where x is the distance from the center of the bridge and $f(x)$ is the height of the cables above the roadway. Use this model to approximate the height of the cables at a distance of **a.** 1000 ft from the center of the bridge and **b.** 1500 ft from the center of the bridge. Round to the nearest tenth.

67. Explain the similarities and differences among the graphs of $f(x) = x^2$, $g(x) = (x - 3)^2$, and $h(x) = x^2 - 3$.

 Multiplication of Monomials

OBJECTIVE **1** **Multiply monomials**

Recall that in the exponential expression x^5, x is the base and 5 is the exponent. The exponent indicates the number of times the base occurs as a factor.

The product of exponential expressions with the *same* base can be simplified by writing each expression in factored form and writing the result with an exponent.

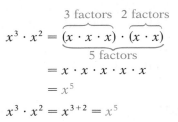

$$x^3 \cdot x^2 = \underbrace{(x \cdot x \cdot x)}_{\text{3 factors}} \cdot \underbrace{(x \cdot x)}_{\text{2 factors}}$$
$$\underbrace{}_{\text{5 factors}}$$
$$= x \cdot x \cdot x \cdot x \cdot x$$
$$= x^5$$

Adding the exponents results in the same product.

$$x^3 \cdot x^2 = x^{3+2} = x^5$$

Rule for Multiplying Exponential Expressions

If m and n are integers, then $x^m \cdot x^n = x^{m+n}$.

For example, in the expression $a^2 \cdot a^6 \cdot a$, the bases are the same. The expression can be simplified by adding the exponents. Recall that $a = a^1$.

$$a^2 \cdot a^6 \cdot a = a^{2+6+1} = a^9$$

Take Note

The Rule for Multiplying Exponential Expressions requires that the bases be the same. The expression $x^3 y^2$ cannot be simplified.

EXAMPLE 1 Multiply: $(2xy)(3x^2y)$

Solution $(2xy)(3x^2y)$
$= (2 \cdot 3)(x \cdot x^2)(y \cdot y)$

• Use the Commutative and Associative Properties of Multiplication to rearrange and group factors.

$= 6x^{1+2}y^{1+1}$

• Multiply variables with the same base by adding the exponents.

$= 6x^3y^2$

PROBLEM 1 Multiply: $(3x^2)(6x^3)$

Solution See page S18.

EXAMPLE 2 Multiply: $(2x^2y)(-5xy^4)$

Solution $(2x^2y)(-5xy^4)$
$= [2(-5)](x^2 \cdot x)(y \cdot y^4)$

• Use the Properties of Multiplication to rearrange and group factors.

$= -10x^3y^5$

• Multiply variables with the same base by adding the exponents.

> PROBLEM 2 Multiply: $(-3xy^2)(-4x^2y^3)$
>
> Solution See page S18.

OBJECTIVE 2 Simplify powers of monomials

Point of Interest

One of the first symbolic representations of powers was given by Diophantus (ca. A.D. 250) in his book Arithmetica. *He used Δ^Y for x^2 and k^Y for x^3. The symbol Δ^Y was the first two letters of the Greek words* dunamis, *meaning "power"; k^Y was from the Greek word* kubos, *meaning "cube." He also combined these symbols to denote higher powers. For instance, Δk^Y was the symbol for x^5.*

A power of a monomial can be simplified by rewriting the expression in factored form and then using the Rule for Multiplying Exponential Expressions.

$$\begin{aligned}(x^2)^3 &= x^2 \cdot x^2 \cdot x^2 \\ &= x^{2+2+2} \\ &= x^6\end{aligned}$$

$$\begin{aligned}(x^4y^3)^2 &= (x^4y^3)(x^4y^3) \\ &= x^4 \cdot y^3 \cdot x^4 \cdot y^3 \\ &= (x^4 \cdot x^4)(y^3 \cdot y^3) \\ &= x^{4+4}y^{3+3} \\ &= x^8y^6\end{aligned}$$

Note that multiplying each exponent inside the parentheses by the exponent outside the parentheses gives the same result.

$$(x^2)^3 = x^{2 \cdot 3} = x^6 \qquad (x^4y^3)^2 = x^{4 \cdot 2}y^{3 \cdot 2} = x^8y^6$$

Rule for Simplifying Powers of Exponential Expressions

If m and n are integers, then $(x^m)^n = x^{mn}$.

Simplify: $(x^5)^2$

Multiply the exponents.

$$(x^5)^2 = x^{5 \cdot 2} = x^{10}$$

Rule for Simplifying Powers of Products

If m, n, and p are integers, then $(x^m y^n)^p = x^{mp}y^{np}$.

Simplify: $(c^5d^3)^6$

Multiply each exponent inside the parentheses by the exponent outside the parentheses.

$$\begin{aligned}(c^5d^3)^6 &= c^{5 \cdot 6}d^{3 \cdot 6} \\ &= c^{30}d^{18}\end{aligned}$$

Simplify: $(3a^2b)^3$

Multiply each exponent inside the parentheses by the exponent outside the parentheses.

$$\begin{aligned}(3a^2b)^3 &= 3^{1 \cdot 3}a^{2 \cdot 3}b^{1 \cdot 3} \\ &= 3^3a^6b^3 \\ &= 27a^6b^3\end{aligned}$$

EXAMPLE 3 Simplify: $(-2x)(-3xy^2)^3$

Solution $(-2x)(-3xy^2)^3$

$= (-2x)(-3)^3 x^3 y^6$ • Multiply each exponent in $-3xy^2$ by the exponent outside the parentheses.

$= (-2x)(-27)x^3 y^6$ • Simplify $(-3)^3$.

$= [-2(-27)](x \cdot x^3)y^6$ • Use the Properties of Multiplication to rearrange and group factors.

$= 54x^4 y^6$ • Multiply variable expressions with the same base by adding the exponents.

PROBLEM 3 Simplify: $(3x)(2x^2y)^3$

Solution See page S18.

5.2 Concept Review

Determine whether the statement is always true, sometimes true, or never true.

1. In the expression $(6x)^4$, x is the base and 4 is the exponent.

2. To multiply $x^m \cdot x^n$, multiply the exponents.

3. The expression "a power of a monomial" means the monomial is the base of an exponential expression.

4. $x^6 \cdot x \cdot x^8 = x^{14}$

5. $(4x^3y^5)^2 = 4x^6y^{10}$

6. $a^2 \cdot b^7 = ab^9$

5.2 Exercises

1 Multiply monomials

1. Explain how to multiply two exponential expressions with the same base. Provide an example.

2. Use the Rule for Multiplying Exponential Expressions to multiply: $(x^7)(x^2) = x^{\underline{?}+\underline{?}} = \underline{?}$.

3. Multiply: $(5a^6b)(2a^5b^3)$

$(5a^6b)(2a^5b^3)$

$= (5 \cdot \underline{\quad?\quad})(a^6 \cdot \underline{\quad?\quad})(b \cdot \underline{\quad?\quad})$ • Use the Commutative and Associative Properties of Multiplication to rearrange and group factors.

$= (\underline{\quad?\quad})(a^{\underline{?}+\underline{?}})(b^{\underline{?}+\underline{?}})$ • Multiply variables with the same base by $\underline{\quad?\quad}$ the exponents.

$= 10a^{\underline{?}}b^{\underline{?}}$ • Simplify.

4. State whether the expression can be simplified using the Rule for Multiplying Exponential Expressions.
 a. $x^4 + x^5$ **b.** $x^4 x^5$ **c.** $x^4 y^4$ **d.** $x^4 + x^4$

Multiply.

5. $(x)(2x)$

6. $(-3y)(y)$

7. $(3x)(4x)$

8. $(7y^3)(7y^2)$

9. $(-2a^3)(-3a^4)$

10. $(5a^6)(-2a^5)$

11. $(x^2y)(xy^4)$

12. $(x^2y^4)(xy^7)$

13. $(-2x^4)(5x^5y)$

14. $(-3a^3)(2a^2b^4)$

15. $(x^2y^4)(x^5y^4)$

16. $(a^2b^4)(ab^3)$

17. $(2xy)(-3x^2y^4)$

18. $(-3a^2b)(-2ab^3)$

19. $(x^2yz)(x^2y^4)$

20. $(-ab^2c)(a^2b^5)$

21. $(a^2b^3)(ab^2c^4)$

22. $(x^2y^3z)(x^3y^4)$

23. $(-a^2b^2)(a^3b^6)$

24. $(xy^4)(-xy^3)$

25. $(-6a^3)(a^2b)$

26. $(2a^2b^3)(-4ab^2)$

27. $(-5y^4z)(-8y^6z^5)$

28. $(3x^2y)(-4xy^2)$

29. $(10ab^2)(-2ab)$

30. $(x^2y)(yz)(xyz)$

31. $(xy^2z)(x^2y)(z^2y^2)$

32. $(-2x^2y^3)(3xy)(-5x^3y^4)$

33. $(4a^2b)(-3a^3b^4)(a^5b^2)$

34. $(3ab^2)(-2abc)(4ac^2)$

2 Simplify powers of monomials

35. Explain how to simplify a power of an exponential expression. Provide an example.

State whether the expression is the product of two exponential expressions or a power of an exponential expression.

36. a. $b^4 \cdot b^8$ **b.** $(b^4)^8$

37. a. $(2z)^2$ **b.** $2z \cdot z$

38. a. $(3a^4)^3$ **b.** $(3a^4)(5a)$

39. a. $x(-xy^4)$ **b.** $(-xy)^4$

40. a. Use the Rule for Simplifying Powers of Exponential Expressions to simplify: $(x^4)^7 = x^{(\underline{\ ?\ })(\underline{\ ?\ })} = \underline{\ ?\ }$.
 b. Use the Rule for Simplifying Powers of Products to simplify:
 $(x^4y^2)^6 = x^{(\underline{\ ?\ })(\underline{\ ?\ })}y^{(\underline{\ ?\ })(\underline{\ ?\ })} = \underline{\ ?\ }$.

41. State whether or not the expression can be simplified using the Rule for Simplifying Powers of Products.
 a. $(xy)^3$ **b.** $(x + y)^3$ **c.** $(a^3 + b^4)^2$ **d.** $(a^3 b^4)^2$

42. a. True or false? $a^2b^5 = ab^{10}$
 b. True or false? $a^2b^5 = ab^7$
 c. True or false? $(a^2b^5)^2 = a^4b^{25}$
 d. True or false? $(a^2 + b^5)^2 = a^4 + b^{10}$

Simplify.

43. $(2^2)^3$

44. $(3^2)^2$

45. $(-2)^2$

46. $(-3)^3$

47. $(-2^2)^3$

48. $(-2^3)^3$

49. $(x^3)^3$

50. $(y^4)^2$

51. $(x^7)^2$

52. $(y^5)^3$

53. $(-x^2)^2$

54. $(-x^2)^3$

55. $(2x)^2$

56. $(3y)^3$

57. $(-2x^2)^3$

58. $(-3y^3)^2$

59. $(x^2y^3)^2$

60. $(x^3y^4)^5$

61. $(3x^2y)^2$

62. $(-2ab^3)^4$

63. $(a^2)(3a^2)^3$

64. $(b^2)(2a^3)^4$

65. $(-2x)(2x^3)^2$

66. $(2y)(-3y^4)^3$

67. $(x^2y)(x^2y)^3$

68. $(a^3b)(ab)^3$

69. $(ab^2)^2(ab)^2$

70. $(x^2y)^2(x^3y)^3$

71. $(-2x)(-2x^3y)^3$

72. $(-3y)(-4x^2y^3)^3$

73. $(-2x)(-3xy^2)^2$

74. $(-3y)(-2x^2y)^3$

75. $(ab^2)(-2a^2b)^3$

76. $(a^2b^2)(-3ab^4)^2$

77. $(-2a^3)(3a^2b)^3$

78. $(-3b^2)(2ab^2)^3$

79. $(-3ab)^2(-2ab)^3$

80. $(-3a^2b)^3(-3ab)^3$

5.2 Applying Concepts

Simplify.

81. $(6x)(2x^2) + (4x^2)(5x)$

82. $(2a^7)(7a^2) - (6a^3)(5a^6)$

83. $(3a^2b^2)(2ab) - (9ab^2)(a^2b)$

84. $(3x^2y^2)^2 - (2xy)^4$

85. $(5xy^3)(3x^4y^2) - (2x^3y)(x^2y^4)$

86. $a^2(ab^2)^3 - a^3(ab^3)^2$

87. $4a^2(2ab)^3 - 5b^2(a^5b)$

88. $9x^3(3x^2y)^2 - x(x^3y)^2$

89. $-2xy(x^2y)^3 - 3x^5(xy^2)^2$

90. $5a^2b(ab^2)^2 + b^3(2a^2b)^2$

91. $a^n \cdot a^n$

92. $(a^n)^2$

93. $(a^2)^n$

94. $a^2 \cdot a^n$

For Exercises 95 to 98, answer true or false. If the answer is false, correct the right side of the equation.

95. $(-a)^5 = -a^5$

96. $(-b)^8 = b^8$

97. $(x^2)^5 = x^{2+5} = x^7$

98. $x^3 + x^3 = 2x^{3+3} = 2x^6$

99. Evaluate $(2^3)^2$ and $2^{(3^2)}$. Are the results the same? If not, which expression has the larger value?

100. What is the Order of Operations for the expression x^{m^n}?

101. If n is a positive integer and $x^n = y^n$, when is $x = y$?

102. Geometry The length of a rectangle is $4ab$. The width is $2ab$. Find the perimeter of the rectangle in terms of ab.

4ab

2ab

103. The distance a rock will fall in t seconds is $16t^2$ ft (neglecting air resistance). Find other examples of quantities that can be expressed in terms of an exponential expression, and explain where the expression is used.

5.3 Multiplication of Polynomials

OBJECTIVE 1 Multiply a polynomial by a monomial

To multiply a polynomial by a monomial, use the Distributive Property and the Rule for Multiplying Exponential Expressions.

Simplify: $-2x(x^2 - 4x - 3)$

$-2x(x^2 - 4x - 3)$
$= -2x(x^2) - (-2x)(4x) - (-2x)(3)$ • Use the Distributive Property.
$= -2x^3 + 8x^2 + 6x$ • Use the Rule for Multiplying Exponential Expressions.

EXAMPLE 1 Multiply. **A.** $(5x + 4)(-2x)$ **B.** $x^3(2x^2 - 3x + 2)$

Solution **A.** $(5x + 4)(-2x)$
$= 5x(-2x) + 4(-2x)$ • Use the Distributive Property.
$= -10x^2 - 8x$ • Use the Rule for Multiplying Exponential Expressions.

B. $x^3(2x^2 - 3x + 2)$
$= 2x^5 - 3x^4 + 2x^3$

PROBLEM 1 Multiply. **A.** $(-2y + 3)(-4y)$ **B.** $-a^2(3a^2 + 2a - 7)$

Solution See page S19.

OBJECTIVE 2 **Multiply two polynomials**

Multiplication of two polynomials requires the repeated application of the Distributive Property.

$$(y - 2)(y^2 + 3y + 1) = (y - 2)(y^2) + (y - 2)(3y) + (y - 2)(1)$$
$$= y^3 - 2y^2 + 3y^2 - 6y + y - 2$$
$$= y^3 + y^2 - 5y - 2$$

A convenient method of multiplying two polynomials is to use a vertical format similar to that used for multiplication of whole numbers.

$$
\begin{array}{r}
y^2 + 3y + 1 \\
y - 2 \\
\hline
-2y^2 - 6y - 2 \\
y^3 + 3y^2 + y \\
\hline
y^3 + y^2 - 5y - 2
\end{array}
$$

Multiply each term in the trinomial by -2.
Multiply each term in the trinomial by y.
Like terms must be in the same column.
Add the terms in each column.

EXAMPLE 2 Multiply: $(2b^3 - b + 1)(2b + 3)$

Solution

$$
\begin{array}{r}
2b^3 - b + 1 \\
2b + 3 \\
\hline
6b^3 - 3b + 3 \\
4b^4 - 2b^2 + 2b \\
\hline
4b^4 + 6b^3 - 2b^2 - b + 3
\end{array}
$$

- Multiply $2b^3 - b + 1$ by 3.
- Multiply $2b^3 - b + 1$ by $2b$. Arrange the terms in descending order.
- Add the terms in each column.

PROBLEM 2 Multiply: $(2y^3 + 2y^2 - 3)(3y - 1)$

Solution See page S19.

EXAMPLE 3 Multiply: $(4a^3 - 5a - 2)(3a - 2)$

Solution

$$
\begin{array}{r}
4a^3 - 5a - 2 \\
3a - 2 \\
\hline
-8a^3 + 10a + 4 \\
12a^4 - 15a^2 - 6a \\
\hline
12a^4 - 8a^3 - 15a^2 + 4a + 4
\end{array}
$$

- Multiply $4a^3 - 5a - 2$ by -2.
- Multiply $4a^3 - 5a - 2$ by $3a$.
- Add the terms in each column.

PROBLEM 3 Multiply: $(3x^3 - 2x^2 + x - 3)(2x + 5)$

Solution See page S19.

OBJECTIVE 3 **Multiply two binomials**

It is often necessary to find the product of two binomials. The product can be found using a method called **FOIL**, which is based on the Distributive Property. The letters of FOIL stand for **F**irst, **O**uter, **I**nner, and **L**ast.

Multiply: $(2x + 3)(x + 5)$

Multiply the First terms. $\quad (2x + 3)(x + 5) \qquad 2x \cdot x = 2x^2$

Multiply the Outer terms. $\quad (2x + 3)(x + 5) \qquad 2x \cdot 5 = 10x$

Multiply the Inner terms. $\quad (2x + 3)(x + 5) \qquad 3 \cdot x = 3x$

Take Note

FOIL is not really a different way of multiplying. It is based on the Distributive Property.

$(2x + 3)(x + 5)$
$\quad = 2x(x + 5) + 3(x + 5)$
$\qquad\; \text{F}\quad\; \text{O}\quad\; \text{I}\quad\; \text{L}$
$\quad = 2x^2 + 10x + 3x + 15$
$\quad = 2x^2 + 13x + 15$

Multiply the Last terms.	$(2x + 3)(x + 5)$	$3 \cdot 5\; = 15$

$\qquad\qquad\qquad\qquad\qquad\qquad\qquad\qquad\;\; \textbf{F}\quad\;\; \textbf{O}\quad\;\; \textbf{I}\quad\; \textbf{L}$

Add the products. $(2x + 3)(x + 5)\quad = 2x^2 + 10x + 3x + 15$
Combine like terms. $\qquad\qquad\qquad\qquad\qquad = 2x^2 + 13x + 15$

EXAMPLE 4 Multiply: $(4x - 3)(3x - 2)$

Solution $(4x - 3)(3x - 2)$
$\quad = 4x(3x) + 4x(-2) + (-3)(3x) + (-3)(-2)$ • Use the FOIL method.
$\quad = 12x^2 - 8x - 9x + 6$
$\quad = 12x^2 - 17x + 6$ • Combine like terms.

PROBLEM 4 Multiply: $(4y - 5)(3y - 3)$

Solution See page S19.

EXAMPLE 5 Multiply: $(3x - 2y)(x + 4y)$

Solution $(3x - 2y)(x + 4y)$
$\quad = 3x(x) + 3x(4y) + (-2y)(x) + (-2y)(4y)$ • Use the FOIL method.
$\quad = 3x^2 + 12xy - 2xy - 8y^2$
$\quad = 3x^2 + 10xy - 8y^2$ • Combine like terms.

PROBLEM 5 Multiply: $(3a + 2b)(3a - 5b)$

Solution See page S19.

OBJECTIVE **4** ## Multiply binomials that have special products

The expression $(a + b)(a - b)$ is the product of the sum and difference of two terms. The first binomial in the expression is a sum; the second is a difference. The two terms are *a* and *b*. The first term in each binomial is a. The second term in each binomial is b.

The expression $(a + b)^2$ is the square of a binomial. The first term in the binomial is a. The second term in the binomial is b.

Using FOIL, it is possible to find a pattern for the product of the sum and difference of two terms and for the square of a binomial.

The Sum and Difference of Two Terms

$$(\boldsymbol{a} + \boldsymbol{b})(\boldsymbol{a} - \boldsymbol{b}) = a^2 - ab + ab - b^2$$
$$= \boldsymbol{a^2} - \boldsymbol{b^2}$$

Square of first term
Square of second term

The Square of a Binomial

$$(a + b)^2 = (a + b)(a + b) = a^2 + ab + ab + b^2$$
$$= a^2 + 2ab + b^2$$

Square of first term

Twice the product of the two terms

Square of last term

$$(a - b)^2 = (a - b)(a - b) = a^2 - ab - ab + b^2$$
$$= a^2 - 2ab + b^2$$

Square of first term

Twice the product of the two terms

Square of last term

EXAMPLE 6 Multiply: $(2x + 3)(2x - 3)$

Solution $(2x + 3)(2x - 3)$ • **$(2x + 3)(2x - 3)$ is the product of the sum and difference of two terms.**

$= (2x)^2 - 3^2$ • **Square the first term. Square the second term.**
$= 4x^2 - 9$ • **Simplify.**

PROBLEM 6 Multiply: $(2a + 5c)(2a - 5c)$

Solution See page S19.

EXAMPLE 7 Multiply: $(4c + 5d)^2$

Solution $(4c + 5d)^2$ • **$(4c + 5d)^2$ is the square of a binomial.**

$= (4c)^2 + 2(4c)(5d) + (5d)^2$ • **Square the first term. Find twice the product of the two terms. Square the second term.**

$= 16c^2 + 40cd + 25d^2$ • **Simplify.**

PROBLEM 7 Multiply: $(3x + 2y)^2$

Solution See page S19.

EXAMPLE 8 Multiply: $(3x - 2)^2$

Solution $(3x - 2)^2$ • **$(3x - 2)^2$ is the square of a binomial.**

$= (3x)^2 + 2(3x)(-2) + (-2)^2$ • **Square the first term. Find twice the product of the two terms. Square the last term.**

$= 9x^2 - 12x + 4$ • **Simplify.**

PROBLEM 8 Multiply: $(6x - y)^2$

Solution See page S19.

OBJECTIVE **5** **Application problems**

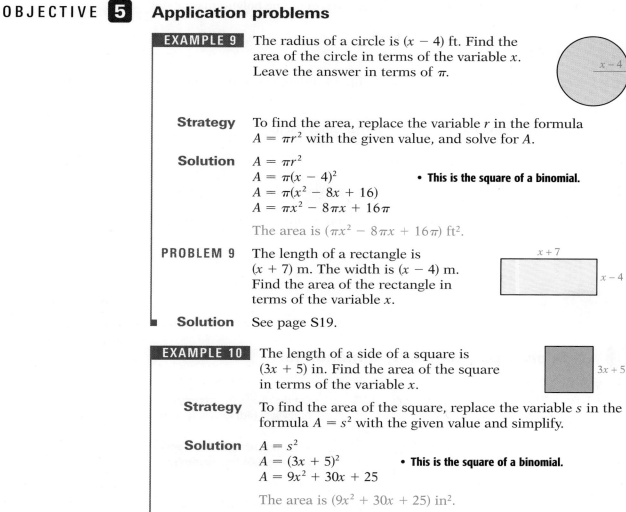

EXAMPLE 9 The radius of a circle is $(x - 4)$ ft. Find the area of the circle in terms of the variable x. Leave the answer in terms of π.

Strategy To find the area, replace the variable r in the formula $A = \pi r^2$ with the given value, and solve for A.

Solution $A = \pi r^2$
$A = \pi(x - 4)^2$ • **This is the square of a binomial.**
$A = \pi(x^2 - 8x + 16)$
$A = \pi x^2 - 8\pi x + 16\pi$

The area is $(\pi x^2 - 8\pi x + 16\pi)$ ft².

PROBLEM 9 The length of a rectangle is $(x + 7)$ m. The width is $(x - 4)$ m. Find the area of the rectangle in terms of the variable x.

Solution See page S19.

EXAMPLE 10 The length of a side of a square is $(3x + 5)$ in. Find the area of the square in terms of the variable x.

Strategy To find the area of the square, replace the variable s in the formula $A = s^2$ with the given value and simplify.

Solution $A = s^2$
$A = (3x + 5)^2$ • **This is the square of a binomial.**
$A = 9x^2 + 30x + 25$

The area is $(9x^2 + 30x + 25)$ in².

PROBLEM 10 The base of a triangle is $(x + 3)$ cm and the height is $(4x - 6)$ cm. Find the area of the triangle in terms of the variable x.

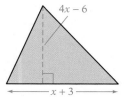

Solution See page S19.

5.3 **Concept Review**

Determine whether the statement is always true, sometimes true, or never true.

1. To multiply a monomial times a polynomial, use the Distributive Property to multiply each term of the polynomial by the monomial.

2. To multiply two polynomials, multiply each term of one polynomial by the other polynomial.

3. A binomial is a polynomial of degree 2.

4. $(x + 7)(x - 7)$ is the product of the sum and difference of the same two terms.

5. To square a binomial means to multiply it times itself.

6. The square of a binomial is a trinomial.

7. The FOIL method is used to multiply two polynomials.

8. Using the FOIL method, the terms $3x$ and 5 are the "First" terms in $(3x + 5)(2x + 7)$.

5.3 Exercises

1 Multiply a polynomial by a monomial

1. Multiply: $-3y(y + 7)$

$-3y(y + 7) = \underline{\quad?\quad}(y) + (\underline{\quad?\quad})(7)$ • Use the $\underline{\quad?\quad}$ **Property** to multiply each term of $(y + 7)$ by $\underline{\quad?\quad}$.

$= \underline{\quad?\quad} - \underline{\quad?\quad}$ • In the first term, the bases are the same. Add the exponents.

2. Multiply: $5x^3(x^2 + 2x - 10)$

$5x^3(x^2 + 2x - 10)$
$= (\underline{\quad?\quad})(x^2) + (\underline{\quad?\quad})(2x) - (\underline{\quad?\quad})(10)$ • Use the Distributive Property to multiply each term of $(\underline{\quad?\quad})$ by $5x^3$.

$= 5x^{\underline{?}} + 10x^{\underline{?}} - 50x^{\underline{?}}$ • Use the Rule for Multiplying Exponential Expressions.

Multiply.

3. $x(x - 2)$ **4.** $y(3 - y)$ **5.** $-x(x + 7)$

6. $-y(7 - y)$ **7.** $3a^2(a - 2)$ **8.** $4b^2(b + 8)$

9. $-5x^2(x^2 - x)$ **10.** $-6y^2(y + 2y^2)$ **11.** $-x^3(3x^2 - 7)$

12. $-y^4(2y^2 - y^6)$ **13.** $2x(6x^2 - 3x)$ **14.** $3y(4y - y^2)$

15. $(2x - 4)3x$

16. $(2x + 1)2x$

17. $-xy(x^2 - y^2)$

18. $-x^2y(2xy - y^2)$

19. $x(2x^3 - 3x + 2)$

20. $y(-3y^2 - 2y + 6)$

21. $-a(-2a^2 - 3a - 2)$

22. $-b(5b^2 + 7b - 35)$

23. $x^2(3x^4 - 3x^2 - 2)$

24. $y^3(-4y^3 - 6y + 7)$

25. $2y^2(-3y^2 - 6y + 7)$

26. $4x^2(3x^2 - 2x + 6)$

27. $(a^2 + 3a - 4)(-2a)$

28. $(b^3 - 2b + 2)(-5b)$

29. $-3y^2(-2y^2 + y - 2)$

30. $-5x^2(3x^2 - 3x - 7)$

31. $xy(x^2 - 3xy + y^2)$

32. $ab(2a^2 - 4ab - 6b^2)$

33. Which expressions are equivalent to $4x - x(3x - 1)$?
 a. $4x - 3x^2 - x$ **b.** $-3x^2 + 5x$ **c.** $4x - 3x^2 + x$
 d. $9x^2 - 3x$ **e.** $3x(3x - 1)$

34. Which expression is equivalent to $b^3(b^3 - b^2)$?
 a. $b^9 - b^6$ **b.** $b^{27} - b^8$ **c.** b^3 **d.** $b^6 - b^5$ **e.** b

2 Multiply two polynomials

Multiply.

35. $(x^2 + 3x + 2)(x + 1)$

36. $(x^2 - 2x + 7)(x - 2)$

37. $(a - 3)(a^2 - 3a + 4)$

38. $(2x - 3)(x^2 - 3x + 5)$

39. $(-2b^2 - 3b + 4)(b - 5)$

40. $(-a^2 + 3a - 2)(2a - 1)$

41. $(3x - 5)(-2x^2 + 7x - 2)$

42. $(2a - 1)(-a^2 - 2a + 3)$

43. $(x^3 - 3x + 2)(x - 4)$

44. $(y^3 + 4y^2 - 8)(2y - 1)$

45. $(3y - 8)(5y^2 + 8y - 2)$

46. $(4y - 3)(3y^2 + 3y - 5)$

47. $(5a^3 - 15a + 2)(a - 4)$

48. $(3b^3 - 5b^2 + 7)(6b - 1)$

49. $(y + 2)(y^3 + 2y^2 - 3y + 1)$

50. $(2a - 3)(2a^3 - 3a^2 + 2a - 1)$

For Exercises 51 and 52, use this product: $(ax^3 + bx + c)(dx + e)$, where a, b, c, d, and e are nonzero real numbers.

51. **a.** What is the degree of the simplified product?
 b. At most, how many terms will the simplified product have?

52. Does the simplified product have an x^2 term? If so, what is its coefficient?

3 Multiply two binomials

53. For the product $(4x - 3)(x + 5)$:
The **First** terms are ___?___ and ___?___.
The **Outer** terms are ___?___ and ___?___.
The **Inner** terms are ___?___ and ___?___.
The **Last** terms are ___?___ and ___?___.

For the product of the binomials, name **a.** the first terms, **b.** the outer terms, **c.** the inner terms, and **d.** the last terms.

54. $(y - 8)(2y + 3)$ **55.** $(3d + 4)(d - 1)$

56. Use the FOIL method to multiply $(x + 2)(8x - 3)$.
The product of the **First** terms is $x \cdot 8x =$ ___?___.
The product of the **Outer** terms is $x \cdot (-3) =$ ___?___.
The product of the **Inner** terms is $2 \cdot 8x =$ ___?___.
The product of the **Last** terms is $2 \cdot (-3) =$ ___?___.
The sum of these four products is ___?___.

Multiply.

57. $(x + 1)(x + 3)$ **58.** $(y + 2)(y + 5)$ **59.** $(a - 3)(a + 4)$

60. $(b - 6)(b + 3)$ **61.** $(y + 3)(y - 8)$ **62.** $(x + 10)(x - 5)$

63. $(y - 7)(y - 3)$ **64.** $(a - 8)(a - 9)$ **65.** $(2x + 1)(x + 7)$

66. $(y + 2)(5y + 1)$ **67.** $(3x - 1)(x + 4)$ **68.** $(7x - 2)(x + 4)$

69. $(4x - 3)(x - 7)$ **70.** $(2x - 3)(4x - 7)$ **71.** $(3y - 8)(y + 2)$

72. $(5y - 9)(y + 5)$ **73.** $(3x + 7)(3x + 11)$ **74.** $(5a + 6)(6a + 5)$

75. $(7a - 16)(3a - 5)$ **76.** $(5a - 12)(3a - 7)$ **77.** $(3b + 13)(5b - 6)$

78. $(x + y)(2x + y)$ **79.** $(2a + b)(a + 3b)$ **80.** $(3x - 4y)(x - 2y)$

81. $(2a - b)(3a + 2b)$ **82.** $(5a - 3b)(2a + 4b)$ **83.** $(2x + y)(x - 2y)$

84. $(3x - 7y)(3x + 5y)$ **85.** $(2x + 3y)(5x + 7y)$ **86.** $(5x + 3y)(7x + 2y)$

87. $(3a - 2b)(2a - 7b)$ **88.** $(5a - b)(7a - b)$ **89.** $(a - 9b)(2a + 7b)$

90. $(2a + 5b)(7a - 2b)$ **91.** $(10a - 3b)(10a - 7b)$ **92.** $(12a - 5b)(3a - 4b)$

93. $(5x + 12y)(3x + 4y)$ **94.** $(11x + 2y)(3x + 7y)$ **95.** $(2x - 15y)(7x + 4y)$

96. $(5x + 2y)(2x - 5y)$ **97.** $(8x - 3y)(7x - 5y)$ **98.** $(2x - 9y)(8x - 3y)$

99. Suppose a and b are positive numbers such that $a > b$. Will the coefficient of the x term of the product $(ax - b)(x + 1)$ be positive or negative?

100. If the constant terms of two binomials are both negative, is the constant term of the product of the two binomials also negative?

4 Multiply two binomials that have special products

101. The product $(a + b)(a - b) = a^2 - b^2$, so
$(5x + 1)(5x - 1) = (\underline{\quad ? \quad})^2 - (\underline{\quad ? \quad})^2 = \underline{\quad ? \quad} - \underline{\quad ? \quad}$.

102. The product $(a + b)^2 = a^2 + 2ab + b^2$, so
$(3x + 7)^2 = (\underline{\quad ? \quad})^2 + 2(\underline{\quad ? \quad})(\underline{\quad ? \quad}) + (\underline{\quad ? \quad})^2$
$= \underline{\quad ? \quad} + \underline{\quad ? \quad} + \underline{\quad ? \quad}$.

Multiply.

103. $(y - 5)(y + 5)$ **104.** $(y + 6)(y - 6)$ **105.** $(2x + 3)(2x - 3)$

106. $(4x - 7)(4x + 7)$ **107.** $(x + 1)^2$ **108.** $(y - 3)^2$

109. $(3a - 5)^2$ **110.** $(6x - 5)^2$ **111.** $(3x - 7)(3x + 7)$

112. $(9x - 2)(9x + 2)$ **113.** $(2a + b)^2$ **114.** $(x + 3y)^2$

115. $(x - 2y)^2$ **116.** $(2x - 3y)^2$ **117.** $(4 - 3y)(4 + 3y)$

118. $(4x - 9y)(4x + 9y)$ **119.** $(5x + 2y)^2$ **120.** $(2a - 9b)^2$

For Exercises 121 to 124, state whether the coefficient of the x term of the product is positive, negative, or zero.

121. $(ax + b)(ax - b)$, where $a > 0$ and $b > 0$

122. $(ax + b)(ax + b)$, where $a > 0$ and $b < 0$

123. $(ax + b)^2$, where $a > 0$ and $b > 0$

124. $(ax + b)^2$, where $a < 0$ and $b < 0$

5 Application problems

Geometry Problems

125. The length of a rectangle is $(2x + 5)$ ft and the width of the rectangle is $(2x - 5)$ ft. The expression that represents the area of the rectangle is $(\underline{\quad ? \quad})(\underline{\quad ? \quad}) = \underline{\quad ? \quad}$.

126. The length of the side of a square is $(3a - 1)$. The expression that represents the area of the square is $(\underline{\quad ? \quad})^2 = \underline{\quad ? \quad}$.

127. The length of a rectangle is $5x$ ft. The width is $(2x - 7)$ ft. Find the area of the rectangle in terms of the variable x.

128. The width of a rectangle is $(x - 6)$ m. The length is $(2x + 3)$ m. Find the area of the rectangle in terms of the variable x.

129. The width of a rectangle is $(3x + 1)$ in. The length of the rectangle is twice the width. Find the area of the rectangle in terms of the variable x.

130. The width of a rectangle is $(4x - 3)$ cm. The length of the rectangle is twice the width. Find the area of the rectangle in terms of the variable x.

131. The length of a side of a square is $(2x + 1)$ km. Find the area of the square in terms of the variable x.

132. The length of a side of a square is $(2x - 3)$ yd. Find the area of the square in terms of the variable x.

133. The base of a triangle is $4x$ m and the height is $(2x + 5)$ m. Find the area of the triangle in terms of the variable x.

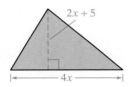

134. The base of a triangle is $(2x + 6)$ in. and the height is $(x - 8)$ in. Find the area of the triangle in terms of the variable x.

135. The radius of a circle is $(x + 4)$ cm. Find the area of the circle in terms of the variable x. Leave the answer in terms of π.

136. The radius of a circle is $(x - 3)$ ft. Find the area of the circle in terms of the variable x. Leave the answer in terms of π.

137. **Sports** A softball diamond has dimensions 45 ft by 45 ft. A base path border x feet wide lies on both the first-base side and the third-base side of the diamond. Express the total area of the softball diamond and the base path in terms of the variable x.

138. Sports An athletic field has dimensions 30 yd by 100 yd. An end zone that is w yards wide borders each end of the field. Express the total area of the field and the end zones in terms of the variable w.

5.3 Applying Concepts

Simplify.

139. $(a + b)^2 - (a - b)^2$

140. $(x + 3y)^2 + (x + 3y)(x - 3y)$

141. $(3a^2 - 4a + 2)^2$

142. $(x + 4)^3$

143. $3x^2(2x^3 + 4x - 1) - 6x^3(x^2 - 2)$

144. $(3b + 2)(b - 6) + (4 + 2b)(3 - b)$

145. $x^n(x^n + 1)$

146. $(x^n + 1)(x^n - 1)$

147. $(x^n + 1)(x^n + 1)$

148. $(x^n - 1)^2$

149. $(x + 1)(x - 1)$

150. $(x + 1)(-x^2 + x - 1)$

151. $(x + 1)(x^3 - x^2 + x - 1)$

152. $(x + 1)(-x^4 + x^3 - x^2 + x - 1)$

Use the pattern of the answers to Exercises 149 to 152 to write the product.

153. $(x + 1)(x^5 - x^4 + x^3 - x^2 + x - 1)$

154. $(x + 1)(-x^6 + x^5 - x^4 + x^3 - x^2 + x - 1)$

Solve.

155. Find $(4n^3)^2$ if $2n - 3 = 4n - 7$.

156. What polynomial has quotient $x^2 + 2x - 1$ when divided by $x + 3$?

157. What polynomial has quotient $3x - 4$ when divided by $4x + 5$?

158. Subtract $4x^2 - x - 5$ from the product of $x^2 + x + 3$ and $x - 4$.

159. Add $x^2 + 2x - 3$ to the product of $2x - 5$ and $3x + 1$.

160. If a polynomial of degree 3 is multiplied by a polynomial of degree 2, what is the degree of the resulting polynomial?

161. Is it possible to multiply a polynomial of degree 2 by a polynomial of degree 2 and have the product be a polynomial of degree 3? If so, give an example. If not, explain why not.

Integer Exponents and Scientific Notation

OBJECTIVE 1

Integer exponents

The quotient of two exponential expressions with the *same* base can be simplified by writing each expression in factored form, dividing by the common factors, and then writing the result with an exponent.

$$\frac{x^5}{x^2} = \frac{\overset{1}{\cancel{x}} \cdot \overset{1}{\cancel{x}} \cdot x \cdot x \cdot x}{\underset{1}{\cancel{x}} \cdot \underset{1}{\cancel{x}}} = x^3$$

Note that subtracting the exponents results in the same quotient.

$$\frac{x^5}{x^2} = x^{5-2} = x^3$$

To divide two monomials with the same base, subtract the exponents of the like bases.

Simplify: $\dfrac{a^7}{a^3}$

The bases are the same. Subtract the exponent in the denominator from the exponent in the numerator.

$$\frac{a^7}{a^3} = a^{7-3} = a^4$$

Simplify: $\dfrac{r^8 s^6}{r^7 s}$

Subtract the exponents of the like bases.

$$\frac{r^8 s^6}{r^7 s} = r^{8-7} s^{6-1} = r s^5$$

Recall that for any number a, $a \neq 0$, $\dfrac{a}{a} = 1$. This property is true for exponential expressions as well. For example, for $x \neq 0$, $\dfrac{x^4}{x^4} = 1$.

This expression also can be simplified using the rule for dividing exponential expressions with the same base.

$$\frac{x^4}{x^4} = x^{4-4} = x^0$$

Because $\dfrac{x^4}{x^4} = 1$ and $\dfrac{x^4}{x^4} = x^{4-4} = x^0$, the following definition of zero as an exponent is used.

Zero as an Exponent

If $x \neq 0$, then $x^0 = 1$. The expression 0^0 is not defined.

Simplify: $(12a^3)^0$, $a \neq 0$

Any nonzero expression to the zero power is 1. $(12a^3)^0 = 1$

Simplify: $-(xy^4)^0$, $x \neq 0$, $y \neq 0$

Any nonzero expression to the zero power is 1. Because the negative sign is outside the parentheses, the answer is -1. $-(xy^4)^0 = -(1) = -1$

The meaning of a negative exponent can be developed by examining the quotient $\dfrac{x^4}{x^6}$.

The expression can be simplified by writing the numerator and denominator in factored form, dividing by the common factors, and then writing the result with an exponent.

$$\frac{x^4}{x^6} = \frac{\cancel{x}\cdot\cancel{x}\cdot\cancel{x}\cdot\cancel{x}}{\cancel{x}\cdot\cancel{x}\cdot\cancel{x}\cdot\cancel{x}\cdot x \cdot x} = \frac{1}{x^2}$$

Now simplify the same expression by subtracting the exponents of the like bases.

$$\frac{x^4}{x^6} = x^{4-6} = x^{-2}$$

Because $\dfrac{x^4}{x^6} = \dfrac{1}{x^2}$ and $\dfrac{x^4}{x^6} = x^{-2}$, the expressions $\dfrac{1}{x^2}$ and x^{-2} must be equal. The following definition of a negative exponent is used.

Definition of Negative Exponents

If n is a positive integer and $x \neq 0$, then $x^{-n} = \dfrac{1}{x^n}$ and $\dfrac{1}{x^{-n}} = x^n$.

Evaluate 2^{-4}.

Write the expression with a positive exponent. Then simplify. $2^{-4} = \dfrac{1}{2^4} = \dfrac{1}{16}$

Now that negative exponents have been defined, the Rule for Dividing Exponential Expressions can be stated.

Rule for Dividing Exponential Expressions

If m and n are integers and $x \neq 0$, then $\dfrac{x^m}{x^n} = x^{m-n}$.

EXAMPLE 1 Write $\dfrac{3^{-3}}{3^2}$ with a positive exponent. Then evaluate.

Solution $\dfrac{3^{-3}}{3^2} = 3^{-3-2}$ • 3^{-3} and 3^2 have the same base. Subtract the exponents.

$= 3^{-5}$

$= \dfrac{1}{3^5}$ • Use the Definition of Negative Exponents to write the expression with a positive exponent.

$= \dfrac{1}{243}$ • Evaluate.

PROBLEM 1 Write $\dfrac{2^{-2}}{2^3}$ with a positive exponent. Then evaluate.

Solution See page S19.

The rules for simplifying exponential expressions and powers of exponential expressions are true for all integers. These rules are restated here.

Rules for Exponents

If m, n, and p are integers, then

$$x^m \cdot x^n = x^{m+n} \qquad (x^m)^n = x^{mn} \qquad (x^m y^n)^p = x^{mp} y^{np}$$

$$\dfrac{x^m}{x^n} = x^{m-n},\; x \neq 0 \qquad x^{-n} = \dfrac{1}{x^n},\; x \neq 0 \qquad x^0 = 1,\; x \neq 0$$

An exponential expression is in simplest form when it is written with only positive exponents.

Simplify: $a^{-7}b^3$

Rewrite a^{-7} with a positive exponent. $a^{-7}b^3 = \dfrac{b^3}{a^7}$

EXAMPLE 2 Simplify: $\dfrac{x^{-4}y^6}{xy^2}$

Solution $\dfrac{x^{-4}y^6}{xy^2} = x^{-4-1}y^{6-2}$ • Divide variables with the same base by subtracting the exponents.

$= x^{-5}y^4$

$= \dfrac{y^4}{x^5}$ • Write the expression with only positive exponents.

PROBLEM 2 Simplify: $\dfrac{b^8}{a^{-5}b^6}$

Solution See page S19.

Simplify: $6d^{-4}$, $d \neq 0$

$6d^{-4} = 6 \cdot \dfrac{1}{d^4} = \dfrac{6}{d^4}$ • Use the Definition of Negative Exponents to rewrite the expression with a positive exponent.

EXAMPLE 3 Simplify: **A.** $\dfrac{-35a^6b^{-2}}{25a^{-2}b^5}$ **B.** $(-2x)(3x^{-2})^{-3}$

Solution **A.** $\dfrac{-35a^6b^{-2}}{25a^{-2}b^5} = -\dfrac{35a^6b^{-2}}{25a^{-2}b^5}$ • A negative sign is placed in front of a fraction.

$$= -\dfrac{\overset{1}{\cancel{5}} \cdot 7a^{6-(-2)}b^{-2-5}}{\underset{1}{\cancel{5} \cdot 5}}$$ • Factor the coefficients. Divide by the common factors. Divide variables with the same base by subtracting the exponents.

$$= -\dfrac{7a^8b^{-7}}{5}$$

$$= -\dfrac{7a^8}{5b^7}$$ • Write the expression with only positive exponents.

B. $(-2x)(3x^{-2})^{-3} = (-2x)(3^{-3}x^6)$ • Use the Rule for Simplifying Powers of Products.

$$= \dfrac{-2x \cdot x^6}{3^3}$$ • Write the expression with positive exponents.

$$= -\dfrac{2x^7}{27}$$ • Use the Rule for Multiplying Exponential Expressions, and simplify the numerical exponential expression.

PROBLEM 3 Simplify: **A.** $\dfrac{12x^{-8}y^4}{-16xy^{-3}}$ **B.** $(-3ab)(2a^3b^{-2})^{-3}$

Solution See page S19.

OBJECTIVE **2** ## Scientific notation

Very large and very small numbers are encountered in the fields of science and engineering. For example, the charge of an electron is 0.00000000000000000160 coulomb. These numbers can be written more easily in scientific notation. In **scientific notation,** a number is expressed as a product of two factors, one a number between 1 and 10 and the other a power of 10.

To change a number written in decimal notation to scientific notation, write it in the form $a \times 10^n$, where a is a number between 1 and 10 and n is an integer.

For numbers greater than 10, move the decimal point to the right of the first digit. The exponent n is positive and equal to the number of places the decimal point has been moved.

$240,000 = 2.4 \times 10^5$

$93,000,000 = 9.3 \times 10^7$

Take Note

There are two steps to writing a number in scientific notation:
(1) determine the number between 1 and 10, and
(2) determine the exponent on 10.

For numbers less than 1, move the decimal point to the right of the first nonzero digit. The exponent n is negative. The absolute value of the exponent is equal to the number of places the decimal point has been moved.

$0.00030 = 3.0 \times 10^{-4}$

$0.0000832 = 8.32 \times 10^{-5}$

Look at the last example above: $0.0000832 = 8.32 \times 10^{-5}$. Using the Definition of Negative Exponents,

$$10^{-5} = \frac{1}{10^5} = \frac{1}{100,000} = 0.00001$$

Because $10^{-5} = 0.00001$, we can write

$$8.32 \times 10^{-5} = 8.32 \times 0.00001 = 0.0000832$$

which is the number we started with. We have not changed the value of the number; we have just written it in another form.

EXAMPLE 4 Write the number in scientific notation.
 A. 824,300,000,000 **B.** 0.000000961

Solution **A.** $824,300,000,000 = 8.243 \times 10^{11}$ • **Move the decimal point 11 places to the left. The exponent on 10 is 11.**

 B. $0.000000961 = 9.61 \times 10^{-7}$ • **Move the decimal point 7 places to the right. The exponent on 10 is −7.**

PROBLEM 4 Write the number in scientific notation.
 A. 57,000,000,000 **B.** 0.000000017

Solution See page S19.

Changing a number written in scientific notation to decimal notation also requires moving the decimal point.

When the exponent on 10 is positive, move the decimal point to the right the same number of places as the exponent.

$3.45 \times 10^9 = 3,450,000,000$

$2.3 \times 10^8 = 230,000,000$

When the exponent on 10 is negative, move the decimal point to the left the same number of places as the absolute value of the exponent.

$8.1 \times 10^{-3} = 0.0081$

$6.34 \times 10^{-6} = 0.00000634$

EXAMPLE 5 Write the number in decimal notation.
 A. 7.329×10^6 **B.** 6.8×10^{-10}

Solution **A.** $7.329 \times 10^6 = 7,329,000$ • **The exponent on 10 is positive. Move the decimal point 6 places to the right.**

 B. $6.8 \times 10^{-10} = 0.00000000068$ • **The exponent on 10 is negative. Move the decimal point 10 places to the left.**

PROBLEM 5 Write the number in decimal notation.
 A. 5×10^{12} **B.** 4.0162×10^{-9}

Solution See page S19.

Numerical calculations involving numbers that have more digits than a hand-held calculator is able to handle can be performed using scientific notation.

EXAMPLE 6 Simplify: $\dfrac{2{,}400{,}000{,}000 \times 0.0000063}{0.00009 \times 480}$

Solution $\dfrac{2{,}400{,}000{,}000 \times 0.0000063}{0.00009 \times 480}$

$= \dfrac{2.4 \times 10^9 \times 6.3 \times 10^{-6}}{9 \times 10^{-5} \times 4.8 \times 10^2}$ • Write the numbers in scientific notation.

$= \dfrac{(2.4)(6.3) \times 10^{9+(-6)-(-5)-2}}{(9)(4.8)}$ • Simplify.

$= 0.35 \times 10^6 = 3.5 \times 10^5$ • Write in scientific notation.

PROBLEM 6 Simplify: $\dfrac{5{,}600{,}000 \times 0.000000081}{900 \times 0.000000028}$

Solution See page S19.

OBJECTIVE **3** **Application problems**

EXAMPLE 7 How many miles does light travel in 1 day? The speed of light is 186,000 mi/s. Write the answer in scientific notation.

Strategy To find the distance traveled:

▶ Write the speed of light in scientific notation.

▶ Write the number of seconds in 1 day in scientific notation.

▶ Use the equation $d = rt$, where r is the speed of light and t is the number of seconds in 1 day.

Solution $186{,}000 = 1.86 \times 10^5$

$24 \cdot 60 \cdot 60 = 86{,}400 = 8.64 \times 10^4$

$d = rt$
$d = (1.86 \times 10^5)(8.64 \times 10^4)$
$d = 1.86 \times 8.64 \times 10^9$
$d = 16.0704 \times 10^9$
$d = 1.60704 \times 10^{10}$

Light travels 1.60704×10^{10} mi in one day.

PROBLEM 7 A computer can do an arithmetic operation in 1×10^{-7} s. How many arithmetic operations can the computer perform in 1 min? Write the answer in scientific notation.

Solution See pages S19–S20.

5.4 **Concept Review**

Determine whether the statement is always true, sometimes true, or never true.

1. The expression $\dfrac{x^5}{y^3}$ can be simplified by subtracting the exponents.

2. The rules of exponents can be applied to expressions that contain an exponent of zero or contain negative exponents.

3. The expression 3^{-2} represents the reciprocal of 3^2.

4. $5x^0 = 0$

5. The expression 4^{-3} represents a negative number.

6. To be in simplest form, an exponential expression cannot contain any negative exponents.

7. $2x^{-5} = \dfrac{1}{2x^5}$

8. $x^4y^{-6} = \dfrac{1}{x^4y^6}$

5.4 Exercises

1 Integer exponents

Write with a positive or zero exponent. Then evaluate.

1. 5^{-2}

2. 3^{-3}

3. $\dfrac{1}{8^{-2}}$

4. $\dfrac{1}{12^{-1}}$

5. $\dfrac{3^{-2}}{3}$

6. $\dfrac{5^{-3}}{5}$

7. $\dfrac{2^3}{2^3}$

8. $\dfrac{3^{-2}}{3^{-2}}$

9. Explain how to rewrite a variable that has a negative exponent as an expression with a positive exponent.

10. As long as x is not zero, x^0 is defined to be equal to ___?___. Using this definition, $3^0 =$ ___?___, $(7x^3)^0 =$ ___?___, and $-2x^0 =$ ___?___.

11. Simplify: $\dfrac{8x^{10}}{2x^3}$

$\dfrac{8x^{10}}{2x^3} = ($___?___$)x^{\underline{?}\,-\,\underline{?}}$ • Divide the coefficients by the common factor 2. Divide the variable parts by subtracting the exponents.

$= $ ___?___ • Simplify.

12. Simplify: $\dfrac{a^3b^7}{a^5b}$

$\dfrac{a^3b^7}{a^5b} = a^{\underline{?}\,-\,\underline{?}}b^{\underline{?}\,-\,\underline{?}}$ • Divide variables with the same base by subtracting the exponents.

$= a^{\underline{?}}b^{\underline{?}}$ • Simplify.

$= \dfrac{b^6}{\underline{\quad?\quad}}$ • Write the expression with only positive exponents.

Simplify.

13. $\dfrac{y^7}{y^3}$

14. $\dfrac{z^9}{z^2}$

15. $\dfrac{a^8}{a^5}$

16. $\dfrac{c^{12}}{c^5}$

17. $\dfrac{p^5}{p}$

18. $\dfrac{w^9}{w}$

19. $\dfrac{4x^8}{2x^5}$

20. $\dfrac{12z^7}{4z^3}$

21. $\dfrac{22k^5}{11k^4}$

22. $\dfrac{14m^{11}}{7m^{10}}$

23. $\dfrac{m^9n^7}{m^4n^5}$

24. $\dfrac{y^5z^6}{yz^3}$

25. $\dfrac{6r^4}{4r^2}$

26. $\dfrac{8x^9}{12x^6}$

27. $\dfrac{-16a^7}{24a^6}$

28. $\dfrac{-18b^5}{27b^4}$

29. x^{-2}

30. y^{-10}

31. $\dfrac{1}{a^{-6}}$

32. $\dfrac{1}{b^{-4}}$

33. $4x^{-7}$

34. $-6y^{-1}$

35. $\dfrac{5}{b^{-8}}$

36. $\dfrac{-3}{v^{-3}}$

37. $\dfrac{1}{3x^{-2}}$

38. $\dfrac{2}{5c^{-6}}$

39. $(ab^5)^0$

40. $(32x^3y^4)^0$

41. $\dfrac{y^3}{y^8}$

42. $\dfrac{z^4}{z^6}$

43. $\dfrac{a^5}{a^{11}}$

44. $\dfrac{m}{m^7}$

45. $\dfrac{4x^2}{12x^5}$

46. $\dfrac{6y^8}{8y^9}$

47. $\dfrac{-12x}{-18x^6}$

48. $\dfrac{-24c^2}{-36c^{11}}$

49. $\dfrac{x^6y^5}{x^8y}$

50. $\dfrac{a^3b^2}{a^2b^3}$

51. $\dfrac{2m^6n^2}{5m^9n^{10}}$

52. $\dfrac{5r^3t^7}{6r^5t^7}$

53. $\dfrac{pq^3}{p^4q^4}$

54. $\dfrac{a^4b^5}{a^5b^6}$

55. $\dfrac{3x^4y^5}{6x^4y^8}$

56. $\dfrac{14a^3b^6}{21a^5b^6}$

57. $\dfrac{14x^4y^6z^2}{16x^3y^9z}$

58. $\dfrac{24a^2b^7c^9}{36a^7b^5c}$

59. $\dfrac{15mn^9p^3}{30m^4n^9p}$

60. $\dfrac{25x^4y^7z^2}{20x^5y^9z^{11}}$

61. $(-2xy^{-2})^3$

62. $(-3x^{-1}y^2)^2$

63. $(3x^{-1}y^{-2})^2$

64. $(5xy^{-3})^{-2}$

65. $(2x^{-1})(x^{-3})$

66. $(-2x^{-5})x^7$

67. $(-5a^2)(a^{-5})^2$

68. $(2a^{-3})(a^7b^{-1})^3$

69. $(-2ab^{-2})(4a^{-2}b)^{-2}$

70. $(3ab^{-2})(2a^{-1}b)^{-3}$

71. $(-5x^{-2}y)(-2x^{-2}y^2)$

72. $\dfrac{a^{-3}b^{-4}}{a^2b}$

73. $\dfrac{3x^{-2}y^2}{6xy^2}$

74. $\dfrac{2x^{-2}y}{8xy}$

75. $\dfrac{3x^{-2}y}{xy}$

76. $\dfrac{2x^{-1}y^4}{x^2y^3}$

77. $\dfrac{2x^{-1}y^{-4}}{4xy^2}$

78. $\dfrac{12a^2b^3}{-27a^2b^2}$

79. $\dfrac{-16xy^4}{96x^4y^4}$

80. $\dfrac{-8x^2y^4}{44y^2z^5}$

81. True or false?

a. $\dfrac{a^{4n}}{a^n} = a^4$ b. $\dfrac{1}{a^{m-n}} = a^{n-m}$

82. True or false?

a. $a^{-n}a^n = 1$ b. $\dfrac{a^n}{b^m} = \left(\dfrac{a}{b}\right)^{n-m}$

2 Scientific notation

83. Why might a number be written in scientific notation instead of decimal notation?

84. A number is written in scientific notation if it is written as the product of a number between ____?____ and ____?____ and a power of ____?____.

85. To write the number 354,000,000 in scientific notation, move the decimal point ____?____ places to the ____?____. The exponent on 10 is ____?____.

86. To write the number 0.0000000086 in scientific notation, move the decimal point ____?____ places to the ____?____. The exponent on 10 is ____?____.

Determine whether the number is written in scientific notation. If not, explain why not.

87. 39.4×10^3

88. 0.8×10^{-6}

89. $7.1 \times 10^{2.4}$

90. 5.8×10^{-132}

Write the number in scientific notation.

91. 2,370,000

92. 75,000

93. 0.00045

94. 0.000076

95. 309,000

96. 819,000,000

97. 0.000000601

98. 0.00000000096

Write the number in decimal notation.

99. 7.1×10^5

100. 2.3×10^7

101. 4.3×10^{-5}

102. 9.21×10^{-7}

103. 6.71×10^8

104. 5.75×10^9

105. 7.13×10^{-6}

106. 3.54×10^{-8}

107. Astronomy Astrophysicists estimate that the radius of the Milky Way galaxy is 1,000,000,000,000,000,000,000 m. Write this number in scientific notation.

108. Geology The mass of Earth is 5,980,000,000,000,000,000,000,000 kg. Write this number in scientific notation.

109. Physics Carbon nanotubes are cylinders of carbon atoms that have remarkable properties of strength. Some nanotubes with a diameter of 0.0000000004 m have been created. Write this number in scientific notation.

110. Biology The weight of a single *E. coli* bacterium is 0.00000000000665 g. Write this number in scientific notation.

111. Archeology The weight of the Great Pyramid of Cheops is estimated to be 12,000,000,000 lb. Write this number in scientific notation.

112. Biophysics Biologists and physicists are working together to measure the mass of a virus. Currently, a virus with a mass of 0.000000000000000000039 g can be measured. Write this number in scientific notation.

113. Food Science The frequency (oscillations per second) of a microwave generated by a microwave oven is approximately 2,450,000,000 hertz. (1 hertz is one oscillation in one second.) Write this number in scientific notation.

114. Astronomy One light-year is the distance traveled by light in one year. One light-year is 5,880,000,000,000 mi. Write this number in scientific notation.

115. $a \times 10^n$ and $a \times 10^{-m}$ are numbers written in scientific notation, where n and m are positive integers such that $n < m$. Is $(a \times 10^n)(a \times 10^{-m})$ greater than or less than 1?

116. $a \times 10^{-n}$ and $a \times 10^{-m}$ are numbers written in scientific notation, where n and m are positive integers such that $n > m$. Is $\dfrac{a \times 10^{-m}}{a \times 10^{-n}}$ greater than or less than 1?

Multiply or divide.

117. $(1.9 \times 10^{12})(3.5 \times 10^7)$

118. $(4.2 \times 10^7)(1.8 \times 10^{-5})$

119. $(2.3 \times 10^{-8})(1.4 \times 10^{-6})$

120. $(3 \times 10^{-20})(2.4 \times 10^9)$

121. $\dfrac{6.12 \times 10^{14}}{1.7 \times 10^9}$

122. $\dfrac{6 \times 10^{-8}}{2.5 \times 10^{-2}}$

123. $\dfrac{5.58 \times 10^{-7}}{3.1 \times 10^{11}}$

124. $\dfrac{9.03 \times 10^6}{4.3 \times 10^{-5}}$

3 Application problems

125. Write the number of seconds in five hours in scientific notation:

$5 \text{ h} = (5 \text{ h})(\underline{\quad ? \quad} \text{ min/h})(\underline{\quad ? \quad} \text{ s/min})$

$= \underline{\quad ? \quad} \text{ s}$

$= 1.8 \times 10^{\underline{\;?\;}}$

126. The speed of light is 3×10^5 km/s. Use your answer to Exercise 125 to find how many kilometers light travels in 5 h:

$d = (3 \times 10^5)(1.8 \times 10^4) = \underline{\quad ? \quad} \times 10^{\underline{\;?\;}} \text{ km}$

Solve. Write the answer in scientific notation.

127. Computer Science A computer can do an arithmetic operation in 5×10^{-7} s. How many arithmetic operations can the computer perform in 1 h?

128. Computer Science A computer can do an arithmetic operation in 2×10^{-9} s. How many arithmetic operations can the computer perform in 1 min?

129. Physics How many meters does light travel in 8 h? The speed of light is 3×10^8 m/s.

130. Physics How many meters does light travel in 1 day? The speed of light is 3×10^8 m/s.

131. Physics A high-speed centrifuge makes 4×10^8 revolutions each minute. Find the time in seconds for the centrifuge to make one revolution.

132. Physics The mass of an electron is 9.109×10^{-31} kg. The mass of a proton is 1.673×10^{-27} kg. How many times heavier is a proton than an electron?

133. Geology The mass of Earth is 5.9×10^{24} kg. The mass of the sun is 2×10^{30} kg. How many times heavier is the sun than Earth?

134. Astronomy It took 11 min for the commands from a computer on Earth to travel to the rover Sojourner on Mars, a distance of 119 million miles. How fast did the signals from Earth to Mars travel? Write the answer in scientific notation.

135. Astronomy How long does it take light to travel to Earth from the sun? The sun is 9.3×10^7 mi from Earth, and light travels 1.86×10^5 mi/s.

136. Measurement The weight of 31 million orchid seeds is 1 oz. Find the weight of one orchid seed.

137. The Federal Government In 2005, the gross national debt was approximately 8.1×10^{12} dollars. How much would each American have to pay in order to pay off the debt? Use 3×10^8 as the number of citizens.

138. Astronomy The distance to Saturn is 8.86×10^8 mi. A satellite leaves Earth traveling at a constant rate of 1×10^5 mph. How long does it take for the satellite to reach Saturn?

139. Astronomy The diameter of Neptune is 3×10^4 mi. Use the formula $SA = 4\pi r^2$ to find the surface area of Neptune in square miles.

140. Biology The radius of a cell is 1.5×10^{-4} mm. Use the formula $V = \frac{4}{3}\pi r^3$ to find the volume of the cell.

141. Chemistry One gram of hydrogen contains 6.023×10^{23} atoms. Find the weight of one atom of hydrogen.

142. Astronomy Our galaxy is estimated to be 5.6×10^{19} mi across. How long (in hours) would it take a space ship to cross the galaxy traveling at 25,000 mph?

5.4 Applying Concepts

Evaluate.

143. $8^{-2} + 2^{-5}$

144. $9^{-2} + 3^{-3}$

145. Evaluate 2^x and 2^{-x} when $x = -2, -1, 0, 1$, and 2.

146. Evaluate 3^x and 3^{-x} when $x = -2, -1, 0, 1$, and 2.

Write in decimal notation.

147. 2^{-4}

148. 25^{-2}

Simplify.

149. $\left(\frac{9x^2y^4}{3xy^2}\right) - \left(\frac{12x^5y^6}{6x^4y^4}\right)$

150. $\left(\frac{6x^2 + 9x}{3x}\right) + \left(\frac{8xy^2 + 4y^2}{4y^2}\right)$

151. $\left(\frac{6x^4yz^3}{2x^2y^3}\right)\left(\frac{2x^2z^3}{4y^2z}\right) \div \left(\frac{6x^2y^3}{x^4y^2z}\right)$

152. $\left(\frac{5x^2yz^3}{3x^4yz^3}\right) \div \left(\frac{10x^2y^5z^4}{2y^3z}\right) \div \left(\frac{5y^4z^2}{x^2y^6z}\right)$

Complete.

153. If $m = n$ and $a \neq 0$, then $\frac{a^m}{a^n} = $ _____.

154. If $m = n + 1$ and $a \neq 0$, then $\frac{a^m}{a^n} = $ _____.

Solve.

155. $(-4.8)^x = 1$

156. $-6.3^x = -1$

Determine whether each equation is true or false. If the equation is false, change the right side of the equation to make a true equation.

157. $(2a)^{-3} = \dfrac{2}{a^3}$

158. $((a^{-1})^{-1})^{-1} = \dfrac{1}{a}$

159. $(2 + 3)^{-1} = 2^{-1} + 3^{-1}$

160. If $x \neq \dfrac{1}{3}$, then $(3x - 1)^0 = (1 - 3x)^0$.

161. Why is the condition $x \neq \dfrac{1}{3}$ given in Exercise 160?

162. If x is a nonzero real number, is x^{-2} always positive, always negative, or positive or negative depending on whether x is positive or negative? Explain your answer.

163. If x is a nonzero real number, is x^{-3} always positive, always negative, or positive or negative depending on whether x is positive or negative? Explain your answer.

5.5 Division of Polynomials

OBJECTIVE 1 ### Divide a polynomial by a monomial

Note that $\dfrac{8 + 4}{2}$ can be simplified by first adding the terms in the numerator and then dividing the result. It can also be simplified by first dividing each term in the numerator by the denominator and then adding the result.

$$\frac{8 + 4}{2} = \frac{12}{2} = 6$$

$$\frac{8 + 4}{2} = \frac{8}{2} + \frac{4}{2} = 4 + 2 = 6$$

To divide a polynomial by a monomial, divide each term in the numerator by the denominator, and write the sum of the quotients.

$$\frac{a + b}{c} = \frac{a}{c} + \frac{b}{c}$$

Divide: $\dfrac{6x^2 + 4x}{2x}$

Divide each term of the polynomial $6x^2 + 4x$ by the monomial $2x$. Simplify.

$$\frac{6x^2 + 4x}{2x} = \frac{6x^2}{2x} + \frac{4x}{2x}$$
$$= 3x + 2$$

EXAMPLE 1 Divide: $\dfrac{6x^3 - 3x^2 + 9x}{3x}$

Solution $\dfrac{6x^3 - 3x^2 + 9x}{3x} = \dfrac{6x^3}{3x} - \dfrac{3x^2}{3x} + \dfrac{9x}{3x}$

$= 2x^2 - x + 3$

• **Divide each term of the polynomial by the monomial 3x.**
• **Simplify each expression.**

PROBLEM 1 Divide: $\dfrac{4x^3y + 8x^2y^2 - 4xy^3}{2xy}$

Solution See page S20.

EXAMPLE 2 Divide: $\dfrac{12x^2y - 6xy + 4x^2}{2xy}$

Solution $\dfrac{12x^2y - 6xy + 4x^2}{2xy}$

$= \dfrac{12x^2y}{2xy} - \dfrac{6xy}{2xy} + \dfrac{4x^2}{2xy}$

$= 6x - 3 + \dfrac{2x}{y}$

• **Divide each term of the polynomial by the monomial 2xy.**
• **Simplify each expression.**

PROBLEM 2 Divide: $\dfrac{24x^2y^2 - 18xy + 6y}{6xy}$

Solution See page S20.

OBJECTIVE 2 **Divide polynomials**

The division method illustrated in Objective 1 is appropriate only when the divisor is a monomial. To divide two polynomials in which the divisor is not a monomial, use a method similar to that used for division of whole numbers.

To check division of polynomials, use

(Quotient × Divisor) + Remainder = Dividend

For example:

$$\begin{array}{r} 3 \\ 5\overline{)17} \\ -15 \\ \hline 2 \end{array}$$

(Quotient × Divisor) + Remainder = Dividend

(3×5) + 2 = 17

Divide: $(x^2 - 5x + 8) \div (x - 3)$

Step 1

$$\begin{array}{r} x \\ x - 3 \overline{)x^2 - 5x + 8} \\ \underline{x^2 - 3x} \\ -2x + 8 \end{array}$$

Think: $x\overline{)x^2} = \dfrac{x^2}{x} = x$

Multiply: $x(x - 3) = x^2 - 3x$

Subtract: $(x^2 - 5x) - (x^2 - 3x) = -2x$

Bring down $+ 8$.

Step 2

$$\begin{array}{r} x - 2 \\ x - 3\overline{)x^2 - 5x + 8} \\ \underline{x^2 - 3x} \end{array}$$

$-2x + 8$ Think: $x\overline{)-2x} = \dfrac{-2x}{x} = -2$

$\underline{-2x + 6}$ Multiply: $-2(x - 3) = -2x + 6$

$\qquad 2$ Subtract: $(-2x + 8) - (-2x + 6) = 2$
 The remainder is 2.

Check:

Quotient × Divisor + Remainder = Dividend

$(x - 2)(x - 3) + 2 = x^2 - 3x - 2x + 6 + 2 = x^2 - 5x + 8$

$(x^2 - 5x + 8) \div (x - 3) = x - 2 + \dfrac{2}{x - 3}$

If a term is missing in the dividend, insert the term with zero as its coefficient. This helps keep like terms in the same column. This is illustrated below.

Divide: $\dfrac{6 - 6x^2 + 4x^3}{2x + 3}$

Arrange the terms in descending order. Note that there is no term of x in $4x^3 - 6x^2 + 6$. Insert a zero as $0x$ for the missing term so that like terms will be in the same columns.

$$\begin{array}{r} 2x^2 - 6x + 9 \\ 2x + 3\overline{)4x^3 - 6x^2 + 0x + 6} \\ \underline{4x^3 + 6x^2} \\ -12x^2 + 0x \\ \underline{-12x^2 - 18x} \\ 18x + 6 \\ \underline{18x + 27} \\ -21 \end{array}$$

$$\dfrac{4x^3 - 6x^2 + 6}{2x + 3} = 2x^2 - 6x + 9 - \dfrac{21}{2x + 3}$$

EXAMPLE 3 Divide. **A.** $\dfrac{12x^2 - 11x + 10}{4x - 5}$ **B.** $\dfrac{x^3 + 1}{x + 1}$

Solution **A.**

$$\begin{array}{r} 3x + 1 \\ 4x - 5\overline{)12x^2 - 11x + 10} \\ \underline{12x^2 - 15x} \\ 4x + 10 \\ \underline{4x - 5} \\ 15 \end{array}$$

$$\dfrac{12x^2 - 11x + 10}{4x - 5} = 3x + 1 + \dfrac{15}{4x - 5}$$

B.

$$\frac{x^3 + 1}{x + 1} = x^2 - x + 1$$

PROBLEM 3 Divide. **A.** $\dfrac{15x^2 + 17x - 20}{3x + 4}$ **B.** $\dfrac{3x^3 + 8x^2 - 6x + 2}{3x - 1}$

■ **Solution** See page S20.

OBJECTIVE 3 **Synthetic division**

Synthetic division is a shorter method of dividing a polynomial by a binomial of the form $x - a$. This method of dividing uses only the coefficients of the variable term and the constant term.

Both long division and synthetic division are used below to simplify the expression $(3x^2 - 10x + 7) \div (x - 2)$.

LONG DIVISION

Compare the coefficients in this problem worked by long division with the coefficients in the same problem worked by synthetic division.

$$\begin{array}{r} 3x - 4 \\ x - 2 \overline{)3x^2 - 10x + 7} \\ \underline{3x^2 - 6x} \\ -4x + 7 \\ \underline{-4x + 8} \\ -1 \end{array}$$

$$(3x^2 - 10x + 7) \div (x - 2) = 3x - 4 - \frac{1}{x - 2}$$

SYNTHETIC DIVISION

$x - a = x - 2; a = 2$ Value of a Coefficients of the dividend

Bring down the 3.

Multiply $2 \cdot 3$ and add the product (6) to -10.
The result is -4.

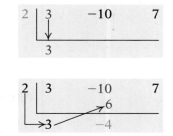

Multiply $2(-4)$ and add the product (-8) to 7. The result is -1.

$$
\begin{array}{r|rrr}
2 & 3 & -10 & 7 \\
 & & 6 & -8 \\
\hline
 & 3 & -4 & -1
\end{array}
$$

$\underbrace{}_{\text{Coefficients of the quotient}}$ $\underbrace{}_{\text{Remainder}}$

The degree of the first term of the quotient is one degree less than the degree of the first term of the dividend.

$(3x^2 - 10x + 7) \div (x - 2) = 3x - 4 - \dfrac{1}{x - 2}$

Check: $(3x - 4)(x - 2) - 1$
$= 3x^2 - 6x - 4x + 8 - 1$
$= 3x^2 - 10x + 7$

Divide: $(2x^3 + 3x^2 - 4x + 8) \div (x + 3)$

Write down the value of a and the coefficients of the dividend.

$$
\begin{array}{r|rrrr}
-3 & 2 & 3 & -4 & 8 \\
 & & -6 & 9 & -15 \\
\hline
 & 2 & -3 & 5 & -7
\end{array}
$$

$\underbrace{}_{\text{Coefficients of the quotient}}$ $\underbrace{}_{\text{Remainder}}$

$x - a = x + 3 = x - (-3);$
$a = -3$

Bring down the 2. Multiply $-3 \cdot 2$ and add the product (-6) to 3. Continue until all the coefficients have been used.

Write the quotient. The degree of the quotient is one degree less than the degree of the dividend.

$(2x^3 + 3x^2 - 4x + 8) \div (x + 3)$
$= 2x^2 - 3x + 5 - \dfrac{7}{x + 3}$

EXAMPLE 4 Divide: **A.** $(5x^2 - 3x + 7) \div (x - 1)$
B. $(3x^4 - 8x^2 + 2x + 1) \div (x + 2)$

Solution **A.**
$$
\begin{array}{r|rrr}
1 & 5 & -3 & 7 \\
 & & 5 & 2 \\
\hline
 & 5 & 2 & 9
\end{array}
$$

• $x - a = x - 1; a = 1$

$(5x^2 - 3x + 7) \div (x - 1)$
$= 5x + 2 + \dfrac{9}{x - 1}$

B.
$$
\begin{array}{r|rrrrr}
-2 & 3 & 0 & -8 & 2 & 1 \\
 & & -6 & 12 & -8 & 12 \\
\hline
 & 3 & -6 & 4 & -6 & 13
\end{array}
$$

• **Insert a zero for the missing term.**

$x - a = x + 2; a = -2$

$(3x^4 - 8x^2 + 2x + 1) \div (x + 2)$
$= 3x^3 - 6x^2 + 4x - 6 + \dfrac{13}{x + 2}$

PROBLEM 4 Divide: **A.** $(6x^2 + 8x - 5) \div (x + 2)$
B. $(2x^4 - 3x^3 - 8x^2 - 2) \div (x - 3)$

Solution See page S20.

OBJECTIVE **4** **Evaluate a polynomial using synthetic division**

A polynomial can be evaluated by using synthetic division. Consider the polynomial $P(x) = 2x^4 - 3x^3 + 4x^2 - 5x + 1$. One way to evaluate the polynomial when $x = 2$ is to replace x by 2 and then simplify the numerical expression.

$$P(x) = 2x^4 - 3x^3 + 4x^2 - 5x + 1$$
$$P(2) = 2(2)^4 - 3(2)^3 + 4(2)^2 - 5(2) + 1$$
$$P(2) = 2(16) - 3(8) + 4(4) - 5(2) + 1$$
$$P(2) = 32 - 24 + 16 - 10 + 1 = 15$$

Now use synthetic division to divide $(2x^4 - 3x^3 + 4x^2 - 5x + 1) \div (x - 2)$.

$$
\begin{array}{r|rrrrr}
2 & 2 & -3 & 4 & -5 & 1 \\
 & & 4 & 2 & 12 & 14 \\
\hline
 & 2 & 1 & 6 & 7 & 15 \\
\end{array}
$$

Coefficients of Remainder
the quotient

Note that the remainder is 15, which is the same value as $P(2)$. This is not a coincidence. The following theorem states that this situation is always true.

> **Remainder Theorem**
>
> If the polynomial $P(x)$ is divided by $x - a$, the remainder is $P(a)$.

Evaluate $P(x) = x^4 - 3x^2 + 4x - 5$ when $x = -2$ by using the Remainder Theorem.

The value at which the polynomial is evaluated

$$
\begin{array}{r|rrrrr}
-2 & 1 & 0 & -3 & 4 & -5 \\
 & & -2 & 4 & -2 & -4 \\
\hline
 & 1 & -2 & 1 & 2 & -9 \\
\end{array}
$$ A 0 is inserted for the x^3 term.

← The remainder

$P(-2) = -9$

EXAMPLE 5 Use the Remainder Theorem to evaluate $P(-2)$ when $P(x) = x^3 - 3x^2 + x + 3$.

Solution Use synthetic division with $a = -2$.

$$
\begin{array}{r|rrrr}
-2 & 1 & -3 & 1 & 3 \\
 & & -2 & 10 & -22 \\
\hline
 & 1 & -5 & 11 & -19 \\
\end{array}
$$

By the Remainder Theorem, $P(-2) = -19$.

PROBLEM 5 Use the Remainder Theorem to evaluate $P(3)$ when $P(x) = 2x^3 - 4x - 5$.

Solution See page S20.

Concept Review

Determine whether the statement is always true, sometimes true, or never true.

1. For $c \neq 0$, $\dfrac{a - b}{c} = \dfrac{a}{c} - \dfrac{b}{c}$.

2. For $x \neq 0$, $\dfrac{9x^2 + 6x}{3x} = 3x + 6x$.

3. Given $(6x^2 + x - 6) \div (3x + 2) = 2x - 1 - \dfrac{4}{3x + 2}$, then

$(3x + 2)(2x - 1) - \dfrac{4}{3x + 2} = 6x^2 + x - 6$.

5.5 Exercises

1 Divide a polynomial by a monomial

1. Replace each question mark to make a true statement.

$$\dfrac{18y^5 + 3y}{3y} = \dfrac{18y^5}{?} + \dfrac{3y}{?}$$
$$= \underline{} + \underline{}$$

2. Replace each question mark to make a true statement.

$$\dfrac{12x^3 - 8x^2}{4x^2} = \dfrac{12x^3}{?} - \dfrac{8x^2}{?}$$
$$= \underline{} - \underline{}$$

Divide.

3. $\dfrac{2x + 2}{2}$

4. $\dfrac{5y + 5}{5}$

5. $\dfrac{10a - 25}{5}$

6. $\dfrac{16b - 40}{8}$

7. $\dfrac{3a^2 + 2a}{a}$

8. $\dfrac{6y^2 + 4y}{y}$

9. $\dfrac{4b^3 - 3b}{b}$

10. $\dfrac{12x^2 - 7x}{x}$

11. $\dfrac{3x^2 - 6x}{3x}$

12. $\dfrac{10y^2 - 6y}{2y}$

13. $\dfrac{5x^2 - 10x}{-5x}$

14. $\dfrac{3y^2 - 27y}{-3y}$

15. $\dfrac{x^3 + 3x^2 - 5x}{x}$

16. $\dfrac{a^3 - 5a^2 + 7a}{a}$

17. $\dfrac{x^6 - 3x^4 - x^2}{x^2}$

18. $\dfrac{a^8 - 5a^5 - 3a^3}{a^2}$

19. $\dfrac{5x^2y^2 + 10xy}{5xy}$

20. $\dfrac{8x^2y^2 - 24xy}{8xy}$

21. $\dfrac{9y^6 - 15y^3}{-3y^3}$

22. $\dfrac{4x^4 - 6x^2}{-2x^2}$

23. $\dfrac{3x^2 - 2x + 1}{x}$

24. $\dfrac{8y^2 + 2y - 3}{y}$

25. $\dfrac{-3x^2 + 7x - 6}{x}$

26. $\dfrac{2y^2 - 6y + 9}{y}$

27. $\dfrac{16a^2b - 20ab + 24ab^2}{4ab}$

28. $\dfrac{22a^2b + 11ab - 33ab^2}{11ab}$

29. $\dfrac{9x^2y + 6xy - 3xy^2}{xy}$

30. Every division equation has a related multiplication equation. For instance, $\dfrac{16}{2} = 8$ means that $16 = 8 \cdot 2$. What is the related multiplication equation for $\dfrac{15x^2 + 12x}{3x} = 5x + 4$?

31. How can multiplication be used to check that $\dfrac{8x^3 - 12x^2 - 4x}{4x} = 2x^2 - 3x - 1$?

2 Divide polynomials

32. Let P and Q be polynomials. To find the quotient $P \div Q$ using long division, place ____?____ inside the long-division symbol and ____?____ outside the long-division symbol.

33. Use this long division:

$$
\begin{array}{r}
3x \phantom{{}+ 7} \\
x - 2 \overline{\smash{)}3x^2 - 4x + 7} \\
\underline{3x^2 - 6x} \phantom{{}+ 7} \\
2x + 7
\end{array}
$$

a. The term $3x$ in the quotient came from dividing ____?____ by ____?____.

b. The polynomial $3x^2 - 6x$ came from finding the product of ____?____ and ____?____.

c. The term $2x$ in the last line came from subtracting ____?____ − (____?____).

d. Finish the division. The quotient is ____?____ with a remainder of ____?____.

34. Explain how you can check the result of dividing two polynomials.

Divide using long division.

35. $(x^2 + 3x - 40) \div (x - 5)$

36. $(x^2 - 14x + 24) \div (x - 2)$

37. $(x^3 - 3x^2 + 2) \div (x - 3)$

38. $(x^3 + 4x^2 - 8) \div (x + 4)$

39. $(6x^2 + 13x + 8) \div (2x + 1)$

40. $(12x^2 + 13x - 14) \div (3x - 2)$

41. $(10x^2 + 9x - 5) \div (2x - 1)$

42. $(18x^2 - 3x + 2) \div (3x + 2)$

43. $(8x^3 - 9) \div (2x - 3)$

44. $(64x^3 + 4) \div (4x + 2)$

45. $(6x^4 - 13x^2 - 4) \div (2x^2 - 5)$

46. $(12x^4 - 11x^2 + 10) \div (3x^2 + 1)$

47. $\dfrac{3x^3 - 8x^2 - 33x - 10}{3x + 1}$

48. $\dfrac{8x^3 - 38x^2 + 49x - 10}{4x - 1}$

49. $\dfrac{4 - 7x + 5x^2 - x^3}{x - 3}$

50. $\dfrac{4 + 6x - 3x^2 + 2x^3}{2x + 1}$

51. $\dfrac{16x^2 - 13x^3 + 2x^4 - 9x + 20}{x - 5}$

52. $\dfrac{x + 3x^4 - x^2 + 5x^3 - 2}{x + 2}$

53. $\dfrac{x^3 - 4x^2 + 2x - 1}{x^2 + 1}$

54. $\dfrac{3x^3 - 2x^2 - 8}{x^2 + 5}$

55. $\dfrac{2x^3 - x + 4 - 3x^2}{x^2 - 1}$

56. $\dfrac{2 - 3x^2 + 5x^3}{x^2 + 3}$

57. $\dfrac{6x^3 + 2x^2 + x + 4}{2x^2 - 3}$

58. $\dfrac{9x^3 + 6x^2 + 2x + 1}{3x^2 + 2}$

59. 🗣 True or false? When a tenth-degree polynomial is divided by a second-degree polynomial, the quotient is a fifth-degree polynomial.

60. 🗣 True or false? When a polynomial of degree $3n$ is divided by a polynomial of degree n, the degree of the quotient polynomial is $2n$.

3 Synthetic division

61. Synthetic division is a shorter method of dividing a polynomial by a _____?_____ of the form _____?_____.

62. Use this display of the synthetic division of two polynomials in x:

$$
\begin{array}{r|rrr}
-2 & 4 & -1 & 3 \\
 & & -8 & 18 \\
\hline
 & 4 & -9 & 21
\end{array}
$$

a. The dividend is the polynomial __?__ .
b. The divisor is $x -$ __?__ $=$ __?__ .
c. The -8 comes from multiplying __?__ and __?__ .
d. The -9 comes from adding __?__ and __?__ .
e. The quotient is the polynomial __?__ plus the remainder __?__ .

Divide using synthetic division.

63. $(2x^2 - 6x - 8) \div (x + 1)$

64. $(3x^2 + 19x + 20) \div (x + 5)$

65. $(3x^2 - 4) \div (x - 1)$

66. $(4x^2 - 8) \div (x - 2)$

67. $(x^2 - 9) \div (x + 4)$

68. $(x^2 - 49) \div (x + 5)$

69. $(2x^2 + 24) \div (2x + 4)$

70. $(3x^2 - 15) \div (x + 3)$

71. $(2x^3 - x^2 + 6x + 9) \div (x + 1)$

72. $(3x^3 + 10x^2 + 6x - 4) \div (x + 2)$

73. $(x^3 - 6x^2 + 11x - 6) \div (x - 3)$

74. $(x^3 - 4x^2 + x + 6) \div (x + 1)$

75. $(6x - 3x^2 + x^3 - 9) \div (x + 2)$

76. $(5 - 5x + 4x^2 + x^3) \div (x - 3)$

77. $(x^3 + x - 2) \div (x + 1)$

78. $(x^3 + 2x + 5) \div (x - 2)$

79. $(18 + x - 4x^3) \div (2 - x)$

80. $(12 - 3x^2 + x^3) \div (3 + x)$

81. $\dfrac{16x^2 - 13x^3 + 2x^4 - 9x + 20}{x - 5}$

82. $\dfrac{2x^3 - x^2 - 10x + 15 + x^4}{x - 2}$

83. $\dfrac{5 + 5x - 8x^2 + 4x^3 - 3x^4}{2 - x}$

84. $\dfrac{3 - 13x - 5x^2 + 9x^3 - 2x^4}{3 - x}$

85. $\dfrac{3x^4 + 3x^3 - x^2 + 3x + 2}{x + 1}$

86. $\dfrac{4x^4 + 12x^3 - x^2 - x + 2}{x + 3}$

87. $\dfrac{2x^4 - x^2 + 2}{x - 3}$

88. $\dfrac{x^4 - 3x^3 - 30}{x + 2}$

89. $\dfrac{x^3 + 125}{x + 5}$

90. $\dfrac{x^3 + 343}{x + 7}$

91. Suppose you know the number to the left of the vertical line in a synthetic division display of $P(x) \div Q(x)$, and you know the numbers below the horizontal line. Do you have enough information to determine $P(x)$?

92. Suppose you know the number to the left of the vertical line in a synthetic division display of $P(x) \div Q(x)$, and you know the numbers in the row above the horizontal line. Do you have enough information to determine $P(x)$?

4 Evaluate a polynomial using synthetic division

93. Finish the synthetic division of $P(x) = 3x^2 - 2x^2 + x - 9$ by $x - 2$.

$$
\begin{array}{r|rrrr}
2 & 3 & -2 & 1 & -9 \\
 & & 6 & ? & ? \\
\hline
 & 3 & 4 & ? & ? \\
\end{array}
$$

The value of $P(2)$ is _____?_____ .

94. Use the synthetic division display to state one point that you know is on the graph of $y = x^3 + x^2 - 9x - 9$.

$$
\begin{array}{r|rrrr}
1 & 1 & 1 & -9 & -9 \\
 & & 1 & 2 & -7 \\
\hline
 & 1 & 2 & -7 & -16 \\
\end{array}
$$

Use the Remainder Theorem to evaluate the polynomial.

95. $P(x) = 2x^2 - 3x - 1$; $P(3)$

96. $Q(x) = 3x^2 - 5x - 1$; $Q(2)$

97. $R(x) = x^3 - 2x^2 + 3x - 1$; $R(4)$

98. $F(x) = x^3 + 4x^2 - 3x + 2$; $F(3)$

99. $P(z) = 2z^3 - 4z^2 + 3z - 1$; $P(-2)$

100. $R(t) = 3t^3 + t^2 - 4t + 2$; $R(-3)$

101. $Q(x) = x^4 + 3x^3 - 2x^2 + 4x - 9$; $Q(2)$

102. $Y(z) = z^4 - 2z^3 - 3z^2 - z + 7$; $Y(3)$

103. $F(x) = 2x^4 - x^3 - 2x - 5$; $F(-3)$

104. $Q(x) = x^4 - 2x^3 + 4x - 2$; $Q(-2)$

105. $P(x) = x^3 - 3$; $P(5)$

106. $S(t) = 4t^3 + 5$; $S(-4)$

107. $R(t) = 4t^4 - 3t^2 + 5$; $R(-3)$

108. $P(z) = 2z^4 + z^2 - 3$; $P(-4)$

109. $Q(x) = x^5 - 4x^3 - 2x^2 + 5x - 2$; $Q(2)$

110. $T(x) = 2x^5 + 4x^4 - x^2 + 4$; $T(3)$

5.5 Applying Concepts

For what value of k will the remainder be zero?

111. $(x^3 - 3x^2 - x + k) \div (x - 3)$

112. $(x^3 - 2x^2 + x + k) \div (x - 2)$

113. $(x^2 + kx - 6) \div (x - 3)$

114. $(x^3 + kx + k - 1) \div (x - 1)$

In Exercises 115 and 116, replace the question marks with expressions to make a true statement.

115. If $\dfrac{x^2 - x - 6}{x - 3} = x + 2$, then $x^2 - x - 6 = (?)(?)$.

116. If $\dfrac{x^2 + 2x - 3}{x - 2} = x + 4 + \dfrac{5}{x - 2}$, then $x^2 + 2x - 3 = (?)(?) + ?$.

Solve.

117. The product of a monomial and $4b$ is $12a^2b$. Find the monomial.

118. The product of a monomial and $6x$ is $24xy^2$. Find the monomial.

119. The quotient of a polynomial and $2x + 1$ is $2x - 4 + \dfrac{7}{2x + 1}$. Find the polynomial.

120. The quotient of a polynomial and $x - 3$ is $x^2 - x + 8 + \dfrac{22}{x - 3}$. Find the polynomial.

Focus on Problem Solving

Dimensional Analysis In solving application problems, it may be useful to include units in order to organize the problem so that the answer is in the proper units. Using units to organize and check the correctness of an application is called **dimensional analysis.** We use the operations of multiplying units and dividing units in applying dimensional analysis to application problems.

The Rule for Multiplying Exponential Expressions states that we multiply two expressions with the same base by adding the exponents.

$$x^4 \cdot x^6 = x^{4+6} = x^{10}$$

In calculations that involve quantities, the units are operated on algebraically.

> A rectangle measures 3 m by 5 m. Find the area of the rectangle.
>
> $$A = LW = (3 \text{ m})(5 \text{ m}) = (3 \cdot 5)(\text{m} \cdot \text{m}) = 15 \text{ m}^2$$
>
> The area of the rectangle is 15 m^2 (**square meters**).

5 m

3 m

> A box measures 10 cm by 5 cm by 3 cm. Find the volume of the box.
>
> $$V = LWH = (10 \text{ cm})(5 \text{ cm})(3 \text{ cm}) = (10 \cdot 5 \cdot 3)(\text{cm} \cdot \text{cm} \cdot \text{cm}) = 150 \text{ cm}^3$$
>
> The volume of the box is 150 cm^3 (**cubic centimeters**).

3 cm

5 cm

10 cm

> Find the area of a square whose side measures $(3x + 5)$ in.
>
> $$A = s^2 = [(3x + 5) \text{ in.}]^2 = (3x + 5)^2 \text{ in}^2 = (9x^2 + 30x + 25) \text{ in}^2$$
>
> The area of the square is $(9x^2 + 30x + 25) \text{ in}^2$ (**square inches**).

$(3x + 5)$ in.

Dimensional analysis is used in the conversion of units.

$\dfrac{1 \text{ mi}}{5280 \text{ ft}}$

The following example converts the unit miles to feet. The equivalent measures 1 mi and 5280 ft are used to form the following rates, which are called conversion factors: $\dfrac{1 \text{ mi}}{5280 \text{ ft}}$ and $\dfrac{5280 \text{ ft}}{1 \text{ mi}}$. Because 1 mi = 5280 ft, both of the conversion factors $\dfrac{1 \text{ mi}}{5280 \text{ ft}}$ and $\dfrac{5280 \text{ ft}}{1 \text{ mi}}$ are equal to 1.

To convert 3 mi to feet, multiply 3 mi by the conversion factor $\dfrac{5280 \text{ ft}}{1 \text{ mi}}$.

3 mi
15,840 ft

$$3 \text{ mi} = 3 \text{ mi} \cdot 1 = \dfrac{3 \text{ mi}}{1} \cdot \dfrac{5280 \text{ ft}}{1 \text{ mi}} = \dfrac{3 \text{ mi} \cdot 5280 \text{ ft}}{1 \text{ mi}} = 3 \cdot 5280 \text{ ft} = 15{,}840 \text{ ft}$$

There are two important points in the above illustration. First, you can think of dividing the numerator and denominator by the common unit "mile" just as you would divide the numerator and denominator of a fraction by a common factor. Second, the conversion factor $\dfrac{5280 \text{ ft}}{1 \text{ mi}}$ is equal to 1, and multiplying an expression by 1 does not change the value of the expression.

In the application problem that follows, the units are kept in the problem while the problem is worked.

In 2005, a horse named Afleet Alex won the Preakness, covering the 1.1875-mile course in 1.92 min. Find Afleet Alex's speed in miles per hour. Round to the nearest tenth.

Strategy To find the average speed, use the formula $r = \dfrac{d}{t}$, where r is the speed, d is the distance, and t is the time. Use the conversion factor $\dfrac{60 \text{ min}}{1 \text{ h}}$.

Afleet Alex

Solution $r = \dfrac{d}{t} = \dfrac{1.1875 \text{ mi}}{1.92 \text{ min}} = \dfrac{1.1875 \text{ mi}}{1.92 \text{ min}} \cdot \dfrac{60 \text{ min}}{1 \text{ h}}$

$\phantom{\text{Solution}\quad r} = \dfrac{71.25 \text{ mi}}{1.92 \text{ h}} \approx 37.1 \text{ mph}$

Afleet Alex's average speed was 37.1 mph.

Try each of the following problems. Round to the nearest tenth or to the nearest cent.

1. Convert 88 ft/s to miles per hour.

2. Convert 8 m/s to kilometers per hour (1 km = 1000 m).

3. A carpet is to be placed in a meeting hall that is 36 ft wide and 80 ft long. At $21.50 per square yard, how much will it cost to carpet the meeting hall?

4. A carpet is to be placed in a room that is 20 ft wide and 30 ft long. At $22.25 per square yard, how much will it cost to carpet the area?

5. Find the number of gallons of water in a fish tank that is 36 in. long and 24 in. wide and is filled to a depth of 16 in. (1 gal = 231 in³).

6. Find the number of gallons of water in a fish tank that is 24 in. long and 18 in. wide and is filled to a depth of 12 in. (1 gal = 231 in³).

7. A $\frac{1}{4}$-acre commercial lot is on sale for $2.15 per square foot. Find the sale price of the commercial lot (1 acre = 43,560 ft²).

8. A 0.75-acre industrial parcel was sold for $98,010. Find the parcel's price per square foot (1 acre = 43,560 ft²).

9. A new driveway will require 800 ft³ of concrete. Concrete is ordered by the cubic yard. How much concrete should be ordered?

10. In 2005, Catherine Ndereba won her fourth Boston Marathon. She ran the 26-mile course in 2.42 h. Find Catherine Ndereba's speed in feet per second. Round to the nearest tenth.

11. The Marianas Trench in the Pacific Ocean is the deepest part of the ocean. Its depth is 6.85 mi. The speed of sound under water is 4700 ft/s. Find the time it takes sound to travel from the surface to the bottom of the Marianas Trench and back.

Projects & Group Activities

Population and Land Allocation*

Texas

Rhode Island

In this project, you are asked to determine hypothetical land allocation for the world's population today. Use the figure 6×10^9 for the current world population and the figure 2.9×10^8 for the current U.S. population. One square mile is approximately 2.8×10^7 ft².

1. If every person in the world moved to Texas and each person were given an equal amount of land, how many square feet of land would each person have? The area of Texas is 2.619×10^5 mi².

2. If every person in the United States moved to Rhode Island and each person were given an equal amount of land, how many square feet of land would each person have? The area of Rhode Island is 1.0×10^3 mi². Round to the nearest whole number.

3. Suppose every person in the world were given a plot of land the size of a two-car garage (22 ft × 22 ft).
 a. How many people would fit in a square mile? Round to the nearest hundred.
 b. How many square miles would be required to accommodate the entire world population? Round to the nearest hundred.

4. If the total land area of Earth were divided equally, how many acres of land would each person be allocated? Use a figure of 5.7×10^7 mi² for the land area of Earth. One acre is 43,560 ft². Round to the nearest tenth.

5. If every person on Earth were given a plot of land the size of a two-car garage, what would be the carrying capacity of Earth? Round to the nearest hundred billion.

**Source:* www.infoplease.com

Chapter 5 Summary

Key Words	**Examples**
A **monomial** is a number, a variable, or a product of numbers and variables. [5.1.1, p. 368]	5 is a number. y is a variable. $8a^2b^2$ is a product of numbers and variables. 5, y, and $8a^2b^2$ are monomials.
The **degree of a monomial** is the sum of the exponents of the variables. [5.1.1, p. 368]	The degree of $3x^4y^2$ is 7.
A **polynomial** is a variable expression in which the terms are monomials. A polynomial of one term is a **monomial.** A polynomial of two terms is a **binomial.** A polynomial of three terms is a **trinomial.** [5.1.1, p. 369]	5, y, and $8a^2b^2$ are monomials. $x + 9$, $y^2 - 3$, and $6a + 7b$ are binomials. $x^2 + 2x - 1$ is a trinomial.
The **degree of a polynomial** is the greatest of the degrees of any of its terms. [5.1.1, p. 369]	The degree of the polynomial $x^3 + 3x^2y^2 - 4xy - 3$ is 4.
The terms of a polynomial in one variable are usually arranged so that the exponents of the variable decrease from left to right. This is called **descending order.** [5.1.1, p. 369]	$7y^4 + 5y^3 - y^2 + 6y - 8$ is written in descending order.
A **polynomial** function is an expression whose terms are monomials. The **leading coefficient** of a polynomial function is the coefficient of the variable with the largest exponent. The **constant term** is the term without a variable. [5.1.1, pp. 369–370]	For the polynomial function $P(x) = 6x^3 - 5x^2 + 4x - 7$, the leading coefficient is 6 and the constant term is -7.
The **opposite of a polynomial** is the polynomial with the sign of every term changed to its opposite. [5.1.2, p. 373]	The opposite of the polynomial $x^2 - 3x + 4$ is $-x^2 + 3x - 4$.
Synthetic division is a shorter method of dividing a polynomial by a binomial of the form $x - a$. This method uses only the coefficients of the variable terms. [5.5.3, p. 411]	

Essential Rules and Procedures

To add polynomials, combine like terms. [5.1.2, p. 373]	$(8x^2 + 2x - 9) + (-3x^2 + 5x - 7)$ $= (8x^2 - 3x^2) + (2x + 5x) + (-9 - 7)$ $= 5x^2 + 7x - 16$
To subtract two polynomials, add the additive inverse of the second polynomial to the first polynomial. [5.1.2, p. 373]	$(3y^2 - 8y + 6) - (-y^2 + 4y - 5)$ $= (3y^2 - 8y + 6) + (y^2 - 4y + 5)$ $= 4y^2 - 12y + 11$
Rule for Multiplying Exponential Expressions [5.2.1, p. 380] $x^m \cdot x^n = x^{m+n}$	$b^5 \cdot b^4 = b^{5+4} = b^9$

Rule for Simplifying Powers of Exponential Expressions
$(x^m)^n = x^{mn}$ [5.2.2, p. 381]

$(y^3)^7 = y^{3(7)} = y^{21}$

Rule for Simplifying Powers of Products
$(x^m y^n)^p = x^{mp} y^{np}$ [5.2.2, p. 381]

$(x^6 y^4 z^5)^2 = x^{6(2)} y^{4(2)} z^{5(2)} = x^{12} y^8 z^{10}$

The FOIL Method
To find the product of two binomials, add the products of the **F**irst terms, the **O**uter terms, the **I**nner terms, and the **L**ast terms. [5.3.3, pp. 386–387]

$(4x + 3)(2x - 5)$
$= (4x)(2x) + (4x)(-5) + (3)(2x) + (3)(-5)$
$= 8x^2 - 20x + 6x - 15$
$= 8x^2 - 14x - 15$

The Sum and Difference of Two Terms
$(a + b)(a - b) = a^2 - b^2$ [5.3.4, p. 387]

$(5x + 3)(5x - 3) = (5x)^2 - 3^2$
$= 25x^2 - 9$

The Square of a Binomial
$(a + b)^2 = a^2 + 2ab + b^2$
$(a - b)^2 = a^2 - 2ab + b^2$ [5.3.4, p. 388]

$(6x - 5)^2 = (6x)^2 + 2(6x)(-5) + (-5)^2$
$= 36x^2 - 60x + 25$

Zero as an Exponent
For $x \neq 0$, $x^0 = 1$. [5.4.1, p. 396]

$17^0 = 1$ $(5y)^0 = 1, y \neq 0$

Definition of Negative Exponents
For $x \neq 0$, $x^{-n} = \dfrac{1}{x^n}$ and $\dfrac{1}{x^{-n}} = x^n$. [5.4.1, p. 397]

$x^{-6} = \dfrac{1}{x^6}$ and $\dfrac{1}{x^{-6}} = x^6$

Rule for Dividing Exponential Expressions
For $x \neq 0$, $\dfrac{x^m}{x^n} = x^{m-n}$. [5.4.1, p. 397]

$\dfrac{y^8}{y^3} = y^{8-3} = y^5$

Scientific Notation
To express a number in scientific notation, write it in the form $a \times 10^n$, where a is a number between 1 and 10, and n is an integer. If the number is greater than 10, the exponent on 10 will be positive. If the number is less than 1, the exponent on 10 will be negative.

$367{,}000{,}000 = 3.67 \times 10^8$

$0.0000059 = 5.9 \times 10^{-6}$

To change a number written in scientific notation to decimal notation, move the decimal point to the right if the exponent on 10 is positive and to the left if the exponent on 10 is negative. Move the decimal point the same number of places as the absolute value of the exponent on 10. [5.4.2, pp. 399–400]

$2.418 \times 10^7 = 24{,}180{,}000$

$9.06 \times 10^{-5} = 0.0000906$

To divide a polynomial by a monomial, divide each term in the numerator by the denominator and write the sum of the quotients. [5.5.1, p. 408]

$\dfrac{8xy^3 - 4y^2 + 12y}{4y}$

$= \dfrac{8xy^3}{4y} - \dfrac{4y^2}{4y} + \dfrac{12y}{4y}$

$= 2xy^2 - y + 3$

To check polynomial division, use the same equation used to check division of whole numbers:

$$(\text{Quotient} \times \text{divisor}) + \text{remainder} = \text{dividend}$$

[5.5.2, p. 409]

$$
\begin{array}{r}
x - 4 \\
x + 3 \overline{)x^2 - x - 10} \\
\underline{x^2 + 3x} \\
-4x - 10 \\
\underline{-4x - 12} \\
2
\end{array}
$$

Check:

$$(x - 4)(x + 3) + 2 = x^2 - x - 12 + 2$$
$$= x^2 - x - 10$$
$$(x^2 - x - 10) \div (x + 3)$$
$$= x - 4 + \frac{2}{x + 3}$$

Chapter 5 Review Exercises

1. Add: $(3x^2 - 2x - 6) + (-x^2 - 3x + 4)$

2. Subtract: $(5x^2 - 8xy + 2y^2) - (x^2 - 3y^2)$

3. Multiply: $(5x^2yz^4)(2xy^3z^{-1})(7x^{-2}y^{-2}z^3)$

4. Multiply: $(2x^{-1}y^2z^5)^4(-3x^3yz^{-3})$

5. Simplify: $\dfrac{3x^4yz^{-1}}{-12xy^3z^2}$

6. Simplify: $\dfrac{(2a^4b^{-3}c^2)^3}{(2a^3b^2c^{-1})^4}$

7. Write 93,000,000 in scientific notation.

8. Write 2.54×10^{-3} in decimal notation.

9. Simplify: $\dfrac{3 \times 10^{-3}}{15 \times 10^2}$

10. Given $P(x) = 2x^3 - x + 7$, evaluate $P(-2)$.

11. Graph: $y = x^2 + 1$

12. Identify **a.** the leading coefficient, **b.** the constant term, and **c.** the degree of the polynomial $P(x) = 3x^5 - 3x^2 + 7x + 8$.

13. Use the Remainder Theorem to evaluate $P(x) = -2x^3 + 2x^2 - 4$ when $x = -3$.

14. Divide: $\dfrac{15x^2 + 2x - 2}{3x - 2}$

15. Divide: $\dfrac{12x^2 - 16x - 7}{6x + 1}$

16. Divide: $\dfrac{4x^3 + 27x^2 + 10x + 2}{x + 6}$

17. Divide: $\dfrac{x^4 - 4}{x - 4}$

18. Multiply: $4x^2y(3x^3y^2 + 2xy - 7y^3)$

19. Multiply: $a^{2n+3}(a^n - 5a + 2)$

20. Simplify: $5x^2 - 4x[x - (3x + 2) + x]$

21. Multiply: $(x^{2n} - x)(x^{n+1} - 3)$

22. Multiply: $(x + 6)(x^3 - 3x^2 - 5x + 1)$

23. Multiply: $(x - 4)(3x + 2)(2x - 3)$

24. Multiply: $(5a + 2b)(5a - 2b)$

25. Simplify: $(4x - 3y)^2$

26. Astronomy The most distant object visible from Earth without the aid of a telescope is the Great Galaxy of Andromeda. It takes light from the Great Galaxy of Andromeda 2.2×10^6 years to travel to Earth. Light travels about 5.9×10^{12} mph. How far from Earth is the Great Galaxy of Andromeda?

27. Power Light from the sun supplies Earth with 2.4×10^{14} horsepower. Earth receives only 2.2×10^{-7} of the power generated by the sun. How much power is generated by the sun?

28. Geometry The length of a rectangle is $(5x + 3)$ cm. The width is $(2x - 7)$ cm. Find the area of the rectangle in terms of the variable x.

29. Geometry The length of the side of a cube is $(3x - 1)$ ft. Find the volume of the cube in terms of the variable x.

30. Geometry Find the area of the figure shown at the right. All dimensions given are in inches.

x

x

$x + 4$

$3x - 2$

Chapter 5 Test

1. Subtract:
$(6x^3 - 7x^2 + 6x - 7) - (4x^3 - 3x^2 + 7)$

2. Simplify: $(-4a^2b)^3(-ab^4)$

3. Simplify: $\dfrac{(2a^{-4}b^2)^3}{4a^{-2}b^{-1}}$

4. Write the number 0.000000501 in scientific notation.

5. Write the number of seconds in 1 week in scientific notation.

6. Simplify: $(2x^{-3}y)^{-4}$

7. Simplify: $-5x[3 - 2(2x - 4) - 3x]$

8. Multiply: $(3a + 4b)(2a - 7b)$

9. Multiply: $(3t^3 - 4t^2 + 1)(2t^2 - 5)$

10. Simplify: $(3z - 5)^2$

11. Divide: $(4x^3 + x - 15) \div (2x - 3)$

12. Divide: $(x^3 - 5x^2 + 5x + 5) \div (x - 3)$

13. Given $P(x) = 3x^2 - 8x + 1$, evaluate $P(2)$.

14. Use the Remainder Theorem to evaluate $P(x) = -x^3 + 4x - 8$ when $x = -2$.

15. Add: $(3x^3 - 2x^2 - 4) + (8x^2 - 8x + 7)$

16. Multiply: $2x(2x^2 - 3x)$

17. Simplify: $\dfrac{12x^2}{-3x^{-4}}$

18. Subtract: $(3a^2 - 2a - 7) - (5a^3 + 2a - 10)$

19. Divide: $\dfrac{16x^5 - 8x^3 + 20x}{4x}$

20. Multiply: $(-2xy^2)(3x^2y^4)$

21. Simplify: $\dfrac{27xy^3}{3x^4y^3}$

22. Multiply: $(2x - 7y)(5x - 4y)$

23. Simplify: $(a^2b^{-3})^2$

24. **Geometry** The length of a rectangle is $(5x + 1)$ ft. The width is $(2x - 1)$ ft. Find the area of the rectangle in terms of the variable x.

5x + 1

2x – 1

25. **Space Travel** A space vehicle travels 2.4×10^5 mi from Earth to the moon at an average velocity of 2×10^4 mph. How long does it take the space vehicle to reach the moon?

Cumulative Review Exercises

1. Simplify: $8 - 2[-3 - (-1)]^2 + 4$

2. Evaluate $\dfrac{2a - b}{b - c}$ when $a = 4$, $b = -2$, and $c = 6$.

3. Identify the property that justifies the statement $2x + (-2x) = 0$.

4. Simplify: $2x - 4[x - 2(3 - 2x) + 4]$

5. Solve: $\dfrac{2}{3} - y = \dfrac{5}{6}$

6. Solve: $8x - 3 - x = -6 + 3x - 8$

7. Simplify: $\dfrac{x^3 - 3}{x - 3}$

8. Solve: $3 - |2 - 3x| = -2$

9. Given $P(x) = 3x^2 - 2x + 2$, evaluate $P(-2)$.

10. What values of x are excluded from the domain of the function $f(x) = \dfrac{x + 1}{x + 2}$?

11. Find the range of the function given by $F(x) = 3x^2 - 4$ if the domain is $\{-2, -1, 0, 1, 2\}$.

12. Find the slope of the line that contains the points $(-2, 3)$ and $(4, 2)$.

13. Find the equation of the line that contains the point $(-1, 2)$ and has slope $-\dfrac{3}{2}$.

14. Find the equation of the line that contains the point $(-2, 4)$ and is perpendicular to the line $3x + 2y = 4$.

15. Solve by using Cramer's Rule: $2x - 3y = 2$
$x + y = -3$

16. Solve by the addition method:
$x - y + z = 0$
$2x + y - 3z = -7$
$-x + 2y + 2z = 5$

17. Graph $3x - 4y = 12$ by using the x- and y-intercepts.

18. Graph the solution set of $-3x + 2y < 6$.

19. Solve by graphing: $x - 2y = 3$
$-2x + y = -3$

20. Graph the solution set: $2x + y < 3$
$-6x + 3y \geq 4$

21. Simplify: $(4a^{-2}b^3)(2a^3b^{-1})^{-2}$

22. Simplify: $\dfrac{(5x^3y^{-3}z)^{-2}}{y^4z^{-2}}$

23. Simplify: $3 - (3 - 3^{-1})^{-1}$

24. Multiply: $(2x + 3)(2x^2 - 3x + 1)$

25. Divide: $(8x^2 + 4x - 3) \div (2x - 3)$

26. Multiply: $(a^2b^5)(ab^2)$

27. Multiply: $(a - 3b)(a + 4b)$

28. Write 3.5×10^{-8} in decimal notation.

29. Integer Problem The sum of two integers is twenty-four. The difference between four times the smaller integer and nine is three less than twice the larger integer. Find the integers.

30. Metallurgy How many ounces of pure gold that costs $360 per ounce must be mixed with 80 oz of an alloy that costs $120 per ounce to make a mixture that costs $200 per ounce?

31. Uniform Motion Two bicycles are 25 mi apart and are traveling toward each other. One cyclist is traveling at $\frac{2}{3}$ the rate of the other cyclist. They pass in 2 h. Find the rate of each cyclist.

32. Investments If $3000 is invested at an annual simple interest rate of 7.5%, how much additional money must be invested at an annual simple interest rate of 10% so that the total interest earned is 9% of the total investment?

33. Uniform Motion The graph below shows the relationship between the distance traveled in miles and the time of travel in hours. Find the slope of the line between the two points on the graph. Write a sentence that states the meaning of the slope.

6

Factoring

DVD SSM

Online Study Center

Need help? For online student resources, such as section quizzes, visit Online Study Center at college.hmco.com/pic/aufmannABI2e.

In 2008, Iowa State will host the first and second rounds of the NCAA Division I women's basketball tournament in Des Moines, Iowa. The teams who will compete in this tournament will be determined by their rankings after the 2008 regular season. Imagine that you have just gotten a job as the manager for NCAA Division I women's basketball, and that you have to organize the game schedule for the upcoming season. How would you go about creating this schedule? **Exercise 88 on page 472** shows you how to use a formula to determine the number of league games that must be scheduled if each team is to play each of the other teams once.

Prep Test

1. Write 30 as a product of prime numbers.

2. Simplify: $-3(4y - 5)$

3. Simplify: $-(a - b)$

4. Simplify: $2(a - b) - 5(a - b)$

5. Solve: $4x = 0$

6. Solve: $2x + 1 = 0$

7. Multiply: $(x + 4)(x - 6)$

8. Multiply: $(2x - 5)(3x + 2)$

9. Simplify: $\dfrac{x^5}{x^2}$

10. Simplify: $\dfrac{6x^4y^3}{2xy^2}$

6.1 Common Factors

OBJECTIVE 1

Factor a monomial from a polynomial

Take Note

12 is the GCF of 24 and 60 because 12 is the largest integer that divides evenly into both 24 and 60.

The **greatest common factor (GCF)** of two or more integers is the greatest integer that is a factor of all the integers.

$$24 = 2 \cdot 2 \cdot 2 \cdot 3$$
$$60 = 2 \cdot 2 \cdot 3 \cdot 5$$
$$\text{GCF} = 2 \cdot 2 \cdot 3 = 12$$

The GCF of two or more monomials is the product of the GCF of the coefficients and the common variable factors.

$$6x^3y = 2 \cdot 3 \cdot x \cdot x \cdot x \cdot y$$
$$8x^2y^2 = 2 \cdot 2 \cdot 2 \cdot x \cdot x \cdot y \cdot y$$
$$\text{GCF} = 2 \cdot x \cdot x \cdot y = 2x^2y$$

Note that the exponent of each variable in the GCF is the same as the *smallest* exponent of that variable in either of the monomials.

The GCF of $6x^3y$ and $8x^2y^2$ is $2x^2y$.

EXAMPLE 1 Find the GCF of $12a^4b$ and $18a^2b^2c$.

Solution $12a^4b = 2 \cdot 2 \cdot 3 \cdot a^4 \cdot b$ • **Factor each monomial.**
$18a^2b^2c = 2 \cdot 3 \cdot 3 \cdot a^2 \cdot b^2 \cdot c$

$\text{GCF} = 2 \cdot 3 \cdot a^2 \cdot b = 6a^2b$ • **The common variable factors are a^2 and b. c is not a common factor.**

PROBLEM 1 Find the GCF of $4x^6y$ and $18x^2y^6$.

Solution See page S20.

The Distributive Property is used to multiply factors of a polynomial. To **factor a polynomial** means to write the polynomial as a product of other polynomials.

In the example above, $2x$ is the GCF of the terms $2x^2$ and $10x$. It is a **common monomial factor** of the terms. $x + 5$ is a **binomial factor** of $2x^2 + 10x$.

EXAMPLE 2 Factor. **A.** $5x^3 - 35x^2 + 10x$ **B.** $16x^2y + 8x^4y^2 - 12x^4y^5$

Solution **A.** The GCF is $5x$.
- **Find the GCF of the terms of the polynomial.**

$$\frac{5x^3}{5x} = x^2, \frac{-35x^2}{5x} = -7x, \frac{10x}{5x} = 2$$
- **Divide each term of the polynomial by the GCF.**

$5x^3 - 35x^2 + 10x$
- **Use the quotients to rewrite the polynomial, expressing each term as a product with the GCF as one of the factors.**

$$= 5x(x^2) + 5x(-7x) + 5x(2)$$

$$= 5x(x^2 - 7x + 2)$$
- **Use the Distributive Property to write the polynomial as a product of factors.**

B. $16x^2y = 2 \cdot 2 \cdot 2 \cdot 2 \cdot x^2 \cdot y$
$8x^4y^2 = 2 \cdot 2 \cdot 2 \cdot x^4 \cdot y^2$
$12x^4y^5 = 2 \cdot 2 \cdot 3 \cdot x^4 \cdot y^5$
The GCF is $4x^2y$.

$16x^2y + 8x^4y^2 - 12x^4y^5$
$= 4x^2y(4) + 4x^2y(2x^2y) + 4x^2y(-3x^2y^4)$
$= 4x^2y(4 + 2x^2y - 3x^2y^4)$

PROBLEM 2 Factor. **A.** $14a^2 - 21a^4b$ **B.** $6x^4y^2 - 9x^3y^2 + 12x^2y^4$

Solution See page S20.

OBJECTIVE 2 **Factor by grouping**

In the examples below, the binomials in parentheses are called **binomial factors**.

$$2a(a + b)$$
$$3xy(x - y)$$

The Distributive Property is used to factor a common binomial factor from an expression.

In the expression at the right, the common binomial factor is $y - 3$. The Distributive Property is used to write the expression as a product of factors.

$x(y - 3) + 4(y - 3)$
$= (y - 3)(x + 4)$

EXAMPLE 3 Factor: $y(x + 2) + 3(x + 2)$

Solution $y(x + 2) + 3(x + 2)$
$= (x + 2)(y + 3)$
- **The common binomial factor is $x + 2$.**

> **PROBLEM 3** Factor: $a(b - 7) + b(b - 7)$
>
> ■ **Solution** See page S20.

Sometimes a binomial factor must be rewritten before a common binomial factor can be found.

Factor: $a(a - b) + 5(b - a)$

$a - b$ and $b - a$ are different binomials.

Note that $(b - a) = (-a + b) = -(a - b)$.

Rewrite $(b - a)$ as $-(a - b)$ so that there is a common factor.

$$a(a - b) + 5(b - a) = a(a - b) + 5[-(a - b)]$$
$$= a(a - b) - 5(a - b)$$
$$= (a - b)(a - 5)$$

EXAMPLE 4 Factor: $2x(x - 5) + y(5 - x)$

Solution $2x(x - 5) + y(5 - x)$
$\quad = 2x(x - 5) - y(x - 5)$ • Rewrite $5 - x$ as $-(x - 5)$ so that there is a common factor.
$\quad = (x - 5)(2x - y)$ • Write the expression as a product of factors.

PROBLEM 4 Factor: $3y(5x - 2) - 4(2 - 5x)$

■ **Solution** See page S20.

Some polynomials can be factored by grouping the terms so that a common binomial factor is found.

Factor: $2x^3 - 3x^2 + 4x - 6$

$$2x^3 - 3x^2 + 4x - 6$$

Group the first two terms and the last two terms. $= (2x^3 - 3x^2) + (4x - 6)$

Factor out the GCF from each group. $= x^2(2x - 3) + 2(2x - 3)$

Write the expression as a product of factors. $= (2x - 3)(x^2 + 2)$

EXAMPLE 5 Factor: $3y^3 - 4y^2 - 6y + 8$

Solution $3y^3 - 4y^2 - 6y + 8$
$\quad = (3y^3 - 4y^2) - (6y - 8)$ • Group the first two terms and the last two terms. Note that $-6y + 8 = -(6y - 8)$.
$\quad = y^2(3y - 4) - 2(3y - 4)$ • Factor out the GCF from each group.
$\quad = (3y - 4)(y^2 - 2)$ • Write the expression as a product of factors.

PROBLEM 5 Factor: $y^5 - 5y^3 + 4y^2 - 20$

■ **Solution** See page S20.

6.1 Concept Review

Determine whether the statement is always true, sometimes true, or never true.

1. To factor a polynomial means to rewrite it as multiplication.

2. The expression $3x(x + 5)$ is a sum, and the expression $3x^2 + 15x$ is a product.

3. The greatest common factor of two numbers is the largest number that divides evenly into both numbers.

4. x is a monomial factor of $x(2x - 1)$. $2x - 1$ is a binomial factor of $x(2x - 1)$.

5. $y(y + 4) + 6$ is in factored form.

6. A common monomial factor is a factor of every term of a polynomial.

7. $a^2 - 8a - ab + 8b = a(a - 8) - b(a + 8)$

8. The binomial $d + 9$ is a common binomial factor of $2c(d + 9) - 7(d + 9)$.

6.1 Exercises

1 Factor a monomial from a polynomial

1. Explain the meaning of "a factor of a polynomial" and the meaning of "to factor a polynomial."

2. Explain why the statement is true.

 a. The terms of the binomial $3x - 6$ have a common factor.
 b. The expression $3x^2 + 15$ is not in factored form.
 c. $5y - 7$ is a factor of $y(5y - 7)$.

3. Use the three monomials $9x^3y^2$, $3xy^3z$, and $6x^2y^2z^2$.
 a. Of the variable factors x, y, and z, the one that is *not* a factor of the GCF of $9x^3y^2$, $3xy^3z$, and $6x^2y^2z^2$ is ___?___.
 b. The GCF of the coefficients 9, 3, and 6 is ___?___.
 c. The GCF of the factors x^3, x, and x^2 is ___?___.
 d. The GCF of the factors y^2, y^3, and y^2 is ___?___.
 e. Use your answers to parts a to d to write the GCF of $9x^3y^2$, $3xy^3z$, and $6x^2y^2z^2$: ___?___

4. a. Of the two expressions $25a^2 - 10a$ and $5a(5a - 2)$, the one that is written in *factored form* is ____?____.

b. Use the Distributive Property to confirm that the two expressions in part a are equivalent: $5a(5a - 2) = ($____?____$)(5a) - ($____?____$)(2) = $ ____?____ $-$ ____?____.

c. The polynomial $25a^2 - 10a$ has a monomial factor and a binomial factor. The monomial factor is ____?____ and the binomial factor is ____?____.

Find the greatest common factor.

5. x^7, x^3

6. y^6, y^{12}

7. x^2y^4, xy^6

8. a^5b^3, a^3b^8

9. $x^2y^4z^6, xy^8z^2$

10. ab^2c^3, a^3b^2c

11. $14a^3, 49a^7$

12. $12y^2, 27y^4$

13. $3x^2y^2, 5ab^2$

14. $8x^2y^3, 7ab^4$

15. $9a^2b^4, 24a^4b^2$

16. $15a^4b^2, 9ab^5$

17. $ab^3, 4a^2b, 12a^2b^3$

18. $12x^2y, x^4y, 16x$

19. $2x^2y, 4xy, 8x$

20. $16x^2, 8x^4y^2, 12xy$

21. $3x^2y^2, 6x, 9x^3y^3$

22. $4a^2b^3, 8a^3, 12ab^4$

Factor.

23. $5a + 5$

24. $7b - 7$

25. $16 - 8a^2$

26. $12 + 12y^2$

27. $8x + 12$

28. $16a - 24$

29. $30a - 6$

30. $20b + 5$

31. $7x^2 - 3x$

32. $12y^2 - 5y$

33. $3a^2 + 5a^5$

34. $9x - 5x^2$

35. $14y^2 + 11y$

36. $6b^3 - 5b^2$

37. $2x^4 - 4x$

38. $3y^4 - 9y$

39. $10x^4 - 12x^2$

40. $12a^5 - 32a^2$

41. $x^2y - xy^3$

42. $a^2b + a^4b^2$

43. $2a^5b + 3xy^3$

44. $5x^2y - 7ab^3$

45. $6a^2b^3 - 12b^2$

46. $8x^2y^3 - 4x^2$

47. $6a^2bc + 4ab^2c$

48. $10x^2yz^2 + 15xy^3z$

49. $18x^2y^2 - 9a^2b^2$

50. $9a^2x - 27a^3x^3$

51. $6x^3y^3 - 12x^6y^6$

52. $3a^2b^2 - 12a^5b^5$

53. $x^3 - 3x^2 - x$

54. $a^3 + 4a^2 + 8a$

55. $2x^2 + 8x - 12$

56. $a^3 - 3a^2 + 5a$

57. $b^3 - 5b^2 - 7b$

58. $5x^2 - 15x + 35$

59. $8y^2 - 12y + 32$

60. $3x^3 + 6x^2 + 9x$

61. $5y^3 - 20y^2 + 10y$

62. $2x^4 - 4x^3 + 6x^2$

63. $3y^4 - 9y^3 - 6y^2$

64. $2x^3 + 6x^2 - 14x$

65. $3y^3 - 9y^2 + 24y$

66. $2y^5 - 3y^4 + 7y^3$

67. $6a^5 - 3a^3 - 2a^2$

68. $x^3y - 3x^2y^2 + 7xy^3$

69. $8x^2y^2 - 4x^2y + x^2$

70. $x^4y^4 - 3x^3y^3 + 6x^2y^2$

71. $4x^5y^5 - 8x^4y^4 + x^3y^3$

72. $16x^2y - 8x^3y^4 - 48x^2y^2$

For Exercises 73 and 74, use the polynomial $x^a + x^b + x^c$, where a, b, and c are all positive integers such that $a < b < c$.

73. What is the GCF of the terms of the polynomial $x^a + x^b + x^c$?

74. Suppose $b = a + 1$ and $c = a + 2$. Write $x^a + x^b + x^c$ in factored form.

2 Factor by grouping

75. a. The expression $5x(3x - 2) + 4(3x - 2)$ is the sum of two terms. One term is $5x(3x - 2)$ and the other term is ___?___. The terms have the common binomial factor ___?___.

b. To write the expression $5x(3x - 2) + 4(3x - 2)$ in factored form, factor out the common binomial factor:
$5x(3x - 2) + 4(3x - 2) = (3x - 2)(\underline{} + \underline{})$.

76. To factor $7a^2 - 7a + 6a - 6$ by grouping, start by factoring $7a$ from the first two terms and 6 from the second two terms:
$7a^2 - 7a + 6a - 6 = (7a^2 - 7a) + (6a - 6) = 7a(\underline{} - \underline{}) +$
$6(\underline{} - \underline{}) = (\underline{})(7a + 6)$

Factor.

77. $x(a + b) + 2(a + b)$

78. $a(x + y) + 4(x + y)$

79. $x(b + 2) - y(b + 2)$

80. $d(z - 8) + 5(z - 8)$

81. $a(y - 4) - b(y - 4)$

82. $c(x - 6) - 7(x - 6)$

Rewrite the expression so that there is a common binomial factor.

83. $a(x - 2) - b(2 - x)$

84. $a(x - 2) + b(2 - x)$

85. $b(a - 7) + 3(7 - a)$

86. $b(a - 7) - 3(7 - a)$

87. $x(a - 2b) - y(2b - a)$

88. $x(a - 2b) + y(2b - a)$

Factor.

89. $a(x - 2) + 5(2 - x)$

90. $a(x - 7) + b(7 - x)$

91. $b(y - 3) + 3(3 - y)$

92. $c(a - 2) - b(2 - a)$

93. $a(x - y) - 2(y - x)$

94. $3(a - b) - x(b - a)$

95. $z(c + 5) - 8(5 + c)$

96. $y(ab + 4) - 7(4 + ab)$

97. $w(3x - 4) - (4 - 3x)$

98. $d(6y - 1) - (1 - 6y)$

99. $x^3 + 4x^2 + 3x + 12$

100. $x^3 - 4x^2 - 3x + 12$

101. $2y^3 + 4y^2 + 3y + 6$

102. $3y^3 - 12y^2 + y - 4$

103. $ab + 3b - 2a - 6$

104. $yz + 6z - 3y - 18$

105. $x^2a - 2x^2 - 3a + 6$

106. $x^2y + 4x^2 + 3y + 12$

107. $3ax - 3bx - 2ay + 2by$

108. $8 + 2c + 4a^2 + a^2c$

109. $x^2 - 3x + 4ax - 12a$

110. $t^2 + 4t - st - 4s$

111. $xy - 5y - 2x + 10$

112. $2y^2 - 10y + 7xy - 35x$

113. $21x^2 + 6xy - 49x - 14y$

114. $4a^2 + 5ab - 10b - 8a$

115. $2ra + a^2 - 2r - a$

116. $2ab - 3b^2 - 3b + 2a$

117. $4x^2 + 3xy - 12y - 16x$

118. $8s + 12r - 6s^2 - 9rs$

119. $10xy^2 - 15xy + 6y - 9$

120. $10a^2b - 15ab - 4a + 6$

121. Not all four-term expressions can be factored by grouping. Which of the following expressions can be factored by grouping?
I. $ab - 3a - 2b - 6$ **II.** $ab + 3a - 2b - 6$ **III.** $ab + 3a + 2b - 6$

122. **a.** Which expressions are equivalent to $x^2 - 5x + 6$?

I. $x^2 - 15x + 10x + 6$ **II.** $x^2 - x - 4x + 6$ **III.** $x^2 - 2x - 3x + 6$
b. Which expression in part a can be factored by grouping?

6.1 Applying Concepts

Factor by grouping.

123. a. $2x^2 + 6x + 5x + 15$ **b.** $2x^2 + 5x + 6x + 15$

124. a. $3x^2 + 3xy - xy - y^2$ **b.** $3x^2 - xy + 3xy - y^2$

125. a. $2a^2 - 2ab - 3ab + 3b^2$ **b.** $2a^2 - 3ab - 2ab + 3b^2$

Compare your answers to parts a and b of Exercises 123 to 125 in order to answer Exercise 126.

126. Do different groupings of the terms in a polynomial affect the binomial factoring?

A whole number is a perfect number if it is the sum of all of its factors that are less than itself. For example, 6 is a perfect number because all the factors of 6 that are less than 6 are 1, 2, and 3, and $1 + 2 + 3 = 6$.

127. Find a perfect number between 20 and 30. **128.** Find a perfect number between 490 and 500.

Solve.

129. Geometry In the equation $P = 2L + 2W$, what is the effect on P when the quantity $L + W$ doubles?

130. Geometry Write an expression in factored form for the shaded portion in the diagram.

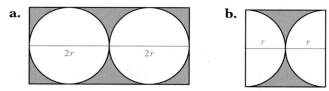

a. $2r$ $2r$

b. r r

<div style="border:1px solid;">6.2</div>

Factoring Polynomials of the Form $x^2 + bx + c$

OBJECTIVE **1** ## Factor trinomials of the form $x^2 + bx + c$

Trinomials of the form $x^2 + bx + c$, where b and c are integers, are shown at the right.

$x^2 + 9x + 14$, $b = 9$, $c = 14$
$x^2 - x - 12$, $b = -1$, $c = -12$
$x^2 - 2x - 15$, $b = -2$, $c = -15$

To factor a trinomial of this form means to express the trinomial as the product of two binomials. Some trinomials expressed as the product of binomials (factored form) are shown at the right.

Trinomial	Factored Form
$x^2 + 9x + 14$	$= (x + 2)(x + 7)$
$x^2 - x - 12$	$= (x + 3)(x - 4)$
$x^2 - 2x - 15$	$= (x + 3)(x - 5)$

The method by which the factors of a trinomial are found is based on FOIL. Consider the following binomial products, noting the relationship between the constant terms of the binomials and the terms of the trinomials.

Signs in the binomials are the same

$(x + 6)(x + 2) = x^2 + 2x + 6x + (6)(2) = x^2 + 8x + 12$
Sum of 6 and 2
Product of 6 and 2

$(x - 3)(x - 4) = x^2 - 4x - 3x + (-3)(-4) = x^2 - 7x + 12$
Sum of -3 and -4
Product of -3 and -4

Signs in the binomials are opposite

$(x + 3)(x - 5) = x^2 - 5x + 3x + (3)(-5) = x^2 - 2x - 15$
Sum of 3 and -5
Product of 3 and -5

$(x - 4)(x + 6) = x^2 + 6x - 4x + (-4)(6) = x^2 + 2x - 24$
Sum of -4 and 6
Product of -4 and 6

Points to Remember in Factoring $x^2 + bx + c$

1. In the trinomial, the coefficient of x is the sum of the constant terms of the binomials.

2. In the trinomial, the constant term is the product of the constant terms of the binomials.

3. When the constant term of the trinomial is positive, the constant terms of the binomials have the same sign as the coefficient of x in the trinomial.

4. When the constant term of the trinomial is negative, the constant terms of the binomials have opposite signs.

Success at factoring a trinomial depends on remembering these four points.

For example, to factor $x^2 - 2x - 24$, find two numbers whose sum is -2 and whose product is -24 [Points 1 and 2]. Because the constant term of the trinomial is negative (-24), the numbers will have opposite signs [Point 4].

A systematic method of finding these numbers involves listing the factors of the constant term of the trinomial and the sum of those factors.

Factors of -24	Sum of the Factors
1, -24	$1 + (-24) = -23$
-1, 24	$-1 + 24 = 23$
2, -12	$2 + (-12) = -10$
-2, 12	$-2 + 12 = 10$
3, -8	$3 + (-8) = -5$
-3, 8	$-3 + 8 = 5$
4, -6	**$4 + (-6) = -2$**
-4, 6	$-4 + 6 = 2$

<div style="float:left">

Take Note

Always check your proposed factorization to ensure accuracy.

</div>

4 and -6 are two numbers whose sum is -2 and whose product is -24. Write the binomial factors of the trinomial.

$$x^2 - 2x - 24 = (x + 4)(x - 6)$$

Check: $(x + 4)(x - 6) = x^2 - 6x + 4x - 24 = x^2 - 2x - 24$

By the Commutative Property of Multiplication, the binomial factors can also be written as

$$x^2 - 2x - 24 = (x - 6)(x + 4)$$

EXAMPLE 1 Factor: $x^2 + 18x + 32$

Solution

Factors of 32	Sum
1, 32	33
2, 16	**18**
4, 8	12

$x^2 + 18x + 32 = (x + 2)(x + 16)$

• Try only positive factors of 32 [Point 3].

• Once the correct pair is found, the other factors need not be tried.

• Write the factors of the trinomial.

Check $(x + 2)(x + 16) = x^2 + 16x + 2x + 32$
$= x^2 + 18x + 32$

PROBLEM 1 Factor: $x^2 - 8x + 15$

Solution See page S20.

EXAMPLE 2 Factor: $x^2 - 6x - 16$

Solution

Factors of -16	Sum
1, -16	-15
-1, 16	15
2, -8	**-6**
-2, 8	6
4, -4	0

• The factors must be of opposite signs [Point 4].

$x^2 - 6x - 16 = (x + 2)(x - 8)$

• Write the factors of the trinomial.

Check $(x + 2)(x - 8) = x^2 - 8x + 2x - 16$
$$= x^2 - 6x - 16$$

PROBLEM 2 Factor: $x^2 + 3x - 18$

Solution See page S21.

Not all trinomials can be factored when using only integers. Consider the trinomial $x^2 - 6x - 8$.

Factors of -8	Sum
1, -8	-7
-1, 8	7
2, -4	-2
-2, 4	2

Because none of the pairs of factors of -8 have a sum of -6, the trinomial is not factorable using integers. The trinomial is said to be **nonfactorable over the integers.**

OBJECTIVE 2 **Factor completely**

A polynomial is **factored completely** when it is written as a product of factors that are nonfactorable over the integers.

Take Note

The first step in *any* factoring problem is to determine whether the terms of the polynomial have a *common factor*. If they do, factor it out first.

EXAMPLE 3 Factor: $3x^3 + 15x^2 + 18x$

Solution The GCF of $3x^3$, $15x^2$, and $18x$ is $3x$.
• Find the GCF of the terms of the polynomial.

$3x^3 + 15x^2 + 18x$
$= 3x(x^2) + 3x(5x) + 3x(6)$
$= 3x(x^2 + 5x + 6)$
• Factor out the GCF.
• Write the polynomial as a product of factors.

Factors of 6	Sum
1, 6	7
2, 3	5

• Factor the trinomial $x^2 + 5x + 6$. Try only positive factors of 6.

$3x^3 + 15x^2 + 18x = 3x(x + 2)(x + 3)$

Check $3x(x + 2)(x + 3) = 3x(x^2 + 3x + 2x + 6)$
$$= 3x(x^2 + 5x + 6)$$
$$= 3x^3 + 15x^2 + 18x$$

PROBLEM 3 Factor: $3a^2b - 18ab - 81b$

Solution See page S21.

EXAMPLE 4 Factor: $x^2 + 9xy + 20y^2$

Solution There is no common factor.

Factors of 20	Sum
1, 20	21
2, 10	12
4, 5	9

• **Try only positive factors of 20.**

$x^2 + 9xy + 20y^2 = (x + 4y)(x + 5y)$

Check $(x + 4y)(x + 5y) = x^2 + 5xy + 4xy + 20y^2$
$= x^2 + 9xy + 20y^2$

PROBLEM 4 Factor: $4x^2 - 40xy + 84y^2$

Solution See page S21.

Take Note

$4y$ and $5y$ are placed in the binomials. This is necessary so that the middle term contains xy and the last term contains y^2.

6.2 Concept Review

Determine whether the statement is always true, sometimes true, or never true.

1. The value of b in the trinomial $x^2 - 8x + 7$ is 8.

2. The factored form of $x^2 - 5x - 14$ is $(x - 7)(x + 2)$.

3. To factor a trinomial of the form $x^2 + bx + c$ means to rewrite the polynomial as a product of two binomials.

4. In the factoring of a trinomial, if the constant term is positive, then the signs in both binomial factors will be the same.

5. In the factoring of a trinomial, if the constant term is negative, then the signs in both binomial factors will be negative.

6. The first step in factoring a trinomial is to determine whether the terms of the trinomial have a common factor.

6.2 Exercises

1 Factor trinomials of the form $x^2 + bx + c$

Complete the table by listing the pairs of factors of the number and the sum of each pair.

1.

Factors of 18	Sum

2.

Factors of 12	Sum

3.

Factors of −21	Sum

4.

Factors of −10	Sum

5.

Factors of −28	Sum

6.

Factors of −32	Sum

Factor.

7. $x^2 + 3x + 2$

8. $x^2 + 5x + 6$

9. $x^2 - x - 2$

10. $x^2 + x - 6$

11. $a^2 + a - 12$

12. $a^2 - 2a - 35$

13. $a^2 - 3a + 2$

14. $a^2 - 5a + 4$

15. $a^2 + a - 2$

16. $a^2 - 2a - 3$

17. $b^2 - 6b + 9$

18. $b^2 + 8b + 16$

19. $b^2 + 7b - 8$

20. $y^2 - y - 6$

21. $y^2 + 6y - 55$

22. $z^2 - 4z - 45$

23. $y^2 - 5y + 6$

24. $y^2 - 8y + 15$

25. $z^2 - 14z + 45$

26. $z^2 - 14z + 49$

27. $z^2 - 12z - 160$

28. $p^2 + 2p - 35$

29. $p^2 + 12p + 27$

30. $p^2 - 6p + 8$

31. $x^2 + 20x + 100$

32. $x^2 + 18x + 81$

33. $b^2 + 9b + 20$

34. $b^2 + 13b + 40$

35. $x^2 - 11x - 42$

36. $x^2 + 9x - 70$

37. $b^2 - b - 20$

38. $b^2 + 3b - 40$

39. $y^2 - 14y - 51$

40. $y^2 - y - 72$

41. $p^2 - 4p - 21$

42. $p^2 + 16p + 39$

43. $y^2 - 8y + 32$

44. $y^2 - 9y + 81$

45. $x^2 - 20x + 75$

46. $p^2 + 24p + 63$

47. $x^2 - 15x + 56$

48. $x^2 + 21x + 38$

49. $x^2 + x - 56$

50. $x^2 + 5x - 36$

51. $a^2 - 21a - 72$ **52.** $a^2 - 7a - 44$ **53.** $a^2 - 15a + 36$ **54.** $a^2 - 21a + 54$

55. $z^2 - 9z - 136$ **56.** $z^2 + 14z - 147$ **57.** $c^2 - c - 90$ **58.** $c^2 - 3c - 180$

59. Suppose b and c are nonzero and n and m are positive constants such that $x^2 + bx + c = (x + n)(x + m)$.
 a. Is c positive or negative?
 b. Is b positive or negative?

60. Suppose b and c are nonzero and n and m are positive constants such that $x^2 + bx + c = (x - n)(x - m)$.
 a. Is c positive or negative?
 b. Is b positive or negative?

2 Factor completely

61. Use the trinomial $5x^2 - 10x - 40$.
 a. The first step in factoring the trinomial is to look for a ___?___ factor of the three terms.
 b. The GCF of the terms of $5x^2 - 10x - 40$ is ___?___.
 c. Factor the GCF from the terms of the trinomial:
 $5x^2 - 10x - 40 = 5(\underline{}?\underline{})$
 d. Factor the trinomial completely:
 $5x^2 - 10x - 40 = 5(x^2 - 2x - 8) = 5(\underline{}?\underline{})(\underline{}?\underline{})$

62. To factor $x^2 + 2xy - 24y^2$, find binomials of the form $(x + ny)(x + my)$, where n and m are numbers whose product is ___?___ and whose sum is ___?___. Two factors of -24 whose sum is 2 are ___?___ and ___?___. The trinomial $x^2 + 2x - 24y^2$ factors as $(\underline{}?\underline{})(\underline{}?\underline{})$.

Factor.

63. $2x^2 + 6x + 4$ **64.** $3x^2 + 15x + 18$ **65.** $3a^2 + 3a - 18$

66. $4x^2 - 4x - 8$ **67.** $ab^2 + 2ab - 15a$ **68.** $ab^2 + 7ab - 8a$

69. $xy^2 - 5xy + 6x$ **70.** $xy^2 + 8xy + 15x$ **71.** $z^3 - 7z^2 + 12z$

72. $2a^3 + 6a^2 + 4a$ **73.** $3y^3 - 15y^2 + 18y$ **74.** $4y^3 + 12y^2 - 72y$

75. $3x^2 + 3x - 36$ **76.** $2x^3 - 2x^2 - 4x$ **77.** $5z^2 - 15z - 140$

78. $6z^2 + 12z - 90$ **79.** $2a^3 + 8a^2 - 64a$ **80.** $3a^3 - 9a^2 - 54a$

81. $x^2 - 5xy + 6y^2$ **82.** $x^2 + 4xy - 21y^2$ **83.** $a^2 - 9ab + 20b^2$

84. $a^2 - 15ab + 50b^2$

85. $x^2 - 3xy - 28y^2$

86. $s^2 + 2st - 48t^2$

87. $y^2 - 15yz - 41z^2$

88. $y^2 + 85yz + 36z^2$

89. $z^4 - 12z^3 + 35z^2$

90. $z^4 + 2z^3 - 80z^2$

91. $b^4 - 22b^3 + 120b^2$

92. $b^4 - 3b^3 - 10b^2$

93. $2y^4 - 26y^3 - 96y^2$

94. $3y^4 + 54y^3 + 135y^2$

95. $x^4 + 7x^3 - 8x^2$

96. $x^4 - 11x^3 - 12x^2$

97. $4x^2y + 20xy - 56y$

98. $3x^2y - 6xy - 45y$

99. $8y^2 - 32y + 24$

100. $10y^2 - 100y + 90$

101. $c^3 + 13c^2 + 30c$

102. $c^3 + 18c^2 - 40c$

103. $3x^3 - 36x^2 + 81x$

104. $4x^3 + 4x^2 - 24x$

105. $x^2 - 8xy + 15y^2$

106. $y^2 - 7xy - 8x^2$

107. $a^2 - 13ab + 42b^2$

108. $y^2 + 4yz - 21z^2$

109. $y^2 + 8yz + 7z^2$

110. $y^2 - 16yz + 15z^2$

111. $3x^2y + 60xy - 63y$

112. $4x^2y - 68xy - 72y$

113. $3x^3 + 3x^2 - 36x$

114. $4x^3 + 12x^2 - 160x$

115. $4z^3 + 32z^2 - 132z$

116. $5z^3 - 50z^2 - 120z$

117. State whether the trinomial has a factor of $(x + 3)$.

 a. $3x^2 - 3x - 36$ **b.** $x^2y - xy - 12y$

118. State whether the trinomial has a factor of $(x + y)$.

 a. $2x^2 - 2xy - 4y^2$ **b.** $2x^2y - 4xy - 4y$

6.2 Applying Concepts

Factor.

119. $20 + c^2 + 9c$

120. $x^2y - 54y - 3xy$

121. $45a^2 + a^2b^2 - 14a^2b$

122. $12p^2 - 96p + 3p^3$

Find all integers k such that the trinomial can be factored over the integers.

123. $x^2 + kx + 35$

124. $x^2 + kx + 18$

125. $x^2 - kx + 21$

126. $x^2 - kx + 14$

Determine the positive integer values of k for which the following polynomials are factorable over the integers.

127. $y^2 + 4y + k$

128. $z^2 + 7z + k$

129. $a^2 - 6a + k$

130. $c^2 - 7c + k$

131. $x^2 - 3x + k$

132. $y^2 + 5y + k$

133. Exercises 127 to 132 included the requirement that $k > 0$. If k is allowed to be any integer, how many different values of k are possible for each polynomial? Explain your answer.

6.3 Factoring Polynomials of the Form $ax^2 + bx + c$

OBJECTIVE 1

Factor trinomials of the form $ax^2 + bx + c$ using trial factors

Trinomials of the form $ax^2 + bx + c$, where a, b, and c are integers and $a \neq 0$, are shown at the right.

$3x^2 - x + 4$, $a = 3$, $b = -1$, $c = 4$
$4x^2 + 5x - 8$, $a = 4$, $b = 5$, $c = -8$

These trinomials differ from those in the previous section in that the coefficient of x^2 is not 1. There are various methods of factoring these trinomials. The method described in this objective is factoring trinomials using trial factors.

To **factor a trinomial of the form $ax^2 + bx + c$** means to express the polynomial as the product of two binomials. Factoring such polynomials by trial and error may require testing many trial factors. To reduce the number of trial factors, remember the following points.

Points to Remember in Factoring $ax^2 + bx + c$

1. If the terms of the trinomial do not have a common factor, then the terms of a binomial factor cannot have a common factor.

2. When the constant term of the trinomial is positive, the constant terms of the binomials have the same sign as the coefficient of x in the trinomial.

3. When the constant term of the trinomial is negative, the constant terms of the binomials have opposite signs.

Factor: $10x^2 - x - 3$

The terms of the trinomial do not have a common factor; therefore, a binomial factor will not have a common factor [Point 1].

Because the constant term, c, of the trinomial is negative (-3), the constant terms of the binomial factors will have opposite signs [Point 3].

Find the factors of a (10) and the factors of c (−3).

Factors of 10	Factors of −3
1, 10	1, −3
2, 5	−1, 3

Using these factors, write trial factors, and use the Outer and Inner products of FOIL to check the middle term.

Trial Factors	Middle Term
$(x + 1)(10x - 3)$	$-3x + 10x = 7x$
$(x - 1)(10x + 3)$	$3x - 10x = -7x$
$\mathbf{(2x + 1)(5x - 3)}$	$\mathbf{-6x + 5x = -x}$
$(2x - 1)(5x + 3)$	$6x - 5x = x$
$(10x + 1)(x - 3)$	$-30x + x = -29x$
$(10x - 1)(x + 3)$	$30x - x = 29x$
$(5x + 1)(2x - 3)$	$-15x + 2x = -13x$
$(5x - 1)(2x + 3)$	$15x - 2x = 13x$

From the list of trial factors, $10x^2 - x - 3 = (2x + 1)(5x - 3)$.

Check: $(2x + 1)(5x - 3) = 10x^2 - 6x + 5x - 3$
$$= 10x^2 - x - 3$$

All the trial factors for this trinomial were listed in this example. However, once the correct binomial factors are found, it is not necessary to continue checking the remaining trial factors.

Factor: $4x^2 - 27x + 18$

The terms of the trinomial do not have a common factor; therefore, a binomial factor will not have a common factor [Point 1].

Because the constant term, c, of the trinomial is positive (18), the constant terms of the binomial factors will have the same sign as the coefficient of x. Because the coefficient of x is −27, both signs will be negative [Point 2].

Find the factors of a (4) and the negative factors of c (18).

Factors of 4	Factors of 18
1, 4	−1, −18
2, 2	−2, −9
	−3, −6

Using these factors, write trial factors, and use the Outer and Inner products of FOIL to check the middle term.

Trial Factors	Middle Term
$(x - 1)(4x - 18)$	Common factor
$(x - 2)(4x - 9)$	$-9x - 8x = -17x$
$(x - 3)(4x - 6)$	Common factor
$(2x - 1)(2x - 18)$	Common factor
$(2x - 2)(2x - 9)$	Common factor
$(2x - 3)(2x - 6)$	Common factor
$(4x - 1)(x - 18)$	$-72x - x = -73x$
$(4x - 2)(x - 9)$	Common factor
$\mathbf{(4x - 3)(x - 6)}$	$\mathbf{-24x - 3x = -27x}$

The correct factors have been found.

$$4x^2 - 27x + 18 = (4x - 3)(x - 6)$$

This last example illustrates that many of the trial factors may have common factors and thus need not be tried. For the remainder of this chapter, the trial factors with a common factor will not be listed.

EXAMPLE 1 Factor: $3x^2 + 20x + 12$

Solution

Factors of 3	Factors of 12
1, 3	1, 12
	2, 6
	3, 4

• Because 20 is positive, only the positive factors of 12 need be tried.

Trial Factors	Middle Term
$(x + 3)(3x + 4)$	$4x + 9x = 13x$
$(3x + 1)(x + 12)$	$36x + x = 37x$
$(3x + 2)(x + 6)$	$18x + 2x = 20x$

• Write the trial factors. Use FOIL to check the middle term.

$$3x^2 + 20x + 12 = (3x + 2)(x + 6)$$

Check $(3x + 2)(x + 6) = 3x^2 + 18x + 2x + 12$
$$= 3x^2 + 20x + 12$$

PROBLEM 1 Factor: $6x^2 - 11x + 5$

Solution See page S21.

EXAMPLE 2 Factor: $6x^2 - 5x - 6$

Solution

Factors of 6	Factors of −6
1, 6	1, −6
2, 3	−1, 6
	2, −3
	−2, 3

• Find the factors of a (6) and the factors of c (−6).

Trial Factors	Middle Term
$(x - 6)(6x + 1)$	$x - 36x = -35x$
$(x + 6)(6x - 1)$	$-x + 36x = 35x$
$(2x - 3)(3x + 2)$	$4x - 9x = -5x$

• Write the trial factors. Use FOIL to check the middle term.

$$6x^2 - 5x - 6 = (2x - 3)(3x + 2)$$

Check $(2x - 3)(3x + 2) = 6x^2 + 4x - 9x - 6$
$$= 6x^2 - 5x - 6$$

PROBLEM 2 Factor: $8x^2 + 14x - 15$

Solution See page S21.

EXAMPLE 3 Factor: $15 - 2x - x^2$

Solution

Factors of 15	Factors of -1
1, 15	1, -1
3, 5	

- The terms have no common factors. The coefficient of x^2 is -1.

Trial Factors	Middle Term
$(1 + x)(15 - x)$	$-x + 15x = 14x$
$(1 - x)(15 + x)$	$x - 15x = -14x$
$(3 + x)(5 - x)$	$-3x + 5x = 2x$
$(3 - x)(5 + x)$	$3x - 5x = -2x$

- Write the trial factors. Use FOIL to check the middle term.

$$15 - 2x - x^2 = (3 - x)(5 + x)$$

Check $(3 - x)(5 + x) = 15 + 3x - 5x - x^2$
$$= 15 - 2x - x^2$$

PROBLEM 3 Factor: $24 - 2y - y^2$

Solution See page S21.

The first step in factoring a trinomial is to determine whether there is a common factor. If there is a common factor, factor out the GCF of the terms.

EXAMPLE 4 Factor: $3x^3 - 23x^2 + 14x$

Solution The GCF of $3x^3$, $23x^2$, and $14x$ is x.

- Find the GCF of the terms of the polynomial.

$$3x^3 - 23x^2 + 14x = x(3x^2 - 23x + 14)$$

- Factor out the GCF.

Factors of 3	Factors of 14
1, 3	$-1, -14$
	$-2, -7$

- Factor the trinomial $3x^2 - 23x + 14$.

Trial Factors	Middle Term
$(x - 1)(3x - 14)$	$-14x - 3x = -17x$
$(x - 14)(3x - 1)$	$-x - 42x = -43x$
$(x - 2)(3x - 7)$	$-7x - 6x = -13x$
$(x - 7)(3x - 2)$	$-2x - 21x = -23x$

$$3x^3 - 23x^2 + 14x = x(x - 7)(3x - 2)$$

Check $x(x - 7)(3x - 2) = x(3x^2 - 2x - 21x + 14)$
$$= x(3x^2 - 23x + 14)$$
$$= 3x^3 - 23x^2 + 14x$$

PROBLEM 4 Factor: $4a^2b^2 - 30a^2b + 14a^2$

Solution See page S21.

OBJECTIVE **2**

Factor trinomials of the form $ax^2 + bx + c$ by grouping

Take Note

In this objective, we are using the skills taught in Objective 2 of Section 1 of this chapter. You may want to review that material before studying this objective.

In the previous objective, trinomials of the form $ax^2 + bx + c$ were factored using trial factors. In this objective, factoring by grouping is used.

To factor $ax^2 + bx + c$, first find two factors of $a \cdot c$ whose sum is b. Use the two factors to rewrite the middle term of the trinomial as the sum of two terms. Then use factoring by grouping to write the factorization of the trinomial.

Factor: $2x^2 + 13x + 15$

$a = 2, c = 15, a \cdot c = 2 \cdot 15 = 30$

Find two positive factors of 30 whose sum is 13.

Positive Factors of 30	Sum
1, 30	31
2, 15	17
3, 10	13
5, 6	11

The factors are 3 and 10.

Use the factors 3 and 10 to rewrite $13x$ as $3x + 10x$.

$$2x^2 + \quad 13x \quad + 15$$
$$= 2x^2 + 3x + 10x + 15$$

Factor by grouping.

$$= (2x^2 + 3x) + (10x + 15)$$
$$= x(2x + 3) + 5(2x + 3)$$
$$= (2x + 3)(x + 5)$$

Check: $(2x + 3)(x + 5) = 2x^2 + 10x + 3x + 15$
$$= 2x^2 + 13x + 15$$

Factor: $6x^2 - 11x - 10$

$a = 6, c = -10, a \cdot c = 6(-10) = -60$

Find two factors of -60 whose sum is -11.

Factors of -60	Sum
1, -60	-59
-1, 60	59
2, -30	-28
-2, 30	28
3, -20	-17
-3, 20	17
4, -15	-11

The required sum has been found. The remaining factors need not be checked. The factors are 4 and -15.

Use the factors 4 and -15 to rewrite $-11x$ as $4x - 15x$.

$$6x^2 - \quad 11x \quad - 10$$
$$= 6x^2 + 4x - 15x - 10$$

Factor by grouping.

Note: $-15x - 10 = -(15x + 10)$

$$= (6x^2 + 4x) - (15x + 10)$$
$$= 2x(3x + 2) - 5(3x + 2)$$
$$= (3x + 2)(2x - 5)$$

Check: $(3x + 2)(2x - 5) = 6x^2 - 15x + 4x - 10$
$$= 6x^2 - 11x - 10$$

Factor: $3x^2 - 2x - 4$

$a = 3, c = -4, a \cdot c = 3(-4) = -12$

Find two factors of -12 whose sum is -2.

Factors of -12	Sum
1, -12	-11
-1, 12	11
2, -6	-4
-2, 6	4
3, -4	-1
-3, 4	1

No integer factors of -12 have a sum of -2. Therefore, $3x^2 - 2x - 4$ is nonfactorable over the integers.

EXAMPLE 5 Factor: $2x^2 + 19x - 10$

Solution $a \cdot c = 2(-10) = -20$ • Find $a \cdot c$.

$-1(20) = -20$
$-1 + 20 = 19$ • Find two numbers whose product is -20 and whose sum is 19.

$2x^2 + 19x - 10$
$= 2x^2 - x + 20x - 10$ • Rewrite $19x$ as $-x + 20x$.
$= (2x^2 - x) + (20x - 10)$ • Factor by grouping.
$= x(2x - 1) + 10(2x - 1)$
$= (2x - 1)(x + 10)$

PROBLEM 5 Factor: $2a^2 + 13a - 7$

Solution See page S22.

EXAMPLE 6 Factor: $8y^2 - 10y - 3$

Solution $a \cdot c = 8(-3) = -24$ • Find $a \cdot c$.

$2(-12) = -24$
$2 + (-12) = -10$ • Find two numbers whose product is -24 and whose sum is -10.

$8y^2 - 10y - 3$
$= 8y^2 + 2y - 12y - 3$ • Rewrite $-10y$ as $2y - 12y$.
$= (8y^2 + 2y) - (12y + 3)$ • Factor by grouping.
$= 2y(4y + 1) - 3(4y + 1)$
$= (4y + 1)(2y - 3)$

PROBLEM 6 Factor: $4a^2 - 11a - 3$

Solution See page S22.

Remember that the first step in factoring a trinomial is to determine whether there is a common factor. If there is a common factor, factor out the GCF of the terms.

EXAMPLE 7 Factor: $24x^2y - 76xy + 40y$

Solution $24x^2y - 76xy + 40y$
$= 4y(6x^2 - 19x + 10)$

$a \cdot c = 6(10) = 60$

$-4(-15) = 60$
$-4 + (-15) = -19$

$6x^2 - 19x + 10$
$= 6x^2 - 4x - 15x + 10$
$= (6x^2 - 4x) - (15x - 10)$
$= 2x(3x - 2) - 5(3x - 2)$
$= (3x - 2)(2x - 5)$

$24x^2y - 76xy + 40y$
$= 4y(6x^2 - 19x + 10)$
$= 4y(3x - 2)(2x - 5)$

- The terms of the polynomial have a common factor, $4y$. Factor out the GCF.
- To factor $6x^2 - 19x + 10$, first find $a \cdot c$.
- Find two numbers whose product is 60 and whose sum is -19.

- Rewrite $-19x$ as $-4x - 15x$.
- Factor by grouping.

- Write the complete factorization of the given polynomial.

PROBLEM 7 Factor: $15x^3 + 40x^2 - 80x$

Solution See page S22.

6.3 Concept Review

Determine whether the statement is always true, sometimes true, or never true.

1. The value of a in the trinomial $3x^2 + 8x - 7$ is -7.

2. The factored form of $2x^2 - 7x - 15$ is $(2x + 3)(x - 5)$.

3. To check the factorization of a trinomial of the form $ax^2 + bx + c$, multiply the two binomial factors.

4. The terms of the binomial $3x - 9$ have a common factor.

5. If the terms of a trinomial of the form $ax^2 + bx + c$ do not have a common factor, then one of its binomial factors cannot have a common factor.

6. To factor a trinomial of the form $ax^2 + bx + c$ by grouping, first find two numbers whose product is ac and whose sum is b.

6.3 Exercises

1 Factor trinomials of the form $ax^2 + bx + c$
using trial factors

1. Use trial factors to factor $2x^2 - 7x - 15$.
 a. Complete the table of the factors of 2 and -15.

Factors of 2	Factors of -15
1, ___?___	1, ___?___
	-1, ___?___
	3, ___?___
	-3, ___?___

 b. Complete the table by finding the middle term for each pair of trial
 factors.

Trial Factors	Middle Term
$(x - 15)(2x + 1)$	___?___
$(x + 15)(2x - 1)$	___?___
$(x - 5)(2x + 3)$	___?___
$(x + 5)(2x - 3)$	___?___

 c. Use the result of part b to write the factored form of $2x^2 - 7x - 15$:
 (___?___)(___?___).

2. Complete the factorization of $8x^2 - 14x + 3$:
 $8x^2 - 14x + 3 = (4x - 1)(\underline{\quad ? \quad})$

Factor by the method of using trial factors.

3. $2x^2 + 3x + 1$ **4.** $5x^2 + 6x + 1$ **5.** $2y^2 + 7y + 3$ **6.** $3y^2 + 7y + 2$

7. $2a^2 - 3a + 1$ **8.** $3a^2 - 4a + 1$ **9.** $2b^2 - 11b + 5$ **10.** $3b^2 - 13b + 4$

11. $2x^2 + x - 1$ **12.** $4x^2 - 3x - 1$ **13.** $2x^2 - 5x - 3$ **14.** $3x^2 + 5x - 2$

15. $6z^2 - 7z + 3$ **16.** $9z^2 + 3z + 2$ **17.** $6t^2 - 11t + 4$ **18.** $10t^2 + 11t + 3$

19. $8x^2 + 33x + 4$ **20.** $7x^2 + 50x + 7$ **21.** $3b^2 - 16b + 16$ **22.** $6b^2 - 19b + 15$

23. $2z^2 - 27z - 14$ **24.** $4z^2 + 5z - 6$ **25.** $3p^2 + 22p - 16$ **26.** $7p^2 + 19p + 10$

27. $6x^2 - 17x + 12$ **28.** $15x^2 - 19x + 6$ **29.** $5b^2 + 33b - 14$ **30.** $8x^2 - 30x + 25$

31. $6a^2 + 7a - 24$ **32.** $14a^2 + 15a - 9$ **33.** $18t^2 - 9t - 5$ **34.** $12t^2 + 28t - 5$

35. $15a^2 + 26a - 21$ **36.** $6a^2 + 23a + 21$ **37.** $8y^2 - 26y + 15$ **38.** $18y^2 - 27y + 4$

39. $8z^2 + 2z - 15$ **40.** $10z^2 + 3z - 4$ **41.** $3x^2 + 14x - 5$ **42.** $15a^2 - 22a + 8$

43. $12x^2 + 25x + 12$ **44.** $10b^2 + 43b - 9$ **45.** $3z^2 + 95z + 10$ **46.** $8z^2 - 36z + 1$

47. $3x^2 + xy - 2y^2$ **48.** $6x^2 + 10xy + 4y^2$ **49.** $28 + 3z - z^2$ **50.** $15 - 2z - z^2$

51. $8 - 7x - x^2$ **52.** $12 + 11x - x^2$ **53.** $9x^2 + 33x - 60$ **54.** $16x^2 - 16x - 12$

55. $24x^2 - 52x + 24$ **56.** $60x^2 + 95x + 20$ **57.** $35a^4 + 9a^3 - 2a^2$ **58.** $15a^4 + 26a^3 + 7a^2$

59. $15b^2 - 115b + 70$ **60.** $25b^2 + 35b - 30$ **61.** $10x^3 + 12x^2 + 2x$ **62.** $9x^3 - 39x^2 + 12x$

63. $10y^3 - 44y^2 + 16y$ **64.** $14y^3 + 94y^2 - 28y$ **65.** $4yz^3 + 5yz^2 - 6yz$ **66.** $2yz^3 - 17yz^2 + 8yz$

67. $20b^4 + 41b^3 + 20b^2$ **68.** $6b^4 - 13b^3 + 6b^2$ **69.** $9x^3y + 12x^2y + 4xy$ **70.** $9a^3b - 9a^2b^2 - 10ab^3$

For Exercises 71 and 72, let $(nx + p)$ and $(mx + q)$ be factors of the trinomial $ax^2 + bx + c$, in which a and c are even and b is odd.

71. If n is even, which of the numbers p, m, and q must be odd?

72. If p is even, which of the numbers n, m, and q must be odd?

2 Factor trinomials of the form $ax^2 + bx + c$ by grouping

73. Complete the table of pairs of factors of -18.

Factors of -18	Sum
1, __?__	__?__
-1, __?__	__?__
2, __?__	__?__
-2, __?__	__?__
3, __?__	__?__
-3, __?__	__?__

74. a. To factor $2x^2 - 3x - 9$ by grouping, first find two integers whose product is $2(-9) = $ __?__ and whose sum is the coefficient of the middle term, __?__. These two integers are __?__ and __?__.

 b. Complete the factoring by grouping:

$$2x^2 - 3x - 9 = 2x^2 - 6x + \underline{\quad ? \quad} - 9$$
$$= (\underline{\quad ? \quad}) + (\underline{\quad ? \quad})$$
$$= 2x(\underline{\quad ? \quad} - \underline{\quad ? \quad}) + 3(\underline{\quad ? \quad} - \underline{\quad ? \quad})$$
$$= (\underline{\quad ? \quad})(\underline{\quad ? \quad})$$

Factor by grouping.

75. $2t^2 - t - 10$ **76.** $2t^2 + 5t - 12$ **77.** $3p^2 - 16p + 5$ **78.** $6p^2 + 5p + 1$

79. $12y^2 - 7y + 1$ **80.** $6y^2 - 5y + 1$ **81.** $5x^2 - 62x - 7$ **82.** $9x^2 - 13x - 4$

83. $12y^2 + 19y + 5$ **84.** $5y^2 - 22y + 8$ **85.** $7a^2 + 47a - 14$ **86.** $11a^2 - 54a - 5$

87. $4z^2 + 11z + 6$ **88.** $6z^2 - 25z + 14$ **89.** $22p^2 + 51p - 10$ **90.** $14p^2 - 41p + 15$

91. $8y^2 + 17y + 9$ **92.** $12y^2 - 145y + 12$ **93.** $6b^2 - 13b + 6$ **94.** $20b^2 + 37b + 15$

95. $33b^2 + 34b - 35$ **96.** $15b^2 - 43b + 22$ **97.** $18y^2 - 39y + 20$ **98.** $24y^2 + 41y + 12$

99. $15x^2 - 82x + 24$ **100.** $13z^2 + 49z - 8$ **101.** $10z^2 - 29z + 10$ **102.** $15z^2 - 44z + 32$

103. $36z^2 + 72z + 35$ **104.** $16z^2 + 8z - 35$ **105.** $14y^2 - 29y + 12$ **106.** $8y^2 + 30y + 25$

107. $6x^2 + 35x - 6$ **108.** $4x^2 + 6x + 2$ **109.** $12x^2 + 33x - 9$ **110.** $15y^2 - 50y + 35$

111. $30y^2 + 10y - 20$ **112.** $2x^3 - 11x^2 + 5x$ **113.** $2x^3 - 3x^2 - 5x$ **114.** $3a^2 + 5ab - 2b^2$

115. $2a^2 - 9ab + 9b^2$ **116.** $4y^2 - 11yz + 6z^2$ **117.** $2y^2 + 7yz + 5z^2$ **118.** $12 - x - x^2$

119. $18 + 17x - x^2$ **120.** $21 - 20x - x^2$ **121.** $360y^2 + 4y - 4$ **122.** $10t^2 - 5t - 50$

123. $16t^2 + 40t - 96$ **124.** $3p^3 - 16p^2 + 5p$ **125.** $6p^3 + 5p^2 + p$ **126.** $26z^2 + 98z - 24$

127. $30z^2 - 87z + 30$ **128.** $12a^3 + 14a^2 - 48a$ **129.** $42a^3 + 45a^2 - 27a$ **130.** $36p^2 - 9p^3 - p^4$

131. $9x^2y - 30xy^2 + 25y^3$ **132.** $8x^2y - 38xy^2 + 35y^3$

133. $9x^3y - 24x^2y^2 + 16xy^3$ **134.** $45a^3b - 78a^2b^2 + 24ab^3$

In Exercises 135 to 138, information is given about the signs of b and c in $ax^2 + bx + c$, where $a > 0$. If you want to factor $ax^2 + bx + c$ by grouping, you look for factors of ac whose sum is b. In each case, state whether the factors of ac should be two positive numbers, two negative numbers, or one positive and one negative number.

135. $b > 0$ and $c > 0$

136. $b < 0$ and $c < 0$

137. $b < 0$ and $c > 0$

138. $b > 0$ and $c < 0$

6.3 Applying Concepts

Factor.

139. $6y + 8y^3 - 26y^2$

140. $22p^2 - 3p^3 + 16p$

141. $a^3b - 24ab - 2a^2b$

142. $3xy^2 - 14xy + 2xy^3$

143. $25t^2 + 60t - 10t^3$

144. $3xy^3 + 2x^3y - 7x^2y^2$

Factor.

145. $2(y + 2)^2 - (y + 2) - 3$

146. $3(a + 2)^2 - (a + 2) - 4$

147. $10(x + 1)^2 - 11(x + 1) - 6$

148. $4(y - 1)^2 - 7(y - 1) - 2$

Find all integers k such that the trinomial can be factored over the integers.

149. $2x^2 + kx + 3$

150. $2x^2 + kx - 3$

151. $3x^2 + kx + 2$

152. $3x^2 + kx - 2$

153. $2x^2 + kx + 5$

154. $2x^2 + kx - 5$

155. Given that $x + 2$ is a factor of $x^3 - 2x^2 - 5x + 6$, factor $x^3 - 2x^2 - 5x + 6$ completely.

156. In your own words, explain how the signs of the last terms of the two binomial factors of a trinomial are determined.

157. Geometry The area of a rectangle is $(3x^2 + x - 2)$ ft². Find the dimensions of the rectangle in terms of the variable x. Given that $x > 0$, specify the dimension that is the length and the dimension that is the width. Can x be negative? Can $x = 0$? Explain your answers.

$A = 3x^2 + x - 2$

158. Geometry Write the area of the shaded region in factored form.

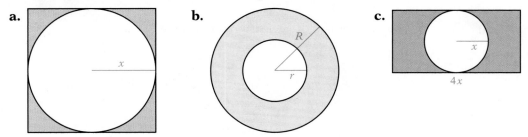

a.

b.

c.

6.4 Special Factoring

OBJECTIVE ▮**1**

Factor the difference of two perfect squares and factor perfect-square trinomials

The product of a term and itself is called a **perfect square.** The exponents on variables of perfect squares are always even numbers.

Term		Perfect Square
5	$5 \cdot 5 =$	25
x	$x \cdot x =$	x^2
$3y^4$	$3y^4 \cdot 3y^4 =$	$9y^8$
x^n	$x^n \cdot x^n =$	x^{2n}

The **square root of a perfect square** is one of the two equal factors of the perfect square. $\sqrt{}$ is the symbol for square root. To find the exponent of the square root of a variable term, divide the exponent by 2. For the examples at the right, assume that x and y represent positive numbers.

$\sqrt{25} = 5$
$\sqrt{x^2} = x$
$\sqrt{9y^8} = 3y^4$
$\sqrt{x^{2n}} = x^n$

Take Note

$(a + b)$ is the sum of the two terms a and b. $(a - b)$ is the difference of the two terms a and b. a and b are the square roots of a^2 and b^2.

The factors of the **difference of two perfect squares** are the sum and difference of the square roots of the perfect squares.

Factors of the Difference of Two Perfect Squares

$$a^2 - b^2 = (a + b)(a - b)$$

The *sum* of two perfect squares, $a^2 + b^2$, is nonfactorable over the integers.

EXAMPLE 1 Factor: $25x^2 - 1$

Solution $25x^2 - 1 = (5x)^2 - 1^2$

• Write the binomial as the difference of two perfect squares.

$= (5x + 1)(5x - 1)$

• The factors are the sum and difference of the square roots of the perfect squares.

PROBLEM 1 Factor: $x^2 - 36y^4$

Solution See page S22.

A perfect-square trinomial is the square of a binomial.

Factors of a Perfect-Square Trinomial

$$a^2 + 2ab + b^2 = (a + b)^2$$
$$a^2 - 2ab + b^2 = (a - b)^2$$

In factoring a **perfect-square trinomial,** remember that the terms of the binomial are the square roots of the perfect squares of the trinomial. The sign in the binomial is the sign of the middle term of the trinomial.

Factor: $4x^2 - 12x + 9$

Because $4x^2$ is a perfect square and 9 is a perfect square, try factoring $4x^2 - 12x + 9$ as the square of a binomial.

$$4x^2 - 12x + 9 \stackrel{?}{=} (2x - 3)^2$$

Check:
$$(2x - 3)^2 = (2x - 3)(2x - 3) = 4x^2 - 6x - 6x + 9 = 4x^2 - 12x + 9$$

The factorization checks. Thus $4x^2 - 12x + 9 = (2x - 3)^2$.

It is important to check a proposed factorization as we did here. The next example illustrates the importance of this check.

Factor: $x^2 + 13x + 36$

Because x^2 is a perfect square and 36 is a perfect square, try factoring $x^2 + 13x + 36$ as the square of a binomial.

$$x^2 + 13x + 36 \stackrel{?}{=} (x + 6)^2$$

Check: $(x + 6)^2 = (x + 6)(x + 6) = x^2 + 6x + 6x + 36 = x^2 + 12x + 36$

In this case, the proposed factorization of $x^2 + 13x + 36$ does *not* check. Try another factorization. The numbers 4 and 9 are factors of 36 whose sum is 13.

$$x^2 + 13x + 36 = (x + 4)(x + 9)$$

EXAMPLE 2 Factor: $4x^2 - 20x + 25$

Solution $4x^2 - 20x + 25 = (2x - 5)^2$

PROBLEM 2 Factor: $9x^2 + 12x + 4$

Solution See page S22.

OBJECTIVE **2** **Factor the sum or the difference of two cubes**

The product of the same three factors is called a **perfect cube.** The exponents on variables of perfect cubes are always divisible by 3.

Term		Perfect Cube
2	$2 \cdot 2 \cdot 2 =$	8
$3y$	$3y \cdot 3y \cdot 3y =$	$27y^3$
x^2	$x^2 \cdot x^2 \cdot x^2 =$	x^6
x^n	$x^n \cdot x^n \cdot x^n =$	x^{3n}

The **cube root** of a perfect cube is one of the three equal factors of the perfect cube. $\sqrt[3]{}$ is the symbol for cube root. To find the exponent of the cube root of a perfect-cube variable term, divide the exponent by 3.

$$\sqrt[3]{8} = 2$$
$$\sqrt[3]{27y^3} = 3y$$
$$\sqrt[3]{x^6} = x^2$$
$$\sqrt[3]{x^{3n}} = x^n$$

The following rules are used to factor the sum or difference of two perfect cubes.

Factoring the Sum or Difference of Two Perfect Cubes

$$a^3 + b^3 = (a + b)(a^2 - ab + b^2)$$
$$a^3 - b^3 = (a - b)(a^2 + ab + b^2)$$

To factor $8x^3 - 27$:

Write the binomial as the difference of two perfect cubes.

$$8x^3 - 27 = (2x)^3 - 3^3$$

The terms of the binomial factor are the cube roots of the perfect cubes. The sign of the binomial factor is the same sign as in the given binomial. The trinomial factor is obtained from the binomial factor.

$$= (2x - 3)(4x^2 + 6x + 9)$$

Square of the first term

Opposite of the product of the two terms

Square of the last term

EXAMPLE 3 Factor. **A.** $x^3y^3 - 1$ **B.** $(x + y)^3 - x^3$

Solution **A.** $x^3y^3 - 1$
$$= (xy)^3 - 1^3$$
$$= (xy - 1)(x^2y^2 + xy + 1)$$

• **Write the binomial as the difference of two perfect cubes.**

B. $(x + y)^3 - x^3$
$$= [(x + y) - x][(x + y)^2 + x(x + y) + x^2]$$

• **This is a difference of two perfect cubes.**

$$= y(x^2 + 2xy + y^2 + x^2 + xy + x^2)$$
$$= y(3x^2 + 3xy + y^2)$$

• **Simplify.**

PROBLEM 3 Factor. **A.** $8x^3 + y^3z^3$ **B.** $(x - y)^3 + (x + y)^3$

Solution See page S22.

OBJECTIVE 3 ## Factor trinomials that are quadratic in form

Certain trinomials can be expressed as quadratic trinomials by making suitable variable substitutions. A trinomial is **quadratic in form** if it can be written as

$$au^2 + bu + c$$

If we let $x^2 = u$, the trinomial $x^4 + 5x^2 + 6$ can be written as shown at the right.

$$x^4 + 5x^2 + 6$$
$$= (x^2)^2 + 5(x^2) + 6$$
$$= u^2 + 5u + 6$$

The trinomial is quadratic in form.

If we let $xy = u$, the trinomial $2x^2y^2 + 3xy - 9$ can be written as shown at the right.

$$2x^2y^2 + 3xy - 9$$
$$= 2(xy)^2 + 3(xy) - 9$$
$$= 2u^2 + 3u - 9$$

The trinomial is quadratic in form.

When a trinomial that is quadratic in form is factored, the variable part of the first term in each binomial factor will be u. For example, because $x^4 + 5x^2 + 6$ is quadratic in form when $x^2 = u$, the first term in each binomial factor will be x^2.

$$x^4 + 5x^2 + 6 = (x^2)^2 + 5(x^2) + 6$$
$$= (x^2 + 2)(x^2 + 3)$$

The trinomial $x^2y^2 - 2xy - 15$ is quadratic in form when $xy = u$. The first term in each binomial factor will be xy.

$$x^2y^2 - 2xy - 15 = (xy)^2 - 2(xy) - 15$$
$$= (xy + 3)(xy - 5)$$

EXAMPLE 4 Factor. **A.** $6x^2y^2 - xy - 12$ **B.** $2x^4 + 5x^2 - 12$

Solution **A.** $6x^2y^2 - xy - 12$
$= (3xy + 4)(2xy - 3)$

• The trinomial is quadratic in form when $xy = u$.

B. $2x^4 + 5x^2 - 12$
$= (x^2 + 4)(2x^2 - 3)$

• The trinomial is quadratic in form when $x^2 = u$.

PROBLEM 4 Factor. **A.** $6x^2y^2 - 19xy + 10$ **B.** $3x^4 + 4x^2 - 4$

Solution See page S22.

OBJECTIVE 4 **Factor completely**

> *When factoring a polynomial completely, ask the following questions about the polynomial:*
>
> 1. Is there a common factor? If so, factor out the GCF.
>
> 2. If the polynomial is a binomial, is it the difference of two perfect squares, the sum of two perfect cubes, or the difference of two perfect cubes? If so, factor.
>
> 3. If the polynomial is a trinomial, is it a perfect-square trinomial or the product of two binomials? If so, factor.
>
> 4. If the polynomial has four terms, can it be factored by grouping? If so, factor.
>
> 5. Is each factor nonfactorable over the integers? If not, factor.

Take Note

Remember that you may have to factor more than once in order to write the polynomial as a product of *prime* factors.

EXAMPLE 5 Factor. **A.** $x^2y + 2x^2 - y - 2$ **B.** $x^{4n} - y^{4n}$

Solution **A.** $x^2y + 2x^2 - y - 2$
$= (x^2y + 2x^2) - (y + 2)$ • Factor by grouping.
$= x^2(y + 2) - (y + 2)$
$= (y + 2)(x^2 - 1)$
$= (y + 2)(x + 1)(x - 1)$ • Factor the difference of
 two perfect squares.

B. $x^{4n} - y^{4n}$
$= (x^{2n})^2 - (y^{2n})^2$ • Factor the difference of
$= (x^{2n} + y^{2n})(x^{2n} - y^{2n})$ two perfect squares.
$= (x^{2n} + y^{2n})[(x^n)^2 - (y^n)^2]$ • Factor the difference of
$= (x^{2n} + y^{2n})(x^n + y^n)(x^n - y^n)$ two perfect squares.

PROBLEM 5 Factor. **A.** $4x - 4y - x^3 + x^2y$ **B.** $x^{4n} - x^{2n}y^{2n}$

Solution See page S22.

 Concept Review

Determine whether the statement is always true, sometimes true, or never true.

1. $a^2b^2c^2$ is an example of a perfect-square trinomial.

2. The factors of $a^3 + b^3$ are $(a + b)(a + b)(a + b)$.

3. The difference of two perfect cubes is not factorable.

4. Factoring out the GCF is the first thing to try in factoring polynomials.

5. The polynomial $b^3 + 1$ is not factorable.

6.4 Exercises

1 **Factor the difference of two perfect squares
and factor perfect-square trinomials**

Which of the expressions are perfect squares?

1. 4; 8; $25x^6$; $12y^{10}$; $100x^4y^4$ **2.** 9; 18; $15a^8$; $49b^{12}$; $64a^{16}b^2$

Name the square root of the expression.

3. $16z^8$ **4.** $36d^{10}$ **5.** $81a^4b^6$ **6.** $25m^2n^{12}$

7. a. The binomial $16x^2 - 1$ is in the form $a^2 - b^2$, where $a = \underline{\quad?\quad}$ and $b = \underline{\quad?\quad}$.

 b. Use the formula $a^2 - b^2 = (a + b)(a - b)$ to factor $16x^2 - 1$:
$16x^2 - 1 = (\underline{\quad?\quad})(\underline{\quad?\quad})$.

8. a. The trinomial $25y^2 - 20y + 4$ is in the form $a^2 - 2ab + b^2$, where $a = \underline{\quad?\quad}$ and $b = \underline{\quad?\quad}$.

 b. Use the formula $a^2 - 2ab + b^2 = (a - b)^2$ to factor $25y^2 - 20y + 4$:
$25y^2 - 20y + 4 = (\underline{\quad?\quad})^2$.

9. Explain how to factor the difference of two perfect squares.

10. What is a perfect-square trinomial?

Factor.

11. $x^2 - 16$

12. $y^2 - 49$

13. $4x^2 - 1$

14. $81x^2 - 4$

15. $b^2 - 2b + 1$

16. $a^2 + 14a + 49$

17. $16x^2 - 40x + 25$

18. $49x^2 + 28x + 4$

19. $x^2y^2 - 100$

20. $a^2b^2 - 25$

21. $x^2 + 4$

22. $a^2 + 16$

23. $x^2 + 6xy + 9y^2$

24. $4x^2y^2 + 12xy + 9$

25. $4x^2 - y^2$

26. $49a^2 - 16b^4$

27. $a^{2n} - 1$

28. $b^{2n} - 16$

29. $a^2 + 4a + 4$

30. $b^2 - 18b + 81$

31. $x^2 - 12x + 36$

32. $y^2 - 6y + 9$

33. $16x^2 - 121$

34. $49y^2 - 36$

35. $1 - 9a^2$

36. $16 - 81y^2$

37. $4a^2 + 4a - 1$

38. $9x^2 + 12x - 4$

39. $b^2 + 7b + 14$

40. $y^2 - 5y + 25$

41. $25 - a^2b^2$

42. $64 - x^2y^2$

43. $25a^2 - 40ab + 16b^2$

44. $4a^2 - 36ab + 81b^2$

45. $x^{2n} + 6x^n + 9$

46. $y^{2n} - 16y^n + 64$

47. Which expressions are equivalent to $x^{16} - 81$?

 a. $(x^4 + 9)(x^4 - 9)$ **b.** $(x^8 + 9)(x^8 - 9)$ **c.** $(x^4 + 9)(x^2 + 3)(x^2 - 3)$
 d. $(x^8 + 9)(x^4 + 3)(x^4 - 3)$

48. Given that a and b are positive numbers, which expressions can be factored?

 a. $a^2x^2 - 2abx + b^2$ **b.** $-a^2x^2 - 2abx - b^2$ **c.** $-a^2x^2 + 2abx - b^2$
 d. $a^2x^2 - 2abx - b^2$

2 Factor the sum or the difference of two cubes

Which of the expressions are perfect cubes?

49. $4; 8; x^9; a^8b^8; 27c^{15}d^{18}$

50. $9; 27; y^{12}; m^3n^6; 64mn^9$

Name the cube root of the expression.

51. $8x^9$

52. $27y^{15}$

53. $64a^6b^{18}$

54. $125c^{12}d^3$

55. a. The binomial $125x^3 + 8$ is in the form $a^3 + b^3$, where $a = $ ___?___ and $b = $ ___?___.

b. Use the formula $a^3 + b^3 = (a + b)(a^2 - ab + b^2)$ to factor $125x^3 + 8$: $125x^3 + 8 = ($ ___?___ $)($ ___?___ $)$.

56. a and b are positive numbers. Which expressions can be factored?

a. $a^3x^3 - b^3y^3$ **b.** $a^9x^3 + b^9y^3$ **c.** $a^8x^3 + b^8y^3$ **d.** $a^6x^3 - b^3y^6$

Factor.

57. $x^3 - 27$

58. $y^3 + 125$

59. $8x^3 - 1$

60. $64a^3 + 27$

61. $x^3 - y^3$

62. $x^3 - 8y^3$

63. $m^3 + n^3$

64. $27a^3 + b^3$

65. $64x^3 + 1$

66. $1 - 125b^3$

67. $27x^3 - 8y^3$

68. $64x^3 + 27y^3$

69. $x^3y^3 + 64$

70. $8x^3y^3 + 27$

71. $16x^3 - y^3$

72. $27x^3 - 8y^3$

73. $8x^3 - 9y^3$

74. $27a^3 - 16$

75. $(a - b)^3 - b^3$

76. $a^3 + (a + b)^3$

77. $x^{6n} + y^{3n}$

78. $x^{3n} + y^{3n}$

3 Factor trinomials that are quadratic in form

79. The trinomial $5x^4y^2 - 17x^2y - 12$ can be written in the quadratic form $5u^2 - 17u - 12$, where $u = $ ___?___.

80. The trinomial $8x^{4n} + 2x^{2n} - 3$ can be written in the quadratic form $8u^2 + 2u - 3$, where $u = $ ___?___.

81. What does it mean for a polynomial to be quadratic in form?

82. Do all polynomials that are quadratic in form factor? If not, give an example of one that does not factor.

Factor.

83. $x^2y^2 - 8xy + 15$

84. $x^2y^2 - 8xy - 33$

85. $x^2y^2 - 17xy + 60$

86. $a^2b^2 + 10ab + 24$

87. $x^4 - 9x^2 + 18$

88. $y^4 - 6y^2 - 16$

89. $b^4 - 13b^2 - 90$

90. $a^4 + 14a^2 + 45$

91. $x^4y^4 - 8x^2y^2 + 12$

92. $a^4b^4 + 11a^2b^2 - 26$

93. $x^{2n} + 3x^n + 2$

94. $a^{2n} - a^n - 12$

95. $3x^2y^2 - 14xy + 15$

96. $5x^2y^2 - 59xy + 44$

97. $6a^2b^2 - 23ab + 21$

98. $10a^2b^2 + 3ab - 7$

99. $2x^4 - 13x^2 - 15$

100. $3x^4 + 20x^2 + 32$

101. $2x^{2n} - 7x^n + 3$

102. $4x^{2n} + 8x^n - 5$

103. $6a^{2n} + 19a^n + 10$

4 Factor completely

104. When factoring a polynomial, you should look first for a ____?____ factor.

105. To factor a polynomial with four terms, try factoring by ____?____.

Factor.

106. $5x^2 + 10x + 5$

107. $12x^2 - 36x + 27$

108. $3x^4 - 81x$

109. $27a^4 - a$

110. $7x^2 - 28$

111. $20x^2 - 5$

112. $y^4 - 10y^3 + 21y^2$

113. $y^5 + 6y^4 - 55y^3$

114. $x^4 - 16$

115. $16x^4 - 81$

116. $8x^5 - 98x^3$

117. $16a - 2a^4$

118. $x^3y^3 - x^3$

119. $x^3 + 2x^2 - x - 2$

120. $2x^3 - 3x^2 - 8x + 12$

121. $2x^3 + 4x^2 - 3x - 6$

122. $3x^3 - 3x^2 + 4x - 4$

123. $x^3 + x^2 - 16x - 16$

124. $4x^3 + 8x^2 - 9x - 18$

125. $a^3b^6 - b^3$

126. $x^6y^6 - x^3y^3$

127. $x^4 - 2x^3 - 35x^2$

128. $x^4 + 15x^3 - 56x^2$

129. $4x^2 + 4x - 1$

130. $8x^4 - 40x^3 + 50x^2$

131. $6x^5 + 74x^4 + 24x^3$

132. $x^4 - y^4$

133. $16a^4 - b^4$

134. $x^6 + y^6$

135. $x^4 - 5x^2 - 4$

136. $a^4 - 25a^2 - 144$

137. $3b^5 - 24b^2$

138. $16a^4 - 2a$

139. $x^4y^2 - 5x^3y^3 + 6x^2y^4$

140. $a^4b^2 - 8a^3b^3 - 48a^2b^4$

141. $16x^3y + 4x^2y^2 - 42xy^3$

142. $24a^2b^2 - 14ab^3 - 90b^4$

143. $x^3 - 2x^2 - x + 2$

144. $x^3 - 2x^2 - 4x + 8$

145. $8xb - 8x - 4b + 4$

146. $4xy + 8x + 4y + 8$

147. $4x^2y^2 - 4x^2 - 9y^2 + 9$

148. $4x^4 - x^2 - 4x^2y^2 + y^2$ **149.** $x^5 - 4x^3 - 8x^2 + 32$ **150.** $x^6y^3 + x^3 - x^3y^3 - 1$

151. $a^{2n+2} - 6a^{n+2} + 9a^2$ **152.** $x^{2n+1} + 2x^{n+1} + x$

153. $2x^{n+2} - 7x^{n+1} + 3x^n$ **154.** $3b^{n+2} + 4b^{n+1} - 4b^n$

155. The factored form of a polynomial $P(x)$ is $x^n(x - a)(x + b)$. What is the degree of the polynomial P?

156. The factored form of a polynomial $P(x)$ is $x^{2n}(x^n + a)(x^n + b)$. What is the degree of the polynomial P?

6.4 Applying Concepts

Find all integers k such that the trinomial is a perfect-square trinomial.

157. $4x^2 - kx + 25$ **158.** $9x^2 - kx + 1$

159. $16x^2 + kxy + y^2$ **160.** $49x^2 + kxy + 64y^2$

Factor.

161. $ax^3 + b - bx^3 - a$ **162.** $xy^2 - 2b - x + 2by^2$

163. $y^{8n} - 2y^{4n} + 1$ **164.** $x^{6n} - 1$

Solve.

165. Number Problem The product of two numbers is 63. One of the two numbers is a perfect square. The other is a prime number. Find the sum of the two numbers.

166. Number Problem What is the smallest whole number by which 250 can be multiplied so that the product will be a perfect square?

167. Number Problem Palindromic numbers are natural numbers that remain unchanged when their digits are written in reverse order. Find all perfect squares less than 500 that are palindromic numbers.

168. Food Science A large circular cookie is cut from a square piece of dough. The diameter of the cookie is x cm. The piece of dough is x cm on a side and 1 cm deep. In terms of x, how many cubic centimeters of dough are left over? Use the approximation 3.14 for π.

169. Factor $x^4 + 64$. (*Suggestion:* Add and subtract $16x^2$ so that the expression becomes $(x^4 + 16x^2 + 64) - 16x^2$. Now factor the difference of two squares.)

170. Using the strategy of Exercise 169, factor $x^4 + 4$. (*Suggestion:* Add and subtract $4x^2$.)

171. Can a third-degree polynomial have factors $x - 1$, $x + 1$, $x - 3$, and $x + 4$? Why or why not?

6.5 Solving Equations by Factoring

OBJECTIVE 1 Solve equations by factoring

Consider the equation $ab = 0$. If a is not zero, then b must be zero. Conversely, if b is not zero, then a must be zero. This is summarized in the Principle of Zero Products.

> **Principle of Zero Products**
>
> If the product of two factors is zero, then at least one of the factors must be zero.
>
> If $ab = 0$, then $a = 0$ or $b = 0$.

The Principle of Zero Products is used to solve equations.

Solve: $(x - 4)(x + 2) = 0$

By the Principle of Zero Products, if $(x - 4)(x + 2) = 0$, then $x - 4 = 0$ or $x + 2 = 0$.

$$x - 4 = 0 \qquad\qquad x + 2 = 0$$
$$x = 4 \qquad\qquad x = -2$$

Check:

$(x - 4)(x + 2) = 0$		$(x - 4)(x + 2) = 0$	
$(4 - 4)(4 + 2)$	0	$(-2 - 4)(-2 + 2)$	0
$0 \cdot 6$	0	$-6 \cdot 0$	0
	$0 = 0$		$0 = 0$

-2 and 4 check as solutions. The solutions are -2 and 4.

An equation of the form $ax^2 + bx + c = 0$, $a \neq 0$, is a **quadratic equation.** A quadratic equation is in **standard form** when the polynomial is written in descending order and is equal to zero.

Some quadratic equations can be solved by factoring and then using the Principle of Zero Products.

Solve: $2x^2 - x = 1$

	$2x^2 - x = 1$
Write the equation in standard form.	$2x - x - 1 = 0$
Factor the trinomial.	$(2x + 1)(x - 1) = 0$
Use the Principle of Zero Products.	$2x + 1 = 0 \qquad x - 1 = 0$
Solve each equation for x.	$2x = -1 \qquad\quad x = 1$
	$x = -\dfrac{1}{2}$
The solutions check.	The solutions are $-\dfrac{1}{2}$ and 1.

The Principle of Zero Products can be extended to more than two factors. For example, if $abc = 0$, then $a = 0$, $b = 0$, or $c = 0$.

Solve: $x^3 - x^2 - 4x + 4 = 0$

Factor by grouping.
$$x^3 - x^2 - 4x + 4 = 0$$
$$(x^3 - x^2) - (4x - 4) = 0$$
$$x^2(x - 1) - 4(x - 1) = 0$$
$$(x - 1)(x^2 - 4) = 0$$
$$(x - 1)(x + 2)(x - 2) = 0$$

Use the Principle of Zero Products.
$$x - 1 = 0 \qquad x + 2 = 0 \qquad x - 2 = 0$$

Solve each equation for x.
$$x = 1 \qquad\quad x = -2 \qquad\quad x = 2$$

The solutions are -2, 1, and 2.

EXAMPLE 1 Solve.

A. $3x^2 + 5x = 2$ **B.** $(x + 4)(x - 3) = 8$

C. $x^3 - x^2 - 25x + 25 = 0$

Solution **A.**
$$3x^2 + 5x = 2$$
$$3x^2 + 5x - 2 = 0$$
• Write the equation in standard form.

$$(3x - 1)(x + 2) = 0$$
• Factor the trinomial.

$$3x - 1 = 0 \qquad x + 2 = 0$$
• Let each factor equal zero (the Principle of Zero Products).

$$3x = 1 \qquad\qquad x = -2$$
• Solve each equation for x.

$$x = \frac{1}{3}$$

The solutions are $\frac{1}{3}$ and -2. • Write the solutions.

B. $(x + 4)(x - 3) = 8$

$x^2 + x - 12 = 8$ • Write the equation in standard form by first multiplying the binomials.

$x^2 + x - 20 = 0$ • Subtract 8 from each side of the equation.

$(x + 5)(x - 4) = 0$ • Factor the trinomial.

$x + 5 = 0 \qquad x - 4 = 0$ • Let each factor equal zero.

$x = -5 \qquad x = 4$ • Solve each equation for x.

The solutions are -5 and 4. • Write the solutions.

C. $x^3 - x^2 - 25x + 25 = 0$

$(x^3 - x^2) - (25x - 25) = 0$ • Factor by grouping.

$x^2(x - 1) - 25(x - 1) = 0$

$(x^2 - 25)(x - 1) = 0$

$(x + 5)(x - 5)(x - 1) = 0$

$x + 5 = 0 \qquad x - 5 = 0 \quad x - 1 = 0$ • Let each factor equal zero.

$x = -5 \qquad x = 5 \qquad x = 1$ • Solve each equation for x.

The solutions are -5, 1, and 5. • Write the solutions.

PROBLEM 1 Solve.

A. $4x^2 + 11x = 3$ **B.** $(x - 2)(x + 5) = 8$

C. $x^3 + 4x^2 - 9x - 36 = 0$

Solution See page S22.

The Principle of Zero Products is used to find elements in the domain of a quadratic function that correspond to a given element in the range.

EXAMPLE 2 Given that -1 is in the range of the function defined by $f(x) = x^2 - 3x - 5$, find two values of c for which $f(c) = -1$.

Solution

$f(c) = -1$

$c^2 - 3c - 5 = -1$ • $f(c) = c^2 - 3c - 5$

$c^2 - 3c - 4 = 0$ • Write in standard form.

$(c - 4)(c + 1) = 0$ • Solve for c by factoring.

$c - 4 = 0 \qquad c + 1 = 0$

$c = 4 \qquad c = -1$

The values of c are -1 and 4.

PROBLEM 2 Given that 4 is in the range of the function defined by $s(t) = t^2 - t - 2$, find two values of c for which $s(c) = 4$.

Solution See page S22.

In Example 2, there are two values in the domain that can be paired with the range element -1. The two values are -1 and 4. Two ordered pairs that belong to the function are $(-1, -1)$ and $(4, -1)$. *Remember:* A function can have different first elements paired with the same second element. A function cannot have the same first element paired with different second elements.

The graph of $f(x) = x^2 - 3x - 5$ is shown at the left.

Take Note

You can check the values of c by evaluating the function.

$f(x) = x^2 - 3x - 5$
$f(4) = 4^2 - 3(4) - 5$
$\quad = -1$ s

$f(-1) = (-1)^2 - 3(-1) - 5$
$\quad = -1$ s

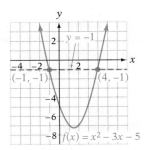

Recall that a *zero* of a function f is a number c for which $f(c) = 0$. The zeros of some quadratic functions can be found by using the Principle of Zero Products. Finding the zeros is similar to Example 2, where the range element is 0.

> **EXAMPLE 3** Find the zeros of $f(x) = x^2 - 2x - 8$.
>
> **Solution** We must find the values of c for which $f(c) = 0$.
>
> $$f(c) = 0$$
> $$c^2 - 2c - 8 = 0$$ • $f(x) = x^2 - 2x - 8$. Thus $f(c) = c^2 - 2c - 8$.
> $$(c + 2)(c - 4) = 0$$ • **Factor and use the Principle of Zero Products.**
>
> $$c + 2 = 0 \qquad c - 4 = 0$$
> $$c = -2 \qquad\quad c = 4$$
>
> The zeros of f are -2 and 4.
>
> As mentioned earlier in the text, the x-coordinates of the x-intercepts of the graph of a function are zeros of the function. See the graph at the left.

PROBLEM 3 Find the zeros of $s(t) = t^2 + 5t - 14$.

Solution See page S23.

OBJECTIVE **2** **Application problems**

> **EXAMPLE 4** An architect wants to design a fountain to be placed on the front lawn of a new art museum. The base of the fountain is to be 30 ft by 40 ft, and there is to be a uniform brick walkway around the fountain. The total area of the fountain and the walkway is 2576 ft². How wide is the walkway?
>
> **Strategy** Let x represent the width of the brick walkway. Then the width of the fountain and walkway is $30 + 2x$, and the length of the fountain and walkway is $40 + 2x$. Use the formula $A = LW$, where $A = 2576$, to write an equation.
>
> **Solution**
> $$A = LW$$
> $$2576 = (40 + 2x)(30 + 2x)$$ • **Replace A, L, and W by their values.**
> $$2576 = 1200 + 140x + 4x^2$$ • **Multiply.**
> $$0 = 4x^2 + 140x - 1376$$
> $$0 = x^2 + 35x - 344$$ • **Divide each side by 4.**
> $$0 = (x - 8)(x + 43)$$ • **Factor and use the Principle of Zero Products.**
>
> $$x - 8 = 0 \qquad x + 43 = 0$$
> $$x = 8 \qquad\qquad x = -43$$
>
> A walkway -43 ft wide would not make sense.
>
> The walkway must be 8 ft wide.

PROBLEM 4 A diagonal of a polygon is a line from one vertex of the polygon to a nonadjacent vertex. The number of diagonals, D, of a polygon with n sides is given by $D = \frac{n(n-3)}{2}$. Find the number of sides for a polygon that has 20 diagonals.

■ **Solution** See page S23.

6.5 Concept Review

Determine whether the statement is always true, sometimes true, or never true.

1. The Principle of Zero Products can be stated as follows: If $ab = 0$, then $a = 0$ or $b = 0$.

2. One side of an equation must be zero in order for us to solve the equation by factoring.

3. Two solutions of an equation are found when solving an equation by the Principle of Zero Products.

4. The Principle of Zero Products can be used to solve a second-degree equation.

5. If n is an integer, then $n + 1$ and $n + 3$ represent odd integers.

6. By the Principle of Zero Products, if $(x + 2)(x + 3) = 4$, then $x + 2 = 4$ or $x + 3 = 4$.

6.5 Exercises

1 Solve equations by factoring

1. What is a quadratic equation? How does it differ from a linear equation? Give an example of each type of equation.

2. The Principle of Zero Products states that if the product of two factors is ___?___, then at least one of the factors is ___?___.

3. Of the two quadratic equations $0 = 5x^2 + 2x - 3$ and $x^2 - 4x = 3$, the one that is in standard form is ___?___.

4. Solve: $3x^2 - 5x - 2 = 0$

$$3x^2 - 5x - 2 = 0$$
$$(x - \underline{\;?\;})(3x + \underline{\;?\;}) = 0$$
$$x - 2 = \underline{\;?\;} \qquad 3x + 1 = \underline{\;?\;}$$
$$x = \underline{\;?\;} \qquad\qquad x = \underline{\;?\;}$$

- The equation is in _____?_____ form.
- Factor the trinomial.
- Use the Principle of Zero Products to set each factor equal to 0.
- Solve each equation for x.

5. Let $f(x) = x^2 + x - 15$. To find two values of c for which $f(c) = 5$, solve the equation $x^2 + x - 15 = \underline{\;?\;}$.

6. How is the Principle of Zero Products used to solve some quadratic equations?

Solve.

7. $(y + 4)(y + 6) = 0$

8. $(a - 5)(a - 2) = 0$

9. $x(x - 7) = 0$

10. $b(b + 8) = 0$

11. $3z(2z + 5) = 0$

12. $4y(3y - 2) = 0$

13. $(2x + 3)(x - 7) = 0$

14. $(4a - 1)(a + 9) = 0$

15. $b^2 - 49 = 0$

16. $4z^2 - 1 = 0$

17. $9t^2 - 16 = 0$

18. $x^2 + x - 6 = 0$

19. $y^2 + 4y - 5 = 0$

20. $a^2 - 8a + 16 = 0$

21. $2b^2 - 5b - 12 = 0$

22. $t^2 - 8t = 0$

23. $x^2 - 9x = 0$

24. $2y^2 - 10y = 0$

25. $3a^2 - 12a = 0$

26. $b^2 - 4b = 32$

27. $z^2 - 3z = 28$

28. $2x^2 - 5x = 12$

29. $3t^2 + 13t = 10$

30. $4y^2 - 19y = 5$

31. $5b^2 - 17b = -6$

32. $6a^2 + a = 2$

33. $8x^2 - 10x = 3$

34. $z(z - 1) = 20$

35. $y(y - 2) = 35$

36. $t(t + 1) = 42$

37. $x(x - 12) = -27$

38. $x(2x - 5) = 12$

39. $y(3y - 2) = 8$

40. $2b^2 - 6b = b - 3$

41. $3a^2 - 4a = 20 - 15a$

42. $2t^2 + 5t = 6t + 15$

43. $(y + 5)(y - 7) = -20$

44. $(x + 2)(x - 6) = 20$

45. $(b + 5)(b + 10) = 6$

46. $(a - 9)(a - 1) = -7$

47. $(t - 3)^2 = 1$

48. $(y - 4)^2 = 4$

49. $(3 - x)^2 + x^2 = 5$

50. $(2 - b)^2 + b^2 = 10$

51. $(a - 1)^2 = 3a - 5$

52. $2x^3 + x^2 - 8x - 4 = 0$

53. $x^3 + 4x^2 - x - 4 = 0$

54. $12x^3 - 8x^2 - 3x + 2 = 0$

Find the values c in the domain of f for which $f(c)$ is the indicated value.

55. $f(x) = x^2 - 3x + 3; f(c) = 1$

56. $f(x) = x^2 + 4x - 2; f(c) = 3$

57. $f(x) = 2x^2 - x - 5; f(c) = -4$

58. $f(x) = 6x^2 - 5x - 9; f(c) = -3$

59. $f(x) = 4x^2 - 4x + 3; f(c) = 2$

60. $f(x) = x^2 - 6x + 12; f(c) = 3$

Find the zero(s) of the function.

61. $f(x) = x^2 + 3x - 4$

62. $g(x) = x^2 + 2x - 15$

63. $s(t) = t^2 - 4t - 12$

64. $f(t) = t^2 + 3t - 18$

65. $f(x) = 2x^2 - 5x + 3$

66. $g(x) = 3x^2 - 7x - 6$

67. $p(x) = x^2 + 6x + 9$

68. $f(x) = 4x^2 - 4x + 1$

2 Application problems

69. Let x represent a positive integer. Then the next consecutive positive integer is ___?___, and an expression that represents the sum of the squares of the two integers is ___?___.

70. Suppose you are given the following problem: "The sum of the squares of two consecutive positive integers is 61. Find the two integers." The equation you solve to determine the answer has solutions -6 and 5. Which solution should be eliminated and why?

71. Number Problem The sum of a number and its square is 90. Find the number.

72. Number Problem The sum of a number and its square is 132. Find the number.

73. Number Problem The sum of the squares of two consecutive positive integers is equal to 145. Find the two integers.

74. Number Problem The sum of the squares of two consecutive odd integers is equal to 290. Find the two integers.

75. Number Problem The sum of the cube of a number and the product of the number and twelve is equal to seven times the square of the number. Find the number.

76. Number Problem The sum of the cube of a number and the product of the number and seven is equal to eight times the square of the number. Find the number.

77. Geometry The length of a rectangle is 5 in. more than twice the width. The area is 168 in². Find the width and length of the rectangle.

78. Geometry The width of a rectangle is 5 ft less than the length. The area of the rectangle is 300 ft². Find the length and width of the rectangle.

79. Construction A trough is made from a rectangular piece of metal by folding up the sides as shown in the figure at the right. What should the value of x be so that the volume is 72 m³?

80. Geometry The number of possible diagonals, D, in a polygon with n sides is given by $D = \frac{n(n - 3)}{2}$. Find the number of sides for a polygon with 54 diagonals.

81. Space Science A small rocket is launched with an acceleration of 10 m/s². The velocity, v, of the rocket after it has traveled s meters is given by $v^2 = 20s$. Find the velocity of this rocket after it has traveled 500 m.

82. Construction A pole is being secured with two guy wires, as shown in the figure at the right. What is the value of x so that the 20-foot guy wire will attach as shown?

83. Geometry The length of the base of a triangle is three times the height. The area of the triangle is 24 cm². Find the base and height of the triangle.

84. Geometry The height of a triangle is 4 cm more than twice the length of the base. The area of the triangle is 35 cm². Find the height of the triangle.

85. Physics An object is thrown downward, with an initial speed of 16 ft/s, from the top of a building 480 ft high. How many seconds later will the object hit the ground? Use the equation $d = vt + 16t^2$, where d is the distance in feet, v is the initial speed in feet per second, and t is the time in seconds.

86. Physics A stone is thrown into an empty well with an initial speed of 8 ft/s. The well is 624 ft deep. How many seconds later will the stone hit the bottom of the well? Use the equation $d = vt + 16t^2$, where d is the distance in feet, v is the initial speed in feet per second, and t is the time in seconds.

87. **Geometry** The length of a rectangle is 6 cm, and the width is 3 cm. If both the length and the width are increased by equal amounts, the area of the rectangle is increased by 70 cm². Find the length and width of the larger rectangle.

88. **Sports** A league has 28 games scheduled. How many teams are in the league if each team plays every other team once? Use the formula $N = \dfrac{t^2 - t}{2}$, where N is the number of basketball games that must be scheduled in a league with t teams if each team is to play every other team once.

6.5 Applying Concepts

89. Find $3n^2 + 2n - 1$ if $n(n + 6) = 16$.

90. Find $4y^2 - y + 3$ if $(y - 1)(y + 2) = 4$.

91. Find $-a^2 + 5a - 2$ if $6a(a - 1) = 36$.

Find the values c in the domain of f for which $f(c)$ is the indicated value.

92. $f(x) = x^3 + 9x^2 - x - 14; f(c) = -5$

93. $f(x) = x^3 + 3x^2 - 4x - 11; f(c) = 1$

94. **Geometry** The perimeter of a rectangular garden is 44 m. The area of the garden is 120 m². Find the length and width of the garden.

95. **Construction** A rectangular piece of cardboard is 10 in. longer than it is wide. Squares 2 in. on a side are to be cut from each corner and then the sides folded up to make an open box with a volume of 112 in³. Find the length and width of the piece of cardboard.

96. **Geometry** The sides of a rectangular box have areas of 16 cm², 20 cm², and 80 cm². Find the volume of the box.

97. Write an equation whose solutions are 3, 2, and −1.

98. The following seems to show that $1 = 2$. Explain the error.

$$a = b$$
$$a^2 = ab \qquad \text{• Multiply each side of the equation by } a.$$
$$a^2 - b^2 = ab - b^2 \qquad \text{• Subtract } b^2 \text{ from each side of the equation.}$$
$$(a - b)(a + b) = b(a - b) \qquad \text{• Factor.}$$
$$a + b = b \qquad \text{• Divide each side by } a - b.$$
$$b + b = b \qquad \text{• Because } a = b, \text{ substitute } b \text{ for } a.$$
$$2b = b$$
$$2 = 1 \qquad \text{• Divide both sides by } b.$$

Focus on Problem Solving

The Trial-and-Error Method

A topic in Section 3 of this chapter is factoring trinomials using trial factors. This method involves writing trial factors and then using the FOIL method to determine which pair of factors is the correct one. The **trial-and-error method** of arriving at a solution to a problem involves performing repeated tests or experiments until a satisfactory conclusion is reached. However, not all problems solved using the trial-and-error method have a strategy by which to determine the answer. Here is an example:

Explain how you could cut through a cube so that the face of the resulting solid is **a.** a square, **b.** an equilateral triangle, **c.** a trapezoid, and **d.** a hexagon.

There is no formula to apply to this problem; there is no computation to perform. This problem requires picturing a cube and the results after cutting through it at different places on its surface and at different angles. For part a, cutting perpendicular to the top and bottom of the cube and parallel to two of its sides will result in a square. The other shapes may prove more difficult.

When solving problems of this type, keep an open mind. Sometimes when using the trial-and-error method, we are hampered by narrowness of vision; we cannot expand our thinking to include other possibilities. Then when we see someone else's solution, it appears so obvious! For example, for the question above, it is necessary to conceive of cutting through the cube at places other than the top surface; we need to be open to the idea of beginning the cut at one of the corner points of the cube.

Look at the letter A printed above at the left. If the letter were folded along line ℓ, the two sides of the letter would match exactly. This letter has **symmetry** with respect to line ℓ. Line ℓ is called the **axis of symmetry.** Now consider the letter H printed below at the left. Both lines ℓ_1 and ℓ_2 are axes of symmetry for this letter; the letter could be folded along either line and the two sides would match exactly. Does the letter A have more than one axis of symmetry? Find axes of symmetry for other capital letters of the alphabet. Which lower-case letters have one axis of symmetry? Do any of the lower-case letters have more than one axis of symmetry?

How many axes of symmetry does a square have? In determining lines of symmetry for a square, begin by drawing a square as shown at the right. The horizontal line of symmetry and the vertical line of symmetry may be immediately obvious to you. But there are two others. Do you see that a line drawn through opposite corners of the square is also a line of symmetry?

Find the number of axes of symmetry for each of the following figures: equilateral triangle, rectangle, regular pentagon, regular hexagon, and circle.

Many of the questions in this text that require an answer of "always true, sometimes true, or never true" are best solved by the trial-and-error method. For example, consider the following statement:

If two rectangles have the same area, then they have the same perimeter.

Try some numbers. Each of two rectangles, one measuring 6 units by 2 units and another measuring 4 units by 3 units, has an area of 12 square units, but the perimeter of the first is 16 units and the perimeter of the second is 14 units. Thus the answer "always true" has been eliminated. We still need to determine whether there is a case where the statement is true. After experimenting with a lot of numbers, you may come to realize that we are trying to determine whether it is possible for two different pairs of factors of a number to have the same sum. Is it?

Don't be afraid to make many experiments, and remember that *errors*, or tests that "don't work," are a part of the trial-and-*error* process.

Projects & Group Activities

Prime and Composite Numbers

A **prime number** is a natural number greater than 1 whose only natural-number factors are itself and 1. The number 11 is a prime number because the only natural-number factors of 11 are 11 and 1.

Eratosthenes, a Greek philosopher and astronomer who lived from 270 to 190 B.C., devised a method of identifying prime numbers. It is called the **Sieve of Eratosthenes.** The procedure is illustrated in the following paragraphs.

1	②	③	4	⑤	6	⑦	8	9	10
⑪	12	⑬	14	15	16	⑰	18	⑲	20
21	22	㉓	24	25	26	27	28	㉙	30
㉛	32	33	34	35	36	㊲	38	39	40
㊶	42	㊸	44	45	46	㊼	48	49	50
51	52	㊿	54	55	56	57	58	㊾	60
61	62	63	64	65	66	67	68	69	70
71	72	73	74	75	76	77	78	79	80
81	82	83	84	85	86	87	88	89	90
91	92	93	94	95	96	97	98	99	100

List all the natural numbers from 1 to 100. Cross out the number 1, because it is not a prime number. The number 2 is prime; circle it. Cross out all the other multiples of 2 (4, 6, 8, . . .), because they are not prime. The number 3 is prime; circle it. Cross out all the other multiples of 3 (6, 9, 12, . . .) that are not already crossed out. The number 4, the next consecutive number in the list, has already been crossed out. The number 5 is prime; circle it. Cross out all the other multiples of 5 that are not already crossed out. Continue in this manner until all the prime numbers less than 100 are circled.

A **composite number** is a natural number greater than 1 that has a natural-number factor other than itself and 1. The number 21 is a composite number because it has factors of 3 and 7. All the numbers crossed out in the table above, except the number 1, are composite numbers.

Solve.

1. Use the Sieve of Eratosthenes to find the prime numbers between 100 and 200.

2. How many prime numbers are even numbers?

3. Find the "twin primes" between 1 and 200. Twin primes are two prime numbers whose difference is 2. For instance, 3 and 5 are twin primes; 5 and 7 are also twin primes.

4. **a.** List two prime numbers that are consecutive natural numbers.

 b. Can there be any other pairs of prime numbers that are consecutive natural numbers?

5. Some primes are the sum of a square and 1. For example, $5 = 2^2 + 1$. Find another prime p such that $p = n^2 + 1$, where n is a natural number.

6. Find a prime p such that $p = n^2 - 1$, where n is a natural number.

7. **a.** 4! (which is read "4 factorial") is equal to $4 \cdot 3 \cdot 2 \cdot 1$. Show that $4! + 2$, $4! + 3$, and $4! + 4$ are all composite numbers.

 b. 5! (which is read "5 factorial") is equal to $5 \cdot 4 \cdot 3 \cdot 2 \cdot 1$. Will $5! + 2$, $5! + 3, 5! + 4$, and $5! + 5$ generate four consecutive composite numbers?

 c. Use the notation 6! to represent a list of five consecutive composite numbers.

Chapter 6 Summary

Key Words

The **greatest common factor** (GCF) of two or more integers is the greatest integer that is a factor of all the integers. [6.1.1, p. 430]

To **factor a polynomial** means to write the polynomial as a product of other polynomials. [6.1.1, p. 431]

Examples

The greatest common factor of 12 and 18 is 6 because 6 is the greatest integer that divides evenly into both 12 and 18.

To factor $8x + 12$ means to write it as the product $4(2x + 3)$. The expression $8x + 12$ is a sum. The expression $4(2x + 3)$ is a product (the polynomial $2x + 3$ is multiplied by 4).

A polynomial that does not factor using only integers is **nonfactorable over the integers.** [6.2.1, p. 439]	The trinomial $x^2 + x + 2$ is nonfactorable over the integers. There are no two integers whose product is 2 and whose sum is 1.
To **factor a trinomial of the form $ax^2 + bx + c$** means to express the trinomial as the product of two binomials. [6.3.1, p. 444]	To factor $3x^2 + 7x + 2$ means to write it as the product $(3x + 1)(x + 2)$.
The product of a term and itself is a **perfect square.** [6.4.1, p. 455]	$(5x)(5x) = 25x^2$; $25x^2$ is a perfect square.
The **square root of a perfect square** is one of the two equal factors of the perfect square. [6.4.1, p. 455]	$\sqrt{25x^2} = 5x$
The product of three factors is called a **perfect cube.** [6.4.2, p. 456]	$(2x)(2x)(2x) = 8x^3$; $8x^3$ is a perfect cube.
The **cube root** of a perfect cube is one of the three equal factors of the perfect cube. [6.4.2, p. 457]	$\sqrt[3]{8x^3} = 2x$
A trinomial is **quadratic in form** if it can be written as $au^2 + bu + c$. [6.4.3, p. 457]	Let $x^2 = u$. $x^4 + 7x^2 + 12 = u^2 + 7u + 12$ $x^4 + 7x^2 + 12$ is quadratic in form.
An equation of the form $ax^2 + bx + c = 0$, $a \neq 0$, is a **quadratic equation.** A quadratic equation is in **standard form** when the polynomial is in descending order and equal to zero. [6.5.1, p. 464]	The equation $2x^2 + 7x + 3 = 0$ is a quadratic equation in standard form.

Essential Rules and Procedures

Factoring the Difference of Two Squares [6.4.1, p. 455]
The difference of two squares is the product of the sum and difference of two terms.
$a^2 - b^2 = (a + b)(a - b)$

$y^2 - 81 = (y + 9)(y - 9)$

Factoring a Perfect-Square Trinomial [6.4.1, p. 455]
A perfect-square trinomial is the square of a binomial.
$a^2 + 2ab + b^2 = (a + b)^2$
$a^2 - 2ab + b^2 = (a - b)^2$

$x^2 + 10x + 25 = (x + 5)^2$
$x^2 - 10x + 25 = (x - 5)^2$

Factoring the Sum or Difference of Two Cubes [6.4.2, p. 457]
$a^3 + b^3 = (a + b)(a^2 - ab + b^2)$
$a^3 - b^3 = (a - b)(a^2 + ab + b^2)$

$x^3 + 64 = (x + 4)(x^2 - 4x + 16)$
$8b^3 - 1 = (2b - 1)(4b^2 + 2b + 1)$

General Factoring Strategy [6.4.4, p. 458]

1. Is there a common factor? If so, factor out the GCF.

2. If the polynomial is a binomial, is it the difference of two perfect squares, the sum of two perfect cubes, or the difference of two perfect cubes? If so, factor.

3. If the polynomial is a trinomial, is it a perfect-square trinomial or the product of two binomials? If so, factor.

4. If the polynomial has four terms, can it be factored by grouping? If so, factor.

5. Is each factor nonfactorable over the integers? If not, factor.

$$24x + 6 = 6(4x + 1)$$
$$2x^2y + 6xy + 8y = 2y(x^2 + 3x + 4)$$
$$4x^2 - 49 = (2x + 7)(2x - 7)$$

$$9x^2 + 6x + 1 = (3x + 1)^2$$
$$2x^2 + 7x + 5 = (2x + 5)(x + 1)$$
$$x^3 - 3x^2 + 2x - 6 = (x^3 - 3x^2) + (2x - 6)$$
$$= x^2(x - 3) + 2(x - 3)$$
$$= (x - 3)(x^2 + 2)$$
$$x^4 - 16 = (x^2 + 4)(x^2 - 4)$$
$$= (x^2 + 4)(x + 2)(x - 2)$$

The Principle of Zero Products [6.5.1, p. 464]

If the product of two factors is zero, then at least one of the factors must be zero.

If $a \cdot b = 0$, then $a = 0$ or $b = 0$.

The Principle of Zero Products is used to solve a quadratic equation by factoring.

$$x^2 + x = 12$$
$$x^2 + x - 12 = 0$$
$$(x + 4)(x - 3) = 0$$

$$x + 4 = 0 \qquad x - 3 = 0$$
$$x = -4 \qquad\quad x = 3$$

Chapter 6 Review Exercises

1. Factor: $14y^9 - 49y^6 + 7y^3$

2. Factor: $3a^2 - 12a + ab - 4b$

3. Factor: $c^2 + 8c + 12$

4. Factor: $a^3 - 5a^2 + 6a$

5. Factor: $6x^2 - 29x + 28$

6. Factor: $3y^2 + 16y - 12$

7. Factor: $18a^2 - 3a - 10$

8. Factor: $a^2b^2 - 1$

9. Factor: $4y^2 - 16y + 16$

10. Solve: $a(5a + 1) = 0$

11. Factor: $12a^2b + 3ab^2$

12. Factor: $b^2 - 13b + 30$

13. Factor: $10x^2 + 25x + 4xy + 10y$

14. Factor: $3a^2 - 15a - 42$

15. Factor: $n^4 - 2n^3 - 3n^2$

16. Factor: $2x^2 - 5x + 6$

17. Factor: $6x^2 - 7x + 2$

18. Factor: $16x^2 + 49$

19. Solve: $(x - 2)(2x - 3) = 0$

20. Factor: $7x^2 - 7$

21. Factor: $3x^5 - 9x^4 - 4x^3$

22. Factor: $4x(x - 3) - 5(3 - x)$

23. Factor: $a^2 + 5a - 14$

24. Factor: $y^2 + 5y - 36$

25. Factor: $5x^2 - 50x - 120$

26. Solve: $(x + 1)(x - 5) = 16$

27. Factor: $5x^{n+5} + x^{n+3} + 4x^2$

28. Factor: $x(y - 3) + 4(3 - y)$

29. Factor: $2ax + 4bx - 3ay - 6by$

30. Factor: $x^2 + 12x + 35$

31. Factor: $12 + x - x^2$

32. Find the zeros of $f(x) = x^2 - 5x - 6$.

33. Factor: $6x^2 - 31x + 18$

34. Factor: $24x^2 + 61x - 8$

35. Factor: $x^2y^2 - 9$

36. Factor: $4x^2 + 12xy + 9y^2$

37. Factor: $x^{2n} - 12x^n + 36$

38. Factor: $36 - a^{2n}$

39. Factor: $64a^3 - 27b^3$

40. Factor: $8 - y^{3n}$

41. Factor: $15x^4 + x^2 - 6$

42. Factor: $36x^8 - 36x^4 + 5$

43. Factor: $21x^4y^4 + 23x^2y^2 + 6$

44. Factor: $3a^6 - 15a^4 - 18a^2$

45. Factor: $x^{4n} - 8x^{2n} + 16$

46. Factor: $3a^4b - 3ab^4$

47. Solve: $x^3 - x^2 - 6x = 0$

48. Solve: $6x^2 + 60 = 39x$

49. Solve: $x^3 - 16x = 0$

50. Solve: $y^3 + y^2 - 36y - 36 = 0$

51. Factor: $16x^2 - 94x + 33$

52. Factor: $9x^2 - 30x + 25$

53. Factor: $12y^2 + 16y - 3$

54. Factor: $3x^2 + 36x + 108$

55. Sports The length of a playing field is twice the width. The area is 5000 yd². Find the length and width of the playing field.

56. Sports The length of a hockey field is 20 yd less than twice the width. The area of the field is 6000 yd². Find the length and width of the hockey field.

57. Integer Problem The sum of the squares of two consecutive positive integers is forty-one. Find the two integers.

58. Motion Pictures The size of a motion picture on the screen is given by the equation $S = d^2$, where d is the distance between the projector and the screen. Find the distance between the projector and the screen when the size of the picture is 400 ft².

59. Gardening A rectangular garden plot has dimensions 15 ft by 12 ft. A uniform path around the garden increases the total area to 270 ft². What is the width of the larger rectangle?

60. Geometry The length of each side of a square is extended 4 ft. The area of the resulting square is 576 ft². Find the length of a side of the original square.

Chapter 6 Test

1. Factor: $6x^2y^2 + 9xy^2 + 12y^2$

2. Factor: $6x^3 - 8x^2 + 10x$

3. Factor: $p^2 + 5p + 6$

4. Factor: $a(x - 2) + b(2 - x)$

5. Solve: $(2a - 3)(a + 7) = 0$

6. Factor: $a^2 - 19a + 48$

7. Factor: $x^3 + 2x^2 - 15x$

8. Factor: $8x^2 + 20x - 48$

9. Factor: $ab + 6a - 3b - 18$

10. Solve: $4x^2 - 1 = 0$

11. Factor: $6x^2 + 19x + 8$

12. Factor: $6a^4 - 13a^2 - 5$

13. Factor: $2b^2 - 32$

14. Factor: $4a^2 - 12ab + 9b^2$

15. Find the zeros of $g(x) = 2x^2 - 5x - 12$.

16. Factor: $27x^3 - 8$

17. Factor: $2x^2 + 4x - 5$

18. Factor: $4x^2 - 49y^2$

19. Solve: $x(x - 8) = -15$

20. Factor: $p^2 + 12p + 36$

21. Factor: $18x^2 - 48xy + 32y^2$

22. Factor: $2y^4 - 14y^3 - 16y^2$

23. **Geometry** The length of a rectangle is 3 cm more than twice the width. The area of the rectangle is 90 cm². Find the length and width of the rectangle.

24. **Geometry** The length of the base of a triangle is three times the height. The area of the triangle is 24 in². Find the length of the base of the triangle.

25. **Integer Problem** The product of two consecutive negative integers is one hundred fifty-six. Find the two integers.

Cumulative Review Exercises

1. Subtract: $4 - (-5) - 6 - 11$

2. Divide: $0.372 \div (-0.046)$
Round to the nearest tenth.

3. Simplify: $(3 - 7)^2 \div (-2) - 3 \cdot (-4)$

4. Evaluate $-2a^2 \div (2b) - c$ when $a = -4$, $b = 2$, and $c = -1$.

5. Identify the property that justifies the statement.
$(3 + 8) + 7 = 3 + (8 + 7)$

6. Multiply: $-\dfrac{3}{4}(-24x^2)$

7. Simplify: $-2[3x - 4(3 - 2x) - 8x]$

8. Solve: $-\dfrac{5}{7}x = -\dfrac{10}{21}$

9. Solve: $4 + 3(x - 2) = 13$

10. Solve: $3x - 2 = 12 - 5x$

11. Solve: $-2 + 4[3x - 2(4 - x) - 3] = 4x + 2$

12. 120% of what number is 42?

13. Solve: $-4x - 2 \geq 10$

14. Solve: $9 - 2(4x - 5) < 3(7 - 6x)$

15. Graph: $y = \dfrac{3}{4}x - 2$

16. Graph: $f(x) = -3x - 3$

17. Find the domain and range of the relation $\{(-5, -4), (-3, -2), (-1, 0), (1, 2), (3, 4)\}$. Is the relation a function?

18. Evaluate $f(x) = 6x - 5$ at $x = 11$.

19. Graph the solution set of $x + 3y > 2$.

20. Solve by substitution: $6x + y = 7$
$\qquad\qquad\qquad\qquad\qquad x - 3y = 17$

21. Solve by the addition method: $2x - 3y = -4$
$\qquad\qquad\qquad\qquad\qquad\qquad\qquad 5x + y = 7$

22. Add: $(3y^3 - 5y^2 - 6) + (2y^2 - 8y + 1)$

23. Simplify: $(-3a^4b^2)^3$

24. Multiply: $(x + 2)(x^2 - 5x + 4)$

25. Divide: $(8x^2 + 4x - 3) \div (2x - 3)$

26. Simplify: $(x^{-4}y^2)^3$

27. Factor: $3a - 3b - ax + bx$

28. Factor: $x^2 + 3xy - 10y^2$

29. Factor: $6a^4 + 22a^3 + 12a^2$

30. Factor: $25a^2 - 36b^2$

31. Factor: $12x^2 - 36xy + 27y^2$

32. Solve: $3x^2 + 11x - 20 = 0$

33. Find the range of the function given by the equation $f(x) = \dfrac{4}{5}x - 3$ if the domain is $\{-10, -5, 0, 5, 10\}$.

34. **Temperature** The daily high temperatures, in degrees Celsius, during one week were recorded as follows: $-4°$, $-7°$, $2°$, $0°$, $-1°$, $-6°$, $-5°$. Find the average daily high temperature for the week.

35. **Geometry** The width of a rectangle is 40% of the length. The perimeter of the rectangle is 42 cm. Find the length and width of the rectangle.

36. **Carpentry** A board 10 ft long is cut into two pieces. Four times the length of the shorter piece is 2 ft less than three times the length of the longer piece. Find the length of each piece.

37. **Rental Cars** Company A rents cars for \$6 a day and 25¢ for every mile driven. Company B rents cars for \$15 a day and 10¢ per mile driven. You want to rent a car for 6 days. What is the maximum number of miles you can drive a Company A car if it is to cost you less than a Company B car?

38. **Investments** An investment of \$4000 is made at an annual simple interest rate of 8%. How much additional money must be invested at an annual simple interest rate of 11% so that the total interest earned is \$1035?

39. **Discounts** A stereo that regularly sells for \$165 is on sale for \$99. Find the discount rate.

40. **Integer Problem** Find three consecutive even integers such that five times the middle integer is twelve more than twice the sum of the first and third.

7

Rational Expressions

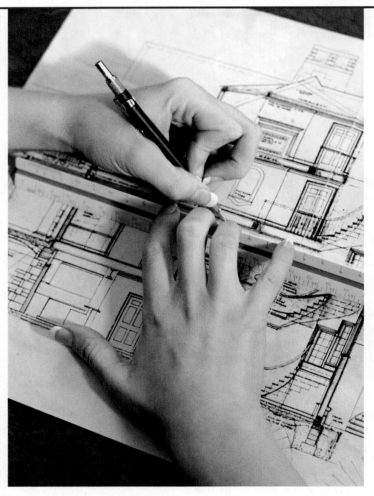

Designing architectural plans such as these requires a great deal of skill and concentration. This architect is creating a scale drawing, in which each dimension is carefully produced using ratios and proportions. These drawings must be very accurate and consistent, since contractors will depend on them to create the real building. You will have a chance to work with scale drawings in **Exercise 19 on page 528,** which asks you to determine the dimensions of a room if one-quarter inch on paper represents one foot in real life.

Online Study Center

DVD SSM

Need help? For online student resources, such as section quizzes, visit the Online Study Center at college.hmco.com/pic/aufmannABI2e.

Prep Test

1. Find the LCM of 10 and 25.

For Exercises 2 to 5, add, subtract, multiply, or divide.

2. $-\dfrac{3}{8} \cdot \dfrac{4}{9}$

3. $-\dfrac{4}{5} \div \dfrac{8}{15}$

4. $-\dfrac{5}{6} + \dfrac{7}{8}$

5. $-\dfrac{3}{8} - \left(-\dfrac{7}{12}\right)$

6. Simplify: $\dfrac{\dfrac{2}{3} - \dfrac{1}{4}}{\dfrac{1}{8} - 2}$

7. Evaluate $\dfrac{2x - 3}{x^2 - x + 1}$ for $x = 2$.

8. Solve: $4(2x + 1) = 3(x - 2)$

9. Solve: $10\left(\dfrac{t}{2} + \dfrac{t}{5}\right) = 10(1)$

10. Two planes start from the same point and fly in opposite directions. The first plane is flying 20 mph slower than the second plane. In 2 h, the planes are 480 mi apart. Find the rate of each plane.

7.1 Introduction to Rational Functions

OBJECTIVE 1 Find the domain of a rational function

An expression in which the numerator and denominator are polynomials is called a **rational expression.** Examples of rational expressions are shown at the right.

$$\dfrac{9}{z} \qquad \dfrac{3x + 4}{2x^2 + 1} \qquad \dfrac{x^3 - x + 1}{x^2 - 3x - 5}$$

The expression $\dfrac{\sqrt{x} + 3}{x}$ is not a rational expression because $\sqrt{x} + 3$ is not a polynomial.

A function that is written in terms of a rational expression is a **rational function.** Each of the following equations represents a rational function.

$$f(x) = \dfrac{x^2 + 3}{2x - 1} \qquad g(t) = \dfrac{3}{t^2 - 4} \qquad R(z) = \dfrac{z^2 + 3z - 1}{z^2 + z - 12}$$

To evaluate a rational function, replace the variable by its value. Then simplify.

EXAMPLE 1 Given $f(x) = \dfrac{3x - 4}{x^2 - 2x + 1}$, find $f(-3)$.

Solution $f(x) = \dfrac{3x - 4}{x^2 - 2x + 1}$

$f(-3) = \dfrac{3(-3) - 4}{(-3)^2 - 2(-3) + 1}$ • **Substitute −3 for x.**

$f(-3) = \dfrac{-9 - 4}{9 + 6 + 1}$

$f(-3) = \dfrac{-13}{16}$

$f(-3) = -\dfrac{13}{16}$

PROBLEM 1 Given $f(x) = \dfrac{3 - 5x}{x^2 + 5x + 6}$, find $f(2)$.

Solution See page S23.

The graph of $f(x) = \dfrac{1}{x - 2}$ is shown at the left. Note that the graph never inter-
sects the line $x = 2$ (shown as a dashed line). This value of x is excluded from
the domain of $f(x) = \dfrac{1}{x - 2}$.

Because division by zero is not defined, the domain of a rational function
must exclude those numbers for which the value of the polynomial in the
denominator is zero.

Determine the domain of $g(x) = \dfrac{x^2 + 4}{3x - 6}$.

The domain of g must exclude values of x for which the denominator
is zero.

Set the denominator equal to zero. $3x - 6 = 0$

Solve for x. $3x = 6$

This value must be *excluded* from the domain. $x = 2$

The domain of g is $\{x \mid x \neq 2\}$.

EXAMPLE 2 Find the domain of $f(x) = \dfrac{2x - 6}{x^2 - 3x - 4}$.

Solution The domain must exclude values of x for which
$x^2 - 3x - 4 = 0$. Solve this equation for x. Because
$x^2 - 3x - 4$ factors over the integers, solve the equation
by using the Principle of Zero Products.

Take Note

In Chapter 8, we will discuss how to solve this problem when the polynomial does not factor.

$$x^2 - 3x - 4 = 0$$
$$(x + 1)(x - 4) = 0$$

$$x + 1 = 0 \qquad x - 4 = 0$$
$$x = -1 \qquad\quad x = 4$$

When $x = -1$ and $x = 4$, the value of the denominator is zero. Therefore, these values must be excluded from the domain of f.

The domain is $\{x \,|\, x \neq -1, 4\}$.

PROBLEM 2 Find the domain of $g(x) = \dfrac{5 - x}{x^2 - 4}$.

■ **Solution** See page S23.

The graph of the function in Example 2 is shown at the left. Note that the graph never intersects the lines $x = -1$ and $x = 4$ (shown as dashed lines). These are the two values of x excluded from the domain of f.

$f(x) = \dfrac{2x - 6}{x^2 - 3x - 4}$

EXAMPLE 3 Find the domain of $f(x) = \dfrac{3x + 2}{x^2 + 1}$.

Solution The domain must exclude values of x for which $x^2 + 1 = 0$. It is not possible for $x^2 + 1 = 0$, because $x^2 \geq 0$, and a positive number added to a number equal to or greater than zero cannot equal zero. Therefore, there are no real numbers that must be excluded from the domain of f.

The domain is $\{x \,|\, x \in \text{real numbers}\}$.

PROBLEM 3 Find the domain of $p(x) = \dfrac{6x}{x^2 + 4}$.

■ **Solution** See page S23.

The graph of the function in Example 3 is shown at the left. Because the domain of f is all real numbers, there are points on the graph for all values of x.

$f(x) = \dfrac{3x + 2}{x^2 + 1}$

OBJECTIVE **2** Simplify rational expressions

A rational expression is in **simplest form** when the numerator and denominator have no common factors other than 1. The Multiplication Property of One is used to write a rational expression in simplest form.

Simplify: $\dfrac{x^2 - 25}{x^2 + 13x + 40}$

$$\frac{x^2 - 25}{x^2 + 13x + 40} = \frac{(x - 5)(x + 5)}{(x + 8)(x + 5)} = \frac{(x - 5)}{(x + 8)} \cdot \frac{\boxed{(x + 5)}}{\boxed{(x + 5)}} = \frac{x - 5}{x + 8} \cdot 1 = \frac{x - 5}{x + 8},$$

$$x \neq -8, -5$$

The requirement $x \neq -8, -5$ must be included because division by 0 is not allowed.

The simplification above is usually shown with slashes to indicate that a common factor has been removed:

$$\frac{x^2 - 25}{x^2 + 13x + 40} = \frac{(x - 5)\cancel{(x + 5)}^{1}}{(x + 8)\cancel{(x + 5)}_{1}} = \frac{x - 5}{x + 8}, x \neq -8, -5$$

We will show a simplification with slashes. We will also omit the restrictions that prevent division by zero. Nonetheless, those restrictions *always* are implied.

EXAMPLE 4 Simplify.

A. $\dfrac{x^2 - 16}{x^2 + 11x + 28}$ **B.** $\dfrac{12 + 5x - 2x^2}{2x^2 - 3x - 20}$ **C.** $\dfrac{x^{2n} + x^n - 2}{x^{2n} - 1}$

Solution **A.** $\dfrac{x^2 - 16}{x^2 + 11x + 28} = \dfrac{(x + 4)(x - 4)}{(x + 4)(x + 7)}$
- Factor the numerator and the denominator.

$= \dfrac{\cancel{(x + 4)}^{1}(x - 4)}{\cancel{(x + 4)}_{1}(x + 7)}$
- Divide by the common factors.

$= \dfrac{x - 4}{x + 7}$
- Write the answer in simplest form.

B. $\dfrac{12 + 5x - 2x^2}{2x^2 - 3x - 20} = \dfrac{(4 - x)(3 + 2x)}{(x - 4)(2x + 5)}$
- Factor the numerator and the denominator.

$= \dfrac{\cancel{(4 - x)}^{-1}(3 + 2x)}{\cancel{(x - 4)}_{1}(2x + 5)}$
- Divide by the common factors. Remember that $4 - x = -(x - 4)$. Therefore, $\dfrac{4 - x}{x - 4} = \dfrac{-(x - 4)}{x - 4} = \dfrac{-1}{1} = -1$.

$= -\dfrac{2x + 3}{2x + 5}$
- Write the answer in simplest form.

C. $\dfrac{x^{2n} + x^n - 2}{x^{2n} - 1} = \dfrac{(x^n - 1)(x^n + 2)}{(x^n - 1)(x^n + 1)}$
- Factor the numerator and the denominator.

$= \dfrac{\cancel{(x^n - 1)}^{1}(x^n + 2)}{\cancel{(x^n - 1)}_{1}(x^n + 1)}$
- Divide by the common factors.

$= \dfrac{x^n + 2}{x^n + 1}$

PROBLEM 4 Simplify.

A. $\dfrac{6x^4 - 24x^3}{12x^3 - 48x^2}$ **B.** $\dfrac{20x - 15x^2}{15x^3 - 5x^2 - 20x}$

C. $\dfrac{x^{2n} + x^n - 12}{x^{2n} - 3x^n}$

Solution See page S23.

Take Note

Recall that
$$(b - a) = -(a - b)$$
Therefore,
$$(4 - x) = -(x - 4)$$
In general,
$$\frac{b - a}{a - b} = \frac{\cancel{-(a - b)}^{1}}{\cancel{a - b}_{1}} = -1$$

7.1 Concept Review

Determine whether the following statements are always true, sometimes true, or never true.

1. $\dfrac{x^2 + 2x + 1}{x^{\frac{1}{2}} - 2x + 4}$ is a rational expression.

2. A rational expression with a variable in the denominator will have restrictions on the value of x.

3. When simplifying rational expressions, divide the numerator and denominator by the GCF of the numerator and denominator.

4. If $x \neq y$, the rational expression $\dfrac{x - y}{y - x}$ can be reduced to -1.

5. $\dfrac{a + 4}{a^2 + 6a + 8} = \dfrac{a + 4}{(a + 4)(a + 2)} = \dfrac{\cancel{a + 4}}{\cancel{(a + 4)}(a + 2)} = \dfrac{0}{a + 2} = 0$

6. For all values of a, $\dfrac{a(a + 4)}{a} = a + 4$.

7.1 Exercises

1 Find the domain of a rational function

1. What is a rational function? Give an example of a rational function.

2. What values are excluded from the domain of a rational function?

Complete Exercises 3 and 4 by replacing the question marks. Use the rational function $f(x) = \dfrac{x + 2}{1 - 2x}$.

3. To find $f(-3)$, substitute -3 for x: $f(-3) = \dfrac{\underline{\quad ? \quad} + 2}{1 - 2(\underline{\quad ? \quad})} = \dfrac{\underline{\quad ? \quad}}{1 + \underline{\quad ? \quad}}$

 $= \underline{\quad ? \quad}$.

4. Find the domain of the function. The domain must exclude values of x for which the denominator is $\underline{\quad ? \quad}$.

 $1 - 2x = \underline{\quad ? \quad}$. • Set the denominator equal to zero.

 $-2x = \underline{\quad ? \quad}$. • Subtract 1 from each side of the equation.

 $x = \underline{\quad ? \quad}$. • Divide each side of the equation by -2.

 The domain must exclude the value $\underline{\quad ? \quad}$, so the domain is $\{x \mid x \neq \underline{\quad ? \quad}\}$.

5. Given $f(x) = \dfrac{2}{x - 3}$, find $f(4)$.

6. Given $f(x) = \dfrac{-7}{5 - x}$, find $f(-2)$.

7. Given $f(x) = \dfrac{x - 2}{x + 4}$, find $f(-2)$.

8. Given $f(x) = \dfrac{x - 3}{2x - 1}$, find $f(3)$.

9. Given $f(x) = \dfrac{1}{x^2 - 2x + 1}$, find $f(-2)$.

10. Given $f(x) = \dfrac{-3}{x^2 - 4x + 2}$, find $f(-1)$.

11. Given $f(x) = \dfrac{x - 2}{2x^2 + 3x + 8}$, find $f(3)$.

12. Given $f(x) = \dfrac{x^2}{3x^2 - 3x + 5}$, find $f(4)$.

13. Given $f(x) = \dfrac{x^2 - 2x}{x^3 - x + 4}$, find $f(-1)$.

14. Given $f(x) = \dfrac{8 - x^2}{x^3 - x^2 + 4}$, find $f(-3)$.

Find the domain of the function.

15. $H(x) = \dfrac{4}{x - 3}$

16. $G(x) = \dfrac{-2}{x + 2}$

17. $f(x) = \dfrac{x}{x + 4}$

18. $g(x) = \dfrac{3x}{x - 5}$

19. $R(x) = \dfrac{5x}{3x + 9}$

20. $p(x) = \dfrac{-2x}{6 - 2x}$

21. $q(x) = \dfrac{4 - x}{(x - 4)(x + 2)}$

22. $h(x) = \dfrac{2x + 1}{(x + 1)(x + 5)}$

23. $V(x) = \dfrac{x^2}{(2x + 5)(3x - 6)}$

24. $F(x) = \dfrac{x^2 - 1}{(4x + 8)(3x - 1)}$

25. $f(x) = \dfrac{x^2 + 1}{x}$

26. $g(x) = \dfrac{2x^3 - x - 1}{x^2}$

27. $k(x) = \dfrac{x + 1}{x^2 + 1}$

28. $P(x) = \dfrac{2x + 3}{2x^2 + 3}$

29. $f(x) = \dfrac{2x - 1}{x^2 + x - 6}$

30. $G(x) = \dfrac{3 - 4x}{x^2 + 4x - 5}$

31. $A(x) = \dfrac{5x + 2}{x^2 + 2x - 24}$

32. $h(x) = \dfrac{3x}{x^2 - 4}$

33. $f(x) = \dfrac{4x - 7}{3x^2 + 12}$

34. $g(x) = \dfrac{x^2 + x + 1}{5x^2 + 1}$

35. $G(x) = \dfrac{x^2 + 1}{6x^2 - 13x + 6}$

36. $A(x) = \dfrac{5x - 7}{x(x - 2)(x - 3)}$

37. $f(x) = \dfrac{x^2 + 8x + 4}{2x^3 + 9x^2 - 5x}$

38. $H(x) = \dfrac{x^4 - 1}{2x^3 + 2x^2 - 24x}$

39. The graph of the function $f(x) = -\dfrac{1}{x + 1}$ never crosses the vertical line $x = -1$. Is the portion of the graph that is to the right of the vertical line $x = -1$ above the x-axis or below the x-axis?

40. The graph of the function $f(x) = \dfrac{1}{(x-1)(x+2)}$ never crosses the vertical lines $x = 1$ and $x = -2$. Is the portion of the graph that is in between the vertical lines $x = 1$ and $x = -2$ above the x-axis or below the x-axis?

2 Simplify rational expressions

41. When is a rational expression in simplest form?

42. Are the rational expressions $\dfrac{x(x-2)}{2(x-2)}$ and $\dfrac{x}{2}$ equal for all values of x? Why or why not?

43. To simplify a rational expression, first write the numerator and denominator in ___?__ form. Then simplify the rational expression by dividing its numerator and denominator by any ___?__ factors.

44. Simplify: $\dfrac{x^2 - 1}{3x^2 + 3x}$

$\dfrac{x^2 - 1}{3x^2 + 3x} = \dfrac{(x - \underline{})(x + \underline{})}{(\underline{})(x + 1)}$ • Factor the numerator and denominator.

$= \dfrac{\underline{}}{\underline{}}$ • Divide the numerator and denominator by the common factor ___?__.

Simplify.

45. $\dfrac{4 - 8x}{4}$

46. $\dfrac{8y + 2}{2}$

47. $\dfrac{6x^2 - 2x}{2x}$

48. $\dfrac{3y - 12y^2}{3y}$

49. $\dfrac{8x^2(x - 3)}{4x(x - 3)}$

50. $\dfrac{16y^4(y + 8)}{12y^3(y + 8)}$

51. $\dfrac{2x - 6}{3x - x^2}$

52. $\dfrac{3a^2 - 6a}{12 - 6a}$

53. $\dfrac{6x^3 - 15x^2}{12x^2 - 30x}$

54. $\dfrac{-36a^2 - 48a}{18a^3 + 24a^2}$

55. $\dfrac{a^2 + 4a}{4a - 16}$

56. $\dfrac{3x - 6}{x^2 + 2x}$

57. $\dfrac{16x^3 - 8x^2 + 12x}{4x}$

58. $\dfrac{3x^3y^3 - 12x^2y^2 + 15xy}{3xy}$

59. $\dfrac{-10a^4 - 20a^3 + 30a^2}{-10a^2}$

60. $\dfrac{-7a^5 - 14a^4 + 21a^3}{-7a^3}$

61. $\dfrac{3x^{3n} - 9x^{2n}}{12x^{2n}}$

62. $\dfrac{8a^n}{4a^{2n} - 8a^n}$

63. $\dfrac{x^2 - 7x + 12}{x^2 - 9x + 20}$

64. $\dfrac{x^2 - x - 20}{x^2 - 2x - 15}$

65. $\dfrac{x^2 - xy - 2y^2}{x^2 - 3xy + 2y^2}$

66. $\dfrac{2x^2 + 7xy - 4y^2}{4x^2 - 4xy + y^2}$

67. $\dfrac{6 - x - x^2}{3x^2 - 10x + 8}$

68. $\dfrac{3x^2 + 10x - 8}{8 - 14x + 3x^2}$

69. $\dfrac{14 - 19x - 3x^2}{3x^2 - 23x + 14}$

70. $\dfrac{x^2 + x - 12}{x^2 - x - 12}$

71. $\dfrac{a^2 - 7a + 10}{a^2 + 9a + 14}$

72. $\dfrac{x^2 - 2x}{x^2 + 2x}$

73. $\dfrac{a^2 - b^2}{a^3 + b^3}$

74. $\dfrac{x^4 - y^4}{x^2 + y^2}$

75. $\dfrac{x^3 + y^3}{3x^3 - 3x^2y + 3xy^2}$

76. $\dfrac{3x^3 + 3x^2 + 3x}{9x^3 - 9}$

77. $\dfrac{x^3 - 4xy^2}{3x^3 - 2x^2y - 8xy^2}$

78. $\dfrac{4a^2 - 8ab + 4b^2}{4a^2 - 4b^2}$

79. $\dfrac{4x^3 - 14x^2 + 12x}{24x + 4x^2 - 8x^3}$

80. $\dfrac{6x^3 - 15x^2 - 75x}{150x + 30x^2 - 12x^3}$

81. $\dfrac{x^2 - 4}{a(x + 2) - b(x + 2)}$

82. $\dfrac{x^2(a - 2) - a + 2}{ax^2 - ax}$

83. $\dfrac{x^4 + 3x^2 + 2}{x^4 - 1}$

84. $\dfrac{x^4 - 2x^2 - 3}{x^4 + 2x^2 + 1}$

85. $\dfrac{x^2y^2 + 4xy - 21}{x^2y^2 - 10xy + 21}$

86. $\dfrac{6x^2y^2 + 11xy + 4}{9x^2y^2 + 9xy - 4}$

87. $\dfrac{a^{2n} - a^n - 2}{a^{2n} + 3a^n + 2}$

88. $\dfrac{a^{2n} + a^n - 12}{a^{2n} - 2a^n - 3}$

89. True or false? $\dfrac{4x^2 - 8x}{16 - 8x} = \dfrac{x^2}{4}$

90. True or false? $\dfrac{4x + 1}{8x + 2} = \dfrac{1}{2}$

7.1 Applying Concepts

91. Evaluate $h(x) = \dfrac{x + 2}{x - 3}$ when $x = 2.9, 2.99, 2.999,$ and 2.9999. On the basis of your evaluations, complete the following sentence. As x becomes closer to 3, the values of $h(x)$ _____ .

92. Evaluate $h(x) = \frac{x + 2}{x - 3}$ when $x = 3.1, 3.01, 3.001,$ and 3.0001. On the basis of your evaluations, complete the following sentence. As x becomes closer to 3, the values of $h(x)$ _____.

93. Photography The focal length of a camera lens is the distance from the lens to the point where parallel rays of light come to a focus.

—Focal length—

The relationship among the focal length (F), the distance between the object and the lens (x, in meters), and the distance between the lens and the film (y, in millimeters) is given by $\frac{1}{F} = \frac{1}{x} + \frac{1}{y}$. A camera used by a professional photographer has a dial that allows the focal length to be set at a constant value. Suppose a photographer chooses a focal length of 50 mm. Substituting this value into the equation, solving for y, and using $y = f(x)$ notation yield $f(x) = \frac{50x}{x - 50}$.

a. Graph this equation for $50 < x \leq 6000$.

b. The point whose coordinates are $(2000, 51)$, to the nearest integer, is on the graph of the function. Give an interpretation of this ordered pair.

c. Give a reason for choosing the domain such that $x > 50$.

d. Photographers refer to depth of field as a range of distances in which an object remains in focus. Use the graph to explain why the depth of field is larger for objects that are far from the lens than for objects that are close to the lens.

94. Suppose that $F(x) = \frac{g(x)}{h(x)}$ and that, for some real number a, $g(a) = 0$ and $h(a) = 0$. Is $F(x)$ in simplest form? Explain your answer.

95. Why can the numerator and denominator of a rational expression be divided by their common factors? What conditions must be placed on the value of the variables when a rational expression is simplified?

 Operations on Rational Expressions

OBJECTIVE **1** **Multiply and divide rational expressions**

The product of two fractions is a fraction whose numerator is the product of the numerators of the two fractions and whose denominator is the product of the denominators of the two fractions.

$$\frac{a}{b} \cdot \frac{c}{d} = \frac{ac}{bd}$$

$$\frac{5}{a+2} \cdot \frac{b-3}{3} = \frac{5(b-3)}{(a+2)3} = \frac{5b-15}{3a+6}$$

The product of two rational expressions can often be simplified by factoring the numerator and the denominator.

Simplify: $\dfrac{x^2 - 2x}{2x^2 + x - 15} \cdot \dfrac{2x^2 - x - 10}{x^2 - 4}$

$$\frac{x^2 - 2x}{2x^2 + x - 15} \cdot \frac{2x^2 - x - 10}{x^2 - 4}$$

Factor the numerator and denominator of each fraction.

$$= \frac{x(x-2)}{(x+3)(2x-5)} \cdot \frac{(x+2)(2x-5)}{(x+2)(x-2)}$$

Multiply.

$$= \frac{x(x-2)(x+2)(2x-5)}{(x+3)(2x-5)(x+2)(x-2)}$$

Divide by the common factors.

$$= \frac{x\cancel{(x-2)}\cancel{(x+2)}\cancel{(2x-5)}}{(x+3)\cancel{(2x-5)}\cancel{(x+2)}\cancel{(x-2)}}$$

Write the answer in simplest form.

$$= \frac{x}{x+3}$$

EXAMPLE 1 Multiply.

A. $\dfrac{2x^2 - 6x}{3x - 6} \cdot \dfrac{6x - 12}{8x^3 - 12x^2}$ **B.** $\dfrac{6x^2 + x - 2}{6x^2 + 7x + 2} \cdot \dfrac{2x^2 + 9x + 4}{4 - 7x - 2x^2}$

Solution **A.** $\dfrac{2x^2 - 6x}{3x - 6} \cdot \dfrac{6x - 12}{8x^3 - 12x^2} = \dfrac{2x(x-3)}{3(x-2)} \cdot \dfrac{6(x-2)}{4x^2(2x-3)}$

$$= \frac{12x(x-3)\cancel{(x-2)}}{12x^2\cancel{(x-2)}(2x-3)} = \frac{x-3}{x(2x-3)}$$

B. $\dfrac{6x^2 + x - 2}{6x^2 + 7x + 2} \cdot \dfrac{2x^2 + 9x + 4}{4 - 7x - 2x^2}$

$$= \frac{(2x-1)(3x+2)}{(3x+2)(2x+1)} \cdot \frac{(2x+1)(x+4)}{(1-2x)(4+x)}$$

$$= \frac{\overset{-1}{\cancel{(2x-1)}}\overset{1}{\cancel{(3x+2)}}\overset{1}{\cancel{(2x+1)}}\overset{1}{\cancel{(x+4)}}}{\underset{1}{\cancel{(3x+2)}}\underset{1}{\cancel{(2x+1)}}\underset{1}{\cancel{(1-2x)}}\underset{1}{\cancel{(x+4)}}} = -1$$

PROBLEM 1 Multiply.

A. $\dfrac{12 + 5x - 3x^2}{x^2 + 2x - 15} \cdot \dfrac{2x^2 + x - 45}{3x^2 + 4x}$

B. $\dfrac{2x^2 - 13x + 20}{x^2 - 16} \cdot \dfrac{2x^2 + 9x + 4}{6x^2 - 7x - 5}$

■ **Solution** See page S23.

The **reciprocal of a rational expression** is the rational expression with the numerator and denominator interchanged.

$$\text{Rational expression} \left\{ \begin{array}{cc} \dfrac{a}{b} & \dfrac{b}{a} \\[2ex] \dfrac{a^2 - 2y}{4} & \dfrac{4}{a^2 - 2y} \end{array} \right\} \text{Reciprocal}$$

To divide two rational expressions, multiply by the reciprocal of the divisor.

$$\frac{a}{b} \div \frac{c}{d} = \frac{a}{b} \cdot \frac{d}{c} = \frac{ad}{bc}$$

$$\frac{2}{a} \div \frac{5}{b} = \frac{2}{a} \cdot \frac{b}{5} = \frac{2b}{5a}$$

$$\frac{x + y}{2} \div \frac{x - y}{5} = \frac{x + y}{2} \cdot \frac{5}{x - y} = \frac{(x + y)5}{2(x - y)} = \frac{5x + 5y}{2x - 2y}$$

EXAMPLE 2 Divide.

A. $\dfrac{12x^2y^2 - 24xy^2}{5z^2} \div \dfrac{4x^3y - 8x^2y}{3z^4}$

B. $\dfrac{3y^2 - 10y + 8}{3y^2 + 8y - 16} \div \dfrac{2y^2 - 7y + 6}{2y^2 + 5y - 12}$

Solution **A.** $\dfrac{12x^2y^2 - 24xy^2}{5z^2} \div \dfrac{4x^3y - 8x^2y}{3z^4}$

$$= \frac{12x^2y^2 - 24xy^2}{5z^2} \cdot \frac{3z^4}{4x^3y - 8x^2y}$$

$$= \frac{12xy^2(x - 2)}{5z^2} \cdot \frac{3z^4}{4x^2y(x - 2)}$$

$$= \frac{36xy^2z^4\overset{1}{\cancel{(x - 2)}}}{20x^2yz^2\underset{1}{\cancel{(x - 2)}}} = \frac{9yz^2}{5x}$$

B. $\dfrac{3y^2 - 10y + 8}{3y^2 + 8y - 16} \div \dfrac{2y^2 - 7y + 6}{2y^2 + 5y - 12}$

$$= \frac{3y^2 - 10y + 8}{3y^2 + 8y - 16} \cdot \frac{2y^2 + 5y - 12}{2y^2 - 7y + 6}$$

$$= \frac{(y-2)(3y-4)}{(3y-4)(y+4)} \cdot \frac{(y+4)(2y-3)}{(y-2)(2y-3)}$$

$$= \frac{\overset{1}{\cancel{(y-2)}}\,\overset{1}{\cancel{(3y-4)}}\,\overset{1}{\cancel{(y+4)}}\,\overset{1}{\cancel{(2y-3)}}}{\underset{1}{\cancel{(3y-4)}}\,\underset{1}{\cancel{(y+4)}}\,\underset{1}{\cancel{(y-2)}}\,\underset{1}{\cancel{(2y-3)}}} = 1$$

PROBLEM 2 Divide.

A. $\dfrac{6x^2 - 3xy}{10ab^4} \div \dfrac{16x^2y^2 - 8xy^3}{15a^2b^2}$

B. $\dfrac{6x^2 - 7x + 2}{3x^2 + x - 2} \div \dfrac{4x^2 - 8x + 3}{5x^2 + x - 4}$

■ **Solution** See page S24.

OBJECTIVE **2** **Add and subtract rational expressions**

When adding rational expressions in which the denominators are the same, add the numerators. The denominator of the sum is the common denominator. Write the answer in simplest form.

$$\frac{a}{c} + \frac{b}{c} = \frac{a+b}{c}$$

$$\frac{4x}{15} + \frac{8x}{15} = \frac{4x+8x}{15} = \frac{12x}{15} = \frac{4x}{5}$$

$$\frac{a}{a^2 - b^2} + \frac{b}{a^2 - b^2} = \frac{a+b}{a^2 - b^2} = \frac{a+b}{(a-b)(a+b)} = \frac{\overset{1}{\cancel{(a+b)}}}{(a-b)\cancel{(a+b)}} = \frac{1}{a-b}$$

When subtracting rational expressions in which the denominators are the same, subtract the numerators. The denominator of the difference is the common denominator. Write the answer in simplest form.

$$\frac{y}{y-3} - \frac{3}{y-3} = \frac{y-3}{y-3} = \frac{\overset{1}{\cancel{(y-3)}}}{\underset{1}{\cancel{(y-3)}}} = 1$$

<aside>
Take Note

Be careful with signs when subtracting algebraic fractions. Note that we must subtract the entire numerator.

$(7x - 12) - (3x - 6) =$
$\qquad 7x - 12 - 3x + 6$
</aside>

$$\frac{7x - 12}{2x^2 + 5x - 12} - \frac{3x - 6}{2x^2 + 5x - 12} = \frac{(7x - 12) - (3x - 6)}{2x^2 + 5x - 12}$$

$$= \frac{7x - 12 - 3x + 6}{2x^2 + 5x - 12}$$

$$= \frac{4x - 6}{2x^2 + 5x - 12}$$

$$= \frac{2(2x - 3)}{(2x - 3)(x + 4)}$$

$$= \frac{2\overset{1}{\cancel{(2x-3)}}}{\cancel{(2x-3)}(x+4)} = \frac{2}{x+4}$$

Before two rational expressions with different denominators can be added or subtracted, both rational expressions must be expressed in terms of a

common denominator. This common denominator is the LCM of the denominators of the rational expressions.

The LCM of two or more polynomials is the simplest polynomial that contains the factors of each polynomial. To find the LCM, first factor each polynomial completely. The LCM is the product of each factor the greatest number of times it occurs in any one factorization.

To find the LCM of $3x^2 + 15x$ and $6x^4 + 24x^3 - 30x^2$, factor each polynomial.

$$3x^2 + 15x = 3x(x + 5)$$
$$6x^4 + 24x^3 - 30x^2 = 6x^2(x^2 + 4x - 5) = 6x^2(x - 1)(x + 5)$$

The LCM is the product of the LCM of the numerical coefficients and each variable factor the greatest number of times it occurs in any one factorization.

$$\text{LCM} = 6x^2(x - 1)(x + 5)$$

Take Note

$x^2 - 2x = x(x - 2)$

$3x - 6 = 3(x - 2)$

The LCM of $x(x - 2)$ and $3(x - 2)$ is $3x(x - 2)$.

Write the fractions $\dfrac{x + 2}{x^2 - 2x}$ and $\dfrac{5x}{3x - 6}$ in terms of the LCM of the denominators.

Find the LCM of the denominators.

The LCM is $3x(x - 2)$.

For each fraction, multiply the numerator and denominator by the factor whose product with the denominator is the LCM.

$$\frac{x + 2}{x^2 - 2x} = \frac{x + 2}{x(x - 2)} \cdot \frac{3}{3} = \frac{3x + 6}{3x(x - 2)}$$

$$\frac{5x}{3x - 6} = \frac{5x}{3(x - 2)} \cdot \frac{x}{x} = \frac{5x^2}{3x(x - 2)}$$

Take Note

Note the steps involved in adding or subtracting rational expressions:

1. Find the LCM of the denominators.

2. Rewrite each fraction in terms of the common denominator.

3. Add or subtract the rational expressions.

4. Simplify the resulting sum or difference.

Add: $\dfrac{3x}{2x - 3} + \dfrac{3x + 6}{2x^2 + x - 6}$

The LCM of $2x - 3$ and $2x^2 + x - 6$ is $(2x - 3)(x + 2)$.

$$\frac{3x}{2x - 3} + \frac{3x + 6}{2x^2 + x - 6}$$

Rewrite each fraction in terms of the LCM of the denominators.

$$= \frac{3x}{2x - 3} \cdot \frac{x + 2}{x + 2} + \frac{3x + 6}{(2x - 3)(x + 2)}$$

$$= \frac{3x^2 + 6x}{(2x - 3)(x + 2)} + \frac{3x + 6}{(2x - 3)(x + 2)}$$

Add the fractions.

$$= \frac{(3x^2 + 6x) + (3x + 6)}{(2x - 3)(x + 2)}$$

$$= \frac{3x^2 + 9x + 6}{(2x - 3)(x + 2)}$$

$$= \frac{3(x^2 + 3x + 2)}{(2x - 3)(x + 2)}$$

Factor the numerator to determine whether there are common factors in the numerator and denominator.

$$= \frac{3(x + 2)(x + 1)}{(2x - 3)(x + 2)}$$

$$= \frac{3\overset{1}{\cancel{(x + 2)}}(x + 1)}{(2x - 3)\underset{1}{\cancel{(x + 2)}}} = \frac{3(x + 1)}{2x - 3}$$

EXAMPLE 3 Subtract: $\dfrac{x}{2x-4} - \dfrac{4-x}{x^2-2x}$

Solution $\dfrac{x}{2x-4} - \dfrac{4-x}{x^2-2x}$

$$= \dfrac{x}{2(x-2)} \cdot \dfrac{x}{x} - \dfrac{4-x}{x(x-2)} \cdot \dfrac{2}{2}$$

• Write each fraction in terms of the LCM. The LCM is $2x(x-2)$.

$$= \dfrac{x^2}{2x(x-2)} - \dfrac{8-2x}{2x(x-2)}$$

• Subtract the fractions.

$$= \dfrac{x^2-(8-2x)}{2x(x-2)}$$

$$= \dfrac{x^2+2x-8}{2x(x-2)}$$

$$= \dfrac{(x+4)(x-2)}{2x(x-2)}$$

$$= \dfrac{(x+4)\cancel{(x-2)}^1}{2x\cancel{(x-2)}_1} = \dfrac{x+4}{2x}$$

• Divide by the common factors.

PROBLEM 3 Add: $\dfrac{a-3}{a^2-5a} + \dfrac{a-9}{a^2-25}$

Solution See page S24.

EXAMPLE 4 Simplify: $\dfrac{6x-23}{2x^2+x-6} + \dfrac{3x}{2x-3} - \dfrac{5}{x+2}$

Solution $\dfrac{6x-23}{2x^2+x-6} + \dfrac{3x}{2x-3} - \dfrac{5}{x+2}$

$$= \dfrac{6x-23}{(2x-3)(x+2)} + \dfrac{3x}{2x-3} \cdot \dfrac{x+2}{x+2} - \dfrac{5}{x+2} \cdot \dfrac{2x-3}{2x-3}$$

$$= \dfrac{6x-23}{(2x-3)(x+2)} + \dfrac{3x^2+6x}{(2x-3)(x+2)} - \dfrac{10x-15}{(2x-3)(x+2)}$$

$$= \dfrac{(6x-23)+(3x^2+6x)-(10x-15)}{(2x-3)(x+2)}$$

$$= \dfrac{6x-23+3x^2+6x-10x+15}{(2x-3)(x+2)}$$

$$= \dfrac{3x^2+2x-8}{(2x-3)(x+2)} = \dfrac{(3x-4)(x+2)}{(2x-3)(x+2)} = \dfrac{3x-4}{2x-3}$$

PROBLEM 4 Simplify: $\dfrac{x-1}{x-2} - \dfrac{7-6x}{2x^2-7x+6} + \dfrac{4}{2x-3}$

Solution See page S24.

7.2 Concept Review

Determine whether the following statements are always true, sometimes true, or never true.

1. To divide an expression by a nonzero rational number, multiply the expression by the reciprocal of the divisor.

2. To multiply rational expressions, multiply the numerators and multiply the denominators.

3. $\dfrac{3}{x+y} + \dfrac{2}{x+y} = \dfrac{3+2}{(x+y)+(x+y)} = \dfrac{5}{2x+2y}$

4. $\dfrac{2x}{x(x+4)} - \dfrac{2}{x+4} = \dfrac{2\cancel{x}}{\cancel{x}(x+4)} - \dfrac{2}{x+4} = \dfrac{2}{x+4} - \dfrac{2}{x+4} = 0$

7.2 Exercises

1 Multiply and divide rational expressions

1. Explain how to multiply two rational expressions.

2. Explain how to divide two rational expressions.

Multiply or divide.

3. $\dfrac{27a^2b^5}{16xy^2} \cdot \dfrac{20x^2y^3}{9a^2b}$

4. $\dfrac{15x^2y^4}{24ab^3} \cdot \dfrac{28a^2b^4}{35xy^4}$

5. $\dfrac{3x-15}{4x^2-2x} \cdot \dfrac{20x^2-10x}{15x-75}$

6. $\dfrac{2x^2+4x}{8x^2-40x} \cdot \dfrac{6x^3-30x^2}{3x^2+6x}$

7. $\dfrac{x^2y^3}{x^2-4x-5} \cdot \dfrac{2x^2-13x+15}{x^4y^3}$

8. $\dfrac{2x^2-5x+3}{x^6y^3} \cdot \dfrac{x^4y^4}{2x^2-x-3}$

9. $\dfrac{x^2-3x+2}{x^2-8x+15} \cdot \dfrac{x^2+x-12}{8-2x-x^2}$

10. $\dfrac{x^2+x-6}{12+x-x^2} \cdot \dfrac{x^2+x-20}{x^2-4x+4}$

11. $\dfrac{x^{n+1}+2x^n}{4x^2-6x} \cdot \dfrac{8x^2-12x}{x^{n+1}-x^n}$

12. $\dfrac{x^{2n}+2x^n}{x^{n+1}+2x} \cdot \dfrac{x^2-3x}{x^{n+1}-3x^n}$

13. $\dfrac{12+x-6x^2}{6x^2+29x+28} \cdot \dfrac{2x^2+x-21}{4x^2-9}$

14. $\dfrac{x^2+5x+4}{4+x-3x^2} \cdot \dfrac{3x^2+2x-8}{x^2+4x}$

15. $\dfrac{x^{2n} - x^n - 6}{x^{2n} + x^n - 2} \cdot \dfrac{x^{2n} - 5x^n - 6}{x^{2n} - 2x^n - 3}$

16. $\dfrac{x^{2n} + 3x^n + 2}{x^{2n} - x^n - 6} \cdot \dfrac{x^{2n} + x^n - 12}{x^{2n} - 1}$

17. $\dfrac{x^3 - y^3}{2x^2 + xy - 3y^2} \cdot \dfrac{2x^2 + 5xy + 3y^2}{x^2 + xy + y^2}$

18. $\dfrac{x^4 - 5x^2 + 4}{3x^2 - 4x - 4} \cdot \dfrac{3x^2 - 10x - 8}{x^2 - 4}$

19. $\dfrac{6x^2y^4}{35a^2b^5} \div \dfrac{12x^3y^3}{7a^4b^5}$

20. $\dfrac{12a^4b^7}{13x^2y^2} \div \dfrac{18a^5b^6}{26xy^3}$

21. $\dfrac{2x - 6}{6x^2 - 15x} \div \dfrac{4x^2 - 12x}{18x^3 - 45x^2}$

22. $\dfrac{4x^2 - 4y^2}{6x^2y^2} \div \dfrac{3x^2 + 3xy}{2x^2y - 2xy^2}$

23. $\dfrac{2x^2 - 2y^2}{14x^2y^4} \div \dfrac{x^2 + 2xy + y^2}{35xy^3}$

24. $\dfrac{8x^3 + 12x^2y}{4x^2 - 9y^2} \div \dfrac{16x^2y^2}{4x^2 - 12xy + 9y^2}$

25. $\dfrac{2x^2 - 5x - 3}{2x^2 + 7x + 3} \div \dfrac{2x^2 - 3x - 20}{2x^2 - x - 15}$

26. $\dfrac{3x^2 - 10x - 8}{6x^2 + 13x + 6} \div \dfrac{2x^2 - 9x + 10}{4x^2 - 4x - 15}$

27. $\dfrac{x^2 - 8x + 15}{x^2 + 2x - 35} \div \dfrac{15 - 2x - x^2}{x^2 + 9x + 14}$

28. $\dfrac{2x^2 + 13x + 20}{8 - 10x - 3x^2} \div \dfrac{6x^2 - 13x - 5}{9x^2 - 3x - 2}$

29. $\dfrac{x^{2n} + x^n}{2x - 2} \div \dfrac{4x^n + 4}{x^{n+1} - x^n}$

30. $\dfrac{x^{2n} - 4}{4x^n + 8} \div \dfrac{x^{n+1} - 2x}{4x^3 - 12x^2}$

31. $\dfrac{2x^2 - 13x + 21}{2x^2 + 11x + 15} \div \dfrac{2x^2 + x - 28}{3x^2 + 4x - 15}$

32. $\dfrac{2x^2 - 13x + 15}{2x^2 - 3x - 35} \div \dfrac{6x^2 + x - 12}{6x^2 + 13x - 28}$

33. $\dfrac{14 + 17x - 6x^2}{3x^2 + 14x + 8} \div \dfrac{4x^2 - 49}{2x^2 + 15x + 28}$

34. $\dfrac{16x^2 - 9}{6 - 5x - 4x^2} \div \dfrac{16x^2 + 24x + 9}{4x^2 + 11x + 6}$

35. $\dfrac{2x^{2n} - x^n - 6}{x^{2n} - x^n - 2} \div \dfrac{2x^{2n} + x^n - 3}{x^{2n} - 1}$

36. $\dfrac{x^{4n} - 1}{x^{2n} + x^n - 2} \div \dfrac{x^{2n} + 1}{x^{2n} + 3x^n + 2}$

37. $\dfrac{6x^2 + 6x}{3x + 6x^2 + 3x^3} \div \dfrac{x^2 - 1}{1 - x^3}$

38. $\dfrac{x^3 + y^3}{2x^3 + 2x^2y} \div \dfrac{3x^3 - 3x^2y + 3xy^2}{6x^2 - 6y^2}$

39. Which expression is *not* equivalent to $\dfrac{x^2 + 4x - 5}{x^2 - x - 6}$?

 a. $\dfrac{x - 1}{x + 2} \div \dfrac{x - 3}{x + 5}$ **b.** $\dfrac{x - 1}{x - 3} \cdot \dfrac{x + 5}{x + 2}$

 c. $\dfrac{x - 3}{x - 1} \div \dfrac{x + 5}{x + 2}$ **d.** $\dfrac{x + 5}{2x - 6} \cdot \dfrac{2x - 2}{x + 2}$

40. The simplified form of the product $\left(\dfrac{x - 1}{x}\right)^8 \cdot \left(\dfrac{x^5}{x - 5}\right)^2$ has a factor in the form x^n in the numerator. What is n?

2 Add and subtract rational expressions

41. To find the LCM of $x^2 - 6x + 9$ and $x^2 - 9$, first factor each polynomial:

$x^2 - 6x + 9 = $ ___?___ • **$x^2 - 6x + 9$ factors as a perfect-square trinomial.**

$x^2 - 9 = $ ___?___ • **$x^2 - 9$ factors as the difference of two perfect squares.**

The LCM will use the factor $x - 3$ ___?___ time(s) and the factor $x + 3$ ___?___ time(s). The LCM is ___?___.

42. True or false? The LCM of $(xy)^3$ and xy^3 is $(xy)^3$.

43. How many factors of $x - 4$ are in the LCM of each pair of expressions?

a. $x^2 - x - 12$ and $x^2 - 8x + 16$ **b.** $x^2 + x - 12$ and $x^2 + 8x + 16$
c. $x^2 - 16$ and $x^2 - x - 12$

44. How many factors of a are in the LCM of $(a^2b)^3$ and a^4b^4? How many factors of b?

Write each fraction in terms of the LCM of the denominators.

45. $\dfrac{3}{4x^2y}$, $\dfrac{17}{12xy^4}$

46. $\dfrac{5}{16a^3b^3}$, $\dfrac{7}{30a^5b}$

47. $\dfrac{3x}{2x - 3}$, $\dfrac{5x}{2x + 3}$

48. $\dfrac{2}{7y - 3}$, $\dfrac{-3}{7y + 3}$

49. $\dfrac{2x}{x^2 - 9}$, $\dfrac{x + 1}{x - 3}$

50. $\dfrac{3x}{x^2 - 16}$, $\dfrac{2x}{4x - 16}$

51. $\dfrac{3x}{x^2 - 1}$, $\dfrac{5x}{x^2 - 2x + 1}$

52. $\dfrac{2x}{x^2 + x - 6}$, $\dfrac{-4x}{x^2 + 5x + 6}$

53. Subtract: $\dfrac{5}{x - 4} - \dfrac{2}{x + 3}$

$\dfrac{5}{x - 4} - \dfrac{2}{x + 3}$ • **The LCM of the denominators is ___?___.**

$= \dfrac{5}{x - 4} \cdot \dfrac{x + 3}{x + 3} - \dfrac{2}{x + 3} \cdot \dfrac{x - 4}{x - 4}$ • **Multiply the numerator and denominator of each fraction by the factor whose product with the denominator is the LCM.**

$= \dfrac{?}{(x - 4)(x + 3)} - \dfrac{?}{(x - 4)(x + 3)}$

$= \dfrac{(\ ?\) - (\ ?\)}{(x - 4)(x + 3)} = \dfrac{5x + 15 - \ ?\ + \ ?\ }{(x - 4)(x + 3)}$

$= \dfrac{\ ?\ + \ ?\ }{(x - 4)(x + 3)}$

54. True or false? $\dfrac{1}{x-3} + \dfrac{1}{3-x} = 0$

Add or subtract.

55. $\dfrac{3}{2xy} - \dfrac{7}{2xy} - \dfrac{9}{2xy}$

56. $-\dfrac{3}{4x^2} + \dfrac{8}{4x^2} - \dfrac{3}{4x^2}$

57. $\dfrac{x}{x^2 - 3x + 2} - \dfrac{2}{x^2 - 3x + 2}$

58. $\dfrac{3x}{3x^2 + x - 10} - \dfrac{5}{3x^2 + x - 10}$

59. $\dfrac{3}{2x^2y} - \dfrac{8}{5x} - \dfrac{9}{10xy}$

60. $\dfrac{2}{5ab} - \dfrac{3}{10a^2b} + \dfrac{4}{15ab^2}$

61. $\dfrac{2}{3x} - \dfrac{3}{2xy} + \dfrac{4}{5xy} - \dfrac{5}{6x}$

62. $\dfrac{3}{4ab} - \dfrac{2}{5a} + \dfrac{3}{10b} - \dfrac{5}{8ab}$

63. $\dfrac{2x-1}{12x} - \dfrac{3x+4}{9x}$

64. $\dfrac{3x-4}{6x} - \dfrac{2x-5}{4x}$

65. $\dfrac{3x+2}{4x^2y} - \dfrac{y-5}{6xy^2}$

66. $\dfrac{2y-4}{5xy^2} + \dfrac{3-2x}{10x^2y}$

67. $\dfrac{2x}{x-3} - \dfrac{3x}{x-5}$

68. $\dfrac{3a}{a-2} - \dfrac{5a}{a+1}$

69. $\dfrac{3}{2a-3} + \dfrac{2a}{3-2a}$

70. $\dfrac{x}{2x-5} - \dfrac{2}{5x-2}$

71. $\dfrac{3}{x+5} + \dfrac{2x+7}{x^2-25}$

72. $\dfrac{x}{4-x} - \dfrac{4}{x^2-16}$

73. $\dfrac{2}{x} - 3 - \dfrac{10}{x-4}$

74. $\dfrac{6a}{a-3} - 5 + \dfrac{3}{a}$

75. $\dfrac{1}{2x-3} - \dfrac{5}{2x} + 1$

76. $\dfrac{5}{x} - \dfrac{5x}{5-6x} + 2$

77. $\dfrac{3}{x^2-1} + \dfrac{2x}{x^2+2x+1}$

78. $\dfrac{1}{x^2-6x+9} - \dfrac{1}{x^2-9}$

79. $\dfrac{x}{x+3} - \dfrac{3-x}{x^2-9}$

80. $\dfrac{1}{x+2} - \dfrac{3x}{x^2+4x+4}$

81. $\dfrac{2x-3}{x+5} - \dfrac{x^2-4x-19}{x^2+8x+15}$

82. $\dfrac{-3x^2+8x+2}{x^2+2x-8} - \dfrac{2x-5}{x+4}$

83. $\dfrac{x^n}{x^{2n} - 1} - \dfrac{2}{x^n + 1}$

84. $\dfrac{2}{x^n - 1} + \dfrac{x^n}{x^{2n} - 1}$

85. $\dfrac{2x - 2}{4x^2 - 9} - \dfrac{5}{3 - 2x}$

86. $\dfrac{x^2 + 4}{4x^2 - 36} - \dfrac{13}{x + 3}$

87. $\dfrac{x - 2}{x + 1} - \dfrac{3 - 12x}{2x^2 - x - 3}$

88. $\dfrac{3x - 4}{4x + 1} + \dfrac{3x + 6}{4x^2 + 9x + 2}$

89. $\dfrac{x + 1}{x^2 + x - 6} - \dfrac{x + 2}{x^2 + 4x + 3}$

90. $\dfrac{x + 1}{x^2 + x - 12} - \dfrac{x - 3}{x^2 + 7x + 12}$

91. $\dfrac{x - 1}{2x^2 + 11x + 12} + \dfrac{2x}{2x^2 - 3x - 9}$

92. $\dfrac{x - 2}{4x^2 + 4x - 3} + \dfrac{3 - 2x}{6x^2 + x - 2}$

93. $\dfrac{x}{x - 3} - \dfrac{2}{x + 4} - \dfrac{14}{x^2 + x - 12}$

94. $\dfrac{x^2}{x^2 + x - 2} + \dfrac{3}{x - 1} - \dfrac{4}{x + 2}$

95. $\dfrac{x^2 + 6x}{x^2 + 3x - 18} - \dfrac{2x - 1}{x + 6} + \dfrac{x - 2}{3 - x}$

96. $\dfrac{2x^2 - 2x}{x^2 - 2x - 15} - \dfrac{2}{x + 3} + \dfrac{x}{5 - x}$

97. $\dfrac{4 - 20x}{6x^2 + 11x - 10} - \dfrac{4}{2 - 3x} + \dfrac{x}{2x + 5}$

98. $\dfrac{x}{4x - 1} + \dfrac{2}{2x + 1} + \dfrac{6}{8x^2 + 2x - 1}$

99. $\dfrac{2x^2}{x^4 - 1} - \dfrac{1}{x^2 - 1} + \dfrac{1}{x^2 + 1}$

100. $\dfrac{x^2 - 12}{x^4 - 16} + \dfrac{1}{x^2 - 4} - \dfrac{1}{x^2 + 4}$

7.2 Applying Concepts

Simplify.

101. $\dfrac{(x + 1)^2}{1 - 2x} \cdot \dfrac{2x - 1}{x + 1}$

102. $\left(\dfrac{2m}{3}\right)^2 \div \left(\dfrac{m^2}{6} + \dfrac{m}{2}\right)$

103. $\left(\dfrac{y - 2}{x^2}\right)^3 \cdot \left(\dfrac{x}{2 - y}\right)^2$

104. $\dfrac{b + 3}{b - 1} \div \dfrac{b + 3}{b - 2} \cdot \dfrac{b - 1}{b + 4}$

105. $\left(\dfrac{y+1}{y-1}\right)^2 - 1$

106. $\left(\dfrac{1}{3} - \dfrac{2}{a}\right) \div \left(\dfrac{3}{a} - 2 + \dfrac{a}{4}\right)$

107. $\dfrac{3x^2 + 6x}{4x^2 - 16} \cdot \dfrac{2x + 8}{x^2 + 2x} \div \dfrac{3x - 9}{5x - 20}$

108. $\dfrac{5y^2 - 20}{3y^2 - 12y} \cdot \dfrac{9y^3 + 6y^2}{2y^2 - 4y} \div \dfrac{y^3 + 2y^2}{2y^2 - 8y}$

109. $\dfrac{a^2 + a - 6}{4 + 11a - 3a^2} \cdot \dfrac{15a^2 - a - 2}{4a^2 + 7a - 2} \div \dfrac{6a^2 - 7a - 3}{4 - 17a + 4a^2}$

110. $\dfrac{25x - x^3}{x^4 - 1} \cdot \dfrac{3 - x - 4x^2}{2x^2 + 7x - 15} \div \dfrac{4x^3 - 23x^2 + 15x}{3 - 5x + 2x^2}$

111. $\left(\dfrac{x+1}{2x-1} - \dfrac{x-1}{2x+1}\right) \cdot \left(\dfrac{2x-1}{x} - \dfrac{2x-1}{x^2}\right)$

112. $\left(\dfrac{y-2}{3y+1} - \dfrac{y+2}{3y-1}\right) \cdot \left(\dfrac{3y+1}{y} - \dfrac{3y-1}{y^2}\right)$

Solve.

113. Use $x = 3$ and $y = 5$ to show that $\dfrac{1}{x} + \dfrac{1}{y} \neq \dfrac{1}{x+y}$.

114. Use $x = 3$ and $y = 5$ to show that $\dfrac{1}{x} - \dfrac{1}{y} \neq \dfrac{1}{x-y}$.

Rewrite the expression as the sum of two fractions in simplest form.

115. $\dfrac{3x + 6y}{xy}$

116. $\dfrac{5a + 8b}{ab}$

117. $\dfrac{4a^2 + 3ab}{a^2b^2}$

118. **Manufacturing** Manufacturers who package their product in cans would like to design the can so that the minimum amount of aluminum is needed. If a soft drink can contains 12 oz (355 cm³), the function that relates the surface area of the can (the amount of aluminum needed) to the radius of the bottom of the can is given by the equation $f(r) = 2\pi r^2 + \dfrac{710}{r}$, where r is measured in centimeters.

a. Express the right side of this equation with a common denominator.

b. Graph the equation for $0 < r \le 7.5$.

c. The point whose coordinates are (7, 409), to the nearest integer, is on the graph of f. Write a sentence that gives an interpretation of this ordered pair.

d. Use a graphing utility to determine the radius of the can that has a minimum surface area. Round to the nearest tenth.

e. The height of the can is determined from $h = \dfrac{355}{\pi r^2}$. Use the answer to part d to determine the height of the can that has a minimum surface area. Round to the nearest tenth.

f. Determine the minimum surface area. Round to the nearest tenth.

119. **Manufacturing** A manufacturer wants to make square tissues and package them in a box. The manufacturer has determined that to be competitive, the box needs to hold 175 tissues, which means that the volume of the box will be 132 in³. The amount of cardboard (surface area) that will be necessary to build this box is given by $f(x) = 2x^2 + \dfrac{528}{x}$, where x is the height of the box in inches.

a. Express the right side of this equation with a common denominator.

b. Graph the equation for $1.5 < x \le 10$.

c. The point whose coordinates are (4, 164) is on the graph. Write a sentence that explains the meaning of this point.

d. With a graphing utility, determine, to the nearest tenth, the height of the box that uses the minimum amount of cardboard.

e. Determine the minimum amount of cardboard. Round to the nearest tenth.

120. For adding or subtracting fractions, any common denominator will do. Explain the advantages and disadvantages of using the LCM of the denominators.

121. A student incorrectly tried to add the fractions $\frac{1}{5}$ and $\frac{2}{3}$ by adding the the numerators and the denominators. The procedure was shown as

$$\frac{1}{5} + \frac{2}{3} = \frac{1 + 2}{5 + 3} = \frac{3}{8}$$

Write the fractions $\frac{1}{5}, \frac{2}{3}$, and $\frac{3}{8}$ in order from smallest to largest. Now take any two other fractions, add the numerators and the denominators, and then write the fractions in order from smallest to largest. Do you see a pattern? If so, explain it. If not, try a few more examples until you find a pattern.

Complex Fractions

OBJECTIVE **1** **Simplify complex fractions**

A **complex fraction** is a fraction whose numerator or denominator contains one or more fractions. Examples of complex fractions are shown below.

$$\dfrac{5}{2+\dfrac{1}{2}} \qquad \dfrac{5+\dfrac{1}{y}}{5-\dfrac{1}{y}} \qquad \dfrac{x+4+\dfrac{1}{x+2}}{x-2+\dfrac{1}{x+2}}$$

Simplify: $\dfrac{\dfrac{1}{x}+\dfrac{1}{y}}{\dfrac{1}{x}-\dfrac{1}{y}}$

Multiply the numerator and denominator of the complex fraction by the LCM of the denominators. The LCM of x and y is xy.

$$\dfrac{\dfrac{1}{x}+\dfrac{1}{y}}{\dfrac{1}{x}-\dfrac{1}{y}} = \dfrac{\dfrac{1}{x}+\dfrac{1}{y}}{\dfrac{1}{x}-\dfrac{1}{y}} \cdot \dfrac{xy}{xy} = \dfrac{\dfrac{1}{x}\cdot xy + \dfrac{1}{y}\cdot xy}{\dfrac{1}{x}\cdot xy - \dfrac{1}{y}\cdot xy} = \dfrac{y+x}{y-x}$$

Note that after the numerator and denominator of the complex fraction have been multiplied by the LCM of the denominators, no fraction remains in the numerator or denominator.

The method shown above of simplifying a complex fraction by multiplying the numerator and denominator by the LCM of the denominators is used in Example 1. However, a different approach is to rewrite the numerator and denominator of the complex fraction as single fractions and then divide the numerator by the denominator. Here the example shown above is simplified by using this alternative method.

Rewrite the numerator and denominator of the complex fraction as single fractions.

$$\dfrac{\dfrac{1}{x}+\dfrac{1}{y}}{\dfrac{1}{x}-\dfrac{1}{y}} = \dfrac{\dfrac{1}{x}\cdot\dfrac{y}{y}+\dfrac{1}{y}\cdot\dfrac{x}{x}}{\dfrac{1}{x}\cdot\dfrac{y}{y}-\dfrac{1}{y}\cdot\dfrac{x}{x}} = \dfrac{\dfrac{y}{xy}+\dfrac{x}{xy}}{\dfrac{y}{xy}-\dfrac{x}{xy}} = \dfrac{\dfrac{y+x}{xy}}{\dfrac{y-x}{xy}}$$

Take Note

Recall that the fraction bar can be read "divided by."

Divide the numerator of the complex fraction by the denominator.

$$= \dfrac{y+x}{xy} \div \dfrac{y-x}{xy} = \dfrac{y+x}{xy} \cdot \dfrac{xy}{y-x}$$

Multiply the fractions. Simplify.

$$= \dfrac{(y+x)xy}{xy(y-x)} = \dfrac{y+x}{y-x}$$

Note that this is the same result we got before.

EXAMPLE 1 Simplify. **A.** $\dfrac{2 - \dfrac{11}{x} + \dfrac{15}{x^2}}{3 - \dfrac{5}{x} - \dfrac{12}{x^2}}$ **B.** $\dfrac{2x - 1 + \dfrac{7}{x + 4}}{3x - 8 + \dfrac{17}{x + 4}}$

Solution **A.** $\dfrac{2 - \dfrac{11}{x} + \dfrac{15}{x^2}}{3 - \dfrac{5}{x} - \dfrac{12}{x^2}} = \dfrac{2 - \dfrac{11}{x} + \dfrac{15}{x^2}}{3 - \dfrac{5}{x} - \dfrac{12}{x^2}} \cdot \dfrac{x^2}{x^2}$ • The LCM of the denominators is x^2.

$$= \dfrac{2 \cdot x^2 - \dfrac{11}{x} \cdot x^2 + \dfrac{15}{x^2} \cdot x^2}{3 \cdot x^2 - \dfrac{5}{x} \cdot x^2 - \dfrac{12}{x^2} \cdot x^2}$$

$$= \dfrac{2x^2 - 11x + 15}{3x^2 - 5x - 12}$$

$$= \dfrac{(2x - 5)(x - 3)}{(3x + 4)(x - 3)} = \dfrac{2x - 5}{3x + 4}$$

B. $\dfrac{2x - 1 + \dfrac{7}{x + 4}}{3x - 8 + \dfrac{17}{x + 4}} = \dfrac{2x - 1 + \dfrac{7}{x + 4}}{3x - 8 + \dfrac{17}{x + 4}} \cdot \dfrac{x + 4}{x + 4}$ • The LCM of the denominators is $x + 4$.

$$= \dfrac{(2x - 1)(x + 4) + \dfrac{7}{x + 4}(x + 4)}{(3x - 8)(x + 4) + \dfrac{17}{x + 4}(x + 4)}$$

$$= \dfrac{2x^2 + 7x - 4 + 7}{3x^2 + 4x - 32 + 17}$$

$$= \dfrac{2x^2 + 7x + 3}{3x^2 + 4x - 15}$$

$$= \dfrac{(2x + 1)(x + 3)}{(3x - 5)(x + 3)} = \dfrac{2x + 1}{3x - 5}$$

PROBLEM 1 **A.** $\dfrac{3 + \dfrac{16}{x} + \dfrac{16}{x^2}}{6 + \dfrac{5}{x} - \dfrac{4}{x^2}}$ **B.** $\dfrac{2x + 5 + \dfrac{14}{x - 3}}{4x + 16 + \dfrac{49}{x - 3}}$

Solution See page S24.

7.3 Concept Review

Determine whether the following statements are always true, sometimes true, or never true.

1. The numerator of a complex fraction contains one or more fractions.

2. We can simplify a complex fraction by multiplying the numerator and denominator of the complex fraction by the LCM of the denominators.

3. The LCM of the denominators in the complex fraction $\dfrac{1 - \dfrac{3}{x-3}}{\dfrac{2}{x} + 1}$ is $x(x-3)$.

4. $\dfrac{c^{-1}}{a^{-1} + b^{-1}} = \dfrac{a+b}{c}$

7.3 Exercises

1 Simplify complex fractions

1. What is a complex fraction?

2. What is the general goal of simplifying a complex fraction?

For Exercises 3 to 6, use the complex fraction $\dfrac{1 - \dfrac{2}{a}}{1 + \dfrac{1}{a^2}}$.

3. To simplify the complex fraction, multiply the numerator and denominator of the complex fraction by the LCM of the denominators of the fractions ___?___ and ___?___. The LCM is ___?___.

4. When you multiply the numerator of the complex fraction by a^2, the numerator of the complex fraction simplifies to ___?___.

5. When you multiply the denominator of the complex fraction by a^2, the denominator of the complex fraction simplifies to ___?___.

6. The simplified form of the complex fraction is ___?___.

Simplify.

7. $\dfrac{2 - \dfrac{1}{3}}{4 + \dfrac{11}{3}}$

8. $\dfrac{3 + \dfrac{5}{2}}{8 - \dfrac{7}{2}}$

9. $\dfrac{3 - \dfrac{2}{3}}{5 + \dfrac{5}{6}}$

10. $\dfrac{1 + \dfrac{1}{x}}{1 - \dfrac{1}{x^2}}$

11. $\dfrac{\dfrac{1}{y^2} - 1}{1 + \dfrac{1}{y}}$

12. $\dfrac{a - 2}{\dfrac{4}{a} - a}$

13. $\dfrac{\dfrac{25}{a} - a}{5 + a}$

14. $\dfrac{\dfrac{1}{a^2} - \dfrac{1}{a}}{\dfrac{1}{a^2} + \dfrac{1}{a}}$

15. $\dfrac{\dfrac{1}{b} + \dfrac{1}{2}}{\dfrac{4}{b^2} - 1}$

16. $\dfrac{2 - \dfrac{4}{x + 2}}{5 - \dfrac{10}{x + 2}}$

17. $\dfrac{4 + \dfrac{12}{2x - 3}}{5 + \dfrac{15}{2x - 3}}$

18. $\dfrac{\dfrac{3}{2a - 3} + 2}{\dfrac{-6}{2a - 3} - 4}$

19. $\dfrac{\dfrac{-5}{b - 5} - 3}{\dfrac{10}{b - 5} + 6}$

20. $\dfrac{\dfrac{x}{x + 1} - \dfrac{1}{x}}{\dfrac{x}{x + 1} + \dfrac{1}{x}}$

21. $\dfrac{\dfrac{2a}{a - 1} - \dfrac{3}{a}}{\dfrac{1}{a - 1} + \dfrac{2}{a}}$

22. $\dfrac{\dfrac{3}{x}}{\dfrac{9}{x^2}}$

23. $\dfrac{\dfrac{x}{3x - 2}}{\dfrac{x}{9x^2 - 4}}$

24. $\dfrac{\dfrac{a^2 - b^2}{4a^2b}}{\dfrac{a + b}{16ab^2}}$

25. $\dfrac{1 - \dfrac{1}{x} - \dfrac{6}{x^2}}{1 - \dfrac{4}{x} + \dfrac{3}{x^2}}$

26. $\dfrac{1 - \dfrac{3}{x} - \dfrac{10}{x^2}}{1 + \dfrac{11}{x} + \dfrac{18}{x^2}}$

27. $\dfrac{\dfrac{15}{x^2} - \dfrac{2}{x} - 1}{\dfrac{4}{x^2} - \dfrac{5}{x} + 4}$

28. $\dfrac{1 - \dfrac{2x}{3x - 4}}{x - \dfrac{32}{3x - 4}}$

29. $\dfrac{1 - \dfrac{12}{3x + 10}}{x - \dfrac{8}{3x + 10}}$

30. $\dfrac{x - 1 + \dfrac{2}{x - 4}}{x + 3 + \dfrac{6}{x - 4}}$

31. $\dfrac{x - 5 - \dfrac{18}{x + 2}}{x + 7 + \dfrac{6}{x + 2}}$

32. $\dfrac{x - 4 + \dfrac{9}{2x + 3}}{x + 3 - \dfrac{5}{2x + 3}}$

33. $\dfrac{\dfrac{1}{a} - \dfrac{3}{a - 2}}{\dfrac{2}{a} + \dfrac{5}{a - 2}}$

34. $\dfrac{\dfrac{2}{b} - \dfrac{5}{b + 3}}{\dfrac{3}{b} + \dfrac{3}{b + 3}}$

35. $\dfrac{\dfrac{1}{y^2} - \dfrac{1}{xy} - \dfrac{2}{x^2}}{\dfrac{1}{y^2} - \dfrac{3}{xy} + \dfrac{2}{x^2}}$

36. $\dfrac{\dfrac{2}{b^2} - \dfrac{5}{ab} - \dfrac{3}{a^2}}{\dfrac{2}{b^2} + \dfrac{7}{ab} + \dfrac{3}{a^2}}$

37. $\dfrac{\dfrac{x-1}{x+1} - \dfrac{x+1}{x-1}}{\dfrac{x-1}{x+1} + \dfrac{x+1}{x-1}}$

38. $\dfrac{\dfrac{y}{y+2} - \dfrac{y}{y-2}}{\dfrac{y}{y+2} + \dfrac{y}{y-2}}$

39. $4 - \dfrac{2}{2 - \dfrac{3}{x}}$

40. Is the expression the reciprocal of the complex fraction $\dfrac{1}{1 - \dfrac{1}{a}}$?

 a. $\dfrac{a}{a-1}$ **b.** $\dfrac{a-1}{a}$ **c.** $1 - \dfrac{1}{a}$

41. The denominator of a certain complex fraction is the reciprocal of its numerator. Which of the following is the simplified form of this complex fraction?

 a. 1
 b. the square of the complex fraction
 c. the square of the numerator of the complex fraction
 d. the reciprocal of the complex fraction

7.3 Applying Concepts

Simplify.

42. $\dfrac{x^{-1} + y^{-1}}{x^{-1} - y^{-1}}$

43. $\dfrac{x^{-1}}{y^{-1}} + \dfrac{y}{x}$

44. $\dfrac{x^{-1} + y}{x^{-1} - y}$

45. $\dfrac{x - \dfrac{1}{x}}{1 + \dfrac{1}{x}}$

46. $1 - \dfrac{1}{1 - \dfrac{1}{b-2}}$

47. $2 - \dfrac{2}{2 - \dfrac{2}{c-1}}$

48. $3 - \dfrac{2}{1 - \dfrac{2}{3 - \dfrac{2}{x}}}$

49. $a + \dfrac{a}{2 + \dfrac{1}{1 - \dfrac{2}{a}}}$

50. $a - \dfrac{1}{2 - \dfrac{2}{2 - \dfrac{2}{a}}}$

51. $\dfrac{\dfrac{1}{x+h} - \dfrac{1}{x}}{h}$

52. $\dfrac{\dfrac{1}{(x+h)^2} - \dfrac{1}{x^2}}{h}$

53. Find the sum of the reciprocals of three consecutive even integers.

54. **Electronics** The total resistance, R, of three resistors in parallel is given by the formula $R = \dfrac{1}{\dfrac{1}{R_1} + \dfrac{1}{R_2} + \dfrac{1}{R_3}}$. Find the total parallel resistance when

 $R_1 = 2$ ohms, $R_2 = 4$ ohms, and $R_3 = 8$ ohms.

55. Car Loans The interest rate on a loan to purchase a car affects the monthly payment. The function that relates the monthly payment for a 5-year loan (60-month loan) to the monthly interest rate is given by

$$P(x) = \dfrac{Cx}{\left[1 - \dfrac{1}{(x+1)^{60}}\right]}$$

where x is the monthly interest rate (as a decimal), C is the loan amount for the car, and $P(x)$ is the monthly payment.
a. Simplify the complex fraction.
b. Graph this equation for $0 < x \le 0.025$ and $C = \$20,000$.
c. What is the interval of *annual* interest rates for the domain in part b?
d. The point whose coordinates are approximately (0.005, 386.66) is on the graph of this equation. Write a sentence that gives an interpretation of this ordered pair.
e. Use a graphing utility to determine the monthly payment for a car with a loan amount of \$20,000 and an annual interest rate of 8%. Round to the nearest dollar.

7.4 Rational Equations

OBJECTIVE 1 Solve fractional equations

To solve an equation containing fractions, **clear denominators** by multiplying each side of the equation by the LCM of the denominators. Then solve for the variable.

Solve: $\dfrac{3x}{x-5} = 5 - \dfrac{5}{x-5}$

$$\dfrac{3x}{x-5} = 5 - \dfrac{5}{x-5}$$

Multiply each side of the equation by the LCM of the denominators.

$$(x-5)\left(\dfrac{3x}{x-5}\right) = (x-5)\left(5 - \dfrac{5}{x-5}\right)$$

$$3x = (x-5)5 - (x-5)\left(\dfrac{5}{x-5}\right)$$

Simplify.

$$3x = 5x - 25 - 5$$
$$3x = 5x - 30$$

Solve the equation for x.

$$-2x = -30$$
$$x = 15$$

15 checks as a solution.
The solution is 15.

Occasionally, a value of the variable that appears to be a solution will make one of the denominators zero. In this case, the equation has no solution for that value of the variable.

Solve: $\dfrac{3x}{x - 3} = 2 + \dfrac{9}{x - 3}$

$$\dfrac{3x}{x - 3} = 2 + \dfrac{9}{x - 3}$$

Multiply each side of the equation by the LCM of the denominators.

$$(x - 3)\left(\dfrac{3x}{x - 3}\right) = (x - 3)\left(2 + \dfrac{9}{x - 3}\right)$$

Use the Distributive Property.

$$3x = (x - 3)2 + (x - 3)\left(\dfrac{9}{x - 3}\right)$$

$$3x = 2x - 6 + 9$$
$$3x = 2x + 3$$
$$x = 3$$

Substituting 3 into the equation results in division by zero. Because division by zero is not defined, the equation has no solution.

$$\dfrac{3x}{x - 3} = 2 + \dfrac{9}{x - 3}$$

$$\dfrac{3(3)}{3 - 3} = 2 + \dfrac{9}{3 - 3}$$

$$\dfrac{9}{0} = 2 + \dfrac{9}{0}$$

> ### Take Note
> If each side of an equation is multiplied by a variable expression, it is essential that the solutions be checked. As this example shows, a proposed solution to an equation may not check when it is substituted into the original equation.

Multiplying each side of an equation by a variable expression may produce an equation with different solutions from the original equation. Thus, any time you multiply each side of an equation by a variable expression, you must check the resulting solution.

EXAMPLE 1 Solve. **A.** $\dfrac{1}{4} = \dfrac{5}{x + 5}$ **B.** $\dfrac{2x}{x - 2} = \dfrac{1}{3x - 4} + 2$

Solution **A.** $\dfrac{1}{4} = \dfrac{5}{x + 5}$

$$4(x + 5)\dfrac{1}{4} = 4(x + 5)\dfrac{5}{x + 5}$$ • **Multiply each side of the equation by the LCM of the denominators.**

$$x + 5 = 4(5)$$
$$x + 5 = 20$$
$$x = 15$$

15 checks as a solution.
The solution is 15.

B. $\dfrac{2x}{x - 2} = \dfrac{1}{3x - 4} + 2$

$$(x - 2)(3x - 4)\dfrac{2x}{x - 2} = (x - 2)(3x - 4)\left(\dfrac{1}{3x - 4} + 2\right)$$

$$(3x - 4)2x = (x - 2)(3x - 4)\left(\dfrac{1}{3x - 4}\right) + (x - 2)(3x - 4)2$$

$$6x^2 - 8x = x - 2 + 6x^2 - 20x + 16$$
$$6x^2 - 8x = 6x^2 - 19x + 14$$
$$11x = 14$$
$$x = \frac{14}{11}$$

$\frac{14}{11}$ checks as a solution.

The solution is $\frac{14}{11}$.

PROBLEM 1 Solve. **A.** $\dfrac{5}{2x - 3} = \dfrac{-2}{x + 1}$ **B.** $\dfrac{4x + 1}{2x - 1} = 2 + \dfrac{3}{x - 3}$

Solution See page S25.

OBJECTIVE **2**

Point of Interest

The following problem was recorded in the Jiuzhang, a Chinese text that dates to the Han dynasty (about 200 B.C. to A.D. 200). "A reservoir has 5 channels bringing water to it. The first channel can fill the reservoir in $\frac{1}{3}$ day, the second in 1 day, the third in $2\frac{1}{2}$ days, the fourth in 3 days, and the fifth in 5 days. If all channels are open, how long does it take to fill the reservoir?" This problem is the earliest known work problem.

Work problems

Rate of work is that part of a task that is completed in one unit of time. If a mason can build a retaining wall in 12 h, then in 1 h the mason can build $\frac{1}{12}$ of the wall. The mason's rate of work is $\frac{1}{12}$ of the wall each hour. If an apprentice can build the wall in x hours, the rate of work for the apprentice is $\frac{1}{x}$ of the wall each hour.

In solving a work problem, the goal is to determine the time it takes to complete a task. The basic equation that is used to solve work problems is

Rate of work × Time worked = Part of task completed

For example, if a pipe can fill a tank in 5 h, then in 2 h the pipe will fill $\frac{1}{5} \times 2 = \frac{2}{5}$ of the tank. In t hours, the pipe will fill $\frac{1}{5} \times t = \frac{t}{5}$ of the tank.

Solve: A mason can build a wall in 10 h. An apprentice can build a wall in 15 h. How long would it take them to build the wall if they worked together?

STRATEGY for solving a work problem
▶ For each person or machine, write a numerical or variable expression for the rate of work, the time worked, and the part of the task completed. The results can be recorded in a table.

Unknown time to build the wall working together: t

	Rate of work	·	Time worked	=	Part of task completed
Mason	$\dfrac{1}{10}$	·	t	=	$\dfrac{t}{10}$
Apprentice	$\dfrac{1}{15}$	·	t	=	$\dfrac{t}{15}$

> ▶ Determine how the parts of the task completed are related. Use the fact that the sum of the parts of the task completed must equal 1, the complete task.

The sum of the part of the task completed by the mason and the part of the task completed by the apprentice is 1.

$$\frac{t}{10} + \frac{t}{15} = 1$$

$$30\left(\frac{t}{10} + \frac{t}{15}\right) = 30(1)$$

$$3t + 2t = 30$$
$$5t = 30$$
$$t = 6$$

Working together, they would build the wall in 6 h.

EXAMPLE 2　An electrician requires 12 h to wire a house. The electrician's apprentice can wire a house in 16 h. After working alone on one job for 4 h, the electrician quits, and the apprentice completes the task. How long does it take the apprentice to finish wiring the house?

Strategy　▶ Time required for the apprentice to finish wiring the house: t

	Rate	Time	Part
Electrician	$\dfrac{1}{12}$	4	$\dfrac{4}{12}$
Apprentice	$\dfrac{1}{16}$	t	$\dfrac{t}{16}$

▶ The sum of the part of the task completed by the electrician and the part of the task completed by the apprentice is 1.

Solution
$$\frac{4}{12} + \frac{t}{16} = 1$$

$$\frac{1}{3} + \frac{t}{16} = 1$$

$$48\left(\frac{1}{3} + \frac{t}{16}\right) = 48(1)$$

$$16 + 3t = 48$$
$$3t = 32$$
$$t = \frac{32}{3}$$

It takes the apprentice $10\frac{2}{3}$ h to finish wiring the house.

Fills tank
in *x* hours

Fills tank
in 9 hours

Fills $\frac{1}{x}$ of the Fills $\frac{1}{9}$ of the
tank each hour tank each hour

PROBLEM 2 Two water pipes can fill a tank with water in 6 h. The larger
pipe working alone can fill the tank in 9 h. How long would
it take the smaller pipe, working alone, to fill the tank?

Solution See page S25.

OBJECTIVE **3** ## Uniform motion problems

A car that travels constantly in a straight line at 55 mph is in uniform motion.
Uniform motion means that the speed of an object does not change.

The basic equation used to solve uniform motion problems is

Distance = Rate × Time

An alternative form of this equation can be written by solving the equation for
time. This form of the equation is used to solve the following problem.

$$\frac{\textbf{Distance}}{\textbf{Rate}} = \textbf{Time}$$

Solve: A motorist drove 150 mi on country roads before driving 50 mi on
mountain roads. The rate of speed on the country roads was three times the
rate on the mountain roads. The time spent traveling the 200 mi was 5 h. Find
the rate of the motorist on the country roads.

50 mi
r

150 mi
3*r*

STRATEGY **for solving a uniform motion problem**

▶ For each object, write a numerical or variable expression for the
distance, rate, and time. The results can be recorded in a table.

The unknown rate of speed on the mountain roads: *r*
Rate of speed on the country roads: 3*r*

	Distance	÷	Rate	=	Time
Country roads	150	÷	3*r*	=	$\dfrac{150}{3r}$
Mountain roads	50	÷	*r*	=	$\dfrac{50}{r}$

▶ Determine how the times traveled by each object are related. For
example, it may be known that the times are equal, or the total time
may be known.

The total time for the trip is 5 h.

$$\frac{150}{3r} + \frac{50}{r} = 5$$

$$\frac{50}{r} + \frac{50}{r} = 5$$

$$r\left(\frac{50}{r} + \frac{50}{r}\right) = r(5)$$

$$50 + 50 = 5r$$

$$100 = 5r$$

$$20 = r$$

The rate of speed on the country roads was $3r$. Replace r with 20 and evaluate.

$$3r = 3(20) = 60$$

The rate of speed on the country roads was 60 mph.

EXAMPLE 3 A marketing executive traveled 810 mi on a corporate jet in the same amount of time that it took to travel an additional 162 mi by helicopter. The rate of the jet was 360 mph greater than the rate of the helicopter. Find the rate of the jet.

Strategy ▶ Rate of the helicopter: r
Rate of the jet: $r + 360$

$\overset{r}{\underset{162 \text{ mi}}{\longmapsto}}$

$\overset{r + 360}{\underset{810 \text{ mi}}{\longmapsto}}$

	Distance	Rate	Time
Jet	810	$r + 360$	$\dfrac{810}{r + 360}$
Helicopter	162	r	$\dfrac{162}{r}$

▶ The time traveled by jet is equal to the time traveled by helicopter.

Solution
$$\frac{810}{r + 360} = \frac{162}{r}$$

$$r(r + 360)\left(\frac{810}{r + 360}\right) = r(r + 360)\left(\frac{162}{r}\right)$$

$$810r = (r + 360)162$$

$$810r = 162r + 58{,}320$$

$$648r = 58{,}320$$

$$r = 90 \qquad \bullet \text{ The rate of the helicopter}$$
$$\text{was 90 mph.}$$

$$r + 360 = 90 + 360 = 450 \qquad \bullet \text{ Substitute the value of } r$$
$$\text{into the variable expression}$$
$$\text{for the rate of the jet.}$$

The rate of the jet was 450 mph.

PROBLEM 3 A plane can fly at a rate of 150 mph in calm air. Traveling with the wind, the plane flew 700 mi in the same amount of time it took to fly 500 mi against the wind. Find the rate of the wind.

■ **Solution** See page S25.

7.4 Concept Review

Determine whether the following statements are always true, sometimes true, or never true.

1. $\frac{3x}{x - 2} + \frac{4}{x} - 3$ is a rational equation.

2. Multiplying both sides of an equation by the same variable expression produces an equivalent equation.

3. -2 and 5 cannot be solutions of the equation $\frac{1}{x + 2} - 3 = \frac{x - 2}{x - 5}$.

4. An equation that contains a variable in a denominator can be simplified to an equivalent equation with no variables in the denominator.

5. If it takes 3 h to fill a tub with water, then $\frac{1}{3}$ of the tub can be filled in 1 h.

7.4 Exercises

 Solve fractional equations

For Exercises 1 to 4, use the equation $\frac{8}{x} = \frac{5}{x + 3}$.

1. The first step in solving the equation is to clear denominators by multiplying each side of the equation by the LCM of the denominators _____?_____ and _____?_____. The LCM is _____?_____.

2. **a.** When you multiply the left side of the equation by the LCM $x(x + 3)$, the left side simplifies to _____?_____.
 b. When you multiply the right side of the equation by the LCM $x(x + 3)$, the right side simplifies to _____?_____.

3. Use your answers to Exercise 2 to write the equation that results from clearing denominators: _____?_____ = _____?_____. The solution of this equation is _____?_____.

4. The solution of a rational equation must be checked in the original equation. To check the solution you found in Exercise 3 in the original equation, substitute ___?___ for x and simplify:

$$\frac{8}{-8} = \frac{5}{-8+3} \text{ simplifies to } \underline{\quad?\quad} = \underline{\quad?\quad}, \text{ which shows that the}$$

solution checks.

Solve.

5. $\dfrac{x}{2} + \dfrac{5}{6} = \dfrac{x}{3}$

6. $\dfrac{x}{5} - \dfrac{2}{9} = \dfrac{x}{15}$

7. $1 - \dfrac{3}{y} = 4$

8. $7 + \dfrac{6}{y} = 5$

9. $\dfrac{8}{2x-1} = 2$

10. $3 = \dfrac{18}{3x-4}$

11. $\dfrac{4}{x-4} = \dfrac{2}{x-2}$

12. $\dfrac{x}{3} = \dfrac{x+1}{7}$

13. $\dfrac{x-2}{5} = \dfrac{1}{x+2}$

14. $\dfrac{x+4}{10} = \dfrac{6}{x-3}$

15. $\dfrac{3}{x-2} = \dfrac{4}{x}$

16. $\dfrac{5}{x} = \dfrac{2}{x+3}$

17. $\dfrac{3}{x-4} + 2 = \dfrac{5}{x-4}$

18. $\dfrac{5}{y+3} - 2 = \dfrac{7}{y+3}$

19. $\dfrac{8}{x-5} = \dfrac{3}{x}$

20. $\dfrac{16}{2-x} = \dfrac{4}{x}$

21. $5 + \dfrac{8}{a-2} = \dfrac{4a}{a-2}$

22. $\dfrac{-4}{a-4} = 3 - \dfrac{a}{a-4}$

23. $\dfrac{x}{2} + \dfrac{20}{x} = 7$

24. $3x = \dfrac{4}{x} - \dfrac{13}{2}$

25. $\dfrac{6}{x-5} = \dfrac{1}{x}$

26. $\dfrac{8}{x-2} = \dfrac{4}{x+1}$

27. $\dfrac{x}{x+2} = \dfrac{6}{x+5}$

28. $\dfrac{x}{x-2} = \dfrac{3}{x-4}$

29. $-\dfrac{5}{x+7} + 1 = \dfrac{4}{x+7}$

30. $5 - \dfrac{2}{2x-5} = \dfrac{3}{2x-5}$

31. $\dfrac{2}{4y^2-9} + \dfrac{1}{2y-3} = \dfrac{3}{2y+3}$

32. $\dfrac{5}{x-2} - \dfrac{2}{x+2} = \dfrac{3}{x^2-4}$

33. $\dfrac{5}{x^2-7x+12} = \dfrac{2}{x-3} + \dfrac{5}{x-4}$

34. $\dfrac{9}{x^2+7x+10} = \dfrac{5}{x+2} - \dfrac{3}{x+5}$

When a proposed solution of a rational equation does not check in the original equation, it is because it results in an expression involving division by zero. For each equation, state the values of x that, when substituted into the original equation, would result in division by zero.

35. $\dfrac{x^2+24}{x^2+x-12} + \dfrac{x}{4} = \dfrac{2x}{x-3}$

36. $\dfrac{1}{x-2} - 1 = \dfrac{2}{x^2-2x}$

2 Work problems

37. ✎ If one person can complete a task in 2 h and another person can complete the same task in 3 h, will it take more or less than 2 h to complete the task when both people are working? Explain your answer.

38. If a gardener can mow a lawn in 20 min, what portion of the lawn can the gardener mow in 1 min?

39. It takes the hot water hose 6 min to fill a washing machine. It takes the cold water hose 4 min to fill the machine. Let t represent the amount of time the hot and cold water hoses together take to fill the machine with warm water. Complete the following table.

	Rate of work	·	Time worked	=	Part of task completed
Hot water hose	?	·	?	=	?
Cold water hose	?	·	?	=	?

40. Refer to Exercise 39. When the washing machine is full, the "part of task completed" is the whole task, so the sum of the parts completed by the hot water hose and by the cold water hose is ___?___. Use this fact and the expressions in the table in Exercise 39 to write an equation that can be solved to find the amount of time it takes to fill the washer with warm water: ___?___ + ___?___ = ___?___.

41. A large biotech firm uses two computers to process the daily results of its research studies. One computer can process the data in 2 h; the other computer takes 3 h to do the same job. How long would it take to process the data if both computers were used?

42. Two college students have started their own business building computers from kits. Working alone, one student can build a computer in 20 h. When the second student helps, they can build a computer in 7.5 h. How long would it take the second student, working alone, to build the computer?

43. One solar heating panel can raise the temperature of water 1 degree in 30 min. A second solar heating panel can raise the temperature 1 degree in 45 min. How long would it take to raise the temperature of the water 1 degree with both solar panels operating?

44. One member of a gardening team can landscape a new lawn in 36 h. The other member of the team can do the job in 45 h. How long would it take to landscape a lawn if both gardeners worked together?

45. One member of a telephone crew can wire new telephone lines in 5 h. It takes 7.5 h for the other member of the crew to do the job. How long would it take to wire new telephone lines if both members of the crew worked together?

46. As the spring flood waters began to recede in Texas, a young family was faced with pumping the water from their basement. One pump they were using could dispose of 9000 gal in 3 h. A second pump could dispose of the same number of gallons in 4.5 h. How many hours would it take to dispose of 9000 gal if both pumps were working?

47. A new machine can package transistors four times as fast as an older machine. Working together, the machines can package the transistors in 8 h. How long would it take the new machine, working alone, to package the transistors?

48. A ski resort can "manufacture" enough snow to open its steepest run in 12 h, whereas naturally falling snow would have to last for 36 h to provide enough snow. If the resort makes snow at the same time as it is snowing naturally, how long will it be before the run can be opened?

49. The larger of two printers being used to print the payroll for a major corporation requires 40 min to print the payroll. After both printers have been operating for 10 min, the larger printer malfunctions. The smaller printer requires 50 more minutes to complete the payroll. How long would it take the smaller printer, working alone, to print the payroll?

50. An experienced bricklayer can work twice as fast as an apprentice bricklayer. After they worked together on a job for 8 h, the experienced bricklayer quit. The apprentice required 12 more hours to finish the job. How long would it take the experienced bricklayer, working alone, to do the job?

51. A roofer requires 12 h to shingle a roof. After the roofer and an apprentice work on a roof for 3 h, the roofer moves on to another job. The apprentice requires 12 more hours to finish the job. How long would it take the apprentice, working alone, to do the job?

52. A welder requires 25 h to do a job. After the welder and an apprentice work on a job for 10 h, the welder quits. The apprentice finishes the job in 17 h. How long would it take the apprentice, working alone, to do the job?

53. Two clerks are addressing envelopes for a mayoral candidate's political campaign. One clerk can address one envelope every 30 s, whereas it takes 40 s for the second clerk to address one envelope. How many minutes will it take them, working together, to address 140 envelopes?

54. Three machines are filling water bottles. The machines can fill the daily quota of water bottles in 12 h, 15 h, and 20 h, respectively. How long would it take to fill the daily quota of water bottles with all three machines working?

55. With both hot and cold water running, a bathtub can be filled in 10 min. The drain will empty the tub in 15 min. A child turns both faucets on and leaves the drain open. How long will it be before the bathtub starts to overflow?

56. The inlet pipe can fill a water tank in 30 min. The outlet pipe can empty the tank in 20 min. How long would it take to empty a full tank with both pipes open?

57. An oil tank has two inlet pipes and one outlet pipe. One inlet pipe can fill the tank in 12 h, and the other inlet pipe can fill the tank in 20 h. The outlet pipe can empty the tank in 10 h. How long would it take to fill the tank with all three pipes open?

58. Water from a tank is being used for irrigation at the same time as the tank is being filled. The two inlet pipes can fill the tank in 6 h and 12 h, respectively. The outlet pipe can empty the tank in 24 h. How long would it take to fill the tank with all three pipes open?

59. It takes Jane n minutes to weed a row of the garden, and it takes Paul m minutes to weed a row of the garden, where $m > n$. Let t be the amount of time it takes Jane and Paul to weed a row of the garden together. Is t less than n, between n and m, or greater than m?

60. Ernie and Mike painted a fence together in a hours. It would have taken Ernie b hours to paint the fence by himself. What fraction of the fence did Ernie paint? What fraction of the fence did Mike paint?

3 Uniform motion problems

61. a. A plane can fly 470 mph in calm air. In the time it takes the plane to fly 1320 mi against a headwind, it could fly 1500 mi if it were flying with the wind. Let w represent the rate of the wind. Complete the following table.

	Distance	÷	Rate	=	Time
Against the wind	___?___	÷	___?___	=	___?___
With the wind	___?___	÷	___?___	=	___?___

 b. Use the relationship between the expressions in the last column of the table to write an equation that can be solved to find the rate of the wind: ___?___ = ___?___ .

62. Use the equation from part b of Exercise 61. The first step in solving this equation is to multiply each side of the equation by ___?___ .

63. Two skaters take off for an afternoon of rollerblading in Central Park. The first skater can cover 15 mi in the same time it takes the second skater, traveling 3 mph slower than the first skater, to cover 12 mi. Find the rate of each rollerblader.

64. A commercial jet travels 1620 mi in the same amount of time it takes a corporate jet to travel 1260 mi. The rate of the commercial jet is 120 mph greater than the rate of the corporate jet. Find the rate of each jet.

Bethesda Fountain in Central Park

65. A passenger train travels 295 mi in the same amount of time it takes a freight train to travel 225 mi. The rate of the passenger train is 14 mph greater than the rate of the freight train. Find the rate of each train.

66. The rate of a bicyclist is 7 mph more than the rate of a long-distance runner. The bicyclist travels 30 mi in the same amount of time it takes the runner to travel 16 mi. Find the rate of the runner.

67. A cyclist rode 40 mi before having a flat tire and then walking 5 mi to a service station. The cycling rate was four times the walking rate. The time spent cycling and walking was 5 h. Find the rate at which the cyclist was riding.

68. A sales executive traveled 32 mi by car and then an additional 576 mi by plane. The rate of the plane was nine times the rate of the car. The total time of the trip was 3 h. Find the rate of the plane.

69. A motorist drove 72 mi before running out of gas and then walking 4 mi to a gas station. The driving rate of the motorist was twelve times the walking rate. The time spent driving and walking was 2.5 h. Find the rate at which the motorist walks.

70. An insurance representative traveled 735 mi by commercial jet and then an additional 105 mi by helicopter. The rate of the jet was four times the rate of the helicopter. The entire trip took 2.2 h. Find the rate of the jet.

71. A business executive can travel the 480 ft between two terminals of an airport by walking on a moving sidewalk in the same time required to walk 360 ft without using the moving sidewalk. If the rate of the moving sidewalk is 2 ft/s, find the rate at which the executive can walk.

72. A cyclist and a jogger start from a town at the same time and head for a destination 18 mi away. The rate of the cyclist is twice the rate of the jogger. The cyclist arrives 1.5 h before the jogger. Find the rate of the cyclist.

73. A single-engine plane and a commercial jet leave an airport at 10 A.M. and head for an airport 960 mi away. The rate of the jet is four times the rate of the single-engine plane. The single-engine plane arrives 4 h after the jet. Find the rate of each plane.

74. Marlys can row a boat 3 mph faster than she can swim. She is able to row 10 mi in the same time it takes her to swim 4 mi. Find Marlys's rate swimming.

75. A cruise ship can sail 28 mph in calm water. Sailing with the Gulf Stream, the ship can sail 170 mi in the same amount of time as it takes to sail 110 mi against the Gulf Stream. Find the rate of the Gulf Stream.

76. A commercial jet can fly 500 mph in calm air. Traveling with the jet stream, the plane flew 2420 mi in the same amount of time as it takes to fly 1580 mi against the jet stream. Find the rate of the jet stream.

77. A tour boat used for river excursions can travel 7 mph in calm water. The amount of time it takes to travel 20 mi with the current is the same amount of time it takes to travel 8 mi against the current. Find the rate of the current.

78. A canoe can travel 8 mph in still water. Traveling with the current of a river, the canoe can travel 15 mi in the same amount of time it takes to travel 9 mi against the current. Find the rate of the current.

7.4 **Applying Concepts**

79. Number Problem The denominator of a fraction is 4 more than the numerator. If both the numerator and the denominator of the fraction are increased by 3, the new fraction is $\frac{5}{6}$. Find the original fraction.

80. Work Problem One pipe can fill a tank in 3 h, a second pipe can fill the tank in 4 h, and a third pipe can fill the tank in 6 h. How long would it take to fill the tank with all three pipes operating?

81. Work Problem One printer can print a company's paychecks in 24 min, a second printer can print the checks in 16 min, and a third printer can do the job in 12 min. How long would it take to print the checks with all three printers operating?

82. Uniform Motion Problem By increasing your speed by 10 mph, you can drive the 200-mile trip to your hometown in 40 min less time than the trip usually takes you. How fast do you usually drive?

83. Uniform Motion Problem Because of weather conditions, a bus driver reduced the usual speed along a 165-mile bus route by 5 mph. The bus arrived only 15 min later than its usual arrival time. How fast does the bus usually travel?

84. Work Problem If a pump can fill a pool in A hours and a second pump can fill the pool in B hours, find a formula, in terms of A and B, for the time it takes both pumps, working together, to fill the pool.

85. Uniform Motion Problem If a parade is 1 mi long and is proceeding at 3 mph, how long will it take a runner, jogging at 5 mph, to run from the beginning of the parade to the end and then back to the beginning?

86. Write a report on the Rhind Papyrus.

87. What is a unit fraction? What role did unit fractions play in early Egyptian mathematics?

Proportions and Variation

Proportions

Quantities such as 3 feet, 5 liters, and 2 miles are number quantities written with units. In these examples, the units are feet, liters, and miles.

A **ratio** is the quotient of two quantities that have the same unit.

The weekly wages of a painter are $800. The painter spends $150 a week for food. The ratio of wages spent for food to the total weekly wages is written as shown below.

$$\frac{\$150}{\$800} = \frac{150}{800} = \frac{3}{16}$$ A ratio is in simplest form when the two numbers do not have a common factor. The units are not written.

A **rate** is the quotient of two quantities that have different units.

A car travels 120 mi on 3 gal of gas. The miles-to-gallon rate is written as shown below.

$$\frac{120 \text{ mi}}{3 \text{ gal}} = \frac{40 \text{ mi}}{1 \text{ gal}}$$ A rate is in simplest form when the two numbers do not have a common factor. The units are written as part of the rate.

A **proportion** is an equation that states that two ratios or rates are equal. For example, $\frac{90 \text{ km}}{4 \text{ L}} = \frac{45 \text{ km}}{2 \text{ L}}$ and $\frac{3}{4} = \frac{x + 2}{16}$ are proportions.

Note that a proportion is a special kind of fractional equation. Many application problems can be solved by using proportions.

The sales tax on a car that costs $24,000 is $1320. Find the sales tax on a car that costs $29,000.

Write a proportion using x to represent the sales tax.	$$\frac{1320}{24,000} = \frac{x}{29,000}$$
To solve the proportion, multiply each side of the equation by the denominators.	$$\frac{11}{200} = \frac{x}{29,000}$$ $$(200)(29,000)\left(\frac{11}{200}\right) = (200)(29,000)\left(\frac{x}{29,000}\right)$$ $$(29,000)(11) = 200x$$ $$319,000 = 200x$$ $$1595 = x$$

The sales tax on the $29,000 car is $1595.

EXAMPLE 1 A stock investment of 50 shares pays a dividend of $106. At this rate, how many aditional shares are needed to earn a dividend of $424?

Strategy To find the additional number of shares that are required, write and solve a proportion using x to represent the additional number of shares. Then $50 + x$ is the total number of shares of stock.

Solution

$$\frac{106}{50} = \frac{424}{50 + x}$$

$$\frac{53}{25} = \frac{424}{50 + x}$$

$$25(50 + x)\frac{53}{25} = 25(50 + x)\frac{424}{50 + x}$$

$$(50 + x)53 = (25)424$$
$$2650 + 53x = 10{,}600$$
$$53x = 7950$$
$$x = 150$$

An additional 150 shares of stock are required.

PROBLEM 1 Two pounds of cashews cost $5.80. At this rate, how much would 15 lb of cashews cost?

Solution See page S25.

OBJECTIVE 2 Variation problems

A **direct variation** is a special function that can be expressed as the equation $y = kx$, where k is a constant. The equation $y = kx$ is read "y varies directly as x" or "y is proportional to x." The constant k is called the **constant of variation** or the **constant of proportionality.**

The circumference (C) of a circle varies directly as the diameter (d). The direct variation equation is written $C = \pi d$. The constant of variation is π.

A nurse earns $28 per hour. The nurse's total wage (w) is directly proportional to the number of hours (h) worked. The equation of variation is $w = 28h$. The constant of proportionality is 28.

In general, a direct variation equation can be written in the form $y = kx^n$, where n is a positive number. For example, the equation $y = kx^2$ is read "y varies directly as the square of x."

The area (A) of a circle varies directly as the square of the radius (r) of the circle. The direct variation equation is $A = \pi r^2$.

The graph of $V = 5r$

Given that V varies directly as r and that $V = 20$ when $r = 4$, the constant of variation can be found by writing the basic direct variation equation, replacing V and r by the given values, and solving for the constant of variation.

$V = kr$
$20 = k \cdot 4$
$5 = k$

The direct variation equation can then be written by substituting the value of k into the basic direct variation equation.

$V = 5r$

The tension (T) in a spring varies directly as the distance (x) it is stretched. If $T = 8$ lb when $x = 2$ in., find T when $x = 4$ in.

Write the basic direct variation equation.	$T = kx$
Replace T and x by the given values.	$8 = k \cdot 2$
Solve for the constant of variation.	$4 = k$
Write the direct variation equation.	$T = 4x$
To find T when $x = 4$ in., substitute 4 for x in the direct variation equation and solve for T.	$T = 4x$ $T = 4 \cdot 4$ $T = 16$

When $x = 4$ in., the tension is 16 lb.

The graph of $T = 4x$

EXAMPLE 2 The amount (A) of medication prescribed for a person varies directly with the person's weight (W). For a person who weighs 50 kg, 2 ml of medication are prescribed. How many milliliters of medication are required for a person who weighs 75 kg?

Strategy To find the required amount of medication:

▶ Write the basic direct variation equation, replace the variables by the given values, and solve for k.

▶ Write the direct variation equation, replacing k by its value. Substitute 75 for W, and solve for A.

Solution $A = kW$

$2 = k \cdot 50$

$\dfrac{1}{25} = k$

$A = \dfrac{1}{25} W$ • **This is the direct variation equation.**

$A = \dfrac{1}{25} \cdot 75 = 3$ • **Replace W by 75.**

The required amount of medication is 3 ml.

PROBLEM 2 The distance (s) a body falls from rest varies directly as the square of the time (t) of the fall. An object falls 64 ft in 2 s. How far will it fall in 5 s?

Solution See pages S25–S26.

A **joint variation** is a variation in which a variable varies directly as the product of two or more other variables. A joint variation can be expressed as the equation $z = kxy$, where k is a constant. The equation $z = kxy$ is read "z varies jointly as x and y."

The area (A) of a triangle varies jointly as the base (b) and the height (h). The joint variation equation is written $A = \dfrac{1}{2}bh$. The constant of variation is $\dfrac{1}{2}$.

An **inverse variation** is a function that can be expressed as the equation $y = \dfrac{k}{x}$, where k is a constant. The equation $y = \dfrac{k}{x}$ is read "y varies inversely as x" or "y is inversely proportional to x."

In general, an inverse variation equation can be written $y = \dfrac{k}{x^n}$, where n is a positive number. For example, the equation $y = \dfrac{k}{x^2}$ is read "y varies inversely as the square of x."

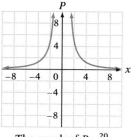

The graph of $P = \dfrac{20}{x^2}$

Given that P varies inversely as the square of x and that $P = 5$ when $x = 2$, the variation constant can be found by writing the basic inverse variation equation, replacing P and x by the given values, and solving for the constant variation.

$$P = \dfrac{k}{x^2}$$
$$5 = \dfrac{k}{2^2}$$
$$5 = \dfrac{k}{4}$$
$$20 = k$$

The inverse variation equation can then be found by substituting the value of k into the basic inverse variation equation.

$$P = \dfrac{20}{x^2}$$

The length (L) of a rectangle with fixed area is inversely proportional to the width (w). If $L = 6$ ft when $w = 2$ ft, find L when $w = 3$ ft.

Write the basic inverse variation equation. $L = \dfrac{k}{w}$

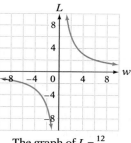

The graph of $L = \dfrac{12}{w}$

Replace L and w by the given values. $6 = \dfrac{k}{2}$

Solve for the constant of variation. $12 = k$

Write the inverse variation equation. $L = \dfrac{12}{w}$

To find L when $w = 3$ ft, substitute 3 for w in the inverse variation equation and solve for L. $L = \dfrac{12}{w}$

$$L = \dfrac{12}{3}$$
$$L = 4$$

When $w = 3$ ft, the length is 4 ft.

EXAMPLE 3 A company that produces personal computers has determined that the number of computers it can sell (s) is inversely proportional to the price (P) of the computer. Two thousand computers can be sold when the price is $2500. How many computers can be sold when the price of a computer is $2000?

Strategy To find the number of computers:

▶ Write the basic inverse variation equation, replace the variables by the given values, and solve for k.

▶ Write the inverse variation equation, replacing k by its value. Substitute 2000 for P, and solve for s.

Solution

$$s = \frac{k}{P}$$

$$2000 = \frac{k}{2500}$$

$$5,000,000 = k$$

$$s = \frac{5,000,000}{P} \qquad \text{• This is the inverse variation equation.}$$

$$s = \frac{5,000,000}{2000} = 2500 \qquad \text{• Replace } P \text{ by 2000.}$$

At a price of \$2000, 2500 computers can be sold.

PROBLEM 3 The resistance (R) to the flow of electric current in a wire of fixed length is inversely proportional to the square of the diameter (d) of the wire. If a wire of diameter 0.01 cm has a resistance of 0.5 ohm, what is the resistance in a wire that is 0.02 cm in diameter?

Solution See page S26.

A **combined variation** is a variation in which two or more types of variation occur at the same time. For example, in chemistry, the volume (V) of a gas varies directly as the temperature (T) and inversely as the pressure (P). This combined variation is written $V = \dfrac{kT}{P}$.

EXAMPLE 4 The pressure (P) of a gas varies directly as the temperature (T) and inversely as the volume (V). When $T = 50°$ and $V = 275$ in^3, $P = 20$ lb/in^2. Find the pressure of a gas when $T = 60°$ and $V = 250$ in^3.

Strategy To find the pressure:

▶ Write the basic combined variation equation, replace the variables by the given values, and solve for k.

▶ Write the combined variation equation, replacing k by its value. Substitute 60 for T and 250 for V, and solve for P.

Solution

$$P = \frac{kT}{V}$$

$$20 = \frac{k(50)}{275}$$

$$110 = k$$

$$P = \frac{110T}{V} \qquad \text{• This is the combined variation equation.}$$

$$P = \frac{110(60)}{250} = 26.4 \qquad \text{• Replace } T \text{ by 60 and } V \text{ by 250.}$$

The pressure is 26.4 lb/in^2.

PROBLEM 4 The strength (s) of a rectangular beam varies jointly as its width (W) and as the square of its depth (d) and inversely as its length (L). If the strength of a beam 2 in. wide, 12 in. deep, and 12 ft long is 1200 lb, find the strength of a beam 4 in. wide, 8 in. deep, and 16 ft long.

■ **Solution** See page S26.

7.5 Concept Review

Determine whether the following statements are always true, sometimes true, or never true.

1. If x varies inversely as y, then when x is doubled, y is doubled.

2. If a varies inversely as b, then ab is a constant.

3. If a varies jointly as b and c, then $a = \dfrac{kb}{c}$.

4. If the length of a rectangle is held constant, then the area of the rectangle varies directly as the width.

5. If the area of a rectangle is held constant, then the length varies directly as the width.

6. The circumference of a circle varies directly as the diameter. If the diameter of a circle is doubled, then the circumference of the circle will be doubled.

7.5 Exercises

1 Proportions

1. How does a ratio differ from a rate?

2. What is a proportion?

3. The scale on a map shows that a distance of 2 cm on the map represents an actual distance of 5 mi. This rate can be expressed as the quotient

$\dfrac{?}{?}$ or as the quotient $\dfrac{?}{?}$.

4. Let x be the distance (in miles) represented by 13 cm on the map described in Exercise 3. Use a proportion to find x.

$$\frac{5}{2} = \frac{?}{\underline{\quad ? \quad}}$$

$$(2)(13)\left(\frac{5}{2}\right) = (2)(13)\left(\frac{x}{13}\right)$$

$$\underline{\quad ? \quad} = (\underline{\quad ? \quad})x$$

$$\underline{\quad ? \quad} = x$$

- Write the proportion so that each side is the ratio of miles to centimeters.
- Multiply each side of the proportion by $\underline{\quad ? \quad}$ and $\underline{\quad ? \quad}$.
- Simplify.
- Divide each side of the equation by $\underline{\quad ? \quad}$.

13 cm on the map represents $\underline{\quad ? \quad}$ mi.

Solve the proportion.

5. $\dfrac{x+1}{10} = \dfrac{2}{5}$

6. $\dfrac{4}{x+2} = \dfrac{3}{4}$

7. $\dfrac{x}{4} = \dfrac{x-2}{8}$

8. $\dfrac{8}{x-5} = \dfrac{3}{x}$

9. $\dfrac{8}{x-2} = \dfrac{4}{x+1}$

10. $\dfrac{4}{x-4} = \dfrac{2}{x-2}$

11. $\dfrac{8}{3x-2} = \dfrac{2}{2x+1}$

12. $\dfrac{x-2}{x-5} = \dfrac{2x}{2x+5}$

13. True or false? If $\dfrac{a}{b} = \dfrac{c}{d}$, then $\dfrac{b}{a} = \dfrac{d}{c}$.

14. True or false? If $\dfrac{a}{b} = \dfrac{c}{d}$, then $\dfrac{a}{c} = \dfrac{b}{d}$.

15. Environmental Science In a wildlife preserve, 60 ducks are captured, tagged, and then released. Later, 200 ducks are examined, and three of the 200 ducks are found to have tags. Estimate the number of ducks in the preserve.

16. Political Science A pre-election survey showed that 7 out of every 12 voters would vote in an election. At this rate, how many people would be expected to vote in a city of 210,000?

17. Food Service A caterer estimates that 2 gal of fruit punch will serve 30 people. How much additional punch is necessary to serve 75 people?

18. Nutrition If a 56-gram serving of pasta contains 7 g of protein, how many grams of protein are in a 454-gram box of the pasta?

19. Architecture On an architectural drawing, $\frac{1}{4}$ in. represents 1 ft. Find the dimensions of a room that measures $4\frac{1}{4}$ in. by $5\frac{1}{2}$ in. on the drawing.

20. Construction A contractor estimated that 15 ft² of window space will be allowed for every 160 ft² of floor space. Using this estimate, how much window space will be allowed for 3200 ft² of floor space?

21. Energy Walking 4 mi in 2 h will use 650 calories. Walking at the same rate, how many miles would a person need to walk to lose 1 pound? (The burning of 3500 calories is equivalent to the loss of 1 pound.) Round to the nearest hundredth.

22. Construction To tile 24 ft² of area, 120 ceramic tiles are needed. At this rate, how many tiles are needed to tile 300 ft²?

23. Health Science Three-fourths of an ounce of a medication are required for a 120-pound adult. At the same rate, how many additional ounces of medication are required for a 200-pound adult?

24. Physics An object that weighs 100 lb on Earth would weigh 90.5 lb on Venus. At this rate, what is the weight on Venus of an object that weighs 150 lb on Earth?

25. Environmental Science Six ounces of an insecticide are mixed with 15 gal of water to make a spray for spraying an orange grove. At the same rate, how much additional insecticide is required to be mixed with 100 gal of water?

26. Ecology In an attempt to estimate the number of Siberian tigers in a preserve, a wildlife management team captured, tagged, and released 50 Siberian tigers. A few months later, they captured 150 Siberian tigers of which 30 had tags. Estimate the number of Siberian tigers in the region.

27. Internet Speeds A certain wireless Internet connection will download a 4-megabyte file in 60 s. At this rate, how many seconds would it take to download a 5-megabyte file?

28. Construction A contractor estimated that 30 ft³ of cement is required to make a concrete floor that measures 90 ft². Using this estimate, find how many additional cubic feet of cement would be required to make a concrete floor that measures 120 ft².

29. Graphic Arts The accompanying picture of a whale uses a scale of 1 in. represents 48 ft. Estimate the actual length of the whale.

30. Graphic Arts The accompanying picture of an elephant uses a scale of 1 in. represents 12 ft. Estimate the actual height of the elephant.

2 **Variation problems**

31. What is a direct variation?

32. What is an inverse variation?

33. A ball whose circumference is 30 cm weighs 250 g. An 800-gram ball made of the same material has a circumference of 96 cm. Is the relationship between the circumference of a ball made of this material and its weight a direct variation or an indirect variation?

34. A person who weighs 150 lb on Earth weighs only 149.25 lb at 10 mi above Earth and 148.5 lb at 20 mi above Earth. Is the relationship between weight and distance above Earth a direct variation or an indirect variation?

35. Scuba Diving The pressure (p) on a diver in the water varies directly as the depth (d). If the pressure is 3.6 lb/in² when the depth is 8 ft, what is the pressure when the depth is 30 ft?

36. Physics The distance (d) a spring will stretch varies directly as the force (f) applied to the spring. If a force of 5 lb is required to stretch a spring 2 in., what force is required to stretch the spring 5 in.?

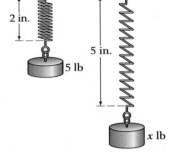

37. Business The profit (P) realized by a company varies directly as the number of products it sells (s). If a company makes a profit of $2500 on the sale of 20 products, what is the profit when the company sells 300 products?

38. Farming The number of bushels of wheat (b) produced by a farm is directly proportional to the number of acres (A) planted in wheat. If a 25-acre farm yields 1125 bushels of wheat, what is the yield of a farm that has 220 acres of wheat?

39. Art Leonardo da Vinci observed that the length of a person's face varies directly as the length of the person's chin. If a person whose face length is 9 in. has a chin length of 1.5 in., what is the length of a person's face whose chin length is 1.7 in.?

40. Art Leonardo da Vinci observed that a person's height varies directly as the width of the person's shoulders. If a person 70 in. tall has a shoulder width of 17.5 in., what is the height of a person whose shoulder width is 16 in.?

41. Optics The distance (d) a person can see to the horizon from a point above the surface of Earth varies directly as the square root of the height (H). If, for a height of 500 ft, the horizon is 19 mi away, how far is the horizon from a point that is 800 ft high? Round to the nearest hundredth.

42. Physics The period (p) of a pendulum, or the time it takes for the pendulum to make one complete swing, varies directly as the square root of the length (L) of the pendulum. If the period of a pendulum is 1.5 s when the length is 2 ft, find the period when the length is 4.5 ft. Round to the nearest hundredth.

43. Physics The distance (s) a ball will roll down an inclined plane is directly proportional to the square of the time (t). If the ball rolls 5 ft in 1 s, how far will it roll in 4 s?

44. Automotive Technology The stopping distance (*s*) of a car varies directly as the square of its speed (*v*). If a car traveling 30 mph requires 60 ft to stop, find the stopping distance for a car traveling 55 mph. Round to the nearest tenth.

45. Chemistry For a constant temperature, the pressure (*P*) of a gas varies inversely as the volume (*V*). If the pressure is 25 lb/in^2 when the volume is 400 ft^3, find the pressure when the volume is 150 ft^3.

46. Physics The pressure (*p*) in a liquid varies directly as the product of the depth (*d*) and the density (*D*) of the liquid. If the pressure is 37.5 lb/in^2 when the depth is 100 in. and the density is 1.2 lb/in^2, find the pressure when the density remains the same and the depth is 60 in.

47. Mechanical Engineering The speed (*v*) of a gear varies inversely as the number of teeth (*t*). If a gear that has 48 teeth makes 20 revolutions per minute, how many revolutions per minute will a gear that has 30 teeth make?

48. Electricity The current (*I*) in a wire varies directly as the voltage (*v*) and inversely as the resistance (*r*). If the current is 27.5 amps when the voltage is 110 volts and the resistance is 4 ohms, find the current when the voltage is 195 volts and the resistance is 12 ohms.

49. Magnetism The repulsive force (*f*) between the north poles of two magnets is inversely proportional to the square of the distance (*d*) between them. If the repulsive force is 18 lb when the distance is 3 in., find the repulsive force when the distance is 1.2 in.

50. Physics The intensity (*I*) of a light source is inversely proportional to the square of the distance (*d*) from the source. If the intensity is 8 lumens at a distance of 6 m, what is the intensity when the distance is 4 m?

51. Electricity The resistance (*R*) of a wire varies directly as the length (*L*) of the wire and inversely as the square of the diameter (*d*). If the resistance is 9 ohms in 50 ft of wire that has a diameter of 0.05 in., find the resistance in 50 ft of a similar wire that has a diameter of 0.02 in.

52. Music The frequency of vibration (*f*) of a guitar string varies directly as the square root of the tension (*T*) and inversely as the length (*L*) of the string. If the frequency is 40 vibrations per second when the tension is 25 lb and the length of the string is 3 ft, find the frequency when the tension is 36 lb and the string is 4 ft long.

53. Sailing The wind force (*w*) on a vertical surface varies directly as the product of the area (*A*) of the surface and the square of the wind velocity (*v*). When the wind is blowing at 30 mph, the force on an area of 10 ft^2 is 45 lb. Find the force on this area when the wind is blowing at 60 mph.

54. Electricity The power (*P*) in an electric circuit is directly proportional to the product of the current (*I*) and the square of the resistance (*R*). If the power is 100 watts when the current is 4 amps and the resistance is 5 ohms, find the power when the current is 2 amps and the resistance is 10 ohms.

7.5 Applying Concepts

55. a. Graph $y = kx$ when $k = 2$.
 b. What kind of function does the graph represent?

56. a. Graph $y = kx$ when $k = \dfrac{1}{2}$.
 b. What kind of function does the graph represent?

57. a. Graph $y = \dfrac{k}{x}$ when $k = 2$ and $x > 0$.
 b. Is this the graph of a function?

58. a. Graph $y = \dfrac{k}{x^2}$ when $k = 2$ and $x > 0$.
 b. Is this the graph of a function?

59. In the inverse variation equation $y = \dfrac{k}{x}$, what is the effect on x when y doubles?

60. In the direct variation equation $y = kx$, what is the effect on y when x doubles?

Complete using the word *directly* or *inversely*.

61. If a varies directly as b and inversely as c, then c varies _____ as b and _____ as a.

62. If a varies _____ as b and c, then abc is constant.

63. If the width of a rectangle is held constant, the area of the rectangle varies _____ as the length.

64. If the area of a rectangle is held constant, the length of the rectangle varies _____ as the width.

65. A number h is the harmonic mean of the numbers a and b if the reciprocal of h is equal to the average of the reciprocals of a and b.
 a. Write an expression for the harmonic mean of a and b.
 b. Find the harmonic mean of 10 and 15.

66. Write a report on financial ratios used by stock market analysts. Include the price-earnings ratio, current ratio, quick ratio, percent yield, and ratio of stockholders' equity to liabilities.

7.6 Literal Equations

OBJECTIVE 1 Solve literal equations

A **literal equation** is an equation that contains more than one variable. Examples of literal equations are shown at the right.

$$3x - 2y = 4$$
$$v^2 = v_0^2 + 2as$$

Formulas are used to express relationships among physical quantities. A **formula** is a literal equation that states a rule about measurement. Examples of formulas are shown below.

$$s = vt - 16t^2 \quad \text{(Physics)}$$
$$c^2 = a^2 + b^2 \quad \text{(Geometry)}$$
$$A = P(1 + r)^t \quad \text{(Business)}$$

The Addition and Multiplication Properties of Equations can be used to solve a literal equation for one of the variables. The goal is to rewrite the equation so that the variable being solved for is alone on one side of the equation, and all the other numbers and variables are on the other side.

Solve $C = \dfrac{5}{9}(F - 32)$ for F.

$$C = \frac{5}{9}(F - 32)$$

Use the Distributive Property to remove parentheses.

$$C = \frac{5}{9}F - \frac{160}{9}$$

Multiply each side of the equation by the LCM of the denominators.

$$9C = 5F - 160$$

Add 160 to each side of the equation.

$$9C + 160 = 5F$$

Divide each side of the equation by the coefficient 5.

$$\frac{9C + 160}{5} = F$$

EXAMPLE 1 **A.** Solve $A = P + Prt$ for P. **B.** Solve $\dfrac{S}{S - C} = R$ for C.

Solution **A.**
$$A = P + Prt$$
$$A = (1 + rt)P$$
$$\frac{A}{1 + rt} = \frac{(1 + rt)P}{1 + rt}$$
$$\frac{A}{1 + rt} = P$$

- Factor *P* from *P* + *Prt*.
- Divide each side of the equation by 1 + *rt*.

B.
$$\frac{S}{S - C} = R$$
$$(S - C)\frac{S}{S - C} = (S - C)R$$
$$S = SR - CR$$
$$CR + S = SR$$

- Multiply each side of the equation by *S* − *C*.
- Add *CR* to each side of the equation.

$$CR = SR - S$$

- Subtract S from each side of the equation.

$$C = \frac{SR - S}{R}$$

- Divide each side of the equation by R.

PROBLEM 1 **A.** Solve $\dfrac{1}{R_1} + \dfrac{1}{R_2} = \dfrac{1}{R}$ for R. **B.** Solve $t = \dfrac{r}{r + 1}$ for r.

 ■ **Solution** See page S26.

7.6 Concept Review

Determine whether the following statements are always true, sometimes true, or never true.

1. A formula is a literal equation that states a rule about measurement.

2. A literal equation contains two variables.

3. The formula $\dfrac{1}{R_1} + \dfrac{1}{R_2} = \dfrac{1}{R}$ is taken from physics. When we solve for R_2, we get $R_2 = R - R_1$.

4. Consider the formula $P = 2L + 2W$. When we double the perimeter, the length and width will both be doubled.

7.6 Exercises

1 Solve literal equations

1. Solve $P = \dfrac{R - C}{n}$ for C.

$$P = \frac{R - C}{n}$$

$$(\underline{\ \ ?\ \ })P = (\underline{\ \ ?\ \ })\left(\frac{R - C}{n}\right)$$

- Multiply each side of the equation by $\underline{\ \ ?\ \ }$.

$$nP = \underline{\ \ ?\ \ }$$

- Simplify.

$$nP - R = -C$$

- Subtract $\underline{\ \ ?\ \ }$ from each side.

$$\underline{\ \ ?\ \ } = C$$

- Multiply each side by -1.

2. Only two steps are needed to solve $P = \dfrac{R - C}{n}$ for n. The first step is to

____?____ each side of the equation by ___?___. The second step is to

____?____ each side of the equation by ___?___. The result is $n =$ ___?___.

Solve the formula for the variable given.

3. $P = 2L + 2W; W$ (Geometry)

4. $F = \dfrac{9}{5}C + 32; C$ (Temperature conversion)

5. $S = C - rC; C$ (Business)

6. $A = P + Prt; t$ (Business)

7. $PV = nRT; R$ (Chemistry)

8. $A = \dfrac{1}{2}bh; h$ (Geometry)

9. $F = \dfrac{Gm_1m_2}{r^2}; m_2$ (Physics)

10. $\dfrac{P_1V_1}{T_1} = \dfrac{P_2V_2}{T_2}; P_2$ (Chemistry)

11. $I = \dfrac{E}{R + r}; R$ (Physics)

12. $S = V_0t - 16t^2; V_0$ (Physics)

13. $A = \dfrac{1}{2}h(b_1 + b_2); b_2$ (Geometry)

14. $V = \dfrac{1}{3}\pi r^2h; h$ (Geometry)

15. $\dfrac{1}{R} = \dfrac{1}{R_1} + \dfrac{1}{R_2}; R_2$ (Physics)

16. $\dfrac{1}{f} = \dfrac{1}{a} + \dfrac{1}{b}; b$ (Physics)

17. $a_n = a_1 + (n - 1)d; d$ (Mathematics)

18. $P = \dfrac{R - C}{n}; R$ (Business)

19. $S = 2WH + 2WL + 2LH; H$ (Geometry)

20. $S = 2\pi r^2 + 2\pi rH; H$ (Geometry)

Solve for x.

21. $ax + by + c = 0$

22. $x = ax + b$

23. $ax + b = cx + d$

24. $y - y_1 = m(x - x_1)$

25. $\dfrac{a}{x} = \dfrac{b}{c}$

26. $\dfrac{1}{x} + \dfrac{1}{a} = b$

27. $\dfrac{1}{a} + \dfrac{1}{b} = \dfrac{1}{x}$

28. $a(a - x) = b(b - x)$

29. 🖏 When asked to solve $ax + b = cx$ for x, one student's result was $x = \dfrac{b}{c - a}$. Another student's result was $x = -\dfrac{b}{a - c}$. Are their answers equivalent?

30. 🖏 When asked to solve $I = \dfrac{E}{R + r}$ for r, one student's result was $r = \dfrac{E}{I} - R$. Another student's result was $r = \dfrac{E - IR}{I}$. Are their answers equivalent?

7.6 Applying Concepts

Solve for x.

31. $\dfrac{x - y}{y} = \dfrac{x + 5}{2y}$

32. $\dfrac{x - 2}{y} = \dfrac{x + 2}{5y}$

33. $\dfrac{x}{x + y} = \dfrac{2x}{4y}$

34. $\dfrac{2x}{x - 2y} = \dfrac{x}{2y}$

35. $\dfrac{x - y}{2x} = \dfrac{x - 3y}{5y}$

36. $\dfrac{x - y}{x} = \dfrac{2x}{9y}$

Solve for the given variable.

37. $\dfrac{w_1}{w_2} = \dfrac{f_2 - f}{f - f_1}; f$

38. $v = \dfrac{v_1 + v_2}{1 + \dfrac{v_1 v_2}{c^2}}; v_1$

Focus on Problem Solving

Implication Sentences that are constructed using "If..., then..." occur frequently in problem-solving situations. These sentences are called **implications.** The sentence "If it rains, then I will stay home" is an implication. The phrase *it rains* is called the **antecedent** of the implication. The phrase *I will stay home* is the **consequent** of the implication.

The sentence "If $x = 4$, then $x^2 = 16$" is a true sentence. The **contrapositive** of an implication is formed by switching the antecedent and the consequent and then negating each one. The contrapositive of "If $x = 4$, then $x^2 = 16$" is "If $x^2 \neq 16$, then $x \neq 4$." This sentence is also true. It is a principle of logic that an implication and its contrapositive are either both true or both false.

The **converse** of an implication is formed by switching the antecedent and the consequent. The converse of "If $x = 4$, then $x^2 = 16$" is "If $x^2 = 16$, then $x = 4$." Note that the converse is not a true statement, because if $x^2 = 16$, then x could equal -4. The converse of an implication may or may not be true.

Those statements for which the implication and the converse are both true are very important. They can be stated in the form "x if and only if y." For instance, a number is divisible by 5 if and only if the last digit of the number is 0 or 5. The implication "If a number is divisible by 5, then it ends in 0 or 5" and the converse "If a number ends in 0 or 5, then it is divisible by 5" are both true.

For each of the following, state the contrapositive and the converse of the implication. If the converse and the implication are both true statements, write a sentence using the phrase "if and only if."

1. If I live in Chicago, then I live in Illinois.

2. If today is June 1, then yesterday was May 31.

3. If today is not Thursday, then tomorrow is not Friday.

4. If a number is divisible by 8, then it is divisible by 4.

5. If a number is an even number, then it is divisible by 2.

6. If a number is a multiple of 6, then it is a multiple of 3.

7. If $4z = 20$, then $z = 5$.

8. If an angle measures 90°, then it is a right angle.

9. If p is a prime number greater than 2, then p is an odd number.

10. If the equation of a graph is $y = mx + b$, then the graph of the equation is a straight line.

11. If $a = 0$ or $b = 0$, then $ab = 0$.

12. If the coordinates of a point are (5, 0), then the point is on the x-axis.

13. If a quadrilateral is a square, then the quadrilateral has four sides of equal length.

14. If $x = y$, then $x^2 = y^2$.

Projects & Group Activities

Continued Fractions The following complex fraction is called a **continued fraction.**

$$1 + \cfrac{1}{1 + \cfrac{1}{1 + \cfrac{1}{1 + \cfrac{1}{1 + \cdots}}}}$$

The dots indicate that the pattern continues to repeat forever.

A **convergent** for a continued fraction is an approximation of the repeated pattern. For instance,

$$c_2 = 1 + \cfrac{1}{1 + \cfrac{1}{1 + 1}} \qquad c_3 = 1 + \cfrac{1}{1 + \cfrac{1}{1 + \cfrac{1}{1 + 1}}} \qquad c_4 = 1 + \cfrac{1}{1 + \cfrac{1}{1 + \cfrac{1}{1 + \cfrac{1}{1 + 1}}}}$$

1. Calculate c_5 for the continued fraction above.

The fraction above is related to the golden rectangle, which has been used in architectural designs as diverse as the Parthenon in Athens, built around 440 B.C., and the United Nations building. A golden rectangle is one for which

$$\frac{\text{length}}{\text{width}} = \frac{\text{length} + \text{width}}{\text{length}}$$

An example of a golden rectangle is shown below.

Here is another continued fraction, which was discovered by Leonhard Euler (1707–1793). Calculating the convergents of this continued fraction yields approximations that are closer and closer to π.

$$\pi = 3 + \cfrac{1^2}{6 + \cfrac{3^2}{6 + \cfrac{5^2}{6 + \cfrac{7^2}{6 + \cdots}}}}$$

2. Calculate $c_5 = 3 + \cfrac{1^2}{6 + \cfrac{3^2}{6 + \cfrac{5^2}{6 + \cfrac{7^2}{6 + \cfrac{9^2}{6 + 11^2}}}}}$

Chapter 7 Summary

Key Words	*Examples*
A fraction in which the numerator and denominator are polynomials is called a **rational expression**. [7.1.1, p. 483]	$\dfrac{x^2 + 2x + 1}{x - 3}$ is a rational expression.
A rational expression is in **simplest form** when the numerator and denominator have no common factors. [7.1.2, p. 485]	$\dfrac{x^2 + 3x + 2}{(x + 1)^2} = \dfrac{(x + 1)(x + 2)}{(x + 1)(x + 1)} = \dfrac{x + 2}{x + 1}$
The **reciprocal of a rational expression** is the rational expression with the numerator and denominator interchanged. [7.2.1, p. 493]	The reciprocal of $\dfrac{a^2 - b^2}{x - y}$ is $\dfrac{x - y}{a^2 - b^2}$.
The **least common multiple (LCM) of two or more polynomials** is the simplest polynomial that contains the factors of each polynomial. [7.2.2, p. 495]	$4x^2 - 12x = 4x(x - 3)$ $3x^3 - 21x^2 + 36x = 3x(x - 3)(x - 4)$ LCM $= 12x(x - 3)(x - 4)$
A **complex fraction** is a fraction whose numerator or denominator contains one or more fractions. [7.3.1, p. 504]	$\dfrac{\dfrac{1}{x} + \dfrac{1}{y}}{\dfrac{1}{x}}$ is a complex fraction.
A **ratio** is the quotient of two quantities that have the same unit. When a ratio is in simplest form, the units are not written. [7.5.1, p. 522]	$\dfrac{\$35}{\$100}$ written as a ratio in simplest form is $\dfrac{7}{20}$.
A **rate** is the quotient of two quantities that have different units. [7.5.1, p. 522]	$\dfrac{65 \text{ mi}}{2 \text{ gal}}$ is a rate.
A **proportion** is an equation that states the equality of two ratios or rates. [7.5.1, p. 522]	$\dfrac{75}{3} = \dfrac{x}{42}$ is a proportion.
Direct variation is a special function that can be expressed as the equation $y = kx^n$, where k is a constant called the **constant of variation** or the **constant of proportionality**. [7.5.2, p. 523]	$E = mc^2$ is Einstein's formula relating energy and mass. c^2 is the constant of proportionality.

Joint variation is a variation in which a variable varies directly as the product of two or more variables. A joint variation can be expressed as the equation $z = kxy$, where k is a constant. [7.5.2, p. 524]

$C = kAT$ is a formula for the cost of insulation, where A is the area to be insulated and T is the thickness of the insulation.

Inverse variation is a function that can be expressed as the equation $y = \dfrac{k}{x^n}$, where k is a constant. [7.5.2, p. 525]

$I = \dfrac{k}{d^2}$ gives the intensity of a light source at a distance d from the source.

Combined variation is a variation in which two or more types of variation occur at the same time. [7.5.2, p. 526]

$V = \dfrac{kT}{P}$ is a formula that states that the volume of a gas is directly proportional to the temperature and inversely proportional to the pressure.

A formula is a literal equation that states rules about measurements. [7.6.1, p. 533]

$P = 2L + 2W$ is the formula for the perimeter of a rectangle.

Essential Rules and Procedures

To multiply fractions
$\dfrac{a}{b} \cdot \dfrac{c}{d} = \dfrac{ac}{bd}$ [7.2.1, p. 492]

$$\dfrac{x^2 + 2x - 8}{x^2 - 2x} \cdot \dfrac{3x - 12}{x^2 - 16}$$
$$= \dfrac{(x + 4)(x - 2)}{x(x - 2)} \cdot \dfrac{3(x - 4)}{(x + 4)(x - 4)}$$
$$= \dfrac{(x + 4)(x - 2) \cdot 3(x - 4)}{x(x - 2)(x + 4)(x - 4)}$$
$$= \dfrac{3}{x}$$

To divide fractions
$\dfrac{a}{b} \div \dfrac{c}{d} = \dfrac{a}{b} \cdot \dfrac{d}{c}$ [7.2.1, p. 493]

$$\dfrac{3x}{x - 5} \div \dfrac{3}{x + 4} = \dfrac{3x}{x - 5} \cdot \dfrac{x + 4}{3}$$
$$= \dfrac{3x(x + 4)}{3(x - 5)} = \dfrac{x(x + 4)}{x - 5}$$
$$= \dfrac{x^2 + 4x}{x - 5}$$

To add fractions
$\dfrac{a}{c} + \dfrac{b}{c} = \dfrac{a + b}{c}$ [7.2.2, p. 494]

$$\dfrac{2x - 7}{x^2 + 4} + \dfrac{x + 2}{x^2 + 4} = \dfrac{(2x - 7) + (x + 2)}{x^2 + 4}$$
$$= \dfrac{3x - 5}{x^2 + 4}$$

To subtract fractions
$\dfrac{a}{c} - \dfrac{b}{c} = \dfrac{a - b}{c}$ [7.2.2, p. 494]

$$\dfrac{5x - 6}{x + 2} - \dfrac{2x + 4}{x + 2} = \dfrac{(5x - 6) - (2x + 4)}{x + 2}$$
$$= \dfrac{3x - 10}{x + 2}$$

Equation for Work Problems

Rate of work × Time worked = Part of task completed
[7.4.2, p. 511]

A roofer requires 24 h to shingle a roof. An apprentice can shingle the roof in 36 h. How long would it take to shingle the roof if both roofers worked together?

$$\frac{t}{24} + \frac{t}{36} = 1$$

Equation for Uniform Motion Problems

Distance = Rate × Time or $\dfrac{\text{Distance}}{\text{Rate}}$ = Time [7.4.3, p. 513]

A motorcycle travels 195 mi in the same amount of time it takes a car to travel 159 mi. The rate of the motorcycle is 12 mph greater than the rate of the car. Find the rate of the car.

$$\frac{195}{r + 12} = \frac{159}{r}$$

Chapter 7 Review Exercises

1. Given $P(x) = \dfrac{x}{x - 3}$, find $P(4)$.

2. Given $P(x) = \dfrac{x^2 - 2}{3x^2 - 2x + 5}$, find $P(-2)$.

3. Find the domain of $g(x) = \dfrac{2x}{x - 3}$.

4. Find the domain of $f(x) = \dfrac{2x - 7}{3x^2 + 3x - 18}$.

5. Find the domain of $F(x) = \dfrac{x^2 - x}{3x^2 + 4}$.

6. Simplify: $\dfrac{6a^{5n} + 4a^{4n} - 2a^{3n}}{2a^{3n}}$

7. Simplify: $\dfrac{16 - x^2}{x^3 - 2x^2 - 8x}$

8. Simplify: $\dfrac{x^3 - 27}{x^2 - 9}$

9. Multiply: $\dfrac{a^6b^4 + a^4b^6}{a^5b^4 - a^4b^4} \cdot \dfrac{a^2 - b^2}{a^4 - b^4}$

10. Multiply: $\dfrac{x^3 - 8}{x^3 + 2x^2 + 4x} \cdot \dfrac{x^3 + 2x^2}{x^2 - 4}$

11. Multiply: $\dfrac{16 - x^2}{6x - 6} \cdot \dfrac{x^2 + 5x + 6}{x^2 - 8x + 16}$

12. Divide: $\dfrac{x^{2n} - 5x^n + 4}{x^{2n} - 2x^n - 8} \div \dfrac{x^{2n} - 4x^n + 3}{x^{2n} + 8x^n + 12}$

13. Divide: $\dfrac{27x^3 - 8}{9x^3 + 6x^2 + 4x} \div \dfrac{9x^2 - 12x + 4}{9x^2 - 4}$

14. Divide: $\dfrac{3 - x}{x^2 + 3x + 9} \div \dfrac{x^2 - 9}{x^3 - 27}$

15. Add: $\dfrac{5}{3a^2b^3} + \dfrac{7}{8ab^4}$

16. Subtract: $\dfrac{3x^2 + 2}{x^2 - 4} - \dfrac{9x - x^2}{x^2 - 4}$

17. Simplify: $\dfrac{8}{9x^2 - 4} + \dfrac{5}{3x - 2} - \dfrac{4}{3x + 2}$

18. Simplify: $\dfrac{6x}{3x^2 - 7x + 2} - \dfrac{2}{3x - 1} + \dfrac{3x}{x - 2}$

19. Simplify: $\dfrac{x}{x - 3} - 4 - \dfrac{2x - 5}{x + 2}$

20. Simplify: $\dfrac{x - 6 + \dfrac{6}{x - 1}}{x + 3 - \dfrac{12}{x - 1}}$

21. Simplify: $\dfrac{x + \dfrac{3}{x - 4}}{3 + \dfrac{x}{x - 4}}$

22. Solve: $\dfrac{5x}{2x - 3} + 4 = \dfrac{3}{2x - 3}$

23. Solve: $\dfrac{x}{x - 3} = \dfrac{2x + 5}{x + 1}$

24. Solve: $\dfrac{6}{x - 3} - \dfrac{1}{x + 3} = \dfrac{51}{x^2 - 9}$

25. Solve: $\dfrac{30}{x^2 + 5x + 4} + \dfrac{10}{x + 4} = \dfrac{4}{x + 1}$

26. Solve $I = \dfrac{1}{R}V$ for R.

27. Solve $Q = \dfrac{N - S}{N}$ for N.

28. Solve $S = \dfrac{a}{1 - r}$ for r.

29. Fuel Efficiency A car uses 4 tanks of fuel to travel 1800 mi. At this rate, how many tanks of fuel would be required for a trip of 3000 mi?

30. Cartography On a certain map, 2.5 in. represents 10 mi. How many miles would be represented by 12 in.?

31. Work Problem An electrician requires 65 min to install a ceiling fan. The electrician and an apprentice, working together, take 40 min to install the fan. How long would it take the apprentice, working alone, to install the ceiling fan?

32. Work Problem The inlet pipe can fill a tub in 24 min. The drain pipe can empty the tub in 15 min. How long would it take to empty a full tub with both pipes open?

33. Work Problem Three students can paint a dormitory room in 8 h, 16 h, and 16 h, respectively. How long would it take to do the job if all three students worked together?

34. Uniform Motion A canoeist can travel 10 mph in calm water. The amount of time it takes to travel 60 mi with the current is the same amount of time it takes to travel 40 mi against the current. Find the rate of the current.

35. Uniform Motion A bus and a cyclist leave a school at 8 A.M. and head for a stadium 90 mi away. The rate of the bus is three times the rate of the cyclist. The cyclist arrives 4 h after the bus. Find the rate of the bus.

36. Uniform Motion A tractor travels 10 mi in the same amount of time it takes a car to travel 15 mi. The rate of the tractor is 15 mph less than the rate of the car. Find the rate of the tractor.

37. Physics The pressure (p) of wind on a flat surface varies jointly as the area (A) of the surface and the square of the wind's velocity (v). If the pressure on 22 ft^2 is 10 lb when the wind's velocity is 10 mph, find the pressure on the same surface when the wind's velocity is 20 mph.

38. Light The illumination (I) produced by a light varies inversely as the square of the distance (d) from the light. If the illumination produced 10 ft from a light is 12 lumens, find the illumination 2 ft from the light.

39. Electricity The electrical resistance (r) of a cable varies directly as its length (l) and inversely as the square of its diameter (d). If a cable 16,000 ft long and $\frac{1}{4}$ in. in diameter has a resistance of 3.2 ohms, what is the resistance of a cable that is 8000 ft long and $\frac{1}{2}$ in. in diameter?

Chapter 7 Test

1. Simplify: $\dfrac{v^3 - 4v}{2v^2 - 5v + 2}$

2. Simplify: $\dfrac{2a^2 - 8a + 8}{4 + 4a - 3a^2}$

3. Multiply: $\dfrac{3x^2 - 12}{5x - 15} \cdot \dfrac{2x^2 - 18}{x^2 + 5x + 6}$

4. Given $P(x) = \dfrac{3 - x^2}{x^3 - 2x^2 + 4}$, find $P(-1)$.

5. Divide: $\dfrac{2x^2 - x - 3}{2x^2 - 5x + 3} \div \dfrac{3x^2 - x - 4}{x^2 - 1}$

6. Multiply: $\dfrac{x^{2n} - x^n - 2}{x^{2n} + x^n} \cdot \dfrac{x^{2n} - x^n}{x^{2n} - 4}$

7. Simplify: $\dfrac{2}{x^2} + \dfrac{3}{y^2} - \dfrac{5}{2xy}$

8. Simplify: $\dfrac{3x}{x - 2} - 3 + \dfrac{4}{x + 2}$

9. Find the domain of $f(x) = \dfrac{3x^2 - x + 1}{x^2 - 9}$.

10. Subtract: $\dfrac{x + 2}{x^2 + 3x - 4} - \dfrac{2x}{x^2 - 1}$

11. Simplify: $\dfrac{1 - \dfrac{1}{x} - \dfrac{12}{x^2}}{1 + \dfrac{6}{x} + \dfrac{9}{x^2}}$

12. Simplify: $\dfrac{1 - \dfrac{1}{x + 2}}{1 - \dfrac{3}{x + 4}}$

13. Solve: $\dfrac{3}{x + 1} = \dfrac{2}{x}$

14. Solve: $\dfrac{4x}{2x - 1} = 2 - \dfrac{1}{2x - 1}$

15. Solve $ax = bx + c$ for x.

16. Work Problem The inlet pipe can fill a water tank in 48 min. The outlet pipe can empty the tank in 30 min. How long would it take to empty a full tank with both pipes open?

17. Wall Papering An interior designer uses two rolls of wallpaper for every 45 ft^2 of wall space in an office. At this rate, how many rolls of wallpaper are needed for an office that has 315 ft^2 of wall space?

18. Work Problem One landscaper can till the soil for a lawn in 30 min, whereas it takes a second landscaper 15 min to do the same job. How long would it take to till the soil for the lawn if both landscapers worked together?

19. Unform Motion A cyclist travels 20 mi in the same amount of time it takes a hiker to walk 6 mi. The rate of the cyclist is 7 mph greater than the rate of the hiker. Find the rate of the cyclist.

20. Physics The stopping distance (s) of a car varies directly as the square of the speed (v) of the car. For a car traveling at 50 mph, the stopping distance is 170 ft. Find the stopping distance of a car that is traveling at 30 mph.

Cumulative Review Exercises

1. Simplify: $8 - 4[-3 - (-2)]^2 \div 5$

2. Solve: $\dfrac{2x - 3}{6} - \dfrac{x}{9} = \dfrac{x - 4}{3}$

3. Solve: $5 - |x - 4| = 2$

4. Find the domain of $f(x) = \dfrac{x}{x - 3}$.

5. Given $P(x) = \dfrac{x-1}{2x-3}$, find $P(-2)$.

6. Write 0.000000035 in scientific notation.

7. Solve: $\dfrac{x}{x+1} = 1$

8. Solve: $(9x - 1)(x - 4) = 0$

9. Simplify: $\dfrac{(2a^{-2}b^3)}{(4a)^{-1}}$

10. Solve: $x - 3(1 - 2x) \geq 1 - 4(2 - 2x)$

11. Multiply: $(2a^2 - 3a + 1)(-2a^2)$

12. Factor: $2x^{2n} + 3x^n - 2$

13. Factor: $x^3y^3 - 27$

14. Simplify: $\dfrac{x^4 + x^3y - 6x^2y^2}{x^3 - 2x^2y}$

15. Find the equation of the line that contains the point $(-2, -1)$ and is parallel to the line $3x - 2y = 6$.

16. Simplify: $(x - x^{-1})^{-1}$

17. Graph $-3x + 5y = -15$ by using the x- and y-intercepts.

18. Graph the solution set: $\begin{aligned} x + y &\leq 3 \\ -2x + y &> 4 \end{aligned}$

19. Divide: $\dfrac{4x^3 + 2x^2 - 10x + 1}{x - 2}$

20. Divide: $\dfrac{16x^2 - 9y^2}{16x^2y - 12xy^2} \div \dfrac{4x^2 - xy - 3y^2}{12x^2y^2}$

21. Find the domain of $f(x) = \dfrac{x^2 + 6x + 7}{3x^2 + 5}$.

22. Subtract: $\dfrac{5x}{3x^2 - x - 2} - \dfrac{2x}{x^2 - 1}$

23. Evaluate the determinant: $\begin{vmatrix} 6 & 5 \\ 2 & -3 \end{vmatrix}$

24. Simplify: $\dfrac{x - 4 + \dfrac{5}{x+2}}{x + 2 - \dfrac{1}{x+2}}$

25. Solve: $\begin{aligned} x + y + z &= 3 \\ -2x + y + 3z &= 2 \\ 2x - 4y + z &= -1 \end{aligned}$

26. Given $f(x) = x^2 - 3x + 3$ and $f(c) = 1$, find c.

27. Solve the proportion: $\dfrac{2}{x-3} = \dfrac{5}{2x-3}$

28. Solve: $\dfrac{3}{x^2-36} = \dfrac{2}{x-6} - \dfrac{5}{x+6}$

29. Multiply: $(a+5)(a^3 - 3a + 4)$

30. Solve $I = \dfrac{E}{R+r}$ for r.

31. Are the graphs of $4x + 3y = 12$ and $3x + 4y = 16$ perpendicular?

32. Graph: $f(x) = 1 - x + x^3$

33. Integer Problem The sum of two integers is 15. Five times the smaller integer is five more than twice the larger integer. Find the integers.

34. Food Mixture How many pounds of almonds that cost $5.40 per pound must be mixed with 50 lb of peanuts that cost $2.60 per pound to make a mixture that costs $4.00 per pound?

35. Elections A pre-election survey showed that three out of five voters would vote in an election. At this rate, how many people would be expected to vote in a city of 125,000?

36. Work Problem A new computer can work six times as fast as an older computer. Working together, the computers can complete a job in 12 min. How long would it take the new computer, working alone, to do the job?

37. Uniform Motion A plane can fly at a rate of 300 mph in calm air. Traveling with the wind, the plane flew 900 mi in the same amount of time it took to fly 600 mi against the wind. Find the rate of the wind.

38. Music The frequency of vibration (f) in an open pipe organ varies inversely as the length (L) of the pipe. If the air in a pipe 2 m long vibrates 60 times per minute, find the frequency in a pipe that is 1.5 m long.

8 Rational Exponents and Radicals

DVD

SSM

Online Study Center

Need help? For online student resources, such as section quizzes, visit the Online Study Center at college.hmco.com/pic/aufmannABI2e.

Satellite phones, satellite radios, and satellite TV are just three examples of the many ways satellites are now being used in everyday life. Thousands of man-made satellites have been launched into space and are now orbiting Earth. The speed at which a satellite orbits Earth is a function of its height above Earth's surface. The height at which it orbits Earth is dependent on the satellite's purpose. The relationship between a satellite's speed and its height above Earth can be modeled by a radical equation, such as the one used in **Exercise 85 on page 585.**

Prep Test

1. Complete: $48 = ? \cdot 3$

For Exercises 2 to 7, simplify.

2. 2^5

3. $6\left(\dfrac{3}{2}\right)$

4. $\dfrac{1}{2} - \dfrac{2}{3} + \dfrac{1}{4}$

5. $(3 - 7x) - (4 - 2x)$

6. $\dfrac{3x^5y^6}{12x^4y}$

7. $(3x - 2)^2$

For Exercises 8 and 9, multiply.

8. $(2 + 4x)(5 - 3x)$

9. $(6x - 1)(6x + 1)$

10. Solve: $x^2 - 14x - 5 = 10$

8.1 Rational Exponents and Radical Expressions

OBJECTIVE 1 Simplify expressions with rational exponents

Point of Interest

Nicolas Chuquet (ca. 1475), a French physician, wrote an algebra text in which he used a notation for expressions with fractional exponents. He wrote R²6 to mean $6^{1/2}$ and R³15 to mean $15^{1/3}$. This was an improvement over earlier notations that used words for these expressions.

In this section, the definition of an exponent is extended beyond integers so that any rational number can be used as an exponent. The definition is expressed in such a way that the Rules of Exponents hold true for rational exponents.

Consider the expression $\left(a^{\frac{1}{n}}\right)^n$ for $a > 0$ and n a positive integer. Now simplify, assuming that the Rule for Simplifying a Power of an Exponential Expression is true.

$$\left(a^{\frac{1}{n}}\right)^n = a^{\frac{1}{n} \cdot n} = a^1 = a$$

Because $\left(a^{\frac{1}{n}}\right)^n = a$, the number $a^{\frac{1}{n}}$ **is the number whose nth power is a.**

$$25^{\frac{1}{2}} = 5 \text{ because } (5)^2 = 25.$$
$$8^{\frac{1}{3}} = 2 \text{ because } (2)^3 = 8.$$

In the expression $a^{\frac{1}{n}}$, if a is a negative number and n is a positive even integer, then $a^{\frac{1}{n}}$ is not a real number.

$(-4)^{\frac{1}{2}}$ is not a real number, because there is no real number whose second power is -4.

When n is a positive odd integer, a can be a positive or a negative number.

$$(-27)^{\frac{1}{3}} = -3 \text{ because } (-3)^3 = -27.$$

Using the definition of $a^{\frac{1}{n}}$ and the Rules of Exponents, it is possible to define any exponential expression that contains a rational exponent.

Definition of $a^{\frac{m}{n}}$

If m and n are positive integers and $a^{\frac{1}{n}}$ is a real number, then

$$a^{\frac{m}{n}} = \left(a^{\frac{1}{n}}\right)^m$$

Take Note

Although we can simplify an expression by rewriting $a^{\frac{m}{n}}$ in the form $(a^m)^{\frac{1}{n}}$, it is usually easier to simplify $\left(a^{\frac{1}{n}}\right)^m$. For instance, simplifying $\left(27^{\frac{1}{3}}\right)^2$ is easier than simplifying $(27^2)^{\frac{1}{3}}$.

The expression $a^{\frac{m}{n}}$ can also be written $a^{\frac{m}{n}} = a^{m \cdot \frac{1}{n}} = (a^m)^{\frac{1}{n}}$. However, rewriting $a^{\frac{m}{n}}$ as $(a^m)^{\frac{1}{n}}$ is not as useful as rewriting it as $\left(a^{\frac{1}{n}}\right)^m$. See the Take Note at the left.

As shown above, expressions that contain rational exponents do not always represent real numbers when the base of the exponential expression is a negative number. For this reason, all variables in this chapter represent positive numbers unless otherwise stated.

EXAMPLE 1 Simplify. **A.** $27^{\frac{2}{3}}$ **B.** $32^{-\frac{2}{5}}$ **C.** $(-49)^{\frac{3}{2}}$

Solution **A.** $27^{\frac{2}{3}} = (3^3)^{\frac{2}{3}}$ • **Rewrite 27 as 3^3.**

$\qquad\qquad = 3^{3\left(\frac{2}{3}\right)}$ • **Use the Rule for Simplifying a Power of an Exponential Expression.**

$\qquad\qquad = 3^2$

$\qquad\qquad = 9$ • **Simplify.**

B. $32^{-\frac{2}{5}} = (2^5)^{-\frac{2}{5}}$ • **Rewrite 32 as 2^5.**

$\qquad\qquad = 2^{-2}$ • **Use the Rule for Simplifying a Power of an Exponential Expression.**

$\qquad\qquad = \dfrac{1}{2^2}$ • **Use the Rule of Negative Exponents.**

$\qquad\qquad = \dfrac{1}{4}$ • **Simplify.**

C. $(-49)^{\frac{3}{2}}$ • **The base of the exponential expression is a negative number, while the denominator of the exponent is a positive even number.**

$(-49)^{\frac{3}{2}}$ is not a real number.

Take Note

Note that $32^{-\frac{2}{5}} = \dfrac{1}{4}$, a positive number. A negative exponent does not affect the sign of a number.

PROBLEM 1 Simplify. **A.** $64^{\frac{2}{3}}$ **B.** $16^{-\frac{3}{4}}$ **C.** $(-81)^{\frac{3}{4}}$

Solution See page S26.

EXAMPLE 2 Simplify. **A.** $b^{\frac{1}{2}} \cdot b^{\frac{2}{3}} \cdot b^{-\frac{1}{4}}$ **B.** $(x^6 y^4)^{\frac{3}{2}}$ **C.** $\left(\dfrac{8a^3 b^{-4}}{64a^{-9} b^2}\right)^{\frac{2}{3}}$

Solution **A.** $b^{\frac{1}{2}} \cdot b^{\frac{2}{3}} \cdot b^{-\frac{1}{4}} = b^{\frac{1}{2}+\frac{2}{3}-\frac{1}{4}}$ • Use the Rule for Multiplying
$$= b^{\frac{6}{12}+\frac{8}{12}-\frac{3}{12}}$$ Exponential Expressions.
$$= b^{\frac{11}{12}}$$

B. $(x^6 y^4)^{\frac{3}{2}} = x^{6\left(\frac{3}{2}\right)} y^{4\left(\frac{3}{2}\right)}$ • Use the Rule for Simplifying Powers of
$$= x^9 y^6$$ Products.

C. $\left(\dfrac{8a^3 b^{-4}}{64a^{-9} b^2}\right)^{\frac{2}{3}} = \left(\dfrac{a^{12}}{8b^6}\right)^{\frac{2}{3}}$ • Use the Rule for Dividing Exponential
 Expressions.
$$= \left(\dfrac{a^{12}}{2^3 b^6}\right)^{\frac{2}{3}}$$ • Rewrite 8 as 2^3.
$$= \dfrac{a^8}{2^2 b^4}$$ • Use the Rule for Simplifying Powers of
 Quotients.
$$= \dfrac{a^8}{4b^4}$$

PROBLEM 2 Simplify. **A.** $\dfrac{x^{\frac{1}{2}} y^{-\frac{5}{4}}}{x^{-\frac{4}{3}} y^{\frac{1}{3}}}$ **B.** $\left(x^{\frac{3}{4}} y^{\frac{1}{2}} z^{-\frac{2}{3}}\right)^{-\frac{4}{3}}$ **C.** $\left(\dfrac{16a^{-2} b^{\frac{4}{3}}}{9a^4 b^{-\frac{2}{3}}}\right)^{-\frac{1}{2}}$

Solution See page S26.

OBJECTIVE 2 **Write exponential expressions as radical expressions and radical expressions as exponential expressions**

Point of Interest

The radical sign was introduced in 1525 in a book by Christoff Rudolff called Coss. He modified the symbol to indicate square roots, cube roots, and fourth roots. The idea of using an index, as we do in our modern notation, did not occur until some years later.

As stated in the previous objective, $a^{\frac{1}{n}}$ is the number whose nth power is a. This can be stated as $a^{\frac{1}{n}}$ is the **nth root** of a. The expression $\sqrt[n]{a}$ is another symbol for the nth root of a.

Alternative Notation for the nth Root of a

If a is a real number and n is a positive integer, then $\sqrt[n]{a} = a^{\frac{1}{n}}$.

In the expression $\sqrt[n]{a}$, the symbol $\sqrt{}$ is called a **radical,** n is the **index** of the radical, and a is the **radicand.** When $n = 2$, the radical expression represents a square root and the index 2 is usually not written.

An exponential expression with a rational exponent can be written as a radical expression.

Definition of the nth Root of a^m

If $a^{\frac{1}{n}}$ is a real number, then $a^{\frac{m}{n}} = a^{m \cdot \frac{1}{n}} = (a^m)^{\frac{1}{n}} = \sqrt[n]{a^m}$.

The expression $a^{\frac{m}{n}}$ can also be written $a^{\frac{m}{n}} = a^{\frac{1}{n} \cdot m} = \left(\sqrt[n]{a}\right)^m$.

The exponential expression at the right has been written as a radical expression.

$$y^{\frac{2}{3}} = (y^2)^{\frac{1}{3}} = \sqrt[3]{y^2}$$

The radical expressions at the right have been written as exponential expressions.

$$\sqrt[5]{x^6} = (x^6)^{\frac{1}{5}} = x^{\frac{6}{5}}$$

$$\sqrt{17} = (17)^{\frac{1}{2}} = 17^{\frac{1}{2}}$$

Write $(5x)^{\frac{2}{5}}$ as a radical expression.

The denominator of the rational exponent is the index of the radical. The numerator is the power of the radicand.

$$(5x)^{\frac{2}{5}} = \sqrt[5]{(5x)^2}$$

Simplify.

$$= \sqrt[5]{25x^2}$$

Write $\sqrt[3]{x^4}$ as an exponential expression with a rational exponent.

The index of the radical is the denominator of the rational exponent. The power of the radicand is the numerator of the rational exponent.

$$\sqrt[3]{x^4} = (x^4)^{\frac{1}{3}} = x^{\frac{4}{3}}$$

EXAMPLE 3 Rewrite the exponential expression as a radical expression.

A. $(3x)^{\frac{2}{3}}$ **B.** $-2x^{\frac{2}{3}}$

Solution **A.** $(3x)^{\frac{2}{3}} = \sqrt[3]{(3x)^2}$

$= \sqrt[3]{9x^2}$

• **The denominator of the rational exponent is the index of the radical. The numerator is the power of the radicand.**

B. $-2x^{\frac{2}{3}} = -2(x^2)^{\frac{1}{3}}$

$= -2\sqrt[3]{x^2}$

• **The -2 is not raised to the power.**

PROBLEM 3 Rewrite the exponential expression as a radical expression.

A. $(2x^3)^{\frac{3}{4}}$ **B.** $-5a^{\frac{5}{6}}$

Solution See page S27.

EXAMPLE 4 Rewrite the radical expression as an exponential expression.

A. $\sqrt[7]{x^5}$ **B.** $\sqrt[3]{a^3 + b^3}$

Solution **A.** $\sqrt[7]{x^5} = (x^5)^{\frac{1}{7}} = x^{\frac{5}{7}}$

• **The index of the radical is the denominator of the rational exponent. The power of the radicand is the numerator of the rational exponent.**

B. $\sqrt[3]{a^3 + b^3} = (a^3 + b^3)^{\frac{1}{3}}$

• **Note that $(a^3 + b^3)^{\frac{1}{3}} \neq a + b$.**

PROBLEM 4 Rewrite the radical expression as an exponential expression.

A. $\sqrt[3]{3ab}$ **B.** $\sqrt[4]{x^4 + y^4}$

Solution See page S27.

OBJECTIVE **3** **Simplify radical expressions that are roots of perfect powers**

Every positive number has two square roots, one a positive and one a negative number. For example, because $(5)^2 = 25$ and $(-5)^2 = 25$, there are two square roots of 25: 5 and -5.

The symbol $\sqrt{}$ is used to indicate the positive or **principal square root.** To indicate the negative square root of a number, a negative sign is placed in front of the radical.

$$\sqrt{25} = 5$$

$$-\sqrt{25} = -5$$

The square root of zero is zero.

$$\sqrt{0} = 0$$

The square root of a negative number is not a real number, because the square of a real number must be positive.

$\sqrt{-25}$ is not a real number.

The square root of a squared negative number is a positive number.

$$\sqrt{(-5)^2} = \sqrt{25} = 5$$

For any real number a, $\sqrt{a^2} = |a|$ and $-\sqrt{a^2} = -|a|$. If a is a positive real number, then $\sqrt{a^2} = a$ and $\left(\sqrt{a}\right)^2 = a$.

The cube root of a positive number is positive.

$\sqrt[3]{8} = 2$, because $2^3 = 8$.

The cube root of a negative number is negative.

$\sqrt[3]{-8} = -2$, because $(-2)^3 = -8$.

For any real number a, $\sqrt[3]{a^3} = a$.

The following holds true for finding the nth root of a real number.

If n is an even integer, then $\sqrt[n]{a^n} = |a|$ and $-\sqrt[n]{a^n} = -|a|$. If n is an odd integer, then $\sqrt[n]{a^n} = a$.

For example,

$$\sqrt[6]{y^6} = |y| \qquad -\sqrt[12]{x^{12}} = -|x| \qquad \sqrt[5]{b^5} = b$$

For the remainder of this chapter, we will assume that variable expressions inside a radical represent positive numbers. Therefore, it is not necessary to use the absolute value signs.

Take Note

Note that when the index is an even natural number, the nth root requires absolute value symbols.

$\sqrt[6]{y^6} = |y|$ but $\sqrt[5]{y^5} = y$

Because we stated that variables within radicals represent *positive* numbers, we will omit the absolute value symbols when writing an answer.

Simplify: $\sqrt[4]{x^4 y^8}$

The radicand is a perfect fourth power because the exponents on the variables are divisible by 4. Write the radical expression as an exponential expression.

$$\sqrt[4]{x^4 y^8} = (x^4 y^8)^{\frac{1}{4}}$$

Use the Rule for Simplifying Powers of Products.

$$= xy^2$$

Simplify: $\sqrt[3]{125c^9d^6}$

The radicand is a perfect cube because 125 is a perfect cube ($125 = 5^3$) and all the exponents on the variables are divisible by 3.

$$\sqrt[3]{125c^9d^6} = (5^3c^9d^6)^{\frac{1}{3}}$$

Use the Rule for Simplifying Powers of Products.

$$= 5c^3d^2$$

Note that a variable expression is a perfect power if the exponents on the factors are evenly divisible by the index of the radical.

The chart below shows roots of perfect powers. Knowledge of these roots is very helpful in simplifying radical expressions.

Square Roots		Cube Roots	Fourth Roots	Fifth Roots
$\sqrt{1} = 1$	$\sqrt{36} = 6$	$\sqrt[3]{1} = 1$	$\sqrt[4]{1} = 1$	$\sqrt[5]{1} = 1$
$\sqrt{4} = 2$	$\sqrt{49} = 7$	$\sqrt[3]{8} = 2$	$\sqrt[4]{16} = 2$	$\sqrt[5]{32} = 2$
$\sqrt{9} = 3$	$\sqrt{64} = 8$	$\sqrt[3]{27} = 3$	$\sqrt[4]{81} = 3$	$\sqrt[5]{243} = 3$
$\sqrt{16} = 4$	$\sqrt{81} = 9$	$\sqrt[3]{64} = 4$	$\sqrt[4]{256} = 4$	
$\sqrt{25} = 5$	$\sqrt{100} = 10$	$\sqrt[3]{125} = 5$	$\sqrt[4]{625} = 5$	

Take Note

From the chart, $\sqrt[5]{243} = 3$, which means that $3^5 = 243$. From this we know that $(-3)^5 = -243$, which means $\sqrt[5]{-243} = -3$.

Simplify: $\sqrt[5]{-243x^5y^{15}}$

From the chart, 243 is a perfect fifth power, and each exponent is divisible by 5. Therefore, the radicand is a perfect fifth power.

$$\sqrt[5]{-243x^5y^{15}} = -3xy^3$$

EXAMPLE 5 Simplify.

 A. $\sqrt{49x^2y^{12}}$ **B.** $\sqrt[3]{-125a^6b^9}$ **C.** $-\sqrt[4]{16a^4b^8}$

Solution **A.** $\sqrt{49x^2y^{12}} = 7xy^6$

 • The radicand is a perfect square. Divide each exponent by 2.

 B. $\sqrt[3]{-125a^6b^9} = -5a^2b^3$

 • The radicand is a perfect cube. Divide each exponent by 3.

 C. $-\sqrt[4]{16a^4b^8} = -2ab^2$

 • The radicand is a perfect fourth power. Divide each exponent by 4.

PROBLEM 5 Simplify.

 A. $\sqrt{121x^{10}y^4}$ **B.** $\sqrt[3]{-8x^{12}y^3}$ **C.** $-\sqrt[4]{81x^{12}y^8}$

Solution See page S27.

8.1 Concept Review

Determine whether the following statements are always true, sometimes true, or never true.

1. For real numbers, $\sqrt{x^2} = x$.

2. $\sqrt{(-2)^2} = -2$

3. $\sqrt[3]{(-3)^3} = -3$

4. For positive integers a and n, $\sqrt[n]{a} = a^{\frac{1}{n}}$.

5. For positive integers a, m, and n, $(a^m)^{\frac{1}{n}} = a^{\frac{m}{n}}$.

6. The nth root of a negative number is negative.

8.1 Exercises

1 Simplify expressions with rational exponents

1. $125^{\frac{1}{3}}$ is the number whose ____?____ power is ____?____, so $125^{\frac{1}{3}} = $ ____?____.

2. $16^{\frac{1}{4}}$ is the number whose ____?____ power is ____?____, so $16^{\frac{1}{4}} = $ ____?____.

3. Simplify: $64^{-\frac{3}{2}}$

$64^{-\frac{3}{2}} = ($ ____?____ $^2)^{-\frac{3}{2}}$ • Write 64 as ____?____2.

$\quad\quad = 8^{-\underline{}}$ • Use the Rule for Simplifying the Power of an Exponential Expression: multiply the exponents.

$\quad\quad = \dfrac{1}{\underline{}}$ • Use the Definition of a Negative Exponent: $a^{-n} = \dfrac{1}{a^n}$.

$\quad\quad = $ ____?____ • Evaluate.

4. Simplify: $\left(a^{\frac{1}{3}} \cdot a^{\frac{1}{2}}\right)^2$

$\left(a^{\frac{1}{3}} \cdot a^{\frac{1}{2}}\right)^2 = (a^{\underline{}})^2$ • Use the Rule for Multiplying Exponential Expressions: add the exponents.

$\quad\quad = a^{\underline{}}$ • Use the Rule for Simplifying the Power of an Exponential Expression: multiply the exponents.

5. Which of the following is undefined?

a. $-25^{\frac{1}{2}}$ **b.** $25^{-\frac{1}{2}}$ **c.** $-25^{-\frac{1}{2}}$ **d.** $(-25)^{\frac{1}{2}}$

6. Which of the following is (are) equivalent to $16^{-\frac{1}{2}}$?

a. -8 **b.** $\dfrac{1}{4}$ **c.** -4 **d.** 4^{-1}

Simplify.

7. $8^{\frac{1}{3}}$

8. $16^{\frac{1}{2}}$

9. $9^{\frac{3}{2}}$

10. $25^{\frac{3}{2}}$

11. $27^{-\frac{2}{3}}$

12. $64^{-\frac{1}{3}}$

13. $32^{\frac{2}{5}}$

14. $16^{\frac{3}{4}}$

15. $(-25)^{\frac{5}{2}}$

16. $(-36)^{\frac{1}{4}}$

17. $\left(\dfrac{25}{49}\right)^{-\frac{3}{2}}$

18. $\left(\dfrac{8}{27}\right)^{-\frac{2}{3}}$

19. $x^{\frac{1}{2}}x^{\frac{1}{2}}$

20. $a^{\frac{1}{3}}a^{\frac{5}{3}}$

21. $y^{-\frac{1}{4}}y^{\frac{3}{4}}$

22. $x^{\frac{2}{5}} \cdot x^{-\frac{4}{5}}$

23. $x^{-\frac{2}{3}} \cdot x^{\frac{3}{4}}$

24. $x \cdot x^{-\frac{1}{2}}$

25. $a^{\frac{1}{3}} \cdot a^{\frac{3}{4}} \cdot a^{-\frac{1}{2}}$

26. $y^{-\frac{1}{6}} \cdot y^{\frac{2}{3}} \cdot y^{\frac{1}{2}}$

27. $\dfrac{a^{\frac{1}{2}}}{a^{\frac{3}{2}}}$

28. $\dfrac{b^{\frac{1}{3}}}{b^{\frac{4}{3}}}$

29. $\dfrac{y^{-\frac{3}{4}}}{y^{\frac{1}{4}}}$

30. $\dfrac{x^{-\frac{3}{5}}}{x^{\frac{1}{5}}}$

31. $\dfrac{y^{\frac{2}{3}}}{y^{-\frac{5}{6}}}$

32. $\dfrac{b^{\frac{3}{4}}}{b^{-\frac{3}{2}}}$

33. $(x^2)^{-\frac{1}{2}}$

34. $(a^8)^{-\frac{3}{4}}$

35. $\left(x^{-\frac{2}{3}}\right)^6$

36. $\left(y^{-\frac{5}{6}}\right)^{12}$

37. $\left(a^{-\frac{1}{2}}\right)^{-2}$

38. $\left(b^{-\frac{2}{3}}\right)^{-6}$

39. $\left(x^{-\frac{3}{8}}\right)^{-\frac{4}{5}}$

40. $\left(y^{-\frac{3}{2}}\right)^{-\frac{2}{9}}$

41. $\left(a^{\frac{1}{2}} \cdot a\right)^2$

42. $\left(b^{\frac{2}{3}} \cdot b^{\frac{1}{6}}\right)^6$

43. $\left(x^{-\frac{1}{2}} \cdot x^{\frac{3}{4}}\right)^{-2}$

44. $\left(a^{\frac{1}{2}} \cdot a^{-2}\right)^3$

45. $\left(y^{-\frac{1}{2}} \cdot y^{\frac{3}{2}}\right)^{\frac{2}{3}}$

46. $\left(b^{-\frac{2}{3}} \cdot b^{\frac{1}{4}}\right)^{-\frac{4}{3}}$

47. $(x^8y^2)^{\frac{1}{2}}$

48. $(a^3b^9)^{\frac{2}{3}}$

49. $(x^4y^2z^6)^{\frac{3}{2}}$

50. $(a^8b^4c^4)^{\frac{3}{4}}$

51. $(x^{-3}y^6)^{-\frac{1}{3}}$

52. $(a^2b^{-6})^{-\frac{1}{2}}$

53. $\left(x^{-2}y^{\frac{1}{3}}\right)^{-\frac{3}{4}}$

54. $\left(a^{-\frac{2}{3}}b^{\frac{2}{3}}\right)^{\frac{3}{2}}$

55. $\left(\dfrac{x^{\frac{1}{2}}}{y^{-2}}\right)^4$

56. $\left(\dfrac{b^{-\frac{3}{4}}}{a^{-\frac{1}{2}}}\right)^8$

57. $\dfrac{x^{\frac{1}{4}} \cdot x^{-\frac{1}{2}}}{x^{\frac{2}{3}}}$

58. $\dfrac{b^{\frac{1}{2}} \cdot b^{-\frac{3}{4}}}{b^{\frac{1}{4}}}$

59. $\left(\dfrac{y^{\frac{2}{3}} \cdot y^{-\frac{5}{6}}}{y^{\frac{1}{9}}}\right)^9$

60. $\left(\dfrac{a^{\frac{1}{3}} \cdot a^{-\frac{2}{3}}}{a^{\frac{1}{2}}}\right)^4$

61. $\left(\dfrac{b^2 \cdot b^{-\frac{3}{4}}}{b^{-\frac{1}{2}}}\right)^{-\frac{1}{2}}$

62. $\dfrac{\left(x^{-\frac{5}{6}} \cdot x^3\right)^{-\frac{2}{3}}}{x^{\frac{4}{3}}}$

63. $\left(a^{\frac{2}{3}}b^2\right)^6 \left(a^3b^3\right)^{\frac{1}{3}}$

64. $\left(x^3 y^{-\frac{1}{2}}\right)^{-2}\left(x^{-3} y^2\right)^{\frac{1}{6}}$

65. $\left(16 m^{-2} n^4\right)^{-\frac{1}{2}}\left(m n^{\frac{1}{2}}\right)$

66. $\left(27 m^3 n^{-6}\right)^{\frac{1}{3}}\left(m^{-\frac{1}{3}} n^{\frac{5}{6}}\right)^6$

67. $\left(\dfrac{x^{\frac{1}{2}} y^{-\frac{3}{4}}}{y^{\frac{2}{3}}}\right)^{-6}$

68. $\left(\dfrac{x^{\frac{1}{2}} y^{-\frac{5}{4}}}{y^{-\frac{3}{4}}}\right)^{-4}$

69. $\left(\dfrac{2^{-6} b^{-3}}{a^{-\frac{1}{2}}}\right)^{-\frac{2}{3}}$

70. $\left(\dfrac{49 c^{\frac{5}{3}}}{a^{-\frac{1}{4}} b^{\frac{5}{6}}}\right)^{-\frac{3}{2}}$

71. $\dfrac{\left(x^{-2} y^4\right)^{\frac{1}{2}}}{\left(x^{\frac{1}{2}}\right)^4}$

72. $\dfrac{\left(x^{-3}\right)^{\frac{1}{3}}}{\left(x^9 y^6\right)^{\frac{1}{6}}}$

73. $a^{-\frac{1}{4}}\left(a^{\frac{5}{4}} - a^{\frac{9}{4}}\right)$

74. $x^{\frac{4}{3}}\left(x^{\frac{2}{3}} + x^{-\frac{1}{3}}\right)$

75. $y^{\frac{2}{3}}\left(y^{\frac{1}{3}} + y^{-\frac{2}{3}}\right)$

76. $b^{-\frac{2}{5}}\left(b^{-\frac{3}{5}} - b^{\frac{7}{5}}\right)$

77. $a^{\frac{1}{6}}\left(a^{\frac{5}{6}} - a^{-\frac{7}{6}}\right)$

78. $\left(x^{\frac{n}{3}}\right)^{3n}$

79. $\left(a^{\frac{2}{n}}\right)^{-5n}$

80. $x^n \cdot x^{\frac{n}{2}}$

81. $a^{\frac{n}{2}} \cdot a^{-\frac{n}{3}}$

82. $\dfrac{y^{\frac{n}{2}}}{y^{-n}}$

83. $\dfrac{b^{\frac{m}{3}}}{b^m}$

84. $\left(x^{\frac{2}{n}}\right)^n$

85. $\left(x^{5n}\right)^{2n}$

86. $\left(x^{\frac{n}{4}} y^{\frac{n}{8}}\right)^8$

87. $\left(x^{\frac{n}{2}} y^{\frac{n}{3}}\right)^6$

2 Write exponential expressions as radical expressions and radical expressions as exponential expressions

88. In the radical expression $\sqrt[n]{a}$, n is called the ____?____ and a is called the ____?____.

89. To rewrite $\sqrt[5]{a^4}$ as an exponential expression, find the exponent for the base a. The rational exponent will be $\dfrac{\text{power on the radicand}}{\text{index}} = \dfrac{\underline{\ \ ?\ \ }}{\underline{\ \ ?\ \ }} \cdot \sqrt[5]{a^4} = a^{\underline{\ \ ?\ \ }}$.

90. True or false? $8x^{\frac{1}{3}} = 2\sqrt[3]{x}$

91. True or false? $\sqrt[4]{x^3} = \left(\sqrt[4]{x}\right)^3$

Rewrite the exponential expression as a radical expression.

92. $3^{\frac{1}{4}}$

93. $5^{\frac{1}{2}}$

94. $a^{\frac{3}{2}}$

95. $b^{\frac{4}{3}}$

96. $(2t)^{\frac{5}{2}}$

97. $(3x)^{\frac{2}{3}}$

98. $-2x^{\frac{2}{3}}$

99. $-3a^{\frac{2}{5}}$

100. $\left(a^2 b\right)^{\frac{2}{3}}$

101. $(x^2y^3)^{\frac{3}{4}}$

102. $(a^2b^4)^{\frac{3}{5}}$

103. $(a^3b^7)^{\frac{3}{2}}$

104. $(4x + 3)^{\frac{3}{4}}$

105. $(3x - 2)^{\frac{1}{3}}$

106. $x^{-\frac{2}{3}}$

Rewrite the radical expression as an exponential expression.

107. $\sqrt{14}$

108. $\sqrt{7}$

109. $\sqrt[3]{x}$

110. $\sqrt[4]{y}$

111. $\sqrt[3]{x^4}$

112. $\sqrt[4]{a^3}$

113. $\sqrt[5]{b^3}$

114. $\sqrt[4]{b^5}$

115. $\sqrt[3]{2x^2}$

116. $\sqrt[5]{4y^7}$

117. $-\sqrt{3x^5}$

118. $-\sqrt[4]{4x^5}$

119. $3x\sqrt[3]{y^2}$

120. $2y\sqrt{x^3}$

121. $\sqrt{a^2 + 2}$

3 Simplify radical expressions that are roots of perfect powers

122. $16x^{12}y^{20}$ is a perfect fourth power because $16 = (\underline{\ \ ?\ \ })^4$, $x^{12} = (\underline{\ \ ?\ \ })^4$, and $y^{20} = (\underline{\ \ ?\ \ })^4$.

123. $32a^{10}b^{15}$ is a perfect fifth power because $32 = (\underline{\ \ ?\ \ })^5$, $a^{10} = (\underline{\ \ ?\ \ })^5$, and $b^{15} = (\underline{\ \ ?\ \ })^5$.

For Exercises 124 to 127, state whether the expression simplifies to a positive number, a negative number, or a number that is not a real number.

124. $-\sqrt[3]{8x^{15}}$

125. $-\sqrt{9x^8}$

126. $\sqrt{-4x^{12}}$

127. $-\sqrt[3]{-27x^9}$

Simplify.

128. $\sqrt{x^{16}}$

129. $\sqrt{y^{14}}$

130. $-\sqrt{x^8}$

131. $-\sqrt{a^6}$

132. $\sqrt{x^2y^{10}}$

133. $\sqrt{a^{14}b^6}$

134. $\sqrt{25x^6}$

135. $\sqrt{121y^{12}}$

136. $\sqrt[3]{x^3y^9}$

137. $\sqrt[3]{a^6b^{12}}$

138. $-\sqrt[3]{x^{15}y^3}$

139. $-\sqrt[3]{a^9b^9}$

140. $\sqrt[3]{27a^9}$

141. $\sqrt[3]{125b^{15}}$

142. $\sqrt[3]{-8x^3}$

143. $\sqrt[3]{-a^6b^9}$

144. $\sqrt{16a^4b^{12}}$

145. $\sqrt{25x^8y^2}$

146. $\sqrt{-16x^4y^2}$

147. $\sqrt{-9a^6b^8}$

148. $\sqrt[3]{27x^9}$

149. $\sqrt[3]{8a^{21}b^6}$

150. $\sqrt[3]{-64x^9y^{12}}$

151. $\sqrt[3]{-27a^3b^{15}}$

152. $\sqrt[4]{x^{16}}$

153. $\sqrt[4]{y^{12}}$

154. $\sqrt[4]{16x^{12}}$

155. $\sqrt[4]{81a^{20}}$

156. $-\sqrt[4]{x^8y^{12}}$

157. $-\sqrt[4]{a^{16}b^4}$

158. $\sqrt[5]{x^{20}y^{10}}$

159. $\sqrt[5]{a^5b^{25}}$

160. $\sqrt[4]{81x^4y^{20}}$

161. $\sqrt[4]{16a^8b^{20}}$

162. $\sqrt[5]{32a^5b^{10}}$

163. $\sqrt[5]{-32x^{15}y^{20}}$

164. $\sqrt[5]{243x^{10}y^{40}}$

165. $\sqrt{\dfrac{16x^2}{y^{14}}}$

166. $\sqrt{\dfrac{49a^4}{b^{24}}}$

167. $\sqrt[3]{\dfrac{27b^3}{a^9}}$

168. $\sqrt[3]{\dfrac{64x^{15}}{y^6}}$

169. $\sqrt{(2x+3)^2}$

170. $\sqrt{(4x+1)^2}$

171. $\sqrt{x^2+2x+1}$

172. $\sqrt{x^2+4x+4}$

8.1 Applying Concepts

Simplify.

173. $\sqrt[3]{\sqrt{x^6}}$

174. $\sqrt{\sqrt[3]{y^6}}$

175. $\sqrt[5]{\sqrt[3]{b^{15}}}$

176. $\sqrt{\sqrt{16x^{12}}}$

177. $\sqrt[5]{\sqrt{a^{10}b^{20}}}$

178. $\sqrt[3]{\sqrt{64x^{36}y^{30}}}$

For what value of p is the given equation true?

179. $y^p y^{\frac{2}{5}} = y$

180. $\dfrac{y^p}{y^{\frac{3}{4}}} = y^{\frac{1}{2}}$

181. $x^p x^{-\frac{1}{2}} = x^{\frac{1}{4}}$

Operations on Radical Expressions

OBJECTIVE 1 Simplify radical expressions

Point of Interest

The Latin expression for irrational numbers was numerus surdus, which means "inaudible number." A prominent 16th-century mathematician wrote of irrational numbers, "just as an infinite number is not a number, so an irrational number is not a true number, but lies hidden in some sort of cloud of infinity." In 1872, Richard Dedekind wrote a paper that established the first logical treatment of irrational numbers.

If a number is not a perfect power, its root can only be approximated; examples include $\sqrt{5}$ and $\sqrt[3]{3}$. These numbers are **irrational numbers.** Their decimal representations never terminate or repeat.

$$\sqrt{5} = 2.2360679... \qquad \sqrt[3]{3} = 1.4422495...$$

A radical expression is not in simplest form when the radicand contains a factor greater than 1 that is a perfect power. The Product Property of Radicals is used to simplify radical expressions whose radicands are not perfect powers.

Product Property of Radicals

If $\sqrt[n]{a}$ and $\sqrt[n]{b}$ are positive real numbers, then

$$\sqrt[n]{ab} = \sqrt[n]{a} \cdot \sqrt[n]{b} \quad \text{and} \quad \sqrt[n]{a} \cdot \sqrt[n]{b} = \sqrt[n]{ab}$$

Simplify: $\sqrt{48}$

Write the radicand as the product of a perfect square and a factor that does not contain a perfect square.	$\sqrt{48} = \sqrt{16 \cdot 3}$
Use the Product Property of Radicals to write the expression as a product.	$= \sqrt{16}\sqrt{3}$
Simplify $\sqrt{16}$.	$= 4\sqrt{3}$

Note that 48 must be written as the product of a perfect square and *a factor that does not contain a perfect square.* Therefore, it would not be correct to rewrite $\sqrt{48}$ as $\sqrt{4 \cdot 12}$ and simplify the expression as shown at the right. Although 4 is a perfect-square factor of 48, 12 contains a perfect square $(12 = 4 \cdot 3)$ and $\sqrt{12}$ can be simplified. Remember to find the largest perfect power that is a factor of the radicand.

$$\sqrt{48} = \sqrt{4 \cdot 12}$$
$$= \sqrt{4}\sqrt{12}$$
$$= 2\sqrt{12}$$

Not in simplest form

Simplify: $\sqrt[3]{x^7}$

Write the radicand as the product of a perfect cube and a factor that does not contain a perfect cube.	$\sqrt[3]{x^7} = \sqrt[3]{x^6 \cdot x}$
Use the Product Property of Radicals to write the expression as a product.	$= \sqrt[3]{x^6}\sqrt[3]{x}$
Simplify.	$= x^2\sqrt[3]{x}$

EXAMPLE 1 Simplify: $\sqrt[4]{32x^7}$

Solution $\sqrt[4]{32x^7} = \sqrt[4]{16x^4(2x^3)}$ • Write the radicand as the product of a perfect fourth power and factors that do not contain a perfect fourth power.

$= \sqrt[4]{16x^4}\sqrt[4]{2x^3}$ • Use the Product Property of Radicals to write the expression as a product.

$= 2x\sqrt[4]{2x^3}$ • Simplify.

PROBLEM 1 Simplify: $\sqrt[5]{128x^7}$

Solution See page S27.

OBJECTIVE 2 ## Add and subtract radical expressions

The Distributive Property is used to simplify the sum or difference of radical expressions that have the same radicand and the same index. For example,

$$3\sqrt{5} + 8\sqrt{5} = (3 + 8)\sqrt{5} = 11\sqrt{5}$$
$$2\sqrt[3]{3x} - 9\sqrt[3]{3x} = (2 - 9)\sqrt[3]{3x} = -7\sqrt[3]{3x}$$

Radical expressions that are in simplest form and have unlike radicands or different indices cannot be simplified by the Distributive Property. The following expressions cannot be simplified by the Distributive Property.

$$3\sqrt[4]{2} - 6\sqrt[4]{3}$$ • **The radicands are different.**
$$2\sqrt[4]{4x} + 3\sqrt[3]{4x}$$ • **The indices are different.**

Simplify: $3\sqrt{32x^2} - 2x\sqrt{2} + \sqrt{128x^2}$

First simplify each term. Then combine like terms by using the Distributive Property.

$$3\sqrt{32x^2} - 2x\sqrt{2} + \sqrt{128x^2} = 3\sqrt{16x^2}\sqrt{2} - 2x\sqrt{2} + \sqrt{64x^2}\sqrt{2}$$
$$= 3 \cdot 4x\sqrt{2} - 2x\sqrt{2} + 8x\sqrt{2}$$
$$= 12x\sqrt{2} - 2x\sqrt{2} + 8x\sqrt{2}$$
$$= 18x\sqrt{2}$$

EXAMPLE 2 Subtract.

 A. $5b\sqrt[4]{32a^7b^5} - 2a\sqrt[4]{162a^3b^9}$ **B.** $5\sqrt[5]{2x^7y^{11}} - y\sqrt[5]{64^7y^6}$

Solution **A.** $5b\sqrt[4]{32a^7b^5} - 2a\sqrt[4]{162a^3b^9}$
 $= 5b\sqrt[4]{16a^4b^4}\sqrt[4]{2a^3b} - 2a\sqrt[4]{81b^8}\sqrt[4]{2a^3b}$
 $= 5b \cdot 2ab\sqrt[4]{2a^3b} - 2a \cdot 3b^2\sqrt[4]{2a^3b}$
 $= 10ab^2\sqrt[4]{2a^3b} - 6ab^2\sqrt[4]{2a^3b}$
 $= 4ab^2\sqrt[4]{2a^3b}$

 B. $5\sqrt[5]{2x^7y^{11}} - y\sqrt[5]{64^7y^6} = 5\sqrt[5]{x^5y^{10}}\sqrt[5]{2x^2y} - y\sqrt[5]{32x^5y^5}\sqrt[5]{2x^2y}$
 $= 5 \cdot xy^2\sqrt[5]{2x^2y} - y \cdot 2xy\sqrt[5]{2x^2y}$
 $= 5xy^2\sqrt[5]{2x^2y} - 2xy^2\sqrt[5]{2x^2y}$
 $= 3xy^2\sqrt[5]{2x^2y}$

PROBLEM 2 Add or subtract.

 A. $3xy\sqrt[3]{81x^5y} - \sqrt[3]{192x^8y^4}$ **B.** $4a\sqrt[3]{54a^7b^9} + a^2b\sqrt[3]{128a^4b^6}$

Solution See page S27.

OBJECTIVE **3** **Multiply radical expressions**

The Product Property of Radicals is used to multiply radical expressions with the same index.

$$\sqrt{3x} \cdot \sqrt{5y} = \sqrt{3x \cdot 5y} = \sqrt{15xy}$$

Multiply: $\sqrt[3]{2a^5b} \ \sqrt[3]{16a^2b^2}$

Use the Product Property of Radicals to multiply the radicands. Then simplify.

$$\sqrt[3]{2a^5b} \ \sqrt[3]{16a^2b^2} = \sqrt[3]{32a^7b^3}$$
$$= \sqrt[3]{8a^6b^3} \ \sqrt[3]{4a}$$
$$= 2a^2b\sqrt[3]{4a}$$

Multiply: $\sqrt{2x}\left(\sqrt{8x} - \sqrt{3}\right)$

Use the Distributive Property to remove parentheses. Then simplify.

$$\sqrt{2x}\left(\sqrt{8x} - \sqrt{3}\right) = \sqrt{16x^2} - \sqrt{6x}$$
$$= 4x - \sqrt{6x}$$

EXAMPLE 3 Multiply: $\sqrt{3x}\left(\sqrt{27x^2} - \sqrt{3x}\right)$

Solution $\sqrt{3x}\left(\sqrt{27x^2} - \sqrt{3x}\right) = \sqrt{81x^3} - \sqrt{9x^2}$
$$= \sqrt{81x^2}\sqrt{x} - \sqrt{9x^2}$$
$$= 9x\sqrt{x} - 3x$$

PROBLEM 3 Multiply: $\sqrt{5b}\left(\sqrt{3b} - \sqrt{10}\right)$

Solution See page S27.

To multiply $\left(\sqrt[3]{x} - 1\right)\left(\sqrt[3]{x} + 7\right)$, use the FOIL method. Then combine like terms.

$$\left(\sqrt[3]{x} - 1\right)\left(\sqrt[3]{x} + 7\right) = \sqrt[3]{x^2} + 7\sqrt[3]{x} - \sqrt[3]{x} - 7$$
$$= \sqrt[3]{x^2} + 6\sqrt[3]{x} - 7$$

The expressions $a + b$ and $a - b$, which are the sum and difference of two terms, are called **conjugates** of each other. The product of conjugates of the form $(a + b)(a - b)$ is $a^2 - b^2$.

$$\left(\sqrt{x} - 3\right)\left(\sqrt{x} + 3\right) = \left(\sqrt{x}\right)^2 - 3^2 = x - 9$$

Take Note

The concept of conjugate is used in a number of different instances. Make sure you understand this idea.

The conjugate of $\sqrt{3} - 4$ is $\sqrt{3} + 4$.

The conjugate of $\sqrt{3} + 4$ is $\sqrt{3} - 4$.

The conjugate of $\sqrt{5a} + \sqrt{b}$ is $\sqrt{5a} - \sqrt{b}$.

EXAMPLE 4 Multiply.

A. $\left(2\sqrt[3]{x} - 3\right)\left(3\sqrt[3]{x} - 4\right)$ **B.** $\left(\sqrt{xy} - 2\right)\left(\sqrt{xy} + 2\right)$

Solution **A.** $\left(2\sqrt[3]{x} - 3\right)\left(3\sqrt[3]{x} - 4\right)$ • Use the FOIL method.
$$= 6\sqrt[3]{x^2} - 8\sqrt[3]{x} - 9\sqrt[3]{x} + 12$$
$$= 6\sqrt[3]{x^2} - 17\sqrt[3]{x} + 12$$

B. $\left(\sqrt{xy} - 2\right)\left(\sqrt{xy} + 2\right)$ • Use $(a - b)(a + b) = a^2 - b^2$.
$$= \left(\sqrt{xy}\right)^2 - 2^2$$
$$= xy - 4$$

PROBLEM 4 Multiply.

A. $\left(2\sqrt[3]{2x} - 3\right)\left(\sqrt[3]{2x} - 5\right)$ **B.** $\left(2\sqrt{x} - 3\right)\left(2\sqrt{x} + 3\right)$

Solution See page S27.

OBJECTIVE **4** **Divide radical expressions**

The Quotient Property of Radicals is used to divide radical expressions with the same index.

Quotient Property of Radicals

If $\sqrt[n]{a}$ and $\sqrt[n]{b}$ are real numbers, and $b \neq 0$, then

$$\sqrt[n]{\dfrac{a}{b}} = \dfrac{\sqrt[n]{a}}{\sqrt[n]{b}} \quad \text{and} \quad \dfrac{\sqrt[n]{a}}{\sqrt[n]{b}} = \sqrt[n]{\dfrac{a}{b}}$$

Simplify: $\sqrt[3]{\dfrac{81x^5}{y^6}}$

Use the Quotient Property of Radicals.
Then simplify each radical expression.

$$\sqrt[3]{\dfrac{81x^5}{y^6}} = \dfrac{\sqrt[3]{81x^5}}{\sqrt[3]{y^6}}$$

$$= \dfrac{\sqrt[3]{27x^3}\,\sqrt[3]{3x^2}}{\sqrt[3]{y^6}}$$

$$= \dfrac{3x\sqrt[3]{3x^2}}{y^2}$$

Simplify: $\dfrac{\sqrt{5a^4b^7c^2}}{\sqrt{ab^3c}}$

Use the Quotient Property of Radicals.
Then simplify the radicand.

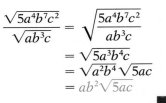

$$\dfrac{\sqrt{5a^4b^7c^2}}{\sqrt{ab^3c}} = \sqrt{\dfrac{5a^4b^7c^2}{ab^3c}}$$

$$= \sqrt{5a^3b^4c}$$

$$= \sqrt{a^2b^4}\,\sqrt{5ac}$$

$$= ab^2\sqrt{5ac}$$

Simplest Form of a Radical Expression

A radical expression is in simplest form when *all* of the following conditions are met.

1. The radicand contains no factor greater than 1 that is a perfect power of the index.

2. There is no fraction under the radical sign.

3. No radical remains in the denominator of the radical expression.

The procedure used to remove a radical from a denominator is called **rationalizing the denominator.** This is illustrated in the next example.

Simplify: $\dfrac{2}{\sqrt{x}}$

Multiply the expression by 1 in the form $\dfrac{\sqrt{x}}{\sqrt{x}}$.

$$\frac{2}{\sqrt{x}} = \frac{2}{\sqrt{x}} \cdot \frac{\sqrt{x}}{\sqrt{x}} = \frac{2\sqrt{x}}{\sqrt{x^2}} = \frac{2\sqrt{x}}{x}$$

Then simplify.

EXAMPLE 5 Simplify. **A.** $\dfrac{5}{\sqrt{5x}}$ **B.** $\dfrac{3x}{\sqrt[3]{4x}}$

Solution **A.** $\dfrac{5}{\sqrt{5x}} = \dfrac{5}{\sqrt{5x}} \cdot \dfrac{\sqrt{5x}}{\sqrt{5x}}$ • Multiply the expression by $\dfrac{\sqrt{5x}}{\sqrt{5x}}$.

$$= \frac{5\sqrt{5x}}{\sqrt{25x^2}}$$

$$= \frac{5\sqrt{5x}}{5x}$$

$$= \frac{\sqrt{5x}}{x}$$

> ### Take Note
> Note that multiplying by $\dfrac{\sqrt[3]{4x}}{\sqrt[3]{4x}}$ will not rationalize the denominator of $\dfrac{3x}{\sqrt[3]{4x}}$.
>
> $$\frac{3x}{\sqrt[3]{4x}} \cdot \frac{\sqrt[3]{4x}}{\sqrt[3]{4x}} = \frac{3x\sqrt[3]{4x}}{\sqrt[3]{16x^2}}$$
>
> Because $16x^2$ is not a perfect cube, the denominator still contains a radical expression.

B. $\dfrac{3x}{\sqrt[3]{4x}} = \dfrac{3x}{\sqrt[3]{4x}} \cdot \dfrac{\sqrt[3]{2x^2}}{\sqrt[3]{2x^2}}$ • Multiply the expression by $\dfrac{\sqrt[3]{2x^2}}{\sqrt[3]{2x^2}}$.

$$= \frac{3x\sqrt[3]{2x^2}}{\sqrt[3]{8x^3}}$$ $\sqrt[3]{4x} \cdot \sqrt[3]{2x^2} = \sqrt[3]{8x^3}$.
$8x^3$ is a perfect cube.

$$= \frac{3x\sqrt[3]{2x^2}}{2x}$$

$$= \frac{3\sqrt[3]{2x^2}}{2}$$

PROBLEM 5 Simplify. **A.** $\dfrac{y}{\sqrt{3y}}$ **B.** $\dfrac{3}{\sqrt[3]{3x^2}}$

Solution See page S27.

To simplify a fraction that has a square-root radical expression with two terms in the denominator, multiply the numerator and denominator by the conjugate of the denominator. Then simplify. This is illustrated in the two examples that follow.

> ### Take Note
> Here are two examples of using a conjugate to simplify a radical expression.

$$\frac{3}{5 - \sqrt{7}} = \frac{3}{5 - \sqrt{7}} \cdot \frac{5 + \sqrt{7}}{5 + \sqrt{7}} = \frac{15 + 3\sqrt{7}}{25 - 7} = \frac{3(5 + \sqrt{7})}{18} = \frac{5 + \sqrt{7}}{6}$$

$$\frac{\sqrt{x} - \sqrt{y}}{\sqrt{x} + \sqrt{y}} = \frac{\sqrt{x} - \sqrt{y}}{\sqrt{x} + \sqrt{y}} \cdot \frac{\sqrt{x} - \sqrt{y}}{\sqrt{x} - \sqrt{y}} = \frac{\sqrt{x^2} - \sqrt{xy} - \sqrt{xy} + \sqrt{y^2}}{(\sqrt{x})^2 - (\sqrt{y})^2}$$

$$= \frac{x - 2\sqrt{xy} + y}{x - y}$$

EXAMPLE 6 Simplify. **A.** $\dfrac{2 - \sqrt{5}}{3 + \sqrt{2}}$ **B.** $\dfrac{\sqrt{3} + \sqrt{y}}{\sqrt{3} - \sqrt{y}}$

Solution **A.** $\dfrac{2 - \sqrt{5}}{3 + \sqrt{2}} = \dfrac{2 - \sqrt{5}}{3 + \sqrt{2}} \cdot \dfrac{3 - \sqrt{2}}{3 - \sqrt{2}} = \dfrac{6 - 2\sqrt{2} - 3\sqrt{5} + \sqrt{10}}{9 - 2}$

$$= \dfrac{6 - 2\sqrt{2} - 3\sqrt{5} + \sqrt{10}}{7}$$

B. $\dfrac{\sqrt{3} + \sqrt{y}}{\sqrt{3} - \sqrt{y}} = \dfrac{\sqrt{3} + \sqrt{y}}{\sqrt{3} - \sqrt{y}} \cdot \dfrac{\sqrt{3} + \sqrt{y}}{\sqrt{3} + \sqrt{y}}$

$$= \dfrac{\sqrt{9} + \sqrt{3y} + \sqrt{3y} + \sqrt{y^2}}{(\sqrt{3})^2 - (\sqrt{y})^2} = \dfrac{3 + 2\sqrt{3y} + y}{3 - y}$$

PROBLEM 6 Simplify. **A.** $\dfrac{4 + \sqrt{2}}{3 - \sqrt{3}}$ **B.** $\dfrac{\sqrt{2} + \sqrt{x}}{\sqrt{2} - \sqrt{x}}$

Solution See page S27.

8.2 Concept Review

Determine whether the following statements are always true, sometimes true, or never true.

1. If a is a real number, then $\sqrt{a}$ represents a real number.

2. If b is a real number, then $\sqrt[3]{b}$ represents a real number.

3. For positive integers x and y, $\sqrt[a]{x} \cdot \sqrt[b]{y} = \sqrt[ab]{xy}$.

4. The conjugate of $\sqrt{a} + \sqrt{b}$ is found by replacing b with its opposite.

5. If $x < 0$, then $\sqrt{x^2} = -x$.

6. For positive integers x and y, $\sqrt[a]{x} \cdot \sqrt[a]{y} = \sqrt[a]{xy}$.

8.2 Exercises

1 Simplify radical expressions

1. Simplify: $\sqrt{50}$

$\sqrt{50} = \sqrt{\underline{} \cdot 2}$

$= \sqrt{\underline{}} \sqrt{\underline{}}$

$= \underline{} \sqrt{2}$

- Write 50 as the product of a perfect-square factor and a factor that does not contain a perfect square.

- Use the Product Property of Radicals to write the expression as the product of two square roots.

- Simplify $\sqrt{25}$.

2. Simplify: $\sqrt[3]{16x^{14}}$

$\sqrt[3]{16x^{14}} = \sqrt[3]{(\underline{\ ?\ })(2x^2)}$ · **Write $16x^4$ as the product of a perfect cube and a factor that does not contain a perfect cube.**

$= \sqrt[3]{\underline{\ ?\ }}\ \sqrt[3]{2x^2}$ · **Use the Product Property of Radicals.**

$= \underline{\ ?\ }\ \sqrt[3]{2x^2}$ · **Simplify.**

3. State whether the number is rational or irrational.
a. $\sqrt{19}$ **b.** $-\sqrt{81}$ **c.** $\sqrt[3]{25}$ **d.** $\sqrt[3]{-27}$

4. State whether the expression is in simplest form.
a. $2\sqrt[5]{x^7}$ **b.** $\sqrt[3]{36y^2}$ **c.** $9a^2\sqrt{54ab}$ **d.** $2b\sqrt[4]{8b^3}$

Explain why the simplification is incorrect.

5. $\sqrt{32} = \sqrt{4 \cdot 8} = \sqrt{4}\sqrt{8} = 2\sqrt{8}$

6. $\sqrt[3]{128} = \sqrt[3]{8 \cdot 16} = \sqrt[3]{8}\sqrt[3]{16} = 2\sqrt[3]{16}$

Simplify.

7. $\sqrt{18}$

8. $\sqrt{40}$

9. $\sqrt{98}$

10. $\sqrt{128}$

11. $\sqrt[3]{72}$

12. $\sqrt[3]{54}$

13. $\sqrt[3]{16}$

14. $\sqrt[3]{128}$

15. $\sqrt{x^4y^3z^5}$

16. $\sqrt{x^3y^6z^9}$

17. $\sqrt{8a^3b^8}$

18. $\sqrt{24a^9b^6}$

19. $\sqrt{45x^2y^3z^5}$

20. $\sqrt{60xy^7z^{12}}$

21. $\sqrt[3]{-125x^2y^4}$

22. $\sqrt[4]{16x^9y^5}$

23. $\sqrt[3]{-216x^5y^9}$

24. $\sqrt[3]{a^8b^{11}c^{15}}$

25. $\sqrt[3]{a^5b^8}$

26. $\sqrt[4]{64x^8y^{10}}$

2 Add and subtract radical expressions

27. The Distributive Property can be used to add or subtract radicals that have the same ___?___ and the same ___?___.

28. Simplify: $5\sqrt[5]{x^{12}} + 2x\sqrt[5]{x^7}$

$5\sqrt[5]{x^{12}} + 2x\sqrt[5]{x^7} = 5\sqrt[5]{\underline{\ ?\ }}\sqrt[5]{x^2} + 2x\sqrt[5]{\underline{\ ?\ }}\ \sqrt[5]{x^2}$

$= 5(\underline{\ ?\ })\ \sqrt[5]{x^2} + 2x(\underline{\ ?\ })\ \sqrt[5]{x^2}$ · **Simplify the perfect fifth roots.**

$= 5x^2\sqrt[5]{x^2} + \underline{\ ?\ }\ \sqrt[5]{x^2}$ · **Multiply.**

$= (\underline{\ ?\ } + \underline{\ ?\ })\ \sqrt[5]{x^2}$ · **Use the Distributive Property.**

$= \underline{\ ?\ }\ \sqrt[5]{x^2}$ · **Add.**

29. State whether the expression can be simplified.
 a. $5x\sqrt{x} + x\sqrt[3]{x}$ **b.** $\sqrt[5]{10y} - \sqrt[5]{y}$
 c. $\sqrt[3]{2+a} + \sqrt[3]{2+a}$ **d.** $\sqrt{8x^2} + \sqrt{2}$

30. Which expressions are equivalent to $\sqrt{ab} + \sqrt{ab}$?
 a. $2\sqrt{ab}$ **b.** $\sqrt{2ab}$ **c.** $2ab$ **d.** $\sqrt{4ab}$

Simplify

31. $\sqrt{2} + \sqrt{2}$

32. $\sqrt{5} - \sqrt{5}$

33. $4\sqrt[3]{7} - \sqrt[3]{7}$

34. $3\sqrt[3]{11} - 8\sqrt[3]{11}$

35. $2\sqrt{x} - 8\sqrt{x}$

36. $3\sqrt{y} + 12\sqrt{y}$

37. $\sqrt{8} - \sqrt{32}$

38. $\sqrt{27} - \sqrt{75}$

39. $\sqrt{128x} - \sqrt{98x}$

40. $\sqrt{48x} + \sqrt{147x}$

41. $\sqrt{27a} - \sqrt{8a}$

42. $\sqrt{18b} + \sqrt{75b}$

43. $2\sqrt{2x^3} + 4x\sqrt{8x}$

44. $5y\sqrt{8y} + 2\sqrt{50y^3}$

45. $x\sqrt{75xy} - \sqrt{27x^3y}$

46. $3\sqrt{8x^2y^3} - 2x\sqrt{32y^3}$

47. $2\sqrt{32x^2y^3} - xy\sqrt{98y}$

48. $6y\sqrt{x^3y} - 2\sqrt{x^3y^3}$

49. $7b\sqrt{a^5b^3} - 2ab\sqrt{a^3b^3}$

50. $2a\sqrt{27ab^5} + 3b\sqrt{3a^3b}$

51. $\sqrt[3]{128} + \sqrt[3]{250}$

52. $\sqrt[3]{16} - \sqrt[3]{54}$

53. $2\sqrt[3]{3a^4} - 3a\sqrt[3]{81a}$

54. $2b\sqrt[3]{16b^2} + \sqrt[3]{128b^5}$

55. $3\sqrt[3]{x^5y^7} - 8xy\sqrt[3]{x^2y^4}$

56. $3\sqrt[4]{32a^5} - a\sqrt[4]{162a}$

57. $2a\sqrt[4]{16ab^5} + 3b\sqrt[4]{256a^5b}$

58. $2\sqrt{50} - 3\sqrt{125} + \sqrt{98}$

59. $3\sqrt{108} - 2\sqrt{18} - 3\sqrt{48}$

60. $\sqrt{9b^3} - \sqrt{25b^3} + \sqrt{49b^3}$

61. $\sqrt{4x^7y^5} + 9x^2\sqrt{x^3y^5} - 5xy\sqrt{x^5y^3}$

62. $2x\sqrt{8xy^2} - 3y\sqrt{32x^3} + \sqrt{8x^3y^2}$

63. $5a\sqrt{3a^3b} + 2a^2\sqrt{27ab} - 4\sqrt{75a^5b}$

64. $\sqrt[3]{54xy^3} - 5\sqrt[3]{2xy^3} + \sqrt[3]{128xy^3}$

3 Multiply radical expressions

65. Multiply: $\sqrt{3}(\sqrt{6} - 6\sqrt{x})$

$\sqrt{3}(\sqrt{6} - 6\sqrt{x}) = \sqrt{3} \cdot \sqrt{6} - \sqrt{3} \cdot 6\sqrt{x}$ • Use the Distributive Property.

$= \sqrt{\underline{\quad?\quad}} - 6\sqrt{\underline{\quad?\quad}}$ • The indices are the same.
Multiply the radicands.

$= \underline{\quad?\quad}\sqrt{2} - 6\sqrt{3x}$ • Simplify $\sqrt{18}$.

66. a. Two expressions in the form $a + b$ and $a - b$ are called ___?___.
　　b. The conjugate of $\sqrt{y} - 5$ is ___?___.
　　c. The product of conjugates of the form $a + b$ and $a - b$ is ___?___.
　　d. $(\sqrt{y} - 5)(\sqrt{y} + 5) = ($ ___?___ $)^2 - ($ ___?___ $)^2 = $ ___?___ $-$ ___?___.

Multiply.

67. $\sqrt{8}\sqrt{32}$

68. $\sqrt{14}\sqrt{35}$

69. $\sqrt[3]{4}\sqrt[3]{8}$

70. $\sqrt[3]{6}\sqrt[3]{36}$

71. $\sqrt{x^2y^5}\sqrt{xy}$

72. $\sqrt{a^3b}\sqrt{ab^4}$

73. $\sqrt{2x^2y}\sqrt{32xy}$

74. $\sqrt{5x^3y}\sqrt{10x^3y^4}$

75. $\sqrt[3]{x^2y}\sqrt[3]{16x^4y^2}$

76. $\sqrt[3]{4a^2b^3}\sqrt[3]{8ab^5}$

77. $\sqrt[4]{12ab^3}\sqrt[4]{4a^5b^2}$

78. $\sqrt[4]{36a^2b^4}\sqrt[4]{12a^5b^3}$

79. $\sqrt{3}\left(\sqrt{27} - \sqrt{3}\right)$

80. $\sqrt{10}\left(\sqrt{10} - \sqrt{5}\right)$

81. $\sqrt{x}\left(\sqrt{x} - \sqrt{2}\right)$

82. $\sqrt{y}\left(\sqrt{y} - \sqrt{5}\right)$

83. $\sqrt{2x}\left(\sqrt{8x} - \sqrt{32}\right)$

84. $\sqrt{3a}\left(\sqrt{27a^2} - \sqrt{a}\right)$

85. $\left(\sqrt{x} - 3\right)^2$

86. $\left(\sqrt{2x} + 4\right)^2$

87. $\left(4\sqrt{5} + 2\right)^2$

88. $2\sqrt{3x^2} \cdot 3\sqrt{12xy^3} \cdot \sqrt{6x^3y}$

89. $2\sqrt{14xy} \cdot 4\sqrt{7x^2y} \cdot 3\sqrt{8xy^2}$

90. $\sqrt[3]{8ab}\sqrt[3]{4a^2b^3}\sqrt[3]{9ab^4}$

91. $\sqrt[3]{2a^2b}\sqrt[3]{4a^3b^2}\sqrt[3]{8a^5b^6}$

92. $\left(\sqrt{2} - 3\right)\left(\sqrt{2} + 4\right)$

93. $\left(\sqrt{5} - 5\right)\left(2\sqrt{5} + 2\right)$

94. $\left(\sqrt{y} - 2\right)\left(\sqrt{y} + 2\right)$

95. $\left(\sqrt{x} - y\right)\left(\sqrt{x} + y\right)$

96. $\left(\sqrt{2x} - 3\sqrt{y}\right)\left(\sqrt{2x} + 3\sqrt{y}\right)$

97. $\left(2\sqrt{3x} - \sqrt{y}\right)\left(2\sqrt{3x} + \sqrt{y}\right)$

98. $\left(\sqrt{a} - 2\right)\left(\sqrt{a} - 3\right)$

99. $\left(\sqrt{x} + 4\right)\left(\sqrt{x} - 7\right)$

100. $\left(\sqrt[3]{a} + 2\right)\left(\sqrt[3]{a} + 3\right)$

101. $\left(\sqrt[3]{x} - 4\right)\left(\sqrt[3]{x} + 5\right)$

102. $\left(2\sqrt{x} - \sqrt{y}\right)\left(3\sqrt{x} + \sqrt{y}\right)$

103. True or false? $\left(\sqrt{a} - 1\right)\left(\sqrt{a} + 1\right) > a$

104. True or false? $\sqrt{a}\left(\sqrt{a} - a\right) = a\left(1 - \sqrt{a}\right)$

4 **Divide radical expressions**

105. Divide: $\dfrac{\sqrt[4]{80x^{10}}}{\sqrt[4]{5x}}$

$\dfrac{\sqrt[4]{80x^{10}}}{\sqrt[4]{5x}} = \sqrt[4]{\dfrac{80x^{10}}{5x}}$ • Use the _____?_____ Property of Radicals.

$= \sqrt[4]{16x^{\underline{?}}}$ • Simplify the radicand.

$= \sqrt[4]{16x^{\underline{?}}}\ \sqrt[4]{x}$

$= \underline{\ ?\ }\ \sqrt[4]{x}$ • Simplify the perfect fourth power.

106. ✎ When is a radical expression in simplest form?

107. State whether the expression is in simplest form.

a. $\dfrac{\sqrt[3]{a^2}}{b}$ b. $\dfrac{\sqrt{a^3}}{b}$ c. $\sqrt[4]{\dfrac{a}{b}}$ d. $\dfrac{\sqrt{a}}{\sqrt{a}+b}$

108. By what expression should the numerator and denominator be multiplied in order to rationalize the denominator?

a. $\dfrac{1}{\sqrt{6}}$ b. $\dfrac{7}{\sqrt[3]{2y^5}}$ c. $\dfrac{8x}{\sqrt[4]{27x}}$ d. $\dfrac{4}{\sqrt{3}-x}$

Simplify.

109. $\dfrac{\sqrt{32x^2}}{\sqrt{2x}}$

110. $\dfrac{\sqrt{60y^4}}{\sqrt{12y}}$

111. $\dfrac{\sqrt{42a^3b^5}}{\sqrt{14a^2b}}$

112. $\dfrac{\sqrt{65ab^4}}{\sqrt{5ab}}$

113. $\dfrac{1}{\sqrt{5}}$

114. $\dfrac{1}{\sqrt{2}}$

115. $\dfrac{1}{\sqrt{2x}}$

116. $\dfrac{2}{\sqrt{3y}}$

117. $\dfrac{5}{\sqrt{5x}}$

118. $\dfrac{9}{\sqrt{3a}}$

119. $\sqrt{\dfrac{x}{5}}$

120. $\sqrt{\dfrac{y}{2}}$

121. $\dfrac{3}{\sqrt[3]{2}}$

122. $\dfrac{5}{\sqrt[3]{9}}$

123. $\dfrac{3}{\sqrt[3]{4x^2}}$

124. $\dfrac{5}{\sqrt[3]{3y}}$

125. $\dfrac{\sqrt{40x^3y^2}}{\sqrt{80x^2y^3}}$

126. $\dfrac{\sqrt{15a^2b^5}}{\sqrt{30a^5b^3}}$

127. $\dfrac{\sqrt{24a^2b}}{\sqrt{18ab^4}}$

128. $\dfrac{\sqrt{12x^3y}}{\sqrt{20x^4y}}$

129. $\dfrac{2}{\sqrt{5}+2}$

130. $\dfrac{5}{2 - \sqrt{7}}$

131. $\dfrac{3}{\sqrt{y} - 2}$

132. $\dfrac{-7}{\sqrt{x} - 3}$

133. $\dfrac{\sqrt{2} - \sqrt{3}}{\sqrt{2} + \sqrt{3}}$

134. $\dfrac{\sqrt{3} + \sqrt{4}}{\sqrt{2} + \sqrt{3}}$

135. $\dfrac{4 - \sqrt{2}}{2 - \sqrt{3}}$

136. $\dfrac{3 - \sqrt{x}}{3 + \sqrt{x}}$

137. $\dfrac{\sqrt{3} - \sqrt{5}}{\sqrt{2} + \sqrt{5}}$

138. $\dfrac{\sqrt{2} + \sqrt{3}}{\sqrt{3} - \sqrt{2}}$

139. $\dfrac{3}{\sqrt[4]{8x^3}}$

140. $\dfrac{-3}{\sqrt[4]{27y^2}}$

141. $\dfrac{4}{\sqrt[5]{16a^2}}$

142. $\dfrac{a}{\sqrt[5]{81a^4}}$

143. $\dfrac{2x}{\sqrt[5]{64x^3}}$

144. $\dfrac{3y}{\sqrt[4]{32y^2}}$

145. $\dfrac{\sqrt{a} + a\sqrt{b}}{\sqrt{a} - a\sqrt{b}}$

146. $\dfrac{\sqrt{3} - 3\sqrt{y}}{\sqrt{3} + 3\sqrt{y}}$

147. $\dfrac{3\sqrt{xy} + 2\sqrt{xy}}{\sqrt{x} - \sqrt{y}}$

148. $\dfrac{2\sqrt{x} + 3\sqrt{y}}{\sqrt{x} - 4\sqrt{y}}$

8.2 Applying Concepts

Simplify.

149. $\left(\sqrt{8} - \sqrt{2}\right)^3$

150. $\left(\sqrt{27} - \sqrt{3}\right)^3$

151. $\left(\sqrt{2} - 3\right)^3$

152. $\left(\sqrt{5} + 2\right)^3$

153. $\dfrac{3}{\sqrt{y + 1} + 1}$

154. $\dfrac{2}{\sqrt{x + 4} + 2}$

Rewrite as an expression with a single radical.

155. $\dfrac{\sqrt[3]{(x + y)^2}}{\sqrt{x + y}}$

156. $\dfrac{\sqrt[4]{(a + b)^3}}{\sqrt{a + b}}$

157. $\sqrt[4]{2y}\,\sqrt{x + 3}$

158. $\sqrt[4]{2x}\,\sqrt{y - 2}$

159. $\sqrt{a}\,\sqrt[3]{a + 3}$

160. $\sqrt{b}\,\sqrt[3]{b - 1}$

Write in exponential form and then simplify.

161. $\sqrt{16^{\frac{1}{2}}}$

162. $\sqrt[3]{4^{\frac{3}{2}}}$

163. $\sqrt[4]{32^{-\frac{4}{5}}}$

164. $\sqrt{243^{-\frac{4}{5}}}$

165. National Debt The approximate national debt, in trillions of dollars, for every other year from 1994 to 2006 is given in the table at the right. (*Source*: www.publicdebt.treas.gov/opd/opdpenny.htm) A function that models this data is $D(x) = 3.06x^{0.3}$, where $x = 2$ corresponds to the year 1992 and $D(x)$ is the national debt in trillions of dollars.

a. Write this function in radical form.
b. Find the value of the national debt predicted by the model for the year 1995.
c. Find the value of the national debt predicted by the model for the year 2003.

166. By what factor must you multiply a number in order to double its square root? to triple its square root? to double its cube root? to triple its cube root? Explain.

Year	National Debt
1994	4.69
1996	5.22
1998	5.53
2000	5.67
2002	6.23
2004	7.38
2006	8.39

8.3 Radical Functions

OBJECTIVE 1 Find the domain of a radical function

A **radical function** is one that contains a fractional exponent or a variable underneath a radical. Examples of radical functions are shown at the right.

$$f(x) = 3^4\sqrt{x^5} - 7$$
$$g(x) = 3x - 2x^{\frac{1}{2}} + 5$$

Note that these are *not* polynomial functions because polynomial functions do not contain variables raised to a fractional power or variable radical expressions.

The domain of a radical function is a set of real numbers for which the radical expression is a real number. For example, -9 is one number that would be excluded from the domain of $f(x) = \sqrt{x + 5}$ because

$$f(-9) = \sqrt{-9 + 5} = \sqrt{-4}, \text{ which is not a real number.}$$

State the domain of $f(x) = \sqrt{x + 5}$ in set-builder notation.

The value of $\sqrt{x + 5}$ is a real number when $x + 5$ is greater than or equal to zero: $x + 5 \geq 0$. Solving this inequality for x results in $x \geq -5$.

The domain of f is $\{x \mid x \geq -5\}$.

State the domain of $F(x) = \sqrt[3]{2x - 6}$.

Because the cube root of a real number is a real number, $\sqrt[3]{2x - 6}$ is a real number for all values of x. [For instance, $F(-1) = \sqrt[3]{2(-1) - 6} = \sqrt[3]{-8} = -2$.] Therefore, the domain of F is all real numbers.

These last two examples suggest the following:

If the index of a radical expression is an even number, the radicand must be greater than or equal to zero to ensure that the value of the expression will be a real number. If the index of a radical expression is an odd number, the radicand may be a positive or a negative number.

EXAMPLE 1 State the domain of each function in set-builder notation.

A. $V(x) = \sqrt[4]{6 - 4x}$ **B.** $R(x) = \sqrt[5]{x + 4}$

Solution **A.** $6 - 4x \geq 0$ • *V* contains an even root. Therefore, the
 $-4x \geq -6$ radicand must be greater than or equal
 $x \leq \dfrac{3}{2}$ to zero.

The domain is $\left\{ x \,\middle|\, x \leq \dfrac{3}{2} \right\}$.

B. $\{x \,|\, x \in \text{real numbers}\}$ • Because *R* contains an odd root, the
 radicand may be positive or negative.

PROBLEM 1 State the domain of each function in interval notation.

A. $Q(x) = \sqrt[3]{6x + 12}$ **B.** $T(x) = (3x + 9)^{\frac{1}{2}}$

Solution See page S27.

OBJECTIVE 2 ## Graph a radical function

The graph of a radical function is produced in the same manner as the graph of any other function. The function is evaluated at several values in the domain of the function, and the resulting ordered pairs are graphed. Ordered pairs must be graphed until an accurate graph can be made.

Take Note

By first determining the domain of the function, you discover the possible values of *x* to use when evaluating the function. Any value of *x* in the domain can be used. We used "nice" values—values of *x* that result in a perfect-square radicand. However, any value of *x* can be used. For instance, if *x* = 3, then

$$f(3) = \sqrt{5} \approx 2.24$$

and the ordered pair (3, 2.24) belongs to the graph. This ordered pair is just a little more difficult to graph.

Graph: $f(x) = \sqrt{x + 2}$

Because *f* contains an even root, the radicand must be positive. To determine the domain of *f*, solve the inequality $x + 2 \geq 0$. The solution is $x \geq -2$, so the domain is $\{x \,|\, x \geq -2\}$. Now determine ordered pairs of the function by choosing values of *x* from the domain. Some possible choices are shown in the following table.

x	$f(x) = \sqrt{x + 2}$	y
-2	$f(-2) = \sqrt{-2 + 2} = \sqrt{0} = 0$	0
-1	$f(-1) = \sqrt{-1 + 2} = \sqrt{1} = 1$	1
2	$f(2) = \sqrt{2 + 2} = \sqrt{4} = 2$	2
7	$f(7) = \sqrt{7 + 2} = \sqrt{9} = 3$	3

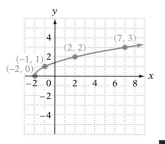

EXAMPLE 2 Graph: $H(x) = \sqrt[3]{x}$

Solution Because H contains only an odd root, the domain of H is all real numbers. Choose some values of x in the domain of H, and evaluate the function for those values. Some possible choices are given in the following table.

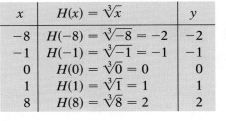

x	$H(x) = \sqrt[3]{x}$	y
-8	$H(-8) = \sqrt[3]{-8} = -2$	-2
-1	$H(-1) = \sqrt[3]{-1} = -1$	-1
0	$H(0) = \sqrt[3]{0} = 0$	0
1	$H(1) = \sqrt[3]{1} = 1$	1
8	$H(8) = \sqrt[3]{8} = 2$	2

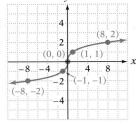

PROBLEM 2 Graph: $F(x) = \sqrt{x - 2}$

Solution See page S27.

The graph of $f(x) = \sqrt{x + 2}$ and $H(x) = \sqrt[3]{x}$ are shown below. Note that each graph satisfies the vertical line test for the graph of a function.

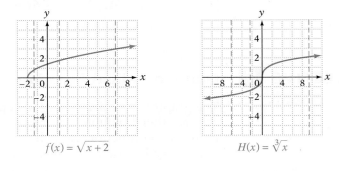

$$f(x) = \sqrt{x + 2} \qquad\qquad H(x) = \sqrt[3]{x}$$

A graphing utility can be used to graph radical functions. See the Appendix for instructions on how to enter a radical function on a graphing utility.

EXAMPLE 3 Graph: $Z(x) = 4 + (8 - 6x)^{\frac{1}{2}}$

Solution Because Z involves an even root, the radicand must be positive. The domain is $\left\{x \mid x \le \dfrac{4}{3}\right\}$.

PROBLEM 3 Graph: $y(x) = 2 - \sqrt[3]{x} - 1$

■ **Solution** See page S27.

8.3 Concept Review

Determine whether the following statements are always true, sometimes true, or never true.

1. The number 5 is not in the domain of $f(x) = \sqrt{4 - x}$.

2. The domain of $f(x) = \sqrt[3]{x - 2}$ is the set of all real numbers.

3. The domain of $f(x) = \dfrac{3}{x^2 + 4}$ is all the real numbers except -2 and 2.

4. If x is a real number, the expression $f(x) = \sqrt{-x^2}$ represents a radical function.

8.3 Exercises

1 Find the domain of a radical function

1. What is a radical function?

2. What is the difference between a radical function and a polynomial function?

3. **a.** When n is an even number, the domain of $f(x) = x^{1/n}$ is $\{x \mid x \underline{\quad ? \quad}\}$.
 b. When n is an odd number, the domain of $f(x) = x^{1/n}$ is $\{x \mid x \underline{\quad ? \quad}\}$.

4. To find the domain of $f(x) = 8 + \sqrt[4]{2x - 1}$, note that the index of the radical is even, so the function is defined only when $\underline{\quad ? \quad} \geq 0$. Solving this inequality for x gives $x \geq \underline{\quad ? \quad}$. In interval notation, the domain of f is $[\underline{\quad ? \quad}, \underline{\quad ? \quad})$.

5. State whether the domain of f is $(-\infty, 0]$, $[0, \infty)$, all real numbers, or the empty set.
 a. $f(x) = \sqrt{-x}$ **b.** $f(x) = \sqrt{-x^2}$
 c. $f(x) = -\sqrt{x^2}$ **d.** $f(x) = \sqrt{(-x)^2}$

6. **a.** True or false? The functions $f(x) = \sqrt[4]{3x} - 4$ and $g(x) = \sqrt[3]{3x} - 4$ have the same domain.

b. True or false? $f(x) = \sqrt{3}\, x^2$ is a radical function.

State the domain of each function in set-builder notation.

7. $f(x) = 2x^{\frac{1}{3}}$

8. $r(x) = -3\sqrt[5]{2x}$

9. $g(x) = -2\sqrt{x + 1}$

10. $h(x) = 3x^{\frac{1}{4}} - 2$

11. $f(x) = 2x\sqrt{x} - 3$

12. $y(x) = -3\sqrt[3]{1 + x}$

13. $C(x) = -3x^{\frac{3}{4}} + 1$

14. $G(x) = 6x^{\frac{2}{5}} + 5$

15. $F(x) = 4(3x - 6)^{\frac{1}{2}}$

State the domain of each function in interval notation.

16. $f(x) = -2(4x - 12)^{\frac{1}{2}}$

17. $g(x) = 2(2x - 10)^{\frac{2}{3}}$

18. $J(x) = 4 - (3x - 3)^{\frac{2}{5}}$

19. $V(x) = x - \sqrt{12 - 4x}$

20. $Y(x) = -6 + \sqrt{6 - x}$

21. $h(x) = 3\sqrt[4]{(x - 2)^3}$

22. $g(x) = \frac{2}{3}\sqrt[4]{(4 - x)^3}$

23. $f(x) = x - (4 - 6x)^{\frac{1}{2}}$

24. $F(x) = (9 + 12x)^{\frac{1}{2}} - 4$

2 Graph a radical function

For Exercises 25 to 28, use the function $f(x) = -\sqrt{x - 5}$.

25. The domain of f is $\{x \mid x\,_\,^?\,_\}$.

26. For what value of x will the radicand be equal to the perfect square 9?

27. $f(14) = -\sqrt{\,_\,^?\,_} - 5 = -\sqrt{\,_\,^?\,_} = \,_\,^?\,_$

28. Use the results of Exercise 27 to state a point that is on the graph of $f(x) = -\sqrt{x - 5}$.

Graph.

29. $F(x) = \sqrt{x}$

30. $G(x) = -\sqrt{x}$

31. $h(x) = -\sqrt[3]{x}$

32. $K(x) = \sqrt[3]{x} + 1$

33. $f(x) = -\sqrt{x + 2}$

34. $g(x) = \sqrt{x - 1}$

35. $S(x) = -\sqrt[4]{x}$

36. $C(x) = (x + 2)^{\frac{1}{4}}$

37. $F(x) = (x - 2)^{\frac{1}{2}}$

38. $f(x) = -(x - 1)^{\frac{1}{2}}$

39. $Q(x) = (x - 3)^{\frac{1}{3}}$

40. $H(x) = (-x)^{\frac{1}{3}}$

Graph.

41. $f(x) = 2x^{\frac{2}{5}} - 1$

42. $h(x) = 3x^{\frac{2}{5}} + 2$

43. $g(x) = 3 - (5 - 2x)^{\frac{1}{2}}$

44. $y(x) = 1 + (4 - 8x)^{\frac{1}{2}}$

45. $V(x) = \sqrt[5]{(x - 2)^2}$

46. $A(x) = x\sqrt{3x - 9}$

47. $F(x) = 2x - 3\sqrt[3]{x} - 1$

48. $f(x) = 2x - \sqrt{x} - 1$

49. $f(x) = 3x\sqrt{4x + 8}$

50. **a.** Which graph below could be the graph of $f(x) = -x^{1/2}$?
b. Which graph below could be the graph of $f(x) = (-x)^{1/2}$?
c. Which graph below could be the graph of $f(x) = -(-x)^{1/2}$?
d. Which graph below could be the graph of $f(x) = x^{1/2}$?

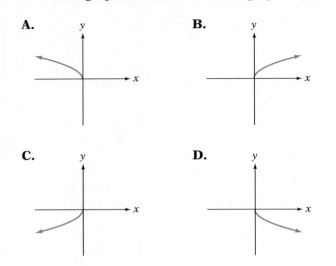

A.

B.

C.

D.

51. State whether the graph of the given function will be above the x-axis, below the x-axis, or both above and below the x-axis.
a. $f(x) = -x^{1/3}$ **b.** $f(x) = -\sqrt{x+2}$ **c.** $f(x) = \sqrt[4]{3x}$
d. $f(x) = (-2x)^{1/2}$

8.3 Applying Concepts

52. **Sports** Many major league baseball parks have a symmetric design as shown in the figure at the right. One decision that the designer must make is the shape of the outfield. One possible design uses the function

$$f(x) = k + (400 - k)\sqrt{1 - \frac{x^2}{a^2}}$$

to determine the shape of the outfield.
a. Graph this equation for $k = 0$, $a = 287$, and $-240 \le x \le 240$.
b. What is the maximum value of this function for the given interval?
c. The equation of the right-field foul line is $y = x$. Where does the foul line intersect the graph of f? That is, find the point on the graph of f for which $y = x$.
d. If the units on the axes are feet, what is the distance from home plate along the foul line to the base of the right-field wall?

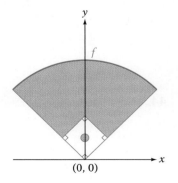

53. **Currency** According to the Bureau of Engraving and Printing, the average life span of different denominations of currency is as shown in the table at the right. The function that approximately models these data is $f(x) = 1.38x^{\frac{2}{5}}$, where x is the denomination of the bill and $f(x)$ is the average life span in years.

Currency	Average Life Span
$1 bill	1.8 years
$5 bill	2 years
$10 bill	3 years
$20 bill	4 years
$50 bill	9 years
$100 bill	9 years

 a. In the context of the problem, what is the domain of the function $f(x) = 1.38x^{\frac{2}{5}}$? Why is 0 not in the domain? Why are negative numbers not in the domain?

 b. If 2 were in the domain of the function, what life span would the model predict for the $2 bill? Round to the nearest tenth. Do you think this estimate is reasonable?

54. Writing the fraction $\frac{2}{4}$ in lowest terms as $\frac{1}{2}$, it appears that $(x^2)^{\frac{1}{4}} = x^{\frac{1}{2}}$.

Using $-10 \le x \le 10$, graph $f(x) = (x^2)^{\frac{1}{4}}$ and then graph $g(x) = x^{\frac{1}{2}}$.

Are the graphs the same? If they are, try graphing $g(x) = x^{\frac{1}{2}}$ first and then $f(x) = (x^2)^{\frac{1}{4}}$. Explain why the graphs are not the same. In your explanation, include why $x^{\frac{2}{4}} = x^{\frac{1}{2}}$ is not always a true statement.

8.4 Solving Equations Containing Radical Expressions

OBJECTIVE 1 **Solve equations containing one or more radical expressions**

An equation that contains a variable expression in a radicand is a **radical equation**.

$$\left.\begin{array}{l} \sqrt[3]{2x - 5} + x = 7 \\ \sqrt{x + 1} - \sqrt{x} = 4 \end{array}\right\} \begin{array}{l}\text{Radical}\\\text{equations}\end{array}$$

The following property is used to solve a radical equation.

Take Note

You might assume that the converse of this is also true. That is, if $a^2 = b^2$, then $a = b$. But this is not the case. For example, if $a = -4$ and $b = 4$, then $(-4)^2 = 4^2$ but $-4 \ne 4$.

Property of Raising Each Side of an Equation to a Power

If two numbers are equal, then the same powers of the numbers are equal.

If $a = b$, then $a^n = b^n$.

Solve: $\sqrt{x-2} - 6 = 0$

Rewrite the equation with the radical on one side of the equation and the constant on the other side.

$\sqrt{x-2} - 6 = 0$
$\sqrt{x-2} = 6$

Square each side of the equation.

$(\sqrt{x-2})^2 = 6^2$

Solve the resulting equation.

$x - 2 = 36$
$x = 38$

Check the solution.

Check: $\sqrt{x-2} - 6 = 0$

$\sqrt{38-2} - 6$	0
$\sqrt{36} - 6$	0
$6 - 6$	0
	$0 = 0$

38 checks as a solution.
The solution is 38.

EXAMPLE 1 Solve. **A.** $\sqrt{3x-2} - 8 = -3$ **B.** $\sqrt[3]{3x-1} = -4$

Solution **A.** $\sqrt{3x-2} - 8 = -3$
$\sqrt{3x-2} = 5$

• Rewrite the equation so that the radical is alone on one side of the equation.

$(\sqrt{3x-2})^2 = 5^2$
$3x - 2 = 25$
$3x = 27$
$x = 9$

• Square each side of the equation.
• Solve the resulting equation.

Check:

$\sqrt{3x-2} - 8 = -3$

$\sqrt{3 \cdot 9 - 2} - 8$	-3
$\sqrt{27 - 2} - 8$	-3
$\sqrt{25} - 8$	-3
$5 - 8$	-3
	$-3 = -3$

• Check the solution.

The solution is 9.

B. $\sqrt[3]{3x-1} = -4$
$(\sqrt[3]{3x-1})^3 = (-4)^3$
$3x - 1 = -64$
$3x = -63$
$x = -21$

• Cube each side of the equation.
• Solve the resulting equation.

Check:

$\sqrt[3]{3x-1} = -4$

$\sqrt[3]{3(-21) - 1}$	-4
$\sqrt[3]{-63 - 1}$	-4
$\sqrt[3]{-64}$	-4
	$-4 = -4$

• Check the solution.

The solution is -21.

> **PROBLEM 1** Solve. **A.** $\sqrt{4x + 5} - 12 = -5$ **B.** $\sqrt[4]{x - 8} = 3$
>
> ■ **Solution** See page S28.

When you raise both sides of an equation to an even power, the resulting equation may have a solution that is not a solution of the original equation. Therefore, it is necessary to check the solution of a radical equation.

> **EXAMPLE 2** Solve. **A.** $x + 2\sqrt{x - 1} = 9$ **B.** $\sqrt{x + 7} = \sqrt{x} + 1$
>
> **Solution** **A.** $x + 2\sqrt{x - 1} = 9$
>
> $$2\sqrt{x - 1} = 9 - x$$
>
> • Rewrite the equation with the radical on one side of the equation.
>
> $$\left(2\sqrt{x - 1}\right)^2 = (9 - x)^2$$
> $$4(x - 1) = 81 - 18x + x^2$$
>
> • Square each side of the equation.
>
> $$4x - 4 = 81 - 18x + x^2$$
> $$0 = x^2 - 22x + 85$$
>
> • Write the quadratic equation in standard form.
>
> $$0 = (x - 5)(x - 17)$$
>
> • Factor.
>
> $$x - 5 = 0 \qquad x - 17 = 0$$
> $$x = 5 \qquad\qquad x = 17$$
>
> • Use the Principle of Zero Products.

> **Check:**
>
$x + 2\sqrt{x - 1} = 9$		$x + 2\sqrt{x - 1} = 9$	
> | $5 + 2\sqrt{5 - 1}$ | 9 | $17 + 2\sqrt{17 - 1}$ | 9 |
> | $5 + 2\sqrt{4}$ | 9 | $17 + 2\sqrt{16}$ | 9 |
> | $5 + 2 \cdot 2$ | 9 | $17 + 2 \cdot 4$ | 9 |
> | $5 + 4$ | 9 | $17 + 8$ | 9 |
> | $9 = 9$ | | $25 \neq 9$ | |

> 17 does not check as a solution. It is called an **extraneous solution** of the equation.
>
> The solution is 5.

> **B.** $\sqrt{x + 7} = \sqrt{x} + 1$
>
> • A radical appears on each side of the equation.
>
> $$\left(\sqrt{x + 7}\right)^2 = \left(\sqrt{x} + 1\right)^2$$
>
> • Square each side of the equation.
>
> $$x + 7 = x + 2\sqrt{x} + 1$$
>
> $$6 = 2\sqrt{x}$$
>
> • Simplify the resulting equation.
>
> $$3 = \sqrt{x}$$
>
> • The equation contains a radical.
>
> $$3^2 = \left(\sqrt{x}\right)^2$$
>
> • Square each side of the equation.
>
> $$9 = x$$

> **Check:**
>
> • Check the solution.
>
$\sqrt{x + 7} = \sqrt{x} + 1$	
> | $\sqrt{9 + 7}$ | $\sqrt{9} + 1$ |
> | $\sqrt{16}$ | $3 + 1$ |
> | $4 = 4$ | |
>
> The solution is 9.

> **PROBLEM 2** Solve. **A.** $x + 3\sqrt{x + 2} = 8$ **B.** $\sqrt{x + 5} = 5 - \sqrt{x}$
>
> ■ **Solution** See page S28

OBJECTIVE **2** **Application problems**

A right triangle contains one 90° angle. The side opposite the 90° angle is called the **hypotenuse.** The other two sides are called **legs.**

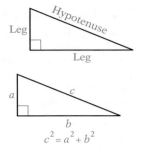

Pythagoras, a Greek mathematician, is credited with the discovery that the square of the hypotenuse of a right triangle is equal to the sum of the squares of the two legs. This is called the Pythagorean Theorem.

Pythagoras
(c. 580 B.C.–529 B.C.)

Pythagorean Theorem

The square of the hypotenuse, c, of a right triangle is equal to the sum of the squares of the two legs, a and b.

$$c^2 = a^2 + b^2$$

EXAMPLE 3 A ladder 20 ft long is leaning against a building. How high on the building will the ladder reach when the bottom of the ladder is 8 ft from the building? Round to the nearest tenth.

20 ft

8 ft

Strategy To find the distance, use the Pythagorean Theorem. The hypotenuse is the length of the ladder. One leg is the distance from the bottom of the ladder to the base of the building. The distance along the building from the ground to the top of the ladder is the unknown leg.

Solution
$$c^2 = a^2 + b^2$$
$$20^2 = 8^2 + b^2$$
$$400 = 64 + b^2$$
$$336 = b^2$$
$$(336)^{\frac{1}{2}} = (b^2)^{\frac{1}{2}}$$
$$\sqrt{336} = b$$
$$18.3 \approx b$$

The distance is 18.3 ft.

PROBLEM 3 Find the diagonal of a rectangle that is 6 cm in length and 3 cm in width. Round to the nearest tenth.

Solution See page S28.

 Concept Review

Determine whether the following statements are always true, sometimes true, or never true.

1. If $a^n = b^n$, then $a = b$.

2. When you raise both sides of an equation to an even power, the resulting equation has a solution that is not a solution of the original equation.

3. If two sides of a triangle are known, the third side can be found by using the Pythagorean Theorem.

4. The solution of $\sqrt[3]{-x} = 8$ is -512.

 Exercises

1 **Solve equations containing one or more radical expressions**

1. Which of the equations $\sqrt{2}\,x + 1 = 7$ and $\sqrt{2x + 1} = 7$ is a radical equation?

2. Solve: $\sqrt[3]{x} - 3 = -4$

$$\sqrt[3]{x} - 3 = -4$$
$$(\sqrt[3]{x} - 3)^{\underline{?}} = (-4)^{\underline{?}}$$ • **Cube each side of the equation.**
$$\underline{\quad?\quad} = \underline{\quad?\quad}$$ • **Simplify.**
$$x = \underline{\quad?\quad}$$ • **Add** $\underline{\quad?\quad}$ **to each side of the equation.**

3. What is the first step when solving $\sqrt{x} + 3 = 9$?

4. Why is it necessary to check the proposed solutions of a radical equation?

Solve.

5. $\sqrt{x} = 5$

6. $\sqrt{y} = 2$

7. $\sqrt[3]{a} = 3$

8. $\sqrt[3]{y} = 5$

9. $\sqrt{3x} = 12$

10. $\sqrt{5x} = 10$

11. $\sqrt[3]{4x} = -2$

12. $\sqrt[3]{6x} = -3$

13. $\sqrt{2x} = -4$

14. $\sqrt{5x} = -5$

15. $\sqrt{3x - 2} = 5$

16. $\sqrt{5x - 4} = 9$

17. $\sqrt{3 - 2x} = 7$

18. $\sqrt{9 - 4x} = 4$

19. $7 = \sqrt{1 - 3x}$

20. $6 = \sqrt{8 - 7x}$

21. $\sqrt[3]{4x - 1} = 2$

22. $\sqrt[3]{5x + 2} = 3$

23. $\sqrt[3]{1 - 2x} = -3$

24. $\sqrt[3]{3 - 2x} = -2$

25. $\sqrt[3]{9x + 1} = 4$

26. $\sqrt{3x + 9} - 12 = 0$

27. $\sqrt{4x - 3} - 5 = 0$

28. $\sqrt{x - 2} = 4$

29. $\sqrt[3]{x - 3} + 5 = 0$

30. $\sqrt[3]{x - 2} = 3$

31. $\sqrt[3]{2x - 6} = 4$

32. $\sqrt[4]{4x + 1} = 2$

33. $\sqrt[4]{2x - 9} = 3$

34. $\sqrt{2x - 3} - 2 = 1$

35. $\sqrt{3x - 5} - 5 = 3$

36. $\sqrt[3]{2x - 3} + 5 = 2$

37. $\sqrt[3]{x - 4} + 7 = 5$

38. $\sqrt{5x - 16} + 1 = 4$

39. $\sqrt{3x - 5} - 2 = 3$

40. $\sqrt{2x - 1} - 8 = -5$

41. $\sqrt{7x + 2} - 10 = -7$

42. $\sqrt[3]{4x - 3} - 2 = 3$

43. $\sqrt[3]{1 - 3x} + 5 = 3$

44. $1 - \sqrt{4x + 3} = -5$

45. $7 - \sqrt{3x + 1} = -1$

46. $\sqrt{x + 1} = 2 - \sqrt{x}$

47. $\sqrt{2x + 4} = 3 - \sqrt{2x}$

48. $\sqrt{x^2 + 3x - 2} - x = 1$

49. $\sqrt{x^2 - 4x - 1} + 3 = x$

50. $\sqrt{x^2 - 3x - 1} = 3$

51. $\sqrt{x^2 - 2x + 1} = 3$

52. $\sqrt{2x + 5} - \sqrt{3x - 2} = 1$

53. $\sqrt{4x + 1} - \sqrt{2x + 4} = 1$

54. $\sqrt{5x - 1} - \sqrt{3x - 2} = 1$

54. $\sqrt{5x + 4} - \sqrt{3x + 1} = 1$

56. $\sqrt[4]{x^2 + 2x + 8} - 2 = 0$

57. $\sqrt[4]{x^2 + x - 1} - 1 = 0$

58. $4\sqrt{x + 1} - x = 1$

59. $3\sqrt{x - 2} + 2 = x$

60. $x + 3\sqrt{x - 2} = 12$

61. $x + 2\sqrt{x + 1} = 7$

62. 🐢 Without actually solving the equations, identify which equation has no solution.

 a. $\sqrt{4x - 8} = -5$ **b.** $\sqrt[3]{4x - 8} = -5$

 c. $\sqrt{4x} - 8 = -5$ **d.** $-\sqrt{4x - 8} = -5$

63. 🐢 How many times will the Property of Raising Each Side of an Equation to a Power be used to solve the equation? Do not solve.

 a. $\sqrt{x - 8} = \sqrt{2x - 1}$ **b.** $\sqrt{x} - 8 = \sqrt{2x} - 1$

2 Application problems

64. The equation $v = 8\sqrt{d}$ gives the speed v, in feet per second, of a dropped object after the object has fallen d feet. A stone is dropped from a bridge. To find the distance the stone has fallen when it reaches a speed of 56 ft/s, replace ____?____ with 56 and solve for ____?____.

65. A 15-foot ladder leans against the side of a building with its bottom d feet from the building. The ladder reaches a height of h feet. Which of the following distances is *not* a possible value for h?
a. 2 ft **b.** 8 ft **c.** 14 ft **d.** 20 ft

66. Sports The equation $s = 16.97\sqrt[9]{n}$ can be used to predict the maximum speed, s (in feet per second), of n rowers on a scull. Find, to the nearest whole number, how many rowers are needed to travel at 20 ft/s. Does doubling the number of rowers double the maximum speed of the scull?

67. Meteorology The sustained wind velocity, v (in meters per second), in a hurricane is given by $v = 6.3\sqrt{1013 - p}$, where p is the air pressure in millibars (mb). If the velocity of the wind in a hurricane is 64 m/s, what is the air pressure? Round to the nearest tenth. What happens to wind speed in a hurricane as air pressure decreases?

68. Astronomy The time, T (in days), that it takes a planet to revolve around the sun can be approximated by the equation $T = 0.407\sqrt{d^3}$, where d is the mean distance of the planet from the sun in millions of miles. It takes Venus approximately 226 days to complete one revolution of the sun. To the nearest million miles, what is the mean distance of Venus from the sun?

69. Astronomy The time, T (in days), that it takes a moon of Saturn to revolve around Saturn can be approximated by the equation $T = 0.373\sqrt{d^3}$, where d is the mean distance of the moon from Saturn in units of 100,000 km. It takes the moon Tethys approximately 1.89 days to complete one revolution of Saturn. To the nearest 1000 km, what is the mean distance of Tethys from Saturn?

70. Construction The maximum velocity of a roller coaster depends on the vertical drop from the top of the highest hill to the bottom of that hill. The formula $v = 8\sqrt{h}$ gives the relationship between maximum velocity, v (in feet per second), and height, h (in feet). The maximum velocity of the Magnum XL-200 roller coaster in Sandusky, Ohio, is approximately 114 ft/s. How tall, to the nearest foot, is the highest hill for this roller coaster?

71. Health Science The number of calories an animal uses per day (called the metabolic rate of the animal) can be approximated by $M = 126.4\sqrt[4]{W^3}$, where M is the metabolic rate and W is the weight of the animal in pounds. Find, to the nearest hundred pounds, the weight of an elephant whose metabolic rate is 60,000 calories per day.

72. Construction A 12-foot ladder is leaning against a building. How high on the building will the ladder reach when the bottom of the ladder is 4 ft from the building? Round to the nearest tenth.

73. Construction A 26-foot ladder is leaning against a building. How far is the bottom of the ladder from the wall when the ladder reaches a height of 24 ft on the building?

74. **Uniform Motion** Two joggers start from the same point and travel in the directions shown in the figure at the right. One jogger is traveling 8 ft/s, and the second jogger is traveling 9 ft/s. What is the distance between the joggers after 1 min? Round to the nearest tenth.

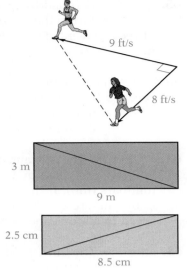

75. **Uniform Motion** Two joggers leave the same point at the same time, one jogging east at 3 m/s and the other jogging south at 3.5 m/s. What is the distance between the joggers after 3 min? Round to the nearest hundredth.

76. **Geometry** The diagonal of a rectangle is a line drawn from one vertex to the opposite vertex. Find the length of the diagonal in the rectangle shown at the right. Round to the nearest tenth.

3 m

9 m

77. **Geometry** The diagonal of a rectangle is a line drawn from one vertex to the opposite vertex. Find the length of the diagonal in the rectangle shown at the right. Round to the nearest tenth.

2.5 cm

8.5 cm

78. **Oceanography** How far would a submarine periscope have to be above the water to locate a ship 3.2 mi away? The equation for the distance in miles that the lookout can see is $d = \sqrt{1.5h}$, where h is the height in feet above the surface of the water. Round to the nearest hundredth.

79. **Oceanography** How far would a submarine periscope have to be above the water to locate a ship 3.5 mi away? The equation for the distance in miles that the lookout can see is $d = \sqrt{1.5h}$, where h is in height in feet above the surface of the water. Round to the nearest hundredth.

80. **Physics** An object is dropped from a bridge. Find the distance the object has fallen when the speed reaches 80 ft/s. Use the equation $v = 8\sqrt{d}$, where v is the speed of the object and d is the distance.

$v = 80$ ft/s

81. **Physics** An object is dropped from a high building. Find the distance the object has fallen when the speed reaches 120 ft/s. Use the equation $v = 8\sqrt{d}$, where v is the speed of the object and d is the distance.

82. **Automotive Technology** Find the distance required for a car to reach a velocity of 60 m/s when the acceleration is 10 m/s². Use the equation $v = \sqrt{2as}$, where v is the velocity, a is the acceleration, and s is the distance.

83. **Automotive Technology** Find the distance required for a car to reach a velocity of 48 ft/s when the acceleration is 12 ft/s². Use the equation $v = \sqrt{2as}$, where v is the velocity, a is the acceleration, and s is the distance.

84. **Clocks** Find the length of a pendulum that makes one swing in 3 s. The equation for the time of one swing of a pendulum is $T = 2\pi\sqrt{\dfrac{L}{32}}$, where T is the time in seconds and L is the length in feet. Round to the nearest hundredth.

2.4 s

85. Satellites At what height above Earth's surface is a satellite in orbit if it is traveling at a speed of 7500 m/s? Use the equation $v = \sqrt{\dfrac{4 \times 10^{14}}{h + 6.4 \times 10^{6}}}$, where v is the speed of the satellite in meters per second and h is the height above Earth's surface in meters. Round to the nearest thousand.

8.4 Applying Concepts

Solve.

86. $x^{\frac{3}{4}} = 8$

87. $x^{\frac{2}{3}} = 9$

88. $x^{\frac{5}{4}} = 32$

Solve the formula for the given variable.

89. $v = \sqrt{64d}$; d

90. $a^2 + b^2 = c^2$; a

91. $V = \pi r^2 h$; r

92. $V = \dfrac{4}{3} \pi r^3$; r

Solve.

93. $\sqrt{3x - 2} = \sqrt{2x - 3} + \sqrt{x - 1}$

94. $\sqrt{2x + 3} + \sqrt{x + 2} = \sqrt{x + 5}$

95. Geometry A box has a base that measures 4 in. by 6 in. The height of the box is 3 in. Find the greatest distance between two corners. Round to the nearest hundredth.

96. Number Problem Find three odd integers a, b, and c such that $a^2 + b^2 = c^2$.

97. Geometry Find the length of the side labeled x.

98. If a and b are nonnegative real numbers, is the equation always true, sometimes true, or never true?

$$\sqrt{a^2 + b^2} = a + b$$

Write a paragraph that supports your answer.

99. If a and b are both positive real numbers and $a > b$, is a^b less than, equal to, or greater than b^a? Does your answer depend on a and b?

 Complex Numbers

OBJECTIVE **1** **Simplify complex numbers**

The radical expression $\sqrt{-4}$ is not a real number because there is no real number whose square is -4. However, the solution of an algebraic equation is sometimes the square root of a negative number.

For example, the equation $x^2 + 1 = 0$ does not have a real number solution, because there is no real number whose square is a negative number.

$$x^2 + 1 = 0$$
$$x^2 = -1$$

During the late 17th century, a new number, called an **imaginary number,** was defined so that a negative number would have a square root. The letter i was chosen to represent the number whose square is -1.

$$i^2 = -1$$

An imaginary number is defined in terms of i.

Definition of $\sqrt{-a}$

If a is a positive real number, then the principal square root of negative a is the imaginary number $i\sqrt{a}$.

$$\sqrt{-a} = i\sqrt{a}$$

Here are some examples:

$$\sqrt{-16} = i\sqrt{16} = 4i$$
$$\sqrt{-12} = i\sqrt{12} = 2i\sqrt{3}$$
$$\sqrt{-21} = i\sqrt{21}$$
$$\sqrt{-1} = i\sqrt{1} = i$$

It is customary to write i in front of a radical to avoid confusing $\sqrt{a}\,i$ with $\sqrt{ai}$.

The real numbers and imaginary numbers make up the complex numbers.

Definition of a Complex Number

A **complex number** is a number of the form $a + bi$, where a and b are real numbers and $i = \sqrt{-1}$. The number a is the **real part** of $a + bi$, and b is the **imaginary part.**

Take Note

The imaginary part of a complex number is a real number. As another example, the imaginary part of $6 - 8i$ is -8.

Examples of complex numbers are shown at the right.

Real part⌐ ⌐Imaginary part

$$a + bi$$
$$3 + 2i$$
$$8 - 10i$$

Complex numbers
$a + bi$
— Real numbers
$a + 0i$

A **real number** is a complex number in which $b = 0$.

— Imaginary numbers
$0 + bi$

An **imaginary number** is a complex number in which $a = 0$.

EXAMPLE 1 Simplify: $\sqrt{-80}$

Solution $\sqrt{-80} = i\sqrt{80} = i\sqrt{16 \cdot 5} = 4i\sqrt{5}$

PROBLEM 1 Simplify: $\sqrt{-45}$

Solution See page S28.

Simplify: $\sqrt{20} - \sqrt{-50}$

Write the complex number in the form $a + bi$.

$\sqrt{20} - \sqrt{-50} = \sqrt{20} - i\sqrt{50}$

Use the Product Property of Radicals to simplify each radical.

$= \sqrt{4 \cdot 5} - i\sqrt{25 \cdot 2}$
$= 2\sqrt{5} - 5i\sqrt{2}$

EXAMPLE 2 Simplify: $\sqrt{25} + \sqrt{-40}$

Solution $\sqrt{25} + \sqrt{-40} = \sqrt{25} + i\sqrt{40} = \sqrt{25} + i\sqrt{4 \cdot 10}$
$= 5 + 2i\sqrt{10}$

PROBLEM 2 Simplify: $\sqrt{98} - \sqrt{-60}$

Solution See page S28.

OBJECTIVE **2** **Add and subtract complex numbers**

Addition and Subtraction of Complex Numbers

To add two complex numbers, add the real parts and add the imaginary parts. To subtract two complex numbers, subtract the real parts and subtract the imaginary parts.

$$(a + bi) + (c + di) = (a + c) + (b + d)i$$
$$(a + bi) - (c + di) = (a - c) + (b - d)i$$

EXAMPLE 3 Add: $(3 + 2i) + (6 - 5i)$

Solution $(3 + 2i) + (6 - 5i)$
$= (3 + 6) + (2 - 5)i$ • Add the real parts and add the imaginary
$= 9 - 3i$ parts.

PROBLEM 3 Subtract: $(-4 + 2i) - (6 - 8i)$

Solution See page S28.

Add: $(3 + \sqrt{-12}) + (7 - \sqrt{-27})$

Write each complex number in the form $a + bi$.

Use the Product Property of Radicals to simplify each radical.

Add the complex numbers.

$(3 + \sqrt{-12}) + (7 - \sqrt{-27})$
$= (3 + i\sqrt{12}) + (7 - i\sqrt{27})$
$= (3 + i\sqrt{4 \cdot 3}) + (7 - i\sqrt{9 \cdot 3})$
$= (3 + 2i\sqrt{3}) + (7 - 3i\sqrt{3})$

$= 10 - i\sqrt{3}$

EXAMPLE 4 Subtract: $(9 - \sqrt{-8}) - (5 + \sqrt{-32})$

Solution $(9 - \sqrt{-8}) - (5 + \sqrt{-32})$
$= (9 - i\sqrt{8}) - (5 + i\sqrt{32})$ • Write each complex number in the form $a + bi$.

$= (9 - i\sqrt{4 \cdot 2}) - (5 + i\sqrt{16 \cdot 2})$ • Simplify each radical.
$= (9 - 2i\sqrt{2}) - (5 + 4i\sqrt{2})$

$= 4 - 6i\sqrt{2}$ • Subtract.

PROBLEM 4 Subtract: $(16 - \sqrt{-45}) - (3 + \sqrt{-20})$

Solution See page S28.

OBJECTIVE 3 ## Multiply complex numbers

When we multiply complex numbers, the term i^2 is frequently a part of the product. Recall that $i^2 = -1$.

Multiply: $2i \cdot 3i$

Multiply the imaginary numbers. $2i \cdot 3i = 6i^2$

Replace i^2 by -1. Then simplify. $= 6(-1) = -6$

When multiplying square roots of negative numbers, first rewrite the radical expressions using i.

Multiply: $\sqrt{-6} \cdot \sqrt{-24}$

Write each radical as the product of a real number and i. $\sqrt{-6} \cdot \sqrt{-24} = i\sqrt{6} \cdot i\sqrt{24}$

Multiply the imaginary numbers. $= i^2\sqrt{144}$

Replace i^2 by -1. $= -\sqrt{144}$

Simplify. $= -12$

Take Note

This example illustrates an important point. When working with a number that has a square root of a negative number, always rewrite the number with i before continuing.

Note from this example that it would have been incorrect to multiply the radicands of the two radical expressions. To illustrate,

$$\sqrt{-6} \cdot \sqrt{-24} = \sqrt{(-6)(-24)} = \sqrt{144} = 12, \textit{ not } -12$$

Multiply: $4i(3 - 2i)$

Use the Distributive Property to remove parentheses.

Replace i^2 by -1.

Write the answer in the form $a + bi$.

$$4i(3 - 2i) = 12i - 8i^2$$
$$= 12i - 8(-1)$$
$$= 8 + 12i$$

EXAMPLE 5 Multiply: $\sqrt{-8}\left(\sqrt{6} - \sqrt{-2}\right)$

Solution $\sqrt{-8}\left(\sqrt{6} - \sqrt{-2}\right)$

$= i\sqrt{8}\left(\sqrt{6} - i\sqrt{2}\right)$ • **Write each complex number in the form $a + bi$.**

$= i\sqrt{48} - i^2\sqrt{16}$ • **Use the Distributive Property.**

$= i\sqrt{16 \cdot 3} - (-1)\sqrt{16}$ • **Simplify each radical. Replace i^2 by -1.**

$= 4i\sqrt{3} + 4$

$= 4 + 4i\sqrt{3}$ • **Write the answer in the form $a + bi$.**

PROBLEM 5 Multiply: $\sqrt{-3}\left(\sqrt{27} - \sqrt{-6}\right)$

Solution See page S28.

The product of two complex numbers can be found by using the FOIL method. For example,

$$(2 + 4i)(3 - 5i) = 6 - 10i + 12i - 20i^2$$
$$= 6 + 2i - 20i^2$$
$$= 6 + 2i - 20(-1)$$
$$= 26 + 2i$$

The conjugate of $a + bi$ is $a - bi$.

The product of conjugates of the form $(a + bi)(a - bi)$ is $a^2 + b^2$.

$$(\boldsymbol{a + bi})(\boldsymbol{a - bi}) = a^2 - b^2i^2 = a^2 - b^2(-1) = a^2 + b^2$$

For example, $(2 + 3i)(2 - 3i) = 2^2 + 3^2 = 4 + 9 = 13$.

Note that the product of a complex number and its conjugate is a real number.

EXAMPLE 6 Multiply.

A. $(3 - 4i)(2 + 5i)$ **B.** $\left(\dfrac{9}{10} + \dfrac{3}{10}i\right)\left(1 - \dfrac{1}{3}i\right)$

C. $(4 + 5i)(4 - 5i)$ **D.** $(6 + i)^2$

Solution **A.** $(3 - 4i)(2 + 5i)$

$= 6 + 15i - 8i - 20i^2$ • **Use the FOIL method.**

$= 6 + 7i - 20i^2$ • **Combine like terms.**

$= 6 + 7i - 20(-1)$ • **Replace i^2 by -1.**

$= 26 + 7i$ • **Write the answer in the form $a + bi$.**

B. $\left(\dfrac{9}{10} + \dfrac{3}{10}i\right)\left(1 - \dfrac{1}{3}i\right)$

$= \dfrac{9}{10} - \dfrac{3}{10}i + \dfrac{3}{10}i - \dfrac{1}{10}i^2$ • Use the FOIL method.

$= \dfrac{9}{10} - \dfrac{1}{10}i^2$ • Combine like terms.

$= \dfrac{9}{10} - \dfrac{1}{10}(-1)$ • Replace i^2 by -1.

$= \dfrac{9}{10} + \dfrac{1}{10} = 1$ • Simplify.

C. $(4 + 5i)(4 - 5i) = 4^2 + 5^2$ • The product of conjugates of
$\qquad\qquad\qquad\quad = 16 + 25$ the form $(a + bi)(a - bi)$ is
$\qquad\qquad\qquad\quad = 41$ $a^2 + b^2$.

D. $(6 + i)^2 = 36 + 12i + i^2$ • $(6 + i)^2 = (6 + i)(6 + i)$
$\qquad\qquad\quad = 36 + 12i + (-1)$
$\qquad\qquad\quad = 35 + 12i$

PROBLEM 6 Multiply.

 A. $(4 - 3i)(2 - i)$ **B.** $(3 - i)\left(\dfrac{3}{10} + \dfrac{1}{10}i\right)$

 C. $(3 + 6i)(3 - 6i)$ **D.** $(1 + 5i)^2$

 ■ **Solution** See pages S28–S29.

OBJECTIVE 4 Divide complex numbers

A fraction containing one or more complex numbers is in simplest form
when no imaginary number remains in the denominator.

Simplify: $\dfrac{4 - 5i}{2i}$

Multiply the expression by 1 in the form $\dfrac{i}{i}$. $\dfrac{4 - 5i}{2i} = \dfrac{4 - 5i}{2i} \cdot \dfrac{i}{i}$

$\qquad\qquad\qquad\qquad\qquad\qquad\qquad\qquad\quad = \dfrac{4i - 5i^2}{2i^2}$

Replace i^2 by -1. $= \dfrac{4i - 5(-1)}{2(-1)}$

Simplify. $= \dfrac{5 + 4i}{-2}$

Write the answer in the form $a + bi$. $= -\dfrac{5}{2} - 2i$

EXAMPLE 7 Simplify: $\dfrac{5 + 4i}{3i}$

Solution $\dfrac{5 + 4i}{3i} = \dfrac{5 + 4i}{3i} \cdot \dfrac{i}{i} = \dfrac{5i + 4i^2}{3i^2} = \dfrac{5i + 4(-1)}{3(-1)} = \dfrac{-4 + 5i}{-3}$

$\qquad\qquad\quad = \dfrac{4}{3} - \dfrac{5}{3}i$

PROBLEM 7 Simplify: $\dfrac{2 - 3i}{4i}$

■ **Solution** See page S29.

Simplify: $\dfrac{3 + 2i}{1 + i}$

To simplify a fraction that has a complex number in the denominator, multiply the numerator and denominator by the conjugate of the complex number. The conjugate of $1 + i$ is $1 - i$.

$\dfrac{3 + 2i}{1 + i} = \dfrac{(3 + 2i)}{(1 + i)} \cdot \dfrac{(1 - i)}{(1 - i)}$

$= \dfrac{3 - 3i + 2i - 2i^2}{1^2 + 1^2}$

$= \dfrac{3 - i - 2(-1)}{2}$

$= \dfrac{5 - i}{2} = \dfrac{5}{2} - \dfrac{1}{2}i$

EXAMPLE 8 Simplify: $\dfrac{5 - 3i}{4 + 2i}$

Solution $\dfrac{5 - 3i}{4 + 2i} = \dfrac{(5 - 3i)}{(4 + 2i)} \cdot \dfrac{(4 - 2i)}{(4 - 2i)} = \dfrac{20 - 10i - 12i + 6i^2}{4^2 + 2^2}$

$= \dfrac{20 - 22i + 6(-1)}{20}$

$= \dfrac{14 - 22i}{20}$

$= \dfrac{7 - 11i}{10}$

$= \dfrac{7}{10} - \dfrac{11}{10}i$

PROBLEM 8 Simplify: $\dfrac{2 + 5i}{3 - 2i}$

■ **Solution** See page S29.

8.5 Concept Review

Determine whether the following statements are always true, sometimes true, or never true.

1. The product of two imaginary numbers is a real number.

2. The sum of two complex numbers is a real number.

3. A complex number is a number of the form $a + bi$, where a and b are real numbers and $\sqrt{-1} = i$.

4. The product of two complex numbers is a real number.

8.5 Exercises

1 Simplify complex numbers

1. What is an imaginary number? What is a complex number?

2. Are all real numbers also complex numbers? Are all complex numbers also real numbers?

3. $\sqrt{-1} = i$, so $i^2 =$ ____?____ .

4. **a.** The real part of the complex number $7 - 9i$ is ____?____ .
 b. The imaginary part of the complex number $7 - 9i$ is ____?____ .

5. Simplify: $\sqrt{-54}$

$\sqrt{-54} =$ ____?____ $\sqrt{54} = i\sqrt{(\underline{\ \ ?\ \ })(\underline{\ \ ?\ \ })} =$ ____?____ $\sqrt{6}$

6. Simplify: $\sqrt{-100x^{16}}$

$\sqrt{-100x^{16}} =$ ____?____ $\sqrt{100x^{16}} =$ ____?____ i

7. Is $10 - 3i$ equivalent to $10 + \sqrt{-9}$ or to $10 - \sqrt{-9}$?

8. Which of the following are equivalent to $\sqrt{25} + \sqrt{-25}$?

 a. $5 - 5i$ **b.** $5 + 5i$ **c.** $\sqrt{25} + i\sqrt{25}$ **d.** $\sqrt{25} - i\sqrt{25}$

Simplify.

9. $\sqrt{-4}$

10. $\sqrt{-64}$

11. $\sqrt{-98}$

12. $\sqrt{-72}$

13. $\sqrt{-27}$

14. $\sqrt{-75}$

15. $\sqrt{16} + \sqrt{-4}$

16. $\sqrt{25} + \sqrt{-9}$

17. $\sqrt{12} - \sqrt{-18}$

18. $\sqrt{60} - \sqrt{-48}$

19. $\sqrt{160} - \sqrt{-147}$

20. $\sqrt{96} - \sqrt{-125}$

21. $\sqrt{-4a^2}$

22. $\sqrt{-16b^6}$

23. $\sqrt{-49x^{12}}$

24. $\sqrt{-32x^3y^2}$

25. $\sqrt{-144a^3b^5}$

26. $\sqrt{-18a^{10}b^9}$

27. $\sqrt{4a} + \sqrt{-12a^2}$

28. $\sqrt{25b} - \sqrt{-48b^2}$

29. $\sqrt{18b^5} - \sqrt{-27b^3}$

30. $\sqrt{a^5b^2} - \sqrt{-a^5b^2}$

31. $\sqrt{-50x^3y^3} + x\sqrt{25x^4y^3}$

32. $\sqrt{-121xy} + \sqrt{60x^2y^2}$

33. $\sqrt{-49a^5b^2} - ab\sqrt{-25a^3}$

34. $\sqrt{-16x^2y} - x\sqrt{-49y}$

35. $\sqrt{12a^3} + \sqrt{-27b^3}$

2 Add and subtract complex numbers

36. Add: $(7 - 9i) + (-3 + 4i)$

$(7 - 9i) + (-3 + 4i)$

$= (7 + \underline{\quad ? \quad}) + (-9 + \underline{\quad ? \quad})i$ • Add the real parts. Add the imaginary parts.

$= \underline{\quad ? \quad}$ • Simplify.

37. Subtract: $12 - (6 - \sqrt{-64})$

$12 - (6 - \sqrt{-64}) = 12 - (6 - \underline{\quad ? \quad}i)$ • Write the complex number in the form $a + bi$.

$= (12 - \underline{\quad ? \quad}) - (\underline{\quad ? \quad})$ • Subtract the real parts. Subtract the imaginary part.

$= \underline{\quad ? \quad} + \underline{\quad ? \quad}$ • Simplify.

Add or subtract.

38. $(2 + 4i) + (6 - 5i)$

39. $(6 - 9i) + (4 + 2i)$

40. $(-2 - 4i) - (6 - 8i)$

41. $(3 - 5i) + (8 - 2i)$

42. $\left(8 - \sqrt{-4}\right) - \left(2 + \sqrt{-16}\right)$

43. $\left(5 - \sqrt{-25}\right) - \left(11 - \sqrt{-36}\right)$

44. $\left(12 - \sqrt{-50}\right) + \left(7 - \sqrt{-8}\right)$

45. $\left(5 - \sqrt{-12}\right) - \left(9 + \sqrt{-108}\right)$

46. $\left(\sqrt{8} + \sqrt{-18}\right) + \left(\sqrt{32} - \sqrt{-72}\right)$

47. $\left(\sqrt{40} - \sqrt{-98}\right) - \left(\sqrt{90} + \sqrt{-32}\right)$

48. $(5 - 3i) + 2i$

49. $(6 - 8i) + 4i$

50. $(7 + 2i) + (-7 - 2i)$

51. $(8 - 3i) + (-8 + 3i)$

52. $(9 + 4i) + 6$

53. $(4 + 6i) + 7$

For Exercises 54 to 57, use the complex numbers $m = a + bi$ and $n = c - di$, where a, b, c, and d are all positive real numbers.

54. True or false? If $a = c$ and $b = d$, then $m - n = 2bi$.

55. True or false? If $m - n$ is a real number, then $b + d = 0$.

56. Suppose $a > c$ and $b > d$. Is the imaginary part of $m + n$ positive or negative?

57. Suppose $a < c$ and $b < d$. Is the real part of $m - n$ positive or negative?

3 Multiply complex numbers

Multiply.

58. $(7i)(-9i)$

59. $(-6i)(-4i)$

60. $\sqrt{-2}\,\sqrt{-8}$

61. $\sqrt{-5}\,\sqrt{-45}$

62. $\sqrt{-3}\,\sqrt{-6}$

63. $\sqrt{-5}\,\sqrt{-10}$

64. $2i(6 + 2i)$

65. $-3i(4 - 5i)$

66. Multiply: $-7i(5 - \sqrt{-64})$

$$-7i(5 - \sqrt{-64}) = -7i(5 - \underline{\quad?\quad}\,i)$$
$$= (\underline{\quad?\quad})(5) - (\underline{\quad?\quad})(8i)$$
$$= \underline{\quad?\quad} + \underline{\quad?\quad}\,i^2$$
$$= -35i + 56(\underline{\quad?\quad})$$
$$= \underline{\quad?\quad}$$

- Rewrite $\sqrt{-64}$ in terms of i.
- Use the Distributive Property.
- Simplify.
- $i^2 = \underline{\quad?\quad}$
- Write the answer in $a + bi$ form.

67. Multiply: $(1 + 9i)(1 - 9i)$

$$(1 + 9i)(1 - 9i) = (\underline{\quad?\quad})^2 + (\underline{\quad?\quad})^2$$
$$= \underline{\quad?\quad} + \underline{\quad?\quad} = \underline{\quad?\quad}$$

- $(a + bi)(a - bi) = a^2 + b^2$
- Simplify.

68. $\sqrt{-2}\left(\sqrt{8} + \sqrt{-2}\right)$

69. $\sqrt{-3}\left(\sqrt{12} - \sqrt{-6}\right)$

70. $(5 - 2i)(3 + i)$

71. $(2 - 4i)(2 - i)$

72. $(6 + 5i)(3 + 2i)$

73. $(4 - 7i)(2 + 3i)$

74. $(1 - i)\left(\dfrac{1}{2} + \dfrac{1}{2}i\right)$

75. $\left(\dfrac{4}{5} - \dfrac{2}{5}i\right)\left(1 + \dfrac{1}{2}i\right)$

76. $\left(\dfrac{6}{5} + \dfrac{3}{5}i\right)\left(\dfrac{2}{3} - \dfrac{1}{3}i\right)$

77. $(2 - i)\left(\dfrac{2}{5} + \dfrac{1}{5}i\right)$

78. $(4 - 3i)(4 + 3i)$

79. $(8 - 5i)(8 + 5i)$

80. $(3 - i)(3 + i)$

81. $(7 - i)(7 + i)$

82. $(6 + i)^2$

83. $(4 - 3i)^2$

84. $(5 - 2i)^2$

85. $(-1 + i)^2$

86. True or false? For all real numbers a and b, the product $(a + bi)(a - bi)$ is a positive real number.

87. True or false? The product of two imaginary numbers is always a negative real number.

4 Divide complex numbers

88. Simplify: $\dfrac{1 - 2i}{-5i}$

$\dfrac{1 - 2i}{-5i} = \dfrac{1 - 2i}{-5i} \cdot \dfrac{?}{?}$ • To remove i from the denominator, multiply the expression by 1 in the form $\dfrac{i}{i}$.

$= \dfrac{i - \underline{\quad ? \quad}}{-5i^2}$ • Multiply the numerators. Multiply the denominators.

$= \dfrac{i - 2(\underline{\quad ? \quad})}{-5(\underline{\quad ? \quad})}$ • Replace i^2 by -1.

$= \dfrac{i + \underline{\quad ? \quad}}{\underline{\quad ? \quad}}$ • Simplify.

$= \underline{\quad ? \quad} + \underline{\quad ? \quad}\, i$ • Write in the form $a + bi$.

89. To simplify $\dfrac{2i}{2 - 3i}$, multiply the numerator and denominator by $\underline{\quad ? \quad}$.

Simplify.

90. $\dfrac{3}{i}$

91. $\dfrac{4}{5i}$

92. $\dfrac{2 - 3i}{-4i}$

93. $\dfrac{16 + 5i}{-3i}$

94. $\dfrac{4}{5 + i}$

95. $\dfrac{6}{5 + 2i}$

96. $\dfrac{2}{2 - i}$

97. $\dfrac{5}{4 - i}$

98. $\dfrac{1 - 3i}{3 + i}$

99. $\dfrac{2 + 12i}{5 + i}$

100. $\dfrac{\sqrt{-10}}{\sqrt{8} - \sqrt{-2}}$

101. $\dfrac{\sqrt{-2}}{\sqrt{12} - \sqrt{-8}}$

102. $\dfrac{2 - 3i}{3 + i}$

103. $\dfrac{3 + 5i}{1 - i}$

104. True or false? The quotient of two imaginary numbers is an imaginary number.

105. True or false? The reciprocal of an imaginary number is an imaginary number.

8.5 Applying Concepts

Note the pattern when successive powers of i are simplified. Use the pattern for Exercises 106 to 114.

$i^1 = i$

$i^2 = -1$

$i^3 = i^2 \cdot i = -i$

$i^4 = i^2 \cdot i^2 = (-1)(-1) = 1$

$i^5 = i \cdot i^4 = i(1) = i$

$i^6 = i^2 \cdot i^4 = -1$

$i^7 = i^3 \cdot i^4 = -i$

$i^8 = i^4 \cdot i^4 = 1$

106. When the exponent on i is a multiple of 4, the power equals _____.

107. i^6

108. i^9

109. i^{57}

110. i^{65}

111. i^{-6}

112. i^{-34}

113. i^{-58}

114. i^{-180}

115. a. Is $3i$ a solution of $2x^2 + 18 = 0$?
b. Is $-3i$ a solution of $2x^2 + 18 = 0$?

116. a. Is $1 + 3i$ a solution of $x^2 - 2x - 10 = 0$?
b. Is $1 - 3i$ a solution of $x^2 - 2x - 10 = 0$?

117. Solve the following equations. Describe the solution by using the following terms: integer, rational number, irrational number, real number, and imaginary number. Note that more than one term may be used to describe the answer.

a. $x^2 + 3 = 7$

b. $x^2 + 1 = 0$

c. $\dfrac{5}{8}x = \dfrac{2}{3}$

d. $x^2 + 1 = 9$

e. $x^{\frac{3}{4}} = 8$

f. $\sqrt[3]{x} = -27$

Focus on Problem Solving

Another Look at Polya's Four-Step Process
Polya's four general steps to follow when attempting to solve a problem are to understand the problem, devise a plan, carry out the plan, and review the solution. (See the Focus on Problem Solving in the chapter titled Review of Real Numbers.) In the process of devising a plan (Step 2), it may be appropriate to write a mathematical expression or an equation. We will illustrate this with the following problem.

Number the letters of the alphabet in sequence from 1 to 26. (See the list on the next page.) Find a word for which the product of the numerical values of the letters of the word equals 1,000,000. We will agree that a "word" is any sequence of letters that contains at least one vowel but is not necessarily in the dictionary.

Understand the Problem

Consider REZB. The product of the letters is $18 \cdot 5 \cdot 26 \cdot 2 = 4680$. This "word" is a sequence of letters with at least one vowel. However, the product of the numerical values of the letters is not 1,000,000. Thus this word does not solve our problem.

Devise a Plan

A = 1	N = 14
B = 2	O = 15
C = 3	P = 16
D = 4	Q = 17
E = 5	R = 18
F = 6	S = 19
G = 7	T = 20
H = 8	U = 21
I = 9	V = 22
J = 10	W = 23
K = 11	X = 24
L = 12	Y = 25
M = 13	Z = 26

Actually, we should have known that the product of the letters in REZB could not equal 1,000,000. The letter R has a factor of 9, and the letter Z has a factor of 13. Neither of these two numbers is a factor of 1,000,000. Consequently, R and Z cannot be letters in the word we are trying to find. This observation leads to an important observation: each of the letters that make up our word must be factors of 1,000,000. To find these letters, consider the prime factorization of 1,000,000.

$$1,000,000 = 2^6 \cdot 5^6$$

Looking at the prime factorization, we note that only letters that contain only 2 or 5 as factors are possible candidates. These letters are B, D, E, H, J, P, T, and Y. One additional point: Because 1 times any number is the number, A can be part of any word we construct.

Our task is now to construct a word from these letters such that the product is 1,000,000. From the prime factorization above, we must have 2 as a factor 6 times and 5 as a factor 6 times.

Carry Out the Plan

We must construct a word with the characteristics described in our plan. Here is a possibility:

THEBEYE

Review the Solution

You should multiply the values of all of the letters and verify that the product is 1,000,000. To ensure that you have an understanding of the problem, try to find other "words" that satisfy the condition of the problem.

Projects & Group Activities

Solving Radical Equations with a Graphing Calculator

The radical equation $\sqrt{x-2} - 6 = 0$ was solved algebraically at the beginning of Section 7.4. To solve $\sqrt{x-2} - 6 = 0$ with a graphing calculator, use the left side of the equation and write an equation in two variables.

$$y = \sqrt{x-2} - 6$$

The graph of $y = \sqrt{x-2} - 6$ is shown on the following page. The solution set of the equation $y = \sqrt{x-2} - 6$ is the set of ordered pairs (x, y) whose coordinates make the equation a true statement. The x-coordinate where $y = 0$ is the solution of the equation $\sqrt{x-2} - 6 = 0$. The solution is the x-intercept of the curve given by the equation $y = \sqrt{x-2} - 6$. The solution is 38.

The solution of the radical equation $\sqrt[3]{x-2} = -0.8$ is 1.488. A graphing calculator can be used to find any rational or irrational solution to a predetermined degree of accuracy. The graph of $y = \sqrt[3]{x-2} + 0.8$ is shown below. The graph intersects the x-axis at $x = 1.488$.

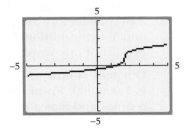

To graph the equation $y = \sqrt[3]{x-2} + 0.8$ on a TI-83/83 Plus or TI-84 Plus, press

| Y= | | MATH | 4 | X, T, θ, n | − | 2 |) | + | .8 | ZOOM | 6 |

Press 2nd CALC 2 to find the x-intercept of the equation.

You were to solve Exercises 5 to 61 in Section 7.4 by algebraic means. Try some of those exercises again, this time using a graphing calculator.

Use a graphing calculator to find the solution of the following equations. Round to the nearest thousandth.

1. $\sqrt{x + 0.3} = 1.3$

2. $\sqrt[3]{x + 1.2} = -1.1$

3. $\sqrt[4]{3x - 1.5} = 1.4$

Chapter 8 Summary

Key Words

The **nth root of a** is $a^{\frac{1}{n}}$. The expression $\sqrt[n]{a}$ is another symbol for the nth root of a. In the expression $\sqrt[n]{a}$, the symbol $\sqrt{}$ is called a **radical**, n is the **index** of the radical, and a is the **radicand**. [8.1.2, p. 550]

Examples

$125^{\frac{1}{3}} = \sqrt[3]{125} = 5$

The index is 3 and the radicand is 125.

If $a^{\frac{1}{n}}$ is a real number, then $a^{\frac{m}{n}} = \sqrt[n]{a^m} = \left(\sqrt[n]{a}\right)^m$. [8.1.2, p. 550]

$$8^{\frac{2}{3}} = \left(\sqrt[3]{8}\right)^2 = 2^2 = 4$$

The symbol $\sqrt{}$ is used to indicate the positive square root, or **principal square root**, of a number. [8.1.3, p. 552]

$$\sqrt{16} = 4$$
$$-\sqrt{16} = -4$$

The expressions $a + b$ and $a - b$ are called **conjugates** of each other. The product of conjugates of the form $(a + b)(a - b)$ is $a^2 - b^2$. [8.2.3, p. 561]

$$(x - 3)(x + 3) = x^2 - 3^2 = x^2 - 9$$

The procedure used to remove a radical from the denominator of a radical expression is called **rationalizing the denominator**. [8.2.4, p. 562]

$$\frac{2}{1 - \sqrt{3}} = \frac{2}{1 - \sqrt{3}} \cdot \frac{1 + \sqrt{3}}{1 + \sqrt{3}}$$
$$= \frac{2(1 + \sqrt{3})}{(1 - \sqrt{3})(1 + \sqrt{3})}$$
$$= \frac{2 + 2\sqrt{3}}{1 - 3} = -1 - \sqrt{3}$$

A **radical equation** is an equation that contains a variable expression in a radicand. [8.4.1, p. 577]

$\sqrt{x - 2} - 3 = 6$ is a radical equation.

A **complex number** is a number of the form $a + bi$, where a and b are real numbers and $i = \sqrt{-1}$. For the complex number $a + bi$, a is the **real part** of the complex number, and b is the **imaginary part** of the complex number. [8.5.1, p. 586]

$3 + 2i$ is a complex number.

Essential Rules and Procedures

Definition of $a^{\frac{m}{n}}$
If m and n are positive integers and $a^{\frac{1}{n}}$ is a real number, then $a^{\frac{m}{n}} = \left(a^{\frac{1}{n}}\right)^m$. [8.1.1, p. 549]

$$8^{\frac{2}{3}} = \left(8^{\frac{1}{3}}\right)^2 = 2^2 = 4$$

Definition of the *n*th root of *a*
If a is a real number, then $a^{\frac{1}{n}} = \sqrt[n]{a}$. [8.1.2, p. 550]

$$x^{\frac{1}{3}} = \sqrt[3]{x}$$

Definition of the *n*th root of a^m
If $a^{\frac{1}{n}}$ is a real number, then $a^{\frac{m}{n}} = \sqrt[n]{a^m}$. [8.1.2, p. 550]

$$b^{\frac{3}{4}} = \sqrt[4]{b^3}$$

Product Property of Radicals
If $\sqrt[n]{a}$ and $\sqrt[n]{b}$ are positive real numbers, then $\sqrt[n]{ab} = \sqrt[n]{a}\,\sqrt[n]{b}$. [8.2.1, p. 559]

$$\sqrt{9 \cdot 7} = \sqrt{9}\,\sqrt{7} = 3\sqrt{7}$$

Quotient Property of Radicals
If $\sqrt[n]{a}$ and $\sqrt[n]{b}$ are positive real numbers, and $b \neq 0$, then $\sqrt[n]{\dfrac{a}{b}} = \dfrac{\sqrt[n]{a}}{\sqrt[n]{b}}$. [8.2.4, p. 562]

$$\sqrt[3]{\frac{5}{27}} = \frac{\sqrt[3]{5}}{\sqrt[3]{27}} = \frac{\sqrt[3]{5}}{3}$$

Property of Raising Both Sides of an Equation to a Power
If a and b are real numbers and $a = b$, then $a^n = b^n$.
[8.4.1, p. 577]

If $x = 4$, then $x^2 = 4^2$.

Pythagorean Theorem
$c^2 = a^2 + b^2$ [8.4.2, p. 580]

$3^2 + 4^2 = 5^2$

Principal Square Root of a Negative Number
If a is a positive real number, then $\sqrt{-a} = i\sqrt{a}$. [8.5.1, p. 586]

$\sqrt{-8} = i\sqrt{8} = 2i\sqrt{2}$

Addition Property of Complex Numbers
If $a + bi$ and $c + di$ are complex numbers, then
$(a + bi) + (c + di) = (a + c) + (b + d)i$. [8.5.2, p. 587]

$(2 + 4i) + (3 + 6i) = (2 + 3) + (4 + 6)i$
$= 5 + 10i$

Subtraction Property of Complex Numbers
If $a + bi$ and $c + di$ are complex numbers, then
$(a + bi) - (c + di) = (a - c) + (b - d)i$. [8.5.2, p. 587]

$(4 + 3i) - (7 + 4i) = (4 - 7) + (3 - 4)i$
$= -3 - i$

Chapter 8 Review Exercises

1. Simplify: $81^{-\frac{1}{4}}$

2. Simplify: $\dfrac{x^{-\frac{3}{2}}}{x^{\frac{7}{2}}}$

3. Simplify: $\left(a^{16}\right)^{-\frac{5}{8}}$

4. Simplify: $(16x^{-4}y^{12})(100x^6y^{-2})^{\frac{1}{2}}$

5. Rewrite $3x^{\frac{3}{4}}$ as a radical expression.

6. Rewrite $7y\sqrt[3]{x^2}$ as an exponential expression.

7. Simplify: $\sqrt[4]{81a^8b^{12}}$

8. Simplify: $-\sqrt{49x^6y^{16}}$

9. Simplify: $\sqrt[3]{-8a^6b^{12}}$

10. Simplify: $\sqrt{18a^3b^6}$

11. Simplify: $\sqrt[5]{-64a^8b^{12}}$

12. Simplify: $\sqrt[4]{x^6y^8z^{10}}$

13. Add: $\sqrt{54} + \sqrt{24}$

14. Subtract: $\sqrt{48x^5y} - x\sqrt{80x^3y}$

15. Subtract: $\sqrt{50a^4b^3} - ab\sqrt{18a^2b}$

16. Simplify: $4x\sqrt{12x^2y} + \sqrt{3x^4y} - x^2\sqrt{27y}$

17. Multiply: $\sqrt{32}\,\sqrt{50}$

18. Multiply: $\sqrt[3]{16x^4y}\,\sqrt[3]{4xy^5}$

19. Multiply: $\sqrt{3x}(3 + \sqrt{3x})$

20. Multiply: $\left(5 - \sqrt{6}\right)^2$

21. Multiply: $\left(\sqrt{3} + 8\right)\left(\sqrt{3} - 2\right)$

22. Simplify: $\dfrac{\sqrt{125x^6}}{\sqrt{5x^3}}$

23. Simplify: $\dfrac{8}{\sqrt{3y}}$

24. Simplify: $\dfrac{x+2}{\sqrt{x}+\sqrt{2}}$

25. Simplify: $\dfrac{\sqrt{x}+\sqrt{y}}{\sqrt{x}-\sqrt{y}}$

26. Simplify: $\sqrt{-36}$

27. Simplify: $\sqrt{-50}$

28. Simplify: $\sqrt{49}-\sqrt{-16}$

29. Simplify: $\sqrt{200}+\sqrt{-12}$

30. Add: $(5+2i)+(4-3i)$

31. Subtract: $(-8+3i)-(4-7i)$

32. Add: $\left(9-\sqrt{-16}\right)+\left(5+\sqrt{-36}\right)$

33. Subtract: $\left(\sqrt{50}+\sqrt{-72}\right)-\left(\sqrt{162}-\sqrt{-8}\right)$

34. Add: $(3-9i)+7$

35. Multiply: $(8i)(2i)$

36. Multiply: $i(3-7i)$

37. Multiply: $\sqrt{-12}\ \sqrt{-6}$

38. Multiply: $(6-5i)(4+3i)$

39. Simplify: $\dfrac{-6}{i}$

40. Simplify: $\dfrac{5+2i}{3i}$

41. Simplify: $\dfrac{7}{2-i}$

42. Simplify: $\dfrac{\sqrt{16}}{\sqrt{4}-\sqrt{-4}}$

43. Simplify: $\dfrac{5+9i}{1-i}$

44. Solve: $\sqrt[3]{9x}=-6$

45. State the domain of $f(x)=\sqrt[4]{3x-2}$ in set-builder notation.

46. State the domain of $f(x)=\sqrt[3]{x-5}$ in set-builder notation.

47. Graph: $g(x)=3x^{\frac{1}{3}}$

48. Graph: $f(x)=\sqrt{x-3}$

49. Solve: $\sqrt[3]{3x-5}=2$

50. Solve: $\sqrt{4x+9}+10=11$

51. **Geometry** Find the width of a rectangle that has a diagonal of 13 in. and a length of 12 in.

52. **Wind Power** The velocity of the wind determines the amount of power generated by a windmill. A typical equation for this relationship is $v=4.05\sqrt[3]{P}$, where v is the velocity in miles per hour and P is the power in watts. Find the amount of power generated by a wind of 20 mph. Round to the nearest whole number.

53. **Physics** Find the distance required for a car to reach a velocity of 88 ft/s when the acceleration is 16 ft/s². Use the equation $v=\sqrt{2as}$, where v is the velocity, a is the acceleration, and s is the distance.

54. **Home Maintenance** A 12-ft ladder is leaning against a building. How far from the building is the bottom of the ladder when the top of the ladder touches the building 10 ft above the ground? Round to the nearest hundredth.

Chapter 8 Test

1. Simplify: $\dfrac{r^{\frac{2}{3}}r^{-1}}{r^{-\frac{1}{2}}}$

2. Simplify: $\dfrac{\left(2x^{\frac{1}{3}}y^{-\frac{2}{3}}\right)^6}{(x^{-4}y^8)^{\frac{1}{4}}}$

3. Simplify: $\left(\dfrac{4a^4}{b^2}\right)^{-\frac{3}{2}}$

4. Rewrite $3y^{\frac{2}{5}}$ as a radical expression.

5. Rewrite $\dfrac{1}{2}\sqrt[4]{x^3}$ as an exponential expression.

6. State the domain of $f(x) = \sqrt{4 - x}$ in set-builder notation.

7. State the domain of $f(x) = (2x - 3)^{\frac{1}{3}}$ in interval notation.

8. Simplify: $\sqrt[3]{27a^4b^3c^7}$

9. Add: $\sqrt{18a^3} + a\sqrt{50a}$

10. Subtract: $\sqrt[3]{54x^7y^3} - x\sqrt[3]{128x^4y^3} - x^2\sqrt[3]{2xy^3}$

11. Multiply: $\sqrt{3x}\left(\sqrt{x} - \sqrt{25x}\right)$

12. Multiply: $\left(2\sqrt{3} + 4\right)\left(3\sqrt{3} - 1\right)$

13. Multiply: $\left(\sqrt{a} - 3\sqrt{b}\right)\left(2\sqrt{a} + 5\sqrt{b}\right)$

14. Multiply: $\left(2\sqrt{x} + \sqrt{y}\right)^2$

15. Simplify: $\dfrac{\sqrt{32x^5y}}{\sqrt{2xy^3}}$

16. Simplify: $\dfrac{4 - 2\sqrt{5}}{2 - \sqrt{5}}$

17. Simplify: $\dfrac{\sqrt{x}}{\sqrt{x} - \sqrt{y}}$

18. Multiply: $\left(\sqrt{-8}\right)\left(\sqrt{-2}\right)$

19. Subtract: $(5 - 2i) - (8 - 4i)$

20. Multiply: $(2 + 5i)(4 - 2i)$

21. Simplify: $\dfrac{2 + 3i}{1 - 2i}$

22. Add: $(2 + i) + (2 - i)$

23. Solve: $\sqrt{x + 12} - \sqrt{x} = 2$

24. Solve: $\sqrt[3]{2x - 2} + 4 = 2$

25. **Guy Wires** A guy wire is attached to a point 30 ft above the ground on a lamppost. The wire is anchored to the ground at a point 6 ft from the base of the post. Find the length of the guy wire. Round to the nearest tenth.

30 ft

6 ft

Cumulative Review Exercises

1. Identify the property that justifies the statement.
$(a + 2)b = ab + 2b$

2. Simplify: $2x - 3[x - 2(x - 4) + 2x]$

3. Find $A \cap B$ given $A = \{2, 4, 6\}$ and $B = \{1, 3, 5\}$.

4. Solve: $\sqrt[3]{2x - 5} + 3 = 6$

5. Solve: $5 - \dfrac{2}{3}x = 4$

6. Solve: $2[4 - 2(3 - 2x)] = 4(1 - x)$

7. Solve: $3x - 4 \le 8x + 1$
Write the solution set in set-builder notation.

8. Solve: $5 < 2x - 3 < 7$
Write the solution set in set-builder notation.

9. Solve: $|7 - 3x| > 1$

10. Factor: $64a^2 - b^2$

11. Factor: $x^5 + 2x^3 - 3x$

12. Solve: $3x^2 + 13x - 10 = 0$

13. Graph: $g(x) = \sqrt{1 - x}$

14. Is the line $x - 2y = 4$ perpendicular to the line $2x + y = 4$?

15. Simplify: $(3^{-1}x^3y^{-5})(3^{-1}y^{-2})^{-2}$

16. Simplify: $\left(\dfrac{x^{-\frac{1}{2}}y^{\frac{3}{4}}}{y^{-\frac{5}{4}}}\right)^4$

17. Subtract: $\sqrt{20x^3} - x\sqrt{45x}$

18. Multiply: $(\sqrt{5} - 3)(\sqrt{5} - 2)$

19. Simplify: $\dfrac{\sqrt[3]{4x^5y^4}}{\sqrt[3]{8x^2y^5}}$

20. Simplify: $\dfrac{3i}{2 - i}$

21. Graph: $\{x \mid x > -1\} \cap \{x \mid x \le 3\}$

22. State the domain of $g(x) = \sqrt[5]{2x + 5}$ in set-builder notation.

23. Given $f(x) = 3x^2 - 2x + 1$, evaluate $f(-3)$.

24. Find an equation of the line that passes through the points $(2, 3)$ and $(-1, 2)$.

25. Evaluate the determinant: $\begin{vmatrix} 1 & 2 & -3 \\ 0 & -1 & 2 \\ 3 & 1 & -2 \end{vmatrix}$

26. Solve by using Cramer's Rule: $\begin{aligned} 2x - y &= 4 \\ -2x + 3y &= 5 \end{aligned}$

27. Find the slope and y-intercept of the equation $3x - 2y = -6$ and then graph the line.

28. Graph the solution set of $3x + 2y \le 4$.

29. Stamps A collection of thirty stamps consists of 13¢ stamps and 18¢ stamps. The total value of the stamps is $4.85. Find the number of 18¢ stamps.

30. Investments An investment of $2500 is made at an annual simple interest rate of 7.2%. How much additional money must be invested at an annual simple interest rate of 8.4% so that the total interest earned is $516?

31. Geometry The width of a rectangle is 6 ft less than the length. The area of the rectangle is 72 ft². Find the length and width of the rectangle.

32. Uniform Motion A sales executive traveled 25 mi by car and then an additional 625 mi by plane. The rate by plane was five times the rate by car. The total time of the trip was 3 h. Find the rate of the plane.

33. Astronomy How long does it take light to travel to Earth from the moon when the moon is 232,500 mi from Earth? Light travels 1.86×10^5 mi/s.

34. Periscopes How far would a submarine periscope have to be above the water to locate a ship 7 mi away? The equation for the distance in miles that the lookout can see is $d = \sqrt{1.5h}$, where h is the height in feet above the surface of the water. Round to the nearest tenth.

35. Investments The graph shows the amount invested and the annual interest income from an investment. Find the slope of the line between the two points shown on the graph. Then write a sentence that states the meaning of the slope.

9

Quadratic Equations and Inequalities

DVD **SSM**

✺ *Online Study Center*

Need help? For online student resources, such as section quizzes, visit the Online Study Center at college.hmco.com/pic/aufmannABI2e.

The H. J. Heinz Company may be known for its ketchup, but it also produces a number of different brands of condiments, sauces, meals, and snacks. Offering a number of different items increases the chance that people will buy at least one Heinz product, thereby increasing the company's profit. Profit is the difference between a company's revenue (the total amount of money the company earns by selling its products or services) and its costs (the total amount of money the company spends in doing business). Quadratic functions are used in **Exercises 29 and 30 on page 672** to determine the maximum profit or revenue a company can earn.

Prep Test

1. Simplify: $\sqrt{18}$

2. Simplify: $\sqrt{-9}$

3. Simplify: $\dfrac{3x - 2}{x - 1} - 1$

4. Evaluate $b^2 - 4ac$ when $a = 2, b = -4$, and $c = 1$.

5. Is $4x^2 + 28x + 49$ a perfect-square trinomial?

6. Factor: $4x^2 - 4x + 1$

7. Factor: $9x^2 - 4$

8. Graph $\{x \mid x < -1\} \cap \{x \mid x < 4\}$

9. Solve: $x(x - 1) = x + 15$

10. Solve: $\dfrac{4}{x - 3} = \dfrac{16}{x}$

9.1 Solving Quadratic Equations by Factoring or by Taking Square Roots

OBJECTIVE 1 Solve quadratic equations by factoring

A **quadratic equation** is an equation of the form $ax^2 + bx + c = 0$, where a and b are coefficients, c is a constant, and $a \neq 0$.

$$3x^2 - x + 2 = 0, \quad a = 3, \quad b = -1, \quad c = 2$$
$$-x^2 + 4 = 0, \quad a = -1, \quad b = 0, \quad c = 4$$
$$6x^2 - 5x = 0, \quad a = 6, \quad b = -5, \quad c = 0$$

A quadratic equation is in **standard form** when the polynomial is in descending order and equal to zero.

Because the degree of the polynomial $ax^2 + bx + c$ is 2, a quadratic equation is also called a **second-degree equation.**

In the chapter entitled "Polynomials and Exponents," the Principle of Zero Products was used to solve some quadratic equations. That procedure is reviewed here.

Principle of Zero Products

If the product of two factors is zero, then at least one of the factors must be zero.

If $ab = 0$, then $a = 0$ or $b = 0$.

The Principle of Zero Products states that if the product of two factors is zero, then at least one of the factors must be zero.

Solve by factoring: $x^2 - 6x = -9$

	$x^2 - 6x = -9$
Write the equation in standard form.	$x^2 - 6x + 9 = 0$
Factor.	$(x - 3)(x - 3) = 0$
Use the Principle of Zero Products.	$x - 3 = 0 \qquad x - 3 = 0$
Solve each equation.	$x = 3 \qquad\qquad x = 3$
	3 checks as a solution.
Write the solutions.	The solution is 3.

When a quadratic equation has two solutions that are the same number, the solution is called a **double root** of the equation. The solution 3 is a double root of the equation $x^2 - 6x = -9$.

EXAMPLE 1 Solve for x by factoring: $x^2 - 4ax - 5a^2 = 0$

Solution $x^2 - 4ax - 5a^2 = 0$ • **This is a literal equation. Solve for**
 x in terms of a.

$(x + a)(x - 5a) = 0$ • **Factor.**

$x + a = 0 \qquad x - 5a = 0$
$\quad x = -a \qquad\quad x = 5a$

The solutions are $-a$ and $5a$.

PROBLEM 1 Solve for x by factoring: $x^2 - 3ax - 4a^2 = 0$

Solution See page S29.

OBJECTIVE 2 Write a quadratic equation given its solutions

As shown below, the solutions of the equation $(x - r_1)(x - r_2) = 0$ are r_1 and r_2.

$(x - r_1)(x - r_2) = 0$ **Check:**

$x - r_1 = 0 \qquad x - r_2 = 0$ $(x - r_1)(x - r_2) = 0 \qquad (x - r_1)(x - r_2) = 0$

$\quad x = r_1 \qquad\qquad x = r_2$ $\overline{(r_1 - r_1)(r_1 - r_2)}\ |\ 0 \qquad \overline{(r_2 - r_1)(r_2 - r_2)}\ |\ 0$

$\qquad\qquad\qquad\qquad\qquad\qquad\qquad 0 \cdot (r_1 - r_2)\ | \qquad\qquad (r_2 - r_1) \cdot 0\ |$

$\qquad\qquad\qquad\qquad\qquad\qquad\qquad\qquad\quad 0 = 0 \qquad\qquad\qquad\qquad\qquad 0 = 0$

Using the equation $(x - r_1)(x - r_2) = 0$ and the fact that r_1 and r_2 are solutions of this equation, it is possible to write a quadratic equation given its solutions.

Write a quadratic equation that has solutions 4 and -5.

	$(x - r_1)(x - r_2) = 0$
Replace r_1 by 4 and r_2 by -5.	$(x - 4)[x - (-5)] = 0$
Simplify.	$(x - 4)(x + 5) = 0$
Multiply.	$x^2 + x - 20 = 0$

EXAMPLE 2 Write a quadratic equation that has integer coefficients and has solutions $\frac{2}{3}$ and $\frac{1}{2}$.

Solution

$$(x - r_1)(x - r_2) = 0$$

$$\left(x - \frac{2}{3}\right)\left(x - \frac{1}{2}\right) = 0 \qquad \bullet \text{ Replace } r_1 \text{ by } \frac{2}{3} \text{ and } r_2 \text{ by } \frac{1}{2}.$$

$$x^2 - \frac{7}{6}x + \frac{1}{3} = 0 \qquad \bullet \text{ Multiply.}$$

$$6\left(x^2 - \frac{7}{6}x + \frac{1}{3}\right) = 6 \cdot 0 \qquad \bullet \text{ Multiply each side of the equation by the LCM of the denominators.}$$

$$6x^2 - 7x + 2 = 0$$

PROBLEM 2 Write a quadratic equation that has integer coefficients and has solutions $-\frac{2}{3}$ and $\frac{1}{6}$.

Solution See page S29.

OBJECTIVE **3** ## Solve quadratic equations by taking square roots

The solution of the quadratic equation $x^2 = 16$ is shown at the right.

$$x^2 = 16$$
$$x^2 - 16 = 0$$
$$(x + 4)(x - 4) = 0$$

Note that the solution is the positive or the negative square root of 16, 4 or -4.

$$x + 4 = 0 \qquad x - 4 = 0$$
$$x = -4 \qquad x = 4$$

The solution can also be found by taking the square root of each side of the equation. Recall that if x is a real number, then $\sqrt{x^2} = |x|$. This fact is used to solve some quadratic equations by taking square roots.

Solve by taking square roots: $x^2 = 16$

$$x^2 = 16$$

Take the square root of each side of the equation.

$$\sqrt{x^2} = \sqrt{16}$$

$\sqrt{x^2} = |x|$. $\sqrt{16} = 4$.

$$|x| = 4$$

Solve the absolute value equation. The notation $x = \pm 4$ means $x = 4$ or $x = -4$.

$$x = \pm 4$$

The solutions are -4 and 4.

Solving a quadratic equation by taking the square root of each side of the equation can lead to solutions that contain radicals.

Solve by taking square roots: $3x^2 = 54$

	$3x^2 = 54$		
Solve for x^2.	$x^2 = 18$		
Take the square root of each side of the equation.	$\sqrt{x^2} = \sqrt{18}$		
Simplify.	$	x	= 3\sqrt{2}$
	$x = \pm 3\sqrt{2}$		

The solutions are $-3\sqrt{2}$ and $3\sqrt{2}$.

This method can also be used when the solutions are complex numbers.

Solve by taking square roots: $2x^2 + 18 = 0$

	$2x^2 + 18 = 0$		
Solve for x^2.	$2x^2 = -18$		
	$x^2 = -9$		
Take the square root of each side of the equation.	$\sqrt{x^2} = \sqrt{-9}$		
Simplify.	$	x	= 3i$
	$x = \pm 3i$		

The solutions are $-3i$ and $3i$.

An equation containing the square of a binomial can be solved by taking square roots.

EXAMPLE 3 Solve by taking square roots: $3(x - 2)^2 + 12 = 0$

Solution
$$3(x - 2)^2 + 12 = 0$$
$$3(x - 2)^2 = -12 \qquad \bullet \text{ Solve for } (x - 2)^2.$$
$$(x - 2)^2 = -4$$
$$\sqrt{(x - 2)^2} = \sqrt{-4} \qquad \bullet \text{ Take the square root of each side of the equation. Then simplify.}$$
$$|x - 2| = 2i$$
$$x - 2 = \pm 2i$$
$$x - 2 = 2i \qquad\qquad x - 2 = -2i \qquad \bullet \text{ Solve for } x.$$
$$x = 2 + 2i \qquad\qquad x = 2 - 2i$$

The solutions are $2 + 2i$ and $2 - 2i$.

PROBLEM 3 Solve by taking square roots: $2(x + 1)^2 + 24 = 0$

Solution See page S29.

9.1 Concept Review

Determine whether the following statements are always true, sometimes true, or never true.

1. A quadratic equation has two real roots.

2. A quadratic equation is also called a second-degree equation.

3. If $(x - 3)(x + 5) = 8$, then $x - 3 = 8$ or $x + 5 = 8$.

4. A quadratic equation has one root.

9.1 Exercises

1 Solve quadratic equations by factoring

1. Explain why the restriction that $a \neq 0$ is given in the definition of a quadratic equation.

2. What does the Principle of Zero Products state? How is it used to solve a quadratic equation?

Write the quadratic equation in standard form and with the coefficient of x^2 positive. Name the values of a, b, and c.

3. $2x^2 - 4x = 5$ **4.** $x^2 = 3x + 1$ **5.** $5x = 4x^2 - 6$ **6.** $3x^2 = 7$

7. Solve: $x^2 - 49 = 0$

$x^2 - 49 = 0$ • The equation is in ___?___ form.

$(\underline{\quad ? \quad})(\underline{\quad ? \quad}) = 0$ • Factor the difference of two perfect squares.

$x + 7 = \underline{\quad ? \quad}$ $x - 7 = \underline{\quad ? \quad}$ • Use the Principle of ___?___ to set each factor equal to 0.

$x = \underline{\quad ? \quad}$ $x = \underline{\quad ? \quad}$ • Solve each equation.

8. Solve for x: $x^2 + ax - 6a^2 = 0$

$x^2 + ax - 6a^2 = 0$ • The equation is a literal equation. Treat a as a constant.

$(x - \underline{\quad ? \quad})(x + \underline{\quad ? \quad}) = 0$ • Factor the trinomial.

$\underline{\quad ? \quad} = 0$ $\underline{\quad ? \quad} = 0$ • Use the Principle of Zero Products to set each factor equal to 0.

$x = \underline{\quad ? \quad}$ $x = \underline{\quad ? \quad}$ • Solve each equation for x in terms of a.

Solve by factoring.

9. $x^2 - 4x = 0$

10. $y^2 + 6y = 0$

11. $t^2 - 25 = 0$

12. $p^2 - 81 = 0$

13. $s^2 - s - 6 = 0$

14. $v^2 + 4v - 5 = 0$

15. $y^2 - 6y + 9 = 0$

16. $x^2 + 10x + 25 = 0$

17. $9z^2 - 18z = 0$

18. $4y^2 + 20y = 0$

19. $r^2 - 3r = 10$

20. $p^2 + 5p = 6$

21. $v^2 + 10 = 7v$

22. $t^2 - 16 = 15t$

23. $2x^2 - 9x - 18 = 0$

24. $3y^2 - 4y - 4 = 0$

25. $4z^2 - 9z + 2 = 0$

26. $2s^2 - 9s + 9 = 0$

27. $3w^2 + 11w = 4$

28. $2r^2 + r = 6$

29. $6x^2 = 23x + 18$

30. $6x^2 = 7x - 2$

31. $4 - 15u - 4u^2 = 0$

32. $3 - 2y - 8y^2 = 0$

33. $x + 18 = x(x - 6)$

34. $t + 24 = t(t + 6)$

35. $4s(s + 3) = s - 6$

36. $3v(v - 2) = 11v + 6$

37. $u^2 - 2u + 4 = (2u - 3)(u + 2)$

38. $(3v - 2)(2v + 1) = 3v^2 - 11v - 10$

39. $(3x - 4)(x + 4) = x^2 - 3x - 28$

Solve for x by factoring.

40. $x^2 + 14ax + 48a^2 = 0$

41. $x^2 - 9bx + 14b^2 = 0$

42. $x^2 + 9xy - 36y^2 = 0$

43. $x^2 - 6cx - 7c^2 = 0$

44. $x^2 - ax - 20a^2 = 0$

45. $2x^2 + 3bx + b^2 = 0$

46. $3x^2 - 4cx + c^2 = 0$

47. $3x^2 - 14ax + 8a^2 = 0$

48. $3x^2 - 11xy + 6y^2 = 0$

49. Given that $ax^2 + bx + c = 0$ is a quadratic equation that can be solved by factoring, state whether the equation has two positive solutions, two negative solutions, or one positive and one negative solution.
a. $a > 0, b < 0, c < 0$ **b.** $a > 0, b > 0, c > 0$ **c.** $a > 0, b < 0, c > 0$

50. Which equation has a double root?

a. $x^2 = a^2$ **b.** $x^2 + 2ax = a^2$
c. $x^2 + 2ax = -a^2$ **d.** $x^2 - 2ax = a^2$

2 **Write a quadratic equation given its solutions**

51. If r is a solution of $ax^2 + bx + c = 0$, then _____?_____ is a factor of $ax^2 + bx + c$.

52. a. The factored form of a quadratic equation that has solutions -4 and 7 is
(___?___)(___?___) $= 0$.
b. The standard form of the equation in part a is _____?_____ $= 0$.

Write a quadratic equation that has integer coefficients and has as solutions the given pair of numbers.

53. 2 and 5

54. 3 and 1

55. -2 and -4

56. -1 and -3

57. 6 and -1

58. -2 and 5

59. 3 and -3

60. 5 and -5

61. 4 and 4

62. 2 and 2

63. 0 and 5

64. 0 and -2

65. 0 and 3

66. 0 and -1

67. 3 and $\dfrac{1}{2}$

68. 2 and $\dfrac{2}{3}$

69. $-\dfrac{3}{4}$ and 2

70. $-\dfrac{1}{2}$ and 5

71. $-\dfrac{5}{3}$ and -2

72. $-\dfrac{3}{2}$ and -1

73. $-\dfrac{2}{3}$ and $\dfrac{2}{3}$

74. $-\dfrac{1}{2}$ and $\dfrac{1}{2}$

75. $\dfrac{1}{2}$ and $\dfrac{1}{3}$

76. $\dfrac{3}{4}$ and $\dfrac{2}{3}$

77. $\dfrac{6}{5}$ and $-\dfrac{1}{2}$

78. $\dfrac{3}{4}$ and $-\dfrac{3}{2}$

79. $-\dfrac{1}{4}$ and $-\dfrac{1}{2}$

80. $-\dfrac{5}{6}$ and $-\dfrac{2}{3}$

81. $\dfrac{3}{5}$ and $-\dfrac{1}{10}$

82. $\dfrac{7}{2}$ and $-\dfrac{1}{4}$

For Exercises 83 to 86, r_1 and r_2 are solutions of the equation $ax^2 + bx + c = 0$, where a is positive. In each case, state whether b is positive or negative, and whether c is positive or negative.

83. $r_1 < 0, r_2 < 0$

84. $r_1 > 0, r_2 < 0, r_1 > |r_2|$

85. $r_1 > 0, r_2 < 0, r_1 < |r_2|$

86. $r_1 > 0, r_2 > 0$

3 Solve quadratic equations by taking square roots

87. The notation $x = \pm 6$ means $x = $ ___?___ or $x = $ ___?___.

88. Solve by taking square roots: $4x^2 + 100 = 0$.

$4x^2 + 100 = 0$

$4x^2 = $ ___?___	• Subtract ___?___ from each side of the equation.		
$x^2 = $ ___?___	• Divide each side of the equation by ___?___.		
$\sqrt{x^2} = \sqrt{-25}$	• Take the ___?___ of each side of the equation.		
$	x	= $ ___?___	• Simplify $\sqrt{-25}$.
$x = \pm$ ___?___	• Solve for x.		

Solve by taking square roots.

89. $y^2 = 49$ **90.** $x^2 = 64$ **91.** $z^2 = -4$ **92.** $v^2 = -16$

93. $s^2 - 4 = 0$ **94.** $r^2 - 36 = 0$ **95.** $4x^2 - 81 = 0$ **96.** $9x^2 - 16 = 0$

97. $y^2 + 49 = 0$ **98.** $z^2 + 16 = 0$ **99.** $v^2 - 48 = 0$ **100.** $s^2 - 32 = 0$

101. $r^2 - 75 = 0$

102. $u^2 - 54 = 0$

103. $z^2 + 18 = 0$

104. $t^2 + 27 = 0$

105. $(x - 1)^2 = 36$

106. $(x + 2)^2 = 25$

107. $3(y + 3)^2 = 27$

108. $4(s - 2)^2 = 36$

109. $5(z + 2)^2 = 125$

110. $(x - 2)^2 = -4$

111. $(x + 5)^2 = -25$

112. $(x - 8)^2 = -64$

113. $3(x - 4)^2 = -12$

114. $5(x + 2)^2 = -125$

115. $3(x - 9)^2 = -27$

116. $2(y - 3)^2 = 18$

117. $\left(v - \dfrac{1}{2}\right)^2 = \dfrac{1}{4}$

118. $\left(r + \dfrac{2}{3}\right)^2 = \dfrac{1}{9}$

119. $\left(x - \dfrac{2}{5}\right)^2 = \dfrac{9}{25}$

120. $\left(y + \dfrac{1}{3}\right)^2 = \dfrac{4}{9}$

121. $\left(a + \dfrac{3}{4}\right)^2 = \dfrac{9}{16}$

122. $4\left(x - \dfrac{1}{2}\right)^2 = 1$

123. $3\left(x - \dfrac{5}{3}\right)^2 = \dfrac{4}{3}$

124. $2\left(x + \dfrac{3}{5}\right)^2 = \dfrac{8}{25}$

125. $(x + 5)^2 - 6 = 0$

126. $(t - 1)^2 - 15 = 0$

127. $(s - 2)^2 - 24 = 0$

128. $(y + 3)^2 - 18 = 0$

129. $(z + 1)^2 + 12 = 0$

130. $(r - 2)^2 + 28 = 0$

131. $(v - 3)^2 + 45 = 0$

132. $(x + 5)^2 + 32 = 0$

133. $\left(u + \dfrac{2}{3}\right)^2 - 18 = 0$

134. $\left(z - \dfrac{1}{2}\right)^2 - 20 = 0$

135. $\left(x + \dfrac{1}{2}\right)^2 + 40 = 0$

136. $\left(r - \dfrac{3}{2}\right)^2 + 48 = 0$

For Exercises 137 to 140, a and b are positive real numbers. In each case, state how many real or complex number solutions the equation has.

137. $(x - a)^2 + b = 0$

138. $x^2 + a = 0$

139. $(x - a)^2 = 0$

140. $(x - a)^2 - b = 0$

9.1 Applying Concepts

Write a quadratic equation that has as solutions the given pair of numbers.

141. $\sqrt{2}$ and $-\sqrt{2}$

142. $\sqrt{5}$ and $-\sqrt{5}$

143. i and $-i$

144. $2i$ and $-2i$

145. $2\sqrt{2}$ and $-2\sqrt{2}$

146. $3\sqrt{2}$ and $-3\sqrt{2}$

147. $2\sqrt{3}$ and $-2\sqrt{3}$

148. $i\sqrt{2}$ and $-i\sqrt{2}$

149. $2i\sqrt{3}$ and $-2i\sqrt{3}$

Solve for x.

150. $2a^2x^2 = 32b^2$

151. $5y^2x^2 = 125z^2$

152. $(x + a)^2 - 4 = 0$

153. $2(x - y)^2 - 8 = 0$

154. $(2x - 1)^2 = (2x + 3)^2$

155. $(x - 4)^2 = (x + 2)^2$

Solve.

156. Show that the solutions of the equation $ax^2 + bx = 0$ are 0 and $-\dfrac{b}{a}$.

157. Show that the solutions of the equation $ax^2 + c = 0$, $a > 0$, $c > 0$, are $\dfrac{\sqrt{ca}}{a}i$ and $-\dfrac{\sqrt{ca}}{a}i$.

158. Chemistry A chemical reaction between carbon monoxide and water vapor is used to increase the ratio of hydrogen gas in certain gas mixtures. In the process, carbon dioxide is also formed. For a certain reaction, the concentration of carbon dioxide, x, in moles per liter, is given by $0.58 = \dfrac{x^2}{(0.02 - x)^2}$. Solve this equation for x. Round to the nearest ten-thousandth.

159. Show that if the complex number $z_1 + z_2i$ is a solution of the equation $ax^2 + bx + c = 0$, where a, b, and c are real numbers and $a \neq 0$, then $z_1 - z_2i$, the complex conjugate, is also a solution of the equation.

Solving Quadratic Equations by Completing the Square and by Using the Quadratic Formula

O B J E C T I V E 1 **Solve quadratic equations by completing the square**

Recall that a perfect-square trinomial is the square of a binomial. Some examples of perfect-square trinomials follow.

Perfect-square trinomial		Square of a binomial
$x^2 + 8x + 16$	$=$	$(x + 4)^2$
$x^2 - 10x + 25$	$=$	$(x - 5)^2$
$x^2 + 2ax + a^2$	$=$	$(x + a)^2$

For each perfect-square trinomial, the square of $\frac{1}{2}$ the coefficient of x equals the constant term.

$$\left(\frac{1}{2} \text{ coefficient of } x\right)^2 = \textbf{Constant term}$$

$$x^2 + 8x + 16, \qquad \left(\frac{1}{2} \cdot 8\right)^2 = 16$$

$$x^2 - 10x + 25, \qquad \left[\frac{1}{2}(-10)\right]^2 = 25$$

$$x^2 + 2ax + a^2, \qquad \left(\frac{1}{2} \cdot 2a\right)^2 = a^2$$

To complete the square on $x^2 + bx$, add $\left(\frac{1}{2}b\right)^2$ to $x^2 + bx$.

Complete the square on $x^2 - 12x$. Write the resulting perfect-square trinomial as the square of a binomial.

Find the constant term.

$$\left[\frac{1}{2}(-12)\right]^2 = (-6)^2 = 36$$

Complete the square on $x^2 - 12x$ by adding the constant term.

$$x^2 - 12x + 36$$

Write the resulting perfect-square trinomial as the square of a binomial.

$$x^2 - 12x + 36 = (x - 6)^2$$

Complete the square on $z^2 + 3z$. Write the resulting perfect-square trinomial as the square of a binomial.

Find the constant term.

$$\left(\frac{1}{2} \cdot 3\right)^2 = \left(\frac{3}{2}\right)^2 = \frac{9}{4}$$

Point of Interest

Early attempts to solve quadratic equations were primarily geometric. The Persian mathematician al-Khwarizmi (c. A.D. 800) essentially completed a square of $x^2 + 12x$ as shown below.

Complete the square on $z^2 + 3z$ by adding the constant term.

$$z^2 + 3z + \frac{9}{4}$$

Write the resulting perfect-square trinomial as the square of a binomial.

$$z^2 + 3z + \frac{9}{4} = \left(z + \frac{3}{2}\right)^2$$

Not all quadratic equations can be solved by factoring, but any quadratic equation can be solved by completing the square. Add to each side of the equation the term that completes the square. Rewrite the equation in the form $(x + a)^2 = b$. Then take the square root of each side of the equation.

Solve by completing the square: $x^2 - 4x - 14 = 0$

$$x^2 - 4x - 14 = 0$$

Add 14 to each side of the equation.

$$x^2 - 4x = 14$$

Add the constant term that completes the square on $x^2 - 4x$ to each side of the equation.

$$x^2 - 4x + 4 = 14 + 4$$

$$\left[\frac{1}{2}(-4)\right]^2 = 4$$

Factor the perfect-square trinomial.

$$(x - 2)^2 = 18$$

Take the square root of each side of the equation.

$$\sqrt{(x - 2)^2} = \sqrt{18}$$

Simplify.

$$|x - 2| = 3\sqrt{2}$$

Solve for x.

$$x - 2 = \pm 3\sqrt{2}$$

$$x - 2 = 3\sqrt{2} \qquad x - 2 = -3\sqrt{2}$$
$$x = 2 + 3\sqrt{2} \qquad x = 2 - 3\sqrt{2}$$

Check:

$$x^2 - 4x - 14 = 0$$

$(2 + 3\sqrt{2})^2 - 4(2 + 3\sqrt{2}) - 14$	0
$4 + 12\sqrt{2} + 18 - 8 - 12\sqrt{2} - 14$	0
$0 = 0$	

$$x^2 - 4x - 14 = 0$$

$(2 - 3\sqrt{2})^2 - 4(2 - 3\sqrt{2}) - 14$	0
$4 - 12\sqrt{2} + 18 - 8 + 12\sqrt{2} - 14$	0
$0 = 0$	

Write the solutions.

The solutions are $2 + 3\sqrt{2}$ and $2 - 3\sqrt{2}$.

As shown on the previous page, the solutions of the equation $x^2 - 4x - 14 = 0$ are $2 + 3\sqrt{2}$ and $2 - 3\sqrt{2}$. These are the exact solutions. However, in some situations it may be preferable to have decimal approximations of the solutions to a quadratic equation. Approximate solutions can be found by using a calculator and then rounding to the desired degree of accuracy.

$$2 + 3\sqrt{2} \approx 6.243 \text{ and } 2 - 3\sqrt{2} \approx -2.243$$

To the nearest thousandth, the approximate solutions of the equation $x^2 - 4x - 14 = 0$ are 6.243 and −2.243.

When a, the coefficient of the x^2 term, is not 1, divide each side of the equation by a before completing the square.

■ Solve $2x^2 - x - 2 = 0$. Find the exact solutions and approximate the solutions to the nearest thousandth.

$$2x^2 - x - 2 = 0$$
$$2x^2 - x = 2$$

Divide each side of the equation by the coefficient of x^2.

$$\frac{2x^2 - x}{2} = \frac{2}{2}$$

The coefficient of the x^2 term is now 1.

$$x^2 - \frac{1}{2}x = 1$$

Add the term that completes the square on $x^2 - \frac{1}{2}x$ to each side of the equation. Factor the perfect-square trinomial.

$$x^2 - \frac{1}{2}x + \frac{1}{16} = 1 + \frac{1}{16}$$
$$\left(x - \frac{1}{4}\right)^2 = \frac{17}{16}$$

Take the square root of each side of the equation.

$$\sqrt{\left(x - \frac{1}{4}\right)^2} = \sqrt{\frac{17}{16}}$$

Simplify.

$$\left|x - \frac{1}{4}\right| = \frac{\sqrt{17}}{4}$$

$$x - \frac{1}{4} = \pm\frac{\sqrt{17}}{4}$$

Solve for x.

$$x - \frac{1}{4} = \frac{\sqrt{17}}{4} \qquad x - \frac{1}{4} = -\frac{\sqrt{17}}{4}$$
$$x = \frac{1}{4} + \frac{\sqrt{17}}{4} \qquad x = \frac{1}{4} - \frac{\sqrt{17}}{4}$$

Check the solutions.

$\dfrac{1 + \sqrt{17}}{4}$ and $\dfrac{1 - \sqrt{17}}{4}$ check as solutions.

$$\frac{1 + \sqrt{17}}{4} \approx 1.281 \qquad \frac{1 - \sqrt{17}}{4} \approx -0.781$$

Write the solutions. The exact solutions are $\frac{1 + \sqrt{17}}{4}$ and $\frac{1 - \sqrt{17}}{4}$. To the nearest thousandth, the solutions are 1.281 and -0.781.

EXAMPLE 1 Solve by completing the square.

A. $4x^2 - 8x + 1 = 0$ **B.** $x^2 + 4x + 5 = 0$

Solution **A.** $4x^2 - 8x + 1 = 0$

$$4x^2 - 8x = -1$$

- Subtract 1 from each side of the equation.

$$\frac{4x^2 - 8x}{4} = \frac{-1}{4}$$

$$x^2 - 2x = -\frac{1}{4}$$

- The coefficient of the x^2 term must be 1. Divide each side of the equation by 4.

$$x^2 - 2x + 1 = -\frac{1}{4} + 1$$

- Complete the square. $\left[\frac{1}{2}(-2)\right]^2 = 1$

$$(x - 1)^2 = \frac{3}{4}$$

- Factor the perfect-square trinomial.

$$\sqrt{(x - 1)^2} = \sqrt{\frac{3}{4}}$$

- Take the square root of each side of the equation.

$$|x - 1| = \frac{\sqrt{3}}{2}$$

$$x - 1 = \pm\frac{\sqrt{3}}{2}$$

- Simplify.

$$x - 1 = \frac{\sqrt{3}}{2} \qquad x - 1 = -\frac{\sqrt{3}}{2}$$

- Solve for x.

$$x = 1 + \frac{\sqrt{3}}{2} \qquad x = 1 - \frac{\sqrt{3}}{2}$$

$$x = \frac{2 + \sqrt{3}}{2} \qquad x = \frac{2 - \sqrt{3}}{2}$$

The solutions are $\frac{2 + \sqrt{3}}{2}$ and $\frac{2 - \sqrt{3}}{2}$.

B. $x^2 + 4x + 5 = 0$

$$x^2 + 4x = -5$$

- Subtract 5 from each side of the equation.

$$x^2 + 4x + 4 = -5 + 4$$
$$(x + 2)^2 = -1$$

- Complete the square.
- Factor the perfect-square trinomial.

$$\sqrt{(x + 2)^2} = \sqrt{-1}$$
$$|x + 2| = i$$
$$x + 2 = \pm i$$

- Take the square root of each side of the equation. Simplify.

$$x + 2 = i \qquad x + 2 = -i$$
$$x = -2 + i \qquad x = -2 - i$$

- Solve for x.

The solutions are $-2 + i$ and $-2 - i$.

PROBLEM 1 Solve by completing the square.

 A. $4x^2 - 4x - 1 = 0$ **B.** $2x^2 + x - 5 = 0$

Solution See pages S29–S30.

OBJECTIVE **2** ## Solve quadratic equations by using the quadratic formula

A general formula known as the **quadratic formula** can be derived by applying the method of completing the square to the standard form of a quadratic equation. This formula can be used to solve any quadratic equation.

The solution of the equation $ax^2 + bx + c = 0$ by completing the square is shown below.

$$ax^2 + bx + c = 0$$

Subtract the constant term from each side of the equation.

$$ax^2 + bx + c - c = 0 - c$$
$$ax^2 + bx = -c$$

Divide each side of the equation by a, the coefficient of x^2.

$$\frac{ax^2 + bx}{a} = \frac{-c}{a}$$
$$x^2 + \frac{b}{a}x = -\frac{c}{a}$$

Complete the square by adding $\left(\frac{1}{2} \cdot \frac{b}{a}\right)^2$ to each side of the equation.

$$x^2 + \frac{b}{a}x + \left(\frac{1}{2} \cdot \frac{b}{a}\right)^2 = \left(\frac{1}{2} \cdot \frac{b}{a}\right)^2 - \frac{c}{a}$$
$$x^2 + \frac{b}{a}x + \frac{b^2}{4a^2} = \frac{b^2}{4a^2} - \frac{c}{a}$$

Simplify the right side of the equation.

$$x^2 + \frac{b}{a}x + \frac{b^2}{4a^2} = \frac{b^2}{4a^2} - \left(\frac{c}{a} \cdot \frac{4a}{4a}\right)$$
$$x^2 + \frac{b}{a}x + \frac{b^2}{4a^2} = \frac{b^2}{4a^2} - \frac{4ac}{4a^2}$$
$$x^2 + \frac{b}{a}x + \frac{b^2}{4a^2} = \frac{b^2 - 4ac}{4a^2}$$

Factor the perfect-square trinomial on the left side of the equation.

$$\left(x + \frac{b}{2a}\right)^2 = \frac{b^2 - 4ac}{4a^2}$$

Take the square root of each side of the equation.

$$\sqrt{\left(x + \frac{b}{2a}\right)^2} = \sqrt{\frac{b^2 - 4ac}{4a^2}}$$
$$\left| x + \frac{b}{2a} \right| = \frac{\sqrt{b^2 - 4ac}}{2a}$$
$$x + \frac{b}{2a} = \pm\frac{\sqrt{b^2 - 4ac}}{2a}$$

Solve for x.

$$x + \frac{b}{2a} = \frac{\sqrt{b^2 - 4ac}}{2a} \qquad\qquad x + \frac{b}{2a} = -\frac{\sqrt{b^2 - 4ac}}{2a}$$
$$x = -\frac{b}{2a} + \frac{\sqrt{b^2 - 4ac}}{2a} \qquad\qquad x = -\frac{b}{2a} - \frac{\sqrt{b^2 - 4ac}}{2a}$$
$$= \frac{-b + \sqrt{b^2 - 4ac}}{2a} \qquad\qquad = \frac{-b - \sqrt{b^2 - 4ac}}{2a}$$

Point of Interest

Although mathematicians have studied quadratic equations since around 500 B.C., it was not until the 18th century that the formula was written as it is today. Of further note, the word quadratic *has the same Latin root as does the word* square.

Quadratic Formula

The solutions of $ax^2 + bx + c = 0$, $a \neq 0$, are

$$\frac{-b + \sqrt{b^2 - 4ac}}{2a} \quad \text{and} \quad \frac{-b - \sqrt{b^2 - 4ac}}{2a}$$

The quadratic formula is frequently written in the form

$$x = \frac{-b \pm \sqrt{b^2 - 4ac}}{2a}$$

Solve by using the quadratic formula: $2x^2 + 5x + 3 = 0$

The equation $2x^2 + 5x + 3 = 0$ is in standard form. $a = 2$, $b = 5$, $c = 3$
Replace a, b, and c in the quadratic formula with these values.

$$x = \frac{-b \pm \sqrt{b^2 - 4ac}}{2a}$$

$$= \frac{-(5) \pm \sqrt{(5)^2 - 4(2)(3)}}{2(2)}$$

$$= \frac{-5 \pm \sqrt{25 - 24}}{4}$$

$$= \frac{-5 \pm \sqrt{1}}{4} = \frac{-5 \pm 1}{4}$$

$$x = \frac{-5 + 1}{4} = \frac{-4}{4} = -1 \qquad x = \frac{-5 - 1}{4} = \frac{-6}{4} = -\frac{3}{2}$$

The solutions are -1 and $-\frac{3}{2}$.

Take Note

The solutions of this quadratic equation are rational numbers. When this happens, the equation could have been solved by factoring and using the Principle of Zero Products. This may be easier than using the quadratic formula.

Solve $3x^2 = 4x + 6$ by using the quadratic formula. Find the exact solutions, and approximate the solutions to the nearest thousandth.

Write the equation in standard form by subtracting $4x$ and 6 from each side of the equation.

$a = 3$, $b = -4$, $c = -6$

Replace a, b, and c in the quadratic formula with these values.

$$3x^2 = 4x + 6$$

$$3x^2 - 4x - 6 = 0$$

$$x = \frac{-b \pm \sqrt{b^2 - 4ac}}{2a}$$

$$= \frac{-(-4) \pm \sqrt{(-4)^2 - 4(3)(-6)}}{2(3)}$$

$$= \frac{4 \pm \sqrt{16 - (-72)}}{6}$$

$$= \frac{4 \pm \sqrt{88}}{6} = \frac{4 \pm 2\sqrt{22}}{6}$$

$$= \frac{\cancel{2}(2 \pm \sqrt{22})}{\cancel{2} \cdot 3} = \frac{2 \pm \sqrt{22}}{3}$$

Check:

$3x^2 = 4x + 6$		$3x^2 = 4x + 6$	
$3\left(\dfrac{2 + \sqrt{22}}{3}\right)^2$	$4\left(\dfrac{2 + \sqrt{22}}{3}\right) + 6$	$3\left(\dfrac{2 - \sqrt{22}}{3}\right)^2$	$4\left(\dfrac{2 - \sqrt{22}}{3}\right) + 6$
$3\left(\dfrac{4 + 4\sqrt{22} + 22}{9}\right)$	$\dfrac{8}{3} + \dfrac{4\sqrt{22}}{3} + \dfrac{18}{3}$	$3\left(\dfrac{4 - 4\sqrt{22} + 22}{9}\right)$	$\dfrac{8}{3} - \dfrac{4\sqrt{22}}{3} + \dfrac{18}{3}$
$3\left(\dfrac{26 + 4\sqrt{22}}{9}\right)$	$\dfrac{26}{3} + \dfrac{4\sqrt{22}}{3}$	$3\left(\dfrac{26 - 4\sqrt{22}}{9}\right)$	$\dfrac{26}{3} - \dfrac{4\sqrt{22}}{3}$
$\dfrac{26 + 4\sqrt{22}}{3} = \dfrac{26 + 4\sqrt{22}}{3}$		$\dfrac{26 - 4\sqrt{22}}{3} = \dfrac{26 - 4\sqrt{22}}{3}$	

The exact solutions are $\dfrac{2 + \sqrt{22}}{3}$ and $\dfrac{2 - \sqrt{22}}{3}$.

$$\dfrac{2 + \sqrt{22}}{3} \approx 2.230 \qquad \dfrac{2 - \sqrt{22}}{3} \approx -0.897$$

To the nearest thousandth, the solutions are 2.230 and −0.897.

Solve by using the quadratic formula: $4x^2 = 8x - 13$

Write the equation in standard form.

$$4x^2 = 8x - 13$$
$$4x^2 - 8x + 13 = 0$$
$$a = 4, b = -8, c = 13$$

Replace a, b, and c in the quadratic formula by their values.

$$x = \dfrac{-b \pm \sqrt{b^2 - 4ac}}{2a}$$
$$= \dfrac{-(-8) \pm \sqrt{(-8)^2 - 4 \cdot 4 \cdot 13}}{2 \cdot 4}$$

Simplify.

$$= \dfrac{8 \pm \sqrt{64 - 208}}{8}$$
$$= \dfrac{8 \pm \sqrt{-144}}{8}$$

Write the answer in the form $a + bi$.

$$= \dfrac{8 \pm 12i}{8} = \dfrac{2 \pm 3i}{2} = 1 \pm \dfrac{3}{2}i$$

Check:

$4x^2 = 8x - 13$		$4x^2 = 8x - 13$	
$4\left(1 + \dfrac{3}{2}i\right)^2$	$8\left(1 + \dfrac{3}{2}i\right) - 13$	$4\left(1 - \dfrac{3}{2}i\right)^2$	$8\left(1 - \dfrac{3}{2}i\right) - 13$
$4\left(1 + 3i - \dfrac{9}{4}\right)$	$8 + 12i - 13$	$4\left(1 - 3i - \dfrac{9}{4}\right)$	$8 - 12i - 13$
$4\left(-\dfrac{5}{4} + 3i\right)$	$-5 + 12i$	$4\left(-\dfrac{5}{4} - 3i\right)$	$-5 - 12i$
$-5 + 12i = -5 + 12i$		$-5 - 12i = -5 - 12i$	

The solutions are $1 + \dfrac{3}{2}i$ and $1 - \dfrac{3}{2}i$.

EXAMPLE 2 Solve by using the quadratic formula.

 A. $4x^2 + 12x + 9 = 0$ **B.** $2x^2 - x + 5 = 0$

Solution **A.** $4x^2 + 12x + 9 = 0$ • $a = 4, b = 12, c = 9$

$$x = \frac{-b \pm \sqrt{b^2 - 4ac}}{2a}$$ • **Replace a, b, and c in the quadratic formula by their values. Then simplify.**

$$= \frac{-12 \pm \sqrt{12^2 - 4 \cdot 4 \cdot 9}}{2 \cdot 4}$$

$$= \frac{-12 \pm \sqrt{144 - 144}}{8}$$

$$= \frac{-12 \pm \sqrt{0}}{8} = \frac{-12}{8} = -\frac{3}{2}$$ • **The equation has a double root.**

The solution is $-\frac{3}{2}$.

B. $2x^2 - x + 5 = 0$ • $a = 2, b = -1, c = 5$

$$x = \frac{-b \pm \sqrt{b^2 - 4ac}}{2a}$$ • **Replace a, b, and c in the quadratic formula by their values. Then simplify.**

$$= \frac{-(-1) \pm \sqrt{(-1)^2 - 4 \cdot 2 \cdot 5}}{2 \cdot 2}$$

$$= \frac{1 \pm \sqrt{1 - 40}}{4}$$

$$= \frac{1 \pm \sqrt{-39}}{4} = \frac{1 \pm i\sqrt{39}}{4}$$

The solutions are $\frac{1}{4} + \frac{\sqrt{39}}{4}i$ and $\frac{1}{4} - \frac{\sqrt{39}}{4}i$.

PROBLEM 2 Solve by using the quadratic formula.

 A. $x^2 + 6x - 9 = 0$ **B.** $4x^2 = 4x - 1$

Solution See page S30.

In Example 2A, the solution of the equation is a double root, and in Example 2B, the solutions are complex numbers.

In the quadratic formula, the quantity $b^2 - 4ac$ is called the **discriminant.** When a, b, and c are real numbers, the discriminant determines whether a quadratic equation will have a double root, two real number solutions that are not equal, or two complex number solutions.

Effect of the Discriminant on the Solutions of a Quadratic Equation

1. If $b^2 - 4ac = 0$, the equation has one real number solution, a double root.

2. If $b^2 - 4ac > 0$, the equation has two real number solutions that are not equal.

3. If $b^2 - 4ac < 0$, the equation has two complex number solutions.

The equation $x^2 - 4x - 5 = 0$ has two real number solutions because the discriminant is greater than zero.

$a = 1, b = -4, c = -5$

$b^2 - 4ac = (-4)^2 - 4(1)(-5)$
$= 16 + 20$
$= 36$

$36 > 0$

EXAMPLE 3 Use the discriminant to determine whether $4x^2 - 2x + 5 = 0$ has one real number solution, two real number solutions, or two complex number solutions.

Solution $b^2 - 4ac = (-2)^2 - 4(4)(5)$ • $a = 4, b = -2, c = 5$
$= 4 - 80$
$= -76$

$-76 < 0$ • **The discriminant is less than 0.**

The equation has two complex number solutions.

PROBLEM 3 Use the discriminant to determine whether $3x^2 - x - 1 = 0$ has one real number solution, two real number solutions, or two complex number solutions.

Solution See page S30.

9.2 Concept Review

Determine whether the following statements are always true, sometimes true, or never true.

1. A quadratic equation can be solved by completing the square.

2. The expression $b^2 - 4ac$ is called the discriminant.

3. If $b^2 > 4ac$, the quadratic equation has two complex number solutions.

4. The quadratic formula can be used to solve quadratic equations that have irrational roots.

5. $x^2 + 5x + 25$ is a perfect-square trinomial.

6. To complete the square on $3x^2 + 6x$, add 9.

9.2 Exercises

1 Solve quadratic equations by completing the square

1. a. To complete the square on $x^2 + 14x$, find the constant c that makes $x^2 + 14x + c$ a ____?____ trinomial.

The constant term c will be the square of half the coefficient of

_____?_____, so $c = \left[\dfrac{1}{2}(\underline{\quad ? \quad})\right]^2 = (\underline{\quad ? \quad})$.

b. Complete the square on $x^2 + 14x$ and write the result as the square of a binomial:

$x^2 + 14x + \underline{\quad ? \quad} = (\underline{\quad ? \quad})^2$

2. Rewrite the equation $3x^2 - 6x + 2 = 0$ in the form $(x - m)^2 = n$.

$3x^2 - 6x + 2 = 0$

$3x^2 - 6x = \underline{\quad ? \quad}$ • Subtract 2 from each side of the equation.

$x^2 - (\underline{\quad ? \quad})x = \underline{\quad ? \quad}$ • To make the coefficient of x^2 equal to 1, divide each side of the equation by 3.

$x^2 - 2x + \underline{\quad ? \quad} = -\dfrac{2}{3} + \underline{\quad ? \quad}$ • Add $\left[\dfrac{1}{2}(\underline{\quad ? \quad})\right]^2$ to each side of the equation.

$(x - \underline{\quad ? \quad})^2 = \underline{\quad ? \quad}$ • Factor the left side of the equation. Simplify the right side of the equation.

3. What is the next step when using completing the square to solve $x^2 + 6x = 4$?

4. Are there some quadratic equations that cannot be solved by completing the square? Why or why not?

Solve by completing the square.

5. $x^2 - 4x - 5 = 0$ **6.** $y^2 + 6y + 5 = 0$ **7.** $v^2 + 8v - 9 = 0$

8. $w^2 - 2w - 24 = 0$ **9.** $z^2 - 6z + 9 = 0$ **10.** $u^2 + 10u + 25 = 0$

11. $r^2 + 4r - 7 = 0$ **12.** $s^2 + 6s - 1 = 0$ **13.** $x^2 - 6x + 7 = 0$

14. $y^2 + 8y + 13 = 0$ **15.** $z^2 - 2z + 2 = 0$ **16.** $t^2 - 4t + 8 = 0$

First try to solve by factoring. If you are unable to solve the equation by factoring, solve the equation by completing the square.

17. $t^2 - t - 1 = 0$ **18.** $u^2 - u - 7 = 0$ **19.** $y^2 - 6y = 4$

20. $w^2 + 4w = 2$ **21.** $x^2 = 8x - 15$ **22.** $z^2 = 4z - 3$

23. $v^2 = 4v - 13$ **24.** $x^2 = 2x - 17$ **25.** $p^2 + 6p = -13$

26. $x^2 + 4x = -20$ **27.** $y^2 - 2y = 17$ **28.** $x^2 + 10x = 7$

29. $z^2 = z + 4$ **30.** $r^2 = 3r - 1$ **31.** $x^2 + 13 = 2x$

32. $x^2 + 27 = 6x$

33. $2y^2 + 3y + 1 = 0$

34. $2t^2 + 5t - 3 = 0$

35. $4r^2 - 8r = -3$

36. $4u^2 - 20u = -9$

37. $6y^2 - 5y = 4$

38. $6v^2 - 7v = 3$

39. $4x^2 - 4x + 5 = 0$

40. $4t^2 - 4t + 17 = 0$

41. $9x^2 - 6x + 2 = 0$

42. $9y^2 - 12y + 13 = 0$

43. $2s^2 = 4s + 5$

44. $3u^2 = 6u + 1$

45. $2r^2 = 3 - r$

46. $2x^2 = 12 - 5x$

47. $y - 2 = (y - 3)(y + 2)$

48. $8s - 11 = (s - 4)(s - 2)$

49. $6t - 2 = (2t - 3)(t - 1)$

50. $2z + 9 = (2z + 3)(z + 2)$

51. $(x - 4)(x + 1) = x - 3$

52. $(y - 3)^2 = 2y + 10$

Solve by completing the square. Approximate the solutions to the nearest thousandth.

53. $z^2 + 2z = 4$

54. $t^2 - 4t = 7$

55. $2x^2 = 4x - 1$

56. $3y^2 = 5y - 1$

57. $4z^2 + 2z - 1 = 0$

58. $4w^2 - 8w = 3$

59. True or false? If $c > 4$, the solutions of $x^2 + 4x + c = 0$ are real numbers.

60. True or false? If $c = b^2$, $x^2 + 2bx = -c$ has one real solution.

2 Solve quadratic equations by using the quadratic formula

61. Write the quadratic formula. What does each variable in the formula represent?

62. Write the expression that appears under the radical symbol in the quadratic formula. What is this quantity called? What can it be used to determine?

63. Suppose you must solve the quadratic equation $x^2 = 3x + 5$. Does it matter whether you rewrite the equation as $x^2 - 3x - 5 = 0$ or as $0 = -x^2 + 3x + 5$ before you begin?

64. Which method of solving a quadratic equation do you prefer: completing the square or using the quadratic formula? Why?

65. To write the equation $x^2 = 6x - 10$ in standard form, subtract ___?___ from and add ___?___ to each side of the equation. The resulting equation is ___?___ $= 0$. Then $a = $ ___?___ , $b = $ ___?___ , $c = $ ___?___ .

66. Use the values of a, b, and c that you found in Exercise 65 to evaluate the discriminant for the equation $x^2 = 6x - 10$: $b^2 - 4ac = ($ ___?___ $)^2 - 4($ ___?___ $)($ ___?___ $) = $ ___?___ $- $ ___?___ $= $ ___?___ .

67. Based on the value of the discriminate you found in Exercise 66, the equation $x^2 = 6x - 10$ must have two ___?___ number solutions.

68. Using the quadratic formula to solve $x^2 = 6x - 10$ leads to the result $\dfrac{6 \pm \sqrt{-4}}{2}$. Simplify this result.

$$\dfrac{6 \pm \sqrt{-4}}{2} = \dfrac{6 \pm \underline{\ \ ?\ \ }}{2}$$ • Rewrite $\sqrt{-4}$ as the imaginary number ___?___.

$$= \underline{\ \ ?\ \ } \pm \underline{\ \ ?\ \ }$$ • Divide each term of the numerator by the denominator.

Solve by using the quadratic formula.

69. $x^2 - 3x - 10 = 0$

70. $z^2 - 4z - 8 = 0$

71. $y^2 + 5y - 36 = 0$

72. $z^2 - 3z - 40 = 0$

73. $w^2 = 8w + 72$

74. $t^2 = 2t + 35$

75. $v^2 = 24 - 5v$

76. $x^2 = 18 - 7x$

77. $2y^2 + 5y - 3 = 0$

78. $4p^2 - 7p + 3 = 0$

79. $8s^2 = 10s + 3$

80. $12t^2 = 5t + 2$

First try to solve by factoring. If you are unable to solve the equation by factoring, solve the equation by using the quadratic formula.

81. $v^2 - 2v - 7 = 0$

82. $t^2 - 2t - 11 = 0$

83. $y^2 - 8y - 20 = 0$

84. $x^2 = 14x - 24$

85. $v^2 = 12v - 24$

86. $2z^2 - 2z - 1 = 0$

87. $4x^2 - 4x - 7 = 0$

88. $2p^2 - 8p + 5 = 0$

89. $2s^2 - 3s + 1 = 0$

90. $4w^2 - 4w - 1 = 0$

91. $3x^2 + 10x + 6 = 0$

92. $3v^2 = 6v - 2$

93. $6w^2 = 19w - 10$

94. $z^2 + 2z + 2 = 0$

95. $p^2 - 4p + 5 = 0$

96. $y^2 - 2y + 5 = 0$

97. $x^2 + 6x + 13 = 0$

98. $s^2 - 4s + 13 = 0$

99. $t^2 - 6t + 10 = 0$

100. $2w^2 - 2w + 5 = 0$

101. $4v^2 + 8v + 3 = 0$

102. $2x^2 + 6x + 5 = 0$

103. $2y^2 + 2y + 13 = 0$

104. $4t^2 - 6t + 9 = 0$

105. $3v^2 + 6v + 1 = 0$

106. $2r^2 = 4r - 11$

107. $3y^2 = 6y - 5$

108. $2x(x - 2) = x + 12$

109. $10y(y + 4) = 15y - 15$

110. $(3s - 2)(s + 1) = 2$

111. $(2t + 1)(t - 3) = 9$

Solve by using the quadratic formula. Approximate the solutions to the nearest thousandth.

112. $x^2 - 6x - 6 = 0$

113. $p^2 - 8p + 3 = 0$

114. $r^2 - 2r = 4$

115. $w^2 + 4w = 1$

116. $3t^2 = 7t + 1$

117. $2y^2 = y + 5$

Use the discriminant to determine whether the quadratic equation has one real number solution, two real number solutions, or two complex number solutions.

118. $2z^2 - z + 5 = 0$

119. $3y^2 + y + 1 = 0$

120. $9x^2 - 12x + 4 = 0$

121. $4x^2 + 20x + 25 = 0$

122. $2v^2 - 3v - 1 = 0$

123. $3w^2 + 3w - 2 = 0$

124. $2p^2 + 5p + 1 = 0$

125. $2t^2 + 9t + 3 = 0$

126. $5z^2 + 2 = 0$

127. True or false? If $a > 0$ and $c < 0$, the solutions of $ax^2 + bx + c = 0$ are real numbers.

128. True or false? If $a > 0$, $b = 0$, and $c > 0$, the solutions of $ax^2 + bx + c = 0$ are real numbers.

Classify each equation as linear or quadratic.

129. $2x^2 + 3x = 1$

130. $2x = 3x - 1$

131. $4x - 2 = 5$

132. $4x^2 - 2x = 5$

133. $6x(x - 2) = 7$

134. $6x - 2 = 7$

9.2 Applying Concepts

Solve.

135. $\sqrt{2}y^2 + 3y - 2\sqrt{2} = 0$

136. $\sqrt{3}z^2 + 10z - 3\sqrt{3} = 0$

137. $\sqrt{2}x^2 + 5x - 3\sqrt{2} = 0$

138. $\sqrt{3}w^2 + w - 2\sqrt{3} = 0$

139. $t^2 - t\sqrt{3} + 1 = 0$

140. $y^2 + y\sqrt{7} + 2 = 0$

Solve for x.

141. $x^2 - ax - 2a^2 = 0$

142. $x^2 - ax - 6a^2 = 0$

143. $2x^2 + 3ax - 2a^2 = 0$

144. $2x^2 - 7ax + 3a^2 = 0$

145. $x^2 - 2x - y = 0$

146. $x^2 - 4xy - 4 = 0$

For what values of p does the quadratic equation have two real number solutions that are not equal? Write the answer in set-builder notation.

147. $x^2 - 6x + p = 0$

148. $x^2 + 10x + p = 0$

For what values of p does the quadratic equation have two complex number solutions? Write the answer in set-builder notation.

149. $x^2 - 2x + p = 0$

150. $x^2 + 4x + p = 0$

Solve.

151. Find all values of x that satisfy the equation $x^2 + ix + 2 = 0$.

152. Show that the equation $2x^2 + bx - 2 = 0$ always has real number solutions regardless of the value of b.

153. Sports The height, h (in feet), of a baseball above the ground t seconds after it is hit can be approximated by the equation $h = -16t^2 + 70t + 4$. Using this equation, determine when the ball will hit the ground. Round to the nearest hundredth. (*Hint:* The ball hits the ground when $h = 0$.)

154. Sports After a baseball is hit, there are two equations that can be considered. One gives the height, h (in feet), the ball is above the ground t seconds after it is hit. The second is the distance, s (in feet), the ball is from home plate t seconds after it is hit. A model of this situation is given by $h = -16t^2 + 70t + 4$ and $s = 44.5t$. Using this model, determine whether the ball will clear a fence 325 ft from home plate. Round to the nearest tenth.

155. Conservation The National Forest Management Act of 1976 specifies that harvesting timber in national forests must be accomplished in conjunction with environmental considerations. One such consideration is providing a habitat for the spotted owl. One model of the survival of the spotted owl requires the solution of the equation $x^2 - s_a x - s_j s_s f = 0$ for x. Different values of s_a, s_j, s_s, and f are given in the table at the right. (*Source:* Charles Biles and Barry Noon, "The Spotted Owl!" from *The Journal of Undergraduate Mathematics and Its Application*, Vol. 11, No. 2, 1990. Reprinted with permission.) The values are particularly important because they relate to the survival of the owl. If $x > 1$, then the model predicts a growth in the population; if $x = 1$, the population remains steady; if $x < 1$, the population decreases. The important solution of the equation is the larger of the two roots of the equation.

	U.S. Forest Service	Lande
s_j	0.34	0.11
s_s	0.97	0.71
s_a	0.97	0.94
f	0.24	0.24

a. Determine the larger root of this equation for values provided by the U.S. Forest Service. Round to the nearest hundredth. Does it predict that the population will increase, remain steady, or decrease?

b. Determine the larger root of this equation for the values provided by R. Lande in *Oecologia* (Vol. 75, 1988). Round to the nearest hundredth. Does it predict that the population will increase, remain steady, or decrease?

156. Agriculture An agricultural research group determined that the amount of grass harvested from a pasture could be modeled by the equation $A = -0.001532x^2 + 0.3358x - 7.5323$, where x is the amount, in 100,000 kg per hectare, of a fertilizer that is applied to the pasture. According to this model, how many kilograms per hectare of fertilizer must be applied to a pasture to produce 800,000 kg per hectare of grass? Round to the nearest thousand.

157. Here is an outline for an alternative method for deriving the quadratic formula. Fill in the details of this derivation. Begin with the equation $ax^2 + bx + c = 0$. Subtract c from each side of the equation. Multiply each side of the equation by $4a$. Now add b^2 to each side. Supply the remaining steps until you have reached the quadratic formula.

158. Explain how to complete the square of $x^2 + bx$.

9.3 Equations That Are Reducible to Quadratic Equations

OBJECTIVE 1 Equations that are quadratic in form

Certain equations that are not quadratic equations can be expressed in quadratic form by making suitable substitutions. An equation is **quadratic in form** if it can be written as $au^2 + bu + c = 0$.

To see that the equation at the right is quadratic in form, let $x^2 = u$. Replace x^2 by u. The equation is quadratic in form.

$$x^4 - 4x^2 - 5 = 0$$
$$(x^2)^2 - 4(x^2) - 5 = 0$$
$$u^2 - 4u - 5 = 0$$

To see that the equation at the right is quadratic in form, let $y^{\frac{1}{2}} = u$. Replace $y^{\frac{1}{2}}$ by u. The equation is quadratic in form.

$$y - y^{\frac{1}{2}} - 6 = 0$$
$$\left(y^{\frac{1}{2}}\right)^2 - \left(y^{\frac{1}{2}}\right) - 6 = 0$$
$$u^2 - u - 6 = 0$$

The key to recognizing equations that are quadratic in form is that when the equation is written in standard form, the exponent on one variable term is $\frac{1}{2}$ the exponent on the other variable term.

Solve: $z + 7z^{\frac{1}{2}} - 18 = 0$

The equation $z + 7z^{\frac{1}{2}} - 18 = 0$ is quadratic in form.

$$z + 7z^{\frac{1}{2}} - 18 = 0$$
$$\left(z^{\frac{1}{2}}\right)^2 + 7\left(z^{\frac{1}{2}}\right) - 18 = 0$$

To solve this equation, let $z^{\frac{1}{2}} = u$.

$$u^2 + 7u - 18 = 0$$

Solve for u by factoring.

$$(u - 2)(u + 9) = 0$$

$$u - 2 = 0 \qquad u + 9 = 0$$
$$u = 2 \qquad u = -9$$

Replace u by $z^{\frac{1}{2}}$.

$$z^{\frac{1}{2}} = 2 \qquad z^{\frac{1}{2}} = -9$$

Solve for z by squaring each side of the equation.

$$\left(z^{\frac{1}{2}}\right)^2 = 2^2 \qquad \left(z^{\frac{1}{2}}\right)^2 = (-9)^2$$
$$z = 4 \qquad z = 81$$

Check the solution. When each side of the equation has been squared, the resulting equation may have an extraneous solution.

Check:
$$\begin{array}{c|c} z + 7z^{\frac{1}{2}} - 18 = 0 & \\ \hline 4 + 7(4)^{\frac{1}{2}} - 18 & 0 \\ 4 + 7 \cdot 2 - 18 & \\ 4 + 14 - 18 & \\ 0 = 0 & \end{array}$$

$$\frac{z + 7z^{\frac{1}{2}} - 18 = 0}{81 + 7(81)^{\frac{1}{2}} - 18 \;\Big|\; 0}$$

$$81 + 7 \cdot 9 - 18$$
$$81 + 63 - 18$$
$$126 \neq 0$$

4 checks as a solution, but 81 does not.

Write the solution. The solution is 4.

EXAMPLE 1 Solve: **A.** $x^4 + x^2 - 12 = 0$ **B.** $x^{\frac{2}{3}} - 2x^{\frac{1}{3}} - 3 = 0$

Solution **A.** $x^4 + x^2 - 12 = 0$ • The equation is quadratic
 $(x^2)^2 + (x^2) - 12 = 0$ in form.
 $u^2 + u - 12 = 0$ • Let $x^2 = u$.
 $(u - 3)(u + 4) = 0$ • Solve for u by factoring.

$$u - 3 = 0 \qquad u + 4 = 0$$
$$u = 3 \qquad u = -4$$

$$x^2 = 3 \qquad\qquad x^2 = -4$$ • Replace u by x^2.
$$\sqrt{x^2} = \sqrt{3} \qquad \sqrt{x^2} = \sqrt{-4}$$ • Solve for x by taking
$$x = \pm\sqrt{3} \qquad\quad x = \pm 2i$$ square roots.

The solutions are $\sqrt{3}$, $-\sqrt{3}$, $2i$, and $-2i$.

B. $x^{\frac{2}{3}} - 2x^{\frac{1}{3}} - 3 = 0$ • The equation is quadratic
 $\left(x^{\frac{1}{3}}\right)^2 - 2\left(x^{\frac{1}{3}}\right) - 3 = 0$ in form.
 $u^2 - 2u - 3 = 0$ • Let $x^{\frac{1}{3}} = u$.
 $(u - 3)(u + 1) = 0$ • Solve for u by factoring.

$$u - 3 = 0 \qquad u + 1 = 0$$
$$u = 3 \qquad u = -1$$

$$x^{\frac{1}{3}} = 3 \qquad\qquad x^{\frac{1}{3}} = -1$$ • Replace u by $x^{\frac{1}{3}}$.

$$\left(x^{\frac{1}{3}}\right)^3 = 3^3 \qquad \left(x^{\frac{1}{3}}\right)^3 = (-1)^3$$ • Solve for x by cubing both
$$x = 27 \qquad\qquad x = -1$$ sides of the equation.

The solutions are 27 and -1.

PROBLEM 1 Solve. **A.** $x - 5x^{\frac{1}{2}} + 6 = 0$ **B.** $4x^4 + 35x^2 - 9 = 0$

Solution See page S30.

OBJECTIVE **2** **Radical equations**

Certain equations containing a radical can be solved by first solving the equation for the radical expression and then squaring each side of the equation.

Remember that when each side of an equation has been squared, the resulting equation may have an extraneous solution. Therefore, the solutions of a radical equation must be checked.

Solve: $\sqrt{x + 2} + 4 = x$

Solve for the radical expression.	$\sqrt{x + 2} + 4 = x$ $\sqrt{x + 2} = x - 4$
Square each side of the equation.	$\left(\sqrt{x + 2}\right)^2 = (x - 4)^2$
Simplify.	$x + 2 = x^2 - 8x + 16$
Write the equation in standard form.	$0 = x^2 - 9x + 14$
Solve for x by factoring.	$0 = (x - 7)(x - 2)$ $x - 7 = 0 \qquad x - 2 = 0$ $x = 7 \qquad\quad x = 2$

Check the solution.

Check:

$$\begin{array}{c|c} \sqrt{x + 2} + 4 = x \\ \hline \sqrt{7 + 2} + 4 & 7 \\ \sqrt{9} + 4 \\ 3 + 4 \\ 7 = 7 \end{array} \qquad \begin{array}{c|c} \sqrt{x + 2} + 4 = x \\ \hline \sqrt{2 + 2} + 4 & 2 \\ \sqrt{4} + 4 \\ 2 + 4 \\ 6 \neq 2 \end{array}$$

2 does not check as a solution.

Write the solution. The solution is 7.

EXAMPLE 2 Solve: $\sqrt{7y - 3} + 3 = 2y$

Solution

$\sqrt{7y - 3} + 3 = 2y$
$\sqrt{7y - 3} = 2y - 3$
$\left(\sqrt{7y - 3}\right)^2 = (2y - 3)^2$
$7y - 3 = 4y^2 - 12y + 9$
$0 = 4y^2 - 19y + 12$

$0 = (4y - 3)(y - 4)$

$4y - 3 = 0 \qquad y - 4 = 0$
$4y = 3 \qquad\quad y = 4$
$y = \dfrac{3}{4}$

- Solve for the radical expression.
- Square each side of the equation.
- Write the equation in standard form.
- Solve for y by factoring.

- 4 checks as a solution.
- $\frac{3}{4}$ does not check as a solution.

The solution is 4.

PROBLEM 2 Solve: $\sqrt{2x + 1} + x = 7$

Solution See page S30.

If an equation contains more than one radical, the procedure of solving for the radical expression and squaring each side of the equation may have to be repeated.

EXAMPLE 3 Solve: $\sqrt{2y + 1} - \sqrt{y} = 1$

Solution

$\sqrt{2y + 1} - \sqrt{y} = 1$
$\sqrt{2y + 1} = \sqrt{y} + 1$
$\left(\sqrt{2y + 1}\right)^2 = \left(\sqrt{y} + 1\right)^2$

- Solve for one of the radical expressions.
- Square each side of the equation.

$$2y + 1 = y + 2\sqrt{y} + 1$$
$$y = 2\sqrt{y}$$
$$y^2 = (2\sqrt{y})^2$$
$$y^2 = 4y$$
$$y^2 - 4y = 0$$
$$y(y - 4) = 0$$

- Solve for the radical expression.
- Square each side of the equation.

$$y = 0 \qquad y - 4 = 0$$
$$y = 4$$

- Write the equation in standard form.

- 0 and 4 check as solutions.

The solutions are 0 and 4.

PROBLEM 3 Solve: $\sqrt{2x - 1} + \sqrt{x} = 2$

Solution See page S31.

OBJECTIVE 3 **Fractional equations**

After each side of a fractional equation has been multiplied by the LCM of the denominators, the resulting equation is sometimes a quadratic equation. The solutions to the resulting equation must be checked, because multiplying each side of an equation by a variable expression may produce an equation that has a solution that is not a solution of the original equation.

Solve: $\dfrac{1}{r} + \dfrac{1}{r + 1} = \dfrac{3}{2}$

$$\frac{1}{r} + \frac{1}{r + 1} = \frac{3}{2}$$

Multiply each side of the equation by the LCM of the denominators.

$$2r(r + 1)\left(\frac{1}{r} + \frac{1}{r + 1}\right) = 2r(r + 1) \cdot \frac{3}{2}$$
$$2(r + 1) + 2r = r(r + 1) \cdot 3$$
$$2r + 2 + 2r = (r^2 + r) \cdot 3$$
$$4r + 2 = 3r^2 + 3r$$

Write the equation in standard form.

$$0 = 3r^2 - r - 2$$
$$0 = (3r + 2)(r - 1)$$

Solve for r by factoring.

$$3r + 2 = 0 \qquad r - 1 = 0$$
$$3r = -2 \qquad r = 1$$
$$r = -\frac{2}{3}$$

$-\dfrac{2}{3}$ and 1 check as solutions.

Write the solutions.

The solutions are $-\dfrac{2}{3}$ and 1.

EXAMPLE 4 Solve: $\dfrac{18}{2a - 1} + 3a = 17$

Solution

$$\frac{18}{2a - 1} + 3a = 17$$

- The LCM of the denominators is $2a - 1$.

$$(2a - 1)\left(\frac{18}{2a - 1} + 3a\right) = (2a - 1)17$$

$$(2a - 1)\frac{18}{2a - 1} + (2a - 1)(3a) = (2a - 1)17$$

$$18 + 6a^2 - 3a = 34a - 17$$ • Write the
$$6a^2 - 37a + 35 = 0$$ equation in
 standard form.

$$(6a - 7)(a - 5) = 0$$ • Solve for *a* by
 factoring.

$$6a - 7 = 0 \qquad a - 5 = 0$$
$$6a = 7 \qquad a = 5$$
$$a = \frac{7}{6}$$

The solutions are $\frac{7}{6}$ and 5.

PROBLEM 4 Solve: $3y + \dfrac{25}{3y - 2} = -8$

■ **Solution** See page S31.

9.3 Concept Review

Determine whether the following statements are always true, sometimes true, or never true.

1. An equation that is quadratic in form can be solved by the quadratic formula.

2. Squaring both sides of a radical equation produces an extraneous root.

3. $x + 3\sqrt{x} - 8 = 0$ is quadratic in form.

4. $\sqrt[4]{x} + 2\sqrt{3x} - 8 = 0$ is quadratic in form.

9.3 Exercises

1 Equations that are quadratic in form

1. The equation $x^8 + 4x^4 - 5 = 0$ is equivalent to the quadratic equation $u^2 + 4u - 5 = 0$, where $u = $ ___?___ .

2. The equation $p - 2p^{\frac{1}{2}} - 8 = 0$ is equivalent to the quadratic equation $u^2 - 2u - 8 = 0$, where $u = $ ___?___ .

3. ✎ What does it mean for an equation to be quadratic in form?

4. ✎ Explain how to show that $x^4 - 2x^2 - 3 = 0$ is quadratic in form.

Solve.

5. $x^4 - 13x^2 + 36 = 0$

6. $y^4 - 5y^2 + 4 = 0$

7. $z^4 - 6z^2 + 8 = 0$

8. $t^4 - 12t^2 + 27 = 0$

9. $p - 3p^{\frac{1}{2}} + 2 = 0$

10. $v - 7v^{\frac{1}{2}} + 12 = 0$

11. $x - x^{\frac{1}{2}} - 12 = 0$

12. $w - 2w^{\frac{1}{2}} - 15 = 0$

13. $z^4 + 3z^2 - 4 = 0$

14. $y^4 + 5y^2 - 36 = 0$

15. $x^4 + 12x^2 - 64 = 0$

16. $x^4 - 81 = 0$

17. $p + 2p^{\frac{1}{2}} - 24 = 0$

18. $v + 3v^{\frac{1}{2}} - 4 = 0$

19. $y^{\frac{2}{3}} - 9y^{\frac{1}{3}} + 8 = 0$

20. $z^{\frac{2}{3}} - z^{\frac{1}{3}} - 6 = 0$

21. $x^6 - 9x^3 + 8 = 0$

22. $y^6 + 9y^3 + 8 = 0$

23. $z^8 - 17z^4 + 16 = 0$

24. $v^4 - 15v^2 - 16 = 0$

25. $p^{\frac{2}{3}} + 2p^{\frac{1}{3}} - 8 = 0$

26. $w^{\frac{2}{3}} + 3w^{\frac{1}{3}} - 10 = 0$

27. $2x - 3x^{\frac{1}{2}} + 1 = 0$

28. $3y - 5y^{\frac{1}{2}} - 2 = 0$

29. Which equations can be solved by solving the quadratic equation $u^2 - 8u - 20 = 0$?
a. $x^{10} - 8x^5 - 20 = 0$ **b.** $x^{16} - 8x^4 - 20 = 0$
c. $x^{1/10} - 8x^{1/5} - 20 = 0$ **d.** $x^{2/5} - 8x^{1/5} - 20 = 0$

30. Each equation can be solved by solving the quadratic equation $u^2 + 5u + 6 = 0$. Which equations could possibly end up with extraneous solutions?
a. $x^{1/4} + 5x^{1/8} + 6 = 0$ **b.** $x^8 + 5x^4 + 6 = 0$
c. $x + 5x^{1/2} + 6 = 0$ **d.** $x^{2/3} + 5x^{1/3} + 6 = 0$

2 Radical equations

31. The first step in the solution of the equation $\sqrt{a + 1} + 7 = a + 2$ is to solve for ___?___ by subtracting ___?___ from each side of the equation. The second step is to ___?___ each side of the equation.

32. Solving the equation given in Exercise 31 yields $a = 8$ and $a = 3$. Checking these solutions in the original equation shows that the solution ___?___ is extraneous.

33. Look at the equations to be solved in Exercises 35 to 52. List the exercise numbers of the equations for which the first step of the solution will be to square both sides of the equation.

34. Look at the equations to be solved in Exercises 35 to 52. List the exercise numbers of the equations whose solutions will require squaring both sides two separate times.

Solve.

35. $\sqrt{x+1} + x = 5$

36. $\sqrt{x-4} + x = 6$

37. $x = \sqrt{x} + 6$

38. $\sqrt{2y-1} = y - 2$

39. $\sqrt{3w+3} = w + 1$

40. $\sqrt{2s+1} = s - 1$

41. $\sqrt{4y+1} - y = 1$

42. $\sqrt{3s+4} + 2s = 12$

43. $\sqrt{10x+5} - 2x = 1$

44. $\sqrt{t+8} = 2t + 1$

45. $\sqrt{p+11} = 1 - p$

46. $x - 7 = \sqrt{x-5}$

47. $\sqrt{x-1} - \sqrt{x} = -1$

48. $\sqrt{y} + 1 = \sqrt{y+5}$

49. $\sqrt{2x-1} = 1 - \sqrt{x-1}$

50. $\sqrt{x+6} + \sqrt{x+2} = 2$

51. $\sqrt{t+3} + \sqrt{2t+7} = 1$

52. $\sqrt{5-2x} = \sqrt{2-x} + 1$

3 Fractional equations

53. Use the equation $\dfrac{4}{y} = \dfrac{y-6}{y-4}$.

 a. The first step in solving the equation is to clear denominators by multiplying each side of the equation by the LCM of the denominators ___?___ and ___?___. The LCM is ___?___.

 b. When you multiply the left side of the equation by the LCM $y(y-4)$, the left side simplifies to ___?___.

 c. When you multiply the right side of the equation by the LCM $y(y-4)$, the right side simplifies to ___?___.

 d. Use your answers to parts b and c to write the equation that results from clearing denominators: ___?___ = ___?___.

54. a. Using the results of Exercise 53, we know we can solve the rational equation $\dfrac{4}{y} = \dfrac{y-6}{y-4}$ by solving the quadratic equation $4y - 16 = y^2 - 6y$. Written in standard form, this equation is $0 = $ ___?___.

 b. Solve $0 = y^2 - 10y + 16$ by factoring: $0 = ($___?___$)($___?___$)$, so $y = $ ___?___ or $y = $ ___?___.

 c. Both solutions from part b check in the original equation, because neither solution makes one of the denominators equal to ___?___.

Solve.

55. $x = \dfrac{10}{x-9}$

56. $z = \dfrac{5}{z-4}$

57. $\dfrac{t}{t+1} = \dfrac{-2}{t-1}$

58. $\dfrac{2v}{v-1} = \dfrac{5}{v+2}$

59. $\dfrac{y-1}{y+2} + y = 1$

60. $\dfrac{2p-1}{p-2} + p = 8$

61. $\dfrac{3r + 2}{r + 2} - 2r = 1$

62. $\dfrac{2v + 3}{v + 4} + 3v = 4$

63. $\dfrac{2}{2x + 1} + \dfrac{1}{x} = 3$

64. $\dfrac{3}{s} - \dfrac{2}{2s - 1} = 1$

65. $\dfrac{16}{z - 2} + \dfrac{16}{z + 2} = 6$

66. $\dfrac{2}{y + 1} + \dfrac{1}{y - 1} = 1$

67. $\dfrac{t}{t - 2} + \dfrac{2}{t - 1} = 4$

68. $\dfrac{4t + 1}{t + 4} + \dfrac{3t - 1}{t + 1} = 2$

69. $\dfrac{5}{2p - 1} + \dfrac{4}{p + 1} = 2$

70. $\dfrac{3w}{2w + 3} + \dfrac{2}{w + 2} = 1$

71. $\dfrac{2v}{v + 2} + \dfrac{3}{v + 4} = 1$

72. $\dfrac{x + 3}{x + 1} - \dfrac{x - 2}{x + 3} = 5$

9.3 Applying Concepts

73. $\dfrac{x^2}{4} + \dfrac{x}{2} = 6$

74. $3\left(\dfrac{x + 1}{2}\right)^2 = 54$

75. $\dfrac{x + 2}{3} + \dfrac{2}{x - 2} = 3$

76. $\dfrac{x^4}{4} + 1 = \dfrac{5x^2}{4}$

77. $\dfrac{x^4}{3} - \dfrac{8x^2}{3} = 3$

78. $\dfrac{x^2}{4} + \dfrac{x}{2} + \dfrac{1}{8} = 0$

79. $\dfrac{x^4}{8} + \dfrac{x^2}{4} = 3$

80. $\sqrt{x^4 - 2} = x$

81. $\sqrt{x^4 + 4} = 2x$

82. $\left(\sqrt{x} - 2\right)^2 - 5\sqrt{x} + 14 = 0$ (*Hint:* Let $u = \sqrt{x} - 2$.)

83. $\left(\sqrt{x} + 3\right)^2 - 4\sqrt{x} - 17 = 0$ (*Hint:* $u = \sqrt{x} + 3$.)

84. **Sports** According to the Compton's Interactive Encyclopedia, the minimum dimensions of a football used in the National Football Association games are 10.875 in. long and 20.75 in. in circumference at the center. A possible model for the cross section of a football is given by $y = \pm 3.3041\sqrt{1 - \dfrac{x^2}{29.7366}}$, where x is the distance from the center of the football and y is the radius of the football at x.

a. What is the domain of the equation?

b. Graph $y = 3.3041\sqrt{1 - \dfrac{x^2}{29.7366}}$ and $y = -3.3041\sqrt{1 - \dfrac{x^2}{29.7366}}$ on the same coordinate axes. Explain why the $\pm$ symbol occurs in the equation.

c. Determine the radius of the football when x is 3 in. Round to the nearest ten-thousandth.

9.4 Applications of Quadratic Equations

OBJECTIVE 1 **Application problems**

The application problems in this section are similar to those problems that were solved earlier in the text. Each of the strategies for the problems in this section will result in a quadratic equation.

Solve: A small pipe takes 16 min longer to empty a tank than does a larger pipe. Working together, the pipes can empty the tank in 6 min. How long would it take each pipe, working alone, to empty the tank?

STRATEGY **for solving an application problem**

▶ Determine the type of problem. Is it a uniform motion problem, a geometry problem, an integer problem, or a work problem?

The problem is a work problem.

▶ Choose a variable to represent the unknown quantity. Write numerical or variable expressions for all the remaining quantities. These results can be recorded in a table.

The unknown time of the larger pipe: t
The unknown time of the smaller pipe: $t + 16$

	Rate of work	·	Time worked	=	Part of task completed
Larger pipe	$\dfrac{1}{t}$	·	6	=	$\dfrac{6}{t}$
Smaller pipe	$\dfrac{1}{t + 16}$	·	6	=	$\dfrac{6}{t + 16}$

▶ Determine how the quantities are related. If necessary, review the strategies presented in earlier chapters.

The sum of the parts of the task completed must equal 1.

$$\frac{6}{t} + \frac{6}{t + 16} = 1$$

$$t(t + 16)\left(\frac{6}{t} + \frac{6}{t + 16}\right) = t(t + 16) \cdot 1$$

$$(t + 16)6 + 6t = t^2 + 16t$$

$$6t + 96 + 6t = t^2 + 16t$$

$$0 = t^2 + 4t - 96$$
$$0 = (t + 12)(t - 8)$$

$$t + 12 = 0 \qquad t - 8 = 0$$
$$t = -12 \qquad t = 8$$

The solution $t = -12$ is not possible because time cannot be a negative number.

The time for the smaller pipe is $t + 16$. Replace t by 8 and evaluate.

$$t + 16 = 8 + 16 = 24$$

The larger pipe requires 8 min to empty the tank.
The smaller pipe requires 24 min to empty the tank.

EXAMPLE 1 In 8 h, two campers rowed 15 mi down a river and then rowed back to their campsite. The rate of the river's current was 1 mph. Find the rate at which the campers row in calm water.

Strategy ▶ This is a uniform motion problem.

▶ Unknown rowing rate of the campers: r

	Distance	÷	Rate	=	Time
Down river	15	÷	$r + 1$	=	$\dfrac{15}{r + 1}$
Up river	15	÷	$r - 1$	=	$\dfrac{15}{r - 1}$

▶ The total time of the trip was 8 h.

Solution
$$\frac{15}{r + 1} + \frac{15}{r - 1} = 8$$
$$(r + 1)(r - 1)\left(\frac{15}{r + 1} + \frac{15}{r - 1}\right) = (r + 1)(r - 1)8$$
$$(r - 1)15 + (r + 1)15 = (r^2 - 1)8$$
$$15r - 15 + 15r + 15 = 8r^2 - 8$$
$$30r = 8r^2 - 8$$
$$0 = 8r^2 - 30r - 8$$
$$0 = 2(4r^2 - 15r - 4)$$
$$0 = 2(4r + 1)(r - 4)$$

$$4r + 1 = 0 \qquad r - 4 = 0$$
$$4r = -1 \qquad r = 4$$
$$r = -\frac{1}{4}$$

• The solution $r = -\frac{1}{4}$ is not possible because the rate cannot be a negative number.

The rowing rate is 4 mph.

PROBLEM 1 The length of a rectangle is 3 m more than the width. The area is 54 m². Find the length of the rectangle.

Solution See page S31.

Determine whether the following statements are always true, sometimes true, or never true.

1. Let x be an integer. Then x^2 and $(x + 2)^2$ represent the squares of two consecutive odd integers.

2. Let x be an integer. Then x^2, $(x + 1)^2$, and $(x + 2)^2$ represent the squares of three consecutive integers.

3. If t is the time to complete an amount of work, then $\frac{1}{t}$ is the rate of doing work.

4. Let r be the rate of rowing a boat in calm water. If the river has a current of 2 mph, then the rate of rowing down the river is $r + 2$, and the rate of rowing up the river is $r - 2$.

9.4 Exercises

1 Application problems

1. The length of a rectangle is 4 ft less than twice the width. Let w be the width of the rectangle.
 a. In terms of w, the length of the rectangle is ___?___ ft.
 b. In terms of w, the area of the rectangle is ___?___ ft².
 c. The area of the rectangle is 96 ft². Use your answer to part b and the fact that the area of the rectangle is 96 ft² to write an equation that can be solved to find the width of the rectangle: ___?___ = ___?___.

2. a. Solve the equation from part c of Exercise 1.

 $$w(2w - 4) = 96$$
 $$\underline{\quad ? \quad} - \underline{\quad ? \quad} = 96 \qquad \text{• Use the Distributive Property.}$$
 $$2w^2 - 4w - 96 = 0 \qquad \text{• Subtract } \underline{\quad ? \quad} \text{ from each side of the equation.}$$
 $$w^2 - 2w - 48 = 0 \qquad \text{• Divide each side of the equation by } \underline{\quad ? \quad}.$$
 $$(w - \underline{\quad ? \quad})(w + \underline{\quad ? \quad}) = 0 \qquad \text{• Factor the left side.}$$
 $$\underline{\quad ? \quad} = 0 \qquad \underline{\quad ? \quad} = 0 \qquad \text{• Use the Principle of Zero Products.}$$
 $$w = \underline{\quad ? \quad} \qquad w = \underline{\quad ? \quad} \qquad \text{• Solve for } w.$$

 b. Only one solution from part a makes sense as the width of the rectangle described in Exercise 1. The width of the rectangle is ___?___. The length of the rectangle is ___?___.

Complete Exercises 3 and 4 *without* writing and solving an equation. Use this situation: A small pipe takes 6 h longer to fill a tank than a large pipe. Working together, the pipes can fill the tank in 2 h.

3. True or false? The amount of time that it takes the large pipe, working alone to fill the tank is less than 2 h.

4. True or false? The amount of time that it takes the small pipe, working alone to fill the tank is greater than 8 h.

Solve.

5. **Farming** A rectangular corral is constructed in a pasture. The length of the rectangle is 8 ft more than twice the width. The area of the rectangle is to be 640 ft². What are the length and width of the rectangle?

6. **Manufacturing** The surface area of the ice cream cone shown at the right is given by $A = \pi r^2 + \pi rs$, where r is the radius of the circular top of the ice cream cone and s is the slant height of the cone. If the area of the cone is 11.25π in² and the slant height of the cone is 6 in., find the radius of the cone.

$s = 6$ in.

7. **Geometry** The length of a rectangle is 2 ft less than three times the width of the rectangle. The area of the rectangle is 65 ft². Find the length and width of the rectangle.

8. **Geometry** The length of a rectangle is 2 cm less than twice the width. The area of the rectangle is 180 cm². Find the length and width of the rectangle.

9. **Geometry** A square piece of cardboard is formed into a box by cutting 10-centimeter squares from each of the four corners and then folding up the sides, as shown in the figure. If the volume of the box is to be 49,000 cm³, what size square piece of cardboard is needed?

10 cm

10 cm

10. **Geography** The state of Colorado is almost perfectly rectangular, with its north border 111 mi longer than its west border. The state encompasses 104,000 mi². Estimate the dimensions of Colorado. Round to the nearest mile.

Colorado

11. **Transportation** An 18-wheeler left a grain depot with a load of wheat and traveled 550 mi to deliver the wheat before returning to the depot. Because of the lighter load on the return trip, the average speed of the truck returning was 5 mph faster than its average speed going. Find the rate returning if the entire trip, not counting unloading time or rest stops, was 21 h.

12. **Racing** The *Tour de France* is a bicycle race that begins in Dublin, Ireland but is primarily a race through France. On each day, racers cover a certain distance depending on the steepness of the terrain. Suppose that on one particular day, the racers must complete 210 mi. One cyclist, traveling 10 mph faster than a second cyclist, covers this distance in 2.4 h less time than the second cyclist. Find the rate of the first cyclist.

13. Physics The height of a projectile fired upward is given by the formula $s = v_0 t - 16t^2$, where s is the height, v_0 is the initial velocity, and t is the time. Find the time for a projectile to return to Earth if it has an initial velocity of 200 ft/s.

14. Physics The height of a projectile fired upward is given by the formula $s = v_0 t - 16t^2$, where s is the height, v_0 is the initial velocity, and t is the time. Find the times at which a projectile with an initial velocity of 128 ft/s will be 64 ft above the ground. Round to the nearest hundredth of a second.

15. Automotive Technology In Germany, there are no speed limits on some portions of the autobahn (highway). Other portions have a speed limit of 180 km/h (approximately 112 mph). The distance, d (in meters), required to stop a car traveling at v kilometers per hour is $d = 0.019v^2 + 0.69v$. Approximate, to the nearest tenth, the maximum speed a driver can be going and still be able to stop within 150 m.

German Autobahn System

16. Sports A penalty kick in soccer is made from a penalty mark that is 36 ft from a goal that is 8 ft high. A possible equation for the flight of a penalty kick is $h = -0.002x^2 + 0.35x$, where h is the height (in feet) of the ball x feet from the penalty mark. Assuming that the flight of the kick is toward the goal and that it is not touched by the goalie, will the ball land in the net?

17. Aeronautics A model rocket is launched with an initial velocity of 200 ft/s. The height, h, of the rocket t seconds after the launch is given by $h = -16t^2 + 200t$. How many seconds after launch will the rocket be 300 ft above the ground? Round to the nearest hundredth of a second.

18. Sports A diver jumps from a platform that is 10 m high. The height of the diver t seconds after the beginning of the dive is given by $h = -4.9t^2 + 3.2t + 10.5$. To the nearest hundredth of a second, how long after the beginning of the dive will the diver enter the water?

19. Petroleum Engineering A small pipe can fill an oil tank in 6 min more time than it takes a larger pipe to fill the same tank. Working together, both pipes can fill the tank in 4 min. How long would it take each pipe, working alone, to fill the tank?

20. Metallurgy A small heating unit takes 8 h longer to melt a piece of iron than does a larger unit. Working together, the heating units can melt the iron in 3 h. How long would it take each heating unit, working alone, to melt the iron?

21. Recreation A cruise ship made a trip of 100 mi in 8 h. The ship traveled the first 40 mi at a constant rate before increasing its speed by 5 mph. Another 60 mi was traveled at the increased speed. Find the rate of the cruise ship for the first 40 mi.

22. Sports A cyclist traveled 60 mi at a constant rate before reducing the speed by 2 mph. Another 40 mi was traveled at the reduced speed. The total time for the 100-mile trip was 9 h. Find the rate during the first 60 mi.

23. **Aeronautics** The rate of a single-engine plane in calm air is 100 mph. Flying with the wind, the plane can fly 240 mi in 1 h less time than is required to make the return trip of 240 mi. Find the rate of the wind.

24. **Transportation** A car travels 120 mi. A second car, traveling 10 mph faster than the first car, makes the same trip in 1 h less time. Find the speed of each car.

25. **Sports** To prepare for an upcoming race, a sculling crew rowed 16 mi down a river and back in 6 h. If the rate of the river's current is 2 mph, find the sculling crew's rate of rowing in calm water.

26. **Sports** A fishing boat traveled 30 mi down a river and then returned. The total time for the round trip was 4 h, and the rate of the river's current was 4 mph. Find the rate of the boat in still water.

9.4 **Applying Concepts**

Solve.

27. **Number Problem** The numerator of a fraction is 3 less than the denominator. The sum of the fraction and four times its reciprocal is $\frac{17}{2}$. Find the fraction.

28. **Integer Problem** Find two consecutive even integers whose cubes differ by 488.

29. **Geometry** An open box is formed from a rectangular piece of cardboard whose length is 8 cm more than its width by cutting squares whose sides are 2 cm in length from each corner and then folding up the sides. Find the dimensions of the box if its volume is 256 cm^3.

30. **Construction** The height of an arch is given by the equation

$$h(x) = -\frac{3}{64}x^2 + 27, \; -24 \le x \le 24$$

where $|x|$ is the distance in feet from the center of the arch.
a. What is the maximum height of the arch?
b. What is the height of the arch 8 ft to the right of center?
c. How far from the center is the arch 8 ft tall?

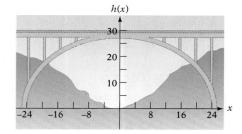

31. **Physics** Using Torricelli's Principle, it can be shown that the depth d, in centimeters, of a liquid in a bottle with a hole of area 0.5 cm^2 in its side can be approximated by $d = 0.0034t^2 - 0.52518t + 20$, where t is the time, in seconds, since a stopper was removed from the hole. When will the depth be 10 cm? Round to the nearest tenth.

32. Geometry The volumes of two spheres differ by 372π cm^3. The radius of the larger sphere is 3 cm more than the radius of the smaller sphere. Find the radius of the larger sphere. $\left(Hint:\text{ The formula for the volume of a sphere is } V = \frac{4}{3}\pi r^3.\right)$

33. Geometry A perfectly spherical scoop of mint chocolate chip ice cream is placed in a cone as shown at the right. How far is the bottom of the scoop of ice cream from the bottom of the cone? (*Hint:* A line segment from the center of the scoop of ice cream to the point at which the ice cream touches the cone is perpendicular to the edge of the cone.)

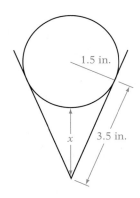

9.5 Nonlinear Inequalities

OBJECTIVE 1 Solve nonlinear inequalities

A **quadratic inequality in one variable** is one that can be written in the form $ax^2 + bx + c < 0$ or $ax^2 + bx + c > 0$, where $a \neq 0$. The symbols $\leq$ and $\geq$ can also be used.

Quadratic inequalities can be solved by algebraic means. However, it is often easier to use a graphical method to solve these inequalities. The graphical method is used in the example that follows.

Solve and graph the solution set of $x^2 - x - 6 < 0$.

Factor the trinomial.

$$x^2 - x - 6 < 0$$
$$(x - 3)(x + 2) < 0$$

On a number line, draw vertical lines at the numbers that make each factor equal to zero.

$$x - 3 = 0 \qquad x + 2 = 0$$
$$x = 3 \qquad\quad x = -2$$

> **Take Note**
> For each factor, choose a number in each region. For example, when $x = -4$, $x - 3$ is negative; when $x = 1$, $x - 3$ is negative; and when $x = 4$, $x - 3$ is positive. When $x = -4$, $x + 2$ is negative; when $x = 1$, $x + 2$ is positive; and when $x = 4$, $x + 2$ is positive.

For each factor, place plus signs above the number line for those regions where the factor is positive and place negative signs where the factor is negative. $x - 3$ is positive for $x > 3$, and $x + 2$ is positive for $x > -2$.

Because $x^2 - x - 6 < 0$, the solution set will be the regions where one factor is positive and the other factor is negative.

Write the solution set.

$\{x \mid -2 < x < 3\}$

The graph of the solution set of the inequality $x^2 - x - 6 < 0$ is shown at the right.

This method of solving quadratic inequalities can be used on any polynomial that can be factored into linear factors.

Solve and graph the solution set of $x^3 - 4x^2 - 4x + 16 > 0$.

Factor the polynomial by grouping.

$$x^3 - 4x^2 - 4x + 16 > 0$$
$$x^2(x - 4) - 4(x - 4) > 0$$
$$(x^2 - 4)(x - 4) > 0$$
$$(x - 2)(x + 2)(x - 4) > 0$$

On a number line, identify for each factor the regions where the factor is positive and where the factor is negative.

There are two regions where the product of the three factors is positive.

Write the solution set.

$\{x \mid -2 < x < 2 \text{ or } x > 4\}$

The graph of the solution set of the inequality $x^3 - 4x^2 - 4x + 16 > 0$ is shown at the right.

EXAMPLE 1 Solve and graph the solution set of $2x^2 - x - 3 \geq 0$.

Solution
$$2x^2 - x - 3 \geq 0$$
$$(2x - 3)(x + 1) \geq 0$$

$$\left\{ x \mid x \leq -1 \text{ or } x \geq \frac{3}{2} \right\}$$

PROBLEM 1 Solve and graph the solution set of $2x^2 - x - 10 \leq 0$.

Solution See page S31.

The graphical method can be used to solve rational inequalities.

Solve: $\dfrac{2x - 5}{x - 4} \le 1$

$$\dfrac{2x - 5}{x - 4} \le 1$$

Rewrite the inequality so that 0 appears on the right side of the inequality.

$$\dfrac{2x - 5}{x - 4} - 1 \le 0$$

Then simplify.

$$\dfrac{2x - 5}{x - 4} - \dfrac{x - 4}{x - 4} \le 0$$

$$\dfrac{x - 1}{x - 4} \le 0$$

On a number line, identify for each factor of the numerator and each factor of the denominator the regions where the factor is positive and where the factor is negative.

$x - 1$ ---------------- | ++++++ | +++
$x - 4$ ---------------- | ------ | +++

$$-5 \ -4 \ -3 \ -2 \ -1 \ \ 0 \ \ 1 \ \ 2 \ \ 3 \ \ 4 \ \ 5$$

The region where the quotient of the two factors is negative is between 1 and 4.

Write the solution set. $\{x \mid 1 \le x < 4\}$

Note that 1 is part of the solution set, but 4 is not part of the solution set because the denominator of the rational expression is zero when $x = 4$.

EXAMPLE 2 Solve and graph the solution set of $\dfrac{x + 4}{x - 3} \ge 0$.

Solution $\dfrac{x + 4}{x - 3} \ge 0$

$x + 4$ -- | +++++++++++++++++ | +++++
$x - 3$ -- | ----------------- | +++++

$$-5 \ -4 \ -3 \ -2 \ -1 \ \ 0 \ \ 1 \ \ 2 \ \ 3 \ \ 4 \ \ 5$$

$\{x \mid x > 3 \text{ or } x \le -4\}$

$$-5 \ -4 \ -3 \ -2 \ -1 \ \ 0 \ \ 1 \ \ 2 \ \ 3 \ \ 4 \ \ 5$$

PROBLEM 2 Solve and graph the solution set of $\dfrac{x}{x - 2} \le 0$.

Solution See page S31.

9.5 Concept Review

Determine whether the following statements are always true, sometimes true, or never true.

1. The endpoints of a solution set of a quadratic inequality are not included in the solution set.

2. The solution set of $(x - 3)(x - 2)(x + 2) > 0$ is 3, 2, and -2.

3. The solution set of a quadratic inequality is an infinite set.

4. The solution set of $\dfrac{1}{x - 1} > 0$ is $\{x \mid x > 1\}$.

9.5 Exercises

1 Solve nonlinear inequalities

1. For the nonlinear inequality $\dfrac{x - 2}{x - 3} \geq 1$, is 3 a possible element of the solution set? Why or why not?

2. To use a number line diagram to solve the inequality $(x + 3)(x - 8) > 0$, begin by drawing vertical lines through ___?___ and ___?___.

3. When x is between -4 and 7, the numerator of $\dfrac{x - 7}{x + 4}$ is negative and the denominator is ___?___. The quotient of a negative number and a positive number is ___?___, so the expression $\dfrac{x - 7}{x + 4}$ is ___?___ for all values of x between -4 and 7.

4. If $(x - 3)(x - 5) > 0$, what must be true of the values of $x - 3$ and $x - 5$?

Solve and graph the solution set.

5. $(x - 4)(x + 2) > 0$

6. $(x + 1)(x - 3) > 0$

7. $x^2 - 3x + 2 \geq 0$

8. $x^2 + 5x + 6 > 0$

9. $x^2 - x - 12 < 0$

10. $x^2 + x - 20 < 0$

11. $(x - 1)(x + 2)(x - 3) < 0$

12. $(x + 4)(x - 2)(x - 1) \geq 0$

13. $\dfrac{x - 4}{x + 2} > 0$

14. $\dfrac{x + 2}{x - 3} > 0$

15. $\dfrac{x - 3}{x + 1} \leq 0$

16. $\dfrac{x - 1}{x} > 0$

17. $\dfrac{(x - 1)(x + 2)}{x - 3} \leq 0$

18. $\dfrac{(x + 3)(x - 1)}{x - 2} \geq 0$

19. Which inequalities can be solved using this diagram?

```
x - 4  ------- | --------- | ---------- | +++
   x   ------- | ---------- | +++++++++++| +++
 x + 3 ------- | +++++++ | ++++++++++ | +++
        +--+--+--+--+--+--+--+--+--+--+--+--
        -5 -4 -3 -2 -1  0  1  2  3  4  5
```

a. $x(x + 3)(x - 4) > 0$ **b.** $x + 3 < x(x - 4)$

c. $0 > \dfrac{x}{3 - x}$ **d.** $\dfrac{x + 3}{x(x - 4)} < 0$

20. a, b, and c are positive numbers such that $a < b < c$. In which of the following intervals is the product $(x - a)(x - b)(x - c)$ always negative?
a. $\{x \mid a < x < c\}$ **b.** $\{x \mid x > c\}$
c. $\{x \mid x < a\}$ **d.** $\{x \mid b < x < c\}$

Solve.

21. $x^2 - 16 > 0$

22. $x^2 - 4 \geq 0$

23. $x^2 - 4x + 4 > 0$

24. $x^2 + 6x + 9 > 0$

25. $x^2 - 9x \leq 36$

26. $x^2 + 4x > 21$

27. $2x^2 - 5x + 2 \geq 0$

28. $4x^2 - 9x + 2 < 0$

29. $4x^2 - 8x + 3 < 0$

30. $2x^2 + 11x + 12 \geq 0$

31. $(x - 6)(x + 3)(x - 2) \leq 0$

32. $(x + 5)(x - 2)(x - 3) > 0$

33. $(2x - 1)(x - 4)(2x + 3) > 0$

34. $(x - 2)(3x - 1)(x + 2) \leq 0$

35. $x^3 + 3x^2 - x - 3 \leq 0$

36. $x^3 + x^2 - 9x - 9 < 0$

37. $x^3 - x^2 - 4x + 4 \geq 0$

38. $2x^3 + 3x^2 - 8x - 12 \geq 0$

39. $\dfrac{3x}{x - 2} > 1$

40. $\dfrac{2x}{x + 1} < 1$

41. $\dfrac{2}{x + 1} \geq 2$

42. $\dfrac{3}{x - 1} < 2$

43. $\dfrac{x}{(x - 1)(x + 2)} \geq 0$

44. $\dfrac{x - 2}{(x + 1)(x - 1)} \leq 0$

45. $\dfrac{1}{x} < 2$

46. $\dfrac{x}{2x - 1} \geq 1$

9.5 Applying Concepts

Graph the solution set.

47. $(x + 2)(x - 3)(x + 1)(x + 4) > 0$

48. $(x - 1)(x + 3)(x - 2)(x - 4) \geq 0$

49. $(x^2 + 2x - 8)(x^2 - 2x - 3) < 0$

50. $(x^2 + 2x - 3)(x^2 + 3x + 2) \geq 0$

51. $(x^2 + 1)(x^2 - 3x + 2) > 0$

52. $(x^2 - 9)(x^2 + 5x + 6) \leq 0$

53. $\dfrac{x^2(3 - x)(2x + 1)}{(x + 4)(x + 2)} \geq 0$

54. $\dfrac{1}{x} + x > 2$

55. $3x - \dfrac{1}{x} \leq 2$

56. $x^2 - x < \dfrac{1 - x}{x}$

Properties of Quadratic Functions

OBJECTIVE 1 **Graph a quadratic function**

Recall that a linear function is one that can be expressed by the equation $f(x) = mx + b$. The graph of a linear function has certain characteristics. It is a straight line with slope m and y-intercept $(0, b)$. A **quadratic function** is one that can be expressed by the equation $f(x) = ax^2 + bx + c, a \neq 0$. The graph of this function, which is called a **parabola,** also has certain characteristics. The graph of a quadratic function can be drawn by finding ordered pairs that belong to the function.

Take Note

Sometimes the value of the independent variable is called the **input** because it is *put in* place of the independent variable. The result of evaluating the function is called the **output.**

An **input/output table** shows the results of evaluating a function for various values of the independent variable. An input/output table for $f(x) = x^2 - 2x - 3$ is shown at the right.

■ Graph: $f(x) = x^2 - 2x - 3$

By evaluating the function for various values of x, find enough ordered pairs to determine the shape of the graph.

x	$f(x) = x^2 - 2x - 3$	$f(x)$	(x, y)
-2	$f(-2) = (-2)^2 - 2(-2) - 3$	5	$(-2, 5)$
-1	$f(-1) = (-1)^2 - 2(-1) - 3$	0	$(-1, 0)$
0	$f(0) = (0)^2 - 2(0) - 3$	-3	$(0, -3)$
1	$f(1) = (1)^2 - 2(1) - 3$	-4	$(1, -4)$
2	$f(2) = (2)^2 - 2(2) - 3$	-3	$(2, -3)$
3	$f(3) = (3)^2 - 2(3) - 3$	0	$(3, 0)$
4	$f(4) = (4)^2 - 2(4) - 3$	5	$(4, 5)$

Because the value of $f(x) = x^2 - 2x - 3$ is a real number for all values of x, the domain of f is all real numbers. From the graph, it appears that no value of y is less than -4. Thus the range is $\{y \mid y \geq -4\}$. The range can also be determined algebraically, as shown below, by completing the square.

Take Note

In completing the square, 1 is both added and subtracted. Because $1 - 1 = 0$, the expression $x^2 - 2x - 3$ is not changed. Note that

$(x - 1)^2 - 4$
$= (x^2 - 2x + 1) - 4$
$= x^2 - 2x - 3$

which is the original expression.

$$f(x) = x^2 - 2x - 3$$

Group the variable terms. $f(x) = (x^2 - 2x) - 3$

Complete the square of $x^2 - 2x$. Add and subtract $\left[\frac{1}{2}(-2)\right]^2 = 1$ to $x^2 - 2x$. $f(x) = (x^2 - 2x + 1) - 1 - 3$

Factor and combine like terms. $f(x) = (x - 1)^2 - 4$

The square of a number is always positive. $(x - 1)^2 \geq 0$

Subtract 4 from each side of the inequality. $(x - 1)^2 - 4 \geq -4$

Replace $(x - 1)^2 - 4$ with $f(x)$. $f(x) \geq -4$
 $y \geq -4$

From the last inequality, the range is $\{y \mid y \geq -4\}$.

In general, the graph of $f(x) = ax^2 + bx + c, a \neq 0$, resembles a "cup" shape as shown on the next page. The parabola opens up when $a > 0$ and opens down

Take Note

The suspension cables for some bridges, such as the Golden Gate bridge, have the shape of a parabola. Parabolic shapes are also used for the mirrors in telescopes and in the design of certain antennas.

when $a < 0$. The **vertex** of the parabola is the point with the smallest y-coordinate when $a > 0$ and is the point with the largest y-coordinate when $a < 0$.

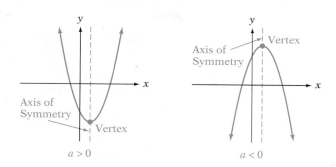

The **axis of symmetry** is a line that passes through the vertex of the parabola and is parallel to the y-axis. To understand the axis of symmetry, think of folding the graph along that line. The two portions of the graph will match up.

The vertex and axis of symmetry of a parabola can be found by completing the square.

Find the vertex and the axis of symmetry of the graph of $F(x) = x^2 + 4x + 3$.

To find the coordinates of the vertex, complete the square. Group the variable terms.

$F(x) = x^2 + 4x + 3$

$F(x) = (x^2 + 4x) + 3$

Complete the square of $x^2 + 4x$. Add and subtract $\left[\frac{1}{2}(4)\right]^2 = 4$ to and from $x^2 + 4x$.

$F(x) = (x^2 + 4x + 4) - 4 + 3$

Factor and combine like terms.

$F(x) = (x + 2)^2 - 1$

Because a, the coefficient of x^2, is positive ($a = 1$), the parabola opens up and the vertex is the point with the least y-coordinate. Because $(x + 2)^2 \geq 0$ for all values of x, the least y-coordinate occurs when $(x + 2)^2 = 0$. When $x = -2$, this expression equals zero.

The x-coordinate of the vertex is -2.

To find the y-coordinate of the vertex, evaluate the function at $x = -2$.

$F(x) = (x + 2)^2 - 1$
$F(-2) = (-2 + 2)^2 - 1 = -1$

The y-coordinate of the vertex is -1.

From these results, the coordinates of the vertex are $(-2, -1)$. The axis of symmetry is the line whose equation is $x = -2$.

By following the process illustrated in the last example and completing the square of $f(x) = ax^2 + bx + c$, we can find a formula for the coordinates of the vertex of a parabola.

Vertex and Axis of Symmetry of a Parabola

Let $f(x) = ax^2 + bx + c$ be the equation of a parabola. The coordinates of the vertex are $\left(-\dfrac{b}{2a}, f\left(-\dfrac{b}{2a}\right)\right)$. The equation of the axis of symmetry is $x = -\dfrac{b}{2a}$.

The coordinates of the vertex and the axis of symmetry can be used to graph a parabola.

Graph $y = x^2 + 2x - 3$ using the vertex and the axis of symmetry.

Find the x-coordinate of the vertex. $a = 1$ and $b = 2$.
$$x = -\frac{b}{2a} = -\frac{2}{2(1)} = -1$$

Find the y-coordinate of the vertex by replacing x with -1 and solving for y.

$$y = x^2 + 2x - 3$$
$$y = (-1)^2 + 2(-1) - 3$$
$$y = 1 - 2 - 3$$
$$y = -4$$

The vertex is $(-1, -4)$.

The axis of symmetry is the line $x = -1$.

Find some ordered-pair solutions of the equation and record these in a table. Because the graph is symmetric to the line $x = -1$, choose values of x greater than -1.

x	y
0	-3
1	0
2	5

Graph the ordered-pair solutions on a rectangular coordinate system. Use symmetry to locate points of the graph on the other side of the axis of symmetry. Remember that corresponding points on the graph are the same distance from the axis of symmetry.

Draw a parabola through the points.

Using a graphing utility, graph the equation $y = x^2 + 2x - 3$ with a domain of $[-4.7, 4.7]$. Trace along the graph and verify that $(-1, -4)$ are the coordinates of the vertex.

EXAMPLE 1 Find the vertex and the axis of symmetry of the parabola whose equation is $y = -3x^2 + 6x + 1$. Then sketch its graph.

Solution $x = -\dfrac{b}{2a} = -\dfrac{6}{2(-3)} = 1$ • **Find the x-coordinate of the vertex.**
 $a = -3$ and $b = 6$.

$y = -3x^2 + 6x + 1$ • **Find the y-coordinate of the vertex by**
$y = -3(1)^2 + 6(1) + 1$ **replacing x by 1 and solving for y.**
$y = 4$

The vertex is $(1, 4)$.

The axis of symmetry • **The axis of symmetry is the line**
is the line $x = 1$. $x = -\dfrac{b}{2a}$.

• **Because a is negative, the parabola opens down. Find a few ordered pairs, and use symmetry to sketch the graph.**

PROBLEM 1 Find the vertex and the axis of symmetry of the parabola whose equation is $y = x^2 - 2$. Then sketch its graph.

Solution See page S31.

EXAMPLE 2 Graph $f(x) = -2x^2 - 4x + 3$. State the domain and range of the function.

Solution Because a is negative ($a = -2$), the graph of f will open down. The x-coordinate of the vertex is

$x = -\dfrac{b}{2a} = -\dfrac{-4}{2(-2)} = -1.$

The y-coordinate of the vertex is
$f(-1) = -2(-1)^2 - 4(-1) + 3 = 5.$ The
vertex is $(-1, 5)$.

Evaluate $f(x)$ for various values of x, and use symmetry to draw the graph.

Because $f(x) = -2x^2 - 4x + 3$ is a real number for all values of x, the domain of the function is $\{x \mid x \in \text{real numbers}\}$. The vertex of the parabola is the highest point on the graph. Because the y-coordinate at that point is 5, the range is $\{y \mid y \leq 5\}$.

PROBLEM 2 Graph $g(x) = x^2 + 4x - 2$. State the domain and range of the function.

Solution See page S31.

> **Take Note**
>
> Once the coordinates of the vertex are found, the range of the quadratic function can be determined.

OBJECTIVE 2 **Find the _x_-intercepts of a parabola**

Recall that a point at which a graph crosses the _x_- or _y_-axis is called an _intercept_ of the graph. The _x_-intercepts of the graph of an equation occur when $y = 0$; the _y_-intercepts occur when $x = 0$.

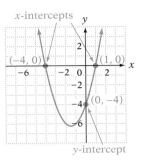

The graph of $y = x^2 + 3x - 4$ is shown at the right. The points whose coordinates are $(-4, 0)$ and $(1, 0)$ are _x_-intercepts of the graph.

Find the _x_-intercepts for the parabola whose equation is $y = 4x^2 - 4x + 1$.

To find _x_-intercepts, let $y = 0$.

$$y = 4x^2 - 4x + 1$$
$$0 = 4x^2 - 4x + 1$$

Solve for _x_ by factoring and using the Principle of Zero Products.

$$0 = (2x - 1)(2x - 1)$$

$$2x - 1 = 0 \qquad 2x - 1 = 0$$
$$2x = 1 \qquad\qquad 2x = 1$$
$$x = \frac{1}{2} \qquad\qquad x = \frac{1}{2}$$

The _x_-intercept is $\left(\frac{1}{2}, 0\right)$.

In this last example, the parabola has only one _x_-intercept. In this case, the parabola is said to be _tangent_ to the _x_-axis at $x = \frac{1}{2}$.

Find the _x_-intercepts of $y = 2x^2 - x - 6$.

To find the _x_-intercepts, let $y = 0$.

$$y = 2x^2 - x - 6$$
$$0 = 2x^2 - x - 6$$

Solve for _x_ by factoring and using the Principle of Zero Products.

$$0 = (2x + 3)(x - 2)$$

$$2x + 3 = 0 \qquad x - 2 = 0$$
$$x = -\frac{3}{2} \qquad\qquad x = 2$$

The _x_-intercepts are $\left(-\frac{3}{2}, 0\right)$ and $(2, 0)$.

Take Note

We are being careful in this discussion to say _real zero_, by which we mean a zero that is a real number. Later in this section, we will discuss complex number zeros.

If the equation $y = 2x^2 - x - 6$ were written in function notation as $f(x) = 2x^2 - x - 6$, then to find the _x_-intercepts you would let $f(x) = 0$ and solve for _x_. Recall that a value of _x_ for which $f(x) = 0$ is a _zero_ of the function.

Thus, $-\frac{3}{2}$ and 2 are zeros of $f(x) = 2x^2 - x - 6$.

Find the zeros of $f(x) = x^2 - 2x - 1$.

To find the zeros, let $f(x) = 0$ and solve for x.

Because $x^2 - 2x - 1$ does not easily factor, use the quadratic formula to solve for x.

$a = 1$, $b = -2$, and $c = -1$.

$$f(x) = x^2 - 2x - 1$$
$$0 = x^2 - 2x - 1$$

$$x = \frac{-b \pm \sqrt{b^2 - 4ac}}{2a}$$

$$= \frac{-(-2) \pm \sqrt{(-2)^2 - 4(1)(-1)}}{2(1)}$$

$$= \frac{2 \pm \sqrt{4 + 4}}{2}$$

$$= \frac{2 \pm \sqrt{8}}{2}$$

$$= \frac{2 \pm 2\sqrt{2}}{2}$$

$$= 1 \pm \sqrt{2}$$

The zeros of the function are $1 - \sqrt{2}$ and $1 + \sqrt{2}$.

The graph of $f(x) = x^2 - 2x - 1$ is shown at the right. Note that the zeros are the x-intercepts of the graph of f.

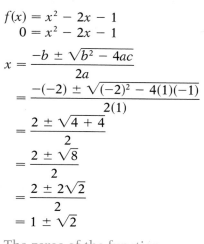

The preceding example demonstrates that the x-coordinate of an x-intercept of the graph of a function is a real zero of the function, and a real zero of a function is the x-coordinate of an x-intercept of the graph of the function.

As mentioned in the Take Note on page 654, a function may have complex number zeros.

Find the zeros of $f(x) = -x^2 + 2x - 2$.

To find the zeros, let $f(x) = 0$ and solve for x.

Because $-x^2 + 2x - 1$ does not easily factor, use the quadratic formula to solve for x.

$a = -1$, $b = 2$, $c = -2$

$$f(x) = -x^2 + 2x - 2$$
$$0 = -x^2 + 2x - 2$$

$$x = \frac{-b \pm \sqrt{b^2 - 4ac}}{2a}$$

$$= \frac{-2 \pm \sqrt{2^2 - 4(-1)(-2)}}{2(-1)}$$

$$= \frac{-2 \pm \sqrt{4 - 8}}{-2}$$

$$= \frac{-2 \pm \sqrt{-4}}{-2} = \frac{-2 \pm 2i}{-2}$$

$$= 1 \pm i$$

The complex zeros of the function are $1 - i$ and $1 + i$.

The graph of $f(x) = -x^2 + 2x - 2$ is shown at the right. Note that the graph does not cross the x-axis.

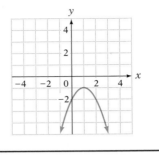

This example shows that a function can have a zero without crossing the x-axis. The zero of the function, in this case, is a complex number. If the graph of a quadratic function crosses the x-axis, it has a real zero. If the graph never crosses the x-axis, it has only complex zeros. See the Applying Concept exercises for functions that have both real and complex zeros.

EXAMPLE 3 Find the x-intercepts of the parabola given by each equation.

 A. $y = x^2 + 2x - 2$ **B.** $y = 4x^2 + 4x + 1$

Solution **A.** $y = x^2 + 2x - 2$

$0 = x^2 + 2x - 2$ • Let $y = 0$.

$$x = \frac{-b \pm \sqrt{b^2 - 4ac}}{2a}$$

• The equation is nonfactorable over the integers. Use the quadratic formula to solve for x.

$$= \frac{-(2) \pm \sqrt{(2)^2 - 4(1)(-2)}}{2 \cdot 1}$$

$$= \frac{-2 \pm \sqrt{4 + 8}}{2}$$

$$= \frac{-2 \pm \sqrt{12}}{2}$$

$$= \frac{-2 \pm 2\sqrt{3}}{2} = -1 \pm \sqrt{3}$$

The x-intercepts are $\left(-1 + \sqrt{3}, 0\right)$ and $\left(-1 - \sqrt{3}, 0\right)$

B. $y = 4x^2 + 4x + 1$

$0 = 4x^2 + 4x + 1$ • Let $y = 0$.

$0 = (2x + 1)(2x + 1)$ • Solve for x by factoring.

$2x + 1 = 0$ $2x + 1 = 0$

$2x = -1$ $2x = -1$

$x = -\dfrac{1}{2}$ $x = -\dfrac{1}{2}$ • The equation has a double root.

The x-intercept is $\left(-\dfrac{1}{2}, 0\right)$.

PROBLEM 3 Find the x-intercepts of the parabola given by each equation.

 A. $y = 2x^2 - 5x + 2$ **B.** $y = x^2 + 4x + 4$

Solution See page S32.

 The zeros of a quadratic function can always be determined exactly by solving a quadratic equation. However, it is possible to approximate the real zeros by graphing the function and determining the x-intercepts.

EXAMPLE 4 Graph $f(x) = 2x^2 - x - 5$. Estimate the real zeros of the function to the nearest tenth.

Solution Graph the function. Then use the features of a graphing utility to estimate the zeros.

The approximate values of the zeros are -1.4 and 1.9.

PROBLEM 4 Graph $f(x) = -x^2 - 4x + 6$. Estimate the real zeros of the function to the nearest tenth.

Solution See page S32.

In Examples 3 and 4, we found values of x in the domain of a function f for which $f(x) = 0$. We can extend this to finding values in the domain for which $f(x) = c$, where c is a real number.

EXAMPLE 5 Find two values of x in the domain of $f(x) = x^2 + 2x - 1$ for which $f(x) = 3$.

Solution

$f(x) = x^2 + 2x - 1$
$3 = x^2 + 2x - 1$
$0 = x^2 + 2x - 4$

• We are trying to find x for which $f(x) = 3$. Replace $f(x)$ by 3 and solve for x.

$$x = \frac{-b \pm \sqrt{b^2 - 4ac}}{2a}$$

$$= \frac{-2 \pm \sqrt{2^2 - 4(1)(-4)}}{2(1)}$$

$$= \frac{-2 \pm \sqrt{4 + 16}}{2}$$

$$= \frac{-2 \pm \sqrt{20}}{2}$$

$$= \frac{-2 \pm 2\sqrt{5}}{2}$$

$$= -1 \pm \sqrt{5}$$

• Because $x^2 + 2x - 4$ does not easily factor, use the quadratic formula to solve for x. $a = 1, b = 2, c = -4$.

The values of x for which $f(x) = 3$ are $-1 - \sqrt{5}$ and $-1 + \sqrt{5}$. The graph of $f(x) = x^2 + 2x - 1$ is shown above. Note that the graph crosses the line $y = 3$ when $x = -1 - \sqrt{5}$ and $x = -1 + \sqrt{5}$.

PROBLEM 5 Find two values of x in the domain of $f(x) = -x^2 - 4x + 2$ for which $f(x) = -4$.

Solution See page S32.

Recall that the discriminant of $ax^2 + bx + c$ is the expression $b^2 - 4ac$ and that this expression can be used to determine whether $ax^2 + bx + c = 0$ has zero, one, or two real number solutions. Because there is a connection between the solutions of $ax^2 + bx + c = 0$ and the x-intercepts of the graph of $y = ax^2 + bx + c$, the discriminant can be used to determine the number of x-intercepts of a parabola.

Effect of the Discriminant on the Number of x-Intercepts of a Parabola

1. If $b^2 - 4ac = 0$, the parabola has one x-intercept.

2. If $b^2 - 4ac > 0$, the parabola has two x-intercepts.

3. If $b^2 - 4ac < 0$, the parabola has no x-intercepts.

Use the discriminant to determine the number of x-intercepts of the parabola whose equation is $y = 2x^2 - x + 2$.

$$a = 2, b = -1, c = 2$$

Evaluate the discriminant.

$$b^2 - 4ac = (-1)^2 - 4(2)(2)$$
$$= 1 - 16$$
$$= -15$$

$$-15 < 0$$

The discriminant is less than zero, so the parabola has no x-intercepts.

EXAMPLE 6 Use the discriminant to determine the number of x-intercepts of the parabola whose equation is $y = x^2 - 6x + 9$.

Solution $b^2 - 4ac = (-6)^2 - 4(1)(9)$ • $a = 1, b = -6, c = 9$
$= 36 - 36$
$= 0$ • The discriminant is equal to zero.

The parabola has one x-intercept.

PROBLEM 6 Use the discriminant to determine the number of x-intercepts of the parabola whose equation is $y = x^2 - x - 6$.

Solution See page S32.

9.6 Concept Review

Determine whether the following statements are always true, sometimes true, or never true.

1. The axis of symmetry of a parabola goes through the origin.

2. The vertex is the lowest point on the graph of a quadratic function.

3. A parabola has two x-intercepts.

4. If the discriminant for a quadratic function is zero, the graph of the function has one x-intercept.

5. The graph of a quadratic function has one y-intercept.

6. The domain of a quadratic function is the set of real numbers.

9.6 Exercises

1 Graph a quadratic function

1. What is a quadratic function?

2. What are the vertex and axis of symmetry of the graph of a parabola?

3. The axis of symmetry of a parabola is the line $x = -5$. What is the x-coordinate of the vertex of the parabola?

4. The axis of symmetry of a parabola is the line $x = 8$. What is the x-coordinate of the vertex of the parabola?

5. The vertex of a parabola is $(7, -9)$. What is the equation of the axis of symmetry of the parabola?

6. The vertex of a parabola is $(-4, 10)$. What is the equation of the axis of symmetry of the parabola?

7. Use the quadratic function $f(x) = 3x^2 - 1$.
 a. The function is in the standard form $f(x) = ax^2 + bx + c$, where
 $a = \underline{}$, $b = \underline{}$, and $c = \underline{}$.
 b. The graph of the function is a $\underline{}$. Because $a > 0$, the graph
 opens $\underline{}$.
 c. The x-coordinate of the vertex is $x = -\dfrac{b}{2a} = -\dfrac{?}{2(\underline{})} = \underline{}$.
 d. The y-coordinate of the vertex is $f(0) = 3(\underline{})^2 - 1 = \underline{}$.
 The vertex of the parabola is $(\underline{}, \underline{})$.

8. Use the results of Exercise 7.
 a. The axis of symmetry of the graph of $f(x) = 3x^2 - 1$ is the vertical line
 $x = \underline{}$.
 b. One point on the graph of $f(x) = 3x^2 - 1$ is $(2, 11)$. This point is
 $\underline{}$ units to the right of the axis of symmetry. The corresponding
 point 2 units to the left of the axis of the symmetry is $(\underline{}, \underline{})$.

Find the vertex and axis of symmetry of the parabola given by each equation.
Then sketch its graph.

9. $y = x^2$

10. $y = -x^2$

11. $y = x^2 - 2$

12. $y = x^2 + 2$

13. $y = -x^2 + 3$

14. $y = -x^2 - 1$

15. $y = \dfrac{1}{2}x^2$

16. $y = 2x^2$

17. $y = 2x^2 - 1$

18. $y = -\dfrac{1}{2}x^2 + 2$

19. $y = x^2 - 2x$

20. $y = x^2 + 2x$

21. $y = -2x^2 + 4x$

22. $y = \dfrac{1}{2}x^2 - x$

23. $y = x^2 - x - 2$

24. $y = x^2 - 3x + 2$ **25.** $y = 2x^2 - x - 5$ **26.** $y = 2x^2 - x - 3$

Graph the function. State the domain and range of the function.

27. $f(x) = 2x^2 - 4x - 5$ **28.** $f(x) = 2x^2 + 8x + 3$

29. $f(x) = -2x^2 - 3x + 2$ **30.** $f(x) = 2x^2 - 7x + 3$

31. $f(x) = x^2 - 4x + 4$ **32.** $f(x) = -x^2 + 6x - 9$

33. $f(x) = x^2 + 4x - 3$ **34.** $f(x) = x^2 - 2x - 2$

35. $f(x) = -x^2 - 4x - 5$ **36.** $f(x) = -x^2 + 4x + 1$

For Exercises 37 to 40, use the quadratic function $f(x) = ax^2 + bx + c$.

37. True or false? If $a < 0$, then the range of f is $\left\{ y \mid y \geq f\left(-\dfrac{b}{2a}\right) \right\}$.

38. True or false? If $b = 0$, then the axis of symmetry is the y-axis.

39. If $a > 0$ and $b < 0$, is the axis of symmetry is to the left or to the right of the y-axis?

40. True or false? If $b = 2a$, then the point with x-coordinate -2 has the same y-coordinate as the point with x-coordinate 0.

2 Find the *x*-intercepts of a parabola

41. If 3 is a zero of the function $y = f(x)$, then $f(\underline{?}) = \underline{?}$, and $(\underline{?}, \underline{?})$ is an $\underline{?}$-intercept of the graph of $y = f(x)$.

42. Use the equation $y = 2x^2 - 3x + 5$. The discriminant is $b^2 - 4ac = (\underline{?})^2 - 4(\underline{?})(\underline{?}) = \underline{?} - \underline{?} = \underline{?}$. Because the discriminant is negative, the graph of $y = 2x^2 - 3x + 5$ has $\underline{?}$ x-intercepts.

43. How can you find the x-intercepts for the graph of a quadratic function?

44. How can the discriminant be used to determine the number of x-intercepts for the graph of a quadratic function?

Find the x-intercepts of the parabola given by each equation.

45. $y = x^2 - 4$ **46.** $y = x^2 - 9$ **47.** $y = 2x^2 - 4x$

48. $y = 3x^2 + 6x$ **49.** $y = x^2 - x - 2$ **50.** $y = x^2 - 2x - 8$

51. $y = 2x^2 - 5x - 3$ **52.** $y = 4x^2 + 11x + 6$

53. $y = 3x^2 - 19x - 14$ **54.** $y = 6x^2 + 7x + 2$

55. $y = 3x^2 - 19x + 20$ **56.** $y = 3x^2 + 19x + 28$

57. $y = 9x^2 - 12x + 4$ **58.** $y = x^2 - 2$

59. $y = 9x^2 - 2$ **60.** $y = 2x^2 - x - 1$

61. $y = 4x^2 - 4x - 15$ **62.** $y = x^2 + 2x - 1$

63. $y = x^2 + 4x - 3$ **64.** $y = x^2 + 6x + 10$

65. $y = -x^2 - 4x - 5$ **66.** $y = x^2 - 2x - 2$

67. $y = -x^2 - 2x + 1$ **68.** $y = -x^2 + 4x + 1$

Find the zeros of the function.

69. $f(x) = x^2 + 3x + 2$ **70.** $f(x) = x^2 - 6x + 9$ **71.** $f(x) = -x^2 + 4x - 5$

72. $f(x) = -x^2 + 3x + 8$

73. $f(x) = 2x^2 - 3x$

74. $f(x) = -3x^2 + 4x$

75. $f(x) = 2x^2 - 4$

76. $f(x) = 3x^2 + 6$

77. $f(x) = 2x^2 + 3x + 2$

78. $f(x) = 3x^2 - x + 4$

79. $f(x) = -3x^2 + 4x - 1$

80. $f(x) = -2x^2 + x + 5$

81. The zeros of the function $f(x) = x^2 - 2x - 3$ are -1 and 3. What are the x-intercepts of the graph of the equation $y = x^2 - 2x - 3$?

82. The zeros of the function $f(x) = x^2 - x - 20$ are -4 and 5. What are the x-intercepts of the graph of the equation $y = x^2 - x - 20$?

Graph the function. Estimate the real zeros of the function to the nearest tenth.

83. $f(x) = x^2 + 3x - 1$

84. $f(x) = x^2 - 2x - 4$

85. $f(x) = 2x^2 - 3x - 7$

86. $f(x) = -2x^2 - x + 2$

87. $f(x) = x^2 + 6x + 12$

88. $f(x) = x^2 - 3x + 9$

89. Find two values of x in the domain of $f(x) = x^2 - 10x - 5$ for which $f(x) = 2$.

90. Find two values of x in the domain of $f(x) = x^2 - 4x - 20$ for which $f(x) = 3$.

91. Find two values of x in the domain of $f(x) = x^2 - 2x - 10$ for which $f(x) = 7$.

92. Find two values of x in the domain of $f(x) = x^2 - 2x + 2$ for which $f(x) = 6$.

93. Show that there are no values of x in the domain of $f(x) = x^2 - 2x + 11$ for which $f(x) = 2$.

94. Show that there are no values of x in the domain of $f(x) = x^2 - 4x + 5$ for which $f(x) = -17$.

Use the discriminant to determine the number of x-intercepts of the graph of the parabola.

95. $y = 2x^2 + x + 1$

96. $y = 2x^2 + 2x - 1$

97. $y = -x^2 - x + 3$

98. $y = -2x^2 + x + 1$

99. $y = x^2 - 8x + 16$

100. $y = x^2 - 10x + 25$

101. $y = -3x^2 - x - 2$

102. $y = -2x^2 + x - 1$

103. $y = 4x^2 - x - 2$

104. $y = 2x^2 + x + 4$

105. $y = -2x^2 - x - 5$

106. $y = -3x^2 + 4x - 5$

107. $y = x^2 + 8x + 16$

108. $y = x^2 - 12x + 36$

109. $y = x^2 + x - 3$

For Exercises 110 to 113, $f(x) = ax^2 + bx + c$ is a quadratic function whose graph is a parabola that has only one x-intercept, $(n, 0)$.

110. True or false? $(n, 0)$ is also the vertex of the parabola.

111. True or false? $ax^2 + bx + c$ is a perfect-square trinomial.

112. True or false? If $a > 0$, then for all real numbers x, $f(x) \geq 0$.

113. True or false? The point with x-coordinate $n - m$ has the same y-coordinate as the point with x-coordinate $n + m$.

9.6 Applying Concepts

Find the value of k such that the graph of the equation contains the given point.

114. $y = x^2 - 3x + k$; $(2, 5)$

115. $y = x^2 + 2x + k$; $(-3, 1)$

116. $y = 2x^2 + kx - 3$; $(4, -3)$

117. $y = 3x^2 + kx - 6$; $(-2, 4)$

For Exercises 118 to 121, find the zeros of the function. Graph the equation and verify that the x-coordinates of the x-intercepts are the real zeros of the function. These functions are examples of functions with both real and complex zeros.

118. $f(x) = x^3 - 3x^2 + x - 3$

119. $f(x) = x^3 + x^2 + 4x + 4$

120. $f(x) = x^4 - 3x^2 - 4$

121. $f(x) = x^4 + 3x^2 - 4$

Solve.

122. One root of the quadratic equation $2x^2 - 5x + k = 0$ is 4. What is the other root?

123. What is the value of k if the vertex of the parabola $y = x^2 - 8x + k$ is a point on the x-axis?

124. The zeros of the function $f(x) = mx^2 + nx + 6$ are -2 and 3. What are the zeros of the function $g(x) = nx^2 + mx - 6$?

125. The point (x_1, y_1) lies in quadrant I and is a solution of the equation $y = 3x^2 - 2x - 1$. Given $y_1 = 5$, find x_1.

126. The point (x_1, y_1) lies in quadrant II and is a solution of the equation $y = 2x^2 + 5x - 3$. Given $y_1 = 9$, find x_1.

127. Match each equation with its graph.

 i. $f(x) = x^2$
 ii. $f(x) = x^2 - 1$
 iii. $f(x) = -x^2$
 iv. $f(x) = -x^2 + 3$
 v. $f(x) = x^2 + 3x - 2$
 vi. $f(x) = -x^2 - 3x + 2$

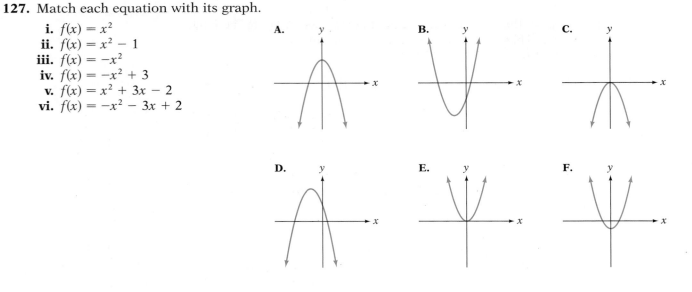

Physics The drag force F, in pounds, on a certain bicyclist is given by $F(v) = 0.151v^2$, where v is the velocity of the bicyclist in miles per hour.

128. Find the vertex of the parabola given by $F(v) = 0.151v^2$, and state what the vertex means in the context of this problem.

129. Find the drag force on the bicyclist when she or he is traveling 15 mph.

130. Graph the equation for the interval $[0, 30]$.

An equation of the form $y = ax^2 + bx + c$ can be written in the form $y = a(x - h)^2 + k$, where (h, k) are the coordinates of the vertex of the parabola. Use the process of completing the square to rewrite the equation in the form $y = a(x - h)^2 + k$. Find the vertex. (*Hint:* Review the example at the bottom of page 650.)

131. $y = x^2 - 4x + 7$ **132.** $y = x^2 - 2x - 2$

133. $y = x^2 + x + 2$ **134.** $y = x^2 - x - 3$

Using $y = a(x - h)^2 + k$ as the equation of a parabola with the vertex at (h, k), find the equation of the parabola that has the given characteristics. Write the final equation in the form $y = ax^2 + bx + c$.

135. Vertex $(1, 2)$; the graph passes through $P(2, 5)$

136. Vertex $(0, -3)$; the graph passes through $P(3, -2)$

9.7 Applications of Quadratic Functions

OBJECTIVE 1

Minimum and maximum problems

The graph of $f(x) = x^2 - 2x + 3$ is shown at the right. Because a is positive, the parabola opens up. The vertex of the parabola is the lowest point on the parabola. It is the point that has the minimum y-coordinate. Therefore, the value of the function at this point is a **minimum.**

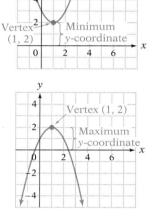

The graph of $f(x) = -x^2 + 2x + 1$ is shown at the right. Because a is negative, the parabola opens down. The vertex of the parabola is the highest point on the parabola. It is the point that has the maximum y-coordinate. Therefore, the value of the function at this point is a **maximum.**

To find the minimum or maximum value of a quadratic function, first find the x-coordinate of the vertex. Then evaluate the function at that value.

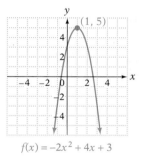

$f(x) = -2x^2 + 4x + 3$

EXAMPLE 1 Find the maximum or minimum value of the function $f(x) = -2x^2 + 4x + 3$.

Solution $x = -\dfrac{b}{2a} = -\dfrac{4}{2(-2)} = 1$

- Find the x-coordinate of the vertex. $a = -2$, $b = 4$.

$f(x) = -2x^2 + 4x + 3$
$f(1) = -2(1)^2 + 4(1) + 3$
$f(1) = 5$

- Evaluate the function at $x = 1$.

Because $a < 0$, the graph of f opens down. Therefore, the function has a maximum value.

The maximum value of the function is 5.

PROBLEM 1 Find the maximum or minimum value of $f(x) = 2x^2 - 3x + 1$.

Solution See page S32.

OBJECTIVE 2

Applications of minimum and maximum

A mason is forming a rectangular floor for a storage shed. The perimeter of the rectangle is 44 ft. What dimensions of the rectangle would give the floor a maximum area? What is the maximum area?

Point of Interest

Calculus is a branch of mathematics that demonstrates, among other things, how to find the maximum or minimum of functions other than quadratic functions. These are very important problems in applied mathematics. For instance, an automotive engineer wants to design a car whose shape will minimize the effect of air flow. The same engineer tries to maximize the efficiency of a car's engine. Similarly, an economist may try to determine what business practices will minimize cost and maximize profit.

We are given the perimeter of the rectangle, and we want to find the dimensions of the rectangle that will yield the maximum area for the floor. Use the equation for the perimeter of a rectangle.

$$P = 2L + 2W$$

$P = 44$

Divide both sides of the equation by 2.

Solve the equation for W.

$$44 = 2L + 2W$$
$$22 = L + W$$
$$22 - L = W$$

Now use the equation for the area of a rectangle. Use substitution to express the area in terms of L.

$$A = LW$$

From the equation above, $W = 22 - L$.
Substitute $22 - L$ for W.

$$A = L(22 - L)$$
$$A = 22L - L^2$$

The area of the rectangle is $22L - L^2$. To find the length of the rectangle, find the L-coordinate of the vertex of the function $f(L) = -L^2 + 22L$.

For the function $f(L) = -L^2 + 22L$,

$$L = -\frac{b}{2a} = -\frac{22}{2(-1)} = 11$$

$a = -1$ and $b = 22$.

The length of the rectangle is 11 ft.

To find the width, replace L in $22 - L$ by the L-coordinate of the vertex and evaluate.

$$W = 22 - L$$
$$W = 22 - 11 = 11$$

The width of the rectangle is 11 ft.

The dimensions of the rectangle that would give the floor a maximum area are 11 ft by 11 ft.

To find the maximum area of the floor, evaluate the function $f(L) = -L^2 + 22L$ at the L-coordinate of the vertex.

$$f(L) = -L^2 + 22L$$
$$f(11) = -(11)^2 + 22(11)$$
$$= -121 + 242 = 121$$

The maximum area of the floor is 121 ft².

The graph of the function $f(L) = -L^2 + 22L$ is shown at the right. Note that the vertex of the parabola is (11, 121). For any value of L less than 11, the area of the floor will be less than 121 ft². For any value of L greater than 11, the area of the floor will be less than 121 ft². 121 is the maximum value of the function, and the maximum value occurs when $L = 11$.

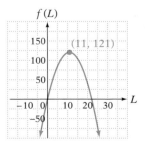

EXAMPLE 2 A mining company has determined that the cost in dollars (c) per ton of mining a mineral is given by

$$c(x) = 0.2x^2 - 2x + 12$$

where x is the number of tons of the mineral that is mined. Find the number of tons of the mineral that should be mined to minimize the cost. What is the minimum cost?

Strategy ▶ To find the number of tons that will minimize the cost, find the x-coordinate of the vertex.

▶ To find the minimum cost, evaluate the function at the x-coordinate of the vertex.

Solution $x = -\dfrac{b}{2a} = -\dfrac{-2}{2(0.2)} = 5$

To minimize the cost, 5 tons should be mined.

$c(x) = 0.2x^2 - 2x + 12$
$c(5) = 0.2(5)^2 - 2(5) + 12 = 5 - 10 + 12 = 7$

The minimum cost per ton is $7.

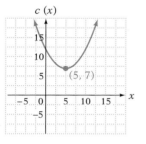

Note: The graph of the function $c(x) = 0.2x^2 - 2x + 12$ is shown at the left. The vertex of the parabola is (5, 7). For any value of x less than 5, the cost per ton is greater than $7. For any value of x greater than 5, the cost per ton is greater than $7. 7 is the minimum value of the function, and the minimum value occurs when $x = 5$.

PROBLEM 2 The height in feet (s) of a ball thrown straight up is given by $s(t) = -16t^2 + 64t$, where t is the time in seconds. Find the time it takes the ball to reach its maximum height. What is the maximum height?

Solution See page S32.

EXAMPLE 3 Find two numbers whose difference is 10 and whose product is a minimum.

Strategy Let x represent one number. Because the difference between the two numbers is 10, $x + 10$ represents the other number. Then their product is represented by $x(x + 10) = x^2 + 10x$.

▶ To find one of the two numbers, find the x-coordinate of the vertex of $f(x) = x^2 + 10x$.

▶ To find the other number, replace x in $x + 10$ by the x-coordinate of the vertex and evaluate.

Solution $x = -\dfrac{b}{2a} = -\dfrac{10}{2(1)} = -5$
$x + 10 = -5 + 10 = 5$

The numbers are -5 and 5.

PROBLEM 3 A rectangular fence is being constructed along a stream to enclose a picnic area. If 100 ft of fence is available, what dimensions of the rectangle will produce the maximum area for picnicking?

Solution See page S32.

9.7 Concept Review

Determine whether the following statements are always true, sometimes true, or never true.

1. A quadratic function has a minimum value.

2. The y-coordinate of the vertex of a quadratic function is the minimum value of the function.

3. The maximum or minimum of a quadratic function is the value of the function evaluated at the x-coordinate of the vertex.

4. The x-value of the vertex of $y = ax^2 + bx + c$, $a \neq 0$, is given by $-\dfrac{b}{2a}$.

5. The maximum value of a quadratic function is a positive number.

6. The function $f(x) = -x^2 + 3x - 2$ will have a maximum value.

9.7 Exercises

1 Minimum and maximum problems

1. What is the minimum value or the maximum value of a quadratic function?

2. Describe how to find the minimum or maximum value of a quadratic function.

For Exercises 3 and 4, use the quadratic function $f(x) = -8x^2 + 8x$.

3. Because $a = $ ___?___ , the graph of f opens ___?___ . The vertex is the highest point on the graph, so its ___?___ -coordinate is the maximum value of $f(x)$.

4. The vertex of the graph of f occurs at $x = -\dfrac{b}{2a} = -\dfrac{?}{2(\underline{\quad?\quad})} = \underline{\quad?\quad}$.

The maximum value of $f(x)$ is $f(\underline{\quad?\quad}) = -8(\underline{\quad?\quad})^2 + 8(\underline{\quad?\quad})$

$= \underline{\quad?\quad} + \underline{\quad?\quad} = \underline{\quad?\quad}$.

5. Does the function have a minimum or a maximum value?
a. $f(x) = -x^2 + 6x - 1$ **b.** $f(x) = 2x^2 - 4$ **c.** $f(x) = -5x^2 + x$

6. Does the function have a minimum or a maximum value?
a. $f(x) = 3x^2 - 2x + 4$ **b.** $f(x) = -x^2 + 9$ **c.** $f(x) = 6x^2 - 3x$

Find the minimum or maximum value of the quadratic function.

7. $f(x) = x^2 - 2x + 3$

8. $f(x) = 2x^2 + 4x$

9. $f(x) = -2x^2 + 4x - 3$

10. $f(x) = -2x^2 + 4x - 5$

11. $f(x) = -2x^2 - 3x + 4$

12. $f(x) = -2x^2 - 3x$

13. $f(x) = 2x^2 + 3x - 8$

14. $f(x) = 3x^2 + 3x - 2$

15. $f(x) = -3x^2 + x - 6$

16. $f(x) = -x^2 - x + 2$

17. $f(x) = x^2 - 5x + 3$

18. $f(x) = 3x^2 + 5x + 2$

19. Which of the following parabolas has the highest minimum value?
 a. $y = x^2 - 2x - 3$ **b.** $y = x^2 - 10x + 20$ **c.** $y = 3x^2 - 6$

20. Which of the following parabolas has the highest maximum value?
 a. $y = -2x^2 + 2x - 1$ **b.** $y = -x^2 + 8x - 2$ **c.** $y = -4x^2 + 3$

21. The vertex of a parabola that opens up is $(-4, 7)$. Does the function have a maximum or a minimum value? What is the maximum or minimum value of the function?

22. The vertex of a parabola that opens down is $(3, -5)$. Does the function have a maximum or a minimum value? What is the maximum or minimum value of the function?

Complete Exercises 23 and 24 given that the vertex of the graph of the function $f(x) = x^2 + bx + c$ is (n, m) and the vertex of the graph of the function $g(x) = -x^2 + dx + e$ is (p, q).

23. True or false? If $m > q$, then for all values of x, $f(x) > g(x)$.

24. True or false? If $m < q$, then for all values of x, $f(x) < g(x)$.

2 Applications of minimum and maximum

Complete Exercises 25 and 26 given that the approximate height h, in meters, of a ball t seconds after it is thrown vertically upward from a height of 2 m, at an initial speed of 30 m/s, is given by the function $h(t) = -5t^2 + 30t + 2$.

25. To find out after how many seconds the ball reaches its maximum height, find the ___?___-coordinate of the vertex of the graph of h:
$t = -\dfrac{b}{2a} = \dfrac{?}{-2(\underline{\quad?\quad})} = \underline{\quad?\quad}$. The ball reaches its maximum height after ___?___ s.

26. Use the result of Exercise 25. To find the maximum height of the ball, evaluate $h(\underline{\quad?\quad}) = -5(\underline{\quad?\quad})^2 + 30(\underline{\quad?\quad}) + 2 = \underline{\quad?\quad} + \underline{\quad?\quad} + 2 = \underline{\quad?\quad}$. The maximum height of the ball is ___?___ m.

Solve.

27. Physics The height s, in feet, of a rock thrown upward at an initial speed of 64 ft/s from a cliff 50 ft above an ocean beach is given by the function $s(t) = -16t^2 + 64t + 50$, where t is the time in seconds. Find the maximum height above the beach that the rock will attain.

28. **Sports** An event in the Summer Olympics is 10-meter springboard diving. In this event, the height s, in meters, of a diver above the water t seconds after jumping is given by $s(t) = -4.9t^2 + 7.8t + 10$. What is the maximum height that the diver will be above the water? Round to the nearest tenth.

29. Business A tour operator believes that the profit P, in dollars, from selling x tickets is given by $P(x) = 40x - 0.25x^2$. Using this model, find the maximum profit that the tour operator can expect.

30. Manufacturing A manufacturer of microwave ovens believes that the revenue R, in dollars, that the company receives is related to the price P, in dollars, of an oven by the function $R(P) = 125P - 0.25P^2$. What price will give the maximum revenue?

31. Chemistry A pool is treated with a chemical to reduce the amount of algae. The amount of algae in the pool t days after the treatment can be approximated by the function $A(t) = 40t^2 - 400t + 500$. How many days after treatment will the pool have the least amount of algae?

32. Structural Engineering The suspension cable that supports a small footbridge hangs in the shape of a parabola. The height h, in feet, of the cable above the bridge is given by the function $h(x) = 0.25x^2 - 0.8x + 25$, where x is the distance in feet from one end of the bridge. What is the minimum height of the cable above the bridge?

33. **Telescopes** An equation of the thickness h, in inches, of the mirror at the Palomar Mountain Observatory is given by

$$h(x) = 0.000379x^2 - 0.0758x + 24,$$

where x is measured from the edge of the mirror. Find the minimum thickness of the mirror.

34. **Art** The Buckingham Fountain in Chicago shoots water from a nozzle at the base of the fountain. The height h, in feet, of the water above the ground t seconds after it leaves the nozzle is given by the function $h(t) = -16t^2 + 90t + 15$. What is the maximum height of the water? Round to the nearest tenth.

35. Fire Science The height s, in feet, of water squirting from a fire hose nozzle is given by the equation $s(x) = -\frac{1}{30}x^2 + 2x + 5$, where x is the horizontal distance from the nozzle. How high on a building 40 ft from the fire hose will the water land?

24 in. x h
200 in.
Not to scale

40 ft
s
5 ft

36. Automotive Technology On wet concrete, the stopping distance s, in feet, of a car traveling v miles per hour is given by $s(v) = 0.055v^2 + 1.1v$. At what speed could a car be traveling and still stop at a stop sign 44 ft away?

37. Number Problem Find two numbers whose sum is 20 and whose product is a maximum.

38. Number Problem Find two numbers whose difference is 14 and whose product is a minimum.

39. Ranching A rancher has 200 ft of fencing to build a rectangular corral alongside an existing fence. Determine the dimensions of the corral that will maximize the enclosed area.

9.7 Applying Concepts

40. Show that the minimum value of
$$S(x) = (2 - x)^2 + (5 - x)^2 + (4 - x)^2 + (7 - x)^2$$
occurs when x is the average of the numbers 2, 5, 4, and 7.

Use a graphing utility to find the minimum or maximum value of the function. Round to the nearest tenth. See the Graphing Calculator Appendix for assistance.

41. $f(x) = x^4 - 2x^2 + 4$

42. $f(x) = x^4 + 2x^3 + 1$

43. $f(x) = -x^6 + x^4 - x^3 + x$

44. $f(x) = -x^8 + x^6 - x^4 + 5x^2 + 7$

45. On the basis of Exercises 41–44, make a conjecture about the relationship between the sign of the leading coefficient of a polynomial function of even degree and whether that function will have a minimum or a maximum value.

Solve.

46. Traffic Control Traffic engineers try to determine the effect a traffic light has at an intersection. By gathering data about the intersection, engineers can determine the approximate number of cars that enter the intersection in the horizontal direction and those that enter in the vertical direction. The engineers would also collect information on the time it takes a stopped car to regain the normal posted speed limit. One model of this situation is $T = \left(\frac{H + V}{2}\right)R^2 + (0.08H - 1.08V)R + 0.58V$, where H is the number of cars arriving at the intersection from the horizontal direction, V is the number of cars arriving at the intersection from the vertical direction, and R is the percent of time the light is red in the horizontal direction. T is the total delay time for all cars and is measured as the number of times the traffic light changes from red to green and back to red.
a. Graph this equation for $H = 100$, $V = 150$, and $0 \le R \le 1$.
b. Write a sentence that explains why the graph is drawn only for $0 \le R \le 1$.
c. What percent of the time should the traffic light remain red in the horizontal direction to minimize T? Round to the nearest whole percent.

47. Write a paper on the history of higher-degree equations. Include discoveries made by del Ferro, Ferrari, and Ruffini.

Focus on Problem Solving

Inductive and Deductive Reasoning

Consider the following sums of odd positive integers.

$1 + 3 = 4 = 2^2$ Sum of the first two odd numbers is 2^2.

$1 + 3 + 5 = 9 = 3^2$ Sum of the first three odd numbers is 3^2.

$1 + 3 + 5 + 7 = 16 = 4^2$ Sum of the first four odd numbers is 4^2.

$1 + 3 + 5 + 7 + 9 = 25 = 5^2$ Sum of the first five odd numbers is 5^2.

1. Make a conjecture about the value of the sum

$$1 + 3 + 5 + 7 + 9 + 11 + 13 + 15$$

without adding the numbers.

If the pattern continues, a possible conjecture is that the sum of the first eight odd positive integers is $8^2 = 64$. By adding the numbers, you can verify that this is correct. Inferring that the pattern established by a few cases is valid for all cases is an example of **inductive reasoning.**

2. Use inductive reasoning to find the next figure in the sequence below.

The fact that a pattern appears to be true does not *prove* that it is true for all cases. For example, consider the polynomial $n^2 - n + 41$. If we begin substituting positive integer values for n and evaluating the expression, we produce the following table.

n	$n^2 - n + 41$	
1	$1^2 - 1 + 41 = 41$	41 is a prime number.
2	$2^2 - 2 + 41 = 43$	43 is a prime number.
3	$3^2 - 3 + 41 = 47$	47 is a prime number.
4	$4^2 - 4 + 41 = 53$	53 is a prime number.

Even if we try a number like 30, we have $30^2 - 30 + 41 = 911$, and 911 is a prime number. Thus it appears that the value of this polynomial is a prime number for any value of n. However, the conjecture is not true. For instance, if $n = 41$, we have $41^2 - 41 + 41 = 41^2 = 1681$, which is not a prime number because it is divisible by 41. This illustrates that inductive reasoning may lead to incorrect conclusions and that an inductive proof must be available to prove conjectures. There is such a proof, called *mathematical induction*, that you may study in a future math course.

Now consider the true statement that the sum of the measures of the interior angles of a triangle is 180°. The figure on the left is a triangle; therefore, the sum of the measures of the interior angles must be 180°. This is an example of *deductive reasoning*. **Deductive reasoning** uses a rule or statement of fact to reach a conclusion.

$\angle A + \angle B + \angle C = 180°$

3. Use deductive reasoning to complete the following: "All even numbers are divisible by 2. Because 41,386 is an even number, . . ."

4. a. Use the pattern given below to find the three missing terms in __ __ __ 64.

 A 2 3 4 B 6 7 8 C 10 11 12 D 14 15 16 ...

 b. Is this inductive or deductive reasoning?

5. a. Draw the next figure in the sequence:

 b. Is this inductive or deductive reasoning?

6. a. If ◊◊◊◊◊ = ‡‡‡ and ‡‡‡ = ∇∇∇∇, how many ◊'s equal ∇∇∇∇∇∇∇∇?

 b. Is this inductive or deductive reasoning?

7. a. If ♠♠ = ♦♦♦♦♦♦, and ♦♦♦ = ♣♣, and ♣♣♣♣ = ♥, how many ♥'s equal ♠♠♠♠♠♠?

 b. Is this inductive or deductive reasoning?

For Exercises 8 to 11, determine whether inductive or deductive reasoning is being used.

8. The Atlanta Braves have won eight games in a row. Therefore, the Atlanta Braves will win their next game.

9. All quadrilaterals have four sides. A square is a quadrilateral. Therefore, a square has four sides.

10. Every English setter likes to hunt. Duke is an English setter, so Duke likes to hunt.

11. All kitchen remodeling jobs cost more than the contractor's estimate. The contractor estimated the cost of remodeling my kitchen at $35,000. Therefore, it will cost more than $35,000 to have my kitchen remodeled.

Projects & Group Activities

Using a Graphing Calculator to Solve a Quadratic Equation

Recall that an x-intercept of the graph of an equation is a point at which the graph crosses the x-axis. For the graph in Figure 1 below, the x-intercepts are $(-2, 0)$ and $(3, 0)$.

Recall also that to find the x-intercept of a graph, set $y = 0$ and then solve for x. For the equation in Figure 1, if we set $y = 0$, the resulting equation is $0 = x^2 - x - 6$, which is a quadratic equation. Solving this equation by factoring, we have

$$0 = x^2 - x - 6$$
$$0 = (x + 2)(x - 3)$$

$$x + 2 = 0 \qquad x - 3 = 0$$
$$x = -2 \qquad x = 3$$

Figure 1

Thus the solutions of the equation are the x-coordinates of the x-intercepts of the graph.

This connection between the solutions of an equation and the x-intercepts of a graph allows us to find approximations of the real number solutions of an equation graphically. For example, to use a TI-83/83 Plus or TI-84 Plus to approximate the solutions of $2x^2 + 5x - 1 = 0$ graphically, use the following keystrokes to graph $y = 2x^2 + 5x - 1$ and find the decimal approximations of the x-coordinates of the x-intercepts. Use the window shown in the graph in Figure 2.

Use the arrow keys to move to the left of the leftmost x-intercept. Press ENTER. Use the arrow keys to move to a point just to the right of the leftmost x-intercept. Press ENTER twice. The x-coordinate at the bottom of the screen is the approximation of one solution of the equation. To find the other solution, use the same procedure but move the cursor first to the left and then to the right of the rightmost x-intercept.

Figure 2

Attempting to find the solutions of an equation graphically will not necessarily yield all the solutions. Because the x-coordinates of the x-intercepts of a graph are *real* numbers, only real number solutions can be found. For instance, consider the equation $x^2 + 4x + 5 = 0$. The graph of $y = x^2 + 4x + 5$ is shown in Figure 3. Note that the graph has no x-intercepts and that consequently it has no real number solutions. However, $x^2 + 4x + 5 = 0$ does have complex number solutions that can be obtained by using the quadratic formula. They are $-2 + i$ and $-2 - i$.

Figure 3

Use a graphing calculator to approximate the solutions of the equation.

1. $x^2 + 6x + 1 = 0$

2. $-x^2 + x + 3 = 0$

3. $2x^2 + 4x - 1 = 0$

4. $x^2 + 4x + 5 = 0$

Finding the Maximum or Minimum of a Function Using a Graphing Calculator

Recall that for a quadratic function, the maximum or minimum value of the function is the y-coordinate of the vertex of the graph of the function. We determined algebraically that the coordinates of the vertex are $\left(-\dfrac{b}{2a}, f\left(-\dfrac{b}{2a}\right)\right)$. It is also possible to use a graphing calculator to approximate a maximum or minimum value. The method used to find the value depends on the calculator model. However, it is generally true for most calculators that the point at which the function has a maximum or minimum must be shown on the screen. If it is not, you must select a different viewing window.

$f(x) = 4x^2 + 4x + 1$

Minimum
X=-.5 Y=0

For a TI-83/83 Plus or TI-84 Plus graphing calculator, press 2nd CALC 3 to determine a minimum value or 2nd CALC 4 to determine a maximum value. Move the cursor to a point on the curve that is to the left of the minimum (or maximum). Press ENTER. Move the cursor to a point on the curve that is to the right of the minimum (or maximum). Press ENTER twice. The minimum (or maximum) value is shown as the y-coordinate at the bottom of the screen.

1. Find the maximum or minimum value of $f(x) = -x^2 - 3x + 2$ by using a graphing calculator.

2. Find the maximum or minimum value of $f(x) = x^2 - 4x - 2$ by using a graphing calculator.

3. Find the maximum or minimum value of $f(x) = -\sqrt{2}x^2 - \pi x + 2$ by using a graphing calculator.

4. Find the maximum or minimum value of $f(x) = \dfrac{x^2}{\sqrt{5}} - \sqrt{7}x - \sqrt{11}$ by using a graphing calculator.

5. **Manufacturing** A manufacturer of camera lenses estimated that the average monthly cost C of production is given by the function $C(x) = 0.1x^2 - 20x + 2000$, where x is the number of lenses produced each month. Find the number of lenses the company should produce in order to minimize the average cost.

6. **Income** The net annual income I, in dollars, of a family physician can be modeled by the equation $I(x) = -290(x - 48)^2 + 148,000$, where x is the age of the physician and $27 \leq x \leq 70$. Find (a) the age at which the physician's income will be a maximum and (b) the maximum income.

Chapter 9 Summary

Key Words

Examples

A **quadratic equation** is an equation of the form $ax^2 + bx + c = 0$, where $a \neq 0$. A quadratic equation is also called a **second-degree equation**. [9.1.1, p. 606]

$3x^2 + 4x - 7 = 0$ and $x^2 - 1 = 0$ are quadratic equations.

A quadratic equation is in **standard form** when the polynomial is in descending order and equal to zero. [9.1.1, p. 606]

$x^2 - 5x + 6 = 0$ is a quadratic equation in standard form.

When a quadratic equation has two solutions that are the same number, the solution is called a **double root** of the equation. [9.1.1, p. 607]

$$x^2 - 4x + 4 = 0$$
$$(x - 2)(x - 2) = 0$$

$$x - 2 = 0 \qquad x - 2 = 0$$
$$x = 2 \qquad\quad x = 2$$

2 is a double root.

For an equation of the form $ax^2 + bx + c = 0$, the quantity $b^2 - 4ac$ is called the **discriminant**. [9.2.2, p. 623]

$$2x^2 - 3x + 2 = 0$$
$$a = 2, b = -3, c = 2$$
$$b^2 - 4ac = (-3)^2 - 4(2)(2)$$
$$= 9 - 16 = -7$$

A **quadratic inequality in one variable** is one that can be written in the form $ax^2 + bx + c > 0$ or $ax^2 + bx + c < 0$, where $a \neq 0$. The symbols $\leq$ and $\geq$ can also be used. [9.5.1, p. 644]

$3x^2 + 5x - 8 \leq 0$ is a quadratic inequality.

The **x-coordinate of the vertex of a parabola** is given by $-\dfrac{b}{2a}$. [9.6.1, p. 652]

$y = 2x^2 - 8x + 5 \quad a = 2, b = -8$
The x-coordinate of the vertex is

$$-\frac{-8}{2(2)} = 2.$$

The **x-intercepts** of the graph of a parabola occur when $y = 0$. [9.6.2, p. 654]

$$x^2 - 5x + 6 = 0$$
$$(x - 3)(x - 2) = 0$$

$$x - 3 = 0 \qquad x - 2 = 0$$
$$x = 3 \qquad\quad x = 2$$

The x-intercepts are $(3, 0)$ and $(2, 0)$.

The graph of a **quadratic function** $f(x) = ax^2 + bx = c$ is a **parabola** that opens up when $a > 0$ and down when $a < 0$. The **axis of symmetry** is a line that passes through the vertex of the parabola. [9.6.1, pp. 650–651]

The graph of $f(x) = ax^2 + bx + c$ has a **minimum** value if $a > 0$ and a **maximum** value if $a < 0$. [9.7.1, p. 667]

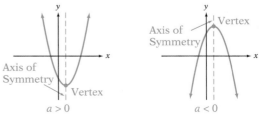

Essential Rules and Procedures

Principle of Zero Products
If $ab = 0$, then $a = 0$ or $b = 0$. [9.1.1, p. 606]

$$(x - 3)(x + 4) = 0$$

$$x - 3 = 0 \qquad x + 4 = 0$$
$$x = 3 \qquad\qquad x = -4$$

A quadratic equation can be solved by **taking the square root of each side of the equation**. [9.1.3, p. 608]

$$(x + 2)^2 - 9 = 0$$
$$(x + 2)^2 = 9$$
$$\sqrt{(x + 2)^2} = \sqrt{9}$$

$$x + 2 = \pm\sqrt{9}$$
$$x + 2 = \pm 3$$

$$x + 2 = 3 \qquad x + 2 = -3$$
$$x = 1 \qquad\qquad x = -5$$

A quadratic equation can be solved by **completing the square**. [9.2.1, p. 617]

$$x^2 + 4x - 1 = 0$$
$$x^2 + 4x = 1$$
$$x^2 + 4x + 4 = 1 + 4$$
$$(x + 2)^2 = 5$$
$$\sqrt{(x + 2)^2} = \sqrt{5}$$
$$x + 2 = \pm\sqrt{5}$$

$$x + 2 = \sqrt{5} \qquad x + 2 = -\sqrt{5}$$
$$x = -2 + \sqrt{5} \qquad x = -2 - \sqrt{5}$$

Quadratic Formula
$$x = \frac{-b \pm \sqrt{b^2 - 4ac}}{2a} \quad [9.2.2,\ p.\ 621]$$

$$2x^2 - 3x + 4 = 0$$
$$a = 2, b = -3, c = 4$$
$$x = \frac{-(-3) \pm \sqrt{(-3)^2 - 4(2)(4)}}{2(2)}$$

$$= \frac{3 \pm \sqrt{9 - 32}}{4} = \frac{3 \pm \sqrt{-23}}{4}$$

$$= \frac{3 \pm i\sqrt{23}}{4} = \frac{3}{4} \pm \frac{\sqrt{23}}{4}i$$

Vertex of a parabola
The coordinates of the vertex of a parabola are $\left(-\frac{b}{2a}, f\left(-\frac{b}{2a}\right)\right)$.
[9.6.1, p. 652]

$$f(x) = x^2 - 2x - 4 \quad a = 1, b = -2$$
$$-\frac{b}{2a} = -\frac{-2}{2(1)} = 1$$
$$f(1) = 1^2 - 2(1) - 4 = -5$$

The coordinates of the vertex are $(1, -5)$.

Discriminant [9.6.2, p. 658]

If $b^2 - 4ac = 0$, the equation has a double root.

$x^2 + 8x + 16 = 0$ has a double root because

$$b^2 - 4ac = 8^2 - 4(1)(16) = 0$$

If $b^2 - 4ac > 0$, the equation has two real number solutions that are not equal.

$2x^2 + 3x - 5 = 0$ has two unequal real number solutions because

$$b^2 - 4ac = 3^2 - 4(2)(-5) = 49$$

If $b^2 - 4ac < 0$, the equation has two complex number solutions.

$3x^2 + 2x + 4 = 0$ has two complex number solutions because

$$b^2 - 4ac = 2^2 - 4(3)(4) = -44$$

Chapter 9 Review Exercises

1. Solve: $2x^2 - 3x = 0$

2. Solve for x: $6x^2 + 9xc = 6c^2$

3. Solve: $x^2 = 48$

4. Solve: $\left(x + \dfrac{1}{2}\right)^2 + 4 = 0$

5. Find the minimum value of the function $f(x) = x^2 - 7x + 8$.

6. Find the maximum value of the function $f(x) = -2x^2 + 4x + 1$.

7. Write a quadratic equation that has integer coefficients and has solutions $\dfrac{1}{3}$ and -3.

8. Solve: $2x^2 + 9x = 5$

9. Solve: $2(x + 1)^2 - 36 = 0$

10. Solve: $x^2 + 6x + 10 = 0$

11. Solve: $\dfrac{2}{x - 4} + 3 = \dfrac{x}{2x - 3}$

12. Solve: $x^4 - 6x^2 + 8 = 0$

13. Solve: $\sqrt{2x - 1} + \sqrt{2x} = 3$

14. Solve: $2x^{\frac{2}{3}} + 3x^{\frac{1}{3}} - 2 = 0$

15. Solve: $\sqrt{3x - 2} + 4 = 3x$

16. Solve: $x^2 - 6x - 2 = 0$

17. Solve: $\dfrac{2x}{x - 4} + \dfrac{6}{x + 1} = 11$

18. Solve: $2x^2 - 2x = 1$

19. Solve: $2x = 4 - 3\sqrt{x - 1}$

20. Solve: $3x = \dfrac{9}{x - 2}$

21. Solve: $\dfrac{3x + 7}{x + 2} + x = 3$

22. Solve: $\dfrac{x - 2}{2x + 3} - \dfrac{x - 4}{x} = 2$

23. Solve: $1 - \dfrac{x + 4}{2 - x} = \dfrac{x - 3}{x + 2}$

24. Find the axis of symmetry of the parabola whose equation is $y = -x^2 + 6x - 5$.

25. Find the vertex of the parabola whose equation is $y = -x^2 + 3x - 2$.

26. Find two values of x in the domain of $f(x) = x^2 - 8x - 4$ for which $f(x) = 1$.

27. Use the discriminant to determine the number of x-intercepts of the parabola whose equation is $y = 3x^2 - 2x - 4$.

28. Find the x-intercepts of the parabola whose equation is $y = 4x^2 + 12x + 4$.

29. Find the x-intercepts of the parabola whose equation is $y = -2x^2 - 3x + 2$.

30. Find the zeros of $f(x) = 3x^2 + 2x + 2$.

31. Solve: $(x + 3)(2x - 5) < 0$

32. Solve: $(x - 2)(x + 4)(2x + 3) \le 0$

33. Solve and graph the solution set of $\dfrac{x - 2}{2x - 3} \ge 0$.

34. Solve and graph the solution set of $\dfrac{(2x - 1)(x + 3)}{x - 4} \le 0$.

35. Graph $f(x) = x^2 + 2x - 4$. State the domain and range.

36. Find the vertex and axis of symmetry of the parabola whose equation is $y = x^2 - 2x + 3$. Then graph the equation.

37. Integer Problem The sum of the squares of three consecutive even integers is fifty-six. Find the three integers.

38. Geometry The length of a rectangle is 2 cm more than twice the width. The area of the rectangle is 60 cm². Find the length and width of the rectangle.

39. Work Problem An older computer requires 12 min longer to print the payroll than does a newer computer. Together the computers can print the payroll in 8 min. Find the time for the newer computer, working alone, to complete the payroll.

40. Uniform Motion A car travels 200 mi. A second car, traveling 10 mph faster than the first car, makes the same trip in 1 h less time. Find the speed of each car.

Chapter 9 Test

1. Solve: $2x^2 + x = 6$

2. Solve: $12x^2 + 7x - 12 = 0$

3. Find the maximum value of the function $f(x) = -x^2 + 8x - 7$.

4. Write the quadratic equation that has integer coefficients and has solutions $-\frac{1}{3}$ and 3.

5. Solve: $2(x + 3)^2 - 36 = 0$

6. Solve: $x^2 + 4x - 1 = 0$

7. Find the zeros of $g(x) = x^2 + 3x - 8$.

8. Solve: $3x^2 - x + 8 = 0$

9. Solve: $\dfrac{2x}{x - 1} + \dfrac{3}{x + 2} = 1$

10. Solve: $2x + 7x^{\frac{1}{2}} - 4 = 0$

11. Solve: $x^4 - 11x^2 + 18 = 0$

12. Solve: $\sqrt{2x + 1} + 5 = 2x$

13. Find two values of x in the domain of $f(x) = x^2 - 4x - 2$ for which $f(x) = 4$.

14. Use the discriminant to determine the number of x-intercepts of the parabola $y = 3x^2 + 2x - 4$.

15. Find the x-intercepts of the parabola whose equation is $y = 2x^2 + 5x - 12$.

16. Find the axis of symmetry of the parabola whose equation is $y = 2x^2 + 6x + 3$.

17. Graph $y = \frac{1}{2}x^2 + x - 4$. State the domain and range.

18. Solve and graph the solution set of $\dfrac{2x - 3}{x + 4} \le 0$.

19. **Geometry** The base of a triangle is 3 ft more than three times the height. The area of the triangle is 30 ft^2. Find the base and height of the triangle.

20. **Uniform Motion** The rate of a river's current is 2 mph. A canoe was rowed 6 mi down the river and back in 4 h. Find the rowing rate in calm water.

Cumulative Review Exercises

1. Evaluate $2a^2 - b^2 \div c^2$ when $a = 3$, $b = -4$, and $c = -2$.

2. Solve: $\dfrac{2x - 3}{4} - \dfrac{x + 4}{6} = \dfrac{3x - 2}{8}$

3. Find the slope of the line that contains the points $(3, -4)$ and $(-1, 2)$.

4. Find the equation of the line that contains the point $(1, 2)$ and is parallel to the line $x - y = 1$.

5. Factor: $-3x^3y + 6x^2y^2 - 9xy^3$

6. Factor: $6x^2 - 7x - 20$

7. Factor: $a^nx + a^ny - 2x - 2y$

8. Divide: $(3x^3 - 13x^2 + 10) \div (3x - 4)$

9. Multiply: $\dfrac{x^2 + 2x + 1}{8x^2 + 8x} \cdot \dfrac{4x^3 - 4x^2}{x^2 - 1}$

10. Find the distance between the points $(-2, 3)$ and $(2, 5)$.

11. Solve $S = \dfrac{n}{2}(a + b)$ for b.

12. Multiply: $-2i(7 - 4i)$

13. Simplify: $a^{-\frac{1}{2}}\left(a^{\frac{1}{2}} - a^{\frac{3}{2}}\right)$

14. Simplify: $\dfrac{\sqrt[3]{8x^4y^5}}{\sqrt[3]{16xy^6}}$

15. Solve: $\dfrac{x}{x + 2} - \dfrac{4x}{x + 3} = 1$

16. Solve: $\dfrac{x}{2x + 3} - \dfrac{3}{4x^2 - 9} = \dfrac{x}{2x - 3}$

17. Solve: $x^4 - 6x^2 + 8 = 0$

18. Solve: $\sqrt{3x + 1} - 1 = x$

19. Solve: $|3x - 2| < 8$

20. Find the x- and y-intercepts of the graph of $6x - 5y = 15$.

21. Graph the solution set: $x + y \le 3$
$\ 2x - y < 4$

22. Solve the system of equations:
$$x + y + z = 2$$
$$-x + 2y - 3z = -9$$
$$x - 2y - 2z = -1$$

23. Given $f(x) = \dfrac{2x - 3}{x^2 - 1}$ find $f(-2)$.

24. Find the domain of the function
$$f(x) = \dfrac{x - 2}{x^2 - 2x - 15}.$$

25. Solve and graph the solution set of $x^3 + x^2 - 6x < 0$.

26. Solve and graph the solution set of
$$\dfrac{(x - 1)(x - 5)}{x + 3} \ge 0.$$

27. **Forestry** The table at the right is based on data from a forestry study that looked at the diameters and heights of trees over a period of time. Between the ages of 2 years and 8 years, what is the average annual rate of change in the height of the trees? Round to the nearest tenth.

Age (in years)	Height (in feet)	Diameter (in inches)
2	7	0.8
3	12	2.0
4	20	3.8
5	22	4.2
6	30	6.9
7	37	7.1
8	42	8.1

28. **Mechanics** A piston rod for an automobile is $9\frac{3}{8}$ in. with a tolerance of $\frac{1}{64}$ in. Find the lower and upper limits of the length of the piston rod.

29. **Geometry** The base of a triangle is $(x + 8)$ ft. The height is $(2x - 4)$ ft. Find the area of the triangle in terms of the variable x.

30. **Depreciation** The graph shows the relationship between the cost of a building and the depreciation allowed for income tax purposes. Find the slope of the line between the two points shown on the graph. Write a sentence that states the meaning of the slope.

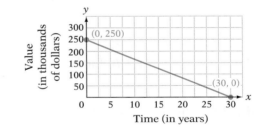

$\mathbf{10}$ Functions and Relations

OBJECTIVES

OBJECTIVES

10.1 1 Graph by using translations

10.2 1 Perform operations on functions
2 Find the composition of two functions

10.3 1 Determine whether a function is one-to-one
2 Find the inverse of a function

DVD **SSM** *Online Study Center*

Need help? For online student resources, such as section quizzes, visit the Online Study Center at college.hmco.com/pic/aufmannABI2e.

Have you ever found yourself wondering why certain products are so expensive? Manufacturing something as complex as a computer requires a lot of high-tech equipment and a lot of money. Usually the price at which a computer is sold can be determined by how much it costs to make the computer plus a certain markup. For a store to calculate the cost of each product, they must use a function such as the one discussed in the **Projects and Group Activities on page 713.**

Prep Test

1. Evaluate $|x + 4|$ for $x = -6$.

2. Given $y = -x^2 + 2x + 1$, find the value of y when $x = -2$.

3. Given $f(x) = x^2 - 3x + 2$, find $f(-4)$.

4. Evaluate $p(r) = r^2 - 5$ when $r = 2 + h$.

5. Solve $x = 2y + 4$ for y.

6. Find the domain and range of the relation $\{(-2, 4), (3, 5), (4, 6), (6, 5)\}$. Is the relation a function?

7. What value is excluded from the domain of $f(x) = \dfrac{3}{x - 8}$?

8. Graph: $f(x) = |x|$ **9.** Graph: $f(x) = x^2 - 2$

10.1 Translations of Graphs

OBJECTIVE 1 **Graph by using translations**

The graphs of $f(x) = |x|$ and $g(x) = |x| + 2$ are shown in Figure 1. Note that for a given x-coordinate, the y-coordinate on the graph of g is 2 units higher than that on the graph of f. The graph of g is said to be a **vertical translation,** or **vertical shift,** of the graph of f.

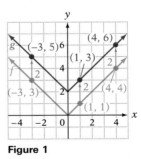

Figure 1

Take Note

Remember that $g(x)$ and $f(x)$ are y-coordinates of an ordered pair. The fact that $g(x)$ is 2 units more than $f(x)$ means that, for a given value of x, the y values for g are 2 units higher than the y values for f.

Note that because $f(x) = |x|$,

$$g(x) = |x| + 2$$
$$= f(x) + 2$$

Thus $g(x)$ is 2 units more than $f(x)$.

Now consider the graphs of the functions $f(x) = |x|$ and $h(x) = |x| - 3$, shown in Figure 2. Note that for a given x-coordinate, the y-coordinate on the graph of h is 3 units lower than that on the graph of f. The graph of h is a vertical translation of the graph of f.

In Figure 1, the graph of $g(x) = |x| + 2$ is the graph of f shifted *up* 2 units, whereas in Figure 2, the graph of $h(x) = |x| - 3$ is the graph of f shifted *down* 3 units.

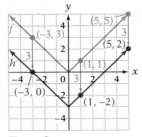

Figure 2

> ### Vertical Translation
>
> If f is a function and c is a positive constant, then
>
> $y = f(x) + c$ is the graph of $y = f(x)$ shifted up c units.
>
> $y = f(x) - c$ is the graph of $y = f(x)$ shifted down c units.

EXAMPLE 1 Given the graph of the function $f(x) = \sqrt[3]{x}$ shown at the right in blue, graph $g(x) = \sqrt[3]{x} + 3$ by using a vertical translation.

Solution The graph of $g(x) = \sqrt[3]{x} + 3$ is the graph of $f(x) = \sqrt[3]{x}$ shifted 3 units up. This graph is shown in red at the right.

PROBLEM 1 Given the graph of the function $f(x) = x^2 - 2x$ shown at the right, graph $g(x) = x^2 - 2x - 3$ by using a vertical translation.

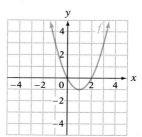

Solution See page S33.

The graphs of $f(x) = |x|$ and $g(x) = |x - 2|$ are shown in Figure 3. Note that the graph of g is the graph of f shifted 2 units to the right. The graph of g is a **horizontal translation,** or **horizontal shift,** of the graph of f. In this situation, each x-coordinate is moved to the right 2 units, but the y-coordinates are unchanged.

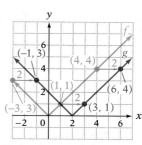

Figure 3

Take Note

Recall that evaluation of a function is accomplished by replacing a variable by another quantity.

$$f(x) = |x|$$
$$f(-3) = |-3|$$
$$f(x - 2) = |x - 2| = g(x)$$
$$f(x + 3) = |x + 3| = h(x)$$

The graphs of $f(x) = |x|$ and $h(x) = |x + 3|$ are shown in Figure 4. In this case, the graph of f is translated 3 units to the left.

For horizontal translations, note that when $f(x) = |x|$,

$$f(x - 2) = |x - 2| = g(x)$$

is a translation to the right and

$$f(x + 3) = |x + 3| = h(x)$$

is a translation to the left.

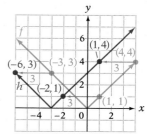

Figure 4

One way to determine the direction of the horizontal shift is to ask, "For what value of x is $y = 0$?" For instance, the horizontal shift of g in Figure 3 is 2 units to the right because $y = 0$ when $x = 2$.

$$g(x) = |x - 2|$$
$$g(2) = |2 - 2| = 0$$

Horizontal Translation

If f is a function and c is a positive constant, then

$y = f(x - c)$ is the graph of $y = f(x)$ shifted to the right c units.

$y = f(x + c)$ is the graph of $y = f(x)$ shifted to the left c units.

Point of Interest

Translations of a figure or pattern may be applied in the design of wallpaper, in the background of a Web page, and in the use of shadows to create an illusion of a letter raised off a page.

EXAMPLE 2 Given the graph of the function $F(x) = x^3$ shown at the right in blue, graph $G(x) = (x + 2)^3$ by using a horizontal translation.

Solution The graph of G (shown in red) is the graph of F shifted horizontally 2 units to the left. Note that if

$$(x + 2)^3 = 0, \text{ then } x = -2.$$

The result, -2, is one confirmation that the graph should be shifted 2 units to the left.

PROBLEM 2 Given the graph of the function $f(x) = \sqrt{x}$ shown at the right, graph $g(x) = \sqrt{x - 3}$ by using a horizontal translation.

Solution See page S33.

It is possible for a graph to involve both a horizontal and a vertical translation.

EXAMPLE 3 Given the graph of the function $f(x) = |x|$ shown at the left in blue, graph $A(x) = |x + 1| - 3$ by using both a horizontal and a vertical translation.

Solution The graph of A includes a horizontal shift of 1 unit to the left ($|x + 1| = 0$ means that $x = -1$) and a vertical shift of 3 units down. The graph of A is shown in red at the left.

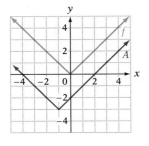

PROBLEM 3 Given the graph of the function
$g(x) = x^3$ shown at the right, graph
$V(x) = (x - 2)^3 + 1$ by using both
a horizontal and a vertical
translation.

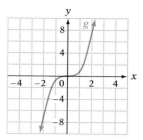

■ **Solution** See page S33.

10.1 Concept Review

Determine whether the following statements are always true, sometimes true,
or never true.

1. If c is a positive constant, the graph of $y = f(x) + c$ is the graph of $y = f(x)$
 shifted upward c units.

2. If c is a positive constant, the graph of $y = f(x + c)$ is the graph of $y = f(x)$
 translated c units to the right.

3. The graph of $y = f(x - 2)$ is the graph of $y = f(x)$ translated 2 units to
 the right.

4. The graph of $y = |x| + 2$ is the graph of $y = |x|$ translated 2 units to the left.

5. If c_1 and c_2 are positive constants, the graph of $y = f(x + c_1) + c_2$ is the
 graph of $y = f(x)$ translated c_1 units to the right and shifted up c_2 units.

6. The graph of $y = (x - 2)^3 + 3$ is the graph of $y = x^3$ translated 2 units to
 the right and shifted 3 units up.

10.1 Exercises

1 Graph by using translations

Complete Exercises 1 to 4 by using the word *up*, *down*, *right*, or *left*.

1. The graph of $g(x) = x^2 - 5$ is the graph of $f(x) = x^2$ shifted 5 units
 ____?____.

2. The graph of $g(x) = (x - 3)^3$ is the graph of $f(x) = x^3$ shifted 3 units
 ____?____.

3. The graph of $g(x) = |x + 7|$ is the graph of $f(x) = |x|$ shifted 7 units ___?___.

4. The graph of $g(x) = (x - 2)^2 + 4$ is the graph of $f(x) = x^2$ shifted 2 units ___?___ and 4 units ___?___.

In Exercises 5 to 30, use translations to draw the graphs. Many of these exercises refer to Figures 1 to 3 below.

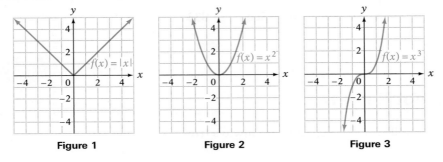

Figure 1 Figure 2 Figure 3

5. Given the graph of $f(x) = |x|$ shown in Figure 1 above, graph $g(x) = |x| + 4$.

6. Given the graph of $f(x) = |x|$ shown in Figure 1 above, graph $g(x) = |x| - 3$.

7. Given the graph of $f(x) = x^2$ shown in Figure 2 above, graph $g(x) = x^2 - 2$.

8. Given the graph of $f(x) = x^2$ shown in Figure 2 above, graph $g(x) = x^2 + 1$.

9. Given the graph of $f(x) = x^3$ shown in Figure 3 above, graph $g(x) = x^3 + 1$.

10. Given the graph of $f(x) = x^3$ shown in Figure 3 above, graph $g(x) = x^3 - 2$.

11. Given the graph of $f(x) = x^2$ shown in Figure 2 above, graph $g(x) = (x - 3)^2$.

12. Given the graph of $f(x) = x^2$ shown in Figure 2 above, graph $g(x) = (x + 2)^2$.

13. Given the graph of $f(x) = |x|$ shown in Figure 1 above, graph $g(x) = |x - 1|$.

14. Given the graph of $f(x) = |x|$ shown in Figure 1 above, graph $g(x) = |x + 4|$.

15. Given the graph of $f(x) = x^3$ shown in Figure 3 on page 690, graph $g(x) = (x - 4)^3$.

16. Given the graph of $f(x) = x^3$ shown in Figure 3 on page 690, graph $g(x) = (x + 3)^3$.

17. Given the graph of $f(x) = -x^2$ shown below, graph $g(x) = -x^2 + 2$.

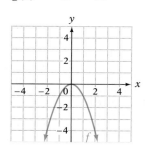

18. Given the graph of $f(x) = -|x|$ shown below, graph $g(x) = -|x| - 2$.

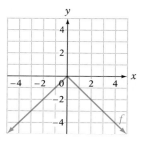

19. Given the graph of $f(x) = x^3$ shown in Figure 3 on page 690, graph $g(x) = (x + 1)^3 - 2$.

20. Given the graph of $f(x) = x^3$ shown in Figure 3 on page 690, graph $g(x) = (x - 2)^3 + 1$.

21. Given the graph of $f(x) = x^2$ shown in Figure 2 on page 690, graph $g(x) = (x - 3)^2 - 4$.

22. Given the graph of $f(x) = x^2$ shown in Figure 2 on page 690, graph $g(x) = (x + 2)^2 + 4$.

23. Given the graph of $f(x) = \sqrt{x}$ shown below, graph $g(x) = \sqrt{x - 1} + 2$.

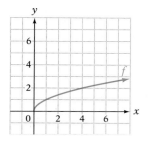

24. Given the graph of $f(x) = -\sqrt{x}$ shown below, graph $g(x) = -\sqrt{x + 1} - 3$.

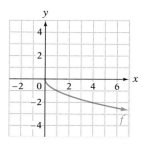

25. Given the graph of $f(x) = x^2 - 2x$ shown below, graph $g(x) = x^2 - 2x - 1$.

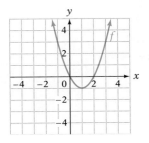

26. Given the graph of $f(x) = x^3 - 4x$ shown below, graph $g(x) = x^3 - 4x + 2$.

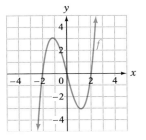

27. Given the graph of $f(x) = -x^2$ shown in Exercise 17, graph $g(x) = -(x + 1)^2 - 2$.

28. Given the graph of $f(x) = -x^2$ shown in Exercise 17, graph $g(x) = -(x - 1)^2 + 1$.

29. Given the graph of $f(x) = -\sqrt{x}$ shown in Exercise 24, graph $g(x) = -\sqrt{x - 2} + 1$.

30. Given the graph of $f(x) = \sqrt{x}$ shown in Exercise 23, graph $g(x) = \sqrt{x + 2} - 3$.

31. **a.** If $(0, 7)$ is the y-intercept of $y = f(x) - 2$, then what point is the y-intercept of $y = f(x)$?
b. If $(5, 0)$ is an x-intercept of $y = f(x)$, then what point is an x-intercept of $y = f(x - 3)$?

32. The graph of $y = f(x)$ is shown at the right. Is graph **a, b,** or **c** the graph of $y = f(x + 2) - 3$?

a.

b.

c.

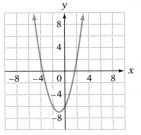

10.1 Applying Concepts

In Exercises 33 to 40, use translations to draw the graph. These exercises refer to Figures 4 to 6 below. No equations are given for the graphs.

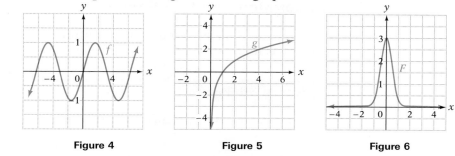

Figure 4 Figure 5 Figure 6

33. Given the graph of $y = f(x)$ shown in Figure 4 above, graph $y = f(x) + 1$.

34. Given the graph of $y = f(x)$ shown in Figure 4 above, graph $y = f(x) - 2$.

35. Given the graph of $y = g(x)$ shown in Figure 5 above, graph $y = g(x) - 3$.

36. Given the graph of $y = g(x)$ shown in Figure 5 above, graph $y = g(x) + 1$.

37. Given the graph of $y = f(x)$ shown in Figure 4 above, graph $y = f(x - 2)$.

38. Given the graph of $y = f(x)$ shown in Figure 4 above, graph $y = f(x + 1)$.

39. Given the graph of $y = F(x)$ shown in Figure 6 above, graph $y = F(x + 3)$.

40. Given the graph of $y = F(x)$ shown in Figure 6 above, graph $y = F(x - 1)$.

41. Given the triangle shown below, draw the triangle for which every value of x is translated 3 units to the right and every value of y is translated 2 units up.

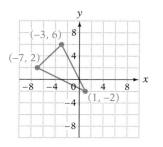

42. Given the rectangle shown below, draw the rectangle for which every value of x is translated 2 units to the right and every value of y is translated 1 unit up.

 Algebra of Functions

OBJECTIVE **1** **Perform operations on functions**

The operations of addition, subtraction, multiplication, and division of functions are defined as follows:

> **Operations on Functions**
>
> If f and g are functions and x is an element of the domain of each function, then
>
> $$(f + g)(x) = f(x) + g(x)$$
> $$(f - g)(x) = f(x) - g(x)$$
> $$(f \cdot g)(x) = f(x) \cdot g(x)$$
> $$\left(\frac{f}{g}\right)(x) = \frac{f(x)}{g(x)}, \; g(x) \neq 0$$

Given $f(x) = x^2 + 1$ and $g(x) = 3x - 2$, find $(f + g)(3)$ and $(f \cdot g)(-1)$.

$$
\begin{aligned}
(f + g)(3) &= f(3) + g(3) \\
&= [(3)^2 + 1] + [3(3) - 2] \\
&= 10 + 7 = 17
\end{aligned}
$$

$$
\begin{aligned}
(f \cdot g)(-1) &= f(-1) \cdot g(-1) \\
&= [(-1)^2 + 1] \cdot [3(-1) - 2] \\
&= 2 \cdot (-5) = -10
\end{aligned}
$$

Consider the functions f and g above. Find $S(x)$, the sum of the two functions.

Use the definition of addition.

$$
\begin{aligned}
S(x) &= (f + g)(x) \\
&= f(x) + g(x)
\end{aligned}
$$

Substitute $x^2 + 1$ for $f(x)$ and $3x - 2$ for $g(x)$.

$$
\begin{aligned}
&= (x^2 + 1) + (3x - 2) \\
S(x) &= x^2 + 3x - 1
\end{aligned}
$$

Now evaluate $S(3)$.

$$
\begin{aligned}
S(3) &= (3)^2 + 3(3) - 1 \\
&= 9 + 9 - 1 \\
&= 17 = (f + g)(3)
\end{aligned}
$$

Note that $S(3) = 17$ and $(f + g)(3) = 17$. This shows that adding $f(x) + g(x)$ and then evaluating is the same as evaluating $f(x)$ and $g(x)$ and then adding. The same is true for the other operations on functions. For instance, let $P(x)$ be the product of the functions f and g. Then

$$
\begin{aligned}
P(x) &= (f \cdot g)(x) \\
&= f(x) \cdot g(x) \\
&= (x^2 + 1)(3x - 2) \\
&= 3x^3 - 2x^2 + 3x - 2
\end{aligned}
$$

$$P(-1) = 3(-1)^3 - 2(-1)^2 + 3(-1) - 2$$
$$= -3 - 2 - 3 - 2$$
$$= -10 = (f \cdot g)(-1)$$

Given $f(x) = 2x^2 - 5x + 3$ and $g(x) = x^2 - 1$, find $\left(\dfrac{f}{g}\right)(1)$.

$$\left(\frac{f}{g}\right)(1) = \frac{f(1)}{g(1)}$$
$$= \frac{2(1)^2 - 5(1) + 3}{(1)^2 - 1}$$
$$= \frac{0}{0} \leftarrow \text{This is not a real number.}$$

Because $\dfrac{0}{0}$ is not defined, the expression $\left(\dfrac{f}{g}\right)(1)$ cannot be evaluated.

EXAMPLE 1 Given $f(x) = x^2 - x + 1$ and $g(x) = x^3 - 4$, find $(f - g)(3)$.

Solution $(f - g)(3) = f(3) - g(3)$
$$= (3^2 - 3 + 1) - (3^3 - 4)$$
$$= 7 - 23$$
$$= -16$$

$(f - g)(3) = -16$

PROBLEM 1 Given $f(x) = x^2 + 2x$ and $g(x) = 5x - 2$, find $(f + g)(-2)$.

Solution See page S33.

EXAMPLE 2 Given $f(x) = x^2 + 2$ and $g(x) = 2x + 3$, find $(f \cdot g)(-2)$.

Solution $(f \cdot g)(-2) = f(-2) \cdot g(-2)$
$$= [(-2)^2 + 2] \cdot [2(-2) + 3]$$
$$= 6 \cdot [-1]$$
$$= -6$$

$(f \cdot g)(-2) = -6$

PROBLEM 2 Given $f(x) = 4 - x^2$ and $g(x) = 3x - 4$, find $(f \cdot g)(3)$.

Solution See page S33.

EXAMPLE 3 Given $f(x) = x^2 + 4x + 4$ and $g(x) = x^3 - 2$, find $\left(\dfrac{f}{g}\right)(3)$.

Solution $\left(\dfrac{f}{g}\right)(3) = \dfrac{f(3)}{g(3)}$
$$= \frac{3^2 + 4(3) + 4}{3^3 - 2}$$
$$= \frac{25}{25}$$
$$= 1$$

$\left(\dfrac{f}{g}\right)(3) = 1$

> **PROBLEM 3** Given $f(x) = x^2 - 4$ and $g(x) = x^2 + 2x + 1$, find $\left(\dfrac{f}{g}\right)(4)$.
>
> **Solution** See page S33.

OBJECTIVE **2** Find the composition of two functions

Composition of functions is another way in which functions can be combined. This method of combining functions uses the output of one function as the input for a second function.

Suppose that a propane heater is used to heat the air in a hot-air balloon and that the shape of the balloon can be approximated by a sphere. The propane expands the air in the balloon in such a way that the radius of the balloon, in feet, is given by $r(t) = 0.3t$, where t is the time, in minutes, that the propane heater has been running.

The volume of the balloon depends on its radius and is given by $V(r) = \dfrac{4}{3}\pi r^3$. To find the volume of the balloon 20 min after the heater is turned on, we first find the radius of the balloon and then use that number to find the volume of the balloon.

$$r(t) = 0.3t \qquad\qquad\qquad V(r) = \frac{4}{3}\pi r^3$$

$$r(20) = 0.3(20) \quad \bullet\ t = 20\ \textbf{min} \qquad V(6) = \frac{4}{3}\pi(6^3) \quad \bullet\ r = 6\ \textbf{ft}$$

$$= 6 \qquad\qquad\qquad\qquad = 288\pi$$

The radius is 6 ft. $\qquad\qquad\qquad\qquad \approx 904.78$

The volume is approximately 904.78 ft³.

There is an alternative way to solve this problem. Because the volume of the balloon depends on the radius and the radius depends on the time, there is a relationship between the volume and the time. We can determine that relationship by evaluating the formula for the volume using $r(t) = 0.3t$. This will give the volume of the balloon as a function of time.

$$V(r) = \frac{4}{3}\pi r^3$$

$$V[r(t)] = \frac{4}{3}\pi[r(t)]^3 \qquad \bullet\ \textbf{Replace } r \textbf{ by } r(t).$$

$$= \frac{4}{3}\pi[0.3t]^3 \qquad \bullet\ r(t) = 0.3t$$

$$= 0.036\pi t^3$$

The volume of the balloon as a function of time is $V(t) = 0.036\pi t^3$. To find the volume of the balloon 20 min after the heater is turned on, evaluate this function at $t = 20$.

$$V(t) = 0.036\pi t^3$$

$$V(20) = 0.036\pi(20)^3 \qquad \bullet\ t = 20\ \textbf{min}$$

$$= 288\pi$$

$$\approx 904.78$$

This is exactly the same result we calculated earlier.

The function $V(t) = 0.036\pi t^3$ is referred to as the *composition* of V with r. The notation $V \circ r$ is used to denote this composition of functions. That is,

$$(V \circ r)(t) = 0.036\pi t^3$$

Definition of the Composition of Two Functions

Let f and g be two functions such that $g(x)$ is in the domain of f for all x in the domain of g. Then the **composition** of the two functions, denoted by $f \circ g$, is the function whose value at x is given by $(f \circ g)(x) = f[g(x)]$.

The function defined by $(f \circ g)(x)$ is called the **composite** of f and g. We read $(f \circ g)(x)$ as "f circle g of x," and we read $f[g(x)]$ as "f of g of x."

Consider the functions $f(x) = 2x$ and $g(x) = x^2$. Then $(f \circ g)(3) = f[g(3)]$ means to evaluate the function f at $g(3)$.

$$g(3) = 3^2 = 9$$

$$f[g(3)] = f(9) = 2(9) = 18$$

The function machine that is illustrated at the right shows the composition of $g(x) = x^2$ and $f(x) = 2x$. Note that the composite function, $(f \circ g)(x) = f[g(x)]$, uses the output of the square function $g(x) = x^2$ as the input of the double function $f(x) = 2x$.

$(f \circ g)(x) = f[g(x)]$

The requirement in the definition of the composition of two functions that $g(x)$ be in the domain of f for all x in the domain of g is important. For instance, let

$$f(x) = \frac{1}{x - 1} \quad \text{and} \quad g(x) = 3x - 5$$

When $x = 2$,

$$g(2) = 3(2) - 5 = 1$$

$$f[g(2)] = f(1) = \frac{1}{1 - 1} = \frac{1}{0} \leftarrow \text{This is not a real number.}$$

In this case, $g(2)$ is not in the domain of f. Thus the composition is not defined at 2.

Given $f(x) = x^3 - x + 1$ and $g(x) = 2x^2 - 10$, evaluate $(g \circ f)(2)$.

$$f(2) = (2)^3 - (2) + 1$$
$$= 7$$

$$\begin{aligned}(g \circ f)(2) &= g[f(2)] \\ &= g(7) \\ &= 2(7)^2 - 10 = 88\end{aligned}$$

$$(g \circ f)(2) = 88$$

Given $f(x) = 3x - 2$ and $g(x) = x^2 - 2x$, evaluate $(f \circ g)(x)$.

$$\begin{aligned}(f \circ g)(x) &= f[g(x)] = 3(x^2 - 2x) - 2 \\ &= 3x^2 - 6x - 2\end{aligned}$$

When we evaluate compositions of functions, the order in which the functions are applied is important. In the two diagrams that follow, the *square function,* $g(x) = x^2$, and the *double function,* $f(x) = 2x$, are interchanged. Note that the final outputs are different. Therefore, $(f \circ g)(x) \neq (g \circ f)(x).$

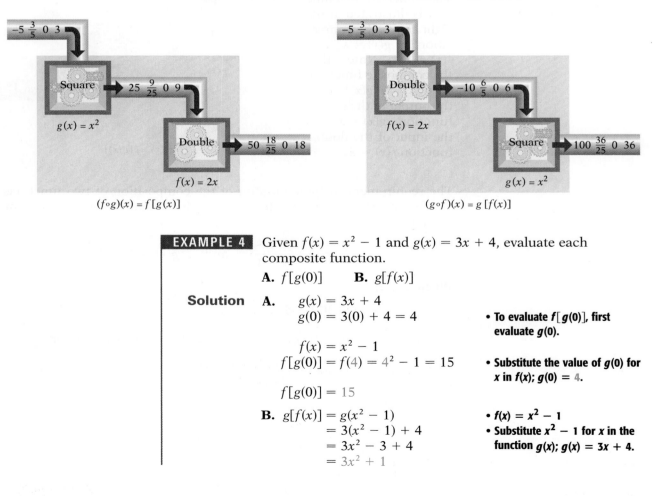

$(f \circ g)(x) = f[g(x)]$ $(g \circ f)(x) = g[f(x)]$

EXAMPLE 4 Given $f(x) = x^2 - 1$ and $g(x) = 3x + 4$, evaluate each composite function.

A. $f[g(0)]$ **B.** $g[f(x)]$

Solution **A.** $g(x) = 3x + 4$
$g(0) = 3(0) + 4 = 4$ • To evaluate $f[g(0)]$, first evaluate $g(0)$.

$f(x) = x^2 - 1$
$f[g(0)] = f(4) = 4^2 - 1 = 15$ • Substitute the value of $g(0)$ for x in $f(x)$; $g(0) = 4$.

$f[g(0)] = 15$

B. $\begin{aligned}g[f(x)] &= g(x^2 - 1) \\ &= 3(x^2 - 1) + 4 \\ &= 3x^2 - 3 + 4 \\ &= 3x^2 + 1\end{aligned}$ • $f(x) = x^2 - 1$
• Substitute $x^2 - 1$ for x in the function $g(x)$; $g(x) = 3x + 4$.

PROBLEM 4 Given $g(x) = 3x - 2$ and $h(x) = x^2 + 1$, evaluate each composite function.

A. $g[h(0)]$ **B.** $h[g(x)]$

■ **Solution** See page S33.

10.2 Concept Review

Determine whether the following statements are always true, sometimes true, or never true.

1. $(f \circ g)(x) = f[g(x)]$

2. If $f(x) = 2x$ and $g(x) = \frac{1}{2}x$, then $(f \circ g)(x) = (g \circ f)(x)$.

3. If $f(x) = 2x + 1$ and $g(x) = x + 4$, then $(f \circ g)(x) = (2x + 1)(x + 4)$.

4. If $f(x) = x^2 - 4$ and $g(x) = x^3$, then $(f \circ g)(x) = (x^2 - 4)^3$.

10.2 Exercises

1 Perform operations on functions

1. Given $f(x) = x^2 + 2x$ and $g(x) = 3x - 1$, find $(f - g)(-2)$.

$(f - g)(-2) = f(-2) - g(-2)$

- For all x in the domains of f and g, $(f - g)(x) = f(x) - g(x)$.

$= [(\underline{})^2 + 2(\underline{})] - [3(\underline{}) - 1]$

- In each function, replace x with -2.

$= [\underline{} - \underline{}] - [\underline{} - 1]$

- Evaluate the expressions inside the grouping symbols.

$= \underline{} - (\underline{})$

$= \underline{}$

- Subtract.

2. Given $f(x) = 2x^3 - x$ and $g(x) = x^2 + 1$, find $(f \cdot g)(-1)$.

$(f \cdot g)(-1) = f(-1) \cdot g(-1)$

- For all x in the domains of f and g, $(f \cdot g)(x) = f(x) \cdot g(x)$.

$= [2(\underline{})^3 - (\underline{})] \cdot [(\underline{})^2 + 1]$

- In each function, replace x with -1.

$= [\underline{} + \underline{}] \cdot [\underline{} + 1]$

- Evaluate the expressions inside the grouping symbols.

$= \underline{} \cdot \underline{}$

$= \underline{}$

- Multiply.

Given $f(x) = 2x^2 - 3$ and $g(x) = -2x + 4$, find:

3. $(f - g)(2)$ **4.** $(f - g)(3)$ **5.** $(f + g)(0)$

6. $(f + g)(1)$

7. $(f \cdot g)(2)$

8. $(f \cdot g)(-1)$

9. $\left(\dfrac{f}{g}\right)(4)$

10. $\left(\dfrac{f}{g}\right)(-1)$

11. $\left(\dfrac{g}{f}\right)(-3)$

Given $f(x) = 2x^2 + 3x - 1$ and $g(x) = 2x - 4$, find:

12. $(f + g)(-3)$

13. $(f + g)(1)$

14. $(f + g)(-2)$

15. $(f - g)(4)$

16. $(f \cdot g)(-2)$

17. $(f \cdot g)(1)$

18. $\left(\dfrac{f}{g}\right)(2)$

19. $\left(\dfrac{f}{g}\right)(-3)$

20. $(f \cdot g)\left(\dfrac{1}{2}\right)$

Given $f(x) = x^2 + 3x - 5$ and $g(x) = x^3 - 2x + 3$, find:

21. $(f - g)(2)$

22. $(f \cdot g)(-3)$

23. $\left(\dfrac{f}{g}\right)(-2)$

For Exercises 24 to 27, use the functions $f(x) = \sqrt{x}$, $g(x) = \sqrt{x^2}$, and $h(x) = \sqrt{a - x}$, where $a > 0$. State whether the given expression is *defined* or *undefined*.

24. $(f \cdot g)(a)$

25. $(f \cdot g)(-a)$

26. $\left(\dfrac{f}{h}\right)(a)$

27. $\left(\dfrac{h}{g}\right)(a)$

2 Find the composition of two functions

For Exercises 28 and 29, use the functions $f(x) = x^2 - 2x$ and $g(x) = 4x$.

28. Evaluate $g[f(-1)]$.

$f(-1) = (\underline{}\,?\,)^2 - 2(\underline{}\,?\,) = \underline{}\,?$ • First evaluate $f(-1)$.

$g(f(-1)) = g(\underline{}\,?\,)$ • Replace $g[f(-1)]$ with its value, 3.

$ = 4(\underline{}\,?\,)$ • Replace x with 3 in $g(x) = 4x$.

$ = \underline{}\,?$ • Evaluate.

29. Find $f[g(x)]$.

$f[g(x)] = f(4x)$ • $g(x) = \underline{}\,?$.

$ = (\underline{}\,?\,)^2 - 2(\underline{}\,?\,)$ • Replace x in $f(x) = x^2 - 2x$ with $4x$.

$ = \underline{}\,? - \underline{}\,?$ • Simplify.

30. Explain the meaning of the notation $f[g(2)]$.

31. What is the meaning of the notation $(f \circ g)(x)$?

Given $f(x) = 2x - 3$ and $g(x) = 4x - 1$, evaluate the composite function.

32. $f[g(0)]$ **33.** $g[f(0)]$ **34.** $f[g(2)]$ **35.** $g[f(-2)]$ **36.** $f[g(x)]$ **37.** $g[f(x)]$

Given $g(x) = x^2 + 3$ and $h(x) = x - 2$, evaluate the composite function.

38. $g[h(0)]$ **39.** $h[g(0)]$ **40.** $g[h(4)]$ **41.** $h[g(-2)]$ **42.** $g[h(x)]$ **43.** $h[g(x)]$

Given $f(x) = x^2 + x + 1$ and $h(x) = 3x + 2$, evaluate the composite function.

44. $f[h(0)]$ **45.** $h[f(0)]$ **46.** $f[h(-1)]$ **47.** $h[f(-2)]$ **48.** $f[h(x)]$ **49.** $h[f(x)]$

Given $f(x) = x - 2$ and $g(x) = x^3$, evaluate the composite function.

50. $f[g(2)]$ **51.** $f[g(-1)]$ **52.** $g[f(2)]$ **53.** $g[f(-1)]$ **54.** $f[g(x)]$ **55.** $g[f(x)]$

For Exercises 56 to 59, state whether or not the given pair of functions will produce the composite function $(f \circ g)(x) = 3x^2 - 3$.

56. $f(x) = x^2 - 1$ and $g(x) = 3x$ **57.** $f(x) = x^2$ and $g(x) = 3x - 3$

58. $f(x) = 3x - 3$ and $g(x) = x^2$ **59.** $f(x) = 3x$ and $g(x) = x^2 - 1$

10.2 Applying Concepts

Given $g(x) = x^2 - 1$, find:

60. $g(2 + h)$ **61.** $g(3 + h) - g(3)$ **62.** $g(-1 + h) - g(-1)$

63. $\dfrac{g(1 + h) - g(1)}{h}$ **64.** $\dfrac{g(-2 + h) - g(-2)}{h}$ **65.** $\dfrac{g(a + h) - g(a)}{h}$

Given $f(x) = 2x$, $g(x) = 3x - 1$, and $h(x) = x - 2$, find:

66. $f(g[h(2)])$ **67.** $g(h[f(1)])$ **68.** $h(g[f(-1)])$

69. $f(h[g(0)])$ **70.** $f(g[h(x)])$ **71.** $g(f[h(x)])$

The graphs of the functions f and g are shown at the right. Use these graphs to determine the values of the following composite functions.

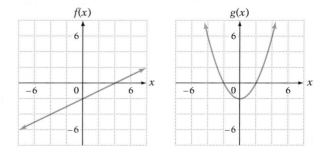

72. $f[g(0)]$ **73.** $f[g(2)]$

74. $f[g(4)]$ **75.** $f[g(-2)]$

76. $f[g(-4)]$ **77.** $g[f(0)]$

78. $g[f(4)]$ **79.** $g[f(-4)]$

One-to-One and Inverse Functions

OBJECTIVE **1**

Determine whether a function is one-to-one

Recall that a function is a set of ordered pairs in which no two ordered pairs that have the same first component have different second components. This means that given any x, there is only one y that can be paired with that x. A **one-to-one function** satisfies the additional condition that given any y, there is only one x that can be paired with the given y. One-to-one functions are commonly expressed by writing 1–1.

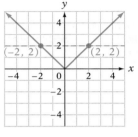

The function given by the equation $y = |x|$ is not a 1–1 function since, given $y = 2$, there are two possible values of x, 2 and -2, that can be paired with the given y-value. The graph at the left illustrates that a horizontal line intersects the graph more than once.

Just as the vertical line test can be used to determine whether a graph represents a function, a **horizontal line test** can be used to determine whether the graph of a function represents a 1–1 function.

> **Horizontal Line Test**
>
> A graph of a function is the graph of a 1–1 function if any horizontal line intersects the graph at no more than one point.

Since any vertical line will intersect the graph at the right at no more than one point, the graph is the graph of a function. Since any horizontal line will intersect the graph at no more than one point, the graph is the graph of a 1–1 function.

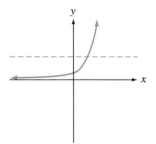

The graph of a quadratic function is shown at the right. Note that a horizontal line can intersect the graph at more than one point. Therefore, this graph is not the graph of a 1–1 function. In general, $f(x) = ax^2 + bx + c$, $a \neq 0$, is not a 1–1 function.

EXAMPLE 1 Determine whether or not the graph represents the graph of a 1–1 function.

A.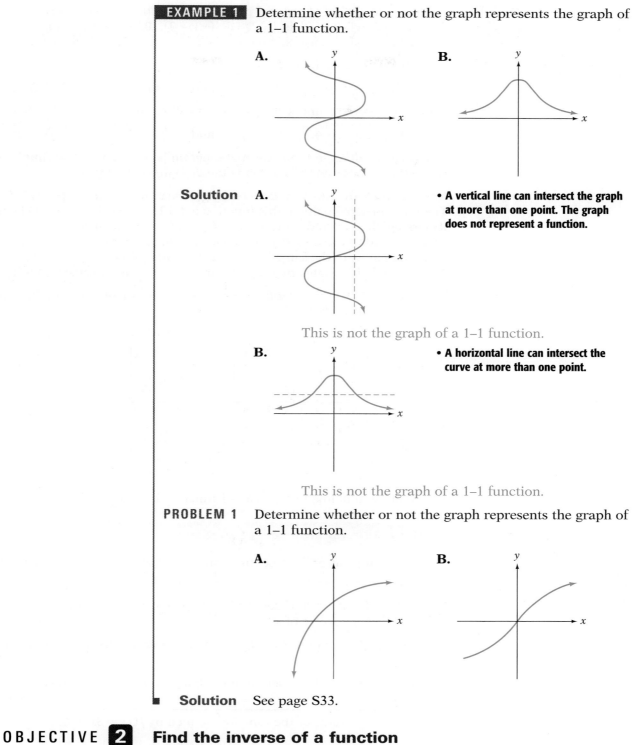

B.

Solution A.

• **A vertical line can intersect the graph at more than one point. The graph does not represent a function.**

This is not the graph of a 1–1 function.

B.

• **A horizontal line can intersect the curve at more than one point.**

This is not the graph of a 1–1 function.

PROBLEM 1 Determine whether or not the graph represents the graph of a 1–1 function.

A.

B.

■ Solution See page S33.

OBJECTIVE **2** **Find the inverse of a function**

The **inverse of a function** is the set of ordered pairs formed by reversing the coordinates of each ordered pair of the function.

For example, the set of ordered pairs of the function defined by $f(x) = 2x$ with domain $\{-2, -1, 0, 1, 2\}$ is $\{(-2, -4), (-1, -2), (0, 0), (1, 2), (2, 4)\}$. The set of ordered pairs of the inverse function is $\{(-4, -2), (-2, -1), (0, 0), (2, 1), (4, 2)\}$.

From the ordered pairs of f, we have

$$\text{Domain} = \{-2, -1, 0, 1, 2\} \quad \text{and} \quad \text{Range} = \{-4, -2, 0, 2, 4\}$$

From the ordered pairs of the inverse function, we have

$$\text{Domain} = \{-4, -2, 0, 2, 4\} \quad \text{and} \quad \text{Range} = \{-2, -1, 0, 1, 2\}$$

Note that the domain of the inverse function is the range of the function, and the range of the inverse function is the domain of the function.

Now consider the function defined by $g(x) = x^2$ with domain $\{-2, -1, 0, 1, 2\}$. The set of ordered pairs of this function is $\{(-2, 4), (-1, 1), (0, 0), (1, 1), (2, 4)\}$. Reversing the ordered pairs gives $\{(4, -2), (1, -1), (0, 0), (1, 1), (4, 2)\}$. These ordered pairs do not satisfy the condition of a function, because there are ordered pairs with the same first coordinate and different second coordinates. This example illustrates that not all functions have an inverse function.

The graphs of $f(x) = 2x$ and $g(x) = x^2$ with the set of real numbers as the domain are shown below.

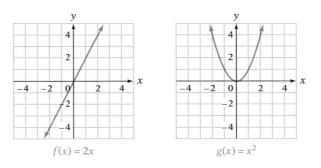

$f(x) = 2x$ $g(x) = x^2$

By the horizontal line test, f is a 1–1 function but g is not.

Condition for an Inverse Function

A function f has an inverse function if and only if f is a 1–1 function.

The symbol f^{-1} is used to denote the inverse of a 1–1 function f. The symbol $f^{-1}(x)$ is read "f inverse of x."

$f^{-1}(x)$ is *not* the reciprocal of $f(x)$ but is the notation for the inverse of a 1–1 function.

To find the inverse of a function, interchange x and y. Then solve for y.

Find the inverse of the function defined by $f(x) = 3x + 6$.

$$f(x) = 3x + 6$$

Replace $f(x)$ by y. $y = 3x + 6$

Interchange x and y. $x = 3y + 6$

Take Note

It is important to note that f^{-1} is the symbol for the *inverse* function and does not mean reciprocal.

$$f^{-1}(x) \neq \frac{1}{f(x)}$$

Take Note

If the ordered pairs of f are given by (x, y), then the ordered pairs of f^{-1} are given by (y, x). That is, x and y are interchanged. This is the reason for Step 2 at the right.

Solve for y. $x - 6 = 3y$

$$\frac{1}{3}x - 2 = y$$

Replace y by $f^{-1}(x)$. $f^{-1}(x) = \frac{1}{3}x - 2$

The inverse of the function is given by $f^{-1}(x) = \frac{1}{3}x - 2$.

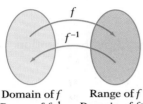

Domain of f Range of f
Range of f^{-1} Domain of f^{-1}

The fact that the ordered pairs of the inverse of a function are the reverse of those of the function has a graphical interpretation. The function shown at the left below includes the points $(-2, 0)$, $(-1, 2)$, $(1, 4)$, and $(5, 6)$. In the graph in the middle, the points with these coordinates reversed are plotted. The inverse function is graphed by drawing a smooth curve through those points, as shown in the figure at the right.

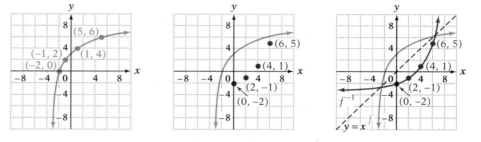

Note that the dashed graph of $y = x$ is shown in the figure at the right. If two functions are inverses of each other, then their graphs are mirror images with respect to the graph of the line $y = x$.

The composition of a function and its inverse have a special property.

Composition of Inverse Functions Property

$$f^{-1}[f(x)] = x \quad \text{and} \quad f[f^{-1}(x)] = x$$

This property can be used to determine whether two functions are inverses of each other.

Are $f(x) = 2x - 4$ and $g(x) = \frac{1}{2}x + 2$ inverses of each other?

To determine whether the functions are inverses, use the Composition of Inverse Functions Property.

$$f[g(x)] = 2\left(\frac{1}{2}x + 2\right) - 4 \qquad g[f(x)] = \frac{1}{2}(2x - 4) + 2$$
$$= x + 4 - 4 \qquad\qquad\qquad = x - 2 + 2$$
$$= x \qquad\qquad\qquad\qquad = x$$

Because $f[g(x)] = x$ and $g[f(x)] = x$, the functions are inverses of each other.

EXAMPLE 2 Find the inverse of the function defined by the equation $f(x) = 2x - 4$.

Solution
$$f(x) = 2x - 4$$
$$y = 2x - 4$$ • Think of the function as the equation $y = 2x - 4$.
$$x = 2y - 4$$ • Interchange x and y.
$$2y = x + 4$$ • Solve for y.
$$y = \frac{1}{2}x + 2$$

$$f^{-1}(x) = \frac{1}{2}x + 2$$

PROBLEM 2 Find the inverse of the function defined by the equation $f(x) = 4x + 2$.

Solution See page S33.

EXAMPLE 3 Are the functions defined by the equations $f(x) = -2x + 3$ and $g(x) = -\frac{1}{2}x + \frac{3}{2}$ inverses of each other?

Solution
$$f[g(x)] = f\left(-\frac{1}{2}x + \frac{3}{2}\right)$$ • Use the property that, for inverses, $f[f^{-1}(x)] = f^{-1}[f(x)] = x$.
$$= -2\left(-\frac{1}{2}x + \frac{3}{2}\right) + 3$$
$$= x - 3 + 3$$
$$= x$$ • $f[g(x)] = x$

$$g[f(x)] = g(-2x + 3)$$
$$= -\frac{1}{2}(-2x + 3) + \frac{3}{2}$$
$$= x - \frac{3}{2} + \frac{3}{2}$$
$$= x$$ • $g[f(x)] = x$

The functions are inverses of each other.

PROBLEM 3 Are the functions defined by the equations $h(x) = 4x + 2$ and $g(x) = \frac{1}{4}x - \frac{1}{2}$ inverses of each other?

Solution See page S33.

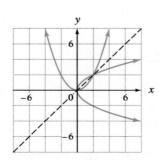

The function given by the equation $f(x) = \frac{1}{2}x^2$ does not have an inverse that is a function. Two of the ordered-pair solutions of this function are $(4, 8)$ and $(-4, 8)$.

The graph of $f(x) = \frac{1}{2}x^2$ is shown at the left. This graph does not pass the horizontal line test for the graph of a 1–1 function. The mirror image of the graph with respect to the line $y = x$ is also shown. This graph does not pass the vertical line test for the graph of a function.

A quadratic function with domain the real numbers does not have an inverse function.

10.3 Concept Review

Determine whether the following statements are always true, sometimes true, or never true.

1. The graph of a function represents the graph of a 1–1 function if any horizontal line intersects the graph at no more than one point.

2. A function has an inverse if and only if it is a 1–1 function.

3. The inverse of the function {(2, 3), (4, 5), (6, 3)} is the function {(3, 2), (5, 4), (3, 6)}.

4. The inverse of a function is a relation.

5. The inverse of a function is a function.

6. The graph of $y = x^2$ is the graph of a 1–1 function.

10.3 Exercises

1 Determine whether a function is one-to-one

1. What is a 1–1 function?

2. What is the horizontal line test?

3. The function $y = x^2$ is not a 1–1 function because, for example, the x values 2 and ____?____ are both paired with the same y value, ____?____

4. Suppose the horizontal line $y = b$ intersects the graph of the function $y = f(x)$ in two different points (a, b) and (c, b). The function $y = f(x)$ is not a 1–1 function because two different x values, ____?____ and ____?____, are paired with the same y value, ____?____

Determine whether the graph represents the graph of a 1–1 function.

5.

6.

7.

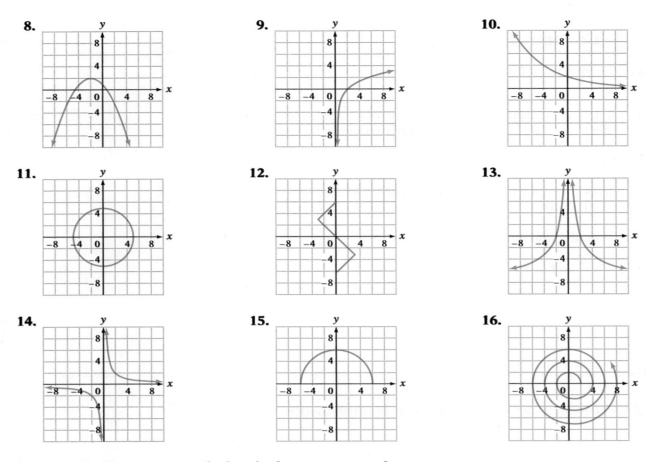

8.

9.

10.

11.

12.

13.

14.

15.

16.

For Exercises 17 to 20, state whether the function is a 1–1 function.

17. $f(x) = x$

18. $f(x) = a$, where a is any real number

19. $f(x) = |x| + a$, where a is any real number

20. $f(x) = x^3$

2 Find the inverse of a function

21. If $(2, -3)$ is a point on the graph of the function $y = f(x)$, then the point
(___?___, ___?___) is a point on the graph of the inverse of $y = f(x)$.

22. a. The only functions whose inverses are also functions are ___?___
functions.
 b. The notation used to denote the inverse function of the 1–1 function
 f is ___?___.

Find the inverse of the function. If the function does not have an inverse, write
"No inverse."

23. $\{(1, 0), (2, 3), (3, 8), (4, 15)\}$

24. $\{(1, 0), (2, 1), (-1, 0), (-2, 1)\}$

25. $\{(3, 5), (-3, -5), (2, 5), (-2, -5)\}$

26. $\{(-5, -5), (-3, -1), (-1, 3), (1, 7)\}$

27. $f(x) = 4x - 8$

28. $f(x) = 3x + 15$

29. $f(x) = x^2 - 1$

30. $f(x) = 2x + 4$

31. $f(x) = x - 5$

32. $f(x) = \dfrac{1}{2}x - 1$

33. $f(x) = \dfrac{1}{3}x + 2$

34. $f(x) = -2x + 2$

35. $f(x) = -3x - 9$

36. $f(x) = 2x^2 + 2$

37. $f(x) = \dfrac{2}{3}x + 4$

38. $f(x) = \dfrac{3}{4}x - 4$

39. $f(x) = -\dfrac{1}{3}x + 1$

40. $f(x) = -\dfrac{1}{2}x + 2$

41. $f(x) = 2x - 5$

42. $f(x) = 3x + 4$

43. $f(x) = x^2 + 3$

44. $f(x) = 5x - 2$

Given $f(x) = 3x - 5$, find:

45. $f^{-1}(0)$ **46.** $f^{-1}(2)$ **47.** $f^{-1}(4)$

State whether the graph is the graph of a function. If it is the graph of a function, does it have an inverse?

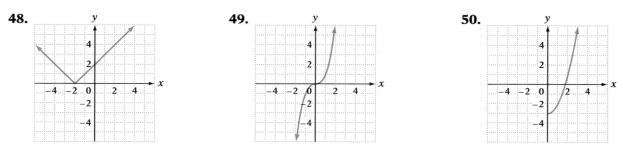

48. **49.** **50.**

Use the Composition of Inverse Functions Property to determine whether the functions are inverses of each other.

51. $f(x) = 4x;\ g(x) = \dfrac{x}{4}$

52. $g(x) = x + 5;\ h(x) = x - 5$

53. $f(x) = 3x;\ h(x) = \dfrac{1}{3x}$

54. $h(x) = x + 2;\ g(x) = 2 - x$

55. $g(x) = 3x + 2;\ f(x) = \dfrac{1}{3}x - \dfrac{2}{3}$

56. $h(x) = 4x - 1;\ f(x) = \dfrac{1}{4}x + \dfrac{1}{4}$

57. $f(x) = \dfrac{1}{2}x - \dfrac{3}{2};\ g(x) = 2x + 3$

58. $g(x) = -\dfrac{1}{2}x - \dfrac{1}{2};\ h(x) = -2x + 1$

Complete.

59. The domain of the inverse function f^{-1} is the _____ of f.

60. The range of the inverse function f^{-1} is the _____ of f.

61. For any function f and its inverse f^{-1}, $f[f^{-1}(3)] = $ _____.

62. For any function f and its inverse f^{-1}, $f^{-1}[f(-4)] = $ _____.

If f is a 1–1 function and $f(0) = 5$, $f(1) = 7$, $f(2) = 9$, and $f(3) = 11$, find:

63. $f^{-1}(5)$

64. $f^{-1}(11)$

65. $f^{-1}(9)$

66. $f^{-1}(7)$

For Exercises 67 to 70, f is a 1–1 function such that $f(a) = b$, $f(b) = c$, and $f(c) = a$.

67. Find $f^{-1}(a)$.

68. Find $f^{-1}(f^{-1}(a))$.

69. Find $f(f^{-1}(a))$.

70. Find $f^{-1}(f^{-1}(b))$.

71. Given $f(x) = 3x + 2$ and $g(x) = 5x + 4$, show that $f[g(x)] = g[f(x)]$. Are f and g inverse functions?

72. Given $f(x) = 7x - 4$ and $g(x) = -2x + 2$, show that $f[g(x)] = g[f(x)]$. Are f and g inverse functions?

10.3 Applying Concepts

Given the graph of the 1–1 function, draw the graph of the inverse of the function by using the technique shown in this section.

73.

74.

75.

76.

Each of the following tables defines a function. Is the inverse of the function a function? Explain your answer.

79.

Grading Scale Table

Score	Grade
90 – 100	A
80 – 89	B
70 – 79	C
60 – 69	D
0 – 59	F

80.

First-Class Postage

Weight (in oz)	Cost
$0 < w \le 1$	$.39
$1 < w \le 2$	$.63
$2 < w \le 3$	$.87
$3 < w \le 4$	$1.11

Exercises 81 and 82 demonstrate the essence of secure communication over the Internet. In practice, the coding functions are very complicated so that it is extremely unlikely that anyone will find the inverse and break the code.

81. Internet Commerce Inverse functions are used to secure business transactions over the Internet. Here is a very simple example of how it can be done. Let $A = 10$, $B = 11$, $C = 12$, ..., $Z = 35$. For instance, **NAME** would be coded as 23102214. Now choose a function that has an inverse function, say $f(x) = 2x - 4$. The function f is used to code **NAME** by evaluating the function at 23102214:

$$f(23102214) = 2(23102214) - 4 = 46204428 - 4 = 46204424$$

The number 46204424 would be sent over the Internet. The computer receiving the code would use the inverse of f to find the original word.

a. Find f^{-1} and show that $f^{-1}(46204424) = 23102214$.

b. Suppose that 44205830 is received by a computer. What word was sent?

82. Cryptography Friends send each other coded messages by using the scheme in Exercise 81. They decide to use the coding function $f(x) = 2x + 1$. If one person receives the message 58285659, what word was sent?

83. Unit Conversion The function given by $f(x) = 3x$ converts yards into feet. Find f^{-1} and explain what it does.

84. Unit Conversion The function given by $f(x) = 16x$ converts pounds into ounces. Find f^{-1} and explain what it does.

85. Geometry The formula for the perimeter of a square is $P(x) = 4x$. Find $P^{-1}(x)$ and explain what it does.

Focus on Problem Solving

Proof in Mathematics

Mathematics is based on the concept of proof, which in turn is based on principles of logic. It was Aristotle (384 B.C.–322 B.C.) who took the logical principles that had developed over time and set them in a structured form. From this structured form, several methods of proof were developed.

A **deductive proof** is one that proceeds in a carefully ordered sequence of steps. It is the type of proof you may have seen in a geometry class. Deductive proofs have a model as shown at the left.

$$\text{A true} \xrightarrow{\text{implies}} \text{B true} \xrightarrow{\text{implies}} \text{C true}$$
$$\text{A true} \xrightarrow{\text{implies}} \text{C true}$$

Here is an example of a deductive proof. It proves that the product of a negative and a positive number is a negative number. The proof is based on the fact that every real number has exactly one additive inverse. For instance, the additive inverse of 4 is -4 and the additive inverse of $-\frac{2}{3}$ is $\frac{2}{3}$.

Theorem: If a and b are positive real numbers, then $(-a)b = -(ab)$.

Proof:
$(-a)b + ab = (-a + a)b$	• **Distributive Property**
$= 0 \cdot b$	• **Inverse Property of Addition**
$= 0$	• **Multiplication Property of Zero**

Since $(-a)b + ab = 0$, $(-a)b$ is the additive inverse of ab. But $-(ab)$ is also the additive inverse of ab, and because every number has exactly one additive inverse, $(-a)b = -(ab)$.

Here is another example of a deductive proof. It is a proof of the Pythagorean Theorem.

Theorem: If a and b are the lengths of the legs of a right triangle and c is the length of the hypotenuse, then $a^2 + b^2 = c^2$.

Proof: Let $\triangle DBA$ be a right triangle and let BC be the altitude to the hypotenuse of $\triangle DBA$, as shown at the left. Then $\triangle BCD$ and $\triangle BCA$ are right triangles.

1. $\triangle DBA$ is a right triangle with height BC.

 Given.

2. $\triangle DBC$ is similar to $\triangle BCA$ and $\triangle DBA$ is similar to $\triangle BCA$.

 Theorem from geometry: the altitude to the hypotenuse of a right triangle forms triangles that are similar to each other.

3. $\dfrac{a}{b} = \dfrac{x}{a}$

 The ratios of corresponding sides of similar triangles are equal.

4. $a^2 = bx$

 Multiply each side by ab.

5. $\dfrac{x + b}{c} = \dfrac{c}{b}$

 The ratios of corresponding sides of similar triangles are equal.

6. $bx + b^2 = c^2$

 Multiply each side by cb.

7. $a^2 + b^2 = c^2$

 Substitution Property from (4).

Because the first statement of the proof is true and each subsequent statement of the proof is true, the first statement means the last statement is true. Therefore we have proved: If a and b are the lengths of the legs of a right triangle and c is the length of the hypotenuse, then $a^2 + b^2 = c^2$.

Projects & Group Activities

Applications of the Composition of Functions Suppose a company's manufacturing cost, in dollars, per computer monitor is given by

$$M(x) = \frac{180x + 2600}{x}$$

where x is the number of computer monitors to be manufactured. An electronics outlet agrees to sell the computer monitors by marking up the manufacturing cost per monitor, $M(x)$, by 30%. Note that the selling price S will be a function of $M(x)$. We can write this as

$$S[M(x)] = 1.30(M(x))$$

Simplifying $S[M(x)]$ yields

$$S[M(x)] = 1.30\left(\frac{180x + 2600}{x}\right)$$

$$= 1.30(180) + 1.30\left(\frac{2600}{x}\right)$$

$$= 234 + \left(\frac{3380}{x}\right)$$

The function produced in this manner is referred to as the composition of M by S. The notation $S \circ M$ is used to denote this composition of functions. That is,

$$(S \circ M)(x) = 234 + \left(\frac{3380}{x}\right)$$

Suppose 10,000 computer monitors are manufactured. Then

$$(S \circ M)(10,000) = 234 + \frac{3380}{10,000}$$

$$= 234.338$$

If 10,000 computer monitors are manufactured, the selling price per computer monitor is $234.34.

Solve.

1. Let x be the number of compact disk players to be manufactured. The manufacturing cost, in dollars, per compact disk player is given by the function

$$M(x) = \frac{60x + 34,000}{x}$$

A computer store will sell the disk players by marking up the manufacturing cost per disk player, $M(x)$, by 45%. Thus the selling price S is a function of $M(x)$ given by the equation $S[M(x)] = 1.45[M(x)]$.

 a. Express the selling price as a function of the number of disk players to be manufactured. That is, find $S \circ M$.

 b. Find $(S \circ M)(24,650)$. What does this represent?

2. The number of bookcases b that a factory can produce per day is a function of the number of hours h it operates.

$$b(h) = 40h \quad \text{for } 0 \le h \le 12$$

The daily cost c to manufacture b bookcases is given by the function

$$c(b) = 0.1b^2 + 90b + 800$$

Evaluate each of the following and interpret your answers.

 a. $b(5)$ **b.** $c(5)$ **c.** $(c \circ b)(h)$ **d.** $(c \circ b)(10)$

3. A car dealership offers a \$1500 rebate and a 10% discount off the price of a new car. Let p be the sticker price of a new car on the dealer's lot, let r be the rebate, and let d be the discount. Then $r(p) = p - 1500$ and $d(p) = 0.90p$.

 a. Write a composite function for the dealer taking the rebate first and then the discount.

 b. Write a composite function for the dealer taking the discount first and then the rebate.

 c. Which composite function would you prefer the dealer use if you were buying a new car?

4. a. Write the area of a circle as a function of r.

 b. Write the radius of a circle as a function of the circumference.

 c. Write the area of a circle as a function of the circumference.

Chapter 10 Summary

Key Words

The **inverse of a function** is the set of ordered pairs formed by reversing the coordinates of each ordered pair of the function. [10.3.2, p. 703]

Examples

Function: {(2, 4), (3, 5), (6, 7), (8, 9)}
Inverse: {(4, 2), (5, 3), (7, 6), (9, 8)}

Essential Rules and Procedures

Vertical Translation [10.1.1, p. 687]
The graph of $y = f(x) + c$, $c > 0$, is the graph of $y = f(x)$ shifted up c units.

The graph $g(x) = x^2 + 1$ is the graph of $f(x) = x^2$ shifted up 1 unit.

The graph of $y = f(x) - c$, $c > 0$, is the graph of $y = f(x)$ shifted down c units.

The graph of $G(x) = x^3 - 3$ is the graph of $F(x) = x^3$ shifted down 3 units.

Horizontal Translation [10.1.1, p. 688]

The graph of $y = f(x - c)$, $c > 0$, is the graph of $y = f(x)$ shifted horizontally c units to the right.

The graph of $y = f(x + c)$, $c > 0$, is the graph of $y = f(x)$ shifted horizontally c units to the left.

The graph $g(x) = (x - 2)^2$ is the graph of $f(x) = x^2$ shifted 2 units to the right.

The graph of $G(x) = (x + 4)^3$ is the graph of $F(x) = x^3$ shifted 4 units to the left.

Operations on Functions [10.2.1, p. 694]

If f and g are functions and x is an element of the domain of each function, then

$(f + g)(x) = f(x) + g(x)$

$(f - g)(x) = f(x) - g(x)$

$(f \cdot g)(x) = f(x) \cdot g(x)$

$\left(\dfrac{f}{g}\right)(x) = \dfrac{f(x)}{g(x)}$, $g(x) \neq 0$

Given $f(x) = x + 2$ and $g(x) = 2x$, then

$$(f + g)(4) = f(4) + g(4)$$
$$= (4 + 2) + 2(4) = 6 + 8$$
$$= 14$$

$$(f - g)(4) = f(4) - g(4)$$
$$= (4 + 2) - 2(4) = 6 - 8$$
$$= -2$$

$$(f \cdot g)(4) = f(4) \cdot g(4)$$
$$= (4 + 2) \cdot (2)(4) = 6 \cdot 8$$
$$= 48$$

$$\left(\frac{f}{g}\right)(4) = \frac{f(4)}{g(4)} = \frac{4 + 2}{2(4)} = \frac{6}{8} = \frac{3}{4}$$

Composition of Two Functions

$(f \circ g)(x) = f[g(x)]$ [10.2.2, p. 697]

Given $f(x) = x - 4$ and $g(x) = 4x$, then

$$(f \circ g)(2) = f[g(2)]$$
$$= f(8) \qquad \text{because } g(2) = 8$$
$$= 8 - 4 = 4$$

Horizontal Line Test

A function f is a 1–1 function if any horizontal line intersects the graph at no more than one point. [10.3.1, p. 702]

A 1–1 function Not a 1–1 function

Condition for an Inverse Function

A function f has an inverse if and only if f is a 1–1 function. [10.3.2, p. 704]

The function $f(x) = x^2$ does not have an inverse function. When $y = 4$, $x = 2$ or -2; therefore, the function $f(x)$ is not a 1–1 function.

Composition of Inverse Functions Property

$f^{-1}[f(x)] = x$ and $f[f^{-1}(x)] = x$ [10.3.2, p. 705]

$$f(x) = 2x - 3 \qquad f^{-1}(x) = \frac{1}{2}x + \frac{3}{2}$$

$$f^{-1}[f(x)] = \frac{1}{2}(2x - 3) + \frac{3}{2} = x$$

$$f[f^{-1}(x)] = 2\left(\frac{1}{2}x + \frac{3}{2}\right) - 3 = x$$

Chapter 10 Review Exercises

1. Use the graph of $f(x) = x^2$ shown below to graph $g(x) = (x - 1)^2$.

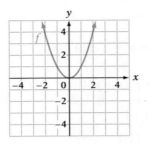

2. Use the graph of $f(x) = x^3$ shown below to graph $g(x) = x^3 + 3$.

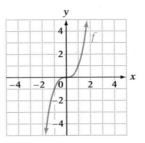

3. Use the graph of $f(x) = |x|$ shown below to graph $g(x) = |x - 1| + 2$.

4. Use the graph of $f(x) = \sqrt{x}$ shown below to graph $g(x) = \sqrt{x + 1} - 2$.

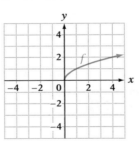

In Exercises 5 to 8, given $f(x) = x^2 + 2x - 3$ and $g(x) = x^2 - 2$, find:

5. $(f + g)(2)$

6. $(f - g)(-4)$

7. $(f \cdot g)(-4)$

8. $\left(\dfrac{f}{g}\right)(3)$

9. Given $f(x) = 3x^2 - 4$ and $g(x) = 2x + 1$, find $f[g(x)]$.

10. Given $f(x) = x^2 + 4$ and $g(x) = 4x - 1$, find $f[g(0)]$.

11. Given $f(x) = 6x + 8$ and $g(x) = 4x + 2$, find $g[f(-1)]$.

12. Given $f(x) = 2x^2 + x - 5$ and $g(x) = 3x - 1$, find $g[f(x)]$.

13. Is the set of ordered pairs a 1–1 function?
$\{(-3, 0), (-2, 0), (-1, 1), (0, 1)\}$

14. Find the inverse of the function
$\{(-2, 1), (2, 3), (5, -4), (7, 9)\}$.

15. Determine whether the graph is the graph of a 1–1 function.

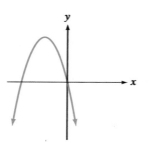

16. Determine whether the graph is the graph of a 1–1 function.

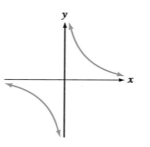

17. Find the inverse of the function $f(x) = \frac{1}{2}x + 8$.

18. Find the inverse of the function $f(x) = -6x + 4$.

19. Find the inverse of the function $f(x) = \frac{2}{3}x - 12$.

20. Are the functions $f(x) = -\frac{1}{4}x + \frac{5}{4}$ and $g(x) = -4x + 5$ inverses of each other?

Chapter 10 Test

1. Determine whether the graph represents the graph of a 1–1 function.

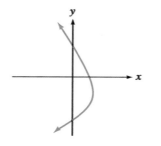

2. Given $f(x) = x^2 + 2x - 3$ and $g(x) = x^3 - 1$, find $(f - g)(2)$.

3. Given $f(x) = x^3 + 1$ and $g(x) = 2x - 3$ find $(f \cdot g)(-3)$.

4. Given $f(x) = 4x - 5$ and $g(x) = x^2 + 3x + 4$, find $\left(\frac{f}{g}\right)(-2)$.

5. Given $f(x) = x^2 + 4$ and $g(x) = 2x^2 + 2x + 1$, find $(f - g)(-4)$.

6. Given $f(x) = 4x + 2$ and $g(x) = \frac{x}{x + 1}$, find $f[g(3)]$.

7. Use the graph of $f(x) = -x^3$ shown below to graph $g(x) = -x^3 - 2$.

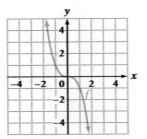

8. Use the graph of $f(x) = -|x|$ shown below to graph $g(x) = -|x| + 3$.

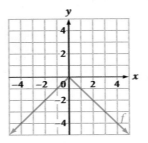

9. If $f(3) = 7$ and $g(x) = x^2 - 2$, find $g[f(3)]$.

10. Use the graph of $f(x) = |x|$ shown below to graph $g(x) = |x + 2| - 3$.

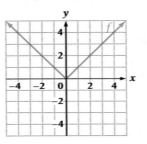

11. Use the graph of $f(x) = -x^2$ shown below to graph $g(x) = -(x - 2)^2 + 2$.

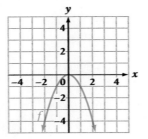

12. Find the inverse of the function $f(x) = \frac{1}{4}x - 4$.

13. Find the inverse of the function $\{(2, 6), (3, 5),(4, 4), (5, 3)\}$.

14. Are the functions $f(x) = \frac{1}{2}x + 2$ and $g(x) = 2x - 4$ inverses of each other?

15. Given $f(x) = 2x^2 - 7$ and $g(x) = x - 1$, find $f[g(x)]$.

16. Find the inverse of the function $f(x) = \frac{1}{2}x - 3$.

17. Are the functions $f(x) = \frac{2}{3}x + 3$ and $g(x) = \frac{3}{2}x - 3$ inverses of each other?

18. Determine whether the graph is the graph of a 1–1 function.

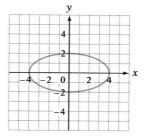

Cumulative Review Exercises

1. Evaluate $-3a + \left| \frac{3b - ab}{3b - c} \right|$ when $a = 2$, $b = 2$, and $c = -2$.

2. Graph: $\{x \mid x < -3\} \cap \{x \mid x > -4\}$

3. Solve: $\frac{3x - 1}{6} - \frac{5 - x}{4} = \frac{5}{6}$

4. Solve: $4x - 2 < -10$ or $3x - 1 > 8$

5. Graph $f(x) = \frac{1}{4}x^2$. Find the vertex and axis of symmetry.

6. Graph the solution set of $3x - 4y \geq 8$.

7. Solve: $|8 - 2x| \geq 0$

8. Simplify: $\left(\frac{3a^3b}{2a} \right)^2 \left(\frac{a^2}{-3b^2} \right)^3$

9. Multiply: $(x - 4)(2x^2 + 4x - 1)$

10. Factor: $a^4 - 2a^2 - 8$

11. Factor: $x^3y + x^2y^2 - 6xy^3$

12. Solve: $(b + 2)(b - 5) = 2b + 14$

13. Solve: $x^2 - 2x > 15$

14. Subtract: $\frac{x^2 + 4x - 5}{2x^2 - 3x + 1} - \frac{x}{2x - 1}$

15. Solve: $\frac{5}{x^2 + 7x + 12} = \frac{9}{x + 4} - \frac{2}{x + 3}$

16. Simplify: $\frac{4 - 6i}{2i}$

17. Find the equation of the line that contains the points $(-3, 4)$ and $(2, -6)$.

18. Find the equation of the line that contains the point $(-3, 1)$ and is perpendicular to the line $2x - 3y = 6$.

19. Solve: $3x^2 = 3x - 1$

20. Solve: $\sqrt{8x + 1} = 2x - 1$

21. Find the minimum value of the function $f(x) = 2x^2 - 3$.

22. Find the range of $f(x) = |3x - 4|$ if the domain is $\{0, 1, 2, 3\}$.

23. Is this set of ordered pairs a function?

$\{(-3, 0), (-2, 0), (-1, 1), (0, 1)\}$

24. Solve: $\sqrt[3]{5x - 2} = 2$

25. Given $g(x) = 3x - 5$ and $h(x) = \frac{1}{2}x + 4$, find $g[h(2)]$.

26. Find the inverse of the function given by $f(x) = -3x + 9$.

27. **Food Mixtures** Find the cost per pound of a tea mixture made from 30 lb of tea costing \$4.50 per pound and 45 lb of tea costing \$3.60 per pound.

28. **Metallurgy** How many pounds of an 80% copper alloy must be mixed with 50 lb of a 20% copper alloy to make an alloy that is 40% copper?

29. **Mixture Problem** Six ounces of an insecticide are mixed with 16 gal of water to make a spray for spraying an orange grove. How much additional insecticide is required if it is to be mixed with 28 gal of water?

30. **Work Problem** A large pipe can fill a tank in 8 min less time than it takes a smaller pipe to fill the same tank. Working together, both pipes can fill the tank in 3 min. How long would it take the larger pipe, working alone, to fill the tank?

31. **Physics** The distance (d) that a spring stretches varies directly as the force (f) used to stretch the spring. If a force of 50 lb can stretch the spring 30 in., how far can a force of 40 lb stretch the spring?

32. **Wind Power** The table at the right shows actual and projected gigawatts (billions of watts) of energy generated in the United States by wind power plants. If the projections are accurate, what will be the average rate of change in the production of energy, in gigawatts per year, from wind power plants between 2003 and 2009? Round to the nearest hundredth.

Year	Energy Generated (in gigawatts)
2003	11.19
2004	14.15
2005	23.22
2006	35.25
2007	48.62
2008	49.50
2009	50.20

Source: Energy Information Administration

33. **Music** The frequency of vibration (f) in an open pipe organ varies inversely as the length (L) of the pipe. If the air in a pipe 2 m long vibrates 60 times per minute, find the frequency in a pipe that is 1.5 m long.

11 Exponential and Logarithmic Functions

DVD SSM

Online Study Center

Need help? For online student resources, such as section quizzes, visit the Online Study Center at college.hmco.com/pic/aufmannABI2e.

Having a credit card can be very helpful when making big purchases, but forgetting to pay your bills can cause major problems later on. When applying for things such as student loans, the lender generally wants to know your past credit history. Usually, they will turn to a credit-reporting agency for a full credit report and a credit score, which helps them determine the likelihood that you will repay the loan in a timely manner. One of the most widely used credit scores is the FICO® score. To learn more about credit reports and FICO scores, see the **Projects and Group Activities on page 776.**

Prep Test

1. Simplify: 3^{-2}

2. Simplify: $\left(\dfrac{1}{2}\right)^{-4}$

3. Complete: $\dfrac{1}{8} = 2^{?}$

4. Evaluate $f(x) = x^4 + x^3$ for $x = -1$ and $x = 3$.

5. Solve: $3x + 7 = x - 5$

6. Solve: $16 = x^2 - 6x$

7. Evaluate $A(1 + i)^n$ for $A = 5000$, $i = 0.04$, and $n = 6$. Round to the nearest hundredth.

8. Graph: $f(x) = x^2 - 1$

11.1 Exponential Functions

OBJECTIVE 1 ### Evaluate exponential functions

Deoxyribonucleic acid (DNA) is the basic material of all living organisms. DNA resembles a twisted rope ladder with the ropes made up of sugar and phosphate molecules. The rungs of the ladder are base pairs, adenine–thymine and cytosine–guanine. The order of these base pairs determines whether an organism is, for example, a flower, a frog, or a human. Molecular biologists are trying to determine the sequences for various organisms. The growth in the number of sequences that have been identified is shown in the graph at the right. This graph is an example of exponential growth.

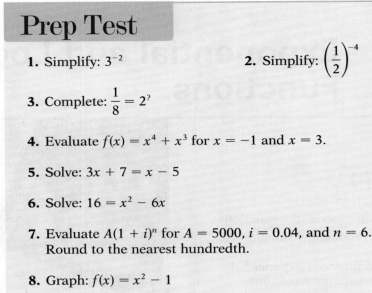

Nuclear medicine physicians use radioisotopes for the diagnosis and treatment of certain diseases. One of the most widely used isotopes is technetium-99m. One use of this isotope is in the diagnosis of cardiovascular disease. The graph at the right shows the

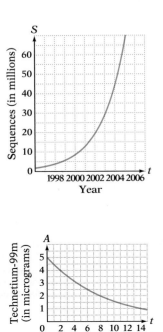

amount of technetium-99m in a patient after its injection into the patient. This graph is an example of exponential decay.

Definition of an Exponential Function

The **exponential function** with base b is defined by

$$f(x) = b^x$$

where $b > 0$, $b \neq 1$, and x is any real number.

In the definition of an exponential function, b, the base, is required to be positive. If the base were a negative number, the value of the function would be a complex number for some values of x. For instance, the value of $f(x) = (-4)^x$ when $x = \frac{1}{2}$ is $f\left(\frac{1}{2}\right) = (-4)^{\frac{1}{2}} = \sqrt{-4} = 2i$. To avoid complex number values of a function, the base of the exponential function is a positive number.

Evaluate $f(x) = 2^x$ at $x = 3$ and $x = -2$.

Substitute 3 for x and simplify. $f(3) = 2^3 = 8$

Substitute -2 for x and simplify. $f(-2) = 2^{-2} = \dfrac{1}{2^2} = \dfrac{1}{4}$

To evaluate an exponential expression for an irrational number such as $\sqrt{2}$, we obtain an approximation to the value of the function by approximating the irrational number. For instance, the value of $f(x) = 4^x$ when $x = \sqrt{2}$ can be approximated by using an approximation of $\sqrt{2}$.

$$f\left(\sqrt{2}\right) = 4^{\sqrt{2}} \approx 4^{1.4142} \approx 7.1029$$

Because $f(x) = b^x$ $(b > 0, b \neq 1)$ can be evaluated at both rational and irrational numbers, the domain of f is all real numbers. And because $b^x > 0$ for all values of x, the range of f is the positive real numbers.

EXAMPLE 1 Evaluate $f(x) = \left(\dfrac{1}{2}\right)^x$ at $x = 2$ and $x = -3$.

Solution $f(x) = \left(\dfrac{1}{2}\right)^x$

$f(2) = \left(\dfrac{1}{2}\right)^2 = \dfrac{1}{4}$ $f(-3) = \left(\dfrac{1}{2}\right)^{-3} = 2^3 = 8$

PROBLEM 1 Evaluate $f(x) = \left(\dfrac{2}{3}\right)^x$ at $x = 3$ and $x = -2$.

Solution See pages S33–S34.

EXAMPLE 2 Evaluate $f(x) = 2^{3x-1}$ at $x = 1$ and $x = -1$.

Solution $f(x) = 2^{3x-1}$

$$f(1) = 2^{3(1)-1} = 2^2 = 4 \qquad f(-1) = 2^{3(-1)-1} = 2^{-4} = \frac{1}{2^4} = \frac{1}{16}$$

PROBLEM 2 Evaluate $f(x) = 2^{2x+1}$ at $x = 0$ and $x = -2$.

Solution See page S34.

EXAMPLE 3 Evaluate $f(x) = \left(\sqrt{5}\right)^x$ at $x = 4$, $x = -2.1$, and $x = \pi$. Round to the nearest ten-thousandth.

Solution $f(x) = \left(\sqrt{5}\right)^x$

$$f(4) = \left(\sqrt{5}\right)^4 \qquad f(-2.1) = \left(\sqrt{5}\right)^{-2.1} \qquad f(\pi) = \left(\sqrt{5}\right)^{\pi}$$
$$= 25 \qquad\qquad \approx 0.1845 \qquad\qquad \approx 12.5297$$

PROBLEM 3 Evaluate $f(x) = \pi^x$ at $x = 3$, $x = -2$, and $x = \pi$. Round to the nearest ten-thousandth.

Solution See page S34.

A frequently used base in applications of exponential functions is an irrational number designated by e. The number e is approximately 2.71828183. It is an irrational number, so it has a nonterminating, nonrepeating decimal representation.

> **Natural Exponential Function**
>
> The function defined by $f(x) = e^x$ is called the **natural exponential function.**

The $\boxed{e^x}$ key on a calculator can be used to evaluate the natural exponential function. The graph of $y = e^x$ is shown at the left.

EXAMPLE 4 Evaluate $f(x) = e^x$ at $x = 2$, $x = -3$, and $x = \pi$. Round to the nearest ten-thousandth.

Solution $f(x) = e^x$
$$f(2) = e^2 \qquad f(-3) = e^{-3} \qquad f(\pi) = e^{\pi}$$
$$\approx 7.3891 \qquad\quad \approx 0.0498 \qquad\quad \approx 23.1407$$

PROBLEM 4 Evaluate $f(x) = e^x$ at $x = 1.2$, $x = -2.5$, and $x = e$. Round to the nearest ten-thousandth.

Solution See page S34.

OBJECTIVE **2** **Graph exponential functions**

Some of the properties of an exponential function can be seen by considering its graph.

Point of Interest

The natural exponential function is an extremely important function. It is used extensively in applied problems in virtually all disciplines from archeology to zoology. Leonhard Euler (1707–1783) was the first to use the letter e as the base of the natural exponential function.

To graph $f(x) = 2^x$, think of the function as the equation $y = 2^x$.

Choose values of x, and find the corresponding values of y. The results can be recorded in a table.

Graph the ordered pairs on a rectangular coordinate system.

Connect the points with a smooth curve.

x	$f(x) = y$
-2	$2^{-2} = \dfrac{1}{4}$
-1	$2^{-1} = \dfrac{1}{2}$
0	$2^0 = 1$
1	$2^1 = 2$
2	$2^2 = 4$
3	$2^3 = 8$

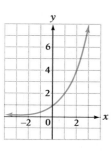

Note that a vertical line would intersect the graph at only one point. Therefore, by the vertical line test, $f(x) = 2^x$ is the graph of a function. Also note that a horizontal line would intersect the graph at only one point. Therefore, $f(x) = 2^x$ is the graph of a 1–1 function.

Graph: $f(x) = \left(\dfrac{1}{2}\right)^x$

Think of the function as the equation $y = \left(\dfrac{1}{2}\right)^x$.

Choose values of x, and find the corresponding values of y.

Graph the ordered pairs on a rectangular coordinate system.

Connect the points with a smooth curve.

x	$f(x) = y$
-3	$\left(\dfrac{1}{2}\right)^{-3} = 8$
-2	$\left(\dfrac{1}{2}\right)^{-2} = 4$
-1	$\left(\dfrac{1}{2}\right)^{-1} = 2$
0	$\left(\dfrac{1}{2}\right)^{0} = 1$
1	$\left(\dfrac{1}{2}\right)^{1} = \dfrac{1}{2}$
2	$\left(\dfrac{1}{2}\right)^{2} = \dfrac{1}{4}$

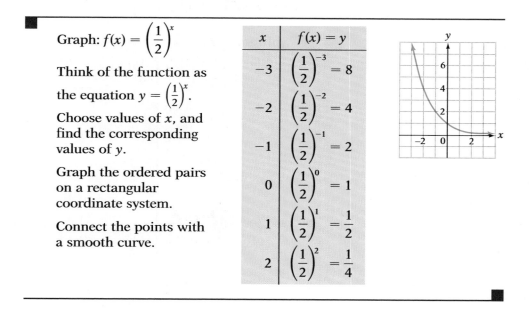

Applying the vertical and horizontal line tests reveals that $f(x) = \left(\dfrac{1}{2}\right)^x$ is also the graph of a 1–1 function.

Exponential Functions Are 1–1

The exponential function defined by $f(x) = b^x$, $b > 0$, $b \neq 1$, is a 1–1 function.

Graph $f(x) = 2^{-x}$.

Think of this as the equation $y = 2^{-x}$.

Choose values of x and find the corresponding values of y.

Graph the ordered pairs on a rectangular coordinate system.

Connect the points with a smooth curve.

x	y
-3	8
-2	4
-1	2
0	1
1	$\frac{1}{2}$
2	$\frac{1}{4}$

EXAMPLE 5 Graph. **A.** $f(x) = 3^{\frac{1}{2}x - 1}$ **B.** $f(x) = 2^x - 1$

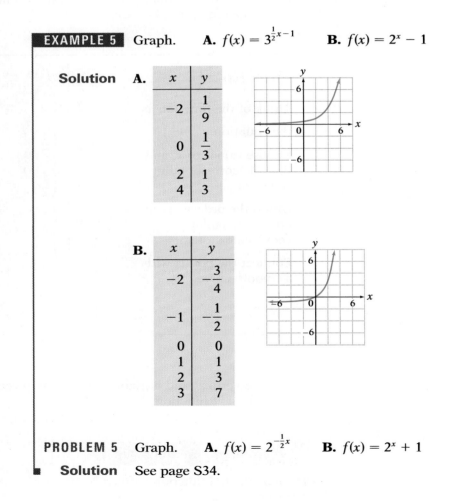

Solution **A.**

x	y
-2	$\frac{1}{9}$
0	$\frac{1}{3}$
2	1
4	3

B.

x	y
-2	$-\frac{3}{4}$
-1	$-\frac{1}{2}$
0	0
1	1
2	3
3	7

PROBLEM 5 Graph. **A.** $f(x) = 2^{-\frac{1}{2}x}$ **B.** $f(x) = 2^x + 1$

Solution See page S34.

EXAMPLE 6 Graph. **A.** $f(x) = 3^{x+1}$ **B.** $f(x) = e^x - 1$

Solution **A.**

x	y
-3	$\frac{1}{9}$
-2	$\frac{1}{3}$
-1	1
0	3
1	9

B.

x	y
-2	-0.86
-1	-0.63
0	0
1	1.72
2	6.39

PROBLEM 6 Graph. **A.** $f(x) = 2^{-x} + 2$ **B.** $f(x) = e^{-2x} - 4$

■ Solution See page S34.

EXAMPLE 7 Graph $f(x) = -\frac{1}{3}e^{2x} + 2$ and approximate the zero of f to the nearest tenth.

Solution

• Recall that a zero of f is a value of x for which $f(x) = 0$. Use the features of a graphing utility to determine the x-intercept of the graph, which is the zero of f.

To the nearest tenth, the zero of f is 0.9.

PROBLEM 7 Graph $f(x) = 2\left(\frac{3}{4}\right)^x - 3$ and approximate, to the nearest tenth, the value of x for which $f(x) = 1$.

■ Solution See page S34.

11.1 Concept Review

Determine whether the following statements are always true, sometimes true, or never true.

1. The domain of an exponential function $f(x) = b^x$, $b > 0$, $b \neq 1$, is the set of positive numbers.

2. An exponential function $f(x) = b^x$, $b > 0$, $b \neq 1$, is a 1–1 function.

3. The graph of the exponential function $f(x) = b^x$, $b > 0$, $b \neq 1$, passes through the point $(0, 0)$.

4. For the function $f(x) = b^x$, $b > 0$, $b \neq 1$, the base b is a positive integer.

5. An exponential function $f(x) = b^x$, $b > 0$, $b \neq 1$, has two x-intercepts.

11.1 Exercises

1 Evaluate exponential functions

1. What is an exponential function?

2. What is the natural exponential function?

3. Which of the following cannot be the base of an exponential function?

 a. 7 **b.** $\dfrac{1}{4}$ **c.** -5 **d.** 0.01

4. Which of the following cannot be the base of an exponential function?

 a. 0.9 **b.** 476 **c.** 8 **d.** $-\dfrac{1}{2}$

5. Given $f(x) = 4^{x-1}$, evaluate $f(1)$.

$$f(1) = 4^{\underline{}-1} = 4^{\underline{}} = \underline{}$$

6. Given $g(x) = e^{3x}$, evaluate $g\left(\dfrac{1}{6}\right)$. Round to the nearest ten-thousandth.

 $$g\left(\frac{1}{6}\right) = e^{3(\underline{})}$$ • Replace x with $\dfrac{1}{6}$.

 $$= e^{\underline{}}$$ • Simplify the exponent.

 $$\approx \underline{}$$ • Use the e^x key on a calculator.

7. Given $f(x) = 3^x$, evaluate:
 a. $f(2)$ **b.** $f(0)$ **c.** $f(-2)$

8. Given $H(x) = 2^x$, evaluate:
 a. $H(-3)$ **b.** $H(0)$ **c.** $H(2)$

9. Given $g(x) = 2^{x+1}$, evaluate:
 a. $g(3)$ **b.** $g(1)$ **c.** $g(-3)$

10. Given $F(x) = 3^{x-2}$, evaluate:
 a. $F(-4)$ **b.** $F(-1)$ **c.** $F(0)$

11. Given $P(x) = \left(\dfrac{1}{2}\right)^{2x}$, evaluate:
 a. $P(0)$ **b.** $P\left(\dfrac{3}{2}\right)$ **c.** $P(-2)$

12. Given $R(t) = \left(\dfrac{1}{3}\right)^{3t}$, evaluate:
 a. $R\left(-\dfrac{1}{3}\right)$ **b.** $R(1)$ **c.** $R(-2)$

13. Given $G(x) = e^{\frac{x}{2}}$, evaluate the following. Round to the nearest ten-thousandth.
 a. $G(4)$ **b.** $G(-2)$ **c.** $G\left(\dfrac{1}{2}\right)$

14. Given $f(x) = e^{2x}$, evaluate the following. Round to the nearest ten-thousandth.
 a. $f(-2)$ **b.** $f\left(-\dfrac{2}{3}\right)$ **c.** $f(2)$

15. Given $H(r) = e^{-r+3}$, evaluate the following. Round to the nearest ten-thousandth.
 a. $H(-1)$ **b.** $H(3)$ **c.** $H(5)$

16. Given $P(t) = e^{-\frac{1}{2}t}$, evaluate the following. Round to the nearest ten-thousandth.
 a. $P(-3)$ **b.** $P(4)$ **c.** $P\left(\dfrac{1}{2}\right)$

17. Given $F(x) = 2^{x^2}$, evaluate:
 a. $F(2)$ **b.** $F(-2)$ **c.** $F\left(\dfrac{3}{4}\right)$

18. Given $Q(x) = 2^{-x^2}$, evaluate:
 a. $Q(3)$ **b.** $Q(-1)$ **c.** $Q(-2)$

19. Given $f(x) = e^{-\frac{x^2}{2}}$, evaluate the following. Round to the nearest ten-thousandth.
 a. $f(-2)$ **b.** $f(2)$ **c.** $f(-3)$

20. Given $f(x) = e^{-2x} + 1$, evaluate the following. Round to the nearest ten-thousandth.
 a. $f(-1)$ **b.** $f(3)$ **c.** $f(-2)$

Complete Exercises 21 to 24 without using a calculator.

21. Which number is a rational number, $(-3)^{1/2}$ or $\left(\dfrac{1}{2}\right)^{-3}$?

22. For which number can you give only an approximate decimal value, $(\sqrt{7})^4$ or $\sqrt[4]{7}$?

23. Is e^3 greater than or less than 27?

24. Is e^{-1} greater than or less than $\dfrac{1}{2}$?

2 **Graph exponential functions**

25. To find the y-coordinate of the point on the graph of $f(x) = 3^x$ that has an x-coordinate of -3, evaluate $f(-3) = 3^{\underline{\ ?\ }} = \dfrac{1}{\underline{\ ?\ }} = \underline{\ ?\ }$. The point on the graph of $f(x) = 3^x$ with an x-coordinate of -3 is $(\underline{\ ?\ }, \underline{\ ?\ })$.

26. To find the y-coordinate of the point on the graph of $g(x) = 2^{x+1}$ that has an x-coordinate of -1, evaluate $g(-1) = 2^{\underline{}+1} = 2^{\underline{}} = \underline{}$. The point on the graph of $g(x) = 2^{x+1}$ with an x-coordinate of -1 is $(\underline{}, \underline{})$.

Graph.

27. $f(x) = 3^x$

28. $f(x) = 3^{-x}$

29. $f(x) = 2^{x+1}$

30. $f(x) = 2^{x-1}$

31. $f(x) = \left(\dfrac{1}{3}\right)^x$

32. $f(x) = \left(\dfrac{2}{3}\right)^x$

33. $f(x) = 2^{-x} + 1$

34. $f(x) = 2^x - 3$

35. $f(x) = \left(\dfrac{1}{3}\right)^{-x}$

36. $f(x) = \left(\dfrac{3}{2}\right)^{-x}$

37. $f(x) = \left(\dfrac{1}{2}\right)^{-x} + 2$

38. $f(x) = \left(\dfrac{1}{2}\right)^x - 1$

39. Which of the following functions have the same graph?

 a. $f(x) = 3^x$ **b.** $f(x) = \left(\dfrac{1}{3}\right)^x$ **c.** $f(x) = x^3$ **d.** $f(x) = 3^{-x}$

40. Which of the following functions have the same graph?

 a. $f(x) = x^4$ **b.** $f(x) = 4^{-x}$ **c.** $f(x) = 4^x$ **d.** $f(x) = \left(\dfrac{1}{4}\right)^x$

41. Graph $f(x) = 3^x$ and $f(x) = 3^{-x}$ and find the point of intersection of the two graphs.

42. Graph $f(x) = 2^{x+1}$ and $f(x) = 2^{-x+1}$ and find the point of intersection of the two graphs.

43. Graph $f(x) = \left(\dfrac{1}{3}\right)^x$. What are the x- and y-intercepts of the graph of the function?

44. Graph $f(x) = \left(\dfrac{1}{3}\right)^{-x}$. What are the x- and y-intercepts of the graph of the function?

45. If the graph of $f(x) = b^x$ rises from left to right, is b greater than 1 or less than 1?

46. Which function has a graph that falls from left to right, $f(x) = 10^x$ or $g(x) = 10^{-x}$?

47. Both a and b are positive and $b \neq 1$. Which function has an x-intercept, $f(x) = b^{x-a}$ or $g(x) = b^x - a$?

48. Is the y-intercept of $f(x) = 5^x$ greater than, equal to, or less than the y-intercept of $f(x) = 5^{-x}$?

Use a graphing calculator to solve Exercises 49 to 55.

49. Graph $f(x) = 2^x - 3$ and approximate the zero of f to the nearest tenth.

50. Graph $f(x) = 5 - 3^x$ and approximate the zero of f to the nearest tenth.

51. Graph $f(x) = e^x$ and approximate, to the nearest tenth, the value of x for which $f(x) = 3$.

52. Graph $f(x) = e^{-2x-3}$ and approximate, to the nearest tenth, the value of x for which $f(x) = 2$.

53. **Investments** The exponential function given by $F(n) = 500(1.00021918)^{365n}$ gives the value in n years of a \$500 investment in a certificate of deposit that earns 8% annual interest compounded daily. Graph F and determine in how many years the investment will be worth \$1000.

54. **Chemistry** The number of grams of radioactive cesium that remain after t years from an original sample of 30 g is given by $N(t) = 30(2^{-0.0322t})$. Graph N and determine in how many years there will be 20 g of cesium remaining.

55. **Oceanography** The percent of light that reaches m meters below the surface of the ocean is given by the equation $P(m) = 100e^{-1.38m}$. Graph P and determine the depth to which 50% of the light will reach.

11.1 Applying Concepts

56. Evaluate $\left(1 + \dfrac{1}{n}\right)^n$ for $n = 100, 1000, 10{,}000,$ and $100{,}000$ and compare the results with the value of e, the base of the natural exponential function. On the basis of your evaluation, complete the following sentence:

As n increases, $\left(1 + \dfrac{1}{n}\right)^n$ becomes closer to _____ .

57. Physics If air resistance is ignored, the speed v, in feet per second, of an object t seconds after it has been dropped is given by the equation $v = 32t$. This is true regardless of the mass of the object. However, if air resistance is considered, then the speed depends on the mass (and on other things). For a certain mass, the speed t seconds after it has been dropped is given by $v = 64\left(1 - e^{-\frac{t}{2}}\right)$.

 a. [graph icon] Graph this equation. *Suggestion:* Use Xmin $= -0.5$, Xmax $= 10$, Xscl $= 1$, Ymin $= -0.5$, Ymax $= 70$, and Yscl $= 10$.

 b. The point whose approximate coordinates are $(4, 55.3)$ is on the graph. Write a sentence that explains the meaning of these coordinates.

58. [pencil icon] Graph $f(x) = e^x$, $g(x) = e^{x-2}$, and $h(x) = e^{x+2}$ on the same rectangular coordinate system. Explain how the graphs are similar and how they are different.

59. [pencil icon] Graph $f(x) = 2^x$, $g(x) = 2^x - 2$, and $h(x) = 2^x + 2$ on the same rectangular coordinate system. Explain how the graphs are similar and how they are different.

11.2 Introduction to Logarithms

OBJECTIVE 1 Write equivalent exponential and logarithmic equations

In Michael Crichton's novel *The Andromeda Strain*, we read, "A single cell of the bacterium *E. coli* would, under ideal circumstances, divide every twenty minutes." If a colony of *E. coli* originally contained 1000 bacteria, then a modified version of Crichton's statement, where the bacteria divide once per hour, can be modeled by the exponential growth equation $A = 1000(2^t)$, where A is the number of *E. coli* present after t hours. The table below shows the number of *E. coli* for selected times.

Time, t (in hours)	$1000(2^t) = A$
3 h	$1000(2^3) = 8000$
6 h	$1000(2^6) = 64{,}000$
8 h	$1000(2^8) = 256{,}000$
12 h	$1000(2^{12}) = 4{,}096{,}000$

Take Note

Crichton goes on to say, in the same paragraph, "In this way it can be shown that in a single day, one cell of *E. coli* could produce a super-colony equal in size and weight to the entire planet Earth." In the model we are using, where the bacteria divide once an hour (instead of three times an hour, as in *The Andromeda Strain*), it would take the colony 3 days to become the size of Earth.

This model would be true if uncontrolled biological growth were possible, but it is not. Other factors, such as food to sustain growth, would not be sufficient to support continued exponential growth.

To determine when there would be 32,000 bacteria in the culture, we would solve the exponential equation $32{,}000 = 1000(2^t)$ for t. Because $A = 1000(2^t)$ is a growth equation and 32,000 is between 8000 and 64,000, t must be between 3 h and 6 h. By trial and error, we find that $t = 5$ h.

$$A = 1000(2^t)$$
$$A = 1000(2^5) \qquad \text{• Replace } t \text{ by } 5.$$
$$A = 1000(32) = 32{,}000$$

Now suppose we want to find how long it would take for the colony to reach 1,000,000 bacteria. From the table, 1,000,000 is between 256,000 and 4,096,000, so this will happen sometime between 8 h and 12 h. If we try 10 h, halfway between 8 and 12, we have

$$A = 1000(2^t)$$
$$A = 1000(2^{10}) \qquad \text{• Replace } t \text{ by } 10.$$
$$A = 1000(1024) = 1{,}024{,}000$$

Because 1,000,000 is less than 1,024,000, the actual value of t is less than 10 h.

We could continue to use trial and error to find t, but it would be more efficient if we could just solve the equation $1{,}000{,}000 = 1000(2^t)$ for t. Using methods described earlier in the text, we have

$$1{,}000{,}000 = 1000(2^t)$$
$$1000 = 2^t \qquad \text{• Divide each side of the equation by 1000.}$$

To solve this equation, it would be helpful to have a function that would give the power of 2 that produces 1000. Around the mid-16th century, mathematicians created such a function, which we now call a *logarithmic function*. We write the solution of $1000 = 2^t$ as $t = \log_2 1000$. This is read "t equals the logarithm base 2 of 1000," and it means "t equals the power of 2 that produces 1000."

When logarithms were first introduced, tables were used to find the numerical value of t. Today, a calculator is used. Using a calculator, we can approximate the value of t as 9.97. Thus $9.97 \approx \log_2 1000$ and $1000 \approx 2^{9.97}$.

The ideas presented above are related to the concept of inverse function discussed in the last chapter. Because the exponential function given by $y = b^x$ is a 1–1 function, it has an inverse function. To find that function, we follow the same procedure we used to find the inverse of other functions.

$$y = b^x$$
$$x = b^y \qquad \text{• Interchange } x \text{ and } y.$$

The equation $x = b^y$ says that y equals the power of b that produces x. That is, y is the logarithm base b of x. The inverse of the exponential function $y = b^x$ is a logarithmic function.

Here is a general definition of logarithm.

Take Note

Recall that for the equation $2^t = 1000$, we said that "t equals the power of 2 that produces 1000" means that "t is the logarithm base 2 of 1000." We are doing the same with $x = b^y$.

Definition of Logarithm

If $x > 0$ and b is a positive constant not equal to 1, then $y = \log_b x$ is equivalent to $b^y = x$.

Read $\log_b x$ as "the logarithm base b of x" or "the log base b of x."

The following table shows equivalent statements written in both exponential and logarithmic form.

Exponential Form		Logarithmic Form
$2^4 = 16$	$\Longleftrightarrow$	$\log_2 16 = 4$
$\left(\dfrac{2}{3}\right)^2 = \dfrac{4}{9}$	$\Longleftrightarrow$	$\log_{\frac{2}{3}}\left(\dfrac{4}{9}\right) = 2$
$10^{-1} = 0.1$	$\Longleftrightarrow$	$\log_{10}(0.1) = -1$
$b^a = x$	$\Longleftrightarrow$	$\log_b x = a$

EXAMPLE 1 Write $4^5 = 1024$ in logarithmic form.

Solution $4^5 = 1024$ is equivalent to $\log_4 1024 = 5$.

PROBLEM 1 Write $3^{-4} = \dfrac{1}{81}$ in logarithmic form.

Solution See page S34.

EXAMPLE 2 Write $\log_7 343 = 3$ in exponential form.

Solution $\log_7 343 = 3$ is equivalent to $7^3 = 343$.

PROBLEM 2 Write $\log_{10} 0.0001 = -4$ in exponential form.

Solution See page S34.

Recalling the equations $y = \log_b x$ and $x = b^y$ from the definition of a logarithm, note that because $b^y > 0$ for all values of y, x is always a positive number. Therefore, in the equation $y = \log_b x$, x is a positive number. The logarithm of a negative number is not a real number.

The 1–1 property of exponential functions can be used to evaluate some logarithms.

Equality of Exponents Property

For $b > 0$, $b \neq 1$, if $b^u = b^v$, then $u = v$.

Evaluate: $\log_2 8$

Write an equation.	$\log_2 8 = x$
Write the equation in its equivalent exponential form.	$8 = 2^x$
Write 8 in exponential form using 2 as the base.	$2^3 = 2^x$
Use the Equality of Exponents Property.	$3 = x$
	$\log_2 8 = 3$

EXAMPLE 3 Evaluate: $\log_3\left(\dfrac{1}{9}\right)$

Solution $\log_3\left(\dfrac{1}{9}\right) = x$ • Write an equation.

$\dfrac{1}{9} = 3^x$ • Write the equation in its equivalent exponential form.

$3^{-2} = 3^x$ • Write $\dfrac{1}{9}$ in exponential form using 3 as the base.

$-2 = x$ • Solve for x using the Equality of Exponents Property.

$\log_3\left(\dfrac{1}{9}\right) = -2$

PROBLEM 3 Evaluate: $\log_4 64$

Solution See page S34.

Solve $\log_4 x = -2$ for x. $\log_4 x = -2$

Write the equation in its equivalent exponential form. $4^{-2} = x$

Solve for x. $\dfrac{1}{16} = x$

The solution is $\dfrac{1}{16}$.

In this example, $\dfrac{1}{16}$ is called the antilogarithm base 4 of negative 2. In general, if $\log_b M = N$, then M is the antilogarithm base b of N. The antilogarithm of a number can be determined by rewriting the logarithmic expression in exponential form. For instance, the exponential form of $\log_5 x = 3$ is $5^3 = x$; x is the antilogarithm base 5 of 3.

Definition of Antilogarithm

If $\log_b M = N$, the **antilogarithm** base b of N is M. In exponential form,

$$M = b^N$$

EXAMPLE 4 Solve $\log_6 x = 2$ for x.

Solution $\log_6 x = 2$
$6^2 = x$ • Write $\log_6 x = 2$ in its equivalent exponential form.
$36 = x$
The solution is 36.

PROBLEM 4 Solve $\log_2 x = -4$ for x.

Solution See page S34.

Logarithms base 10 are called **common logarithms.** We usually omit the base, 10, when writing the common logarithm of a number. Therefore, $\log_{10} x$ is written $\log x$. To find the common logarithm of most numbers, a calculator is necessary. A calculator was used to find the value of $\log 384$, shown below.

$$\log 384 \approx 2.584331224$$

When e (the base of the natural exponential function) is used as a base of a logarithm, the logarithm is referred to as the **natural logarithm** and is abbreviated $\ln x$. This is read "el en x." Using a calculator, we find that

$$\ln 23 \approx 3.135494216$$

EXAMPLE 5 Solve $\ln x = -1$ for x. Round to the nearest ten-thousandth.

Solution
$$\ln x = -1$$
$$e^{-1} = x$$
$$0.3679 \approx x$$

- $\ln x$ is the abbreviation for $\log_e x$.
- Write the equation in its equivalent exponential form. Evaluate.

The solution is 0.3679.

PROBLEM 5 Solve $\log x = 1.5$ for x. Round to the nearest ten-thousandth.

Solution See page S34.

OBJECTIVE 2 ## The properties of logarithms

Because a logarithm is an exponent, the properties of logarithms are similar to the properties of exponents.

The table at the right shows some powers of 2 and the equivalent logarithmic form of each.

The table can be used to show that $\log_2 4 + \log_2 8$ equals $\log_2 32$.

$2^0 = 1$	$\log_2 1 = 0$
$2^1 = 2$	$\log_2 2 = 1$
$2^2 = 4$	$\log_2 4 = 2$
$2^3 = 8$	$\log_2 8 = 3$
$2^4 = 16$	$\log_2 16 = 4$
$2^5 = 32$	$\log_2 32 = 5$

$$\log_2 4 + \log_2 8 = 2 + 3 = 5$$
$$\log_2 32 = 5$$
$$\log_2 4 + \log_2 8 = \log_2 32$$

Note that $\log_2 32 = \log_2(4 \times 8) = \log_2 4 + \log_2 8$.

The property of logarithms that states that the logarithm of the product of two numbers equals the sum of the logarithms of the two numbers is similar to the property of exponents that states that to multiply two exponential expressions with the same base, we add the exponents.

Product Property of Logarithms

For any positive real numbers x, y, and b, $b \neq 1$,

$$\log_b(xy) = \log_b x + \log_b y$$

Proof:
Let $\log_b x = m$ and $\log_b y = n$.

| Write each equation in its equivalent exponential form. | $x = b^m \qquad y = b^n$ |

| Use substitution and the properties of exponents. | $xy = b^m b^n$
 $xy = b^{m+n}$ |

| Write the equation in its equivalent logarithmic form. | $\log_b xy = m + n$ |

| Substitute $\log_b x$ for m and $\log_b y$ for n. | $\log_b xy = \log_b x + \log_b y$ |

The Product Property of Logarithms is used to rewrite logarithmic expressions.

The $\log_b 6z$ is written in **expanded form** as $\log_b 6 + \log_b z$.

$$\log_b 6z = \log_b 6 + \log_b z$$

The $\log_b 12 + \log_b r$ is written as a single logarithm as $\log_b 12r$.

$$\log_b 12 + \log_b r = \log_b 12r$$

The Product Property of Logarithms can be extended to include the logarithm of the product of more than two factors. For example,

$$\log_b xyz = \log_b(xy)z = \log_b xy + \log_b z = \log_b x + \log_b y + \log_b z$$

To write $\log_b 5st$ in expanded form, use the Product Property of Logarithms.

$$\log_b 5st = \log_b 5 + \log_b s + \log_b t$$

A second property of logarithms involves the logarithm of the quotient of two numbers. This property of logarithms is also based on the fact that a logarithm is an exponent and that to divide two exponential expressions with the same base, we subtract the exponents.

Quotient Property of Logarithms

For any positive real numbers x, y, and b, $b \neq 1$, $\log_b \dfrac{x}{y} = \log_b x - \log_b y$.

Take Note

This theorem is used to rewrite expressions such as

$$\log_5\left(\frac{m}{8}\right) = \log_5 m - \log_5 8$$

It does *not* state any relationship that involves $\dfrac{\log_b x}{\log_b y}$.

This expression cannot be simplified.

Proof:
Let $\log_b x = m$ and $\log_b y = n$.

| Write each equation in its equivalent exponential form. | $x = b^m \qquad y = b^n$ |

| Use substitution and the properties of exponents. | $\dfrac{x}{y} = \dfrac{b^m}{b^n}$
 $\dfrac{x}{y} = b^{m-n}$ |

| Write the equation in its equivalent logarithmic form. | $\log_b \dfrac{x}{y} = m - n$ |

| Substitute $\log_b x$ for m and $\log_b y$ for n. | $\log_b \dfrac{x}{y} = \log_b x - \log_b y$ |

The Quotient Property of Logarithms is used to rewrite logarithmic expressions.

The $\log_b \frac{p}{8}$ is written in expanded form as $\log_b p - \log_b 8$.

$$\log_b \frac{p}{8} = \log_b p - \log_b 8$$

The $\log_b y - \log_b v$ is written as a single logarithm as $\log_b \frac{y}{v}$.

$$\log_b y - \log_b v = \log_b \frac{y}{v}$$

A third property of logarithms, useful especially in the computation of a power of a number, is based on the fact that a logarithm is an exponent and that the power of an exponential expression is found by multiplying the exponents.

The table of the powers of 2 shown earlier can be used to show that $\log_2 2^3$ equals $3 \log_2 2$.

$$\log_2 2^3 = \log_2 8 = 3$$
$$3 \log_2 2 = 3 \cdot 1 = 3$$
$$\log_2 2^3 = 3 \log_2 2$$

Point of Interest

Logarithms were developed independently by Jobst Burgi (1552–1632) and John Napier (1550–1617) as a means of simplifying the calculations of astronomers. The idea was to devise a method by which two numbers could be multiplied by performing additions. Napier is usually given credit for logarithms because he published his result first.

John Napier

In Napier's original work, the logarithm of 10,000,000 was 0. After this work was published, Napier, in discussions with Henry Briggs (1561–1631), decided that tables of logarithms would be easier to use if the logarithm of 1 was 0. Napier died before new tables could be determined, and Briggs took on the task. His table consisted of logarithms accurate to 30 decimal places, all accomplished without a calculator!

The logarithms Briggs calculated are the common logarithms mentioned earlier.

Power Property of Logarithms

For any positive real numbers x and b, $b \neq 1$, and for any real number r, $\log_b x^r = r \log_b x$.

Proof:
Let $\log_b x = m$.

Write the equation in its equivalent exponential form. $x = b^m$

Raise each side to the r power. $x^r = (b^m)^r$
 $x^r = b^{mr}$

Write the equation in its equivalent logarithmic form. $\log_b x^r = mr$

Substitute $\log_b x$ for m. $\log_b x^r = r \log_b x$

The Power Property of Logarithms is used to rewrite logarithmic expressions.

The $\log_b x^3$ is written in terms of $\log_b x$ as $3 \log_b x$.

$$\log_b x^3 = 3 \log_b x$$

$\frac{2}{3} \log_4 z$ is written with a coefficient of 1 as $\log_4 z^{\frac{2}{3}}$.

$$\frac{2}{3} \log_4 z = \log_4 z^{\frac{2}{3}}$$

The properties of logarithms can be used in combination to simplify expressions that contain logarithms.

EXAMPLE 6 Write the logarithm in expanded form.

 A. $\log_b \dfrac{xy}{z}$ **B.** $\ln \dfrac{x^2}{y^3}$ **C.** $\log_8 \sqrt{x^3 y}$

Solution **A.** $\log_b \dfrac{xy}{z} = \log_b(xy) - \log_b z$ • Use the Quotient Property of Logarithms.

 $= \log_b x + \log_b y - \log_b z$ • Use the Product Property of Logarithms.

 B. $\ln \dfrac{x^2}{y^3} = \ln x^2 - \ln y^3$ • Use the Quotient Property of Logarithms.

 $= 2 \ln x - 3 \ln y$ • Use the Power Property of Logarithms.

 C. $\log_8 \sqrt{x^3 y} = \log_8 (x^3 y)^{\frac{1}{2}}$ • Write the radical expression as an exponential expression.

 $= \dfrac{1}{2} \log_8 x^3 y$ • Use the Power Property of Logarithms.

 $= \dfrac{1}{2} (\log_8 x^3 + \log_8 y)$ • Use the Product Property of Logarithms.

 $= \dfrac{1}{2} (3 \log_8 x + \log_8 y)$ • Use the Power Property of Logarithms.

 $= \dfrac{3}{2} \log_8 x + \dfrac{1}{2} \log_8 y$ • Use the Distributive Property.

PROBLEM 6 Write the logarithm in expanded form.

 A. $\log_b \dfrac{x^2}{y}$ **B.** $\ln y^{\frac{1}{3}} z^3$ **C.** $\log_8 \sqrt[3]{xy^2}$

Solution See page S34.

EXAMPLE 7 Express as a single logarithm with a coefficient of 1.

 A. $3 \log_5 x + \log_5 y - 2 \log_5 z$

 B. $\dfrac{1}{2} (\log_3 x - 3 \log_3 y + \log_3 z)$

 C. $\dfrac{1}{3} (2 \ln x - 4 \ln y)$

Solution **A.** $3 \log_5 x + \log_5 y - 2 \log_5 z$

 $= \log_5 x^3 + \log_5 y - \log_5 z^2$ • Use the Power Property of Logarithms.

 $= \log_5 x^3 y - \log_5 z^2$ • Use the Product Property of Logarithms.

 $= \log_5 \dfrac{x^3 y}{z^2}$ • Use the Quotient Property of Logarithms.

B. $\dfrac{1}{2}(\log_3 x - 3\log_3 y + \log_3 z)$

$= \dfrac{1}{2}(\log_3 x - \log_3 y^3 + \log_3 z)$
• Use the Power Property of Logarithms.

$= \dfrac{1}{2}\left(\log_3 \dfrac{x}{y^3} + \log_3 z\right)$
• Use the Quotient Property of Logarithms.

$= \dfrac{1}{2}\left(\log_3 \dfrac{xz}{y^3}\right)$
• Use the Product Property of Logarithms.

$= \log_3\left(\dfrac{xz}{y^3}\right)^{\frac{1}{2}} = \log_3 \sqrt{\dfrac{xz}{y^3}}$
• Use the Power Property of Logarithms. Write the exponential expression as a radical expression.

C. $\dfrac{1}{3}(2\ln x - 4\ln y)$

$= \dfrac{1}{3}(\ln x^2 - \ln y^4)$
• Use the Power Property of Logarithms.

$= \dfrac{1}{3}\left(\ln \dfrac{x^2}{y^4}\right)$
• Use the Quotient Property of Logarithms.

$= \ln\left(\dfrac{x^2}{y^4}\right)^{\frac{1}{3}} = \ln \sqrt[3]{\dfrac{x^2}{y^4}}$
• Use the Power Property of Logarithms. Write the exponential expression as a radical expression.

PROBLEM 7 Express as a single logarithm with a coefficient of 1.

A. $2\log_b x - 3\log_b y - \log_b z$

B. $\dfrac{1}{3}(\log_4 x - 2\log_4 y + \log_4 z)$

C. $\dfrac{1}{2}(2\ln x - 5\ln y)$

■ **Solution** See pages S34–S35.

There are three other properties of logarithms that are useful in simplifying logarithmic expressions.

Properties of Logarithms

Logarithmic Property of One
For any positive real number b, $b \neq 1$, $\log_b 1 = 0$.

Inverse Property of Logarithms
For any positive real numbers x and b, $b \neq 1$, $\log_b b^x = x$.

1–1 Property of Logarithms
For any positive real numbers x, y, and b, $b \neq 1$, if $\log_b x = \log_b y$, then $x = y$.

EXAMPLE 8 Simplify. **A.** $8 \log_4 4$ **B.** $\log_8 1$

Solution **A.** $8 \log_4 4 = \log_4 4^8$ • Use the Power Property of Logarithms.
$\phantom{\textbf{A.} \; 8 \log_4 4} = 8$ • Use the Inverse Property of Logarithms.

B. $\log_8 1 = 0$ • Use the Logarithmic Property of One.

PROBLEM 8 Simplify. **A.** $\log_{16} 1$ **B.** $12 \log_3 3$

Solution See page S35.

Although only common logarithms and natural logarithms are programmed into a calculator, the logarithms for other positive bases can be found.

Evaluate $\log_5 22$. Round to the nearest ten-thousandth.

Write an equation.	$\log_5 22 = x$
Write the equation in its equivalent exponential form.	$5^x = 22$
Apply the common logarithm to each side of the equation.	$\log 5^x = \log 22$
Use the Power Property of Logarithms.	$x \log 5 = \log 22$
This is an exact answer.	$x = \dfrac{\log 22}{\log 5}$
This is an approximate answer.	$x \approx 1.9206$

$$\log_5 22 \approx 1.9206$$

In the third step above, the natural logarithm, instead of the common logarithm, could have been applied to each side of the equation. The same result would have been obtained.

Using a procedure similar to the one used to evaluate $\log_5 22$, a formula for changing bases can be derived.

Change-of-Base Formula

$$\log_a N = \frac{\log_b N}{\log_b a}$$

EXAMPLE 9 Evaluate $\log_7 32$. Round to the nearest ten-thousandth.

Solution $\log_7 32 = \dfrac{\ln 32}{\ln 7} \approx 1.7810$ • Use the Change-of-Base Formula.
$N = 32, a = 7, b = e$

PROBLEM 9 Evaluate $\log_4 2.4$. Round to the nearest ten-thousandth.

Solution See page S35.

EXAMPLE 10 Rewrite $f(x) = -3 \log_7(2x - 5)$ in terms of natural logarithms.

Solution $f(x) = -3 \log_7(2x - 5)$

$$= -3 \cdot \frac{\ln(2x - 5)}{\ln 7}$$

• **Use the Change-of-Base Formula to rewrite** $\log_7(2x - 5)$ **as** $\frac{\ln(2x - 5)}{\ln 7}$.

$$= -\frac{3}{\ln 7} \ln(2x - 5)$$

PROBLEM 10 Rewrite $f(x) = 4 \log_8(3x + 4)$ in terms of common logarithms.

Solution See page S35.

In Example 10, it is important to understand that $-\frac{3}{\ln 7} \ln(2x - 5)$ and $-3 \log_7(2x - 5)$ are *exactly* equal. If common logarithms had been used, the result would have been $f(x) = -\frac{3}{\log 7} \log(2x - 5)$. The expressions $-\frac{3}{\log 7} \log(2x - 5)$ and $-3 \log_7(2x - 5)$ are also *exactly* equal.

If you are working in a base other than base 10 or base e, the Change-of-Base Formula will enable you to calculate the value of the logarithm in that base just as though that base were programmed into the calculator.

11.2 Concept Review

Determine whether the following statements are always true, sometimes true, or never true.

1. For $x > 0$, $b > 0$, $b \neq 1$, $y = \log_b x$ is equivalent to $b^y = x$.

2. $\log_b \dfrac{x}{y} = \dfrac{\log_b x}{\log_b y}$

3. $\dfrac{\log x}{\log y} = \dfrac{x}{y}$

4. $\log(x + y) = \log x + \log y$

5. $\log_b \sqrt{x} = \dfrac{1}{2} \log_b x$

6. $\log(x^{-1}) = \dfrac{1}{\log x}$

7. The inverse of an exponential function is a logarithmic function.

8. If x and y are positive real numbers, $x < y$, and $b > 0$, $b \neq 1$, then $\log_b x < \log_b y$.

11.2 Exercises

1 Write equivalent exponential and logarithmic equations

1. **a.** What is a common logarithm?
 b. How is the common logarithm of $4z$ written?

2. **a.** What is a natural logarithm?
 b. How is the natural logarithm of $3x$ written?

3. The notation "$\log_b x$" can be read as "log ____?____ of ____?____."

4. A logarithm is an exponent. For example, $\log_2 8$ is the exponent used on ____?____ to give a result of ____?____, so $\log_2 8$ is ____?____.

Write the exponential equation in logarithmic form.

5. $5^2 = 25$

6. $10^3 = 1000$

7. $4^{-2} = \dfrac{1}{16}$

8. $3^{-3} = \dfrac{1}{27}$

9. $10^y = x$

10. $e^y = x$

11. $a^x = w$

12. $b^y = c$

Write the logarithmic equation in exponential form.

13. $\log_3 9 = 2$

14. $\log_2 32 = 5$

15. $\log 0.01 = -2$

16. $\log_5 \dfrac{1}{5} = -1$

17. $\ln x = y$

18. $\log x = y$

19. $\log_b u = v$

20. $\log_c x = y$

21. Complete the Equality of Exponents Property: For $b > 0$, $b \neq 1$, if $b^u = b^v$, then ____?____.

22. Evaluate $\log_4 64$.

$\log_4 64 = x$ • Write an equation.
$\underline{\ \ ?\ \ } = 4^{\overline{\ ?\ }}$ • Write the equation in its equivalent exponential form.
$4^{\underline{\ ?\ }} = 4^x$ • Write 64 as a power of 4.
$\underline{\ \ ?\ \ } = x$ • Solve for x using the Equality of Exponents Property.

Evaluate.

23. $\log_3 81$

24. $\log_7 49$

25. $\log_2 128$

26. $\log_5 125$

27. $\log 100$

28. $\log 0.001$

29. $\ln e^3$

30. $\ln e^2$

31. $\log_8 1$

32. $\log_3 243$

33. $\log_5 625$

34. $\log_2 64$

Solve for x.

35. $\log_3 x = 2$ **36.** $\log_5 x = 1$ **37.** $\log_4 x = 3$ **38.** $\log_2 x = 6$

39. $\log_7 x = -1$ **40.** $\log_8 x = -2$ **41.** $\log_6 x = 0$ **42.** $\log_4 x = 0$

Solve for x. Round to the nearest hundredth.

43. $\log x = 2.5$ **44.** $\log x = 3.2$ **45.** $\log x = -1.75$ **46.** $\log x = -2.1$

47. $\ln x = 2$ **48.** $\ln x = 4$ **49.** $\ln x = -\dfrac{1}{2}$ **50.** $\ln x = -1.7$

51. $b > 1$ and m and n are positive numbers such that $m > n$. Is $\log_b m$ less than, equal to, or greater than $\log_b n$?

52. $0 < b < 1$ and $m > 1$. Is $\log_b m$ less than, equal to, or greater than zero?

2 The properties of logarithms

53. What is the Product Property of Logarithms?

54. What is the Quotient Property of Logarithms?

55. Complete the Power Property of Logarithms: For any positive real numbers x and b, $b \neq 1$, and any real number r, $\log_b x^r = \underline{\quad ? \quad}$.

56. Write $\log_4 x^8 y$ in expanded form.

$$\log_4 x^8 y = \log_4(\underline{\quad ? \quad}) + \log_4(\underline{\quad ? \quad})$$

- By the Product Property of Logarithms, the log of a product expands into the sum of the logs of the factors.

$$= (\underline{\quad ? \quad})\log_4 x + \log_4 y$$

- Use the Power Property of Logarithms.

Write the logarithm in expanded form.

57. $\log_8(xz)$ **58.** $\log_7(4y)$ **59.** $\log_3 x^5$

60. $\log_2 y^7$ **61.** $\ln\left(\dfrac{r}{s}\right)$ **62.** $\ln\left(\dfrac{z}{4}\right)$

63. $\log_3(x^2 y^6)$ **64.** $\log_4(t^4 u^2)$ **65.** $\log_7\left(\dfrac{u^3}{v^4}\right)$

66. $\log\left(\dfrac{s^5}{t^2}\right)$ **67.** $\log_2(rs)^2$ **68.** $\log_3(x^2 y)^3$

69. $\log_9 x^2 yz$

70. $\log_6 xy^2 z^3$

71. $\ln\left(\dfrac{xy^2}{z^4}\right)$

72. $\ln\left(\dfrac{r^2 s}{t^3}\right)$

73. $\log_8\left(\dfrac{x^2}{yz^2}\right)$

74. $\log_9\left(\dfrac{x}{y^2 z^3}\right)$

75. $\log_7\sqrt{xy}$

76. $\log_8\sqrt[3]{xz}$

77. $\log_2\sqrt{\dfrac{x}{y}}$

78. $\log_3\sqrt[3]{\dfrac{r}{s}}$

79. $\ln\sqrt{x^3 y}$

80. $\ln\sqrt{x^5 y^3}$

81. $\log_7\sqrt{\dfrac{x^3}{y}}$

82. $\log_b\sqrt[3]{\dfrac{r^2}{t}}$

Express as a single logarithm with a coefficient of 1.

83. $\log_3 x^3 - \log_3 y$

84. $\log_7 t + \log_7 v^2$

85. $\log_8 x^4 + \log_8 y^2$

86. $\log_2 r^2 + \log_2 s^3$

87. $3\ln x$

88. $4\ln y$

89. $3\log_5 x + 4\log_5 y$

90. $2\log_6 x + 5\log_6 y$

91. $-2\log_4 x$

92. $-3\log_2 y$

93. $2\log_3 x - \log_3 y + 2\log_3 z$

94. $4\log_5 r - 3\log_5 s + \log_5 t$

95. $\log_b x - (2\log_b y + \log_b z)$

96. $2\log_2 x - (3\log_2 y + \log_2 z)$

97. $2(\ln x + \ln y)$

98. $3(\ln r + \ln t)$

99. $\dfrac{1}{2}(\log_6 x - \log_6 y)$

100. $\dfrac{1}{3}(\log_8 x - \log_8 y)$

101. $2(\log_4 s - 2\log_4 t + \log_4 r)$

102. $3(\log_9 x + 2\log_9 y - 2\log_9 z)$

103. $\log_5 x - 2(\log_5 y + \log_5 z)$

104. $\log_4 t - 3(\log_4 u + \log_4 v)$

105. $3\ln t - 2(\ln r - \ln v)$

106. $2\ln x - 3(\ln y - \ln z)$

107. $\dfrac{1}{2}(3\log_4 x - 2\log_4 y + \log_4 z)$

108. $\dfrac{1}{3}(4\log_5 t - 3\log_5 u - 3\log_5 v)$

Complete Exercises 109 to 112 without using a calculator.

109. True or false? $\log_b 10 + \log_b 5 = \log_b 15$

110. True or false? $\log_b 10 - \log_b 5 = \log_b 2$

111. True or false? $\log_b b^{10} - \log_b 1 = 10$

112. True or false? $\dfrac{\log_b 10}{\log_b 2} = \log_b 8$

Evaluate. Round to the nearest ten-thousandth.

113. $\ln 4$

114. $\ln 6$

115. $\ln\left(\dfrac{17}{6}\right)$

116. $\ln\left(\dfrac{13}{17}\right)$

117. $\log_8 6$

118. $\log_4 8$

119. $\log_5 30$

120. $\log_6 28$

121. $\log_3(0.5)$

122. $\log_5(0.6)$

123. $\log_7(1.7)$

124. $\log_6(3.2)$

125. $\log_5 15$

126. $\log_3 25$

127. $\log_{12} 120$

128. $\log_9 90$

Rewrite each function in terms of common logarithms.

129. $f(x) = \log_3(3x - 2)$

130. $f(x) = \log_5(x^2 + 4)$

131. $f(x) = \log_8(4 - 9x)$

132. $f(x) = \log_7(3x^2)$

133. $f(x) = 5 \log_9(6x + 7)$

134. $f(x) = 3 \log_2(2x^2 - x)$

Rewrite each function in terms of natural logarithms.

135. $f(x) = \log_2(x + 5)$

136. $f(x) = \log_4(3x + 4)$

137. $f(x) = \log_3(x^2 + 9)$

138. $f(x) = \log_7(9 - x^2)$

139. $f(x) = 7 \log_8(10x - 7)$

140. $f(x) = 7 \log_3(2x^2 - x)$

11.2 Applying Concepts

Use the properties of logarithms to solve for x.

141. $\log_8 x = 3 \log_8 2$

142. $\log_5 x = 2 \log_5 3$

143. $\log_4 x = \log_4 2 + \log_4 3$

144. $\log_3 x = \log_3 4 + \log_3 7$

145. $\log_6 x = 3 \log_6 2 - \log_6 4$

146. $\log_9 x = 5 \log_9 2 - \log_9 8$

147. $\log x = \dfrac{1}{3} \log 27$

148. $\log_2 x = \dfrac{3}{2} \log_2 4$

149. Complete the statement using the equation $\log_a b = c$.
$a^c = $ _____

150. Given $f(x) = 3 \log_6(2x - 1)$, determine $f(7)$ to the nearest hundredth.

151. Given $S(t) = 8 \log_5(6t + 2)$, determine $S(2)$ to the nearest hundredth.

152. Given $P(v) = -3 \log_6(4 - 2v)$, determine $P(-4)$ to the nearest hundredth.

153. Given $G(x) = -5 \log_7(2x + 19)$, determine $G(-3)$ to the nearest hundredth.

154. Solve for x: $\log_2(\log_2 x) = 3$

155. Solve for x: $\ln(\ln x) = 1$

156. **Biology** To discuss the variety of species that live in a certain environment, a biologist needs a precise definition of *diversity*. Let $p_1, p_2, \ldots, p_n$ be the proportions of n species that live in an environment. The biological diversity, D, of this system is

$$D = -(p_1 \log_2 p_1 + p_2 \log_2 p_2 + \cdots + p_n \log_2 p_n)$$

The larger the value of D, the greater the diversity of the system. Suppose an ecosystem has exactly five different varieties of grass: rye (R), Bermuda (B), blue (L), fescue (F), and St. Augustine (A).
a. Calculate the diversity of this ecosystem if the proportions are as shown in Table 1.
b. Because Bermuda and St. Augustine are virulent grasses, after a time the proportions are as shown in Table 2. Does this system have more or less diversity than the one given in Table 1?
c. After an even longer period, the Bermuda and St. Augustine completely overrun the environment, and the proportions are as in Table 3. Calculate the diversity of this system. (*Note:* For purposes of the diversity definition, $0 \log_2 0 = 0$.) Does it have more or less diversity than the system given in Table 2?
d. Finally, the St. Augustine overruns the Bermuda, and the proportions are as in Table 4. Calculate the diversity of this system. Write a sentence that describes your answer.

Table 1

R	B	L	F	A
$\dfrac{1}{5}$	$\dfrac{1}{5}$	$\dfrac{1}{5}$	$\dfrac{1}{5}$	$\dfrac{1}{5}$

Table 2

R	B	L	F	A
$\dfrac{1}{8}$	$\dfrac{3}{8}$	$\dfrac{1}{16}$	$\dfrac{1}{8}$	$\dfrac{5}{16}$

Table 3

R	B	L	F	A
0	$\dfrac{1}{4}$	0	0	$\dfrac{3}{4}$

Table 4

R	B	L	F	A
0	0	0	0	1

11.3 Graphs of Logarithmic Functions

OBJECTIVE 1 ### Graph logarithmic functions

The graph of a logarithmic function can be drawn by using the relationship between the exponential and logarithmic functions.

To graph $f(x) = \log_2 x$, think of the function as the equation $y = \log_2 x$.

$$f(x) = \log_2 x$$
$$y = \log_2 x$$

Point of Interest

Although logarithms were originally developed to assist with computations, logarithmic functions have a much broader use today. These functions occur in geology, acoustics, chemistry, and economics, for example.

Write the equivalent exponential equation.

$$x = 2^y$$

Because the equation is solved for x in terms of y, it is easier to choose values of y and find the corresponding values of x. The results can be recorded in a table.

Graph the ordered pairs on a rectangular coordinate system.

Connect the points with a smooth curve.

x	y
$\dfrac{1}{4}$	-2
$\dfrac{1}{2}$	-1
1	0
2	1
4	2

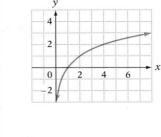

Applying the vertical line and horizontal line tests reveals that $f(x) = \log_2 x$ is the graph of a 1–1 function.

Recall that the graph of the inverse of a function f is the mirror image of the graph of f with respect to the line $y = x$. The graph of $f(x) = 2^x$ was shown earlier. Because $g(x) = \log_2 x$ is the inverse of $f(x) = 2^x$, the graphs of these functions are mirror images of each other with respect to the line $y = x$.

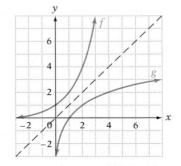

Graph: $f(x) = \log_2 x + 1$

Think of the function as the equation $y = \log_2 x + 1$.

$$f(x) = \log_2 x + 1$$
$$y = \log_2 x + 1$$

Solve the equation for $\log_2 x$.

$$y - 1 = \log_2 x$$

Write the equivalent exponential equation.

$$2^{y-1} = x$$

Choose values of y, and find the corresponding values of x.

Graph the ordered pairs on a rectangular coordinate system.

Connect the points with a smooth curve.

x	y
$\dfrac{1}{4}$	-1
$\dfrac{1}{2}$	0
$1\dfrac{1}{2}$	2
4	3

EXAMPLE 1 Graph. **A.** $f(x) = \log_3 x$ **B.** $f(x) = 2 \log_3 x$

Solution **A.** $f(x) = \log_3 x$
$$y = \log_3 x$$
$$x = 3^y$$

x	y
$\dfrac{1}{9}$	-2
$\dfrac{1}{3}$	-1
1	0
3	1

• Substitute y for $f(x)$.
• Write the equivalent exponential equation.
• Choose values of y, and find the corresponding values of x. Graph the ordered pairs on a rectangular coordinate system. Connect the points with a smooth curve.

B. $f(x) = 2 \log_3 x$
$$y = 2 \log_3 x$$
$$\frac{y}{2} = \log_3 x$$
$$x = 3^{\frac{y}{2}}$$

x	y
$\dfrac{1}{9}$	-4
$\dfrac{1}{3}$	-2
1	0
3	2

• Substitute y for $f(x)$.
• Solve the equation for $\log_3 x$.
• Write the equivalent exponential equation.
• Choose values of y, and find the corresponding values of x. Graph the ordered pairs on a rectangular coordinate system. Connect the points with a smooth curve.

PROBLEM 1 Graph. **A.** $f(x) = \log_2(x - 1)$ **B.** $f(x) = \log_3 2x$

Solution See page S35.

EXAMPLE 2 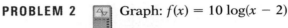 Graph: $f(x) = 2 \ln x + 3$

Solution The graph is shown below. To verify the accuracy of the graph, evaluate $f(x)$ for a few values of x, and compare the results to values found by using the TRACE feature of the graphing utility. For example, $f(1) = 2 \ln(1) + 3 = 3$, so $(1, 3)$ is a point on the graph. When $x = 3$, $f(3) = 2 \ln(3) + 3 \approx 5.2$, so $(3, 5.2)$ is a point on the graph.

PROBLEM 2 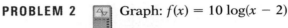 Graph: $f(x) = 10 \log(x - 2)$

Solution See page S35.

The graphs of logarithmic functions to other than base e or base 10 can be drawn with a graphing utility by first using the Change-of-Base Formula $\log_a N = \dfrac{\log_b N}{\log_b a}$ to rewrite the logarithmic function in terms of base e or base 10.

Graph: $f(x) = \log_3 x$

Use the Change-of-Base Formula to rewrite $\log_3 x$ in terms of $\log x$ or $\ln x$. The natural logarithm function $\ln x$ is used here.

$$\log_3 x = \frac{\ln x}{\ln 3}$$

To graph $f(x) = \log_3 x$ using a graphing utility, use the equivalent form $f(x) = \dfrac{\ln x}{\ln 3}$.

The graph is shown at the right.

The graph of $f(x) = \log_3 x$ could have been drawn by rewriting $\log_3 x$ in terms of $\log x$, $\log_3 x = \dfrac{\log x}{\log 3}$. The graph of $f(x) = \log_3 x$ is identical to the graph of $f(x) = \dfrac{\log x}{\log 3}$.

The example that follows was graphed by rewriting the logarithmic function in terms of the natural logarithmic function. The common logarithmic function could also have been used.

EXAMPLE 3 Graph: $f(x) = -3 \log_2 x$

Solution $f(x) = -3 \log_2 x$

$$= -3 \cdot \frac{\ln x}{\ln 2} = -\frac{3}{\ln 2} \ln x$$

- Rewrite $\log_2 x$ in terms of $\ln x$.
- The graph of $f(x) = -3 \log_2 x$ is the same as the graph of

$$f(x) = -\frac{3}{\ln 2} \ln x.$$

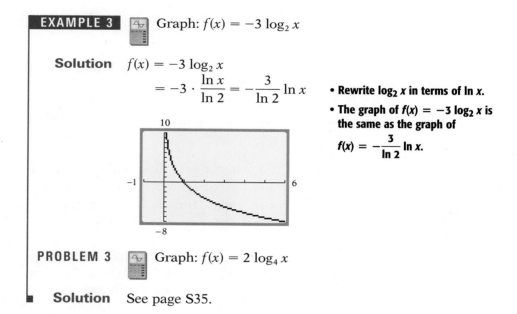

PROBLEM 3 Graph: $f(x) = 2 \log_4 x$

Solution See page S35.

11.3 Concept Review

Determine whether the following statements are always true, sometimes true, or never true.

1. The domain of $f(x) = \log_b x$, $b > 0$, $b \neq 1$, is the real numbers.

2. The range of $f(x) = \log_b x$, $b > 0$, $b \neq 1$, is the real numbers.

3. The graphs of $x = 2^y$ and $y = \log_2 x$ are identical graphs.

4. The range of $f(x) = \ln x$ is the real numbers greater than zero.

11.3 Exercises

1 Graph logarithmic functions

1. Is the function $f(x) = \log x$ a 1–1 function? Why or why not?

2. Name two characteristics of the graph of $y = \log_b x$, $b > 1$.

3. What is the relationship between the graph of $x = 3^y$ and that of $y = \log_3 x$?

4. What is the relationship between the graph of $y = 3^x$ and that of $y = \log_3 x$?

5. To find points on the graph of $f(x) = \log_3(x - 1)$, first replace $f(x)$ with ___?___ and write the equation in ___?___ form: $x - 1 = 3^y$. Solve the equation for x: $x =$ ___?___. Choose values for ___?___ and find the corresponding values for ___?___.

6. Use the results of Exercise 5. The x-coordinate of the point on the graph of $f(x) = \log_3(x - 1)$ that has y-coordinate 2 is $3^{\underline{?}} + 1 =$ ___?___. The point on the graph of $f(x) = \log_3(x - 1)$ with a y-coordinate of 2 is (___?___, ___?___).

Graph.

7. $f(x) = \log_4 x$

8. $f(x) = \log_2(x + 1)$

9. $f(x) = \log_3(2x - 1)$

10. $f(x) = \log_2\left(\dfrac{1}{2}x\right)$

11. $f(x) = 3 \log_2 x$

12. $f(x) = \dfrac{1}{2} \log_2 x$

13. $f(x) = -\log_2 x$

14. $f(x) = -\log_3 x$

15. $f(x) = \log_2(x - 1)$

16. $f(x) = \log_3(2 - x)$

17. $f(x) = -\log_2(x - 1)$

18. $f(x) = -\log_2(1 - x)$

Use a graphing utility to graph the following.

19. $f(x) = \log_2 x - 3$

20. $f(x) = \log_3 x + 2$

21. $f(x) = -\log_2 x + 2$

22. $f(x) = -\dfrac{1}{2}\log_2 x - 1$

23. $f(x) = x - \log_2(1 - x)$

24. $f(x) = x + \log_3(2 - x)$

Graph the functions on the same rectangular coordinate system.

25. $f(x) = 4^x;\ g(x) = \log_4 x$

26. $f(x) = 3^x;\ g(x) = \log_3 x$

27. $f(x) = \left(\dfrac{1}{2}\right)^x;\ g(x) = \log_{\frac{1}{2}} x$

28. $f(x) = 10^x;\ g(x) = \log_{10} x$

29. Which function(s) will have the same graph as the graph of the function $f(x) = \log_3 3x$?
a. $g(x) = 3\log_3 x$ **b.** $h(x) = 1 + \log_3 x$
c. $F(x) = \log_3 x^3$ **d.** $G(x) = \dfrac{\log 3x}{\log 3}$

30. Which function(s) will have the same graph as the graph of the function $f(x) = -\log_b x$?
a. $g(x) = \log_b(-x)$ **b.** $h(x) = \dfrac{1}{\log_b x}$
c. $F(x) = \log_b \dfrac{1}{x}$ **d.** $G(x) = \dfrac{\log b}{\log x}$

11.3 Applying Concepts

31. **Employment** The proficiency of a typist decreases (without practice) over time. An equation that approximates this decrease is given by $S = 60 - 7 \ln(t + 1)$, where S is the typing speed in words per minute and t is the number of months without typing.

 a. Graph the equation.

 b. The point whose approximate coordinates are (4, 49) is on the graph. Write a sentence that describes the meaning of this ordered pair.

32. **Astronomy** Astronomers use the *distance modulus* of a star as a method of determining the star's distance from Earth. The formula is $M = 5 \log s - 5$, where M is the distance modulus and s is the star's distance from Earth in parsecs. (One parsec $\approx 2.1 \times 10^{13}$ miles.)

 a. Graph the equation.

 b. The point whose approximate coordinates are (25.1, 2) is on the graph. Write a sentence that describes the meaning of this ordered pair.

33. **Energy** The Energy Information Administration forecasts energy production for the United States. Based on data from that agency, the equation $y = 4.1596 + 1.2296 \ln t$ models the estimated quadrillion BTU, y, of energy that will be produced in the United States by renewable energy sources such as wind energy, solar power, biomass energy, and geothermal energy, during year t, where t is the number of years after 2002.

 a. Graph the equation.

 b. According to this model, in what year will energy production from renewable sources first exceed 6 quadrillion BTU?

 c. According to this model, what will be the energy production from renewable energy sources in 2010? Round to the nearest tenth of a quadrillion.

Determine whether the ordered pairs in the table belong to a linear function, a quadratic function, an exponential function, or a logarithmic function.

34.

x	y
0	5
1	3
2	1
3	−1
4	−3

35.

x	y
−1	6
0	1
1	2
2	9
3	22

36.

x	y
0	1
1	4
2	16
3	64
4	256

37.

x	y
0	−3
2	−2
4	−1
6	0
8	1

38.

x	y
0	-3
1	-7
2	-9
3	-9
4	-7

39.

x	y
0	160
1	80
2	40
3	20
4	10

40.

x	y
$\dfrac{1}{2}$	-1
1	0
2	1
4	2
8	3

41.

x	y
$\dfrac{1}{4}$	-2
1	0
4	2
16	4
256	8

42. Using the Power Property of Logarithms, $\ln x^2 = 2 \ln x$. Graph the equations $f(x) = \ln x^2$ and $g(x) = 2 \ln x$ on the same rectangular coordinate system. Are the graphs the same? Why or why not?

43. Because $f(x) = e^x$ and $g(x) = \ln x$ are inverse functions of each other, $f[g(x)] = x$ and $g[f(x)] = x$. Graph $f[g(x)] = e^{\ln x}$ and $g[f(x)] = \ln e^x$. Explain why the graphs are different even though $f[g(x)] = g[f(x)]$.

11.4 Exponential and Logarithmic Equations

OBJECTIVE 1 Solve exponential equations

An **exponential equation** is one in which a variable occurs in the exponent. The examples at the right are exponential equations.

$$6^{2x+1} = 6^{3x-2}$$
$$4^x = 3$$
$$2^{x+1} = 7$$

An exponential equation in which each side of the equation can be expressed in terms of the same base can be solved by using the Equality of Exponents Property. Recall that the Equality of Exponents Property states that

$$\text{If } b^u = b^v, \text{ then } u = v.$$

In Example 1 below, this property is used to solve an exponential equation.

EXAMPLE 1 Solve and check: $9^{x+1} = 27^{x-1}$

Solution

$$9^{x+1} = 27^{x-1}$$
$$(3^2)^{x+1} = (3^3)^{x-1}$$ • Rewrite each side of the equation using the same base.
$$3^{2x+2} = 3^{3x-3}$$

$$2x + 2 = 3x - 3$$ • Use the Equality of Exponents Property to equate the exponents.

$$2 = x - 3$$ • Solve the resulting equation.
$$5 = x$$

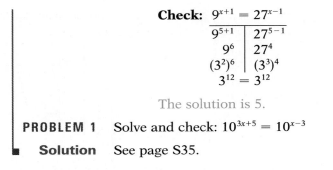

Check: $9^{x+1} = 27^{x-1}$

$$
\begin{array}{c|c}
9^{5+1} & 27^{5-1} \\
9^6 & 27^4 \\
(3^2)^6 & (3^3)^4 \\
3^{12} = {} & 3^{12}
\end{array}
$$

The solution is 5.

PROBLEM 1 Solve and check: $10^{3x+5} = 10^{x-3}$

■ **Solution** See page S35.

When both sides of an exponential equation cannot easily be expressed in terms of the same base, logarithms are used to solve the exponential equation.

EXAMPLE 2 Solve for x. Round to the nearest ten-thousandth.

A. $4^x = 7$ **B.** $3^{2x} = 4$

Solution **A.** $4^x = 7$

$\log 4^x = \log 7$ • Take the common logarithm of each side of the equation.

$x \log 4 = \log 7$ • Rewrite using the properties of logarithms.

$x = \dfrac{\log 7}{\log 4}$ • Solve for x.

$x \approx 1.4037$

The solution is 1.4037.

B. $3^{2x} = 4$

$\log 3^{2x} = \log 4$ • Take the common logarithm of each side of the equation.

$2x \log 3 = \log 4$ • Rewrite using the properties of logarithms.

$2x = \dfrac{\log 4}{\log 3}$ • Solve for x.

$x = \dfrac{\log 4}{2 \log 3}$

$x \approx 0.6309$

The solution is 0.6309.

PROBLEM 2 Solve for x. Round to the nearest ten-thousandth.

A. $4^{3x} = 25$ **B.** $(1.06)^x = 1.5$

■ **Solution** See page S35.

Take Note

To evaluate $\dfrac{\log 7}{\log 4}$ on a scientific calculator, use the keystrokes

7 [log] [÷] 4 [log] [=]

The display should read 1.4036775.

The equations in Example 2 can be solved by graphing. For Example 2A, by subtracting 7 from each side of the equation $4^x = 7$, the equation can be written as $4^x - 7 = 0$. The graph of $f(x) = 4^x - 7$ is shown on the next page. The values of x for which $f(x) = 0$ are the solutions of the equation $4^x = 7$. These values of x are the zeros of the function. By using the features of a graphing

utility, it is possible to determine a very accurate solution. The solution to the nearest tenth is shown in the graph.

EXAMPLE 3 Solve $e^x = 2x + 1$ for x. Round to the nearest hundredth.

Solution Rewrite the equation by subtracting $2x + 1$ from each side and writing the equation as $e^x - 2x - 1 = 0$. The zeros of $f(x) = e^x - 2x - 1$ are the solutions of $e^x = 2x + 1$. Graph f and use the features of a graphing utility to estimate the solutions to the nearest hundredth.

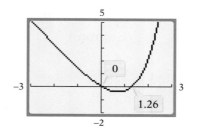

The solutions are 0 and 1.26.

PROBLEM 3 Solve $e^x = x$ for x. Round to the nearest hundredth.

Solution See page S36.

OBJECTIVE **2** ## Solve logarithmic equations

A logarithmic equation can be solved by using the properties of logarithms. Here is an example.

Solve: $\log_9 x + \log_9 (x - 8) = 1$

Use the Product Property of Logarithms to rewrite the left side of the equation.

$$\log_9 x + \log_9 (x - 8) = 1$$
$$\log_9 x(x - 8) = 1$$

Write the equation in exponential form.

$$9^1 = x(x - 8)$$

Simplify and solve for x.

$$9 = x^2 - 8x$$
$$0 = x^2 - 8x - 9$$
$$0 = (x - 9)(x + 1)$$

$$x - 9 = 0 \qquad x + 1 = 0$$
$$x = 9 \qquad\quad x = -1$$

When x is replaced by 9 in the original equation, 9 checks as a solution. When x is replaced by -1, the original equation contains the expression $\log_9(-1)$. Because the logarithm of a negative number is not a real number, -1 does not check as a solution.

The solution of the equation is 9.

Some logarithmic equations can be solved by using the 1–1 Property of Logarithms, which states that for $x > 0$, $y > 0$, $b > 0$, $b \neq 1$,

If $\log_b x = \log_b y$, then $x = y$.

This property is used in Example 4B.

EXAMPLE 4 Solve for x.

A. $\log_3(2x - 1) = 2$

B. $\log_2 x - \log_2(x - 1) = \log_2 2$

Solution **A.** $\log_3(2x - 1) = 2$

$$3^2 = 2x - 1 \qquad \bullet \text{ Rewrite in exponential form.}$$
$$9 = 2x - 1 \qquad \bullet \text{ Solve for } x.$$
$$10 = 2x$$
$$5 = x$$

The solution is 5.

B. $\log_2 x - \log_2(x - 1) = \log_2 2$

$$\log_2\left(\frac{x}{x - 1}\right) = \log_2 2 \qquad \bullet \text{ Use the Quotient Property of Logarithms.}$$

$$\frac{x}{x - 1} = 2 \qquad \bullet \text{ Use the 1–1 Property of Logarithms.}$$

$$(x - 1)\left(\frac{x}{x - 1}\right) = (x - 1)2 \qquad \bullet \text{ Solve for } x.$$
$$x = 2x - 2$$
$$-x = -2$$
$$x = 2$$

The solution is 2.

PROBLEM 4 Solve for x.

A. $\log_4(x^2 - 3x) = 1$

B. $\log_3 x + \log_3(x + 3) = \log_3 4$

Solution See page S36.

Some logarithmic equations cannot be solved algebraically. In these cases, a graphical approach may be appropriate.

EXAMPLE 5 Solve $\ln(2x + 4) = x^2$ for x. Round to the nearest hundredth.

Solution Rewrite the equation by subtracting x^2 from each side and writing the equation as $\ln(2x + 4) - x^2 = 0$. The zeros of the function defined by $f(x) = \ln(2x + 4) - x^2$ are the solutions of the equation. Graph f and then use the features of a graphing utility to estimate the solutions to the nearest hundredth.

The solutions are -0.89 and 1.38.

PROBLEM 5 Solve $\log(3x - 2) = -2x$ for x. Round to the nearest hundredth.

Solution See page S36.

Recall the equation $\log_9 x + \log_9(x - 8) = 1$ from the beginning of this objective. The algebraic solution showed that although -1 and 9 may be solutions, only 9 satisfies the equation. The extraneous solution was introduced at the second step. The Product Property of Logarithms, $\log_b(xy) = \log_b x + \log_b y$, applies only when both x and y are positive numbers. This occurs when $x > 8$. Therefore, a solution to this equation must be greater than 8. The graphs of $f(x) = \log_9 x + \log_9(x - 8) - 1$ and $g(x) = \log_9 x(x - 8) - 1$ are shown below.

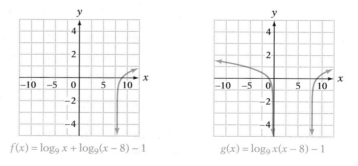

$f(x) = \log_9 x + \log_9(x - 8) - 1$ $g(x) = \log_9 x(x - 8) - 1$

Note that the only zero of f is 9, whereas the zeros of g are -1 and 9.

11.4 Concept Review

Determine whether the following are always true, sometimes true, or never true.

1. Suppose $b > 0$, $b \neq 1$. If $b^x = b^y$, then $x = y$.

2. $\log x + \log(x + 2) = \log(2x + 2)$

3. $\log_2(-4) = -2$

4. $\log(2x - 2) - \log x = 4$ is equivalent to $\log \dfrac{2x - 2}{x} = 4$.

5. $\log(2x) + \log 4 = 6$ is equivalent to $\log(8x) = 6$.

6. $\log_{10} 10^4 = 4$

11.4 Exercises

1 Solve exponential equations

1. What is an exponential equation?

2. **a.** What does the Equality of Exponents Property state?
 b. Describe when you would use this property.

3. Solve: $3^{3x-1} = 9^{2x}$.

$3^{3x-1} = 9^{2x}$	• Both bases are powers of ___?___.
$3^{3x-1} = (3^{\underline{?}})^{2x}$	• Write 9 as a power of 3.
$3^{3x-1} = 3^{\underline{?}}$	• Simplify the exponent.
$\underline{\quad?\quad} = \underline{\quad?\quad}$	• Use the Equality of Exponents Property to equate the exponents.
$x = \underline{\quad?\quad}$	• Solve for x.

4. Solve: $5^x = 18$

$5^x = 18$	• The bases are not powers of the same number.
$\log 5^x = \log 18$	• Take the ___?___ of each side of the equation.
$(\underline{\quad?\quad}) \log 5 = \log 18$	• Use the Power Property of Logarithms.
$x = \dfrac{\underline{\quad?\quad}}{\underline{\quad?\quad}}$	• Solve for x by dividing each side of the equation by ___?___.
$x \approx \underline{\quad?\quad}$	• Use a calculator to evaluate x to the nearest ten-thousandth.

Solve for x. Round to the nearest ten-thousandth.

5. $5^{4x-1} = 5^{x+2}$ **6.** $7^{4x-3} = 7^{2x+1}$ **7.** $8^{x-4} = 8^{5x+8}$

8. $10^{4x-5} = 10^{x+4}$ **9.** $5^x = 6$ **10.** $7^x = 10$

11. $12^x = 6$ **12.** $10^x = 5$ **13.** $\left(\dfrac{1}{2}\right)^x = 3$

14. $\left(\dfrac{1}{3}\right)^x = 2$ **15.** $(1.5)^x = 2$ **16.** $(2.7)^x = 3$

17. $10^x = 21$

18. $10^x = 37$

19. $2^{-x} = 7$

20. $3^{-x} = 14$

21. $2^{x-1} = 6$

22. $4^{x+1} = 9$

23. $3^{2x-1} = 4$

24. $4^{-x+2} = 12$

25. $9^x = 3^{x+1}$

26. $2^{x-1} = 4^x$

27. $8^{x+2} = 16^x$

28. $9^{3x} = 81^{x-4}$

29. $5^{x^2} = 21$

30. $3^{x^2} = 40$

31. $2^{4x-2} = 20$

32. $4^{3x+8} = 12$

33. $3^{-x+2} = 18$

34. $5^{-x+1} = 15$

35. $4^{2x} = 100$

36. $3^{3x} = 1000$

37. $2.5^{-x} = 4$

38. $3.25^{x+1} = 4.2$

39. $0.25^x = 0.125$

40. $0.1^{5x} = 10^{-2}$

41. Given that a is any real number, which equations have the same solution?
 a. $9^{x-a} = 81^x$ **b.** $5^{2x-2a} = 25^{2x}$ **c.** $49^{(x-a)/4} = 7^x$ **d.** $9^x = 81^{x-a}$

42. Use the graphs shown below. Which graph(s) can be used to solve the equation $e^x - 5 = x$? How many positive solutions does the equation $e^x - 5 = x$ have?
 a. $y = e^x + x + 5$ **b.** $y = e^x - x - 5$

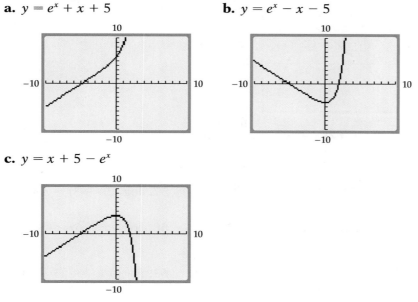

 c. $y = x + 5 - e^x$

Solve for x by graphing. Round to the nearest hundredth.

43. $3^x = 2$

44. $5^x = 9$

45. $2^x = 2x + 4$

46. $3^x = -x - 1$

47. $e^x = -2x - 2$

48. $e^x = 3x + 4$

2 Solve logarithmic equations

49. What is a logarithmic equation?

50. What does the 1–1 Property of Logarithms state?

51. A logarithmic equation in the form $\log_b$ (expression) $= n$, where n is a real number, can be solved by rewriting the equation in ___?___.

52. A logarithmic equation in the form $\log_b$(expression) $= \log_b$(expression), can be solved by using the ___?___ Property of Logarithms to set the two expressions ___?___ to each other.

53. Look at the equation in each exercise. State whether you will use the method described in Exercise 51 or the method described in Exercises 52 to solve the equation.
a. Exercise 63 **b.** Exercise 65 **c.** Exercise 67 **d.** Exercise 69

54. Use the graphs shown below. Which equation has no solution, $\log(x - 1) = -2x + 3$ or $\log(x - 1) = 2x - 3$?

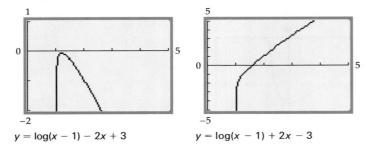

$y = \log(x - 1) - 2x + 3$ $y = \log(x - 1) + 2x - 3$

Solve for x.

55. $\log_3(x + 1) = 2$

56. $\log_5(x - 1) = 1$

57. $\log_2(2x - 3) = 3$

58. $\log_4(3x + 1) = 2$

59. $\log_2(x^2 + 2x) = 3$

60. $\log_3(x^2 + 6x) = 3$

61. $\log_5\left(\dfrac{2x}{x - 1}\right) = 1$

62. $\log_6 \dfrac{3x}{x + 1} = 1$

63. $\log_7 x = \log_7(1 - x)$

64. $\dfrac{3}{4} \log x = 3$

65. $\dfrac{2}{3} \log x = 6$

66. $\log(x - 2) - \log x = 3$

67. $\log_2(x - 3) + \log_2(x + 4) = 3$

68. $\log x - 2 = \log(x - 4)$

69. $\log_3 x + \log_3(x - 1) = \log_3 6$

70. $\log_4 x + \log_4(x - 2) = \log_4 15$

71. $\log_2(8x) - \log_2(x^2 - 1) = \log_2 3$

72. $\log_5(3x) - \log_5(x^2 - 1) = \log_5 2$

73. $\log_9 x + \log_9(2x - 3) = \log_9 2$

74. $\log_6 x + \log_6(3x - 5) = \log_6 2$

75. $\log_8(6x) = \log_8 2 + \log_8(x - 4)$

76. $\log_7(5x) = \log_7 3 + \log_7(2x + 1)$

77. $\log_9(7x) = \log_9 2 + \log_9(x^2 - 2)$

78. $\log_3 x = \log_3 2 + \log_3(x^2 - 3)$

79. $\log(x^2 + 3) - \log(x + 1) = \log 5$

80. $\log(x + 3) + \log(2x - 4) = \log 3$

Solve for x by graphing. Round to the nearest hundredth.

81. $\log x = -x + 2$

82. $\log x = -2x$

83. $\log(2x - 1) = -x + 3$

84. $\log(x + 4) = -2x + 1$

85. $\ln(x + 2) = x^2 - 3$

86. $\ln x = -x^2 + 1$

11.4 Applying Concepts

Solve for x. Round to the nearest ten-thousandth.

87. $8^{\frac{x}{2}} = 6$

88. $4^{\frac{x}{3}} = 2$

89. $5^{\frac{3x}{2}} = 7$

90. $9^{\frac{2x}{3}} = 8$

91. $1.2^{\frac{x}{2} - 1} = 1.4$

92. $5.6^{\frac{x}{3} + 1} = 7.8$

Solve.

93. If $4^x = 7$, find the value of $2^{(6x + 3)}$.

94. **Physics** A model for the distance s (in feet) that an object experiencing air resistance will fall in t seconds is given by $s = 78 \ln\left(\dfrac{e^{0.8t} + e^{-0.8t}}{2}\right)$.

 a. Graph this equation. *Suggestion:* Use Xmin = 0, Xmax = 4.5, Xscl = 0.5, Ymin = 0, Ymax = 140, and Yscl = 20.

 b. Determine, to the nearest hundredth of a second, the time it takes the object to travel 125 ft.

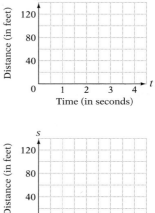

95. **Physics** A model for the distance s (in feet) that an object experiencing air resistance will fall in t seconds is given by $s = 312.5 \ln\left(\dfrac{e^{0.32t} + e^{-0.32t}}{2}\right)$.

 a. Graph this equation. *Suggestion:* Use Xmin = 0, Xmax = 4.5, Xscl = 0.5, Ymin = 0, Ymax = 140, and Yscl = 20.

 b. Determine, to the nearest hundredth of a second, the time it takes the object to travel 100 ft.

96. The following "proof" appears to show that $0.04 < 0.008$. Explain the error.

$$2 < 3$$
$$2 \log 0.2 < 3 \log 0.2$$
$$\log(0.2)^2 < \log(0.2)^3$$
$$(0.2)^2 < (0.2)^3$$
$$0.04 < 0.008$$

11.5 Applications of Exponential and Logarithmic Functions

OBJECTIVE 1 Application problems

A biologist places one single-celled bacterium in a culture, and each hour that particular species of bacterium divides into two bacteria. After one hour there will be two bacteria. After two hours, each of the two bacteria will divide and there will be four bacteria. After three hours, each of the four bacteria will divide and there will be eight bacteria.

Time, t	Number of Bacteria, N
0	1
1	2
2	4
3	8
4	16

The table at the left shows the number of bacteria in the culture after various intervals of time, t, in hours. Values in this table could also be found by using the exponential equation $N = 2^t$.

The equation $N = 2^t$ is an example of an **exponential growth equation.** In general, any equation that can be written in the form $A = A_0 b^{kt}$, where A is the size at time t, A_0 is the initial size, $b > 1$, and k is a positive real number, is an exponential growth equation. These equations are important not only in population growth studies but also in physics, chemistry, psychology, and economics.

Recall that interest is the amount of money one pays (or receives) when borrowing (or investing) money. **Compound interest** is interest that is computed not only on the original principal but also on the interest already earned. The compound interest formula is an exponential growth equation.

The **compound interest formula** is $P = A(1 + i)^n$, where A is the original value of an investment, i is the interest rate per compounding period, n is the total number of compounding periods, and P is the value of the investment after n periods.

An investment broker deposits $1000 into an account that earns 12% annual interest compounded quarterly. What is the value of the investment after 2 years? Round to the nearest dollar.

Find i, the interest rate per quarter. The quarterly rate is the annual rate divided by 4, the number of quarters in 1 year.

$$i = \frac{12\%}{4} = \frac{0.12}{4} = 0.03$$

Find n, the number of compounding periods. The investment is compounded quarterly, 4 times a year, for 2 years.

$$n = 4 \cdot 2 = 8$$

Use the compound interest formula. $P = A(1 + i)^n$

Replace A, i, and n by their values. $P = 1000(1 + 0.03)^8$

Solve for P. $P \approx 1267$

The value of the investment after 2 years is $1267.

Exponential decay can also be modeled by an exponential equation. One of the most common illustrations of exponential decay is the decay of a radioactive substance.

A radioactive isotope of cobalt has a half-life of approximately 5 years. This means that one-half of any given amount of the cobalt isotope will disintegrate in 5 years. Suppose you begin with 10 mg of a cobalt isotope. The table at the left indicates the amount of the initial 10 mg of the cobalt isotope that remains after various intervals of time, t, in years. Values in this table could also be found by using the exponential equation $A = 10\left(\frac{1}{2}\right)^{\frac{t}{5}}$.

Time, t	Amount, A
0	10
5	5
10	2.5
15	1.25
20	0.625

The equation $A = 10\left(\frac{1}{2}\right)^{\frac{t}{5}}$ is an example of an **exponential decay equation.**

Compare this equation to the exponential growth equation, and note that for exponential growth, the base of the exponential expression is greater than 1, whereas for exponential decay, the base is between 0 and 1.

A method by which an archeologist can measure the age of a bone is called **carbon dating.** Carbon dating is based on a radioactive isotope of carbon called carbon-14, which has a half-life of approximately 5570 years. The exponential decay equation is given by $A = A_0\left(\frac{1}{2}\right)^{\frac{t}{5570}}$, where A_0 is the original amount of carbon-14 present in the bone, t is the age of the bone, and A is the amount of carbon-14 present after t years.

A bone that originally contained 100 mg of carbon-14 now has 70 mg of carbon-14. What is the approximate age of the bone? Round to the nearest year.

Use the exponential decay equation.

$$A = A_0\left(\frac{1}{2}\right)^{\frac{t}{5570}}$$

Replace A_0 and A by their given values, and solve for t.

$$70 = 100\left(\frac{1}{2}\right)^{\frac{t}{5570}}$$

$$70 = 100(0.5)^{\frac{t}{5570}}$$

Divide each side of the equation by 100.

$$0.7 = (0.5)^{\frac{t}{5570}}$$

Take the common logarithm of each side of the equation. Then simplify.

$$\log 0.7 = \log(0.5)^{\frac{t}{5570}}$$

$$\log 0.7 = \frac{t}{5570} \log 0.5$$

$$\frac{5570 \log 0.7}{\log 0.5} = t$$

$$2866 \approx t$$

The age of the bone is approximately 2866 years.

| | EXAMPLE 1 | The number of words per minute a student can type will increase with practice and can be approximated by the equation $N = 100[1 - (0.9)^t]$, where N is the number of words typed per minute after t days of practice. In how many days will the student be able to type 60 words per minute? |

Strategy To find the number of days, replace N by 60 and solve for t.

Solution

$$N = 100[1 - (0.9)^t]$$
$$60 = 100[1 - (0.9)^t]$$
$$0.6 = 1 - (0.9)^t$$

- **Divide each side of the equation by 100.**

$$-0.4 = -(0.9)^t$$
$$0.4 = (0.9)^t$$

- **Subtract 1 from each side of the equation.**

$$\log 0.4 = \log(0.9)^t$$
$$\log 0.4 = t \log 0.9$$

$$t = \frac{\log 0.4}{\log 0.9} \approx 8.6967184$$

After approximately 9 days the student will type 60 words per minute.

The graph at the left is the graph of $N = 100[1 - (0.9)^t]$. Note that t is approximately 8.7 when $N = 60$.

N

Words per minute

80
60 (8.7, 60)
40
20

t

0 10 20

Number of days

PROBLEM 1 An investment of \$3000 is placed into an account that earns 12% annual interest compounded monthly. In approximately how many years will the investment be worth twice the original amount? Round to the nearest whole number.

Solution See page S36.

The first applications of logarithms (and the main reason why they were developed) were to reduce computational drudgery. Today, with the widespread use of calculators and computers, the computational uses of logarithms have diminished. However, a number of other applications of logarithms have emerged.

A chemist measures the acidity or alkalinity of a solution by the formula **pH = −log(H⁺),** where H^+ is the concentration of hydrogen ions in the solution. A neutral solution such as distilled water has a pH of 7, acids have a pH less than 7, and alkaline solutions (also called basic solutions) have a pH greater than 7.

Point of Interest

Søren Sørensen

The pH scale was created by the Danish biochemist Søren Sørensen in 1909 to measure the acidity of water used in the brewing of beer. pH is an abbreviation for pondus hydrogenii, *which translates as "potential hydrogen."*

Find the pH of vinegar for which $H^+ = 1.26 \times 10^{-3}$. Round to the nearest tenth.

Use the pH equation.
$H^+ = 1.26 \times 10^{-3}$.

$$pH = -\log(H^+)$$
$$= -\log(1.26 \times 10^{-3})$$
$$= -(\log 1.26 + \log 10^{-3})$$
$$\approx -[0.1004 + (-3)] = 2.8996$$

The pH of vinegar is approximately 2.9.

Logarithmic functions are used to scale very large or very small numbers into numbers that are easier to comprehend. For instance, the *Richter scale* magnitude of an earthquake uses a logarithmic function to convert the intensity of shock waves I into a number M, which for most earthquakes is in the range of 0 to 10. The intensity I of an earthquake is often given in terms of the constant I_0, where I_0 is the intensity of the smallest earthquake, called a **zero-level earthquake,** that can be measured on a seismograph near the earthquake's epicenter. An earthquake with an intensity I_0 has a Richter scale magnitude of $M = \log \dfrac{I}{I_0}$, where I_0 is the measure of a zero-level earthquake.

The earthquake with the largest intensity ever recorded in the continental United States since 1900 occurred in San Francisco, California, in 1906. The intensity I of the quake was approximately $50{,}118{,}000 I_0$. Find the Richter scale magnitude of the earthquake. Round to the nearest tenth.

$$M = \log \dfrac{I}{I_0}$$

$I = 50{,}118{,}000 I_0.$

$$M = \log \dfrac{50{,}118{,}000\, I_0}{I_0}$$

Divide the numerator and denominator by I_0.

$$M = \log 50{,}118{,}000$$

Evaluate $\log 50{,}118{,}000$.

$$M \approx 7.7$$

The San Francisco earthquake had a Richter scale magnitude of 7.7.

> **Take Note**
>
> Note that we do not need to know the value of I_0 in order to determine the Richter scale magnitude of the quake.

If you know the Richter scale magnitude of an earthquake, you can determine the intensity of the earthquake.

In 1964, Prince William Sound, Alaska, experienced what was then the earthquake with the largest magnitude ever to have occurred in the United States. The magnitude of the earthquake was 9.2 on the Richter scale. Find the intensity of this earthquake in terms of I_0. Round to the nearest thousand.

Replace M in $M = \log \dfrac{I}{I_0}$ by 9.2.

$$9.2 = \log \dfrac{I}{I_0}$$

Write the equation in its equivalent exponential form.

$$10^{9.2} = \dfrac{I}{I_0}$$

Multiply by I_0.

$$10^{9.2} I_0 = I$$

Evaluate $10^{9.2}$.

$$1{,}584{,}893{,}000 I_0 \approx I$$

The Prince William Sound earthquake had an intensity that was approximately 1,584,893,000 times the intensity of a zero-level earthquake.

Point of Interest

The Richter scale was created by seismologist Charles F. Richter in 1935. Note that a tenfold increase in the intensity level of an earthquake increases the Richter scale magnitude of the earthquake by only 1.

The percent of light that will pass through a substance is given by the equation **log P = −kd,** where P is the percent of light, as a decimal, passing through the substance, k is a constant that depends on the substance, and d is the thickness of the substance in centimeters.

Find the percent of light that will pass through translucent glass for which $k = 0.4$ and d is 0.5 cm.

Replace k and d in the equation by their given values, and solve for P.

$$\log P = -kd$$
$$\log P = -(0.4)(0.5)$$
$$\log P = -0.2$$

Use the relationship between the logarithmic and exponential functions.

$$P = 10^{-0.2}$$
$$P \approx 0.6310$$

Approximately 63.1% of the light will pass through the glass.

EXAMPLE 2 Find the hydrogen ion concentration, H^+, of orange juice that has a pH of 3.6.

Strategy To find the hydrogen ion concentration, replace pH by 3.6 in the equation $pH = -\log(H^+)$ and solve for H^+.

Solution
$$pH = -\log(H^+)$$
$$3.6 = -\log(H^+)$$
$$-3.6 = \log(H^+)$$
$$10^{-3.6} = H^+$$
$$0.00025 \approx H^+$$

The hydrogen ion concentration is approximately 0.00025.

PROBLEM 2 On September 3, 2000, an earthquake measuring 5.2 on the Richter scale struck the Napa Valley, 50 mi north of San Francisco. Find the intensity of the quake in terms of I_0.

Solution See page S36.

11.5 Concept Review

Determine whether the following statements are always true, sometimes true, or never true.

1. In the exponential growth equation $A = A_0 b^{kt}$, A_0 is the initial size of the growth material.

2. An investor deposits $1000 into an account that earns 8% compounded quarterly. The investor receives interest of $20 every quarter.

3. The equation $A = A_0 e^{kt}$ is an exponential growth equation. An equivalent equation to this exponential growth equation is $t = \dfrac{1}{k} \ln \dfrac{A}{A_0}$.

4. The hydrogen ion concentration of a solution is given by the equation pH $= -\log(H^+)$. Solving this equation for H^+ gives the equivalent equation $H^+ = 10^{-pH}$.

11.5 Exercises

1 Application Problems

1. What is meant by the phrase *exponential decay*?

2. Explain how compound interest differs from simple interest. Is compound interest an example of exponential decay or exponential growth?

Finance For Exercises 3 to 8, use the compound interest formula $P = A(1 + i)^n$, where A is the original value of an investment, i is the interest rate per compounding period, n is the total number of compounding periods, and P is the value of the investment after n periods.

Replace the question marks in Exercises 3 and 4 with the correct number from the problem situation or with the word "unknown."

3. Problem Situation: You invest $3000 in an account that earns 3% annual interest compounded monthly. In approximately how many years will your investment be worth $4000?

In the formula $P = A(1 + i)^n$, $P = $ ____?____, $A = $ ____?____, $i = $ ____?____, and $n = $ ____?____.

4. Problem Situation: A sample of a radioactive isotope with a half-life of 4 h contains 40 mg. How much of the radioactive isotope is present after 24 h?

In the formula $A = A_0\left(\dfrac{1}{2}\right)^{t/k}$, $A = $ ____?____, $A_0 = $ ____?____, $t = $ ____?____, and $k = $ ____?____.

5. An investment broker deposits $1000 into an account that earns 8% annual interest compounded quarterly. What is the value of the investment after 2 years? Round to the nearest dollar.

6. A financial advisor recommends that a client deposit $2500 into a fund that earns 7.5% annual interest compounded monthly. What will be the value of the investment after 3 years? Round to the nearest cent.

7. To save for college tuition, the parents of a preschooler invest $5000 in a bond fund that earns 6% annual interest compounded monthly. In approximately how many years will the investment be worth $15,000? Round to the nearest whole number.

8. A hospital administrator deposits $10,000 into an account that earns 9% annual interest compounded monthly. In approximately how many years will the investment be worth $15,000? Round to the nearest whole number.

Biology For Exercises 9 to 12, use the exponential decay equation $A = A_0\left(\dfrac{1}{2}\right)^{t/k}$,

where A is the amount of a radioactive material present after time t, k is the half-life of the radioactive material, and A_0 is the original amount of radioactive substance. Round to the nearest tenth.

9. An isotope of technetium is used to prepare images of internal body organs. This isotope has a half-life of approximately 6 h. A patient is injected with 30 mg of this isotope.
 a. What will be the amount of technetium in the patient after 3 h?
 b. How long (in hours) will it take for the amount of technetium in the patient to reach 20 mg?

10. Iodine-131 is an isotope that is used to study the functioning of the thyroid gland. This isotope has a half-life of approximately 8 days. A patient is given an injection that contains 8 micrograms of iodine-131.
 a. What will be the amount of iodine in the patient after 5 days?
 b. How long (in days) will it take for the amount of iodine in the patient to reach 5 micrograms?

11. A sample of promethium-147 (used in some luminous paints) contains 25 mg. One year later, the sample contains 18.95 mg. What is the half-life of promethium-147, in years?

12. Francium-223 is a very rare radioactive isotope discovered in 1939 by Marguerite Percy. A 3-microgram sample of francium-223 decays to 2.54 micrograms in 5 min. What is the half-life of francium-223, in minutes?

13. After 6 years, $\dfrac{1}{8}$ of the original amount of a radioactive isotope is left. What is the half-life of the isotope?

14. After 12 h, $\dfrac{1}{16}$ of the original amount of a radioactive isotope is left. What is the half-life of the isotope?

15. Atmospheric Pressure Earth's atmospheric pressure changes as you rise above the surface. At an altitude of h kilometers, where $0 < h < 80$, the pressure P in newtons per square centimeter is approximately modeled by the equation $P(h) = 10.13e^{-0.116h}$.
 a. What is the approximate pressure at 40 km above Earth?
 b. What is the approximate pressure on Earth's surface?
 c. Does atmospheric pressure increase or decrease as you rise above Earth's surface?

16. Demography The number of centenarians (people aged 100 or older) in the United States has been increasing dramatically. Based on data from the U.S. Census Bureau, and using $x = 0$ to correspond to the year 2000, the equation $y = 66.77(1.0532^x)$ approximates the projected U.S. centenarian population through 2050, where y is in thousands.

 a. According to the model, what is the predicted population of centenarians in 2015? Round to the nearest thousand.

 b. In what year does the model predict that the U.S. centenarian population will first exceed 250,000?

Chemistry For Exercises 17 and 18, use the equation $pH = -\log(H^+)$, where H^+ is the hydrogen ion concentration of a solution. Round to the nearest hundredth.

17. Find the pH of milk, for which the hydrogen ion concentration is 3.97×10^{-7}.

18. Find the pH of a baking soda solution, for which the hydrogen ion concentration is 3.98×10^{-9}.

Optics For Exercises 19 and 20, use the equation $\log P = -kd$, which gives the relationship between the percent P, as a decimal, of light passing through a substance of thickness d.

19. The value of k for a swimming pool is approximately 0.05. At what depth, in meters, will the percent of light be 75% of the light at the surface of the pool?

20. The constant k for a piece of blue stained glass is 20. What percent of light will pass through a piece of this glass that is 0.005 m thick?

Acoustics The number of decibels, D, of a sound can be given by the equation $D = 10(\log I + 16)$, where I is the power of a sound measured in watts. Use this equation for Exercises 21 and 22. Round to the nearest whole number.

21. Find the number of decibels of normal conversation. The power of the sound of normal conversation is approximately 3.2×10^{-10} watts.

22. The loudest sound made by any animal is made by the blue whale and can be heard over 500 mi away. The power of the sound is 630 watts. Find the number of decibels of sound emitted by the blue whale.

23. E. coli The actual equation for the population growth of E. coli bacteria mentioned in the novel *The Andromeda Strain* by Michael Crichton is given by $y = 2^{t/3}$, where y is the size of the population in t hours. The mass of Earth is 5.98×10^{27} g, and the mass of one E. coli bacterium is approximately 6.7×10^{-15} g. Assuming that a colony of E. coli bacteria continues to grow as predicted by the equation, in how many hours would the mass of the colony be equal to the mass of Earth? Round to the nearest hour. Is this more or less time than predicted in the novel? (See the Take Note on page 623.)

24. **Electricity** The current, in amperes, in a electric circuit is given by $I = 6(1 - e^{-2t})$, where t is the time in seconds. Using this model, after how many seconds is the current 4 amperes? Round to the nearest hundredth.

25. **Medicine** The intensity I of an x-ray after it has passed through a material that is x centimeters thick is given by $I = I_0 e^{-kx}$, where I_0 is the initial intensity and k is a number that depends on the material. The constant k for copper is 3.2. Find the thickness of copper that is needed so that the intensity of the x-ray after passing through the copper is 25% of the original intensity. Round to the nearest tenth.

26. **Energy** One model for the time it will take for the world's oil supply to be depleted is given by the equation $T = 14.29 \ln (0.00411r + 1)$, where r is the estimated world oil reserves in billions of barrels and T is the time before that amount of oil is depleted. Use this equation to determine how many barrels of oil are necessary to last 20 years. Round to the nearest tenth.

Geology For Exercises 27 to 30, use the Richter scale equation $M = \log \dfrac{I}{I_0}$, where M is the magnitude of an earthquake, I is the intensity of the shock waves, and I_0 is the measure of a zero-level earthquake.

27. The largest earthquake ever recorded occurred on May 22, 1960, off the coast of Chile. The earthquake had an intensity of approximately $I = 3{,}162{,}277{,}000 I_0$. Find the Richter scale magnitude of the earthquake. Round to the nearest tenth.

28. One of the largest earthquakes ever recorded in the United States occurred on February 3, 1965, in the Rat Islands of Alaska. The earthquake had an intensity of approximately $I = 501{,}187{,}000 I_0$. Find the Richter scale magnitude of the earthquake. Round to the nearest tenth.

29. The Kamchatka Peninsula is located off the northeastern coast of Russia. This peninsula has experienced two of the ten largest earthquakes ever recorded. The smaller of the two Kamchatka earthquakes measured 8.5 on the Richter scale. Find the intensity of this earthquake in terms of I_0. Round to the nearest thousand.

30. The magnitude of the Sumatra–Andaman earthquake that occurred on December 26, 2004, off the northern coast of Sumatra was 9.15 on the Richter scale. Find the intensity of the earthquake in terms of I_0.

Geology Shown at the right is a **seismogram,** which is used to measure the magnitude of an earthquake. The magnitude is determined by the amplitude A of the wave and the time t between the occurrence of two types of waves, called primary waves and secondary waves. As you can see on the graph, a primary wave is abbreviated p-wave, and a secondary wave is abbreviated s-wave. The amplitude A of a wave is one-half the difference

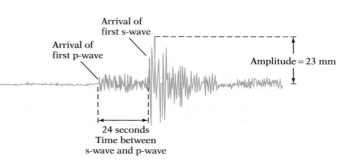

between its highest and lowest points. For this graph, A is 23 millimeters. The equation of the graph is $M = \log A + 3 \log 8t - 2.92$.

31. Determine the magnitude of the earthquake for the seismogram given in the graph. Round to the nearest tenth.

32. Find the magnitude of an earthquake that has a seismogram with an amplitude of 30 mm and for which t is 21 s.

33. Find the magnitude of an earthquake that has a seismogram with an amplitude of 28 mm and for which t is 28 s.

11.5 Applying Concepts

34. Biology At 9 A.M., a culture of bacteria had a population of 1.5×10^6. At noon, the population was 3.0×10^6. If the population is growing exponentially, at what time will the population be 9×10^6? Round to the nearest hour.

35. Inflation If the average annual rate of inflation is 5%, in how many years will prices double? Round to the nearest whole number.

36. Investments The value of an investment in an account that earns an annual interest rate of 10% compounded daily grows according to the equation $A + A_0\left(1 + \dfrac{0.10}{365}\right)^{365t}$. Find the time for the investment to double in value. Round to the nearest whole number.

37. Investments Some banks now use continuous compounding of an amount invested. In this case, the equation that relates the value of an initial investment of A dollars after t years at an annual interest rate of r% is $P = Ae^{rt}$. Using this equation, find the value after 5 years of an investment of $2500 in an account that earns 5% annual interest.

Rocketry When a rock is tossed into the air, the mass of the rock remains constant and a reasonable model for the height of the rock can be given by a quadratic function. However, when a rocket is launched straight up from Earth's surface, the rocket is burning fuel, so the mass of the rocket is always changing. The height of the rocket above Earth can be approximated by the equation

$$y(t) = At - 16t^2 + \frac{A}{k}(M + m - kt)\ln\left(1 - \frac{k}{M + m}t\right)$$

where M is the mass of the rocket, m is the mass of the fuel, A is the rate at which fuel is ejected from the engines, k is the rate at which fuel is burned, t is the time in seconds, and $y(t)$ is the height in feet after t seconds.

38. During the development of the V-2 rocket program in the United States, approximate values for a V-2 rocket were $M = 8000$ lb, $m = 16{,}000$ lb, $A = 8000$ ft/s, and $k = 250$ lb/s.

a. Use a graphing utility to estimate, to the nearest second, the time required for the rocket to reach a height of 1 mi (5280 ft).

b. Use $v(t) = -32t + A \ln\left(\dfrac{M + m}{M + m - kt}\right)$, and the answer to part (a), to determine the velocity of the rocket. Round to the nearest whole number.

c. Determine the domain of the velocity function.

39. Mortgages When you purchase a car or home and make monthly payments on the loan, you are amortizing the loan. Part of each monthly payment is interest on the loan, and the remaining part of the payment is a repayment of the loan amount. The amount remaining to be repaid on the loan after x months is given by $y = A(1 + i)^x + B$, where y is the amount of the loan to be repaid. In this equation, $A = \dfrac{Pi - M}{i}$ and $B = \dfrac{M}{i}$, where P is the original loan amount, i is the monthly interest rate $\left(\dfrac{\text{annual interest rate}}{12}\right)$, and M is the monthly payment. For a 30-year home mortgage of $100,000 with an annual interest rate of 8%, $i = 0.00667$ and $M = 733.76$.

a. How many months are required to reduce the loan amount to $90,000? Round to the nearest month.

b. How many months are required to reduce the loan amount to one-half the original amount? Round to the nearest month.

c. The total amount of interest that is paid after x months is given by $I = Mx + A(1 + i)^x + B - P$. Determine the month in which the total interest paid exceeds $100,000. Round to the nearest month.

40. One scientific study suggested that the *carrying capacity* of Earth is around 10 billion people. What is meant by "carrying capacity"? Find the current world population, and project when Earth's population would reach 10 billion, assuming population growth rates of 1%, 2%, 3%, 4%, and 5%. Find the current rate of world population growth and use that number to determine when, according to your model, the population will reach 10 billion.

41. Radioactive carbon dating can be used only in certain situations. Other methods of radioactive dating, such as the rubidium-strontium method, are appropriate in other situations. Write a report on radioactive dating methods. Include in your report a description of when each form of dating is appropriate.

Focus on Problem Solving

Proof by Contradiction
The four-step plan for solving problems that we have used before is restated here.

1. Understand the problem.

2. Devise a plan.

3. Carry out the plan.

4. Review the solution.

One of the techniques that can be used in the second step is a method called *proof by contradiction.* In this method, you assume that the conditions of the problem you are trying to solve can be met and then show that your assumption leads to a condition you already know is not true.

To illustrate this method, suppose we try to prove that $\sqrt{2}$ is a rational number. We begin by recalling that a rational number is one that can be written as the quotient of integers. Therefore, let a and b be two integers with *no common factors* other than 1. If $\sqrt{2}$ is a rational number, then

$$\sqrt{2} = \frac{a}{b}$$

$$(\sqrt{2})^2 = \left(\frac{a}{b}\right)^2 \qquad \bullet \textbf{ Square each side.}$$

$$2 = \frac{a^2}{b^2}$$

$$2b^2 = a^2 \qquad \bullet \textbf{ Multiply both sides by } b^2.$$

The last equation states that a^2 is an even number. Because a^2 is an even number, a is an even number. Now divide each side of the last equation by 2.

$$2b^2 = a^2$$
$$2b^2 = a \cdot a$$
$$b^2 = a \cdot x \qquad \bullet x = \frac{a}{2}$$

Because a is an even number, $a \cdot x$ is an even number. Because $a \cdot x$ is an even number, b^2 is an even number, and this in turn means that b is an even number. This result, however, contradicts the assumption that a and b are two integers with *no common factors* other than 1. Because this assumption is now known not to be true, we conclude that our original assumption, that $\sqrt{2}$ is a rational number, is false. This proves that $\sqrt{2}$ is an irrational number.

Try a proof by contradiction for the following problem: "Is it possible to write numbers using each of the digits 0, 1, 2, 3, 4, 5, 6, 7, 8, and 9 exactly once so that the sum of the numbers is exactly 100?"[1] Some suggestions for the proof are outlined on the next page.

[1]G. Polya, *How to Solve It: A New Aspect of Mathematical Method,* copyright © 1945 by G. Polya. renewed 1973 Princeton University Press. Reprinted by permission of Princeton University Press.

First note that the sum of the ten digits is 45. This means that some of the digits must be used as tens digits. Let x be the sum of those digits.

a. What is the sum of the remaining units digits?
b. Express "the sum of the units digits and the tens digits equals 100" as an equation. (*Note:* Writing an equation is one of the possible problem-solving strategies.)
c. Solve the equation for x.
d. Explain why this means that it is impossible to satisfy both conditions of the problem.

Projects & Group Activities

Credit Reports and FICO® Scores[2]

When a consumer applies for a loan, the lender generally wants to know the consumer's credit history. For this, the lender turns to a credit reporting agency. These agencies maintain files on millions of borrowers. They provide the lender with a credit report, which lists information such as the types of credit the consumer uses, the length of time the consumer's credit accounts have been open, the amounts owed by the consumer, and whether the consumer has paid his or her bills on time. Along with the credit report, the lender can buy a credit score based on the information in the report. The credit score gives the lender a quick measure of the consumer's credit risk, or how likely the consumer is to repay a debt. It answers the lender's question: "If I lend this person money (or give this person a credit card), what is the probability that I will be paid back in a timely manner?"

One of the most widely used credit bureau scores is the FICO® score, so named because scores are produced from software developed by Fair, Isaac and Company (FICO). FICO scores range from 300 to 850. The higher the score, the lower the predicted credit risk for lenders.

The following graph shows the delinquency rate, or credit risk, associated with ranges of FICO scores. For example, the delinquency rate of consumers in the 550–599 range of scores is 51%. This means that within the next 2 years, for every 100 borrowers in this range, approximately 51 will default on a loan, file for bankruptcy, or fail to pay a credit card bill within 90 days of the due date.

[2]Source: Information provided by Fair, Isaac and Company, Inc. For more information on credit scoring, please visit www.myfico.com. Copyright © 2002 Fair, Isaac and Company, Inc. All Rights Reserved. FAIR, ISAAC, and FICO are registered trademarks of Fair, Isaac and Company, Inc.

Because the delinquency rate depends on the FICO score, we can let the independent variable x represent the FICO score and the dependent variable y represent the delinquency rate. We will use the middle of each range for the values of x. The resulting ordered pairs are recorded in the table at the right.

x	y
400	87
525	71
575	51
625	31
675	15
725	5
775	2
825	1

1. Use your graphing calculator to find a logarithmic equation that approximates these data. Here are instructions from the TI-83, the TI-83 Plus, and the TI-84 Plus:

 Press STAT ENTER. This will bring up a table into which you can enter data. Enter the values of x in L1. Enter the values of y in L2.

 Press STAT again, press the right arrow key to highlight CALC, and arrow down to LnReg (ln for the natural logarithm). Press ENTER twice. The values for a and b in the equation $y = a + b \ln x$ will appear on the screen.

2. Use your equation to predict the delinquency rate of a consumer with a score of 500. Round to the nearest whole number.

3. The equation pairs a delinquency rate of 36% with what score?

4. What is the highest FICO score that will result in a delinquency rate of 71% or more?

5. What is the lowest FICO score that will result in a delinquency rate of 5% or lower?

6. Use the Internet to find the name of at least one of the major credit reporting agencies.

Chapter 11 Summary

Key Words	**Examples**
A function of the form $f(x) = b^x$, where b is a positive real number not equal to 1, is an **exponential function**. The number b is the **base** of the exponential function. [11.1.1, p. 723]	$f(x) = 3^x$; 3 is the base of the function.
The function $f(x) = e^x$ is called the **natural exponential function**. [11.1.1, p. 724]	$f(x) = e^x$; e is an irrational number approximately equal to 2.71828183.
For $x > 0$, $b > 0$, $b \neq 1$, $y = \log_b x$ is equivalent to $x = b^y$. Read $\log_b x$ as "the **logarithm** of x, base b." [11.2.1, p. 733]	$\log_2 8 = 3$ is equivalent to $8 = 2^3$.
Common logarithms are logarithms to the base 10. [11.2.1, p. 736]	$\log_{10} 100 = 2$ is usually written as $\log 100 = 2$.
Natural logarithms are logarithms to the base e. [11.2.1, p. 736]	$\log_e 100 \approx 4.61$ is usually written as $\ln 100 \approx 4.61$.
An **exponential equation** is an equation in which a variable occurs in the exponent. [11.4.1, p. 755]	$2^x = 12$ is an exponential equation.

Essential Rules and Procedures

Equality of Exponents Property For $b > 0$, $b \neq 1$, if $b^u = b^v$, then $u = v$. [11.2.1, p. 734]	If $b^x = b^5$, then $x = 5$.
Definition of Antilogarithm If $\log_b M = N$, the antilogarithm base b of N is M. In exponential form, $M = b^N$. [11.2.1, p. 735]	$\log_4 x = 2$ $4^2 = x$ $16 = x$ The antilogarithm base 4 of 2 is 16.
Product Property of Logarithms For any positive real numbers x, y, and b, $b \neq 1$, $\log_b(xy) = \log_b x + \log_b y$. [11.2.2, p. 736]	$\log_b(3x) = \log_b 3 + \log_b x$
Quotient Property of Logarithms For any positive real numbers x, y, and b, $b \neq 1$, $\log_b \dfrac{x}{y} = \log_b x - \log_b y$. [11.2.2, p. 737]	$\log_b \dfrac{x}{20} = \log_b x - \log_b 20$
Power Property of Logarithms For any positive real numbers x and b, $b \neq 1$, and for any real number r, $\log_b x^r = r \log_b x$. [11.2.2, p. 738]	$\log_b x^5 = 5 \log_b x$

Logarithmic Property of One
For any positive real number b, $b \neq 1$, $\log_b 1 = 0$.
[11.2.2, p. 740]

$\log_6 1 = 0$

Inverse Property of Logarithms
For any positive real numbers x and b, $b \neq 1$, $\log_b b^x = x$.
[11.2.2, p. 740]

$\log_3 3^4 = 4$

1–1 Property of Logarithms
For any positive real numbers x, y, and b, $b \neq 1$, if
$\log_b x = \log_b y$, then $x = y$. [11.2.2, p. 740]

If $\log_5(x - 2) = \log_5 3$, then $x - 2 = 3$.

Change-of-Base Formula
$\log_a N = \dfrac{\log_b N}{\log_b a}$ [11.2.2, p. 741]

$\log_3 12 = \dfrac{\log 12}{\log 3}$ $\log_6 16 = \dfrac{\ln 16}{\ln 6}$

Chapter 11 Review Exercises

1. Evaluate: $\log_4 16$

2. Write $\frac{1}{2}(\log_3 x - \log_3 y)$ as a single logarithm with a coefficient of 1.

3. Evaluate the function $f(x) = e^{x-2}$ at $x = 2$.

4. Solve for x: $8^x = 2^{x-6}$

5. Evaluate $f(x) = \left(\dfrac{2}{3}\right)^x$ at $x = 0$.

6. Solve for x: $\log_3 x = -2$

7. Write $2^5 = 32$ in logarithmic form.

8. Solve for x: $\log x + \log(x - 4) = \log 12$

9. Write $\log_6 \sqrt{xy^3}$ in expanded form.

10. Solve for x: $4^{5x-2} = 4^{3x+2}$

11. Solve for x: $3^{7x+1} = 3^{4x-5}$

12. Evaluate $f(x) = 3^{x+1}$ at $x = -2$.

13. Evaluate: $\log_2 16$

14. Solve for x: $\log_6 2x = \log_6 2 + \log_6(3x - 4)$

15. Evaluate $\log_2 5$. Round to the nearest ten-thousandth.

16. Evaluate $\log_6 22$. Round to the nearest ten-thousandth.

17. Solve for x: $4^x = 8^{x-1}$

18. Evaluate $f(x) = \left(\dfrac{1}{4}\right)^x$ at $x = -1$.

19. Write $\log_5 \sqrt{\dfrac{x}{y}}$ in expanded form.

20. Solve for x: $\log_5 \dfrac{7x + 2}{3x} = 1$

21. Solve for x: $\log_5 x = 3$

22. Solve for x: $\log x + \log(2x + 3) = \log 2$

23. Write $3 \log_b x - 5 \log_b y$ as a single logarithm with a coefficient of 1.

24. Evaluate $f(x) = 2^{-x-1}$ at $x = -3$.

25. Evaluate $\log_3 19$. Round to the nearest ten-thousandth.

26. Solve $3^{x+2} = 5$ for x. Round to the nearest ten-thousandth.

27. Graph: $f(x) = 2^x - 3$

28. Graph: $f(x) = \left(\dfrac{1}{2}\right)^x + 1$

29. Graph: $f(x) = \log_2(2x)$

30. Graph: $f(x) = \log_2 x - 1$

31. Graph $f(x) = \dfrac{\log x}{x}$ and approximate the zeros of f to the nearest tenth.

32. **Radioactivity** Use the exponential decay equation $A = A_0 \left(\dfrac{1}{2}\right)^{\frac{t}{k}}$, where A is the amount of a radioactive material present after time t, k is the half-life, and A_0 is the original amount of radioactive material, to find the half-life of a material that decays from 10 mg to 9 mg in 5 h. Round to the nearest whole number.

33. **Optics** The percent of light that will pass through a translucent material is given by the equation $\log P = -0.5d$, where P is the percent of light, as a decimal, that passes through the material, and d is the thickness of the material in centimeters. How thick must this translucent material be so that only 50% of the light that is incident to the material will pass through it? Round to the nearest thousandth.

Chapter 11 Test

1. Evaluate $f(x) = \left(\dfrac{3}{4}\right)^x$ at $x = 0$.

2. Evaluate $f(x) = 4^{x-1}$ at $x = -2$.

3. Evaluate: $\log_4 64$

4. Solve for x: $\log_4 x = -2$

5. Write $\log_6 \sqrt[3]{x^2 y^5}$ in expanded form.

6. Write $\dfrac{1}{2}(\log_5 x - \log_5 y)$ as a single logarithm with a coefficient of 1.

7. Solve for x: $\log_6 x + \log_6(x - 1) = 1$

8. Evaluate $f(x) = 3^{x+1}$ at $x = -2$.

9. Solve for x: $3^x = 17$

10. Solve for x: $\log_2 x + 3 = \log_2(x^2 - 20)$

11. Solve for x: $5^{6x-2} = 5^{3x+7}$

12. Solve for x: $4^x = 2^{3x+4}$

13. Solve for x: $\log(2x + 1) + \log x = \log 6$

14. Graph: $f(x) = 2^x - 1$

15. Graph: $f(x) = 2^x + 2$

16. Graph: $f(x) = \log_2(3x)$

17. Graph: $f(x) = \log_3(x + 1)$

18. Graph $f(x) = 3 - 2^x$ and approximate the zeros of f to the nearest tenth.

19. Investments Find the value of $10,000 invested for 6 years at 7.5% compounded monthly. Use the compound interest formula $P = A(1 + i)^n$, where A is the original value of the investment, i is the interest rate per compounding period, and n is the number of compounding periods. Round to the nearest dollar.

20. Radioactivity Use the exponential decay equation $A = A_0\left(\dfrac{1}{2}\right)^{\frac{t}{k}}$ where A is the amount of a radioactive material present after time t, k is the half-life, and A_0 is the original amount of radioactive material, to find the half-life of a material that decays from 40 mg to 30 mg in 10 h. Round to the nearest whole number.

Cumulative Review Exercises

1. Solve: $4 - 2[x - 3(2 - 3x) - 4x] = 2x$

2. Solve $S = 2WH + 2WL + 2LH$ for L.

3. Solve: $|2x - 5| \le 3$

4. Factor: $4x^{2n} + 7x^n + 3$

5. Solve: $x^2 + 4x - 5 \le 0$

6. Simplify: $\dfrac{1 - \dfrac{5}{x} + \dfrac{6}{x^2}}{1 + \dfrac{1}{x} - \dfrac{6}{x^2}}$

7. Simplify: $\dfrac{\sqrt{xy}}{\sqrt{x} - \sqrt{y}}$

8. Simplify: $y\sqrt{18x^5y^4} - x\sqrt{98x^3y^6}$

9. Simplify: $\dfrac{i}{2 - i}$

10. Find the equation of the line that contains the point $(2, -2)$ and is parallel to the line $2x - y = 5$.

11. Write a quadratic equation that has integer coefficients and has as solutions $\frac{1}{3}$ and -3.

12. Solve: $x^2 - 4x - 6 = 0$

13. Find the range of $f(x) = x^2 - 3x - 4$ if the domain is $\{-1, 0, 1, 2, 3\}$.

14. Given $f(x) = x^2 + 2x + 1$ and $g(x) = 2x - 3$, find $f[g(0)]$.

15. Solve by the addition method:
$$3x - y + z = 3$$
$$x + y + 4z = 7$$
$$3x - 2y + 3z = 8$$

16. Solve: $y = -2x - 3$
$y = 2x - 1$

17. Evaluate $f(x) = 3^{-x+1}$ at $x = -4$.

18. Solve for x: $\log_4 x = 3$

19. Solve for x: $2^{3x+2} = 4^{x+5}$

20. Solve for x: $\log x + \log(3x + 2) = \log 5$

21. Graph: $\{x \mid x < 0\} \cap \{x \mid x > -4\}$

22. Graph the solution set of $\dfrac{x + 2}{x - 1} \ge 0$.

23. Graph: $y = -x^2 - 2x + 3$

24. Graph: $f(x) = |x| - 2$

25. Graph: $f(x) = \left(\dfrac{1}{2}\right)^x - 1$

26. Graph: $f(x) = \log_2 x + 1$

27. Metallurgy An alloy containing 25% tin is mixed with an alloy containing 50% tin. How much of each were used to make 2000 lb of an alloy containing 40% tin?

28. Compensation An account executive earns $500 per month plus 8% commission on the amount of sales. The executive's goal is to earn a minimum of $3000 per month. What amount of sales will enable the executive to earn $3000 or more per month?

29. Education The table below shows the actual and projected number of high school graduates, in thousands, for selected years. (*Source:* Department of Education, Office of Policy Research and Improvement.) Draw a scatter diagram of these data.

Year, x	High School Graduates (in thousands)
2004	124
2005	132
2006	133
2007	134
2008	135
2009	139
2010	141

30. Work Problem A new printer can print checks three times faster than an older printer. The old printer can print the checks in 30 min. How long would it take to print the checks with both printers operating?

31. Chemistry For a constant temperature, the pressure (*P*) of a gas varies inversely as the volume (*V*). If the pressure is 50 lb/in² when the volume is 250 ft³, find the pressure when the volume is 25 ft³.

32. Purchasing A contractor buys 45 yd of nylon carpet and 30 yd of wool carpet for $2340. A second purchase, at the same prices, includes 25 yd of nylon carpet and 80 yd of wool carpet for $2820. Find the cost per yard of the wool carpet.

33. Investments An investor deposits $10,000 into an account that earns 9% interest compounded monthly. What is the value of the investment after 5 years? Use the compound interest formula $P = A(1 + i)^n$, where *A* is the original value of an investment, *i* is the interest rate per compounding period, *n* is the total number of compounding periods, and *P* is the value of the investment after *n* periods. Round to the nearest dollar.

R

Review of Beginning Algebra Topics

Online Study Center

DVD

SSM

Need help? For online student resources, such as section quizzes, visit the Online Study Center at college.hmco.com/pic/aufmannABI2e.

R.1 Variable Expressions

OBJECTIVE 1 **Evaluate variable expressions**

Whenever an expression contains more than one operation, the operations must be performed in a specified order, as listed below in the Order of Operations Agreement.

1.4.2* Order of Operations Agreement

The Order of Operations Agreement

Step 1 Perform operations inside grouping symbols. Grouping symbols include parentheses (), brackets [], braces { }, the absolute value symbol ||, and fraction bars.

Step 2 Simplify exponential expressions.

Step 3 Do multiplication and division as they occur from left to right.

Step 4 Do addition and subtraction as they occur from left to right.

EXAMPLE 1 Evaluate: $-2(7 - 3)^2 + 4 - 2(5 - 2)$

Solution $-2(7 - 3)^2 + 4 - 2(5 - 2)$

$= -2(4)^2 + 4 - 2(3)$ • **Perform operations inside parentheses.**

$= -2(16) + 4 - 2(3)$ • **Simplify the exponential expression.**

$= -32 + 4 - 2(3)$ • **Do the multiplication and division from left to right.**

$= -32 + 4 - 6$

$= -28 - 6$ • **Do the addition and subtraction from left to right.**

$= -28 + (-6) = -34$

PROBLEM 1 Evaluate: $(-4)(6 - 8)^2 - (-12 \div 4)$

Solution See page S36.

1.5.1 Evaluate variable expressions

A **variable** is a letter that represents a quantity that is unknown or that can change or vary. An expression that contains one or more variables is a **variable expression.** $3x - 4y + 7z$ is a variable expression. It contains the variables x, y, and z.

Replacing a variable in a variable expression by a number and then simplifying the resulting numerical expression is called **evaluating the variable expression.** The number substituted for the variable is called the **value of the variable.** The result is called the **value of the variable expression.**

*Review this objective for more detailed coverage of this topic.

EXAMPLE 2 Evaluate $5ab^3 + 2a^2b^2 - 4$ when $a = 3$ and $b = -2$.

Solution $5ab^3 + 2a^2b^2 - 4$

$5(3)(-2)^3 + 2(3)^2(-2)^2 - 4$ • **Replace a by 3 and b by -2.**

$= 5(3)(-8) + 2(9)(4) - 4$ • **Use the Order of Operations Agreement to simplify the numerical expression.**

$= -120 + 72 - 4$

$= -48 - 4$

$= -48 + (-4)$

$= -52$

PROBLEM 2 Evaluate $3xy^2 - 3x^2y$ when $x = -2$ and $y = 5$.

Solution See page S37.

OBJECTIVE 2 Simplify variable expressions

The Properties of Real Numbers are used to simplify variable expressions.

1.5.3
1.5.4 Simplify variable expressions using the Properties of Addition and Multiplication

EXAMPLE 3 Simplify: $-\dfrac{1}{3}(-3y)$

Solution $-\dfrac{1}{3}(-3y)$

$= \left[-\dfrac{1}{3}(-3)\right]y$ • **Use the Associative Property of Multiplication to regroup factors.**

$= 1y$ • **Use the Inverse Property of Multiplication.**

$= y$ • **Use the Multiplication Property of One.**

PROBLEM 3 Simplify: $-5(-3a)$

Solution See page S37.

A variable expression is shown at the right. The expression can be rewritten by writing subtraction as addition of the opposite. Note that the expression has four addends. The **terms** of a variable expression are the addends of the expression. The expression has four terms.

$3x^2 - 4xy + 5z - 2$

$3x^2 + (-4xy) + 5z + (-2)$

Four terms

$\underbrace{3x^2 - 4xy + 5z}_{\text{Variable terms}} \underbrace{- 2}_{\text{Constant term}}$

The terms $3x^2$, $-4xy$, and $5z$ are **variable terms.** The term -2 is a **constant term,** or simply a **constant.**

Like terms of a variable expression are terms that have the same variable parts. The terms $3x$ and $-7x$ are like terms. Constant terms are also like terms. Thus -6 and 9 are like terms.

Like terms

$$3x^2 + 3x - 6 - 7x + 9$$

Like terms

By using the Commutative Property of Multiplication, we can rewrite the Distributive Property as $ba + ca = (b + c)a$. This form of the Distributive Property is used to **combine like terms** of a variable expression by adding their coefficients. For instance,

$$7x + 9x = (7 + 9)x \quad \text{• Use the Distributive Property: } \textbf{\textit{ba} + \textit{ca} = (\textit{b} + \textit{c})\textit{a}.}$$
$$= 16x$$

EXAMPLE 4 Simplify: $4x^2 + 5x - 6x^2 - 7x$

Solution $4x^2 + 5x - 6x^2 - 7x$

$$= 4x^2 - 6x^2 + 5x - 7x \qquad \text{• Use the Associative and}$$
$$= (4x^2 - 6x^2) + (5x - 7x) \qquad \quad \text{Commutative Properties of}$$
$$\qquad\qquad\qquad\qquad\qquad\qquad \text{Addition to rearrange and}$$
$$\qquad\qquad\qquad\qquad\qquad\qquad \text{group like terms.}$$
$$= -2x^2 + (-2x) \qquad\qquad \text{• Use the Distributive Property}$$
$$\qquad\qquad\qquad\qquad\qquad\qquad \text{to combine like terms.}$$
$$= -2x^2 - 2x$$

PROBLEM 4 Simplify: $2z^2 - 5z - 3z^2 + 6z$

Solution See page S37.

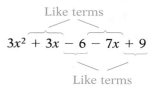

1.5.5 Simplify variable expressions using the Distributive Property

The Distributive Property is also used to remove parentheses from a variable expression. Here is an example.

$$4(2x + 5z) = 4(2x) + 4(5z) \qquad \text{• Use the Distributive Property:}$$
$$\qquad\qquad\qquad\qquad\qquad\qquad \textbf{\textit{a}(\textit{b} + \textit{c}) = \textit{ab} + \textit{ac}}$$
$$= (4 \cdot 2)x + (4 \cdot 5)z \qquad \text{• Use the Associative Property of}$$
$$\qquad\qquad\qquad\qquad\qquad\qquad \text{Multiplication to regroup factors.}$$
$$= 8x + 20z \qquad\qquad \text{• Multiply 4 · 2 and 4 · 5.}$$

The Distributive Property can be extended to expressions containing more than two terms. For instance,

$$4(2x + 3y + 5z) = 4(2x) + 4(3y) + 4(5z)$$
$$= 8x + 12y + 20z$$

EXAMPLE 5 Simplify. **a.** $-3(2x + 4)$ **b.** $6(3x - 4y + z)$

Solution **a.** $-3(2x + 4) = -3(2x) + (-3)(4)$
$$= -6x - 12$$
 b. $6(3x - 4y + z)$
$$= 6(3x) - 6(4y) + 6(z)$$
$$= 18x - 24y + 6z$$

PROBLEM 5 Simplify. **a.** $-3(5y - 2)$ **b.** $-2(4x + 2y - 6z)$

Solution See page S37.

Take Note

Recall that the Distributive Property states that if a, b, and c are real numbers, then

$$a(b + c) = ab + ac$$

To simplify the expression $5 + 3(4x - 2)$, use the Distributive Property to remove the parentheses.

$$5 + 3(4x - 2) = 5 + 3(4x) - 3(2)$$ • Use the Distributive Property.
$$= 5 + 12x - 6$$
$$= 12x - 1$$ • Add the like terms 5 and −6.

EXAMPLE 6 Simplify.
 a. $3(2x - 4) - 5(3x + 2)$
 b. $3a - 2[7a - 2(2a + 1)]$

Solution **a.** $3(2x - 4) - 5(3x + 2)$
 $$= 6x - 12 - 15x - 10$$ • Use the Distributive Property twice.
 $$= -9x - 22$$ • Combine like terms.

 b. $3a - 2[7a - 2(2a + 1)]$
 $$= 3a - 2[7a - 4a - 2]$$ • Use the Distributive Property to remove the parentheses.
 $$= 3a - 2[3a - 2]$$ • Combine like terms.
 $$= 3a - 6a + 4$$ • Use the Distributive Property to remove the brackets.
 $$= -3a + 4$$ • Combine like terms.

PROBLEM 6 Simplify.
 a. $7(-3x - 4y) - 3(3x + y)$
 b. $2y - 3[5 - 3(3 + 2y)]$

Solution See page S37.

R.1 Exercises

1 Evaluate variable expressions

Evaluate the variable expression when $a = 2$, $b = 3$, and $c = -4$.

1. $a - 2c$ **2.** $-3a + 4b$ **3.** $3b - 3c$

4. $-3c + 4$ **5.** $16 \div (2c)$ **6.** $6b \div (-a)$

7. $3b - (a + c)^2$ **8.** $(a - b)^2 + 2c$ **9.** $(b - 3a)^2 + bc$

Evaluate the variable expression when $a = -1$, $b = 3$, $c = -2$, and $d = 4$.

10. $\dfrac{b - a}{d}$ **11.** $\dfrac{d - b}{a}$ **12.** $\dfrac{2d + b}{-a}$

13. $\dfrac{b - d}{c - a}$ **14.** $2(b + c) - 2a$ **15.** $3(b - a) - bc$

16. $\dfrac{-4bc}{2a + c}$

17. $\dfrac{abc}{b - d}$

18. $(d - a)^2 - (b - c)^2$

19. $(-b + d)^2 + (-a + c)^2$

20. $4ab + (2c)^2$

21. $3cd - (4a)^2$

Evaluate the variable expression when $a = 2.7$, $b = -1.6$, and $c = -0.8$.

22. $c^2 - ab$

23. $(a + b)^2 - c$

24. $\dfrac{b^3}{c} - 4a$

2 Simplify variable expressions

Simplify each of the following.

25. $x + 7x$

26. $12y + 9y$

27. $8b - 5b$

28. $4y - 11y$

29. $-12a + 17a$

30. $-15xy + 7xy$

31. $4x + 5x + 2x$

32. $-5x^2 - 10x^2 + x^2$

33. $6x - 2y + 9x$

34. $3x - 7y - 6x + 4x$

35. $5a + 6a - 2a$

36. $2a - 5a + 3a$

37. $12y^2 + 10y^2$

38. $2z^2 - 9z^2$

39. $\dfrac{3}{4}x - \dfrac{1}{4}x$

40. $\dfrac{2}{5}y - \dfrac{3}{5}y$

41. $-4(5x)$

42. $-2(-8y)$

43. $(6a)(-4)$

44. $-5(7x^2)$

45. $\dfrac{1}{4}(4x)$

46. $\dfrac{12x}{5}\left(\dfrac{5}{12}\right)$

47. $\dfrac{1}{3}(21x)$

48. $-\dfrac{5}{8}(24a^2)$

49. $(36y)\left(\dfrac{1}{12}\right)$

50. $-(z + 4)$

51. $-3(a + 5)$

52. $(4 - 3b)9$

53. $(-2x - 6)8$

54. $3(5x^2 + 2x)$

55. $-5(2y^2 - 1)$

56. $4(x^2 - 3x + 5)$

57. $6(3x^2 - 2xy - y^2)$

58. $5a - (4a + 6)$

59. $3 - (10 + 8y)$

60. $12(y - 2) + 3(7 - 4y)$

61. $-5[2x + 3(5 - x)]$

62. $-3[2x - (x + 7)]$

63. $-5a - 2[2a - 4(a + 7)]$

 Equations and Inequalities

OBJECTIVE **1** **First-degree equations in one variable**

An **equation** expresses the equality of two mathematical expressions. Each of the equations that follows is a **first-degree equation in one variable.** *First degree* means that the variable has an exponent of 1.

$$x + 11 = 14$$
$$3a + 5 = 8a$$
$$2(6y - 1) = 3$$

A **solution** of an equation is a number that, when substituted for the variable, results in a true equation.

3 is a solution of the equation $x + 4 = 7$ because $3 + 4 = 7$.

9 is not a solution of the equation $x + 4 = 7$ because $9 + 4 \neq 7$.

To **solve an equation** means to find a solution of the equation. In solving an equation, the goal is to rewrite the given equation with the variable alone on one side of the equation and a constant term on the other side of the equation; the constant term is the solution of the equation. The following properties of equations are used to rewrite equations in this form.

Properties of Equations

Addition Property of Equations

The same number can be added to each side of an equation without changing the solution of the equation. In symbols, the equation $a = b$ has the same solution as the equation $a + c = b + c$.

Multiplication Property of Equations

Each side of an equation can be multiplied by the same nonzero number without changing the solution of the equation. In symbols, if $c \neq 0$, then the equation $a = b$ has the same solution as the equation $ac = bc$.

2.1.2 Solve equations using the Addition and Multiplication
2.1.3 Properties of Equations

Take Note

Subtraction is defined as addition of the opposite.

$$a - b = a + (-b)$$

The Addition Property of Equations is used to remove a term from one side of the equation by adding the opposite of that term to each side of the equation. Because subtraction is defined in terms of addition, the Addition Property of Equations also makes it possible to subtract the same number from each side of an equation without changing the solution of the equation.

For example, to solve the equation $t + 9 = -4$, subtract the constant term (9) from each side of the equation.

$$t + 9 = -4$$
$$t + 9 - 9 = -4 - 9$$
$$t = -13$$

Now the variable is alone on one side of the equation, and a constant term (-13) is on the other side. The solution is the constant. The solution is -13.

2.2.3 Solve general equations

Take Note

Division is defined as multiplication by the reciprocal.

$$a \div b = a \cdot \frac{1}{b}$$

Take Note

When using the Multiplication Property of Equations, multiply each side of the equation by the reciprocal of the coefficient when the coefficient is a fraction. Divide each side of the equation by the coefficient when the coefficient is an integer or a decimal.

The Multiplication Property of Equations is used to remove a coefficient by multiplying each side of the equation by the reciprocal of the coefficient. Because division is defined in terms of multiplication, each side of an equation can be divided by the same nonzero number without changing the solution of the equation.

For example, to solve the equation $-5q = 120$, divide each side of the equation by the coefficient -5.

$$-5q = 120$$
$$\frac{-5q}{-5} = \frac{120}{-5}$$
$$q = -24$$

Now the variable is alone on one side of the equation, and a constant (-24) is on the other side. The solution is the constant. The solution is -24.

In solving more complicated first-degree equations in one variable, use the following sequence of steps.

Steps for Solving a First-Degree Equation in One Variable

1. Use the Distributive Property to remove parentheses.

2. Combine any like terms on the right side of the equation and any like terms on the left side of the equation.

3. Use the Addition Property to rewrite the equation with only one variable term.

4. Use the Addition Property to rewrite the equation with only one constant term.

5. Use the Multiplication Property to rewrite the equation with the variable alone on one side of the equation and a constant on the other side of the equation.

If one of the above steps is not needed to solve a given equation, proceed to the next step.

EXAMPLE 1 Solve.
 a. $5x + 9 = 23 - 2x$
 b. $8x - 3(4x - 5) = -2x + 6$

Solution **a.**
$$5x + 9 = 23 - 2x$$
$$5x + 2x + 9 = 23 - 2x + 2x \quad \bullet \textbf{ Step 3}$$
$$7x + 9 = 23$$
$$7x + 9 - 9 = 23 - 9 \quad \bullet \textbf{ Step 4}$$
$$7x = 14$$
$$\frac{7x}{7} = \frac{14}{7} \quad \bullet \textbf{ Step 5}$$
$$x = 2$$

The solution is 2.

b. $8x - 3(4x - 5) = -2x + 6$

$8x - 12x + 15 = -2x + 6$ • **Step 1**

$-4x + 15 = -2x + 6$ • **Step 2**

$-4x + 2x + 15 = -2x + 2x + 6$ • **Step 3**

$-2x + 15 = 6$

$-2x + 15 - 15 = 6 - 15$ • **Step 4**

$-2x = -9$

$\dfrac{-2x}{-2} = \dfrac{-9}{-2}$ • **Step 5**

$x = \dfrac{9}{2}$

The solution is $\dfrac{9}{2}$.

PROBLEM 1 Solve. **a.** $4x + 3 = 7x + 9$
b. $4 - (5x - 8) = 4x + 3$

Solution See page S37.

OBJECTIVE **2** **Inequalities in one variable**

An **inequality** contains the symbol $>$, $<$, $\geq$, or $\leq$. An inequality expresses the relative order of two mathematical expressions. Here are some examples of inequalities in one variable.

$$\left.\begin{array}{l} 4x \geq 12 \\ 2x + 7 \leq 9 \\ x^2 + 1 > 3x \end{array}\right\} \quad \text{Inequalities in one variable}$$

2.7.2 Solve an inequality in one variable

A **solution of an inequality in one variable** is a number that, when substituted for the variable, results in a true inequality. For the inequality $x < 4$, 3, 0, and -5 are solutions of the inequality because replacing the variable by these numbers results in a true inequality.

$x < 4$	$x < 4$	$x < 4$
$3 < 4$ True	$0 < 4$ True	$-5 < 4$ True

The number 7 is not a solution of the inequality $x < 4$ because $7 < 4$ is a false inequality.

Besides the numbers 3, 0, and -5, there are an infinite number of other solutions of the inequality $x < 4$. Any number less than 4 is a solution; for instance, -5.2, $\dfrac{5}{2}$, π, and 1 are also solutions of the inequality. The set of all the solutions of an inequality is called the **solution set of the inequality.** The solution set of the inequality $x < 4$ is written in set-builder notation as $\{x \,|\, x < 4\}$. This is read "the set of all x such that x is less than 4." It is written in interval notation as $(-\infty, 4)$.

The graph of the solution set of $x < 4$ is shown at the right.

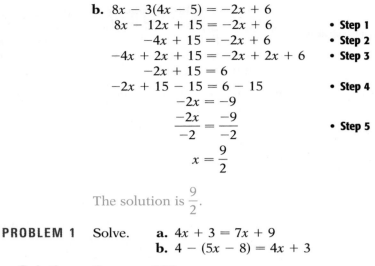

$$\begin{array}{cccccccccccc} & & & & & & & & & & & \\ \hline -5 & -4 & -3 & -2 & -1 & 0 & 1 & 2 & 3 & 4 & 5 \end{array}$$

In solving an inequality, the goal is to rewrite the given inequality in the form

$$variable < constant \quad \text{or} \quad variable > constant$$

The Addition Property of Inequalities is used to rewrite an inequality in this form.

The Addition Property of Inequalities

If $a > b$ and c is a real number, then the inequalities $a > b$ and $a + c > b + c$ have the same solution set.

If $a < b$ and c is a real number, then the inequalities $a < b$ and $a + c < b + c$ have the same solution set.

The Addition Property of Inequalities is also true for the symbols $\leq$ and $\geq$.

The Addition Property of Inequalities is used to remove a term from one side of an inequality by adding the additive inverse of that term to each side of the inequality. Because subtraction is defined in terms of addition, the same number can be subtracted from each side of an inequality without changing the solution set of the inequality.

As shown in the following example, the Addition Property of Inequalities applies to variable terms as well as to constants.

Solve $4x - 5 \leq 3x - 2$. Write the solution set in interval notation.

$$4x - 5 \leq 3x - 2$$
$$4x - 3x - 5 \leq 3x - 3x - 2 \qquad \text{• Subtract } 3x \text{ from each side of the inequality.}$$
$$x - 5 \leq -2 \qquad \text{• Simplify.}$$
$$x - 5 + 5 \leq -2 + 5 \qquad \text{• Add 5 to each side of the inequality.}$$
$$x \leq 3 \qquad \text{• Simplify.}$$

The solution set is $(-\infty, 3]$.

When we multiply or divide an inequality by a number, the inequality symbol may be reversed, depending on whether the number is positive or negative. Look at the following two examples.

$$3 < 5$$
$$2(3) < 2(5) \qquad \text{• Multiply by positive 2.}$$
$$6 < 10 \qquad \text{The inequality symbol remains the same. } 6 < 10 \text{ is a true statement.}$$

$$3 < 5$$
$$-2(3) > -2(5) \qquad \text{• Multiply by negative 2.}$$
$$-6 > -10 \qquad \text{The inequality symbol is reversed in order to make the inequality a true statement.}$$

This is summarized in the Multiplication Property of Inequalities.

The Multiplication Property of Inequalities

Rule 1

If $a > b$ and $c > 0$, then $ac > bc$.

If $a < b$ and $c > 0$, then $ac < bc$.

Rule 2

If $a > b$ and $c < 0$, then $ac < bc$.

If $a < b$ and $c < 0$, then $ac > bc$.

Rule 1 states that when each side of an inequality is multiplied by a positive number, the inequality symbol remains the same. Rule 2 states that when each side of an inequality is multiplied by a negative number, the inequality symbol must be reversed.

Here are a few more examples of this property.

Rule 1		**Rule 2**	
$-4 < -2$	$5 > -3$	$-1 < 7$	$-2 > -6$
$2(-4) < 2(-2)$	$3(5) > 3(-3)$	$-2(-1) > -2(7)$	$-3(-2) < -3(-6)$
$-8 < -4$	$15 > -9$	$2 > -14$	$6 < 18$

Use the Multiplication Property of Inequalities to remove a coefficient other than 1 from one side of an inequality so that the inequality can be rewritten with the variable alone on one side of the inequality and a constant term on the other side. The Multiplication Property of Inequalities is also true for the symbols $\leq$ and $\geq$.

Because division is defined in terms of multiplication, when each side of an inequality is divided by a positive number, the inequality symbol remains the same. When each side of an inequality is divided by a negative number, the inequality symbol must be reversed.

Take Note

Solving inequalities in one variable is similar to solving equations in one variable, except that when you multiply or divide by a negative number, you must reverse the inequality symbol.

Solve $-3x < 9$. Write the solution set in set-builder notation.

$$-3x < 9$$

$$\frac{-3x}{-3} > \frac{9}{-3}$$ • **Divide each side of the inequality by the coefficient -3 and reverse the inequality symbol.**

$$x > -3$$ • **Simplify.**

$$\{x \mid x > -3\}$$ • **Write the answer in set-builder notation.**

EXAMPLE 2 Solve $x + 3 > 4x + 6$. Write the solution set in set-builder notation.

Solution

$$x + 3 > 4x + 6$$

$$x - 4x + 3 > 4x - 4x + 6$$ • **Subtract $4x$ from each side.**

$$-3x + 3 > 6$$

$$-3x + 3 - 3 > 6 - 3$$ • **Subtract 3 from each side.**
$$-3x > 3$$
$$\frac{-3x}{-3} < \frac{3}{-3}$$ • **Divide each side by −3. Reverse the symbol.**
$$x < -1$$
$$\{x \mid x < -1\}$$

PROBLEM 2 Solve $3x - 1 \leq 5x - 7$. Write the solution set in set-builder notation.

■ **Solution** See page S37.

When an inequality contains parentheses, often the first step in solving the inequality is to use the Distributive Property to remove the parentheses.

EXAMPLE 3 Solve $-2(x - 7) > 3 - 4(2x - 3)$. Write the solution set in interval notation.

Solution
$$-2(x - 7) > 3 - 4(2x - 3)$$
$$-2x + 14 > 3 - 8x + 12$$ • **Use the Distributive Property.**
$$-2x + 14 > 15 - 8x$$ • **Combine like terms.**
$$-2x + 8x + 14 > 15 - 8x + 8x$$ • **Add 8x to each side.**
$$6x + 14 > 15$$
$$6x + 14 - 14 > 15 - 14$$ • **Subtract 14 from each side.**
$$6x > 1$$
$$\frac{6x}{6} > \frac{1}{6}$$ • **Divide each side by 6.**
$$x > \frac{1}{6}$$
$$\left(\frac{1}{6}, \infty\right)$$

PROBLEM 3 Solve $3 - 2(3x + 1) \leq 7 - 2x$. Write the solution set in interval notation.

■ **Solution** See page S37.

R.2 Exercises

1 First-degree equations in one variable

Solve.

1. $x + 7 = -5$ **2.** $9 + b = 21$ **3.** $-9 = z - 8$

4. $b - 11 = 11$ **5.** $-48 = 6z$ **6.** $-9a = -108$

7. $-\dfrac{3}{4}x = 15$

8. $\dfrac{5}{2}x = -10$

9. $-\dfrac{x}{4} = -2$

10. $\dfrac{2x}{5} = -8$

11. $4 - 2b = 2 - 4b$

12. $4y - 10 = 6 + 2y$

13. $5x - 3 = 9x - 7$

14. $3m + 5 = 2 - 6m$

15. $6a - 1 = 2 + 2a$

16. $5x + 7 = 8x + 5$

17. $2 - 6y = 5 - 7y$

18. $4b + 15 = 3 - 2b$

19. $2(x + 1) + 5x = 23$

20. $9n - 15 = 3(2n - 1)$

21. $7a - (3a - 4) = 12$

22. $5(3 - 2y) = 3 - 4y$

23. $9 - 7x = 4(1 - 3x)$

24. $2(3b + 5) - 1 = 10b + 1$

25. $2z - 2 = 5 - (9 - 6z)$

26. $4a + 3 = 7 - (5 - 8a)$

27. $5(6 - 2x) = 2(5 - 3x)$

28. $4(3y + 1) = 2(y - 8)$

29. $2(3b - 5) = 4(6b - 2)$

30. $3(x - 4) = 1 - (2x - 7)$

2 Inequalities in one variable

Solve. Write the answer in set-builder notation.

31. $x - 5 > -2$

32. $5 + n \geq 4$

33. $-2 + n \geq 0$

34. $x - 3 < 2$

35. $8x \leq -24$

36. $-4x < 8$

37. $3n > 0$

38. $-2n \leq -8$

39. $2x - 1 > 7$

40. $5x - 2 \leq 8$

41. $4 - 3x < 10$

42. $7 - 2x \geq 1$

43. $3x - 1 > 2x + 2$

44. $6x + 4 \leq 8 + 5x$

45. $8x + 1 \geq 2x + 13$

46. $6x + 3 > 4x - 1$

47. $-3 - 4x > -11$

48. $4x - 2 < x - 11$

Solve. Write the answer using interval notation.

49. $4x - 2 > 3x + 1$

50. $7x + 5 \leq 9 + 6x$

51. $9x + 2 \geq 3x + 14$

52. $8x + 1 > 6x + 3$

53. $-5 - 2x > -13$

54. $5x - 3 < x - 11$

55. $4(2x - 1) > 3x - 2(3x - 5)$

56. $2 - 5(x + 1) \geq 3(x - 1) - 8$

57. $3(4x + 3) \leq 7 - 4(x - 2)$

58. $3 + 2(x + 5) \geq x + 5(x + 1) + 1$

59. $3 - 4(x + 2) \leq 6 + 4(2x + 1)$

60. $12 - 2(3x - 2) \geq 5x - 2(5 - x)$

R.3 Linear Equations in Two Variables

OBJECTIVE 1

Graph points in a rectangular coordinate system

A **rectangular coordinate system** is formed by two number lines, one horizontal and one vertical. The point of intersection is called the **origin.** The two axes are called the **coordinate axes** or simply the **axes.** Generally, the horizontal axis is labeled the *x*-axis, and the vertical axis is labeled the *y*-axis.

 3.1.1 Points in the plane

Each point in the plane can be identified by a pair of numbers called an **ordered pair.** The first number of the ordered pair measures a horizontal change from the *y*-axis and is called the **abscissa,** or **x-coordinate.** The second number of the pair measures a vertical change from the *x*-axis and is called the **ordinate,** or **y-coordinate.** The ordered pair (x, y) associated with a point is also called the **coordinates** of the point.

Horizontal change ————┐ ┌———— Vertical change

Ordered pair → (4, 3)

x-coordinate ————┘ └———— *y*-coordinate

To **graph,** or **plot,** a point means to place a dot at the coordinates of the point. For example, to graph the ordered pair $(4, 3)$, start at the origin. Move 4 units to the right and then 3 units up. Draw a dot. To graph $(-3, -4)$, start at the origin. Move 3 units to the left and then 4 units down. Draw a dot.

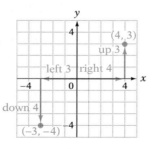

EXAMPLE 1 Graph the ordered pairs $(-4, 2)$, $(4, 3)$, $(0, -1)$, $(2, 0)$, and $(-1, -3)$.

Solution

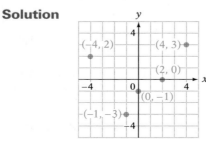

PROBLEM 1 Graph the ordered pairs $(-2, 4)$, $(4, 0)$, $(0, 3)$, $(-3, -4)$ and $(6, -1)$.

Solution See page S37.

OBJECTIVE **2** ## Graph a linear equation in two variables

The equations that follow are examples of equations in two variables.

$$y = 3x - 4$$
$$2x - y = 7$$
$$y = x^2 + 1$$

A **solution of an equation in two variables** is an ordered pair (x, y) whose coordinates make the equation a true statement.

Graph an equation of the form
3.3.1 $y = mx + b$

Is the ordered pair $(-4, 9)$ a solution of the equation $y = -2x + 1$?

$$y = -2x + 1$$

9	$-2(-4) + 1$	• **Replace x by -4 and y by 9.**
9	$8 + 1$	• **Simplify the right side.**
$9 = 9$		• **Compare the results. If the resulting equation is true, the ordered pair is a solution of the equation. If it is not true, the ordered pair is not a solution of the equation.**

Yes, the ordered pair $(-4, 9)$ is a solution of the equation $y = -2x + 1$.

Besides $(-4, 9)$, there are many other ordered pairs that are solutions of the equation $y = -2x + 1$. For example, $(0, 1)$, $(3, -5)$, and $\left(-\frac{3}{2}, 4\right)$ are also solutions. In general, an equation in two variables has an infinite number of solutions. By choosing any value of x and substituting that value into the equation, we can calculate a corresponding value of y.

Find the ordered-pair solution of $y = \frac{2}{5}x - 4$ that corresponds to $x = 5$.

$$y = \frac{2}{5}x - 4$$

$$y = \frac{2}{5}(5) - 4 \quad \text{• We are given that } x = 5. \text{ Replace } x \text{ by 5 in the equation.}$$

$$y = 2 - 4 \quad \text{• Simplify the right side.}$$

$$y = -2 \quad \text{• When } x = 5, y = -2.$$

The ordered-pair solution of $y = \frac{2}{5}x - 4$ in which $x = 5$ is $(5, -2)$.

Solutions of an equation in two variables can be graphed in the rectangular coordinate system.

Graph the ordered-pair solutions of $y = -2x + 1$ for $x = -2, -1, 0, 1,$ and 2.

x	$y = -2x + 1$	y	(x, y)
-2	$-2(-2) + 1$	5	$(-2, 5)$
-1	$-2(-1) + 1$	3	$(-1, 3)$
0	$-2(0) + 1$	1	$(0, 1)$
1	$-2(1) + 1$	-1	$(1, -1)$
2	$-2(2) + 1$	-3	$(2, -3)$

• **Use the given values of x to determine ordered-pair solutions of the equation. It is convenient to record these in a table.**

The ordered-pair solutions of $y = -2x + 1$ for $x = -2, -1, 0, 1,$ and 2 are $(-2, 5)$, $(-1, 3)$, $(0, 1)$, $(1, -1)$, and $(2, -3)$. These are graphed at the right.

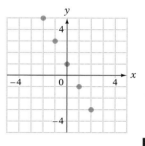

The **graph of an equation in two variables** is a graph of the ordered-pair solutions of the equation. Consider the equation $y = -2x + 1$ above. The ordered-pair solutions $(-2, 5)$, $(-1, 3)$, $(0, 1)$, $(1, -1)$, and $(2, -3)$ are graphed in the figure above. We can choose values of x that are not integers to produce more ordered pairs to graph, such as $\left(\frac{5}{2}, -4\right)$ and $\left(-\frac{1}{2}, 2\right)$. Choosing still other values of x would result in more and more ordered pairs being graphed. The result would be so many dots that the graph would appear as a straight line, as shown at the left. This is the graph of $y = -2x + 1$.

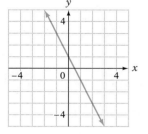

The equation $y = -2x + 1$ is an example of a linear equation because its graph is a straight line. It is also called a first-degree equation in two variables because the exponent on each variable is 1.

Take Note

The equation $y = x^2 + 2x - 4$ is not a linear equation in two variables because there is a term with a variable squared. The equation $y = \dfrac{5}{x - 3}$ is not a linear equation because a variable occurs in the denominator of a fraction.

Linear Equation in Two Variables

An equation of the form $y = mx + b$, where m is the coefficient of x and b is a constant, is a linear equation in two variables. The graph of a linear equation in two variables is a straight line.

Examples of linear equations are shown at the right.

$y = 5x + 3$ $(m = 5, b = 3)$

$y = x - 4$ $(m = 1, b = -4)$

$y = -\dfrac{3}{4}x$ $\left(m = -\dfrac{3}{4}, b = 0\right)$

To graph a linear equation, find ordered-pair solutions of the equation. Do this by choosing any value of x and finding the corresponding value of y. Repeat

this procedure, choosing different values for x, until you have found the number of solutions desired. Because the graph of a linear equation in two variables is a straight line, and a straight line is determined by two points, it is necessary to find only two solutions. However, it is recommended that at least three points be used to ensure accuracy.

Graph: $y = 2x - 3$

x	$y = 2x - 3$	y
0	$2(0) - 3$	-3
2	$2(2) - 3$	1
-1	$2(-1) - 3$	-5

- **Choose any values of x. Then find the corresponding values of y. The numbers 0, 2, and −1 were chosen arbitrarily for x.**

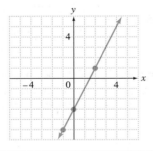

- **Graph the ordered-pair solutions (0, −3), (2, 1), and (−1, −5). Draw a straight line through the points.**

Remember that a graph is a drawing of the ordered-pair solutions of the equation. Therefore, every point on the graph is a solution of the equation, and every solution of the equation is a point on the graph.

When graphing an equation of the form $y = mx + b$, if m is a fraction, choose values of x that will simplify the evaluation. This is illustrated in Example 2. Note that the values of x chosen are multiples of the denominator, 2.

EXAMPLE 2 Graph $y = -\dfrac{3}{2}x - 3$.

Solution

x	y
0	-3
-2	0
-4	3

PROBLEM 2 Graph $y = \dfrac{3}{5}x - 4$.

Solution See pages S37–S38.

An equation of the form $Ax + By = C$ is also a linear equation in two variables. Examples of these equations follow.

$$3x + 4y = 12 \qquad (A = 3, B = 4, C = 12)$$
$$x - 5y = -10 \qquad (A = 1, B = -5, C = -10)$$
$$2x - y = 0 \qquad (A = 2, B = -1, C = 0)$$

One method of graphing an equation of the form $Ax + By = C$ involves first solving the equation for y and then following the same procedure used for graphing an equation of the form $y = mx + b$. To solve the equation for y means to rewrite the equation so that y is alone on one side of the equation and the term containing x and the constant are on the other side of the equation. The Addition and Multiplication Properties of Equations are used to rewrite an equation of the form $Ax + By = C$ in the form $y = mx + b$.

3.3.2 Graph an equation of the form $Ax + By = C$

Graph $3x + 2y = 6$.

$$3x + 2y = 6$$

- **The equation is in the form $Ax + By = C$.**

$$2y = -3x + 6$$

- **Solve the equation for y. Subtract $3x$ from each side of the equation.**

$$y = -\frac{3}{2}x + 3$$

- **Divide each side of the equation by 2. Note that each term on the right side is divided by 2.**

x	y
0	3
2	0
4	-3

- **Find at least three solutions.**

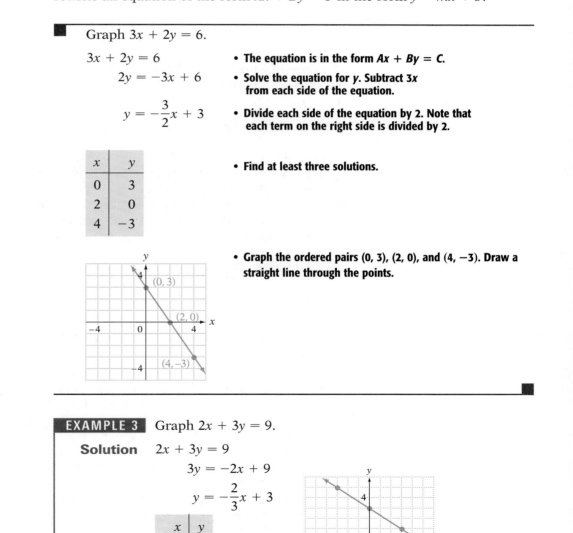

- **Graph the ordered pairs (0, 3), (2, 0), and (4, −3). Draw a straight line through the points.**

EXAMPLE 3 Graph $2x + 3y = 9$.

Solution $2x + 3y = 9$

$$3y = -2x + 9$$

$$y = -\frac{2}{3}x + 3$$

x	y
-3	5
0	3
3	1

PROBLEM 3 Graph $-3x + 2y = 4$.

■ **Solution** See page S38.

See page S38.

3.4.1 Slope

The **slope** of a line is a measure of the slant of the line. The symbol for slope is m.

For an equation of the form $y = mx + b$, m is the slope. Here are a few examples.

> The slope of the line $y = -2x + 5$ is -2.
>
> The slope of the line $y = 8x$ is 8.
>
> The slope of the line $y = \dfrac{3}{4}x - 1$ is $\dfrac{3}{4}$.

The slope of a line containing two points is the ratio of the change in the y values between the two points to the change in the x values.

Take Note

Slope $= m = \dfrac{\text{change in } y}{\text{change in } x}$

Slope Formula

The slope of the line containing the two points $P_1(x_1, y_1)$ and $P_2(x_2, y_2)$ is given by

$$m = \frac{y_2 - y_1}{x_2 - x_1}, \qquad x_1 \neq x_2$$

EXAMPLE 4 Find the slope of the line containing the points $(-4, -3)$ and $(-1, 1)$.

Solution Let $(x_1, y_1) = (-4, -3)$ and $(x_2, y_2) = (-1, 1)$.

$$m = \frac{y_2 - y_1}{x_2 - x_1} = \frac{1 - (-3)}{-1 - (-4)} = \frac{4}{3}$$

The slope is $\dfrac{4}{3}$.

PROBLEM 4 Find the slope of the line containing the points $(-2, 3)$ and $(1, -3)$.

■ **Solution** See page S38.

See page S38.

One important characteristic of the graph of a linear equation is its *intercepts*. An **x-intercept** is a point at which the graph crosses the x-axis. A **y-intercept** is a point at which the graph crosses the y-axis.

3.4.2 Slope-intercept form of a straight line

The graph of the equation $y = \dfrac{1}{2}x - 2$ is shown at the right. The x-intercept of the graph is $(4, 0)$. The y-intercept of the graph is $(0, -2)$.

Note that at an x-intercept, the y-coordinate is 0. At a y-intercept, the x-coordinate is 0.

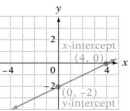

We can find the *y*-intercept of a linear equation by letting $x = 0$.

Find the *y*-intercept of the graph of the equation $y = 3x + 4$.

$$y = 3x + 4$$
$$y = 3(0) + 4 \qquad \bullet \text{ To find the } \textbf{\textit{y}}\textbf{-intercept, let } \textbf{\textit{x}} = \textbf{0.}$$
$$y = 0 + 4$$
$$y = 4$$

The *y*-intercept is (0, 4).

Note that the constant term of $y = 3x + 4$ is the *y*-coordinate of the *y*-intercept.

In general, for any equation of the form $y = mx + b$, the *y*-intercept is (0, *b*).

Because the slope and the *y*-intercept can be determined directly from the equation $y = mx + b$, this equation is called the slope–intercept form of a straight line.

Slope-Intercept Form of a Straight Line

The equation $y = mx + b$ is called the **slope-intercept form of a straight line.** The slope of the line is *m*, the coefficient of *x*. The *y*-intercept is (0, *b*).

The following equations are written in slope-intercept form.

$$y = -4x + 3 \qquad \text{Slope} = -4, y\text{-intercept} = (0, 3)$$

$$y = \frac{2}{5}x - 1 \qquad \text{Slope} = \frac{2}{5}, y\text{-intercept} = (0, -1)$$

$$y = -x \qquad \text{Slope} = -1, y\text{-intercept} = (0, 0)$$

When an equation is in slope-intercept form, it is possible to draw a graph of the function quickly.

Graph $x + 2y = 4$ by using the slope and *y*-intercept.

Solve the equation for *y*.

$$x + 2y = 4$$
$$2y = -x + 4$$
$$y = -\frac{1}{2}x + 2$$

From the equation $y = -\frac{1}{2}x + 2$, the slope is $-\frac{1}{2}$ and the *y*-intercept is (0, 2).

Rewrite the slope $-\dfrac{1}{2}$ as $\dfrac{-1}{2}$.

Beginning at the y-intercept, move right 2 units (change in x) and then down 1 unit (change in y).

The point whose coordinates are (2, 1) is a second point on the graph. Draw a straight line through the points (0, 2) and (2, 1).

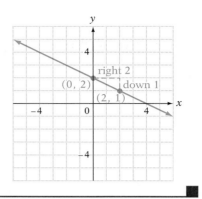

EXAMPLE 5 Graph $y = -\dfrac{3}{2}x + 4$ by using the slope and the y-intercept.

Solution From the equation, the slope is $-\dfrac{3}{2}$ and the y-intercept is (0, 4).

Rewrite the slope $-\dfrac{3}{2}$ as $\dfrac{-3}{2}$.

Place a dot at the y-intercept.

Starting at the y-intercept, move to the right 2 units (the change in x) and move down 3 units (the change in y). Place a dot at that location. Draw a line through the two points.

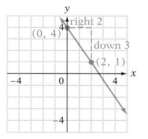

PROBLEM 5 Graph $y = -\dfrac{2}{3}x + 2$ by using the slope and the y-intercept.

Solution See page S38.

OBJECTIVE 3

3.2.1 Evaluate a function

Evaluate a function

A **relation** is a set of ordered pairs. A **function** is a relation in which no two ordered pairs have the same first coordinate and different second coordinates.

The relation {(0, 4), (1, 5), (1, 6), (2, 7)} is not a function because the ordered pairs (1, 5), and (1, 6) have the same first coordinate and different second coordinates.

The relation $[(0, 4), (1, 5), (2, 6), (3, 7)]$ is a function.

The phrase "y is a function of x," or a similar phrase with different variables, is used to describe those equations in two variables that define functions. To emphasize that the equation represents a function, function notation is used. For example, the square function is written in function notation as follows:

$$f(x) = x^2$$

The process of determining $f(x)$ for a given value of x is called **evaluating the function.** For instance, to evaluate $f(x) = x^2$ when $x = 3$, replace x by 3 and simplify.

$$f(x) = x^2$$
$$f(3) = 3^2 = 9$$

The value of the function is 9 when $x = 3$. An ordered pair of the function is $(3, 9)$.

> **Take Note**
>
> The symbol $f(x)$ is read "the value of f at x" or "f of x."

EXAMPLE 6 Evaluate $f(x) = 2x - 4$ when $x = 3$. Use your answer to write an ordered pair of the function.

Solution
$$f(x) = 2x - 4$$
$$f(3) = 2(3) - 4 \qquad \bullet\ \boldsymbol{x = 3}$$
$$f(3) = 6 - 4$$
$$f(3) = 2$$

An ordered pair of the function is $(3, 2)$.

PROBLEM 6 Evaluate the function $f(x) = 4 - 2x$ at $x = -3$. Use your answer to write an ordered pair of the function.

Solution See page S38.

OBJECTIVE 4

> 3.5.1 Find the equation of a line given a point and the slope

Find the equation of a line

When the slope of a line and a point on the line are known, the equation of the line can be determined by using the point-slope formula.

> **Point-Slope Formula**
>
> Let m be the slope of a line, and let (x_1, y_1) be the coordinates of a point on the line. The equation of the line can be found using the point-slope formula:
>
> $$y - y_1 = m(x - x_1)$$

EXAMPLE 7 Find the equation of the line that contains the point $(1, -3)$ and has slope -2.

Solution
$$y - y_1 = m(x - x_1)$$
$$y - (-3) = -2(x - 1) \qquad \bullet\ \boldsymbol{(x_1, y_1) = (1, -3),\ m = -2}$$
$$y + 3 = -2x + 2$$
$$y = -2x - 1$$

The equation of the line is $y = -2x - 1$.

PROBLEM 7 Find the equation of the line that contains the point $(-2, 2)$ and has slope $-\frac{1}{2}$.

■ **Solution** See page S38.

R.3 Exercises

1 Graph points in a rectangular coordinate system

Graph the ordered pairs.

1. $(2, 3)$, $(4, 0)$, $(-4, 1)$, $(-2, -2)$

2. $(0, 2)$, $(-4, -1)$, $(2, 0)$, $(1, -3)$

3. $(-2, 5)$, $(3, 4)$, $(0, 0)$, $(-3, -2)$

2 Graph a linear equation in two variables

Graph.

4. $y = 2x - 4$

5. $y = 2x + 1$

6. $y = -4x + 1$

7. $y = -3x + 4$

8. $y = 2x$

9. $y = 3x$

10. $y = -\dfrac{5}{2}x$

11. $y = -\dfrac{4}{3}x$

12. $y = \dfrac{2}{3}x + 1$

13. $y = \dfrac{3}{2}x - 1$

14. $y = -\dfrac{1}{2}x + 3$

15. $y = -\dfrac{2}{3}x + 1$

16. $2x - y = 3$ **17.** $2x + y = -3$ **18.** $2x + 5y = 10$

19. $x - 4y = 8$ **20.** $3x - y = -2$ **21.** $3x - 2y = 8$

Find the slope of the line containing the given points.

22. $P_1(4, 2), P_2(3, 4)$ **23.** $P_1(2, 1), P_2(3, 4)$ **24.** $P_1(-1, 3), P_2(2, 4)$

25. $P_1(-2, 1), P_2(2, 2)$ **26.** $P_1(2, 4), P_2(4, -1)$ **27.** $P_1(1, 3), P_2(5, -3)$

28. $P_1(3, -4), P_2(3, 5)$ **29.** $P_1(-1, 2), P_2(-1, 3)$ **30.** $P_1(4, -2), P_2(3, -2)$

31. $P_1(5, 1), P_2(-2, 1)$ **32.** $P_1(0, -1), P_2(3, -2)$ **33.** $P_1(3, 0), P_2(2, -1)$

Give the slope and the y-intercept of the graph of the equation.

34. $y = -3x + 7$ **35.** $y = \dfrac{5}{2}x - 4$ **36.** $y = x$

Graph by using the slope and the y-intercept.

37. $y = \dfrac{1}{2}x + 2$ **38.** $y = \dfrac{2}{3}x - 3$ **39.** $y = -\dfrac{3}{2}x$

40. $y = \dfrac{3}{4}x$ **41.** $y = -\dfrac{1}{2}x + 2$ **42.** $y = \dfrac{2}{3}x - 1$

43. $x - 3y = 3$ **44.** $3x + 2y = 8$ **45.** $4x + y = 2$

 Evaluate a function

Evaluate the function for the given value of the variable. Use your answer to write an ordered pair of the function.

46. $f(x) = 2x + 7; x = -2$ **47.** $g(x) = -3x + 1; x = -4$ **48.** $h(x) = 4x - 2; x = 3$

49. $p(x) = 6 - 8x; x = -1$ **50.** $t(x) = 5 - 7x; x = 0$ **51.** $f(t) = t^2 - t - 3; t = 2$

52. $p(n) = n^2 - 4n - 7; n = -3$ **53.** $h(x) = -3x^2 + x - 1; x = -2$ **54.** $g(t) = 4t^3 - 2t; t = -1$

4 **Find the equation of a line**

Find the equation of the line that contains the given point and has the given slope.

55. Point $(-1, 2); m = -3$ **56.** Point $(2, -3); m = 3$ **57.** Point $(4, -5); m = -2$

58. Point $(-3, 5); m = 3$ **59.** Point $(5, -3); m = -\dfrac{3}{5}$ **60.** Point $(3, 1); m = \dfrac{1}{3}$

61. Point $(-3, -2); m = -\dfrac{2}{3}$ **62.** Point $(4, -2); m = \dfrac{3}{4}$ **63.** Point $(2, 3); m = -\dfrac{1}{2}$

 Polynomials

OBJECTIVE **1** **Multiply and divide monomials**

A **monomial** is a number, a variable, or a product of a number and variables. The following rules and definitions are used to multiply and divide monomials and to write monomials in simplest form.

> **Rules of Exponents**
>
> **Rule for Multiplying Exponential Expressions**
>
> If m and n are integers, then $x^m \cdot x^n = x^{m+n}$.
>
> **Rule for Simplifying the Power of an Exponential Expression**
>
> If m and n are integers, then $(x^m)^n = x^{mn}$.
>
> **Rule for Simplifying Powers of Products**
>
> If m, n, and p are integers, then $(x^m y^n)^p = x^{mp} y^{np}$.

5.2.1
5.2.2 Multiply and
5.4.1 divide monomials

Rule for Dividing Exponential Expressions

If m and n are integers and $x \neq 0$, then $\dfrac{x^m}{x^n} = x^{m-n}$.

Rule for Simplifying Powers of Quotients

If m, n, and p are integers and $y \neq 0$, then $\left(\dfrac{x^m}{y^n}\right)^p = \dfrac{x^{mp}}{y^{np}}$.

Definition of Zero as an Exponent

If $x \neq 0$, then $x^0 = 1$. The expression 0^0 is undefined.

Definition of Negative Exponents

If n is a positive integer and $x \neq 0$, then $x^{-n} = \dfrac{1}{x^n}$ and $\dfrac{1}{x^{-n}} = x^n$.

Simplify: $(3x^4)^2(4x^3)$

$$(3x^4)^2(4x^3) = (3^{1 \cdot 2}x^{4 \cdot 2})(4x^3)$$

- **Use the Rule for Simplifying Powers of Products to simplify $(3x^4)^2$.**

$$= (3^2 x^8)(4x^3)$$
$$= (9x^8)(4x^3)$$
$$= (9 \cdot 4)(x^8 \cdot x^3)$$

- **Use the Rule for Multiplying Exponential Expressions.**

$$= 36x^{8+3}$$
$$= 36x^{11}$$

An exponential expression is in simplest form when there are no negative exponents in the expression. For example, the expression y^{-7} is not in simplest form; use the Definition of Negative Exponents to rewrite the expression with a positive exponent: $y^{-7} = \dfrac{1}{y^7}$. The expression $\dfrac{1}{c^{-4}}$ is not in simplest form; use the Definition of Negative Exponents to rewrite the expression with a positive exponent: $\dfrac{1}{c^{-4}} = c^4$.

Simplify: $\dfrac{6x^2}{8x^9}$

$$\frac{6x^2}{8x^9} = \frac{3x^2}{4x^9} = \frac{3x^{2-9}}{4}$$

- **Divide the coefficients by their common factors. Use the Rule for Dividing Exponential Expressions.**

$$= \frac{3x^{-7}}{4} = \frac{3}{4} \cdot \frac{x^{-7}}{1} = \frac{3}{4} \cdot \frac{1}{x^7}$$

- **Rewrite the expression with only positive exponents.**

$$= \frac{3}{4x^7}$$

Simplify: $\left(\dfrac{a^4}{b^3}\right)^{-2}$

$\left(\dfrac{a^4}{b^3}\right)^{-2} = \dfrac{a^{4(-2)}}{b^{3(-2)}} = \dfrac{a^{-8}}{b^{-6}}$

• **Use the Rule for Simplifying Powers of Quotients.**

$= \dfrac{b^6}{a^8}$

• **Rewrite the expression with positive exponents.**

EXAMPLE 1 Simplify.

 a. $(-2x)(3x^{-2})^{-3}$ **b.** $\left(\dfrac{3a^2b^{-1}}{27a^{-3}b^{-4}}\right)^{-2}$

Solution **a.** $(-2x)(3x^{-2})^{-3}$

$= (-2x)(3^{-3}x^6)$

• **Use the Rule for Simplifying Powers of Products.**

$= \dfrac{-2x \cdot x^6}{3^3}$

• **Write the expression with positive exponents.**

$= -\dfrac{2x^7}{27}$

• **Use the Rule for Multiplying Exponential Expressions. Simplify 3^3.**

b. Use the Rule for Simplifying Powers of Quotients. Then simplify the expression and write it with positive exponents.

$\left(\dfrac{3a^2b^{-1}}{27a^{-3}b^{-4}}\right)^{-2} = \left(\dfrac{a^2b^{-1}}{9a^{-3}b^{-4}}\right)^{-2}$

$= \dfrac{a^{2(-2)}b^{(-1)(-2)}}{9^{1(-2)}a^{(-3)(-2)}b^{(-4)(-2)}}$

$= \dfrac{a^{-4}b^2}{9^{-2}a^6b^8}$

$= 9^2a^{-4-6}b^{2-8}$

$= 81a^{-10}b^{-6} = \dfrac{81}{a^{10}b^6}$

PROBLEM 1 Simplify.

 a. $(-2ab)(2a^3b^{-2})^{-3}$ **b.** $\left(\dfrac{2x^2y^{-4}}{4x^{-2}y^{-5}}\right)^{-3}$

Solution See page S38.

OBJECTIVE 2 **Add and subtract polynomials**

A **polynomial** is a variable expression in which the terms are monomials. The polynomial $15t^2 - 2t + 3$ has three terms: $15t^2$, $-2t$, and 3. Note that each of these three terms is a monomial.

A polynomial of *one* term is a **monomial**. $-7x^2$ is a monomial.
A polynomial of *two* terms is a **binomial**. $4y + 3$ is a binomial.
A polynomial of *three* terms is a **trinomial**. $6b^2 + 5b - 8$ is a trinomial.

Polynomials can be added by combining like terms. This is illustrated in Example 2.

5.1.2 Add and subtract polynomials

EXAMPLE 2 Add: $(8x^2 - 4x - 9) + (2x^2 + 9x - 9)$

Solution $(8x^2 - 4x - 9) + (2x^2 + 9x - 9)$
$$= (8x^2 + 2x^2) + (-4x + 9x) + (-9 - 9)$$
$$= 10x^2 + 5x - 18$$

PROBLEM 2 Add: $(-4x^3 + 2x^2 - 8) + (4x^3 + 6x^2 - 7x + 5)$

Solution See page S38.

The additive inverse of the polynomial $(3x^2 - 7x + 8)$ is $-(3x^2 - 7x + 8)$.

To find the additive inverse of a polyno- $-(3x^2 - 7x + 8) = -3x^2 + 7x - 8$
mial, change the sign of each term inside
the parentheses.

To subtract two polynomials, add the additive inverse of the second polyno-
mial to the first.

Simplify: $(5a^2 - a + 2) - (-2a^3 + 3a - 3)$

$(5a^2 - a + 2) - (-2a^3 + 3a - 3)$
$$= (5a^2 - a + 2) + (2a^3 - 3a + 3)$$ • **Rewrite subtraction as addition of the additive inverse.**
$$= 2a^3 + 5a^2 - 4a + 5$$ • **Combine like terms.**

EXAMPLE 3 Subtract: $(7c^2 - 9c - 12) - (9c^2 + 5c - 8)$

Solution $(7c^2 - 9c - 12) - (9c^2 + 5c - 8)$
$$= (7c^2 - 9c - 12) + (-9c^2 - 5c + 8)$$
$$= -2c^2 - 14c - 4$$

PROBLEM 3 Subtract: $(-4w^3 + 8w - 8) - (3w^3 - 4w^2 + 2w - 1)$

Solution See page S38.

OBJECTIVE ❸ **Multiply polynomials**

5.3.1 Multiply a polynomial by a monomial

The Distributive Property is used to multiply a polynomial by a monomial.
Each term of the polynomial is multiplied by the monomial.

Multiply: $3x^3(4x^4 - 2x + 5)$

$3x^3(4x^4 - 2x + 5)$
$$= 3x^3(4x^4) - 3x^3(2x) + 3x^3(5)$$ • **Use the Distributive Property. Multiply each term of the polynomial by $3x^3$.**

$$= 12x^7 - 6x^4 + 15x^3$$ • **Use the Rule for Multiplying Exponential Expressions.**

EXAMPLE 4 Multiply: $2xy(3x^2 - xy + 2y^2)$

Solution $2xy(3x^2 - xy + 2y^2) = 2xy(3x^2) - 2xy(xy) + 2xy(2y^2)$
$$= 6x^3y - 2x^2y^2 + 4xy^3$$

PROBLEM 4 Multiply: $3mn^2(2m^2 - 3mn - 1)$

Solution See page S38.

5.3.2 Multiply two polynomials

A vertical format similar to that used for multiplication of whole numbers is used to multiply two polynomials. The product $(2y - 3)(y^2 + 2y + 5)$ is shown below.

$$
\begin{array}{r}
y^2 + \ 2y + \ 5 \\
2y - \ 3 \\
\hline
-3y^2 - \ 6y - 15 \\
2y^3 + 4y^2 + 10y \\
\hline
2y^3 + \ y^2 + \ 4y - 15
\end{array}
$$

This is $-3(y^2 + 2y + 5)$.

This is $2y(y^2 + 2y + 5)$. Like terms are placed in the same columns.

Add the terms in each column.

EXAMPLE 5 Multiply: $(2a^2 + 4a - 5)(3a + 5)$

Solution
$$
\begin{array}{r}
2a^2 + \ 4a - \ 5 \\
3a + \ 5 \\
\hline
10a^2 + 20a - 25 \\
6a^3 + 12a^2 - 15a \\
\hline
6a^3 + 22a^2 + \ 5a - 25
\end{array}
$$
• **Align like terms in the same column.**

PROBLEM 5 Multiply: $(3c^2 - 4c + 5)(2c - 3)$

Solution See page S39.

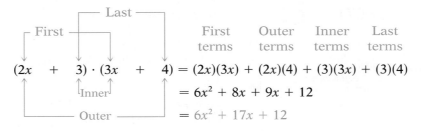
5.3.3 Multiply two binomials

It is frequently necessary to multiply two binomials. The product is computed by using a method called FOIL. The letters of FOIL stand for **F**irst, **O**uter, **I**nner, and **L**ast. The FOIL method is based on the Distributive Property and involves adding the products of the first terms, the outer terms, the inner terms, and the last terms.

The product $(2x + 3)(3x + 4)$ is shown below using FOIL.

$$(2x + 3) \cdot (3x + 4) = (2x)(3x) + (2x)(4) + (3)(3x) + (3)(4)$$

First terms Outer terms Inner terms Last terms

$$= 6x^2 + 8x + 9x + 12$$
$$= 6x^2 + 17x + 12$$

EXAMPLE 6 Multiply: $(4x - 3)(2x + 5)$

Solution $(4x - 3)(2x + 5) = (4x)(2x) + (4x)(5) + (-3)(2x) + (-3)(5)$
$$= 8x^2 + 20x + (-6x) + (-15)$$
$$= 8x^2 + 14x - 15$$

PROBLEM 6 Multiply: $(4y - 7)(3y - 5)$

■ **Solution** See page S39.

OBJECTIVE 4

Divide polynomials

To divide two polynomials, use a method similar to that used for division of whole numbers. To divide $(x^2 - 5x + 8)$ by $(x - 3)$:

Step 1:

$$
\begin{array}{r}
x \phantom{{}- 5x + 8} \\
x - 3 \overline{\smash{)}\, x^2 - 5x + 8} \\
\underline{x^2 - 3x} \quad \downarrow \\
-2x + 8
\end{array}
$$

Think: $x\overline{)x^2} = \dfrac{x^2}{x} = x$

Multiply: $x(x - 3) = x^2 - 3x$

Subtract: $(x^2 - 5x) - (x^2 - 3x) = -2x$
Bring down the $+8$.

Step 2:

$$
\begin{array}{r}
x - 2 \\
x - 3 \overline{\smash{)}\, x^2 - 5x + 8} \\
\underline{x^2 - 3x} \\
-2x + 8 \\
\underline{-2x + 6} \\
2
\end{array}
$$

Think: $x\overline{)-2x} = \dfrac{-2x}{x} = -2$

Multiply: $-2(x - 3) = -2x + 6$

Subtract: $(-2x + 8) - (-2x + 6) = 2$
The remainder is 2.

Check: (Quotient × Divisor) + Remainder = Dividend

$$(x - 2)(x - 3) + 2 = x^2 - 3x - 2x + 6 + 2 = x^2 - 5x + 8$$

$$(x^2 - 5x + 8) \div (x - 3) = x - 2 + \frac{2}{x - 3}$$

EXAMPLE 7 Divide: $(6x + 2x^3 + 26) \div (x + 2)$

Solution Arrange the terms of the dividend in descending order. There is no x^2 term in $2x^3 + 6x + 26$. Insert $0x^2$ for the missing term so that like terms will be in columns.

$$
\begin{array}{r}
2x^2 - 4x + 14 \\
x + 2 \overline{\smash{)}\, 2x^3 + 0x^2 + 6x + 26} \\
\underline{2x^3 + 4x^2} \\
-4x^2 + 6x \\
\underline{-4x^2 - 8x} \\
14x + 26 \\
\underline{14x + 28} \\
-2
\end{array}
$$

Check: $(x + 2)(2x^2 - 4x + 14) - 2$
$= 2x^3 + 6x + 28 - 2 = 2x^3 + 6x + 26$

$$(6x + 2x^3 + 26) \div (x + 2) = 2x^2 - 4x + 14 - \frac{2}{x + 2}$$

PROBLEM 7 Divide: $(x^3 - 7 - 2x) \div (x - 2)$

■ **Solution** See page S39.

OBJECTIVE **5**

Factor trinomials of the form $ax^2 + bx + c$

6.1.1 Factor a monomial from a polynomial

A polynomial is in **factored form** when it is written as a product of other polynomials. Factoring can be thought of as the reverse of multiplication.

Polynomial		Factored Form
$2x^3 + 6x^2 - 10x$	$=$	$2x(x^2 + 3x - 5)$
$x^2 - 3x - 28$	$=$	$(x + 4)(x - 7)$

To factor out a common monomial from the terms of a polynomial, first find the greatest common factor (GCF) of the terms.

The GCF of two or more monomials is the product of the GCF of the coefficients and the common variable factors.

$$10a^3b = \mathbf{2} \cdot 5 \cdot \mathbf{a} \cdot a \cdot a \cdot \mathbf{b}$$
$$4a^2b^2 = \mathbf{2} \cdot 2 \cdot \mathbf{a} \cdot \mathbf{a} \cdot b \cdot b$$
$$\text{GCF} = \mathbf{2} \cdot \mathbf{a} \cdot \mathbf{a} \cdot \mathbf{b} = 2a^2b$$

Note that the exponent of each variable in the GCF is the same as the smallest exponent of that variable in either of the monomials.

The GCF of $10a^3b$ and $4a^2b^2$ is $2a^2b$.

Factor: $5x^2 - 35x^2 + 10x$

The GCF is $5x$.

$\dfrac{5x^3 - 35x^2 + 10x}{5x}$
$= x^2 - 7x + 2$
$5x^3 - 35x^2 + 10x$
$= 5x(x^2 - 7x + 2)$
$5x(x^2 - 7x + 2)$
$= 5x^3 - 35x^2 + 10x$

- **Find the GCF of the terms $5x^3$, $-35x^2$, and $10x$.**

- **Divide each term of the polynomial by the GCF.**

- **Write the polynomial as the product of the GCF and the quotient found above.**
- **Check the factorization by multiplying.**

EXAMPLE 8 Factor: $16x^4y^5 + 8x^4y^2 - 12x^3y$

Solution The GCF is $4x^3y$.

$$16x^4y^5 + 8x^4y^2 - 12x^3y = 4x^3y(4xy^4 + 2xy - 3)$$

PROBLEM 8 Factor: $6x^4y^2 - 9x^3y^2 + 12x^2y^4$

Solution See page S39.

A **quadratic trinomial** is a trinomial of the form $ax^2 + bx + c$, where a and b are coefficients and c is a constant. Examples of quadratic trinomials follow.

$x^2 + 9x + 14$	$a = 1, b = 9, c = 14$
$x^2 - 2x - 15$	$a = 1, b = -2, c = -15$
$3x^2 - x + 4$	$a = 3, b = -1, c = 4$

To **factor a quadratic trinomial** means to express the trinomial as the product of two binomials. For example,

Trinomial		Factored Form
$2x^2 - x - 1$	$=$	$(2x + 1)(x - 1)$
$y^2 - 3y + 2$	$=$	$(y - 1)(y - 2)$

We will begin by factoring trinomials of the form $x^2 + bx + c$, where the coefficient of x^2 is 1.

The method by which factors of a trinomial are found is based on FOIL. Consider the following binomial products, noting the relationship between the constant term of the binomials and the terms of the trinomial.

Sum of the binomial constants

Product of the binomial constants

$$(x + 4)(x + 5) = x \cdot x + 5x + 4x + 4 \cdot 5 \qquad = x^2 + 9x + 20$$
$$(x - 6)(x + 8) = x \cdot x + 8x - 6x + (-6)(8) \qquad = x^2 + 2x - 48$$
$$(x - 3)(x - 2) = x \cdot x - 2x - 3x + (-3)(-2) \quad = x^2 - 5x + 6$$

Factor: $x^2 - 7x + 10$

Find two integers whose product is 10 and whose sum is -7.

Negative Factors of 10	Sum
$-1, -10$	-11
$-2, -5$	-7

• These are the correct factors.

$x^2 - 7x + 10 = (x - 2)(x - 5)$

• Write the trinomial as a product of the factors.

Check: $(x - 2)(x - 5) = x^2 - 5x - 2x + 10$
$$= x^2 - 7x + 10$$

6.2.1 Factor trinomials of the form $x^2 + bx + c$

Take Note

Once the correct factors are found, it is not necessary to try the remaining factors.

EXAMPLE 9 Factor: $x^2 + 6x - 27$

Solution Two factors of -27 whose sum is 6 are -3 and 9.

$$x^2 + 6x - 27 = (x - 3)(x + 9)$$

PROBLEM 9 Factor: $x^2 - 8x + 15$

Solution See page S39.

6.3.1 Factor trinomials of the form $ax^2 + bx + c$ using trial factors

To use the trial factor method to factor a trinomial of the form $ax^2 + bx + c$, where $a \neq 1$, use the factors of a and the factors of c to write all of the possible binomial factors of the trinomial. Then use FOIL to determine the correct factorization. To reduce the number of trial factors that must be considered, remember the following guidelines.

Use the signs of the constant term and the coefficient of x in the trinomial to determine the signs of the binomial factors. If the constant term is positive, the signs of the binomial factors will be the same as the sign of the coefficient of x in the trinomial. If the sign of the constant term is negative, the constant terms in the binomials will have different signs.

Factor: $2x^2 - 7x + 3$

Because the constant term is positive $(+3)$ and the coefficient of x is negative (-7), the binomial constants will be negative.

Positive Factors of 2 ($a = 2$)	Negative Factors of 3 ($c = 3$)
1, 2	$-1, -3$

Write trial factors. Use the **O**uter and **I**nner products of FOIL to determine the middle term, $-7x$, of the trinomial.

Trial Factors	Middle Term	
$(x - 1)(2x - 3)$	$-3x - 2x = -5x$	
$(x - 3)(2x - 1)$	$-x - 6x = -7x$	• $-7x$ is the middle term.

$2x^2 - 7x + 3 = (x - 3)(2x - 1)$

Check: $(x - 3)(2x - 1) = 2x^2 - x - 6x + 3 = 2x^2 - 7x + 3$

Factor: $5x^2 + 22x - 15$

The constant term is negative (-15). The binomial constants will have different signs.

Positive Factors of 5 ($a = 5$)	Factors of -15 ($c = -15$)
1, 5	$-1, 15$
	$1, -15$
	$-3, 5$
	$3, -5$

Write trial factors. Use the **O**uter and **I**nner products of FOIL to determine the middle term, $22x$, of the trinomial.

Take Note

It is not necessary to test trial factors that have a common factor. If the trinomial does not have a common factor, then its factors cannot have a common factor.

Trial Factors	Middle Term	
$(x - 1)(5x + 15)$	common factor	
$(x + 15)(5x - 1)$	$-x + 75x = 74x$	
$(x + 1)(5x - 15)$	common factor	
$(x - 15)(5x + 1)$	$x - 75x = -74x$	
$(x - 3)(5x + 5)$	common factor	
$(x + 5)(5x - 3)$	$-3x + 25x = 22x$	• $22x$ is the middle term.
$(x + 3)(5x - 5)$	common factor	
$(x - 5)(5x + 3)$	$3x - 25x = -22x$	

$5x^2 + 22x - 15 = (x + 5)(5x - 3)$

Check: $(x + 5)(5x - 3) = 5x^2 - 3x + 25x - 15 = 5x^2 + 22x - 15$

EXAMPLE 10 Factor: $2x^2 + 3x - 5$

Solution

Positive factors of 2	Factors of -5
1, 2	1, -5
	-1, 5

Trial Factors	Middle Term
$(x + 1)(2x - 5)$	$-5x + 2x = -3x$
$(x - 5)(2x + 1)$	$x - 10x = -9x$
$\mathbf{(x - 1)(2x + 5)}$	$\mathbf{5x - 2x = 3x}$
$(x + 5)(2x - 1)$	$-x + 10x = 9x$

$2x^2 + 3x - 5 = (x - 1)(2x + 5)$

Check: $(x - 1)(2x + 5) = 2x^2 + 5x - 2x - 5$
$= 2x^2 + 3x - 5$

PROBLEM 10 Factor: $3x^2 - x - 2$

Solution See page S39.

6.4.4 Factor completely

A polynomial is factored completely when it is written as a product of factors that are nonfactorable over the integers.

The first step in any factoring problem is to determine whether the terms of the polynomial have a common factor. If they do, factor it out first.

EXAMPLE 11 Factor: $5x^2y + 60xy + 100y$

Solution There is a common factor, $5y$.

Factor out the GCF.

$5x^2y + 60xy + 100y = 5y(x^2 + 12x + 20)$

Factor $x^2 + 12x + 20$. The two factors of 20 whose sum is 12 are 2 and 10.

$5y(x^2 + 12x + 20) = 5y(x + 2)(x + 10)$
$5x^2y + 60xy + 100y = 5y(x + 2)(x + 10)$

Check: $5y(x + 2)(x + 10) = (5xy + 10y)(x + 10)$
$= 5x^2y + 50xy + 10xy + 100y$
$= 5x^2y + 60xy + 100y$

PROBLEM 11 Factor: $4a^3 - 4a^2 - 24a$

Solution See page S39.

R.4 Exercises

1 Multiply and divide monomials

Simplify.

1. $z^3 \cdot z \cdot z^4$

2. $b \cdot b^2 \cdot b^6$

3. $(x^3)^5$

4. $(b^2)^4$

5. $(x^2y^3)^6$

6. $(m^4n^2)^3$

7. $\dfrac{a^8}{a^2}$

8. $\dfrac{c^{12}}{c^5}$

9. $(-m^3n)(m^6n^2)$

10. $(-r^4t^3)(r^2t^9)$

11. $(-2a^3bc^2)^3$

12. $(-4xy^3z^2)^2$

13. $\dfrac{m^4n^7}{m^3n^5}$

14. $\dfrac{a^5b^6}{a^3b^2}$

15. $\dfrac{-16a^7}{24a^6}$

16. $\dfrac{18b^5}{-45b^4}$

17. $(9mn^4p)(-3mp^2)$

18. $(-3v^2wz)(-4vz^4)$

19. $(-2n^2)(-3n^4)^3$

20. $(-3m^3n)(-2m^2n^3)^3$

21. $\dfrac{14x^4y^6z^2}{16x^3y^9z}$

22. $\dfrac{25x^4y^7z^2}{20x^5y^9z^{11}}$

23. $(-2x^3y^2)^3(-xy^2)^4$

24. $(-m^4n^2)^5(-2m^3n^3)^3$

25. $4x^{-7}$

26. $-6y^{-1}$

27. $d^{-4}d^{-6}$

28. $x^{-3}x^{-5}$

29. $\dfrac{x^{-3}}{x^2}$

30. $\dfrac{x^4}{x^{-5}}$

31. $\dfrac{1}{3x^{-2}}$

32. $\dfrac{2}{5c^{-6}}$

33. $(x^2y^{-4})^3$

34. $(x^3y^5)^{-4}$

35. $(3x^{-1}y^{-2})^2$

36. $(5xy^{-3})^{-2}$

37. $(2x^{-1})(x^{-3})$

38. $(-2x^{-5})(x^7)$

39. $\dfrac{3x^{-2}y^2}{6xy^2}$

40. $\dfrac{2x^{-2}y}{8xy}$

41. $\dfrac{2x^{-1}y^{-4}}{4xy^2}$

42. $\dfrac{3a^{-2}b}{ab}$

43. $(x^{-2}y)^2(xy)^{-2}$

44. $(x^{-1}y^2)^{-3}(x^2y^{-4})^{-3}$

45. $\left(\dfrac{x^2y^{-1}}{xy}\right)^{-4}$

46. $\left(\dfrac{x^{-2}y^{-4}}{x^{-2}y}\right)^{-2}$

47. $\left(\dfrac{4a^{-2}b}{8a^3b^{-4}}\right)^2$

48. $\left(\dfrac{6ab^{-2}}{3a^{-2}b}\right)^{-2}$

2 Add and subtract polynomials

Add or subtract.

49. $(4b^2 - 5b) + (3b^2 + 6b - 4)$

50. $(2c^2 - 4) + (6c^2 - 2c + 4)$

51. $(2a^2 - 7a + 10) + (a^2 + 4a + 7)$

52. $(-6x^2 + 7x + 3) + (3x^2 + x + 3)$

53. $(x^2 - 2x + 1) - (x^2 + 5x + 8)$

54. $(3x^2 + 2x - 2) - (5x^2 - 5x + 6)$

55. $(-2x^3 + x - 1) - (-x^2 + x - 3)$

56. $(2x^2 + 5x - 3) - (3x^3 + 2x - 5)$

57. $(x^3 - 7x + 4) + (2x^2 + x - 10)$

58. $(3y^3 + y^2 + 1) + (-4y^3 - 6y - 3)$

59. $(5x^3 + 7x - 7) + (10x^2 - 8x + 3)$

60. $(3y^3 + 4y + 9) + (2y^2 + 4y - 21)$

61. $(2y^3 + 6y - 2) - (y^3 + y^2 + 4)$

62. $(-2x^2 - x + 4) - (-x^3 + 3x - 2)$

63. $(4y^3 - y - 1) - (2y^2 - 3y + 3)$

64. $(3x^2 - 2x - 3) - (2x^3 - 2x^2 + 4)$

3 Multiply polynomials

Multiply.

65. $4b(3b^3 - 12b^2 - 6)$

66. $-2a^2(3a^2 - 2a + 3)$

67. $3b(3b^4 - 3b^2 + 8)$

68. $-2x^2(2x^2 - 3x - 7)$

69. $-2x^2y(x^2 - 3xy + 2y^2)$

70. $3ab^2(3a^2 - 2ab + 4b^2)$

71. $(x^2 + 3x + 2)(x + 1)$

72. $(x^2 - 2x + 7)(x - 2)$

73. $(a - 3)(a^2 - 3a + 4)$

74. $(2x - 3)(x^2 - 3x + 5)$

75. $(-2b^2 - 3b + 4)(b - 5)$

76. $(-a^2 + 3a - 2)(2a - 1)$

77. $(x^3 - 3x + 2)(x - 4)$

78. $(y^3 + 4y^2 - 8)(2y - 1)$

79. $(y + 2)(y^3 + 2y^2 - 3y + 1)$

80. $(2a - 3)(2a^3 - 3a^2 + 2a - 1)$

81. $(a - 3)(a + 4)$

82. $(b - 6)(b + 3)$

83. $(y - 7)(y - 3)$

84. $(a - 8)(a - 9)$

85. $(2x + 1)(x + 7)$

86. $(y + 2)(5y + 1)$

87. $(3x - 1)(x + 4)$

88. $(7x - 2)(x + 4)$

89. $(4x - 3)(x - 7)$

90. $(2x - 3)(4x - 7)$

91. $(3y - 8)(y + 2)$

92. $(5y - 9)(y + 5)$

93. $(7a - 16)(3a - 5)$

94. $(5a - 12)(3a - 7)$

95. $(x + y)(2x + y)$

96. $(2a + b)(a + 3b)$

97. $(3x - 4y)(x - 2y)$

98. $(2a - b)(3a + 2b)$

99. $(5a - 3b)(2a + 4b)$ **100.** $(2x + 3)(2x - 3)$ **101.** $(4x - 7)(4x + 7)$

4 Divide polynomials

Divide.

102. $(b^2 - 14b + 49) \div (b - 7)$ **103.** $(x^2 - x - 6) \div (x - 3)$ **104.** $(2x^2 + 5x + 2) \div (x + 2)$

105. $(2y^2 - 13y + 21) \div (y - 3)$ **106.** $(x^2 + 1) \div (x - 1)$ **107.** $(x^2 + 4) \div (x + 2)$

108. $(6x^2 - 7x) \div (3x - 2)$ **109.** $(6y^2 + 2y) \div (2y + 4)$ **110.** $(a^2 + 5a + 10) \div (a + 2)$

111. $(b^2 - 8b - 9) \div (b - 3)$ **112.** $(2y^2 - 9y + 8) \div (2y + 3)$ **113.** $(3x^2 + 5x - 4) \div (x - 4)$

114. $(8x + 3 + 4x^2) \div (2x - 1)$ **115.** $(10 + 21y + 10y^2) \div (2y + 3)$

116. $(x^3 + 3x^2 + 5x + 3) \div (x + 1)$ **117.** $(x^3 - 6x^2 + 7x - 2) \div (x - 1)$

118. $(x^4 - x^2 - 6) \div (x^2 + 2)$ **119.** $(x^4 + 3x^2 - 10) \div (x^2 - 2)$

5 Factor trinomials of the form $ax^2 + bx + c$

Factor.

120. $8x + 12$ **121.** $12y^2 - 5y$ **122.** $10x^4 - 12x^2$ **123.** $10x^2yz^2 + 15xy^3z$

124. $x^3 - 3x^2 - x$ **125.** $5x^2 - 15x + 35$ **126.** $3x^3 + 6x^2 + 9x$ **127.** $3y^4 - 9y^3 - 6y^2$

128. $2x^3 + 6x^2 - 14x$ **129.** $x^4y^4 - 3x^3y^3 + 6x^2y^2$

130. $4x^5y^5 - 8x^4y^4 + x^3y^3$ **131.** $16x^2y - 8x^3y^4 - 48x^2y^2$

132. $x^2 + 5x + 6$ **133.** $x^2 + x - 2$ **134.** $x^2 + x - 6$ **135.** $a^2 + a - 12$

136. $a^2 - 2a - 35$ **137.** $a^2 - 3a + 2$ **138.** $a^2 - 5a + 4$ **139.** $b^2 + 7b - 8$

140. $y^2 + 6y - 55$

141. $z^2 - 4z - 45$

142. $y^2 - 8y + 15$

143. $z^2 - 14z + 45$

144. $p^2 + 12p + 27$

145. $b^2 + 9b + 20$

146. $y^2 - 8y + 32$

147. $y^2 - 9y + 81$

148. $p^2 + 24p + 63$

149. $x^2 - 15x + 56$

150. $5x^2 + 6x + 1$

151. $2y^2 + 7y + 3$

152. $2a^2 - 3a + 1$

153. $3a^2 - 4a + 1$

154. $4x^2 - 3x - 1$

155. $2x^2 - 5x - 3$

156. $6t^2 - 11t + 4$

157. $10t^2 + 11t + 3$

158. $8x^2 + 33x + 4$

159. $10z^2 + 3z - 4$

160. $3x^2 + 14x - 5$

161. $3z^2 + 95z + 10$

162. $8z^2 - 36z + 1$

163. $2t^2 - t - 10$

164. $2t^2 + 5t - 12$

165. $12y^2 + 19y + 5$

166. $5y^2 - 22y + 8$

167. $11a^2 - 54a - 5$

168. $4z^2 + 11z + 6$

169. $6b^2 - 13b + 6$

170. $6x^2 + 35x - 6$

171. $3x^2 + 15x + 18$

172. $3a^2 + 3a - 18$

173. $ab^2 + 7ab - 8a$

174. $3y^3 - 15y^2 + 18y$

175. $2y^4 - 26y^3 - 96y^2$

176. $3y^4 + 54y^3 + 135y^2$

177. $2x^3 - 11x^2 + 5x$

178. $2x^3 + 3x^2 - 5x$

179. $10t^2 - 5t - 50$

180. $16t^2 + 40t - 96$

181. $6p^3 + 5p^2 + p$

182. $12x^2y - 36xy + 27y$

Appendix

Table of Properties

Properties of Real Numbers

The Associative Property of Addition

If a, b, and c are real numbers, then
$(a + b) + c = a + (b + c)$.

The Commutative Property of Addition

If a and b are real numbers, then $a + b = b + a$.

The Addition Property of Zero

If a is a real number, then $a + 0 = 0 + a = a$.

The Multiplication Property of Zero

If a is a real number, then $a \cdot 0 = 0 \cdot a = 0$.

The Inverse Property of Addition

If a is a real number, then
$a + (-a) = (-a) + a = 0$.

The Associative Property of Multiplication

If a, b, and c are real numbers, then
$(a \cdot b) \cdot c = a \cdot (b \cdot c)$.

The Commutative Property of Multiplication

If a and b are real numbers, then $a \cdot b = b \cdot a$.

The Multiplication Property of One

If a is a real number, then $a \cdot 1 = 1 \cdot a = a$.

The Inverse Property of Multiplication

If a is a real number and $a \neq 0$, then
$$a \cdot \frac{1}{a} = \frac{1}{a} \cdot a = 1.$$

Distributive Property

If a, b, and c are real numbers, then
$a(b + c) = ab + ac$ or $(b + c)a = ba + ca$.

Properties of Equations

Addition Property of Equations

The same number or variable term can be added to each side of an equation without changing the solution of the equation.

Multiplication Property of Equations

Each side of an equation can be multiplied by the same nonzero number without changing the solution of the equation.

Properties of Inequalities

Addition Property of Inequalities

If $a > b$, then $a + c > b + c$.

If $a < b$, then $a + c < b + c$.

Multiplication Property of Inequalities

If $a > b$ and $c > 0$, then $ac > bc$.

If $a < b$ and $c > 0$, then $ac < bc$.

If $a > b$ and $c < 0$, then $ac < bc$.

If $a < b$ and $c < 0$, then $ac > bc$.

Properties of Exponents

If m and n are integers, then $x^m \cdot x^n = x^{m+n}$.

If m and n are integers, then $(x^m)^n = x^{mn}$.

If $x \neq 0$, then $x^0 = 1$.

If m and n are integers and $x \neq 0$, then
$$\frac{x^m}{x^n} = x^{m-n}.$$

If m, n, and p are integers, then
$(x^m \cdot y^n)^p = x^{mp}y^{np}$.

If n is a positive integer and $x \neq 0$, then
$$x^{-n} = \frac{1}{x^n} \text{ and } \frac{1}{x^{-n}} = x^n.$$

If m, n, and p are integers and $y \neq 0$, then
$$\left(\frac{x^m}{y^n}\right)^p = \frac{x^{mp}}{y^{np}}.$$

823

Table of Properties (*continued*)

Principle of Zero Products

If $a \cdot b = 0$, then $a = 0$ or $b = 0$.

Properties of Radical Expressions

If a and b are positive real numbers, then $\sqrt[n]{ab} = \sqrt[n]{a}\,\sqrt[n]{b}$.

If a and b are positive real numbers, then

$$\sqrt[n]{\frac{a}{b}} = \frac{\sqrt[n]{a}}{\sqrt[n]{b}}.$$

Property of Squaring Both Sides of an Equation

If a and b are real numbers and $a = b$, then $a^2 = b^2$.

Properties of Logarithms

If x, y, and b are positive real numbers and $b \neq 1$, then $\log_b(xy) = \log_b x + \log_b y$.

If x, y, and b are positive real numbers and $b \neq 1$, then $\log_b \frac{x}{y} = \log_b x - \log_b y$.

If x and b are positive real numbers, $b \neq 1$, and r is any real number, then $\log_b x^r = r \log_b x$.

If x and b are positive real numbers and $b \neq 1$, then $\log_b b^x = x$.

Keystroke Guide for the TI-83 and TI-83 Plus/TI-84 Plus

Basic Operations

Numerical calculations are performed on the **home screen.** You can always return to the home screen by pressing [2nd] QUIT. Pressing [CLEAR] erases the home screen.

To evaluate the expression $-2(3 + 5) - 8 \div 4$, use the following keystrokes.

Note: There is a difference between the key to enter a negative number, [(-)], and the key for subtraction, [-]. You cannot use these keys interchangeably.

The [2nd] key is used to access the commands in gold writing above a key. For instance, to evaluate the $\sqrt{49}$, press [2nd] √ 49 [)] [ENTER].

The [ALPHA] key is used to place a letter on the screen. One reason to do this is to store a value of a variable. The following keystrokes give A the value of 5.

5 [STO▸] [ALPHA] A [ENTER]

This value is now available in calculations. For instance, we can find the value of $3a^2$ by using the following keystrokes: 3 [ALPHA] A [x²]. To display the value of the variable on the screen, press [2nd] RCL [ALPHA] A.

Note: When you use the [ALPHA] key, only capital letters are available on the TI calculators.

Complex Numbers

To perform operations on complex numbers, first press [MODE] and then use the arrow keys to select a+bi. Then press [ENTER] [2nd] QUIT.

Addition of complex numbers To add $(3 + 4i) + (2 - 7i)$, use the keystrokes

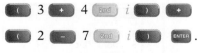

[(] 2 [-] 7 [2nd] *i* [)] [ENTER].

Division of complex numbers To divide $\frac{26 + 2i}{2 + 4i}$, use the keystrokes [(] 26 [+] 2 [2nd] *i* [)] [÷]

[(] 2 [+] 4 [2nd] *i* [)] [ENTER].

Note: Operations for subtraction and multiplication are similar.

Additional operations on complex numbers can be found by selecting CPX under the MATH key.

To find the absolute value of $2 - 5i$, press MATH (scroll to CPX) (scroll to abs) ENTER (2 − 5 2nd i) ENTER .

Evaluating Functions

There are various methods of evaluating a function, but all methods require that the expression be entered as one of the ten functions Y₁ to Y₀. To evaluate $f(x) = \frac{x^2}{x-1}$ when $x = -3$, enter the expression into, for instance, Y₁, and then press VARS ▸ 1 1 ((-) 3) ENTER .

Take Note

Use the down arrow key to scroll past Y₇ to see Y₈, Y₉, and Y₀.

Note: If you try to evaluate a function at a number that is not in the domain of the function, you will get an error message. For instance, 1 is not in the domain of $f(x) = \frac{x^2}{x-1}$. If we try to evaluate the function at 1, the error screen at the right appears.

Evaluating Valuable Expressions

To evaluate a variable expression, first store the values of each variable. Then enter the variable expression. For instance, to evaluate $s^2 + 2sl$ when $s = 4$ and $l = 5$, use the following keystrokes.

4 STO▸ ALPHA S ENTER 5 STO▸ ALPHA L ENTER ALPHA S x^2 + 2 ALPHA S ALPHA L ENTER

Graph

To graph a function, use the Y= key to enter the expression for the function, select a suitable viewing window, and then press GRAPH . For instance, to graph $f(x) = 0.1x^3 - 2x - 1$ in the standard viewing window, use the following keystrokes.

Y= 0.1 X,T,θ,n ^ 3 − 2 X,T,θ,n − 1 ZOOM (scroll to 6) ENTER

Note: For the keystrokes above, you do not have to scroll to 6. Alternatively, use ZOOM 6. This will select the standard viewing window and automatically start the graph. Use the WINDOW key to create a custom window for a graph.

Graphing Inequalities

To illustrate this feature, we will graph $y \leq 2x - 1$. Enter $2x - 1$ into Y₁. Because $y \leq 2x - 1$, we want to shade below the graph. Move the cursor to the left of Y₁ and press ENTER three times. Press GRAPH .

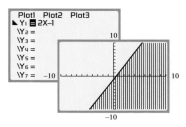

Note: To shade above the graph, move the cursor to the left of Y₁ and press ENTER two times. An inequality with the symbol $\leq$ or $\geq$ should be graphed with a solid line, and an inequality with the symbol $<$ or $>$ should be graphed with a dashed line. However, the graph of a linear inequality on a graphing calculator does not distinguish between a solid line and a dashed line.

To graph the solution set of a system of inequalities, solve each inequality for y and graph each inequality. The solution set is the intersection of the two inequalities. The solution set of $\begin{array}{l} 3x + 2y > 10 \\ 4x - 3y \leq 5 \end{array}$ is shown at the right.

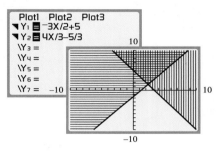

Intersect

The INTERSECT feature is used to solve a system of equations. To illustrate this feature, we will use the system of equations $\begin{array}{l} 2x - 3y = 13 \\ 3x + 4y = -6 \end{array}$.

Note: Some equations can be solved by this method. See the section "Solve an equation" below. Also, this method is used to find a number in the domain of a function for a given number in the range. See the section "Find a domain element."

Solve each of the equations in the system of equations for y. In this case, we have $y = \frac{2}{3}x - \frac{13}{3}$ and $y = -\frac{3}{4}x - \frac{3}{2}$.

Use the Y= editor to enter $\frac{2}{3}x - \frac{13}{3}$ into Y₁ and $-\frac{3}{4}x - \frac{3}{2}$ into Y₂. Graph the two functions in the standard viewing window. (If the window does not show the point of intersection of the two graphs, adjust the window until you can see the point of intersection.)

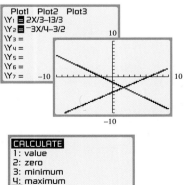

Press 2nd CALC (scroll to 5, intersect) ENTER .

Alternatively, you can just press 2nd CALC 5.

First curve? is shown at the bottom of the screen and identifies one of the two graphs on the screen. Press ENTER.

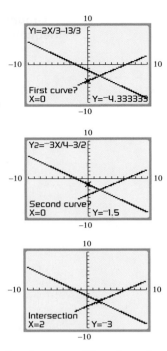

Second curve? is shown at the bottom of the screen and identifies the second of the two graphs on the screen. Press ENTER.

Guess? shown at the bottom of the screen asks you to use the left or right arrow key to move the cursor to the *approximate* location of the point of intersection. (If there are two or more points of intersection, it does not matter which one you choose first.) Press ENTER.

The solution of the system of equations is $(2, -3)$.

Solve an equation To illustrate the steps involved, we will solve the equation $2x + 4 = -3x - 1$. The idea is to write the equation as the system of equations $\begin{aligned} y &= 2x + 4 \\ y &= -3x - 1 \end{aligned}$ and then use the steps for solving a system of equations.

Use the Y= editor to enter the left and right sides of the equation into Y_1 and Y_2. Graph the two functions and then follow the steps for **Intersect**.

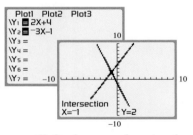

The solution is -1, the x-coordinate of the point of intersection.

Find a domain element For this example, we will find a number in the domain of $f(x) = -\dfrac{2}{3}x + 2$ that corresponds to 4 in the range of the function. This is like solving the system of equations $y = -\dfrac{2}{3}x + 2$ and $y = 4$.

Use the Y= editor to enter the expression for the function in Y_1 and the desired output, 4, in Y_2. Graph the two functions and then follow the steps for **Intersect**.

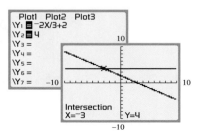

The point of intersection is $(-3, 4)$. The number -3 in the domain of f produces an output of 4 in the range of f.

Math Pressing MATH gives you access to many built-in functions. The following keystrokes will convert 0.125 into a fraction: .125 MATH 1 ENTER .

Additional built-in functions under [MATH] can be found by pressing [MATH] ◖. For instance, to evaluate $-|-25|$, press (-) [MATH] ◖ 1 (-) 25 ▌)▐ [ENTER] .

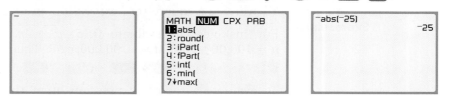

See your owner's manual for assistance with other functions under the [MATH] key.

Min and Max The local minimum and the local maximum values of a function are calculated by accessing the CALC menu. For this demonstration, we will find the minimum value and the maximum value of $f(x) = 0.2x^3 + 0.3x^2 - 3.6x + 2$.

Enter the function into Y₁. Press [2nd] CALC (scroll to 3 for minimum of the function) [ENTER] .

Alternatively, you can just press [2nd] CALC 3.

Left Bound? shown at the bottom of the screen asks you to use the left or right arrow key to move the cursor to the *left* of the minimum. Press [ENTER] .

Right Bound? shown at the bottom of the screen asks you to use the left or right arrow key to move the cursor to the *right* of the minimum. Press [ENTER] .

Guess? shown at the bottom of the screen asks you to use the left or right arrow key to move the cursor to the *approximate* location of the minimum. Press [ENTER] .

The minimum value of the function is the y-coordinate. For this example, the minimum value of the function is -2.4.

The x-coordinate for the minimum is 2. However, because of rounding errors in the calculation, it is shown as a number close to 2.

To find the maximum value of the function, follow the same steps as above, except select maximum under the CALC menu. The screens for this calculation are shown at the top of the next page.

The maximum value of the function is 10.1.

Radical Expressions

To evaluate a square-root expression, press ⟦2nd⟧ √ .

For instance, to evaluate $0.15\sqrt{p^2 + 4p + 10}$ when $p = 100,000$, first store 100,000 in P. Then press 0.15 ⟦2nd⟧ √ ⟦ALPHA⟧ P ⟦x²⟧ ⟦+⟧ 4 ⟦ALPHA⟧ P ⟦+⟧ 10 ⟦)⟧ ⟦ENTER⟧ .

To evaluate a radical expression other than a square root, access $\sqrt[x]{\ }$ by pressing ⟦MATH⟧ . For instance, to evaluate $\sqrt[4]{67}$, press 4 (the index of the radical) ⟦MATH⟧ (scroll to 5) ⟦ENTER⟧ 67 ⟦ENTER⟧ .

Scientific Notation

To enter a number in scientific notation, use ⟦2nd⟧ EE. For instance, to find $\dfrac{3.45 \times 10^{-12}}{1.5 \times 10^{25}}$, press 3.45 ⟦2nd⟧ EE ⟦(-)⟧ 12 ⟦÷⟧ 1.5 ⟦2nd⟧ EE 25 ⟦ENTER⟧ . The answer is 2.3×10^{-37}.

Table

There are three steps in creating an input/output table for a function. First use the ⟦Y=⟧ editor to input the function. The second step is setting up the table, and the third step is displaying the table.

To set up the table, press ⟦2nd⟧ TBLSET. **TblStart** is the first value of the independent variable in the input/output table. △**Tbl** is the difference between successive values. Setting this to 1 means that, for this table, the input values are $-2, -1, 0, 1, 2, \ldots$. If △**Tbl** = 0.5, then the input values are $-2, -1.5, -1, -0.5, 0, 0.5, \ldots$.

Indpnt is the independent variable. When this is set to **Auto**, values of the independent variable are automatically entered into the table. **Depend** is the dependent variable. When this is set to **Auto**, values of the dependent variable are automatically entered into the table.

To display the table, press ⟦2nd⟧ TABLE. An input/output table for $f(x) = x^2 - 1$ is shown at the right.

Once the table is on the screen, the up and down arrow keys can be used to display more values in the table. For the table at the right, we used the up arrow key to move to $x = -7$.

An input/output table for any given input can be created by selecting **Ask** for the independent variable. The table at the right shows an input/output table for $f(x) = \dfrac{4x}{x-2}$ for selected values of x. Note that **ERROR** or **ERR:** appeared when 2 was entered. This occurred because f is not defined when $x = 2$.

Note: Using the table feature in **Ask** mode is the same as evaluating a function for given values of the independent variable. For instance, from the table at the right, we have $f(4) = 8$.

Test The TEST feature has many uses, one of which is to graph the solution set of a linear inequality in one variable. To illustrate this feature, we will graph the solution set of $x - 1 < 4$. Press ⬤**Y=** ⬤**X,T,θ,n** ⬤**−** 1 ⬤**2nd** TEST (scroll to 5) ⬤**ENTER** 4 ⬤**GRAPH** .

Trace Once a graph is drawn, pressing ⬤**TRACE** will place a cursor on the screen, and the coordinates of the point below the cursor are shown at the bottom of the screen. Use the left and right arrow keys to move the cursor along the graph. For the graph at the right, we have $f(4.8) = 3.4592$, where $f(x) = 0.1x^3 - 2x + 2$ is shown at the top left of the screen.

In TRACE mode, you can evaluate a function at any value of the independent variable that is within Xmin and Xmax. To do this, first graph the function. Now press ⬤**TRACE** (the value of x) ⬤**ENTER** . For the graph at the left below, we used $x = -3.5$. If a value of x is chosen outside the window, an error message is displayed.

In the example above where we entered -3.5 for x, the value of the function was calculated as 4.7125. This means that $f(-3.5) = 4.7125$. The keystrokes (2nd) QUIT (VARS) ◊ 11 (MATH) 1 (ENTER) will convert the decimal value into a fraction.

When the TRACE feature is used with two or more graphs, the up and down arrow keys are used to move between the graphs. The graphs below are for the functions $f(x) = 0.1x^3 - 2x + 2$ and $g(x) = 2x - 3$. By using the up and down arrows, we can place the cursor on either graph. The right and left arrows are used to move along the graph.

Window The viewing window for a graph is controlled by pressing (WINDOW). Xmin and Xmax are the minimum value and maximum value, respectively, of the independent variable shown on the graph. Xscl is the distance between tick marks on the x-axis. Ymin and Ymax are the minimum value and maximum value, respectively, of the dependent variable shown on the graph. Yscl is the distance between tick marks on the y-axis. Leave Xres as 1.

Note: In the standard viewing window, the distance between tick marks on the x-axis is different from the distance between tick marks on the y-axis. This will distort a graph. A more accurate picture of a graph can be created by using a square viewing window. See ZOOM.

(Y=) The (Y=) editor is used to enter the expression for a function. There are ten possible functions, labeled Y_1 to Y_0, that can be active at any one time. For instance, to enter $f(x) = x^2 + 3x - 2$ as Y_1, use the following keystrokes.

(Y=) (X,T,θ,n) (x^2) (+) 3 (X,T,θ,n) (−) 2

Note: If an expression is already entered for Y_1, place the cursor anywhere on that expression and press (CLEAR).

To enter $s = \dfrac{2v - 1}{v^3 - 3}$ into Y_2, place the cursor to the right

of the equals sign for Y_2. Then press (2 (X,T,θ,n) (−)

1) (÷) ((X,T,θ,n) (^) 3 (−) 3).

Note: When we enter an equation, the independent variable (*v* in the expression above) is entered using X,T,θ,*n*. The dependent variable (*s* in the expression above) is one of Y₁ to Y₀. Also note the use of parentheses to ensure the correct order of operations.

Observe the black rectangle that covers the equals sign for the two examples we have shown. This rectangle means that the function is "active." If we were to press GRAPH, then the graph of both functions would appear. You can make a function inactive by using the arrow keys to move the cursor over the equals sign of that function and then pressing ENTER. This will remove the black rectangle. We have done that for Y₂, as shown above. Now if GRAPH is pressed, only Y₁ will be graphed.

```
Plot1  Plot2  Plot3
\Y₁ ■ X²+3X−2
\Y₂ = (2X−1)/(X^3−3)
\Y₃ =
\Y₄ =
\Y₅ =
\Y₆ =
\Y₇ =
```

It is also possible to control the appearance of the graph by moving the cursor on the Y= screen to the left of any Y. With the cursor in this position, pressing ENTER will change the appearance of the graph. The options are shown at the right.

```
Plot1  Plot2  Plot3
\Y₁ = Default graph line
 Y₂ = Bold graph line
▜Y₃ = Shade above graph
▙Y₄ = Shade below graph
-₀Y₅ = Draw path of graph
₀Y₆ = Travel path of graph
∶Y₇ = Dashed graph line
```

Zero The ZERO feature of a graphing calculator is used for various calculations: to find the *x*-intercepts of a function, to solve some equations, and to find the zero of a function.

x-intercepts To illustrate the procedure for finding *x*-intercepts, we will use $f(x) = x^2 + x - 2$.

First, use the Y= editor to enter the expression for the function and then graph the function in the standard viewing window. (It may be necessary to adjust this window so that the intercepts are visible.) Once the graph is displayed, use the following keystrokes to find the *x*-intercepts of the graph of the function.

Press 2nd CALC (scroll to 2 for **zero** of the function) ENTER.

Alternatively, you can just press 2nd CALC 2.

```
CALCULATE
1: value
2: zero
3: minimum
4: maximum
5: intersect
6: dy/dx
7: ∫f(x)dx
```

Left Bound? shown at the bottom of the screen asks you to use the left or right arrow key to move the cursor to the *left* of the desired *x*-intercept. Press ENTER.

Right Bound? shown at the bottom of the screen asks you to use the left or right arrow key to move the cursor to the *right* of the desired *x*-intercept. Press ENTER.

Guess? shown at the bottom of the screen asks you to use the left or right arrow key to move the cursor to the *approximate* location of the desired *x*-intercept. Press ENTER .

The *x*-coordinate of an *x*-intercept is −2. Therefore, an *x*-intercept is (−2, 0).

To find the other *x*-intercept, follow the same steps as above. The screens for this calculation are shown below.

A second *x*-intercept is (1, 0).

Solve an equation To use the ZERO feature to solve an equation, first rewrite the equation with all terms on one side. For instance, one way to solve the equation $x^3 - x + 1 = -2x + 3$ is first to rewrite it as $x^3 + x - 2 = 0$. Enter $x^3 + x - 2$ into Y_1 and then follow the steps for finding *x*-intercepts.

Find the real zeros of a function To find the real zeros of a function, follow the steps for finding *x*-intercepts.

Zoom Pressing ZOOM enables you to select some preset viewing windows. This key also gives you access to ZBox, Zoom In, and Zoom Out. These functions enable you to redraw a selected portion of a graph in a new window. Some windows that are used frequently in this text are shown below.

ZOOM MEMORY
1: ZBox
2: Zoom In
3: Zoom Out
4: ZDecimal
5: ZSquare
6: ZStandard
7↓ ZTrig

WINDOW
Xmin = ‾4.7
Xmax = 4.7
Xscl = 1
Ymin = ‾3.1
Ymax = 3.1
Yscl = 1
Xres = 1

ZOOM MEMORY
1: ZBox
2: Zoom In
3: Zoom Out
4: ZDecimal
5: ZSquare
6: ZStandard
7↓ ZTrig

WINDOW
Xmin = ‾15.16129...
Xmax = 15.161290...
Xscl = 1
Ymin = ‾10
Ymax = 10
Yscl = 1
Xres = 1

ZOOM MEMORY
1: ZBox
2: Zoom In
3: Zoom Out
4: ZDecimal
5: ZSquare
6: ZStandard
7↓ ZTrig

WINDOW
Xmin = ‾10
Xmax = 10
Xscl = 1
Ymin = ‾10
Ymax = 10
Yscl = 1
Xres = 1

ZOOM MEMORY
4↑ ZDecimal
5: ZSquare
6: ZStandard
7: ZTrig
8: ZInteger
9: ZoomStat
0: ZoomFit

WINDOW
Xmin = ‾47
Xmax = 47
Xscl = 10
Ymin = ‾31
Ymax = 31
Yscl = 10
Xres = 1

Solutions to Chapter Problems

Solutions to Chapter 1 Problems

SECTION 1.1

Problem 1
$A = \{1, 2, 3, 4\}$

Problem 2
$-5 < -1$
$-1 = -1$
$5 > -1$

The element 5 is greater than -1.

Problem 3
A. 9 **B.** -62

Problem 4
A. $|-5| = 5$ **B.** $-|-9| = -9$

SECTION 1.2

Problem 1
A. $-162 + 98 = -64$

B. $-154 + (-37) = -191$

C. $-36 + 17 + (-21) = -19 + (-21)$
$ = -40$

Problem 2
$-8 - 14 = -8 + (-14)$
$ = -22$

Problem 3
$4 - (-3) - 12 - (-7) - 20$
$= 4 + 3 + (-12) + 7 + (-20)$
$= 7 + (-12) + 7 + (-20)$
$= -5 + 7 + (-20)$
$= 2 + (-20)$
$= -18$

Problem 4
A. $-38 \cdot 51 = -1938$

B. $-7(-8)(9)(-2) = 56(9)(-2)$
$ = 504(-2)$
$ = -1008$

Problem 5
A. $(-135) \div (-9) = 15$

B. $\dfrac{84}{-6} = -14$

C. $-\dfrac{36}{-12} = -(-3) = 3$

Problem 6
Strategy To find the average daily high temperature:
▶ Add the seven temperature readings.
▶ Divide the sum by 7.

Solution $-5 + (-6) + 3 + 0 + (-4) + (-7) + (-2)$
$= -11 + 3 + 0 + (-4) + (-7) + (-2)$
$= -8 + 0 + (-4) + (-7) + (-2)$
$= -8 + (-4) + (-7) + (-2)$
$= -12 + (-7) + (-2)$
$= -19 + (-2)$
$= -21$

$-21 \div 7 = -3$

The average daily high temperature was $-3°C$.

SECTION 1.3

Problem 1

$$\begin{array}{r} 0.16 \\ 25\overline{)4.00} \\ \underline{-2\,5} \\ 1\,50 \\ \underline{-1\,50} \\ 0 \end{array}$$

$\dfrac{4}{25} = 0.16$

Problem 2

$$\begin{array}{r} 0.444 \\ 9\overline{)4.000} \\ \underline{-3\,6} \\ 40 \\ \underline{-36} \\ 40 \\ \underline{-36} \\ 4 \end{array}$$

$\dfrac{4}{9} = 0.\overline{4}$

Problem 3
$$-\frac{7}{12} \cdot \frac{9}{14} = -\frac{7 \cdot 9}{12 \cdot 14} = -\frac{\overset{1}{\cancel{7}} \cdot \overset{1}{\cancel{3}} \cdot 3}{2 \cdot 2 \cdot \underset{1}{\cancel{3}} \cdot 2 \cdot \underset{1}{\cancel{7}}}$$
$$= -\frac{3}{8}$$

Problem 4
$$-\frac{3}{8} \div \left(-\frac{5}{12}\right) = \frac{3}{8} \cdot \frac{12}{5}$$
$$= \frac{3 \cdot 12}{8 \cdot 5} = \frac{3 \cdot \overset{1}{\cancel{2}} \cdot \overset{1}{\cancel{2}} \cdot 3}{\underset{1}{\cancel{2}} \cdot \underset{1}{\cancel{2}} \cdot 2 \cdot 5} = \frac{9}{10}$$

Problem 5

$$
\begin{array}{r}
5.44 \\
\times \quad 3.8 \\
\hline
4352 \\
1632 \\
\hline
20.672
\end{array}
$$

$(-5.44)(3.8) = -20.672$

Problem 6

$$
\begin{array}{r}
0.231 \\
1.7.\overline{)0.3.940} \\
-3\,4 \\
\hline
54 \\
-51 \\
\hline
30 \\
-17 \\
\hline
13
\end{array}
$$

$-0.394 \div 1.7 \approx -0.23$

Problem 7

Prime factorization of 9 and 12:

$9 = 3 \cdot 3 \qquad 12 = 2 \cdot 2 \cdot 3$

$\text{LCM} = 2 \cdot 2 \cdot 3 \cdot 3 = 36$

$$
\frac{5}{9} - \frac{11}{12} = \frac{20}{36} - \frac{33}{36} = \frac{20}{36} + \frac{-33}{36}
$$

$$
= \frac{20 + (-33)}{36} = \frac{-13}{36} = -\frac{13}{36}
$$

Problem 8

$$
-\frac{7}{8} - \frac{5}{6} + \frac{1}{2} = -\frac{21}{24} - \frac{20}{24} + \frac{12}{24}
$$

$$
= \frac{-21}{24} + \frac{-20}{24} + \frac{12}{24}
$$

$$
= \frac{-21 + (-20) + 12}{24} = \frac{-29}{24} = -\frac{29}{24}
$$

Problem 9

$$
\begin{array}{r}
3.097 \\
4.9 \\
+ \; 3.09 \\
\hline
11.087
\end{array}
$$

Problem 10

$$
\begin{array}{r}
67.910 \\
-16.127 \\
\hline
51.783
\end{array}
$$

$16.127 - 67.91 = -51.783$

Problem 11

$$
125\% = 125\left(\frac{1}{100}\right) = \frac{125}{100} = 1\frac{1}{4}
$$

$$
125\% = 125(0.01) = 1.25
$$

Problem 12

$$
16\frac{2}{3}\% = 16\frac{2}{3}\left(\frac{1}{100}\right) = \frac{50}{3}\left(\frac{1}{100}\right) = \frac{1}{6}
$$

Problem 13

$6.08\% = 6.08(0.01) = 0.0608$

Problem 14

A. $0.043 = 0.043(100\%) = 4.3\%$

B. $2.57 = 2.57(100\%) = 257\%$

Problem 15

$$
\frac{5}{9} = \frac{5}{9}(100\%) = \frac{500}{9}\% \approx 55.6\%
$$

Problem 16

$$
\frac{9}{16} = \frac{9}{16}(100\%) = \frac{900}{16}\% = 56\frac{1}{4}\%
$$

SECTION 1.4

Problem 1

$(-5)^3 = (-5)(-5)(-5) = 25(-5) = -125$

$-5^3 = -(5 \cdot 5 \cdot 5) = -(25 \cdot 5) = -125$

Problem 2

$(-3)^3 = (-3)(-3)(-3) = 9(-3) = -27$

$(-3)^4 = (-3)(-3)(-3)(-3)$

$\quad = 9(-3)(-3) = -27(-3) = 81$

Problem 3

$(3^3)(-2)^3 = (3)(3)(3) \cdot (-2)(-2)(-2)$

$\quad = 27 \cdot (-8) = -216$

$$
\left(-\frac{2}{5}\right)^2 = \left(-\frac{2}{5}\right)\left(-\frac{2}{5}\right) = \frac{2 \cdot 2}{5 \cdot 5} = \frac{4}{25}
$$

Problem 4

$36 \div (8 - 5)^2 - (-3)^2 \cdot 2$

$\quad = 36 \div (3)^2 - (-3)^2 \cdot 2$

$\quad = 36 \div 9 - 9 \cdot 2$

$\quad = 4 - 9 \cdot 2$

$\quad = 4 - 18$

$\quad = -14$

Problem 5

$27 \div 3^2 + (-3)^2 \cdot 4$

$\quad = 27 \div 9 + 9 \cdot 4$

$\quad = 3 + 9 \cdot 4$

$\quad = 3 + 36$

$\quad = 39$

Problem 6

$4 - 3[4 - 2(6 - 3)] \div 2$

$\quad = 4 - 3[4 - 2(3)] \div 2$

$\quad = 4 - 3[4 - 6] \div 2$

$\quad = 4 - 3[-2] \div 2$

$\quad = 4 - (-6) \div 2$

$\quad = 4 - (-3)$

$\quad = 7$

SECTION 1.5

Problem 1
-4

Problem 2
$2xy + y^2$
$2(-4)(2) + (2)^2 = 2(-4)(2) + 4$
$\qquad\qquad\qquad = -8(2) + 4$
$\qquad\qquad\qquad = -16 + 4$
$\qquad\qquad\qquad = -12$

Problem 3
$\dfrac{a^2 + b^2}{a + b}$
$\dfrac{(5)^2 + (-3)^2}{5 + (-3)} = \dfrac{25 + 9}{5 + (-3)}$
$\qquad\qquad\quad = \dfrac{34}{2}$
$\qquad\qquad\quad = 17$

Problem 4
$x^3 - 2(x + y) + z^2$
$(2)^3 - 2[2 + (-4)] + (-3)^2$
$\quad = (2)^3 - 2(-2) + (-3)^2$
$\quad = 8 - 2(-2) + 9$
$\quad = 8 + 4 + 9$
$\quad = 12 + 9$
$\quad = 21$

Problem 5
$V = \dfrac{1}{3}\pi r^2 h$
$V = \dfrac{1}{3}\pi(4.5)^2(9.5) \qquad \bullet\, r = \dfrac{1}{2}d = \dfrac{1}{2}(9) = 4.5$
$V = \dfrac{1}{3}\pi(20.25)(9.5)$
$V \approx 201.5$

The volume is approximately 201.5 cm³.

Problem 6
$7 + (-7) = 0$

Problem 7
The Associative Property of Addition

Problem 8
A. $9x + 6x = (9 + 6)x = 15x$
B. $-4y - 7y = [-4 + (-7)]y = -11y$

Problem 9
A. $3a - 2b + 5a = 3a + 5a - 2b$
$\qquad\qquad\qquad = (3a + 5a) - 2b$
$\qquad\qquad\qquad = 8a - 2b$

B. $x^2 - 7 + 9x^2 - 14$
$\quad = x^2 + 9x^2 - 7 - 14$
$\quad = (x^2 + 9x^2) + (-7 - 14)$
$\quad = 10x^2 - 21$

Problem 10
A. $-7(-2a) = [-7(-2)]a$
$\qquad\qquad = 14a$

B. $-\dfrac{5}{6}(-30y^2) = \left[-\dfrac{5}{6}(-30)\right]y^2$
$\qquad\qquad\qquad = 25y^2$

C. $(-5x)(-2) = (-2)(-5x)$
$\qquad\qquad\quad = [-2(-5)]x$
$\qquad\qquad\quad = 10x$

Problem 11
A. $7(4 + 2y) = 7(4) + 7(2y)$
$\qquad\qquad\quad = 28 + 14y$

B. $-(5x - 12) = -1(5x - 12)$
$\qquad\qquad\quad = -1(5x) - (-1)(12)$
$\qquad\qquad\quad = -5x + 12$

C. $(3a - 1)5 = (3a)(5) - (1)(5)$
$\qquad\qquad\quad = 15a - 5$

D. $-3(6a^2 - 8a + 9)$
$\quad = -3(6a^2) - (-3)(8a) + (-3)(9)$
$\quad = -18a^2 + 24a - 27$

Problem 12
$7(x - 2y) - 3(-x - 2y)$
$\quad = 7x - 14y + 3x + 6y$
$\quad = 10x - 8y$

Problem 13
$3y - 2[x - 4(2 - 3y)]$
$\quad = 3y - 2[x - 8 + 12y]$
$\quad = 3y - 2x + 16 - 24y$
$\quad = -2x - 21y + 16$

SECTION 1.6

Problem 1
A. eighteen <u>less than</u> the <u>cube</u> of x
$\quad x^3 - 18$

B. y <u>decreased by</u> the <u>sum</u> of z and nine
$\quad y - (z + 9)$

C. the <u>difference between</u> the <u>square</u> of q
$\quad$ and the <u>sum</u> of r and t
$\quad q^2 - (r + t)$

Problem 2
the unknown number: n
the square of the number: n^2
the product of five and the square of the number: $5n^2$
$5n^2 + n$

Problem 3
the unknown number: n
twice the number: $2n$
the sum of seven and twice the number: $7 + 2n$
$3(7 + 2n)$

Problem 4

the unknown number: n

twice the number: $2n$

the difference between twice the number and 17:

$2n - 17$

$n - (2n - 17) = n - 2n + 17$

$\qquad\qquad\qquad = -n + 17$

Problem 5

the unknown number: n

three-fourths of the number: $\dfrac{3}{4}n$

one-fifth of the number: $\dfrac{1}{5}n$

$\dfrac{3}{4}n + \dfrac{1}{5}n = \dfrac{15}{20}n + \dfrac{4}{20}n$

$\qquad\qquad = \dfrac{19}{20}n$

Problem 6

the time required by the newer model: t

the time required by the older model is twice the time required for the newer model: $2t$

Problem 7

the length of one piece: L

the length of the second piece: $6 - L$

Solutions to Chapter 2 Problems

SECTION 2.1

Problem 1

$$5 - 4x = 8x + 2$$

$$\begin{array}{c|c} 5 - 4\left(\dfrac{1}{4}\right) & 8\left(\dfrac{1}{4}\right) + 2 \\ 5 - 1 & 2 + 2 \\ & 4 = 4 \end{array}$$

Yes, $\dfrac{1}{4}$ is a solution.

Problem 2

$$10x - x^2 = 3x - 10$$

$$\begin{array}{c|c} 10(5) - (5)^2 & 3(5) - 10 \\ 50 - 25 & 15 - 10 \\ & 25 \neq 5 \end{array}$$

No, 5 is not a solution.

Problem 3

$$x - \dfrac{1}{3} = -\dfrac{3}{4}$$

$$x - \dfrac{1}{3} + \dfrac{1}{3} = -\dfrac{3}{4} + \dfrac{1}{3}$$

$$x + 0 = -\dfrac{9}{12} + \dfrac{4}{12}$$

$$x = -\dfrac{5}{12}$$

Check $\qquad x - \dfrac{1}{3} = -\dfrac{3}{4}$

$$\begin{array}{c|c} -\dfrac{5}{12} - \dfrac{1}{3} & -\dfrac{3}{4} \\ -\dfrac{3}{4} & = -\dfrac{3}{4} \end{array}$$

The solution is $-\dfrac{5}{12}$.

Problem 4

$$-8 = 5 + x$$

$$-8 - 5 = 5 - 5 + x$$

$$-13 = x$$

The solution is -13.

Problem 5

$$-\dfrac{2x}{5} = 6$$

$$\left(-\dfrac{5}{2}\right)\left(-\dfrac{2}{5}x\right) = \left(-\dfrac{5}{2}\right)(6)$$

$$1x = -15$$

$$x = -15$$

The solution is -15.

Problem 6

$$6x = 10$$

$$\dfrac{6x}{6} = \dfrac{10}{6}$$

$$x = \dfrac{5}{3}$$

The solution is $\dfrac{5}{3}$.

Problem 7

$$4x - 8x = 16$$

$$-4x = 16$$

$$\dfrac{-4x}{-4} = \dfrac{16}{-4}$$

$$x = -4$$

The solution is -4.

Problem 8

$$PB = A$$

$$P(60) = 27$$

$$60P = 27$$

$$\dfrac{60P}{60} = \dfrac{27}{60}$$

$$P = 0.45$$

27 is 45% of 60.

Problem 9

Strategy To find the percent of the questions answered correctly, solve the basic percent equation using $B = 80$ and $A = 72$. The percent is unknown.

Solution
$$PB = A$$
$$P(80) = 72$$
$$80P = 72$$
$$\frac{80P}{80} = \frac{72}{80}$$
$$P = 0.9$$

90% of the questions were answered correctly.

Problem 10

Strategy To determine the sales tax, solve the basic percent equation using $B = 895$ and $P = 6\% = 0.06$. The amount is unknown.

Solution
$$PB = A$$
$$895(0.06) = A$$
$$53.7 = A$$

The sales tax is $53.70.

Problem 11

Strategy To determine how much she must deposit into the bank account:

▶ Find the amount of interest earned on the municipal bond by solving the equation $I = Prt$ for I using $P = 1000$, $r = 6.4\% = 0.064$, and $t = 1$.

▶ Solve $I = Prt$ for P using the amount of interest earned on the municipal bond for I, $r = 8\% = 0.08$, and $t = 1$.

Solution $I = Prt$
$$I = 1000(0.064)(1)$$
$$I = 64$$

The interest earned on the municipal bond is $64.

$$I = Prt$$
$$64 = P(0.08)(1)$$
$$64 = 0.08P$$
$$\frac{64}{0.08} = \frac{0.08P}{0.08}$$
$$800 = P$$

Clarissa must deposit $800 into the bank account.

Problem 12

Strategy To find the number of ounces of cereal in the bowl, solve the equation $Q = Ar$ for A using $Q = 2$ and $r = 25\% = 0.25$.

Solution
$$Q = Ar$$
$$2 = A(0.25)$$

$$\frac{2}{0.25} = \frac{A(0.25)}{0.25}$$
$$8 = A$$

The cereal bowl contains 8 oz of cereal.

SECTION 2.2

Problem 1

$$5x + 7 = 10$$
$$5x + 7 - 7 = 10 - 7$$
$$5x = 3$$
$$\frac{5x}{5} = \frac{3}{5}$$
$$x = \frac{3}{5}$$

The solution is $\frac{3}{5}$.

Problem 2

$$11 = 11 + 3x$$
$$11 - 11 = 11 - 11 + 3x$$
$$0 = 3x$$
$$\frac{0}{3} = \frac{3x}{3}$$
$$0 = x$$

The solution is 0.

Problem 3

$$\frac{5}{2} - \frac{2}{3}x = \frac{1}{2}$$
$$6\left(\frac{5}{2} - \frac{2}{3}x\right) = 6\left(\frac{1}{2}\right)$$
$$6\left(\frac{5}{2}\right) - 6\left(\frac{2}{3}x\right) = 3$$
$$15 - 4x = 3$$
$$15 - 15 - 4x = 3 - 15$$
$$-4x = -12$$
$$\frac{-4x}{-4} = \frac{-12}{-4}$$
$$x = 3$$

The solution is 3.

Problem 4

$$5x + 4 = 6 + 10x$$
$$5x - 10x + 4 = 6 + 10x - 10x$$
$$-5x + 4 = 6$$
$$-5x + 4 - 4 = 6 - 4$$
$$-5x = 2$$
$$\frac{-5x}{-5} = \frac{2}{-5}$$
$$x = -\frac{2}{5}$$

The solution is $-\frac{2}{5}$.

Problem 5

$$5x - 4(3 - 2x) = 2(3x - 2) + 6$$
$$5x - 12 + 8x = 6x - 4 + 6$$
$$13x - 12 = 6x + 2$$
$$13x - 6x - 12 = 6x - 6x + 2$$
$$7x - 12 = 2$$
$$7x - 12 + 12 = 2 + 12$$
$$7x = 14$$
$$\frac{7x}{7} = \frac{14}{7}$$
$$x = 2$$

The solution is 2.

Problem 6

$$-2[3x - 5(2x - 3)] = 3x - 8$$
$$-2[3x - 10x + 15] = 3x - 8$$
$$-2[-7x + 15] = 3x - 8$$
$$14x - 30 = 3x - 8$$
$$14x - 3x - 30 = 3x - 3x - 8$$
$$11x - 30 = -8$$
$$11x - 30 + 30 = -8 + 30$$
$$11x = 22$$
$$\frac{11x}{11} = \frac{22}{11}$$
$$x = 2$$

The solution is 2.

Problem 7

Strategy To find the number of years, replace V with 10,200 in the given equation and solve for t.

Solution
$$V = 450t + 7500$$
$$10,200 = 450t + 7500$$
$$10,200 - 7500 = 450t + 7500 - 7500$$
$$2700 = 450t$$
$$\frac{2700}{450} = \frac{450t}{450}$$
$$6 = t$$

The value of the investment will be $10,200 in 6 years.

Problem 8

Strategy The lever is 14 ft long, so $d = 14$. One force is 6 ft from the fulcrum, so $x = 6$. The one force is 40 lb, so $F_1 = 40$. To find the other force when the system balances, replace the variables F_1, x, and d in the lever system equation with the given values, and solve for F_2.

Solution
$$F_1 x = F_2(d - x)$$
$$40(6) = F_2(14 - 6)$$
$$240 = F_2(8)$$
$$240 = 8F_2$$
$$\frac{240}{8} = \frac{8F_2}{8}$$
$$30 = F_2$$

A 30-pound force must be applied to the other end.

SECTION 2.3

Problem 1

the unknown number: n

nine less than twice a number	is	five times the sum of the number and twelve

$$2n - 9 = 5(n + 12)$$
$$2n - 9 = 5n + 60$$
$$2n - 5n - 9 = 5n - 5n + 60$$
$$-3n - 9 = 60$$
$$-3n - 9 + 9 = 60 + 9$$
$$-3n = 69$$
$$\frac{-3n}{-3} = \frac{69}{-3}$$
$$n = -23$$

The number is -23.

Problem 2

Strategy To find the number of carbon atoms, write and solve an equation using n to represent the number of carbon atoms.

Solution

eight	represents	twice the number of carbon atoms

$$8 = 2n$$
$$\frac{8}{2} = \frac{2n}{2}$$
$$4 = n$$

There are 4 carbon atoms in a butane molecule.

Problem 3

Strategy To find the number of 10-speed bicycles made each day, write and solve an equation using n to represent the number of 10-speed bicycles and $160 - n$ to represent the number of 3-speed bicycles.

Solution

four times the number of 3-speed bicycles made	equals	30 less than the number of 10-speed bicycles made

$$4(160 - n) = n - 30$$
$$640 - 4n = n - 30$$
$$640 - 4n - n = n - n - 30$$
$$640 - 5n = -30$$
$$640 - 640 - 5n = -30 - 640$$
$$-5n = -670$$
$$\frac{-5n}{-5} = \frac{-670}{-5}$$
$$n = 134$$

There are 134 10-speed bicycles made each day.

Problem 4

Strategy ▶ First consecutive integer: n
Second consecutive integer: $n + 1$
Third consecutive integer: $n + 2$

▶ The sum of the three integers is -12.

Solution
$$n + (n + 1) + (n + 2) = -12$$
$$3n + 3 = -12$$
$$3n = -15$$
$$n = -5$$

$$n + 1 = -5 + 1 = -4$$
$$n + 2 = -5 + 2 = -3$$

The three consecutive integers are -5, -4, and -3.

Problem 5

Strategy ▶ Pounds of $4.00 hamburger: x
Pounds of $2.80 hamburger: $75 - x$

	Amount	Cost	Value
$4.00 hamburger	x	4.00	$4.00x$
$2.80 hamburger	$75 - x$	2.80	$2.80(75 - x)$
Mixture	75	3.20	$75(3.20)$

▶ The sum of the values before mixing equals the value after mixing.

Solution
$$4.00x + 2.80(75 - x) = 75(3.20)$$
$$4x + 210 - 2.80x = 240$$
$$1.2x + 210 = 240$$
$$1.2x = 30$$
$$x = 25$$

$$75 - x = 75 - 25 = 50$$

The mixture must contain 25 lb of the $4.00 hamburger and 50 lb of the $2.80 hamburger.

Problem 6

Strategy The distance is 350 mi. Therefore, $d = 350$. The plane is flying into a headwind, so the rate of the plane is the difference between its rate in calm air (175 mph) and the rate of the wind (35 mph): 175 mph − 35 mph = 140 mph. Therefore, $r = 140$. To find the time, solve the equation $d = rt$ for t.

Solution
$$d = rt$$
$$350 = 140t$$
$$\frac{350}{140} = \frac{140t}{140}$$
$$2.5 = t$$

It takes the plane 2.5 h to travel 350 mi.

Problem 7

Strategy ▶ Rate of the second plane: r
Rate of the first plane: $r + 30$

	Rate	Time	Distance
First plane	$r + 30$	4	$4(r + 30)$
Second plane	r	4	$4r$

▶ The total distance traveled by the two planes is 1160 mi.

Solution
$$4(r + 30) + 4r = 1160$$
$$4r + 120 + 4r = 1160$$
$$8r + 120 = 1160$$
$$8r = 1040$$
$$r = 130$$

$$r + 30 = 130 + 30 = 160$$

The first plane is traveling 160 mph.
The second plane is traveling 130 mph.

SECTION 2.4

Problem 1

Strategy ▶ Each equal side: s

▶ Use the equation for the perimeter of a square.

Solution
$$P = 4s$$
$$52 = 4s$$
$$13 = s$$

The length of each side is 13 ft.

Problem 2

Strategy To find the supplement, let x represent the supplement of a 107° angle. Use the fact that supplementary angles are two angles whose sum is 180° to write an equation. Solve for x.

Solution
$$107° + x = 180°$$
$$x = 73°$$

The supplement of a 107° angle is a 73° angle.

Problem 3

Strategy The angles labeled are adjacent angles of intersecting lines and are therefore supplementary angles. To find x, write an equation and solve for x.

Solution
$$x + (3x + 20°) = 180°$$
$$4x + 20° = 180°$$
$$4x = 160°$$
$$x = 40°$$

Problem 4

Strategy $2x = y$ because alternate exterior angles have the same measure.

$y + (x + 15°) = 180°$ because adjacent angles of intersecting lines are supplementary angles. Substitute $2x$ for y and solve for x.

Solution
$$y + (x + 15°) = 180°$$
$$2x + (x + 15°) = 180°$$
$$3x + 15° = 180°$$
$$3x = 165°$$
$$x = 55°$$

Problem 5

Strategy ▶ To find the measure of $\angle a$, use the fact that $\angle a$ and $\angle y$ are vertical angles.

▶ To find the measure of $\angle b$, use the fact that the sum of the interior angles of a triangle is 180°.

▶ To find the measure of $\angle d$, use the fact that $\angle b$ and $\angle d$ are supplementary angles.

Solution $\angle a = \angle y = 55°$

$$\angle a + \angle b + 90° = 180°$$
$$55° + \angle b + 90° = 180°$$
$$145° + \angle b = 180°$$
$$\angle b = 35°$$

$$\angle b + \angle d = 180°$$
$$35° + \angle d = 180°$$
$$\angle d = 145°$$

Problem 6

Strategy To find the measure of the third angle, use the fact that the sum of the measures of the interior angles of a triangle is 180°. Write an equation using x to represent the measure of the third angle. Solve the equation for x.

Solution $x + 90° + 27° = 180°$
$$x + 117° = 180°$$
$$x = 63°$$

The measure of the third angle is 63°.

SECTION 2.5

Problem 1

Strategy Given: $C = \$120$
$S = \$180$
Unknown markup rate: r
Use the equation $S = C + rC$.

Solution $S = C + rC$
$$180 = 120 + 120r$$
$$60 = 120r$$
$$0.5 = r$$

The markup rate is 50%.

Problem 2

Strategy Given: $r = 40\% = 0.40$
$S = \$266$
Unknown cost: C
Use the equation $S = C + rC$.

Solution $S = C + rC$
$$266 = C + 0.4C$$

$$266 = 1.40C$$
$$190 = C$$

The cost is \$190.

Problem 3

Strategy Given: $R = \$39.80$
$S = \$29.85$
Unknown discount rate: r
Use the equation $S = R - rR$.

Solution $S = R - rR$
$$29.85 = 39.80 - 39.80r$$
$$-9.95 = -39.80r$$
$$0.25 = r$$

The discount rate is 25%.

Problem 4

Strategy Given: $S = \$43.50$
$r = 25\% = 0.25$
Unknown regular price: R
Use the equation $S = R - rR$.

Solution $S = R - rR$
$$43.50 = R - 0.25R$$
$$43.50 = 0.75R$$
$$58 = R$$

The regular price is \$58.

SECTION 2.6

Problem 1

Strategy ▶ Amount invested at 7.5%: x

	Principal	Rate	Interest
Amount at 5.2%	3500	0.052	0.052(3500)
Amount at 7.5%	x	0.075	0.075x

▶ The sum of the interest earned by the two investments equals the total annual interest earned (\$575).

Solution $0.052(3500) + 0.075x = 575$
$$182 + 0.075x = 575$$
$$0.075x = 393$$
$$x = 5240$$

The amount invested at 7.5% is \$5240.

Problem 2

Strategy ▶ Pounds of 22% fat hamburger: x
Pounds of 12% fat hamburger: $80 - x$

	Amount	Percent	Quantity
22%	x	0.22	0.22x
12%	$80 - x$	0.12	0.12(80 − x)
18%	80	0.18	0.18(80)

▶ The sum of the quantities before mixing is equal to the quantity after mixing.

Solution $0.22x + 0.12(80 - x) = 0.18(80)$
$0.22x + 9.6 - 0.12x = 14.4$
$0.10x + 9.6 = 14.4$
$0.10x = 4.8$
$x = 48$

$80 - x = 80 - 48 = 32$

The butcher needs 48 lb of the hamburger that is 22% fat and 32 lb of the hamburger that is 12% fat.

SECTION 2.7

Problem 1
$\{x \,|\, x < 7, x \in \text{real numbers}\}$

Problem 2
$A \cup C = \{-5, -2, -1, 0, 1, 2, 5\}$

Problem 3
$E \cap F = \varnothing$

Problem 4

Problem 5

Problem 6
$[-8, -1)$

Problem 7
$\{x \,|\, x > -12\}$

Problem 8
$2x - 1 < 6x + 7$
$-4x - 1 < 7$
$-4x < 8$
$\dfrac{-4x}{-4} > \dfrac{8}{-4}$
$x > -2$
$\{x \,|\, x > -2\}$

Problem 9
$5x - 2 \le 4 - 3(x - 2)$
$5x - 2 \le 4 - 3x + 6$
$5x - 2 \le 10 - 3x$
$8x - 2 \le 10$
$8x \le 12$
$\dfrac{8x}{8} \le \dfrac{12}{8}$
$x \le \dfrac{3}{2}$
$\left(-\infty, \dfrac{3}{2}\right]$

Problem 10
$-2 \le 5x + 3 \le 13$
$-2 - 3 \le 5x + 3 - 3 \le 13 - 3$
$-5 \le 5x \le 10$

$\dfrac{-5}{5} \le \dfrac{5x}{5} \le \dfrac{10}{5}$
$-1 \le x \le 2$
$[-1, 2]$

Problem 11
$5 - 4x > 1$ and $6 - 5x < 11$
$-4x > -4$ $-5x < 5$
$x < 1$ $x > -1$
$\{x \,|\, x < 1\}$ $\{x \,|\, x > -1\}$
$\{x \,|\, x < 1\} \cap \{x \,|\, x > -1\} = (-1, 1)$

Problem 12
$2 - 3x > 11$ or $5 + 2x > 7$
$-3x > 9$ $2x > 2$
$x < -3$ $x > 1$
$\{x \,|\, x < -3\}$ $\{x \,|\, x > 1\}$
$\{x \,|\, x < -3\} \cup \{x \,|\, x > 1\}$
$= \{x \,|\, x < -3 \text{ or } x > 1\}$

Problem 13
Strategy To find the maximum height, substitute the given values in the inequality $\frac{1}{2}bh < A$ and solve for x.

Solution $\dfrac{1}{2}bh < A$

$\dfrac{1}{2}(12)(x + 2) < 50$

$6(x + 2) < 50$
$6x + 12 < 50$
$6x < 38$
$x < \dfrac{19}{3}$

The largest integer less than $\dfrac{19}{3}$ is 6.

$x + 2 = 6 + 2 = 8$

The maximum height of the triangle is 8 in.

Problem 14
Strategy To find the range of scores, write and solve an inequality using N to represent the score on the fifth exam.

Solution $80 \le \dfrac{72 + 94 + 83 + 70 + N}{5} \le 89$

$80 \le \dfrac{319 + N}{5} \le 89$

$5(80) \le 5\left(\dfrac{319 + N}{5}\right) \le 5(89)$

$400 \le 319 + N \le 445$
$400 - 319 \le 319 - 319 + N \le 445 - 319$
$81 \le N \le 126$

Because 100 is a maximum score, the range of scores that will give the student a B for the course is $81 \le N \le 100$.

SECTION 2.8

Problem 1

A. $|x| = 25$

$x = 25 \qquad x = -25$

The solutions are 25 and −25.

B. $|2x - 3| = 5$

$2x - 3 = 5 \qquad 2x - 3 = -5$

$2x = 8 \qquad\quad 2x = -2$

$x = 4 \qquad\quad x = -1$

The solutions are 4 and −1.

C. $|x - 3| = -2$

The absolute value of a number must be nonnegative.

There is no solution to this equation.

D. $5 - |3x + 5| = 3$

$-|3x + 5| = -2$

$|3x + 5| = 2$

$3x + 5 = 2 \qquad 3x + 5 = -2$

$3x = -3 \qquad\quad 3x = -7$

$x = -1 \qquad\quad x = -\dfrac{7}{3}$

The solutions are −1 and $-\dfrac{7}{3}$.

Problem 2

$|3x + 2| < 8$

$-8 < 3x + 2 < 8$

$-8 - 2 < 3x + 2 - 2 < 8 - 2$

$-10 < 3x < 6$

$\dfrac{-10}{3} < \dfrac{3x}{3} < \dfrac{6}{3}$

$-\dfrac{10}{3} < x < 2$

$\left\{ x \,\middle|\, -\dfrac{10}{3} < x < 2 \right\}$

Problem 3

$|3x - 7| < 0$

The absolute value of a number must be nonnegative. The solution set is the empty set.

$\varnothing$

Problem 4

$|5x + 3| > 8$

$5x + 3 < -8 \quad \text{or} \quad 5x + 3 > 8$

$5x < -11 \qquad\qquad 5x > 5$

$x < -\dfrac{11}{5} \qquad\qquad x > 1$

$\left\{ x \,\middle|\, x < -\dfrac{11}{5} \right\} \qquad \{x \,|\, x > 1\}$

$\left\{ x \,\middle|\, x < -\dfrac{11}{5} \right\} \cup \{x \,|\, x > 1\}$

$= \left\{ x \,\middle|\, x < -\dfrac{11}{5} \ \text{or} \ x > 1 \right\}$

Problem 5

Strategy Let b represent the desired diameter of the bushing, T the tolerance, and d the actual diameter of the bushing. Solve the absolute value inequality $|d - b| \le T$ for d.

Solution

$|d - b| \le T$

$|d - 2.55| \le 0.003$

$-0.003 \le d - 2.55 \le 0.003$

$-0.003 + 2.55 \le d - 2.55 + 2.55 \le 0.003 + 2.55$

$2.547 \le d \le 2.553$

The lower and upper limits of the diameter of the bushing are 2.547 in. and 2.553 in.

Solutions to Chapter 3 Problems

SECTION 3.1

Problem 1

Replace x by −2 and solve for y.

$y = \dfrac{3x}{x + 1} = \dfrac{3(-2)}{-2 + 1} = \dfrac{-6}{-1} = 6$

The ordered-pair solution is (−2, 6).

Problem 2

$y = |x + 1|$

x	y
−3	2
−2	1
−1	0
0	1
1	2

Problem 3

$(x_1, y_1) = (5, -2) \qquad (x_2, y_2) = (-4, 3)$

$d = \sqrt{(x_2 - x_1)^2 + (y_2 - y_1)^2}$

$= \sqrt{(-4 - 5)^2 + (3 - (-2))^2}$

$= \sqrt{(-9)^2 + (5)^2}$

$= \sqrt{81 + 25}$

$= \sqrt{106}$

Problem 4

$(x_1, y_1) = (-3, -5) \qquad (x_2, y_2) = (-2, 3)$

$x_m = \dfrac{x_1 + x_2}{2} \qquad\qquad y_m = \dfrac{y_1 + y_2}{2}$

$= \dfrac{-3 + (-2)}{2} \qquad\qquad = \dfrac{-5 + 3}{2}$

$= -\dfrac{5}{2} \qquad\qquad\qquad = -1$

The midpoint is $\left(-\dfrac{5}{2}, -1 \right)$.

Problem 5

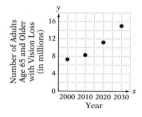

Problem 6

In 1900, the population was 1.5 million: (1900, 1.5)

In 2000, the population was 33.9 million: (2000, 33.9)

Average rate of change

$$= \frac{\text{change in } y}{\text{change in } x}$$

$$= \frac{\text{population in 2000} - \text{population in 1900}}{2000 - 1900}$$

$$= \frac{33.9 - 1.5}{2000 - 1900}$$

$$= \frac{32.4}{100} = 0.324$$

The average rate of change in the population was 324,000 people per year.

Problem 7

Solution A. In 1990, the median salary was 437,500: (1990, 437,500)

In 2005, the median salary was 5,833,334: (2005, 5,833,334)

The average rate of change

$$= \frac{\text{change in } y}{\text{change in } x}$$

$$= \frac{\text{median salary in 2005} - \text{median salary in 1990}}{2005 - 1990}$$

$$= \frac{5,833,334 - 437,500}{2005 - 1990}$$

$$= \frac{5,395,834}{15} \approx 360,000$$

The average rate of change was \$360,000 per year.

B. \$360,000 > \$173,000

The average annual rate of change in the median salary of New York Yankees players between 1990 and 2005 is greater than the average annual rate of change in the median salary of Boston Red Sox players between 1990 and 2005.

SECTION 3.2

Problem 1

$\{(-2, 6), (-1, 3), (0, 0), (1, -3), (2, -6), (3, -9)\}$

Domain $= \{-2, -1, 0, 1, 2, 3\}$
- **The domain of the relation is the set of first coordinates of the ordered pairs.**

Range $= \{-9, -6, -3, 0, 3, 6\}$
- **The range of the relation is the set of second coordinates of the ordered pairs.**

Because there are no two ordered pairs with the same first coordinate and different second coordinates, the relation is a function.

Problem 2

A. $f(z) = z + |z|$
$f(-4) = -4 + |-4| = -4 + 4$
$f(-4) = 0$
- **Replace z by -4. Then simplify.**

B. To find the value of f when $z = 3$ means to evaluate the function when $z = 3$.

$f(z) = z + |z|$
$f(3) = 3 + |3| = 3 + 3$
$f(3) = 6$
- **Replace z by 3. Then simplify.**

The value of the function when $z = 3$ is 6.

Problem 3

$r(s) = 3s - 6$
$r(2a + 3) = 3(2a + 3) - 6$
$= 6a + 9 - 6$
$= 6a + 3$
- **Replace s by $2a + 3$.**
- **Simplify.**

Problem 4

Evaluate the function for each element of the domain. The range includes the values of $f(-3), f(-2), f(-1), f(0), f(1)$, and $f(2)$.

$f(x) = x^3 + x$
$f(-3) = (-3)^3 + (-3) = -30$
$f(-2) = (-2)^3 + (-2) = -10$
$f(-1) = (-1)^3 + (-1) = -2$
$f(0) = (0)^3 + 0 = 0$
$f(1) = (1)^3 + 1 = 2$
$f(2) = (2)^3 + 2 = 10$

The range is $\{-30, -10, -2, 0, 2, 10\}$.

Problem 5

$f(x) = 2x - 5$
$f(c) = 2c - 5$
$-3 = 2c - 5$
$2 = 2c$
$1 = c$
- **Replace x by c.**
- **$f(c) = -3$.**
- **Solve for c.**

The value of c is 1. The corresponding ordered pair is $(1, -3)$.

Problem 6

x	$y = f(x) = -x^2 - 4x + 2$	(x, y)
-5	$f(-5) = -(-5)^2 - 4(-5) + 2$	$(-5, -3)$
-4	$f(-4) = -(-4)^2 - 4(-4) + 2$	$(-4, 2)$
-3	$f(-3) = -(-3)^2 - 4(-3) + 2$	$(-3, 5)$
-2	$f(-2) = -(-2)^2 - 4(-2) + 2$	$(-2, 6)$
-1	$f(-1) = -(-1)^2 - 4(-1) + 2$	$(-1, 5)$
0	$f(0) = -(0)^2 - 4(0) + 2$	$(0, 2)$
1	$f(1) = -(1)^2 - 4(1) + 2$	$(1, -5)$

Problem 7

x	$y = f(x) = \lvert x - 2 \rvert$	(x, y)
-1	$f(-1) = \lvert -1 - 2 \rvert = 3$	$(-1, 3)$
0	$f(0) = \lvert 0 - 2 \rvert = 2$	$(0, 2)$
1	$f(1) = \lvert 1 - 2 \rvert = 1$	$(1, 1)$
2	$f(2) = \lvert 2 - 2 \rvert = 0$	$(2, 0)$
3	$f(3) = \lvert 3 - 2 \rvert = 1$	$(3, 1)$
4	$f(4) = \lvert 4 - 2 \rvert = 2$	$(4, 2)$

Problem 8

A. Any vertical line intersects the graph at most once. The graph is the graph of a function.

B. There are vertical lines that intersect the graph at more than one point. The graph is not the graph of a function.

SECTION 3.3

Problem 1

x	y
-5	-7
0	-4
5	-1

Problem 2

$$-3x + 2y = 4$$
$$2y = 3x + 4$$
$$y = \frac{3}{2}x + 2$$

x	y
-2	-1
0	2
2	5

Problem 3

$$y - 3 = 0$$
$$y = 3$$

Problem 4

x-intercept:
$$3x - y = 2$$
$$3x - 0 = 2$$
$$3x = 2$$
$$x = \frac{2}{3}$$
$$\left(\frac{2}{3}, 0\right)$$

y-intercept:
$$3x - y = 2$$
$$3(0) - y = 2$$
$$-y = 2$$
$$y = -2$$
$$(0, -2)$$

Problem 5

$$f(x) = \frac{2}{3}x + 4$$
$$0 = \frac{2}{3}x + 4$$
$$-4 = \frac{2}{3}x$$
$$-6 = x$$

The zero is -6.

Problem 6

The ordered pair (32, 74) means that a person with a stride of 32 in. is 74 in. tall.

SECTION 3.4

Problem 1

Let $P_1 = (4, -3)$ and $P_2 = (2, 7)$.

$$m = \frac{y_2 - y_1}{x_2 - x_1} = \frac{7 - (-3)}{2 - 4} = \frac{10}{-2} = -5$$

The slope is -5.

Problem 2

$$m = \frac{55{,}000 - 25{,}000}{2 - 5} = \frac{30{,}000}{-3} = -10{,}000$$

A slope of $-10{,}000$ means that the value of the printing press is decreasing by \$10,000 per year.

Problem 3

$$2x + 3y = 6$$
$$3y = -2x + 6$$
$$y = -\frac{2}{3}x + 2$$
$$m = -\frac{2}{3} = \frac{-2}{3}$$

y-intercept $= (0, 2)$

Problem 4

$(x_1, y_1) = (-3, -2)$

$$m = 3 = \frac{3}{1} = \frac{\text{change in } y}{\text{change in } x}$$

SECTION 3.5

Problem 1

$m = -3 \qquad (x_1, y_1) = (4, -3)$

$$y - y_1 = m(x - x_1)$$
$$y - (-3) = -3(x - 4)$$
$$y + 3 = -3x + 12$$
$$y = -3x + 9$$

The equation of the line is $y = -3x + 9$.

Problem 2

Let $(x_1, y_1) = (2, 0)$ and $(x_2, y_2) = (5, 3)$.

$$m = \frac{y_2 - y_1}{x_2 - x_1} = \frac{3 - 0}{5 - 2} = \frac{3}{3} = 1$$

$$y - y_1 = m(x - x_1)$$
$$y - 0 = 1(x - 2)$$
$$y = x - 2$$

The equation of the line is $y = x - 2$.

Problem 3

Strategy ▶ Select the independent and dependent variables. Because the function will predict the Celsius temperature, that quantity is the dependent variable, y. The Fahrenheit temperature is the independent variable.

▶ From the given data, two ordered pairs are $(212, 100)$ and $(32, 0)$. Use these ordered pairs to determine the linear function.

Solution Let $(x_1, y_1) = (32, 0)$ and $(x_2, y_2) = (212, 100)$.

$$m = \frac{y_2 - y_1}{x_2 - x_1} = \frac{100 - 0}{212 - 32} = \frac{100}{180} = \frac{5}{9}$$
$$y - y_1 = m(x - x_1)$$
$$y - 0 = \frac{5}{9}(x - 32)$$
$$y = \frac{5}{9}(x - 32), \text{ or } C = \frac{5}{9}(F - 32).$$

The linear function is $f(F) = \dfrac{5}{9}(F - 32)$.

SECTION 3.6

Problem 1

$$5x + 2y = 2$$
$$2y = -5x + 2$$
$$y = -\frac{5}{2}x + 1$$
$$m_1 = -\frac{5}{2}$$

$$5x + 2y = -6$$
$$2y = -5x - 6$$
$$y = -\frac{5}{2}x - 3$$
$$m_2 = -\frac{5}{2}$$

$$m_1 = m_2 = -\frac{5}{2}$$

Yes, the lines are parallel.

Problem 2

$$y - y_1 = m(x - x_1) \qquad \bullet \text{ Use the point-slope formula.}$$
$$y - 1 = \frac{3}{4}(x - 4) \qquad \bullet \ m = \frac{3}{4}, \text{ the slope of the line}$$
$$\hspace{3cm} y = \frac{3}{4}x - 1. \ (x_1, y_1) = (4, 1), \text{ the given point.}$$
$$y - 1 = \frac{3}{4}x - 3 \qquad \bullet \text{ Use the Distributive Property.}$$
$$y = \frac{3}{4}x - 2 \qquad \bullet \text{ Add 1 to each side of the equation.}$$

The equation of the line is $y = \dfrac{3}{4}x - 2$.

Problem 3

$$x - 4y = 3$$
$$-4y = -x + 3$$
$$y = \frac{1}{4}x - \frac{3}{4}$$
$$m_1 = \frac{1}{4}$$
$$m_1 \cdot m_2 = -1$$
$$\frac{1}{4} \cdot m_2 = -1$$
$$m_2 = -4$$

$$y - y_1 = m(x - x_1)$$
$$y - 3 = -4[x - (-2)]$$
$$y - 3 = -4(x + 2)$$
$$y - 3 = -4x - 8$$
$$y = -4x - 5$$

The equation of the line is $y = -4x - 5$.

SECTION 3.7

Problem 1

$$x + 3y > 6$$
$$3y > -x + 6$$
$$y > -\frac{1}{3}x + 2$$

Problem 2

$$y < 2$$

Solutions to Chapter 4 Problems

SECTION 4.1

Problem 1

The solution is $(-1, 2)$.

Problem 2

A. The lines are parallel and therefore do not intersect. The system of equations has no solution. The system of equations is inconsistent.

B. The two equations represent the same line. The system of equations is dependent. The solutions are the ordered pairs $\left(x, \frac{3}{4}x - 3\right)$.

Problem 3

A. (1) $3x - y = 3$
 (2) $6x + 3y = -4$

Solve equation (1) for y.
$$3x - y = 3$$
$$-y = -3x + 3$$
$$y = 3x - 3$$

Substitute into equation (2).

$$6x + 3y = -4$$
$$6x + 3(3x - 3) = -4$$
$$6x + 9x - 9 = -4$$
$$15x - 9 = -4$$
$$15x = 5$$
$$x = \frac{5}{15} = \frac{1}{3}$$

Substitute into equation (1).
$$3x - y = 3$$
$$3\left(\frac{1}{3}\right) - y = 3$$
$$1 - y = 3$$
$$-y = 2$$
$$y = -2$$

The solution is $\left(\frac{1}{3}, -2\right)$.

B. (1) $6x - 3y = 6$
 (2) $2x - y = 2$

Solve equation (2) for y.
$$2x - y = 2$$
$$-y = -2x + 2$$
$$y = 2x - 2$$

Substitute into equation (1).
$$6x - 3y = 6$$
$$6x - 3(2x - 2) = 6$$
$$6x - 6x + 6 = 6$$
$$6 = 6$$

$6 = 6$ is a true equation. The system of equations is dependent. The solutions are the ordered pairs $(x, 2x - 2)$.

SECTION 4.2

Problem 1

A. (1) $2x + 5y = 6$
 (2) $3x - 2y = 6x + 2$

Write equation (2) in the form $Ax + By = C$.
$$3x - 2y = 6x + 2$$
$$-3x - 2y = 2$$

Solve the system: $2x + 5y = 6$
 $-3x - 2y = 2$

Eliminate y.
$$2(2x + 5y) = 2(6)$$
$$5(-3x - 2y) = 5(2)$$
$$4x + 10y = 12$$
$$-15x - 10y = 10$$

Add the equations.
$$-11x = 22$$
$$x = -2$$

Replace x in equation (1).

$$2x + 5y = 6$$
$$2(-2) + 5y = 6$$
$$-4 + 5y = 6$$
$$5y = 10$$
$$y = 2$$

The solution is $(-2, 2)$.

B. $$2x + y = 5$$
$$4x + 2y = 6$$

Eliminate y.

$$-2(2x + y) = -2(5)$$
$$4x + 2y = 6$$

$$-4x - 2y = -10$$
$$4x + 2y = 6$$

Add the equations.

$$0x + 0y = -4$$
$$0 = -4$$

$0 = -4$ is not a true equation. The system is inconsistent and therefore has no solution.

Problem 2

(1) $x - y + z = 6$
(2) $2x + 3y - z = 1$
(3) $x + 2y + 2z = 5$

Eliminate z. Add equations (1) and (2).

$$x - y + z = 6$$
$$2x + 3y - z = 1$$
(4) $3x + 2y = 7$

Eliminate z. Multiply equation (2) by 2 and add to equation (3).

$$4x + 6y - 2z = 2$$
$$x + 2y + 2z = 5$$
(5) $5x + 8y = 7$

Solve the system of two equations.

(4) $3x + 2y = 7$
(5) $5x + 8y = 7$

Multiply equation (4) by -4 and add to equation (5).

$$-12x - 8y = -28$$
$$5x + 8y = 7$$
$$-7x = -21$$
$$x = 3$$

Replace x by 3 in equation (4).

$$3x + 2y = 7$$
$$3(3) + 2y = 7$$
$$9 + 2y = 7$$
$$2y = -2$$
$$y = -1$$

Replace x by 3 and y by -1 in equation (1).

$$x - y + z = 6$$
$$3 - (-1) + z = 6$$
$$4 + z = 6$$
$$z = 2$$

The solution is $(3, -1, 2)$.

SECTION 4.3

Problem 1

A. $\begin{vmatrix} -1 & -4 \\ 3 & -5 \end{vmatrix} = -1(-5) - 3(-4)$
$$= 5 + 12 = 17$$

The value of the determinant is 17.

B. Expand by cofactors of the first column.

$$\begin{vmatrix} 1 & 4 & -2 \\ 3 & 1 & 1 \\ 0 & -2 & 2 \end{vmatrix} = 1\begin{vmatrix} 1 & 1 \\ -2 & 2 \end{vmatrix} - 3\begin{vmatrix} 4 & -2 \\ -2 & 2 \end{vmatrix} + 0$$
$$= 1(2 + 2) - 3(8 - 4)$$
$$= 4 - 12$$
$$= -8$$

The value of the determinant is -8.

Problem 2

$$D = \begin{vmatrix} 6 & -6 \\ 2 & -10 \end{vmatrix} = -48$$

$$D_x = \begin{vmatrix} 5 & -6 \\ -1 & -10 \end{vmatrix} = -56$$

$$D_y = \begin{vmatrix} 6 & 5 \\ 2 & -1 \end{vmatrix} = -16$$

$$x = \frac{D_x}{D} = \frac{-56}{-48} = \frac{7}{6} \qquad y = \frac{D_y}{D} = \frac{-16}{-48} = \frac{1}{3}$$

The solution is $\left(\dfrac{7}{6}, \dfrac{1}{3}\right)$.

Problem 3

$$D = \begin{vmatrix} 2 & -1 & 1 \\ 3 & 2 & -1 \\ 1 & 3 & 1 \end{vmatrix} = 21$$

$$D_x = \begin{vmatrix} -1 & -1 & 1 \\ 3 & 2 & -1 \\ -2 & 3 & 1 \end{vmatrix} = 9$$

$$D_y = \begin{vmatrix} 2 & -1 & 1 \\ 3 & 3 & -1 \\ 1 & -2 & 1 \end{vmatrix} = -3$$

$$D_z = \begin{vmatrix} 2 & -1 & -1 \\ 3 & 2 & 3 \\ 1 & 3 & -2 \end{vmatrix} = -42$$

$$x = \frac{D_x}{D} = \frac{9}{21} = \frac{3}{7} \qquad y = \frac{D_y}{D} = \frac{-3}{21} = -\frac{1}{7}$$

$$z = \frac{D_z}{D} = \frac{-42}{21} = -2$$

The solution is $\left(\dfrac{3}{7}, -\dfrac{1}{7}, -2\right)$.

Problem 4

$$\begin{bmatrix} 2 & -3 & 1 & | & 4 \\ 1 & 0 & -2 & | & 3 \\ 0 & 1 & 2 & | & -3 \end{bmatrix}$$

$$2x - 3y + z = 4$$
$$x - 2z = 3$$
$$y + 2z = -3$$

Problem 5

A. $\begin{bmatrix} 1 & 8 & -2 & | & 3 \\ 2 & -3 & 4 & | & 1 \\ 3 & 5 & -7 & | & 3 \end{bmatrix}$ $R_2 \leftrightarrow R_3$ $\begin{bmatrix} 1 & 8 & -2 & | & 3 \\ 3 & 5 & -7 & | & 3 \\ 2 & -3 & 4 & | & 1 \end{bmatrix}$

B. $\begin{bmatrix} 1 & 8 & -2 & | & 3 \\ 2 & -3 & 4 & | & 1 \\ 3 & 5 & -7 & | & 3 \end{bmatrix}$ $3R_2 \rightarrow$ $\begin{bmatrix} 1 & 8 & -2 & | & 3 \\ 6 & -9 & 12 & | & 3 \\ 3 & 5 & -7 & | & 3 \end{bmatrix}$

C. $\begin{bmatrix} 1 & 8 & -2 & | & 3 \\ 2 & -3 & 4 & | & 1 \\ 3 & 5 & -7 & | & 3 \end{bmatrix}$ $-3R_1 + R_3 \rightarrow$ $\begin{bmatrix} 1 & 8 & -2 & | & 3 \\ 2 & -3 & 4 & | & 1 \\ 0 & -19 & -1 & | & -6 \end{bmatrix}$

Problem 6

$$\begin{bmatrix} 1 & -3 & 2 & | & 1 \\ -4 & 14 & 0 & | & -2 \\ 2 & -5 & -3 & | & 16 \end{bmatrix} \quad 4R_1 + R_2 \rightarrow \quad \begin{bmatrix} 1 & -3 & 2 & | & 1 \\ 0 & 2 & 8 & | & 2 \\ 2 & -5 & -3 & | & 16 \end{bmatrix}$$

$$\begin{bmatrix} 1 & -3 & 2 & | & 1 \\ 0 & 2 & 8 & | & 2 \\ 2 & -5 & -3 & | & 16 \end{bmatrix} \quad -2R_1 + R_3 \rightarrow \quad \begin{bmatrix} 1 & -3 & 2 & | & 1 \\ 0 & 2 & 8 & | & 2 \\ 0 & 1 & -7 & | & 14 \end{bmatrix}$$

$$\begin{bmatrix} 1 & -3 & 2 & | & 1 \\ 0 & 2 & 8 & | & 2 \\ 0 & 1 & -7 & | & 14 \end{bmatrix} \quad \frac{1}{2}R_2 \rightarrow \quad \begin{bmatrix} 1 & -3 & 2 & | & 1 \\ 0 & 1 & 4 & | & 1 \\ 0 & 1 & -7 & | & 14 \end{bmatrix}$$

$$\begin{bmatrix} 1 & -3 & 2 & | & 1 \\ 0 & 1 & 4 & | & 1 \\ 0 & 1 & -7 & | & 14 \end{bmatrix} \quad -1R_2 + R_3 \rightarrow \quad \begin{bmatrix} 1 & -3 & 2 & | & 1 \\ 0 & 1 & 4 & | & 1 \\ 0 & 0 & -11 & | & 13 \end{bmatrix}$$

$$\begin{bmatrix} 1 & -3 & 2 & | & 1 \\ 0 & 1 & 4 & | & 1 \\ 0 & 0 & -11 & | & 13 \end{bmatrix} \quad -\frac{1}{11}R_3 \rightarrow \quad \begin{bmatrix} 1 & -3 & 2 & | & 1 \\ 0 & 1 & 4 & | & 1 \\ 0 & 0 & 1 & | & -\frac{13}{11} \end{bmatrix}$$

A row echelon form of the matrix is $\begin{bmatrix} 1 & -3 & 2 & | & 1 \\ 0 & 1 & 4 & | & 1 \\ 0 & 0 & 1 & | & -\frac{13}{11} \end{bmatrix}$.

Problem 7

$$4x - 5y = 17$$
$$3x + 2y = 7$$

$$\begin{bmatrix} 4 & -5 & | & 17 \\ 3 & 2 & | & 7 \end{bmatrix} \xrightarrow[\frac{1}{4}R_1]{\text{Change } a_{11} \text{ to 1.}} \begin{bmatrix} 1 & -\frac{5}{4} & | & \frac{17}{4} \\ 3 & 2 & | & 7 \end{bmatrix}$$

• **We multiplied R_1 by the reciprocal of a_{11}.**

$$\begin{bmatrix} 1 & -\frac{5}{4} & | & \frac{17}{4} \\ 3 & 2 & | & 7 \end{bmatrix} \xrightarrow[-3R_1 + R_2]{\text{Change } a_{21} \text{ to 0.}} \begin{bmatrix} 1 & -\frac{5}{4} & | & \frac{17}{4} \\ 0 & \frac{23}{4} & | & -\frac{23}{4} \end{bmatrix}$$

• **We multiplied R_1 by the opposite of a_{21}.**

$$\begin{bmatrix} 1 & -\frac{5}{4} & | & \frac{17}{4} \\ 0 & \frac{23}{4} & | & -\frac{23}{4} \end{bmatrix} \xrightarrow[\frac{4}{23}R_2]{\text{Change } a_{22} \text{ to 1.}} \begin{bmatrix} 1 & -\frac{5}{4} & | & \frac{17}{4} \\ 0 & 1 & | & -1 \end{bmatrix}$$

• **This is row echelon form.**

$$x - \frac{5}{4}y = \frac{17}{4}$$
$$y = -1$$

$$x - \frac{5}{4}(-1) = \frac{17}{4}$$
$$x + \frac{5}{4} = \frac{17}{4}$$
$$x = 3$$

The solution is $(3, -1)$.

Problem 8

$$2x + 3y + 3z = -2$$
$$x + 2y - 3z = 9$$
$$3x - 2y - 4z = 1$$

$$\begin{bmatrix} 2 & 3 & 3 & | & -2 \\ 1 & 2 & -3 & | & 9 \\ 3 & -2 & -4 & | & 1 \end{bmatrix} \quad R_1 \leftrightarrow R_2 \quad \begin{bmatrix} 1 & 2 & -3 & | & 9 \\ 2 & 3 & 3 & | & -2 \\ 3 & -2 & -4 & | & 1 \end{bmatrix}$$

$$\begin{bmatrix} 1 & 2 & -3 & | & 9 \\ 2 & 3 & 3 & | & -2 \\ 3 & -2 & -4 & | & 1 \end{bmatrix} \quad -2R_1 + R_2 \rightarrow \quad \begin{bmatrix} 1 & 2 & -3 & | & 9 \\ 0 & -1 & 9 & | & -20 \\ 3 & -2 & -4 & | & 1 \end{bmatrix}$$

$$\begin{bmatrix} 1 & 2 & -3 & | & 9 \\ 0 & -1 & 9 & | & -20 \\ 3 & -2 & -4 & | & 1 \end{bmatrix} \quad -3R_1 + R_3 \rightarrow \quad \begin{bmatrix} 1 & 2 & -3 & | & 9 \\ 0 & -1 & 9 & | & -20 \\ 0 & -8 & 5 & | & -26 \end{bmatrix}$$

$$\begin{bmatrix} 1 & 2 & -3 & | & 9 \\ 0 & -1 & 9 & | & -20 \\ 0 & -8 & 5 & | & -26 \end{bmatrix} \quad -1R_2 \rightarrow \quad \begin{bmatrix} 1 & 2 & -3 & | & 9 \\ 0 & 1 & -9 & | & 20 \\ 0 & -8 & 5 & | & -26 \end{bmatrix}$$

$$\begin{bmatrix} 1 & 2 & -3 & | & 9 \\ 0 & 1 & -9 & | & 20 \\ 0 & -8 & 5 & | & -26 \end{bmatrix} \quad 8R_2 + R_3 \rightarrow \quad \begin{bmatrix} 1 & 2 & -3 & | & 9 \\ 0 & 1 & -9 & | & 20 \\ 0 & 0 & -67 & | & 134 \end{bmatrix}$$

$$\begin{bmatrix} 1 & 2 & -3 & | & 9 \\ 0 & 1 & -9 & | & 20 \\ 0 & 0 & -67 & | & 134 \end{bmatrix} \quad -\frac{1}{67}R_2 \rightarrow \quad \begin{bmatrix} 1 & 2 & -3 & | & 9 \\ 0 & 1 & -9 & | & 20 \\ 0 & 0 & 1 & | & -2 \end{bmatrix}$$

$$x + 2y - 3z = 9$$
$$y - 9z = 20$$
$$z = -2$$

$$y - 9(-2) = 20$$
$$y + 18 = 20$$
$$y = 2$$

$$x + 2(2) - 3(-2) = 9$$
$$x + 10 = 9$$
$$x = -1$$

The solution is $(-1, 2, -2)$.

SECTION 4.4

Problem 1

Strategy ▶ Rate of the rowing team in calm water: t
Rate of the current: c

	Rate	Time	Distance
With current	$t + c$	2	$2(t + c)$
Against current	$t - c$	2	$2(t - c)$

▶ The distance traveled with the current is 18 mi.
The distance traveled against the current is 10 mi.

Solution
$$2(t + c) = 18 \rightarrow t + c = 9$$
$$2(t - c) = 10 \rightarrow t - c = 5$$
$$2t = 14$$
$$t = 7$$

$$t + c = 9$$
$$7 + c = 9$$
$$c = 2$$

The rate of the rowing team in calm water is 7 mph.
The rate of the current is 2 mph.

Problem 2

Strategy ▶ Cost of an orange tree: x
Cost of a grapefruit tree: y

First purchase:

	Amount	Unit cost	Value
Orange trees	25	x	$25x$
Grapefruit trees	20	y	$20y$

Second purchase:

	Amount	Unit cost	Value
Orange trees	20	x	$20x$
Grapefruit trees	30	y	$30y$

▶ The total of the first purchase was $290.
The total of the second purchase was $330.

Solution
$$25x + 20y = 290$$
$$20x + 30y = 330$$

$$4(25x + 20y) = 4 \cdot 290$$
$$-5(20x + 30y) = -5 \cdot 330$$

$$100x + 80y = 1160$$
$$-100x - 150y = -1650$$
$$-70y = -490$$
$$y = 7$$

$$25x + 20y = 290$$
$$25x + 20(7) = 290$$
$$25x + 140 = 290$$
$$25x = 150$$
$$x = 6$$

The cost of an orange tree is $6.
The cost of a grapefruit tree is $7.

Problem 3

Strategy Number of full-price tickets sold: x
Number of member admission tickets sold: y
Number of student tickets sold: z

There were 750 tickets sold: $x + y + z = 750$
The income was $5400: $10x + 7y + 5z = 5400$
20 more student tickets were sold than full-price tickets: $z = x + 20$

Solution
(1) $\qquad x + y + z = 750$
(2) $\qquad 10x + 7y + 5z = 5400$
(3) $\qquad z = x + 20$

Solve the system of equations by substitution. Replace z in equation (1) and equation (2) by $x + 20$.

$$x + y + (x + 20) = 750$$
$$10x + 7y + 5(x + 20) = 5400$$

(4) $\qquad 2x + y = 730$
(5) $\qquad 15x + 7y = 5300$

Multiply equation (4) by -7 and add to equation (5).

$$-14x - 7y = -5110$$
$$15x + 7y = 5300$$
$$x = 190$$

Substitute the value of x into equation (4) and solve for y.

$$2x + y = 730$$
$$2(190) + y = 730$$
$$380 + y = 730$$
$$y = 350$$

Substitute the values of x and y into equation (1) and solve for z.

$$x + y + z = 750$$
$$190 + 350 + z = 750$$
$$540 + z = 750$$
$$z = 210$$

The museum sold 190 full-priced tickets, 350 member tickets, and 210 student tickets.

SECTION 4.5

Problem 1

A. Shade above the solid line $y = 2x - 3$.
Shade above the dashed line $y = -3x$.

The solution set of the system is the intersection of the solution sets of the individual inequalities.

B. $3x + 4y > 12$
$$4y > -3x + 12$$
$$y > -\frac{3}{4}x + 3$$

Shade above the dashed line $y = -\frac{3}{4}x + 3$.

Shade below the dashed line $y = \frac{3}{4}x - 1$.

The solution set of the system is the intersection of the solution sets of the individual inequalities.

Solutions to Chapter 5 Problems

SECTION 5.1

Problem 1

Strategy Evaluate the function when $r = 0.4$.

Solution $v(r) = 600r^2 - 1000r^3$
$$= 600(0.4)^2 - 1000(0.4)^3$$
$$= 32$$

The velocity will be 32 cm/s.

Problem 2

The leading coefficient is -3, the constant term is -12, and the degree is 4.

Problem 3

A. This is a polynomial function.

B. A polynomial function does not have a variable expression raised to a negative power. This is not a

polynomial function.

C. A polynomial function does not have a variable expression within a radical. This is not a polynomial function.

Problem 4

x	y
-1	3
0	0
1	-1
2	0
3	3

$f(x) = x^2 - 2x$

Problem 5

x	y
-3	28
-2	9
-1	2
0	1
1	0
2	-7
3	-26

$F(x) = -x^3 + 1$

Problem 6

$$
\begin{aligned}
5x^2 + 3x - 1 \\
2x^2 + 4x - 6 \\
\underline{x^2 - 7x + 8} \\
8x^2 \qquad + 1
\end{aligned}
$$

Problem 7

Rewrite the subtraction as the addition of the opposite.

$$(-5x^2 + 2x - 3) + (-6x^2 - 3x + 7)$$

$$
\begin{aligned}
-5x^2 + 2x - 3 \\
\underline{-6x^2 - 3x + 7} \\
-11x^2 - x + 4
\end{aligned}
$$

Problem 8

$$
\begin{aligned}
D(x) &= (4x^3 - 3x^2 + 2) - (-2x^2 + 2x - 3) \\
&= (4x^3 - 3x^2 + 2) + (2x^2 - 2x + 3) \\
&= 4x^3 - x^2 - 2x + 5
\end{aligned}
$$

SECTION 5.2

Problem 1

$$(3x^2)(6x^3) = (3 \cdot 6)(x^2 \cdot x^3) = 18x^5$$

Problem 2

$$
\begin{aligned}
&(-3xy^2)(-4x^2y^3) \\
&= [(-3)(-4)](x \cdot x^2)(y^2 \cdot y^3) \\
&= 12x^3y^5
\end{aligned}
$$

Problem 3

$$
\begin{aligned}
(3x)(2x^2y)^3 &= (3x)(2^3x^6y^3) = (3x)(8x^6y^3) \\
&= (3 \cdot 8)(x \cdot x^6)y^3 = 24x^7y^3
\end{aligned}
$$

SECTION 5.3

Problem 1

A. $(-2y + 3)(-4y) = 8y^2 - 12y$

B. $-a^2(3a^2 + 2a - 7) = -3a^4 - 2a^3 + 7a^2$

Problem 2

$$
\begin{array}{r}
2y^3 + 2y^2 \quad\;\; - 3 \\
3y\; - 1 \\
\hline
-2y^3 - 2y^2 \qquad + 3 \\
6y^4 + 6y^3 \qquad - 9y \\
\hline
6y^4 + 4y^3 - 2y^2 - 9y\; + 3
\end{array}
$$

Problem 3

$$
\begin{array}{r}
3x^3 - 2x^2 + x - 3 \\
2x + 5 \\
\hline
15x^3 - 10x^2 + 5x - 15 \\
6x^4 - 4x^3 + 2x^2 - 6x \\
\hline
6x^4 + 11x^3 - 8x^2 - x - 15
\end{array}
$$

Problem 4

$(4y - 5)(3y - 3)$
$= 12y^2 - 12y - 15y + 15$
$= 12y^2 - 27y + 15$

Problem 5

$(3a + 2b)(3a - 5b)$
$= 9a^2 - 15ab + 6ab - 10b^2$
$= 9a^2 - 9ab - 10b^2$

Problem 6

$(2a + 5c)(2a - 5c) = (2a)^2 - (5c)^2$
$\qquad\qquad\qquad\quad = 4a^2 - 25c^2$

Problem 7

$(3x + 2y)^2 = (3x)^2 + 2(3x)(2y) + (2y)^2$
$\qquad\qquad = 9x^2 + 12xy + 4y^2$

Problem 8

$(6x - y)^2 = (6x)^2 + 2(6x)(-y) + (-y)^2$
$\qquad\qquad = 36x^2 - 12xy + y^2$

Problem 9

Strategy To find the area in terms of x, replace the variables L and W in the equation $A = LW$ with the given values, and solve for A.

Solution $A = LW$
$A = (x + 7)(x - 4)$
$A = x^2 - 4x + 7x - 28$
$A = x^2 + 3x - 28$

The area is $(x^2 + 3x - 28)$ m².

Problem 10

Strategy To find the area of the triangle in terms of x, replace the variables b and h in the equation $A = \frac{1}{2}bh$ with the given values, and solve for A.

Solution $A = \frac{1}{2}bh$

$A = \frac{1}{2}(x + 3)(4x - 6)$

$A = \frac{1}{2}(4x^2 + 6x - 18)$

$A = 2x^2 + 3x - 9$

The area is $(2x^2 + 3x - 9)$ cm².

SECTION 5.4

Problem 1

$$\frac{2^{-2}}{2^3} = 2^{-2-3} = 2^{-5} = \frac{1}{2^5} = \frac{1}{32}$$

Problem 2

$$\frac{b^8}{a^{-5}b^6} = a^5 b^{8-6} = a^5 b^2$$

Problem 3

A. $\dfrac{12x^{-8}y^4}{-16xy^{-3}} = -\dfrac{3x^{-9}y^7}{4} = -\dfrac{3y^7}{4x^9}$

B. $(-3ab)(2a^3b^{-2})^{-3} = (-3ab)(2^{-3}a^{-9}b^6)$
$\qquad\qquad = -\dfrac{3a^{-8}b^7}{2^3} = -\dfrac{3b^7}{8a^8}$

Problem 4

A. $57{,}000{,}000{,}000 = 5.7 \times 10^{10}$

B. $0.000000017 = 1.7 \times 10^{-8}$

Problem 5

A. $5 \times 10^{12} = 5{,}000{,}000{,}000{,}000$

B. $4.0162 \times 10^{-9} = 0.0000000040162$

Problem 6

$\dfrac{5{,}600{,}000 \times 0.000000081}{900 \times 0.000000028}$

$= \dfrac{5.6 \times 10^6 \times 8.1 \times 10^{-8}}{9 \times 10^2 \times 2.8 \times 10^{-8}}$

$= \dfrac{(5.6)(8.1) \times 10^{6+(-8)-2-(-8)}}{(9)(2.8)} = 1.8 \times 10^4$

Problem 7

Strategy To find the number of arithmetic operations:
► Find the reciprocal of 1×10^{-7}, which is the number of operations performed in 1 s.
► Write the number of seconds in 1 min (60) in scientific notation.
► Multiply the number of arithmetic operations per second by the number of seconds in 1 min.

Solution $\dfrac{1}{1 \times 10^{-7}} = 10^7$

$60 = 6 \times 10$

$6 \times 10 \times 10^7 = 6 \times 10^8$

The computer can perform 6×10^8 operations in 1 min.

SECTION 5.5

Problem 1

$$\frac{4x^3y + 8x^2y^2 - 4xy^3}{2xy} = \frac{4x^3y}{2xy} + \frac{8x^2y^2}{2xy} - \frac{4xy^3}{2xy}$$
$$= 2x^2 + 4xy - 2y^2$$

Problem 2

$$\frac{24x^2y^2 - 18xy + 6y}{6xy} = \frac{24x^2y^2}{6xy} - \frac{18xy}{6xy} + \frac{6y}{6xy}$$
$$= 4xy - 3 + \frac{1}{x}$$

Problem 3

A.

$$
\begin{array}{r}
5x - 1 \\
3x + 4 \overline{)15x^2 + 17x - 20} \\
\underline{15x^2 + 20x} \\
-3x - 20 \\
\underline{-3x - 4} \\
-16
\end{array}
$$

$$\frac{15x^2 + 17x - 20}{3x + 4} = 5x - 1 - \frac{16}{3x + 4}$$

B.

$$
\begin{array}{r}
x^2 + 3x - 1 \\
3x - 1 \overline{)3x^3 + 8x^2 - 6x + 2} \\
\underline{3x^3 - x^2} \\
9x^2 - 6x \\
\underline{9x^2 - 3x} \\
-3x + 2 \\
\underline{-3x + 1} \\
1
\end{array}
$$

$$\frac{3x^3 + 8x^2 - 6x + 2}{3x - 1} = x^2 + 3x - 1 + \frac{1}{3x - 1}$$

Problem 4

A.

$$
\begin{array}{r|rrr}
-2 & 6 & 8 & -5 \\
 & & -12 & 8 \\
\hline
 & 6 & -4 & 3
\end{array}
$$

$$(6x^2 + 8x - 5) \div (x + 2) = 6x - 4 + \frac{3}{x + 2}$$

B.

$$
\begin{array}{r|rrrrr}
3 & 2 & -3 & -8 & 0 & -2 \\
 & & 6 & 9 & 3 & 9 \\
\hline
 & 2 & 3 & 1 & 3 & 7
\end{array}
$$

$$(2x^4 - 3x^3 - 8x^2 - 2) \div (x - 3)$$
$$= 2x^3 + 3x^2 + x + 3 + \frac{7}{x - 3}$$

Problem 5

$$
\begin{array}{r|rrrr}
3 & 2 & 0 & -4 & -5 \\
 & & 6 & 18 & 42 \\
\hline
 & 2 & 6 & 14 & 37
\end{array}
$$

By the Remainder Theorem, $P(3) = 37$.

Solutions to Chapter 6 Problems

SECTION 6.1

Problem 1

$4x^6y = 2 \cdot 2 \cdot x^6 \cdot y$
$18x^2y^6 = 2 \cdot 3 \cdot 3 \cdot x^2 \cdot y^6$
GCF $= 2 \cdot x^2 \cdot y = 2x^2y$

Problem 2

A. $14a^2 = 2 \cdot 7 \cdot a^2$
$21a^4b = 3 \cdot 7 \cdot a^4 \cdot b$
The GCF is $7a^2$.

$14a^2 - 21a^4b = 7a^2(2) + 7a^2(-3a^2b)$
$\qquad\qquad\quad = 7a^2(2 - 3a^2b)$

B. $6x^4y^2 = 2 \cdot 3 \cdot x^4 \cdot y^2$
$9x^3y^2 = 3 \cdot 3 \cdot x^3 \cdot y^2$
$12x^2y^4 = 2 \cdot 2 \cdot 3 \cdot x^2 \cdot y^4$
The GCF is $3x^2y^2$.

$6x^4y^2 - 9x^3y^2 + 12x^2y^4$
$= 3x^2y^2(2x^2) + 3x^2y^2(-3x) + 3x^2y^2(4y^2)$
$= 3x^2y^2(2x^2 - 3x + 4y^2)$

Problem 3

$a(b - 7) + b(b - 7) = (b - 7)(a + b)$

Problem 4

$3y(5x - 2) - 4(2 - 5x)$
$= 3y(5x - 2) + 4(5x - 2)$
$= (5x - 2)(3y + 4)$

Problem 5

$y^5 - 5y^3 + 4y^2 - 20$
$= (y^5 - 5y^3) + (4y^2 - 20)$
$= y^3(y^2 - 5) + 4(y^2 - 5)$
$= (y^2 - 5)(y^3 + 4)$

SECTION 6.2

Problem 1

Factors of 15	Sum
$-1, -15$	-16
$-3, \ -5$	-8

$x^2 - 8x + 15 = (x - 3)(x - 5)$

Problem 2

Factors of -18	Sum
1, -18	-17
-1, 18	17
2, -9	-7
-2, 9	7
3, -6	-3
-3, 6	3

$x^2 + 3x - 18 = (x + 6)(x - 3)$

Problem 3

The GCF is $3b$.

$3a^2b - 18ab - 81b = 3b(a^2 - 6a - 27)$

Factors of -27	Sum
1, -27	-26
-1, 27	26
3, -9	-6
-3, 9	6

$3a^2b - 18ab - 81b = 3b(a + 3)(a - 9)$

Problem 4

The GCF is 4.

$4x^2 - 40xy + 84y^2$
$\quad = 4(x^2 - 10xy + 21y^2)$

Factors of 21	Sum
$-1, -21$	-22
$-3, -7$	-10

$4x^2 - 40xy + 84y^2 = 4(x - 3y)(x - 7y)$

SECTION 6.3

Problem 1

Factors of 6	Factors of 5
1, 6	$-1, -5$
2, 3	

Trial Factors	Middle Term
$(x - 1)(6x - 5)$	$-5x - 6x = -11x$

$6x^2 - 11x + 5 = (x - 1)(6x - 5)$

Problem 2

Factors of 8	Factors of -15
1, 8	1, -15
2, 4	-1, 15
	3, -5
	-3, 5

Trial Factors	Middle Term
$(x + 1)(8x - 15)$	$-15x + 8x = -7x$
$(x - 1)(8x + 15)$	$15x - 8x = 7x$
$(x + 3)(8x - 5)$	$-5x + 24x = 19x$
$(x - 3)(8x + 5)$	$5x - 24x = -19x$
$(2x + 1)(4x - 15)$	$-30x + 4x = -26x$
$(2x - 1)(4x + 15)$	$30x - 4x = 26x$
$(2x + 3)(4x - 5)$	$-10x + 12x = 2x$
$(2x - 3)(4x + 5)$	$10x - 12x = -2x$
$(8x + 1)(x - 15)$	$-120x + x = -119x$
$(8x - 1)(x + 15)$	$120x - x = 119x$
$(8x + 3)(x - 5)$	$-40x + 3x = -37x$
$(8x - 3)(x + 5)$	$40x - 3x = 37x$
$(4x + 1)(2x - 15)$	$-60x + 2x = -58x$
$(4x - 1)(2x + 15)$	$60x - 2x = 58x$
$(4x + 3)(2x - 5)$	$-20x + 6x = -14x$
$(4x - 3)(2x + 5)$	$20x - 6x = 14x$

$8x^2 + 14x - 15 = (4x - 3)(2x + 5)$

Problem 3

Factors of 24	Factors of -1
1, 24	1, -1
2, 12	
3, 8	
4, 6	

Trial Factors	Middle Term
$(1 - y)(24 + y)$	$y - 24y = -23y$
$(2 - y)(12 + y)$	$2y - 12y = -10y$
$(3 - y)(8 + y)$	$3y - 8y = -5y$
$(4 - y)(6 + y)$	$4y - 6y = -2y$

$24 - 2y - y^2 = (4 - y)(6 + y)$

Problem 4

The GCF is $2a^2$.

$4a^2b^2 - 30a^2b + 14a^2$
$\quad = 2a^2(2b^2 - 15b + 7)$

Factors of 2	Factors of 7
1, 2	$-1, -7$

Trial Factors	Middle Term
$(b - 1)(2b - 7)$	$-7b - 2b = -9b$
$(b - 7)(2b - 1)$	$-b - 14b = -15b$

$4a^2b^2 - 30a^2b + 14a^2$
$\quad = 2a^2(b - 7)(2b - 1)$

Problem 5

$a \cdot c = -14$
$-1(14) = -14, -1 + 14 = 13$

$2a^2 + 13a - 7$
$= 2a^2 - a + 14a - 7$
$= (2a^2 - a) + (14a - 7)$
$= a(2a - 1) + 7(2a - 1)$
$= (2a - 1)(a + 7)$

Problem 6

$a \cdot c = -12$
$1(-12) = -12, 1 - 12 = -11$

$4a^2 - 11a - 3$
$= 4a^2 + a - 12a - 3$
$= (4a^2 + a) - (12a + 3)$
$= a(4a + 1) - 3(4a + 1)$
$= (4a + 1)(a - 3)$

Problem 7

The GCF is $5x$.

$15x^3 + 40x^2 - 80x = 5x(3x^2 + 8x - 16)$
$3(-16) = -48$
$-4(12) = -48, -4 + 12 = 8$

$3x^2 + 8x - 16$
$= 3x^2 - 4x + 12x - 16$
$= (3x^2 - 4x) + (12x - 16)$
$= x(3x - 4) + 4(3x - 4)$
$= (3x - 4)(x + 4)$

$15x^3 + 40x^2 - 80x = 5x(3x^2 + 8x - 16)$
$\qquad\qquad\qquad\quad = 5x(3x - 4)(x + 4)$

SECTION 6.4

Problem 1

$x^2 - 36y^4 = x^2 - (6y^2)^2$
$\qquad\qquad = (x + 6y^2)(x - 6y^2)$

Problem 2

$9x^2 + 12x + 4 = (3x + 2)^2$

Problem 3

A. $8x^3 + y^3z^3 = (2x)^3 + (yz)^3$
$\qquad\qquad\quad = (2x + yz)(4x^2 - 2xyz + y^2z^2)$

B. $(x - y)^3 + (x + y)^3$
$\quad = [(x - y) + (x + y)]$
$\qquad \times [(x - y)^2 - (x - y)(x + y) + (x + y)^2]$
$\quad = 2x[x^2 - 2xy + y^2 - (x^2 - y^2) + x^2 + 2xy + y^2]$
$\quad = 2x[x^2 - 2xy + y^2 - x^2 + y^2 + x^2 + 2xy + y^2]$
$\quad = 2x(x^2 + 3y^2)$

Problem 4

A. $6x^2y^2 - 19xy + 10$
$\quad = (3xy - 2)(2xy - 5)$

B. $3x^4 + 4x^2 - 4 = (x^2 + 2)(3x^2 - 2)$

Problem 5

A. $4x - 4y - x^3 + x^2y$
$\quad = (4x - 4y) - (x^3 - x^2y)$
$\quad = 4(x - y) - x^2(x - y)$
$\quad = (x - y)(4 - x^2)$
$\quad = (x - y)(2 + x)(2 - x)$

B. $x^{4n} - x^{2n}y^{2n} = x^{2n+2n} - x^{2n}y^{2n}$
$\qquad\qquad\qquad = x^{2n}(x^{2n} - y^{2n})$
$\qquad\qquad\qquad = x^{2n}[(x^n)^2 - (y^n)^2]$
$\qquad\qquad\qquad = x^{2n}(x^n + y^n)(x^n - y^n)$

SECTION 6.5

Problem 1

A. $\qquad 4x^2 + 11x = 3$
$\qquad 4x^2 + 11x - 3 = 0$
$\qquad (4x - 1)(x + 3) = 0$

$\quad 4x - 1 = 0 \quad x + 3 = 0$
$\qquad\quad 4x = 1 \qquad\quad x = -3$
$\qquad\qquad x = \dfrac{1}{4}$

The solutions are $\dfrac{1}{4}$ and -3.

B. $(x - 2)(x + 5) = 8$
$\quad x^2 + 3x - 10 = 8$
$\quad x^2 + 3x - 18 = 0$
$\quad (x + 6)(x - 3) = 0$

$\quad x + 6 = 0 \qquad x - 3 = 0$
$\qquad x = -6 \qquad\quad x = 3$

The solutions are -6 and 3.

C. $\quad x^3 + 4x^2 - 9x - 36 = 0$
$\quad (x^3 + 4x^2) - (9x + 36) = 0$
$\quad x^2(x + 4) - 9(x + 4) = 0$
$\qquad (x + 4)(x^2 - 9) = 0$
$\quad (x + 4)(x + 3)(x - 3) = 0$

$x + 4 = 0 \quad x + 3 = 0 \quad x - 3 = 0$
$\quad x = -4 \qquad x = -3 \qquad x = 3$

The solutions are -4, -3, and 3.

Problem 2

$\qquad\qquad s(c) = 4$
$\quad c^2 - c - 2 = 4$
$\quad c^2 - c - 6 = 0$
$(c + 2)(c - 3) = 0$

$c + 2 = 0 \qquad c - 3 = 0$
$\quad c = -2 \qquad\quad c = 3$

The values of c are -2 and 3.

Problem 3

We must find the values of c for which $f(c) = 0$.

$$f(c) = 0$$
$$c^2 + 5c - 14 = 0 \qquad \text{• } f(x) = x^2 + 5x - 14, \text{ so}$$
$$f(c) = c^2 + 5c - 14.$$
$$(c + 7)(c - 2) = 0 \qquad \text{• Factor and use the Principle of Zero Products.}$$

$$c + 7 = 0 \qquad c - 2 = 0$$
$$c = -7 \qquad c = 2$$

The zeros of f are -7 and 2.

Problem 4

Strategy Replace D by 20 in the equation

$$D = \frac{n(n - 3)}{2} \text{ and then solve for } n.$$

Solution $D = \dfrac{n(n - 3)}{2}$

$$20 = \frac{n(n - 3)}{2}$$
$$40 = n^2 - 3n$$
$$0 = n^2 - 3n - 40$$
$$0 = (n - 8)(n + 5)$$

$$n - 8 = 0 \qquad n + 5 = 0$$
$$n = 8 \qquad n = -5$$

The answer -5 does not make sense in the context of this problem.

The polygon has 8 sides.

Solutions to Chapter 7 Problems

SECTION 7.1

Problem 1

$$f(x) = \frac{3 - 5x}{x^2 + 5x + 6}$$
$$f(2) = \frac{3 - 5(2)}{2^2 + 5(2) + 6}$$
$$f(2) = \frac{3 - 10}{4 + 10 + 6}$$
$$f(2) = \frac{-7}{20}$$
$$f(2) = -\frac{7}{20}$$

Problem 2

$$x^2 - 4 = 0$$
$$(x - 2)(x + 2) = 0$$

$$x - 2 = 0 \qquad x + 2 = 0$$
$$x = 2 \qquad x = -2$$

The domain is $\{x \mid x \neq -2, 2\}$.

Problem 3

$x^2 + 4$ is a positive number.

The domain is $\{x \mid x \in \text{real numbers}\}$.

Problem 4

A. $\dfrac{6x^4 - 24x^3}{12x^3 - 48x^2} = \dfrac{6x^3(x - 4)}{12x^2(x - 4)}$

$$= \frac{\overset{1}{6x^3}\cancel{(x - 4)}}{\underset{1}{12x^2}\cancel{(x - 4)}} = \frac{x}{2}$$

B. $\dfrac{20x - 15x^2}{15x^3 - 5x^2 - 20x} = \dfrac{5x(4 - 3x)}{5x(3x^2 - x - 4)}$

$$= \frac{5x(4 - 3x)}{5x(3x - 4)(x + 1)}$$

$$= \frac{\overset{-1}{5x\cancel{(4 - 3x)}}}{\underset{1}{5x\cancel{(3x - 4)}}(x + 1)}$$

$$= -\frac{1}{x + 1}$$

C. $\dfrac{x^{2n} + x^n - 12}{x^{2n} - 3x^n} = \dfrac{(x^n + 4)(x^n - 3)}{x^n(x^n - 3)}$

$$= \frac{(x^n + 4)\overset{1}{\cancel{(x^n - 3)}}}{x^n\underset{1}{\cancel{(x^n - 3)}}} = \frac{x^n + 4}{x^n}$$

SECTION 7.2

Problem 1

A. $\dfrac{12 + 5x - 3x^2}{x^2 + 2x - 15} \cdot \dfrac{2x^2 + x - 45}{3x^2 + 4x}$

$$= \frac{(4 + 3x)(3 - x)}{(x + 5)(x - 3)} \cdot \frac{(2x - 9)(x + 5)}{x(3x + 4)}$$

$$= \frac{(4 + 3x)(3 - x)(2x - 9)(x + 5)}{(x + 5)(x - 3) \cdot x(3x + 4)}$$

$$= \frac{\overset{1}{\cancel{(4 + 3x)}}\overset{-1}{\cancel{(3 - x)}}(2x - 9)\overset{1}{\cancel{(x + 5)}}}{\underset{1}{\cancel{(x + 5)}}\underset{1}{\cancel{(x - 3)}} \cdot x\underset{1}{\cancel{(3x + 4)}}}$$

$$= -\frac{2x - 9}{x}$$

B. $\dfrac{2x^2 - 13x + 20}{x^2 - 16} \cdot \dfrac{2x^2 + 9x + 4}{6x^2 - 7x - 5}$

$$= \frac{(2x - 5)(x - 4)}{(x - 4)(x + 4)} \cdot \frac{(2x + 1)(x + 4)}{(3x - 5)(2x + 1)}$$

$$= \frac{(2x - 5)(x - 4)(2x + 1)(x + 4)}{(x - 4)(x + 4)(3x - 5)(2x + 1)}$$

$$= \frac{(2x - 5)\overset{1}{\cancel{(x - 4)}}\overset{1}{\cancel{(2x + 1)}}\overset{1}{\cancel{(x + 4)}}}{\underset{1}{\cancel{(x - 4)}}\underset{1}{\cancel{(x + 4)}}(3x - 5)\underset{1}{\cancel{(2x + 1)}}}$$

$$= \frac{2x - 5}{3x - 5}$$

Problem 2

A. $\dfrac{6x^2 - 3xy}{10ab^4} \div \dfrac{16x^2y^2 - 8xy^3}{15a^2b^2}$

$= \dfrac{6x^2 - 3xy}{10ab^4} \cdot \dfrac{15a^2b^2}{16x^2y^2 - 8xy^3}$

$= \dfrac{3x(2x - y)}{10ab^4} \cdot \dfrac{15a^2b^2}{8xy^2(2x - y)}$

$= \dfrac{45a^2b^2x(2x - y)}{80ab^4xy^2(2x - y)} = \dfrac{45a^2b^2x(2x \!-\! \overset{1}{\cancel{y)}}}{80ab^4xy^2(2x \!-\! \underset{1}{\cancel{y)}}}$

$= \dfrac{9a}{16b^2y^2}$

B. $\dfrac{6x^2 - 7x + 2}{3x^2 + x - 2} \div \dfrac{4x^2 - 8x + 3}{5x^2 + x - 4}$

$= \dfrac{6x^2 - 7x + 2}{3x^2 + x - 2} \cdot \dfrac{5x^2 + x - 4}{4x^2 - 8x + 3}$

$= \dfrac{(2x - 1)(3x - 2)}{(x + 1)(3x - 2)} \cdot \dfrac{(x + 1)(5x - 4)}{(2x - 1)(2x - 3)}$

$= \dfrac{(2x - 1)(3x - 2)(x + 1)(5x - 4)}{(x + 1)(3x - 2)(2x - 1)(2x - 3)}$

$= \dfrac{\overset{1}{\cancel{(2x - 1)}}\,\overset{1}{\cancel{(3x - 2)}}\,\overset{1}{\cancel{(x + 1)}}(5x - 4)}{\underset{1}{\cancel{(x + 1)}}\,\underset{1}{\cancel{(3x - 2)}}\,\underset{1}{\cancel{(2x - 1)}}(2x - 3)}$

$= \dfrac{5x - 4}{2x - 3}$

Problem 3

The LCM is $a(a - 5)(a + 5)$.

$\dfrac{a - 3}{a^2 - 5a} + \dfrac{a - 9}{a^2 - 25}$

$= \dfrac{a - 3}{a(a - 5)} \cdot \dfrac{a + 5}{a + 5} + \dfrac{a - 9}{(a - 5)(a + 5)} \cdot \dfrac{a}{a}$

$= \dfrac{a^2 + 2a - 15}{a(a - 5)(a + 5)} + \dfrac{a^2 - 9a}{a(a - 5)(a + 5)}$

$= \dfrac{(a^2 + 2a - 15) + (a^2 - 9a)}{a(a - 5)(a + 5)}$

$= \dfrac{a^2 + 2a - 15 + a^2 - 9a}{a(a - 5)(a + 5)}$

$= \dfrac{2a^2 - 7a - 15}{a(a - 5)(a + 5)} = \dfrac{(2a + 3)(a - 5)}{a(a - 5)(a + 5)}$

$= \dfrac{(2a + 3)\overset{1}{\cancel{(a - 5)}}}{a\underset{1}{\cancel{(a - 5)}}(a + 5)} = \dfrac{2a + 3}{a(a + 5)}$

Problem 4

The LCM is $(x - 2)(2x - 3)$.

$\dfrac{x - 1}{x - 2} - \dfrac{7 - 6x}{2x^2 - 7x + 6} + \dfrac{4}{2x - 3}$

$= \dfrac{x - 1}{x - 2} \cdot \dfrac{2x - 3}{2x - 3} - \dfrac{7 - 6x}{(x - 2)(2x - 3)}$

$\quad + \dfrac{4}{2x - 3} \cdot \dfrac{x - 2}{x - 2}$

$= \dfrac{2x^2 - 5x + 3}{(x - 2)(2x - 3)} - \dfrac{7 - 6x}{(x - 2)(2x - 3)}$

$\quad + \dfrac{4x - 8}{(x - 2)(2x - 3)}$

$= \dfrac{(2x^2 - 5x + 3) - (7 - 6x) + (4x - 8)}{(x - 2)(2x - 3)}$

$= \dfrac{2x^2 - 5x + 3 - 7 + 6x + 4x - 8}{(x - 2)(2x - 3)}$

$= \dfrac{2x^2 + 5x - 12}{(x - 2)(2x - 3)} = \dfrac{(x + 4)(2x - 3)}{(x - 2)(2x - 3)}$

$= \dfrac{(x + 4)\overset{1}{\cancel{(2x - 3)}}}{(x - 2)\underset{1}{\cancel{(2x - 3)}}} = \dfrac{x + 4}{x - 2}$

SECTION 7.3

Problem 1

A. The LCM of x and x^2 is x^2.

$\dfrac{3 + \dfrac{16}{x} + \dfrac{16}{x^2}}{6 + \dfrac{5}{x} - \dfrac{4}{x^2}} = \dfrac{3 + \dfrac{16}{x} + \dfrac{16}{x^2}}{6 + \dfrac{5}{x} - \dfrac{4}{x^2}} \cdot \dfrac{x^2}{x^2}$

$= \dfrac{3 \cdot x^2 + \dfrac{16}{x} \cdot x^2 + \dfrac{16}{x^2} \cdot x^2}{6 \cdot x^2 + \dfrac{5}{x} \cdot x^2 - \dfrac{4}{x^2} \cdot x^2}$

$= \dfrac{3x^2 + 16x + 16}{6x^2 + 5x - 4}$

$= \dfrac{(3x + 4)(x + 4)}{(2x - 1)(3x + 4)}$

$= \dfrac{\overset{1}{\cancel{(3x + 4)}}(x + 4)}{(2x - 1)\underset{1}{\cancel{(3x + 4)}}} = \dfrac{x + 4}{2x - 1}$

B. The LCM is $x - 3$.

$\dfrac{2x + 5 + \dfrac{14}{x - 3}}{4x + 16 + \dfrac{49}{x - 3}}$

$= \dfrac{2x + 5 + \dfrac{14}{x - 3}}{4x + 16 + \dfrac{49}{x - 3}} \cdot \dfrac{x - 3}{x - 3}$

$= \dfrac{(2x + 5)(x - 3) + \dfrac{14}{x - 3}(x - 3)}{(4x + 16)(x - 3) + \dfrac{49}{x - 3}(x - 3)}$

$= \dfrac{2x^2 - x - 15 + 14}{4x^2 + 4x - 48 + 49} = \dfrac{2x^2 - x - 1}{4x^2 + 4x + 1}$

$= \dfrac{(2x + 1)(x - 1)}{(2x + 1)(2x + 1)} = \dfrac{\overset{1}{\cancel{(2x + 1)}}(x - 1)}{\underset{1}{\cancel{(2x + 1)}}(2x + 1)}$

$= \dfrac{x - 1}{2x + 1}$

SECTION 7.4

Problem 1

A.
$$\frac{5}{2x-3} = \frac{-2}{x+1}$$
$$(x+1)(2x-3)\frac{5}{2x-3} = (x+1)(2x-3)\frac{-2}{x+1}$$
$$5(x+1) = -2(2x-3)$$
$$5x+5 = -4x+6$$
$$9x+5 = 6$$
$$9x = 1$$
$$x = \frac{1}{9}$$

The solution is $\frac{1}{9}$.

B.
$$\frac{4x+1}{2x-1} = 2 + \frac{3}{x-3}$$
$$(2x-1)(x-3)\frac{4x+1}{2x-1} = (2x-1)(x-3)\left(2 + \frac{3}{x-3}\right)$$
$$(x-3)(4x+1) = (2x-1)(x-3)2 + (2x-1)3$$
$$4x^2 - 11x - 3 = 4x^2 - 14x + 6 + 6x - 3$$
$$-11x - 3 = -8x + 3$$
$$-3x = 6$$
$$x = -2$$

The solution is -2.

Problem 2

Strategy ▶ Time required for the small pipe to fill the tank: x

	Rate	Time	Part
Large pipe	$\frac{1}{9}$	6	$\frac{6}{9}$
Small pipe	$\frac{1}{x}$	6	$\frac{6}{x}$

▶ The sum of the part of the task completed by the large pipe and the part of the task completed by the small pipe is 1.

Solution
$$\frac{6}{9} + \frac{6}{x} = 1$$
$$\frac{2}{3} + \frac{6}{x} = 1$$
$$3x\left(\frac{2}{3} + \frac{6}{x}\right) = 3x \cdot 1$$
$$2x + 18 = 3x$$
$$18 = x$$

The small pipe working alone will fill the tank in 18 h.

Problem 3

Strategy ▶ Rate of the wind: r

	Distance	Rate	Time
With wind	700	$150 + r$	$\frac{700}{150+r}$
Against wind	500	$150 - r$	$\frac{500}{150-r}$

▶ The time flying with the wind equals the time flying against the wind.

Solution
$$\frac{700}{150+r} = \frac{500}{150-r}$$
$$(150+r)(150-r)\left(\frac{700}{150+r}\right) =$$
$$(150+r)(150-r)\left(\frac{500}{150-r}\right)$$
$$(150-r)700 = (150+r)500$$
$$105{,}000 - 700r = 75{,}000 + 500r$$
$$30{,}000 = 1200r$$
$$25 = r$$

The rate of the wind is 25 mph.

SECTION 7.5

Problem 1

Strategy To find the cost, write and solve a proportion using x to represent the cost.

Solution
$$\frac{2}{5.80} = \frac{15}{x}$$
$$x(5.80)\frac{2}{5.80} = x(5.80)\frac{15}{x}$$
$$2x = 15(5.80)$$
$$2x = 87$$
$$x = 43.5$$

The cost of 15 lb of cashews is $43.50.

Problem 2

Strategy To find the distance:

▶ Write the basic direct variation equation, replace the variables by the given values, and solve for k.

▶ Write the direct variation equation, replacing k by its value. Substitute 5 for t, and solve for s.

Solution $s = kt^2$
$64 = k(2)^2$
$64 = k \cdot 4$
$16 = k$

$s = 16t^2$
$s = 16(5)^2 = 400$

The object will fall 400 ft in 5 s.

Problem 3

Strategy To find the resistance:

▶ Write the basic inverse variation equation, replace the variables by the given values, and solve for k.

▶ Write the inverse variation equation, replacing k by its value. Substitute 0.02 for d, and solve for R.

Solution $R = \dfrac{k}{d^2}$

$0.5 = \dfrac{k}{(0.01)^2}$

$0.5 = \dfrac{k}{0.0001}$

$0.00005 = k$

$R = \dfrac{0.00005}{d^2}$

$R = \dfrac{0.00005}{(0.02)^2} = 0.125$

The resistance is 0.125 ohm.

Problem 4

Strategy To find the strength of the beam:

▶ Write the basic combined variation equation, replace the variables by the given values, and solve for k.

▶ Write the basic combined variation equation, replacing k by its value and substituting 4 for W, 8 for d, and 16 for L. Solve for s.

Solution $s = \dfrac{kWd^2}{L}$ $s = \dfrac{50Wd^2}{L}$

$1200 = \dfrac{k(2)(12)^2}{12}$ $s = \dfrac{50(4)8^2}{16}$

$1200 = \dfrac{k(288)}{12}$ $s = \dfrac{12,800}{16}$

$14,400 = 288k$ $s = 800$
$50 = k$

The strength of the beam is 800 lb.

Problem 1

A. $\dfrac{1}{R_1} + \dfrac{1}{R_2} = \dfrac{1}{R}$

$RR_1R_2\left(\dfrac{1}{R_1} + \dfrac{1}{R_2}\right) = RR_1R_2\left(\dfrac{1}{R}\right)$

$RR_1R_2\left(\dfrac{1}{R_1}\right) + RR_1R_2\left(\dfrac{1}{R_2}\right) = R_1R_2$

$RR_2 + RR_1 = R_1R_2$
$R(R_2 + R_1) = R_1R_2$

$R = \dfrac{R_1R_2}{R_2 + R_1}$

B. $t = \dfrac{r}{r + 1}$

$(r + 1)t = (r + 1)\left(\dfrac{r}{r + 1}\right)$

$tr + t = r$
$t = r - tr$
$t = r(1 - t)$

$\dfrac{t}{1 - t} = r$

Solutions to Chapter 8 Problems

Problem 1

A. $64^{\frac{2}{3}} = (2^6)^{\frac{2}{3}} = 2^4 = 16$

B. $16^{-\frac{3}{4}} = (2^4)^{-\frac{3}{4}} = 2^{-3} = \dfrac{1}{2^3} = \dfrac{1}{8}$

C. $(-81)^{\frac{3}{4}}$

The base of the exponential expression is negative, and the denominator of the exponent is a positive even number.

Therefore, $(-81)^{\frac{3}{4}}$ is not a real number.

Problem 2

A. $\dfrac{x^{\frac{1}{2}}y^{-\frac{5}{4}}}{x^{-\frac{4}{3}}y^{\frac{1}{3}}} = \dfrac{x^{\frac{1}{2}+\frac{4}{3}}}{y^{\frac{1}{3}+\frac{5}{4}}} = \dfrac{x^{\frac{11}{6}}}{y^{\frac{19}{12}}}$

B. $\left(x^{\frac{3}{4}}y^{\frac{1}{2}}z^{-\frac{2}{3}}\right)^{-\frac{4}{3}} = x^{-1}y^{-\frac{2}{3}}z^{\frac{8}{9}} = \dfrac{z^{\frac{8}{9}}}{xy^{\frac{2}{3}}}$

C. $\left(\dfrac{16a^{-2}b^{\frac{4}{3}}}{9a^4b^{-\frac{2}{3}}}\right)^{-\frac{1}{2}} = \left(\dfrac{2^4a^{-6}b^2}{3^2}\right)^{-\frac{1}{2}} = \dfrac{2^{-2}a^3b^{-1}}{3^{-1}}$

$= \dfrac{3a^3}{2^2b} = \dfrac{3a^3}{4b}$

Problem 3

A. $(2x^3)^{\frac{3}{4}} = \sqrt[4]{(2x^3)^3} = \sqrt[4]{8x^9}$

B. $-5a^{\frac{5}{6}} = -5(a^5)^{\frac{1}{6}} = -5\sqrt[6]{a^5}$

Problem 4

A. $\sqrt[3]{3ab} = (3ab)^{\frac{1}{3}}$

B. $\sqrt[4]{x^4 + y^4} = (x^4 + y^4)^{\frac{1}{4}}$

Problem 5

A. $\sqrt{121x^{10}y^4} = 11x^5y^2$

B. $\sqrt[3]{-8x^{12}y^3} = -2x^4y$

C. $-\sqrt[4]{81x^{12}y^8} = -3x^3y^2$

SECTION 8.2

Problem 1

$\sqrt[5]{128x^7} = \sqrt[5]{32x^5(4x^2)}$
$= \sqrt[5]{32x^5}\sqrt[5]{4x^2}$
$= 2x\sqrt[5]{4x^2}$

Problem 2

A. $3xy\sqrt[3]{81x^5y} - \sqrt[3]{192x^8y^4}$
$= 3xy\sqrt[3]{27x^3}\sqrt[3]{3x^2y} - \sqrt[3]{64x^6y^3}\sqrt[3]{3x^2y}$
$= 3xy \cdot 3x\sqrt[3]{3x^2y} - 4x^2y\sqrt[3]{3x^2y}$
$= 9x^2y\sqrt[3]{3x^2y} - 4x^2y\sqrt[3]{3x^2y}$
$= 5x^2y\sqrt[3]{3x^2y}$

B. $4a\sqrt[3]{54a^7b^9} + a^2b\sqrt[3]{128a^4b^6}$
$= 4a\sqrt[3]{27a^6b^9}\sqrt[3]{2a} + a^2b\sqrt[3]{64a^3b^6}\sqrt[3]{2a}$
$= 4a \cdot 3a^2b^3\sqrt[3]{2a} + a^2b \cdot 4ab^2\sqrt[3]{2a}$
$= 12a^3b^3\sqrt[3]{2a} + 4a^3b^3\sqrt[3]{2a}$
$= 16a^3b^3\sqrt[3]{2a}$

Problem 3

$\sqrt{5b}(\sqrt{3b} - \sqrt{10})$
$= \sqrt{15b^2} - \sqrt{50b}$
$= \sqrt{b^2}\sqrt{15} - \sqrt{25}\sqrt{2b}$
$= b\sqrt{15} - 5\sqrt{2b}$

Problem 4

A. $(2\sqrt[3]{2x} - 3)(\sqrt[3]{2x} - 5)$
$= 2\sqrt[3]{4x^2} - 10\sqrt[3]{2x} - 3\sqrt[3]{2x} + 15$
$= 2\sqrt[3]{4x^2} - 13\sqrt[3]{2x} + 15$

B. $(2\sqrt{x} - 3)(2\sqrt{x} + 3)$
$= (2\sqrt{x})^2 - 9 = 4x - 9$

Problem 5

A. $\dfrac{y}{\sqrt{3y}} = \dfrac{y}{\sqrt{3y}} \cdot \dfrac{\sqrt{3y}}{\sqrt{3y}} = \dfrac{y\sqrt{3y}}{\sqrt{9y^2}} = \dfrac{y\sqrt{3y}}{3y}$
$= \dfrac{\sqrt{3y}}{3}$

B. $\dfrac{3}{\sqrt[3]{3x^2}} = \dfrac{3}{\sqrt[3]{3x^2}} \cdot \dfrac{\sqrt[3]{9x}}{\sqrt[3]{9x}} = \dfrac{3\sqrt[3]{9x}}{\sqrt[3]{27x^3}} = \dfrac{3\sqrt[3]{9x}}{3x}$
$= \dfrac{\sqrt[3]{9x}}{x}$

Problem 6

A. $\dfrac{4 + \sqrt{2}}{3 - \sqrt{3}} = \dfrac{4 + \sqrt{2}}{3 - \sqrt{3}} \cdot \dfrac{3 + \sqrt{3}}{3 + \sqrt{3}}$
$= \dfrac{12 + 4\sqrt{3} + 3\sqrt{2} + \sqrt{6}}{9 - (\sqrt{3})^2}$
$= \dfrac{12 + 4\sqrt{3} + 3\sqrt{2} + \sqrt{6}}{6}$

B. $\dfrac{\sqrt{2} + \sqrt{x}}{\sqrt{2} - \sqrt{x}} = \dfrac{\sqrt{2} + \sqrt{x}}{\sqrt{2} - \sqrt{x}} \cdot \dfrac{\sqrt{2} + \sqrt{x}}{\sqrt{2} + \sqrt{x}}$
$= \dfrac{\sqrt{4} + \sqrt{2x} + \sqrt{2x} + \sqrt{x^2}}{(\sqrt{2})^2 - (\sqrt{x})^2}$
$= \dfrac{2 + 2\sqrt{2x} + x}{2 - x}$

SECTION 8.3

Problem 1

A. Because Q contains an odd root, there are no restrictions on the radicand.

The domain is $(-\infty, \infty)$.

B. $3x + 9 \geq 0$
$3x \geq -9$
$x \geq -3$

The domain is $[-3, \infty)$.

Problem 2

Problem 3

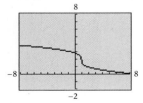

SECTION 8.4

Problem 1

A.
$$\sqrt{4x + 5} - 12 = -5$$
$$\sqrt{4x + 5} = 7$$
$$\left(\sqrt{4x + 5}\right)^2 = 7^2$$
$$4x + 5 = 49$$
$$4x = 44$$
$$x = 11$$

Check:
$$\sqrt{4x + 5} - 12 \;\big|\; -5$$
$$\sqrt{4 \cdot 11 + 5} - 12 \;\big|\; -5$$
$$\sqrt{44 + 5} - 12$$
$$\sqrt{49} - 12$$
$$7 - 12$$
$$-5 = -5$$

The solution is 11.

B.
$$\sqrt[4]{x - 8} = 3$$
$$\left(\sqrt[4]{x - 8}\right)^4 = 3^4$$
$$x - 8 = 81$$
$$x = 89$$

Check:
$$\sqrt[4]{x - 8} = 3$$
$$\sqrt[4]{89 - 8} \;\big|\; 3$$
$$\sqrt[4]{81}$$
$$3 = 3$$

The solution is 89.

Problem 2

A.
$$x + 3\sqrt{x + 2} = 8$$
$$3\sqrt{x + 2} = 8 - x$$
$$\left(3\sqrt{x + 2}\right)^2 = (8 - x)^2$$
$$9(x + 2) = 64 - 16x + x^2$$
$$9x + 18 = 64 - 16x + x^2$$
$$0 = x^2 - 25x + 46$$
$$0 = (x - 2)(x - 23)$$

$$x - 2 = 0 \qquad x - 23 = 0$$
$$x = 2 \qquad\quad x = 23$$

Check:
$$x + 3\sqrt{x + 2} = 8$$
$$2 + 3\sqrt{2 + 2} \;\big|\; 8$$
$$2 + 3\sqrt{4}$$
$$2 + 3 \cdot 2$$
$$2 + 6$$
$$8 = 8$$

$$x + 3\sqrt{x + 2} = 8$$
$$23 + 3\sqrt{23 + 2} \;\big|\; 8$$
$$23 + 3\sqrt{25}$$
$$23 + 3 \cdot 5$$
$$23 + 15$$
$$38 \neq 8$$

23 does not check as a solution. The solution is 2.

B.
$$\sqrt{x + 5} = 5 - \sqrt{x}$$
$$\left(\sqrt{x + 5}\right)^2 = \left(5 - \sqrt{x}\right)^2$$
$$x + 5 = 25 - 10\sqrt{x} + x$$

$$-20 = -10\sqrt{x}$$
$$2 = \sqrt{x}$$
$$2^2 = \left(\sqrt{x}\right)^2$$
$$4 = x$$

Check:
$$\sqrt{x + 5} = 5 - \sqrt{x}$$
$$\sqrt{4 + 5} \;\big|\; 5 - \sqrt{4}$$
$$\sqrt{9} \;\big|\; 5 - 2$$
$$3 = 3$$

The solution is 4.

Problem 3

Strategy To find the diagonal, use the Pythagorean Theorem. One leg is the length of the rectangle. The second leg is the width of the rectangle. The hypotenuse is the diagonal of the rectangle.

Solution
$$c^2 = a^2 + b^2$$
$$c^2 = (6)^2 + (3)^2$$
$$c^2 = 36 + 9$$
$$c^2 = 45$$
$$(c^2)^{\frac{1}{2}} = (45)^{\frac{1}{2}}$$
$$c = \sqrt{45}$$
$$c \approx 6.7$$

The diagonal is 6.7 cm.

SECTION 8.5

Problem 1
$$\sqrt{-45} = i\sqrt{45} = i\sqrt{9 \cdot 5} = 3i\sqrt{5}$$

Problem 2
$$\sqrt{98} - \sqrt{-60} = \sqrt{98} - i\sqrt{60}$$
$$= \sqrt{49 \cdot 2} - i\sqrt{4 \cdot 15}$$
$$= 7\sqrt{2} - 2i\sqrt{15}$$

Problem 3
$$(-4 + 2i) - (6 - 8i) = -10 + 10i$$

Problem 4
$$\left(16 - \sqrt{-45}\right) - \left(3 + \sqrt{-20}\right)$$
$$= \left(16 - i\sqrt{45}\right) - \left(3 + i\sqrt{20}\right)$$
$$= \left(16 - i\sqrt{9 \cdot 5}\right) - \left(3 + i\sqrt{4 \cdot 5}\right)$$
$$= \left(16 - 3i\sqrt{5}\right) - \left(3 + 2i\sqrt{5}\right)$$
$$= 13 - 5i\sqrt{5}$$

Problem 5
$$\sqrt{-3}\left(\sqrt{27} - \sqrt{-6}\right)$$
$$= i\sqrt{3}\left(\sqrt{27} - i\sqrt{6}\right)$$
$$= i\sqrt{81} - i^2\sqrt{18}$$
$$= i\sqrt{81} - (-1)\sqrt{9 \cdot 2}$$
$$= 9i + 3\sqrt{2}$$
$$= 3\sqrt{2} + 9i$$

Problem 6

A.
$$(4 - 3i)(2 - i) = 8 - 4i - 6i + 3i^2$$
$$= 8 - 10i + 3i^2 = 8 - 10i + 3(-1)$$
$$= 5 - 10i$$

B. $(3 - i)\left(\dfrac{3}{10} + \dfrac{1}{10}i\right)$

$= \dfrac{9}{10} + \dfrac{3}{10}i - \dfrac{3}{10}i - \dfrac{1}{10}i^2 = \dfrac{9}{10} - \dfrac{1}{10}i^2$

$= \dfrac{9}{10} - \dfrac{1}{10}(-1) = \dfrac{9}{10} + \dfrac{1}{10} = 1$

C. $(3 + 6i)(3 - 6i) = 3^2 + 6^2 = 9 + 36$
 $= 45$

D. $(1 + 5i)^2 = 1 + 10i + 25i^2$
 $= 1 + 10i + 25(-1)$
 $= 1 + 10i - 25$
 $= -24 + 10i$

Problem 7

$\dfrac{2 - 3i}{4i} = \dfrac{2 - 3i}{4i} \cdot \dfrac{i}{i} = \dfrac{2i - 3i^2}{4i^2} = \dfrac{2i - 3(-1)}{4(-1)}$

$= \dfrac{3 + 2i}{-4} = -\dfrac{3}{4} - \dfrac{1}{2}i$

Problem 8

$\dfrac{2 + 5i}{3 - 2i} = \dfrac{(2 + 5i)}{(3 - 2i)} \cdot \dfrac{(3 + 2i)}{(3 + 2i)}$

$= \dfrac{6 + 4i + 15i + 10i^2}{3^2 + 2^2} = \dfrac{6 + 19i + 10(-1)}{13}$

$= \dfrac{-4 + 19i}{13} = -\dfrac{4}{13} + \dfrac{19}{13}i$

Solutions to Chapter 9 Problems

SECTION 9.1

Problem 1

$x^2 - 3ax - 4a^2 = 0$
$(x + a)(x - 4a) = 0$

$x + a = 0 \qquad x - 4a = 0$
$\quad x = -a \qquad\quad x = 4a$

The solutions are $-a$ and $4a$.

Problem 2

$$(x - r_1)(x - r_2) = 0$$

$$\left[x - \left(-\dfrac{2}{3}\right)\right]\left(x - \dfrac{1}{6}\right) = 0$$

$$\left(x + \dfrac{2}{3}\right)\left(x - \dfrac{1}{6}\right) = 0$$

$$x^2 + \dfrac{3}{6}x - \dfrac{2}{18} = 0$$

$$18\left(x^2 + \dfrac{3}{6}x - \dfrac{2}{18}\right) = 0$$

$$18x^2 + 9x - 2 = 0$$

Problem 3

$2(x + 1)^2 + 24 = 0$
$\quad 2(x + 1)^2 = -24$
$\quad\quad (x + 1)^2 = -12$
$\quad \sqrt{(x + 1)^2} = \sqrt{-12}$
$\quad\quad |x + 1| = 2i\sqrt{3}$
$\quad\quad x + 1 = \pm 2i\sqrt{3}$

$x + 1 = 2i\sqrt{3} \qquad\qquad x + 1 = -2i\sqrt{3}$
$\quad x = -1 + 2i\sqrt{3} \qquad\qquad x = -1 - 2i\sqrt{3}$

The solutions are $-1 + 2i\sqrt{3}$ and $-1 - 2i\sqrt{3}$.

SECTION 9.2

Problem 1

A. $4x^2 - 4x - 1 = 0$
$\quad 4x^2 - 4x = 1$
$\quad \dfrac{4x^2 - 4x}{4} = \dfrac{1}{4}$

$\quad\quad x^2 - x = \dfrac{1}{4}$

Complete the square.

$x^2 - x + \dfrac{1}{4} = \dfrac{1}{4} + \dfrac{1}{4}$

$\left(x - \dfrac{1}{2}\right)^2 = \dfrac{2}{4}$

$\sqrt{\left(x - \dfrac{1}{2}\right)^2} = \sqrt{\dfrac{2}{4}}$

$\left|x - \dfrac{1}{2}\right| = \dfrac{\sqrt{2}}{2}$

$x - \dfrac{1}{2} = \pm\dfrac{\sqrt{2}}{2}$

$x - \dfrac{1}{2} = \dfrac{\sqrt{2}}{2} \qquad x - \dfrac{1}{2} = -\dfrac{\sqrt{2}}{2}$

$x = \dfrac{1}{2} + \dfrac{\sqrt{2}}{2} \qquad x = \dfrac{1}{2} - \dfrac{\sqrt{2}}{2}$

The solutions are $\dfrac{1 + \sqrt{2}}{2}$ and $\dfrac{1 - \sqrt{2}}{2}$.

B. $2x^2 + x - 5 = 0$
$\quad 2x^2 + x = 5$
$\quad \dfrac{2x^2 + x}{2} = \dfrac{5}{2}$

$\quad\quad x^2 + \dfrac{1}{2}x = \dfrac{5}{2}$

Complete the square.

$x^2 + \dfrac{1}{2}x + \dfrac{1}{16} = \dfrac{5}{2} + \dfrac{1}{16}$

$\left(x + \dfrac{1}{4}\right)^2 = \dfrac{41}{16}$

$$\sqrt{\left(x + \frac{1}{4}\right)^2} = \sqrt{\frac{41}{16}}$$

$$\left|x + \frac{1}{4}\right| = \frac{\sqrt{41}}{4}$$

$$x + \frac{1}{4} = \pm\frac{\sqrt{41}}{4}$$

$$x + \frac{1}{4} = \frac{\sqrt{41}}{4} \qquad x + \frac{1}{4} = -\frac{\sqrt{41}}{4}$$

$$x = -\frac{1}{4} + \frac{\sqrt{41}}{4} \qquad x = -\frac{1}{4} - \frac{\sqrt{41}}{4}$$

The solutions are $\dfrac{-1 + \sqrt{41}}{4}$ and $\dfrac{-1 - \sqrt{41}}{4}$.

Problem 2

A. $x^2 + 6x - 9 = 0$

$a = 1, b = 6, c = -9$

$$x = \frac{-b \pm \sqrt{b^2 - 4ac}}{2a}$$

$$= \frac{-6 \pm \sqrt{6^2 - 4(1)(-9)}}{2 \cdot 1}$$

$$= \frac{-6 \pm \sqrt{36 + 36}}{2}$$

$$= \frac{-6 \pm \sqrt{72}}{2}$$

$$= \frac{-6 \pm 6\sqrt{2}}{2} = -3 \pm 3\sqrt{2}$$

The solutions are $-3 + 3\sqrt{2}$ and $-3 - 3\sqrt{2}$.

B. $\qquad 4x^2 = 4x - 1$

$4x^2 - 4x + 1 = 0$

$a = 4, b = -4, c = 1$

$$x = \frac{-b \pm \sqrt{b^2 - 4ac}}{2a}$$

$$= \frac{-(-4) \pm \sqrt{(-4)^2 - 4(4)(1)}}{2 \cdot 4}$$

$$= \frac{4 \pm \sqrt{16 - 16}}{8}$$

$$= \frac{4 \pm \sqrt{0}}{8} = \frac{4}{8} = \frac{1}{2}$$

The solution is $\dfrac{1}{2}$.

Problem 3

$3x^2 - x - 1 = 0$

$a = 3, b = -1, c = -1$

$b^2 - 4ac = (-1)^2 - 4(3)(-1)$

$\quad = 1 + 12 = 13$

$13 > 0$

Because the discriminant is greater than zero, the equation has two real number solutions.

SECTION 9.3

Problem 1

A. $\qquad x - 5x^{\frac{1}{2}} + 6 = 0$

$$\left(x^{\frac{1}{2}}\right)^2 - 5\left(x^{\frac{1}{2}}\right) + 6 = 0$$

$$u^2 - 5u + 6 = 0$$

$$(u - 2)(u - 3) = 0$$

$$u - 2 = 0 \qquad u - 3 = 0$$

$$u = 2 \qquad u = 3$$

Replace u with $x^{\frac{1}{2}}$.

$$x^{\frac{1}{2}} = 2 \qquad\qquad x^{\frac{1}{2}} = 3$$

$$\left(x^{\frac{1}{2}}\right)^2 = 2^2 \qquad \left(x^{\frac{1}{2}}\right)^2 = 3^2$$

$$x = 4 \qquad\qquad x = 9$$

4 and 9 check as solutions.

The solutions are 4 and 9.

B. $\qquad 4x^4 + 35x^2 - 9 = 0$

$$4(x^2)^2 + 35(x^2) - 9 = 0$$

$$4u^2 + 35u - 9 = 0$$

$$(4u - 1)(u + 9) = 0$$

$$4u - 1 = 0 \qquad u + 9 = 0$$

$$4u = 1 \qquad\quad u = -9$$

$$u = \frac{1}{4}$$

Replace u with x^2.

$$x^2 = \frac{1}{4} \qquad\qquad x^2 = -9$$

$$\sqrt{x^2} = \sqrt{\frac{1}{4}} \qquad \sqrt{x^2} = \sqrt{-9}$$

$$|x| = \frac{1}{2} \qquad\qquad |x| = 3i$$

$$x = \pm\frac{1}{2} \qquad\qquad x = \pm 3i$$

The solutions are $\dfrac{1}{2}$, $-\dfrac{1}{2}$, $3i$, and $-3i$.

Problem 2

$$\sqrt{2x + 1} + x = 7$$

$$\sqrt{2x + 1} = 7 - x$$

$$\left(\sqrt{2x + 1}\right)^2 = (7 - x)^2$$

$$2x + 1 = 49 - 14x + x^2$$

$$0 = x^2 - 16x + 48$$

$$0 = (x - 4)(x - 12)$$

$$x - 4 = 0 \qquad x - 12 = 0$$

$$x = 4 \qquad\quad x = 12$$

4 checks as a solution, but 12 does not check as a solution.

The solution is 4.

Problem 3

$\sqrt{2x - 1} + \sqrt{x} = 2$

Solve for one of the radical expressions.

$$\sqrt{2x - 1} = 2 - \sqrt{x}$$
$$(\sqrt{2x - 1})^2 = (2 - \sqrt{x})^2$$
$$2x - 1 = 4 - 4\sqrt{x} + x$$
$$x - 5 = -4\sqrt{x}$$

Square each side of the equation.

$$(x - 5)^2 = (-4\sqrt{x})^2$$
$$x^2 - 10x + 25 = 16x$$
$$x^2 - 26x + 25 = 0$$
$$(x - 1)(x - 25) = 0$$

$$x - 1 = 0 \qquad x - 25 = 0$$
$$x = 1 \qquad\qquad x = 25$$

1 checks as a solution, but 25 does not check as a solution.

The solution is 1.

Problem 4

$$3y + \frac{25}{3y - 2} = -8$$

$$(3y - 2)\left(3y + \frac{25}{3y - 2}\right) = (3y - 2)(-8)$$

$$(3y - 2)(3y) + (3y - 2)\left(\frac{25}{3y - 2}\right) = (3y - 2)(-8)$$

$$9y^2 - 6y + 25 = -24y + 16$$
$$9y^2 + 18y + 9 = 0$$
$$9(y^2 + 2y + 1) = 0$$
$$9(y + 1)(y + 1) = 0$$

$$y + 1 = 0 \qquad y + 1 = 0$$
$$y = -1 \qquad\quad y = -1$$

The solution is -1.

SECTION 9.4

Problem 1

Strategy ▶ This is a geometry problem.
▶ Width of the rectangle: W
 Length of the rectangle: $W + 3$
▶ Use the equation $A = L \cdot W$.

Solution
$$A = L \cdot W$$
$$54 = (W + 3)(W)$$
$$54 = W^2 + 3W$$
$$0 = W^2 + 3W - 54$$
$$0 = (W + 9)(W - 6)$$

$$W + 9 = 0 \qquad W - 6 = 0$$
$$W = -9 \qquad\quad W = 6$$

The solution -9 is not possible.

Length $= W + 3 = 6 + 3 = 9$

The length is 9 m.

SECTION 9.5

Problem 1

$$2x^2 - x - 10 \le 0$$
$$(2x - 5)(x + 2) \le 0$$

$2x - 5$ $---\ |\ --------\ |\ +++$
$x + 2$ $---\ |\ +++++++\ |\ +++$

$\begin{array}{ccccccc} -3 & -2 & -1 & 0 & 1 & 2 & 3 \end{array}$

$$\left\{x \mid -2 \le x \le \frac{5}{2}\right\}$$

$\begin{array}{ccccccccccc} -5 & -4 & -3 & -2 & -1 & 0 & 1 & 2 & 3 & 4 & 5 \end{array}$

Problem 2

$$\frac{x}{x - 2} \le 0$$

x $----\ |\ +++\ |\ ++++++$
$x - 2$ $----\ |\ ---\ |\ ++++++$

$\begin{array}{ccccccc} -1 & 0 & 1 & 2 & 3 & 4 \end{array}$

$$\{x \mid 0 \le x < 2\}$$

$\begin{array}{ccccccccccc} -5 & -4 & -3 & -2 & -1 & 0 & 1 & 2 & 3 & 4 & 5 \end{array}$

SECTION 9.6

Problem 1

x-coordinate: $-\dfrac{b}{2a} = -\dfrac{0}{2(1)} = 0$

$$y = x^2 - 2$$
$$= 0^2 - 2$$
$$= -2$$

The vertex is $(0, -2)$.

The axis of symmetry is the line $x = 0$.

Problem 2

The domain is $\{x \mid x \in \text{real numbers}\}$. The range is $\{y \mid y \ge -6\}$.

Problem 3

A. $y = 2x^2 - 5x + 2$

$0 = 2x^2 - 5x + 2$

$0 = (2x - 1)(x - 2)$

$2x - 1 = 0$	$x - 2 = 0$
$2x = 1$	$x = 2$

$x = \dfrac{1}{2}$

The x-intercepts are $\left(\dfrac{1}{2}, 0\right)$ and $(2, 0)$.

B. $y = x^2 + 4x + 4$

$0 = x^2 + 4x + 4$

$0 = (x + 2)(x + 2)$

$x + 2 = 0$	$x + 2 = 0$
$x = -2$	$x = -2$

The x-intercept is $(-2, 0)$.

Problem 4

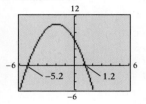

The approximate values of the zeros are -5.2 and 1.2.

Problem 5

$f(x) = -x^2 - 4x + 2$

$-4 = -x^2 - 4x + 2$

$0 = -x^2 - 4x + 6$

$x = \dfrac{-b \pm \sqrt{b^2 - 4ac}}{2a}$

$= \dfrac{4 \pm \sqrt{-(-4)^2 - 4(-1)(6)}}{2(-1)}$

$= \dfrac{4 \pm \sqrt{16 + 24}}{-2}$

$= \dfrac{4 \pm \sqrt{40}}{-2}$

$= \dfrac{4 \pm 2\sqrt{10}}{-2}$

$= -2 \pm \sqrt{10}$

The values of x for which $f(x) = -4$ are $-2 + \sqrt{10}$ and $-2 - \sqrt{10}$.

Problem 6

$y = x^2 - x - 6$

$a = 1, b = -1, c = -6$

$b^2 - 4ac = (-1)^2 - 4(1)(-6)$

$\qquad = 1 + 24 = 25$

Because the discriminant is greater than zero, the parabola has two x-intercepts.

SECTION 9.7

Problem 1

$x = -\dfrac{b}{2a} = -\dfrac{-3}{2(2)} = \dfrac{3}{4}$

$f(x) = 2x^2 - 3x + 1$

$f\left(\dfrac{3}{4}\right) = 2\left(\dfrac{3}{4}\right)^2 - 3\left(\dfrac{3}{4}\right) + 1 = \dfrac{9}{8} - \dfrac{9}{4} + 1$

$\qquad = -\dfrac{1}{8}$

Because a is positive, the function has a minimum value. The minimum value of the function is $-\dfrac{1}{8}$.

Problem 2

Strategy ▶ To find the time it takes the ball to reach its maximum height, find the t-coordinate of the vertex.

▶ To find the maximum height, evaluate the function at the t-coordinate of the vertex.

Solution $t = -\dfrac{b}{2a} = -\dfrac{64}{2(-16)} = 2$

The ball reaches its maximum height in 2 s.

$s(t) = -16t^2 + 64t$

$s(2) = -16(2)^2 + 64(2) = -64 + 128 = 64$

The maximum height is 64 ft.

Problem 3

Strategy

$P = 2x + y$

$100 = 2x + y$ • $P = 100$

$100 - 2x = y$ • Solve for y.

Express the area of the rectangle in terms of x.

$A = xy$

$A = x(100 - 2x)$ • $y = 100 - 2x$

$A = -2x^2 + 100x$

▶ To find the width, find the x-coordinate of the vertex of $f(x) = -2x^2 + 100x$.

▶ To find the length, replace x in $y = 100 - 2x$ by the x-coordinate of the vertex.

Solution $x = -\dfrac{b}{2a} = -\dfrac{100}{2(-2)} = 25$

The width is 25 ft.

$100 - 2x = 100 - 2(25)$

$\qquad = 100 - 50 = 50$

The length is 50 ft.

Solutions to Chapter 10 Problems

SECTION 10.1

Problem 1

The graph of $g(x) = x^2 - 2x - 3$ is the graph of $f(x) = x^2 - 2x$ shifted 3 units down.

Problem 2

The graph of $g(x) = \sqrt{x - 3}$ is the graph of $f(x) = \sqrt{x}$ shifted 3 units to the right.

Problem 3

The graph of $V(x) = (x - 2)^3 + 1$ is the graph of $g(x) = x^3$ shifted up 1 unit and shifted 2 units to the right.

SECTION 10.2

Problem 1

$$\begin{aligned}
(f + g)(-2) &= f(-2) + g(-2) \\
&= [(-2)^2 + 2(-2)] + [5(-2) - 2] \\
&= (4 - 4) + (-10 - 2) \\
&= -12
\end{aligned}$$

$(f + g)(-2) = -12$

Problem 2

$$\begin{aligned}
(f \cdot g)(3) &= f(3) \cdot g(3) \\
&= (4 - 3^2) \cdot [3(3) - 4] \\
&= (4 - 9) \cdot (9 - 4) \\
&= (-5)(5) \\
&= -25
\end{aligned}$$

$(f \cdot g)(3) = -25$

Problem 3

$$\begin{aligned}
\left(\frac{f}{g}\right)(4) &= \frac{f(4)}{g(4)} \\
&= \frac{4^2 - 4}{4^2 + 2 \cdot 4 + 1} \\
&= \frac{16 - 4}{16 + 8 + 1} \\
&= \frac{12}{25}
\end{aligned}$$

$$\left(\frac{f}{g}\right)(4) = \frac{12}{25}$$

Problem 4

A. $\quad h(x) = x^2 + 1$
$\quad h(0) = 0 + 1 = 1$

$\quad g(x) = 3x - 2$
$\quad g(1) = 3(1) - 2 = 1$

$\quad g[h(0)] = 1$

B. $h[g(x)] = h(3x - 2)$
$\qquad = (3x - 2)^2 + 1$
$\qquad = 9x^2 - 12x + 4 + 1$
$\qquad = 9x^2 - 12x + 5$

SECTION 10.3

Problem 1

A. Because any vertical line will intersect the graph at no more than one point, and any horizontal line will intersect the graph at no more than one point, the graph is the graph of a 1–1 function.

B. Because any vertical line will intersect the graph at no more than one point, and any horizontal line will intersect the graph at no more than one point, the graph is the graph of a 1–1 function.

Problem 2

$$\begin{aligned}
f(x) &= 4x + 2 \\
y &= 4x + 2 \\
x &= 4y + 2 \\
4y &= x - 2 \\
y &= \frac{1}{4}x - \frac{1}{2}
\end{aligned}$$

$$f^{-1}(x) = \frac{1}{4}x - \frac{1}{2}$$

Problem 3

$$\begin{aligned}
h[g(x)] &= 4\left(\frac{1}{4}x - \frac{1}{2}\right) + 2 \\
&= x - 2 + 2 = x
\end{aligned}$$

$$\begin{aligned}
g[h(x)] &= \frac{1}{4}(4x + 2) - \frac{1}{2} \\
&= x + \frac{1}{2} - \frac{1}{2} = x
\end{aligned}$$

The functions are inverses of each other.

Solutions to Chapter 11 Problems

SECTION 11.1

Problem 1

$$f(3) = \left(\frac{2}{3}\right)^3 = \frac{8}{27}$$

$$f(-2) = \left(\frac{2}{3}\right)^{-2} = \left(\frac{3}{2}\right)^2 = \frac{9}{4}$$

Problem 2

$$f(x) = 2^{2x+1}$$
$$f(0) = 2^{2(0)+1} = 2^1 = 2$$
$$f(-2) = 2^{2(-2)+1} = 2^{-3} = \frac{1}{2^3} = \frac{1}{8}$$

Problem 3

$$f(x) = \pi^x$$
$$f(3) = \pi^3$$
$$\approx 31.0063$$
$$f(-2) = \pi^{-2}$$
$$\approx 0.1013$$
$$f(\pi) = \pi^\pi$$
$$\approx 36.4622$$

Problem 4

$$f(x) = e^x$$
$$f(1.2) = e^{1.2}$$
$$\approx 3.3201$$
$$f(-2.5) = e^{-2.5}$$
$$\approx 0.0821$$
$$f(e) = e^e$$
$$\approx 15.1543$$

Problem 5

A.

B.

Problem 6

A.

B.

Problem 7

−2.4

SECTION 11.2

Problem 1

$3^{-4} = \dfrac{1}{81}$ is equivalent to $\log_3 \dfrac{1}{81} = -4$.

Problem 2

$\log_{10} 0.0001 = -4$ is equivalent to $10^{-4} = 0.0001$.

Problem 3

$$\log_4 64 = x$$
$$64 = 4^x$$
$$4^3 = 4^x$$
$$3 = x$$
$$\log_4 64 = 3$$

Problem 4

$$\log_2 x = -4$$
$$2^{-4} = x$$
$$\frac{1}{2^4} = x$$
$$\frac{1}{16} = x$$

The solution is $\dfrac{1}{16}$.

Problem 5

$$\log x = 1.5$$
$$10^{1.5} = x$$
$$31.6228 \approx x$$

The solution is 31.6228.

Problem 6

A. $\log_b \dfrac{x^2}{y} = \log_b x^2 - \log_b y$

$$= 2 \log_b x - \log_b y$$

B. $\ln y^{\frac{1}{3}} z^3 = \ln y^{\frac{1}{3}} + \ln z^3$

$$= \frac{1}{3} \ln y + 3 \ln z$$

C. $\log_8 \sqrt[3]{xy^2} = \log_8 (xy^2)^{\frac{1}{3}}$

$$= \frac{1}{3} \log_8 xy^2$$
$$= \frac{1}{3} (\log_8 x + \log_8 y^2)$$
$$= \frac{1}{3} (\log_8 x + 2 \log_8 y)$$
$$= \frac{1}{3} \log_8 x + \frac{2}{3} \log_8 y$$

Problem 7

A. $2 \log_b x - 3 \log_b y - \log_b z = \log_b x^2 - \log_b y^3 - \log_b z$

$$= \log_b \frac{x^2}{y^3} - \log_b z$$
$$= \log_b \frac{x^2}{y^3 z}$$

B. $\dfrac{1}{3}(\log_4 x - 2\log_4 y + \log_4 z)$

$$= \dfrac{1}{3}(\log_4 x - \log_4 y^2 + \log_4 z)$$

$$= \dfrac{1}{3}\left(\log_4 \dfrac{x}{y^2} + \log_4 z\right)$$

$$= \dfrac{1}{3}\left(\log_4 \dfrac{xz}{y^2}\right)$$

$$= \log_4 \left(\dfrac{xz}{y^2}\right)^{\frac{1}{3}} = \log_4 \sqrt[3]{\dfrac{xz}{y^2}}$$

C. $\dfrac{1}{2}(2\ln x - 5\ln y) = \dfrac{1}{2}(\ln x^2 - \ln y^5)$

$$= \dfrac{1}{2}\left(\ln \dfrac{x^2}{y^5}\right)$$

$$= \ln \left(\dfrac{x^2}{y^5}\right)^{\frac{1}{2}}$$

$$= \ln \sqrt{\dfrac{x^2}{y^5}}$$

Problem 8

A. $\log_{16} 1 = 0$

B. $12\log_3 3 = \log_3 3^{12} = 12$

Problem 9

$$\log_4 2.4 = \dfrac{\ln 2.4}{\ln 4} \approx 0.6315$$

Problem 10

$$f(x) = 4\log_8(3x + 4) = 4 \cdot \dfrac{\log(3x + 4)}{\log 8}$$

$$= \dfrac{4}{\log 8}\log(3x + 4)$$

SECTION 11.3

Problem 1

A. $f(x) = \log_2(x - 1)$
$\quad y = \log_2(x - 1)$

$\quad y = \log_2(x - 1)$ is equivalent to $2^y = x - 1$.

$\quad 2^y + 1 = x$

B. $f(x) = \log_3 2x$
$\quad y = \log_3 2x$

$\quad y = \log_3 2x$ is equivalent to $3^y = 2x$.

$\quad \dfrac{3^y}{2} = x$

Problem 2

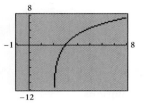

Problem 3

$f(x) = 2\log_4 x$

$$= 2 \cdot \dfrac{\ln x}{\ln 4}$$

$$= \dfrac{2}{\ln 4}\ln x$$

SECTION 11.4

Problem 1

$$10^{3x+5} = 10^{x-3}$$

$$3x + 5 = x - 3$$
$$2x + 5 = -3$$
$$2x = -8$$
$$x = -4$$

Check:

10^{3x+5}	$= 10^{x-3}$
$10^{3(-4)+5}$	10^{-4-3}
10^{-12+5}	10^{-7}
10^{-7}	$= 10^{-7}$

The solution is -4.

Problem 2

A.
$$4^{3x} = 25$$
$$\log 4^{3x} = \log 25$$
$$3x\log 4 = \log 25$$
$$3x = \dfrac{\log 25}{\log 4}$$
$$x = \dfrac{\log 25}{3\log 4}$$
$$x \approx 0.7740$$

The solution is 0.7740.

B.
$$(1.06)^x = 1.5$$
$$\log(1.06)^x = \log 1.5$$
$$x\log 1.06 = \log 1.5$$
$$x = \dfrac{\log 1.5}{\log 1.06}$$
$$x \approx 6.9585$$

The solution is 6.9585.

Problem 3

$$e^x = x$$
$$e^x - x = 0$$

Graph $f(x) = e^x - x$.

The equation has no real number solutions.

Problem 4

A. $\log_4(x^2 - 3x) = 1$

Rewrite in exponential form.

$$4^1 = x^2 - 3x$$
$$4 = x^2 - 3x$$
$$0 = x^2 - 3x - 4$$
$$0 = (x + 1)(x - 4)$$

$$x + 1 = 0 \qquad x - 4 = 0$$
$$x = -1 \qquad x = 4$$

The solutions are -1 and 4.

B. $\log_3 x + \log_3(x + 3) = \log_3 4$
$$\log_3[x(x + 3)] = \log_3 4$$

Use the One-to-One Property of Logarithms.

$$x(x + 3) = 4$$
$$x^2 + 3x = 4$$
$$x^2 + 3x - 4 = 0$$
$$(x + 4)(x - 1) = 0$$

$$x + 4 = 0 \qquad x - 1 = 0$$
$$x = -4 \qquad x = 1$$

-4 does not check as a solution. The solution is 1.

Problem 5

$$\log(3x - 2) = -2x$$
$$\log(3x - 2) + 2x = 0$$

Graph $f(x) = \log(3x - 2) + 2x$.

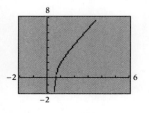

The solution is 0.68.

SECTION 11.5

Problem 1

Strategy ▶ To find the time, solve the compound interest formula for n. Use $P = 6000$, $A = 3000$, and $i = \dfrac{12\%}{12} = \dfrac{0.12}{12} = 0.01$.

Solution
$$P = A(1 + i)^n$$
$$6000 = 3000(1 + 0.01)^n$$
$$6000 = 3000(1.01)^n$$
$$2 = (1.01)^n$$
$$\log 2 = \log(1.01)^n$$
$$\log 2 = n \log 1.01$$
$$\frac{\log 2}{\log 1.01} = n$$
$$70 \approx n$$

70 months $\div\ 12 \approx 5.8$ years

In approximately 6 years, the investment will be worth $6000.

Problem 2

Strategy ▶ To find the intensity, use the equation for the Richter scale magnitude of an earthquake, $M = \log \dfrac{I}{I_0}$. Replace M by 5.2 and solve for I.

Solution
$$M = \log \frac{I}{I_0}$$
$$5.2 = \log \frac{I}{I_0} \qquad \bullet \textbf{ Replace } \textbf{\textit{M}} \textbf{ by 5.2.}$$
$$10^{5.2} = \frac{I}{I_0} \qquad \bullet \textbf{ Write in exponential form.}$$
$$10^{5.2}I_0 = I$$
$$158{,}489 I_0 \approx I$$

The earthquake had an intensity that was approximately 158,489 times the intensity of a zero-level earthquake.

Solutions to Chapter R Problems

SECTION R.1

Problem 1

$$(-4)(6 - 8)^2 - (-12 \div 4) = -4(-2)^2 - (-3)$$
$$= -4(4) - (-3)$$
$$= -16 - (-3)$$
$$= -16 + 3$$
$$= -13$$

Problem 2

$3xy^2 - 3x^2y$

$3(-2)(5)^2 - 3(-2)^2(5) = 3(-2)(25) - 3(4)(5)$

$$= -6(25) - 3(4)(5)$$
$$= -150 - 3(4)(5)$$
$$= -150 - 12(5)$$
$$= -150 - 60$$
$$= -150 + (-60)$$
$$= -210$$

Problem 3

$-5(-3a) = [-5(-3)]a$

$$= 15a$$

Problem 4

$2z^2 - 5z - 3z^2 + 6z = 2z^2 - 3z^2 - 5z + 6z$

$$= (2z^2 - 3z^2) + (-5z + 6z)$$
$$= -1z^2 + z$$
$$= -z^2 + z$$

Problem 5

a. $-3(5y - 2) = -3(5y) - (-3)(2)$

$$= -15y + 6$$

b. $-2(4x + 2y - 6z) = -2(4x) + (-2)(2y) - (-2)(6z)$

$$= -8x - 4y + 12z$$

Problem 6

a. $7(-3x - 4y) - 3(3x + y) = -21x - 28y - 9x - 3y$

$$= -30x - 31y$$

b. $2y - 3[5 - 3(3 + 2y)] = 2y - 3[5 - 9 - 6y]$

$$= 2y - 3[-4 - 6y]$$
$$= 2y + 12 + 18y$$
$$= 20y + 12$$

SECTION R.2

Problem 1

a.

$4x + 3 = 7x + 9$

$4x - 7x + 3 = 7x - 7x + 9$ • Subtract 7x.

$-3x + 3 = 9$

$-3x + 3 - 3 = 9 - 3$ • Subtract 3.

$-3x = 6$

$\dfrac{-3x}{-3} = \dfrac{6}{-3}$ • Divide by −3.

$x = -2$

The solution is −2.

b.

$4 - (5x - 8) = 4x + 3$

$4 - 5x + 8 = 4x + 3$ • **Distributive Property**

$-5x + 12 = 4x + 3$ • **Combine like terms.**

$-5x - 4x + 12 = 4x - 4x + 3$ • **Subtract 4x.**

$-9x + 12 = 3$

$-9x + 12 - 12 = 3 - 12$ • **Subtract 12.**

$-9x = -9$

$\dfrac{-9x}{-9} = \dfrac{-9}{-9}$ • **Divide by −9.**

$x = 1$

The solution is 1.

Problem 2

$3x - 1 \le 5x - 7$

$3x - 5x - 1 \le 5x - 5x - 7$ • **Subtract 5x.**

$-2x - 1 \le -7$

$-2x - 1 + 1 \le -7 + 1$ • **Add 1.**

$-2x \le -6$

$\dfrac{-2x}{-2} \ge \dfrac{-6}{-2}$ • **Divide by −2.**

$x \ge 3$

$\{x | x \ge 3\}$

Problem 3

$3 - 2(3x + 1) \le 7 - 2x$

$3 - 6x - 2 \le 7 - 2x$ • **Distributive Property**

$1 - 6x \le 7 - 2x$ • **Combine like terms.**

$1 - 6x + 2x \le 7 - 2x + 2x$ • **Add 2x.**

$1 - 4x \le 7$ • **Subtract 1.**

$1 - 1 - 4x \le 7 - 1$ • **Divide by −4.**

$-4x \le 6$

$\dfrac{-4x}{-4} \ge \dfrac{6}{-4}$

$x \ge -\dfrac{3}{2}$

$\left[-\dfrac{3}{2}, \infty\right)$

SECTION R.3

Problem 1

Problem 2

$y = \dfrac{3}{5}x - 4$

x	y
0	-4
-5	-7
5	-1

Problem 3

$$-3x + 2y = 4$$
$$2y = 3x + 4$$
$$y = \frac{3}{2}x + 2$$

x	y
0	2
-2	-1
2	5

Problem 4

Let $(x_1, y_1) = (-2, 3)$ and $(x_2, y_2) = (1, -3)$.

$$m = \frac{y_2 - y_1}{x_2 - x_1} = \frac{-3 - 3}{1 - (-2)} = \frac{-6}{3} = -2$$

The slope is -2.

Problem 5

$$y = -\frac{2}{3}x + 2$$

$$m = -\frac{2}{3} = \frac{-2}{3}$$

y-intercept $= (0, 2)$

Place a dot at the y-intercept.

Starting at the y-intercept, move to the right 3 units (the change in x) and down 2 units (the change in y).

Place a dot at that location.

Draw a line through the two points.

Problem 6

$$f(x) = 4 - 2x$$
$$f(-3) = 4 - 2(-3) \qquad \bullet\ x = -3$$
$$f(-3) = 4 + 6$$
$$f(-3) = 10$$

An ordered pair of the function is $(-3, 10)$.

Problem 7

$$(x_1, y_1) = (-2, 2),\ m = -\frac{1}{2}$$

$$y - y_1 = m(x - x_1)$$

$$y - 2 = -\frac{1}{2}[x - (-2)]$$

$$y - 2 = -\frac{1}{2}(x + 2)$$

$$y - 2 = -\frac{1}{2}x - 1$$

$$y = -\frac{1}{2}x + 1$$

SECTION R.4

Problem 1

a. $(-2ab)(2a^3b^{-2})^{-3} = (-2ab)(2^{-3}a^{-9}b^6)$
$$= -(2^{-2})a^{-8}b^7$$
$$= -\frac{b^7}{2^2a^8}$$
$$= -\frac{b^7}{4a^8}$$

b. $\left(\dfrac{2x^2y^{-4}}{4x^{-2}y^{-5}}\right)^{-3} = \left(\dfrac{x^2y^{-4}}{2x^{-2}y^{-5}}\right)^{-3}$
$$= \frac{x^{2(-3)}y^{(-4)(-3)}}{2^{1(-3)}x^{(-2)(-3)}y^{(-5)(-3)}}$$
$$= \frac{x^{-6}y^{12}}{2^{-3}x^6y^{15}}$$
$$= 2^3x^{-6-6}y^{12-15}$$
$$= 2^3x^{-12}y^{-3}$$
$$= \frac{8}{x^{12}y^3}$$

Problem 2

$$(-4x^3 + 2x^2 - 8) + (4x^3 + 6x^2 - 7x + 5)$$
$$= (-4x^3 + 4x^3) + (2x^2 + 6x^2) + (-7x) + (-8 + 5)$$
$$= 8x^2 - 7x - 3$$

Problem 3

$$(-4w^3 + 8w - 8) - (3w^3 - 4w^2 + 2w - 1)$$
$$= (-4w^3 + 8w - 8) + (-3w^3 + 4w^2 - 2w + 1)$$
$$= (-4w^3 - 3w^3) + 4w^2 + (8w - 2w) + (-8 + 1)$$
$$= -7w^3 + 4w^2 + 6w - 7$$

Problem 4

$$3mn^2(2m^2 - 3mn - 1)$$
$$= 3mn^2(2m^2) - 3mn^2(3mn) - 3mn^2(1)$$
$$= 6m^3n^2 - 9m^2n^3 - 3mn^2$$

Problem 5

$$3c^2 - 4c + 5$$
$$2c - 3$$
$$\overline{-9c^2 + 12c - 15}$$
$$6c^3 - 8c^2 + 10c$$
$$\overline{6c^3 - 17c^2 + 22c - 15}$$

Problem 6

$(4y - 7)(3y - 5)$
$= (4y)(3y) + (4y)(-5) + (-7)(3y) + (-7)(-5)$
$= 12y^2 - 20y - 21y + 35$
$= 12y^2 - 41y + 35$

Problem 7

$(x^3 - 7 - 2x) \div (x - 2)$

$$\begin{array}{r} x^2 + 2x + 2 \\ x - 2\overline{)x^3 + 0x^2 - 2x - 7} \\ \underline{x^3 - 2x^2} \\ 2x^2 - 2x \\ \underline{2x^2 - 4x} \\ 2x - 7 \\ \underline{2x - 4} \\ -3 \end{array}$$

Check: $(x - 2)(x^2 + 2x + 2) - 3$
$= x^3 - 2x - 4 - 3$
$= x^3 - 2x - 7$
$(x^3 - 7 - 2x) \div (x - 2)$

$= x^2 + 2x + 2 - \dfrac{3}{x - 2}$

Problem 8

The GCF is $3x^2y^2$.
$6x^4y^2 - 9x^3y^2 + 12x^2y^4$
$= 3x^2y^2(2x^2 - 3x + 4y^2)$

Problem 9

Two factors of 15 whose sum is -8 are -3 and -5.
$x^2 - 8x + 15 = (x - 3)(x - 5)$

Problem 10

$3x^2 - x - 2$

Positive Factors of 3	Factors of -2
1, 3	1, -2
	-1, 2

Trial Factors	Middle Term
$(x + 1)(3x - 2)$	$-2x + 3x = x$
$(x - 2)(3x + 1)$	$x - 6x = -5x$
$(x - 1)(3x + 2)$	$2x - 3x = -x$
$(x + 2)(3x - 1)$	$-x + 6x = 5x$

$3x^2 - x - 2 = (x - 1)(3x + 2)$

Check: $(x - 1)(3x + 2) = 3x^2 + 2x - 3x - 2$
$= 3x^2 - x - 2$

Problem 11

$4a^3 - 4a^2 - 24a$

There is a common factor, $4a$.
Factor out the GCF.
$$4a^3 - 4a^2 - 24a = 4a(a^2 - a - 6)$$
Factor $a^2 - a - 6$. The two factors of -6 whose sum is -1 are 2 and -3.
$4a(a^2 - a - 6) = 4a(a + 2)(a - 3)$
$4a^3 - 4a^2 - 24a = 4a(a + 2)(a - 3)$
Check: $4a(a + 2)(a - 3) = (4a^2 + 8a)(a - 3)$
$= 4a^3 - 12a^2 + 8a^2 - 24a$
$= 4a^3 - 4a^2 - 24a$

Answers to Selected Exercises

Answers to Chapter 1 Selected Exercises

PREP TEST

1. 127.16 **2.** 49,743 **3.** 4517 **4.** 11,396 **5.** 24 **6.** 24 **7.** 4 **8.** $3 \cdot 7$
9. $\frac{2}{5}$

CONCEPT REVIEW 1.1

1. Sometimes true **3.** Never true **5.** Sometimes true **7.** Always true **9.** Always true

SECTION 1.1

3. natural **5.** is greater than or equal to **7.** $>$ **9.** $<$ **11.** $>$ **13.** $>$ **15.** $<$
17. $>$ **19.** $>$ **21.** $<$ **23.** a **25.** Yes **27.** $\{1, 2, 3, 4, 5, 6, 7, 8\}$
29. $\{1, 2, 3, 4, 5, 6, 7, 8\}$ **31.** $\{-6, -5, -4, -3, -2, -1\}$ **33.** 5 **35.** $-23, -18$ **37.** 21, 37
39. $-52, -46, 0$ **41.** $-17, 0, 4, 29$ **43.** 5, 6, 7, 8, 9 **45.** $-10, -9, -8, -7, -6, -5$ **49.** positive
51. absolute value **53.** -22 **55.** 31 **57.** 168 **59.** -630 **61.** 18 **63.** -49
65. 16 **67.** 12 **69.** -29 **71.** -14 **73.** 0 **75.** -34 **77. a.** 8, 5, 2, -1, -3
b. 8, 5, 2, 1, 3 **79.** True **81.** $>$ **83.** $<$ **85.** $>$ **87.** $<$ **89.** $-19, -|-8|, |-5|, 6$
91. $-22, -(-3), |-14|, |-25|$ **93. a.** 5°F with a 20 mph wind feels colder. **b.** -25°F with a 10 mph wind
feels colder. **95.** $-4, 4$ **97.** $-3, 11$ **99.** Negative **101.** 0 **103.** True

CONCEPT REVIEW 1.2

1. Sometimes true **3.** Always true **5.** Always true **7.** Never true **9.** Never true

SECTION 1.2

3. 8; -3; 5 **5.** -11 **7.** -9 **9.** -3 **11.** 1 **13.** -5 **15.** -30 **17.** 9
19. 1 **21.** -10 **23.** -28 **25.** -392 **27.** Positive **31.** (-4); -14 **33.** 8
35. -7 **37.** 7 **39.** -2 **41.** -28 **43.** -13 **45.** 6 **47.** -9 **49.** 2
51. -138 **53.** -8 **55.** Negative **61.** -10; 7; -70 **63.** positive; 48 **65.** 42
67. -20 **69.** -16 **71.** 25 **73.** 0 **75.** -72 **77.** -102 **79.** 140 **81.** -70
83. 162 **85.** 120 **87.** 36 **89.** 192 **91.** -108 **93.** 0 **95.** Negative
97. -15; 3 **99.** $3(-12) = -36$ **101.** $-5(11) = -55$ **103.** -2 **105.** 8 **107.** 0
109. -9 **111.** -9 **113.** 9 **115.** -24 **117.** -12 **119.** -13 **121.** -18
123. 19 **125.** 26 **127.** 11 **129.** -13 **131.** 13 **133.** Positive **135.** b
137. The temperature is 3°C. **139.** The difference is 14°C. **141.** The difference is 399°C.
143. The difference in elevation is 5662 m. **145.** The difference in elevation is 6028 m. **147.** The
difference in elevation is 9244 m. **149.** The average daily high temperature is -3°C. **151.** The temperature
dropped 100°F. **153.** The average temperature at 30,000 ft is 36°F colder than at 20,000 ft.
155. The student's score is 93. **157.** The 5-day moving average for the Disney stock is $-24, -18, -20, -2, 2, 8$.
159. 17 **161.** 3 **163.** $-4, -9, -14$ **165.** $-16, 4, -1$ **167.** 5436 **169.** b **171.** a
173. No. $10 - (-8) = 18$

CONCEPT REVIEW 1.3

1. Never true **3.** Sometimes true **5.** Always true **7.** Never true **9.** Always true

SECTION 1.3

1. 2; 3; repeating **3.** $0.\overline{3}$ **5.** 0.25 **7.** 0.4 **9.** $0.1\overline{6}$ **11.** 0.125 **13.** $0.\overline{2}$
15. $0.\overline{45}$ **17.** $0.58\overline{3}$ **19.** $0.2\overline{6}$ **21.** 0.4375 **23.** 0.24 **25.** 0.225 **27.** $0.68\overline{1}$
29. $0.458\overline{3}$ **31.** $0.1\overline{5}$ **33.** No **35.** four **37.** $-\dfrac{3}{8}$ **39.** $\dfrac{1}{10}$ **41.** $\dfrac{15}{64}$ **43.** $\dfrac{3}{2}$
45. $-\dfrac{8}{9}$ **47.** $\dfrac{2}{3}$ **49.** 4.164 **51.** 4.347 **53.** −4.028 **55.** −2.22 **57.** 12.26448
59. a. Less than one **b.** Greater than one **61.** 0.75 **63.** −2060.55 **65.** 0.09
67. 6 **69.** $\dfrac{1}{2}$ **71.** −1 **73.** $-\dfrac{25}{18}$ **75.** $\dfrac{1}{24}$ **77.** $\dfrac{17}{18}$ **79.** $-\dfrac{47}{48}$ **81.** $\dfrac{3}{8}$
83. $-\dfrac{7}{60}$ **85.** $-\dfrac{1}{16}$ **87.** $-\dfrac{7}{24}$ **89.** 7.29 **91.** −3.049 **93.** −1.06 **95.** −23.845
97. −10.7893 **99.** 11.56 **101.** −60.03 **103.** −34.99 **105.** 239.2351 **107. a.** 2
b. 0 **c.** 1 **d.** −1 **111.** $\dfrac{1}{100}; \dfrac{1}{100}; \dfrac{4}{5}$ **113.** 100%; 100%; 30% **115.** $\dfrac{3}{4}$, 0.75
117. $\dfrac{1}{2}$, 0.5 **119.** $\dfrac{16}{25}$, 0.64 **121.** $1\dfrac{3}{4}$, 1.75 **123.** $\dfrac{19}{100}$, 0.19 **125.** $\dfrac{1}{20}$, 0.05
127. $4\dfrac{1}{2}$, 4.5 **129.** $\dfrac{2}{25}$, 0.08 **131.** $\dfrac{1}{9}$ **133.** $\dfrac{5}{16}$ **135.** $\dfrac{1}{200}$ **137.** $\dfrac{1}{16}$ **139.** 0.073
141. 0.158 **143.** 0.0915 **145.** 0.1823 **147.** 15% **149.** 5% **151.** 17.5%
153. 115% **155.** 0.8% **157.** 6.5% **159.** 54% **161.** 33.3% **163.** 44.4%
165. 250% **167.** $37\dfrac{1}{2}\%$ **169.** $35\dfrac{5}{7}\%$ **171.** 125% **173.** $155\dfrac{5}{9}\%$ **175.** Greater
177. $\dfrac{2}{5}$ of the respondents found their most recent jobs on the Internet. **179.** Fewer than one-quarter of the
respondents found their most recent jobs through a newspaper ad. **181.** Natural number, integer, positive integer,
rational number, real number **183.** Rational number, real number **185.** Irrational number, real number
187. The cost is $1.35. **189.** The difference is $138.499 billion. **191.** The deficit in 1985 was 4 times
greater than in 1975. **193.** The average deficit for the years 1995 through 2000 was −$4,579 million.
195. The new fraction is greater than $\dfrac{2}{5}$. **197.** 1.06x **199.** 53¢ **201.** Yes. For example, $c = \dfrac{a+b}{2}$.

CONCEPT REVIEW 1.4

1. Never true **3.** Never true **5.** Always true **7.** Never true **9.** Never true

SECTION 1.4

1. base; exponent; −5; −5; 25 **3.** 36 **5.** −49 **7.** 9 **9.** 81 **11.** $\dfrac{1}{4}$ **13.** 0.09
15. 12 **17.** 0.216 **19.** −12 **21.** 16 **23.** −864 **25.** −1008 **27.** 3
29. −77,760 **31.** Positive **35.** −2; 4; 20; −17 **37.** 9 **39.** 12 **41.** 1 **43.** 7
45. −36 **47.** 13 **49.** 4 **51.** 15 **53.** −1 **55.** 4 **57.** 1.7 **59.** $\dfrac{17}{48}$
61. b **63.** < **65.** > **67.** 100 **69.** 120 **71.** It will take the computer 17 s.
73. 6 **75.** 9

CONCEPT REVIEW 1.5

1. Never true **3.** Always true **5.** Always true **7.** Never true

SECTION 1.5

1. $-4x$; -7; $3x^2$; $-4x$; -7 **5.** -7 **7.** 13 **9.** -15 **11.** 41 **13.** 57 **15.** 5
17. 12 **19.** 6 **21.** 10 **23.** 24 **25.** 4.96 **27.** -5.68 **29. a.** No, V cannot be a whole number. **b.** No, volume is not measured in square units. **31.** The volume is 25.8 in³.
33. The area is 93.7 cm². **35.** The volume is 484.9 m³. **37.** 41 **39.** Commutative
41. reciprocal (or multiplicative inverse) **43.** 2 **45.** 5 **47.** 6 **49.** -8 **51.** The Inverse Property of Addition **53.** The Commutative Property of Addition **55.** The Commutative Property of Multiplication **57.** The Associative Property of Multiplication **61.** $3x$, $5x$ **63.** $-8a$; Distributive; -3
65. $5a$ **67.** $-2ab$ **69.** $5xy$ **71.** 0 **73.** $-\dfrac{5}{6}x$ **75.** $-\dfrac{1}{24}x^2$ **77.** $11x$ **79.** $7a$
81. $-14x^2$ **83.** $-x + 3y$ **85.** $17x - 3y$ **87.** $-2a - 6b$ **89.** $-3x - 8y$ **91.** $-4x^2 - 2x$
93. a. Negative **b.** Positive **c.** 0 **95.** Associative; $-48x$ **97.** $12x$ **99.** $-21a$
101. $6y$ **103.** $8x$ **105.** $-6a$ **107.** $12b$ **109.** $-15x^2$ **111.** x^2 **113.** x
115. n **117.** x **119.** n **121.** $2x$ **123.** $-2x$ **125.** $-15a^2$ **127.** $6y$
129. Less than one **131.** Distributive; -24; 16 **133.** $-x - 2$ **135.** $8x - 6$ **137.** $-2a - 14$
139. $-6y + 24$ **141.** $35 - 21b$ **143.** $-9 + 15x$ **145.** $15x^2 + 6x$ **147.** $2y - 18$
149. $-15x - 30$ **151.** $-6x^2 - 28$ **153.** $-6y^2 + 21$ **155.** $-6a^2 + 7b^2$ **157.** $4x^2 - 12x + 20$
159. $-3y^2 + 9y + 21$ **161.** $-12a^2 - 20a + 28$ **163.** 3 **165.** m; 3 **167.** $a - 7$
169. $-11x + 13$ **171.** $-4y - 4$ **173.** $-2x - 16$ **175.** $18y - 51$ **177.** $a + 7b$
179. $6x + 28$ **181.** $5x - 75$ **183.** $4x - 4$ **185.** $38x - 63$ **187.** b **189.** $20x + y$
191. **b** **193.** $-3x + y$ **195. a.** 0 **b.** 1 or -1

CONCEPT REVIEW 1.6

1. Never true **3.** Never true **5.** Always true **7.** Never true

SECTION 1.6

1. sum, times **3.** total, divided by **5.** $19 - d$ **7.** $r - 12$ **9.** $28a$ **11.** $5(n - 7)$
13. $y - 3y$ **15.** $-6b$ **17.** $\dfrac{4}{p - 6}$ **19.** $\dfrac{x - 9}{2x}$ **21.** $-4s - 21$ **23.** $\dfrac{d + 8}{d}$
25. $\dfrac{3}{8}(t + 15)$ **27.** $w + \dfrac{7}{w}$ **29.** $d + (16d - 3)$ **31.** $\dfrac{n}{19}$ **33.** $n + 40$ **35.** $(n - 90)^2$
37. $\dfrac{4}{9}n + 20$ **39.** $n(n + 10)$ **41.** $7n + 14$ **43.** $\dfrac{12}{n + 2}$ **45.** $\dfrac{2}{n + 1}$ **47.** $60 - \dfrac{n}{50}$
49. $n^2 + 3n$ **51.** $(n + 3) + n^3$ **53.** $n - \dfrac{1}{4}n^2$ **55.** $2n - \dfrac{7}{n}$ **57.** $n^3 - 12n$ **59.** **c**
61. $\dfrac{1}{2}$; $\dfrac{3}{4}$; $\dfrac{5}{4}n$ **63.** $n + (n + 10)$; $2n + 10$ **65.** $n - (9 - n)$; $2n - 9$ **67.** $\dfrac{1}{5}n - \dfrac{3}{8}n$; $-\dfrac{7}{40}n$
69. $(n + 9) + 4$; $n + 13$ **71.** $2(3n + 40)$; $6n + 80$ **73.** $7(5n)$; $35n$ **75.** $17n + 2n$; $19n$
77. $n + 12n$; $13n$ **79.** $3(n^2 + 4)$; $3n^2 + 12$ **81.** $\dfrac{3}{4}(16n + 4)$; $12n + 3$ **83.** $16 - (n + 9)$; $-n + 7$
85. $\dfrac{4n}{2} + 5$; $2n + 5$ **87.** $6(n + 8)$; $6n + 48$ **89.** $7 - (n + 2)$; $-n + 5$ **91.** $\dfrac{1}{3}(n + 6n)$; $\dfrac{7n}{3}$
93. $(8 + n^3) + 2n^3$; $3n^3 + 8$ **95.** $(n - 6) + (n + 12)$; $2n + 6$ **97.** $(n - 20) + (n + 9)$; $2n - 11$

99. $14 + (n - 3)10$; $10n - 16$ **101.** $25 - x$ **103.** $\frac{1}{2}L$ **105.** Let x be one number; x and $18 - x$

107. Let G be the number of genes in the roundworm genome; $G + 11{,}000$ **109.** Let N be the total number of Americans; $\frac{3}{4}N$ **111.** Let s be the number of points awarded for a safety; $3s$ **113.** Let d be the diameter of Dione; $d + 253$ **115.** Let p be the pounds of pecans produced in Texas; $\frac{1}{2}p$ **117.** Let p be the world population in 1980; $2p$ **119.** Let L be the measure of the largest angle; $\frac{1}{2}L - 10$ **121.** Let n be either the number of nickels or the number of dimes; n and $35 - n$ **123.** Let h be the number of hours of overtime worked; $720 + 27h$ **125.** Let m be the number of minutes of phone calls; $19.95 + 0.08m$

127. $\frac{1}{4}x$

CHAPTER 1 REVIEW EXERCISES*

1. $\{1, 2, 3, 4, 5, 6\}$ [1.1.1] **2.** 62.5% [1.3.4] **3.** -4 [1.1.2] **4.** 4 [1.2.2] **5.** 17 [1.2.4]

6. $0.\overline{7}$ [1.3.1] **7.** -5.3578 [1.3.2] **8.** 8 [1.4.2] **9.** 67.2% [1.3.4] **10.** $\frac{159}{200}$ [1.3.4]

11. 90 [1.2.3] **12.** -6.881 [1.3.3] **13.** $-5, -3$ [1.1.1] **14.** 0.07 [1.3.4] **15.** 12 [1.4.1]

16. $>$ [1.1.1] **17.** $\frac{1}{15}$ [1.3.3] **18.** -108 [1.4.1] **19.** $-\frac{7}{6}$ [1.3.2] **20.** 12 [1.4.2]

21. $288\frac{8}{9}\%$ [1.3.4] **22. a.** $12, 8, 1, -7$ **b.** $12, 8, 1, 7$ [1.1.2] **23.** -11.5 [1.3.2]

24. 8 [1.4.2] **25.** -8 [1.2.1] **26.** $-\frac{11}{24}$ [1.3.3] **27.** 3.561 [1.3.3] **28.** -27 [1.2.2]

29. $-\frac{1}{10}$ [1.3.2] **30.** $<$ [1.1.1] **31.** y^2 [1.5.3] **32.** $3x$ [1.5.4] **33.** $7x + 38$ [1.5.6]

34. 16 [1.5.1] **35.** 9 [1.5.2] **36.** $-3x + 21y$ [1.5.6] **37.** $15 - 35b$ [1.5.5] **38.** $-7x + 33$ [1.5.6] **39.** The Commutative Property of Multiplication [1.5.2] **40.** $24 - 6x$ [1.5.5]

41. 3 [1.5.1] **42.** $36y$ [1.5.4] **43.** $5x - 43$ [1.5.6] **44.** $2x$ [1.5.4] **45.** $-6x^2 + 21y^2$ [1.5.5] **46.** 6 [1.5.1] **47.** $-x + 6$ [1.5.6] **48.** $-5a - 2b$ [1.5.3] **49.** $-10x^2 + 15x - 30$ [1.5.5] **50.** $-9x - 7y$ [1.5.3] **51.** $-2x - 5y$ [1.5.3] **52.** $30b$ [1.5.4] **53.** 21 [1.5.2]

54. $-2x^2 + 4x$ [1.5.3] **55.** $b - 7b$ [1.6.1] **56.** $\frac{10}{y - 2}$ [1.6.1] **57.** The temperature is 8°C. [1.2.5] **58.** The average low temperature was -2°C. [1.2.5] **59.** The difference is 106°F. [1.2.5]

60. The temperature is -6°C. [1.2.5] **61.** The difference is 714°C. [1.2.5] **62.** $8\left(\frac{2n}{16}\right)$; n [1.6.2]

63. $4(2 + 5n)$; $8 + 20n$ [1.6.2] **64.** Let h be the height of the triangle; $h + 15$ [1.6.3]
65. Let b be the amount of either espresso beans or mocha java beans; b and $20 - b$ [1.6.3]

CHAPTER 1 TEST

1. $\frac{11}{20}$ [1.3.4] **2.** $-6, -8$ [1.1.1] **3.** -3 [1.2.2] **4.** 0.15 [1.3.1] **5.** $-\frac{1}{14}$ [1.3.2]

6. -15 [1.2.4] **7.** $-\frac{8}{3}$ [1.4.1] **8.** 2 [1.2.1] **9.** $\{1, 2, 3, 4, 5, 6\}$ [1.1.1] **10.** 159% [1.3.4]

*The numbers in brackets following the answers in the Chapter Review Exercises refer to the objective that corresponds to that problem. For example, the reference [1.2.1] stands for Section 1.2, Objective 1. This notation will be used for all Prep Tests, Chapter Review Exercises, Chapter Tests, and Cumulative Review Exercises throughout the text.

11. 29 [1.1.2] **12.** > [1.1.1] **13.** $-\dfrac{23}{18}$ [1.3.3] **14.** $\dfrac{5}{6}$ [1.3.2] **15.** 23.1% [1.3.4]

16. 0.062 [1.3.4] **17.** 14 [1.2.2] **18.** $0.4\overline{3}$ [1.3.1] **19.** −2.43 [1.3.2] **20.** −12 [1.2.1]

21. a. 17, 6, −5, −9 **b.** 17, 6, 5, 9 [1.1.2] **22.** −11.384 [1.3.3] **23.** 160 [1.2.3]

24. $36y$ [1.5.4] **25.** $4x - 3y$ [1.5.3] **26.** $10n - 6$ [1.5.6] **27.** The Multiplication Property of

One [1.5.2] **28.** $4x - 40$ [1.5.5] **29.** $\dfrac{1}{12}x^2$ [1.5.3] **30.** $4x$ [1.5.4] **31.** $-24y^2 + 48$

[1.5.5] **32.** 19 [1.5.2] **33.** 6 [1.5.1] **34.** $-3x + 13y$ [1.5.6] **35.** $3x^2 - 15x + 12$ [1.5.5]

36. −32 [1.5.1] **37.** $37x - 5y$ [1.5.6] **38.** −2 [1.4.2] **39.** The temperature is 4°C. [1.2.5]

40. The average is −48°F. [1.2.5] **41.** $(a + b) - b^2$ [1.6.1] **42.** $20(n + 9)$; $20n + 180$ [1.6.2]

43. $(n - 3) + (n + 2)$; $2n - 1$ [1.6.2] **44.** Let d be the distance from Earth to the sun; $30d$ [1.6.3]

45. Let L be the length of one piece; L and $9 - L$ [1.6.3]

Answers to Chapter 2 Selected Exercises

PREP TEST

1. −4 [1.2.2] **2.** 1 [1.3.2] **3.** −10 [1.3.2] **4.** 0.9 [1.3.4] **5.** 75% [1.3.4]

6. 63 [1.5.1] **7.** $10x - 5$ [1.5.3] **8.** −9 [1.5.3] **9.** $9x - 18$ [1.5.6]

CONCEPT REVIEW 2.1

1. Sometimes true **3.** Always true **5.** Always true **7.** Always true

SECTION 2.1

1. a. equation **b.** expression **3.** Yes **5.** No **7.** No **9.** Yes **11.** Yes

13. Yes **15.** No **17.** No **19.** No **21.** Yes **23.** No **25.** No **29.** 30; 87

33. 2 **35.** 15 **37.** 6 **39.** −6 **41.** 3 **43.** 0 **45.** −2 **47.** −7 **49.** −7

51. −12 **53.** 2 **55.** −5 **57.** 15 **59.** 9 **61.** 14 **63.** −1 **65.** 1

67. $-\dfrac{1}{2}$ **69.** $-\dfrac{3}{4}$ **71.** $\dfrac{1}{12}$ **73.** $-\dfrac{7}{12}$ **75.** $-\dfrac{5}{7}$ **77.** $-\dfrac{1}{3}$ **79.** 1.869 **81.** 0.884

85. divide; 7 **89.** 7 **91.** −7 **93.** −4 **95.** 5 **97.** 9 **99.** −8 **101.** 0

103. −7 **105.** 8 **107.** 12 **109.** 12 **111.** −18 **113.** −18 **115.** 6

117. −12 **119.** 32 **121.** −12 **123.** 15 **125.** 25 **127.** $\dfrac{8}{7}$ **129.** 2 **131.** 3

133. −3 **135.** −22.62 **137.** 1.05 **139.** −3.4 **141.** No **145.** 24% **147.** 7.2

149. 400 **151.** 9 **153.** 25% **155.** 5 **157.** 200% **159.** 400 **161.** 7.7

163. 40% of 80 is equal to 80% of 40. **165.** unknown; 30; 24 **167.** unknown; 32; 0.1 **169.** 250 seats

are reserved for wheelchair accessibility. **171.** In the average single-family home, 74 gal of water are used per

person per day. **173.** 82.1% of the vacation cost is charged on a credit card. **175.** There is insufficient

information to answer the question. **177.** 12% of the deaths were not attributed to motor vehicle accidents.

179. In this country 96.1 billion kilowatt-hours of electricity are used for home lighting per year.

181. The annual simple interest rate is 9%. **183.** Sal earned $240 in interest from the two accounts.

185. The interest rate on the combined investment is between 6% and 9%. **187.** The percent concentration of

the hydrogen peroxide is 2%. **189.** Apple Dan's has the greater concentration of apple juice. **191.** 12.5 g of

the cream are not glycerine. **193.** The percent concentration of salt in the remaining solution is 12.5%.

195. −21 **197.** −27 **199.** 21 **201.** The cost of the dinner was $80. **203.** 2

205. One possible answer is $5x = -10$.

CONCEPT REVIEW 2.2

1. Always true **3.** Never true **5.** Never true **7.** Sometimes true

SECTION 2.2

1. 5; 8 **3.** Negative **5.** 3 **7.** 6 **9.** -1 **11.** 1 **13.** -9 **15.** 3 **17.** 3

19. 3 **21.** 2 **23.** 2 **25.** 4 **27.** $\dfrac{6}{7}$ **29.** $\dfrac{2}{3}$ **31.** -1 **33.** $\dfrac{3}{4}$ **35.** $\dfrac{1}{3}$

37. $-\dfrac{3}{4}$ **39.** $\dfrac{1}{3}$ **41.** $-\dfrac{1}{6}$ **43.** 1 **45.** 0 **47.** $\dfrac{2}{5}$ **49.** $-\dfrac{3}{2}$ **51.** 18 **53.** 8

55. -16 **57.** 25 **59.** 21 **61.** 15 **63.** -16 **65.** $\dfrac{7}{6}$ **67.** $\dfrac{1}{8}$ **69.** $\dfrac{15}{2}$

71. $-\dfrac{18}{5}$ **73.** 2 **75.** 3 **77.** 1 **79.** 2 **81.** 3.95 **83.** -0.8 **85.** -11

87. 0 **89.** 12; 12; Distributive; 2; 84; $2x$; $2x$; 7; Divide; 7; -12 **91.** -2 **93.** 3 **95.** 8
97. 2 **99.** -2 **101.** -3 **103.** 2 **105.** -2 **107.** -2 **109.** -7 **111.** 0

113. -2 **115.** -2 **117.** 4 **119.** 10 **121.** 3 **123.** $\dfrac{3}{4}$ **125.** $\dfrac{2}{7}$ **127.** $-\dfrac{3}{4}$

129. 3 **131.** 10 **133.** $\dfrac{4}{3}$ **135.** 3 **137.** -14 **139.** 7 **141.** $3x$; 18

143. **a** and **c** **145.** 3 **147.** 4 **149.** 2 **151.** 2 **153.** 2 **155.** -7 **157.** $\dfrac{4}{7}$

159. $\dfrac{1}{2}$ **161.** $-\dfrac{1}{3}$ **163.** $\dfrac{10}{3}$ **165.** $-\dfrac{1}{4}$ **167.** 0.5 **169.** 0 **171.** -1

173. The car will slide 168 ft. **175.** The depth of the diver is 40 ft. **177.** The initial velocity is 8 ft/s.
179. An adult whose humerus measures 36 cm is approximately 182.5 cm tall. **181.** The passenger was driven
6 mi. **183.** He made 8 errors. **185.** The population is approximately 51,000 people. **187.** The
break-even point is 350 television sets. **191.** 5 ft **193.** No **195.** No, the seesaw is not balanced.
197. A force of 25 lb must be applied to the other end of the lever. **199.** The fulcrum must be placed 10 ft
from the 128-pound acrobat. **201.** No solution **203.** $-\dfrac{11}{4}$ **205.** 62 **207.** One possible
answer is $3x - 6 = 2x - 2$.

CONCEPT REVIEW 2.3

1. Always true **3.** Always true **5.** Always true **7.** Never true **9.** Always true

SECTION 2.3

1. equals **3.** $n - 15 = 7; n = 22$ **5.** $7n = -21; n = -3$ **7.** $3n - 4 = 5; n = 3$
9. $4(2n + 3) = 12; n = 0$ **11.** $12 = 6(n - 3); n = 5$ **13.** $2(n - 25) = 3n; n = -50$
15. $3n = 2(20 - n)$; 8 and 12 **17.** $3n + 2(18 - n) = 44$; 8 and 10 **19.** Answers may vary. For example:
$n + n + 10 = 14; n + n - 10 = 14$ **21.** The original value was $16,000. **23.** There are 58 calories in a
medium-size orange. **25.** The amount of mulch is 15 lb. **27.** The intensity of the sound is 140 decibels.
29. The length is 80 ft; the width is 50 ft. **31.** To replace the water pump, 5 h of labor were required.
33. The area of Greenland is 840,000 mi². **35.** consecutive **37.** The integers are 17, 18, and 19.
39. The integers are 26, 28, and 30. **41.** The integers are 17, 19, and 21. **43.** The integers are 8 and 10.
45. The integers are 7 and 9. **47.** The integers are -9, -8, and -7. **49.** The integers are 10, 12, and 14.
51. No solution **53.** **b** **55.** **a.** $8.40 **b.** $.70 **57.** The vegetable medley costs $1.66 per
pound. **59.** There were 320 adult tickets sold. **61.** The mixture must contain 225 L of imitation maple syrup.

63. The instructor used 8 lb of peanuts. **65.** The cost of the mixture is $4.04 per pound.
67. The mixture must contain 37.5 gal of cranberry juice. **69.** **d** and **f** **71.** **a.** $12t$; $15t$
b. 400; 360 **73.** **a.** Less than **b.** Equal to **c.** 3 mi **75.** The student drives at an average rate of 40 mph. **77.** It would take a Boeing 737-800 jet 3.2 h to make the same trip. **79.** The two cyclists will meet at 2:30 P.M. **81.** The bicyclist overtakes the in-line skater 11.25 mi from the starting point.
83. The speed of the first plane is 420 mph. The speed of the second plane is 500 mph. **85.** The distance to the island is 43.2 mi. **87.** The rate of the first plane is 255 mph. The rate of the second plane is 305 mph.
89. The distance between the student's home and the bicycle shop is 2.8 mi. **91.** The two trains will pass each other in 2 h. **93.** At 3:39 P.M., $\frac{1}{3}$ of the container is filled. **95.** The integers are -15, -13, -11, and -9.
97. For any four consecutive integers, the sum of the first and fourth integers equals the sum of the second and third integers. **99.** It will take the runner $\frac{2}{3}$ h. **101.** The grass seed is 18 ft from the 40-foot tower.
103. **a.** Rutan and Yeager landed at Edwards Air Force Base on December 23. **b.** The average speed was 116 mph. **c.** The approximate distance of the flight above Earth was 14 mi.

CONCEPT REVIEW 2.4

1. Always true **3.** Always true **5.** Never true

SECTION 2.4

3. $2x - 3$ **5.** True **7.** The sides measure 50 ft, 50 ft, and 25 ft. **9.** The length is 13 m. The width is 8 m. **11.** The length is 40 ft. The width is 20 ft. **13.** The sides measure 40 cm, 20 cm, and 50 cm.
15. The length is 130 ft. The width is 39 ft. **17.** The width is 12 ft. **19.** Each side measures 12 in.
21. **a.** 90° **b.** 180° **c.** 360° **d.** between 0° and 90° **e.** between 90° and 180° **f.** 90° **g.** 180°
23. 90; 37; 37° **25.** The complement of a 28° angle is 62°. **27.** The supplement of a 73° angle is 107°.
29. 35° **31.** 20° **33.** 53° **35.** 121° **37.** 15° **39.** 18° **41.** 45° **43.** 49°
45. 12° **47.** False **49.** True **51.** $\angle a = 122°$, $\angle b = 58°$ **53.** $\angle a = 44°$, $\angle b = 136°$
55. 20° **57.** 40° **59.** 128° **61.** Vertical; 180°; 50°; 85°; 135°; 45° **63.** $\angle x = 160°$, $\angle y = 145°$
65. $\angle a = 40°$, $\angle b = 140°$ **67.** $75° - x$ **69.** The measure of the third angle is 45°.
71. The measure of the third angle is 73°. **73.** The measure of the third angle is 43°.
75. The angles measure 38°, 38°, and 104°. **77.** The angles measure 60°, 30°, and 90°. **79.** True
81. True **83.** The length is 9 cm. The width is 4 cm.

CONCEPT REVIEW 2.5

1. Always true **3.** Never true **5.** Always true

SECTION 2.5

3. 40; unknown; 0.25 **5.** The selling price is $56. **7.** The selling price is $565.64. **9.** The markup rate is 75%. **11.** The markup rate is 44.4%. **13.** The markup rate is 60%. **15.** The cost of the compact disc player is $120. **17.** The cost of the basketball is $59. **19.** The cost of the computer is $1750.
21. True **25.** 318.75; 375; unknown **27.** The sale price is $101.25. **29.** The discount price is $218.50. **31.** The discount rate is 25%. **33.** The discount rate is 23.2%. **35.** The markdown rate is 38%. **37.** The regular price is $510. **39.** The regular price is $300. **41.** The regular price is $275.
43. True **45.** The markup is $18. **47.** The markup rate is 40%. **49.** The cost of the camera is $230.
51. The regular price is $80. **53.** No. The single discount that would give the same sale price is 28%.

CONCEPT REVIEW 2.6

1. Never true **3.** Never true

SECTION 2.6

3. $0.0625x$; $0.06(x - 500)$; 115 **5.** Joseph will earn $327.50 on the two accounts. **7.** Deon must invest $1600 in the account that earns 8% interest. **9.** The amount invested in the CD is $15,000. **11.** An additional $2000 must be invested at an annual simple interest rate of 10.5%. **13.** The amount invested at 8.5% is $3000. The amount invested at 6.4% is $5000. **15.** She must invest an additional $3000 at an annual interest rate of 10%. **17.** The amount that should be invested at 4.2% is $8000. The amount that should be invested at 6% is $5600. **19. a.** 5.5% and 7.2% **b.** $6000 **21.** 0.10; 3.2 **23.** A 40-ounce bottle of Orange Ade contains 10 oz of orange juice. **25.** The 850-milliliter solution contains 12.5 more milliliters of hydrogen peroxide. **27.** The resulting alloy is 30% silver. **29.** The resulting alloy is $33\frac{1}{3}$% silver. **31.** 500 lb of the 12% aluminum alloy is needed. **33.** 25 L of the 65% solution and 25 L of the 15% solution were used. **35.** 3 qt of water must be added. **37.** 30 oz of pure water must be added. **39.** The result is 6% real fruit juice. **41.** The resulting alloy is 31.1% copper. **43. b** **45.** The amount invested at 9% was $5000. The amount invested at 8% was $6000. The amount invested at 9.5% was $9000. **47.** The cost per pound of the tea mixture is $4.84. **49.** 60 g of pure water were in the beaker before the acid was added. **51. a.** 2000 **b.** 2005

CONCEPT REVIEW 2.7

1. Always true **3.** Sometimes true **5.** Sometimes true

SECTION 2.7

3. roster; set-builder **5.** infinity **7.** $\{x \mid x > 4, x \in \text{integers}\}$ **9.** $\{x \mid x \geq -2\}$ **11.** $\{x \mid 0 < x < 1\}$ **13.** $\{x \mid 1 \leq x \leq 4\}$ **15.** $\{1, 2, 4, 6, 9\}$ **17.** $\{2, 3, 5, 8, 9, 10\}$ **19.** $\{-4, -2, 0, 2, 4, 8\}$ **21.** $\{1, 2, 3, 4, 5\}$ **23.** $\{6\}$ **25.** $\{5, 10, 20\}$ **27.** $\varnothing$ **29.** $\{4, 6\}$

31. **33.** **35.**

37. **39.** **41.**

43. **45.** **47.** $\{x \mid 0 < x < 8\}$

49. $\{x \mid -5 \leq x \leq 7\}$ **51.** $\{x \mid -3 \leq x < 6\}$ **53.** $\{x \mid x \leq 4\}$ **55.** $\{x \mid x > 5\}$ **57.** $(-2, 4)$ **59.** $[-1, 5]$ **61.** $(-\infty, 1)$ **63.** $[-2, 6)$ **65.** $(-\infty, \infty)$ **67. c** **71. a, c** **73.** remains the same **75.** is reversed **77.** $\{x \mid x < 5\}$ **79.** $\{x \mid x \leq 2\}$ **81.** $\{x \mid x < -4\}$ **83.** $\{x \mid x > 3\}$ **85.** $\{x \mid x > 4\}$ **87.** $\{x \mid x > -2\}$ **89.** $\{x \mid x \geq 2\}$ **91.** $\{x \mid x \leq 3\}$ **93.** $\{x \mid x < -3\}$

95. $(-\infty, 5]$ **97.** $[1, \infty)$ **99.** $(-\infty, -5)$ **101.** $\left(-\infty, \frac{23}{16}\right)$ **103.** $\left[\frac{8}{3}, \infty\right)$ **105.** $(-\infty, 1)$

107. $(-\infty, 3)$ **109.** Only positive numbers **111.** Only negative numbers **115.** 4; 2; 1; 6 **117.** One interval of real numbers **119.** Empty set **121.** $(-1, 2)$ **123.** $(-\infty, 1] \cup [3, \infty)$ **125.** $(-2, 4)$ **127.** $(-\infty, -3) \cup (0, \infty)$ **129.** $[3, \infty)$ **131.** $\varnothing$ **133.** $\varnothing$

135. $(-\infty, 1) \cup (3, \infty)$ **137.** $\{x \mid -3 < x < 4\}$ **139.** $\{x \mid 3 < x < 5\}$ **141.** $\left\{x \mid x > 3 \text{ or } x \leq -\frac{5}{2}\right\}$ **143.** $\{x \mid x > 4\}$ **145.** $\varnothing$ **147.** $\{x \mid x \text{ is a real number}\}$ **149.** $\{x \mid -4 < x < -2\}$ **151.** $\{x \mid x < -4 \text{ or } x > 3\}$ **153.** $\{x \mid -4 \leq x < 1\}$ **155.** $\left\{x \mid x > \frac{27}{2} \text{ or } x < \frac{5}{2}\right\}$ **157.** $\left\{x \mid -10 \leq x \leq \frac{11}{2}\right\}$

159. $n \geq 40$ **161.** The smallest integer is 3. **163.** The maximum width of the rectangle is 2 ft. **165.** The four integers are 15, 16, 17, 18; or 16, 17, 18, 19; or 17, 18, 19, 20. **167.** The length of the second side could be 5 in., 6 in., or 7 in. **169.** A customer can use a cellular phone 60 min before the charges exceed the

first option.　　**171.**　The AirTouch Plan is less expensive for more than 460 messages.　　**173.**　A call must be 7 min or less if it is to be cheaper to pay with coins.　　**175.**　If a business chooses the Glendale Federal Bank, then the business writes more than 200 checks per month.　　**177.**　The car can travel between 429 mi and 536.25 mi on a full tank of gasoline.　　**179.**　{1, 2}　　**181.**　{1, 2, 3, 4, 5, 6}　　**183.**　{1, 2, 3, 4}　　**185.**　{1, 2} **187.**　The temperature is between 25°C and 30°C.　　**189.**　The largest whole number of minutes the call would last is 10 min.

CONCEPT REVIEW 2.8

1.　Always true　　**3.**　Never true　　**5.**　Never true

SECTION 2.8

1.　Yes　　**3.**　Yes　　**5.**　8, −8　　**7.**　−7 and 7　　**9.**　−3 and 3　　**11.**　No solution　　**13.**　−5 and 1

15.　8 and 2　　**17.**　2　　**19.**　No solution　　**21.**　$\frac{1}{2}$ and $\frac{9}{2}$　　**23.**　0 and $\frac{4}{5}$　　**25.**　−1

27.　No solution　　**29.**　4 and 14　　**31.**　4 and 12　　**33.**　$\frac{7}{4}$　　**35.**　No solution　　**37.**　−3 and $\frac{3}{2}$

39.　$\frac{4}{5}$ and 0　　**41.**　No solution　　**43.**　No solution　　**45.**　−2 and 1　　**47.**　$-\frac{3}{5}$　　**49.**　Two positive solutions　　**51.**　Two negative solutions　　**53.**　<; >　　**55.**　{x | x > 3 or x < −3}　　**57.**　{x | x > 1 or x < −3}　　**59.**　{x | 4 ≤ x ≤ 6}　　**61.**　{x | x ≤ −1 or x ≥ 5}　　**63.**　{x | −3 < x < 2}

65.　$\left\{x \mid x > 2 \text{ or } x < -\frac{14}{5}\right\}$　　**67.**　∅　　**69.**　{x | x ∈ real numbers}　　**71.**　$\left\{x \mid x \leq -\frac{1}{3} \text{ or } x \geq 3\right\}$

73.　$\left\{x \mid -2 \leq x \leq \frac{9}{2}\right\}$　　**75.**　{x | x = 2}　　**77.**　$\left\{x \mid x < -2 \text{ or } x > \frac{22}{9}\right\}$　　**79.**　All negative solutions

81.　tolerance; 5.5; upper; 49.5; lower　　**83.**　The lower and upper limits of the diameter of the bushing are 1.742 in. and 1.758 in.　　**85.**　The lower and upper limits of the amount of medicine to be given to the patient are 2.3 cc and 2.7 cc.　　**87.**　The lower and upper limits of voltage to the computer are 93.5 volts and 126.5 volts.

89.　The lower and upper limits of the length of the piston rod are $10\frac{11}{32}$ in. and $10\frac{13}{32}$ in.　　**91.**　5 in.; 0.01 in.

93.　The lower and upper limits of the resistor are 13,500 ohms and 16,500 ohms.　　**95.**　The lower and upper limits of the resistor are 53.2 ohms and 58.8 ohms.　　**97.**　2, $-\frac{2}{3}$　　**99.**　{x | −7 ≤ x ≤ 8}　　**101.**　{y | y ≥ −6}

103.　{b | b ≤ 7}　　**105.**　|x − 2| < 5　　**107. a.**　≤　　**b.**　≥　　**c.**　≤　　**d.**　=　　**e.**　=

CHAPTER 2 REVIEW EXERCISES

1.　No　[2.1.1]　　**2.**　20　[2.1.2]　　**3.**　−7　[2.1.3]　　**4.**　7　[2.2.1]　　**5.**　4　[2.2.2]

6.　$-\frac{1}{5}$　[2.2.3]　　**7.**　405　[2.1.4]　　**8.**　25　[2.1.4]　　**9.**　67.5%　[2.1.4]　　**10.**　$-\frac{9}{19}$　[2.2.3]

11.　{2, 3}　[2.7.1]　　**12.**　 −5 −4 −3 −2 −1　0　1　2　3　4　5　　[2.7.1]　　**13.**　$\left(\frac{5}{3}, \infty\right)$　[2.7.2]　　**14.**　(−4, ∞) [2.7.2]　　**15.**　(−1, 2)　[2.7.3]　　**16.**　(−∞, −2) ∪ (2, ∞)　[2.7.3]　　**17.**　$\left\{x \mid -3 < x < \frac{4}{3}\right\}$　[2.7.3]

18.　{x | x ∈ real numbers}　[2.7.3]　　**19.**　$-\frac{5}{2}, \frac{11}{2}$　[2.8.1]　　**20.**　$-\frac{8}{5}$　[2.8.1]　　**21.**　No solution　[2.8.1]

22.　{x | 1 ≤ x ≤ 4}　[2.8.2]　　**23.**　$\left\{x \mid x \leq \frac{1}{2} \text{ or } x \geq 2\right\}$　[2.8.2]　　**24.**　∅　[2.8.2]　　**25.**　A force of 24 lb must be applied to the other end of the lever.　[2.2.4]　　**26.**　Approximately 11,065 plants and animals are at risk

of extinction on Earth. [2.3.1] **27.** The depth is 80 ft. [2.2.4] **28.** $2x - 7 = x$; 7 [2.3.1]
29. The length of the shorter piece is 14 in. [2.3.1] **30.** The two numbers are 8 and 13. [2.3.1]
31. The discount rate is $33\frac{1}{3}\%$ [2.5.2] **32.** The angles measure 65°, 65°, and 50°. [2.4.3] **33.** The
markup rate is 80%. [2.5.1] **34.** The concentration of butterfat is 14%. [2.6.2] **35.** The length is 80 ft.
The width is 20 ft. [2.4.1] **36.** The integers are -17, -15, and -13. [2.3.2] **37.** $\angle a = 138°$, $\angle b = 42°$
[2.4.2] **38.** The lower and upper limits of the amount of medicine to be given are 1.75 cc and 2.25 cc. [2.8.3]
39. The integers are -1, 0, and 1. [2.3.2] **40.** The mixture costs $4.25 per ounce. [2.3.3] **41.** The
mixture must contain 52 gal of apple juice. [2.3.3] **42.** The speed of the first plane is 440 mph. The speed of
the second plane is 520 mph. [2.3.4] **43.** The amount invested at 10.5% was $3000. The amount invested at
6.4% was $5000. [2.6.1] **44.** 375 lb of the 30% tin alloy and 125 lb of the 70% tin alloy were used. [2.6.2]
45. The range of scores to receive a B grade is $82 \le N \le 100$. [2.7.4]

CHAPTER 2 TEST

1. -12 [2.1.3] **2.** $-\frac{1}{2}$ [2.2.2] **3.** -3 [2.2.1] **4.** No [2.1.1] **5.** $-\frac{1}{3}$ [2.2.3] **6.** -5

[2.1.2] **7.** 40 [2.1.4] **8.** $(-1, \infty)$ [2.7.2] **9.** $\{x \mid x > -2\}$ [2.7.3] **10.** 3 and $-\frac{9}{5}$ [2.8.1]

11. $\{x \mid x > 2 \text{ or } x < -1\}$ [2.8.2] **12.** $3n - 15 = 27$; $n = 14$ [2.3.1] **13.** The cost is $200. [2.5.1]
14. The number of calculators produced was 200. [2.2.4] **15.** The integers are 5, 7, and 9. [2.3.2]
16. The measures of the three angles are 48°, 33°, and 99°. [2.4.3] **17.** The minimum length is 24 ft. [2.7.4]
18. The sale price is $79.20. [2.5.2] **19.** The measures of the three other angles are 45°, 135°, and 45°.
[2.4.2] **20.** It costs less to rent from agency A if the car is driven less than 120 mi. [2.7.4] **21.** The lower
and upper limits of the amount of medication to be given are 2.9 cc and 3.1 cc. [2.8.3] **22.** The price of the
hamburger mixture is $3.20/lb. [2.3.3] **23.** The jogger ran a total distance of 12 mi. [2.3.4] **24.** The
amount invested at 7.8% was $5000. The amount invested at 9% was $7000. [2.6.1] **25.** 100 oz of pure water
must be added. [2.6.2]

CUMULATIVE REVIEW EXERCISES

1. 6 [1.2.2] **2.** -48 [1.2.3] **3.** $-\frac{19}{48}$ [1.3.3] **4.** -2 [1.3.2] **5.** 54 [1.4.1]

6. 24 [1.4.2] **7.** 6 [1.5.1] **8.** $-17x$ [1.5.3] **9.** $-5a - 2b$ [1.5.3] **10.** $2x$ [1.5.4]
11. $36y$ [1.5.4] **12.** $2x^2 + 6x - 4$ [1.5.5] **13.** $-4x + 14$ [1.5.6] **14.** $6x - 34$ [1.5.6]
15. $\{-7, -6, -5, -4, -3, -2, -1\}$ [1.1.1] **16.** $87\frac{1}{2}\%$ [1.3.4] **17.** 3.42 [1.3.4] **18.** $\frac{5}{8}$ [1.3.4]
19. Yes [2.1.1] **20.** -5 [2.1.2] **21.** -25 [2.1.3] **22.** 3 [2.2.1] **23.** -3 [2.2.2]
24. 13 [2.2.3] **25.** $x < -\frac{8}{9}$ [2.7.2] **26.** $x \ge 12$ [2.7.2] **27.** $x > 9$ [2.7.2] **28.** $8 - \frac{n}{12}$
[2.3.1] **29.** $n + (n + 2)$; $2n + 2$ [2.3.1] **30.** A force of 24 lb must be applied to the other end of the lever.
[2.2.4] **31.** The first integer is 1. [2.3.2] **32.** 48 adult tickets were sold. [2.3.3] **33.** The speed of
the faster plane is 340 mph. [2.3.4] **34.** 3 L of 12% acid solution must be in the mixture. [2.6.2]
35. $6500 was invested at 9.8%. [2.6.1]

Answers to Chapter 3 Selected Exercises

PREP TEST

1. $-4x + 12$ [1.5.5] **2.** 10 [1.4.2] **3.** -2 [1.4.2] **4.** 11 [1.5.1] **5.** 2.5 [1.5.1]
6. 5 [1.5.1] **7.** 1 [1.5.1] **8.** 4 [2.1.3]

CONCEPT REVIEW 3.1

1. Always true **3.** Never true **5.** Never true

SECTION 3.1

1. right; down **3.** 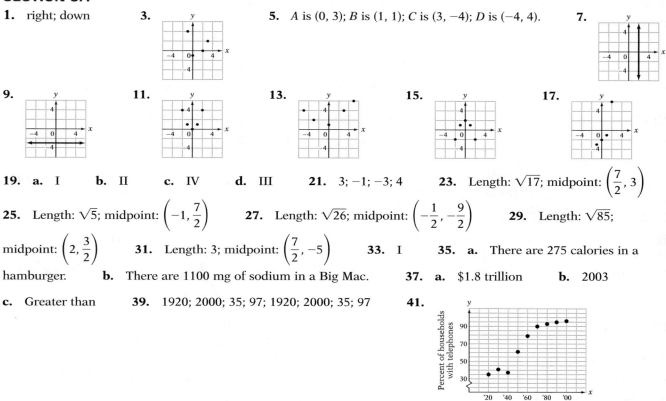 **5.** A is (0, 3); B is (1, 1); C is (3, -4); D is (-4, 4). **7.**

9. **11.** **13.** **15.** **17.**

19. a. I **b.** II **c.** IV **d.** III **21.** 3; -1; -3; 4 **23.** Length: $\sqrt{17}$; midpoint: $\left(\dfrac{7}{2}, 3\right)$

25. Length: $\sqrt{5}$; midpoint: $\left(-1, \dfrac{7}{2}\right)$ **27.** Length: $\sqrt{26}$; midpoint: $\left(-\dfrac{1}{2}, -\dfrac{9}{2}\right)$ **29.** Length: $\sqrt{85}$;

midpoint: $\left(2, \dfrac{3}{2}\right)$ **31.** Length: 3; midpoint: $\left(\dfrac{7}{2}, -5\right)$ **33.** I **35. a.** There are 275 calories in a

hamburger. **b.** There are 1100 mg of sodium in a Big Mac. **37. a.** \$1.8 trillion **b.** 2003

c. Greater than **39.** 1920; 2000; 35; 97; 1920; 2000; 35; 97 **41.**

43. a. The average rate of change in the population was 179,000 people per year. **b.** The average annual rate of change in the population from 1900 to 1950 was less than the average annual rate of change from 1950 to 2000. **c.** The average annual rate of change in the population of Texas from 1980 to 1990 was less than the average annual rate of change in the population of California from 1980 to 1990. **45.** 1970s **47. a.** The average annual rate of change in the number of pedestrian fatalities from 1990 to 2000 was -175.5 pedestrians per year. **b.** The answer to part a can be considered encouraging because it reflects the fact that the number of fatalities is decreasing. **49. a.** The average rate of change in the number of applications from 1990 to 2000 was 25,500 applications per year. **b.** The average annual rate of change in the number of applications from 1995 to 2000 was 29,000 applications per year greater than that from 1990 to 1995. **51.** **53.**

CONCEPT REVIEW 3.2

1. Sometimes true **3.** Sometimes true **5.** Always true

SECTION 3.2

3. exactly one **5.** Function; D = {−3, −2, 1, 4}; R = {−7, 1, 2, 5} **7.** Function; D = {1, 2, 3, 4, 5}; R = {5}

9. Function; D = {−2, −1, 1, 2, 3}; R = $\left\{-1, -\frac{1}{2}, \frac{1}{3}, \frac{1}{2}, 1\right\}$ **11.** Not a function; D = {2, 4, 6, 8}; R = {3, 5, 7, 8, 9}

13. Yes, each element of the domain is paired with exactly one element of the range. **15.** No, 1 and 4 in the domain are paired with more than one element in range. **17.** Yes, each element of the domain is paired with exactly one element of the range. **19. a.** Yes **b.** $29.25 **23.** True **25.** 3; 3; 8

27. a. 13 **b.** −3 **c.** 13 **29. a.** −3 **b.** 12 **c.** 8 **31. a.** −8 **b.** 1

c. −4 **33. a.** 1 **b.** −1 **c.** $-\frac{1}{2}$ **35. a.** 4a + 11 **b.** 8a − 1 **37. a.** 17

b. 7 **c.** −1 **39. a.** 15 **b.** 2 **c.** $3w^2 − 4w − 5$ **41. a.** 2 **b.** 2

c. $a^2 + 2a − 1$ **43. a.** 35 **b.** 0 **c.** $4a^2 + 8a + 3$ **45. a.** 4 **b.** 2 **c.** 0

47. a. 7 **b.** 10 **c.** −2 **49. a.** 5 **b.** −10 **c.** −10 **51.** 0 **53.** 13

55. 1 **57.** {−14, −11, −8, −5, −2, 1} **59.** $\left\{-\frac{3}{2}, -1, -\frac{1}{2}, 0, \frac{1}{2}, 1\right\}$ **61.** {3, 4, 7, 12}

63. {−26, −8, −2, 4} **65.** 4; (4, 5) **67.** 4; (4, −7) **69.** 3; (3, 0) **71.** A, C **73.** No

75. Yes **77.** **79.** **81.** **83.**

85. **87.** **89.** **91.** **93.**

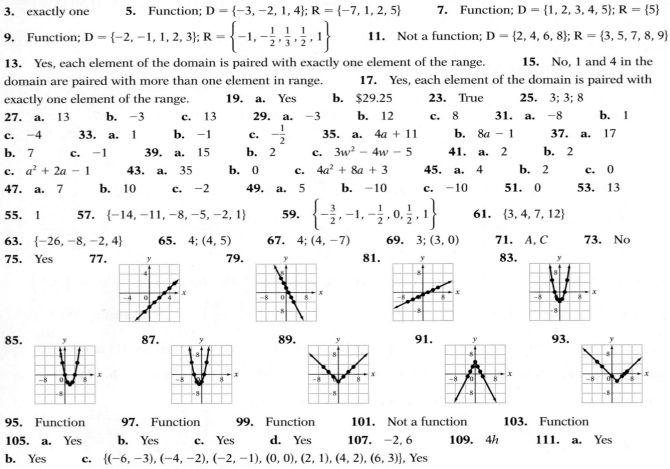

95. Function **97.** Function **99.** Function **101.** Not a function **103.** Function

105. a. Yes **b.** Yes **c.** Yes **d.** Yes **107.** −2, 6 **109.** 4h **111. a.** Yes

b. Yes **c.** {(−6, −3), (−4, −2), (−2, −1), (0, 0), (2, 1), (4, 2), (6, 3)}, Yes

d. {(9, −3), (4, −2), (1, −1), (0, 0), (1, 1), (4, 2), (9, 3)}, No **e.** No **113.** 61.2 ft **117. a.** 20 ft/s

b. 28 ft/s **119. a.** 60°F **b.** 52°F

CONCEPT REVIEW 3.3

1. Never true **3.** Always true **5.** Sometimes true

SECTION 3.3

3. a. −7; −1; −7 **b.** −5; 0; −5 **c.** −3; 1; −3 **5.** **7.**

9. **11.** **13.** **15.** **17.**

19. 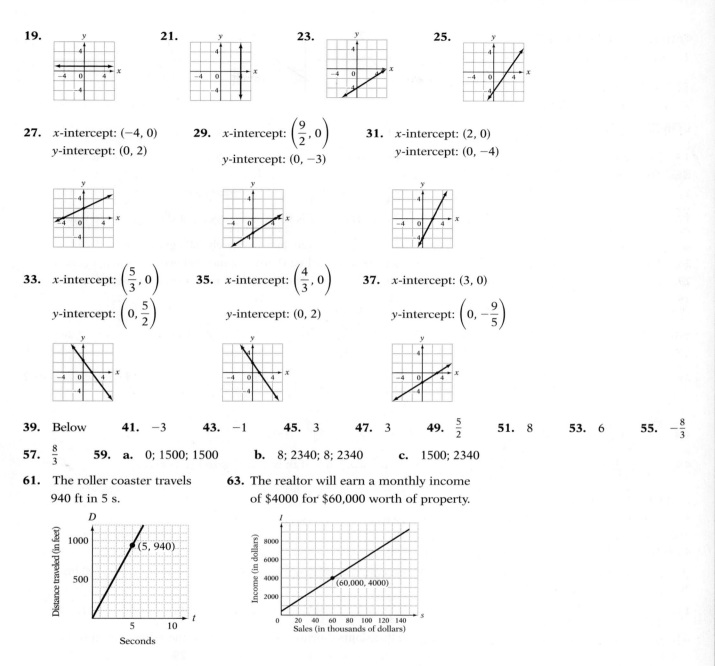 **21.** **23.** **25.**

27. *x*-intercept: (−4, 0) **29.** *x*-intercept: $\left(\dfrac{9}{2}, 0\right)$ **31.** *x*-intercept: (2, 0)

 y-intercept: (0, 2) *y*-intercept: (0, −3) *y*-intercept: (0, −4)

33. *x*-intercept: $\left(\dfrac{5}{3}, 0\right)$ **35.** *x*-intercept: $\left(\dfrac{4}{3}, 0\right)$ **37.** *x*-intercept: (3, 0)

 y-intercept: $\left(0, \dfrac{5}{2}\right)$ *y*-intercept: (0, 2) *y*-intercept: $\left(0, -\dfrac{9}{5}\right)$

39. Below **41.** −3 **43.** −1 **45.** 3 **47.** 3 **49.** $\dfrac{5}{2}$ **51.** 8 **53.** 6 **55.** $-\dfrac{8}{3}$

57. $\dfrac{8}{3}$ **59. a.** 0; 1500; 1500 **b.** 8; 2340; 8; 2340 **c.** 1500; 2340

61. The roller coaster travels **63.** The realtor will earn a monthly income
940 ft in 5 s. of $4000 for $60,000 worth of property.

65. The caterer will charge $614
for 120 hot appetizers.

67. The sale price is $160. **69.** **71.** **73.**

CONCEPT REVIEW 3.4

1. Always true **3.** Never true

SECTION 3.4

1. -5; -4; 3; 2 **3.** -1 **5.** $\dfrac{1}{3}$ **7.** $-\dfrac{2}{3}$ **9.** $-\dfrac{3}{4}$ **11.** Undefined **13.** $\dfrac{7}{5}$ **15.** 0

17. $-\dfrac{1}{2}$ **19.** Undefined **21.** The slope is 40. The slope is the average speed of the motorist in miles per hour. **23.** The slope is -0.05. For each mile the car is driven, approximately 0.05 gal of fuel is used.

25. Line A represents Lois's distance; line B, Tanya's distance; and line C, the distance between Lois and Tanya.

27. a. The ramp does not meet the ANSI requirements. **b.** The ramp does meet the ANSI requirements.

29. 5; $(0, -3)$ **31.** **33.** **35.** **37.**

39. **41.** Below; upward **43.** **45.** **47.** increases by 2

49. increases by $\dfrac{1}{2}$ **51.** 10 **53.** i and D; ii and C; iii and B; iv and F; v and E; vi and A

CONCEPT REVIEW 3.5

1. Always true **3.** Never true **5.** Never true **7.** Never true

SECTION 3.5

1. $-\dfrac{6}{5}$; 2; $-\dfrac{6}{5}x + 2$ **3.** $y = 2x + 5$ **5.** $y = \dfrac{1}{2}x + 2$ **7.** $y = -\dfrac{5}{3}x + 5$ **9.** $y = -3x + 4$

11. $y = \dfrac{1}{2}x$ **13.** $y = 3x - 9$ **15.** $y = -\dfrac{2}{3}x + 7$ **17.** $y = -x - 3$ **19.** $x = 3$ **21.** $y = -3$

23. $y = -2x + 3$ **25.** $x = -5$ **27.** The line slants downward to the right, so the value of m must be negative. The line appears to cross the y-axis at $(0, 2)$, so the value of b cannot be 6. **29.** $y = x + 2$

31. $y = -2x - 3$ **33.** $y = \dfrac{1}{3}x + \dfrac{10}{3}$ **35.** $y = -\dfrac{3}{2}x + 3$ **37.** $y = x - 1$ **39.** $y = \dfrac{2}{3}x + \dfrac{5}{3}$

41. $y = \dfrac{1}{2}x - 1$ **43.** $y = -4$ **45.** $y = \dfrac{3}{4}x$ **47.** $x = -2$ **49.** $y = x - 1$ **51.** $y = -x + 3$

53. 1000; 500 **55.** $y = 1200x$; the model predicts the plane will be 13,200 ft above sea level in 11 min.

57. $y = 0.59x + 4.95$; the model predicts that using a cell phone for 13 min in one month will cost $12.62.

59. $y = 63x$; the model predicts that there are 315 calories in a 5-ounce serving of lean hamburger.

61. $y = -0.431x + 864$; the model predicts that in 1967 a plane could travel from New York to Paris in 16.2 h.

63. $y = -0.032x + 16$; the model predicts that this car would contain 11.2 gal of gas after driving 150 mi.

65. $\left(-\dfrac{b}{m}, 0\right)$ **67.** 0 **69.** Changing b moves the graph of the line up or down.

CONCEPT REVIEW 3.6

1. Sometimes true **3.** Never true **5.** Always true **7.** Always true

SECTION 3.6

3. -5 **5.** $-\dfrac{1}{4}$ **7.** $-\dfrac{1}{3}$ **9.** Yes **11.** No **13.** No **15.** Yes **17.** Yes

19. Yes **21.** No **23.** Yes **25.** Yes **27.** $y = \dfrac{2}{3}x - \dfrac{8}{3}$ **29.** $y = \dfrac{1}{3}x - \dfrac{1}{3}$

31. $y = -\dfrac{5}{3}x - \dfrac{14}{3}$ **33.** Downward to the right **35.** $\dfrac{A_1}{B_1} = \dfrac{A_2}{B_2}$ **37.** $y = -2x + 15$

39. Any equation of the form $y = 2x + b$, where $b \neq -13$, or of the form $y = -\dfrac{3}{2}x + c$, where $c \neq 8$

CONCEPT REVIEW 3.7

1. Never true **3.** Always true **5.** Always true

SECTION 3.7

3. Yes **5.** No **7.** **9.** **11.** **13.**

15. 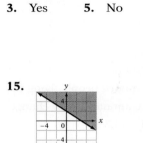 **17.** **19.** **21.** **23.**

25. Negative

CHAPTER 3 REVIEW EXERCISES

1. $(4, 2)$ [3.1.1] **2.** Midpoint: $\left(\dfrac{1}{2}, \dfrac{9}{2}\right)$; length: $\sqrt{26}$ [3.1.2] **3.** [3.2.2]

4. [3.3.2] **5.** $P(-2) = -2$; $P(a) = 3a + 4$ [3.2.1] **6.** Domain: $\{-1, 0, 1, 2, 5\}$; range: $\{0, 2, 3\}$; Yes [3.2.1]

7. Range: $\{-2, -1, 2\}$ [3.2.1] **8.** 2; $(2, 5)$ [3.2.1] **9.** x-intercept: $(3, 0)$ [3.3.2]

10. 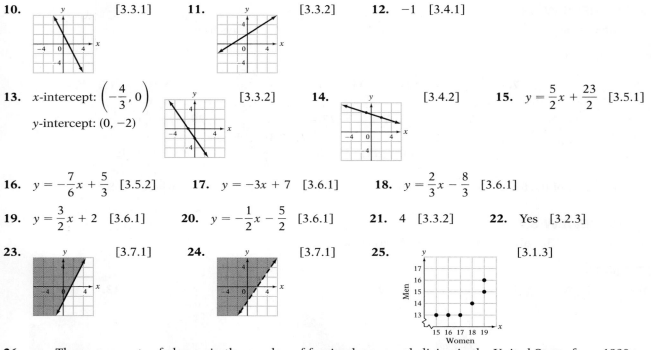 [3.3.1] **11.** [3.3.2] **12.** -1 [3.4.1]

13. x-intercept: $\left(-\dfrac{4}{3}, 0\right)$ [3.3.2] **14.** [3.4.2] **15.** $y = \dfrac{5}{2}x + \dfrac{23}{2}$ [3.5.1]
 y-intercept: $(0, -2)$

16. $y = -\dfrac{7}{6}x + \dfrac{5}{3}$ [3.5.2] **17.** $y = -3x + 7$ [3.6.1] **18.** $y = \dfrac{2}{3}x - \dfrac{8}{3}$ [3.6.1]

19. $y = \dfrac{3}{2}x + 2$ [3.6.1] **20.** $y = -\dfrac{1}{2}x - \dfrac{5}{2}$ [3.6.1] **21.** 4 [3.3.2] **22.** Yes [3.2.3]

23. [3.7.1] **24.** [3.7.1] **25.** [3.1.3]

26. a. The average rate of change in the number of foreign-born people living in the United States from 1980 to 2000 was 715,000 people per year. **b.** The average rate of change in the number of foreign-born people living in the United States from 1970 to 1980 was 450,000 people per year. This was less than the average annual rate of change from 1980 to 1990. [3.1.4] **27.** After 4 h, the car has traveled 220 mi. [3.3.3]

28. The slope is 20. The cost of manufacturing one calculator is $20. [3.4.1] **29.** $y = 80x + 25,000$; this model predicts that the cost to build a house that contains 2000 ft² will be $185,000. [3.5.3]

CHAPTER 3 TEST

1. 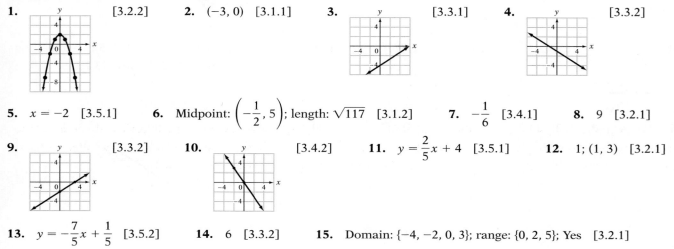 [3.2.2] **2.** $(-3, 0)$ [3.1.1] **3.** [3.3.1] **4.** [3.3.2]

5. $x = -2$ [3.5.1] **6.** Midpoint: $\left(-\dfrac{1}{2}, 5\right)$; length: $\sqrt{117}$ [3.1.2] **7.** $-\dfrac{1}{6}$ [3.4.1] **8.** 9 [3.2.1]

9. [3.3.2] **10.** [3.4.2] **11.** $y = \dfrac{2}{5}x + 4$ [3.5.1] **12.** 1; $(1, 3)$ [3.2.1]

13. $y = -\dfrac{7}{5}x + \dfrac{1}{5}$ [3.5.2] **14.** 6 [3.3.2] **15.** Domain: $\{-4, -2, 0, 3\}$; range: $\{0, 2, 5\}$; Yes [3.2.1]

16. $y = -\dfrac{3}{2}x + \dfrac{7}{2}$ [3.6.1] **17.** $y = 2x + 1$ [3.6.1] **18.** [3.7.1] **19.** No [3.2.3]

20. $y = -\dfrac{3}{10}x + 175$; this model predicts that 85 students will enroll when the tuition is $300. [3.5.3]

21. The slope is $-\dfrac{10,000}{3}$. The value of the house decreases by $3333.33 each year. [3.4.1]

22. [3.1.3]

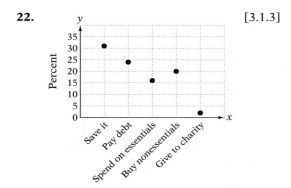

CUMULATIVE REVIEW EXERCISES

1. Commutative Property of Multiplication [1.5.2] **2.** $\dfrac{9}{2}$ [2.2.3] **3.** $\dfrac{8}{9}$ [2.2.3] **4.** $-\dfrac{1}{14}$ [2.2.3]

5. $\left\{ x \mid x < -1 \text{ or } x > \dfrac{1}{2} \right\}$ [2.7.3] **6.** $\dfrac{5}{2}$ and $-\dfrac{3}{2}$ [2.8.1] **7.** $\left\{ x \mid 0 < x < \dfrac{10}{3} \right\}$ [2.8.2] **8.** 8 [1.4.2]

9. -18 [1.5.1] **10.** [2.7.1] **11.** $\varnothing$ [2.7.3] **12.** 14 [3.2.1]

13. $(-8, 13)$ [3.1.1] **14.** $-\dfrac{7}{4}$ [3.4.1] **15.** $y = \dfrac{3}{2}x + \dfrac{13}{2}$ [3.5.1] **16.** $y = -\dfrac{5}{4}x + 3$ [3.5.2]

17. $y = -\dfrac{3}{2}x + 7$ [3.6.1] **18.** $y = -\dfrac{2}{3}x + \dfrac{8}{3}$ [3.6.1] **19.** [3.3.2]

20. [3.4.2] **21.** [3.7.1]

22. There are 12 nickels in the coin purse. [2.3.3] **23.** The rate of the first plane is 200 mph, and the rate of the second plane is 400 mph. [2.3.4] **24.** The mixture consists of 32 lb of $8 coffee and 48 lb of $3 coffee. [2.3.3]
25. $y = -2500x + 15,000$; the value of the truck decreases by $2500 each year. [3.4.1, 3.5.2]

Answers to Chapter 4 Selected Exercises

PREP TEST

1. $6x + 5y$ [1.5.5] **2.** 7 [1.5.1] **3.** 0 [2.2.1] **4.** -3 [2.2.3] **5.** 1000 [2.2.3]
6. [3.3.1] **7.** [3.3.2] **8.** [3.7.1]

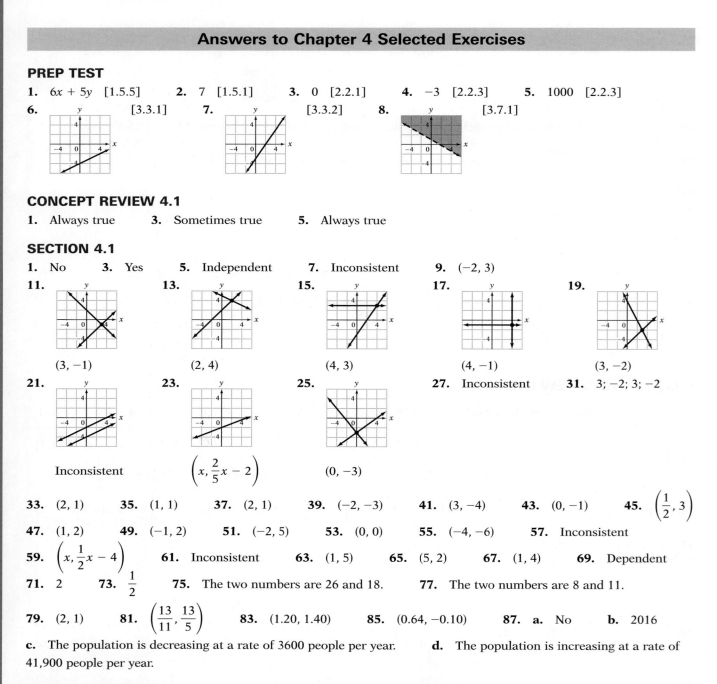

CONCEPT REVIEW 4.1

1. Always true **3.** Sometimes true **5.** Always true

SECTION 4.1

1. No **3.** Yes **5.** Independent **7.** Inconsistent **9.** $(-2, 3)$
11. $(3, -1)$ **13.** $(2, 4)$ **15.** $(4, 3)$ **17.** $(4, -1)$ **19.** $(3, -2)$
21. Inconsistent **23.** $\left(x, \frac{2}{5}x - 2\right)$ **25.** $(0, -3)$ **27.** Inconsistent **31.** $3; -2; 3; -2$

33. $(2, 1)$ **35.** $(1, 1)$ **37.** $(2, 1)$ **39.** $(-2, -3)$ **41.** $(3, -4)$ **43.** $(0, -1)$ **45.** $\left(\frac{1}{2}, 3\right)$
47. $(1, 2)$ **49.** $(-1, 2)$ **51.** $(-2, 5)$ **53.** $(0, 0)$ **55.** $(-4, -6)$ **57.** Inconsistent
59. $\left(x, \frac{1}{2}x - 4\right)$ **61.** Inconsistent **63.** $(1, 5)$ **65.** $(5, 2)$ **67.** $(1, 4)$ **69.** Dependent
71. 2 **73.** $\frac{1}{2}$ **75.** The two numbers are 26 and 18. **77.** The two numbers are 8 and 11.

79. $(2, 1)$ **81.** $\left(\frac{13}{11}, \frac{13}{5}\right)$ **83.** $(1.20, 1.40)$ **85.** $(0.64, -0.10)$ **87. a.** No **b.** 2016
c. The population is decreasing at a rate of 3600 people per year. **d.** The population is increasing at a rate of 41,900 people per year.

CONCEPT REVIEW 4.2

1. Always true **3.** Sometimes true **5.** Never true

SECTION 4.2

3. $(6, 1)$ **5.** $(1, 1)$ **7.** $(2, 1)$ **9.** $(x, 3x - 4)$ **11.** $\left(-\frac{1}{2}, 2\right)$ **13.** Inconsistent
15. $(-1, -2)$ **17.** $(2, 5)$ **19.** $\left(\frac{1}{2}, \frac{3}{4}\right)$ **21.** $(0, 0)$ **23.** $\left(\frac{2}{3}, -\frac{2}{3}\right)$ **25.** $(1, -1)$

27. $\left(x, \frac{4}{3}x - 6\right)$ **29.** (5, 3) **31.** $\left(\frac{1}{3}, -1\right)$ **33.** $\left(\frac{5}{3}, \frac{1}{3}\right)$ **35.** Inconsistent

39. Dependent **41.** The first and second equations, the second and third equations **43.** (−1, 2, 1)

45. (6, 2, 4) **47.** (4, 1, 5) **49.** (3, 1, 0) **51.** (−1, −2, 2) **53.** Inconsistent **55.** (2, 1, −3)

57. (2, −1, 3) **59.** (6, −2, 2) **61.** (0, −2, 0) **63.** (2, 3, 1) **65.** (1, 1, 3) **67.** (2, 1)

69. (2, −1) **71.** (2, 0, −1) **73.** $A = 2, B = 3, C = -3$ **75.** The number of nickels is $3z - 5$, the number of dimes is $-4z + 35$, and the number of quarters is z, when $z = 2, 3, 4, 5, 6, 7,$ or 8. **77.** (1, 1)

79. (3, 2)

CONCEPT REVIEW 4.3

1. Always true **3.** Sometimes true **5.** Always true

SECTION 4.3

1. 5 **3.** square **5.** 8; −1; 13 **7.** 11 **9.** 18 **11.** 0 **13.** 15 **15.** −30

17. 0 **19. a.** 2; −3; −5; 4 **b.** −7 **21. a.** 2; 5; −5; −2 **b.** 21 **23.** (3, −4)

25. (4, −1) **27.** $\left(\frac{11}{14}, \frac{17}{21}\right)$ **29.** $\left(\frac{1}{2}, 1\right)$ **31.** Not possible by Cramer's Rule **33.** (−1, 0)

35. (1, −1, 2) **37.** (2, −2, 3) **39.** Not possible by Cramer's Rule **41.** $\left(\frac{68}{25}, \frac{56}{25}, -\frac{8}{25}\right)$

43. 1; −3; 2 **45.** 9 **47.** −3; 6; −9; −6 **49.** $x - 2y + 3z = 2; y - 5z = -3; z = 1$

51. $\begin{bmatrix} 1 & -2 & \bigm| & \frac{1}{2} \\ 0 & 1 & \bigm| & -\frac{5}{2} \end{bmatrix}$ **53.** $\begin{bmatrix} 1 & -\frac{2}{5} & \bigm| & \frac{3}{5} \\ 0 & 1 & \bigm| & 26 \end{bmatrix}$ **55.** $\begin{bmatrix} 1 & 4 & 1 & \bigm| & -2 \\ 0 & 1 & 4 & \bigm| & -8 \\ 0 & 0 & 1 & \bigm| & -\frac{32}{19} \end{bmatrix}$ **57.** $\begin{bmatrix} 1 & -3 & \frac{1}{2} & \bigm| & -\frac{3}{2} \\ 0 & 1 & \frac{3}{2} & \bigm| & \frac{5}{2} \\ 0 & 0 & 1 & \bigm| & 3 \end{bmatrix}$

59. (−7, 4, 3) **61.** (1, 3) **63.** (−1, −3) **65.** (3, 2) **67.** Inconsistent **69.** (−2, 2)

71. (0, 0, −3) **73.** (1, −1, −1) **75.** Inconsistent **77.** $\left(\frac{1}{3}, \frac{1}{2}, 0\right)$ **79.** $\left(\frac{1}{5}, \frac{2}{5}, -\frac{3}{5}\right)$

81. $\left(\frac{1}{4}, 0, -\frac{2}{3}\right)$ **83.** −14 **85.** 0 **87. a.** 0 **b.** 0 **89.** The area of the polygon is 239 ft².

CONCEPT REVIEW 4.4

1. Never true **3.** Always true

SECTION 4.4

1. Row 1: $b + c$; 3; $3(b + c)$; Row 2: $b - c$; 5; $5(b - c)$ **3.** Less than **5.** The rate of the plane in calm air is 150 mph. The rate of the wind is 10 mph. **7.** The rate of the cabin cruiser in calm water is 14 mph. The rate of the current is 2 mph. **9.** The rate of the plane in calm air is 165 mph. The rate of the wind is 15 mph. **11.** The rate of the boat in calm water is 19 km/h. The rate of the current is 3 km/h. **13.** The rate of the plane in calm air is 105 mph. The rate of the wind is 15 mph. **15.** The rate of the boat in calm water is 16.5 mph. The rate of the current is 1.5 mph. **17.** Table 1, Row 1: 200; h; 200h; Row 2: 300; w; 300w; Table 2, Row 1: 350; h; 350h; Row 2: 100; w; 100w **19.** Greater than **21.** The cost of the pine is $.18 per foot. The cost of the redwood is $.30 per foot. **23.** The cost per unit of gas is $.16. **25.** There are 18 quarters in the bank. **27.** The company plans to manufacture 25 mountain bikes during the week. **29.** The pharmacist should use 200 mg of the first powder and 450 mg of the second powder. **31.** The model VI costs $4000. **33.** $8000 was invested at 9%, $6000 at 7%, and $4000 at 5%. **35.** The distances are $d_1 = 6$ in., $d_2 = 3$ in., and $d_3 = 9$ in. **37.** There were 10 nickels, 5 dimes, and 4 quarters. **39.** There was $14,000 invested at 12%, $10,000 invested

at 8%, and $9000 invested at 9%. **41.** The measures of the two angles are 9° and 81°. **43.** There are 25 nickels, 10 dimes, and 5 quarters in the coin bank.

CONCEPT REVIEW 4.5

1. Sometimes true **3.** Always true

SECTION 4.5

1. b **3. a.** A; D **b.** A; B **c.** A **5.** **7.**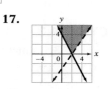

9. **11.** **13.** **15.** **17.**

19. **21.** **23.** Points below the line $x + y = b$ **25.** Region between the parallel lines $x + y = a$ and $x + y = b$

27. **29.** **31.**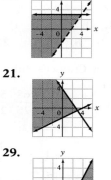

CHAPTER 4 REVIEW EXERCISES

1. Inconsistent [4.1.2] **2.** $\left(x, -\dfrac{1}{4}x + \dfrac{3}{2}\right)$ [4.1.2] **3.** $(-4, 7)$ [4.2.1] **4.** $\left(x, \dfrac{1}{3}x - 2\right)$ [4.2.1]

5. $(5, -2, 3)$ [4.2.2] **6.** $(3, -1, -2)$ [4.2.2] **7.** 28 [4.3.1] **8.** 0 [4.3.1] **9.** $(3, -1)$ [4.3.2]

10. $\left(\dfrac{110}{23}, \dfrac{25}{23}\right)$ [4.3.2] **11.** $(-1, -3, 4)$ [4.3.2] **12.** $(2, 3, -5)$ [4.3.2] **13.** $(1, -1, 4)$ [4.2.2]

14. $\left(\dfrac{8}{5}, \dfrac{7}{5}\right)$ [4.3.2] **15.** $\left(\dfrac{1}{2}, -1, \dfrac{1}{3}\right)$ [4.3.3] **16.** 12 [4.3.1] **17.** $(2, -3)$ [4.3.2]

18. $(2, -3, 1)$ [4.3.3] **19.** [4.1.1] **20.** [4.1.1]

(0, 3) (x, 2x − 4)

21. [4.5.1] **22.** [4.5.1] **23.** The rate of the cabin cruiser in calm water is 16 mph. The rate of the current is 4 mph. [4.4.1]

24. The rate of the plane in calm air is 175 mph. The rate of the wind is 25 mph. [4.4.1] **25.** The number of children attending on Friday was 100. [4.4.2] **26.** There is $10,400 invested at 8%, $5200 invested at 6%, and $9400 invested at 4%. [4.4.2]

CHAPTER 4 TEST

1. $\left(\dfrac{3}{4}, \dfrac{7}{8}\right)$ [4.1.2] 2. $(-3, -4)$ [4.1.2] 3. $(2, -1)$ [4.1.2] 4. $(-2, 1)$ [4.3.3]

5. Inconsistent [4.2.1] 6. $(1, 1)$ [4.2.1] 7. Inconsistent [4.2.2] 8. $(2, -1, -2)$ [4.3.3]

9. 10 [4.3.1] 10. -32 [4.3.1] 11. $\left(-\dfrac{1}{3}, -\dfrac{10}{3}\right)$ [4.3.2] 12. $\left(\dfrac{1}{5}, -\dfrac{6}{5}, \dfrac{3}{5}\right)$ [4.3.2]

13. $(0, -2, 3)$ [4.3.2] 14. [4.1.1] 15. [4.1.1]

(3, 4)

(−5, 0)

16. [4.5.1] 17. [4.5.1] 18. $(-0.14, 2.43)$ [4.1.1]

19. The rate of the plane in calm air is 150 mph. The rate of the wind is 25 mph. [4.4.1] 20. The cost per yard of cotton is \$9. The cost per yard of wool is \$14. [4.4.2]

CUMULATIVE REVIEW EXERCISES

1. $-\dfrac{11}{28}$ [2.2.3] 2. $y = 5x - 11$ [3.5.2] 3. $3x - 24$ [1.5.6] 4. -4 [1.5.1]

5. $\{x \mid x < 6\}$ [2.7.3] 6. $\{x \mid -4 < x < 8\}$ [2.8.2] 7. $\{x \mid x > 4 \text{ or } x < -1\}$ [2.8.2] 8. -98 [3.2.1]

9. $\{0, 1, 5, 8, 16\}$ [3.2.1] 10. 1 [3.2.1] 11. $3h$ [3.2.1] 12. [2.7.1]

13. $y = -\dfrac{2}{3}x + \dfrac{5}{3}$ [3.5.1] 14. $y = -\dfrac{3}{2}x + \dfrac{1}{2}$ [3.6.1] 15. $2\sqrt{10}$ [3.1.2] 16. $\left(-\dfrac{1}{2}, 4\right)$ [3.1.2]

17. [3.4.2] 18. [3.7.1] 19. $(-5, -11)$ [4.1.2] 20. $(1, 0, -1)$ [4.2.2]

21. 3 [4.3.1] 22. [4.1.1] 23. $(2, -3)$ [4.3.2] 24. [4.5.1]

25. There are 16 nickels in the purse. [2.3.3] 26. The amount of water that should be added is 60 ml. [2.6.2]
27. The rate of the wind is 12.5 mph. [4.4.1] 28. The cost per pound of steak is \$5. [4.4.2] 29. The lower and upper limits of the resistor are 10,200 ohms and 13,800 ohms. [2.8.3] 30. The slope is 40. The account executive earns \$40 for each \$1000 of sales. [3.4.1]

Answers to Chapter 5 Selected Exercises

PREP TEST

1. 1 [1.2.2] 2. -18 [1.2.3] 3. $\dfrac{2}{3}$ [1.2.4] 4. 48 [1.5.1] 5. 0 [1.2.4]

6. No [1.5.3] **7.** $5x^2 - 9x - 6$ [1.5.3] **8.** 0 [1.5.3] **9.** $-6x + 24$ [1.5.5]
10. $-7xy + 10y$ [1.5.6]

CONCEPT REVIEW 5.1

1. Always true **3.** Always true **5.** Sometimes true

SECTION 5.1

3. 2; leading; coefficient **5. a.** The leading coefficient is -1. **b.** The constant term is 8.
c. The degree is 2. **7.** The expression is not a polynomial. **9.** The expression is not a polynomial.
11. a. The leading coefficient is 3 **b.** The constant term is π. **c.** The degree is 5. **13. a.** The
leading coefficient is -5. **b.** The constant term is 2. **c.** The degree is 3. **15. a.** The leading
coefficient is 14. **b.** The constant term is 14. **c.** The degree is 0. **17.** 13 **19.** 10
21. -11 **23.** **25.** **27.**

29. The length of the wave is 23.1 m. **31.** There must be 56 games scheduled. **33.** The baked Alaska
needs 260 in^3 of meringue. **35.** A **37.** $-8x^2 - 3x + 5$ **39.** $6x^2 - 6x + 5$ **41.** $-x^2 + 1$
43. $5y^2 - 15y + 2$ **45.** $7a^2 - a + 2$ **47.** $5x^3 - 4x^2 + 4x - 7$ **49.** $9x^n + 5$ **51.** $3x^4 - 8x^2 + 2x$
53. $-2x^3 + 4x^2 + 8$ **55. a.** Positive **b.** Positive **c.** Negative **57.** -2 **59.** $P(x) + Q(x)$
is a fourth-degree polynomial. **61.** The maximum deflection of the beam is 3.125 in. **63.** 2
65. The graph of k is the graph of f moved 2 units down.

CONCEPT REVIEW 5.2

1. Never true **3.** Always true **5.** Never true

SECTION 5.2

3. 2; a^5; b^3; 10; 6; 5; 1; 3; adding; 11; 4 **5.** $2x^2$ **7.** $12x^2$ **9.** $6a^7$ **11.** x^3y^5 **13.** $-10x^9y$
15. x^7y^8 **17.** $-6x^3y^5$ **19.** x^4y^5z **21.** $a^3b^5c^4$ **23.** $-a^5b^8$ **25.** $-6a^5b$ **27.** $40y^{10}z^6$
29. $-20a^2b^3$ **31.** $x^3y^5z^3$ **33.** $-12a^{10}b^7$ **37. a.** Power **b.** Product **39. a.** Product
b. Power **41. a.** Yes **b.** No **c.** No **d.** Yes **43.** 64 **45.** 4 **47.** -64
49. x^9 **51.** x^{14} **53.** x^4 **55.** $4x^2$ **57.** $-8x^6$ **59.** x^4y^6 **61.** $9x^4y^2$ **63.** $27a^8$
65. $-8x^7$ **67.** x^8y^4 **69.** a^4b^6 **71.** $16x^{10}y^3$ **73.** $-18x^3y^4$ **75.** $-8a^7b^5$ **77.** $-54a^9b^3$
79. $-72a^5b^5$ **81.** $32x^3$ **83.** $-3a^3b^3$ **85.** $13x^5y^5$ **87.** $27a^5b^3$ **89.** $-5x^7y^4$ **91.** a^{2n}
93. a^{2n} **95.** True **97.** False; $x^{2\cdot5} = x^{10}$ **99.** No; $2^{(3^2)}$ is larger. $(2^3)^2 = 8^2 = 64$, whereas $2^{(3^2)} = 2^9 = 512$.
101. $x = y$ when n is odd.

CONCEPT REVIEW 5.3

1. Always true **3.** Sometimes true **5.** Always true **7.** Sometimes true

SECTION 5.3

1. $-3y$; $-3y$; Distributive; $-3y$; $-3y^2$; $21y$ **3.** $x^2 - 2x$ **5.** $-x^2 - 7x$ **7.** $3a^3 - 6a^2$
9. $-5x^4 + 5x^3$ **11.** $-3x^5 + 7x^3$ **13.** $12x^3 - 6x^2$ **15.** $6x^2 - 12x$ **17.** $-x^3y + xy^3$

19. $2x^4 - 3x^2 + 2x$ **21.** $2a^3 + 3a^2 + 2a$ **23.** $3x^6 - 3x^4 - 2x^2$ **25.** $-6y^4 - 12y^3 + 14y^2$

27. $-2a^3 - 6a^2 + 8a$ **29.** $6y^4 - 3y^3 + 6y^2$ **31.** $x^3y - 3x^2y^2 + xy^3$ **33. b** and **c**

35. $x^3 + 4x^2 + 5x + 2$ **37.** $a^3 - 6a^2 + 13a - 12$ **39.** $-2b^3 + 7b^2 + 19b - 20$

41. $-6x^3 + 31x^2 - 41x + 10$ **43.** $x^4 - 4x^3 - 3x^2 + 14x - 8$ **45.** $15y^3 - 16y^2 - 70y + 16$

47. $5a^4 - 20a^3 - 15a^2 + 62a - 8$ **49.** $y^4 + 4y^3 + y^2 - 5y + 2$ **51. a.** 4 **b.** 5

53. $4x; x; 4x; 5; -3; x; -3; 5$ **55. a.** $3d, d$ **b.** $3d, -1$ **c.** $4, d$ **d.** $4, -1$ **57.** $x^2 + 4x + 3$

59. $a^2 + a - 12$ **61.** $y^2 - 5y - 24$ **63.** $y^2 - 10y + 21$ **65.** $2x^2 + 15x + 7$ **67.** $3x^2 + 11x - 4$

69. $4x^2 - 31x + 21$ **71.** $3y^2 - 2y - 16$ **73.** $9x^2 + 54x + 77$ **75.** $21a^2 - 83a + 80$

77. $15b^2 + 47b - 78$ **79.** $2a^2 + 7ab + 3b^2$ **81.** $6a^2 + ab - 2b^2$ **83.** $2x^2 - 3xy - 2y^2$

85. $10x^2 + 29xy + 21y^2$ **87.** $6a^2 - 25ab + 14b^2$ **89.** $2a^2 - 11ab - 63b^2$ **91.** $100a^2 - 100ab + 21b^2$

93. $15x^2 + 56xy + 48y^2$ **95.** $14x^2 - 97xy - 60y^2$ **97.** $56x^2 - 61xy + 15y^2$ **99.** Positive

101. $5x; 1; 25x^2; 1$ **103.** $y^2 - 25$ **105.** $4x^2 - 9$ **107.** $x^2 + 2x + 1$ **109.** $9a^2 - 30a + 25$

111. $9x^2 - 49$ **113.** $4a^2 + 4ab + b^2$ **115.** $x^2 - 4xy + 4y^2$ **117.** $16 - 9y^2$

119. $25x^2 + 20xy + 4y^2$ **121.** Zero **123.** Positive **125.** $2x + 5; 2x - 5; 4x^2 - 25$ **127.** The area is $(10x^2 - 35x)$ ft². **129.** The area is $(18x^2 + 12x + 2)$ in². **131.** The area $(4x^2 + 4x + 1)$ km².

133. The area is $(4x^2 + 10x)$ m². **135.** The area is $(\pi x^2 + 8\pi x + 16\pi)$ cm². **137.** The area is $(90x + 2025)$ ft². **139.** $4ab$ **141.** $9a^4 - 24a^3 + 28a^2 - 16a + 4$ **143.** $24x^3 - 3x^2$ **145.** $x^{2n} + x^n$

147. $x^{2n} + 2x^n + 1$ **149.** $x^2 - 1$ **151.** $x^4 - 1$ **153.** $x^6 - 1$ **155.** 1024

157. $12x^2 - x - 20$ **159.** $7x^2 - 11x - 8$

CONCEPT REVIEW 5.4

1. Never true **3.** Always true **5.** Never true **7.** Never true

SECTION 5.4

1. $\dfrac{1}{25}$ **3.** 64 **5.** $\dfrac{1}{27}$ **7.** 1 **11.** $4; 10; 3; 4x^7$ **13.** y^4 **15.** a^3 **17.** p^4

19. $2x^3$ **21.** $2k$ **23.** m^5n^2 **25.** $\dfrac{3r^2}{2}$ **27.** $-\dfrac{2a}{3}$ **29.** $\dfrac{1}{x^2}$ **31.** a^6 **33.** $\dfrac{4}{x^7}$

35. $5b^8$ **37.** $\dfrac{x^2}{3}$ **39.** 1 **41.** $\dfrac{1}{y^5}$ **43.** $\dfrac{1}{a^6}$ **45.** $\dfrac{1}{3x^3}$ **47.** $\dfrac{2}{3x^5}$ **49.** $\dfrac{y^4}{x^2}$

51. $\dfrac{2}{5m^3n^8}$ **53.** $\dfrac{1}{p^3q}$ **55.** $\dfrac{1}{2y^3}$ **57.** $\dfrac{7xz}{8y^3}$ **59.** $\dfrac{p^2}{2m^3}$ **61.** $-\dfrac{8x^3}{y^6}$ **63.** $\dfrac{9}{x^2y^4}$

65. $\dfrac{2}{x^4}$ **67.** $-\dfrac{5}{a^8}$ **69.** $-\dfrac{a^5}{8b^4}$ **71.** $\dfrac{10y^3}{x^4}$ **73.** $\dfrac{1}{2x^3}$ **75.** $\dfrac{3}{x^3}$ **77.** $\dfrac{1}{2x^2y^6}$

79. $-\dfrac{1}{6x^3}$ **81. a.** False **b.** True **85.** 8; left; 8 **87.** No. 39.4 is not between 1 and 10.

89. No. 2.4 is not an integer. **91.** 2.37×10^6 **93.** 4.5×10^{-4} **95.** 3.09×10^5

97. 6.01×10^{-7} **99.** $710,000$ **101.** 0.000043 **103.** $671,000,000$ **105.** 0.00000713

107. 1×10^{21} **109.** 4×10^{-10} **111.** 1.2×10^{10} **113.** 2.45×10^9 **115.** Less than

117. 6.65×10^{19} **119.** 3.22×10^{-14} **121.** 3.6×10^5 **123.** 1.8×10^{-18} **125.** $60; 60; 18,000; 4$

127. The computer can perform 7.2×10^9 operations in 1 h. **129.** Light travels 8.64×10^{12} m in 8 h.

131. The time for the centrifuge to make one revolution is 1.5×10^{-7} s. **133.** The sun is 3.3898305×10^5 times heavier than Earth. **135.** It takes light 5×10^2 s for light to travel to Earth from the sun.

137. Each American would have to pay 2.7×10^4 dollars. **139.** The surface area of Neptune is 2.82743339×10^9 mi². **141.** The weight of one atom of hydrogen is $1.66030217 \times 10^{-24}$ g. **143.** $\dfrac{3}{64}$

145. For 2^x: $\dfrac{1}{4}, \dfrac{1}{2}$, 1, 2, 4. For 2^{-x}: 4, 2, 1, $\dfrac{1}{2}, \dfrac{1}{4}$. **147.** 0.0625 **149.** xy^2 **151.** $\dfrac{x^6z^6}{4y^5}$ **153.** 1

155. 0 **157.** False; $(2a)^{-3} = \dfrac{1}{8a^3}$ **159.** False; $(2+3)^{-1} = 5^{-1} = \dfrac{1}{5}$

CONCEPT REVIEW 5.5

1. Always true **3.** Never true

SECTION 5.5

1. $3y$; $3y$; $6y^4$; 1 **3.** $x + 1$ **5.** $2a - 5$ **7.** $3a + 2$ **9.** $4b^2 - 3$ **11.** $x - 2$
13. $-x + 2$ **15.** $x^2 + 3x - 5$ **17.** $x^4 - 3x^2 - 1$ **19.** $xy + 2$ **21.** $-3y^3 + 5$
23. $3x - 2 + \dfrac{1}{x}$ **25.** $-3x + 7 - \dfrac{6}{x}$ **27.** $4a - 5 + 6b$ **29.** $9x + 6 - 3y$
31. Multiply $4x$ and $2x^2 - 3x - 1$. **33. a.** $3x^2$; x **b.** $3x$; $x - 2$ **c.** $-4x$; $-6x$
d. $3x + 2$; 11 **35.** $x + 8$ **37.** $x^2 + \dfrac{2}{x - 3}$ **39.** $3x + 5 + \dfrac{3}{2x + 1}$ **41.** $5x + 7 + \dfrac{2}{2x - 1}$
43. $4x^2 + 6x + 9 + \dfrac{18}{2x - 3}$ **45.** $3x^2 + 1 + \dfrac{1}{2x^2 - 5}$ **47.** $x^2 - 3x - 10$ **49.** $-x^2 + 2x - 1 + \dfrac{1}{x - 3}$
51. $2x^3 - 3x^2 + x - 4$ **53.** $x - 4 + \dfrac{x + 3}{x^2 + 1}$ **55.** $2x - 3 + \dfrac{1}{x - 1}$ **57.** $3x + 1 + \dfrac{10x + 7}{2x^2 - 3}$
59. False **61.** binomial; $x - a$ **63.** $2x - 8$ **65.** $3x + 3 - \dfrac{1}{x - 1}$ **67.** $x - 4 + \dfrac{7}{x + 4}$
69. $x - 2 + \dfrac{16}{x + 2}$ **71.** $2x^2 - 3x + 9$ **73.** $x^2 - 3x + 2$ **75.** $x^2 - 5x + 16 - \dfrac{41}{x + 2}$
77. $x^2 - x + 2 - \dfrac{4}{x + 1}$ **79.** $4x^2 + 8x + 15 + \dfrac{12}{x - 2}$ **81.** $2x^3 - 3x^2 + x - 4$
83. $3x^3 + 2x^2 + 12x + 19 + \dfrac{33}{x - 2}$ **85.** $3x^3 - x + 4 - \dfrac{2}{x + 1}$ **87.** $2x^3 + 6x^2 + 17x + 51 + \dfrac{155}{x - 3}$
89. $x^2 - 5x + 25$ **91.** Yes **93.** 8; 9; 18; 9; 9 **95.** $P(3) = 8$ **97.** $R(4) = 43$
99. $P(-2) = -39$ **101.** $Q(2) = 31$ **103.** $F(-3) = 190$ **105.** $P(5) = 122$ **107.** $R(-3) = 302$
109. $Q(2) = 0$ **111.** 3 **113.** -1 **115.** $(x + 2), (x - 3)$ **117.** $3a^2$ **119.** $4x^2 - 6x + 3$

CHAPTER 5 REVIEW EXERCISES

1. $2x^2 - 5x - 2$ [5.1.2] **2.** $4x^2 - 8xy + 5y^2$ [5.1.2] **3.** $70xy^2z^6$ [5.4.1] **4.** $-\dfrac{48y^9z^{17}}{x}$ [5.4.1]

5. $-\dfrac{x^3}{4y^2z^3}$ [5.4.1] **6.** $\dfrac{c^{10}}{2b^{17}}$ [5.4.1] **7.** 9.3×10^7 [5.4.2] **8.** 0.00254 [5.4.2]

9. 2×10^{-6} [5.4.2] **10.** -7 [5.1.1] **11.** [5.1.1] **12. a.** 3 **b.** 8

c. 5 [5.1.1] **13.** 68 [5.5.4] **14.** $5x + 4 + \dfrac{6}{3x - 2}$ [5.5.2] **15.** $2x - 3 - \dfrac{4}{6x + 1}$ [5.5.2]

16. $4x^2 + 3x - 8 + \dfrac{50}{x + 6}$ [5.5.2/5.5.3] **17.** $x^3 + 4x^2 + 16x + 64 + \dfrac{252}{x - 4}$ [5.5.2/5.5.3]
18. $12x^5y^3 + 8x^3y^2 - 28x^2y^4$ [5.3.1] **19.** $a^{3n+3} - 5a^{2n+4} + 2a^{2n+3}$ [5.3.1] **20.** $9x^2 + 8x$ [5.3.1]

21. $x^{3n+1} - 3x^{2n} - x^{n+2} + 3x$ [5.3.3] **22.** $x^4 + 3x^3 - 23x^2 - 29x + 6$ [5.3.2] **23.** $6x^3 - 29x^2 + 14x + 24$ [5.3.2] **24.** $25a^2 - 4b^2$ [5.3.4] **25.** $16x^2 - 24xy + 9y^2$ [5.3.4] **26.** The distance from Earth to the Great Galaxy of Andromeda is 1.137048×10^{23} mi. [5.4.3] **27.** The sun generates 1.09×10^{21} horsepower. [5.4.3] **28.** The area is $(10x^2 - 29x - 21)$ cm². [5.3.5] **29.** The volume is $(27x^3 - 27x^2 + 9x - 1)$ ft³. [5.3.5] **30.** The area is $(5x^2 + 8x - 8)$ in². [5.3.5]

CHAPTER 5 TEST

1. $2x^3 - 4x^2 + 6x - 14$ [5.1.2] **2.** $64a^7b^7$ [5.2.1] **3.** $\dfrac{2b^7}{a^{10}}$ [5.4.1] **4.** 5.01×10^{-7} [5.4.2]

5. 6.048×10^5 [5.4.2] **6.** $\dfrac{x^{12}}{16y^4}$ [5.4.1] **7.** $35x^2 - 55x$ [5.3.1] **8.** $6a^2 - 13ab - 28b^2$ [5.3.3]

9. $6t^5 - 8t^4 - 15t^3 + 22t^2 - 5$ [5.3.2] **10.** $9z^2 - 30z + 25$ [5.3.4] **11.** $2x^2 + 3x + 5$ [5.5.2]

12. $x^2 - 2x - 1 + \dfrac{2}{x-3}$ [5.5.2/5.5.3] **13.** -3 [5.1.1] **14.** -8 [5.5.4] **15.** $3x^3 + 6x^2 - 8x + 3$ [5.1.2] **16.** $4x^3 - 6x^2$ [5.3.1] **17.** $-4x^6$ [5.4.1] **18.** $-5a^3 + 3a^2 - 4a + 3$ [5.1.2]

19. $4x^4 - 2x^2 + 5$ [5.5.1] **20.** $-6x^3y^6$ [5.2.1] **21.** $\dfrac{9}{x^3}$ [5.4.1] **22.** $10x^2 - 43xy + 28y^2$ [5.3.3]

23. $\dfrac{a^4}{b^6}$ [5.4.1] **24.** The area of the rectangle is $(10x^2 - 3x - 1)$ ft². [5.3.5] **25.** It takes 12 h for the space vehicle to reach the moon. [5.4.3]

CUMULATIVE REVIEW EXERCISES

1. 4 [1.4.2] **2.** $-\dfrac{5}{4}$ [1.5.1] **3.** Inverse Property of Addition [1.5.2] **4.** $-18x + 8$ [1.5.6]

5. $-\dfrac{1}{6}$ [2.1.2] **6.** $-\dfrac{11}{4}$ [2.2.2] **7.** $x^2 + 3x + 9 + \dfrac{24}{x-3}$ [5.5.3] **8.** -1 and $\dfrac{7}{3}$ [2.8.1]

9. 18 [3.2.1] **10.** -2 [3.2.1] **11.** $\{-4, -1, 8\}$ [3.2.1] **12.** $-\dfrac{1}{6}$ [3.4.1]

13. $y = -\dfrac{3}{2}x + \dfrac{1}{2}$ [3.5.1] **14.** $y = \dfrac{2}{3}x + \dfrac{16}{3}$ [3.6.1] **15.** $\left(-\dfrac{7}{5}, -\dfrac{8}{5}\right)$ [4.3.2]

16. $\left(-\dfrac{9}{7}, \dfrac{2}{7}, \dfrac{11}{7}\right)$ [4.2.2] **17.** [3.3.2] **18.** [3.7.1]

19. $(1, -1)$ [4.1.1] **20.** [4.5.1] **21.** $\dfrac{b^5}{a^8}$ [5.4.1] **22.** $\dfrac{y^2}{25x^6}$ [5.4.1]

23. $\dfrac{21}{8}$ [5.4.1] **24.** $4x^3 - 7x + 3$ [5.3.2] **25.** $4x + 8 + \dfrac{21}{2x-3}$ [5.5.2] **26.** a^3b^7 [5.2.1]

27. $a^2 + ab - 12b^2$ [5.3.3] **28.** 0.000000035 [5.4.2] **29.** The integers are 9 and 15. [2.3.2]

30. 40 oz of pure gold must be mixed with the alloy. [2.3.3] **31.** The slower cyclist travels at 5 mph, the faster cyclist at 7.5 mph. [2.3.4] **32.** The additional investment is $4500. [2.6.1] **33.** $m = 50$; a slope of 50 means the average speed was 50 mph. [3.4.1]

Answers to Chapter 6 Selected Exercises

PREP TEST

1. $2 \cdot 3 \cdot 5$ [1.2.3] **2.** $-12y + 15$ [1.5.5] **3.** $-a + b$ [1.5.5] **4.** $-3a + 3b$ [1.5.6]

5. 0 [2.1.3] **6.** $-\dfrac{1}{2}$ [2.2.1] **7.** $x^2 - 2x - 24$ [5.3.3] **8.** $6x^2 - 11x - 10$ [5.3.3]

9. x^3 [5.4.1] **10.** $3x^3y$ [5.4.1]

CONCEPT REVIEW 6.1

1. Always true **3.** Always true **5.** Never true **7.** Never true

SECTION 6.1

3. a. z **b.** 3 **c.** x **d.** y^2 **e.** $3xy^2$ **5.** x^3 **7.** xy^4 **9.** xy^4z^2 **11.** $7a^3$
13. 1 **15.** $3a^2b^2$ **17.** ab **19.** $2x$ **21.** $3x$ **23.** $5(a + 1)$ **25.** $8(2 - a^2)$
27. $4(2x + 3)$ **29.** $6(5a - 1)$ **31.** $x(7x - 3)$ **33.** $a^2(3 + 5a^3)$ **35.** $y(14y + 11)$
37. $2x(x^3 - 2)$ **39.** $2x^2(5x^2 - 6)$ **41.** $xy(x - y^2)$ **43.** $2a^5b + 3xy^3$ **45.** $6b^2(a^2b - 2)$
47. $2abc(3a + 2b)$ **49.** $9(2x^2y^2 - a^2b^2)$ **51.** $6x^3y^3(1 - 2x^3y^3)$ **53.** $x(x^2 - 3x - 1)$
55. $2(x^2 + 4x - 6)$ **57.** $b(b^2 - 5b - 7)$ **59.** $4(2y^2 - 3y + 8)$ **61.** $5y(y^2 - 4y + 2)$
63. $3y^2(y^2 - 3y - 2)$ **65.** $3y(y^2 - 3y + 8)$ **67.** $a^2(6a^3 - 3a - 2)$ **69.** $x^2(8y^2 - 4y + 1)$
71. $x^3y^3(4x^2y^2 - 8xy + 1)$ **73.** x^a **75. a.** $4(3x - 2); 3x - 2$ **b.** $5x; 4$ **77.** $(a + b)(x + 2)$
79. $(b + 2)(x - y)$ **81.** $(y - 4)(a - b)$ **83.** $a(x - 2) + b(x - 2)$ **85.** $b(a - 7) - 3(a - 7)$
87. $x(a - 2b) + y(a - 2b)$ **89.** $(x - 2)(a - 5)$ **91.** $(y - 3)(b - 3)$ **93.** $(x - y)(a + 2)$
95. $(c + 5)(z - 8)$ **97.** $(3x - 4)(w + 1)$ **99.** $(x + 4)(x^2 + 3)$ **101.** $(y + 2)(2y^2 + 3)$
103. $(a + 3)(b - 2)$ **105.** $(a - 2)(x^2 - 3)$ **107.** $(a - b)(3x - 2y)$ **109.** $(x - 3)(x + 4a)$
111. $(x - 5)(y - 2)$ **113.** $(7x + 2y)(3x - 7)$ **115.** $(2r + a)(a - 1)$ **117.** $(4x + 3y)(x - 4)$
119. $(2y - 3)(5xy + 3)$ **121.** II **123. a.** $(x + 3)(2x + 5)$ **b.** $(2x + 5)(x + 3)$
125. a. $(a - b)(2a - 3b)$ **b.** $(2a - 3b)(a - b)$ **127.** 28 **129.** P doubles

CONCEPT REVIEW 6.2

1. Never true **3.** Always true **5.** Never true

SECTION 6.2

1.

Factors of 18	Sum
1, 18	19
2, 9	11
3, 6	9

3.

Factors of −21	Sum
1, −21	−20
−1, 21	20
3, −7	−4
−3, 7	4

5.

Factors of −28	Sum
1, −28	−27
−1, 28	27
2, −14	−12
−2, 14	12
4, −7	−3
−4, 7	3

7. $(x + 1)(x + 2)$ **9.** $(x + 1)(x - 2)$ **11.** $(a + 4)(a - 3)$ **13.** $(a - 1)(a - 2)$ **15.** $(a + 2)(a - 1)$
17. $(b - 3)(b - 3)$ **19.** $(b + 8)(b - 1)$ **21.** $(y + 11)(y - 5)$ **23.** $(y - 2)(y - 3)$
25. $(z - 5)(z - 9)$ **27.** $(z + 8)(z - 20)$ **29.** $(p + 3)(p + 9)$ **31.** $(x + 10)(x + 10)$
33. $(b + 4)(b + 5)$ **35.** $(x + 3)(x - 14)$ **37.** $(b + 4)(b - 5)$ **39.** $(y + 3)(y - 17)$
41. $(p + 3)(p - 7)$ **43.** Nonfactorable over the integers **45.** $(x - 5)(x - 15)$ **47.** $(x - 7)(x - 8)$

49. $(x + 8)(x - 7)$ **51.** $(a + 3)(a - 24)$ **53.** $(a - 3)(a - 12)$ **55.** $(z + 8)(z - 17)$
57. $(c + 9)(c - 10)$ **59. a.** Positive **b.** Positive **61. a.** common **b.** 5
c. $x^2 - 2x - 8$ **d.** $x + 2; x - 4$ **63.** $2(x + 1)(x + 2)$ **65.** $3(a + 3)(a - 2)$ **67.** $a(b + 5)(b - 3)$
69. $x(y - 2)(y - 3)$ **71.** $z(z - 3)(z - 4)$ **73.** $3y(y - 2)(y - 3)$ **75.** $3(x + 4)(x - 3)$
77. $5(z + 4)(z - 7)$ **79.** $2a(a + 8)(a - 4)$ **81.** $(x - 2y)(x - 3y)$ **83.** $(a - 4b)(a - 5b)$
85. $(x + 4y)(x - 7y)$ **87.** Nonfactorable over the integers **89.** $z^2(z - 5)(z - 7)$
91. $b^2(b - 10)(b - 12)$ **93.** $2y^2(y + 3)(y - 16)$ **95.** $x^2(x + 8)(x - 1)$ **97.** $4y(x + 7)(x - 2)$
99. $8(y - 1)(y - 3)$ **101.** $c(c + 3)(c + 10)$ **103.** $3x(x - 3)(x - 9)$ **105.** $(x - 3y)(x - 5y)$
107. $(a - 6b)(a - 7b)$ **109.** $(y + z)(y + 7z)$ **111.** $3y(x + 21)(x - 1)$ **113.** $3x(x + 4)(x - 3)$
115. $4z(z + 11)(z - 3)$ **117. a.** Yes **b.** Yes **119.** $(c + 4)(c + 5)$ **121.** $a^2(b - 5)(b - 9)$
123. $36, 12, -12, -36$ **125.** $22, 10, -10, -22$ **127.** $3, 4$ **129.** $5, 8, 9$ **131.** 2

CONCEPT REVIEW 6.3

1. Never true **3.** Sometimes true **5.** Always true

SECTION 6.3

1. a. $2; -15; 15; -5; 5$ **b.** $-29x; 29x; -7x; 7x$ **c.** $x - 5; 2x + 3$ **3.** $(x + 1)(2x + 1)$
5. $(y + 3)(2y + 1)$ **7.** $(a - 1)(2a - 1)$ **9.** $(b - 5)(2b - 1)$ **11.** $(x + 1)(2x - 1)$
13. $(x - 3)(2x + 1)$ **15.** Nonfactorable over the integers **17.** $(2t - 1)(3t - 4)$ **19.** $(x + 4)(8x + 1)$
21. $(b - 4)(3b - 4)$ **23.** $(z - 14)(2z + 1)$ **25.** $(p + 8)(3p - 2)$ **27.** $(2x - 3)(3x - 4)$
29. $(b + 7)(5b - 2)$ **31.** $(2a - 3)(3a + 8)$ **33.** $(3t + 1)(6t - 5)$ **35.** $(3a + 7)(5a - 3)$
37. $(2y - 5)(4y - 3)$ **39.** $(2z + 3)(4z - 5)$ **41.** $(x + 5)(3x - 1)$ **43.** $(3x + 4)(4x + 3)$
45. Nonfactorable over the integers **47.** $(x + y)(3x - 2y)$ **49.** $(4 + z)(7 - z)$ **51.** $(1 - x)(8 + x)$
53. $3(x + 5)(3x - 4)$ **55.** $4(2x - 3)(3x - 2)$ **57.** $a^2(5a + 2)(7a - 1)$ **59.** $5(b - 7)(3b - 2)$
61. $2x(x + 1)(5x + 1)$ **63.** $2y(y - 4)(5y - 2)$ **65.** $yz(z + 2)(4z - 3)$ **67.** $b^2(4b + 5)(5b + 4)$
69. $xy(3x + 2)(3x + 2)$ **71.** m and p **73.** $-18; -17; 18; 17; -9; -7; 9; 7; -6; -3; 6; 3$
75. $(t + 2)(2t - 5)$ **77.** $(p - 5)(3p - 1)$ **79.** $(3y - 1)(4y - 1)$ **81.** Nonfactorable over the integers
83. $(3y + 1)(4y + 5)$ **85.** $(a + 7)(7a - 2)$ **87.** $(z + 2)(4z + 3)$ **89.** $(2p + 5)(11p - 2)$
91. $(y + 1)(8y + 9)$ **93.** $(2b - 3)(3b - 2)$ **95.** $(3b + 5)(11b - 7)$ **97.** $(3y - 4)(6y - 5)$
99. Nonfactorable over the integers **101.** $(2z - 5)(5z - 2)$ **103.** $(6z + 5)(6z + 7)$
105. $(2y - 3)(7y - 4)$ **107.** $(x + 6)(6x - 1)$ **109.** $3(x + 3)(4x - 1)$ **111.** $10(y + 1)(3y - 2)$
113. $x(x + 1)(2x - 5)$ **115.** $(a - 3b)(2a - 3b)$ **117.** $(y + z)(2y + 5z)$ **119.** $(1 + x)(18 - x)$
121. $4(9y + 1)(10y - 1)$ **123.** $8(t + 4)(2t - 3)$ **125.** $p(2p + 1)(3p + 1)$ **127.** $3(2z - 5)(5z - 2)$
129. $3a(2a + 3)(7a - 3)$ **131.** $y(3x - 5y)(3x - 5y)$ **133.** $xy(3x - 4y)(3x - 4y)$ **135.** Two positive
137. Two negative **139.** $2y(y - 3)(4y - 1)$ **141.** $ab(a + 4)(a - 6)$ **143.** $5t(3 + 2t)(4 - t)$
145. $(y + 3)(2y + 1)$ **147.** $(2x - 1)(5x + 7)$ **149.** $7, -7, 5, -5$ **151.** $7, -7, 5, -5$
153. $11, -11, 7, -7$ **155.** $(x + 2)(x - 3)(x - 1)$

CONCEPT REVIEW 6.4

1. Never true **3.** Never true **5.** Never true

SECTION 6.4

1. $4; 25x^6; 100x^4y^4$ **3.** $4z^4$ **5.** $9a^2b^3$ **7. a.** $4x; 1$ **b.** $4x + 1; 4x - 1$
11. $(x + 4)(x - 4)$ **13.** $(2x + 1)(2x - 1)$ **15.** $(b - 1)^2$ **17.** $(4x - 5)^2$ **19.** $(xy + 10)(xy - 10)$

21. Nonfactorable **23.** $(x + 3y)^2$ **25.** $(2x + y)(2x - y)$ **27.** $(a^n + 1)(a^n - 1)$ **29.** $(a + 2)^2$
31. $(x - 6)^2$ **33.** $(4x + 11)(4x - 11)$ **35.** $(1 + 3a)(1 - 3a)$ **37.** Nonfactorable
39. Nonfactorable **41.** $(5 + ab)(5 - ab)$ **43.** $(5a - 4b)^2$ **45.** $(x^n + 3)^2$ **47.** **b** and **d**
49. $8; x^9; 27c^{15}d^{18}$ **51.** $2x^3$ **53.** $4a^2b^6$ **55.** **a.** $5x; 2$ **b.** $5x + 2; 25x^2 - 10x + 4$
57. $(x - 3)(x^2 + 3x + 9)$ **59.** $(2x - 1)(4x^2 + 2x + 1)$ **61.** $(x - y)(x^2 + xy + y^2)$
63. $(m + n)(m^2 - mn + n^2)$ **65.** $(4x + 1)(16x^2 - 4x + 1)$ **67.** $(3x - 2y)(9x^2 + 6xy + 4y^2)$
69. $(xy + 4)(x^2y^2 - 4xy + 16)$ **71.** Nonfactorable **73.** Nonfactorable **75.** $(a - 2b)(a^2 - ab + b^2)$
77. $(x^{2n} + y^n)(x^{4n} - x^{2n}y^n + y^{2n})$ **79.** x^2y **83.** $(xy - 3)(xy - 5)$ **85.** $(xy - 5)(xy - 12)$
87. $(x^2 - 6)(x^2 - 3)$ **89.** $(b^2 - 18)(b^2 + 5)$ **91.** $(x^2y^2 - 6)(x^2y^2 - 2)$ **93.** $(x^n + 2)(x^n + 1)$
95. $(3xy - 5)(xy - 3)$ **97.** $(2ab - 3)(3ab - 7)$ **99.** $(2x^2 - 15)(x^2 + 1)$ **101.** $(2x^n - 1)(x^n - 3)$
103. $(2a^n + 5)(3a^n + 2)$ **105.** grouping **107.** $3(2x - 3)^2$ **109.** $a(3a - 1)(9a^2 + 3a + 1)$
111. $5(2x + 1)(2x - 1)$ **113.** $y^3(y + 11)(y - 5)$ **115.** $(4x^2 + 9)(2x + 3)(2x - 3)$
117. $2a(2 - a)(4 + 2a + a^2)$ **119.** $(x + 2)(x + 1)(x - 1)$ **121.** $(x + 2)(2x^2 - 3)$
123. $(x + 1)(x + 4)(x - 4)$ **125.** $b^3(ab - 1)(a^2b^2 + ab + 1)$ **127.** $x^2(x - 7)(x + 5)$
129. Nonfactorable **131.** $2x^3(3x + 1)(x + 12)$ **133.** $(4a^2 + b^2)(2a + b)(2a - b)$ **135.** Nonfactorable
137. $3b^2(b - 2)(b^2 + 2b + 4)$ **139.** $x^2y^2(x - 3y)(x - 2y)$ **141.** $2xy(4x + 7y)(2x - 3y)$
143. $(x - 2)(x + 1)(x - 1)$ **145.** $4(b - 1)(2x - 1)$ **147.** $(y + 1)(y - 1)(2x + 3)(2x - 3)$
149. $(x + 2)(x - 2)^2(x^2 + 2x + 4)$ **151.** $a^2(a^n - 3)^2$ **153.** $x^n(2x - 1)(x - 3)$ **155.** $n + 2$
157. $20, -20$ **159.** $8, -8$ **161.** $(x - 1)(x^2 + x + 1)(a - b)$ **163.** $(y^{2n} + 1)^2(y^n - 1)^2(y^n + 1)^2$
165. The sum of the numbers is 16. **167.** The palindromic perfect squares less than 500 are 1, 4, 9, 121, and 484.
169. $(x^2 - 4x + 8)(x^2 + 4x + 8)$

CONCEPT REVIEW 6.5

1. Always true **3.** Sometimes true **5.** Sometimes true

SECTION 6.5

3. $0 = 5x^2 + 2x - 3$ **5.** 5 **7.** -4 and -6 **9.** 0 and 7 **11.** 0 and $-\dfrac{5}{2}$ **13.** $-\dfrac{3}{2}$ and 7
15. 7 and -7 **17.** $\dfrac{4}{3}$ and $-\dfrac{4}{3}$ **19.** -5 and 1 **21.** $-\dfrac{3}{2}$ and 4 **23.** 0 and 9 **25.** 0 and 4
27. 7 and -4 **29.** $\dfrac{2}{3}$ and -5 **31.** $\dfrac{2}{5}$ and 3 **33.** $\dfrac{3}{2}$ and $-\dfrac{1}{4}$ **35.** 7 and -5 **37.** 9 and 3
39. $-\dfrac{4}{3}$ and 2 **41.** $\dfrac{4}{3}$ and -5 **43.** 5 and -3 **45.** -11 and -4 **47.** 4 and 2 **49.** 2 and 1
51. 2 and 3 **53.** $-4, -1$, and 1 **55.** 1 and 2 **57.** $-\dfrac{1}{2}$ and 1 **59.** $\dfrac{1}{2}$ **61.** -4 and 1
63. -2 and 6 **65.** 1 and $\dfrac{3}{2}$ **67.** -3 **69.** $x + 1; x^2 + (x + 1)^2$ **71.** The number is -10 or 9.
73. The integers are 8 and 9. **75.** The number is 0, 3, or 4. **77.** The width is 8 in. The length is 21 in.
79. The value of x is 2 m or 3 m. **81.** The velocity of the rocket is 100 m/s. **83.** The height is 4 cm.
The base is 12 cm. **85.** The object will hit the ground in 5 s. **87.** The length of the larger rectangle
is 11 cm. The width is 8 cm. **89.** 175 or 15 **91.** -16 or 4 **93.** $-3, -2$, and 2 **95.** The length is
18 in. The width is 8 in. **97.** For example, $x^3 - 4x^2 + x + 6 = 0$

CHAPTER 6 REVIEW EXERCISES

1. $7y^3(2y^6 - 7y^3 + 1)$ [6.1.1] **2.** $(a - 4)(3a + b)$ [6.1.2] **3.** $(c + 2)(c + 6)$ [6.2.1]

4. $a(a - 2)(a - 3)$ [6.2.2] **5.** $(2x - 7)(3x - 4)$ [6.3.1/6.3.2] **6.** $(y + 6)(3y - 2)$ [6.3.1/6.3.2]

7. $(3a + 2)(6a - 5)$ [6.3.1/6.3.2] **8.** $(ab + 1)(ab - 1)$ [6.4.1] **9.** $4(y - 2)^2$ [6.4.4]

10. $0, -\dfrac{1}{5}$ [6.5.1] **11.** $3ab(4a + b)$ [6.1.1] **12.** $(b - 3)(b - 10)$ [6.2.1]

13. $(2x + 5)(5x + 2y)$ [6.1.2] **14.** $3(a + 2)(a - 7)$ [6.2.2] **15.** $n^2(n + 1)(n - 3)$ [6.2.2]

16. Nonfactorable over the integers [6.3.1/6.3.2] **17.** $(2x - 1)(3x - 2)$ [6.3.1/6.3.2]

18. Nonfactorable over the integers [6.4.1] **19.** $2, \dfrac{3}{2}$ [6.5.1] **20.** $7(x + 1)(x - 1)$ [6.4.4]

21. $x^3(3x^2 - 9x - 4)$ [6.1.1] **22.** $(x - 3)(4x + 5)$ [6.1.2] **23.** $(a + 7)(a - 2)$ [6.2.1]

24. $(y + 9)(y - 4)$ [6.2.1] **25.** $5(x - 12)(x + 2)$ [6.2.2] **26.** $-3, 7$ [6.5.1]

27. $x^2(5x^{n+3} + x^{n+1} + 4)$ [6.1.1] **28.** $(y - 3)(x - 4)$ [6.1.2] **29.** $(a + 2b)(2x - 3y)$ [6.1.2]

30. $(x + 5)(x + 7)$ [6.2.1] **31.** $(3 + x)(4 - x)$ [6.3.1] **32.** -1 and 6 [6.5.1]

33. $(3x - 2)(2x - 9)$ [6.3.1/6.3.2] **34.** $(8x - 1)(3x + 8)$ [6.3.1/6.3.2] **35.** $(xy + 3)(xy - 3)$ [6.4.1]

36. $(2x + 3y)^2$ [6.4.1] **37.** $(x^n - 6)^2$ [6.4.1] **38.** $(6 + a^n)(6 - a^n)$ [6.4.1]

39. $(4a - 3b)(16a^2 + 12ab + 9b^2)$ [6.4.2] **40.** $(2 - y^n)(4 + 2y^n + y^{2n})$ [6.4.2]

41. $(3x^2 + 2)(5x^2 - 3)$ [6.4.3] **42.** $(6x^4 - 5)(6x^4 - 1)$ [6.4.3]

43. $(3x^2y^2 + 2)(7x^2y^2 + 3)$ [6.4.3] **44.** $3a^2(a^2 - 6)(a^2 + 1)$ [6.4.4] **45.** $(x^n + 2)^2(x^n - 2)^2$ [6.4.4]

46. $3ab(a - b)(a^2 + ab + b^2)$ [6.4.4] **47.** $-2, 0,$ and 3 [6.5.1] **48.** $\dfrac{5}{2}$ and 4 [6.5.1] **49.** $-4, 0,$

and 4 [6.5.1] **50.** $-1, -6,$ and 6 [6.5.1] **51.** $(2x - 11)(8x - 3)$ [6.3.1/6.3.2]

52. $(3x - 5)^2$ [6.4.1] **53.** $(2y + 3)(6y - 1)$ [6.3.1/6.3.2] **54.** $3(x + 6)^2$ [6.4.4] **55.** The length is

100 yd. The width is 50 yd. [6.5.2] **56.** The length is 100 yd. The width is 60 yd. [6.5.2] **57.** The two

integers are 4 and 5. [6.5.2] **58.** The distance is 20 ft. [6.5.2] **59.** The width of the larger rectangle is

15 ft. [6.5.2] **60.** The length of a side of the original square was 20 ft. [6.5.2]

CHAPTER 6 TEST

1. $3y^2(2x^2 + 3x + 4)$ [6.1.1] **2.** $2x(3x^2 - 4x + 5)$ [6.1.1] **3.** $(p + 2)(p + 3)$ [6.2.1]

4. $(x - 2)(a - b)$ [6.1.2] **5.** $\dfrac{3}{2}, -7$ [6.5.1] **6.** $(a - 3)(a - 16)$ [6.2.1] **7.** $x(x + 5)(x - 3)$ [6.2.2]

8. $4(x + 4)(2x - 3)$ [6.3.1/6.3.2] **9.** $(b + 6)(a - 3)$ [6.1.2] **10.** $\dfrac{1}{2}, -\dfrac{1}{2}$ [6.5.1]

11. $(2x + 1)(3x + 8)$ [6.3.1/6.3.2] **12.** $(2a^2 - 5)(3a^2 + 1)$ [6.4.3] **13.** $2(b + 4)(b - 4)$ [6.4.4]

14. $(2a - 3b)^2$ [6.4.1] **15.** $-\dfrac{3}{2}$ and 4 [6.5.1] **16.** $(3x - 2)(9x^2 + 6x + 4)$ [6.4.2]

17. Nonfactorable over the integers [6.3.1/6.3.2] **18.** $(2x + 7y)(2x - 7y)$ [6.4.1] **19.** $3, 5$ [6.5.1]

20. $(p + 6)^2$ [6.4.1] **21.** $2(3x - 4y)^2$ [6.4.4] **22.** $2y^2(y + 1)(y - 8)$ [6.2.2]

23. The length is 15 cm. The width is 6 cm. [6.5.2] **24.** The length of the base is 12 in. [6.5.2]

25. The two integers are -13 and -12. [6.5.2]

CUMULATIVE REVIEW EXERCISES

1. -8 [1.2.2] **2.** -8.1 [1.3.2] **3.** 4 [1.4.2] **4.** -7 [1.5.1]

5. The Associative Property of Addition [1.5.2] **6.** $18x^2$ [1.5.4] **7.** $-6x + 24$ [1.5.6]

8. $\dfrac{2}{3}$ [2.1.3] **9.** 5 [2.2.3] **10.** $\dfrac{7}{4}$ [2.2.2] **11.** 3 [2.2.3] **12.** 35 [2.1.4]

13. $x \le -3$ [2.7.2] **14.** $x < \dfrac{1}{5}$ [2.7.2] **15.** [3.3.1] **16.** [3.3.1]

17. Domain: $\{-5, -3, -1, 1, 3\}$; range: $\{-4, -2, 0, 2, 4\}$; yes [3.2.1] **18.** 61 [3.2.1]

19. [3.7.1] **20.** $(2, -5)$ [4.1.2] **21.** $(1, 2)$ [4.2.1] **22.** $3y^3 - 3y^2 - 8y - 5$ [5.1.2]

23. $-27a^{12}b^6$ [5.2.1] **24.** $x^3 - 3x^2 - 6x + 8$ [5.3.2] **25.** $4x + 8 + \dfrac{21}{2x - 3}$ [5.5.2]

26. $\dfrac{y^6}{x^{12}}$ [5.4.1] **27.** $(a - b)(3 - x)$ [6.1.2] **28.** $(x + 5y)(x - 2y)$ [6.2.1]

29. $2a^2(3a + 2)(a + 3)$ [6.4.4] **30.** $(5a + 6b)(5a - 6b)$ [6.4.1] **31.** $3(2x - 3y)^2$ [6.4.4]

32. $-5, \dfrac{4}{3}$ [6.5.1] **33.** $\{-11, -7, -3, 1, 5\}$ [3.2.1] **34.** The average high temperature for the week was $-3°$C. [1.2.5] **35.** The length of the rectangle is 15 cm. The width is 6 cm. [2.4.1] **36.** The pieces measure 4 ft and 6 ft. [2.2.4] **37.** You can drive a Company C car a maximum of 359 mi. [2.7.4] **38.** An additional \$6500 must be invested at an annual simple interest rate of 11%. [2.6.1] **39.** The discount rate is 40%. [2.5.2] **40.** The integers are 10, 12, and 14. [2.3.2]

Answers to Chapter 7 Selected Exercises

PREP TEST

1. 50 [1.3.3] **2.** $-\dfrac{1}{6}$ [1.3.2] **3.** $-\dfrac{3}{2}$ [1.3.2] **4.** $\dfrac{1}{24}$ [1.3.3] **5.** $\dfrac{5}{24}$ [1.3.3]

6. $-\dfrac{2}{9}$ [1.4.2] **7.** $\dfrac{1}{3}$ [1.5.1] **8.** -2 [2.2.3] **9.** $\dfrac{10}{7}$ [2.2.3] **10.** 110 mph, 130 mph [2.3.4]

CONCEPT REVIEW 7.1

1. Never true **3.** Always true **5.** Never true

SECTION 7.1

3. $-3; -3; -1; 6; -\dfrac{1}{7}$ **5.** $f(4) = 2$ **7.** $f(-2) = -2$ **9.** $f(-2) = \dfrac{1}{9}$ **11.** $f(3) = \dfrac{1}{35}$

13. $f(-1) = \dfrac{3}{4}$ **15.** $\{x \mid x \ne 3\}$ **17.** $\{x \mid x \ne -4\}$ **19.** $\{x \mid x \ne -3\}$ **21.** $\{x \mid x \ne -2, 4\}$

23. $\left\{x \mid x \ne -\dfrac{5}{2}, 2\right\}$ **25.** $\{x \mid x \ne 0\}$ **27.** $\{x \mid x \in \text{real numbers}\}$ **29.** $\{x \mid x \ne -3, 2\}$

31. $\{x \mid x \ne -6, 4\}$ **33.** $\{x \mid x \in \text{real numbers}\}$ **35.** $\left\{x \mid x \ne \dfrac{2}{3}, \dfrac{3}{2}\right\}$ **37.** $\left\{x \mid x \ne 0, \dfrac{1}{2}, -5\right\}$

39. Below **43.** factored; common **45.** $1 - 2x$ **47.** $3x - 1$ **49.** $2x$ **51.** $-\dfrac{2}{x}$

53. $\dfrac{x}{2}$ **55.** The expression is in simplest form. **57.** $4x^2 - 2x + 3$ **59.** $a^2 + 2a - 3$

61. $\dfrac{x^n - 3}{4}$ **63.** $\dfrac{x - 3}{x - 5}$ **65.** $\dfrac{x + y}{x - y}$ **67.** $-\dfrac{x + 3}{3x - 4}$ **69.** $-\dfrac{x + 7}{x - 7}$ **71.** The expression is in

simplest form. **73.** $\dfrac{a - b}{a^2 - ab + b^2}$ **75.** $\dfrac{x + y}{3x}$ **77.** $\dfrac{x + 2y}{3x + 4y}$ **79.** $-\dfrac{2x - 3}{2(2x + 3)}$ **81.** $\dfrac{x - 2}{a - b}$

83. $\dfrac{x^2 + 2}{(x + 1)(x - 1)}$ **85.** $\dfrac{xy + 7}{xy - 7}$ **87.** $\dfrac{a^n - 2}{a^n + 2}$ **89.** False **91.** decrease

93. **a.**

b. The ordered pair (2000, 51) means that when the distance between the object and the lens is 2000 m, the distance

between the lens and the film is 51 mm. **c.** For $x = 50$, the expression $\dfrac{50x}{x - 50}$ is undefined. For $0 < x < 50$, $f(x)$

is negative, and distance cannot be negative. Therefore, the domain is $x > 50$. **d.** For $x > 1000$, $f(x)$ changes
very little for large changes in x.

CONCEPT REVIEW 7.2

1. Always true **3.** Never true

SECTION 7.2

3. $\dfrac{15b^4xy}{4}$ **5.** 1 **7.** $\dfrac{2x - 3}{x^2(x + 1)}$ **9.** $-\dfrac{x - 1}{x - 5}$ **11.** $\dfrac{2(x + 2)}{x - 1}$ **13.** $-\dfrac{x - 3}{2x + 3}$ **15.** $\dfrac{x^n - 6}{x^n - 1}$

17. $x + y$ **19.** $\dfrac{a^2y}{10x}$ **21.** $\dfrac{3}{2}$ **23.** $\dfrac{5(x - y)}{xy(x + y)}$ **25.** $\dfrac{(x - 3)^2}{(x + 3)(x - 4)}$ **27.** $-\dfrac{x + 2}{x + 5}$

29. $\dfrac{x^{2n}}{8}$ **31.** $\dfrac{(x - 3)(3x - 5)}{(2x + 5)(x + 4)}$ **33.** -1 **35.** 1 **37.** $-\dfrac{2(x^2 + x + 1)}{(x + 1)^2}$ **39.** c

41. $(x - 3)^2$; $(x + 3)(x - 3)$; two; one; $(x - 3)^2(x + 3)$ **43.** **a.** Two **b.** Zero **c.** One

45. $\dfrac{9y^3}{12x^2y^4}, \dfrac{17x}{12x^2y^4}$ **47.** $\dfrac{6x^2 + 9x}{(2x - 3)(2x + 3)}, \dfrac{10x^2 - 15x}{(2x - 3)(2x + 3)}$ **49.** $\dfrac{2x}{(x + 3)(x - 3)}, \dfrac{x^2 + 4x + 3}{(x + 3)(x - 3)}$

51. $\dfrac{3x^2 - 3x}{(x + 1)(x - 1)^2}, \dfrac{5x^2 + 5x}{(x + 1)(x - 1)^2}$ **53.** $(x - 4)(x + 3)$; $5x + 15$; $2x - 8$; $5x + 15$; $2x - 8$; $2x$; 8; $3x$; 23

55. $-\dfrac{13}{2xy}$ **57.** $\dfrac{1}{x - 1}$ **59.** $\dfrac{15 - 16xy - 9x}{10x^2y}$ **61.** $-\dfrac{5y + 21}{30xy}$ **63.** $-\dfrac{6x + 19}{36x}$

65. $\dfrac{10x + 6y + 7xy}{12x^2y^2}$ **67.** $-\dfrac{x^2 + x}{(x - 3)(x - 5)}$ **69.** -1 **71.** $\dfrac{5x - 8}{(x + 5)(x - 5)}$ **73.** $-\dfrac{3x^2 - 4x + 8}{x(x - 4)}$

75. $\dfrac{4x^2 - 14x + 15}{2x(2x - 3)}$ **77.** $\dfrac{2x^2 + x + 3}{(x + 1)^2(x - 1)}$ **79.** $\dfrac{x + 1}{x + 3}$ **81.** $\dfrac{x + 2}{x + 3}$ **83.** $-\dfrac{x^n - 2}{(x^n + 1)(x^n - 1)}$

85. $\dfrac{12x + 13}{(2x + 3)(2x - 3)}$ **87.** $\dfrac{2x + 3}{2x - 3}$ **89.** $\dfrac{2x + 5}{(x + 3)(x - 2)(x + 1)}$ **91.** $\dfrac{3x^2 + 4x + 3}{(2x + 3)(x + 4)(x - 3)}$

93. $\dfrac{x - 2}{x - 3}$ **95.** $-\dfrac{2x^2 - 9x - 9}{(x + 6)(x - 3)}$ **97.** $\dfrac{3x^2 - 14x + 24}{(3x - 2)(2x + 5)}$ **99.** $\dfrac{2}{x^2 + 1}$ **101.** $-x - 1$

103. $\dfrac{y - 2}{x^4}$ **105.** $\dfrac{4y}{(y - 1)^2}$ **107.** $\dfrac{5(x + 4)(x - 4)}{2(x + 2)(x - 2)(x - 3)}$ **109.** $-\dfrac{(a + 3)(a - 2)(5a - 2)}{(3a + 1)(a + 2)(2a - 3)}$

111. $\dfrac{6(x - 1)}{x(2x + 1)}$ **113.** $\dfrac{8}{15} \neq \dfrac{1}{8}$ **115.** $\dfrac{3}{y} + \dfrac{6}{x}$ **117.** $\dfrac{4}{b^2} + \dfrac{3}{ab}$ **119** **a.** $f(x) = \dfrac{2x^3 + 528}{x}$

b.

c. When the height of the box is 4 in., 164 in² of cardboard is needed.

d. The height of the box is 5.1 in.

e. The minimum amount of cardboard is 155.5 in².

CONCEPT REVIEW 7.3

1. Sometimes true **3.** Always true

SECTION 7.3

3. $\dfrac{2}{a}; \dfrac{1}{a^2}; a^2$ **5.** $a^2 + 1$ **7.** $\dfrac{5}{23}$ **9.** $\dfrac{2}{5}$ **11.** $\dfrac{1-y}{y}$ **13.** $\dfrac{5-a}{a}$ **15.** $\dfrac{b}{2(2-b)}$

17. $\dfrac{4}{5}$ **19.** $-\dfrac{1}{2}$ **21.** $\dfrac{2a^2 - 3a + 3}{3a - 2}$ **23.** $3x + 2$ **25.** $\dfrac{x+2}{x-1}$ **27.** $-\dfrac{(x-3)(x+5)}{4x^2 - 5x + 4}$

29. $\dfrac{1}{x+4}$ **31.** $\dfrac{x-7}{x+5}$ **33.** $-\dfrac{2(a+1)}{7a - 4}$ **35.** $\dfrac{x+y}{x-y}$ **37.** $-\dfrac{2x}{x^2 + 1}$ **39.** $\dfrac{6(x-2)}{2x - 3}$

41. c **43.** $\dfrac{2y}{x}$ **45.** $x - 1$ **47.** $\dfrac{c-3}{c-2}$ **49.** $\dfrac{2a(2a-3)}{3a - 4}$ **51.** $-\dfrac{1}{x(x+h)}$

53. $\dfrac{3n^2 + 12n + 8}{n(n+2)(n+4)}$ **55. a.** $P(x) = \dfrac{Cx(x+1)^{60}}{(x+1)^{60} - 1}$

b.

c. The interval of annual interest rates is 0% to 30%.

d. The ordered pair (0.005, 386.66) means that when the monthly interest rate on a car loan is 0.5%, the monthly payment on the loan is $386.66.

e. The monthly payment with a loan amount of $20,000 and an annual interest rate of 8% is $406.

CONCEPT REVIEW 7.4

1. Never true **3.** Always true **5.** Always true

SECTION 7.4

1. $x; x + 3; x(x + 3)$ **3.** $8(x + 3); 5x; -8$ **5.** -5 **7.** -1 **9.** $\dfrac{5}{2}$ **11.** 0 **13.** -3 and 3

15. 8 **17.** 5 **19.** -3 **21.** No solution **23.** 4 and 10 **25.** -1 **27.** -3 and 4

29. 2 **31.** $\dfrac{7}{2}$ **33.** No solution **35.** $3; -4$ **39.** Row 1: $\dfrac{1}{6}; t; \dfrac{t}{6}$; Row 2: $\dfrac{1}{4}; t; \dfrac{t}{4}$

41. It would take 1.2 h to process the data with both computers working. **43.** With both panels working, it would take 18 min to raise the temperature 1 degree. **45.** With both members working together, it would take 3 h to wire the telephone lines. **47.** Working alone, the new machine would take 10 h to package the transistors. **49.** Working alone, the smaller printer would take 80 min to print the payroll. **51.** It would take the apprentice 20 h to shingle the roof working alone. **53.** It will take them 40 min to address 140 envelopes. **55.** In 30 min, the bathtub will start to overflow. **57.** With all three pipes open, it would take 30 h to fill the tank. **59.** Less than n **61. a.** Row 1: $1320; 470 - w; \dfrac{1320}{470 - w}$; Row 2: $1500; 470 + w; \dfrac{1500}{470 + w}$

b. $\dfrac{1320}{470-w}$; $\dfrac{1500}{470+w}$ **63.** The rate of the first skater is 15 mph; the rate of the second skater is 12 mph.

65. The rate of the freight train is 45 mph. The rate of the passenger train is 59 mph. **67.** The cyclist was riding at the rate of 12 mph. **69.** The motorist walks at the rate of 4 mph. **71.** The business executive can walk at the rate of 6 ft/s. **73.** The rate of the single-engine plane is 180 mph. The rate of the jet is 720 mph.

75. The rate of the Gulf Stream is 6 mph. **77.** The rate of the current is 3 mph. **79.** The original fraction is $\dfrac{17}{21}$. **81.** With all three printers operating, it would take $5\frac{1}{3}$ min to print the checks.

83. The bus usually travels at 60 mph. **85.** The total time is $\dfrac{5}{8}$ h.

CONCEPT REVIEW 7.5

1. Never true **3.** Never true **5.** Never true

SECTION 7.5

3. 5 mi; 2 cm; 2 cm; 5 mi **5.** 3 **7.** -2 **9.** -4 **11.** $-\dfrac{6}{5}$ **13.** True **15.** There are 4000 ducks in the preserve. **17.** The caterer will need an additional 3 gal of punch. **19.** The dimensions of the room are 17 ft by 22 ft. **21.** To lose 1 pound, a person would need to walk 21.54 mi.

23. An additional 0.5 oz of medication is required. **25.** An additional 34 oz of insecticide is required.

27. It would take 75 s to download a 5-megabyte file. **29.** The length of the whale is 96 ft.

33. Direct variation **35.** The pressure is 13.5 lb/in². **37.** When the company sells 300 products, the profit is \$37,500. **39.** The length of the person's face is 10.2 in. **41.** The horizon is 24.03 mi from a point that is 800 ft high. **43.** The ball will roll 80 ft in 4 s. **45.** When the volume is 150 ft³, the pressure is $66\frac{2}{3}$ lb/in². **47.** A gear that has 30 teeth will make 32 revolutions per minute. **49.** The repulsive force is 112.5 lb when the distance is 1.2 in. **51.** The resistance is 56.25 ohms. **53.** The wind force is 180 lb.

55. a. **b.** A linear function **57. a.** **b.** Yes **59.** x is halved.

61. directly; inversely **63.** directly **65. a.** $\dfrac{2ab}{a+b}$ **b.** 12

CONCEPT REVIEW 7.6

1. Always true **3.** Never true

SECTION 7.6

1. n; n; n; $R-C$; R; $R-nP$ **3.** $W=\dfrac{P-2L}{2}$ **5.** $C=\dfrac{S}{1-r}$ **7.** $R=\dfrac{PV}{nT}$ **9.** $m_2=\dfrac{Fr^2}{Gm_1}$

11. $R=\dfrac{E-Ir}{I}$ **13.** $b_2=\dfrac{2A-hb_1}{h}$ **15.** $R_2=\dfrac{RR_1}{R_1-R}$ **17.** $d=\dfrac{a_n-a_1}{n-1}$ **19.** $H=\dfrac{S-2WL}{2W+2L}$

21. $x=\dfrac{-by-c}{a}$ **23.** $x=\dfrac{d-b}{a-c}$ **25.** $x=\dfrac{ac}{b}$ **27.** $x=\dfrac{ab}{a+b}$ **29.** Yes **31.** $5+2y$

33. $0, y$ **35.** $\dfrac{y}{2}, 5y$ **37.** $f=\dfrac{w_2f_2+w_1f_1}{w_1+w_2}$

CHAPTER 7 REVIEW EXERCISES

1. 4 [7.1.1] **2.** $\dfrac{2}{21}$ [7.1.1] **3.** $\{x \mid x \neq 3\}$ [7.1.1] **4.** $\{x \mid x \neq -3, 2\}$ [7.1.1]

5. $\{x \mid x \in \text{real numbers}\}$ [7.1.1] **6.** $3a^{2n} + 2a^n - 1$ [7.1.2] **7.** $-\dfrac{x+4}{x(x+2)}$ [7.1.2]

8. $\dfrac{x^2 + 3x + 9}{x + 3}$ [7.1.2] **9.** $\dfrac{1}{a-1}$ [7.2.1] **10.** x [7.2.1] **11.** $-\dfrac{(x+4)(x+3)(x+2)}{6(x-1)(x-4)}$ [7.2.1]

12. $\dfrac{x^n + 6}{x^n - 3}$ [7.2.1] **13.** $\dfrac{3x+2}{x}$ [7.2.1] **14.** $-\dfrac{x-3}{x+3}$ [7.2.1] **15.** $\dfrac{21a + 40b}{24a^2 b^4}$ [7.2.2]

16. $\dfrac{4x-1}{x+2}$ [7.2.2] **17.** $\dfrac{3x+26}{(3x-2)(3x+2)}$ [7.2.2] **18.** $\dfrac{9x^2 + x + 4}{(3x-1)(x-2)}$ [7.2.2]

19. $-\dfrac{5x^2 - 17x - 9}{(x-3)(x+2)}$ [7.2.2] **20.** $\dfrac{x-4}{x+5}$ [7.3.1] **21.** $\dfrac{x-1}{4}$ [7.3.1] **22.** $\dfrac{15}{13}$ [7.4.1]

23. -3 and 5 [7.5.1] **24.** 6 [7.4.1] **25.** No solution [7.4.1] **26.** $R = \dfrac{V}{I}$ [7.6.1]

27. $N = \dfrac{S}{1-Q}$ [7.6.1] **28.** $r = \dfrac{S-a}{S}$ [7.6.1] **29.** The number of tanks of fuel is $6\dfrac{2}{3}$. [7.5.1]

30. The number of miles is 48. [7.5.1] **31.** Working alone, the apprentice would take 104 min to install the ceiling fan. [7.4.2] **32.** With both pipes open, it would take 40 min to empty the tub. [7.4.2] **33.** It would take 4 h for the three painters to paint the dormitory together. [7.4.2] **34.** The rate of the current is 2 mph. [7.4.3] **35.** The rate of the bus is 45 mph. [7.4.3] **36.** The rate of the tractor is 30 mph. [7.4.3]
37. The pressure is 40 lb. [7.5.2] **38.** The illumination is 300 lumens 2 ft from the light. [7.5.2]
39. The resistance of the cable is 0.4 ohm. [7.5.2]

CHAPTER 7 TEST

1. $\dfrac{v(v+2)}{2v-1}$ [7.1.2] **2.** $-\dfrac{2(a-2)}{3a+2}$ [7.1.2] **3.** $\dfrac{6(x-2)}{5}$ [7.2.1] **4.** 2 [7.1.1]

5. $\dfrac{x+1}{3x-4}$ [7.2.1] **6.** $\dfrac{x^n - 1}{x^n + 2}$ [7.2.1] **7.** $\dfrac{4y^2 + 6x^2 - 5xy}{2x^2 y^2}$ [7.2.2] **8.** $\dfrac{2(5x+2)}{(x-2)(x+2)}$ [7.2.2]

9. $\{x \mid x \neq -3, 3\}$ [7.1.1] **10.** $-\dfrac{x^2 + 5x - 2}{(x+1)(x+4)(x-1)}$ [7.2.2] **11.** $\dfrac{x-4}{x+3}$ [7.3.1]

12. $\dfrac{x+4}{x+2}$ [7.3.1] **13.** 2 [7.5.1] **14.** No solution [7.4.1] **15.** $x = \dfrac{c}{a-b}$ [7.6.1]

16. It would take 80 min to empty the full tank with both pipes open. [7.4.2] **17.** The office requires 14 rolls of wallpaper. [7.5.1] **18.** Working together, the landscapers would complete the task in 10 min. [7.4.2]
19. The rate of the cyclist is 10 mph. [7.4.3] **20.** The stopping distance of the car is 61.2 ft. [7.5.2]

CUMULATIVE REVIEW EXERCISES

1. $\dfrac{36}{5}$ [1.4.2] **2.** $\dfrac{15}{2}$ [2.2.3] **3.** 7 and 1 [2.8.1] **4.** $\{x \mid x \neq 3\}$ [7.1.1] **5.** $\dfrac{3}{7}$ [7.1.1]

6. 3.5×10^{-8} [5.4.2] **7.** No solution [7.4.1] **8.** $\dfrac{1}{9}$ and 4 [6.5.1] **9.** $\dfrac{8b^3}{a}$ [5.4.1]

10. $\{x \mid x \leq 4\}$ [2.7.2] **11.** $-4a^4 + 6a^3 - 2a^2$ [5.3.1] **12.** $(2x^n - 1)(x^n + 2)$ [6.4.3]

13. $(xy - 3)(x^2 y^2 + 3xy + 9)$ [6.4.2] **14.** $x + 3y$ [7.1.2] **15.** $y = \dfrac{3}{2}x + 2$ [3.6.1]

16. $\dfrac{x}{x^2 - 1}$ [5.4.1] **17.** [3.3.2] **18.** [4.5.1]

19. $4x^2 + 10x + 10 + \dfrac{21}{x - 2}$ [5.5.2/5.5.3] **20.** $\dfrac{3xy}{x - y}$ [7.2.1] **21.** $\{x \mid x \in \text{real numbers}\}$ [7.1.1]

22. $-\dfrac{x}{(3x + 2)(x + 1)}$ [7.2.2] **23.** -28 [4.3.1] **24.** $\dfrac{x - 3}{x + 3}$ [7.3.1] **25.** $(1, 1, 1)$ [4.2.2 or 4.3.2]

26. $1, 2$ [6.5.1] **27.** 9 [7.5.1] **28.** 13 [7.4.1] **29.** $a^4 + 5a^3 - 3a^2 - 11a + 20$ [5.3.2]

30. $r = \dfrac{E - IR}{I}$ [7.6.1] **31.** No [3.6.1] **32.** [5.1.1]

33. The smaller integer is 5 and the larger integer is 10. [2.3.1] **34.** The number of pounds of almonds is 50. [2.3.3] **35.** The number of people expected to vote is 75,000. [7.5.1] **36.** It would take the new computer 14 min to do the job working alone. [7.4.2] **37.** The rate of the wind is 60 mph. [7.4.3] **38.** The frequency of vibration of the open pipe organ is 80 vibrations/min. [7.5.2]

Answers to Chapter 8 Selected Exercises

PREP TEST

1. 16 [1.2.4] **2.** 32 [1.4.1] **3.** 9 [1.3.2] **4.** $\dfrac{1}{12}$ [1.3.3] **5.** $-5x - 1$ [1.5.6]

6. $\dfrac{xy^5}{4}$ [5.4.1] **7.** $9x^2 - 12x + 4$ [5.3.4] **8.** $-12x^2 + 14x + 10$ [5.3.2] **9.** $36x^2 - 1$ [5.3.4]

10. $-1, 15$ [6.5.1]

CONCEPT REVIEW 8.1

1. Sometimes true **3.** Always true **5.** Always true

SECTION 8.1

1. third; 125; 5 **3.** $8; 8; -3; 8^3; \dfrac{1}{512}$ **5.** d **7.** 2 **9.** 27 **11.** $\dfrac{1}{9}$ **13.** 4

15. Not a real number **17.** $\dfrac{343}{125}$ **19.** x **21.** $y^{\frac{1}{2}}$ **23.** $x^{\frac{1}{12}}$ **25.** $a^{\frac{7}{12}}$ **27.** $\dfrac{1}{a}$

29. $\dfrac{1}{y}$ **31.** $y^{\frac{3}{2}}$ **33.** $\dfrac{1}{x}$ **35.** $\dfrac{1}{x^4}$ **37.** a **39.** $x^{\frac{3}{10}}$ **41.** a^3 **43.** $\dfrac{1}{x^{\frac{1}{2}}}$ **45.** $y^{\frac{2}{3}}$

47. x^4y **49.** $x^6y^3z^9$ **51.** $\dfrac{x}{y^2}$ **53.** $\dfrac{x^{\frac{3}{2}}}{y^{\frac{1}{4}}}$ **55.** x^2y^8 **57.** $\dfrac{1}{x^{\frac{11}{12}}}$ **59.** $\dfrac{1}{y^{\frac{5}{2}}}$ **61.** $\dfrac{1}{b^{\frac{7}{8}}}$

63. a^5b^{13} **65.** $\dfrac{m^2}{4n^{\frac{3}{2}}}$ **67.** $\dfrac{y^{\frac{17}{2}}}{x^3}$ **69.** $\dfrac{16b^2}{a^{\frac{1}{3}}}$ **71.** $\dfrac{y^2}{x^3}$ **73.** $a - a^2$ **75.** $y + 1$

77. $a - \dfrac{1}{a}$ **79.** $\dfrac{1}{a^{10}}$ **81.** $a^{\frac{n}{6}}$ **83.** $\dfrac{1}{b^{\frac{2m}{3}}}$ **85.** x^{10n^2} **87.** $x^{3n}y^{2n}$ **89.** $4; 5; \dfrac{4}{5}$

91. True **93.** $\sqrt{5}$ **95.** $\sqrt[3]{b^4}$ **97.** $\sqrt[3]{9x^2}$ **99.** $-3\sqrt[5]{a^2}$ **101.** $\sqrt[4]{x^6y^9}$ **103.** $\sqrt{a^9b^{21}}$

105. $\sqrt[3]{3x-2}$ **107.** $14^{\frac{1}{2}}$ **109.** $x^{\frac{1}{3}}$ **111.** $x^{\frac{4}{3}}$ **113.** $b^{\frac{3}{5}}$ **115.** $(2x^2)^{\frac{1}{3}}$ **117.** $-(3x^5)^{\frac{1}{2}}$

119. $3xy^{\frac{2}{3}}$ **121.** $(a^2+2)^{\frac{1}{2}}$ **123.** $2;a^2;b^3$ **125.** Negative **127.** Positive **129.** y^7

131. $-a^3$ **133.** a^7b^3 **135.** $11y^6$ **137.** a^2b^4 **139.** $-a^3b^3$ **141.** $5b^5$ **143.** $-a^2b^3$

145. $5x^4y$ **147.** Not a real number **149.** $2a^7b^2$ **151.** $-3ab^5$ **153.** y^3 **155.** $3a^5$

157. $-a^4b$ **159.** ab^5 **161.** $2a^2b^5$ **163.** $-2x^3y^4$ **165.** $\dfrac{4x}{y^7}$ **167.** $\dfrac{3b}{a^3}$ **169.** $2x+3$

171. $x+1$ **173.** x **175.** b **177.** ab^2 **179.** $\dfrac{3}{5}$ **181.** $\dfrac{3}{4}$

CONCEPT REVIEW 8.2

1. Sometimes true **3.** Never true **5.** Always true

SECTION 8.2

1. $25;25;2;5$ **3. a.** Irrational **b.** Rational **c.** Irrational **d.** Rational **7.** $3\sqrt{2}$

9. $7\sqrt{2}$ **11.** $2\sqrt[3]{9}$ **13.** $2\sqrt[3]{2}$ **15.** $x^2yz^2\sqrt{yz}$ **17.** $2ab^4\sqrt{2a}$ **19.** $3xyz^2\sqrt{5yz}$

21. $-5y\sqrt[3]{x^2y}$ **23.** $-6xy^3\sqrt[3]{x^2}$ **25.** $ab^2\sqrt[3]{a^2b^2}$ **27.** radicand; index **29. a.** No **b.** No

c. Yes **d.** Yes **31.** $2\sqrt{2}$ **33.** $3\sqrt[3]{7}$ **35.** $-6\sqrt{x}$ **37.** $-2\sqrt{2}$ **39.** $\sqrt{2x}$

41. $3\sqrt{3a}-2\sqrt{2a}$ **43.** $10x\sqrt{2x}$ **45.** $2x\sqrt{3xy}$ **47.** $xy\sqrt{2y}$ **49.** $5a^2b^2\sqrt{ab}$ **51.** $9\sqrt[3]{2}$

53. $-7a\sqrt[3]{3a}$ **55.** $-5xy^2\sqrt[3]{x^2y}$ **57.** $16ab\sqrt[4]{ab}$ **59.** $6\sqrt{3}-6\sqrt{2}$ **61.** $6x^3y^2\sqrt{xy}$

63. $-9a^2\sqrt{3ab}$ **65.** $18;3x;3$ **67.** 16 **69.** $2\sqrt[3]{4}$ **71.** $xy^3\sqrt{x}$ **73.** $8xy\sqrt{x}$

75. $2x^2y\sqrt[3]{2}$ **77.** $2ab\sqrt[4]{3a^2b}$ **79.** 6 **81.** $x-\sqrt{2x}$ **83.** $4x-8\sqrt{x}$ **85.** $x-6\sqrt{x}+9$

87. $84+16\sqrt{5}$ **89.** $672x^2y^2$ **91.** $4a^3b^3\sqrt[3]{a}$ **93.** $-8\sqrt{5}$ **95.** $x-y^2$ **97.** $12x-y$

99. $x-3\sqrt{x}-28$ **101.** $\sqrt[3]{x^2}+\sqrt[3]{x}-20$ **103.** False **105.** Quotient; $9;8;2x^2$

107. a. Yes **b.** No **c.** No **d.** No **109.** $4\sqrt{x}$ **111.** $b^2\sqrt{3a}$ **113.** $\dfrac{\sqrt{5}}{5}$

115. $\dfrac{\sqrt{2x}}{2x}$ **117.** $\dfrac{\sqrt{5x}}{x}$ **119.** $\dfrac{\sqrt{5x}}{5}$ **121.** $\dfrac{3\sqrt[3]{4}}{2}$ **123.** $\dfrac{3\sqrt[3]{2x}}{2x}$ **125.** $\dfrac{\sqrt{2xy}}{2y}$

127. $\dfrac{2\sqrt{3ab}}{3b^2}$ **129.** $2\sqrt{5}-4$ **131.** $\dfrac{3\sqrt{y}+6}{y-4}$ **133.** $-5+2\sqrt{6}$ **135.** $8+4\sqrt{3}-2\sqrt{2}-\sqrt{6}$

137. $\dfrac{\sqrt{15}+\sqrt{10}-\sqrt{6}-5}{3}$ **139.** $\dfrac{3\sqrt[4]{2x}}{2x}$ **141.** $\dfrac{2\sqrt[5]{2a^3}}{a}$ **143.** $\dfrac{\sqrt[5]{16x^2}}{2}$ **145.** $\dfrac{1+2\sqrt{ab}+ab}{1-ab}$

147. $\dfrac{5x\sqrt{y}+5y\sqrt{x}}{x-y}$ **149.** $2\sqrt{2}$ **151.** $29\sqrt{2}-45$ **153.** $\dfrac{3\sqrt{y+1}-3}{y}$ **155.** $\sqrt[6]{x+y}$

157. $\sqrt[4]{2y(x+3)^2}$ **159.** $\sqrt[6]{a^3(a+3)^2}$ **161.** $16^{\frac{1}{4}},2$ **163.** $32^{-\frac{1}{5}},\dfrac{1}{2}$ **165. a.** $D(x)=3.06\sqrt[10]{x^3}$

b. The model predicts a national debt of \$4.96 trillion. **c.** The model predicts a national debt of \$6.61 trillion.

CONCEPT REVIEW 8.3

1. Always true **3.** Never true

SECTION 8.3

3. a. ≥ 0 **b.** $\in$ real numbers **5. a.** $(-\infty,0]$ **b.** Empty set **c.** $[0,\infty)$ **d.** All real numbers **7.** $\{x\,|\,x\in\text{real numbers}\}$ **9.** $\{x\,|\,x\geq-1\}$ **11.** $\{x\,|\,x\geq 0\}$ **13.** $\{x\,|\,x\geq 0\}$

15. $\{x\,|\,x\geq 2\}$ **17.** $(-\infty,\infty)$ **19.** $(-\infty,3]$ **21.** $[2,\infty)$ **23.** $\left(-\infty,\dfrac{2}{3}\right]$ **25.** ≥ 5

27. 14; 9; −3 **29.** **31.** **33.** **35.**

37. **39.** **41.** **43.** **45.**

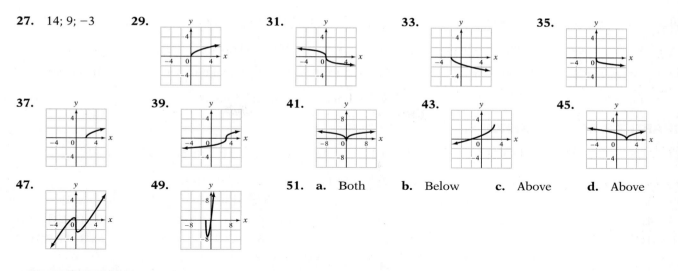

47. **49.** **51. a.** Both **b.** Below **c.** Above **d.** Above

CONCEPT REVIEW 8.4

1. Sometimes true **3.** Sometimes true

SECTION 8.4

1. $\sqrt{2x+1}=7$ **5.** 25 **7.** 27 **9.** 48 **11.** −2 **13.** No solution **15.** 9

17. −23 **19.** −16 **21.** $\dfrac{9}{4}$ **23.** 14 **25.** 7 **27.** 7 **29.** −122 **31.** 35

33. 45 **35.** 23 **37.** −4 **39.** 10 **41.** 1 **43.** 3 **45.** 21 **47.** $\dfrac{25}{72}$ **49.** 5

51. −2, 4 **53.** 6 **55.** 0, 1 **57.** −2, 1 **59.** 2, 11 **61.** 3 **63. a.** One

b. Two **65. d** **67.** The air pressure is 909.8 mb. The wind speed increases. **69.** The mean

distance is 295,000 km. **71.** The elephant weighs 3700 lb. **73.** The distance is 10 ft. **75.** The

distance between the joggers after 3 min is 829.76 m. **77.** The length of the diagonal is 8.9 cm. **79.** The

periscope must be 8.17 ft above the water. **81.** The distance is 225 ft. **83.** The distance is 96 ft.

85. The satellite is 711,000 m above Earth's surface. **87.** 27 **89.** $d=\dfrac{v^2}{64}$ **91.** $r=\dfrac{\sqrt{V\pi h}}{\pi h}$

93. 2 **95.** The distance between the corners is 7.81 in. **97.** $\sqrt{6}$

CONCEPT REVIEW 8.5

1. Always true **3.** Always true

SECTION 8.5

3. −1 **5.** i; 9; 6; 3i **7.** $10-\sqrt{-9}$ **9.** 2i **11.** 7$i\sqrt{2}$ **13.** 3$i\sqrt{3}$ **15.** 4 + 2i

17. $2\sqrt{3}-3i\sqrt{2}$ **19.** $4\sqrt{10}-7i\sqrt{3}$ **21.** 2ai **23.** $7x^6i$ **25.** $12ab^2i\sqrt{ab}$

27. $2\sqrt{a}+2ai\sqrt{3}$ **29.** $3b^2\sqrt{2b}-3bi\sqrt{3b}$ **31.** $5x^3y\sqrt{y}+5xyi\sqrt{2xy}$ **33.** $2a^2bi\sqrt{a}$

35. $2a\sqrt{3a}+3bi\sqrt{3b}$ **37.** 8; 6; −8i; 6; 8i **39.** $10-7i$ **41.** $11-7i$ **43.** $-6+i$

45. $-4-8i\sqrt{3}$ **47.** $-\sqrt{10}-11i\sqrt{2}$ **49.** $6-4i$ **51.** 0 **53.** $11+6i$ **55.** True

57. Negative **59.** −24 **61.** −15 **63.** $-5\sqrt{2}$ **65.** $-15-12i$ **67.** 1; 9; 1; 81; 82

69. $3\sqrt{2}+6i$ **71.** $-10i$ **73.** $29-2i$ **75.** 1 **77.** 1 **79.** 89 **81.** 50

83. $7-24i$ **85.** $-2i$ **87.** False **89.** $2+3i$ **91.** $-\dfrac{4}{5}i$ **93.** $-\dfrac{5}{3}+\dfrac{16}{3}i$

95. $\dfrac{30}{29}-\dfrac{12}{29}i$ **97.** $\dfrac{20}{17}+\dfrac{5}{17}i$ **99.** $\dfrac{11}{13}+\dfrac{29}{13}i$ **101.** $-\dfrac{1}{5}+\dfrac{\sqrt{6}}{10}i$ **103.** $-1+4i$ **105.** True

107. -1 **109.** i **111.** -1 **113.** -1 **115. a.** Yes **b.** Yes

117. a. $2, -2$; integer, rational number, real number **b.** $i, -i$; imaginary

c. $\dfrac{16}{15}$; rational number, real number **d.** $2\sqrt{2}, -2\sqrt{2}$; irrational number, real number **e.** 16; integer, rational number, real number **f.** $-19{,}683$; integer, rational number, real number

CHAPTER 8 REVIEW EXERCISES

1. $\dfrac{1}{3}$ [8.1.1] **2.** $\dfrac{1}{x^5}$ [8.1.1] **3.** $\dfrac{1}{a^{10}}$ [8.1.1] **4.** $\dfrac{160y^{11}}{x}$ [8.1.1] **5.** $3\sqrt[4]{x^3}$ [8.1.2]

6. $7x^{\frac{2}{3}}y$ [8.1.2] **7.** $3a^2b^3$ [8.1.3] **8.** $-7x^3y^8$ [8.1.3] **9.** $-2a^2b^4$ [8.1.3]

10. $3ab^3\sqrt{2a}$ [8.2.1] **11.** $-2ab^2\sqrt[5]{2a^3b^2}$ [8.2.1] **12.** $xy^2z^2\sqrt[4]{x^2z^2}$ [8.2.1] **13.** $5\sqrt{6}$ [8.2.2]

14. $4x^2\sqrt{3xy} - 4x^2\sqrt{5xy}$ [8.2.2] **15.** $2a^2b\sqrt{2b}$ [8.2.2] **16.** $6x^2\sqrt{3y}$ [8.2.2] **17.** 40 [8.2.3]

18. $4xy^2\sqrt[3]{x^2}$ [8.2.3] **19.** $3x + 3\sqrt{3x}$ [8.2.3] **20.** $31 - 10\sqrt{6}$ [8.2.3] **21.** $-13 + 6\sqrt{3}$ [8.2.3]

22. $5x\sqrt{x}$ [8.2.4] **23.** $\dfrac{8\sqrt{3y}}{3y}$ [8.2.4] **24.** $\dfrac{x\sqrt{x} - x\sqrt{2} + 2\sqrt{x} - 2\sqrt{2}}{x - 2}$ [8.2.4]

25. $\dfrac{x + 2\sqrt{xy} + y}{x - y}$ [8.2.4] **26.** $6i$ [8.5.1] **27.** $5i\sqrt{2}$ [8.5.1] **28.** $7 - 4i$ [8.5.1]

29. $10\sqrt{2} + 2i\sqrt{3}$ [8.5.1] **30.** $9 - i$ [8.5.2] **31.** $-12 + 10i$ [8.5.2] **32.** $14 + 2i$ [8.5.2]

33. $-4\sqrt{2} + 8i\sqrt{2}$ [8.5.2] **34.** $10 - 9i$ [8.5.2] **35.** -16 [8.5.3] **36.** $7 + 3i$ [8.5.3]

37. $-6\sqrt{2}$ [8.5.3] **38.** $39 - 2i$ [8.5.3] **39.** $6i$ [8.5.4] **40.** $\dfrac{2}{3} - \dfrac{5}{3}i$ [8.5.4]

41. $\dfrac{14}{5} + \dfrac{7}{5}i$ [8.5.4] **42.** $1 + i$ [8.5.4] **43.** $-2 + 7i$ [8.5.4] **44.** -24 [8.4.1]

45. $\left\{x \,\middle|\, x \geq \dfrac{2}{3}\right\}$ [8.3.1] **46.** $\{x \,|\, x \in \text{real numbers}\}$ [8.3.1] **47.** [8.3.2]

48. [8.3.2] **49.** $\dfrac{13}{3}$ [8.4.1] **50.** -2 [8.4.1] **51.** The width is 5 in. [8.4.2]

52. The amount of power is 120 watts. [8.4.2] **53.** The distance required is 242 ft. [8.4.2]

54. The distance is 6.63 ft. [8.4.2]

CHAPTER 8 TEST

1. $r^{\frac{1}{6}}$ [8.1.1] **2.** $\dfrac{64x^3}{y^6}$ [8.1.1] **3.** $\dfrac{b^3}{8a^6}$ [8.1.1] **4.** $3\sqrt[5]{y^2}$ [8.1.2] **5.** $\dfrac{1}{2}x^{\frac{3}{4}}$ [8.1.2]

6. $\{x \,|\, x \leq 4\}$ [8.3.1] **7.** $(-\infty, \infty)$ [8.3.1] **8.** $3abc^2\sqrt[3]{ac}$ [8.2.1] **9.** $8a\sqrt{2a}$ [8.2.2]

10. $-2x^2y\sqrt[3]{2x}$ [8.2.2] **11.** $-4x\sqrt{3}$ [8.2.3] **12.** $14 + 10\sqrt{3}$ [8.2.3]

13. $2a - \sqrt{ab} - 15b$ [8.2.3] **14.** $4x + 4\sqrt{xy} + y$ [8.2.3] **15.** $\dfrac{4x^2}{y}$ [8.2.4] **16.** 2 [8.2.4]

17. $\dfrac{x + \sqrt{xy}}{x - y}$ [8.2.4] **18.** -4 [8.5.3] **19.** $-3 + 2i$ [8.5.2] **20.** $18 + 16i$ [8.5.3]

21. $-\dfrac{4}{5} + \dfrac{7}{5}i$ [8.5.4] **22.** 4 [8.5.2] **23.** 4 [8.4.1] **24.** -3 [8.4.1] **25.** The length of the guy wire is 30.6 ft. [8.4.2]

CUMULATIVE REVIEW EXERCISES

1. The Distributive Property [1.5.2] **2.** $-x - 24$ [1.5.6] **3.** $\varnothing$ [2.7.1] **4.** 16 [8.4.1]

5. $\dfrac{3}{2}$ [2.2.1] **6.** $\dfrac{2}{3}$ [2.2.3] **7.** $\{x \mid x \geq -1\}$ [2.7.2] **8.** $\{x \mid 4 < x < 5\}$ [2.7.3]

9. $\left\{ x \mid x < 2 \text{ or } x > \dfrac{8}{3} \right\}$ [2.8.2] **10.** $(8a + b)(8a - b)$ [6.4.1] **11.** $x(x^2 + 3)(x + 1)(x - 1)$ [6.4.4]

12. $\dfrac{2}{3}$ and -5 [6.5.1] **13.**

[8.3.2] **14.** Yes [3.6.1] **15.** $\dfrac{3x^3}{y}$ [5.4.1]

16. $\dfrac{y^8}{x^2}$ [8.1.1] **17.** $-x\sqrt{5x}$ [8.2.2] **18.** $11 - 5\sqrt{5}$ [8.2.3] **19.** $\dfrac{x\sqrt[3]{4y^2}}{2y}$ [8.2.4]

20. $-\dfrac{3}{5} + \dfrac{6}{5}i$ [8.5.4] **21.**

[2.7.1] **22.** $\{x \mid x \in \text{real numbers}\}$ [8.3.1]

23. 34 [3.2.1] **24.** $y = \dfrac{1}{3}x + \dfrac{7}{3}$ [3.5.2] **25.** 3 [4.3.1] **26.** $\left(\dfrac{17}{4}, \dfrac{9}{2} \right)$ [4.3.2]

27.

$m = \dfrac{3}{2}, b = 3$ [3.3.2] **28.**

[3.7.1]

29. There are nineteen 18¢ stamps. [2.3.3] **30.** The additional investment must be $4000. [2.6.1]
31. The length is 12 ft. The width is 6 ft. [6.5.2] **32.** The rate of the plane is 250 mph. [7.4.3]
33. The time is 1.25 s. [5.4.2] **34.** The height of the periscope is 32.7 ft. [8.4.2] **35.** The slope is 0.08.
A slope of 0.08 means that the annual interest income is 8% of the investment. [3.4.1]

Answers to Chapter 9 Selected Exercises

PREP TEST

1. $3\sqrt{2}$ [8.2.1] **2.** $3i$ [8.5.1] **3.** $\dfrac{2x - 1}{x - 1}$ [7.2.2] **4.** 8 [1.5.1] **5.** Yes [5.3.4]

6. $(2x - 1)^2$ [6.4.1] **7.** $(3x + 2)(3x - 2)$ [6.4.1] **8.**

[2.7.1]

9. $-3, 5$ [6.5.1] **10.** 4 [7.5.1]

CONCEPT REVIEW 9.1

1. Sometimes true **3.** Never true

SECTION 9.1

3. $2x^2 - 4x - 5 = 0; a = 2, b = -4, c = -5$ **5.** $4x^2 - 5x - 6 = 0; a = 4, b = -5, c = -6$ **7.** standard;
$x + 7; x - 7; 0; 0;$ Zero Products; $-7; 7$ **9.** 0 and 4 **11.** -5 and 5 **13.** -2 and 3 **15.** 3

17. 0 and 2 **19.** -2 and 5 **21.** 2 and 5 **23.** 6 and $-\dfrac{3}{2}$ **25.** $\dfrac{1}{4}$ and 2 **27.** -4 and $\dfrac{1}{3}$

29. $-\dfrac{2}{3}$ and $\dfrac{9}{2}$ **31.** -4 and $\dfrac{1}{4}$ **33.** -2 and 9 **35.** -2 and $-\dfrac{3}{4}$ **37.** 2 and -5 **39.** $-\dfrac{3}{2}$

and -4 **41.** $2b$ and $7b$ **43.** $-c$ and $7c$ **45.** $-\dfrac{b}{2}$ and $-b$ **47.** $\dfrac{2a}{3}$ and $4a$ **49. a.** One

positive and one negative **b.** Two negative **c.** Two positive **51.** $x - r$ **53.** $x^2 - 7x + 10 = 0$
55. $x^2 + 6x + 8 = 0$ **57.** $x^2 - 5x - 6 = 0$ **59.** $x^2 - 9 = 0$ **61.** $x^2 - 8x + 16 = 0$
63. $x^2 - 5x = 0$ **65.** $x^2 - 3x = 0$ **67.** $2x^2 - 7x + 3 = 0$ **69.** $4x^2 - 5x - 6 = 0$
71. $3x^2 + 11x + 10 = 0$ **73.** $9x^2 - 4 = 0$ **75.** $6x^2 - 5x + 1 = 0$ **77.** $10x^2 - 7x - 6 = 0$
79. $8x^2 + 6x + 1 = 0$ **81.** $50x^2 - 25x - 3 = 0$ **83.** b is positive and c is positive. **85.** b is positive

and c is negative. **87.** 6; -6 **89.** 7 and -7 **91.** $2i$ and $-2i$ **93.** 2 and -2 **95.** $\dfrac{9}{2}$ and

$-\dfrac{9}{2}$ **97.** $7i$ and $-7i$ **99.** $4\sqrt{3}$ and $-4\sqrt{3}$ **101.** $5\sqrt{3}$ and $-5\sqrt{3}$ **103.** $3i\sqrt{2}$ and $-3i\sqrt{2}$

105. 7 and -5 **107.** 0 and -6 **109.** -7 and 3 **111.** $-5 + 5i$ and $-5 - 5i$ **113.** $4 + 2i$ and

$4 - 2i$ **115.** $9 + 3i$ and $9 - 3i$ **117.** 0 and 1 **119.** $-\dfrac{1}{5}$ and 1 **121.** $-\dfrac{3}{2}$ and 0 **123.** $\dfrac{7}{3}$

and 1 **125.** $-5 + \sqrt{6}$ and $-5 - \sqrt{6}$ **127.** $2 + 2\sqrt{6}$ and $2 - 2\sqrt{6}$ **129.** $-1 + 2i\sqrt{3}$ and $-1 - 2i\sqrt{3}$

131. $3 + 3i\sqrt{5}$ and $3 - 3i\sqrt{5}$ **133.** $\dfrac{-2 + 9\sqrt{2}}{3}$ and $\dfrac{-2 - 9\sqrt{2}}{3}$ **135.** $-\dfrac{1}{2} + 2i\sqrt{10}$ and $-\dfrac{1}{2} - 2i\sqrt{10}$

137. Two complex solutions **139.** One real solution **141.** $x^2 - 2 = 0$ **143.** $x^2 + 1 = 0$

145. $x^2 - 8 = 0$ **147.** $x^2 - 12 = 0$ **149.** $x^2 + 12 = 0$ **151.** $-\dfrac{5z}{y}$ and $\dfrac{5z}{y}$ **153.** $y + 2$ and

$y - 2$ **155.** 1

CONCEPT REVIEW 9.2

1. Always true **3.** Never true **5.** Never true

SECTION 9.2

1. a. perfect-square; x; 14; 49 **b.** 49; $x + 7$ **5.** 5 and -1 **7.** -9 and 1 **9.** 3

11. $-2 + \sqrt{11}$ and $-2 - \sqrt{11}$ **13.** $3 + \sqrt{2}$ and $3 - \sqrt{2}$ **15.** $1 + i$ and $1 - i$ **17.** $\dfrac{1 + \sqrt{5}}{2}$ and

$\dfrac{1 - \sqrt{5}}{2}$ **19.** $3 + \sqrt{13}$ and $3 - \sqrt{13}$ **21.** 3 and 5 **23.** $2 + 3i$ and $2 - 3i$

25. $-3 + 2i$ and $-3 - 2i$ **27.** $1 + 3\sqrt{2}$ and $1 - 3\sqrt{2}$ **29.** $\dfrac{1 + \sqrt{17}}{2}$ and $\dfrac{1 - \sqrt{17}}{2}$

31. $1 + 2i\sqrt{3}$ and $1 - 2i\sqrt{3}$ **33.** $-\dfrac{1}{2}$ and -1 **35.** $\dfrac{1}{2}$ and $\dfrac{3}{2}$ **37.** $-\dfrac{1}{2}$ and $\dfrac{4}{3}$ **39.** $\dfrac{1}{2} + i$ and

$\dfrac{1}{2} - i$ **41.** $\dfrac{1}{3} + \dfrac{1}{3}i$ and $\dfrac{1}{3} - \dfrac{1}{3}i$ **43.** $\dfrac{2 + \sqrt{14}}{2}$ and $\dfrac{2 - \sqrt{14}}{2}$ **45.** $-\dfrac{3}{2}$ and 1 **47.** $1 + \sqrt{5}$ and

$1 - \sqrt{5}$ **49.** $\dfrac{1}{2}$ and 5 **51.** $2 + \sqrt{5}$ and $2 - \sqrt{5}$ **53.** -3.236 and 1.236 **55.** 1.707 and 0.293

57. 0.309 and -0.809 **59.** False **65.** $6x$; 10; $x^2 - 6x + 10$; 1; -6; 10 **67.** complex

69. 5 and -2 **71.** 4 and -9 **73.** $4 + 2\sqrt{22}$ and $4 - 2\sqrt{22}$ **75.** 3 and -8 **77.** $\dfrac{1}{2}$ and -3

79. $-\dfrac{1}{4}$ and $\dfrac{3}{2}$ **81.** $1 + 2\sqrt{2}$ and $1 - 2\sqrt{2}$ **83.** 10 and -2 **85.** $6 + 2\sqrt{3}$ and $6 - 2\sqrt{3}$

87. $\dfrac{1 + 2\sqrt{2}}{2}$ and $\dfrac{1 - 2\sqrt{2}}{2}$ **89.** $\dfrac{1}{2}$ and 1 **91.** $\dfrac{-5 + \sqrt{7}}{3}$ and $\dfrac{-5 - \sqrt{7}}{3}$ **93.** $\dfrac{2}{3}$ and $\dfrac{5}{2}$

95. $2 + i$ and $2 - i$ **97.** $-3 + 2i$ and $-3 - 2i$ **99.** $3 + i$ and $3 - i$ **101.** $-\dfrac{3}{2}$ and $-\dfrac{1}{2}$

103. $-\dfrac{1}{2} + \dfrac{5}{2}i$ and $-\dfrac{1}{2} - \dfrac{5}{2}i$ **105.** $\dfrac{-3 + \sqrt{6}}{3}$ and $\dfrac{-3 - \sqrt{6}}{3}$ **107.** $1 + \dfrac{\sqrt{6}}{3}i$ and $1 - \dfrac{\sqrt{6}}{3}i$ **109.** $-\dfrac{3}{2}$

and -1 **111.** $-\dfrac{3}{2}$ and 4 **113.** 7.606 and 0.394 **115.** 0.236 and -4.236 **117.** 1.851 and

-1.351 **119.** Two complex **121.** One real **123.** Two real **125.** Two real **127.** True

129. Quadratic **131.** Linear **133.** Quadratic **135.** $\dfrac{\sqrt{2}}{2}$ and $-2\sqrt{2}$ **137.** $-3\sqrt{2}$ and $\dfrac{\sqrt{2}}{2}$

139. $\dfrac{\sqrt{3}}{2} + \dfrac{1}{2}i$ and $\dfrac{\sqrt{3}}{2} - \dfrac{1}{2}i$ **141.** $2a$ and $-a$ **143.** $\dfrac{a}{2}$ and $-2a$ **145.** $1 + \sqrt{y + 1}$ and $1 - \sqrt{y + 1}$

147. $\{p\,|\,p < 9, p \in \text{real numbers}\}$ **149.** $\{p\,|\,p > 1, p \in \text{real numbers}\}$ **151.** $i, -2i$ **153.** The ball
takes 4.43 s to hit the ground. **155. a.** 1.05; The model predicts that the population will increase.
b. 0.96; The model predicts that the population will decrease.

CONCEPT REVIEW 9.3

1. Always true **3.** Always true

SECTION 9.3

1. x^4 **5.** $3, -3, 2, -2$ **7.** $2, -2, \sqrt{2}, -\sqrt{2}$ **9.** 1 and 4 **11.** 16 **13.** $2i, -2i, 1, -1$

15. $4i, -4i, 2, -2$ **17.** 16 **19.** 512 and 1 **21.** $2, 1, -1 + i\sqrt{3}, -1 - i\sqrt{3}, -\dfrac{1}{2} + \dfrac{\sqrt{3}}{2}i, -\dfrac{1}{2} - \dfrac{\sqrt{3}}{2}i$

23. $1, -1, 2, -2, i, -i, 2i, -2i$ **25.** -64 and 8 **27.** $\dfrac{1}{4}$ and 1 **29.** a and d

31. $\sqrt{a + 1}$; 7; square **33.** 38, 39, 40, 44, 45, 46, 48, 49, 52 **35.** 3 **37.** 9 **39.** 2 and -1

41. 0 and 2 **43.** $-\dfrac{1}{2}$ and 2 **45.** -2 **47.** 1 **49.** 1 **51.** -3

53. a. $y; y - 4; y(y - 4)$ **b.** $4y - 16$ **c.** $y^2 - 6y$ **d.** $4y - 16; y^2 - 6y$ **55.** 10 and -1

57. $-\dfrac{1}{2} + \dfrac{\sqrt{7}}{2}i$ and $-\dfrac{1}{2} - \dfrac{\sqrt{7}}{2}i$ **59.** 1 and -3 **61.** 0 and -1 **63.** $\dfrac{1}{2}$ and $-\dfrac{1}{3}$

65. 6 and $-\dfrac{2}{3}$ **67.** $\dfrac{4}{3}$ and 3 **69.** $-\dfrac{1}{4}$ and 3 **71.** $\dfrac{-5 + \sqrt{33}}{2}$ and $\dfrac{-5 - \sqrt{33}}{2}$ **73.** 4 and -6

75. 5 and 4 **77.** $3, -3, i, -i$ **79.** $i\sqrt{6}, -i\sqrt{6}, 2, -2$ **81.** $\sqrt{2}$ **83.** 4

CONCEPT REVIEW 9.4

1. Sometimes true **3.** Always true

SECTION 9.4

1. a. $2w - 4$ **b.** $w(2w - 4)$ **c.** $w(2w - 4)$; 96 **3.** False **5.** The length is 40 ft. The width is
16 ft. **7.** The length is 13 ft. The width is 5 ft. **9.** The cardboard needs to be 90 cm by 90 cm.
11. The rate returning was 55 mph. **13.** The projectile takes 12.5 s to return to Earth. **15.** The
maximum speed is 72.5 km/h. **17.** The rocket will be 300 ft above the ground after 1.74 s and after 10.76 s.
19. Working alone, it would take the smaller pipe 12 min to fill the tank. It would take the larger pipe 6 min.
21. The rate of the cruise ship for the first 40 mi was 10 mph. **23.** The rate of the wind is 20 mph.

25. The rate of the crew in calm water is 6 mph. **27.** The fraction is $\dfrac{3}{6}$. **29.** The width is 8 cm. The

length is 16 cm. The height is 2 cm. **31.** The depth will be 10 cm after 22.2 s. **33.** The bottom of the
scoop of ice cream is 2.3 in. from the bottom of the cone.

CONCEPT REVIEW 9.5

1. Sometimes true **3.** Sometimes true

SECTION 9.5

3. positive; negative; negative **5.** $\{x \mid x < -2 \text{ or } x > 4\}$ **7.** $\{x \mid x \le 1 \text{ or } x \ge 2\}$

9. $\{x \mid -3 < x < 4\}$ **11.** $\{x \mid x < -2 \text{ or } 1 < x < 3\}$

13. $\{x \mid x < -2 \text{ or } x > 4\}$ **15.** $\{x \mid -1 < x \le 3\}$

17. $\{x \mid x \le -2 \text{ or } 1 \le x < 3\}$

19. a and d **21.** $\{x \mid x < -4 \text{ or } x > 4\}$ **23.** $\{x \mid x < 2 \text{ or } x > 2\}$ **25.** $\{x \mid -3 \le x \le 12\}$

27. $\left\{x \mid x \le \dfrac{1}{2} \text{ or } x \ge 2\right\}$ **29.** $\left\{x \mid \dfrac{1}{2} < x < \dfrac{3}{2}\right\}$ **31.** $\{x \mid x \le -3 \text{ or } 2 \le x \le 6\}$

33. $\left\{x \mid -\dfrac{3}{2} < x < \dfrac{1}{2} \text{ or } x > 4\right\}$ **35.** $\{x \mid x \le -3 \text{ or } -1 \le x \le 1\}$ **37.** $\{x \mid -2 \le x \le 1 \text{ or } x \ge 2\}$

39. $\{x \mid x < -1 \text{ or } x > 2\}$ **41.** $\{x \mid -1 < x \le 0\}$ **43.** $\{x \mid -2 < x \le 0 \text{ or } x > 1\}$ **45.** $\left\{x \mid x < 0 \text{ or } x > \dfrac{1}{2}\right\}$

47. **49.** **51.**

53. **55.**

CONCEPT REVIEW 9.6

1. Sometimes true **3.** Sometimes true **5.** Always true

SECTION 9.6

3. -5 **5.** $x = 7$ **7. a.** 3; 0; -1 **b.** parabola; up **c.** 0; 3; 0 **d.** 0; -1; 0; -1

9. Vertex: (0, 0) **11.** Vertex: (0, -2)
Axis of symmetry: $x = 0$ Axis of symmetry: $x = 0$

13. Vertex: (0, 3) **15.** Vertex: (0, 0) **17.** Vertex: (0, -1)
Axis of symmetry: $x = 0$ Axis of symmetry: $x = 0$ Axis of symmetry: $x = 0$

19. Vertex: (1, -1) **21.** Vertex: (1, 2) **23.** Vertex: $\left(\dfrac{1}{2}, -\dfrac{9}{4}\right)$
Axis of symmetry: $x = 1$ Axis of symmetry: $x = 1$
Axis of symmetry: $x = \dfrac{1}{2}$

25. Vertex: $\left(\dfrac{1}{4}, -\dfrac{41}{8}\right)$

Axis of symmetry: $x = \dfrac{1}{4}$

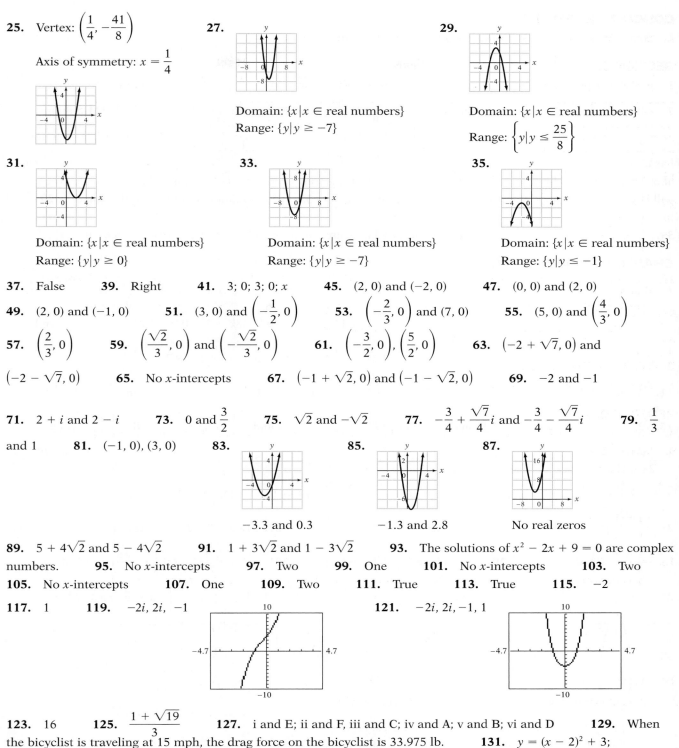

27.

Domain: $\{x \mid x \in \text{real numbers}\}$
Range: $\{y \mid y \geq -7\}$

29.

Domain: $\{x \mid x \in \text{real numbers}\}$
Range: $\left\{y \mid y \leq \dfrac{25}{8}\right\}$

31.

Domain: $\{x \mid x \in \text{real numbers}\}$
Range: $\{y \mid y \geq 0\}$

33.

Domain: $\{x \mid x \in \text{real numbers}\}$
Range: $\{y \mid y \geq -7\}$

35.

Domain: $\{x \mid x \in \text{real numbers}\}$
Range: $\{y \mid y \leq -1\}$

37. False **39.** Right **41.** 3; 0; 3; 0; x **45.** $(2, 0)$ and $(-2, 0)$ **47.** $(0, 0)$ and $(2, 0)$

49. $(2, 0)$ and $(-1, 0)$ **51.** $(3, 0)$ and $\left(-\dfrac{1}{2}, 0\right)$ **53.** $\left(-\dfrac{2}{3}, 0\right)$ and $(7, 0)$ **55.** $(5, 0)$ and $\left(\dfrac{4}{3}, 0\right)$

57. $\left(\dfrac{2}{3}, 0\right)$ **59.** $\left(\dfrac{\sqrt{2}}{3}, 0\right)$ and $\left(-\dfrac{\sqrt{2}}{3}, 0\right)$ **61.** $\left(-\dfrac{3}{2}, 0\right), \left(\dfrac{5}{2}, 0\right)$ **63.** $(-2 + \sqrt{7}, 0)$ and

$(-2 - \sqrt{7}, 0)$ **65.** No x-intercepts **67.** $(-1 + \sqrt{2}, 0)$ and $(-1 - \sqrt{2}, 0)$ **69.** -2 and -1

71. $2 + i$ and $2 - i$ **73.** 0 and $\dfrac{3}{2}$ **75.** $\sqrt{2}$ and $-\sqrt{2}$ **77.** $-\dfrac{3}{4} + \dfrac{\sqrt{7}}{4}i$ and $-\dfrac{3}{4} - \dfrac{\sqrt{7}}{4}i$ **79.** $\dfrac{1}{3}$

and 1 **81.** $(-1, 0), (3, 0)$ **83.**

85.

87.

-3.3 and 0.3

-1.3 and 2.8

No real zeros

89. $5 + 4\sqrt{2}$ and $5 - 4\sqrt{2}$ **91.** $1 + 3\sqrt{2}$ and $1 - 3\sqrt{2}$ **93.** The solutions of $x^2 - 2x + 9 = 0$ are complex
numbers. **95.** No x-intercepts **97.** Two **99.** One **101.** No x-intercepts **103.** Two
105. No x-intercepts **107.** One **109.** Two **111.** True **113.** True **115.** -2

117. 1 **119.** $-2i, 2i, -1$

121. $-2i, 2i, -1, 1$

123. 16 **125.** $\dfrac{1 + \sqrt{19}}{3}$ **127.** i and E; ii and F; iii and C; iv and A; v and B; vi and D **129.** When
the bicyclist is traveling at 15 mph, the drag force on the bicyclist is 33.975 lb. **131.** $y = (x - 2)^2 + 3$;

vertex: $(2, 3)$ **133.** $y = \left(x + \dfrac{1}{2}\right)^2 + \dfrac{7}{4}$; vertex: $\left(-\dfrac{1}{2}, \dfrac{7}{4}\right)$ **135.** $y = 3x^2 - 6x + 5$

CONCEPT REVIEW 9.7

1. Sometimes true **3.** Always true **5.** Sometimes true

SECTION 9.7

3. -8; down; y **5. a.** Maximum value **b.** Minimum value **c.** Maximum value

7. Minimum value at 2 **9.** Maximum value at -1 **11.** Maximum value at $\dfrac{41}{8}$ **13.** Minimum value

at $-\dfrac{73}{8}$ **15.** Maximum value at $-\dfrac{71}{12}$ **17.** Minimum value at $-\dfrac{13}{4}$ **19.** a **21.** The function

has a minimum value at 7. **23.** True **25.** t; 30; -5; 3; 3 **27.** The rock will attain a maximum height
above the beach of 114 ft. **29.** The maximum profit the tour operator can expect is $1600. **31.** The pool
will have the least amount of algae 5 days after treatment. **33.** The minimum thickness of the mirror is 20.21
in. **35.** The water will land $31\frac{2}{3}$ ft up on the building. **37.** The two numbers are 10 and 10.
39. The dimensions that will maximize the enclosed area are 100 ft by 50 ft. **41.** 3.0 **43.** 0.5

CHAPTER 9 REVIEW EXERCISES

1. 0 and $\dfrac{3}{2}$ [9.1.1] **2.** $-2c$ and $\dfrac{c}{2}$ [9.1.1] **3.** $-4\sqrt{3}$ and $4\sqrt{3}$ [9.1.3] **4.** $-\dfrac{1}{2} - 2i$ and

$-\dfrac{1}{2} + 2i$ [9.1.3] **5.** $-\dfrac{17}{4}$ [9.7.1] **6.** 3 [9.7.1] **7.** $3x^2 + 8x - 3 = 0$ [9.1.2] **8.** -5 and

$\dfrac{1}{2}$ [9.1.1] **9.** $-1 - 3\sqrt{2}$ and $-1 + 3\sqrt{2}$ [9.1.3] **10.** $-3 - i$ and $-3 + i$ [9.2.1]

11. 2 and 3 [9.3.3] **12.** $-\sqrt{2}, \sqrt{2}, -2, 2$ [9.3.1] **13.** $\dfrac{25}{18}$ [9.3.2] **14.** -8 and $\dfrac{1}{8}$ [9.3.1]

15. 2 [9.3.2] **16.** $3 - \sqrt{11}$ and $3 + \sqrt{11}$ [9.2.2] **17.** 5 and $-\dfrac{4}{9}$ [9.3.3] **18.** $\dfrac{1 + \sqrt{3}}{2}$ and

$\dfrac{1 - \sqrt{3}}{2}$ [9.2.2] **19.** $\dfrac{5}{4}$ [9.3.2] **20.** 3 and -1 [9.3.3] **21.** -1 [9.3.3] **22.** $\dfrac{-3 + \sqrt{249}}{10}$ and

$\dfrac{-3 - \sqrt{249}}{10}$ [9.3.3] **23.** $\dfrac{-11 - \sqrt{129}}{2}$ and $\dfrac{-11 + \sqrt{129}}{2}$ [9.3.3] **24.** $x = 3$ [9.6.1]

25. $\left(\dfrac{3}{2}, \dfrac{1}{4}\right)$ [9.6.1] **26.** $4 + \sqrt{21}$ and $4 - \sqrt{21}$ [9.6.2] **27.** Two [9.6.2] **28.** $\left(\dfrac{-3 - \sqrt{5}}{2}, 0\right)$ and

$\left(\dfrac{-3 + \sqrt{5}}{2}, 0\right)$ [9.6.2] **29.** $(-2, 0)$ and $\left(\dfrac{1}{2}, 0\right)$ [9.6.2] **30.** $-\dfrac{1}{3} + \dfrac{\sqrt{5}}{3}i$ and $-\dfrac{1}{3} - \dfrac{\sqrt{5}}{3}i$ [9.6.2]

31. $\left\{x \mid -3 < x < \dfrac{5}{2}\right\}$ [9.5.1] **32.** $\left\{x \mid x \le -4 \text{ or } -\dfrac{3}{2} \le x \le 2\right\}$ [9.5.1] **33.** $\left\{x \mid x < \dfrac{3}{2} \text{ or } x \ge 2\right\}$

⟨number line: -5 -4 -3 -2 -1 0 1 2 3 4 5⟩ [9.5.1] **34.** $\left\{x \mid x \le -3 \text{ or } \dfrac{1}{2} \le x < 4\right\}$ ⟨number line: -5 -4 -3 -2 -1 0 1 2 3 4 5⟩ [9.5.1]

35. [9.6.1]

Domain: $\{x \mid x \in \text{real numbers}\}$
Range: $\{y \mid y \ge -5\}$

36. [9.6.1]

Vertex: $(1, 2)$
Axis of symmetry: $x = 1$

37. The integers are 2, 4, and 6 or -6, -4, and -2. [9.4.1] **38.** The width of the rectangle is 5 cm. The length
of the rectangle is 12 cm. [9.4.1] **39.** Working alone, the new computer can print the payroll in 12 min. [9.4.1]
40. The rate of the first car is 40 mph. The rate of the second car is 50 mph. [9.4.1]

CHAPTER 9 TEST

1. $\dfrac{3}{2}$ and -2 [9.1.1] **2.** $\dfrac{3}{4}$ and $-\dfrac{4}{3}$ [9.1.1] **3.** 9 [9.7.1] **4.** $3x^2 - 8x - 3 = 0$ [9.1.2]

5. $-3 + 3\sqrt{2}$ and $-3 - 3\sqrt{2}$ [9.1.3] **6.** $-2 + \sqrt{5}$ and $-2 - \sqrt{5}$ [9.2.2] **7.** $\dfrac{-3 + \sqrt{41}}{2}$ and

$\dfrac{-3 - \sqrt{41}}{2}$ [9.6.2] **8.** $\dfrac{1}{6} - \dfrac{\sqrt{95}}{6}i$ and $\dfrac{1}{6} + \dfrac{\sqrt{95}}{6}i$ [9.2.2] **9.** $-3 - \sqrt{10}$ and $-3 + \sqrt{10}$ [9.3.3]

10. $\dfrac{1}{4}$ [9.3.1] **11.** $-3, 3, -\sqrt{2}, \sqrt{2}$ [9.3.1] **12.** 4 [9.3.2] **13.** $2 + \sqrt{10}$ and $2 - \sqrt{10}$ [9.6.2]

14. Two [9.6.2] **15.** $(-4, 0)$ and $\left(\dfrac{3}{2}, 0\right)$ [9.6.2] **16.** $x = -\dfrac{3}{2}$ [9.6.1]

17.

[9.6.1] **18.** $\left\{x \mid -4 < x \le \dfrac{3}{2}\right\}$

[9.5.1]

Domain: $\{x \mid x \in \text{real numbers}\}$
Range: $\{y \mid y \ge -4.5\}$

19. The height of the triangle is 4 ft. The base of the triangle is 15 ft. [9.4.1] **20.** The rate of the canoe in calm water is 4 mph. [9.4.1]

CUMULATIVE REVIEW EXERCISES

1. 14 [1.5.1] **2.** -28 [2.2.3] **3.** $-\dfrac{3}{2}$ [3.4.1] **4.** $y = x + 1$ [3.6.1]

5. $-3xy(x^2 - 2xy + 3y^2)$ [6.1.1] **6.** $(2x - 5)(3x + 4)$ [6.3.1/6.3.2] **7.** $(x + y)(a^n - 2)$ [6.1.2]

8. $x^2 - 3x - 4 - \dfrac{6}{3x - 4}$ [5.5.2] **9.** $\dfrac{x}{2}$ [7.2.1] **10.** $2\sqrt{5}$ [3.1.2] **11.** $b = \dfrac{2S - an}{n}$ [7.6.1]

12. $-8 - 14i$ [8.5.3] **13.** $1 - a$ [8.1.1] **14.** $\dfrac{x\sqrt[3]{4y^2}}{2y}$ [8.2.4] **15.** $-\dfrac{3}{2}$ and -1 [9.3.3]

16. $-\dfrac{1}{2}$ [9.3.3] **17.** $2, -2, \sqrt{2}$, and $-\sqrt{2}$ [9.3.1] **18.** 0 and 1 [9.3.2]

19. $\left\{x \mid -2 < x < \dfrac{10}{3}\right\}$ [2.8.2] **20.** $\left(\dfrac{5}{2}, 0\right)$ and $(0, -3)$ [3.3.2] **21.**

[4.5.1]

22. $(1, -1, 2)$ [4.2.2] **23.** $-\dfrac{7}{3}$ [7.1.1] **24.** $\{x \mid x \ne 5, -3\}$ [7.1.1]

25. $\{x \mid x < -3 \text{ or } 0 < x < 2\}$

[9.5.1]

26. $\{x \mid -3 < x \le 1 \text{ or } x \ge 5\}$

[9.5.1] **27.** The average annual rate of change is 5.8 ft. [3.1.4] **28.** The lower and upper limits of the length of the piston rod are $9\dfrac{23}{64}$ in. and

$9\dfrac{25}{64}$ in. [2.8.3] **29.** The area of the triangle is $(x^2 + 6x - 16)$ ft^2. [5.3.5] **30.** $m = -\dfrac{25,000}{3}$; a slope of

$-\dfrac{25,000}{3}$ means that the value of the building decreases $8333.33 each year. [3.4.1]

Answers to Chapter 10 Selected Exercises

PREP TEST

1. 2 [1.5.1] **2.** −7 [1.5.1] **3.** 30 [3.2.1] **4.** $h^2 + 4h − 1$ [3.2.1] **5.** $y = \dfrac{1}{2}x − 2$ [3.3.2]

6. D: {−2, 3, 4, 6}; R: {4, 5, 6}; Yes [3.2.1] **7.** 8 [7.1.1]

8. 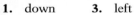 [3.2.2] **9.** [9.6.1]

CONCEPT REVIEW 10.1

1. Always true **3.** Always true **5.** Never true

SECTION 10.1

1. down **3.** left

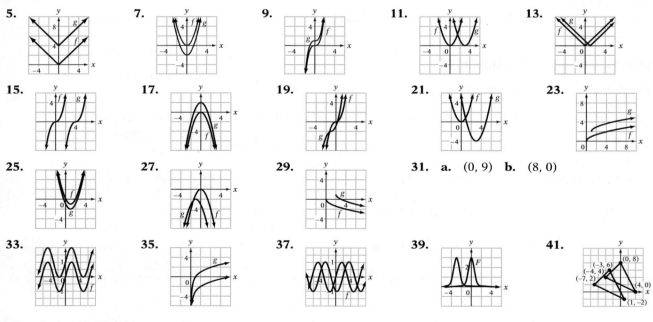

31. a. (0, 9) **b.** (8, 0)

CONCEPT REVIEW 10.2

1. Always true **3.** Never true

SECTION 10.2

1. −2; −2; −2; 4; 4; −6; 0; −7; 7 **3.** 5 **5.** 1 **7.** 0 **9.** $-\dfrac{29}{4}$ **11.** $\dfrac{2}{3}$ **13.** 2

15. 39 **17.** −8 **19.** $-\dfrac{4}{5}$ **21.** −2 **23.** 7 **25.** Undefined **27.** Defined

29. $4x$; $4x$; $4x$; $16x^2$; $8x$ **33.** −13 **35.** −29 **37.** $8x − 13$ **39.** 1 **41.** 5 **43.** $x^2 + 1$

45. 5 **47.** 11 **49.** $3x^2 + 3x + 5$ **51.** −3 **53.** −27 **55.** $x^3 − 6x^2 + 12x − 8$

57. No **59.** Yes **61.** $6h + h^2$ **63.** $2 + h$ **65.** $h + 2a$ **67.** −1 **69.** −6

71. $6x − 13$ **73.** −2 **75.** −2 **77.** 0 **79.** 6

CONCEPT REVIEW 10.3

1. Always true **3.** Never true **5.** Sometimes true

SECTION 10.3

3. $-2; 4$ **5.** Yes **7.** No **9.** Yes **11.** No **13.** No **15.** No **17.** Yes

19. No **21.** $-3; 2$ **23.** $\{(0, 1), (3, 2), (8, 3), (15, 4)\}$ **25.** No inverse

27. $f^{-1}(x) = \dfrac{1}{4}x + 2$ **29.** No inverse **31.** $f^{-1}(x) = x + 5$ **33.** $f^{-1}(x) = 3x - 6$

35. $f^{-1}(x) = -\dfrac{1}{3}x - 3$ **37.** $f^{-1}(x) = \dfrac{3}{2}x - 6$ **39.** $f^{-1}(x) = -3x + 3$ **41.** $f^{-1}(x) = \dfrac{1}{2}x + \dfrac{5}{2}$

43. No inverse **45.** $\dfrac{5}{3}$ **47.** 3 **49.** Yes; yes **51.** Yes **53.** No **55.** Yes

57. Yes **59.** Range **61.** 3 **63.** 0 **65.** 2 **67.** c **69.** a

71. $f[g(x)] = f(5x + 4) = 3(5x + 4) + 2 = 15x + 14$; $g[f(x)] = g(3x + 2) = 5(3x + 2) + 4 = 15x + 14$; No

73. **75.** The function and its inverse have the same graph.

77. **81. a.** $f^{-1} = \dfrac{1}{2}x + 2$ **b.** MATH **83.** $f^{-1}(x) = \dfrac{1}{3}x$. The inverse function converts feet into yards.

85. $P^{-1}(x) = \dfrac{1}{4}x$. The inverse function gives the length of a side of a square given its perimeter.

CHAPTER 10 REVIEW EXERCISES

1. [10.1.1] **2.** [10.1.1] **3.** [10.1.1] **4.** [10.1.1]

5. 7 [10.2.1] **6.** -9 [10.2.1] **7.** 70 [10.2.1] **8.** $\dfrac{12}{7}$ [10.2.1] **9.** $12x^2 + 12x - 1$ [10.2.2]

10. 5 [10.2.2] **11.** 10 [10.2.2] **12.** $6x^2 + 3x - 16$ [10.2.2] **13.** No [10.3.1]

14. $\{(1, -2), (3, 2), (-4, 5), (9, 7)\}$ [10.3.2] **15.** No [10.3.1] **16.** Yes [10.3.1]

17. $f^{-1}(x) = 2x - 16$ [10.3.2] **18.** $f^{-1}(x) = -\dfrac{1}{6}x + \dfrac{2}{3}$ [10.3.2] **19.** $f^{-1}(x) = \dfrac{3}{2}x + 18$ [10.3.2]

20. Yes [10.3.2]

CHAPTER 10 TEST

1. No [10.3.1] **2.** -2 [10.2.1] **3.** 234 [10.2.1] **4.** $-\dfrac{13}{2}$ [10.2.1] **5.** -5 [10.2.1]

6. 5 [10.2.2] **7.** [10.1.1] **8.** [10.1.1] **9.** 47 [10.2.2]

10. [10.1.1] **11.** [10.1.1] **12.** $f^{-1}(x) = 4x + 16$ [10.3.2]

13. $\{(6, 2), (5, 3), (4, 4), (3, 5)\}$ [10.3.2] **14.** Yes [10.3.2] **15.** $2x^2 - 4x - 5$ [10.2.2]

16. $f^{-1}(x) = 2x + 6$ [10.3.2] **17.** No [10.3.2] **18.** No [10.3.1]

CUMULATIVE REVIEW EXERCISES

1. $-\dfrac{23}{4}$ [1.5.1] **2.** [2.7.1] **3.** 3 [2.2.3]

4. $\{x \mid x < -2 \text{ or } x > 3\}$ [2.7.3] **5.** [9.6.1] **6.** [3.7.1]

Vertex: (0, 0)

Axis of symmetry: $x = 0$

7. $\{x \mid x \in \text{real numbers}\}$ [2.8.2] **8.** $-\dfrac{a^{10}}{12b^4}$ [5.2.1] **9.** $2x^3 - 4x^2 - 17x + 4$ [5.3.2]

10. $(a + 2)(a - 2)(a^2 + 2)$ [6.4.4] **11.** $xy(x - 2y)(x + 3y)$ [6.4.4] **12.** -3 and 8 [6.5.1]

13. $\{x \mid x < -3 \text{ or } x > 5\}$ [9.5.1] **14.** $\dfrac{5}{2x - 1}$ [7.2.2] **15.** -2 [7.4.1] **16.** $-3 - 2i$ [8.5.4]

17. $y = -2x - 2$ [3.5.2] **18.** $y = -\dfrac{3}{2}x - \dfrac{7}{2}$ [3.6.1] **19.** $\dfrac{1}{2} + \dfrac{\sqrt{3}}{6}i$ and $\dfrac{1}{2} - \dfrac{\sqrt{3}}{6}i$ [9.2.2]

20. 3 [8.4.1] **21.** -3 [9.7.1] **22.** $\{1, 2, 4, 5\}$ [3.2.1] **23.** Yes [3.2.1] **24.** 2 [8.4.1]

25. $g[h(2)] = 10$ [10.2.2] **26.** $f^{-1}(x) = -\dfrac{1}{3}x + 3$ [10.3.2] **27.** The cost per pound of the mixture is

\$3.96. [2.3.3] **28.** 25 lb of the 80% copper alloy must be used. [2.6.2] **29.** An additional 4.5 oz of

insecticide is required. [7.5.1] **30.** It would take the larger pipe 4 min to fill the tank. [7.4.2]

31. A force of 40 lb will stretch the spring 24 in. [7.5.2] **32.** The average rate of change in the production of

gigawatts of energy is 6.50 gigawatts per year. [3.1.4] **33.** The frequency is 80 vibrations per minute. [7.5.2]

Answers to Chapter 11 Selected Exercises

PREP TEST

1. $\dfrac{1}{9}$ [5.4.1] **2.** 16 [5.4.1] **3.** -3 [5.4.1] **4.** 0; 108 [3.2.1] **5.** -6 [2.2.2]

6. $-2, 8$ [9.1.1] **7.** 6326.60 [1.5.1] **8.** [3.2.2]

CONCEPT REVIEW 11.1

1. Never true **3.** Never true **5.** Never true

SECTION 11.1

3. c **5.** 1; 0; 1 **7. a.** 9 **b.** 1 **c.** $\dfrac{1}{9}$ **9. a.** 16 **b.** 4 **c.** $\dfrac{1}{4}$ **11. a.** 1

b. $\dfrac{1}{8}$ **c.** 16 **13. a.** 7.3891 **b.** 0.3679 **c.** 1.2840 **15. a.** 54.5982 **b.** 1

c. 0.1353 **17. a.** 16 **b.** 16 **c.** 1.4768 **19. a.** 0.1353 **b.** 0.1353 **c.** 0.0111

21. $\left(\dfrac{1}{2}\right)^{-3}$ **23.** Less than **25.** $-3; 3^3; \dfrac{1}{27}; -3; \dfrac{1}{27}$ **27.** **29.**

31. **33.** **35.** **37.** **39.** b and d

41. (0, 1) **43.** There is no x-intercept. The y-intercept is (0, 1). **45.** Greater than **47.** $g(x) = b^x - a$

49. 1.6 **51.** 1.1 **53.** The investment will be worth $1000 in 9 years.

55. 50% of the light will reach to a depth of 0.5 m. **57. a.** **b.** The point (4, 55.3) means that after 4 s, the object is moving at a speed of 55.3 ft/s.

CONCEPT REVIEW 11.2

1. Always true **3.** Never true **5.** Always true **7.** Always true

SECTION 11.2

3. base b; x **5.** $\log_5 25 = 2$ **7.** $\log_4 \dfrac{1}{16} = -2$ **9.** $\log_{10} x = y$ **11.** $\log_a w = x$ **13.** $3^2 = 9$

15. $10^{-2} = 0.01$ **17.** $e^y = x$ **19.** $b^v = u$ **21.** $u = v$ **23.** 4 **25.** 7 **27.** 2

29. 3 **31.** 0 **33.** 4 **35.** 9 **37.** 64 **39.** $\dfrac{1}{7}$ **41.** 1 **43.** 316.23

45. 0.02 **47.** 7.39 **49.** 0.61 **51.** Greater than **55.** $r\log_b x$ **57.** $\log_8 x + \log_8 z$

59. $5 \log_3 x$ **61.** $\ln r - \ln s$ **63.** $2 \log_3 x + 6 \log_3 y$ **65.** $3 \log_7 u - 4 \log_7 v$

67. $2 \log_2 r + 2 \log_2 s$ **69.** $2 \log_9 x + \log_9 y + \log_9 z$ **71.** $\ln x + 2 \ln y - 4 \ln z$

73. $2 \log_8 x - \log_8 y - 2 \log_8 z$ **75.** $\dfrac{1}{2} \log_7 x + \dfrac{1}{2} \log_7 y$ **77.** $\dfrac{1}{2} \log_2 x - \dfrac{1}{2} \log_2 y$ **79.** $\dfrac{3}{2} \ln x + \dfrac{1}{2} \ln y$

81. $\dfrac{3}{2} \log_7 x - \dfrac{1}{2} \log_7 y$ **83.** $\log_3 \dfrac{x^3}{y}$ **85.** $\log_8 x^4 y^2$ **87.** $\ln x^3$ **89.** $\log_5 x^3 y^4$ **91.** $\log_4 \dfrac{1}{x^2}$

93. $\log_3 \dfrac{x^2 z^2}{y}$ **95.** $\log_b \dfrac{x}{y^2 z}$ **97.** $\ln x^2 y^2$ **99.** $\log_6 \sqrt{\dfrac{x}{y}}$ **101.** $\log_4 \dfrac{s^2 r^2}{t^4}$ **103.** $\log_5 \dfrac{x}{y^2 z^2}$

105. $\ln \dfrac{t^3 v^2}{r^2}$ **107.** $\log_4 \sqrt{\dfrac{x^3 z}{y^2}}$ **109.** False **111.** True **113.** 1.3863 **115.** 1.0415

117. 0.8617 **119.** 2.1133 **121.** −0.6309 **123.** 0.2727 **125.** 1.6826 **127.** 1.9266

129. $\dfrac{\log(3x - 2)}{\log 3}$ **131.** $\dfrac{\log(4 - 9x)}{\log 8}$ **133.** $\dfrac{5}{\log 9} \log(6x + 7)$ **135.** $\dfrac{\ln(x + 5)}{\ln 2}$ **137.** $\dfrac{\ln(x^2 + 9)}{\ln 3}$

139. $\dfrac{7}{\ln 8} \ln(10x - 7)$ **141.** 8 **143.** 6 **145.** 2 **147.** 3 **149.** b **151.** 13.12

153. −6.59 **155.** 15.1543

CONCEPT REVIEW 11.3

1. Never true **3.** Always true

SECTION 11.3

5. y; exponential; $3^y + 1$; y; x **7.** **9.** **11.**

13. **15.** **17.** **19.** **21.**

23. **25.** **27.** **29.** **b** and **d**

31. **a.**

b. The point (4, 49) means that after 4 months, the typist's proficiency has dropped to 49 words per minute.

33. **a.**

b. According to the model, energy production will first exceed 6 quadrillion BTU in 2006.

c. According to the model, energy production in 2010 will be 6.7 quadrillion BTU.

35. A quadratic function **37.** A linear function **39.** An exponential function

41. A logarithmic function

CONCEPT REVIEW 11.4

1. Always true **3.** Never true **5.** Always true

SECTION 11.4

3. 3; 2; 4x; 3x − 1; 4x; −1 **5.** 1 **7.** −3 **9.** 1.1133 **11.** 0.7211 **13.** −1.5850
15. 1.7095 **17.** 1.3222 **19.** −2.8074 **21.** 3.5850 **23.** 1.1309 **25.** 1 **27.** 6
29. 1.3754 and −1.3754 **31.** 1.5805 **33.** −0.6309 **35.** 1.6610 **37.** −1.5129 **39.** 1.5
41. **a, b,** and **c** **43.** 0.63 **45.** −1.86 and 3.44 **47.** −1.16 **51.** exponential form
53. **a.** Exercise 52 **b.** Exercise 51 **c.** Exercise 51 **d.** Exercise 52 **55.** 8 **57.** $\dfrac{11}{2}$
59. 2 and −4 **61.** $\dfrac{5}{3}$ **63.** $\dfrac{1}{2}$ **65.** 1,000,000,000 **67.** 4 **69.** 3 **71.** 3 **73.** 2
75. No solution **77.** 4 **79.** $\dfrac{5 \pm \sqrt{33}}{2}$ **81.** 1.76 **83.** 2.42 **85.** −1.51 and 2.10
87. 1.7233 **89.** 0.8060 **91.** 5.6910 **93.** 2774

95. **a.**

b. It will take the object approximately 2.64 s to fall 100 ft.

CONCEPT REVIEW 11.5

1. Always true **3.** Always true

SECTION 11.5

3. 4000; 3000; 0.0025; unknown **5.** The value of the investment after 2 years is $1172. **7.** The
investment will be worth $15,000 in 18 years. **9.** **a.** The amount of technetium will be 21.2 mg after 3 h.
b. It takes 3.5 h for the amount of technetium to reach 20 mg. **11.** The half-life of promethium-147 is
2.5 years. **13.** 2 years **15.** **a.** The approximate pressure at 40 km above Earth is 0.098 newtons per
square centimeter. **b.** The approximate pressure on Earth's surface is 10.13 newtons per square centimeter.
c. The atmospheric pressure decreases as you rise above Earth's surface. **17.** The pH of milk is 6.4.
19. At a depth of 2.5 m, the percent of light will be 75% of the light at the surface of the pool. **21.** Normal
conversation is 65 decibels. **23.** The mass of the colony would be equal to the mass of Earth in 46 h. This is
more time than predicted in the novel. **25.** A thickness of 0.4 cm is needed. **27.** The Richter scale
magnitude of the earthquake was 9.5. **29.** The intensity of the earthquake was 316,228,000I_0.
31. The magnitude of the earthquake is 5.3. **33.** The magnitude of the earthquake is 5.6. **35.** Prices will
double in 14 years. **37.** The value of the investment after 5 years is $3210.06. **39.** **a.** 104 months are
required to reduce the loan amount to $90,000. **b.** 269 months are required to reduce the loan amount to one-
half the original amount. **c.** The total interest paid exceeds $100,000 in month 163.

CHAPTER 11 REVIEW EXERCISES

1. 2 [11.2.1] **2.** $\log_3 \sqrt{\dfrac{x}{y}}$ [11.2.2] **3.** 1 [11.1.1] **4.** −3 [11.4.1] **5.** 1 [11.1.1]

6. $\dfrac{1}{9}$ [11.2.1] **7.** $\log_2 32 = 5$ [11.2.1] **8.** 6 [11.4.2] **9.** $\dfrac{1}{2}\log_6 x + \dfrac{3}{2}\log_6 y$ [11.2.2]

10. 2 [11.4.1] **11.** -2 [11.4.1] **12.** $\dfrac{1}{3}$ [11.1.1] **13.** 4 [11.2.1] **14.** 2 [11.4.2]

15. 2.3219 [11.2.2] **16.** 1.7251 [11.2.2] **17.** 3 [11.4.1] **18.** 4 [11.1.1]

19. $\dfrac{1}{2}\log_5 x - \dfrac{1}{2}\log_5 y$ [11.2.2] **20.** $\dfrac{1}{4}$ [11.4.2] **21.** 125 [11.2.1] **22.** $\dfrac{1}{2}$ [11.4.2]

23. $\log_b \dfrac{x^3}{y^5}$ [11.2.2] **24.** 4 [11.1.1] **25.** 2.6801 [11.2.2] **26.** -0.5350 [11.4.1]

27. [11.1.2] **28.** [11.1.2] **29.** [11.3.1]

30. [11.3.1] **31.** 1.0 [11.3.1] **32.** The half-life is 33 h. [11.5.1]

33. The material must be 0.602 cm thick. [11.5.1]

CHAPTER 11 TEST

1. 1 [11.1.1] **2.** $\dfrac{1}{64}$ [11.1.1] **3.** 3 [11.2.1] **4.** $\dfrac{1}{16}$ [11.2.1]

5. $\dfrac{2}{3}\log_6 x + \dfrac{5}{3}\log_6 y$ [11.2.2] **6.** $\log_5 \sqrt{\dfrac{x}{y}}$ [11.2.2] **7.** 3 [11.4.2] **8.** $\dfrac{1}{3}$ [11.1.1]

9. 2.5789 [11.4.1] **10.** 10 [11.4.2] **11.** 3 [11.4.1] **12.** -4 [11.4.1] **13.** $\dfrac{3}{2}$ [11.4.2]

14. [11.1.2] **15.** [11.1.2] **16.** [11.3.1]

17. [11.3.1] **18.** 1.6 [11.1.2] **19.** The value of the investment after 6 years is \$15,661. [11.5.1]

20. The half-life is 24 h. [11.5.1]

CUMULATIVE REVIEW EXERCISES

1. $\dfrac{8}{7}$ [2.2.3] **2.** $L = \dfrac{S - 2WH}{2W + 2H}$ [7.6.1] **3.** $\{x \mid 1 \le x \le 4\}$ [2.8.2] **4.** $(4x^n + 3)(x^n + 1)$ [6.4.3]

5. $\{x \mid -5 \le x \le 1\}$ [9.5.1] **6.** $\dfrac{x - 3}{x + 3}$ [7.3.1] **7.** $\dfrac{x\sqrt{y} + y\sqrt{x}}{x - y}$ [8.2.4] **8.** $-4x^2 y^3 \sqrt{2x}$ [8.2.2]

9. $-\dfrac{1}{5} + \dfrac{2}{5}i$ [8.5.4] **10.** $y = 2x - 6$ [3.6.1] **11.** $3x^2 + 8x - 3 = 0$ [9.1.2] **12.** $2 + \sqrt{10}$ and $2 - \sqrt{10}$ [9.2.1, 9.2.2] **13.** $\{-6, -4, 0\}$ [3.2.1] **14.** 4 [10.2.2] **15.** $(0, -1, 2)$ [4.2.2]

16. $\left(-\dfrac{1}{2}, -2\right)$ [4.1.2] **17.** 243 [11.1.1] **18.** 64 [11.2.1] **19.** 8 [11.4.1] **20.** 1 [11.4.2]

21. [2.7.1] **22.** [9.5.1] **23.** [9.6.1]

24. [3.2.2] **25.** [11.1.2] **26.** [11.3.1]

27. 800 lb of the alloy containing 25% tin and 1200 lb of the alloy containing 50% tin were used. [2.6.2]

28. To earn $3000 or more a month, the sales executive must make sales amounting to $31,250 or more. [2.7.4]

29. [3.1.3] **30.** With both printers operating, it would take 7.5 min to print the checks. [7.4.2]

31. When the volume is 25 ft^3, the pressure is 500 lb/in^2. [7.5.2] **32.** The cost per yard of the wool carpet is $24. [4.4.2] **33.** The value of the investment after 5 years is $15,657. [11.5.1]

Answers to Chapter R Selected Exercises

SECTION R.1

1. 10 **3.** 21 **5.** −2 **7.** 5 **9.** −3 **11.** −1 **13.** 1 **15.** 18 **17.** −6
19. 2 **21.** −40 **23.** 2.01 **25.** 8x **27.** 3b **29.** 5a **31.** 11x **33.** 15x − 2y
35. 9a **37.** 22y^2 **39.** $\frac{1}{2}x$ **41.** −20x **43.** −24a **45.** x **47.** 7x **49.** 3y
51. −3a − 15 **53.** −16x − 48 **55.** −10y^2 + 5 **57.** 18x^2 − 12xy − 6y^2 **59.** −8y − 7
61. 5x − 75 **63.** −a + 56

SECTION R.2

1. −12 **3.** −1 **5.** −8 **7.** −20 **9.** 8 **11.** −1 **13.** 1 **15.** $\frac{3}{4}$ **17.** 3
19. 3 **21.** 2 **23.** −1 **25.** $\frac{1}{2}$ **27.** 5 **29.** $-\frac{1}{9}$ **31.** {x|x > 3} **33.** {n|n ≥ 2}
35. {x|x ≤ −3} **37.** {n|n > 0} **39.** {x|x > 4} **41.** {x|x > −2} **43.** {x|x > 3}
45. {x|x ≥ 2} **47.** {x|x < 2} **49.** (3, ∞) **51.** [2, ∞) **53.** (−∞, 4) **55.** $\left(\frac{14}{11}, \infty\right)$
57. $\left(-\infty, \frac{3}{8}\right]$ **59.** $\left[-\frac{5}{4}, \infty\right)$

SECTION R.3

1.

3.

5.

7.

9.

11.

13.

15.

17.

19.

21.

23. 3 **25.** $\frac{1}{4}$ **27.** $-\frac{3}{2}$ **29.** undefined **31.** 0 **33.** 1

35. $m = \frac{5}{2}, b = (0, -4)$ **37.**

39.

41.

43.

45.

47. 13; $(-4, 13)$ **49.** 14; $(-1, 14)$ **51.** -1; $(2, -1)$

53. -15; $(-2, -15)$ **55.** $y = -3x - 1$ **57.** $y = -2x + 3$ **59.** $y = -\frac{3}{5}x$ **61.** $y = -\frac{2}{3}x - 4$

63. $y = -\frac{1}{2}x + 4$

SECTION R.4

1. z^8 **3.** x^{15} **5.** $x^{12}y^{18}$ **7.** a^6 **9.** $-m^9n^3$ **11.** $-8a^9b^3c^6$ **13.** mn^2 **15.** $-\frac{2a}{3}$

17. $-27m^2n^4p^3$ **19.** $54n^{14}$ **21.** $\frac{7xz}{8y^3}$ **23.** $-8x^{13}y^{14}$ **25.** $\frac{4}{x^7}$ **27.** $\frac{1}{d^{10}}$ **29.** $\frac{1}{x^5}$

31. $\frac{x^2}{3}$ **33.** $\frac{x^6}{y^{12}}$ **35.** $\frac{9}{x^2y^4}$ **37.** $\frac{2}{x^4}$ **39.** $\frac{1}{2x^3}$ **41.** $\frac{1}{2x^2y^6}$ **43.** $\frac{1}{x^6}$ **45.** $\frac{y^8}{x^4}$

47. $\frac{b^{10}}{4a^{10}}$ **49.** $7b^2 + b - 4$ **51.** $3a^2 - 3a + 17$ **53.** $-7x - 7$ **55.** $-2x^3 + x^2 + 2$

57. $x^3 + 2x^2 - 6x - 6$ **59.** $5x^3 + 10x^2 - x - 4$ **61.** $y^3 - y^2 + 6y - 6$ **63.** $4y^3 - 2y^2 + 2y - 4$

65. $12b^4 - 48b^3 - 24b$ **67.** $9b^5 - 9b^3 + 24b$ **69.** $-2x^4y + 6x^3y^2 - 4x^2y^3$ **71.** $x^3 + 4x^2 + 5x + 2$

73. $a^3 - 6a^2 + 13a - 12$ **75.** $-2b^3 + 7b^2 + 19b - 20$ **77.** $x^4 - 4x^3 - 3x^2 + 14x - 8$

79. $y^4 + 4y^3 + y^2 - 5y + 2$ **81.** $a^2 + a - 12$ **83.** $y^2 - 10y + 21$ **85.** $2x^2 + 15x + 7$

87. $3x^2 + 11x - 4$ **89.** $4x^2 - 31x + 21$ **91.** $3y^2 - 2y - 16$ **93.** $21a^2 - 83a + 80$

95. $2x^2 + 3xy + y^2$ **97.** $3x^2 - 10xy + 8y^2$ **99.** $10a^2 + 14ab - 12b^2$ **101.** $16x^2 - 49$

103. $x + 2$ **105.** $2y - 7$ **107.** $x - 2 + \frac{8}{x + 2}$ **109.** $3y - 5 + \frac{10}{y + 2}$ **111.** $b - 5 - \frac{24}{b - 3}$

113. $3x + 17 + \dfrac{64}{x - 4}$ **115.** $5y + 3 + \dfrac{1}{2y + 3}$ **117.** $x^2 - 5x + 2$ **119.** $x^2 + 5$ **121.** $y(12y - 5)$

123. $5xyz(2xz + 3y^2)$ **125.** $5(x^2 - 3x + 7)$ **127.** $3y^2(y^2 - 3y - 2)$ **129.** $x^2y^2(x^2y^2 - 3xy + 6)$

131. $8x^2y(2 - xy^3 - 6y)$ **133.** $(x + 2)(x - 1)$ **135.** $(a + 4)(a - 3)$ **137.** $(a - 1)(a - 2)$

139. $(b + 8)(b - 1)$ **141.** $(z + 5)(z - 9)$ **143.** $(z - 5)(z - 9)$ **145.** $(b + 4)(b + 5)$

147. Nonfactorable **149.** $(x - 7)(x - 8)$ **151.** $(y + 3)(2y + 1)$ **153.** $(a - 1)(3a - 1)$

155. $(x - 3)(2x + 1)$ **157.** $(2t + 1)(5t + 3)$ **159.** $(2z - 1)(5z + 4)$ **161.** Nonfactorable

163. $(t + 2)(2t - 5)$ **165.** $(3y + 1)(4y + 5)$ **167.** $(a - 5)(11a + 1)$ **169.** $(2b - 3)(3b - 2)$

171. $3(x + 2)(x + 3)$ **173.** $a(b + 8)(b - 1)$ **175.** $2y^2(y + 3)(y - 16)$ **177.** $x(x - 5)(2x - 1)$

179. $5(t + 2)(2t - 5)$ **181.** $p(2p + 1)(3p + 1)$

Glossary

abscissa The first number in an ordered pair; it measures a horizontal distance and is also called the first coordinate of an ordered pair. (Sec. 3.1)

absolute value equation An equation that contains the absolute value symbol. (Sec. 2.8)

absolute value of a number The distance of the number from zero on the number line. (Sec. 1.1)

acute angle An angle whose measure is between $0°$ and $90°$. (Sec. 2.4)

addend In addition, a number being added. (Sec. 1.2)

addition The process of finding the total of two numbers. (Sec. 1.2)

addition method An algebraic method of finding an exact solution of a system of linear equations. (Sec. 4.2)

additive inverse of a polynomial The polynomial with the sign of every term changed. (Sec. 5.1)

additive inverses Numbers that are the same distance from zero on the number line but lie on different sides of zero; also called opposites. (Sec. 1.1, 1.5)

adjacent angles Two angles that share a common side. (Sec. 2.4)

alternate exterior angles Two nonadjacent angles that are on opposite sides of the transversal and outside the parallel lines. (Sec. 2.4)

alternate interior angles Two nonadjacent angles that are on opposite sides of the transversal and between the parallel lines. (Sec. 2.4)

analytic geometry Geometry in which a coordinate system is used to study relationships between variables. (Sec. 3.1)

antilogarithm If $\log_b M = N$, then the antilogarithm, base b, of N is M. (Sec. 11.2)

arithmetic mean of values Average determined by calculating the sum of the values and then dividing that result by the number of values. (Sec. 1.2)

average of a set of numbers The sum of the numbers divided by the number of numbers in the set. (Sec. 1.2)

axes The two number lines that form a rectangular coordinate system; also called coordinate axes. (Sec. 3.1)

axis of symmetry of a parabola A line of symmetry that passes through the vertex of the parabola and is parallel to the y-axis for an equation of the form $y = ax^2 + bx + c$. (Sec. 9.6)

base In an exponential expression, the number that is taken as a factor as many times as indicated by the exponent. (Sec. 1.4)

binomial A polynomial of two terms. (Sec. 5.1)

binomial factor A factor that has two terms. (Sec. 6.1)

clear denominators Remove denominators from an equation that contains fractions by multiplying each side of the equation by the LCM of the denominators. (Sec. 2.1, 7.4)

closed interval In interval notation, an interval that contains its endpoints. (Sec. 2.7)

cofactor of an element of a matrix $(-1)^{i+j}$ times the minor of that element of the matrix, where i is the row number of the element and j is its column number. (Sec. 4.3)

combined variation A variation in which two or more types of variation occur at the same time. (Sec. 7.5)

combine like terms Use the Distributive Property to add the coefficients of like variable terms; add like terms of a variable expression. (Sec. 1.5)

common logarithms Logarithms to the base 10. (Sec. 11.2)

common monomial factor A monomial factor that is a factor of the terms in a polynomial. (Sec. 6.1)

complementary angles Two angles whose sum is $90°$. (Sec. 2.4)

complete the square Add to a binomial the constant term that makes it a perfect-square trinomial. (Sec. 9.2)

complex fraction A fraction whose numerator or denominator contains one or more fractions. (Sec. 7.3)

complex number A number of the form $a + bi$, where a and b are real numbers and $i = \sqrt{-1}$. (Sec. 8.5)

composition of functions The operation on two functions f and g denoted by $f \circ g$. The value of the composition of f and g is given by $(f \circ g)(x) = f[g(x)]$. (Sec. 10.2)

compound inequality Two inequalities joined with a connective word such as "and" or "or." (Sec. 2.7)

compound interest Interest that is computed not only on the original principal but also on the interest already earned. (Sec. 11.5)

conjugates Binomial expressions that differ only in the sign of a term. The expressions $a + b$ and $a - b$ are conjugates. (Sec. 8.2)

consecutive even integers Even integers that follow one another in order. (Sec. 2.3)

consecutive integers Integers that follow one another in order. (Sec. 2.3)

consecutive odd integers Odd integers that follow one another in order. (Sec. 2.3)

constant function A function given by $f(x) = b$, where b is a constant. (Sec. 3.3)

constant of proportionality k in a variation equation; also called the constant of variation. (Sec. 7.5)

constant of variation k in a variation equation; also called the constant of proportionality. (Sec. 7.5)

constant term A term that contains no variable part. (Sec. 1.5)

coordinate axes The two number lines that form a rectangular coordinate system; also called simply axes. (Sec. 3.1)

coordinates of a point The numbers in the ordered pair that is associated with the point. (Sec. 3.1)

corresponding angles Two angles that are on the same side of the transversal and are both acute angles or are both obtuse angles. (Sec. 2.4)

cost The price that a business pays for a product. (Sec. 2.5)

cube root of a perfect cube One of the three equal factors of the perfect cube. (Sec. 6.4)

cubic function A third-degree polynomial function. (Sec. 5.1)

decimal notation Notation in which a number consists of a whole-number part, a decimal point, and a decimal part. (Sec. 1.3)

degree A unit used to measure angles. (Sec. 2.4)

degree of a monomial The sum of the exponents of the variables in a monomial. (Sec. 5.1)

degree of a polynomial The greatest of the degrees of any of the terms of a polynomial. (Sec. 5.1)

dependent system of equations A system of equations whose graphs coincide. (Sec. 4.1)

dependent variable A variable whose value depends on that of another variable, which is known as the independent variable. (Sec. 3.2)

descending order The terms of a polynomial in one variable arranged so that the exponents of the variable decrease from left to right. (Sec. 5.1)

determinant The number associated with a square matrix. (Sec. 4.3)

difference of two perfect cubes A polynomial of the form $a^3 - b^3$. (Sec. 6.4)

difference of two perfect squares A polynomial of the form $a^2 - b^2$. (Sec. 6.4)

direct variation A special function that can be expressed as the equation $y = kx$, where k is a constant called the constant of variation or the constant of proportionality. (Sec. 7.5)

discount The amount by which a retailer reduces the regular price of a product for a promotional sale. (Sec. 2.5)

discount rate The percent of the regular price that the discount represents. (Sec. 2.5)

discriminant For an equation of the form $ax^2 + bx + c = 0$, the quantity $b^2 - 4ac$. (Sec. 9.2)

domain The set of the first coordinates of all the ordered pairs of a function. (Sec. 3.2)

double root The two equal roots of a quadratic equation. (Sec. 9.1)

element of a matrix A number in the matrix. (Sec. 4.3)

elements of a set The objects in the set. (Sec. 1.1)

empty set The set that contains no elements. (Sec. 2.7)

equation A statement of the equality of two mathematical expressions. (Sec. 2.1)

equilateral triangle A triangle in which all three sides are of equal length. (Sec. 2.4)

evaluate a function Determine $f(x)$ for a given value of x. (Sec. 5.1)

evaluate a variable expression Replace the variable in a variable expression by a numerical value and then simplify the resulting expression. (Sec. 1.5)

even integer An integer that is divisible by 2. (Sec. 2.3)

expanding by cofactors A technique for finding the value of a 3×3 or larger determinant. (Sec. 4.3)

exponent In an exponential expression, the raised number that indicates how many times the factor, or base, occurs in the multiplication. (Sec. 1.4)

exponential equation An equation in which the variable occurs in the exponent. (Sec. 11.4)

exponential form The expression 2^6 is in exponential form. Compare *factored form*. (Sec. 1.4)

exponential function The function with base b defined by $f(x) = b^x$, where b is a positive real number not equal to 1. (Sec. 11.1)

exterior angle of a triangle An angle adjacent to an interior angle of a triangle. (Sec. 2.4)

extraneous solution When each side of an equation is raised to an even power, the resulting equation may have a solution that is not a solution of the original equation. Such a solution is called an extraneous solution. (Sec. 8.4)

factor In multiplication, a number being multiplied. (Sec. 1.2)

factored form The multiplication $2 \cdot 2 \cdot 2 \cdot 2 \cdot 2$ is in factored form. Compare *exponential form*. (Sec. 1.4)

factor by grouping Group and factor terms in a polynomial in such a way that a common binomial factor is found. (Sec. 6.1)

factor a polynomial Write the polynomial as a product of other polynomials. (Sec. 6.1)

finite set A set in which all the elements can be listed. (Sec. 2.7)

first coordinate of an ordered pair The first number of the ordered pair; it measures a horizontal distance and is also called the abscissa. (Sec. 3.1)

FOIL method A method of finding the product of two binomials in which the sum of the products of the first terms, of the outer terms, of the inner terms, and of the last terms is found. (Sec. 5.3)

formula A literal equation that states a rule about measurement. (Sec. 7.6)

function A relation in which no two ordered pairs that have the same first coordinate have different second coordinates. (Sec. 3.2)

function notation Notation used for those equations that define functions. The letter f is commonly used to name a function. (Sec. 3.2)

graph of a function A graph of the ordered pairs that belong to the function. (Sec. 3.2)

graph of an ordered pair The dot drawn at the coordinates of the point in the plane. (Sec. 3.1)

graph of a real number A heavy dot placed directly above the number on the number line. (Sec. 1.1)

graph of $x = a$ A vertical line passing through the point whose coordinates are $(a, 0)$. (Sec. 5.2)

graph of $y = b$ A horizontal line passing through the point whose coordinates are $(0, b)$. (Sec. 5.2)

greatest common factor of two numbers The greatest number that is a factor of the two numbers. (Sec. 6.1)

grouping symbols Parentheses (), brackets [], braces { }, the absolute value symbol | |, and the fraction bar. (Sec. 1.4)

half-plane The solution set of a linear inequality in two variables. (Sec. 3.7)

horizontal shift The displacement of a graph to the right or left on the coordinate axes; also known as a horizontal translation. (Sec. 10.1)

horizontal translation The displacement of a graph to the right or left on the coordinate axes; also known as a horizontal shift. (Sec. 10.1)

hypotenuse In a right triangle, the side opposite the 90° angle. (Sec. 3.1, 8.4)

imaginary number A number of the form ai, where a is a real number and $i = \sqrt{-1}$. (Sec. 8.5)

imaginary part of a complex number For the complex number $a + bi$, b is the imaginary part. (Sec. 8.5)

inconsistent system of equations A system of equations that has no solution. (Sec. 4.1)

independent system of equations A system of equations that has exactly one solution. (Sec. 4.1)

independent variable A variable whose value determines that of another variable, which is known as the dependent variable. (Sec. 3.2)

index In the expression $\sqrt[n]{a}$, n is the index of the radical. (Sec. 8.1)

infinite set A set in which all the elements cannot be listed. (Sec. 2.7)

integers The numbers $\ldots, -3, -2, -1, 0, 1, 2, 3, \ldots$. (Sec. 1.1)

interior angle of a triangle An angle within the region enclosed by a triangle. (Sec. 2.4)

intersection of two sets The set that contains all elements that are common to both of the sets. (Sec. 2.7)

interval notation A type of set-builder notation in which the property that distinguishes the elements of the set is their location within a specified interval. (Sec. 2.7)

inverse of a function The set of ordered pairs formed by reversing the coordinates of each ordered pair of the function. (Sec. 10.3)

inverse variation A function that can be expressed as the equation $y = k/x$, where k is a constant. (Sec. 7.5)

irrational number A number that cannot be written in the form a/b, where a and b are integers and b is not equal to zero. (Sec. 1.3)

isosceles triangle A triangle that has two equal angles and two equal sides. (Sec. 2.4)

joint variation A variation in which a variable varies directly as the product of two or more variables. A joint variation can be expressed as the equation $z = kxy$, where k is a constant. (Sec. 7.5)

leading coefficient In a polynomial, the coefficient of the variable with the largest exponent. (Sec. 5.1)

least common multiple of two polynomials The polynomial of least degree that contains the factors of each polynomial. (Sec. 7.2)

leg of a right triangle In a right triangle, one of the two sides that are not opposite the 90° angle. (Sec. 3.1, 8.4)

like terms Terms of a variable expression that have the same variable part. Having no variable part, constant terms are like terms. (Sec. 1.5)

linear equation in three variables An equation of the form $Ax + By + Cz = D$, where A, B, and C are coefficients of the variables and D is a constant. (Sec. 4.2)

linear equation in two variables An equation of the form $y = mx + b$, where m is the coefficient of x and b is a constant, or of the form $Ax + By = C$. Also called a linear function. (Sec. 3.3)

linear function An equation of the form $y = mx + b$, where m is the coefficient of x and b is a constant; also called a linear equation in two variables. Its graph is a straight line. (Sec. 3.3)

linear inequality in two variables An inequality of the form $y > mx + b$, or $Ax + By > C$. The symbol $>$ could be replaced by $\geq$, $<$, or $\leq$. (Sec. 3.7)

literal equation An equation that contains more than one variable. (Sec. 3.3)

logarithm The mathematical operation that is the inverse of exponentiation. The logarithm of a number x in base b is the number y such that $b^y = x$, written $\log_b x = y$. (Sec. 11.2)

lower limit In a tolerance, the lowest acceptable value. (Sec. 2.8)

main diagonal The elements $a_{11}, a_{22}, a_{33}, \ldots, a_{nm}$ of a matrix. (Sec. 4.3)

markup The difference between selling price and cost. (Sec. 2.5)

markup rate The percent of a retailer's cost that the markup represents. (Sec. 2.5)

matrix A rectangular array of numbers. (Sec. 4.3)

maximum value of a function The greatest value that the function can have. (Sec. 9.7)

midpoint of a line segment The point on a line segment that is equidistant from its endpoints. (Sec. 3.1)

minimum value of a function The least value that the function can have. (Sec. 9.7)

minor of an element The minor of an element in a 3×3 determinant is the 2×2 determinant obtained by eliminating the row and column that contain the element. (Sec. 4.3)

monomial A number, a variable, or a product of a number and variables; a polynomial of one term. (Sec. 5.1)

multiplication The process of finding the product of two numbers. (Sec. 1.2)

multiplicative inverse The multiplicative inverse of a nonzero real number a is $1/a$; also called the reciprocal of a. (Sec. 1.5)

natural exponential function The function defined by $f(x) = e^x$, where $e \approx 2.71828$. (Sec. 11.1)

natural logarithm Logarithm to the base e (the base of the natural exponential function), abbreviated $\ln x$. (Sec. 11.2)

natural numbers The numbers $1, 2, 3, \ldots$; also called the positive integers. (Sec. 1.1)

negative integers The numbers $\ldots, -3, -2, -1$. (Sec. 1.1)

negative slope The slope of a line that slants downward to the right. (Sec. 3.4)

nonfactorable over the integers A description of a polynomial that does not factor using only integers. (Sec. 6.2)

null set A set that contains no elements; also called an empty set. (Sec. 2.7)

numerical coefficient The number part of a variable term. (Sec. 1.5)

obtuse angle An angle whose measure is between 90° and 180°. (Sec. 2.4)

odd integer An integer that is not divisible by 2. (Sec. 2.3)

one-to-one function A function in which, given any y, there is only one x that can be paired with the given y. (Sec. 10.3)

open interval In interval notation, an interval that does not contain its endpoints. (Sec. 2.7)

opposites Numbers that are the same distance from zero on the number line but lie on different sides of zero; also called additive inverses. (Sec. 1.1)

order $m \times n$ A matrix of m rows and n columns. (Sec. 4.3)

Order of Operations Agreement Rules that specify the order in which operations are performed in simplifying numerical expressions. (Sec. 1.4)

ordered pair A pair of numbers expressed in the form (a, b) and used to locate a point in a rectangular coordinate system. (Sec. 3.1)

ordered triple Three numbers expressed in the form (x, y, z) and used to locate a point in an xyz-coordinate system. (Sec. 4.2)

ordinate The second number of an ordered pair; it measures a vertical distance and is also called the second coordinate of an ordered pair. (Sec. 3.1)

origin The point of intersection of the two number lines that form a rectangular coordinate system. (Sec. 3.1)

parabola The name given to the graph of a quadratic function. (Sec. 9.6)

parallel lines Lines that never meet; the distance between them is always the same. (Sec. 2.4)

percent Parts of 100. (Sec. 1.3)

percent mixture problem A problem that involves combining two solutions or alloys that have different concentrations of the same substance. (Sec. 2.6)

perfect cube The product of the same three factors. (Sec. 6.4)

perfect square The product of a term and itself.(Sec. 6.4)

perfect-square trinomial The square of a binomial. (Sec. 6.4)

perimeter The distance around a plane geometric figure. (Sec. 2.4)

perpendicular lines Intersecting lines that form right angles. (Sec. 2.4)

plane The infinitely extending, two-dimensional space in which a rectangular coordinate system lies; may be pictured as a large, flat piece of paper. (Sec. 3.1)

polynomial A variable expression in which the terms are monomials. (Sec. 5.1)

positive integers The numbers $1, 2, 3, \ldots$; also called the natural numbers. (Sec. 1.1)

positive slope The slope of a line that slants upward to the right. (Sec. 3.4)

principal square root The positive square root of a number. (Sec. 8.1)

product In multiplication, the result of multiplying two numbers. (Sec. 1.2)

proportion An equation that states the equality of two ratios or rates. (Sec. 7.5)

quadrant One of the four regions into which a rectangular coordinate system divides the plane. (Sec. 3.1)

quadratic equation An equation that can be written in the form $ax^2 + bx + c = 0$, where a and b are coefficients, c is a constant, and a 0; also called a second-degree equation. (Sec. 6.5, 9.1)

quadratic function A function that can be expressed by the equation $f(x) = ax^2 + bx + c$, where a is not equal to zero. (Sec. 5.1, 9.6)

quadratic inequality in one variable An inequality that can be written in the form $ax^2 + bx + c < 0$ or $ax^2 + bx + c > 0$, where a is not equal to zero. The symbols $\leq$ and $\geq$ can also be used. (Sec. 9.5)

radical In a radical expression, the symbol $\sqrt{}$. (Sec. 8.1)

radical equation An equation that contains a variable expression in a radical. (Sec. 8.4)

radical function A function that contains a variable expression in a radical. (Sec. 7.1)

radicand In a radical expression, the expression within the radical. (Sec. 8.1)

range The set of the second coordinates of all the ordered pairs of a function. (Sec. 3.2)

rate The quotient of two quantities that have different units. (Sec. 7.5)

rate of work The part of a task that is completed in one unit of time. (Sec. 7.5)

ratio The quotient of two quantities that have the same unit. (Sec. 7.5)

rational expression A fraction in which the numerator and denominator are polynomials. (Sec. 7.1)

rational function A function that is written in terms of a rational expression. (Sec. 7.1)

rationalize the denominator Remove a radical from the denominator of a fraction. (Sec. 8.2)

rational number A number of the form a/b, where a and b are integers and b is not equal to zero. (Sec. 1.3)

real numbers The rational numbers and the irrational numbers taken together. (Sec. 1.3)

real part of a complex number For the complex number $a + bi$, a is the real part. (Sec. 8.5)

reciprocal of a fraction The fraction that results when the numerator and denominator of a fraction are interchanged. (Sec. 1.3, 1.5)

reciprocal of a rational expression The rational expression that results when the numerator and denominator of a rational expression are interchanged. (Sec. 7.2)

rectangular coordinate system A coordinate system formed by two number lines, one horizontal and one vertical, that intersect at the zero point of each line. (Sec. 3.1)

relation A set of ordered pairs. (Sec. 3.2)

repeating decimal A decimal formed when dividing the numerator of a fraction by its denominator, in which a digit or a sequence of digits in the decimal repeats infinitely. (Sec. 1.3)

right angle An angle whose measure is 90°. (Sec. 2.4)

right triangle A triangle that contains a 90° angle. (Sec. 3.1)

roster method A method of designating a set by enclosing a list of its elements in braces. (Sec. 1.1)

scatter diagram A graph of ordered-pair data. (Sec. 3.1)

scientific notation Notation in which each number is expressed as the product of a number between 1 and 10 and a power of 10. (Sec. 5.4)

second coordinate of an ordered pair The second number of the ordered pair; it measures a vertical distance and is also called the ordinate. (Sec. 3.1)

second-degree equation An equation of the form $ax^2 + bx + c = 0$, where a and b are coefficients, c is a constant, and $a \neq 0$; also called a quadratic equation. (Sec. 9.1)

selling price The price for which a business sells a product to a customer. (Sec. 2.5)

set A collection of objects. (Sec. 1.1)

set-builder notation A method of designating a set that makes use of a variable and a certain property that only elements of that set possess. (Sec. 2.7)

slope A measure of the slant, or tilt, of a line. (Sec. 3.4)

slope-intercept form of a straight line The equation $y = mx + b$, where m is the slope of the line and $(0, b)$ is the y-intercept. (Sec. 3.4)

solution(s) of an equation The replacement value(s) of the variable that will make the equation true; also called the root(s) of the equation. (Sec. 2.1)

solution of an equation in two variables An ordered pair whose coordinates make the equation a true statement. (Sec. 3.1)

solution of a system of equations in three variables An ordered triple that is a solution of each equation of the system. (Sec. 4.2)

solution of a system of equations in two variables An ordered pair that is a solution of each equation of the system. (Sec. 4.1)

solution set of an inequality A set of numbers, each element of which, when substituted for the variable, results in a true inequality. (Sec. 2.7)

solution set of a system of inequalities The intersection of the solution sets of the individual inequalities. (Sec. 4.5)

solve an equation Find the root(s), or solution(s), of the equation. (Sec. 2.1)

square of a binomial A polynomial that can be expressed in the form $(a + b)^2$. (Sec. 5.3)

square matrix A matrix that has the same number of rows as columns. (Sec. 4.3)

square root of a perfect square One of the two equal factors of the perfect square. (Sec. 6.4)

standard form of a quadratic equation A quadratic equation written with the polynomial in descending order and equal to zero. $ax^2 + bx + c = 0$ is in standard form. (Sec. 6.5, 9.1)

straight angle An angle whose measure is 180°. (Sec. 2.4)

substitution method An algebraic method of finding a solution of a system of linear equations. (Sec. 4.1)

subtraction The process of finding the difference between two numbers. (Sec. 1.2)

sum In addition, the total of two or more numbers. (Sec. 1.2)

supplementary angles Two angles whose sum is 180°. (Sec. 2.4)

synthetic division A shorter method of dividing a polynomial by a binomial of the form $x - a$. This method uses only the coefficients of the variable terms. (Sec. 5.5)

system of equations Two or more equations considered together. (Sec. 4.1)

system of inequalities Two or more inequalities considered together. (Sec. 4.5)

terms of a variable expression The addends of the variable expression. (Sec. 1.5)

tolerance of a component The acceptable amount by which the component may vary from a given measurement. (Sec. 2.8)

transversal A line intersecting two other lines at two different points. (Sec. 2.4)

triangle A three-sided closed figure. (Sec. 2.4)

trinomial A polynomial of three terms. (Sec. 5.1)

undefined slope The slope of a vertical line is undefined. (Sec. 3.4)

uniform motion The motion of an object whose speed and direction do not change. (Sec. 2.3)

union of two sets The set that contains all elements that belong to either of the sets. (Sec. 2.7)

upper limit In a tolerance, the greatest acceptable value. (Sec. 2.8)

value mixture problem A problem that involves combining, into a single blend, two ingredients that have different prices. (Sec. 2.3)

value of a function The value of the dependent variable for a given value of the independent variable. (Sec. 3.2)

variable A letter of the alphabet used to stand for a number that is unknown or that can change. (Sec. 1.1, 1.5)

variable expression An expression that contains one or more variables. (Sec. 1.5)

variable part In a variable term, the variable or variables and their exponents. (Sec. 1.5)

variable term A term composed of a numerical coefficient and a variable part. When the numerical coefficient is 1 or -1, the 1 is not usually written. (Sec. 1.5)

vertex of a parabola The point on the parabola with the smallest y-coordinate or the largest y-coordinate. (Sec. 9.6)

vertical angles Two angles that are on opposite sides of the intersection of two lines. (Sec. 2.4)

vertical shift The displacement of a graph up or down on the coordinate axes; also known as a vertical translation. (Sec. 10.1)

vertical translation The displacement of a graph up or down on the coordinate axes; also known as a vertical shift. (Sec. 10.1)

whole numbers The numbers 0, 1, 2, 3, The whole numbers include the natural numbers and zero. (Sec. 1.1)

x-coordinate The abscissa in an ordered pair. (Sec. 3.1)

x-intercept The point at which a graph crosses the x-axis. (Sec. 3.3)

xy-coordinate system A rectangular coordinate system in which the horizontal axis is labeled x and the vertical axis is labeled y. (Sec. 3.1)

xyz-coordinate system A three-dimensional coordinate system formed when a third coordinate axis (the z-axis) is located perpendicular to the xy-plane. (Sec. 4.2)

y-coordinate The ordinate in an ordered pair. (Sec. 3.1)

y-intercept The point at which a graph crosses the y-axis. (Sec. 3.3)

zero of a function A value of x for which $f(x) = 0$. (Sec. 3.3)

zero slope The slope of a horizontal line. (Sec. 3.4)

Index

Table of Equations and Formulas

Slope of a Straight Line

$$\text{Slope} = m = \frac{y_2 - y_1}{x_2 - x_1}, \; x_1 \neq x_2$$

Slope-intercept Form of a Straight Line

$$y = mx + b$$

Point-slope Formula for a Line

$$y - y_1 = m(x - x_1)$$

Quadratic Formula

$$x = \frac{-b \pm \sqrt{b^2 - 4ac}}{2a}$$

Perimeter and Area of a Triangle, and Sum of the Measures of the Angles

$$P = a + b + c$$
$$A = \frac{1}{2}bh$$
$$\angle A + \angle B + \angle C = 180°$$

Pythagorean Theorem

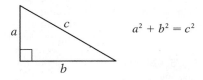

$$a^2 + b^2 = c^2$$

Perimeter and Area of a Rectangle

$$P = 2L + 2W$$
$$A = LW$$

Perimeter and Area of a Square

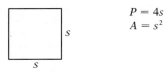

$$P = 4s$$
$$A = s^2$$

Circumference and Area of a Circle

$$C = 2\pi r \quad \text{or} \quad C = \pi d$$
$$A = \pi r^2$$

Area of a Trapezoid

$$A = \frac{1}{2}h(b_1 + b_2)$$

Volume and Surface Area of a Rectangular Solid

$$V = LWH$$
$$SA = 2LW + 2LH + 2WH$$

Volume and Surface Area of a Cube

$$V = s^3$$
$$SA = 6s^2$$

Volume and Surface Area of a Right Circular Cylinder

$$V = \pi r^2 h$$
$$SA = 2\pi r^2 + 2\pi rh$$

Volume and Surface Area of a Right Circular Cone

$$V = \frac{1}{3}\pi r^2 h$$
$$SA = \pi r^2 + \pi rl$$

Volume and Surface Area of a Sphere

$$V = \frac{4}{3}\pi r^3$$
$$SA = 4\pi r^2$$

Table of Measurement Abbreviations

U.S. Customary System

Length		Capacity		Weight		Area	
in.	inches	oz	fluid ounces	oz	ounces	in^2	square inches
ft	feet	c	cups	lb	pounds	ft^2	square feet
yd	yards	qt	quarts				
mi	miles	gal	gallons				

Metric System

Length		Capacity		Weight/Mass		Area	
mm	millimeters (0.001 m)	ml	milliliters (0.001 L)	mg	milligrams (0.001 g)	cm^2	square
cm	centimeters (0.01 m)	cl	centiliters (0.01 L)	cg	centigrams (0.01 g)		centimeters
dm	decimeters (0.1 m)	dl	deciliters (0.1 L)	dg	decigrams (0.1 g)	m^2	square meters
m	meters	L	liters	g	grams		
dam	decameters (10 m)	dal	decaliters (10 L)	dag	decagrams (10 g)		
hm	hectometers (100 m)	hl	hectoliters (100 L)	hg	hectograms (100 g)		
km	kilometers (1000 m)	kl	kiloliters (1000 L)	kg	kilograms (1000 g)		

Time

h	hours	min	minutes	s	seconds

Table of Symbols

Symbol	Meaning		
+	add		
−	subtract		
$\cdot$, $(a)(b)$	multiply		
$\frac{a}{b}$, ÷	divide		
()	parentheses, a grouping symbol		
[]	brackets, a grouping symbol		
π	pi, a number approximately equal to $\frac{22}{7}$ or 3.14		
$-a$	the opposite, or additive inverse, of a		
$\frac{1}{a}$	the reciprocal, or multiplicative inverse, of a		
=	is equal to		
≈	is approximately equal to		
≠	is not equal to		
<	is less than		
≤	is less than or equal to		
>	is greater than		
≥	is greater than or equal to		
(a, b)	an ordered pair whose first component is a and whose second component is b		
°	degree (for angles and temperature)		
$\sqrt{a}$	the principal square root of a		
$	a	$	the absolute value of a

TI-30X IIS

6 [Ab/c] 2 [Ab/c] 3 [+] 3 [Ab/c] 4 [ENTER] — Access operations in blue

Operations on fractions
$6\frac{2}{3} + \frac{3}{4} = 7\frac{5}{12}$

```
6⌐2⌐3+3⌐4
              7⌐5/12
```

The value of π
```
π
3.141592654
```

Power of a number (See Note 1 below.)
13 [∧] 4 [ENTER]
```
13⁴
              28561
```

Square root of a number
[2nd] √36 [)] [ENTER]
```
√(36)
               6
```

Square a number
7 [x²] [ENTER]
```
7²
              49
```

Photo courtesy of Texas Instruments Incorporated

.4 [2nd] F◊D [ENTER] **Change decimal to fraction or fraction to decimal**
```
.4▸F◊D
              2/5
```

3 [+] 2 [(] 10 [−] 6 [)] [ENTER] **Operations with parentheses**
```
3+2(10−6)
              11
```

11 [×] 25 [2nd] % [ENTER] **Operations with percent**
```
11*25%
              2.75
```

Used to complete an operation

[(−)] 12 [÷] 6 [ENTER] **Enter a negative number (See Note 2 below.)**
```
−12/6
              −2
```

fx-300MS

√ 36 [=] **Square root of a number**
```
√36
               6
```

Access operations in gold

6 [ab/c] 2 [ab/c] 3 [+] 3 [ab/c] 4 [=] **Operations on fractions**
$6\frac{2}{3} + \frac{3}{4} = 7\frac{5}{12}$
```
6⌐2⌐3+3⌐4
              7⌐5⌐12
```

7 [x²] [=] **Square a number**
```
7²
              49
```

[(−)] 12 [÷] 6 [=] **Enter a negative number (See Note 2 below.)**
```
−12÷6
              −2
```

Photo courtesy of Casio, Inc.

.4 [=] d/c **Change decimal to fraction**
```
.4
              2⌐5
```

13 [∧] 4 [=] **Power of a number (See Note 1 below.)**
```
13⁴
              28561
```

3 [+] 2 [(] 10 [−] 6 [)] [=] **Operations with parentheses**
```
3+2(10−6)
              11
```

11 [×] 25 % [=] **Operations with percent**
```
11x25%
              2.75
```

Used to complete an operation

π [=] **The value of π**
```
π
3.141592654
```

NOTE 1: Some calculators use the [yˣ] key to calculate a power.
For those calculators, enter 13 [yˣ] 4 [=] to evaluate 13^4.

NOTE 2: Some calculators use the [+/−] key to enter a negative number.
For those calculators, enter 12 [+/−] [÷] 6 [=] to calculate $-12 \div 6$.